The Living Record of Science

《自然》学科经典系列

总顾问：李政道（Tsung-Dao Lee）

英方总主编：Sir John Maddox
Sir Philip Campbell

中方总主编：路甬祥

生命科学的进程 V

PROGRESS IN LIFE SCIENCES V

（英汉对照）

主编：许智宏

外语教学与研究出版社 · 麦克米伦教育 · 《自然》旗下期刊与服务集合

FOREIGN LANGUAGE TEACHING AND RESEARCH PRESS · MACMILLAN EDUCATION · NATURE PORTFOLIO

北京 BEIJING

图书在版编目（CIP）数据

生命科学的进程．V：英汉对照／许智宏主编．-- 北京：外语教学与研究出版社，2022.3
（《自然》学科经典系列／路甬祥等总主编）
ISBN 978-7-5213-3358-9

Ⅰ．①生… Ⅱ．①许… Ⅲ．①生命科学－文集－英、汉 Ⅳ．①Q1-53

中国版本图书馆 CIP 数据核字（2022）第 035009 号

地图审图号：GS（2021）7155 号
书中地图系原文插附地图

出 版 人　王　芳
项目统筹　章思英
项目负责　刘晓楠　顾海成
责任编辑　刘晓楠
责任校对　王　菲　夏洁媛
封面设计　孙莉明　高　蕾
版式设计　孙莉明
出版发行　外语教学与研究出版社
社　　址　北京市西三环北路 19 号（100089）
网　　址　http://www.fltrp.com
印　　刷　北京华联印刷有限公司
开　　本　787×1092　1/16
印　　张　61.5
版　　次　2022 年 3 月第 1 版　2022 年 3 月第 1 次印刷
书　　号　ISBN 978-7-5213-3358-9
定　　价　568.00 元

购书咨询：（010）88819926　电子邮箱：club@fltrp.com
外研书店：https://waiyants.tmall.com
凡印刷、装订质量问题，请联系我社印制部
联系电话：（010）61207896　电子邮箱：zhijian@fltrp.com
凡侵权、盗版书籍线索，请联系我社法律事务部
举报电话：（010）88817519　电子邮箱：banquan@fltrp.com
物料号：333580001

《自然》学科经典系列

（英汉对照）

生命科学的进程

（英汉对照）

主编：许智宏

审稿专家（以姓氏笔画为序）

于　军	王　宇	王晓良	邢　松	巩克瑞	肖景发	汪筱林
张颖奇	陈继征	陈捷胤	赵凌霞	胡松年	徐文堪	姬书安
常　江	彭小忠	曾长青	解彬彬	裴端卿	潘　雷	

翻译工作组稿人（以姓氏笔画为序）

王丽霞	王晓蕾	王耀杨	刘　明	关秀清	李　琦	何　铭
蔡　迪						

翻译人员（以姓氏笔画为序）

王海纳	毛晨晖	田晓光	吕　静	任　奕	刘振明	刘皓芳
李　平	李　梅	李　辉	杨　晶	张玉光	张瑶楠	高如丽
董培智						

校对人员（以姓氏笔画为序）

王　菲	王丽霞	王晓蕾	田晓阳	任　奕	刘立云	刘雨佳
刘琰璐	许静静	李　龙	李　平	李　景	李　婷	李照涛
李霄霞	杨　晶	吴　茜	张玉光	张亚盟	张茜楠	张瑶楠
陈思原	周　晔	周少贞	赵凤轩	洪雅强	贺舒雅	夏洁媛
顾海成	黄小斌	第文龙	焦晓林	潘卫东	薛　陕	

Contents
目　　录

2002

2003

2004

2005

2006

2007

Volume V

An Exceptionally Well-preserved Theropod Dinosaur from the Yixian Formation of China

Pei-ji Chen *et al.*

Editor's Note

The first sign that the palaeontological world was about to be turned upside down at the 1996 meeting of the Society of Vertebrate Paleontology in New York came when Pei-Ji Chen showed snapshots of a dinosaur from the hitherto obscure Yixian Formation of Liaoning Province, PRC. A dinosaur—with feathers. The relationship between dinosaurs and birds had long been suspected. Indeed, *Archaeopteryx*, the earliest known bird, looked like a feathered dinosaur. However, there had been no hard evidence for feathers in non-flying dinosaurs. To be sure, Chen's *Sinosauropteryx* had more of a fringe of feather-like fibrils than true feathers, but more was to come, and soon. Fossils of feathered dinosaurs began to pour from Liaoning to capture the world's imagination.

Two spectacular fossilized dinosaur skeletons were recently discovered in Liaoning in northeastern China. Here we describe the two nearly complete skeletons of a small theropod that represent a species closely related to *Compsognathus*. *Sinosauropteryx* has the longest tail of any known theropod, and a three-fingered hand dominated by the first finger, which is longer and thicker than either of the bones of the forearm. Both specimens have interesting integumentary structures that could provide information about the origin of feathers. The larger individual also has stomach contents, and a pair of eggs in the abdomen.

THE Jehol biota[1] was widely distributed in eastern Asia during latest Jurassic and Early Cretaceous times. These freshwater and terrestrial fossils include macroplants, palynomorphs, charophytes, flagellates, conchostracans, ostracods, shrimps, insects, bivalves, gastropods, fish, turtles, lizards, pterosaurs, crocodiles, dinosaurs, birds and mammals. In recent years, the Jehol biota has become famous as an abundant source of remains of early birds[2,3]. Dinosaurs are less common in the lacustrine beds, but the specimens described here consist of two nearly complete skeletons of a small theropod discovered by farmers in Liaoning. The skeletons are from the basal part of the Yixian Formation, from the same horizon as the fossil birds *Confuciusornis* and *Liaoningornis*[3]. Both are remarkably well preserved, and include fossilized integument, organ pigmentation and abdominal contents. One of the two was split into part and counterpart, and the sections were deposited in two different institutions. One side (in the National Geological Museum of China, Beijing) became the holotype of *Sinosauropteryx prima*, a supposed bird[4]. The

在中国义县组发现的一具保存异常完美的兽脚类恐龙

陈丕基等

编者按

当陈丕基在1996年纽约举办的古脊椎动物学会会议上展示了来自中国辽宁省义县组（年代还没有研究透彻）的一只恐龙的照片时，古生物学世界被震撼了。一只带羽毛的恐龙！恐龙与鸟类之间的关联一直被怀疑。的确，始祖鸟——已知最早的鸟类——看上去非常像一只带羽毛的恐龙。然而，之前并未发现不会飞的恐龙长有羽毛的确凿证据。诚然，陈丕基命名的中华龙鸟具有的更像是羽毛状的纤维，而不是真的羽毛，但很快就有更多的发现。带羽毛恐龙化石开始在辽宁大量出现，吸引了全世界的目光。

最近在中国东北部的辽宁省发现了两具非常吸引人的恐龙骨架化石。在本文中我们描述了这两具近乎完整的小型兽脚类恐龙骨架，它们代表了一种与美颌龙具有密切关系的恐龙。中华龙鸟是已知的兽脚类恐龙中尾巴最长的，前肢具有三指，由第一指主导运动，第一指要比前肢的任何骨骼都长且粗壮。两件标本都保存了吸引人的皮肤衍生物结构，这些结构能够为羽毛的起源提供信息。较大个体的腹腔中还保存着胃容物以及一对蛋化石。

热河生物群[1]广泛地分布在东亚地区的侏罗纪末到早白垩世地层中，这里发现的淡水陆相化石包括大型植物、孢粉类、轮藻类、鞭毛虫、叶肢介、介形类、虾、昆虫、双壳类、腹足类、鱼类、龟鳖类、蜥蜴、翼龙、鳄类、恐龙、鸟和哺乳动物。在最近几年里，热河生物群成了著名的早期鸟类化石的丰富产地[2,3]。恐龙化石在湖相地层较少，但是本文记述的两具接近完整的小型兽脚类恐龙骨架是由在辽宁的农民发现的。这两具标本发现自义县组的基部，与孔子鸟和辽宁鸟[3]层位相同。这两件标本都保存完好，并且包括了石化的皮肤衍生物结构、器官色素沉着和腹容物。两者中有一件标本分离成为正体和负体两块，这两块标本保存在两个不同的机构，一面作为原始中华龙鸟（曾被误认为是鸟类[4]）的正模标本（保存于中国地质博物馆，北京），而负体则和第二件较大的标本一起保存于中国科学院南京地质古

counterpart and the second larger specimen are in the collections of the Nanjing Institute of Geology and Palaeontology.

The Yixian Formation is mainly composed of andesites, andesite-breccia, agglomerates and basalts, but has four fossil-bearing sedimentary intercalations that are rich in tuffaceous materials. The Jianshangou (formerly Jianshan[5,6]) intercalated bed (60 m thick) is the basal part of this volcanic sedimentary formation, and is made up of greyish–white, greyish–yellow and greyish–black sandstones, siltstones, mudstones and shales. These sediments are rich in fossils of mixed Jurassic–Cretaceous character. The primitive nature of the fossil birds of the Jianshangou fossil group has led to suggestions that the beds could be as early as Tithonian in age[2]. But although *Confuciusornis* and the other birds[3] are more advanced than *Archaeopteryx* in a number of significant features, we can only conclude that the beds that the fossils came from are probably younger than the Solnhofen Lithographic Limestones (Early Tithonian). The presence of *Psittacosaurus* in the same beds is more consistent with an Early Cretaceous age[7], as are the palynomorphs[8] and a recent radiometric date of the formation[9], but other radiometric dating attempts have indicated older ages[10].

Dinosauria OWEN 1842

Theropoda MARSH 1881

Coelurosauria VON HUENE 1914

Compsognathidae MARSH 1882

Sinosauropteryx prima JI and JI 1996

Holotype. Part (National Geological Museum of China, GMV 2123) and counterpart (Nanjing Institute of Geology and Palaeontology, NIGP 127586) slabs of a complete skeleton.

Referred specimen. Nanjing Institute of Geology and Palaeontology NIGP 127587. Nearly complete skeleton, lacking only the distal half of the tail.

Locality and horizon. Jianshangou-Sihetun area of Beipiao, Liaoning, People's Republic of China. Yixian Formation, Jehol Group, Upper Jurassic or Lower Cretaceous (Fig. 1; Fig. 1 has been omitted in this edited version).

Diagnosis. Compsognathid with longest tail known for any theropod (64 caudals). Skull 15% longer than femur, and forelimb (humerus plus radius) only 30% length of leg (femur plus tibia), in contrast with *Compsognathus* where skull is same length as femur, and forelimb length is 40% leg length. Within the Compsognathidae, forelimb length (compared to femur length) is shorter in *Sinosauropteryx* (61–65%) than it is in *Compsognathus* (90–99%). In contrast with all other theropods, ungual phalanx II–2 is longer than the radius. Haemal spines simple and spatulate, whereas those of *Compsognathus* taper distally.

生物研究所。

义县组的主要岩性组成为安山岩、安山角砾岩、集块岩和玄武岩，但是有四个富含凝灰质的含化石沉积岩夹层。尖山沟（以前称“尖山”[5,6]）夹层（60 米厚）位于火山岩沉积构造的基部，组成层序是：灰白色、灰黄色和灰黑色的砂岩、粉砂岩、泥岩和页岩。这段沉积富含混合侏罗纪–白垩纪特征的化石。尖山沟化石群含有的鸟类化石的原始特征表明该层位可能与提塘阶年代相当[2]。尽管孔子鸟和其他鸟类[3]在许多重要特征上比始祖鸟更加进步，我们仅能推断该化石层位可能要比德国索伦霍芬石印灰岩的层位（早提塘阶）靠上。发现于相同层位的鹦鹉嘴龙生存时代为早白垩世[7]，这与孢粉测年[8]和最近对该地层进行的放射性同位素测年[9]的结果相一致，但是其他的放射性测年显示该地层的年代更古老[10]。

恐龙总目 Dinosauria OWEN 1842
兽脚亚目 Theropoda MARSH 1881
虚骨龙类 Coelurosauria VON HUENE 1914
美颌龙科 Compsognathidae MARSH 1882
原始中华龙鸟 *Sinosauropteryx prima* JI and JI 1996

正模标本 正体（中国地质博物馆，标本编号 GMV 2123）和负体（南京地质古生物研究所，NIGP 127586）完整骨骼的板状化石。

归入标本 南京地质古生物研究所，NIGP 127587。近乎完整的骨骼，仅缺少尾部的远端一半。

产地与层位 中国辽宁省北票市尖山沟–四合屯地区。上侏罗统或下白垩统，热河群，义县组（图 1；该图显示中华龙鸟和孔子鸟产地，此版本中省略）。

鉴别特征 一种美颌龙科恐龙，具有已知兽脚类中最长的尾（64 个尾椎）。头骨比股骨长 15%，前肢（肱骨 + 桡骨）仅是后肢（股骨 + 胫骨）长度的 30%，而美颌龙头骨与股骨等长，前肢长是后肢长的 40%。在美颌龙科成员中，中华龙鸟的前肢与股骨的长度比例为 61% ~ 65%，短于美颌龙（90% ~ 99%）。不同于其他兽脚类恐龙，中华龙鸟的第二指爪比桡骨长。脉弧简单且呈片状，而美颌龙的则是锥形。

Description

Sinosauropteryx is comparable in size and morphology to known specimens of *Compsognathus*[11,12] from Germany and France. The smaller Chinese specimen (Fig. 2) is 0.68 m long (snout to end of tail) and has a femur length of 53.2 mm, whereas the second specimen (Fig. 3) has a femur length of 86.4 mm. The former is smaller than the type specimen of *Compsognathus longipes* (femur length about 67 mm) and the latter is smaller than the second specimen of *Compsognathus* from Canjuers (France), which has a femur length of 110 mm and an estimated length of 1.25 m. Although size and body proportions indicate that the smaller specimen was younger when it died, well-ossified limb joints and tarsals suggest that it was approaching maturity.

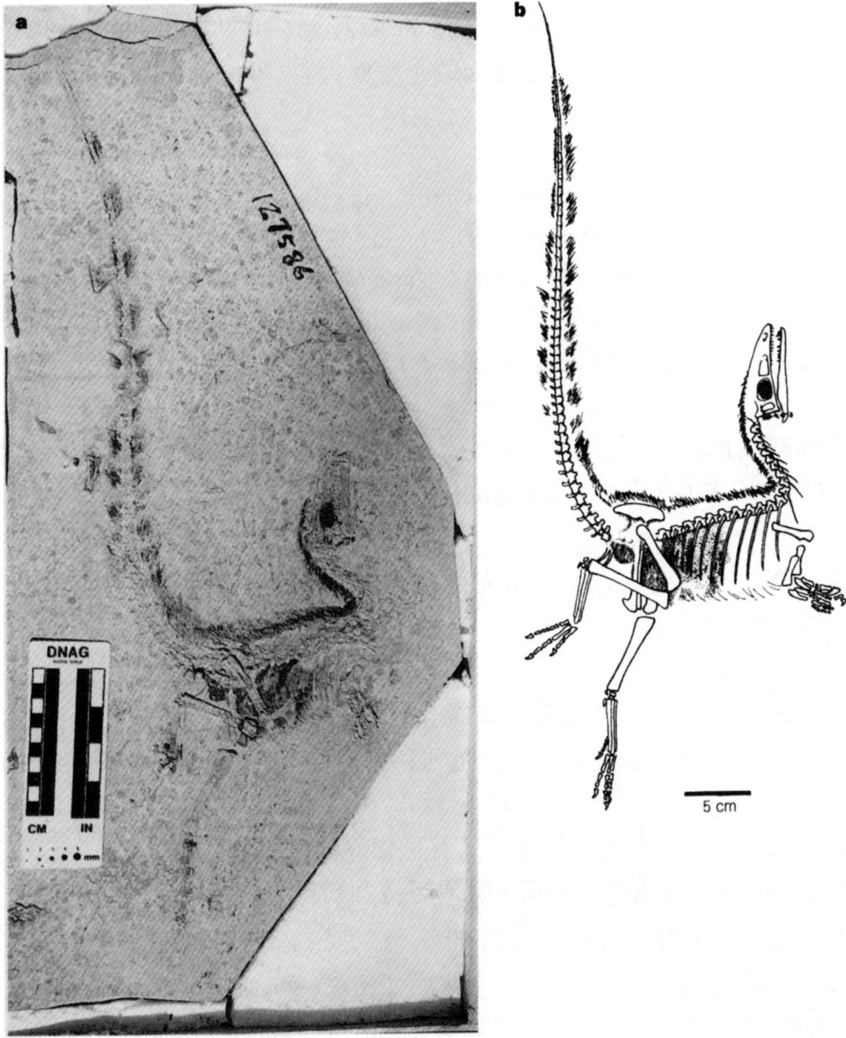

Fig. 2. *Sinosauropteryx prima* Ji and Ji. **a**, NIGP 127586, the counterpart of holotype (GMV 2123). **b**, Skeletal reconstruction of NIGP 127586. The integumentary structures are along the dorsal side and tail and dark pigmentation in the abdominal region might be some soft tissues of viscera.

描　述

中华龙鸟在大小和形态上同德国、法国已知的美颌龙标本[11,12]相近。中国标本中较小的一块（图 2）为 0.68 米长（从吻到尾巴末端），股骨长为 53.2 毫米，而第二件标本（图 3）的股骨长为 86.4 毫米。前者比模式种长足美颌龙（股骨长约 67 毫米）小，而后者比另一件来自法国 Canjuers 的美颌龙（股骨长为 110 毫米，估计全长 1.25 米）小。尽管大小和身体比例反映出小的标本死时比较年轻，但它死亡后保存下来的完全骨化的肢骨关节和跗骨表明它已经接近成熟。

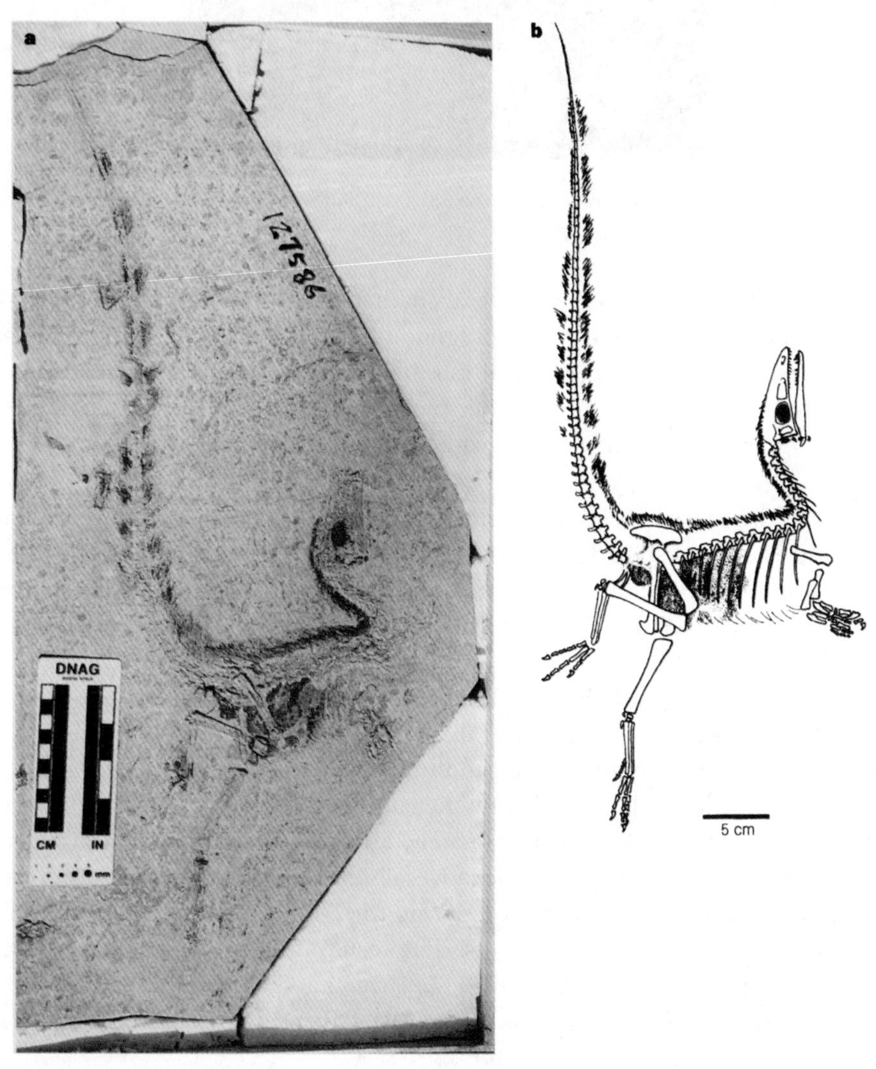

图 2. 原始中华龙鸟。**a**，NIGP 127586，是正模（GMV 2123）的负体。**b**，标本 NIGP 127586 的骨骼复原。皮肤衍生物结构沿着背部到尾部，腹部区域的黑色沉着可能是一些内脏软体组织残留。

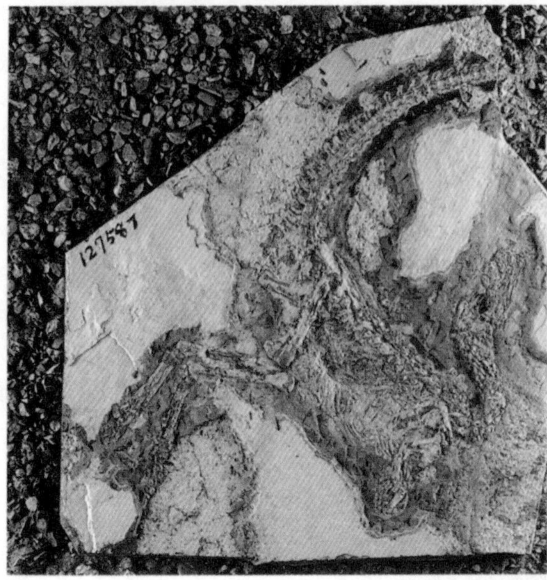

Fig. 3. *Sinosauropteryx prima* Ji and Ji, NIGP 127587, an adult individual from the same locality as holotype.

Sinosauropteryx and *Compsognathus* share several characteristics that indicate close relationship. These can be used to diagnose the Compsognathidae and include unserrated premaxillary but serrated maxillary teeth, a powerful manual phalanx I–1 (shaft diameter is greater than that of the radius), fan-shaped neural spines on the dorsal vertebrae, limited anterior expansion of the pubic boot and a prominent obturator process of the ischium.

Other characteristics were used to diagnose *Compsognathus*, including the presence of a relatively large skull and short forelimbs. In *Compsognathus*, skull length is 30% of that of the presacral vertebral column, whereas this same ratio is 40% in the new specimen NIGP 127586 and 36% in NIGP 127587. Unfortunately, relative skull length is highly variable in theropods. Comparing skull length with femur length, which is less variable than vertebral length, most theropods have skulls 100–119% the length of the femur[13]. The *Compsognathus* skulls are 99–100% and *Sinosauropteryx* skulls are 113–117%. Compsognathids have short forelimbs[11], 40% of the length of the hindlimb in *Compsognathus*. In *Sinosauropteryx*, the lengths of humerus plus radius divided by the sum of femur and tibia lengths produces a figure of less than 30% (Table 1). Unfortunately, such ratios are dependent on the absolute size of the animal, mostly because of negative allometry experienced by the tibia during growth or interspecific size increase. Comparing the lengths of humerus plus radius with femur length produces more useful results. The resulting figures fall within the range of most theropods (60–110%). The abelisaurid *Carnosaurus* and all tyrannosaurs have relatively shorter arms. Within the Compsognathidae, however, arm length is shorter in *Sinosauropteryx* (61–65%) than it is in *Compsognathus* (90–99%).

图 3. 原始中华龙鸟，NIGP 127587，一个成年个体，与正模标本来自同一产地

中华龙鸟和美颌龙具有几个共同的特征，显示出它们之间存在密切的关系。这些特征通常能够用于鉴别美颌龙科，包括无锯齿的前颌骨齿和有锯齿的上颌骨齿，强有力的第一指第一骨（直径要比桡骨大），背椎上扇形的神经棘，耻骨靴状突前端有限的扩展以及坐骨明显的闭孔突。

其他鉴定美颌龙的特征包括相对较大的头骨和短小的前肢。美颌龙头骨的长度占荐前椎总长的 30%，然而在新标本 NIGP 127586 和 NIGP 127587 中，该比率分别为 40% 和 36%。不巧的是，相对头骨长度在兽脚类恐龙中是一个非常易变的变量。头骨和股骨的长度的比值相对于与椎体的长度比值在兽脚类恐龙中变化较小，大部分兽脚类恐龙的头骨是股骨的长度的 100% ~ 119%[13]。其中美颌龙的该值为 99% ~ 100%，中华龙鸟为 113% ~ 117%。美颌龙科具有较短的前肢 [11]，其中美颌龙前肢长度是后肢的 40%。在中华龙鸟中，肱骨加上桡骨的长度与股骨和胫骨的总和相比小于 30%（表 1）。不过，这个比率取决于动物的绝对大小，主要是由生长或种间大小增长的过程中胫骨的异速生长造成的。肱骨加上桡骨的长度与股骨长度的比值更加可靠。该比例在大多数兽脚类的范围内（60% ~ 110%）。在阿贝力龙类和暴龙类中，前肢相对较短。在美颌龙类中，中华龙鸟的前肢长度比值（61% ~ 65%）要比美颌龙的（90% ~ 99%）小。

Table 1. Comparison of size and proportions of *Sinosauropteryx* and *Compsognathus*

Species	Specimen	Skull	Humerus	Radius	Femur	Tibia	Skull/femur	Arm/leg
Compsognathus sp.	BSP ASI	70	39	24.7	71	87.6	0.99	0.40
Compsognathus sp.	MNHN	110	67	42	110		1.00	
Sinosauropteryx primus	NIGP 127586	62.5	20.3	12.4	53.2	61	1.17	0.29
Sinosauropteryx primus	NIGP 127587	97.2	35.5	21	86.4	97	1.13	0.31

Length measurements are given in millimetres. Data about *Compsognathus* are from ref. 11.

Both specimens of *Sinosauropteryx* have 10 cervical and 13 dorsal vertebrae. The posterior cervical vertebrae have biconcave centra. We could not determine the number of sacrals. The tail is extremely long. In the smaller specimen it is almost double the snout–vent length, and there are 59 caudal vertebrae exposed with an estimated five more than have been lost from the middle of the tail of NIGP 127586 (but present in GMV 2123). Only the first 23 vertebrae are preserved in the larger specimen, but this section is longer than the summed lengths of the cervical, dorsal and sacral vertebrae. Neither of the European specimens has a complete tail, but in both cases the tail was clearly longer than the body. When vertebral lengths are normalized (divided by the average lengths of caudal vertebrae 2–5), there are no significant differences between vertebral lengths in any of the four tails. As in *Compsognathus*, the dorsal neural spines are peculiar in that they are anteroposteriorly long but low, and often are fan-shaped.

The caudal centra increase in length over the first six segments, but posteriorly decrease progressively in length and all other dimensions. The first 10 tail vertebrae have neural spines, most of which slope posterodorsally. There are at least four pairs of caudal ribs in NIGP 127586, and more distal caudals have low bumps in this region that could also be interpreted as transverse processes. This could be another way to distinguish the Asian and European compsognathids, because the German specimen of *Compsognathus* apparently lacks caudal ribs and transverse processes[11]. Haemal spines are found on at least the first 47 caudals of NIGP 127586, and the anterior ones are simple spatulate structures that curve gently posteroventrally. The haemal spines are oriented more posteriorly than ventrally, and are more strongly curved.

Both specimens have 13 pairs of dorsal ribs. The ribs indicate a high but narrow body. The distal ends of the first two pairs of ribs are expanded and end in cup-like depressions that suggest the presence of a cartilaginous sternum. The gastralia are well preserved with two gastralia on each side of a segment. The median gastralia cross to form the interconnected "zig-zag" pattern characteristic of all theropods[14] and primitive birds like *Archaeopteryx* and *Confuciusornis*.

The front limb is relatively short and stout. Both NIGP 127586 and NIGP 127587 have articulated hands, something that is lacking in the two European specimens. What has been interpreted by some as the first metacarpal[11] in *Compsognathus* is the first phalanx of

表 1. 中华龙鸟和美颌龙个体大小和比例的相互比较

种类	标本编号	头骨	肱骨	桡骨	股骨	胫骨	头骨/股骨	前肢/后肢
美颌龙未定种	BSP ASI	70	39	24.7	71	87.6	0.99	0.40
美颌龙未定种	MNHN	110	67	42	110		1.00	
原始中华龙鸟	NIGP 127586	62.5	20.3	12.4	53.2	61	1.17	0.29
原始中华龙鸟	NIGP 127587	97.2	35.5	21	86.4	97	1.13	0.31

长度测量的单位是毫米，美颌龙资料来源于参考文献 11。

中华龙鸟的两件标本均有 10 节颈椎和 13 节背椎。后部的颈椎是双凹型椎体。不过我们不能确定荐椎的数量。尾巴特别长。在较小的标本（NIGP 127586）中，尾长差不多是吻肛距的两倍，有 59 节尾椎暴露，估计尾巴中段至少缺失 5 节以上的椎体（但在 GMV 2123 中存在）。在较大的标本中只有前 23 节尾椎椎体保存，但是这一段比颈椎、背椎和荐椎的总长要长。而欧洲的两个标本都没有保存完整的尾巴，但二者的尾巴明显要比身体长。当脊椎的长度进行标准化计算（除以第 2 ~ 5 节尾椎的平均长度）后，这四件标本的尾巴长度没有明显的差异。和美颌龙一样，中华龙鸟背神经棘很奇特，其前后向长但高度略低，且多呈现为扇形。

前六节尾椎的椎体长度递增，而后面尾椎的长度和其他尺寸都逐渐递减。前十节尾椎上有神经棘，多数向后背侧倾斜。在 NIGP 127586 中至少有 4 对尾肋，且很多末端尾椎在肋骨着生的部位有较低的突起，可以看作是横突。这可以作为区分亚洲和欧洲的美颌龙类的另一个方法，因为德国的美颌龙标本明显缺少尾肋和尾椎横突 [11]。NIGP 127586 至少前 47 节尾椎上都具有脉弧，前面的脉弧呈轻微向后腹侧弯曲的简单的片状结构。脉弧向后的幅度比向腹部大，且弯曲得更加强烈。

两件标本均有 13 对背肋。肋骨的形态揭示了该个体的身躯高而窄。前面两对肋骨的最末端膨大，且末端具有杯状的凹陷，揭示了软骨质胸骨的存在。腹膜肋保存较好，在每一节两侧都各有两个腹膜肋。中间的腹膜肋像所有兽脚类恐龙 [14] 和原始鸟类（如始祖鸟和孔子鸟）中一样交叉连通呈“之”字形排列。

前肢相对短而粗壮，NIGP 127586 和 NIGP 127587 两件标本均保存着有关节的手部，这在欧洲的两标本中是缺乏的。美颌龙中被其他人鉴定为是第一掌骨 [11] 的

digit I, as was originally proposed by von Huene[15]. The first metacarpal is short (4.2 mm long in NIGP 127586, and double that length in 127587), and is probably the element identified as a carpal in the French specimen[12]. As is typical of all theropods, the collateral ligament pits of the first phalanx are much closer to the extensor surface of the bone than they are to the flexor surface. Both phalanx I–1 and the ungual that it supports are relatively large, each being as long as the radius, and thicker than the shafts and the distal ends of either the radius or the ulna. This unusual character seems to have been partially developed in at least phalanx I–1 of *Compsognathus*. Relative to the length of the radius, both these elements are longer in *Sinosauropteryx* than in any other known theropod except for *Mononykus*[16]. As indicated by the proposed phylogenetic placement[16] of *Mononykus*, there are too many anatomical differences between compsognathids and *Mononykus* to suggest a close relationship, and the similarities probably represent convergence.

The long (39 mm in NIGP 127586, 67.5 mm in NIGP 127587), low (22.2 mm high at both pubic and ischial peduncles in NIGP 127587) ilium is shallowly convex on the dorsal side in lateral aspect. The pubis, which is 82.8 mm long in NIGP 127587, is oriented anteroventrally, but is closer to vertical than it is in most non-avian theropods. The distal end expands into a pubic boot as in most tetanuran theropods. In the larger specimen, this expansion is 17.7 mm. As in *Archaeopteryx*[17], *Compsognathus*[11] and dromaeosaurs[18], the boot expands posteriorly from the shaft of the pubis, and the anterior expansion is moderate. The lack of the significant anterior expansion of the pubic boot may be correlated with the inclination of the shaft. The ischium is only two-thirds the length of the pubis in NIGP 127587. It tapers distally into a narrow shaft (3.2 mm in diameter), and like *Compsognathus*, there is a slight expansion at the end (6 mm in NIGP 127587). The prominent obturator process is also found in *Compsognathus*.

The shaft of the femur is gently curved. Both tibia and fibula are elongate. The astragalus and calcaneum are present in both specimens, although not clearly seen in either. There are five metatarsals, but as in other theropods and early birds, the first is reduced to a distal articular condyle, and the fifth is reduced to a proximal splint (Fig. 2b). Metatarsals II, III and IV are closely appressed and elongate, but are not co-ossified. The second and fourth metatarsals do not contact each other. Pedal phalanges are conservative in number (2–3–4–5–0) and morphology.

Inclusions within the Body Cavity

Like the German *Compsognathus*, the larger Chinese specimen has stomach contents preserved within the rib cage. This consists of a semi-articulated skeleton of a lizard, complete with skull (Fig. 4a, b). Numerous lizard skeletons have been recovered from these beds, but have yet to be described. Low in the abdomen of NIGP 127587, anterior to and slightly above the pubic boot, lies a pair of small eggs (37 × 26 mm) (Fig. 4a, c), one in front of the other. Additional eggs may lie underneath. Gastralia lie over the exposed surfaces of the eggs, and the left femur protrudes from beneath them, so there can be no doubt that

骨骼实际上是第一指的第一节，与最早由许纳[15]鉴别的一致。第一掌骨较短（NIGP 127586中长4.2毫米，NIGP 127587中长度为前者的两倍），它可能就是法国标本[12]中鉴定为腕骨的成分。像典型的兽脚类恐龙一样，第一指骨的侧韧带凹更靠近伸肌面而非屈肌面。第一指的第一节和其支撑的爪相对巨大，都像桡骨一样长，而且比桡骨或尺骨的骨干和末端都要粗。在美颌龙中，至少第一指的第一节有类似的特征。中华龙鸟第一指第一节及爪的长度与桡骨的长度比要比除单爪龙[16]外其他所有已知兽脚类恐龙都大。之前的系统发育结果表明[16]，美颌龙类和单爪龙之间存在许多解剖学的差异，不可能是相近的类群，二者的相似性可能为趋同所致。

髂骨长而低（NIGP 127586中长39毫米，NIGP 127587中长67.5毫米；NIGP 127587中髂骨在耻骨和坐骨柄处高度均为22.2毫米），侧视上背部略微突起。NIGP 127587的耻骨长82.8毫米，指向前腹侧，但比很多非鸟兽脚类恐龙的更接近垂直。像大多数坚尾龙类兽脚类恐龙一样，耻骨末端膨大呈靴状。在较大的标本中，这个膨大有17.7毫米。类似始祖鸟[17]、美颌龙[11]和驰龙类[18]，靴状突从耻骨柄向后膨大，前部适度膨大。耻骨突缺少明显的前突可能与耻骨柄的倾斜有关。NIGP 127587的坐骨只是耻骨长度的三分之二。坐骨向远端变尖形成一个狭窄的骨干（直径3.2毫米），且跟美颌龙一样，在末端存在微小的膨大（NIGP 127587中为6毫米）。在美颌龙中也发现了明显的闭孔突。

股骨柄略微弯曲。胫骨和腓骨相对较长。两件标本都保存了距骨和跟骨，尽管保存得都不清晰。具有5个跖骨，但和其他兽脚类恐龙和早期鸟类一样，第一跖骨退化为一个末端关节髁，第五跖骨退化成近端的一小块（图2b）。第Ⅱ、Ⅲ、Ⅳ跖骨细长并紧密排列，但没有联合骨化。第Ⅱ、Ⅳ跖骨不互相接触。脚趾骨在数量（2–3–4–5–0）和形态学上相对保守。

体腔内含物

同德国的美颌龙一样，中国发现的中华龙鸟标本中较大的一块的腹腔中有胃容物保存。胃容物包括一具部分关节保存的蜥蜴骨骼，头骨完整（图4a和4b）。发现中华龙鸟化石的地层已经发现过众多的蜥蜴骨骼，但还没有进行过描述研究。在NIGP 127587的腹部下方，耻骨靴状突的前面略上方，有一对小型的蛋化石（37毫米×26毫米）（图4a和4c），一个在前一个在后。其余的蛋可能分布在下面。腹膜肋排

they were within the body cavity. It is possible that the eggs were eaten by the dinosaur. However, given their position in the abdomen behind and below the stomach contents, and the fact that they are in the wrong part of the body cavity for the egg shell to be intact, it is more likely that these were unlaid eggs of the compsognathid. Eggs have also been reported in the holotype of *Compsognathus*[19], but they are more numerous and are only 10 mm in diameter. As they were also found outside the body cavity, their identification as *Compsognathus* eggs has not been widely accepted. The presence of fewer but larger eggs in *Sinosauropteryx* casts additional doubt on this identification.

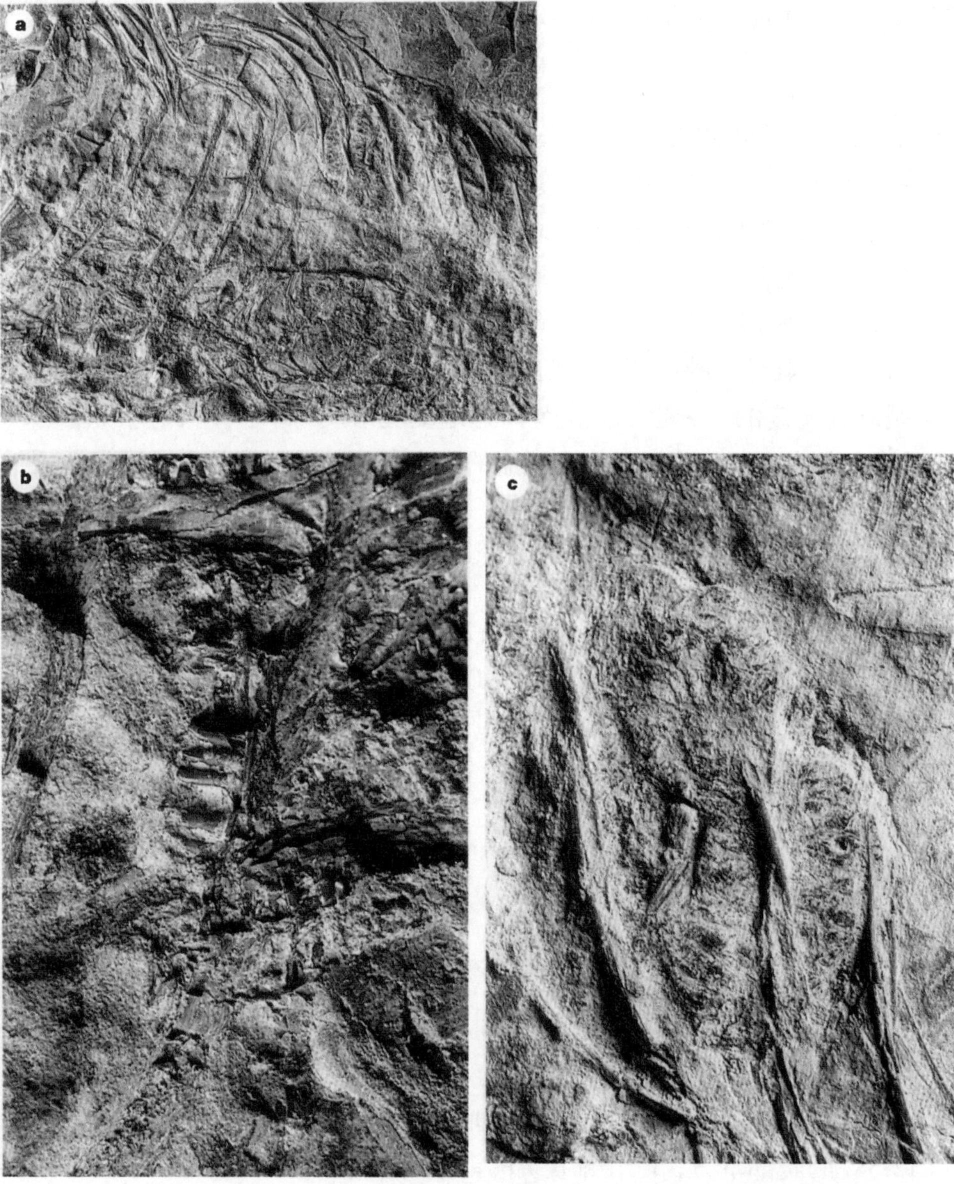

Fig. 4. Body of NIGP 127587. **a**, Stomach contents are preserved within the rib cage, and include a small lizard and a pair of eggs. **b**, A close-up of the lizard skull. **c**, A close-up of a pair of the eggs.

列在蛋化石表面，左股骨从它们下面伸出，因此它们毫无疑问处于体腔之内。蛋有可能是被恐龙吃下去的。但是，考虑到蛋的位置在腹部胃容物后侧和下面，而且如果它们是食物的话，在这个位置蛋壳不可能是完整的，因此这更像是美颌龙类还未产出的蛋。蛋化石也在美颌龙的正模标本中有过报道 [19]，但是蛋的数量很多，直径却仅有 10 毫米。由于这些蛋都在体腔外面发现，把它们鉴定为美颌龙的蛋并没有被广泛接受。中华龙鸟腹腔内少而更大型的蛋对之前美颌龙蛋化石的鉴定提出了新的质疑。

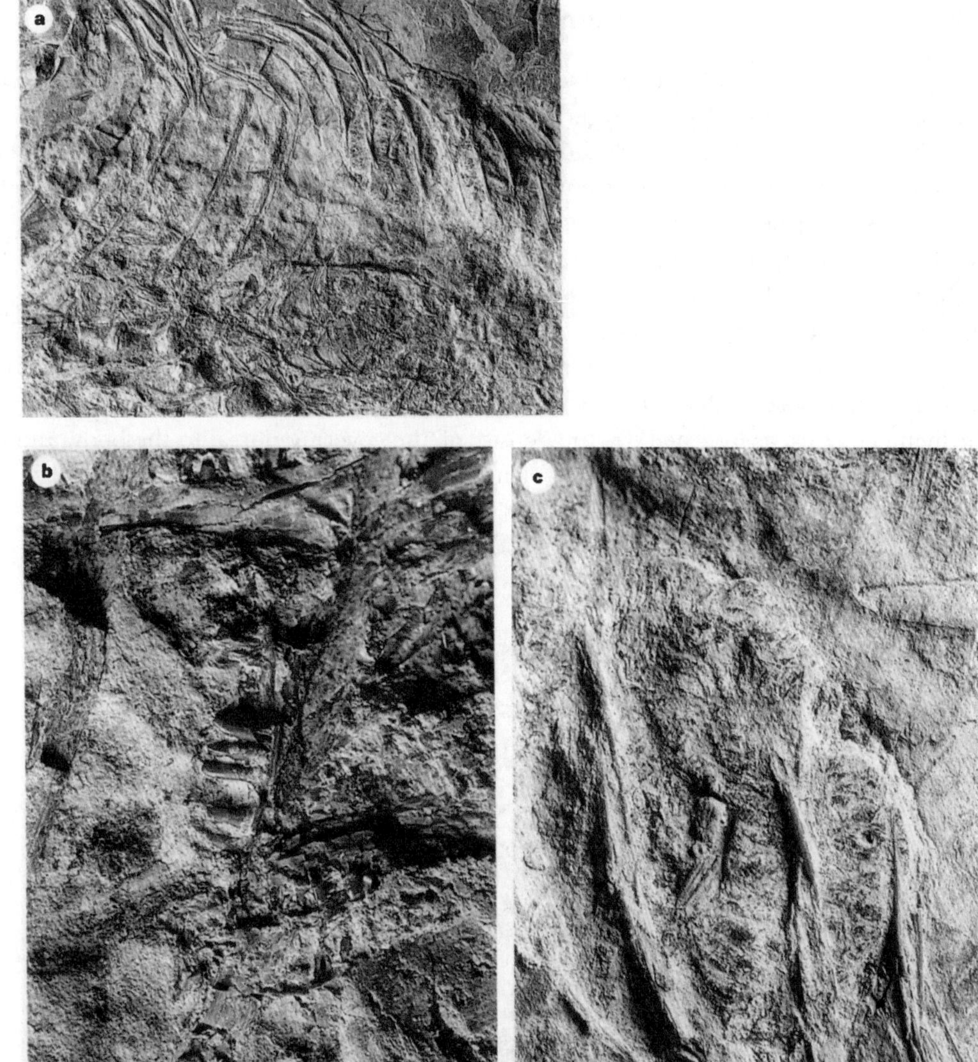

图 4. 标本 NIGP 127587 整体。**a**，保存在腹腔中的胃容物，包括一只小的蜥蜴和一对蛋化石。**b**，蜥蜴头骨的特写。**c**，成对蛋化石的特写。

Although more than two eggs may have been present in the larger specimen of *Sinosauropteryx*, it does not seem as though many could have been held within the abdomen. It may well be that these dinosaurs laid fewer eggs than most (some species are known to have produced in excess of 40)[20]. However, it is more likely that their presence demonstrates paired ovulation, as has been suggested for *Oviraptor*[21], *Troodon*[22] and other theropods. *Sinosauropteryx* therefore probably laid eggs in pairs, with a delay for ovulation between each pair.

Integumentary Structures

One of the most remarkable features of both Chinese specimens is the preservation of integumentary structures. In the larger specimen, these structures can be seen along the dorsal surface of the neck and back, and along the upper and lower margins of the tail, but in the smaller specimen the integumentary structures are clearer (Fig. 5). They cover the top of the back half of the skull, the neck, the back, the hips and the tail. They also extend along the entire ventral margin of the tail. Small patches can be seen on the side of the skull (behind the quadrate and over the articular), behind the right humerus, and in front of the right ulna. With the exception of a small patch outside the left ribs of NIGP 127587 and several areas on the left side of the tail (lateral to the vertebrae), integumentary structures cannot be seen along the sides of the body. The structures were probably present in the living animals, as indicated by the density of the covering dorsal to the body, and by the few random patches of integumentary structures that can be seen elsewhere.

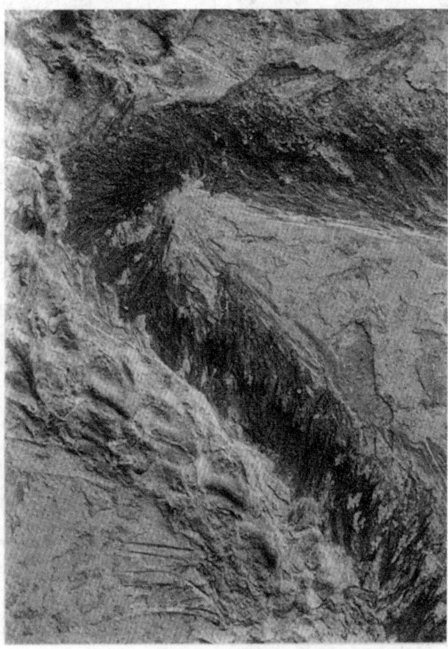

Fig. 5. Integumentary structures in the neck and dorsal sides of NIGP 127586.

尽管在大的中华龙鸟标本中可能存在多于两个蛋化石，但看上去腹腔内蛋化石的数量不会很多。很可能这些恐龙要比多数恐龙产蛋少（一些我们了解的种类产蛋会超过 40 枚）[20]。不过，这些蛋化石的出现再次证实了双排卵现象，如同在窃蛋龙[21]、伤齿龙[22]和其他兽脚类恐龙中发现的那样。因此，中华龙鸟很可能每次产一对卵，每次排卵之间需要延迟一段时间。

皮肤衍生物结构

两件中国化石标本最显著的特点之一就是保存了皮肤衍生物结构。在较大的标本中，这些结构沿着颈部和背部的背面以及尾巴的上下缘分布，而在较小的标本中皮肤衍生物的结构更清晰（图 5）。它们覆盖头骨背侧的后半部、颈部、背部、臀部和尾巴，同时沿着整个尾巴的腹侧扩展。头骨（方骨后和关节骨上）、右肱骨后侧和右尺骨前侧都能够看出小片的皮肤衍生物。而在 NIGP 127587 上，除了左肋外的一小片和尾巴左侧的几个区域（椎体侧面），身体侧面的其他位置并不能观察到皮肤衍生物。这种结构可能在现生动物中也存在，类似身体背部的浓密毛发和身体其他部位可见的少量皮肤衍生物。

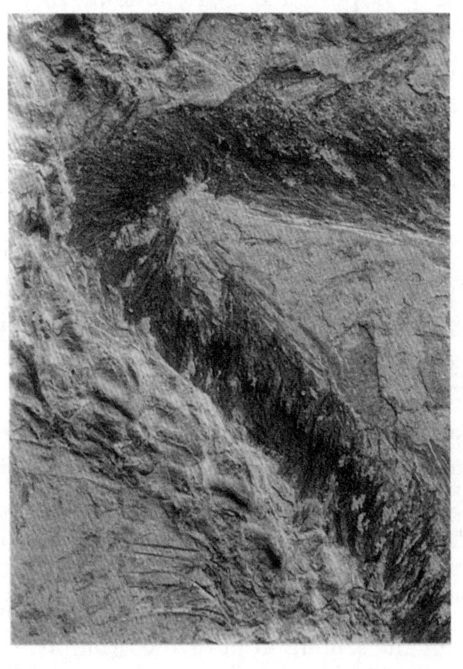

图 5. NIGP 127586 的颈部和背部的皮肤衍生物结构

Fig. 6. Integumentary structures over caudal 27 of NIGP 127587.

In the two theropods, the distances separating the integumentary structures from the underlying bones are directly proportional to the amount of skin and muscle that would have been present. As in modern animals, the integument closely adheres to the tops of the skull and hips, and becomes progressively closer to the caudal vertebrae towards the tip of the tail. In the posterior part of the neck, over the shoulders, and at the base of the tail, the integumentary structures are more distant from the underlying skeletal elements, and in life would have been separated by greater thicknesses of muscle and other soft tissues.

The orientation and frequently sinuous lines of the integumentary structures suggest they were soft and pliable, and semi-independent of each other. They frequently cross each other, and are tangled in some areas. There is an apparent tendency for the integumentary structures to clump along the tail of the smaller specimen, but this is an artefact of the splitting plane between NIGP 127586 (Fig. 2a) and GMV 2123. As both individuals were lying in the water of a lake when they were buried, it is clear that we are not looking at the normal orientation of the integumentary structures in the fossils. Under magnification, the margins of the larger structures are darker along the edges, but lighter medially, which indicates that they might have been hollow. Overall, the integumentary structures are rather coarse for such a small animal, and the thickest strands are much thicker than the hairs of the vast majority of small mammals[23]. In NIGP 127586, integumentary structures are first seen on the dorsal surface of the skull in front of the orbit. The skull is semidisarticulated, and sediment still covers the snout region, so it is possible that the integumentary structures extended more anteriorly. The most rostral integumentary structures are 5.5 mm long, and

图 6. NIGP 127587 的第 27 节尾椎上方的皮肤衍生物结构

在这两个兽脚类恐龙中，皮肤衍生物与下覆骨骼的间距与活着时存在的皮肤和肌肉量呈正比。像现生动物一样，皮肤衍生物紧密黏附在头骨和臀部顶部，并逐渐靠近尾巴末端的尾椎。在颈部后侧、肩部之上和尾巴的基部，皮肤衍生物距离下面的骨骼更远，可能是因为这些区域在动物活着时被非常厚的肌肉和其他软组织与骨骼分隔开。

皮肤衍生物的方向和错综复杂的排列表明它们柔软而易弯曲，相互间较为分离。它们彼此频繁交叉，在某些区域又比较紊乱。在较小的标本中，皮肤衍生物明显的趋向是沿着尾部丛生，但这是 NIGP 127586(图 2a) 和 GMV 2123 之间分割面上的假象。因为两个个体被掩埋时都是在湖水中死亡，因此很显然我们在化石中看到的并非皮肤衍生物的正常方向。当放大一定倍数后，大的结构边缘是暗色的，但是中间是亮的，这显示它们中间可能有空洞。总的来说，这种小型动物皮肤衍生物结构比较粗糙，毛发最浓密部分其浓密程度超过大多数小型哺乳动物的毛发[23]。在 NIGP 127586 上，皮肤衍生物结构首先是在眼眶之前的头骨背部表面看到。头骨是半脱离的，沉积物依然覆盖在吻部，因此皮肤衍生物结构可能会更向前扩展。最靠近吻端的皮肤衍生物有 5.5 毫米长，在头骨上方又延伸了约 4 毫米。它们在肩胛骨远端的上

extend about 4 mm above the skull. They quickly lengthen to at least 21 mm above the distal ends of the scapulae. This axial length seems to stay constant along most of the back, but decreases sharply to 16 mm dorsal to the ilium. The longest integumentary structures seem to have been above the base of the tail, although it is impossible to measure any single structure. More distally along the tail, integumentary structures decrease more rapidly on the lower side of the tail than on the upper. By caudal 47, the ventral structures are 4.2 mm long, about half the length of the dorsal structures in that region.

The size distribution of the integumentary structures of NIGP 127587 follow the same general pattern as in the smaller specimen. Although the integument tends to look thinner on this specimen, it is simply because the integumentary structures are lying closer to the body. Individual measurements are consistently larger than those of NIGP 127586. The integumentary structures are 13 mm long above the skull, 23.5 mm above the fourth cervical, at least 35 mm over the scapulae, at least 40 mm over caudal 27 (Fig. 6), and at least 35 mm below caudal 25. Integumentary structures on the left side of the body are largely covered by ribs, gastralia, stomach contents and matrix, so it is only possible to say that each is more than 5 mm long. Those associated with the right ulna are 14 mm long.

Integumentary structures have also been reported in the theropod *Pelecanimimus*[24] from the Lower Cretaceous of Spain. These consist of subparallel fibres arranged perpendicular to the bones, with a less conspicuous secondary system parallel to them. As described, they seem to be similar to the integumentary structures of *Sinosauropteryx*.

Skin impressions have been found on most main types of dinosaurs, including sauropods, ankylosaurs, ornithopods, stegosaurs, ceratopsians, and several genera of large theropods. In all of these animals, there is no evidence of integumentary structures, and the skin usually has a "pebbly" surface texture. Integumentary structures have been claimed for both specimens of *Compsognathus*[11,25] though the interpretations have been questioned in both cases[11]. In the German specimen, there was supposedly a patch of skin over the abdominal region. The French specimen included some strange markings in the region of the forearm, that were originally identified as a swimming appendage formed either of dermal bone or of thick skin[12], but it is clearly not well enough preserved to be positively identified. The identification of these structures as integumentary is questionable[11], and there is nothing on the Chinese specimens to support the presence of such structures in compsognathids. Evidence of feathers in *Compsognathus* was sought[11] without success, but this lack of evidence on the German specimen of *Compsognathus* does not eliminate the possibility that they might have existed.

Discussion

The integumentary structures of *Sinosauropteryx* are extremely interesting regardless of whether they are referred to as feathers, protofeathers, or some other structure. Unfortunately, they are piled so thick that we have been unable to isolate a single one for

方迅速延长至至少 21 毫米。这个长度似乎在背部大部分区域是恒定的，但是在髂骨背侧迅速递减到 16 毫米。最长的皮肤衍生物可能在尾巴基部的上方，尽管不可能去测量所有单个结构。顺着尾巴向远端延伸，皮肤衍生物结构在尾巴下方要比在上方递减更快。到第 47 节尾椎，腹侧的结构是 4.2 毫米长，约是背部相同区域长度的一半。

NIGP 127587 的皮肤衍生物的大小分布与小的标本遵循相同的普遍模式。这件标本中的皮肤衍生物趋向是逐渐变稀，这仅仅是因为皮肤衍生物更贴近身体。每一个部位皮肤衍生物的测量尺寸都要比标本 NIGP 127586 的大。头骨上的皮肤衍生物结构长 13 毫米，第四颈椎上的长 23.5 毫米，肩胛骨上的至少 35 毫米长，在第 27 节尾椎上的至少 40 毫米长（图 6），在第 25 尾椎下的至少 35 毫米长。身体左侧的皮肤衍生物基本被肋骨、腹膜肋、胃容物和围岩所覆盖，因此，只能推测皮肤衍生物可能长于 5 毫米。附着在右侧尺骨上的皮肤衍生物是 14 毫米长。

从西班牙下白垩统发现的兽脚类恐龙似鹈鹕龙标本 [24] 同样报道过皮肤衍生物。它们是由与骨骼垂直的近似平行排列的纤维组成的，具有不显著的次级结构与之平行排列。正如描述所说，它们似乎与中华龙鸟的皮肤衍生物结构相似。

皮肤印痕在很多主要类型的恐龙中曾被发现，包括蜥脚类、甲龙类、鸟脚类、剑龙类、角龙类和一些大型兽脚类恐龙属。但是在所有的这些恐龙中都没有皮肤衍生物存在的证据，它们的皮肤表面通常具有卵纹状的纹理。虽然关于两件标本的皮肤衍生物的解释都存在质疑 [11]，但皮肤衍生物确实在两个美颌龙标本中都存在 [11,25]。在德国的标本中，推测腹部区域上方有一块皮肤。在法国的标本中，前肢区域包括一些奇怪的痕迹，最初被鉴定为游泳的附肢，可能是膜质骨或者厚的皮肤 [12]，但由于保存得不够完好而无法确切地下结论。这些结构被鉴定为皮肤衍生物是存在问题的 [11]，中国的标本没有为美颌龙存在这样的结构提供支持依据。美颌龙存在羽毛的证据并没有寻找 [11] 成功，但是仅凭德国美颌龙标本缺少羽毛证据这一事实并不能排除美颌龙存在皮肤衍生物的可能性。

讨　论

不论中华龙鸟的皮肤衍生物是羽毛、原始羽毛还是一些其他的结构，它们都是特别引人瞩目的。不巧的是，它们堆积得太厚，使我们不能单独将其分离开来

examination. Comparison with birds from the same locality shows that the same problem exists with identifying individual feathers (other than the flight feathers) and components of feathers in avian specimens. The morphological characteristics that we describe suggest that the integumentary structures seem to resemble most closely the plumules of modern birds, having relatively short quills and long, filamentous barbs. The absence of barbules and hooklets is uncommon in modern birds, but has been noted in Cretaceous specimens[26].

It has been proposed that the feathers of another recently discovered animal from the same locality in Liaoning are structurally intermediate between the integumentary structures of *Sinosauropteryx* and the feathers of *Archaeopteryx*[27]. The clearly preserved feathers of *Protarchaeopteryx robusta* are symmetrical, which indicates that the animal was not capable of flight. This is confirmed by the relatively short length of the forelimb. Both *Sinosauropteryx* and *Protarchaeopteryx* had been identified as birds because of the presence of feathers[27], but much more work needs to be done to prove that the integumentary structures of *Sinosauropteryx* have any structural relationship to feathers, and phylogenetic analysis of the skeleton clearly places compsognathids far from the ancestry of birds. Despite arguments to the contrary[28], cladistic analysis favours the notion that the bird lineage originated within theropod dinosaurs[29,30]. If this phylogenetic framework is accepted, the integumentary structures of *Sinosauropteryx* could shed light on some of the many hypotheses concerning feather origins. Three main functions have been suggested for the initial development of feathers—display, aerodynamics and insulation.

The integumentary structures of *Sinosauropteryx* have no apparent aerodynamic characteristics, but might be representative of what covered the ancestral stock of birds. It is highly unlikely that something as complex as a bird feather could evolve in one step, and many animals glide and fly with much simpler structures. Even birds secondarily simplify feathers when airborne flight ceases to be their main method of locomotion, and produce structures that are intermediate between reptilian scales and feathers[31]. The multi-branched integumentary structures of the Chinese compsognathids are relatively simple, but are suitable for modification into the more complex structures required for flight.

Feathers may have appeared first as display structures[31], but the density, distribution, and relatively short lengths of the integumentary structures of *Sinosauropteryx* suggest that they were not used for display. It is conceivable that both specimens are female, and that the males had more elaborate integumentary structures for display. It is also possible that the integumentary structures were coloured to serve a display function. Therefore, the existing *Sinosauropteryx* specimens do not support the hypothesis that feathers evolved primarily for display, but do not disprove it either.

The dense, pliable integumentary structures of the Chinese compsognathids would not have been appropriate as heat shields to screen and shade the body from the Sun's rays[32]. Although they may have been effective in protecting the body from solar radiation in warm weather, they would also have been effective in preventing an ectothermic

进行检测。和同一产地的鸟类化石相比发现，在鸟类标本中鉴别单独的羽毛(除了翼羽)和羽毛的组成时也存在同样的问题。根据我们描述的形态学特征，该皮肤衍生物结构似乎最接近现代鸟类的绒羽，具有相对较短的羽轴和长的纤维状的羽枝。现代鸟类很少缺失羽小枝和羽小钩，但这种现象在白垩纪的鸟类[26]中是较为常见的。

最近在辽宁同一产地发现的另外一种动物的羽毛被认为是介于中华龙鸟的皮肤衍生物和始祖鸟羽毛之间的中间结构[27]。粗壮原始祖鸟清晰保存的羽毛是对称的，反映出这种动物是不具飞行能力的，这一点同时也根据它的前肢长度相对较短得以证明。中华龙鸟和原始祖鸟都曾经因为羽毛的存在被鉴定为鸟类[27]，但是还有很多工作需要去做来证明中华龙鸟的皮肤衍生物同鸟类羽毛之间存在结构上的联系，而且基于骨骼形态的系统发育分析结果显示美颌龙距离鸟类的祖先很远。尽管存在反对的论断[28]，但系统发育分析仍主张鸟类起源于兽脚类恐龙[29,30]。如果这个系统发生的结果能够被接受的话，那么中华龙鸟的皮肤衍生物结构能阐明关于羽毛起源的很多假说。羽毛的早期起源被认为主要与三个方面的功能有关：展示、空气动力学和保温。

中华龙鸟的皮肤衍生物结构没有明显的空气动力学特征，可能代表鸟类体表覆盖物的原始状态。像鸟类羽毛这种复杂的结构，其演化进程基本上是不可能一步发展到位的，很多动物是利用更为简单的结构进行滑翔和飞行。甚至当一些鸟类不再有飞行的习性时，它们的羽毛会发生次生简化，形成一种介于爬行动物的鳞甲和羽毛中间的结构[31]。中国美颌龙类多分枝的皮肤衍生物结构是相对简单的，但是适合修饰成更复杂的鸟类飞行结构。

羽毛最初出现时可能是作为展示的结构[31]，但中华龙鸟皮肤衍生物的密度、分布和相对短的长度显示它们不是用来展示的。可以想象这两件标本都属于雌性个体，雄性应该具有更为精细的皮肤衍生物用来展示。皮肤衍生物也可能是色彩斑斓的，以适合展示。因此，现存的中华龙鸟标本不支持羽毛原始演化是为了展示的假说，但也不能反驳这种假说。

中国美颌龙类浓密且柔软的皮肤衍生物不适合作为保暖层以及遮蔽太阳辐射[32]。但这些皮肤衍生物在温暖的天气保护身体避免太阳辐射可能是有效的，同时也可以有效预防外温的兽脚类恐龙在太阳直射中吸收外部热量而使体温迅速升高。

theropod from rapidly warming up by basking in the sunshine. If small theropods were endothermic, they would have needed insulation to maintain high body temperatures[33-35]. The presence of dense integumentary structures may suggest that *Sinosauropteryx* was endothermic, and that heat retention was the primary function for the evolution of integumentary structures[36-38]. Recently published histological studies suggest at least some early birds were not truly endothermic[39], although they may have been physiologically intermediate between poikilothermic ectotherms and homeothermic endotherms[40].

The Chinese compsognathids have integumentary structures consisting of vertical fibres running from the base of the head along the back and around the tail extending forwards almost to the legs. There are no structures showing the fundamental morphological features of modern bird feathers, but they could be previously unidentified protofeathers which are not as complex as either down feathers or even the hair-like feathers of secondarily flightless birds. Their simplicity would not have made them ineffective for insulation when wet any more than it negates the insulatory capabilities of mammalian hair. We cannot determine whether or not the integumentary structures were arranged in pterylae, but they were long enough to cover apteria, if they existed, and could therefore still have been effective in thermoregulation. Continuous distribution is not essential to be effective in this function[28], especially if the apteria are part of a mechanism for dispersing excess heat. Finally, the aerodynamic capabilities of bird feathers are not comprised by the previous evolution of less complex protofeathers that had some other function, such as insulation.

In addition to the integumentary structures, there is dark pigmentation over the eyes of both specimens. A second region of dark pigmentation in the abdominal region of the smaller specimen might represent some soft tissues of viscera.

Multidisciplinary and multinational research is just beginning on these unique small theropods. Techniques developed to study fossil feathers[38,41] will be useful research tools as work progresses. In the meantime, the integumentary structures of *Sinosauropteryx* suggest that feathers evolved from simpler, branched structures that evolved in non-avian theropod dinosaurs, possibly for insulation.

(**391**, 147-152; 1998)

Pei-ji Chen*, **Zhi-ming Dong**† **& Shuo-nan Zhen**‡

* Nanjing Institute of Geology and Palaeontology, Academia Sinica, 39 East Beijing Road, Nanjing 210008, People's Republic of China

† Institute of Vertebrate Paleontology and Paleoanthropology, Academia Sinica, PO Box 643, Beijing 100044, People's Republic of China

‡ Beijing Natural History Museum, 126 Tien Qiao Street, Beijing 100050, People's Republic of China

Received 14 January; accepted 18 September 1997.

References:

1. Chen, P. J. Distribution and migration of Jehol fauna with reference to nonmarine Jurassic–Cretaceous boundary in China. *Acta Palaeontol. Sin.* **27**, 659-683 (1988).

如果小型兽脚类恐龙是内温的，它们需要隔绝外界来维持较高的体温[33-35]。稠密的皮肤衍生物表明中华龙鸟可能是内温动物，而保暖性是皮肤衍生物结构演化的最初功能[36-38]。最近发表的组织学研究提出，至少一些早期的鸟类不是真正的内温动物[39]，不过它们可能在生理学上介于变温外温动物与恒温内温动物之间[40]。

中国美颌龙类具有的皮肤衍生物结构由垂直的纤维组成，从头部基部开始沿着背部延伸，包围尾部，同时沿尾部向前延伸，几乎到达腿部。它们没有现代鸟类羽毛的基本形态结构，但是可以被看作是一种未被确认的原始羽毛，其复杂程度低于绒羽甚至次级失去飞行能力的鸟类毛发状的羽毛。它们简单的构造在潮湿情况下能有效隔热，比哺乳动物毛发潮湿情况下的保温能力要强。我们不能确定皮肤衍生物结构是否按羽区排列，但如果它们存在，它们长到足够覆盖裸区，就能有效地进行温度调节。皮肤衍生物连续分布对于这个功能的有效性并不是必不可少的[28]，尤其是如果裸区是用来分散多余热量的机制的一部分。最后，鸟类羽毛的空气动力学功能并不为这些处于演化早期的、结构不很复杂的原始羽毛所拥有，它们有另外的功能，如保暖。

除皮肤衍生物结构外，在两件标本眼部的上方都存在深色色素沉着。此外，在较小标本的腹部也保存有深色色素沉着，可能意味着存在一些内脏的软组织。

对于这些独特的小型兽脚类恐龙进行的多学科和多国合作研究只是刚刚开始。羽毛化石研究的技术发展[38,41]为加快研究进展提供了有益的研究工具。同时，中华龙鸟的皮肤衍生物结构揭示了羽毛的演化是从简单、多分枝结构开始的，这种结构在非鸟兽脚类恐龙中出现可能是用于保温的。

（张玉光 翻译；汪筱林 审稿）

2. Hou, L.-H., Zhang, J.-Y., Martin, L. D. & Feduccia, A. A beaked bird from the Jurassic of China. *Nature* **377,** 616-618 (1995).

3. Hou, L.-H., Martin, L. D., Zhang, J.-Y. & Feduccia, A. Early adaptive radiation of birds: evidence from fossils from northeastern China. *Science* **274,** 1164-1167 (1996).

4. Ji, Q. & Ji, S. A. On discovery of the earliest bird fossil in China and the origin of birds. *Chinese Geol.* **233,** 30-33 (1996).

5. Chen, P. J. *et al.* Studies on the Late Mesozoic continental formations of western Liaoning. *Bull. Nanjing Inst. Geol. Palaeontol.* **1,** 22-25 (1980).

6. Chen, P. J. Nonmarine Jurassic strata of China. *Bull. Mus. N. Arizona* **60,** 395-412 (1996).

7. Dong, Z. M. Early Cretaceous dinosaur faunas in China: an introduction. *Can. J. Earth Sci.* **30,** 2096-2100 (1993).

8. Li, W. B. & Liu, Z. S. The Cretaceous palynofloras and their bearing on stratigraphic correlation in China. *Cretaceous Res.* **15,** 333-365 (1994).

9. Smith, P. E. *et al.* Dates and rates in ancient lakes: ^{40}Ar-^{39}Ar evidence for an Early Cretaceous age for the Jehol Group, northeast China. *Can. J. Earth Sci.* **32,** 1426-1431 (1995).

10. Wang, D. F. & Diao, N. C. Geochronology of Jura-Cretaceous volcanics in west Liaoning, China. *Scientific papers on geology for international exchange* **5,** 1-12 (Geological Publishing House, Beijing, 1984).

11. Ostrom, J. H. The osteology of *Compsognathus longipes* Wagner. *Zitteliana* **4,** 73-118 (1978).

12. Bidar, A., Demay, L. & Thomel, G. *Compsognathus corallestris* nouvelle espèce de dinosaurien théropode du Portlandiend de Canjuers (sud-est de la France). *Ann. Mus. d'Hist. Nat. Nice* **1,** 3-34 (1972).

13. Currie, P. J. & Zhao, X. J. A new large theropod (Dinosauria, Theropoda) from the Jurassic of Xinjiang, People's Republic of China. *Can. J. Earth Sci.* **30,** 2037-2081 (1993).

14. Claesseus, L. Dinosaur gastralia and their function in respiration. *J. Vert. Palaeontol.* **16,** 28A (1996).

15. von Huene, F. The carnivorous Saurischia in the Jura and Cretaceous formations principally in Europe. *Revista Museo Plata* **29,** 35-167 (1926).

16. Perle, A., Chiappe, L. M., Barsbold, R., Clark, J. M. & Norell, M. Skeletal morphology of *Mononykus olecranus* (Theropoda: Avialae) from the Late Cretaceous of Mongolia. *Am. Mus. Novit.* **3105,** 1-29 (1994).

17. Wellnhofer, P. Das siebte Exemplar von *Archaeopteryx* aus den Solnhofener Schichten. *Archaeopteryx* **11,** 1-48 (1993).

18. Barsbold, R. Carnivorous dinosaurs from the Cretaceous of Mongolia. *Sovmestnaya Sovetsko-Mongol'skaya Paleontol. Ekspiditsiya, Trudy* **19,** 5-119 (1983).

19. Griffiths, P. The question of *Compsognathus* eggs. *Rev. Paleobiol.* Spec. issue **7,** 85-94 (1993).

20. Carpenter, K., Hirsch, K. F. & Horner, J. R. *Dinosaur Eggs and Babies* (Cambridge Univ. Press, 1994).

21. Dong, Z. M. & Currie, P. J. On the discovery of an oviraptorid skeleton on a nest of eggs at Bayan Mandahu, Inner Mongolia, People's Republic of China. *Can. J. Earth Sci.* **33,** 631-636 (1996).

22. Varricchio, D. J., Jackson, F., Borkowski, J. J. & Horner, J. R. Nest and egg clutches of the dinosaur *Troodon formosus* and the evolution of avian reproductive traits. *Nature* **385,** 247-250 (1997).

23. Meng, J. & Wyss, A. R. Multituberculate and other mammal hair recovered from Palaeogene excreta. *Nature* **385,** 712-714 (1997).

24. Pérez-Moreno, B. P. *et al.* A unique multitoothed ornithomimosaur dinosaur from the Lower Cretaceous of Spain. *Nature* **370,** 363-367 (1994).

25. von Huene, F. Der Vermuthliche Hautpanzer des *Compsognathus longipes* Wagner. *Neues Jb. F. Min.* **1,** 157-160 (1901).

26. Grimaldi, D. & Case, G. R. A feather in amber from the Upper Cretaceous of New Jersey. *Am. Mus. Novit.* **3126,** 1-6 (1995).

27. Ji, Q. & Ji, S. A. *Protarchaeopteryx*, a new genus of Archaeopterygidae in China. *Chinese Geol.* **238,** 38-41 (1997).

28. Feduccia, A. *The Origin and Evolution of Birds* (Yale Univ. Press, New Haven, 1996).

29. Gauthier, J. in *The Origin of Birds and the Evolution of Flight* (ed. Padian, K.) 1-55 (California Acad. Sci., San Francisco, 1986).

30. Fastovsky, D. E. & Weishampel, D. B. *The Evolution and Extinction of the Dinosaurs* (Cambridge Univ. Press, 1996).

31. McGowan, C. Feather structure in flightless birds and its bearing on the question of the origin of feathers. *J. Zool. (Lond.)* **218,** 537-547 (1989).

32. Paul, G. S. *Predatory Dinosaurs of the World* (Simon and Schuster, New York, 1988).

33. Ewart, J. C. The nestling feathers of the mallard, with observations on the composition, origin, and history of feathers. *Proc. Zool. Soc. Lond.* 609-642 (1921).

34. Van Tyne, J. & Berger, A. J. *Fundamentals of Ornithology* (Wiley, New York, 1976).

35. Young, J. Z. *The Life of Vertebrates* (Oxford Univ. Press, **1950**).

36. Chinsamy, A., Chiappe, L. M. & Dodson, P. Growth rings in Mesozoic birds. *Nature* **368,** 196-197 (1994).

37. Chiappe, L. M. The first 85 million years of avian evolution. *Nature* **378,** 349-355 (1995).

38. Brush, A. H. in *Avian Biology* vol. 9 (eds Farner, D. S., King, J. R. & Parkes, K. C.) 121-162 (Academic, London, 1993).

39. Regal, P. J. The evolutionary origin of feathers. *Quart. Rev. Biol.* **50,** 35-66 (1975).

40. Ostrom, J. H. Reply to "Dinosaurs as reptiles". *Evolution* **28,** 491-493 (1974).

41. Davis, P. G. & Briggs, D. E. G. Fossilization of feathers. *Geology* **23,** 783-786 (1995).

Acknowledgements. This study was supported by NSFC. We thank L.-s. Chen and P. J. Currie (Royal Tyrrell Museum of Palaeontology) for helping to prepare the fossil materials and manuscript; M.-m. Zhang, X.-n. Mu, G. Sun, J. H. Ostram, A. Brush, L. Martin, P. Wellnhofer, N. J. Mateer, E. B. Koppelhus, D. B. Brinkman, D. A. Eberth, J. A. Ruben, L. Chiappe, S. Czerkas, R. O'Brien, D. Rimlinger, M. Vickaryous and D. Unwin for assistance and comments; and L. Mazzatenta and M. Skrepnick for help producing the photographs and drawings.

Correspondence and requests for materials should be addressed to P-j.C. (e-mail: lpsnigp@nanjing.jspta.chinamail.sprint.com).

Potent and Specific Genetic Interference by Double-stranded RNA in *Caenorhabditis elegans*

A. Fire *et al.*

Editor's Note

Here American biologists Andrew Fire and Craig Mello describe a fundamental form of gene regulation called RNA interference, where snippets of double-stranded RNA instruct the cell to destroy genetically identical messenger RNA molecules, and so effectively silence the corresponding gene. The process occurs naturally in plants and animals, where it helps regulate gene expression and is a vital part of the immune response to viruses and other foreign genetic material. It is widely used in the laboratory to study gene function, and holds promise as a therapeutic tool for thwarting viruses and controlling the expression of aberrant, disease-causing genes. This demonstration of RNA interference in the nematode worm earned Fire and Mello a Nobel Prize eight years later.

Experimental introduction of RNA into cells can be used in certain biological systems to interfere with the function of an endogenous gene[1,2]. Such effects have been proposed to result from a simple antisense mechanism that depends on hybridization between the injected RNA and endogenous messenger RNA transcripts. RNA interference has been used in the nematode *Caenorhabditis elegans* to manipulate gene expression[3,4]. Here we investigate the requirements for structure and delivery of the interfering RNA. To our surprise, we found that double-stranded RNA was substantially more effective at producing interference than was either strand individually. After injection into adult animals, purified single strands had at most a modest effect, whereas double-stranded mixtures caused potent and specific interference. The effects of this interference were evident in both the injected animals and their progeny. Only a few molecules of injected double-stranded RNA were required per affected cell, arguing against stochiometric interference with endogenous mRNA and suggesting that there could be a catalytic or amplification component in the interference process.

DESPITE the usefulness of RNA interference in *C. elegans*, two features of the process have been difficult to explain. First, sense and antisense RNA preparations are each sufficient to cause interference[3,4]. Second, interference effects can persist well into the next generation, even though many endogenous RNA transcripts are rapidly degraded in the early embryo[5]. These results indicate a fundamental difference in behaviour between native RNAs (for example, mRNAs) and the molecules responsible for interference. We sought to test the possibility that this contrast reflects an underlying difference in RNA

秀丽隐杆线虫体内双链 RNA 强烈而特异性的基因干扰作用

法厄等

编者按

在本文中，美国生物学家安德鲁·法厄和克雷格·梅洛描述了基因调控的一种基本形式——称为 RNA 干扰，即小片段的双链 RNA 导致细胞内基因（序列）相同的信使 RNA（mRNA）分子被破坏，从而有效地沉默相应靶基因的表达。这个过程在植物和动物体内自然发生，它有助于调控基因表达，并且也是机体对病毒和其他外来遗传物质进行免疫应答的重要组成部分。RNA 干扰在实验室中被广泛应用于基因功能的研究，并且有望成为抗病毒和控制异常基因与致病基因表达的一种治疗工具。本文中对线虫体内 RNA 干扰机制的阐述令法厄和梅洛于 8 年后获得诺贝尔奖。

在特定生物系统中，用实验方法将 RNA 导入细胞可以干扰内源基因的功能[1,2]。研究人员认为这种干扰效应源于一种简单的反义机制，该机制依赖于导入的 RNA 与内源信使 RNA 转录本之间的杂交。RNA 干扰已经被用于操控秀丽隐杆线虫基因的表达[3,4]。在此我们研究了干扰 RNA 结构和释放的必要条件。我们惊奇地发现，双链 RNA 产生的干扰作用比其中的任意一条单链都要强。将纯化的单链导入成年动物体内后最多只产生中等强度的干扰作用，而双链混合物的导入则能产生强烈而特异性的干扰。这种干扰作用在被注射的动物及其后代中都很明显。与内源 mRNA 产生干扰的剂量不同，仅需要几个双链 RNA 分子就可以对细胞产生干扰，这表明，在该干扰过程中可能存在具有催化或放大作用的组分。

虽然 RNA 干扰在秀丽隐杆线虫中非常有效，但是该过程的两个特征却难以得到解释。首先，单独制备的正义 RNA 链或反义 RNA 链都足以产生干扰作用[3,4]。其次，尽管许多内源 RNA 转录本都在胚胎早期迅速地降解了，但是干扰作用却能够持续到下一代[5]。这些结果表明产生干扰效应的 RNA 分子和天然分子（如 mRNA）在性质上具有本质差别。我们认为可能是 RNA 的根本结构差异导致了它们具有不同的

structure. RNA populations to be injected are generally prepared using bacteriophage RNA polymerases[6]. These polymerases, although highly specific, produce some random or ectopic transcripts. DNA transgene arrays also produce a fraction of aberrant RNA products[3]. From these facts, we surmised that the interfering RNA populations might include some molecules with double-stranded character. To test whether double-stranded character might contribute to interference, we further purified single-stranded RNAs and compared interference activities of individual strands with the activity of a deliberately prepared double-stranded hybrid.

The *unc-22* gene was chosen for initial comparisons of activity. *unc-22* encodes an abundant but nonessential myofilament protein[7-9]. Several thousand copies of *unc-22* mRNA are present in each striated muscle cell[3]. Semiquantitative correlations between *unc-22* activity and phenotype of the organism have been described[8]: decreases in *unc-22* activity produce an increasingly severe twitching phenotype, whereas complete loss of function results in the additional appearance of muscle structural defects and impaired motility.

Purified antisense and sense RNAs covering a 742-nucleotide segment of *unc-22* had only marginal interference activity, requiring a very high dose of injected RNA to produce any observable effect (Table 1). In contrast, a sense-antisense mixture produced highly effective interference with endogenous gene activity. The mixture was at least two orders of magnitude more effective than either single strand alone in producing genetic interference. The lowest dose of the sense-antisense mixture that was tested, ~60,000 molecules of each strand per adult, led to twitching phenotypes in an average of 100 progeny. Expression of *unc-22* begins in embryos containing ~500 cells. At this point, the original injected material would be diluted to at most a few molecules per cell.

Table 1. Effects of sense, antisense and mixed RNAs on progeny of injected animals

Gene	segment	Size (kilobases)	Injected RNA	F1 phenotype
unc-22				*unc-22*-null mutants: strong twitchers[7,8]
*unc22A**	Exon 21–22	742	Sense	Wild type
			Antisense	Wild type
			Sense + antisense	Strong twitchers (100%)
unc22B	Exon 27	1,033	Sense	Wild type
			Antisense	Wild type
			Sense + antisense	Strong twitchers (100%)
unc22C	Exon 21–22†	785	Sense + antisense	Strong twitchers (100%)
fem-1				*fem-1*-null mutants: female (no sperm)[13]
fem1A	Exon 10‡	531	Sense	Hermaphrodite (98%)
			Antisense	Hermaphrodite (> 98%)
			Sense + antisense	Female (72%)
fem1B	Intron 8	556	Sense + antisense	Hermaphrodite (> 98%)
unc-54				*unc-54*-null mutants: paralysed[7,11]
unc54A	Exon 6	576	Sense	Wild type (100%)
			Antisense	Wild type (100%)
			Sense + antisense	Paralysed (100%)§

功能，并希望证实这个观点。通常情况下，用于导入的 RNA 分子都是利用噬菌体的 RNA 聚合酶来获得的[6]。这些聚合酶尽管具有很高的特异性，但也会产生随机突变或者异位突变。DNA 转基因阵列同样也会产生一定比例的异常 RNA 分子[3]。基于这些事实，我们总结出了干扰 RNA 分子中可能包含一些有双链特征的分子。为了验证双链分子是否可以导致干扰效应，我们进一步纯化了单链 RNA 分子，并将各个单链的干扰活性与特意准备的杂交双链分子进行了比较。

我们选择 *unc-22* 基因来进行初步的活性比较。*unc-22* 编码一种大量却非必需的肌丝蛋白[7-9]。每个横纹肌细胞中都有数千拷贝的 *unc-22* mRNA[3]。*unc-22* 的活性与机体表型的半定量关系已有报道[8]：*unc-22* 活性降低会加剧肌肉抽搐这一表型，而 *unc-22* 完全丧失功能则进一步导致肌肉结构缺陷且运动能力受损。

包含 *unc-22* 上 742 个核苷酸的正义链和反义链 RNA 纯化片段只有微小的干扰活性，需要注射很高剂量的 RNA 才能产生可见效应（表 1）。相反，正义链–反义链混合物对内源基因活性产生高效的干扰作用。混合物产生的遗传性基因干扰比任何一条单链所产生的至少高两个数量级。实验证明，正义链–反义链混合物的最低剂量（即每个成年动物中每条单链为 60,000 个分子左右）平均可在 100 个后代中导致抽搐表型。*unc-22* 的表达始于包含约 500 个细胞的胚胎。此时，初始的注射物被稀释为每个细胞最多只含有几个分子。

表 1. 动物注射 RNA 分子正义链、反义链及其混合物后对后代的影响

基因	片段	大小（kb）	注射的 RNA 链	子一代表型
unc-22				*unc-22* 无效突变体：剧烈抽搐[7,8]
*unc22A**	外显子 21~22	742	正义链	野生型
			反义链	野生型
			正义链 + 反义链	剧烈抽搐（100%）
unc22B	外显子 27	1,033	正义链	野生型
			反义链	野生型
			正义链 + 反义链	剧烈抽搐（100%）
unc22C	外显子 21~22†	785	正义链 + 反义链	剧烈抽搐（100%）
fem-1				*fem-1* 无效突变体：雌体（无精子）[13]
fem1A	外显子 10‡	531	正义链	雌雄同体（98%）
			反义链	雌雄同体（>98%）
			正义链 + 反义链	雌体（72%）
fem1B	内含子 8	556	正义链 + 反义链	雌雄同体（>98%）
unc-54				*unc-54* 无效突变体：瘫痪[7,11]
unc54A	外显子 6	576	正义链	野生型（100%）
			反义链	野生型（100%）
			正义链 + 反义链	瘫痪（100%）§

Continued

Gene	segment	Size (kilobases)	Injected RNA	F1 phenotype
unc54B	Exon 6	651	Sense	Wild type (100%)
			Antisense	Wild type (100%)
			Sense+antisense	Paralysed (100%)§
unc54C	Exon 1–5	1,015	Sense+antisense	Arrested embryos and larvae (100%)
unc54D	Promoter	567	Sense+antisense	Wild type (100%)
unc54E	Intron 1	369	Sense+antisense	Wild type (100%)
unc54F	Intron 3	386	Sense+antisense	Wild type (100%)
hlh-1				*hlh-1*-null mutants: lumpy-dumpy larvae[16]
hlh1A	Exons1–6	1,033	Sense	Wild type (< 2% lpy-dpy)
			Antisense	Wild type (< 2% lpy-dpy)
			Sense+ antisense	Lpy-dpy larvae (> 90%)‖
hlh1B	Exons1–2	438	Sense+ antisense	Lpy-dpy larvae (> 80%)‖
hlh1C	Exons4–6	299	Sense+ antisense	Lpy-dpy larvae (> 80%)‖
hlh1D	Intron 1	697	Sense+ antisense	Wild type (< 2% lpy-dpy)
myo-3-driven GFP transgenes¶				Makes nuclear GFP in body muscle
myo-3::NLS:: gfp:: lacZ		730	Sense	Nuclear GFP–LacZ pattern of parent strain
gfpG	Exons 2–5		Antisense	Nuclear GFP–LacZ pattern of parent strain
			Sense+antisense	Nuclear GFP–LacZ absent in 98% of cells
lacZL	Exon 12–14	830	Sense+antisense	Nuclear GFP–LacZ absent in > 95% of cells
myo-3::MtLS:: gfp				Makes mitochondrial GFP in body muscle
gfpG	Exons 2–5	730	Sense	Mitochondrial-GFP pattern of parent strain
			Antisense	Mitochondrial-GFP pattern of parent strain
			Sense+antisense	Mitochondrial-GFP absent in 98% of cells
LacZL	Exon 12–14	830	Sense+antisense	Mitochondrial-GFP pattern of parent strain

Each RNA was injected into 6–10 adult hermaphrodites (0.5×10^6–1×10^6 molecules into each gonad arm). After 4–6 h (to clear prefertilized eggs from the uterus), injected animals were transferred and eggs collected for 20–22 h. Progeny phenotypes were scored upon hatching and subsequently at 12–24-h intervals.

* to obtain a semiquantitative assessment of the relationship between RNA dose and phenotypic response, we injected each *unc22A* RNA preparation at a series of different concentrations (see figure in Supplementary information for details). At the highest dose tested (3.6×10^6 molecules per gonad), the individual sense and antisense *unc22A* preparations produced some visible twitching (1% and 11% of progeny, respectively). Comparable doses of double-stranded *unc22A* RNA produced visible twitching in all progeny, whereas a 120-fold lower dose of double-stranded *unc22A* RNA produced visible twitching in 30% of progeny.

† *unc22C* also carries the 43-nucleotide intron between exons 21 and 22.

‡ *fem1A* carries a portion (131 nucleotides) of intron 10.

§ Animals in the first affected broods (layed 4–24 h after injection) showed movement defects indistinguishable from those of *unc-54*-null mutants. A variable fraction of these animals (25%–75%) failed to lay eggs (another phenotype of *unc-54*-null mutants), whereas the remainder of the paralysed animals did lay eggs. This may indicate incomplete interference with *unc-54* activity in vulval muscles. Animals from later broods frequently show a distinct partial loss-of-function phenotype, with contractility in a subset of body-wall muscles.

‖ Phenotypes produced by RNA-mediated interference with *hlh-1* included arrested embryos and partially elongated L1 larvae (the *hlh-1*-null phenotype). These phenotypes were seen in virtually all progeny after injection of double-stranded *hlh1A* and in about half of the affected animals produced after injection of double-stranded *hlh1B* and double-stranded *hlh1C*. A set of less severe defects was seen in the remainder of the animals produced after injection of double-stranded *hlh1B* and double-stranded *hlh1C*. The less severe phenotypes are characteristic of partial loss of function of *hlh-1* (B. Harfe and A.F., unpublished observations).

¶ the host for these injections, strain PD4251, expresses both mitochondrial GFP and nuclear GFP-LacZ (see Methods). This allows simultaneous assay for interference with *gfp* (seen as loss of all fluorescence) and with *lacZ* (loss of nuclear fluorescence). The table describes scoring of animals as L1 larvae. Double-stranded *gfpG* caused a loss of GFP in all but 0–3 of the 85 body muscles in these larvae. As these animals mature to adults, GFP activity was seen in 0–5 additional body-wall muscles and in the 8 vulval muscles. Lpy-dpy, lumpy-dumpy.

续表

基因	片段	大小(kb)	注射的 RNA 链	子一代表型
unc-54B	外显子 6	651	正义链	野生型(100%)
			反义链	野生型(100%)
			正义链 + 反义链	瘫痪(100%)§
unc54C	外显子 1~5	1,015	正义链 + 反义链	幼虫及胚胎发育停滞(100%)
unc54D	启动子	567	正义链 + 反义链	野生型(100%)
unc54E	内含子 1	369	正义链 + 反义链	野生型(100%)
unc54F	内含子 3	386	正义链 + 反义链	野生型(100%)
hlh-1				*hlh-1* 无效突变体：矮胖粗笨幼虫[16]
hlh1A	外显子 1~6	1,033	正义链	野生型(<2% lpy-dpy)
			反义链	野生型(<2% lpy-dpy)
			正义链 + 反义链	Lpy-dpy 幼虫(>90%)‖
hlh1B	外显子 1~2	438	正义链 + 反义链	Lpy-dpy 幼虫(>80%)‖
hlh1C	外显子 4~6	299	正义链 + 反义链	Lpy-dpy 幼虫(>80%)‖
hlh1D	内含子 1	697	正义链 + 反义链	野生型(<2% lpy-dpy)
myo-3 驱动 *GFP* 转移基因 ¶				
myo-3::NLS:: gfp::lacZ				肌肉产生核 GFP
gfpG	外显子 2~5	730	正义链	亲代核 GFP-LacZ 模式
			反义链	亲代核 GFP-LacZ 模式
			正义链 + 反义链	98% 细胞缺失核 GFP-LacZ
lacZL	外显子 12~14	830	正义链 + 反义链	95% 以上细胞缺失核 GFP-LacZ
myo-3::MtLS:: gfp				肌肉产生线粒体 GFP
gfpG	外显子 2~5	730	正义链	亲代线粒体 GFP 模式
			反义链	亲代线粒体 GFP 模式
			正义链 + 反义链	98% 细胞缺失线粒体 GFP
lacZL	外显子 12~14	830	正义链 + 反义链	亲代线粒体 GFP 模式

每种 RNA 分子被注入 6~10 个雌雄同体的成虫中(每个生殖腺臂注入 0.5×10^6~1×10^6 个分子)。4~6 小时后(这段时间用于清除子宫中受精前的卵子)，移出被注射动物并于 20~22 小时收集其卵子。孵育之后即记录后代的表型，且每隔 12~24 小时记录一次。

* 为了测定 RNA 剂量和表型反应之间的半定量关系，我们以一系列的浓度梯度注射 *unc22A* RNA(细节详见补充信息图)。在测试的最高剂量(每个生殖腺 3.6×10^6 个分子)时，单独制备的正义链和反义链 *unc22A* 产生一些可见的抽搐(分别占后代的 1% 和 11%)。相等剂量的双链 *unc22A* RNA 在所有后代中都产生了可见的抽搐，而 1/120 剂量的双链 *unc22A* RNA 在 30% 的后代中产生可见抽搐。

† *unc22C* 同时携带了外显子 21 和 22 之间的含 43 个核苷酸的内含子。

‡ *fem1A* 携带内含子 10 的一部分(131 个核苷酸)。

§ 首先受影响的一批动物(注射后 4~24 小时产卵)表现出的行动缺陷与 *unc-54* 无效突变体难以区分。这些动物中的一部分(25%~75%)不能产卵(*unc-54* 无效突变体的另一个表型)，而剩余的瘫痪动物确实产下了卵。这可能说明在外阴肌肉中对 *unc-54* 活性的干扰不完全。更晚的一批动物表现出明显的部分功能缺失表型，部分体壁肌肉有收缩性。

‖ 通过 RNA 介导的对 *hlh-1* 的干扰，产生的表型包括胚胎发育停滞和 L1 幼虫局部不能延长(*hlh-1* 无效表型)。这些表型在注射双链 *hlh1A* 后的所有后代中及注射双链 *hlh1B* 和双链 *hlh1C* 后受影响的一半动物中都可以见到。其余注射双链 *hlh1B* 和双链 *hlh1C* 的动物中看到不太严重的缺陷。不太严重的缺陷表型源自部分缺失 *hlh-1* 的功能(哈弗和法厄，未发表的观察结果)。

¶ 这些注射的宿主，品系为 PD4251，既表达线粒体 GFP 也表达核 GFP-LacZ(见方法部分)。这允许同时测定对 *gfp*(可见所有荧光丧失)和 *lacZ*(核荧光丧失)的干扰作用。这个表描述了 L1 幼虫的情况。双链 *gfpG* 引起这些幼虫中除 0~3 个以外的 85 个全身肌群中 GFP 缺失。在这些动物长到成虫时，可以在 0~5 条其他的体壁肌肉和 8 条外阴肌肉中观测到 GFP 活性。Lpy-dpy，粗笨–矮胖。

The potent interfering activity of the sense-antisense mixture could reflect the formation of double-stranded RNA (dsRNA) or, conceivably, some other synergy between the strands. Electrophoretic analysis indicated that the injected material was predominantly double-stranded. The dsRNA was gel-purified from the annealed mixture and found to retain potent interfering activity. Although annealing before injection was compatible with interference, it was not necessary. Mixing of sense and antisense RNAs in low-salt concentrations (under conditions of minimal dsRNA formation) or rapid sequential injection of sense and antisense strands were sufficient to allow complete interference. A long interval (> 1 h) between sequential injections of sense and antisense RNA resulted in a dramatic decrease in interfering activity. This suggests that injected single strands may be degraded or otherwise rendered inaccessible in the absence of the opposite strand.

A question of specificity arises when considering known cellular responses to dsRNA. Some organisms have a dsRNA-dependent protein kinase that activates a panic-response mechanism[10]. Conceivably, our sense-antisense synergy might have reflected a nonspecific potentiation of antisense effects by such a panic mechanism. This is not the case: co-injection of dsRNA segments unrelated to *unc-22* did not potentiate the ability of single *unc-22*-RNA strands to mediate inhibition (data not shown). We also investigated whether double-stranded structure could potentiate interference activity when placed in *cis* to a single-stranded segment. No such potentiation was seen: unrelated double-stranded sequences located 5′ or 3′ of a single-stranded *unc-22* segment did not stimulate interference. Thus, we have only observed potentiation of interference when dsRNA sequences exist within the region of homology with the target gene.

The phenotype produced by interference using *unc-22* dsRNA was extremely specific. Progeny of injected animals exhibited behaviour that precisely mimics loss-of-function mutations in *unc-22*. We assessed target specificity of dsRNA effects using three additional genes with well characterized phenotypes (Fig. 1, Table 1). *unc-54* encodes a body-wall-muscle heavy-chain isoform of myosin that is required for full muscle contraction[7,11,12]; *fem-1* encodes an ankyrin-repeat-containing protein that is required in hermaphrodites for sperm production[13,14]; and *hlh-1* encodes a *C. elegans* homologue of myoD-family proteins that is required for proper body shape and motility[15,16]. For each of these genes, injection of related dsRNA produced progeny broods exhibiting the known null-mutant phenotype, whereas the purified single RNA strands produced no significant interference. With one exception, all of the phenotypic consequences of dsRNA injection were those expected from interference with the corresponding gene. The exception (segment *unc54C* which led to an embryonic- and larval-arrest phenotype not seen with *unc-54-null* mutants) was illustrative. This segment covers the highly conserved myosin-motor domain, and might have been expected to interfere with activity of other highly related myosin heavy-chain genes[17]. The *unc54C* segment has been unique in our overall experience to date: effects of 18 other dsRNA segments (Table 1; and our unpublished observations) have all been limited to those expected from previously characterized null mutants.

正义链–反义链混合物的强干扰活性可以反映双链 RNA(dsRNA)的形成，或者可以想象为，反映了链之间的一些协同作用。电泳分析显示注射的物质主要是双链。dsRNA 是从退火的混合物中凝胶纯化得到的，并且保留了强干扰活性。虽然在注射前退火与干扰不矛盾，但并不是必需的。在低盐浓度下混合正义链和反义链的 RNA(此条件下形成的 dsRNA 最少)或快速连续注射正义和反义链都可以产生完全干扰。连续注射正义链和反义链 RNA 之间的长时间间隔(>1 小时)会导致干扰活性显著下降。这表明已注射的单链可能被降解或因缺少互补链而难以实施干扰效应。

当考虑到已知的细胞对 dsRNA 的反应时，就产生了一个特殊问题。一些有机体具有 dsRNA 依赖性蛋白激酶，它活化一个恐慌反应机制[10]。可以想象，我们的正义–反义协同作用可能反映了由这一恐慌机制引起的反义效应的非特异性增强。事实并非如此：共同注射与 *unc-22* 无关的 dsRNA 片段不能增强单链 *unc-22*-RNA 介导的抑制作用(数据未显示)。我们也研究了将 *cis* 放入一个单链片段后双链结构是否可以增强干扰能力。并未观察到这种增强作用：位于一个单链 *unc-22* 片段 5′ 或 3′ 的无关双链序列不能激发干扰。因此，只有 dsRNA 序列位于与目标基因同源的区域内时才能观察到对干扰的增强作用。

用 *unc-22* dsRNA 干扰产生的表型非常特殊。注射动物的后代表现出的行为非常类似于 *unc-22* 功能缺失突变体。我们用表型明确的其他三个基因来评价 dsRNA 效应的目标特异性(图 1 和表 1)。*unc-54* 编码的体壁肌肉重链同工型肌球蛋白是完整肌肉收缩所必需的[7,11,12]；*fem-1* 编码的锚定蛋白重复包含蛋白在雌雄同体动物中是产生精子所必需的[13,14]；*hlh-1* 编码的 myoD 家族蛋白的秀丽隐杆线虫同源物是维持适当的身体形状和运动能力所必需的[15,16]。对于每个基因来说，注射与之相关的 dsRNA 后，后代表现出已知的无效突变体表型，而提纯的单链 RNA 不能产生显著干扰。dsRNA 注射产生的所有表型皆符合对应基因干扰所预期的表型，仅有一个例外。对此例外(*unc54C* 片段导致胚胎及幼虫发育阻碍，这在 *unc-54* 无效突变体中未曾见过)做以下说明：这个片段覆盖高度保守的肌球蛋白分子马达结构域，因此预计它会干扰其他高度相关的肌球蛋白重链基因[17]。该 *unc54C* 片段也是我们所有实验数据中唯一出现的例外，其他 18 个 dsRNA 片段所出现的效应均可见于先前已知的无效突变表型中(表 1 和我们未发表的结果)。

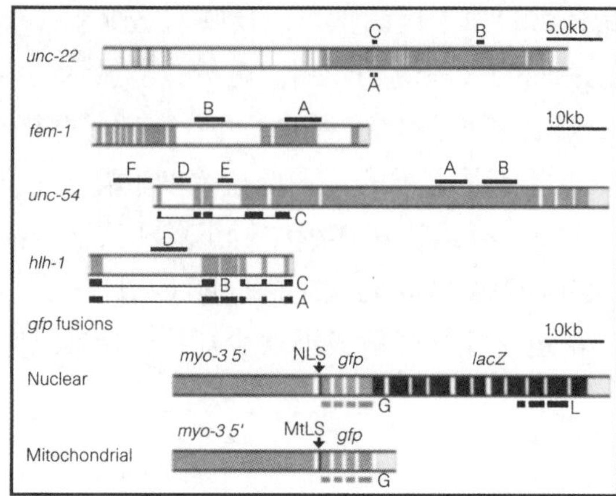

Fig. 1. Genes used to study RNA-mediated genetic interference in *C. elegans*. Intron-exon structure for genes used to test RNA-mediated inhibition are shown (grey and filled boxes, exons; open boxes, introns; patterned and striped boxes, 5′ and 3′ untranslated regions. *unc-22*. ref. 9, *unc*-54, ref. 12, *fem-1*, ref. 14, and *hlh-1*, ref. 15). Each segment of a gene tested for RNA interference is designated with the name of the gene followed by a single letter (for example, *unc22C*). These segments are indicated by bars and uppercase letters above and below each gene. Segments derived from genomic DNA are shown above the gene; segments derived from cDNA are shown below the gene. NLS, nuclear-localization sequence; MtLS, mitochondrial localization sequence.

The pronounced phenotypes seen following dsRNA injection indicate that interference effects are occurring in a high fraction of cells. The phenotypes seen in *unc-54* and *hlh-1* null mutants, in particular, are known to result from many defective muscle cells[11,16]. To examine interference effects of dsRNA at a cellular level, we used a transgenic line expressing two different green fluorescent protein (GFP)-derived fluorescent-reporter proteins in body muscle. Injection of dsRNA directed to *gfp* produced marked decreases in the fraction of fluorescent cells (Fig. 2). Both reporter proteins were absent from the affected cells, whereas the few cells that were fluorescent generally expressed both GFP proteins.

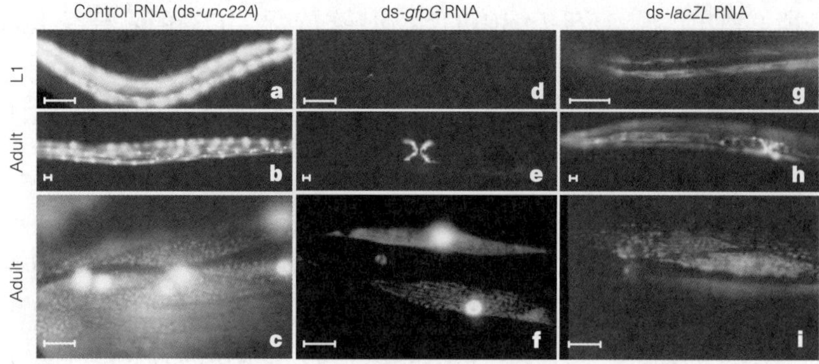

Fig. 2. Analysis of RNA-interference effects in individual cells. Fluorescence micrographs show progeny of injected animals from GFP-reporter strain PD4251. **a–c**, Progeny of animals injected with a control RNA (double-stranded (ds)-*unc22A*). **a**, Young larva, **b**, adult, **c**, adult body wall at high magnification. These GFP patterns appear identical to patterns in the parent strain, with prominent fluorescence in nuclei (nuclear-localized GFP-LacZ) and mitochondria (mitochondrially targeted GFP). **d–f**, Progeny of

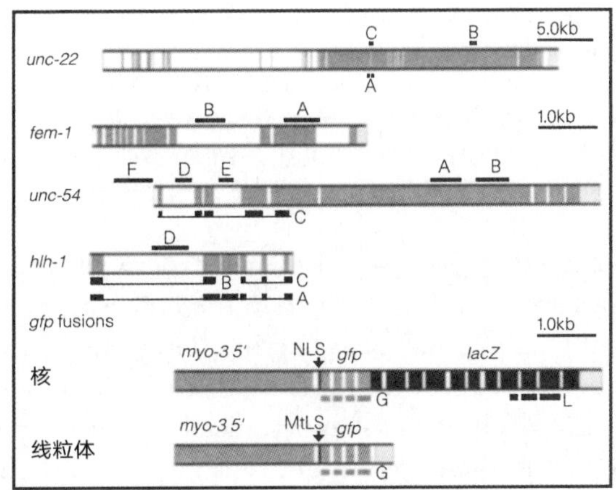

图 1. 用于研究秀丽隐杆线虫中 RNA 介导的遗传性干扰的基因。显示了用于测试 RNA 介导的抑制作用的基因内含子–外显子结构(灰色及填充的方框，外显子；空白的方框，内含子；有图案和条纹的方框，5′ 和 3′ 非翻译区。*unc-22*，参考文献 9；*unc-54*，参考文献 12；*fem-1*，参考文献 14；*hlh-1*，参考文献 15)。用于测试 RNA 干扰的基因片段用基因名加一个字母表示(如 *unc22C*)。这些片段用横条标出，并且在每个基因上部或下部用大写字母表示。从基因组 DNA 得到的片段在基因上方显示；从 cDNA 得到的片段在基因下方显示。NLS，核定位序列；MtLS，线粒体定位序列。

dsRNA 注射后所观察到的明显表型说明在大量细胞中发生了干扰效应。特别是已经知道在 *unc-54* 和 *hlh-1* 无效突变体中见到的表型是源自众多缺陷型肌细胞[11,16]。为了在细胞水平上检验 dsRNA 的干扰效果，我们使用了一个转基因株，它可以在肌肉中表达两种绿色荧光蛋白(GFP)衍生的荧光报告蛋白。注射 dsRNA 直接导致在表达绿色荧光的细胞(图 2)内 *gfp* 荧光强度的显著下降。两个报告蛋白从受影响的细胞中消失，而少量有荧光的细胞基本都表达两种 GFP 蛋白。

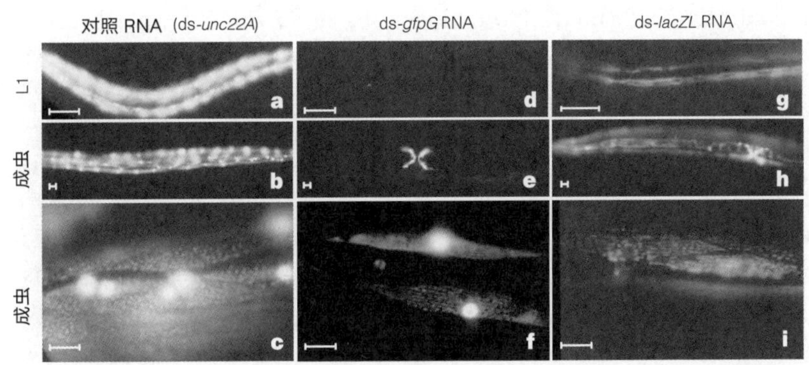

图 2. 在单个细胞中分析 RNA 干扰效应。荧光显微图显示了注射 GFP 报告品系 PD4251 的动物后代。**a~c**，注射对照 RNA(双链(ds)-*unc22A*)的动物后代。**a**，年轻幼虫，**b**，成虫，**c**，高倍放大的成虫体壁。这些子代的 GFP 表达情况与亲本一致，核(核定位 GFP-LacZ)中和线粒体(线粒体靶标 GFP)中均具有显著的荧光。**d~f**，注射 ds-*gfpG* 的动物后代。**d** 中幼虫只可见一个活性细胞，而 **e**、**f** 中的成虫的

animals injected with ds-*gfpG*. Only a single active cell is seen in the larva in **d**, whereas the entire vulval musculature expresses active GFP in the adult animal in **e. f**, Two rare GFP-positive cells in an adult: both cells express both nuclear-targeted GFP-LacZ and mitochondrial GFP. **g–i**, Progeny of animals injected with *ds-lacZL* RNA: mitochondrial-targeted GFP seems unaffected, while the nuclear-targeted GFP-LacZ is absent from almost all cells (for example, see larva in **g**). **h**, A typical adult, with nuclear GFP-LacZ lacking in almost all body-wall muscles but retained in vulval muscles. Scale bars represent 20 μm.

The mosaic pattern observed in the *gfp*-interference experiments was nonrandom. At low doses of dsRNA, we saw frequent interference in the embryonically derived muscle cells that are present when the animal hatches. The interference effect in these differentiated cells persisted throughout larval growth: these cells produced little or no additional GFP as the affected animals grew. The 14 postembryonically derived striated muscles are born during early larval stages and these were more resistant to interference. These cells have come through additional divisions (13–14 divisions versus 8–9 divisions for embryonic muscles[18,19]). At high concentrations of *gfp* dsRNA, we saw interference in virtually all striated body-wall muscles, with occasional lone escaping cells, including cells born during both embryonic and postembryonic development. The non-striated vulval muscles, which are born during late larval development, appeared to be resistant to interference at all tested concentrations of injected dsRNA.

We do not yet know the mechanism of RNA-mediated interference in *C. elegans*. Some observations, however, add to the debate about possible targets and mechanisms.

First, dsRNA segments corresponding to various intron and promoter sequences did not produce detectable interference (Table 1). Although consistent with interference at a post-transcriptional level, these experiments do not rule out interference at the level of the gene.

Second, we found that injection of dsRNA produces a pronounced decrease or elimination of the endogenous mRNA transcript (Fig. 3). For this experiment, we used a target transcript (*mex-3*) that is abundant in the gonad and early embryos[20], in which straightforward *in situ* hybridization can be performed[5]. No endogenous *mex-3* mRNA was observed in animals injected with a dsRNA segment derived from *mex-3*. In contrast, animals into which purified *mex-3* antisense RNA was injected retained substantial endogenous mRNA levels (Fig. 3d).

Third, dsRNA-mediated interference showed a surprising ability to cross cellular boundaries. Injection of dsRNA (for *unc-22*, *gfp* or *lacZ*) into the body cavity of the head or tail produced a specific and robust interference with gene expression in the progeny brood (Table 2). Interference was seen in the progeny of both gonad arms, ruling out the occurrence of a transient "nicking" of the gonad in these injections. dsRNA injected into the body cavity or gonad of young adults also produced gene-specific interference in somatic tissues of the injected animal (Table 2).

整个外阴肌肉都表达活性 GFP，成虫中仅有两个 GFP 活性细胞：两个细胞都表达活性核靶标 GFP-LacZ 和线粒体靶标 GFP。**g~i**，注射 ds-*lacZL* RNA 的动物后代：线粒体靶标 GFP 好像不受影响，而核靶标 GFP-LacZ 在几乎所有细胞中都缺失了(例如，见 **g** 中的幼虫)。**h**，一个典型的成虫，其核靶标 GFP-LacZ 几乎在所有体壁肌肉都缺失但仍存在于外阴肌肉中。比例尺代表 20 μm。

gfp 干扰实验中观察到的镶嵌状图像不是随机的。在低剂量 dsRNA 时，我们经常看到动物孵化时会出现胚胎源的肌肉细胞发生干扰。不同分化细胞中的干扰效应在整个幼虫成长过程中持续发生：这些细胞在受影响动物的生长过程中很少产生或不产生额外的 GFP。14 个源自胚胎后期的横纹肌在幼虫早期阶段产生并且更能抵抗干扰。这些细胞进行额外的分裂(13~14 次分裂，胚胎源肌肉为 8~9 次分裂[18,19])。在高浓度 *gfp* dsRNA 作用时，我们在所有体壁横纹肌中都看到了干扰，只有个别漏网的细胞，包括在胚胎发育和胚胎后期发育过程中出现的细胞。出现于幼虫发育晚期的外阴非横纹肌，似乎在所有检测的 dsRNA 注射浓度下都不发生干扰。

我们还不知道秀丽隐杆线虫中 RNA 介导的干扰机理。然而，一些观察结果增加到对可能靶标和机制方面的讨论中。

第一，对应于各种内含子和启动子序列的 dsRNA 片段没有产生可检测到的干扰效应(表 1)。虽然这支持了干扰发生在转录后水平的观点，但这些实验不能排除基因水平上的干扰。

第二，我们发现注射 dsRNA 导致内源 mRNA 转录本明显减少或消失(图 3)。在这个实验中，我们选用目标转录本(*mex-3*)，它在生殖腺和早期胚胎中含量很丰富[20]，在早期胚胎中可以直接进行原位杂交[5]。在注射了来自 *mex-3* 的 dsRNA 片段的动物体内没有观察到内源性的 *mex-3* mRNA。相反，注射了纯化的 *mex-3* 反义 RNA 的动物维持基本的内源 mRNA 水平(图 3d)。

第三，dsRNA 介导的干扰显示了穿过细胞边界的惊人能力。头部或尾部体腔内注射 dsRNA(*unc-22*、*gfp* 或 *lacZ*)对孵化后代基因表达产生了特异的、稳定的干扰(表 2)。干扰主要出现在具两个性腺的后代中，且排除了在这些注射中发生短暂的“切口”的可能性。将 dsRNA 注射到年轻成虫的体腔或生殖腺后，在被注射动物的体壁组织内也产生了基因特异性干扰(表 2)。

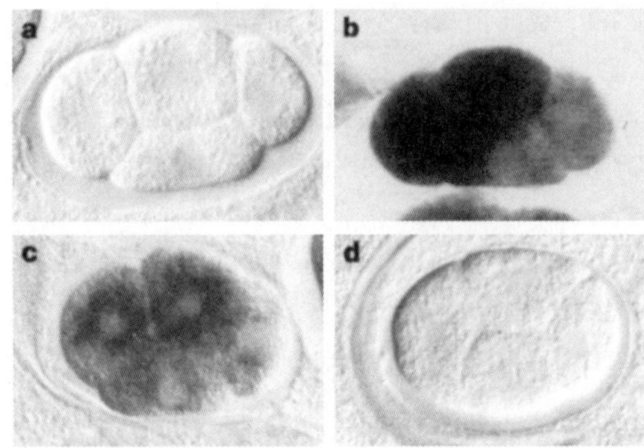

Fig. 3. Effects of *mex-3* RNA interference on levels of the endogenous mRNA. Interference contrast micrographs show *in situ* hybridization in embryos. The 1,262-nt *mex-3* cDNA clone[20] was divided into two segments, *mex-3A* and *mex-3B*, with a short (325-nt) overlap (similar results were obtained in experiments with no overlap between interfering and probe segments). *mex-3B* antisense or dsRNA was injected into the gonads of adult animals, which were fed for 24 h before fixation and *in situ* hybridization (ref. 5; B. Harfe and A.F., unpublished observations). The *mex-3B* dsRNA produced 100% embryonic arrest, whereas >90% of embryos produced after the antisense injections hatched. Antisense probes for the *mex-3A* portion of *mex-3* were used to assay distribution of the endogenous *mex-3* mRNA (dark stain). Four-cell-stage embryos are shown; similar results were observed from the one to eight cell stage and in the germ line of injected adults. **a**, Negative control showing lack of staining in the absence of the hybridization probe. **b**, Embryo from uninjected parent (showing normal pattern of endogenous *mex-3* RNA[20]). **c**, Embryo from a parent injected with purified *mex-3B* antisense RNA. These embryos (and the parent animals) retain the *mex-3* mRNA, although levels may be somewhat less than wild type. **d**, Embryo from a parent injected with dsRNA corresponding to *mex-3B*; no *mex-3* RNA is detected. Each embryo is approximately 50 μm in length.

Table 2. Effect of site of injection on interference in injected animals and their progeny

dsRNA	Site of injection	Injected-animal phenotype	Progeny phenotype
None	Gonad or body cavity	No twitching	No twitching
None	Gonad or body cavity	Strong nuclear and mitochondrial GFP expression	Strong nuclear and mitochondrial GFP expression
unc22B	Gonad	Weak twitchers	Strong twitchers
unc22B	Body-cavity head	Weak twitchers	Strong twitchers
unc22B	Body-cavity tail	Weak twitchers	Strong twitchers
gfpG	Gonad	Lower nuclear and mitochondrial GFP expression	Rare or absent nuclear and mitochondrial GFP expression
gfpG	Body-cavity tail	Lower nuclear and mitochondrial GFP expression	Rare or absent nuclear and mitochondrial GFP expression
lacZL	Gonad	Lower nuclear GFP expression	Rare or absent nuclear-GFP expression
lacZL	Body-cavity tail	Lower nuclear GFP expression	Rare or absent nuclear-GFP expression

The GFP-reporter strain PD4251, which expresses both mitochondrial GFP and nuclear GFP-LacZ, was used for injections. The use of this strain allowed simultaneous assay for interference with *gfp* (fainter overall fluorescence), *lacZ* (loss of nuclear fluorescence) and *unc-22* (twitching). Body-cavity injections into the tail region were carried out to minimize accidental injection of the gonad; equivalent results have been observed with injections into the anterior body cavity. An equivalent set of injections was also performed into a single gonad arm. The entire progeny broods showed phenotypes identical to those described in Table 1. This included progeny of both injected and uninjected gonad arms. Injected animals were scored three days after recovery and showed somewhat less dramatic phenotypes than their progeny. This could be partly due to the persistence of products already present in the injected adult.

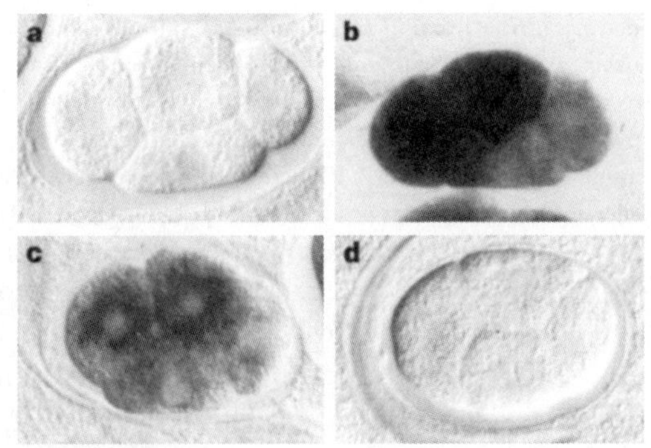

图 3. *mex-3* RNA 对内源性 mRNA 水平的干扰作用。干扰对比显微照片显示胚胎中的原位杂交结果。将 1,262 个核苷酸的 *mex-3* cDNA 克隆[20] 分为两个片段，即 *mex-3A* 和 *mex-3B*，两者有一小段（325 个核苷酸）重叠（在干扰片段和探针片段之间无重叠的实验中也得到相似的结果）。将 *mex-3B* 反义链或 dsRNA 注射到成年动物生殖腺内，并在固定和进行原位杂交前饲养 24 小时（参考文献 5；哈弗和法厄，未发表的观察结果）。*mex-3B* dsRNA 可产生 100% 的胚胎抑制，然而大于 90% 的注射了反义链的胚胎都可以孵化。*mex-3* 的部分片段 *mex-3A* 的反义探针用于分析内源性 *mex-3* mRNA（深染色）的分布。四细胞期的胚胎被显示出来。在单细胞期到八细胞期和注射的成虫生殖系中都观察到相似结果。**a**，阴性对照显示无杂交探针时不染色。**b**，来自未注射亲本的胚胎（显示正常内源 *mex-3* RNA 模式[20]）。**c**，亲本注射了纯化的 *mex-3B* RNA 反义链产生的胚胎。这些胚胎（和亲本动物）保留了 *mex-3* mRNA，但水平相对野生型有一定程度降低。**d**，亲本注射了对应 *mex-3B* 的 dsRNA 产生的胚胎；未检测到 *mex-3* RNA。每个胚胎长度大约为 50 μm。

表 2. 在注射动物及其后代中注射位置对干扰的影响

dsRNA	注射位置	注射动物的表型	后代表型
无	生殖腺或体腔	不抽搐	不抽搐
无	生殖腺或体腔	核 GFP 及线粒体 GFP 的高表达	核 GFP 及线粒体 GFP 的高表达
unc22B	生殖腺	弱抽搐	强烈抽搐
unc22B	头部体腔	弱抽搐	强烈抽搐
unc22B	尾部体腔	弱抽搐	强烈抽搐
gfpG	生殖腺	核 GFP 及线粒体 GFP 的低表达	很少或无核 GFP 及线粒体 GFP 表达
gfpG	尾部体腔	核 GFP 及线粒体 GFP 的低表达	很少或无核 GFP 及线粒体 GFP 表达
lacZL	生殖腺	核 GFP 的低表达	很少或无核 GFP 表达
lacZL	尾部体腔	核 GFP 的低表达	很少或无核 GFP 表达

GFP 报告品系 PD4251 中既可以表达线粒体 GFP 又可以表达核 GFP-LacZ，因此被用于注射。使用这个品系可以同时分析 *gfp*（昏暗的整体荧光），*lacZ*（缺失核荧光）和 *unc-22*（抽搐）的干扰作用。注射到体腔尾部区域以最大限度减少偶然注射到生殖腺的情况；注射到前部体腔也观察到了相同的结果。还有相同的一组只注射到单个生殖腺臂。所有孵化后代呈现的表型与表 1 所描述的相同。包含了注射和未注射生殖腺臂的后代。注射的动物恢复三天后统计的表型一定程度上不如其后代明显。这可以部分归因于被注射的成年动物体内已有产物的持久性。注射双链 *unc22B* 后，部分注射动物在标准饲养条件下产生微弱抽搐（21 个动物中有 10 个）。左旋咪唑的处理导致这些动物发生

After injection of double-stranded *unc22B*, a fraction of the injected animals twitch weakly under standard growth conditions (10 out of 21 animals). Levamisole treatment led to twitching of 100% (21 out of 21) of these animals. Similar effects (not shown) were seen with double-stranded *unc22A*. Injections of double-stranded *gfpG* or double-stranded *lacZL* produced a dramatic decrease (but not elimination) of the corresponding GFP reporters. In some cases, isolated cells or parts of animals retained strong GFP activity. These were most frequently seen in the anterior region and around the vulva. Injections of double-stranded *gfpG* and double-stranded *lacZL* produced no twitching, whereas injections of double-stranded *unc22A* produced no change in the GFP-fluorescence pattern.

The use of dsRNA injection adds to the tools available for studying gene function in *C. elegans*. In particular, it should now be possible functionally to analyse many interesting coding regions[21] for which no specific function has been defined. Although the effects of dsRNA-mediated interference are potent and specific we have observed several limitations that should be taken into account when designing RNA-interference-based experiments. First, a sequence shared between several closely related genes may interfere with several members of the gene family. Second, it is likely that a low level of expression will resist RNA-mediated interference for some or all genes, and that a small number of cells will likewise escape these effects.

Genetic tools are available for only a few organisms. Double-stranded RNA could conceivably mediate interference more generally in other nematodes, in other invertebrates, and, potentially, in vertebrates. RNA interference might also operate in plants: several studies have suggested that inverted-repeat structures or characteristics of dsRNA viruses are involved in transgene-dependent co-suppression in plants[22,23].

There are several possible mechanisms for RNA interference in *C. elegans*. A simple antisense model is not likely: annealing between a few injected RNA molecules and excess endogenous transcripts would not be expected to yield observable phenotypes. RNA-targeted processes cannot, however, be ruled out, as they could include a catalytic component. Alternatively, direct RNA-mediated interference at the level of chromatin structure or transcription could be involved. Interactions between RNA and the genome, combined with propagation of changes along chromatin, have been proposed in mammalian X-chromosome inactivation and plant-gene co-suppression[22,24]. If RNA interference in *C. elegans* works by such a mechanism, it would be new in targeting regions of the template that are present in the final mRNA (as we observed no phenotypic interference using intron or promoter sequences). Whatever their target, the mechanisms underlying RNA interference probably exist for a biological purpose. Genetic interference by dsRNA could be used by the organism for physiological gene silencing. Likewise, the ability of dsRNA to work at a distance from the site of injection, and particularly to move into both germline and muscle cells, suggests that there is an effective RNA-transport mechanism in *C. elegans*.

Methods

RNA synthesis and microinjection. RNA was synthesized from phagemid clones by using T3

100% 的抽搐（21 个中有 21 个）。在双链 *unc22A* 实验中出现了类似结果（未显示）。注射双链 *gfpG* 或双链 *lacZL* 产生的相应 GFP 报告物明显减少（但未消失）。在某些情况下分离的细胞或部分动物体仍有很强的 GFP 活性，这在前部区域和外阴周围最为常见。注射双链 *gfpG* 和双链 *lacZL* 不产生抽搐，而注射双链 *unc22A* 后荧光模式不发生改变。

dsRNA 注射的使用为研究秀丽隐杆线虫基因功能增加了可利用的工具。特别是它可以用于针对许多研究者感兴趣但具体功能尚未确定的编码区的功能分析。虽然 dsRNA 介导的干扰效应很强并具有特异性，但我们也观察到在设计以 RNA 干扰为基础的实验时应考虑的一些局限性。第一，几个紧密相关的基因之间共享的一段序列可以干扰这个基因家族的几个成员。第二，似乎低水平表达可以使部分或全部基因对抗 RNA 介导的干扰，而且少量细胞也可借此逃避 RNA 干扰作用。

遗传工具仅对很少的生物体适用。可以想象双链 RNA 可以更普遍地在其他线虫、无脊椎动物，甚至也可能在脊椎动物中介导干扰效应。RNA 干扰在植物中也可以操作：一些研究提示植物中依赖于转基因的共抑制涉及反向重复结构或 dsRNA 病毒特征 [22,23]。

秀丽隐杆线虫的 RNA 干扰存在几种可能的机制。而简单的反义链模型也似乎不大可能：少量注射的 RNA 分子和过量的内源转录本之间发生退火，估计不能产生可观察到的表型。然而，不能排除以 RNA 为目标的过程中可能包括催化组分。也许，RNA 直接介导的干扰也包括了发生在染色质和转录水平的作用。RNA 和基因组之间的相互作用，伴随着染色质的增殖变化，被认为与哺乳动物 X 染色体失活和植物基因共抑制有关 [22,24]。如果秀丽隐杆线虫研究中的 RNA 干扰是这样的机制，在最终的 mRNA 中表现的应该是新的模板靶标区域（因为我们用内含子和启动子序列时没有观察到表型干扰）。不管它们的目标是什么，RNA 干扰机制的存在可能具有某种生物学目的。dsRNA 产生的遗传干扰可被有机体用于生理基因沉默。同样，dsRNA 能够在距注射位置一定距离处产生作用，尤其是能进入生殖腺和肌肉细胞，这表明秀丽隐杆线虫体内存在有效的 RNA 转移机制。

方　法

RNA 合成和微量注射　使用 T3 和 T7 聚合酶从噬菌粒中合成了 RNA[6]。然后通过两个

and T7 polymerase[6]. Templates were then removed with two sequential DNase treatments. When sense-, antisense-, and mixed-RNA populations were to be compared, RNAs were further purified by electrophoresis on low-gelling-temperature agarose. Gel-purified products appeared to lack many of the minor bands seen in the original "sense" and "antisense" preparations. Nonetheless, RNA species comprising $< 10\%$ of purified RNA preparations would not have been observed. Without gel purification, the "sense" and "antisense" preparations produced notable interference. This interference activity was reduced or eliminated upon gel purification. In contrast, sense-plus-antisense mixtures of gel-purified and non-gel-purified RNA preparations produced identical effects.

Sense/antisense annealing was carried out in injection buffer (ref. 27) at 37 °C for 10–30 min. Formation of predominantly double-stranded material was confirmed by testing migration on a standard (nondenaturing) agarose gel: for each RNA pair, gel mobility was shifted to that expected for dsRNA of the appropriate length. Co-incubation of the two strands in a lower-salt buffer (5 mM Tris-Cl, pH 7.5, 0.5 mM EDTA) was insufficient for visible formation of dsRNA *in vitro*. Non-annealed sense-plus-antisense RNAs for *unc22B* and *gfpG* were tested for RNA interference and found to be much more active than the individual single strands, but twofold to fourfold less active than equivalent preannealed preparations.

After preannealing of the single strands for *unc22A*, the single electrophoretic species, corresponding in size to that expected for the dsRNA, was purified using two rounds of gel electrophoresis. This material retained a high degree of interference activity.

Except where noted, injection mixes were constructed so that animals would receive an average of 0.5×10^6 to 1.0×10^6 RNA molecules. For comparisons of sense, antisense, and double-stranded RNA activity, equal masses of RNA were injected (that is, dsRNA was used at half the molar concentration of the single strands). Numbers of molecules injected per adult are approximate and based on the concentration of RNA in the injected material (estimated from ethidium bromide staining) and the volume of injected material (estimated from visible displacement at the site of injection). It is likely that this volume will vary several-fold between individual animals; this variability would not affect any of the conclusions drawn from this work.

Analysis of phenotypes. Interference with endogenous genes was generally assayed in a wild-type genetic background (N2). Features analysed included movement, feeding, hatching, body shape, sexual identity, and fertility. Interference with *gfp* (ref. 25) and *lacZ* activity was assessed using *C. elegans* strain PD4251. This strain is a stable transgenic strain containing an integrated array (ccIs4251) made up of three plasmids: pSAK4 (*myo-3* promoter driving mitochondrially targeted GFP); pSAK2 (*myo-3* promoter driving a nuclear-targeted GFP-LacZ fusion); and a *dpy-20* subclone[26] as a selectable marker. This strain produces GFP in all body muscles, with a combination of mitochondrial and nuclear localization. The two distinct compartments are easily distinguished in these cells, allowing easy distinction between cells expressing both, either, or neither of the original GFP constructs.

Gonadal injection was done as described[27]. Body-cavity injections followed a similar procedure, with needle insertion into regions of the head and tail beyond the positions of the two gonad arms. Injection into the cytoplasm of intestinal cells is also effective, and may be the least disruptive to the

连续的 DNase 处理将模板除去。当正义链、反义链和双链混合物进行比较时，通过低凝胶温度琼脂糖电泳进一步纯化 RNA。凝胶纯化产物似乎缺少在最初“正义链”和“反义链”制备物中看到的许多小条带。然而，包含 $<10\%$ 的纯化 RNA 制备物的 RNA 物质无法被观察到。不经过凝胶纯化时，“正义链”和“反义链”制备物会产生明显的干扰。这种干扰活性可被凝胶纯化降低或消除。相反，凝胶纯化的和未经凝胶纯化的正义链加反义链混合物能产生相同的作用。

37℃下在注射缓冲液（参考文献 27）中进行正义链/反义链退火 10~30 分钟。通过测试在标准（非变性）琼脂糖凝胶上的迁移来确定主要双链物质的形成：对每对 RNA，凝胶迁移到预期的 dsRNA 的适当位置。在低盐缓冲液（5 mM Tris-Cl，pH 7.5，0.5 mM EDTA）中共同孵育两条链，这不足以在体外形成可见的 dsRNA。用未经退火的 *unc22B* 和 *gfpG* 的有义链和反义链混合 RNA 测试 RNA 干扰，发现比单独用一个链的活性高很多，但比提前退火的制备物活性低 1/2 到 1/4。

unc22A 的单链提前退火后，将对应于预期的 dsRNA 大小的单一的电泳条带通过两轮凝胶电泳纯化。这种物质仍保持高度干扰活性。

除了特别指出的地方，构建的注射混合物能使动物平均接受 0.5×10^6 到 1.0×10^6 个 RNA 分子。为比较正义链、反义链和双链 RNA 活性，进行了等量的 RNA 注射（即 dsRNA 的摩尔浓度为单链 RNA 的一半）。每个成虫注射的分子数量相互接近，分子数量取决于注射物中 RNA 的浓度（通过溴化乙锭染色来确定）和注射物体积（通过注射位置的可见位移来确定）。这个体积在不同个体之间可能有几倍的变化；这一变化不影响本研究中得出的任何结论。

表型分析 内源基因的干扰一般在野生型遗传背景（N2）中分析。特征分析包括运动、进食、孵化、身体形态、性别特征和生殖能力。*gfp*（参考文献 25）和 *lacZ* 的干扰活性用秀丽隐杆线虫品系 PD4251 进行评价。这个品系是稳定的转基因品系，包含一个融合序列（ccIs4251），由三个质粒组成：pSAK4（*myo-3* 启动子启动以线粒体为靶标的 GFP），pSAK2（*myo-3* 启动子启动以核为靶标的 GFP-LacZ 融合子）和一个 *dpy-20* 亚克隆 [26] 作为选择性标记。这个品系在全身肌肉中产生 GFP，并结合线粒体和核定位。这两个清楚的部位在这些细胞中很容易区分，在表达两种、表达其中一种和不表达初始 GFP 结构的细胞中都易于区分。

按照文献描述进行了生殖腺注射 [27]。按相似的步骤进行体腔注射，用针插入位于两个生殖腺臂上方的头部和尾部区域。注射到肠细胞的细胞质内也有效，同时可能对动物造成的

animal. After recovery and transfer to standard solid media, injected animals were transferred to fresh culture plates at 16-h intervals. This yields a series of semisynchronous cohorts in which it was straightforward to identify phenotypic differences. A characteristic temporal pattern of phenotypic severity is observed among progeny. First, there is a short "clearance" interval in which unaffected progeny are produced. These include impermeable fertilized eggs present at the time of injection. Second, after the clearance period, individuals that show the interference phenotype are produced. Third, after injected animals have produced eggs for several days, gonads can in some cases "revert" to produce incompletely affected or phenotypically normal progeny.

(**391**, 806-811; 1998)

Andrew Fire[*], **SiQun Xu**[*], **Mary K. Montgomery**[*], **Steven A. Kostas**[*†], **Samuel E. Driver**[‡] **& Craig C. Mello**[‡]

[*] Carnegie Institution of Washington, Department of Embryology, 115 West University Parkway, Baltimore, Maryland 21210, USA

[†] Biology Graduate Program, Johns Hopkins University, 3400 North Charles Street, Baltimore, Maryland 21218, USA

[‡] Program in Molecular Medicine, Department of Cell Biology, University of Massachusetts Cancer Center, Two Biotech Suite 213, 373 Plantation Street, Worcester, Massachusetts 01605, USA

Received 16 September; accepted 24 November 1997.

References:

1. Izant, J. & Weintraub, H. Inhibition of thymidine kinase gene expression by antisense RNA: a molecular approach to genetic analysis. *Cell* **36**, 1007-1015 (1984).
2. Nellen, W. & Lichtenstein, C. What makes an mRNA anti-sense-itive? *Trends Biochem. Sci.* **18**, 419-423 (1993).
3. Fire, A., Albertson, D., Harrison, S. & Moerman, D. Production of antisense RNA leads to effective and specific inhibition of gene expression in *C. elegans* muscle. *Development* **113**, 503-514 (1991).
4. Guo, S. & Kemphues, K. *par-1*, a gene required for establishing polarity in *C. elegans* embryos, encodes a putative Ser/Thr kinase that is asymmetrically distributed. *Cell* **81**, 611-620 (1995).
5. Seydoux, G. & Fire, A. Soma-germline asymmetry in the distributions of embryonic RNAs in *Caenorhabditis elegans. Development* **120**, 2823-2834 (1994).
6. Ausubel, F. *et al. Current Protocols in Molecular Biology* (Wiley, New York, 1990).
7. Brenner, S. The genetics of *Caenorhabditis elegans. Genetics* **77**, 71-94 (1974).
8. Moerman, D. & Baillie, D. Genetic organization in *Caenorhabditis* elegans: fine structure analysis of the *unc-22 gene. Genetics* **91**, 95-104 (1979).
9. Benian, G., L'Hernault, S. & Morris, M. Additional sequence complexity in the muscle gene, *unc-22*, and its encoded protein, twitchin, of *Caenorhabditis elegans. Genetics* **134**, 1097-1104 (1993).
10. Proud, C. PKR: a new name and new roles. *Trends Biochem. Sci.* **20**, 241-246 (1995).
11. Epstein, H., Waterston, R. & Brenner, S. A mutant affecting the heavy chain of myosin in *C. elegans. J. Mol. Biol.* **90**, 291-300 (1974).
12. Karn, J., Brenner, S. & Barnett, L. Protein structural domains in the *C. elegans unc-54* myosin heavy chain gene are not separated by introns. *Proc. Natl Acad. Sci. USA* **80**, 4253-4257 (1983).
13. Doniach, T. & Hodgkin, J. A. A sex-determining gene, *fem-1*, required for both male and hermaphrodite development in *C. elegans. Dev. Biol.* **106**, 223-235 (1984).
14. Spence, A., Coulson, A. & Hodgkin, J. The product of *fem-1*, a nematode sex-determining gene, contains a motif found in cell cycle control proteins and receptors for cell-cell interactions. *Cell* **60**, 981-990 (1990).
15. Krause, M., Fire, A., Harrison, S., Priess, J. & Weintraub, H. CeMyoD accumulation defines the body wall muscle cell fate during *C . elegans* embryogenesis. *Cell* **63**, 907-919 (1990).
16. Chen, L., Krause, M., Sepanski, M. & Fire, A. The *C. elegans* MyoD homolog *HLH-1* is essential for proper muscle function and complete morphogenesis. *Development* **120**, 1631-1641(1994).
17. Dibb, N. J., Maruyama, I. N., Krause, M. & Karn, J. Sequence analysis of the complete *Caenorhabditis elegans* myosin heavy chain gene family. *J. Mol. Biol.* **205**, 603-613 (1989).
18. Sulston, J., Schierenberg, E., White, J. & Thomson, J. The embryonic cell lineage of the nematode *Caenorhabditis elegans. Dev. Biol.* **100**, 64-119 (1983).
19. Sulston, J. & Horvitz, H. Postembyonic cell lineages of the nematode *Caenorhabiditis elegans. Dev. Biol.* **82**, 41-55 (1977).
20. Draper, B. W., Mello, C. C., Bowerman, B., Hardin, J. & Priess, J. R. *MEX-3* is a KH domain protein that regulates blastomere identity in early *C. elegans* embryos. *Cell* **87**, 205-216 (1996).

损害最小。恢复并转移到标准固体培养基上后，注射的动物每间隔 16 小时转移到新鲜的培养平板上。这样产生的一系列半同步同生群可以直接用于鉴定表型差异。在后代中观察到表型严重的暂时性特征样式。首先，在一个短暂的“间隙”中产生不受影响的后代。这包括在注射时已出现的不透水的受精卵。第二，在这个间隙之后产生显示干扰表型的个体。第三，注射的动物产卵几天后生殖腺在某些情况下又“恢复”产生不完全受影响或表型正常的后代。

（高如丽 李梅 翻译；彭小忠 审稿）

21. Sulston, J. *et al.* The *C. elegans* genome sequencing project: a beginning. *Nature* **356**, 37-41 (1992).

22. Matzke, M. & Matzke, A. How and why do plants inactivate homologous (*trans*) genes? *Plant Physiol.* **107**, 679-685 (1995).

23. Ratcliff, F., Harrison, B. & Baulcombe, D. A similarity between viral defense and gene silencing in plants. *Science* **276**, 1558-1560 (1997).

24. Latham, K. X chromosome imprinting and inactivation in the early mammalian embryo. *Trends Genet.* **12**, 134-138 (1996).

25. Chalfie, M., Tu, Y., Euskirchen, G., Ward, W. & Prasher, D. Green fluorescent protein as a marker for gene expression. *Science* **263**, 802-805 (1994).

26. Clark, D., Suleman, D., Beckenbach, K., Gilchrist, E. & Baillie, D. Molecular cloning and characterization of the *dpy-20* gene of *C. elegans*. *Mol. Gen. Genet.* **247**, 367-378 (1995).

27. Mello, C. & Fire, A. DNA transformation. *Methods Cell Biol.* **48**, 451-482 (1995).

Supplementary information is available on *Nature*'s World-Wide Web site (http://www.nature.com) or as paper copy from Mary Sheehan at the London editorial office of *Nature*.

Acknowledgements. We thank A. Grishok, B. Harfe, M. Hsu, B. Kelly, J. Hsieh, M. Krause, M. Park, W. Sharrock, T. Shin, M. Soto and H. Tabara for discussion. This work was supported by the NIGMS (A.F.) and the NICHD (C.M.), and by fellowship and career awards from the NICHD (M.K.M.), NIGMS (S.K.), PEW charitable trust (C.M.), American Cancer Society (C.M.), and March of Dimes (C.M.).

Correspondence and requests for materials should be addressed to A.F. (e-mail: fire@mail1.ciwemb.edu).

Deciphering the Biology of *Mycobacterium tuberculosis* from the Complete Genome Sequence

S. T. Cole *et al.*

Editor's Note

This paper reveals the complete genome sequence of the tuberculosis-causing bacterium *Mycobacterium tuberculosis*. The genome, which contains over 4 million base pairs and around 4,000 genes, provides a sequence of every potential drug target and every possible antigen that could be used in a vaccine, making it of huge importance. The data, collected by microbial scientist Stewart T. Cole and colleagues, also shed light on the bacterium's basic biology. For example, the genome contains many genes involved in lipid metabolism, thought to encode the bacterium's complex cell wall. And two new families of proteins are thought to represent a possible source of antigenic variation, perhaps explaining how the pathogen alters its surface proteins to evade the host's immune response.

Countless millions of people have died from tuberculosis, a chronic infectious disease caused by the tubercle bacillus. The complete genome sequence of the best-characterized strain of *Mycobacterium tuberculosis*, H37Rv, has been determined and analysed in order to improve our understanding of the biology of this slow-growing pathogen and to help the conception of new prophylactic and therapeutic interventions. The genome comprises 4,411,529 base pairs, contains around 4,000 genes, and has a very high guanine+cytosine content that is reflected in the biased amino-acid content of the proteins. *M. tuberculosis* differs radically from other bacteria in that a very large portion of its coding capacity is devoted to the production of enzymes involved in lipogenesis and lipolysis, and to two new families of glycine-rich proteins with a repetitive structure that may represent a source of antigenic variation.

DESPITE the availability of effective short-course chemotherapy (DOTS) and the Bacille Calmette-Guérin (BCG) vaccine, the tubercle bacillus continues to claim more lives than any other single infectious agent[1]. Recent years have seen increased incidence of tuberculosis in both developing and industrialized countries, the widespread emergence of drug-resistant strains and a deadly synergy with the human immunodeficiency virus (HIV). In 1993, the gravity of the situation led the World Health Organisation (WHO) to declare tuberculosis a global emergency in an attempt to heighten public and political awareness. Radical measures are needed now to prevent the grim predictions of the WHO becoming reality. The combination of genomics and bioinformatics has the potential to generate the information and knowledge that will enable the conception and development of new

根据全基因组序列破译结核分枝杆菌生物学

科尔等

编者按

这篇文章揭示了引起结核病的细菌——结核分枝杆菌的完整基因组序列。该基因组包含超过 400 万个碱基对和大约 4,000 个基因，为每个潜在的药物靶标和每个可能用于生产疫苗的抗原提供了序列，这使得该基因组的序列尤为重要。由微生物学家斯图尔特·科尔及其同事收集的数据也揭示了该细菌的基本生物学特性。例如，该基因组包含很多参与脂类代谢的基因，这些基因被认为编码细菌复杂的细胞壁。两个新蛋白家族被认为代表了抗原变异的可能来源，这也许能解释病原体是如何改变自身表面蛋白来躲避宿主的免疫反应。

数以百万计的人死于肺结核——一种由结核杆菌引起的慢性传染病。目前研究最清楚的结核分枝杆菌菌株 H37Rv 的完整基因组序列已完成测序和分析，有助于增加我们对这种生长缓慢的病原菌的生物学理解，帮助我们研发新的预防和治疗干预措施。该基因组由 4,411,529 个碱基对组成，包含约 4,000 个基因，鸟嘌呤 + 胞嘧啶含量很高，这在蛋白质的氨基酸含量偏倚上有所反映。结核分枝杆菌与其他细菌的根本不同之处在于，它很大一部分编码能力用于编码脂肪合成和分解的酶，以及两个富含甘氨酸且有重复性结构的蛋白新家族，这两个家族可能是抗原变异的来源。

尽管已有有效的短程化疗(DOTS)和卡介苗(BCG)疫苗可用，相比其他单一传染源，结核杆菌仍然夺走更多人的生命[1]。近年来，结核病发病率在发展中国家和工业化国家都呈上升趋势，耐药菌株普遍出现，并且与人类免疫缺陷病毒(HIV)共同形成致命的双重感染。1993 年，由于势态严重，世界卫生组织(WHO)宣布结核病为全球紧急事件，试图提高公众意识和政治意识。现在需要严格的措施防止世界卫生组织残酷的预言成为现实。基因组学和生物信息学的结合，有望产生新信息和新知识，构思和开发针对该空气传播疾病的新疗法及干预措施，并阐明其病原

therapies and interventions needed to treat this airborne disease and to elucidate the unusual biology of its aetiological agent, *Mycobacterium tuberculosis*.

The characteristic features of the tubercle bacillus include its slow growth, dormancy, complex cell envelope, intracellular pathogenesis and genetic homogeneity[2]. The generation time of *M. tuberculosis*, in synthetic medium or infected animals, is typically ~24 hours. This contributes to the chronic nature of the disease, imposes lengthy treatment regimens and represents a formidable obstacle for researchers. The state of dormancy in which the bacillus remains quiescent within infected tissue may reflect metabolic shutdown resulting from the action of a cell-mediated immune response that can contain but not eradicate the infection. As immunity wanes, through ageing or immune suppression, the dormant bacteria reactivate, causing an outbreak of disease often many decades after the initial infection[3]. The molecular basis of dormancy and reactivation remains obscure but is expected to be genetically programmed and to involve intracellular signalling pathways.

The cell envelope of *M. tuberculosis*, a Gram-positive bacterium with a G+C-rich genome, contains an additional layer beyond the peptidoglycan that is exceptionally rich in unusual lipids, glycolipids and polysaccharides[4,5]. Novel biosynthetic pathways generate cell-wall components such as mycolic acids, mycocerosic acid, phenolthiocerol, lipoarabinomannan and arabinogalactan, and several of these may contribute to mycobacterial longevity, trigger inflammatory host reactions and act in pathogenesis. Little is known about the mechanisms involved in life within the macrophage, or the extent and nature of the virulence factors produced by the bacillus and their contribution to disease.

It is thought that the progenitor of the *M. tuberculosis* complex, comprising *M. tuberculosis*, *M. bovis*, *M. bovis* BCG, *M. africanum* and *M. microti*, arose from a soil bacterium and that the human bacillus may have been derived from the bovine form following the domestication of cattle. The complex lacks interstrain genetic diversity, and nucleotide changes are very rare[6]. This is important in terms of immunity and vaccine development as most of the proteins will be identical in all strains and therefore antigenic drift will be restricted. On the basis of the systematic sequence analysis of 26 loci in a large number of independent isolates[6], it was concluded that the genome of *M. tuberculosis* is either unusually inert or that the organism is relatively young in evolutionary terms.

Since its isolation in 1905, the H37Rv strain of *M. tuberculosis* has found extensive, worldwide application in biomedical research because it has retained full virulence in animal models of tuberculosis, unlike some clinical isolates; it is also susceptible to drugs and amenable to genetic manipulation. An integrated map of the 4.4 megabase (Mb) circular chromosome of this slow-growing pathogen had been established previously and ordered libraries of cosmids and bacterial artificial chromosomes (BACs) were available[7,8].

体——结核分枝杆菌独特的生物学机理。

结核杆菌的典型特征包括生长缓慢、休眠、细胞被膜复杂、胞内发病和遗传同质性[2]。在合成培养基或感染的动物中，结核分枝杆菌的代时通常约为 24 小时。这导致这种病本质上是慢性的，必须采取长期的治疗方案，对研究者而言也是一个严重障碍。休眠状态下的结核杆菌在感染组织内保持静态，这可能反映了细胞介导的免疫反应导致了代谢的关闭，这些免疫反应只能抑制感染但不能消除感染。伴随着衰老或免疫抑制引起的免疫力减弱，休眠状态的细菌再次活化，导致疾病通常在最初感染的几十年后爆发[3]。休眠和再次活化的分子基础仍然不清楚，但预计受到遗传调控并且涉及细胞内的信号通路。

结核分枝杆菌是革兰氏阳性细菌，基因组富含 G+C，它的细胞被膜在肽聚糖外还有另外一层，这一层中的罕见脂、糖脂和多糖极其丰富[4,5]。新的生物合成途径产生细胞壁组分，如分枝菌酸、结核蜡酸、苯酚硫代醇、脂阿拉伯甘露聚糖和阿拉伯半乳糖，其中一些可能有助于结核杆菌长期存活，引发宿主炎症反应，并且在致病中发挥作用。目前对它在巨噬细胞内的生存机制了解甚少，对杆菌产生的毒力因子的水平和本质以及它们对疾病的作用也知之甚少。

结核分枝杆菌复合群包含结核分枝杆菌、牛型分枝杆菌、牛型分枝杆菌卡介苗、非洲分枝杆菌和田鼠分枝杆菌，其祖先被认为源自土壤细菌，而人型杆菌可能是在牛驯化后来源于牛型分枝杆菌。复合群菌株缺乏菌株间的遗传多样性，核苷酸变化非常少[6]。就免疫和疫苗研发来说这一点很重要，因为大多数蛋白质在所有菌株中完全相同，所以抗原漂移会受限。基于对大量独立分离株的 26 个位点进行系统序列分析[6]得出的结论是结核分枝杆菌的基因组要么异常不活跃，要么该生物在进化上相对年轻。

自1905 年分离到结核分枝杆菌的 H37Rv 菌株，该菌株已在世界范围的生物医学研究中得到了深入的应用，因为它在肺结核动物模型中保留了全部的毒力，这与有些临床菌株不同；并且它还对药物敏感，易于遗传操作。我们已经建立了该生长缓慢的病原菌的 4.4 Mb 环状染色体的整合图谱，并且规则排序的黏粒和细菌人工染色体（BACs）文库也已建好[7,8]。

Organization and Sequence of the Genome

Sequence analysis. To obtain the contiguous genome sequence, a combined approach was used that involved the systematic sequence analysis of selected large-insert clones (cosmids and BACs) as well as random small-insert clones from a whole-genome shotgun library. This culminated in a composite sequence of 4,411,529 base pairs (bp) (Figs 1, 2; Editorial note: Figure 2, showing a linear map of the chromosome of *M. tuberculosis* H37Rv, was originally included as a fold-out insert. For details, see http://www.sanger.ac.uk/resources/downloads/bacteria/mycobacterium.html), with a G+C content of 65.6%. This represents the second-largest bacterial genome sequence currently available (after that of *Escherichia coli*)[9]. The initiation codon for the *dnaA* gene, a hallmark for the origin of replication, *oriC*, was chosen as the start point for numbering. The genome is rich in repetitive DNA, particularly insertion sequences, and in new multigene families and duplicated housekeeping genes. The G+C content is relatively constant throughout the genome (Fig. 1) indicating that horizontally transferred pathogenicity islands of atypical base composition are probably absent. Several regions showing higher than average G+C content (Fig. 1) were detected; these correspond to sequences belonging to a large gene family that includes the polymorphic G+C-rich sequences (PGRSs).

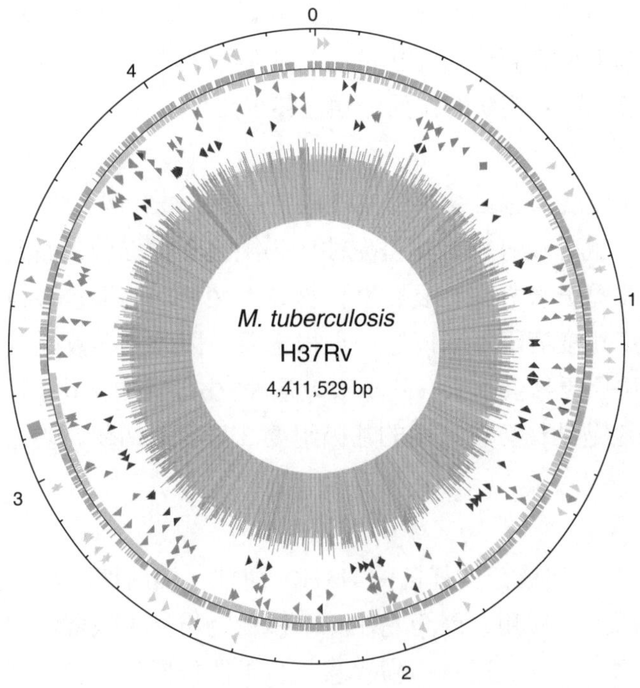

Fig. 1. Circular map of the chromosome of *M. tuberculosis* H37Rv. The outer circle shows the scale in Mb, with 0 representing the origin of replication. The first ring from the exterior denotes the positions of stable RNA genes (tRNAs are blue, others are pink) and the direct repeat region (pink cube); the second ring inwards shows the coding sequence by strand (clockwise, dark green; anticlockwise, light green); the third ring depicts repetitive DNA (insertion sequences, orange; 13E12 REP family, dark pink; prophage, blue);

基因组结构和序列

序列分析 为得到连续的基因组序列，我们采取了一个组合方法，这个方法包括对选定的大插入片段克隆（黏粒和 BACs）和一个全基因组鸟枪文库的随机小插入片段克隆进行系统的序列分析。最终得到 4,411,529 个碱基对（bp）的组合序列（图 1 和图 2；编者注：图 2 显示了结核分枝杆菌 H37Rv 染色体的线性图谱，在原文中是一个折叠插入图。更多的细节请看 http://www.sanger.ac.uk/resources/downloads/bacteria/mycobacterium.html），其中 G+C 含量为 65.6%。这是目前已有的第二大细菌基因组序列（在大肠杆菌之后）[9]。*dnaA* 基因的起始密码子，复制起始位点的标志 *oriC* 被选定为编码的起点。该基因组富含重复 NDA 序列，特别是插入序列，以及新的多基因家族和重复的管家基因。G+C 含量在整个基因组中相对恒定（图 1），这表明可能没有横向转移来的具有非典型碱基组成的致病岛。我们检测到几个 G+C 含量高于平均水平的区域（图 1）；它们对应属于同一个大基因家族的序列，这些序列富含 G+C 且具有多态性（PGRS）。

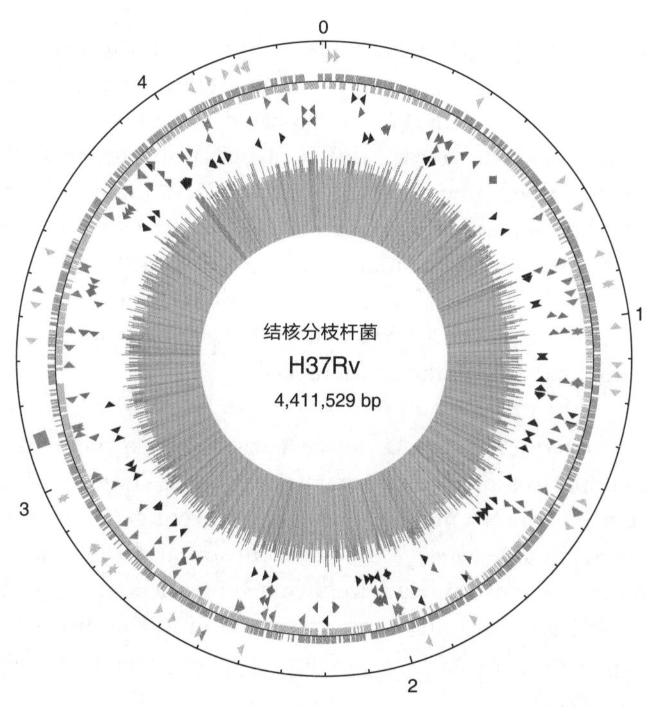

图 1. 结核分枝杆菌 H37Rv 染色体的环形图。外环单位的刻度为 Mb，0 代表复制起点。从外侧数第一环表示稳定 RNA 基因的位置（tRNA 是蓝色的，其他是粉红色）和正向重复区域（粉红条）；向内第二环按链表示编码序列（顺时针，暗绿色；逆时针，浅绿色）；第三环表示重复 DNA（插入序列，橙色；13E12 REP 家族，深粉色；原噬菌体，蓝色）；第四环表示 PPE 家族成员的位置（绿色）；第五环表示

the fourth ring shows the positions of the PPE family members (green); the fifth ring shows the PE family members (purple, excluding PGRS); and the sixth ring shows the positions of the PGRS sequences (dark red). The histogram (centre) represents G+C content, with < 65% G+C in yellow, and > 65% G+C in red. The figure was generated with software from DNASTAR.

Genes for stable RNA. Fifty genes coding for functional RNA molecules were found. These molecules were the three species produced by the unique ribosomal RNA operon, the 10Sa RNA involved in degradation of proteins encoded by abnormal messenger RNA, the RNA component of RNase P, and 45 transfer RNAs. No 4.5S RNA could be detected. The *rrn* operon is situated unusually as it occurs about 1,500 kilobases (kb) from the putative *oriC*; most eubacteria have one or more *rrn* operons near to *oriC* to exploit the gene-dosage effect obtained during replication[10]. This arrangement may be related to the slow growth of *M. tuberculosis*. The genes encoding tRNAs that recognize 43 of the 61 possible sense codons were distributed throughout the genome and, with one exception, none of these uses A in the first position of the anticodon, indicating that extensive wobble occurs during translation. This is consistent with the high G+C content of the genome and the consequent bias in codon usage. Three genes encoding tRNAs for methionine were found; one of these genes (*metV*) is situated in a region that may correspond to the terminus of replication (Figs 1, 2). As *metV* is linked to defective genes for integrase and excisionase, perhaps it was once part of a phage or similar mobile genetic element.

Insertion sequences and prophages. Sixteen copies of the promiscuous insertion sequence IS*6110* and six copies of the more stable element IS*1081* reside within the genome of H37Rv[8]. One copy of IS*1081* is truncated. Scrutiny of the genomic sequence led to the identification of a further 32 different insertion sequence elements, most of which have not been described previously, and of the 13E12 family of repetitive sequences which exhibit some of the characteristics of mobile genetic elements (Fig. 1). The newly discovered insertion sequences belong mainly to the IS*3* and IS*256* families, although six of them define a new group. There is extensive similarity between IS*1561* and IS*1552* with insertion sequence elements found in *Nocardia* and *Rhodococcus* spp., suggesting that they may be widely disseminated among the actinomycetes.

Most of the insertion sequences in *M. tuberculosis* H37Rv appear to have inserted in intergenic or non-coding regions, often near tRNA genes (Fig. 1). Many are clustered, suggesting the existence of insertional hot-spots that prevent genes from being inactivated, as has been described for *Rhizobium*[11]. The chromosomal distribution of the insertion sequences is informative as there appears to have been a selection against insertions in the quadrant encompassing *oriC* and an overrepresentation in the direct repeat region that contains the prototype IS*6110*. This bias was also observed experimentally in a transposon mutagenesis study[12].

At least two prophages have been detected in the genome sequence and their presence may explain why *M. tuberculosis* shows persistent low-level lysis in culture. Prophages phiRv1 and phiRv2 are both ~10 kb in length and are similarly organized, and some of their

PE 家族成员（紫色，除 PGRS 外）；第六环表示 PGRS 序列的位置（暗红色）。柱状图（中心）代表 G+C 含量，< 65%的 G+C 含量为黄色，> 65%的 G+C 含量为红色。此图使用软件 DNASTAR 生成。

编码稳定 RNA 的基因　我们发现了 50 个编码功能性 RNA 分子的基因。这些分子分别是：由独特的核糖体 RNA 操纵子产生的三类分子、参与降解异常信使 RNA 所编码的蛋白质的 10Sa RNA、核糖核酸酶 P(RNase P) 的 RNA 组分和 45 个转运 RNA 分子。没有检测到 4.5S RNA。*rrn* 操纵子的位置异常，因为它距假定的 *oriC* 1,500 kb；大多数真细菌在 *oriC* 附近有一个或多个 *rrn* 操纵子，从而利用复制期间得到的基因剂量效应 [10]。这种排列可能与结核分枝杆菌生长缓慢有关。编码 tRNA 的基因分布在基因组中，能识别分散在基因组中 61 个可能的有义密码子中的 43 个，但有一个例外，所有反密码子的第一位都不用 A，表明翻译过程中摆动广泛存在。这与基因组中 G+C 含量高及其导致的密码子偏好相符。我们发现了三个甲硫氨酸 tRNA 编码基因；其中一个基因 (*metV*) 位于可能对应复制终点的区域（图 1 和图 2）。*metV* 与整合酶和切除酶的缺陷基因相关，或许它曾经是噬菌体或相似的可移动遗传元件的一部分。

插入序列和原噬菌体　H37Rv 基因组中有 16 个拷贝的混杂的 IS*6110* 插入序列和 6 个拷贝的较稳定的 IS*1081* 元件 [8]。其中一个 IS*1081* 拷贝是截短的。仔细查看基因组序列又进一步鉴定了 32 个不同的插入序列元件，其中大部分以前没有描述过，另外还鉴定了 13E12 家族重复序列，该序列具有可移动遗传元件的某些特征（图 1）。新发现的插入序列大多属于 IS*3* 和 IS*256* 家族，但其中 6 个定义了一个新的种类。IS*1561* 和 IS*1552* 与诺卡氏菌属和红球菌属中发现的插入序列元件有广泛的相似性，这表明它们可能在放线菌中广泛传播。

结核分枝杆菌 H37Rv 中的大多数插入序列似乎插到了基因间区或非编码区，通常靠近 tRNA 基因（图 1）。许多插入序列聚集成簇，意味着存在插入热点以防止基因被灭活，这与根瘤菌中所描述的类似 [11]。插入序列在染色体上的分布可以提供有用的信息，因为似乎存在某种选择避免插入到 *oriC* 周围四分之一的区域，并且插入序列似乎在包含 IS*6110* 原型的正向重复区域中出现次数过多。这种倾向在一个研究转座子突变的实验中发现过 [12]。

在基因组序列中至少发现了两个原噬菌体，它们的存在可能解释了为什么结核分枝杆菌在培养中存在持续低水平裂解。原噬菌体 phiRv1 和 phiRv2 长度都约为 10 kb，结构相似，并且它们的一些基因产物与链霉菌和腐生分枝杆菌编码的产物有显著的

gene products show marked similarity to those encoded by certain bacteriophages from *Streptomyces* and saprophytic mycobacteria. The site of insertion of phiRv1 is intriguing as it corresponds to part of a repetitive sequence of the 13E12 family that itself appears to have integrated into the biotin operon. Some strains of *M. tuberculosis* have been described as requiring biotin as a growth supplement, indicating either that phiRv1 has a polar effect on expression of the distal *bio* genes or that aberrant excision, leading to mutation, may occur. During the serial attenuation of *M. bovis* that led to the vaccine strain *M. bovis* BCG, the phiRv1 prophage was lost[13]. In a systematic study of the genomic diversity of prophages and insertion sequences (S.V.G. *et al.*, manuscript in preparation), only IS*1532* exhibited significant variability, indicating that most of the prophages and insertion sequences are currently stable. However, from these combined observations, one can conclude that horizontal transfer of genetic material into the free-living ancestor of the *M. tuberculosis* complex probably occurred in nature before the tubercle bacillus adopted its specialized intracellular niche.

Genes encoding proteins. 3,924 open reading frames were identified in the genome (see Methods), accounting for ~91% of the potential coding capacity (Figs 1, 2). A few of these genes appear to have in-frame stop codons or frameshift mutations (irrespective of the source of the DNA sequenced) and may either use frameshifting during translation or correspond to pseudogenes. Consistent with the high G+C content of the genome, GTG initiation codons (35%) are used more frequently than in *Bacillus subtilis* (9%) and *E. coli* (14%), although ATG (61%) is the most common translational start. There are a few examples of atypical initiation codons, the most notable being the ATC used by *infC*, which begins with ATT in both *B. subtilis* and *E. coli* [9,14]. There is a slight bias in the orientation of the genes (Fig. 1) with respect to the direction of replication as ~59% are transcribed with the same polarity as replication, compared with 75% in *B. subtilis*. In other bacteria, genes transcribed in the same direction as the replication forks are believed to be expressed more efficiently[9,14]. Again, the more even distribution in gene polarity seen in *M. tuberculosis* may reflect the slow growth and infrequent replication cycles. Three genes (*dnaB*, *recA* and Rv1461) have been invaded by sequences encoding inteins (protein introns) and in all three cases their counterparts in *M. leprae* also contain inteins, but at different sites[15] (S.T.C. *et al.*, unpublished observations).

Protein function, composition and duplication. By using various database comparisons, we attributed precise functions to ~40% of the predicted proteins and found some information or similarity for another 44%. The remaining 16% resembled no known proteins and may account for specific mycobacterial functions. Examination of the amino-acid composition of the *M. tuberculosis* proteome by correspondence analysis[16], and comparison with that of other microorganisms whose genome sequences are available, revealed a statistically significant preference for the amino acids Ala, Gly, Pro, Arg and Trp, which are all encoded by G+C-rich codons, and a comparative reduction in the use of amino acids encoded by A+T-rich codons such as Asn, Ile, Lys, Phe and Tyr (Fig. 3). This approach also identified two groups of proteins rich in Asn or Gly that belong to new families, PE and PPE (see below). The fraction of the proteome that has arisen through gene

相似性。phiRv1 的插入位点很有趣，因为它与 13E12 家族的一段重复序列对应，这段序列本身似乎已经整合到生物素操纵子中。有一些结核分枝杆菌菌株被描述为需要生物素作为生长补充，这说明要么 phiRv1 对远端 *bio* 基因的表达有极性效应，要么可能发生了异常切除，导致基因突变。在牛型分枝杆菌连续衰减产生牛型分枝杆菌卡介苗疫苗株的过程中，phiRv1 原噬菌体丢失[13]。对原噬菌体基因组多样性和插入序列的系统研究(戈登等，稿件准备中)显示，只有 IS*1532* 显示出显著变化，这表明大多数原噬菌体和插入序列目前处于稳定状态。但是，综合这些观察结果我们可以得出结论，遗传物质向营自由生活的结核分枝杆菌复合群祖先发生水平转移，实际上很可能发生在结核菌采用它独特的细胞内生活方式之前。

基因编码蛋白 从基因组中鉴定到 3,924 个开放读码框(见方法)，约占潜在编码能力的 91%(图 1 和图 2)。其中少数基因似乎有读码框内终止密码子或移码突变(不考虑测序 DNA 的来源)，这些基因可能在翻译过程中有移码，或是假基因的转录产物。与基因组高 G+C 含量一致，GTG 起始密码子的使用频率(35%)高于枯草芽孢杆菌(9%)和大肠杆菌(14%)，尽管 ATG 在翻译起始最常见(61%)。这里有少数非典型起始密码子的例子，最值得注意的是 *infC* 使用 ATC，而在枯草芽孢杆菌和大肠杆菌中则都使用 ATT[9,14]。相对于复制方向，基因定位有轻微偏好(图 1)，约 59%的转录方向与复制方向相同，而枯草芽孢杆菌中为 75%。在其他细菌中，基因转录与复制叉方向相同并可以更高效地表达[9,14]。同样，结核分枝杆菌基因极性分布更均衡，这可以反映出其生长缓慢和复制周期稀少。三个基因(*dnaB*、*recA* 和 Rv1461)都有内含肽(蛋白内含子)编码序列插入，并且这三个基因在麻风分枝杆菌中的同源基因也都包含内含肽，只是插入位点不同[15](科尔等，未发表的观察结果)。

蛋白质功能、组成和复制 通过使用各种数据库进行比较，我们对约 40%的预测蛋白的功能有了精确的注释，另外也发现了其他 44%的蛋白质的一些信息或相似性。其余的 16%与已知蛋白质没有相似性，可能负责特定的结核杆菌功能。通过对应分析检查结核分枝杆菌的蛋白质组的氨基酸组成[16]，并与其他基因组序列可用的微生物进行比较，统计上显示结核分枝杆菌显著偏好丙氨酸(Ala)、甘氨酸(Gly)、脯氨酸(Pro)、精氨酸(Arg)和色氨酸(Trp)等，这些氨基酸均由富含 G+C 的密码子编码，相对少使用富含 A+T 的密码子编码的氨基酸，如天冬酰胺(Asn)、异亮氨酸(Ile)、赖氨酸(Lys)、苯丙氨酸(Phe)和酪氨酸(Tyr)(见图 3)。这种方法还发现了两组富含天冬酰胺(Asn)或甘氨酸(Gly)的蛋白质新家族，PE 和 PPE(见

duplication is similar to that seen in *E. coli* or *B. subtilis* (~51%; refs 9, 14), except that the level of sequence conservation is considerably higher, indicating that there may be extensive redundancy or differential production of the corresponding polypeptides. The apparent lack of divergence following gene duplication is consistent with the hypothesis that *M. tuberculosis* is of recent descent[6].

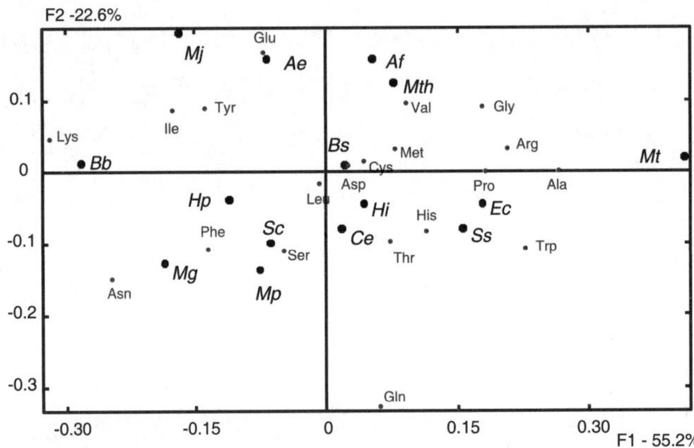

Fig. 3. Correspondence analysis of the proteomes from extensively sequenced organisms as a function of amino-acid composition. Note the extreme position of *M. tuberculosis* and the shift in amino-acid preference reflecting increasing G+C content from left to right. Abbreviations used: *Ae*, *Aquifex aeolicus*; *Af*, *Archaeoglobus fulgidis*; *Bb*, *Borrelia burgdorfei*; *Bs*, *B. subtilis*; *Ce*, *Caenorhabditis elegans*; *Ec*, *E. coli*; *Hi*, *Haemophilus influenzae*; *Hp*, *Helicobacter pylori*; *Mg*, *Mycoplasma genitalium*; *Mj*, *Methanococcus jannaschi*; *Mp*, *Mycoplasma pneumoniae*; *Mt*, *M. tuberculosis*; *Mth*, *Methanobacterium thermoautotrophicum*; *Sc*, *Saccharomyces cerevisiae*; *Ss*, *Synechocystis* sp. strain PCC6803. F1 and F2, first and second factorial axes[16].

General Metabolism, Regulation and Drug Resistance

Metabolic pathways. From the genome sequence, it is clear that the tubercle bacillus has the potential to synthesize all the essential amino acids, vitamins and enzyme co-factors, although some of the pathways involved may differ from those found in other bacteria. *M. tuberculosis* can metabolize a variety of carbohydrates, hydrocarbons, alcohols, ketones and carboxylic acids[2,17]. It is apparent from genome inspection that, in addition to many functions involved in lipid metabolism, the enzymes necessary for glycolysis, the pentose phosphate pathway, and the tricarboxylic acid and glyoxylate cycles are all present. A large number (~200) of oxidoreductases, oxygenases and dehydrogenases is predicted, as well as many oxygenases containing cytochrome P450, that are similar to fungal proteins involved in sterol degradation. Under aerobic growth conditions, ATP will be generated by oxidative phosphorylation from electron transport chains involving a ubiquinone cytochrome *b* reductase complex and cytochrome *c* oxidase. Components of several anaerobic phosphorylative electron transport chains are also present, including genes for nitrate reductase (*narGHJI*), fumarate reductase (*frdABCD*) and possibly nitrite reductase (*nirBD*), as well as a new reductase (*narX*) that results from a rearrangement of a homologue of the *narGHJI* operon. Two genes encoding haemoglobin-like proteins,

下文)。通过基因复制蛋白质组的比例升高，与大肠杆菌或枯草杆菌相似(大约51%；参考文献9和14)，但序列的保守水平要高很多，这表明相应的多肽可能存在大量冗余或差异化表达。基因复制后明显缺少分化，这与结核分枝杆菌在进化上属于近代物种的假说是一致的[6]。

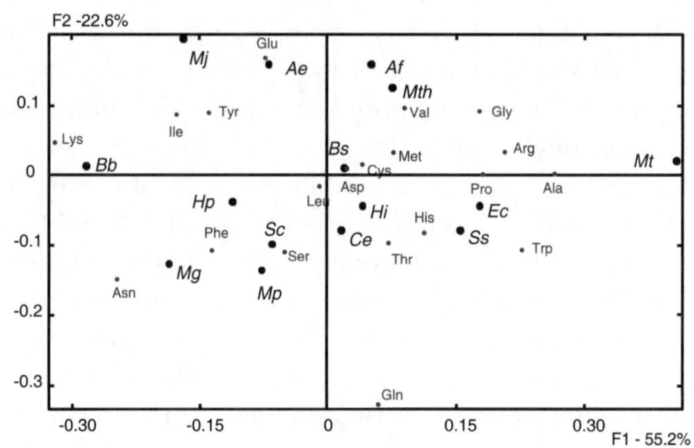

图3. 大量测序的有机蛋白质组作为氨基酸组成的函数的对应分析。注意结核分枝杆菌的极端位置和氨基酸偏好的改变，该改变反映了G+C含量由左至右增加。缩写的使用：*Ae*，风产液菌；*Af*，闪亮古生球菌；*Bb*，伯氏疏螺旋体；*Bs*，枯草芽孢杆菌；*Ce*，秀丽隐杆线虫；*Ec*，大肠杆菌；*Hi*，流感嗜血杆菌；*Hp*，幽门螺杆菌；*Mg*，生殖支原体；*Mj*，詹氏甲烷球菌；*Mp*，肺炎支原体；*Mt*，结核分枝杆菌；*Mth*，嗜热自养甲烷杆菌；*Sc*，酿酒酵母；*Ss*，集胞藻PCC6803菌株。F1和F2，第一和第二因子轴[16]。

基本代谢、调控及耐药性

代谢途径 从基因组序列可以清楚地看到，尽管结核杆菌一些途径可能不同于在其他细菌中发现的途径，但它有能力合成所有必需氨基酸、维生素和辅酶。结核分枝杆菌可以代谢多种碳水化合物、烃、醇、酮和羧酸[2,17]。基因组检测清楚地显示，除了脂代谢的许多功能，糖酵解、戊糖磷酸途径、三羧酸和乙醛酸循环等所必需的酶也都存在。预测有大量(约200种)的氧化还原酶、氧化酶和脱氢酶，还有许多含有细胞色素P450的氧化酶，它们和参与固醇降解的真菌蛋白类似。有氧生长条件下，通过电子传递链包括泛醌–细胞色素*b*还原酶复合体和细胞色素*c*氧化酶，氧化磷酸化产生腺苷三磷酸(ATP)。还存在几个厌氧磷酸化电子传递链组分，这包括硝酸还原酶基因(*narGHJI*)、富马酸还原酶基因(*frdABCD*)，可能还有亚硝酸盐还原酶基因(*nirBD*)，另外还有一个新的还原酶基因(*narX*)，它可能由一个与*narGHJI*操纵子同源的基因重排而来。我们还发现两个血红蛋白类似蛋白的编码基因，可能防

which may protect against oxidative stress or be involved in oxygen capture, were found. The ability of the bacillus to adapt its metabolism to environmental change is significant as it not only has to compete with the lung for oxygen but must also adapt to the microaerophilic/anaerobic environment at the heart of the burgeoning granuloma.

Regulation and signal transduction. Given the complexity of the environmental and metabolic choices facing *M. tuberculosis*, an extensive regulatory repertoire was expected. Thirteen putative sigma factors govern gene expression at the level of transcription initiation, and more than 100 regulatory proteins are predicted (Table 1)(Editorial note: this large table, giving a functional classification of all protein-coding genes in *M. tuberculosis*, has been omitted in this edited version. Note that the original version published in *Nature* contained errors, which were corrected in an erratum in *Nature* **396**, 190–198 (1998)). Unlike *B. subtilis* and *E. coli*, in which there are > 30 copies of different two-component regulatory systems[14], *M. tuberculosis* has only 11 complete pairs of sensor histidine kinases and response regulators, and a few isolated kinase and regulatory genes. This relative paucity in environmental signal transduction pathways is probably offset by the presence of a family of eukaryotic-like serine/threonine protein kinases (STPKs), which function as part of a phosphorelay system[18]. The STPKs probably have two domains: the well-conserved kinase domain at the amino terminus is predicted to be connected by a transmembrane segment to the carboxy-terminal region that may respond to specific stimuli. Several of the predicted envelope lipoproteins, such as that encoded by *lppR* (Rv2403), show extensive similarity to this putative receptor domain of STPKs, suggesting possible interplay. The STPKs probably function in signal transduction pathways and may govern important cellular decisions such as dormancy and cell division, and although their partners are unknown, candidate genes for phosphoprotein phosphatases have been identified.

Drug resistance. *M. tuberculosis* is naturally resistant to many antibiotics, making treatment difficult[19]. This resistance is due mainly to the highly hydrophobic cell envelope acting as a permeability barrier[4], but many potential resistance determinants are also encoded in the genome. These include hydrolytic or drug-modifying enzymes such as β-lactamases and aminoglycoside acetyl transferases, and many potential drug–efflux systems, such as 14 members of the major facilitator family and numerous ABC transporters. Knowledge of these putative resistance mechanisms will promote better use of existing drugs and facilitate the conception of new therapies.

Lipid Metabolism

Very few organisms produce such a diverse array of lipophilic molecules as *M. tuberculosis*. These molecules range from simple fatty acids such as palmitate and tuberculostearate, through isoprenoids, to very-long-chain, highly complex molecules such as mycolic acids and the phenolphthiocerol alcohols that esterify with mycocerosic acid to form the scaffold for attachment of the mycosides. Mycobacteria contain examples of every known lipid and

御氧化应激或参与氧捕获。结核分枝杆菌的代谢适应环境变化的能力很有意义，因为它不仅要和肺竞争氧气，还必须适应迅速生长的肉芽瘤中心的微需氧/厌氧环境。

调控和信号传导 鉴于结核分枝杆菌面临的复杂环境和代谢选择，预计会有一套广泛的调控基因集。13 个假定的 σ 因子在转录起始水平控制基因表达，并且预测有 100 多个调节蛋白（表 1）。（编者注：给出结核分枝杆菌所有蛋白编码基因的大表在此版中被删除了。注意发表在《自然》中的版本含有错误，在《自然》1998 年第 396 卷 190 ~ 198 页的勘误中得以校正。）与枯草芽孢杆菌和大肠杆菌中有 >30 个拷贝的不同的双组分调控系统 [14] 不同，结核分枝杆菌只有完整的 11 对传感器组氨酸激酶和反应调节子，以及少量单独的激酶和调控基因。环境信号传导通路的相对不足可能通过一个真核样丝氨酸/苏氨酸蛋白激酶（STPK）家族来抵消，这个家族是磷酸传递系统的一部分 [18]。STPK 可能有两个结构域：氨基端为高度保守的激酶结构域，羧基端则可能对一些特定的刺激产生反应，据推测，这两个区域通过一个跨膜片段相连。一些预测的细胞被膜脂蛋白，如 *lppR*（Rv2403）编码的蛋白，与 STPK 中假定的受体结构域有广泛的相似性，这暗示可能存在相互作用。STPK 可能在信号传导通路中起作用，并可能控制重要的细胞决策，如休眠和细胞分裂，尽管它们的搭档未知，但已经发现磷蛋白磷酸酶的候选基因。

耐药性 结核分枝杆菌自身能抵抗多种抗生素，这使治疗变得困难 [19]。这种抗性的主要原因是细胞被膜高度疏水成为渗透屏障 [4]，但许多潜在的耐药性决定成分也是由基因组编码的。这些成分包括水解酶或药物修饰酶，如 β–内酰胺酶和氨基糖苷乙酰转移酶，以及许多潜在的药物外排系统，如主要协助转运蛋白家族的 14 个成员和大量 ABC 转运体。关于这些假设的耐药机制的知识将促进对现有药物的更好利用并有助于设计新的治疗方案。

脂 质 代 谢

极少数生物体像结核分枝杆菌一样产生一系列不同的亲脂性分子。这些分子包括简单脂肪酸，如棕榈酸酯和结核硬脂酸，以及类异戊二烯和极长链、高度复杂的分子，如分枝菌酸和苯酚结核菌醇，后者酯化结核蜡酸形成供海藻糖苷附着的支架。结核分枝杆菌包含已知的各类脂质和聚酮生物合成系统，包括通常在哺乳动物和植

polyketide biosynthetic system, including enzymes usually found in mammals and plants as well as the common bacterial systems. The biosynthetic capacity is overshadowed by the even more remarkable radiation of degradative, fatty acid oxidation systems and, in total, there are ~250 distinct enzymes involved in fatty acid metabolism in *M. tuberculosis* compared with only 50 in *E. coli*[20].

Fatty acid degradation. *In vivo*-grown mycobacteria have been suggested to be largely lipolytic, rather than lipogenic, because of the variety and quantity of lipids available within mammalian cells and the tubercle[2] (Fig. 4a). The abundance of genes encoding components of fatty acid oxidation systems found by our genomic approach supports this proposition, as there are 36 acyl-CoA synthases and a family of 36 related enzymes that could catalyse the first step in fatty acid degradation. There are 21 homologous enzymes belonging to the enoyl-CoA hydratase/isomerase superfamily of enzymes, which rehydrate the nascent product of the acyl-CoA dehydrogenase. The four enzymes that convert the 3-hydroxy fatty acid into a 3-keto fatty acid appear less numerous, mainly because they are difficult to distinguish from other members of the short-chain alcohol dehydrogenase family on the basis of primary sequence. The five enzymes that complete the cycle by thiolysis of the β-ketoester, the acetyl-CoA C-acetyltransferases, do indeed appear to be a more limited family. In addition to this extensive set of dissociated degradative enzymes, the genome also encodes the canonical FadA/FadB β-oxidation complex (Rv0859 and Rv0860). Accessory activities are present for the metabolism of odd-chain and multiply unsaturated fatty acids.

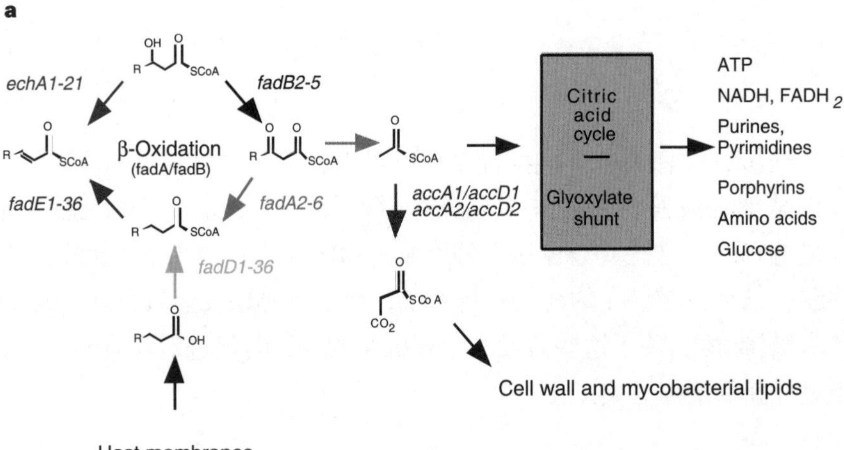

物以及一般细菌系统中发现的酶。与辐射分布更加显著的脂肪酸氧化降解系统相比，生物合成能力甚至相形见绌，总的来说在结核分枝杆菌中共有约 250 种不同的酶参与脂肪酸代谢，相比之下大肠杆菌中只有 50 种[20]。

脂肪酸的降解 研究发现体内生长的分枝杆菌主要参与脂肪降解，而不是脂肪合成，因为哺乳动物细胞和结节中有大量各种可用的脂类[2]（图 4a）。我们用基因组学方法发现大量脂肪酸氧化系统组分的编码基因支持这一观点，有 36 个酰基辅酶 A 合酶和一个含有 36 种酶，能催化脂肪酸降解第一步的酶家族。有 21 个属于烯脂酰辅酶 A 水合酶/异构酶超家族的同源酶，它们对酯酰辅酶 A 脱氢酶的新生产物进行再水合。有四个酶负责将 3–羟基脂肪酸转化为 3–酮脂肪酸，这个数目似乎少了很多，主要是因为在一维序列水平上难以将它们和短链醇脱氢酶家族的其他成员区分开。乙酰辅酶 A C–乙酰转移酶通过硫解 β–酮酸酯完成循环，该家族确实较小，只有五个成员。除了这一系列数目众多的游离降解酶，基因组也编码典型的 FadA/FadB β–氧化复合体（Rv0859 和 Rv0860），还包括代谢奇数链脂肪酸和多不饱和脂肪酸等辅助活性。

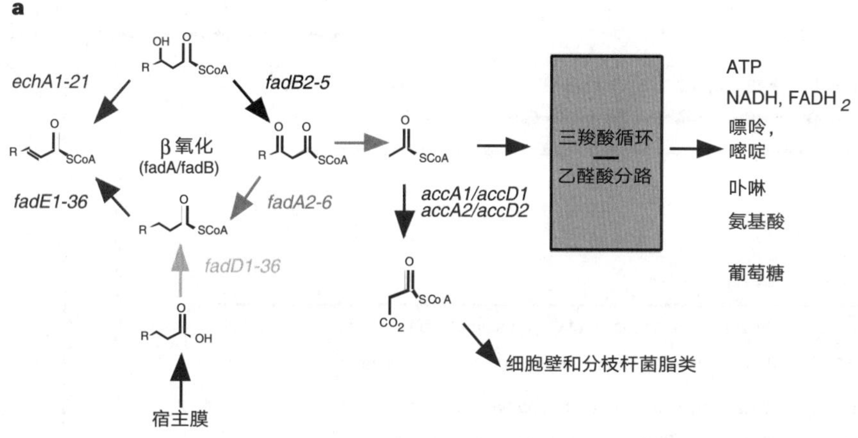

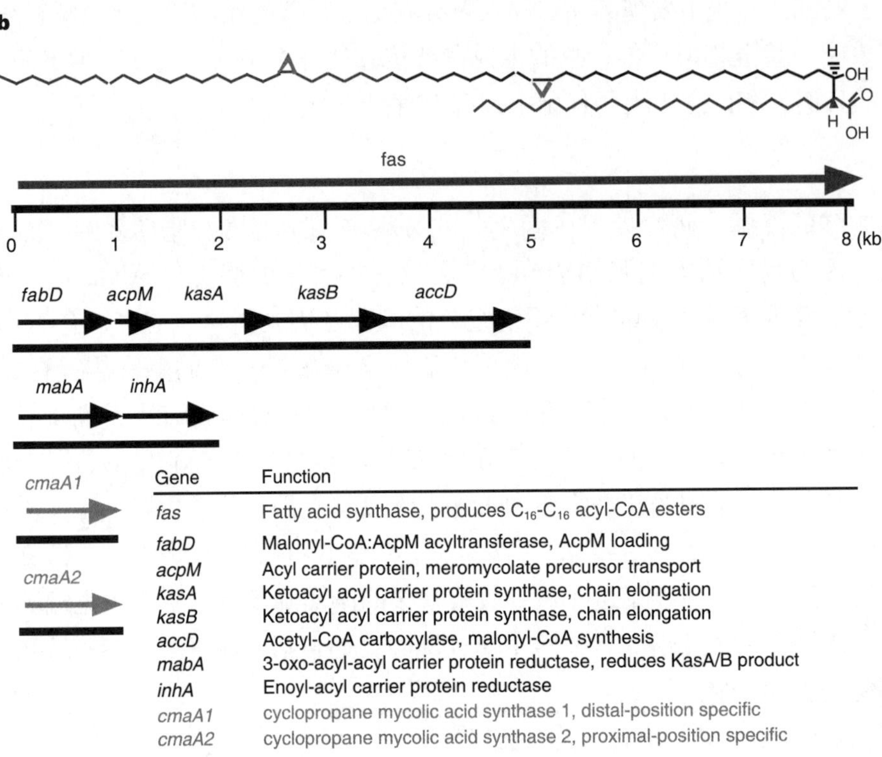

Gene	Function
fas	Fatty acid synthase, produces C_{16}-C_{16} acyl-CoA esters
fabD	Malonyl-CoA:AcpM acyltransferase, AcpM loading
acpM	Acyl carrier protein, meromycolate precursor transport
kasA	Ketoacyl acyl carrier protein synthase, chain elongation
kasB	Ketoacyl acyl carrier protein synthase, chain elongation
accD	Acetyl-CoA carboxylase, malonyl-CoA synthesis
mabA	3-oxo-acyl-acyl carrier protein reductase, reduces KasA/B product
inhA	Enoyl-acyl carrier protein reductase
cmaA1	cyclopropane mycolic acid synthase 1, distal-position specific
cmaA2	cyclopropane mycolic acid synthase 2, proximal-position specific

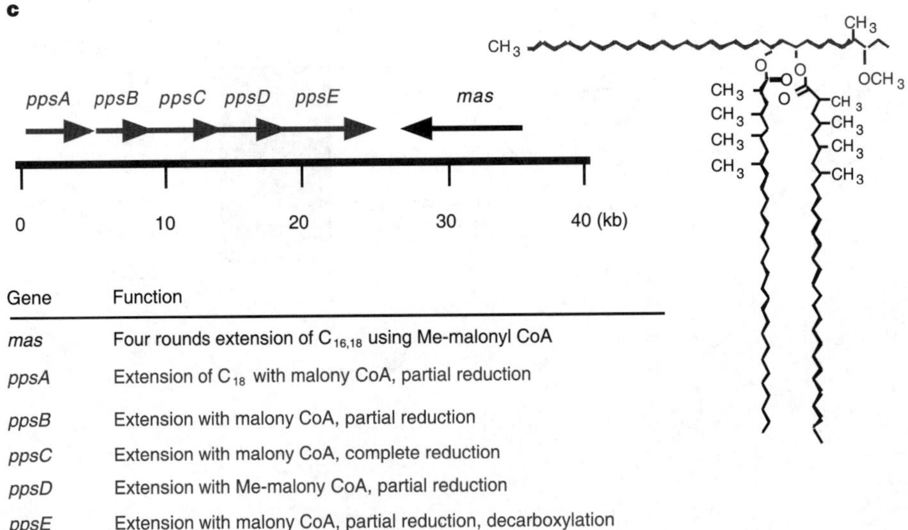

Gene	Function
mas	Four rounds extension of $C_{16,18}$ using Me-malonyl CoA
ppsA	Extension of C_{18} with malony CoA, partial reduction
ppsB	Extension with malony CoA, partial reduction
ppsC	Extension with malony CoA, complete reduction
ppsD	Extension with Me-malony CoA, partial reduction
ppsE	Extension with malony CoA, partial reduction, decarboxylation

Fig. 4. Lipid metabolism. **a**, Degradation of host-cell lipids is vital in the intracellular life of *M. tuberculosis*. Host-cell membranes provide precursors for many metabolic processes, as well as potential precursors of mycobacterial cell-wall constituents, through the actions of a broad family of β-oxidative enzymes encoded by multiple copies in the genome. These enzymes produce acetyl CoA, which can be converted into many different metabolites and fuel for the bacteria through the actions of the enzymes of the citric acid cycle and the glyoxylate shunt of this cycle. **b**, The genes that synthesize mycolic acids, the dominant lipid component of the mycobacterial cell wall, include the type I fatty acid synthase (*fas*) and

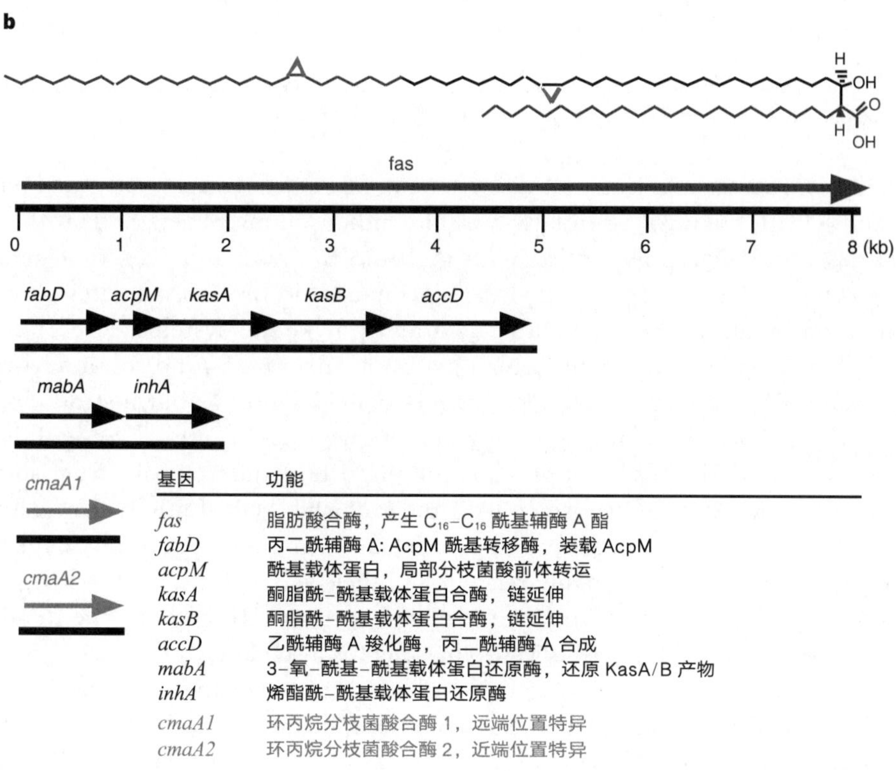

基因	功能
fas	脂肪酸合酶，产生 C_{16}–C_{16} 酰基辅酶 A 酯
fabD	丙二酰辅酶 A: AcpM 酰基转移酶，装载 AcpM
acpM	酰基载体蛋白，局部分枝菌酸前体转运
kasA	酮脂酰–酰基载体蛋白合酶，链延伸
kasB	酮脂酰–酰基载体蛋白合酶，链延伸
accD	乙酰辅酶 A 羧化酶，丙二酰辅酶 A 合成
mabA	3–氧–酰基–酰基载体蛋白还原酶，还原 KasA/B 产物
inhA	烯酯酰–酰基载体蛋白还原酶
cmaA1	环丙烷分枝菌酸合酶 1，远端位置特异
cmaA2	环丙烷分枝菌酸合酶 2，近端位置特异

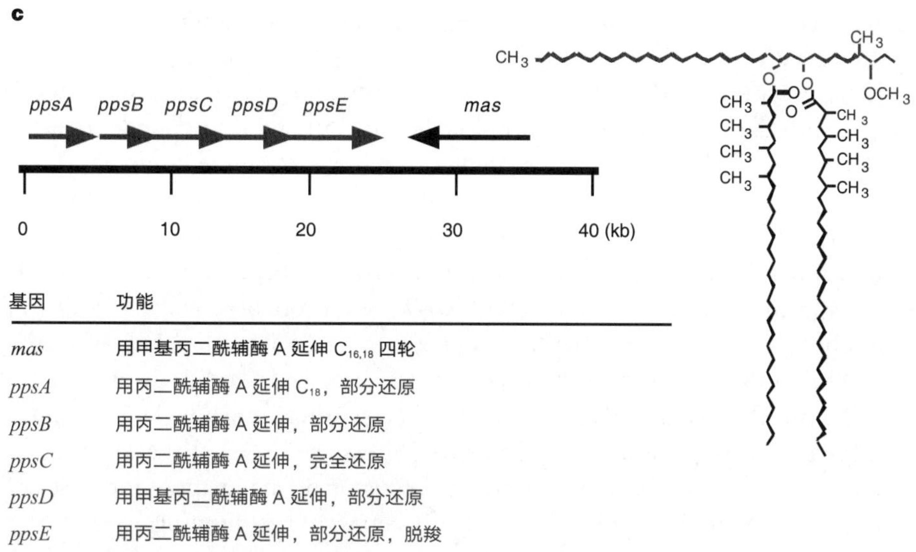

基因	功能
mas	用甲基丙二酰辅酶 A 延伸 $C_{16,18}$ 四轮
ppsA	用丙二酰辅酶 A 延伸 C_{18}，部分还原
ppsB	用丙二酰辅酶 A 延伸，部分还原
ppsC	用丙二酰辅酶 A 延伸，完全还原
ppsD	用甲基丙二酰辅酶 A 延伸，部分还原
ppsE	用丙二酰辅酶 A 延伸，部分还原，脱羧

图 4. 脂质代谢。**a**，宿主细胞脂质的降解对于结核分枝杆菌的胞内生活是至关重要的。在基因组中多拷贝基因编码的β氧化酶大家族的作用下，宿主细胞膜为许多代谢过程提供前体，也为分枝杆菌细胞壁成分提供潜在的前体。这些酶产生乙酰辅酶 A，在柠檬酸循环及乙醛酸旁路循环酶的作用下，乙酰辅酶 A 转换成许多不同的代谢物和供细菌生长用的原料。**b**，合成分枝菌酸（分枝杆菌细胞壁的主要脂分）的基因，包括 I 型脂肪酸合酶（*fas*）和独特的 II 型系统，II 型系统依靠扩展一个结合在酰基载体蛋白的前体

a unique type II system which relies on extension of a precursor bound to an acyl carrier protein to form full-length (~80-carbon) mycolic acids. The *cma* genes are responsible for cyclopropanation. **c**, The genes that produce phthiocerol dimycocerosate form a large operon and represent type I (*mas*) and type II (the *pps* operon) polyketide synthase systems. Functions are colour coordinated.

Fatty acid biosynthesis. At least two discrete types of enzyme system, fatty acid synthase (FAS) I and FAS II, are involved in fatty acid biosynthesis in mycobacteria (Fig. 4b). FAS I (Rv2524, *fas*) is a single polypeptide with multiple catalytic activities that generates several shorter CoA esters from acetyl-CoA primers[5] and probably creates precursors for elongation by all of the other fatty acid and polyketide systems. FAS II consists of dissociable enzyme components which act on a substrate bound to an acyl-carrier protein (ACP). FAS II is incapable of *de novo* fatty acid synthesis but instead elongates palmitoyl-ACP to fatty acids ranging from 24 to 56 carbons in length[17,21]. Several different components of FAS II may be targets for the important tuberculosis drug isoniazid, including the enoyl-ACP reductase InhA[22], the ketoacyl-ACP synthase KasA and the ACP AcpM[21]. Analysis of the genome shows that there are only three potential ketoacyl synthases: KasA and KasB are highly related, and their genes cluster with *acpM*, whereas KasC is a more distant homologue of a ketoacyl synthase III system. The number of ketoacyl synthase and ACP genes indicates that there is a single FAS II system. Its genetic organization, with two clustered ketoacyl synthases, resembles that of type II aromatic polyketide biosynthetic gene clusters, such as those for actinorhodin, tetracycline and tetracenomycin in *Streptomyces* species[23]. InhA seems to be the sole enoyl-ACP reductase and its gene is co-transcribed with a *fabG* homologue, which encodes 3-oxoacyl-ACP reductase. Both of these proteins are probably important in the biosynthesis of mycolic acids.

Fatty acids are synthesized from malonyl-CoA and precursors are generated by the enzymatic carboxylation of acetyl (or propionyl)-CoA by a biotin-dependent carboxylase (Fig. 4b). From study of the genome we predict that there are three complete carboxylase systems, each consisting of an α- and a β-subunit, as well as three β-subunits without an α-counterpart. As a group, all of the carboxylases seem to be more related to the mammalian homologues than to the corresponding bacterial enzymes. Two of these carboxylase systems (*accA1*, *accD1* and *accA2*, *accD2*) are probably involved in degradation of odd-numbered fatty acids, as they are adjacent to genes for other known degradative enzymes. They may convert propionyl-CoA to succinyl-CoA, which can then be incorporated into the tricarboxylic acid cycle. The synthetic carboxylases (*accA3*, *accD3*, *accD4*, *accD5* and *accD6*) are more difficult to understand. The three extra β-subunits might direct carboxylation to the appropriate precursor or may simply increase the total amount of carboxylated precursor available if this step were rate-limiting.

Synthesis of the paraffinic backbone of fatty and mycolic acids in the cell is followed by extensive postsynthetic modifications and unsaturations, particularly in the case of the mycolic acids[24,25]. Unsaturation is catalysed either by a FabA-like β-hydroxyacyl-ACP dehydrase, acting with a specific ketoacyl synthase, or by an aerobic terminal mixed function desaturase that uses both molecular oxygen and NADPH. Inspection of the genome revealed no obvious candidates for the FabA-like activity. However, three potential aerobic

形成全长（约 80 碳）分枝菌酸。*cma* 基因负责形成环丙烷。**c**，产生结核菌醇双结核蜡酸酯的基因形成一个大操纵子，代表 I 型（*mas*）和 II 型（*pps* 操纵子）聚酮合酶系统。功能和颜色相对应。

脂肪酸生物合成 至少有两个不同类型的酶系统，脂肪酸合酶（FAS）I 和 FAS II，参与结核分枝杆菌的脂肪酸生物合成（图 4b）。FAS I（Rv2524，*fas*）是一个具有多种催化活性的单一多肽，它能利用乙酰辅酶 A 引物产生一些较短的辅酶 A 酯[5]，并且可能通过所有其他脂肪酸和聚酮系统产生延伸所需的前体。FAS II 由多个可解离的酶组分组成，这些酶组分都作用于一个结合在酰基载体蛋白（ACP）上的底物。FAS II 不能从头合成脂肪酸，但可以将软脂酰–ACP 延伸为长度为 24 到 56 个碳的脂肪酸[17,21]。FAS II 的几个不同组分可能是重要的抗结核病药物异烟肼的靶标，包括烯脂酰–ACP 还原酶 InhA[22]、酮脂酰–ACP 合酶 KasA 和 ACP AcpM[21]。基因组分析表明只有三个潜在酮脂酰合酶：KasA 和 KasB 高度相关，它们的基因与 *acpM* 聚集成簇，而 KasC 则是酮脂酰合酶 III 系统的一个更远源的同源蛋白。酮脂酰合酶和 ACP 基因的数量表明有单一的 FAS II 系统。该系统的遗传结构有两个聚集成簇的酮脂酰酶，类似于 II 型芳香族聚酮生物合成基因簇，如链霉菌中的放线紫红素、四环素、特曲霉素合成基因簇[23]。InhA 似乎是唯一的烯脂酰–ACP 还原酶，其基因与编码 3–氧酰基–ACP 还原酶的 *fabG* 同源基因进行共转录。这两种蛋白质很可能在分枝菌酸合成中起重要作用。

脂肪酸是由丙二酰辅酶 A 合成的，前体由依赖生物素的羧化酶酶促羧化乙酰（或丙酰）–辅酶 A（图 4b）生成。通过对基因组的研究，我们预测出三个完整的羧化酶系统，每一个都包括一个 α 亚基和一个 β 亚基，另外还预测出三个 β 亚基，没有 α 亚基与之配对。整体而言，所有羧化酶和哺乳动物来源的同源蛋白的亲缘关系似乎高于细菌来源的同源酶。羧化酶系统（*accA1*、*accD1* 和 *accA2*、*accD2*）中的两个很可能参与奇数碳原子脂肪酸的降解，因为在基因组上它们与其他已知降解酶基因邻近。他们可能将丙酰辅酶 A 转换为琥珀酰辅酶 A，然后琥珀酰辅酶 A 被纳入三羧酸循环。具有合成作用的羧化酶（*accA3*、*accD3*、*accD4*、*accD5* 和 *accD6*）更难理解。如果这步是限速步骤的话，这三个额外的 β 亚基可能介导适当的前体羧化或只是简单地增加可用的羧化前体总数。

在细胞内合成脂肪酸和分枝菌酸的石蜡骨架后接着进行大量的合成后修饰和不饱和化，特别是在分枝菌酸的合成中[24,25]。不饱和化由一个 FabA 样的 β–羟脂酰–ACP 脱水酶催化，该酶与特定的酮酯酰合酶共同起作用，或者由一个使用分子氧和 NADPH 的需氧末端混合功能去饱和酶催化。检测基因组没有发现明显具有 FabA 样活性的候选基因。然而，很显然，三个潜在的需氧去饱和酶（由 *desA1*、

desaturases (encoded by *desA1*, *desA2* and *desA3*) were evident that show little similarity to related vertebrate or yeast enzymes (which act on CoA esters) but instead resemble plant desaturases (which use ACP esters). Consequently, the genomic data indicate that unsaturation of the meromycolate chain may occur while the acyl group is bound to AcpM.

Much of the subsequent structural diversity in mycolic acids is generated by a family of *S*-adenosyl-L-methionine-dependent enzymes, which use the unsaturated meromycolic acid as a substrate to generate *cis* and *trans* cyclopropanes and other mycolates. Six members of this family have been identified and characterized[25] and two clustered, convergently transcribed new genes are evident in the genome (*umaA1* and *umaA2*). From the functions of the known family members and the structures of mycolic acids in *M. tuberculosis*, it is tempting to speculate that these new enzymes may introduce the *trans* cyclopropanes into the meromycolate precursor. In addition to these two methyltransferases, there are two other unrelated lipid methyltransferases (Ufa1 and Ufa2) that share homology with cyclopropane fatty acid synthase of *E. coli*[25]. Although cyclopropanation seems to be a relatively common modification of mycolic acids, cyclopropanation of plasma-membrane constituents has not been described in mycobacteria. Tuberculostearic acid is produced by methylation of oleic acid, and may be synthesized by one of these two enzymes.

Condensation of the fully functionalized and preformed meromycolate chain with a 26-carbon α-branch generates full-length mycolic acids that must be transported to their final location for attachment to the cell-wall arabinogalactan. The transfer and subsequent transesterification is mediated by three well-known immunogenic proteins of the antigen 85 complex[26]. The genome encodes a fourth member of this complex, antigen 85C′ (*fbpC2*, Rv0129), which is highly related to antigen 85C. Further studies are needed to show whether the protein possesses mycolytransferase activity and to clarify the reason behind the apparent redundancy.

Polyketide synthesis. Mycobacteria synthesize polyketides by several different mechanisms. A modular type I system, similar to that involved in erythromycin biosynthesis[23], is encoded by a very large operon, *ppsABCDE*, and functions in the production of phenolphthiocerol[5]. The absence of a second type I polyketide synthase suggests that the related lipids phthiocerol A and B, phthiodiolone A and phthiotriol may all be synthesized by the same system, either from alternative primers or by differential postsynthetic modification. It is physiologically significant that the *pps* gene cluster occurs immediately upstream of *mas*, which encodes the multifunctional enzyme mycocerosic acid synthase (MAS), as their products phthiocerol and mycocerosic acid esterify to form the very abundant cell-wall-associated molecule phthiocerol dimycocerosate (Fig. 4c).

Members of another large group of polyketide synthase enzymes are similar to MAS, which also generates the multiply methyl-branched fatty acid components of mycosides and phthiocerol dimycocerosate, abundant cell-wall-associated molecules[5]. Although some of these polyketide synthases may extend type I FAS CoA primers to produce other long-chain methyl-branched fatty acids such as mycolipenic, mycolipodienic and mycolipanolic acids or the phthioceranic and hydroxyphthioceranic acids, or may even show functional overlap[5],

desA2 和 *desA3* 编码）与脊椎动物或酵母中的相关酶（作用于辅酶 A 酯）几乎没有相似性，而与植物中的去饱和酶（使用 ACP 酯）相似。因此，基因组数据表明，当酰基结合到 AcpM 上可能使得局部分枝菌酸酯链去饱和。

分枝菌酸许多后来的结构多样性是由 S-腺苷-L-甲硫氨酸依赖的酶家族产生的，它以不饱和的局部分枝菌酸为底物生成顺式-和反式-环丙烷以及其他分枝菌酸。现在已经鉴定和描述了这个家族的六个成员[25]，两个基因簇集中转录的新基因在基因组中很明显（*umaA1* 和 *umaA2*）。从已知家族成员的功能和结核分枝杆菌分枝菌酸的结构推测，这些新酶可能将反式环丙烷引入到局部分枝菌酸前体中。除了这两个甲基转移酶，还有另外两个无关的脂质甲基转移酶（Ufa1 和 Ufa2）与大肠杆菌的环丙烷脂肪酸合酶有同源性[25]。虽然环丙烷化似乎是分枝菌酸的一种相对常见的修饰，但质膜组分的环丙烷化在分枝杆菌中还未见描述。结核硬脂酸通过油酸的甲基化产生，该合成过程可能由这两个酶中的一个完成。

完全功能的预先形成的局部分枝菌酸链和 26 碳 α 分枝缩合产生全长分枝菌酸，分枝菌酸必须转移到最终的位置以结合细胞壁阿拉伯半乳糖。转移和随后的酯交换反应由抗原 85 复合体中三个众所周知的免疫原性蛋白介导[26]。基因组编码这个复合体的第四个成员，抗原 85C′（*fbpC2*，Rv0129）与抗原 85C 高度相关。这个蛋白是否具有分枝酸转移酶活性，以及其明显冗余背后的原因的阐明，都需要进一步的研究。

聚酮合成　分枝杆菌通过几种不同机制合成聚酮。模块化的 I 型系统，类似于参与红霉素生物合成的系统[23]，由一个非常大的操纵子 *ppsABCDE* 编码，并在苯酚结核菌醇的产生中起作用[5]。基因组中不存在第二个 I 型聚酮合酶说明相关的脂质结核菌醇 A 和 B、结核菌二醇 A 和结核菌三醇都可能由同一系统合成，这要么使用了不同的引物，要么合成后进行了不同的修饰。*pps* 基因簇紧邻 *mas* 上游，*mas* 编码多功能酶分枝杆菌结核蜡酸合酶（MAS），这在生理上很重要，因为它们的产物结核菌醇和结核蜡酸酯化形成了非常丰富的细胞壁关联分子结核菌醇双结核蜡酸酯（图 4c）。

另一个聚酮合酶大家族的成员与 MAS 相似，它也产生丰富的细胞壁关联分子——海藻糖苷和结核菌醇双结核蜡酸酯的多种甲基化分枝脂肪酸组分[5]。虽然其中一些聚酮合酶可能延长 I 型 FAS 辅酶 A 引物，产生其他长链甲基化分枝脂肪酸，如霉脂酸、霉脂二烯酸和霉脂羟基酸或分枝菌蜡酸和羟基分枝菌蜡酸，甚至出现功能重叠[5]，但是这些酶的数目比已知代谢物多得多。因此，可能存在只在特定条件

there are many more of these enzymes than there are known metabolites. Thus there may be new lipid and polyketide metabolites that are expressed only under certain conditions, such as during infection and disease.

A fourth class of polyketide synthases is related to the plant enzyme superfamily that includes chalcone and stilbene synthase[23]. These polyketide synthases are phylogenetically divergent from all other polyketide and fatty acid synthases and generate unreduced polyketides that are typically associated with anthocyanin pigments and flavonoids. The function of these systems, which are often linked to apparent type I modules, is unknown. An example is the gene cluster spanning *pks10*, *pks7*, *pks8* and *pks9*, which includes two of the chalcone-synthase-like enzymes and two modules of an apparent type I system. The unknown metabolites produced by these enzymes are interesting because of the potent biological activities of some polyketides such as the immunosuppressor rapamycin.

Siderophores. Peptides that are not ribosomally synthesized are made by a process that is mechanistically analogous to polyketide synthesis[23,27]. These peptides include the structurally related iron-scavenging siderophores, the mycobactins and the exochelins[2,28], which are derived from salicylate by the addition of serine (or threonine), two lysines and various fatty acids and possible polyketide segments. The *mbt* operon, encoding one apparent salicylate-activating protein, three amino-acid ligases, and a single module of a type I polyketide synthase, may be responsible for the biosynthesis of the mycobacterial siderophores. The presence of only one non-ribosomal peptide-synthesis system indicates that this pathway may generate both siderophores and that subsequent modification of a single ε-amino group of one lysine residue may account for the different physical properties and function of the siderophores[28].

Immunological Aspects and Pathogenicity

Given the scale of the global tuberculosis burden, vaccination is not only a priority but remains the only realistic public health intervention that is likely to affect both the incidence and the prevalence of the disease[29]. Several areas of vaccine development are promising, including DNA vaccination, use of secreted or surface-exposed proteins as immunogens, recombinant forms of BCG and rational attenuation of *M. tuberculosis*[29]. All of these avenues of research will benefit from the genome sequence as its availability will stimulate more focused approaches. Genes encoding ~90 lipoproteins were identified, some of which are enzymes or components of transport systems, and a similar number of genes encoding preproteins (with type I signal peptides) that are probably exported by the Sec-dependent pathway. *M. tuberculosis* seems to have two copies of *secA*. The potent T-cell antigen Esat-6 (ref. 30), which is probably secreted in a Sec-independent manner, is encoded by a member of a multigene family. Examination of the genetic context reveals several similarly organized operons that include genes encoding large ATP-hydrolysing membrane proteins that might act as transporters. One of the surprises of the genome project was the discovery of two extensive families of novel glycine-rich proteins, which may be of immunological significance as they are predicted to be abundant and potentially polymorphic antigens.

下，如在感染和疾病时表达的新的脂和聚酮代谢产物。

第四类聚酮合酶与包含查尔酮和二苯乙烯合酶的植物酶超家族[23]有关。这些聚酮合酶在系统发生上与其他所有聚酮和脂肪酸合酶不同，它们产生通常与花青素和黄酮类化合物相关联的非还原聚酮。这些系统的功能未知，经常与一些明显的I型模块相连接。一个例子是涵盖*pks10*、*pks7*、*pks8*和*pks9*的基因簇，它包括两个查尔酮合酶的类似酶和明显的I型系统的两个模块。因为一些聚酮如免疫抑制剂雷帕霉素具有强大的生物活性，所以这些酶产生的未知代谢产物很有趣。

铁载体　非核糖体合成的肽链是通过与聚酮合成相似的过程产生的[23,27]。这些肽包括结构相关的清除铁的铁载体，分枝杆菌素和胞外螯合素[2,28]，它们是通过向水杨酸盐添加丝氨酸（或苏氨酸）、两个赖氨酸和各种脂肪酸和可能的聚酮片段得到的。*mbt*操纵子可能负责分枝杆菌铁载体的合成，它编码一个明显的水杨酸盐活化蛋白、三个氨基酸连接酶和I型聚酮合酶的一个单一模块。基因组中只有一个非核糖体肽合成系统表明该途径可能产生这两种铁载体，并且一个赖氨酸残基的ε–氨基的后续修饰可能是铁载体具有不同物理性质和功能的原因[28]。

免疫学特性和致病性

鉴于全球结核病的发生规模，接种疫苗不只是一个优先选择，也可能是对结核病的发病率和流行都发挥作用的唯一实际的公共健康干预措施[29]。疫苗开发的几个领域是有前途的，这包括使用分泌蛋白或表面暴露蛋白作为免疫原的DNA疫苗、重组形式的卡介苗和合理减毒的结核分枝杆菌[29]。所有这些研究途径将从基因组序列中受益，基因组序列的获得将促进产生更加聚焦的解决方法。我们鉴定了约90个脂蛋白编码基因，其中有些是酶或转运系统组分，还有数量相当的基因编码蛋白质前体（含I型信号肽），它们很可能通过Sec依赖途径向胞外转运。结核分枝杆菌似乎有两个拷贝的*secA*。有效的T细胞抗原Esat-6（参考文献30）由多基因家族的一个成员编码，很可能以不依赖Sec的方式分泌。对基因组邻近区域的检查发现了几个结构相似的操纵子，其中包括编码具有ATP水解功能的高分子量膜蛋白基因，而这些膜蛋白可能是转运体。本基因组项目的惊喜之一是发现了新的富含甘氨酸的蛋白质，它们分别属于两个大家族，可能有免疫学意义，因为它们被预测是丰富且潜在的多态性抗原。

The PE and PPE multigene families. About 10% of the coding capacity of the genome is devoted to two large unrelated families of acidic, glycine-rich proteins, the PE and PPE families, whose genes are clustered (Figs 1, 2) and are often based on multiple copies of the polymorphic repetitive sequences referred to as PGRSs, and major polymorphic tandem repeats (MPTRs), respectively[31,32]. The names PE and PPE derive from the motifs Pro–Glu (PE) and Pro–Pro–Glu (PPE) found near the N terminus in most cases[33]. The 99 members of the PE protein family all have a highly conserved N-terminal domain of ~110 amino-acid residues that is predicted to have a globular structure, followed by a C-terminal segment that varies in size, sequence and repeat copy number (Fig. 5). Phylogenetic analysis separated the PE family into several subfamilies. The largest of these is the highly repetitive PGRS class, which contains 61 members; members of the other subfamilies, share very limited sequence similarity in their C-terminal domains (Fig. 5). The predicted molecular weights of the PE proteins vary considerably as a few members contain only the N-terminal domain, whereas most have C-terminal extensions ranging in size from 100 to 1,400 residues. The PGRS proteins have a high glycine content (up to 50%), which is the result of multiple tandem repetitions of Gly– Gly–Ala or Gly–Gly–Asn motifs, or variations thereof.

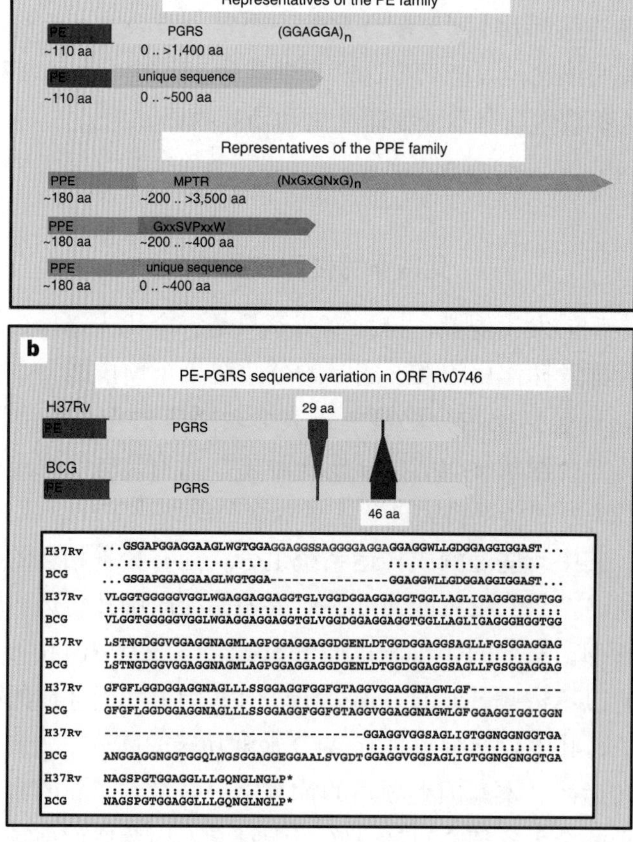

Fig. 5. The PE and PPE protein families. **a**, Classification of the PE and PPE protein families. **b**, Sequence

PE 和 PPE 多基因家族　基因组约 10%的编码能力用于编码酸性富含甘氨酸但相互之间不相关的两大蛋白家族，PE 和 PPE 家族，它们的基因聚集成簇（图 1 和图 2），且通常分别基于多个拷贝的被称为 PGRS 的多态性重复序列和主要多态性串联重复序列（MPTR）[31,32]。命名为 PE 和 PPE 是因为在大多数情况下靠近 N 端有脯氨酸–谷氨酸（PE）和脯氨酸–脯氨酸–谷氨酸（PPE）基序 [33]。PE 蛋白家族的 99 个成员都有一个高度保守的 N 末端结构域，该结构域含有约 110 个氨基酸残基，预测为球状结构，其后是大小、序列和重复拷贝数都不同的 C 端片段（图 5）。系统发育分析将 PE 家族分为几个亚家族。其中最大的是高度重复的 PGRS 类，包含 61 个成员；其他亚家族成员 C 端结构域的序列相似性很有限（图 5）。PE 蛋白的预测分子量差别很大，因为少数成员只包含 N 端结构域，而大多数成员具有大小从 100 到 1,400 个残基不等的 C 端延伸。PGRS 蛋白甘氨酸含量高（达 50%），这是甘氨酸–甘氨酸–丙氨酸或甘氨酸–甘氨酸–天冬酰氨基序或其他变异基序多次串联重复的结果，或者是由于突变导致的。

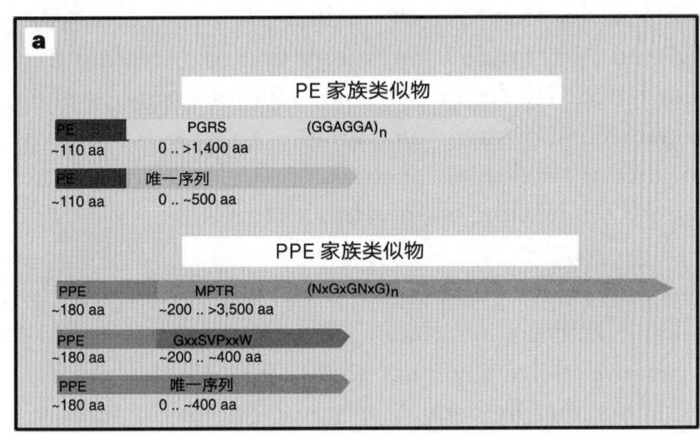

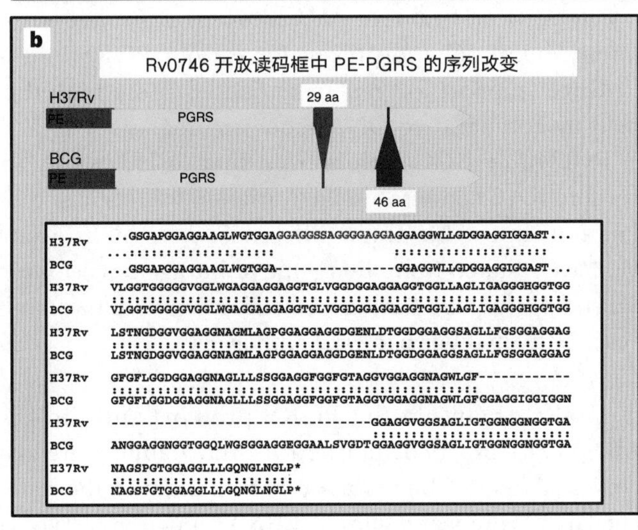

图 5. PE 和 PPE 蛋白家族。**a**，PE 和 PPE 蛋白家族的分类。**b**，结核分枝杆菌 H37Rv 和牛分枝杆菌卡介

variation between *M. tuberculosis* H37Rv and *M. bovis* BCG-Pasteur in the PE-PGRS encoded by open reading frame (ORF) Rv0746.

The 68 members of the PPE protein family (Fig. 5) also have a conserved N-terminal domain that comprises ~180 amino-acid residues, followed by C-terminal segments that vary markedly in sequence and length. These proteins fall into at least three groups, one of which constitutes the MPTR class characterized by the presence of multiple, tandem copies of the motif Asn–X–Gly–X–Gly–Asn–X–Gly. The second subgroup contains a characteristic, well-conserved motif around position 350, whereas the third contains proteins that are unrelated except for the presence of the common 180-residue PPE domain.

The subcellular location of the PE and PPE proteins is unknown and in only one case, that of a lipase (Rv3097), has a function been demonstrated. On examination of the protein database from the extensively sequenced *M. leprae*[15], no PGRS- or MPTR-related polypeptides were detected but a few proteins belonging to the non-MPTR subgroup of the PPE family were found. These proteins include one of the major antigens recognized by leprosy patients, the serine-rich antigen[34]. Although it is too early to attribute biological functions to the PE and PPE families, it is tempting to speculate that they could be of immunological importance. Two interesting possibilities spring to mind. First, they could represent the principal source of antigenic variation in what is otherwise a genetically and antigenically homogeneous bacterium. Second, these glycine-rich proteins might interfere with immune responses by inhibiting antigen processing.

Several observations and results support the possibility of antigenic variation associated with both the PE and the PPE family proteins. The PGRS member Rv1759 is a fibronectin-binding protein of relative molecular mass 55,000 (ref. 35) that elicits a variable antibody response, indicating either that individuals mount different immune responses or that this PGRS protein may vary between strains of *M. tuberculosis*. The latter possibility is supported by restriction fragment length polymorphisms for various PGRS and MPTR sequences in clinical isolates[33]. Direct support for genetic variation within both the PE and the PPE families was obtained by comparative DNA sequence analysis (Fig. 5). The gene for the PE–PGRS protein Rv0746 of BCG differs from that in H37Rv by the deletion of 29 codons and the insertion of 46 codons. Similar variation was seen in the gene for the PPE protein Rv0442 (data not shown). As these differences were all associated with repetitive sequences they could have resulted from intergenic or intragenic recombinational events or, more probably, from strand slippage during replication[32]. These mechanisms are known to generate antigenic variability in other bacterial pathogens[36].

There are several parallels between the PGRS proteins and the Epstein–Barr virus nuclear antigens (EBNAs). Members of both polypeptide families are glycine-rich, contain extensive Gly–Ala repeats, and exhibit variation in the length of the repeat region between different isolates. The Gly–Ala repeat region of EBNA1 functions as a *cis*-acting inhibitor of the ubiquitin/proteasome antigen-processing pathway that generates peptides presented in

苗–巴斯德开放阅读框(ORF)Rv0746 编码的 PE–PGRS 的序列变异。

PPE 蛋白家族的 68 个成员(图 5)也有保守的 N 端结构域，由 180 个氨基酸残基组成，其后是序列和长度都有很大差异的 C 端片段。这些蛋白质至少可分为三组，其中一组组成了 MPTR 类，其特点是含有多个天冬酰胺–X–甘氨酸–X–甘氨酸–天冬酰胺–X–甘氨酸基序的串联拷贝。第二组在位置 350 左右包含一个独特的高保守基序，而第三组包含的蛋白，除了都含有长 180 个氨基酸残基的 PPE 结构域外没有其他相关性。

PE 和 PPE 蛋白的亚细胞定位未知，仅在一个脂酶(Rv3097)的例子中证明是有功能的。对全面测序的麻风分枝杆菌 [15] 蛋白数据库进行检测，没有检测到 PGRS 或 MPTR 相关多肽，但发现了少量非 MPTR 亚组的 PPE 蛋白家族。这些蛋白质包括麻风病患者识别的一个主要抗原——富含丝氨酸的抗原 [34]。尽管认定 PE 和 PPE 家族有生物学功能还为时过早，但对它们可能有免疫学功能的推测是很吸引人的。我们想到两个有趣的可能性。第一，它们可能代表结核分枝杆菌抗原变异的主要来源，否则结核分枝杆菌将是遗传和抗原均一的细菌。第二，这些富含甘氨酸的蛋白质可能会通过抑制抗原加工干扰免疫应答。

少量观察数据和结果支持抗原变异与 PE 和 PPE 家族蛋白存在相关的可能性。PGRS 成员 Rv1759 是纤连蛋白结合蛋白，相对分子量为 55,000(参考文献 35)，诱导不同的抗体反应，这表明要么不同个体产生了不同的免疫应答，要么结核分枝杆菌不同菌株的 PGRS 蛋白是不同的。后者的可能性得到临床分离株中不同的 PGRS 和 MPTR 序列的限制性片段长度多态性的支持 [33]。PE 和 PPE 家族内的遗传变异得到了比较 DNA 序列分析的直接支持(图 5)。BCG 的 PE–PGRS 蛋白 Rv0746 的基因与 H37Rv 中的不同，前者含 29 个密码子的缺失和 46 个密码子的插入。类似的变化在 PPE 蛋白 Rv0442 的基因中也可以看到(数据未显示)。由于这些差异都与重复序列相关，它们可能来自基因间或基因内的重组事件，或者更可能是复制过程中链的滑动引起的 [32]。已知其他病原菌可利用这些机制产生抗原变异 [36]。

PGRS 蛋白和 EB 病毒核抗原(EBNAs)之间有一些相似之处。两个多肽家族成员都富含甘氨酸，含有丰富的甘氨酸–丙氨酸重复序列，不同分离株重复区域的长度不同。EBNA1 的甘氨酸–丙氨酸重复区域是泛素/蛋白酶体抗原加工途径的顺式作用抑制剂，这一途径产生的肽出现在主要组织相容性复合物(MHC)I 类分子存在的

the context of major histocompatibility complex (MHC) class I molecules[37,38]. MHC class I knockout mice are very susceptible to *M. tuberculosis*, underlining the importance of a cytotoxic T-cell response in protection against disease[3,39]. Given the many potential effects of the PPE and PE proteins, it is important that further studies are performed to understand their activity. If extensive antigenic variability or reduced antigen presentation were indeed found, this would be significant for vaccine design and for understanding protective immunity in tuberculosis, and might even explain the varied responses seen in different BCG vaccination programmes[40].

Pathogenicity. Despite intensive research efforts, there is little information about the molecular basis of mycobacterial virulence[41]. However, this situation should now change as the genome sequence will accelerate the study of pathogenesis as never before, because other bacterial factors that may contribute to virulence are becoming apparent. Before the completion of the genome sequence, only three virulence factors had been described[41]: catalase-peroxidase, which protects against reactive oxygen species produced by the phagocyte; *mce*, which encodes macrophage-colonizing factor[42]; and a sigma factor gene, *sigA* (aka *rpoV*), mutations in which can lead to attenuation[41]. In addition to these single-gene virulence factors, the mycobacterial cell wall[4] is also important in pathology, but the complex nature of its biosynthesis makes it difficult to identify critical genes whose inactivation would lead to attenuation.

On inspection of the genome sequence, it was apparent that four copies of *mce* were present and that these were all situated in operons, comprising eight genes, organized in exactly the same manner. In each case, the genes preceding *mce* code for integral membrane proteins, whereas *mce* and the following five genes are all predicted to encode proteins with signal sequences or hydrophobic stretches at the N terminus. These sets of proteins, about which little is known, may well be secreted or surface-exposed; this is consistent with the proposed role of Mce in invasion of host cells[42]. Furthermore, a homologue of *smpB*, which has been implicated in intracellular survival of *Salmonella typhimurium*, has also been identified[43]. Among the other secreted proteins identified from the genome sequence that could act as virulence factors are a series of phospholipases C, lipases and esterases, which might attack cellular or vacuolar membranes, as well as several proteases. One of these phospholipases acts as a contact-dependent haemolysin (N. Stoker, personal communication). The presence of storage proteins in the bacillus, such as the haemoglobin-like oxygen captors described above, points to its ability to stockpile essential growth factors, allowing it to persist in the nutrient-limited environment of the phagosome. In this regard, the ferritin-like proteins, encoded by *bfrA* and *bfrB*, may be important in intracellular survival as the capacity to acquire enough iron in the vacuole is very limited.

Methods

Sequence analysis. Initially, ~3.2 Mb of sequence was generated from cosmids[8] and the remainder was obtained from selected BAC clones[7] and 45,000 whole-genome shotgun clones. Sheared

条件下[37,38]。MHC I 类敲除小鼠非常容易感染结核分枝杆菌，这强调了细胞毒性 T 细胞反应在抗病保护中的重要性[3,39]。鉴于 PPE 和 PE 蛋白质的多种潜在作用，进一步研究了解它们的活性有很重要的意义。如果确实发现大量抗原变异或抗原呈递减少，将对设计疫苗和理解肺结核中保护性免疫有重大意义，甚至可能解释在不同的卡介苗接种方案中看到的不同反应[40]。

致病性 尽管进行了大量研究，但关于分枝杆菌毒性的分子基础仍知之甚少[41]。然而，这种情况现在应该会有所改变，因为基因组序列将前所未有地加快致病机制的研究，其他可能与毒力有关的细菌因子正在变得更加清楚。在完成基因组测序之前，只有三个毒力因子被描述过[41]：过氧化氢酶–过氧化物酶，可以保护其免受巨噬细胞产生的活性氧的伤害；*mce*，编码巨噬细胞集落因子[42]；一个 σ 因子基因，*sigA*(又名 *rpoV*)，该基因的突变可导致减毒[41]。除了这些单基因毒力因子，结核菌的细胞壁[4]在病理学上也很重要，但因其生物合成的复杂性致使鉴定失活导致减毒的关键基因很困难。

检查基因组序列发现，很显然存在 *mec* 的四个拷贝，而且都位于操纵子中，这些操纵子包括 8 个基因，组织形式完全相同。在每个操纵子中，*mec* 前面的基因编码整合膜蛋白，而 *mec* 及其后面的五个基因都被预测编码 N 端有信号肽序列或疏水序列的蛋白质。我们对这些蛋白质知之甚少，它们很可能是分泌或表面暴露的蛋白，这与有人已经提出的 Mce 在侵入宿主细胞中的功能一致[42]。此外，我们还鉴定了 *smpB* 的一个同源基因，该基因与鼠伤寒沙门菌在细胞内的生存密切相关[43]。根据基因组序列鉴定的其他分泌蛋白可能作为毒力因子的蛋白包括一系列磷脂酶 C、脂肪酶和酯酶，它们可能攻击细胞膜或者液泡膜，还包括几个蛋白酶。其中一个磷脂酶发挥接触依赖性溶血素的作用(斯托克，个人交流)。芽孢杆菌中存在储存蛋白，例如上述血红蛋白样氧捕捉蛋白，表明其具备储存必需生长因子的能力，使之能够在吞噬小体养分有限的环境中生存。在这点上，*bfrA* 和 *bfrB* 编码的铁蛋白样蛋白，可能对细胞内生存很重要，因为在液泡中获得足够铁的能力是非常有限的。

方　法

序列分析 起初，黏粒测序获得了约 3.2 Mb 的序列[8]，其余序列是从选定的 BAC 克隆[7]和 45,000 个全基因组鸟枪克隆中得到的。将黏粒和 BAC 的打断片段(1.4 ~ 2.0 kb)克隆

fragments (1.4–2.0 kb) from cosmids and BACs were cloned into M13 vectors, whereas genomic DNA was cloned in pUC18 to obtain both forward and reverse reads. The PGRS genes were grossly underrepresented in pUC18 but better covered in the BAC and cosmid M13 libraries. We used small-insert libraries[44] to sequence regions prone to compression or deletion and, in some cases, obtained sequences from products of the polymerase chain reaction or directly from BACs[7]. All shotgun sequencing was performed with standard dye terminators to minimize compression problems, whereas finishing reactions used dRhodamine or BigDye terminators (http://www.sanger.ac.uk). Problem areas were verified by using dye primers. Thirty differences were found between the genomic shotgun sequences and the cosmids; twenty of which were due to sequencing errors and ten to mutations in cosmids (1 error per 320 kb). Less than 0.1% of the sequence was from areas of single-clone coverage, and ~0.2% was from one strand with only one sequencing chemistry.

Informatics. Sequence assembly involved PHRAP, GAP4 (ref. 45) and a customized perl script that merges sequences from different libraries and generates segments that can be processed by several finishers simultaneously. Sequence analysis and annotation was managed by DIANA (B.G.B. *et al.*, unpublished). Genes encoding proteins were identified by TB-parse[46] using a hidden Markov model trained on known *M. tuberculosis* coding and non-coding regions and translation-initiation signals, with corroboration by positional base preference. Interrogation of the EMBL, TREMBL, SwissProt, PROSITE[47] and in-house databases involved BLASTN, BLASTX[48], DOTTER (http://www.sanger.ac.uk) and FASTA[49]. tRNA genes were located and identified using tRNAscan and tRNAscan-SE[50]. The complete sequence, a list of annotated cosmids and linking regions can be found on our website (http:// www. sanger.ac.uk) and in MycDB (http://www.pasteur.fr/mycdb/).

(**393**, 537-544; 1998)

S. T. Cole*, R. Brosch*, J. Parkhill, T. Garnier*, C. Churcher, D. Harris, S. V. Gordon*, K. Eiglmeier*, S. Gas*, C. E. Barry III†, F. Tekaia‡, K. Badcock, D. Basham, D. Brown, T. Chillingworth, R. Connor, R. Davies, K. Devlin, T. Feltwell, S. Gentles, N. Hamlin, S. Holroyd, T. Hornsby, K. Jagels, A. Krogh§, J. McLean, S. Moule, L. Murphy, K. Oliver, J. Osborne, M. A. Quail, M.-A. Rajandream, J. Rogers, S. Rutter, K. Seeger, J. Skelton, R. Squares, S. Squares, J. E. Sulston, K. Taylor, S. Whitehead & B. G. Barrell

Sanger Centre, Wellcome Trust Genome Campus, Hinxton CB10 1SA, UK

* Unité de Génétique Moléculaire Bactérienne, and ‡ Unité de Génétique Moléculaire des Levures, Institut Pasteur, 28 rue du Docteur Roux, 75724 Paris Cedex 15, France

† Tuberculosis Research Unit, Laboratory of Intracellular Parasites, Rocky Mountain Laboratories, National Institute of Allergy and Infectious Diseases, National Institutes of Health, Hamilton, Montana 59840, USA

§ Center for Biological Sequence Analysis, Technical University of Denmark, Lyngby, Denmark

Received 15 April; accepted 8 May 1998.

References:

1. Snider, D. E. Jr, Raviglione, M. & Kochi, A. in *Tuberculosis: Pathogenesis, Protection, and Control* (ed. Bloom, B. R.) 2-11 (Am. Soc. Microbiol., Washington DC, 1994).

2. Wheeler, P. R. & Ratledge, C. in *Tuberculosis: Pathogenesis, Protection, and Control* (ed. Bloom, B. R.) 353-385 (Am. Soc. Microbiol., Washington DC, 1994).

3. Chan, J. & Kaufmann, S. H. E. in *Tuberculosis: Pathogenesis, Protection, and Control* (ed. Bloom, B. R.) 271-284 (Am. Soc. Microbiol., Washington DC, 1994).

4. Brennan, P. J. & Draper, P. in *Tuberculosis: Pathogenesis, Protection, and Control* (ed. Bloom, B. R.) 271-284 (Am. Soc. Microbiol., Washington DC, 1994).

5. Kolattukudy, P. E., Fernandes, N. D., Azad, A. K., Fitzmaurice, A. M. & Sirakova, T. D. Biochemistry and molecular genetics of cell-wall lipid biosynthesis in mycobacteria. *Mol. Microbiol.* **24,** 263-270 (1997).

到 M13 载体上，将基因组 DNA 克隆到质粒 pUC18 上获得正向和反向可读片段。PGRS 基因在 pUC18 中出现的频率严重低于正常水平，但在 BAC 和黏粒 M13 文库中覆盖得很好。我们使用小插入片段文库[44]对易压缩或缺失的区域测序，在某些情况下，从 PCR 反应产物中或直接从 BACs 中获得序列[7]。所有鸟枪测序采用标准染料终止剂，以尽量减少压缩问题，终止反应用 dRhodamine 或 BigDye 终止剂(http://www.sanger.ac.uk)。存在问题的区域用染料标记引物验证。在基因组鸟枪测序和黏粒测序的序列之间发现了 30 个差异；其中 20 个是由测序错误引起的，10 个是由黏粒中的突变(1 个错误/320 kb)引起的。小于 0.1%的序列来自单个克隆覆盖的区域，大约 0.2%来自只有一个测序化学反应的一条链。

信息学 序列组装使用了 PHRAP、GAP4(参考文献 45)和定制的 perl 脚本，该脚本用于合并不同文库中的序列，并生成可以同时被几个基因组完成工具处理的片段。序列分析和注释由 DIANA(巴雷尔等，未发表)完成。使用 TB-parse[46]鉴定蛋白质编码基因，使用马尔可夫模型分析已知的结核分枝杆菌编码和非编码区域及翻译起始信号，通过位置碱基偏好进行验证。综合使用 EMBL、TREMBL、SwissProt、PROSITE[47]数据库和内部数据库 BLASTN、BLASTX[48]、DOTTER(http://www.sanger.ac.uk)和 FASTA[49]等。使用 tRNAscan 和 tRNAscan-SE[50]定位和鉴定 tRNA 基因。在我们的网站(http:// www.sanger.ac.uk)上和 MycDB(http://www.pasteur.fr/mycdb/)中可找到完整的基因组序列、一系列已注释的黏粒和相连接区域。

(李梅 翻译；解彬彬 审稿)

6. Sreevatsan, S. *et al.* Restricted structural gene polymorphism in the *Mycobacterium tuberculosis* complex indicates evolutionarily recent global dissemination. *Proc. Natl Acad. Sci. USA* **94,** 9869-9874 (1997).
7. Brosch, R. *et al.* Use of a *Mycobacterium tuberculosis* H37Rv bacterial artificial chromosome library for genome mapping, sequencing and comparative genomics. *Infect. Immun.* **66,** 2221-2229 (1998).
8. Philipp, W. J. *et al.* An integrated map of the genome of the tubercle bacillus, *Mycobacterium tuberculosis* H37Rv, and comparison with *Mycobacterium leprae. Proc. Natl Acad. Sci. USA* **93,** 3132-3137 (1996).
9. Blattner, F. R. *et al.* The complete genome sequence of *Escherichia coli* K-12. *Science* **277,** 1453-1462 (1997).
10. Cole, S. T. & Saint-Girons, I. Bacterial genomics. *FEMS Microbiol. Rev.* **14,** 139-160 (1994).
11. Freiberg, C. *et al.* Molecular basis of symbiosis between *Rhizobium* and legumes. *Nature* **387,** 394-401 (1997).
12. Bardarov, S. *et al.* Conditionally replicating mycobacteriophages: a system for transposon delivery to *Mycobacterium tuberculosis. Proc. Natl Acad. Sci. USA* **94,** 10961-10966 (1997).
13. Mahairas, G. G., Sabo, P. J., Hickey, M. J., Singh, D. C. & Stover, C. K. Molecular analysis of genetic differences between *Mycobacterium bovis* BCG and virulent *M. bovis. J. Bacteriol.* **178,** 1274-1282 (1996).
14. Kunst, F. *et al.* The complete genome sequence of the gram-positive bacterium *Bacillus subtilis. Nature* **390,** 249-256 (1997).
15. Smith, D. R. *et al.* Multiplex sequencing of 1.5 Mb of the *Mycobacterium leprae* genome. *Genome Res.* **7,** 802-819 (1997).
16. Greenacre, M. *Theory and Application of Correspondence Analysis* (Academic, London, 1984).
17. Ratledge, C. R. in *The Biology of the Mycobacteria* (eds Ratledge, C. & Stanford, J.) 53-94 (Academic, San Diego, 1982).
18. Av-Gay, Y. & Davies, J. Components of eukaryotic-like protein signaling pathways in *Mycobacterium tuberculosis. Microb. Comp. Genomics* **2,** 63-73 (1997).
19. Cole, S. T. & Telenti, A. Drug resistance in *Mycobacterium tuberculosis. Eur. Resp. Rev.* **8,** 701S-713S (1995).
20. Riley, M. & Labedan, B. in Escherichia coli *and* Salmonella (ed. Neidhardt, F. C.) 2118-2202 (ASM, Washington, 1996).
21. Mdluli, K. *et al.* Inhibition of a *Mycobacterium tuberculosis* β-ketoacyl ACP synthase by isoniazid. *Science* **280,** 1607-1610 (1998).
22. Banerjee, A. *et al. inhA*, a gene encoding a target for isoniazid and ethionamide in *Mycobacterium tuberculosis. Science* **263,** 227-230 (1994).
23. Hopwood, D. A. Genetic contributions to understanding polyketide synthases. *Chem. Rev.* **97,** 2465-2497 (1997).
24. Minnikin, D. E. in *The Biology of the Mycobacteria* (eds Ratledge, C. & Stanford, J.) 95-184 (Academic, London, 1982).
25. Barry, C. E. III *et al.* Mycolic acids: structure, biosynthesis, and physiological functions. *Prog. Lipid Res.* (in the press).
26. Belisle, J. T. *et al.* Role of the major antigen of *Mycobacterium tuberculosis* in cell wall biogenesis. *Science* **276,** 1420-1422 (1997).
27. Marahiel, M. A., Stachelhaus, T. & Mootz, H. D. Modular peptide synthetases involved in nonribosomal peptide synthesis. *Chem. Rev.* **97,** 2651-2673 (1997).
28. Gobin, J. *et al.* Iron acquisition by *Mycobacterium tuberculosis*: isolation and characterization of a family of iron-binding exochelins. *Proc. Natl Acad. Sci. USA* **92,** 5189-5193 (1995).
29. Young, D. B. & Fruth, U. in *New Generation Vaccines* (eds Levine, M., Woodrow, G., Kaper, J. & Cobon, G. S.) 631-645 (Marcel Dekker, New York, 1997).
30. Sorensen, A. L., Nagai, S., Houen, G., Andersen, P. & Anderson, A. B. Purification and characterization of a low-molecular-mass T-cell antigen secreted by *Mycobacterium tuberculosis. Infect. Immun.* **63,** 1710-1717 (1995).
31. Hermans, P. W. M., van Soolingen, D. & van Embden, J. D. A. Characterization of a major polymorphic tandem repeat in *Mycobacterium tuberculosis* and its potential use in the epidemiology of *Mycobacterium kansasii* and *Mycobacterium gordonae. J. Bacteriol.* **174,** 4157-4165 (1992).
32. Poulet, S. & Cole, S. T. Characterisation of the polymorphic GC-rich repetitive sequence (PGRS) present in *Mycobacterium tuberculosis. Arch. Microbiol.* **163,** 87-95 (1995).
33. Cole, S. T. & Barrell, B. G. in *Genetics and Tuberculosis* (eds Chadwick, D. J. & Cardew, G., *Novartis Foundation Symp. 217*) 160-172 (Wiley, Chichester, 1998).
34. Vega-Lopez, F. *et al.* Sequence and immunological characterization of a serine-rich antigen from *Mycobacterium leprae. Infect. Immun.* **61,** 2145-2153 (1993).
35. Abou-Zeid, C. *et al.* Genetic and immunological analysis of *Mycobacterium tuberculosis* fibronectin-binding proteins. *Infect. Immun.* **59,** 2712-2718 (1991).
36. Robertson, B. D. & Meyer, T. F. Genetic variation in pathogenic bacteria. *Trends Genet.* **8,** 422-427 (1992).
37. Levitskaya, J. *et al.* Inhibition of antigen processing by the internal repeat region of the Epstein-Barr virus nuclear antigen-1. *Nature* **375,** 685-688 (1995).
38. Levitskaya, J., Sharipo, A., Leonchiks, A., Ciechanover, A. & Masucci, M. G. Inhibition of ubiquitin/ proteasome-dependent protein degradation by the Gly-Ala repeat domain of the Epstein-Barr virus nuclear antigen 1. *Proc. Natl Acad. Sci. USA* **94,** 12616-12621 (1997).
39. Flynn, J. L., Goldstein, M. A., Treibold, K. J., Koller, B. & Bloom, B. R. Major histocompatibility complex class-I restricted T cells are required for resistance to *Mycobacterium tuberculosis* infection. *Proc. Natl Acad. Sci. USA* **89,** 12013-12017 (1992).
40. Bloom, B. R. & Fine, P. E. M. in *Tuberculosis: Pathogenesis, Protection, and Control* (ed. Bloom, B. R.) 531-557 (Am. Soc. Microbiol., Washington DC, 1994).
41. Collins, D. M. In search of tuberculosis virulence genes. *Trends Microbiol.* **4,** 426-430 (1996).
42. Arruda, S., Bomfim, G., Knights, R., Huima-Byron, T. & Riley, L. W. Cloning of an *M. tuberculosis* DNA fragment associated with entry and survival inside cells. *Science* **261,** 1454-1457 (1993).
43. Baumler, A. J., Kusters, J. G., Stojikovic, I. & Heffron, F. *Salmonella typhimurium* loci involved in survival within macrophages. *Infect. Immun.* **62,** 1623-1630 (1994).
44. McMurray, A. A., Sulston, J. E. & Quail, M. A. Short-insert libraries as a method of problem solving in genome sequencing. *Genome Res.* **8,** 562-566 (1998).
45. Bonfield, J. K., Smith, K. F. & Staden, R. A new DNA sequence assembly program. *Nucleic Acids Res.* **24,** 4992-4999 (1995).
46. Krogh, A., Mian, I. S. & Haussler, D. A hidden Markov model that finds genes in *E. coli* DNA. *Nucleic Acids Res.* **22,** 4768-4778 (1994).
47. Bairoch, A., Bucher, P. & Hofmann, K. The PROSITE database, its status in 1997. *Nucleic Acids Res.* **25,** 217-221 (1997).

48. Altschul, S., Gish, W., Miller, W., Myers, E. & Lipman, D. A basic local alignment search tool. *J. Mol. Biol.* **215,** 403-410 (1990).

49. Pearson, W. & Lipman, D. Improved tools for biological sequence comparisons. *Proc. Natl Acad. USA* **85,** 2444-2448 (1988).

50. Lowe, T. M. & Eddy, S. R. tRNAscan-SE: a program for improved detection of transfer RNA genes in genomic DNA. *Nucleic Acids Res.* **25,** 955-964 (1997).

Acknowledgements. We thank Y. Av-Gay, F.-C. Bange, A. Danchin, B. Dujon, W. R. Jacobs Jr, L. Jones, M. McNeil, I. Moszer, P. Rice and J. Stephenson for advice, reagents and support. This work was supported by the Wellcome Trust. Additional funding was provided by the Association Francaise Raoul Follereau, the World Health Organisation and the Institut Pasteur. S.V.G. received a Wellcome Trust travelling research fellowship.

Correspondence and requests for materials should be addressed to B.G.B. (barrell@sanger.ac.uk) or S.T.C. (stcole@pasteur.fr). The complete sequence has been deposited in EMBL/GenBank/DDBJ as MTBH37RV, accession number AL123456.

Two Feathered Dinosaurs from Northeastern China

Ji Qiang *et al.*

Editor's Note

***Sinosauropteryx* was soon followed by even more spectacular creatures. This report from Ji and colleagues describes two: *Protarchaeopteryx* and *Caudipteryx* which, like *Sinosauropteryx*, came from Liaoning Province in northeastern PRC. Phylogenetic analysis placed these long-legged, ground-living runners close to the origin of birds, although they were both more primitive than *Archaeopteryx* and presumably incapable of flight. Unlike *Sinosauropteryx*, however, there was no doubting that these creatures had feathers. *Protarchaeopteryx* had a switch of feathers on the end of its tail, and *Caudipteryx* appeared to have feathers fringing its forelimbs. These creatures provided a graphic demonstration that feathers appeared in evolution long before dinosaurs became capable of flight.**

Current controversy over the origin and early evolution of birds centres on whether or not they are derived from coelurosaurian theropod dinosaurs. Here we describe two theropods from the Upper Jurassic/Lower Cretaceous Chaomidianzi Formation of Liaoning Province, China. Although both theropods have feathers, it is likely that neither was able to fly. Phylogenetic analysis indicates that they are both more primitive than the earliest known avialan (bird), *Archaeopteryx*. These new fossils represent stages in the evolution of birds from feathered, ground-living, bipedal dinosaurs.

Dinosauria Owen 1842
Theropoda Marsh 1881
Maniraptora Gauthier 1986
Unnamed clade
Protarchaeopteryx robusta Ji & Ji 1997

Holotype. National Geological Museum of China, NGMC 2125 (Figs 1, 2 and 3).

Locality and horizon. Sihetun area near Beipiao City, Liaoning, China. Jiulongsong Member of Chaomidianzi Formation, Jehol Group[1]. This underlies the Yixian Formation, the age of which has been determined to be Late Jurassic to Early Cretaceous[2-4].

Diagnosis. Large straight premaxillary teeth, and short, bulbous maxillary and dentary teeth, all of which are primitively serrated. Rectrices form a fan at the end of the tail.

Description. The skull of *Protarchaeopteryx* is shorter than the femur (Table 1). There are

中国东北地区两类长羽毛的恐龙

季强等

编者按

继中华龙鸟发现之后更多引人注目的生物相继被发现。这篇季强和同事们发表的文章描述了其中两种：原始祖鸟和尾羽龙，它们跟中华龙鸟一样来自中国东北地区的辽宁省。系统发育分析显示这些长腿、陆地生活的擅跑者与鸟类的起源关系密切，尽管它们都比始祖鸟原始，并且可能不会飞行。不过，与中华龙鸟不同的是，它们毫无疑问确实存在羽毛。原始祖鸟尾巴的末端发育一簇羽毛，尾羽龙的前肢边缘也覆盖羽毛。这些生物为羽毛的出现远远早于恐龙变得能够飞行提供了形象的说明。

最近关于鸟类起源及其早期演化的争论集中在鸟类是否起源于兽脚类恐龙的虚骨龙类。本文描述了在中国辽宁省上侏罗统/下白垩统炒米甸子组的两类兽脚类恐龙。虽然这两类都有羽毛，但很可能都不会飞行。系统发育分析表明两者都比已知最早的鸟类——始祖鸟更加原始。这些新的化石代表了从长有羽毛、地面生活和两足行走的恐龙向鸟类演化的阶段。

恐龙总目 Dinosauria Owen 1842

兽脚亚目 Theropoda Marsh 1881

手盗龙类 Maniraptora Gauthier 1986

未命名分类单元 Unnamed clade

粗壮原始祖鸟 *Protarchaeopteryx robusta* Ji & Ji 1997

正模标本 中国地质博物馆，标本编号 NGMC 2125(图 1、2 和 3)。

产地与层位 中国辽宁省北票市四合屯，热河群炒米甸子组九龙松段[1]。位于义县组之下，义县组时代为晚侏罗世至早白垩世[2-4]。

鉴定特征 大且直的前颌骨齿，短而呈球状的上颌骨齿和齿骨齿，所有的牙齿都是原始的锯齿状。尾羽在尾部末端形成扇形。

描述 原始祖鸟的头骨比股骨短(表 1)。有四颗带锯齿的前颌骨齿(图 1c)，齿

four serrated premaxillary teeth (Fig. 1c), with crown heights of up to 12 mm. Premaxillary teeth of coelophysids[5], compsognathids[6,7] and early birds lack serrations, but premaxillary denticles are present in most other theropods. Six maxillary and seven dentary teeth are preserved (Fig. 1), all of which are less than a quarter the height of the premaxillary teeth. They most closely resemble those of *Archaeopteryx*[8] in shape (Figs 1b, c and 2b, c), but have anterior and posterior serrations (7–10 serrations per mm).

Table 1. Lengths of elements in *Protarchaeopteryx* and *Caudipteryx*

Element	NGMC 2125	NGMC 97-4-A	NGMC 97-9-A
Body length	690	890	725
Skull	70	76	79
Sternal plates	25	36	–
Humerus	88	69	70
Arm (humerus to end of phalange II-2)	297	214	220
Ilium	95	101	–
Ischium	–	77	–
Leg (femur to end of phalange III-4)	450	550	540
Femur	122	147	149
Tibia	160	188	182
Metatarsal III	85	115	117

Length measurements are given in millimetres. NGMC 2125, *Protarchaeopteryx*; NGMC 97-4-A and NGMC 97-9-A, *Caudipteryx*.

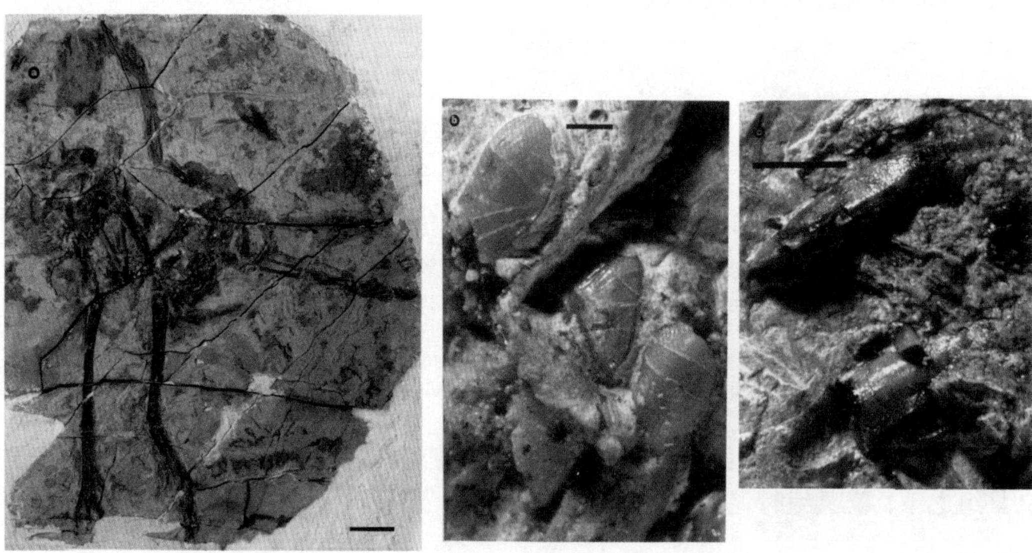

Fig. 1. *Protarchaeopteryx robusta*. **a**, NGMC 2125, holotype. Scale bar, 5 cm. **b**, Fourth to sixth left dentary teeth. Scale bar, 1 mm. **c**, Premaxillary teeth showing small serrations. Scale bar, 5 mm.

冠高达 12 毫米。腔骨龙类[5]、美颌龙类[6,7]和早期鸟类的前颌骨齿缺少锯齿，但前颌骨齿的锯齿出现在大多数其他的兽脚类恐龙中。标本保存了六颗上颌骨齿和七颗齿骨齿（图 1），这些牙齿要短于前颌骨齿高度的四分之一。它们在形态上非常接近始祖鸟的牙齿[8]（图 1b、1c 和图 2b、2c），但其前缘和后缘具有锯齿（每毫米 7 ~ 10 个锯齿）。

表 1. 原始祖鸟和尾羽龙骨骼成分的长度

骨骼成分	NGMC 2125	NGMC 97-4-A	NGMC 97-9-A
身长	690	890	725
头骨	70	76	79
胸骨板	25	36	–
肱骨	88	69	70
前臂（肱骨到第 II 指第 2 指骨末端）	297	214	220
髂骨	95	101	–
坐骨	–	77	–
后肢（股骨到第 III 趾第 4 趾骨末端）	450	550	540
股骨	122	147	149
胫骨	160	188	182
第 III 跖骨	85	115	117

长度测量单位是毫米。NGMC 2125 为原始祖鸟，NGMC 97-4-A 和 NGMC 97-9-A 为尾羽龙。

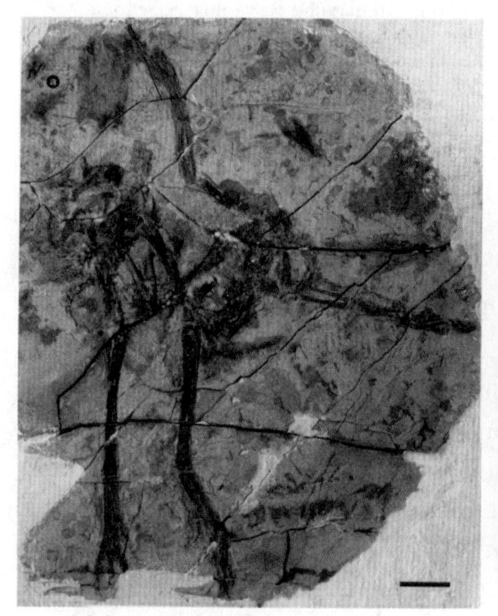

图 1. 粗壮原始祖鸟。**a**，NGMC 2125，正模标本，比例尺为 5 厘米。**b**，左侧第 4 到第 6 齿骨齿，比例尺为 1 毫米。**c**，前颌骨齿的小锯齿，比例尺为 5 毫米。

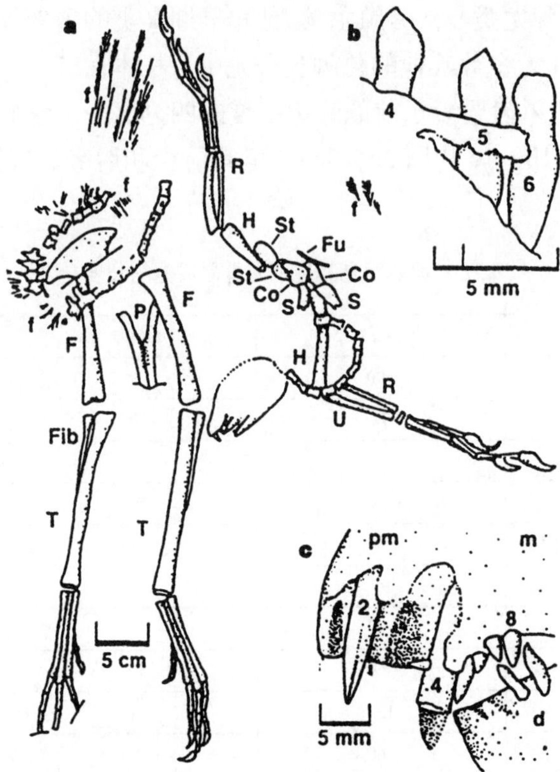

Fig. 2. *Protarchaeopteryx robusta*. **a**, Outline of the specimen shown in Fig. 1a. **b**, Outline of the left dentary teeth shown in Fig. 1b. **c**, Drawing of the front of the jaws, showing the large size of the premaxillary teeth compared with maxillary and dentary ones. Abbreviations: Co, coracoid; d, dentary; F, femur; f, feathers; Fib, fibula; Fu, furcula; H, humerus; m, maxilla; P, pubis; pm, premaxilla; R, radius; S, scapula; St, sternal plate; T, tibia; U, ulna. Numbers represent tooth positions from front to back.

The amphicoelous posterior cervicals are the same length as the posterior dorsals, which have large pleurocoels. If the lengths of missing segments of the tail are accounted for, there were fewer than 28 caudals. Vertebrae increase in length from proximal to mid-caudals, as in most non-avian coelurosaurs.

There are two thin, flat, featureless sternal plates. The clavicles are fused into a broad, U-shaped furcula (interclavicular angle is about 60°) as in *Archaeopteryx*, *Confuciusornis* and many non-avian theropods. The forelimb is shorter than the hindlimb. The arm is shorter (compared to the femur) than it is in birds, but is longer than those of long-armed non-avian coelurosaurs such as dromaeosaurids and oviraptorids (Table 2). The better preserved right wrist of NGMC 2125 has a single semilunate carpal capping the first two metacarpals. The hand has the normal theropod phalangeal formula of 2-3-4-x-x. The manus is longer than either the humerus or radius. Compared to femur length, the hand is more elongate than those of any theropods other than *Archaeopteryx*[9] and *Confuciusornis* (Table 2). More advanced birds such as *Cathayornis* have shorter hands[10]. Phalanges III-1 and III-2 in the hand of *Protarchaeopteryx* are almost the same size, and are about half the length of III-3. The unguals are long and sharp, and keratinous sheaths are preserved on two of them.

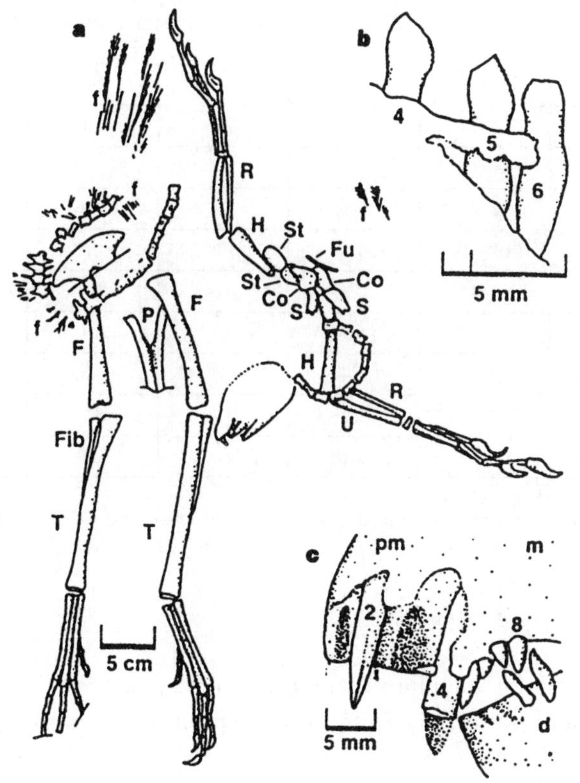

图 2. 粗壮原始祖鸟。**a**，根据图 1a 绘制的标本轮廓。**b**，根据图 1b 绘制的左侧齿骨齿轮廓。**c**，颌部前部素描图，显示与上颌骨齿和齿骨齿相比较大的前颌骨齿。缩写：Co，乌喙骨；d，齿骨；F，股骨；f，羽毛；Fib，腓骨；Fu，叉骨；H，肱骨；m，上颌骨；P，耻骨；pm，前颌骨；R，桡骨；S，肩胛骨；St，胸骨板；T，胫骨；U，尺骨。数字代表牙齿位置的前后顺序。

双凹型的后部颈椎与后部背椎等长，背椎具有大的侧凹。如果算上尾巴缺失的片段，尾椎应该少于 28 节。尾椎的椎体从近端到中部逐渐加长，这与大多数非鸟虚骨龙类一样。

有两块薄而扁平且没有明显特征的胸骨板。锁骨愈合成宽阔的 U 形叉骨（锁骨间夹角为 60 度），类似于始祖鸟、孔子鸟和很多非鸟兽脚类恐龙。前肢比后肢短。前臂（与股骨相比较）要比鸟类的对应部位短，但要比那些非鸟虚骨龙类如驰龙类、窃蛋龙类等较长的前臂长（表 2）。NGMC 2125 的右腕保存很好，单一的半月形腕骨覆盖前两个掌骨。手指指式是兽脚类恐龙通常的 2-3-4-x-x。手部比肱骨和桡骨都长。和股骨长度相比较，手指要比任何兽脚类（除了始祖鸟 [9] 和孔子鸟）的都要长（表 2）。较进步鸟类诸如华夏鸟的手指比较短 [10]。原始祖鸟的第Ⅲ指第 1、2 指骨几乎等长，大约为第 3 指骨长的一半。指爪长而锋利，其中两枚指爪上都保存有角质鞘。

Table 2. Relative proportions of elements in relevant avian and non-avian theropods

Element	Drom	Ov	Tro	Cx	Px	Ax	Con
Arm/F	1.8–2.6	1.5–1.8	1.8	1.5	2.4	3.7	3.9
S/H	0.8	1.0–1.2	–	1.1	–	0.6	0.8
R/H	0.7–0.8	0.8–0.9	0.6–0.7	0.9	0.8	0.9	0.8
Manus/H	0.9–1.2	1.2–1.4	1.3	1.2	1.6	1.2	1.3
Manus/F	1.0	0.7–1.0	0.8	0.6	1.2	1.5	1.6
McI/McII	0.4–0.5	0.4–0.6	0.3	0.4	0.4	0.3	0.4
T/F	1.1–1.4	1.2	1.1–1.2	1.2	1.3	1.4	1.1
Leg/F	3.6	3.3	3.8	3.7	3.7	3.8	3.3
Leg/arm	1.4	1.7	2.1	2.5	1.5	1.1	0.8

All data were collected from original specimens by P.J.C. Ax, *Archaeopteryx*; Con, *Confuciusornis*; Cx, *Caudipteryx*; Drom, dromaeosaurids; F, femur; H, humerus; Mc, metacarpal; Ov, oviraptorids; Px, *Protarchaeopteryx*; R, radius; S, scapula; T, tibia; Tro, troodontids.

The preacetabular blade of the ilium is about the same length as the postacetabular blade. The pubic boot expands posteriorly. Anteriorly, the pubis is not exposed.

The tibia is longer than the femur, as it is in most advanced theropods and early birds. It is not known if the fibula extended to the tarsus.

The metatarsals are separate from each other and the distal tarsals. Metatarsal I is centred halfway up the posteromedial edge of the second metatarsal. In perching birds such as *Sinornis*[9], metatarsal I is positioned near the end of metatarsal II and is retroverted. Its condition in *Archaeopteryx* is intermediate. Pedal unguals are smaller than manual unguals.

A clump of at least six plumulaceous feathers is preserved anterior to the chest, with some showing well-developed vanes (Fig. 3a). Evenly distributed plumulaceous feathers up to 27 mm long are associated with ten proximal caudal vertebrae. Twenty-millimetre plumulaceous feathers are preserved along the lateral side of the right femur and the proximal end of the left femur.

Parts of more than twelve rectrices are preserved[11] attached to the distal caudals. One of the symmetrical tail feathers (Fig. 3b) extends 132 mm from the closest tail vertebra, and has a long tapering rachis with a basal diameter of 1.5 mm. The well-formed pennaceous vanes of *Protarchaeopteryx* show that barbules were present. The vane is 5.3 mm wide on either side of the rachis. At midshaft, five barbs come off the rachis every 5 mm (compared with six in *Archaeopteryx*), and individual barbs are 15 mm long. As in modern rectrices, the barbs at the base of the feather are plumulaceous.

表 2. 相关的鸟类和非鸟兽脚类恐龙的骨骼成分比例

骨骼成分	Drom	Ov	Tro	Cx	Px	Ax	Con
Arm/F	1.8 ~ 2.6	1.5 ~ 1.8	1.8	1.5	2.4	3.7	3.9
S/H	0.8	1.0 ~ 1.2	–	1.1	–	0.6	0.8
R/H	0.7 ~ 0.8	0.8 ~ 0.9	0.6 ~ 0.7	0.9	0.8	0.9	0.8
Manus/H	0.9 ~ 1.2	1.2 ~ 1.4	1.3	1.2	1.6	1.2	1.3
Manus/F	1.0	0.7 ~ 1.0	0.8	0.6	1.2	1.5	1.6
McI/McII	0.4 ~ 0.5	0.4 ~ 0.6	0.3	0.4	0.4	0.3	0.4
T/F	1.1 ~ 1.4	1.2	1.1 ~ 1.2	1.2	1.3	1.4	1.1
Leg/F	3.6	3.3	3.8	3.7	3.7	3.8	3.3
Leg/arm	1.4	1.7	2.1	2.5	1.5	1.1	0.8

所有数据由菲利普 · 柯里从原始标本收集。Arm，前臂；Ax，始祖鸟；Con，孔子鸟；Cx，尾羽龙；Drom，驰龙类；F，股骨；H，肱骨；Leg，后肢；Manus，手部；Mc，掌骨；Ov，窃蛋龙类；Px，原始祖鸟；R，桡骨；S，肩胛骨；T，胫骨；Tro，伤齿龙类。

髂骨的前、后髋臼区域大约等长。耻骨的靴状突向后扩展。耻骨前部没有暴露出来。

胫骨比股骨长，这与大多数进步的兽脚类和早期鸟类一样。但不太清楚腓骨是否延伸到跗骨。

跖骨彼此分离，并与跗骨远端分离。第 I 跖骨位于第 II 跖骨后内侧缘的一半位置。在栖禽类如中国鸟中 [9]，第 I 跖骨的位置接近第 II 跖骨末端并向后倾。始祖鸟的情形介于中间。趾爪比指爪小得多。

一丛至少包括 6 枚绒羽的羽毛保存在胸部前方，有一些显示发育很好的羽片 (图 3a)。均匀分布的长达 27 毫米的绒羽，与近端 10 节尾椎相关联。20 毫米长的绒羽沿着右股骨侧面及左股骨近端保存。

多于 12 根的部分尾羽保存下来，附着在远端尾椎上 [11]。其中一根对称的尾羽 (图 3b) 从最近的尾椎延伸 132 毫米，羽轴长而向末端逐渐变尖，羽轴基部直径为 1.5 毫米。原始祖鸟已经成型的羽片保存有羽小枝。在羽轴任一侧的羽片宽 5.3 毫米。在羽轴中部，每隔 5 毫米有 5 个羽枝从羽轴伸出 (始祖鸟是 6 个)，单个羽枝长 15 毫米。而现生鸟类的尾羽中，羽毛基部的羽枝为似绒羽状。

Fig. 3. *Protarchaeopteryx robusta*, NGMC 2125. **a**, Contour and plumulaceous feathers. Scale bar, 10 mm. **b**, Rectrices. Scale bar, 5 mm.

Maniraptora Gauthier 1986
Unnamed clade

Diagnosis. The derived presence of a short tail (less than 23 caudal vertebrae) and arms with remiges attached to the second digit.

Caudipteryx zoui gen. et sp. nov.

Etymology. "*Caudipteryx*" means "tail feather"; "*zoui*" refers to Zou Jiahua, vice-premier of China and an avid supporter of the scientific work in Liaoning.

Holotype. NGMC 97-4-A (Figs 4 and 5b).

图 3. 粗壮原始祖鸟，NGMC 2125。**a**，廓羽及绒羽，比例尺为 10 毫米。**b**, 尾羽，比例尺为 5 毫米。

手盗龙类 Maniraptora Gauthier 1986

未命名分类单元 Unnamed clade

鉴定特征 具有的进步特征为短尾（尾椎少于 23 节）以及前臂飞羽着生于第 II 指上。

邹氏尾羽龙（新属新种） *Caudipteryx zoui* gen. et sp. nov.

词源 “*Caudipteryx*”意为“尾羽”；“*zoui*”指的是时任中国国务院副总理的邹家华，他大力支持辽宁的化石科学研究。

正模标本 NGMC 97-4-A（图 4 和图 5b）。

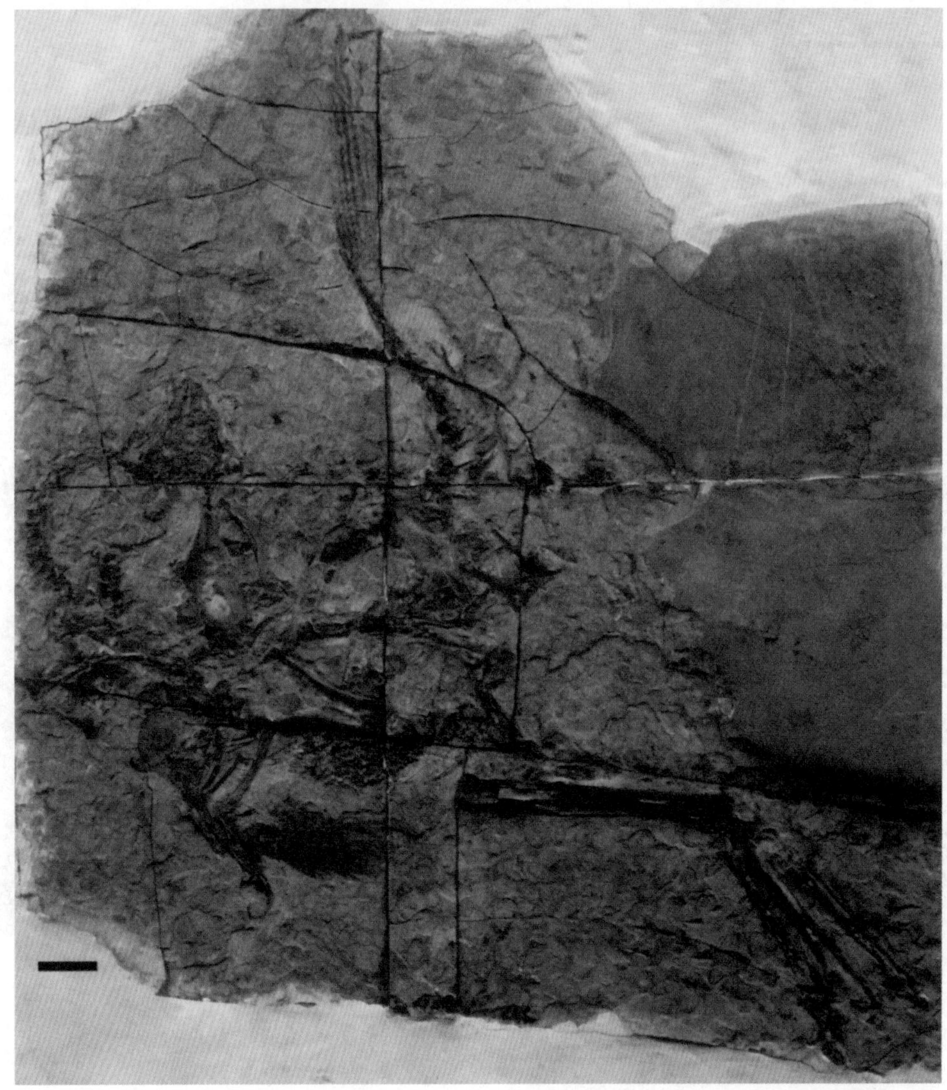

Fig. 4. *Caudipteryx zoui*, holotype, NGMC 97-4-A. Scale bar, 5 cm.

Paratype. NGMC 97-9-A (Fig. 5d).

Locality and horizon. Sihetun area, Liaoning. Jiulongsong Member of the Chaomidianzi Formation.

Diagnosis. Elongate, hooked premaxillary teeth with broad roots; maxilla and dentary edentulous. Tail short (one-quarter of the length of the body). Arm is long for a non-avian theropod; short manual claws. Leg-to-arm ratio, 2.5.

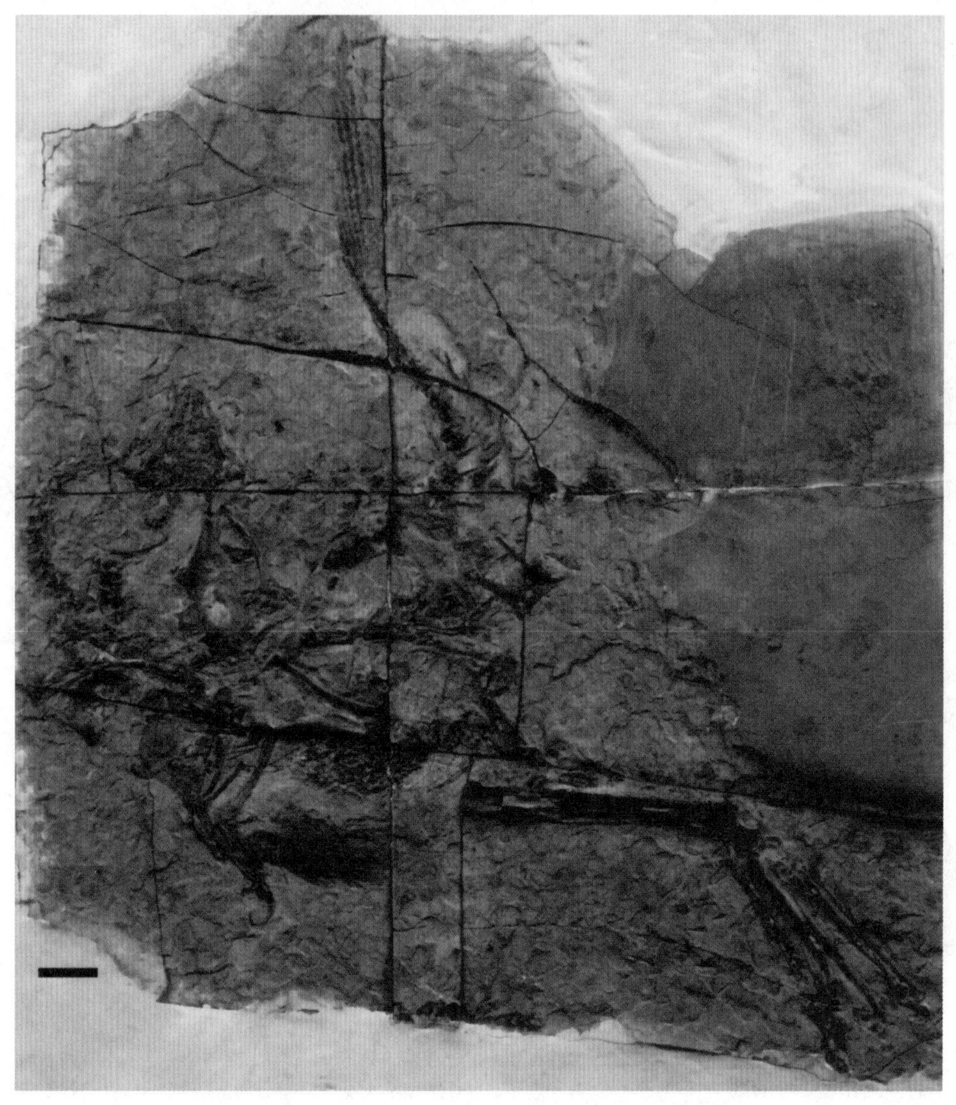

图 4. 邹氏尾羽龙，正模标本，NGMC 97-4-A。比例尺为 5 厘米。

副模标本　NGMC 97-9-A（图 5d）。

产地与层位　辽宁省四合屯，炒米甸子组九龙松段。

鉴定特征　长的钩状前颌骨齿，齿根比较宽；上颌骨和齿骨缺少牙齿。尾巴短（约为身体长度的四分之一）。作为非鸟兽脚类恐龙，其前臂较长；指爪短小。后肢与前臂长度之比为 2.5。

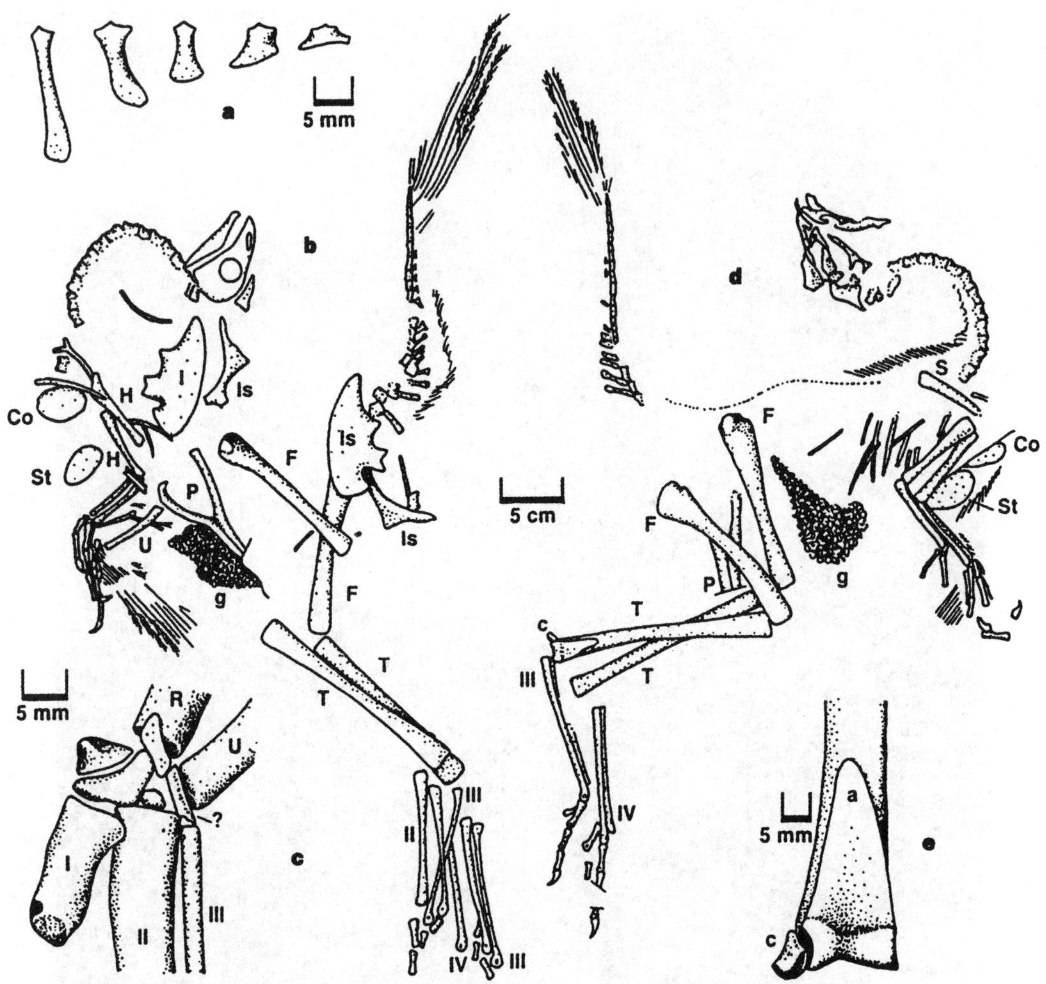

Fig. 5. *Caudipteryx zoui*. **a**, Haemal spines from the fourth, sixth, eighth, eleventh and thirteenth caudal vertebrae (from left to right) of NGMC 97-4-A in left lateral view. **b**, Drawing of the specimen shown in Fig. 4. **c**, Wrist of NGMC 97-4-A. **d**, Drawing of NGMC 97-9-A. **e**, Proximal tarsals of NGMC 97-9-A. Abbreviations: a, astragalus; c, calcaneum; Co, coracoid; F, femur; g, gastroliths; H, humerus; I, ilium; Is, ischium; P, pubis; R, radius; S, scapula; St, sternal plate; T, tibia; U, ulna; ?, possibly fragment of gastralia. Roman numerals represent digit numbers.

Description. The skulls of both specimens of *Caudipteryx* are shorter than the corresponding femora because of a reduction in the length of the antorbital region. The relatively large premaxilla (Figs 6 and 7) borders most of the large external naris. The maxilla and nasal are short, but the frontals and jugals are long. The lacrimal of NGMC 97-4-A is an inverted L-shaped, pneumatic bone. Scleral plates are preserved in the 20-mm-diameter orbits of both specimens. The tall quadratojugal seems to have contacted the squamosal and abutted the lateral surface of the quadrate. The single-headed quadrate is vertical in orientation. The ectopterygoid has a normal theropod hooklike jugal process. There is a broad, beak-like margin at the symphysis of the dentaries. Posteriorly, the dentary bifurcates around a large external mandibular fenestra as in oviraptorids. A well-developed, sliding intramandibular joint is present between dentary and surangular.

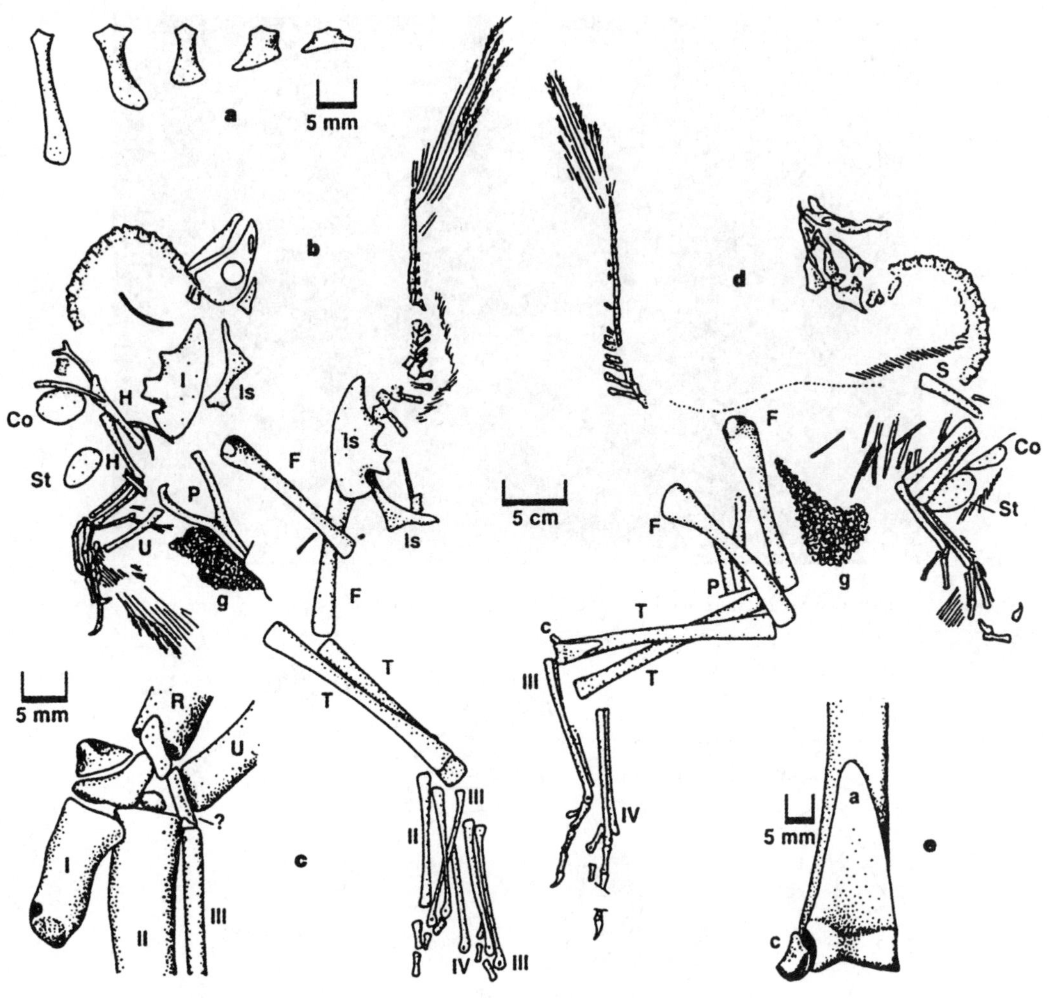

图 5. 邹氏尾羽龙。**a**，NGMC 97-4-A 的第 4、6、8、11、13 尾椎相应的脉弧左侧视（从左至右）。**b**，图 4 中标本的素描图。**c**，NGMC 97-4-A 的腕部。**d**，NGMC 97-9-A 的素描图。**e**，NGMC 97-9-A 的近端跗骨。缩写：a，距骨；c，跟骨；Co，乌喙骨；F，股骨；g，胃石；H，肱骨；I，髂骨；Is，坐骨；P，耻骨；R，桡骨；S，肩胛骨；St，胸骨板；T，胫骨；U，尺骨；?，可能的腹膜肋片段。罗马数字表示指/趾的顺序。

描述 两件尾羽龙标本的头骨长度要比各自的股骨短，这是由眶前区缩短造成的。相对较大的前颌骨（图 6 和图 7）构成较大的外鼻孔的大部分边缘。上颌骨和鼻骨较短，而额骨和轭骨较长。NGMC 97-4-A 标本的泪骨呈倒转的“L”形，为含气骨骼。两件标本的巩膜板均保存在直径 20 毫米的眼眶中。高的方轭骨似乎与鳞骨连接且与方骨的侧面相邻。单头的方骨近于直立。外翼骨具有兽脚类恐龙通常具有的钩状轭骨突。齿骨联合的边缘表面宽阔且呈鸟喙状。像窃蛋龙类一样，齿骨后部围绕着大的外下颌孔分叉。非常发育的、可滑动的颌内关节存在于齿骨与上隅骨之间。

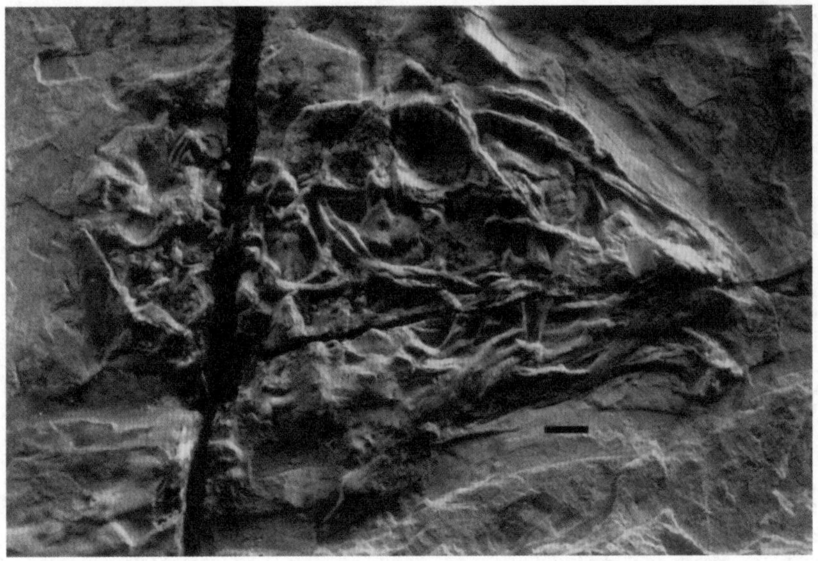

Fig. 6. *Caudipteryx zoui*, skull of NGMC 97-9-A in right lateral view. Scale bar, 1 cm.

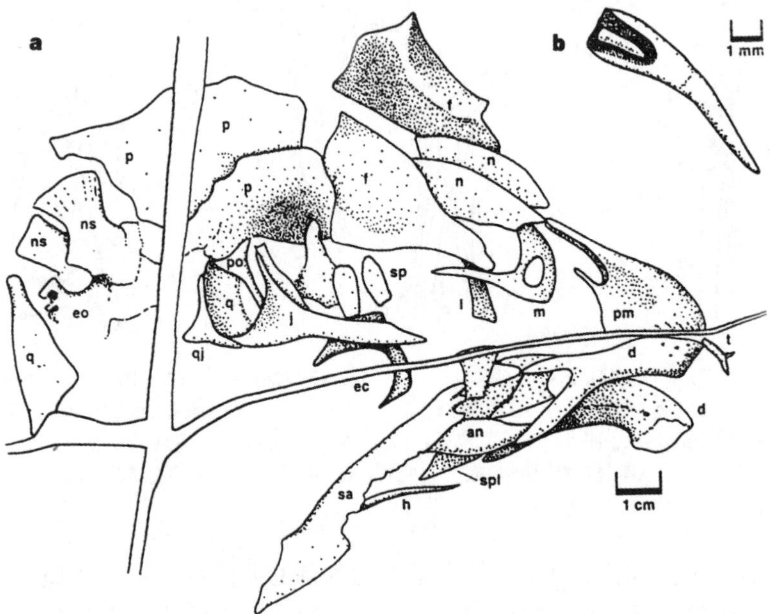

Fig. 7. *Caudipteryx zoui*. **a**, Sketch of skull shown in Fig. 6. **b**, Premaxillary tooth of NGMC 97-4-A, showing resorption pit and germ tooth. Abbreviations: an, angular; d, dentary; ec, ectopterygoid; eo, exoccipital; f, frontal; h, hyoid; j, jugal; l, lacrymal; m, maxilla; n, nasal; ns, neural spine; p, parietal; pm, premaxilla; po, postorbital; q, quadrate; qj, quadratojugal; sa, surangular; sp, scleral plate; spl, splenial; t, premaxillary teeth.

There are four teeth in each premaxilla. They have elongate, needlelike crowns, and the roots are five times wider than the crowns (Fig. 7b). The lingual wall of the root of the third right tooth has been resorbed for the crown of a replacement tooth. The teeth seem

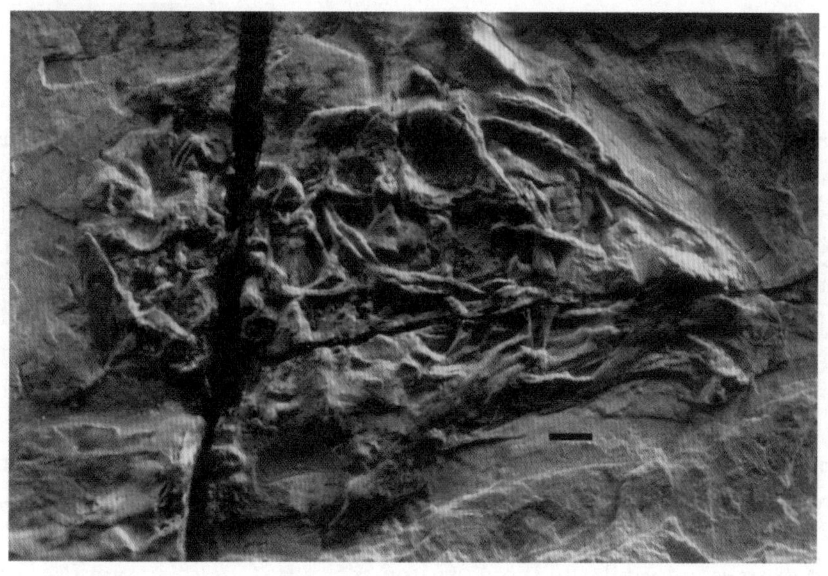

图 6. 邹氏尾羽龙，NGMC 97-9-A 头骨右侧视。比例尺为 1 厘米。

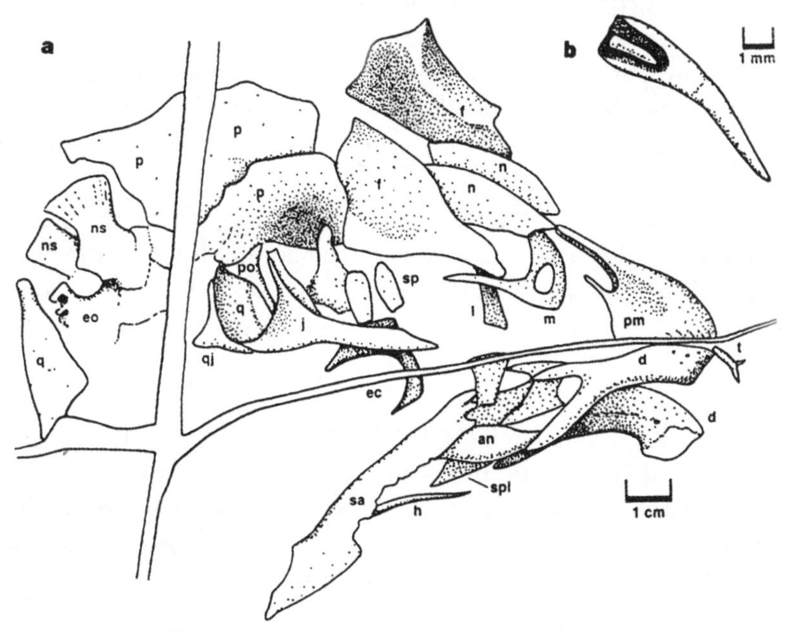

图 7. 邹氏尾羽龙。**a**，图 6 的头骨素描图。**b**，NGMC 97-4-A 的前颌骨齿，显示出再吸收窝和胚齿。缩写：an，隅骨；d，齿骨；ec，外翼骨；eo，外枕骨；f，额骨；h，舌骨；j，轭骨；l，泪骨；m，上颌骨；n，鼻骨；ns，神经棘；p，顶骨；pm，前颌骨；po，眶后骨；q，方骨；qj，方轭骨；sa，上隅骨；sp，巩膜板；spl, 夹板骨；t，前颌骨齿。

每侧前颌骨上各有 4 颗牙齿。它们具有拉长的、像针一样的齿冠，根部是冠部的 5 倍宽（图 7b）。右侧第 3 颗牙齿齿根的舌面被一颗替换齿的齿冠再吸收。牙齿似

to have been procumbent, with an inflection at the gumline. *Caudipteryx* had no maxillary or dentary teeth.

There are ten amphicoelous cervical vertebrae and five sacrals as in most non-avian theropods and *Archaeopteryx*[8,12]. The tail of NGMC 97-4-A is articulated and well-preserved, and includes 22 vertebrae, as in *Archaeopteryx*. It is shorter than the 30-segment tails of oviraptorids. Most other non-avian theropods have much longer tails. Caudals do not become longer posteriorly, as they do in most non-avian theropods and *Archaeopteryx*. Almost two-thirds of the tail of NGMC 97-4-A is preserved as a straight rod, but the vertebrae are not fused. The first six haemal spines are elongate, rodlike structures. More posterior haemal spines decrease in height, but expand anteriorly and posteriorly (Fig. 5a).

Each segment of gastralia is formed by two pairs of slender, tapering rods, as in all non-avian theropods, *Protarchaeopteryx* and early birds[8,9,13,14].

The paired sternals are similar to those of dromaeosaurids and oviraptorids. *Confuciusornis* had a relatively larger, unkeeled sternum. Some short bones with slight expansions at each end are found near the sternal plates of NGMC 97-9-A, and may be sternal ribs.

The scapula is longer than the humerus, whereas the scapula-to-humerus ratio is less than 1.0 in flying birds (Table 2) because of humerus elongation. The clavicles are fused into a broad, U-shaped furcula in NGMC 97-9-A as in *Archaeopteryx*, *Confuciusornis* and many non-avian theropods.

Compared to the humerus, forearm length is similar to that in oviraptorosaurs (Table 2), *Archaeopteryx* and *Protarchaeopteryx*. In more advanced birds[15,16], the radius is longer than the humerus. The external surface of the ulna, as in *Archaeopteryx*[8], lacks any evidence of quill nodes.

There are three carpals preserved in NGMC 97-4-A, including a large semi-lunate one that caps metacarpals I and II as in dromaeosaurids, oviraptorids, troodontids, *Archaeopteryx*, *Confuciusornis* and other birds. Four carpals have been recognized in *Archaeopteryx*[12]. A large triangular radiale sits between the semi-lunate and the radius. A small carpal articulates with the third metacarpal. A thin wedge of bone at the end of the ulna is probably a fragment of gastralia.

The unfused metacarpals and digits of both specimens are well preserved. The third metacarpal is almost as long as the second, but is more slender. The hand has the normal theropod phalangeal formula of 2-3-4-x-x. The manus is longer than either the humerus or the radius, which is a primitive characteristic shared with most non-avian coelurosaurs, *Archaeopteryx*[9] and *Confuciusornis*. In contrast with *Archaeopteryx*, *Confuciusornis*, *Protarchaeopteryx* and many non-avian theropods (ornithomimids, troodontids, dromaeosaurids and oviraptorids), the manus is relatively short compared with the femur.

乎呈平伏状，齿龈有些弯曲。尾羽龙没有上颌骨齿和齿骨齿。

与大多数非鸟兽脚类恐龙和始祖鸟一样[8,12]有10个双凹型的颈椎和5个荐椎。NGMC 97-4-A尾部骨骼相关节且保存完好，同始祖鸟一样共包括22节尾椎。这比窃蛋龙类的30节要短。大多数其他非鸟兽脚类恐龙都具有长得多的尾巴。尾椎没有像大多非鸟兽脚类恐龙和始祖鸟一样向后变长。NGMC 97-4-A几乎三分之二的尾巴呈杆状，但尾椎没有愈合。前6个脉弧是伸长的棒状结构。靠后的脉弧高度减小，但前后向都扩展（图5a）。

每节腹膜肋由两对细长且两端变尖的棒状小骨组成，这与所有非鸟兽脚类恐龙、原始祖鸟以及早期鸟类一样[8,9,13,14]。

成对的胸骨与驰龙类和窃蛋龙类的相似。孔子鸟具有一个相对较大的、无龙骨突的胸骨。NGMC 97-9-A的每个胸骨板附近都有一些短小、两端微弱膨大的骨骼，可能是胸肋。

肩胛骨比肱骨长，但在飞行鸟类中，肱骨的加长使其肩胛骨与肱骨的比率小于1.0（表2）。NGMC 97-9-A的锁骨愈合成宽阔的U形叉骨，就像始祖鸟、孔子鸟和许多非鸟兽脚类恐龙一样。

以肱骨比较，尾羽龙前臂的长度与窃蛋龙类（表2）、始祖鸟和原始祖鸟的是较为相似的。在较进步的鸟类中[15,16]，桡骨比肱骨长。尺骨的外表面比较像始祖鸟[8]，缺乏羽茎节点存在的证据。

NGMC 97-4-A有三个腕骨保存，包括一块大的覆盖第Ⅰ、Ⅱ掌骨的半月形腕骨，这与驰龙类、窃蛋龙类、伤齿龙类、始祖鸟、孔子鸟和其他鸟类一样。始祖鸟保存了四个腕骨[12]。一块大的三角形桡腕骨位于半月形腕骨和桡骨之间，小的腕骨和第Ⅲ掌骨相关节。尺骨末端有块细小的楔形骨，大概是腹膜肋的碎块。

两件标本中未愈合的掌骨和指骨保存很完整。第Ⅲ掌骨和第Ⅱ掌骨差不多一样长，但较纤细。手指指式是兽脚类恐龙通常的2-3-4-x-x。手部比肱骨和桡骨都长，这是与大多数非鸟虚骨龙类、始祖鸟[9]和孔子鸟共有的原始特征。同始祖鸟、孔子鸟和原始祖鸟及许多非鸟兽脚类恐龙（似鸟龙类、伤齿龙类、驰龙类和窃蛋龙类）相反的是，手部相对股骨而言要短。

The curved second manual ungual is about two-thirds the size of the same element in *Protarchaeopteryx*, and is less than 70% the length of the penultimate phalanx.

Pelvic elements are unfused, as they are in all non-avian theropods (except some ceratosaurs) and the most primitive birds[16]. The acetabulum is large, comprising almost a quarter of the length of the ilium (the ratio of acetabulum-to-ilium length is less than 0.11 in birds[17]). It has a deeper, shorter, more squared-off preacetabular region than that of *Protarchaeopteryx*, and closely resembles the ilium of dromaeosaurids[18]. The tapering postacetabular region is lower and longer than the preacetabular. The pubic peduncle is anteroposteriorly elongated, and has a notch (Figs 4 and 5b) in the ventral margin that divides the suture into two surfaces. This notch and the deep pubic peduncle of the ischium are characteristic of opisthopubic pelves. The ischium has no dorsal process such as that found in *Archaeopteryx* and *Confuciusornis*, and the shaft curves down and back. A well-developed ventromedial flange is present, perhaps indicating contact between elements. In general appearance, the ischium most closely resembles those of non-avian coelurosaurs.

The ratio of hindlimb-to-forelimb length is higher than in other coelurosaurs (Table 2) except alvarezsaurids[19], which had exceptionally short arms. The greater trochanter is separated from the lesser trochanter of the femur by a shallow notch, and forms a raised, semi-lunate rim that is similar to the trochanter femoris of birds, troodontids and avimimids.

None of the fibulae is complete, but NGMC 97-9-A has a socket for the distal end of the fibula formed by the calcaneum, astragalus and tibia. The astragalus is not fused to the tibia. The ascending process of NGMC 97-9-A (Fig. 5e) extends 22% of the distance up the front surface of the tibia, compared with 12% in *Archaeopteryx*[12]. As in *Archaeopteryx*[8], *Confuciusornis*[10] and most non-avian theropods, the calcaneum is retained as a separate, disk-like element. Two distal tarsals are positioned over the third and fourth metatarsals, as in *Archaeopteryx*, *Boluochia*[20] and all non-avian theropods that lack fused tarsometatarsals.

The metatarsals of *Caudipteryx* are not fused; this is the plesiomorphic condition expressed in most non-avian theropods. Metatarsal I is centred about a quarter of the way up the posteromedial corner of the second metatarsal. The third is the longest of the metatarsals, and in anterior view completely separates the second and fourth metatarsals, unlike in the arctometatarsalian condition of many theropods[21]. Nevertheless, at midshaft the third metatarsal is thin anteroposteriorly and is triangular in cross-section. The pedal unguals are triangular in cross-section and are about the same size as the manual unguals.

At least fourteen remiges are attached to the second metacarpal, phalanx II-1, and the base of phalanx II-2 of NGMC 97-4-A (Fig. 8a). Each remex has a well-preserved rachis and vane. The most distal remex is less than 30 mm long. The second most distal remex is 63.5 mm long, is symmetrical, and has 6.5-mm-long barbs on either side of the rachis. The fourth most distal primary remex is 95 mm long and is longer than the humerus. Unfortunately, the distal ends of the remaining remiges are not preserved. In flying birds

弯曲的第Ⅱ指爪长度是原始祖鸟同一骨骼成分的三分之二，也比倒数第2指骨长度的70%短。

腰带各骨骼未愈合，类似于所有的非鸟兽脚类恐龙（一些角鼻龙类除外）和大多数原始的鸟类[16]。髋臼很大，几乎占髂骨长度的四分之一（在鸟类中，髋臼与髂骨的比率小于0.11[17]）。它的前髋臼区域比原始祖鸟的深而短，更接近方形，十分接近驰龙类的髂骨[18]。向后端变窄的后髋臼区域要比前髋臼区域低而长。耻骨柄前后延长，在靠近腹侧边缘发育的凹槽（图4和图5b）将缝合线分开为两面。这个凹槽和深的坐骨的耻骨柄是后伸型耻骨具有的特征。坐骨没有如同始祖鸟和孔子鸟一样的背突，坐骨柄向下向后弯曲。存在发育很好的腹中缘，或许表明两骨骼之间的接触。大体上看，坐骨表现出同非鸟虚骨龙类十分相似的特征。

后肢和前肢长度的比率要比其他的虚骨龙类（除了阿瓦拉慈龙类[19]，其前臂异常短小）的高（表2）。股骨的大转子与小转子被一个浅的凹槽分开，并形成了一个突起的、半月形边缘，形态和鸟类、伤齿龙类以及拟鸟龙类的股骨转子相似。

腓骨都保存不完整，但是NGMC 97-9-A的跟骨、距骨和胫骨形成一个腓骨末端的窝。距骨没有和胫骨愈合。NGMC 97-9-A的上升突（图5e）延伸了胫骨的前表面之上距离的22%，而始祖鸟的则是12%[12]。类似于始祖鸟[8]、孔子鸟[10]和大多数非鸟兽脚类恐龙，跟骨仍保留为一分离的圆盘状骨骼。两个远端跗骨的位置超过第Ⅲ、Ⅳ跖骨，像始祖鸟、波罗赤鸟[20]以及所有缺少愈合的跗跖骨的非鸟兽脚类恐龙一样。

尾羽龙的跖骨没有愈合，这是与大多非鸟兽脚类恐龙相似的近祖性状。第Ⅰ跖骨位于第Ⅱ跖骨后内侧角向上四分之一处。第Ⅲ跖骨是最长的跖骨，从前视上看，完全地将第Ⅱ、Ⅳ跖骨分开，这与很多兽脚类恐龙[21]的窄跖型情形不同。然而，第Ⅲ跖骨的骨干前后向窄细，横截面呈三角形。趾爪的横截面是三角形，大约与指爪的尺寸相同。

NGMC 97-4-A标本中，至少有14根飞羽附着在第Ⅱ掌骨和第Ⅱ指的第1指骨上，以及第Ⅱ指第2指骨的基部（图8a）。每根飞羽上很好地保留着羽轴和羽片。最远端的飞羽长不超过30毫米。第二远端的飞羽长63.5毫米，左右对称，羽轴的两侧有6.5毫米长的羽枝。第四远端的初级飞羽长95毫米，比肱骨都长。遗憾的是，

(even *Archaeopteryx*[12]), each remex is longer than they are in *Caudipteryx*, and the most distal remiges are the longest. For example, the remiges of *Archaeopteryx*[22] are more than double the length of the femur. The barbs on either side of the rachis are symmetrical, contrasting with *Archaeopteryx* and modern flying birds[23].

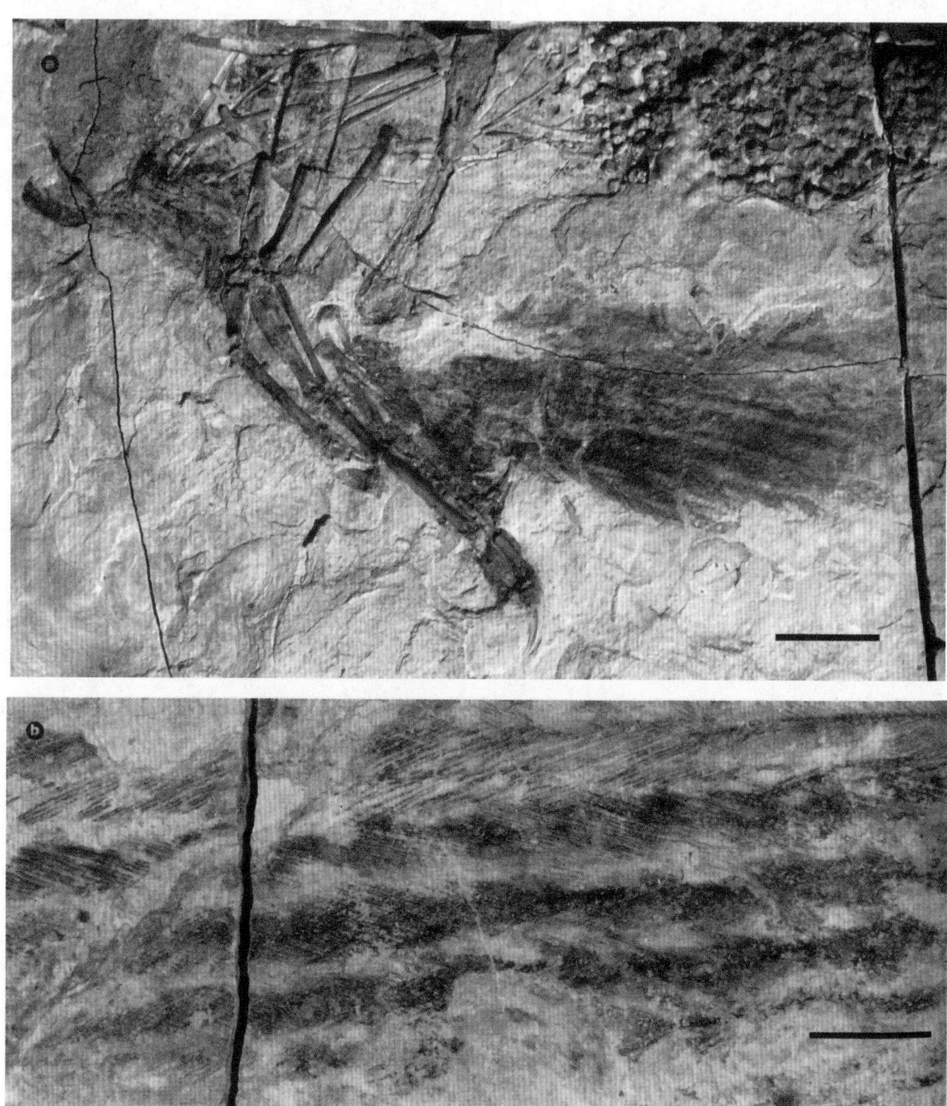

Fig. 8. Feathers of *Caudipteryx zoui*, NGMC 97-4-A. **a**, Remiges of left arm. Scale bar, 1.75 cm. **b**, Rectrices, showing colour banding. Scale bar, 1 cm.

The holotype preserves ten complete and two partial rectrices. Eleven are attached to the left side of the tail, and were probably paired with another eleven feathers on the right side (only the terminal feather is preserved). Two rectrices are attached to each side of the last five or six caudal vertebrae, but not to more anterior ones. NGMC 97-9-A preserves most of nine rectrices. In *Archaeopteryx*, rectrices are associated with all but the first five or

其他飞羽的远端未能保存下来。在飞行鸟类（甚至始祖鸟[12]）中，每个飞羽都要比尾羽龙的飞羽长，且最远端的飞羽是最长的。例如，始祖鸟[22]的飞羽要长于股骨的两倍。与始祖鸟和现生飞行鸟类[23]不同，尾羽龙羽轴两侧的羽枝是左右对称的。

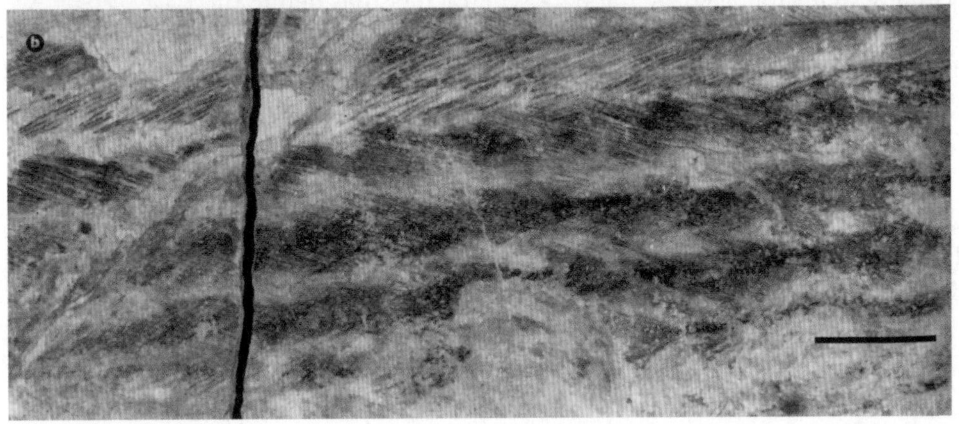

图 8. 邹氏尾羽龙的羽毛，NGMC 97-4-A。**a**，左臂上的飞羽，比例尺为 1.75 厘米。**b**，尾羽，显示出色带，比例尺为 1 厘米。

正模标本保存了 10 根完整的和 2 根局部的尾羽。其中 11 根羽毛附着在尾巴的左侧，可能右侧存在另外 11 根与之成对的羽毛（只有末端的羽毛保留着）。2 根尾羽附着在最后 5 或 6 节尾椎的两侧，但没有更靠前一些。NGMC 97-9-A 保留着 9 根尾羽的大部分。在始祖鸟中，尾羽和所有尾椎（除了前 5 或 6 节）相互关联[12,22]。每个

six caudals[12,22]. Each rachis has a basal diameter of 0.74 mm and tapers distally. All the feathers appear to be symmetrical (Fig. 8b), although in most cases the tips of the barbs of adjacent feathers overlap. The vane of the sixth feather is 6 mm wide on either side of the rachis.

The body of NGMC 97-4-A, especially the hips and the base of the tail, is covered by small, plumulaceous feathers of up to 14 mm long.

Both specimens have concentrations of small polished and rounded pebbles in the stomach region. These gastroliths are up to 4.5 mm in diameter, although most are considerably less than 4 mm wide.

Phylogenetic Analysis

We examined the systematic positions of *Protarchaeopteryx* and *Caudipteryx* by coding these specimens for the 90 characters used in an analysis of avialan phylogeny[24] (for a matrix of these characters, see Supplementary Information). Characters were unordered, and a tree was produced using the branch-and-bound option of PAUP[25]. We rooted the tree with Velociraptorinae[26,27]. A single tree resulted with a length of 110 steps, a retention index of 0.849 and a consistency index of 0.855. Analysis shows *Caudipteryx* to be the sister group to the Avialae, and *Protarchaeopteryx* to be unresolved from the Velociraptorinae root (Fig. 9). The placement of *Protarchaeopteryx* as the sister group to *Caudipteryx*+Avialae, as the sister group to Velociraptorinae, or as the sister group to Velociraptorinae+(*Caudipteryx*+Avialae) are equally well supported by the data. Characters that define the *Caudipteryx*+Avialae clade in the shortest tree include unambiguous (uninfluenced by missing data or optimization) characters 2 and 12 and several more ambiguous ones (characters 4, 5, 10, 11, 15, 19, 24, 37, 85 and 86). *Caudipteryx* is separated from the Avialae by three unambiguous characters (7, 8 and 71) and additional ambiguous ones (characters 5, 6, 9, 10, 11, 18, 24, 39, 40, 56 and 69). The important characteristic of this phylogeny is that the Avialae (not including *Protarchaeopteryx* and *Caudipteryx*) is monophyletic; this placement is supported by the unequivocal presence of a quadratojugal that is joined to the quadrate by a ligament[17] (character 7), the absence of a quadratojugal squamosal contact (character 8) and a reduced or absent process of the ischium (character 71).

羽轴的基部直径是0.74毫米，并且末端越来越细。所有的羽毛都是对称的（图8b），尽管在大多数情况下相邻羽毛的羽枝末梢是相互交叠的。第6根羽毛羽轴两侧的羽片均为6毫米宽。

在NGMC 97-4-A的身体上，特别是臀部和尾巴基部，覆盖着小的绒羽，长度可达14毫米。

两件标本在胃部区域具有密集的磨光的圆形小卵石。这些胃石的直径最大可达4.5毫米，不过大部分胃石的宽度远小于4毫米。

系统发育分析

我们将标本编码到初鸟类包含90个特征的系统发育分析中，检验了原始祖鸟和尾羽龙的系统发育位置[24]（这些特征的矩阵可见补充信息。编者注：本书未收录补充信息）。特征是无序的，系统树是通过PAUP程序[25]的分支界定计算的。我们以疾走龙类作为系统树的根[26,27]。运行110步得出单一的系统树，保留指数是0.849，稠度指数是0.855。分析显示尾羽龙和初鸟类是姊妹群，而原始祖鸟在疾走龙类分支上的位置并未能解决（图9）。原始祖鸟与尾羽龙+初鸟类构成姊妹群，或与疾走龙类构成姊妹群，或与疾走龙类+（尾羽龙+初鸟类）构成姊妹群，这些结果同等程度地被数据所支持。在该最短的树中，定义尾羽龙+初鸟类这一分支的特征包括确定的（未受缺失数据或最优化影响）特征2和12，以及另外一些不甚明确的特征（特征4、5、10、11、15、19、24、37、85和86）。尾羽龙可从初鸟类中划分开来，这被三个确定的特征（7、8和71）和其他不确定的特征（特征5、6、9、10、11、18、24、39、40、56和69）所支持。这一系统关系树的重要特征是初鸟类（不包括原始祖鸟和尾羽龙）是单系的，它的定位被以下特征所支持：确切存在的方轭骨以及其与方骨以韧带连接[17]（特征7），方轭骨和鳞骨不接触（特征8）以及减弱或缺失的坐骨突（特征71）。

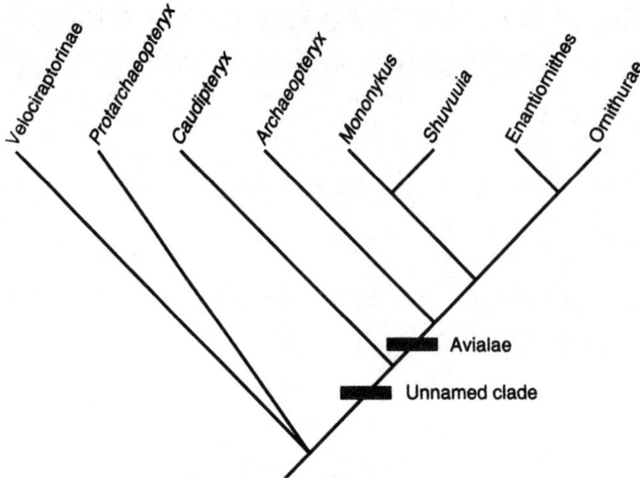

Fig. 9. Cladogram of proposed relationships of *Protoarchaeopteryx* and *Caudipteryx*. This tree is based on 90 characters and has a length of 110 steps.

As characters dealing with feathers cannot be scored relative to outgroup conditions, they were not used in the phylogenetic analysis. However, our analysis indicates that feathers can no longer be used in the diagnosis of the Avialae.

Discussion

The three *Protarchaeopteryx* and *Caudipteryx* individuals were close to maturity at the time of death. The neural spines seem to be fused to cervical and dorsal centra in *Protarchaeopteryx*. Sternal plates ossify late in the ontogeny of non-avian theropods, and are present in both *Caudipteryx* and *Protarchaeopteryx*. Well-ossified sternal ribs, wrist bones and ankle bones in *Caudipteryx* also indicate the maturity of the specimens.

The remiges of *Caudipteryx* and the rectrices of both *Protarchaeopteryx* and *Caudipteryx* have symmetrical veins, whereas even those of *Archaeopteryx* are asymmetrical. Birds with asymmetrical feathers are generally considered to be capable of flight[23], but it is possible that an animal with symmetrical feathers could also fly. Relative arm length of *Protarchaeopteryx* is shorter than that of *Archaeopteryx*, but is longer than in non-avian coelurosaurs. The arms of *Caudipteryx*, in contrast, are shorter than those of most non-avian coelurosaurs; the remiges are only slightly longer than the humerus; and the distal remiges are shorter than more proximal ones. It seems unlikely that this animal was capable of active flight. The relatively long legs of *Protarchaeopteryx* and *Caudipteryx*, both of which have the hallux positioned high and orientated anteromedially, indicate that they were ground-dwelling runners.

Paired rectrices of *Protarchaeopteryx* and *Caudipteryx* are restricted to the end of the tail, whereas in *Archaeopteryx* they extend over more than two-thirds the length of the tail[12].

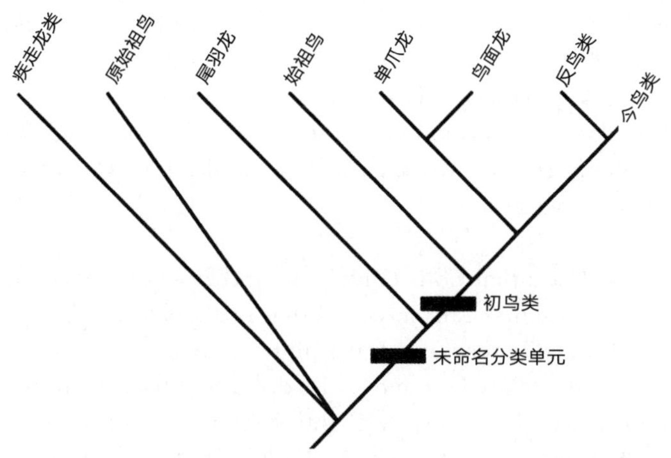

图 9. 原始祖鸟和尾羽龙系统关系的进化分支图。该系统树基于 90 个特征和 110 步得出。

由于羽毛这一特征不能与外类群中的情形作出量化联系，因此羽毛特征没有用于系统发育分析当中。然而，我们的分析指出羽毛不能再作为鉴定初鸟类的特征指标。

讨　论

这三件原始祖鸟和尾羽龙的标本个体在死亡时都比较接近成熟。原始祖鸟的颈椎与背椎椎体似乎与对应的神经棘已经愈合。在非鸟兽脚类恐龙个体发育过程中，胸骨板的骨化相对较晚，但这些在尾羽龙和原始祖鸟中都是存在的。骨化较好的胸肋、腕骨和踝骨同样显示尾羽龙是成熟个体。

尾羽龙的飞羽、原始祖鸟和尾羽龙的尾羽都具有对称的纹理，但始祖鸟的这些羽毛是不对称的。具有不对称羽毛的鸟类一般被认为能够飞行 [23]，但具有对称羽毛的动物也可能会飞。原始祖鸟前臂的相对长度要比始祖鸟的短一些，但还是要比非鸟虚骨龙类的长。而尾羽龙的前臂则相反，要比大多数非鸟虚骨龙类的短；飞羽仅比肱骨长一点点；远端飞羽短于近端飞羽。这些特征都表明这种动物具备主动飞行能力是不太可能的。原始祖鸟和尾羽龙都具相对较长的腿，它们的拇趾位置比较高且指向前内侧，表明它们是陆地生活的奔跑者。

原始祖鸟和尾羽龙成对的尾羽仅局限于尾的末端，而在始祖鸟中尾羽延伸的长度可超过尾部长度的三分之二 [12]。不管是在什么情况保存下来的，我们发现半羽和

Wherever preservation made it possible, we found semi-plumes and down-like feathers around the periphery of the bodies, suggesting that most of the bodies were feather-covered, possibly like *Archaeopteryx*[28]. Feathers found with *Otogornis*[29] were also apparently plumulaceous. Plumulaceous and downy feathers cover the bodies of *Protarchaeopteryx* and *Caudipteryx*, and possibly that of *Sinosauropteryx* as well[7]. This suggests that the original function of feathers was insulation.

Phylogenetic analysis shows that both *Caudipteryx* and *Protarchaeopteryx* lie outside Avialae and are non-avian coelurosaurs. This indicates that feathers are irrelevant in the diagnosis of birds. It can no longer be certain that isolated down and semi-plume feathers[30-33] discovered in Mesozoic rocks belonged to birds rather than to non-avian dinosaurs. Furthermore, the presence of feathers on flightless theropods suggests that the hypothesis that feathers and flight evolved together is incorrect. Finally, the presence of remiges, rectrices and plumulaceous feathers on non-avian theropods provides unambiguous evidence supporting the theory that birds are the direct descendants of theropod dinosaurs.

(**393**, 753-761; 1998)

Ji Qiang*, Philip J. Currie†, Mark A. Norell‡ & Ji Shu-An*

* National Geological Museum of China, Yangrou Hutong 15, Xisi, 100034 Beijing, People's Republic of China

† Royal Tyrrell Museum of Palaeontology, Box 7500, Drumheller, Alberta T0J 0Y0, Canada

‡ American Museum of Natural History, Central Park West at 79th Street, New York, New York 10024-5192, USA

Received 19 January; accepted 27 May 1998.

References:

1. Ji, Q. *et al.* On the sequence and age of the protobird bearing deposits in the Sihetun-Jianshangou area, Beipiao, western Liaoning. *Prof. Pap. Strat. Paleo.* (in the press).

2. Hou, L.-H., Zhou, Z.-H., Martin, L. D. & Feduccia, A. A beaked bird from the Jurassic of China. *Nature* **377,** 616-618 (1995).

3. Smith, P. E. *et al.* Dates and rates in ancient lakes: ^{40}Ar-^{39}Ar evidence for an Early Cretaceous age for the Jehol Group, northeast China. *Can. J. Earth Sci.* **32,** 1426-1431 (1995).

4. Smith, J. B., Hailu, Y. & Dodson, P. in *The Dinofest Symposium, Abstracts* (eds Wolberg, D. L. *et al.*) 55 (Academy of Natural Sciences, Philadelphia, 1998).

5. Colbert, E. H. The Triassic dinosaur *Coelophysis*. *Bull. Mus. N. Arizona* **57,** 1-160 (1989).

6. Ostrom, J. H. The osteology of *Compsognathus longipipes* Wagner. *Zitteliana* **4,** 73-118 (1978).

7. Chen, P.-j., Dong, Z.-m. & Zhen, S.-n. An exceptionally well-preserved theropod dinosaur from the Yixian Formation of China. *Nature* **391,** 147-152 (1998).

8. Wellnhofer, P. A new specimen of *Archaeopteryx* from the Solnhofen Limestone. *Nat. Hist. Mus. Los Angeles County Sci. Ser.* **36,** 3-23 (1992).

9. Sereno, P. C. & Rao, C. G. Early evolution of avian flight and perching: new evidence from the Lower Cretaceous of China. *Science* **255,** 845-848 (1992).

10. Zhou, Z. H. in *Sixth Symposium on Mesozoic Terrestrial Ecosystems and Biota, Short Papers* (eds Sun, A. & Wang, Y.) 209-214 (China Ocean, Beijing, 1995).

11. Ji, Q. & Ji, S. A. Protarchaeopterygid bird (*Protarchaeopteryx* gen. nov.)—fossil remains of archaeopterygids from China. *Chinese Geol.* **238,** 38-41 (1997).

12. Wellnhofer, P. Das fünfte skelettexemplar von *Archaeopteryx*. *Palaeontogr.* A **147,** 169-216 (1974).

13. Wellnhofer, P. Das siebte Examplar von *Archaeopteryx* aus den Solnhofener Schichten. *Archaeopteryx* **11,** 1-48 (1993).

14. Hou, L. H. A carinate bird from the Upper Jurassic of western Liaoning, China. *Chinese Sci. Bull.* **42,** 413-416 (1997).

15. Dong, Z. M. A lower Cretaceous enantiornithine bird from the Ordos Basin of Inner Mongolia, People's Republic of China. *Can. J. Earth Sci.* **30,** 2177-2179 (1993).

16. Forster, C. A., Sampson, S. D., Chiappe, L. M. & Krause, D. W. The theropod ancestry of birds: new evidence from the Late Cretaceous of Madagascar. *Science* **279,** 1915-1919 (1998).

17. Chiappe, L. M., Norell, M. A. & Clark, J. M. The skull of a relative of the stem-group bird *Mononykus*. *Nature* **392,** 275-278 (1998).

18. Norell, M. A. & Makovicky, P. Important features of the dromaeosaur skeleton: information from a new specimen. *Am. Mus. Novit.* **3215,** 1-28 (1997).

19. Perle, A., Chiappe, L. M., Barsbold, R., Clark, J. M. & Norell, M. Skeletal morphology of *Mononykus olecranus* (Theropoda: Avialae) from the Late Cretaceous of Mongolia. *Am. Mus. Novit.* **3105,** 1-29 (1994).

20. Zhou, Z. H. The discovery of Early Cretaceous birds in China. *Courier Forschungsinstitut Senckenberg* **181,** 9-22 (1995).

绒羽状的羽毛围绕在身体周围，初步判断身体表面大部分披有羽毛，这些特点大概有些类似于始祖鸟[28]。在鄂托克鸟[29]上发现的羽毛同样很显然是绒羽。绒羽和绒羽状羽毛覆盖原始祖鸟和尾羽龙的身体，也许中华龙鸟也是这样[7]。这表明羽毛的最初功能是用来保温的。

系统发育分析表明了尾羽龙和原始祖鸟位于初鸟类之外，而属于非鸟虚骨龙类恐龙。这显示羽毛对于定义鸟类是一个毫不相关的特征。发现于中生代岩石中的孤立存在的半羽和绒羽状羽毛[30-33]将不再确切地属于鸟类，而可能为非鸟兽脚类恐龙的羽毛。而且，在不会飞的兽脚类恐龙身上存在羽毛表明羽毛和飞行同步演化的假说是不正确的。最后，非鸟兽脚类恐龙身上的飞羽、尾羽和绒羽的存在，为鸟类是兽脚类恐龙直接后裔的理论提供了确凿的证据。

（张玉光 翻译；姬书安 审稿）

21. Holtz, T. R. Jr The phylogenetic position of the Tyrannosauridae: implications for theropod systematics. *J. Paleontol.* **68,** 1100-1117 (1994).
22. deBeer, G. *Archaeopteryx lithographica. Br. Mus. Nat. Hist.* **244,** 1-68 (1954).
23. Feduccia, A. & Tordoff, H. B. Feathers of *Archaeopteryx*: asymmetric vanes indicate aerodynamic function. *Science* **203,** 1021-1022 (1979).
24. Chiappe, L. M. in *The Encyclopedia of Dinosaurs* (eds Currie, P. J. & Padian, K.) 32-38 (Academic, San Diego, 1997).
25. Swofford, D. & Begle, D. P. *Phylogenetic Analysis Using Parsimony. Version 3.1.1.* (Smithsonian Institution, Washington DC, 1993).
26. Gauthier, J. in *The Origin of Birds and the Evolution of Flight* (ed. Padian, K.) 1-55 (Calif. Acad. Sci., San Francisco, 1986).
27. Holtz, T. R. Jr Phylogenetic taxonomy of the Coelurosauria (Dinosauria: Theropoda). *J. Paleontol.* **70,** 536-538 (1996).
28. Owen, R. On the *Archaeopteryx* of von Meyer, with a description of the fossil remains of a long-tailed species, from the Lithographic Stone of Solenhofen. *Phil. Trans., Lond.* **153,** 33-47 (1863).
29. Hou, L. H. A late Mesozoic bird from Inner Mongolia. *Vert. PalAsiatica* **32,** 258-266 (1994).
30. Kurochkin, E. N. A true carinate bird from Lower Cretaceous deposits in Mongolia and other evidence of Early Cretaceous birds in Asia. *Cretaceous Res.* **6,** 271-278 (1985).
31. Sanz, J. L., Bonapart, J. F. & Lacasa, A. Unusual Early Cretaceous birds from Spain. *Nature* **331,** 433- 435 (1988).
32. Kellner, A. W. A., Maisey, J. G. & Campos, D. A. Fossil down feather from the Lower Cretaceous of Brazil. *Palaeontol.* **37,** 489-492 (1994).
33. Grimaldi, D. & Case, G. R. A feather in amber from the Upper Cretaceous of New Jersey. *Am. Mus. Novit.* **3126,** 1-6 (1995).

Supplementary information is available on *Nature*'s World-Wide Web site (http://www.nature.com) or as paper copy from the London editorial office of *Nature*.

Acknowledgements. We thank A. Brush, B. Creisler, M. Ellison, W.-D. Heinrich, N. Jacobsen, E. and R. Koppelhus, P. Makovicky, A. Milner, G. Olshevsky, J. Ostrom and H.-P. Schultze for advice, access to collections and logistic support; and the National Geographic Society, National Science Foundation (USA), the American Museum of Natural History, National Natural Science Foundation of China and the Ministry of Geology for support. Photographs were taken by O. L. Mazzatenta and K. Aulenback; the latter was also responsible for preliminary preparation of the *Caudipteryx* specimens. Line drawings are by P. J. C.

Correspondence and requests for materials should be addressed to P.J.C. (e-mail: pcurrie@mcd.gov.ab.ca).

Jefferson Fathered Slave's Last Child

E. A. Foster *et al.*

Editor's Note

This paper reports one of the more dramatic examples of the ability, developed over the previous decade or so, to make genetic comparisons between individuals by looking at specific "markers" in their genomes. For many years there had been suggestions that Thomas Jefferson, one of the Founding Fathers of the United States of America and its third president, had fathered one or more children by one of his slaves, named Sally Hemings. The paper reports a genetic analysis of the male-line descendants of Hemings' youngest and eldest sons (Thomas Woodson and Eston Hemings Jefferson, each of whom will have received a Y chromosome in direct paternal succession from these two respective men), which is compared with that of male-line descendants of Jefferson's paternal uncle. The results support the view that Eston, but not Thomas Woodson, was fathered by Thomas Jefferson.

THERE is a long-standing historical controversy over the question of US President Thomas Jefferson's paternity of the children of Sally Hemings, one of his slaves[1-4]. To throw some scientific light on the dispute, we have compared Y-chromosomal DNA haplotypes from male-line descendants of Field Jefferson, a paternal uncle of Thomas Jefferson, with those of male-line descendants of Thomas Woodson, Sally Hemings' putative first son, and of Eston Hemings Jefferson, her last son. The molecular findings fail to support the belief that Thomas Jefferson was Thomas Woodson's father, but provide evidence that he was the biological father of Eston Hemings Jefferson.

In 1802, President Thomas Jefferson was accused of having fathered a child, Tom, by Sally Hemings[5]. Tom was said to have been born in 1790, soon after Jefferson and Sally Hemings returned from France where he had been minister. Present-day members of the African–American Woodson family believe that Thomas Jefferson was the father of Thomas Woodson, whose name comes from his later owner[6]. No known documents support this view.

Sally Hemings had at least four more children. Her last son, Eston (born in 1808), is said to have borne a striking resemblance to Thomas Jefferson, and entered white society in Madison, Wisconsin, as Eston Hemings Jefferson. Although Eston's descendants believe that Thomas Jefferson was Eston's father, most Jefferson scholars give more credence to the oral tradition of the descendants of Martha Jefferson Randolph, the president's daughter. They believe that Sally Hemings' later children, including Eston, were fathered by either

杰斐逊是其奴隶最小孩子的生父

福斯特等

编者按

这篇论文报道了一种非常引人注目的、在过去十年发展起来的科学手段，该手段通过在不同个体基因组中寻找特异性的“标记”来进行遗传比对。许多年以来，一直有人认为托马斯·杰斐逊——美国的开国之父之一以及第三任总统，与他的一个奴隶莎丽·海明斯，生下了一个或者多个孩子。这篇论文报道了海明斯最大的和最小的儿子（托马斯·伍德森和艾斯顿·海明斯·杰斐逊）的男性后裔（他们分别从这两个人直接遗传一条父系Y染色体）的遗传分析，并与托马斯·杰斐逊的伯父的男性后裔的基因进行比较。结果支持托马斯·杰斐逊是艾斯顿而非托马斯·伍德森的父亲这一观点。

关于美国总统托马斯·杰斐逊与其奴隶之一莎丽·海明斯的孩子之间的血缘关系，长期以来一直是历史上争论不休的一个问题[1-4]。从科学的角度来看待和分析这个争论，我们比较了不同来源的Y染色体DNA的单体型。其中一些标本来自托马斯·杰斐逊的伯父菲尔德·杰斐逊的男性后裔，另一些标本来自托马斯·伍德森的男性后裔，他被认定为莎丽·海明斯的第一个儿子，以及来自她最小的儿子艾斯顿·海明斯·杰斐逊的男性后裔的样本。分子层面的比对结果无法支持我们相信托马斯·杰斐逊是托马斯·伍德森的生父，但是提供了证据表明他是艾斯顿·海明斯·杰斐逊生物学上的生父。

1802年，总统托马斯·杰斐逊被指控与莎丽·海明斯生下了一个孩子，名叫汤姆[5]。据说汤姆生于1790年，就在杰斐逊和莎丽·海明斯从法国返回不久，杰斐逊曾经在法国做过部长。现在的非裔美籍伍德森家族成员都相信托马斯·杰斐逊是托马斯·伍德森的生父，后者的名字来自他之后的主人[6]。但是，没有任何已知的文件支持这种观点。

莎丽·海明斯至少有四个孩子。她最小的孩子艾斯顿（生于1808年），据说与托马斯·杰斐逊惊人地相似，他以艾斯顿·海明斯·杰斐逊的身份进入威斯康星州麦迪逊城的白人社会中。尽管艾斯顿的男性后裔相信托马斯·杰斐逊是艾斯顿的生父，但是研究杰斐逊的绝大多数学者更愿意相信总统的女儿玛莎·杰斐逊·伦道夫的后裔的口口相传。他们认为莎丽·海明斯所生的孩子，包括艾斯顿在内，他们的生父

Samuel or Peter Carr, sons of Jefferson's sister, which would explain their resemblance to the president.

Because most of the Y chromosome is passed unchanged from father to son, apart from occasional mutations, DNA analysis of the Y chromosome can reveal whether or not individuals are likely to be male-line relatives. We therefore analysed DNA from the Y chromosomes of: five male-line descendants of two sons of the president's paternal uncle, Field Jefferson; five male-line descendants of two sons of Thomas Woodson; one male-line descendant of Eston Hemings Jefferson; and three male-line descendants of three sons of John Carr, grandfather of Samuel and Peter Carr (Fig. 1a). No Y-chromosome data were available from male-line descendants of President Thomas Jefferson because he had no surviving sons.

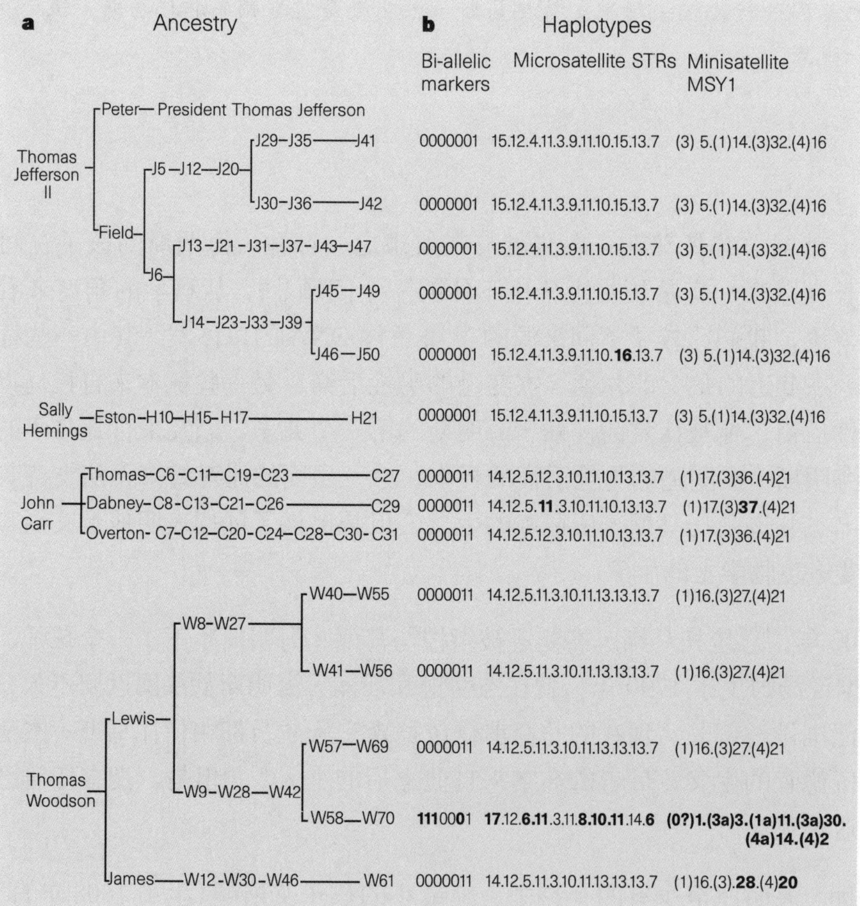

Fig. 1. Male-line ancestry and haplotypes of participants. **a,** Ancestry. Numbers correspond to reference numbers and names in more detailed genealogical charts for each family. **b,** Haplotypes. Entries in bold highlight deviations from the usual patterns for the group of descendants. **Bi-allelic markers**. Order of loci: YAP-SRYm8299-sY81-LLY22g-Tat-92R7-SRYm1532. 0, ancestral state; 1, derived state. **Microsatellite short tandem repeats (STRs)**. Order of loci: 19-388-389A-389B-389C-389D-390-391-

是塞缪尔或者彼得·卡尔——杰斐逊妹妹的儿子，这样也可以解释为什么他们的染色体会和杰斐逊总统之间有着相似之处。

绝大多数的 Y 染色体都会毫无改变地由生父传给儿子，除非偶然发生突变，因此，对 Y 染色体的 DNA 分析可以揭示不同个体之间是否具有某种亲缘关系。我们分析了来自不同人的 Y 染色体上的 DNA：总统的伯父菲尔德·杰斐逊的两个儿子的五个男性后裔；托马斯·伍德森的两个儿子的五个男性后裔；艾斯顿·海明斯·杰斐逊的男性后裔；约翰·卡尔的三个儿子的三个男性后裔，约翰·卡尔是塞缪尔和彼得·卡尔的祖父（见图 1a）。样本中没有来自总统托马斯·杰斐逊的男性后裔的数据，因为他没有活下来的儿子。

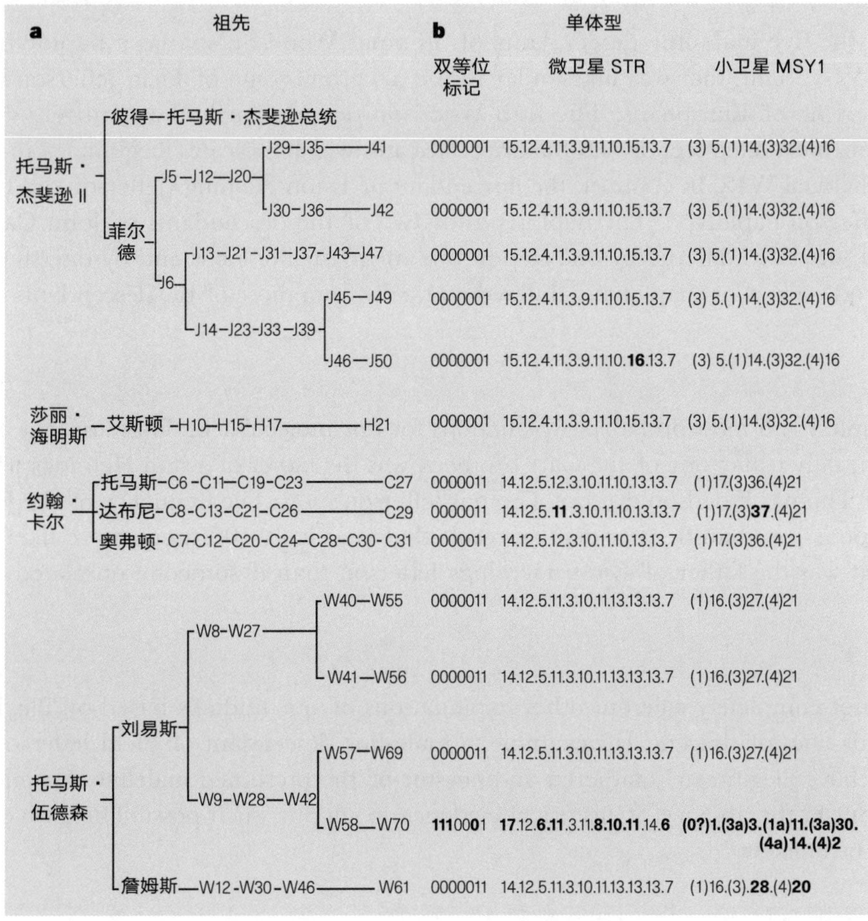

图 1. 男性族谱以及参与者的单体型。**a**，祖先。数字对应相应的人的编码，名字则是更为具体的每个家族的宗族谱图。**b**，单体型。粗体用来标注与整个家族正常遗传情况的差异。**双等位基因标记**。位点的顺序：YAP-SRYm8299-sY81-LLY22g-Tat-92R7-SRYm 1532。0，祖先状态；1，派生状态。**微卫星短串联重复序列（STR）**。位点顺序：19-388-389A-389B-389C-389D-390-391-392-393-dxys156y。在每一个位点

392-393-dxys156y. The number of repeats at each locus is shown. **Minisatellite MSY1.** Each number in brackets represents the sequence type of the repeat unit; the number after it is the number of units with this sequence type. For example, J41 has 5 units of sequence type 3, 14 units of sequence type 1, 32 units of sequence type 3, and 16 units of sequence type 4.

Seven bi-allelic markers (refs 7–12), eleven microsatellites (ref. 13) and the minisatellite MSY1 (ref. 14) were analysed (Fig. 1b). Four of the five descendants of Field Jefferson shared the same haplotype at all loci, and the fifth differed by only a single unit at one microsatellite locus, probably a mutation. This haplotype is rare in the population, where the average frequency of a microsatellite haplotype is about 1.5 per cent. Indeed, it has never been observed outside the Jefferson family, and it has not been found in 670 European men (more than 1,200 worldwide) typed with the microsatellites or 308 European men (690 worldwide) typed with MSY1.

Four of the five male-line descendants of Thomas Woodson shared a haplotype (with one MSY1 variant) that was not similar to the Y chromosome of Field Jefferson but was characteristic of Europeans. The fifth Woodson descendant had an entirely different haplotype, most often seen in sub-Saharan Africans, which indicates illegitimacy in the line after individual W42. In contrast, the descendant of Eston Hemings Jefferson did have the Field Jefferson haplotype. The haplotypes of two of the descendants of John Carr were identical; the third differed by one step at one microsatellite locus and by one step in the MSY1 code. The Carr haplotypes differed markedly from those of the descendants of Field Jefferson.

The simplest and most probable explanations for our molecular findings are that Thomas Jefferson, rather than one of the Carr brothers, was the father of Eston Hemings Jefferson, and that Thomas Woodson was not Thomas Jefferson's son. The frequency of the Jefferson haplotype is less than 0.1 per cent, a result that is at least 100 times more likely if the president was the father of Eston Hemings Jefferson than if someone unrelated was the father.

We cannot completely rule out other explanations of our findings based on illegitimacy in various lines of descent. For example, a male-line descendant of Field Jefferson could possibly have illegitimately fathered an ancestor of the presumed maleline descendant of Eston. But in the absence of historical evidence to support such possibilities, we consider them to be unlikely.

(**396**, 27-28; 1998)

上重复的次数显示在图上。**小卫星 MSY1**。每一个括号中的数字代表重复单元的序列类型；在它之后的数字代表该序列类型单元的数目。例如，J41 个体具有 5 个单元的序列类型 3，14 个单元的序列类型 1，32 个单元的序列类型 3，以及 16 个单元的序列类型 4。

我们对七个双等位基因标记（参考文献 7 ~ 12）、十一个微卫星遗传标记（参考文献 13）以及小卫星 MSY1（参考文献 14）进行了分析研究（图 1b）。来自菲尔德 · 杰斐逊家族五个男性后裔中的四个人在所有位点上都具有同样的单体型，第五个人只是在其中一个微卫星遗传位点上的一个单元上略有不同，这可能是一个突变。这个单体型在人群中出现的概率是很小的，每个微卫星单体型出现的平均概率大概是百分之一点五。事实上，这种单体型从来没有在杰斐逊家族以外被发现过，在微卫星标记的 670 个欧洲男性（在全世界范围内超过 1,200 个）或者 308 个欧洲男性（全世界范围内 690 个男性）中都未曾发现过 MSY1 位点标记。

来自托马斯 · 伍德森五个男性后裔中有四个具有相同的单体型（有一个 MSY1 位点突变），这个单体型与来自菲尔德 · 杰斐逊的 Y 染色体之间并不相似，但却是欧洲人的特征。来自伍德森男性后裔的第五个后裔具有完全不同的单体型，这种单体型在亚撒哈拉非裔中最常见，这说明个体 W42 之后子嗣中存在私生情况。与之相反，艾斯顿 · 海明斯 · 杰斐逊的后代确实具有与菲尔德 · 杰斐逊相同的单体型。来自约翰 · 卡尔的两个后代的单体型是一样的；第三个后代的差别其中之一位于微卫星位点上，另一个位于 MSY1 编码上。来自卡尔男性后裔的单体型与来自菲尔德 · 杰斐逊男性后裔的单体型有着显著的差别。

针对我们分子研究的发现，最简单和最有可能的解释是艾斯顿 · 海明斯 · 杰斐逊的生父是托马斯 · 杰斐逊，而不是卡尔兄弟中的某一个。另外，托马斯 · 伍德森不是托马斯 · 杰斐逊的儿子。杰斐逊单体型出现的频率低于百分之零点一，这个结果说明，总统作为艾斯顿 · 海明斯 · 杰斐逊生父的可能性是一个毫不相干的人作为他的生父的概率的 100 倍。

基于不同谱系的异常情况，我们无法完全排除针对我们发现的其他解释。例如，菲尔德 · 杰斐逊的一个男性后裔有可能非法生下了一个孩子，而他是我们假定的艾斯顿后裔中的祖先。但缺少历史证据来支持这样一种假设，我们认为这是不可能发生的。

（刘振明 翻译；胡松年 审稿）

Eugene A. Foster*, M. A. Jobling†, P. G. Taylor†, P. Donnelly‡, P. de Knijff§, Rene Mieremet§, T. Zerjal¶, C. Tyler-Smith¶
* 6 Gildersleeve Wood, Charlottesville, Virginia 22903, USA e-mail: eafoster@aol.com
† Department of Genetics, University of Leicester, Adrian Building, University Road, Leicester LE1 7RH, UK
‡ Department of Statistics, University of Oxford, South Parks Road, Oxford OX1 3TG, UK
§ MGC Department of Human Genetics, Leiden University, PO Box 9503, 2300 RA Leiden, The Netherlands
¶ Department of Biochemistry, University of Oxford, South Parks Road, Oxford OX1 3QU, UK

References:

1. Peterson, M. D. *The Jefferson Image in the American Mind* 181-187 (Oxford Univ. Press, New York, 1960).
2. Malone, D. *Jefferson the President: First Term, 1801–1805* Appendix II, 494-498 (Little Brown, Boston, MA, 1970).
3. Brodie, F. M. *Thomas Jefferson: An Intimate History* (Norton, New York, 1974).
4. Ellis, J. J. *American Sphinx: The Character of Thomas Jefferson* (Knopf, New York, 1997).
5. Callender, J. T. *Richmond Recorder* 1 September 1802. [Cited in: Gordon-Reed, A. *ThomasJefferson and Sally Hemings: An American Controversy* (Univ. Press of Virginia, Charlottesville, 1997)].
6. Woodson, M. S. *The Woodson Source Book* 2nd edn (Washington, 1984).
7. Hammer, M. F. *Mol. Biol. Evol.* **11,** 749-761 (1994).
8. Whitfield, L. S., Sulston, J. E. & Goodfellow, P. N. *Nature* **378,** 379-380 (1995).
9. Seielstad, M. T. *et al. Hum. Mol. Genet.* **3,** 2159-2161 (1994).
10. Zerjal, T. *et al. Am. J. Hum. Genet.* **60,** 1174-1183 (1997).
11. Mathias, N., Bayes, M. & Tyler-Smith, C. *Hum. Mol. Genet.* **3,** 115-123 (1994).
12. Kwok, C. *et al. J. Med. Genet.* **33,** 465-468 (1996).
13. Kayser, M. *et al. Int. J. Legal Med.* **110,** 125-133 (1997).
14. Jobling, M. A., Bouzekri, N. & Taylor, P. G. *Hum. Mol. Genet.* **7,** 643-653 (1998).

Anticipation of Moving Stimuli by the Retina

M. J. Berry II *et al.*

Editor's Note

Our senses seem to show us the world as it is. But what we see is not just the imprint of photons hitting the retina: it is produced as our neural systems filter and highlight this incoming information. Thus it takes time to generate even the simplest neural message, creating the risk that the real world has changed by the time we perceive it—a risk that crucially applies to fast-moving objects such as predators or prey. But as this paper by Michael J. Berry at Harvard University and coworkers shows, our neural systems can predict where an object in our visual field is likely to be, giving us a better chance, say, of evading or catching it.

A flash of light evokes neural activity in the brain with a delay of 30–100 milliseconds[1], much of which is due to the slow process of visual transduction in photoreceptors[2,3]. A moving object can cover a considerable distance in this time, and should therefore be seen noticeably behind its actual location. As this conflicts with everyday experience, it has been suggested that the visual cortex uses the delayed visual data from the eye to extrapolate the trajectory of a moving object, so that it is perceived at its actual location[4-7]. Here we report that such anticipation of moving stimuli begins in the retina. A moving bar elicits a moving wave of spiking activity in the population of retinal ganglion cells. Rather than lagging behind the visual image, the population activity travels near the leading edge of the moving bar. This response is observed over a wide range of speeds and apparently compensates for the visual response latency. We show how this anticipation follows from known mechanisms of retinal processing.

BECAUSE a moving object often follows a smooth trajectory, one can extrapolate from its past position and velocity to obtain an estimate of its current location. Recent experiments on motion perception[5-7] indicate that the human brain possesses just such a mechanism: Subjects were shown a moving bar sweeping at constant velocity; a second bar was flashed briefly in alignment with the moving bar. When asked what they perceived at the time of the flash, observers reliably reported seeing the flashed bar trailing behind the moving bar. This flash lag effect has been confirmed repeatedly[8-10], and various high-level processes have been invoked to explain it, such as a time delay due to the shift of visual attention. To assess whether processing in the retina contributes to this effect we analysed the "neural image" of these two stimuli at the retinal output. We recorded simultaneously the spike trains of many ganglion cells in the isolated retina of tiger salamander or rabbit. The responses to flashed and moving bars were then analysed by plotting the firing rate in the retinal ganglion-cell population as a function of space and time.

视网膜可对运动物体的刺激进行预测

贝里等

编者按

我们的感官似乎将世界原原本本地呈现给我们。其实我们所看到的世界不仅仅是投射在视网膜上的光子印记：它是输入信息经由神经系统过滤并筛选产生的。因此即使生成最简单的神经信号都需要耗费时间，这就产生以下风险，真实的世界在我们感知到它的时候就已经改变了——这对快速移动的物体诸如捕食者和猎物都是至关重要的。但是如哈佛大学迈克尔·贝里及其同事在文章中所指出的，我们的神经系统能够预测视野中物体的可能位置，从而给我们提供更好的机会去逃避追捕或者抓住猎物。

闪光所引发的脑神经活动会有一个 30~100 毫秒的延迟 [1]，而延迟的大部分原因在于光感受器中迟缓的视觉传导过程 [2,3]。在这段时间里一个运动的物体会移动相当远的距离，因此当物体被看到时，已经显著落后于其实际位置。这与我们的日常体验相矛盾，有一种解释是视觉大脑皮层利用从眼睛中获取滞后的视觉数据来推测运动物体的轨迹，从而察觉其真正的位置 [4-7]。本文报道这种对运动刺激的预测始发于视网膜。移动光柱会在一群视网膜神经节细胞诱发一个随之移动的放电。这种总体的活动紧跟着移动光柱的最前缘行进，而不是滞后于视觉的影像。这种反应在很宽的速度范围内均能观察到，很明显这是在补偿视觉反应的潜伏期。我们在本文揭示这种预测是如何遵循已知视觉处理机制的。

由于运动的物体一般沿着平滑的轨迹移动，人们可以根据其过去的位置和速度推测其目前的大体位置。最近运动知觉的实验表明 [5-7]，人类大脑拥有这样一种机制：一个移动光柱匀速从受试者前掠过；第二个光柱与这个移动光柱重合，但短暂闪烁。当询问受试者在闪烁出现时觉察到什么时，他们肯定地回答看到一个闪烁的光柱尾随在移动光柱后面。这种闪烁滞后效应已经被多次证实 [8-10]，许多高级活动被援引用来解释这种现象，例如视觉注意转换引起时间的滞后。为了确定视网膜的活动是否参与了这种效应，我们分析了这两种刺激在视网膜输出中所产生的“神经影像”。我们同时记录了从虎螈或兔分离出来的视网膜神经节细胞所产生的放电序列。通过把群体放电频率绘制为空间和时间的函数的方法，来分析视网膜神经节细胞对闪烁和移动光柱的反应。

Figure 1 illustrates the responses of individual OFF-type ganglion cells to a dark bar flashed briefly over the receptive-field centre. In both salamander (Fig. 1a) and rabbit (Fig. 1b), the cells remained silent for a latency of ~50 ms, then fired a burst of spikes that lasted another 50 ms. When the bar was swept over the retina at constant speed (Fig. 1c, d), these same cells fired for a more extended period, beginning some time before the bar reached the position at which the flash occurred, and extending for a shorter time thereafter. When the bar was swept in the opposite direction (Fig. 1e, f), it produced a very similar response, showing that these cells had no direction-selective preference.

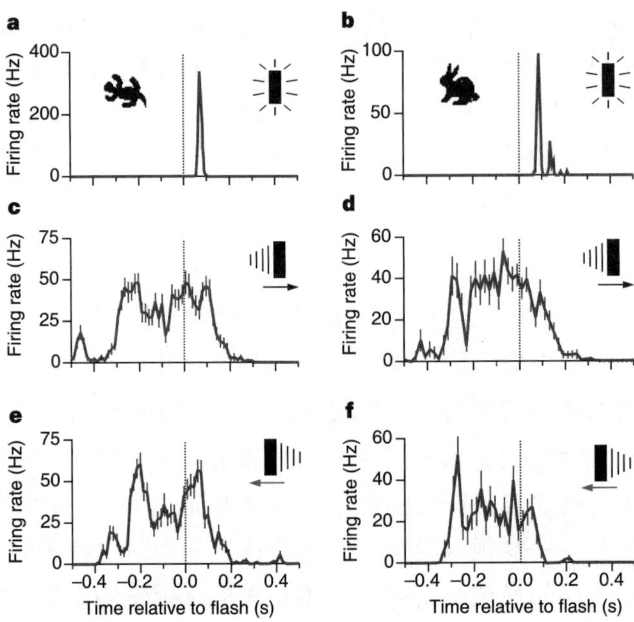

Fig. 1. Responses of two ganglion cells to flashed and moving bars. **a**, **c**, **e** Results from a "fast OFF" ganglion cell in salamander retina; **b**, **d**, **f**, results from a brisk-sustained OFF cell in rabbit retina. **a**, **b**, Firing rate as a function of time after a dark bar (90% contrast,133 μm width) was flashed for 15 ms on the receptive-field centre. **c**, **d**, Firing rate of the same two cells while the dark bar moved continuously across the retina at 0.44 mm s^{-1}. At time zero, the bar was aligned with the position of the flash in **a** and **b**. **e**, **f**, As in **c**, **d**, but with the bar moving in the opposite direction. Error bars denote standard error across repeated presentations of the stimulus.

From many single-unit measurements such as these, we compiled the neural image of the visual stimulus in the population of ganglion cells. This was done by plotting the firing rate of every cell as a function of distance from the bar stimulus and interpolating these points with a smooth line (see Methods). The neural image of the flashed bar among salamander "fast OFF" cells is shown in Fig. 2a. After a latency of 40 ms, a hump of neural activity appears that increases rapidly to a peak at 60 ms, then declines and disappears at 100 ms. As might be expected, the profile is centred on the bar. It has a width on the retina of ~200 μm at half-maximum, close to the size of the receptive-field centre for these neurons[11]. The width increases somewhat during the late phase of the response, creating the impression of an outward "splash"[12].

图 1 显示了单个的 OFF 型神经节细胞对暗光柱在感受域中央短暂闪现的反应。虎螈（图 1a）和兔（图 1b）的细胞均保持大约 50 ms 的静息，然后爆发一段持续 50 ms 的放电。当光柱匀速掠过视网膜时（图 1c 和 1d），同样一群细胞放电持续时间较长，开始于光柱到达闪烁位置之前的某个时刻，然后仍持续较短的一段时间。当光柱以相反的方向扫过时（图 1e 和 1f），会产生非常相似的反应，这表明这些细胞没有方向选择的偏好性。

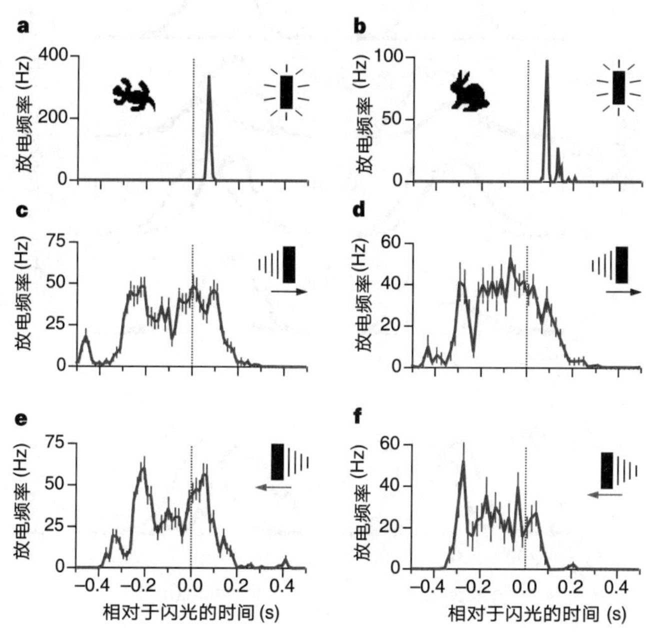

图 1. 两个神经节细胞对闪烁光柱和移动光柱的反应。**a, c, e,** 从虎螈视网膜“快速 OFF”神经节细胞中得来的结果；**b, d, f,** 从兔视网膜“持续激活 OFF”细胞得来的结果。**a, b,** 暗光柱（对比度 90%，宽度 133 μm）在感受域中央闪现时间为 15 ms，放电频率记为时间的函数。**c, d,** 当暗光柱以同样的速度 0.44 mm · s^{-1} 持续不断地经过视网膜时，两个细胞的放电频率。在时间为 0 时，**a** 和 **b** 中移动光柱与闪烁光柱的位置相对应。**e** 和 **f** 与 **c** 和 **d** 一样，但是移动光柱向相反的方向移动。误差线表示刺激重复表现出来的标准偏差。

从许多像这样的单个细胞测量的结果中，我们记录到对一群神经节细胞进行视觉刺激而产生的神经影像数据。我们采用以下方法：将每个细胞的放电频率记为移动光柱距离的函数，并将这些点用平滑的曲线连接（参见实验方法）。图 2a 显示在虎螈“快速 OFF”细胞中闪烁光柱的神经影像。潜伏 40 ms 后，神经活动的高峰期开始出现，在 60 ms 时快速增长到峰值，然后减弱，在 100 ms 时消失。正如预测的那样，这些变化以移动光柱为中心。在峰值一半时，其在视网膜的宽度大约是 200 μm，这个数值接近于这些神经元感受域中央的大小 [11]。这个宽度在反应后期有所增加，给人一种向外“扩散”的感觉 [12]。

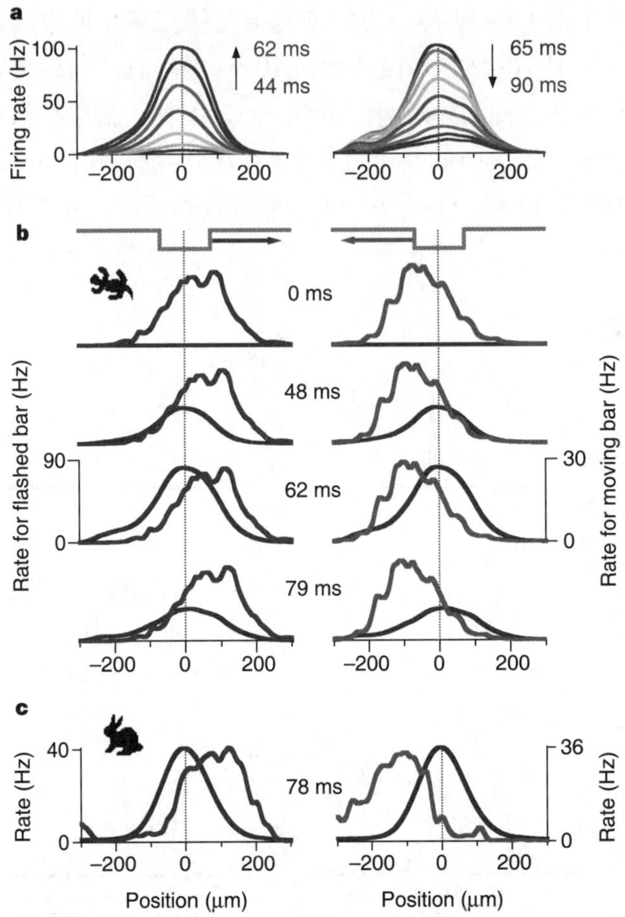

Fig. 2. Population response to flashed and moving bars. **a**, Spatial profile of firing in the population of salamander fast OFF ganglion cells in response to a flashed dark bar (90% contrast, 133 μm width, see stimulus trace in **b**) at a series of times after the flash (colour scale, 3 ms steps). **b**, Profile of the population response at four time points following a flashed bar (red, from **a**), and the same bar travelling at 0.44 mm s^{-1} rightward (blue) and leftward (green). At time 0 ms, the moving bars were aligned with the position of the flash; at 62 ms, the flash response was maximal. Curves in **a** and **b** derived from 15 cells. **c**, As in **b**, for the population of brisk-sustained OFF cells in rabbit retina. At 78 ms, the flash response was maximal. Curves derived from six cells.

The neural image of the moving bar is shown in Fig. 2b. Again, a hump of firing activity is observed, which now sweeps over the retina along with the moving bar. If this response were subject to the same time delay as the flash, one would expect the neural image to trail behind the visual image of the bar. Instead, the hump of firing activity is clearly ahead of the centre of the bar, and the peak firing rate seems to occur near the bar's leading edge. The same response occurs when the bar moves in the opposite direction. By superposing on this the response to a flash that was aligned with the moving bar (from Fig. 2a), we find that the two neural images are clearly separated—at the time when the response to the flash peaks, the response of the moving bar is displaced ~100 μm ahead in the direction of motion. A very similar displacement between the two neural images was observed among

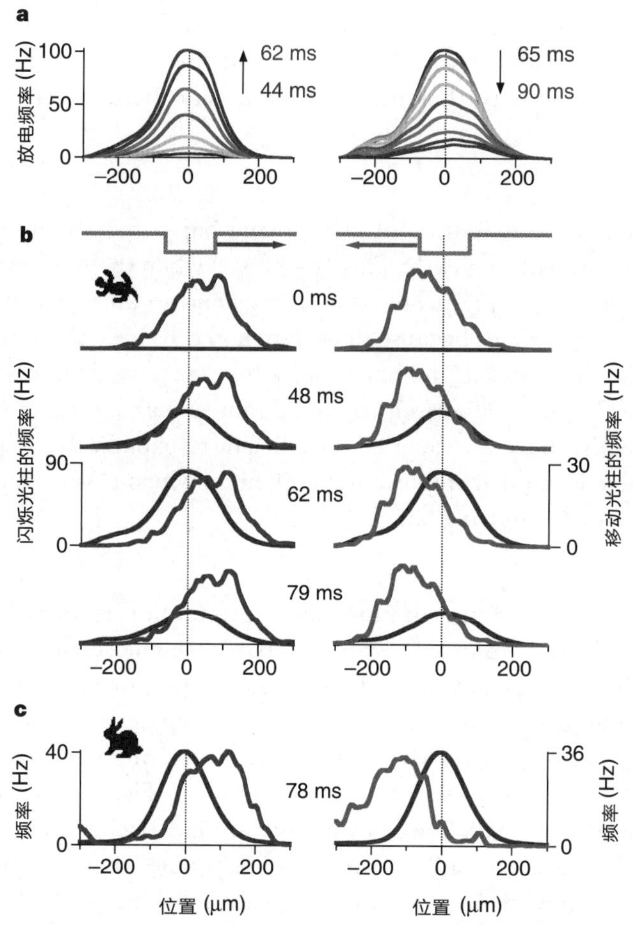

图 2. 对闪烁光柱和移动光柱的群体响应。**a**，虎螈快速 OFF 神经节细胞群体对闪烁暗光柱（对比度 90%，宽度 133 μm，同 **b** 的刺激一样）在放电之后一系列时间点上的放电空间特性（色标，3 ms 时间间距）。**b**，闪烁光柱后四个时间点上的群体反应特性（红色曲线数据来自 **a**），同一移动光柱以 0.44 mm · s^{-1} 速度向右移动（蓝）和向左移动（绿）。在 0 ms 时，移动光柱与闪烁光柱的位置对齐；在 62 ms 时，对闪烁光柱的反应达到最大化。**a** 和 **b** 的曲线数据来源于 15 个细胞。**c** 和 **b** 一样，是兔视网膜持续激活 OFF 细胞群体的结果。在 78 ms 时反应最大。曲线数据采集于 6 个细胞。

图 2b 所示为移动光柱的神经影像。我们再次观察到驼峰状的放电活动，且与移动光柱掠过视网膜的过程一致。如果这种反应与闪烁光柱滞后时间相同，人们应该会看到神经影像尾随在移动光柱的视觉影像之后。相反，放电活动的峰值明显位于移动光柱中心之前，其放电频率的最高点似乎出现在靠近移动光柱最前缘的地方。当移动光柱向相反方向移动时会发生相同的反应。通过将这种反应与一个闪烁光柱（该光柱与移动光柱对齐）所引发的反应进行重叠（图 2a），我们发现两个神经影像完全分开——在对闪烁光柱的反应达峰值时，移动光柱的反应沿着运动方向向前移动大约 100 μm。在兔视网膜持续激活 OFF 细胞中（图 2c）也观察到类似的两个神

brisk-sustained OFF cells in the rabbit retina (Fig. 2c). If subsequent stages of the visual system estimate the location of the flashed bar and the moving bar by the position of these humps of neural activity, they must conclude that the moving bar is ahead of the flashed bar.

How does this apparent anticipation of the moving bar come about? One suspects that cells ahead of the bar start firing early (Figs 1c–f, 2b, c) when the bar begins to invade their receptive-field centre. The firing profile does not extend to an equal distance behind the trailing edge of the bar, perhaps because these ganglion cells have transient responses: They fire while stimulation increases as the bar invades the receptive field, not while stimulation decreases as the bar leaves it. However, we found that spatial and temporal filtering by the ganglion cell's receptive field was by itself insufficient to explain the response profiles (see below). Instead, there is another important component, which was revealed in experiments varying the intensity of the moving bar.

Figure 3a illustrates the response of this neural population to dark bars of increasing contrast relative to the background. As expected, bars of higher contrast produced stronger modulations in firing. The peak firing rate increased in proportion to contrast at first, but then appeared to saturate (Fig. 3a, inset). In addition, the shape of the neural image changed significantly with contrast. At low contrast, the peak in firing occurred behind the bar's leading edge. At high contrast—the same condition as in Fig. 2—the peak of the profile was ahead of the leading edge, followed by a more gradual decline in firing. The saturation of the peak firing rate and the shift in the response profile can be explained if the high-contrast stimulus somehow desensitizes the response of the ganglion cell after a short time delay. In that case, a ganglion cell just ahead of the bar should be strongly excited as the edge begins to enter its receptive-field centre, but then its response gain gets reduced and the firing rate declines even before the edge is half-way across. A well-known component of retinal processing that fits this description is the "contrast-gain control"[13-15].

Following ref. 16, we incorporated this aspect into a quantitative description of a ganglion cell's light response (Fig. 4). In this scenario, the retina integrates the light stimulus over space and time, with a weighting function $k(x,t)$ given by the ganglion cell's receptive field, and the resulting signal determines the neuron's firing rate. If the stimulus provides strong excitation for an extended period of time, a negative feedback loop reduces the gain at the input and consequently the response to subsequent stimulation[17]. With just four free parameters, this model produced a satisfying account of neural responses throughout the entire contrast series (Fig. 3a). It indicates that the retinal gain is modulated as much as fourfold during passage of the high-contrast bar (Fig. 3b), which pushes the response profile towards the leading edge of the bar. Without contrast-gain control the predicted profile always lagged significantly behind (Fig. 3a).

经影像之间的位置偏差。如果视觉系统的后续阶段通过这些神经活动峰值出现的位置来估计闪烁光柱束和移动光柱的位置，他们的结论一定是移动光柱位于闪烁光柱之前。

这种对移动光柱明显的预感是如何产生的呢？有人猜测当移动光柱开始进入神经节细胞的感受域中心时，在移动光柱之前的细胞提早放电（图 1c ~ f、2b 和 2c）。这种放电特性在移动光柱的后缘并未进行等距离的延长，可能因为这些神经节细胞具有瞬态响应：神经节细胞在移动光柱进入感受域、刺激增加时放电；而在移动光柱离开、感受域刺激降低时不放电。但是，我们发现，通过神经节细胞感受域对空间和时间的过滤本身不足以解释反应的特性（见下文）。相反，还有另外一种重要的组分，是通过变化移动光柱的强度发现的。

图 3a 阐明了神经细胞群体对相对背景的对比度逐渐增强的暗光柱的反应。正如所料，具有较高对比度的移动光柱对放电具有更强的调节能力。起初放电的峰值频率与对比度成正比，但是随后出现饱和（图 3a，插入图）。另外，神经影像的形状随对比度的变化而显著变化。在对比度低时，放电的峰值出现在移动光柱前缘之后。当对比度高时——与图 2 条件相同——放电的峰值在移动光柱前缘之前，随后放电逐渐减弱。如果高对比度刺激在短时间之后，从某种程度上降低了神经节细胞反应的敏感性，那么放电频率峰值的饱和与放电特性之间的偏差就可以解释了。在这种情况下，当移动光柱边缘开始进入其感受域的中心时，位于移动光柱之前的神经节细胞应该非常兴奋，但是随后其反应的增益变弱，其放电频率甚至在移动光柱边缘仅通过一半时就开始衰减。符合这种描述的一个著名的视网膜信息处理过程叫作“对比度增益控制”[13-15]。

按照参考文献 16 的方法，我们把这方面纳入神经节细胞对光反应的定量描述中（图 4）。在本情形中，视网膜将空间和时间上的光刺激与神经节细胞感受域得出的权重函数 $k(x,t)$ 相结合，最终信号决定了神经元的放电频率。如果光刺激引发的强烈兴奋超过特定的时间，负反馈环在输入端减少增益，进而降低对随后刺激的响应[17]。仅用这四个独立的参数，此模型会很好地解释贯穿整个对比系列中的神经反应（图 3a）。结果显示在高对比度移动光柱经过时视网膜增益可被调节到 4 倍（图 3b），从而将反应的曲线推向移动光柱的前缘。没有对比度增益控制，预测的曲线总是明显滞后（图 3a）。

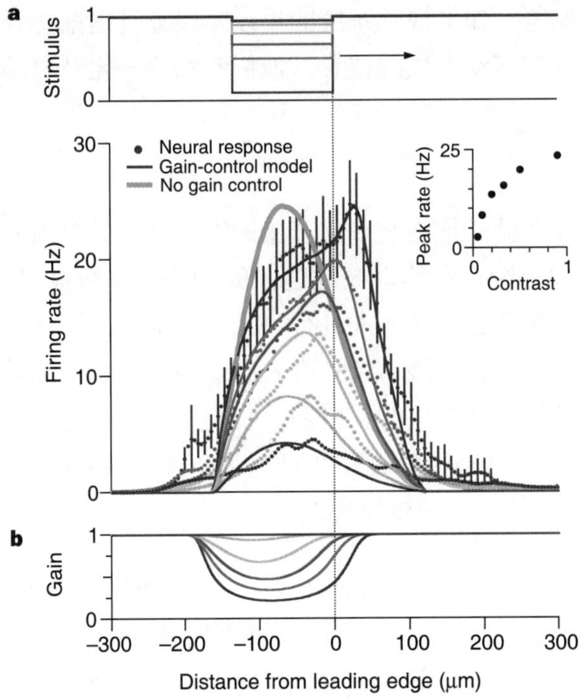

Fig. 3. Dependence of motion extrapolation on contrast. **a**, Stimulation with moving dark bars (133 μm width, 0.44 mm s^{-1} speed) of varying contrast: 5, 10, 20, 33, 50 and 90% (see top stimulus traces). Main panel shows the response profile derived from 15 salamander fast OFF ganglion cells (coloured dots), and the predicted response (coloured lines) from a model incorporating contrast-gain control (Fig. 4). Grey line shows the prediction without contrast-gain control. Inset, the peak firing rate of the response profile as a function of the contrast of the bar. Error bars denote standard error, derived from variation among ganglion cells. **b**, Gain variable g of the gain control model (Fig. 4). Model parameters: $\theta = 0$, $\alpha = 85$ Hz, $B = 45\ s^{-1}$, $\tau = 170$ ms.

This explanation indicates that there will be clear limits to what stimuli can be anticipated. For example, if the bar moves fast enough to cross the receptive field before the contrast-gain control sets in, then the peak of the firing profile should lag behind the leading edge. Figure 5a explores these limits, and shows that up to speeds of about 1 mm s^{-1} on the retina, the shape of the firing profile among ganglion cells remained essentially unchanged, with a peak near or ahead of the leading edge. At higher speeds, however, the response profile began to slip significantly behind the leading edge. This basic relationship was confirmed for several different populations of ganglion cells in both rabbit and salamander (Fig. 5b). The various cell types differed in the extent of anticipation at low speeds, but all began to show a lag in the neural image at speeds of 1–2 mm s^{-1}. In particular, direction selectivity does not play a special role in motion anticipation.

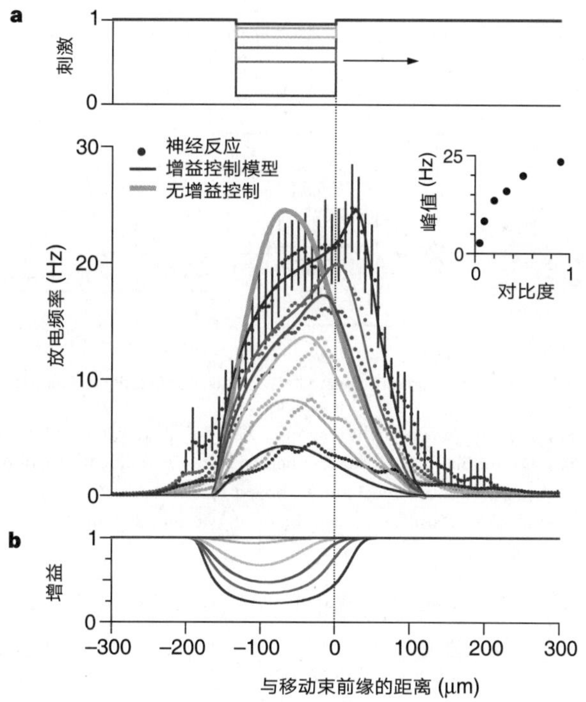

图 3. 运动推测对对比度的依赖性。**a**，移动暗光柱（宽度 133 μm，速度 0.44 mm · s^{-1}）在不同对比度：5%、10%、20%、33%、50% 和 90%（参见顶端的刺激轨迹）下的刺激。主图显示的是 15 个虎螈快速 OFF 神经节细胞采集到的反应特性（彩色的点），和来自混合的对比度增益控制模型中的预测反应特性（彩色的线）（图 4）。灰色的线显示没有对比度增益控制的预测。插图显示反应曲线的放电频率峰值与移动光柱对比度之间的函数关系。误差线表示从不同的神经节细胞变异中得来的标准误差。**b**，增益控制模型的增益变异 g（图 4）。模型参数：$\theta = 0$，$\alpha = 85$ Hz，$B = 45\ s^{-1}$，$\tau = 170$ ms。

这种解释表明细胞对于何种刺激可被预测有着明确的限定。例如，如果一个光柱移动得足够快，在对比度增益控制介入之前经过感受域，那么细胞放电的峰值应该滞后于光束前缘。图 5a 探索了这种限定，结果显示在视网膜上速度高达大约 1 mm · s^{-1}，神经节细胞放电特性本质上保持不变，峰值靠近前缘或在前缘之前。然而，在更高速度下，反应特性开始明显落后于前缘。这种基本关系在几种不同群体的神经节细胞（如兔和虎螈中）中得到验证（图 5b）。在低速情况下，不同类型细胞对运动的预测程度不同，但是所有细胞在速度为 1 ~ 2 mm · s^{-1} 时均开始出现神经影像的滞后。尤其是方向选择性在运动的预测中没有起到特殊作用。

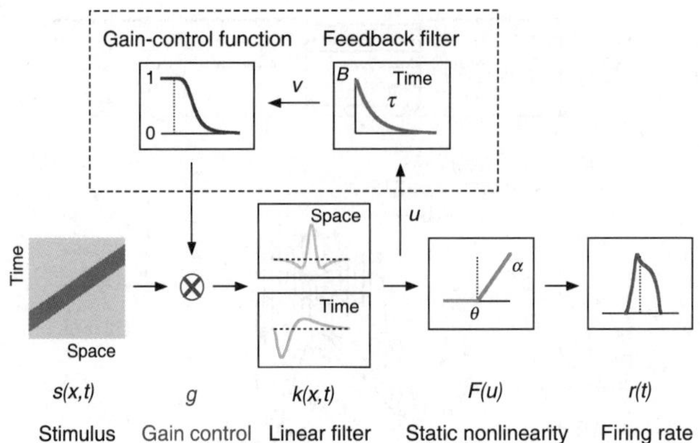

Fig. 4. Cascade model for a ganglion cell's light response. The stimulus $s(x,t)$ is multiplied by a gain factor g, convolved with a spatiotemporal filter $k(x,t)$, and rectified by a static nonlinear function $F(u)$ to produce the firing rate $r(t)$. The contrast-gain control mechanism (boxed region) takes the output of the linear filter u, averages it by exponential filtering in time, and uses the result v to set the gain factor g through a decreasing gain control function $g(v)$. Formally,

$$u(t) = g(v) \int_{-\infty}^{\infty} dx \int_{-\infty}^{t} dt'\, s(x,t')\, k(x,t-t')$$

$$v(t) = \int_{-\infty}^{t} dt'\, u(t')\, B \exp\left(-\frac{t-t'}{\tau}\right)$$

$$g(v) = \begin{cases} 1 & v < 0 \\ 1/(1+v^4) & v > 0 \end{cases}$$

$$F(u) = \begin{cases} 0 & u < 0 \\ \alpha(u-\theta^4) & u > 0 \end{cases}$$

The filter $k(x,t)$ is measured (see Methods), and $g(v)$ is taken from a previous, successful model of contrast-gain control in the salamander retina[17]. Thus, the model has four parameters: the threshold θ and slope α of the rectifier $F(u)$, and the amplitude B and time constant τ of the gain control filter.

In summary, we have shown that the extrapolation of a moving object's trajectory begins in the retina. In the neural image that the eye transmits to the brain, the moving object is clearly ahead of the corresponding flashed object (Fig. 2). According to a successful model for the ganglion cell's light response (Figs 3, 4), motion anticipation in these populations can be explained on the basis of the spatially extended receptive field, the biphasic temporal response and a nonlinear contrast-gain control. There are several indications that this retinal mechanism contributes strongly to human perception of moving stimuli. First, the requisite components of our model are well documented in many species. In the primate retina, a nonlinear contrast-gain control is found specifically in the M-type ganglion cells[15], neurons that feed the central pathways leading to motion perception[18]. Second, retinal motion extrapolation breaks down at speeds above 1 mm s^{-1} (Fig. 5b). This corresponds well with observations on human subjects: At retinal speeds of 0.3–0.9 mm s^{-1}, perceptual motion extrapolation appeared to compensate for the entire visual delay[5], whereas at speeds of ~4 mm s^{-1} only partial extrapolation was observed[10]. Finally, the retina anticipates high-contrast stimuli more than low-contrast stimuli (Fig. 3), a further departure from ideal extrapolation. Again, this effect has been observed in human psychophysics[9].

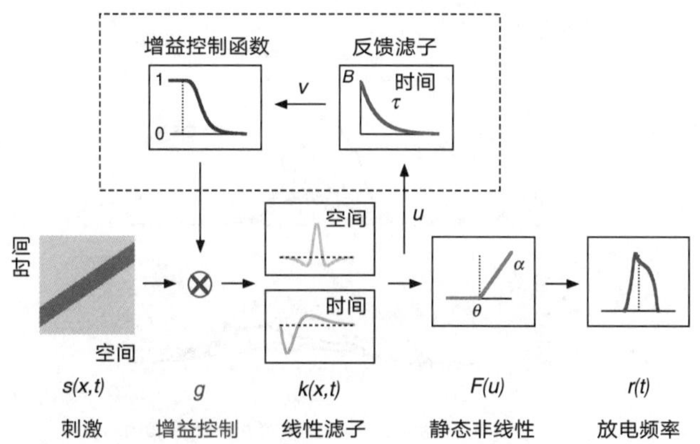

图 4. 一个神经节细胞光反应的级联模型。刺激 $s(x,t)$ 乘以增益因子 g，与时间空间滤子 $k(x,t)$ 卷积，经过静态的非线性函数 $F(u)$ 校正，产生放电频率 $r(t)$。增益控制机理（方框区域）接受线性滤子 u 的输出，通过时间的指数滤子求其平均值，通过减少增益控制函数 $g(v)$，用结果 v 设定增益因子 g。用公式表示为：

$$u(t) = g(v) \int_{-\infty}^{\infty} \mathrm{d}x \int_{-\infty}^{t} \mathrm{d}t' \, s(x,t') \, k(x,t-t')$$

$$v(t) = \int_{-\infty}^{t} \mathrm{d}t' \, u(t') \, B \exp\left(-\frac{t-t'}{\tau}\right)$$

$$g(v) = \begin{cases} 1 & v < 0 \\ 1/(1+v^4) & v > 0 \end{cases}$$

$$F(u) = \begin{cases} 0 & u < 0 \\ \alpha(u-\theta^4) & u > 0 \end{cases}$$

滤子 $k(x,t)$ 可以测量（参见方法），$g(v)$ 从先前成功的虎螈视网膜增益控制模型中得来[17]。因此，模型有 4 个参数：整流器 $F(u)$ 的阈值 θ 和斜率 α，以及振幅 B 和增益控制滤子的时间常数 τ。

总而言之，我们证明对运动物体轨迹的推测始于视网膜。在经眼传输至脑的神经影像中，运动物体明显先于相应的闪烁物体（图 2）。根据神经节细胞光反应的成功模型（图 3 和图 4），这些群体的运动预测可以基于如下解释：感受域空间上的延伸，时间上两阶段反应和非线性对比度增益控制。有些迹象表明这种视网膜机制极大地促成了人对运动刺激的感知。首先，我们模型中的必要组件在许多物种中得到很好的证实。在灵长类视网膜中，非线性的对比度增益控制被发现，尤其存在于 M 型神经节细胞中[15]，满足中枢通路的神经元导致对运动的感知[18]。第二，视网膜对运动的预测在速度大于 1 mm · s^{-1} 时失效（图 5b）。这与人体实验的观察结果很好地对应：视网膜速度为 0.3 ~ 0.9 mm · s^{-1}，知觉运动的预测能够对整个视觉延迟产生补偿[5]，当速度约为 4 mm · s^{-1} 时，只观察到对运动的部分预测[10]。最后，视网膜对高对比度刺激的预测要强于对低对比度刺激的预测（图 3），这更加偏离理想推测。再一次强调，这种效应在人类精神物理学中已经观察到[9]。

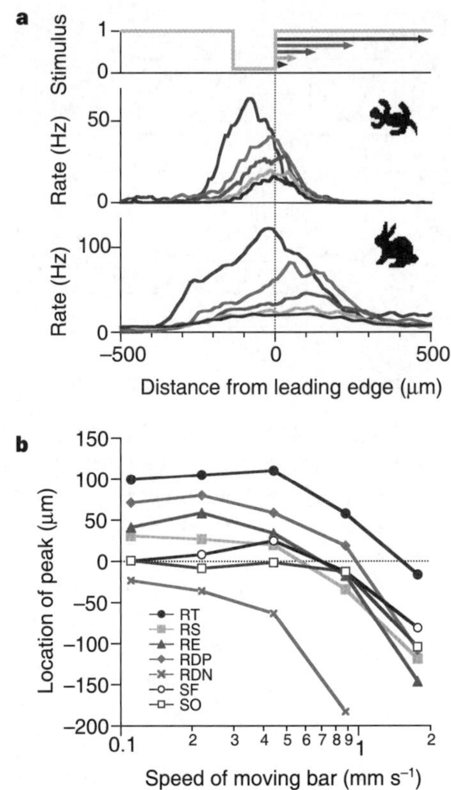

Fig. 5. Dependence of motion extrapolation on speed. **a**, Stimulation with moving dark bars (90% contrast, 133 μm width) of varying speed: 0.11, 0.22, 0.44, 0.88 and 1.76 mm s^{-1} (see top stimulus traces). Firing profiles are plotted for salamander fast OFF ganglion cells (middle panel, 19 cells) and rabbit brisk-transient OFF cells (bottom panel, 3 cells). **b**, Motion extrapolation as a function of speed for various populations of ganglion cells. The distance between the peak in the firing-rate profile and the leading edge of the moving bar is plotted as a function of speed; positive numbers indicate that the response profile peaked ahead of the leading edge. The functional types are: salamander, fast OFF (SF, 19 cells); salamander, other OFF (SO, 16 cells); rabbit, brisk-transient OFF (RT, 3 cells); rabbit, brisk-sustained OFF (RS, 6 cells); rabbit, local edge detectors (RE, 4 cells); rabbit, ON/OFF direction-selective cells (5 cells) probed in the preferred direction (RDP) and the null direction (RDN).

In general, an animal is likely to benefit from anticipating the future position of an object, for example to pounce on it or to evade it. This is particularly urgent when the primary sensory data are delayed. In principle, this delay could be compensated anywhere within the behavioural loop, even within the motor system that executes the response. However, it is advantageous to perform the correction early, before different sensory pathways merge. For example, in many animals the retina projects directly to the tectum or the superior colliculus, where a visual map of space is overlaid with an auditory map[19,20]. Auditory transduction in hair cells incurs a much shorter delay than phototransduction[21]. If the visual and auditory images of a moving object should align on the target map, the compensation for the delay in the visual pathway must occur within the retina.

It is likely that subsequent stages of the visual system continue this process, possibly by using a similar mechanism. Within the visual cortex we can certainly find the requisite

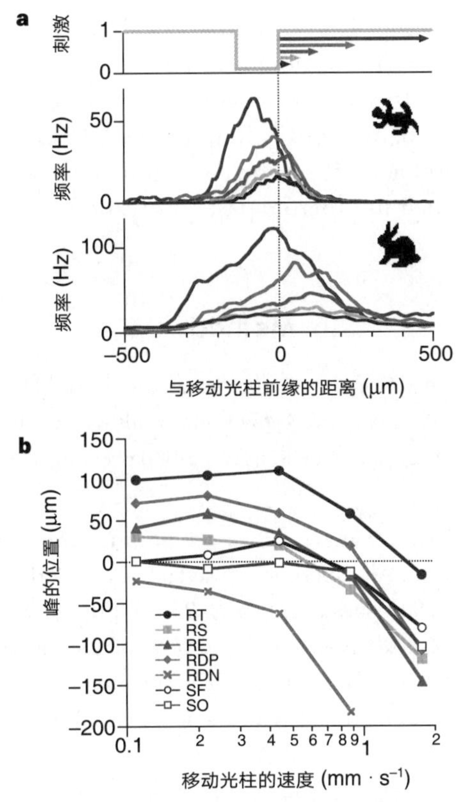

图 5. 运动推测对速度的依赖性。**a**，移动暗光柱（对比度 90%，宽度 133 μm）在不同速度下：0.11、0.22、0.44、0.88 和 1.76 mm · s^{-1}（参见顶部的刺激轨迹）的刺激。放电特性是根据虎螈快速 OFF 神经节细胞（中间图，19 个细胞）和兔临时激活 OFF 细胞（底部图，3 个细胞）的数据绘制而来的。**b**，不同群体的神经节细胞速度函数的运动推测。细胞放电频率峰值和移动光柱前缘的距离被绘制为速度的函数；正数表明反应在光柱前缘之前出现峰值。功能类型包括：虎螈，快速 OFF 细胞（SF，19 个）；虎螈，其他 OFF 细胞（SO，16 个）；兔，临时激活 OFF 细胞（RT，3 个）；兔，持续激活 OFF 细胞（RS，6 个）；兔，局部边缘探测器细胞（RE，4 个）；兔，在首选方向（RDP）和无效方向（RDN）探测的 ON/OFF 方向选择性细胞（5 个）。

通常，动物可能会受益于对运动物体未来位置的预测能力，例如用来袭击或逃脱。如果初级感官数据传输滞后，这一点就显得尤其重要。从原理上来说，这种滞后在行为回路中任何地方均可以补偿，甚至在执行反应的运动系统中。然而，在不同的感觉路径合并之前，提前执行正确的决定是有利的。例如，许多动物的视网膜直接映射到上表皮层或者视上丘，在此处空间的视觉图像与听觉图像叠加 [19,20]。在毛细胞中的听觉传导引发的滞后比视觉传导更短 [21]。如果运动物体的视觉和听觉影像能与目标图像并列，对视觉通路的滞后补偿一定发生在视网膜。

很可能通过相似的机理，视觉系统随后的信息处理阶段会延续这个过程。在视觉皮层我们能够确切地发现图 4 模型中的必要组件：例如，兴奋输入的局部汇集，

components of the model in Fig. 4: for example, local pooling of excitatory inputs, time-delayed inhibition and mechanisms of nonlinear gain control[22,23]. More generally, there are many instances within the cortex where variables relevant to our behaviour are mapped onto two-dimensional sheets of neurons[24]. If the time course of these variables produces a smooth trajectory of neural activity on the cortical map, then a mechanism such as that described here can predict their future from past observations.

Methods

Recording. Retinae were obtained from larval tiger salamanders and Dutch belted rabbits. A piece of isolated retina was placed ganglion-cell-layer-down on a multi-electrode array, which recorded spike trains simultaneously from many ganglion cells, as described previously[11,25].

Stimulation. Visual stimuli were generated on a computer monitor and projected onto the photoreceptor layer, as described[25]. All experiments used a background of white light, with a photopic intensity of $M = 11$ mW m^{-2} . Dark bars of intensity B were presented on this background, and the contrast of a bar is defined as $C=(M-B)/M$. A screen pixel of the monitor measured 6.7 μm on the retina, and each video frame lasted 15 ms. Thus, a bar sweeping at 0.44 mm s^{-1} moved by one pixel every video frame. Flashed bars were presented for a single video frame.

Receptive fields. The spatiotemporal receptive fields of all ganglion cells were measured by reverse correlation to randomly flickering stripes[25], orientated parallel to the bars from other experiments. Each of the contiguous 13-μm wide stripes was randomly turned on or off every 30 ms. From ~60 min of recording, we computed for each ganglion cell the average stimulus sequence in the one second preceding an action potential. This reverse correlation is a measure of how the ganglion cell integrates light over space and time. Its time-reverse is the ganglion cell's linear kernel $k(x,t)$ (ref. 26), which can also be interpreted as the effect of a thin line flashed at distance x on the cell's firing rate at time t after the flash[11]. As expected, $k(x,t)$ had a "Mexican hat" spatial profile, reflecting opposite effects from centre and surround[27], and a biphasic time course (see Fig. 4 and ref. 11).

Cell types. Retinal ganglion cells appear in distinct functional types, and we took care to analyse these subpopulations separately. Salamander cells were classified based on their spatiotemporal receptive fields and on responses to uniform square-wave flashes, as described[28]. Rabbit cells were classified based on the spatiotemporal receptive field and the shape of the spike train's autocorrelation function, following the criteria of ref. 29. Direction-selective cells produced at least tenfold more spikes to one direction of the moving bar than to the opposite direction.

Population activity. The profile of population activity was evaluated along the spatial dimension perpendicular to the bars. Each cell's position was defined as the middle of its receptive field, determined by fitting a spatial gaussian to the centre lobe of the kernel $k(x,t)$. To estimate the population response to a flashed bar, the flash was repeated 75 times in each of 15 locations, separated by 33 μm. For each flash location and each ganglion cell, the firing rate following the flash

时间滞后的抑制，和非线性增益控制的机理 [22,23]。通常，在皮质层有许多这样的例子，在那里与我们行为有关的可变物被映射到神经元的二维层 [24]。如果这些可变物的时间进程在大脑皮层的神经活动中产生一个平稳的轨迹，那么本文描述的机制能够从过去的观察结果中预测可变物的未来。

方　法

记录　视网膜采集于幼年期的虎螈和荷兰条纹兔。将一片分离的视网膜放置在一个多电极阵列上，放置方向为神经节细胞层朝下，按照之前所述的方法 [11,25] 同时从许多神经节细胞记录放电信号。

刺激　视觉刺激由计算机显示器产生，然后按照文献描述的方法投射到光感受器细胞层 [25]。所有实验以白光为背景，光强为 $M=11\ \mathrm{mW\cdot m^{-2}}$。强度为 B 的暗光柱呈现在此背景中，暗光柱的对比度定义为 $C=(M-B)/M$。显示器单个屏幕像素在视网膜上的投射大小为 6.7 μm，每个视频帧持续的时间为 15 ms。因此，移动光柱以 $0.44\ \mathrm{mm\cdot s^{-1}}$ 的速度运动，每个视频帧就是一个像素。闪烁光柱在单个视频帧里出现。

感受域　所有神经节细胞的时空感受域均通过随机闪烁光柱的负相关测量 [25]，方向与其他实验里的移动光柱平行。每一个连续的 13 μm 宽的条纹每隔 30 ms 随机开关一次。在大约 60 分钟的记录中，我们可以为每一个神经节细胞计算出动作电位前一秒钟之内的平均刺激序列。这种负相关是对神经节细胞如何对光进行时空整合的一种测量。其时间的逆转是神经节细胞的线性核 $k(x,t)$（文献 26），也可以解释为一条细线闪现后在 t 时间 x 距离处细胞放电频率所产生的效应 [11]。正如所料，$k(x,t)$ 在空间曲线上有一个“墨西哥帽”，反映了中间和周围的相反效应 [27]，以及双相的时间进程。(参见图 4 和参考文献 11)。

细胞类型　视网膜神经节细胞具有不同的功能类型，我们分别对这些亚种群进行了仔细地分析。正如前人所描述的，虎螈的细胞是根据它们的时空感受域及其对统一的矩形波闪烁光柱的反应来区分的 [28]。兔的细胞是按照参考文献 29 的标准，根据时间和空间感受域以及放电序列自相关函数的形状进行区分的。方向选择性的细胞对移动光柱移动的特定方向产生的放电至少是相反方向的 10 倍。

群体活动　根据垂直于光柱的空间维度来评价种群活动的特性。每个细胞的位置定位在其感受域中央，由空间的高斯曲线拟合核 $k(x,t)$ 的中心突起形状决定。为了估计细胞群体对闪烁光柱的反应，闪烁光柱在间隔 33 μm 的 15 个位置处每处重复 75 次。对于每一个闪烁光柱的位置和每个神经节细胞，其在闪烁光柱之后的放电频率（图 1a 和 1b）用二进制 2 ms

(Fig. 1a, b) was calculated using a time bin of 2 ms. To compose the firing profile at a given time after the flash, each ganglion cell's firing rate was plotted against the cell's position relative to the flashing line. This plot was smoothed by convolution with a gaussian of standard deviation 20 μm. To estimate the population response to a moving bar, the stimulus was repeated 50 times, and each ganglion cell's firing rate computed as for flashes, but with a time bin of 15 ms. Then the firing rate was plotted against the distance from the bar, and averaged over all cells.

(**398**, 334-338; 1999)

Michael J. Berry II, Iman H. Brivanlou, Thomas A. Jordan* & Markus Meister

Department of Molecular and Cellular Biology, Harvard University, 16 Divinity Avenue, Cambridge, Massachusetts 02138, USA

* Present address: Department of Psychiatry and Behavioral Science, Stanford University School of Medicine, Stanford, California 94305, USA.

Received 11 December 1998; accepted 2 February 1999.

References:

1. Maunsell, J. H. & Gibson, J. R. Visual response latencies in striate cortex of the macaque monkey. *J. Neurophysiol.* **68,** 1332-1344 (1992).
2. Lennie, P. The physiological basis of variations in visual latency. *Vision Res.* **21,** 815-824 (1981).
3. Schnapf, J. L., Kraft, T. W. & Baylor, D. A. Spectral sensitivity of human cone photoreceptors. *Nature* **325,** 439-441 (1987).
4. De Valois, R. L. & De Valois, K. K. Vernier acuity with stationary moving Gabors. *Vision Res.* **31,** 1619-1626 (1991).
5. Nijhawan, R. Motion extrapolation in catching. *Nature* **370,** 256-257 (1994).
6. Khurana, B. & Nijhawan, R. Extrapolation or attention shift? *Nature* **378,** 566 (1995).
7. Nijhawan, R. Visual decomposition of colour through motion extrapolation. *Nature* **386,** 66-69 (1997).
8. Baldo, M. V. & Klein, S. A. Extrapolation or attention shift? *Nature* **378,** 565-566 (1995).
9. Purushothaman, G., Patel, S. S., Bedell, H. E. & Ogmen, H. Moving ahead through differential visual latency. *Nature* **396,** 424 (1998).
10. Whitney, D. & Murakami, I. Latency difference, not spatial extrapolation. *Nature Neurosci.* **1,** 656-657 (1998).
11. Smirnakis, S. M., Berry, M. J., Warland, D. K., Bialek, W. & Meister, M. Adaptation of retinal processing to image contrast and spatial scale. *Nature* **386,** 69-73 (1997).
12. Jacobs, A. L. & Werblin, F. S. Spatiotemporal patterns at the retinal output. J. *Neurophysiol.* **80,** 447-451 (1998).
13. Shapley, R. M. & Victor, J. D. The effect of contrast on the transfer properties of cat retinal ganglion cells. *J. Physiol.* **285,** 275-298 (1978).
14. Sakai, H. M., Wang, J. L. & Naka, K. Contrast gain control in the lower vertebrate retinas. *J. Gen. Physiol.* **105,** 815-835 (1995).
15. Benardete, E. A., Kaplan, E. & Knight, B. W. Contrast gain control in the primate retina: P cells are not X-like, some M cells are. *Visual Neurosci.* **8,** 483-486 (1992).
16. Victor, J. D. The dynamics of the cat retinal X cell centre. *J. Physiol.* **386,** 219-246 (1987).
17. Crevier, D. W. & Meister, M. Synchronous period-doubling in flicker vision of salamander and man. *J. Neurophysiol.* **79,** 1869-1878 (1998).
18. Merigan, W. H. & Maunsell, J. H. How parallel are the primate visual pathways? *Annu. Rev. Neurosci.* **16,** 369-402 (1993).
19. Sparks, D. L. Translation of sensory signals into commands for control of saccadic eye movements: role of primate superior colliculus. *Physiol. Rev.* **66,** 118-171 (1986).
20. Knudsen, E. I. Auditory and visual maps of space in the optic tectum of the owl. *J. Neurosci.* **2,** 1177-1194 (1982).
21. Corey, D. P. & Hudspeth, A. J. Response latency of vertebrate hair cells. *Biophys. J.* **26,** 499-506 (1979).
22. Carandini, M., Heeger, D. J. & Movshon, J. A. Linearity and normalization in simple cells of the macaque primary visual cortex. *J. Neurosci.* **17,** 8621-8644 (1997).
23. Abbott, L. F., Varela, J. A., Sen, K. & Nelson, S. B. Synaptic depression and cortical gain control. *Science* **275,** 220-224 (1997).
24. Knudsen, E. I., du Lac, S. & Esterly, S. D. Computational maps in the brain. *Annu. Rev. Neurosci.* **10,** 41-65 (1987).
25. Meister, M., Pine, J. & Baylor, D. A. Multi-neuronal signals from the retina: acquisition and analysis. *J. Neurosci. Methods* **51,** 95-106 (1994).
26. Hunter, I. W. & Korenberg, M. J. The identification of nonlinear biological systems: Wiener and Hammerstein cascade models. *Biol. Cybern.* **55,** 135-144 (1986).
27. Rodieck, R. W. Quantitative analysis of cat retinal ganglion cell response to visual stimuli. *Vision Res.* **5,** 583-601 (1965).
28. Warland, D. K., Reinagel, P. & Meister, M. Decoding visual information from a population of retinal ganglion cells. *J. Neurophysiol.* **78,** 2336-2350 (1997).
29. Devries, S. H. & Baylor, D. A. Mosaic arrangement of ganglion cell receptive fields in rabbit retina. *J. Neurophysiol.* **78,** 2048-2060 (1997).

Acknowledgements. We thank J. Keat for assistance in generating the visual stimulus, and H. Berg and T. Holy for comments on the manuscript. This work was supported by a NRSA to M.B. and a grant from the NIH and a Presidential Faculty Fellowship to M.M.

Correspondence and requests for material should be addressed to M.J.B. (e-mail: berry@biosun.harvard.edu).

的时间计算。为了描绘闪烁光柱之后给定时间的放电特性，每个神经节细胞的放电频率根据相对于闪烁光柱线的细胞位置描绘。该曲线通过与标准偏差为 20 μm 的高斯曲线卷积变得平滑。为了估计对移动光柱的群体反应，每个刺激被重复 50 次，每个神经节细胞对闪光的放电频率都以二进制 15 ms 的时间计算。然后，对神经节细胞的放电频率以及其相对于光柱的距离进行绘图，并对所有的细胞进行平均。

（董培智 翻译；巩克瑞 审稿）

The DNA Sequence of Human Chromosome 22

I. Dunham *et al.*

Editor's Note

This paper, a key landmark in the Human Genome Project, describes the first complete sequence of a human chromosome. This sequence of chromosome 22 offered insights into the way that genes are arranged along chromosomes and into how these genes might be controlled, paving the way for developments in medical diagnostics and therapeutics. The project, led by British-based geneticist Ian Dunham, involved an international consortium of sequencing centres, and the data were made freely available. The study shows that chromosome 22 is made up of 33.5 million "letters" and includes 679 genes, over half of which were previously unknown in humans. The complete human genome sequence was published four years later.

Knowledge of the complete genomic DNA sequence of an organism allows a systematic approach to defining its genetic components. The genomic sequence provides access to the complete structures of all genes, including those without known function, their control elements, and, by inference, the proteins they encode, as well as all other biologically important sequences. Furthermore, the sequence is a rich and permanent source of information for the design of further biological studies of the organism and for the study of evolution through cross-species sequence comparison. The power of this approach has been amply demonstrated by the determination of the sequences of a number of microbial and model organisms. The next step is to obtain the complete sequence of the entire human genome. Here we report the sequence of the euchromatic part of human chromosome 22. The sequence obtained consists of 12 contiguous segments spanning 33.4 megabases, contains at least 545 genes and 134 pseudogenes, and provides the first view of the complex chromosomal landscapes that will be found in the rest of the genome.

TWO alternative approaches have been proposed to determine the human genome sequence. In the clone by clone approach, a map of the genome is constructed using clones of a suitable size (for example, 100–200 kilobases (kb)), and then the sequence is determined for each of a representative set of clones that completely covers the map[1]. Alternatively, a whole genome shotgun[2] requires the sequencing of unmapped genomic clones, typically in a size range of 2–10 kb, followed by a monolithic assembly to produce the entire sequence. Although the merits of these two strategies continue to be debated[3], the public domain human genome sequencing project is following the clone by clone approach[4] because it is modular, allows efficient organization of distributed resources and sequencing capacities, avoids problems arising from distant repeats and results in early completion of significant units of the genome. Here we report the first sequencing landmark of the

人类 22 号染色体的 DNA 序列

邓纳姆等

编者按

本文是人类基因组计划的一个关键里程碑，介绍了单条人类染色体的第一个完整序列。22 号染色体的序列帮助我们了解基因在染色体上的编排方式以及这些基因是如何被调控的，并为医学诊断和治疗的发展奠定了基础。本项目由英国遗传学家伊恩·邓纳姆领衔，多个测序中心组成的国际联盟共同完成，数据可免费获取。本研究表明，22 号染色体是由 3,350 万个“字母”组成，包含 679 个基因，其中半数以上之前在人类中是未知的。完整的人类基因组序列发表于 4 年后。

关于一种生物的完整基因组 DNA 序列的知识使我们可以用系统的方法确定其遗传组分。基因组序列可以提供所有基因的完整结构，包括功能未知的基因及其控制元件，通过推理还可得到它们编码的蛋白质，以及所有其他生物学上的重要序列。此外，对于针对生物体设计进一步的生物学研究和通过跨物种序列比对来进行演化研究而言，该物种的序列都是丰富且永久的信息来源。一系列微生物和模式生物序列的测定已经充分证明了这种方法的强大能力。下一步是获得整个人类基因组的完整序列。在本文中我们报道了人类 22 号染色体常染色质部分的序列。该序列由跨越 33.4 Mb 的 12 个连续片段组成，包含至少 545 个基因和 134 个假基因，并首次展示了在基因组其他部分将会发现的复杂的染色体概况。

目前有两种测定人类基因组序列的方法可供选择。连续克隆法用适当大小(如 100 ~ 200 千碱基(kb))的多个克隆构建基因组图谱，然后测定完全覆盖图谱的一套代表性克隆中每个克隆的序列[1]。另一种是全基因组鸟枪法[2]，需要对未作图的基因组克隆进行测序，大小通常在 2 ~ 10 kb，然后通过整体组装得到整个序列。虽然这两种策略的优点仍然存在争议[3]，但公开的人类基因组测序项目已经采取连续克隆法[4]，因为这种方法是模块化的，可以有效地组织分散的资源及测序能力，避免远距离重复引起的问题，而且能够迅速完成基因组重要单元的测定。在本文中，我们报道了人类基因组计划的第一个测序里程碑：第一条人类染色体常染色质部分的

human genome project, the operationally complete sequence of the euchromatic portion of a human chromosome.

Chromosome 22 is the second smallest of the human autosomes, comprising 1.6–1.8% of the genomic DNA[5]. It is one of five human acrocentric chromosomes, each of which shares substantial sequence similarity in the short arm, which encodes the tandemly repeated ribosomal RNA genes and a series of other tandem repeat sequence arrays. There is no evidence to indicate the presence of any protein coding genes on the short arm of chromosome 22 (22p). In contrast, direct[6] and indirect[7,8] mapping methods suggest that the long arm of the chromosome (22q) is rich in genes compared with other chromosomes. The relatively small size and the existence of a high-resolution framework map of the chromosome[9] suggested to us that sequencing human chromosome 22 would provide an excellent opportunity to show the feasibility of completing the sequence of a substantial unit of the human genome. In addition, alteration of gene dosage on part of 22q is responsible for the aetiology of a number of human congenital anomaly disorders including cat eye syndrome (CES, Mendelian Inheritance in Man (MIM) 115470, http://www.ncbi.nlm.nih.gov/omim/) and velocardiofacial/DiGeorge syndrome (VCFS, MIM 192430; DGS, MIM 188400). Other regions associated with human disease are the schizophrenia susceptibility locus[10,11], and the sequences involved in spinocerebellar ataxia 10 (SCA10)[12]. Making the sequence of human chromosome 22 freely available to the community early in the data collection phase has benefited studies of disease-related and other genes associated with this human chromosome[13-19].

Genomic Sequencing

To identify genomic clones as the substrate for sequencing chromosome 22, extensive clone maps of the chromosome were constructed using cosmids, fosmids, bacterial artificial chromosomes (BACs) and P1-derived artificial chromosomes (PACs). Clones representing parts of chromosome 22 were identified by screening BAC and PAC libraries representing more than 20 genome equivalents using sequence tagged site (STS) markers known to be derived from the chromosome, or by using cosmid and fosmid libraries derived from flow-sorted DNA from chromosome 22. Overlapping clone contigs were assembled on the basis of restriction enzyme fingerprints and STS-content data, and ordered relative to each other using the established framework map of the chromosome[9]. The resulting nascent contigs were extended and joined by iterative cycles of chromosome walking using sequences from the end of each contig. In two places, yeast artificial chromosome (YAC) clones were used to join or extend contigs (AL049708, AL049760). The sequence-ready map covers 22q in 11 clone contigs with 10 gaps and stretches from sequences containing known chromosome 22 centromeric tandem repeats to the 22q telomere[20].

In the final sequence, one additional gap that was intractable to sequencing is found 234 kb from the centromere (see below). The gaps between the clone contigs are located at the two ends of the map, in the 4.3 Mb adjacent to the centromere and in 7.3 Mb at the telomeric

可使用的完整序列。

22 号染色体是人类第二小的常染色体，占基因组 DNA 的 1.6% ~ 1.8%[5]。它是人类 5 条近端着丝粒染色体之一，这 5 条染色体具有序列高度相似的短臂，这些短臂编码串联重复的核糖体 RNA 基因和一系列其他串联重复序列阵列。目前没有证据表明 22 号染色体短臂（22p）上存在任何蛋白质编码基因。相反，直接[6]和间接[7,8]作图方法都表明该染色体的长臂（22q）与其他染色体相比拥有丰富的基因。人类 22 号染色体相对较小，并具有高分辨率框架图[9]，这表明对其进行测序是展示人类基因组大型单元的序列测定可行性的绝佳机会。此外，部分 22q 上基因数量的改变是引起很多人类先天性异常疾病的病因，这些疾病包括猫眼综合征（CES，人类孟德尔遗传数据库（MIM）115470，http://www.ncbi.nlm.nih.gov/omim/）和腭心面/迪格奥尔格综合征（VCFS，MIM 192430；DGS，MIM 188400）。其他与人类疾病相关的区域包括精神分裂症易感性位点[10,11]和涉及脊髓小脑性共济失调 10（SCA10）的序列[12]。我们早在数据收集阶段就将人类 22 号染色体的序列免费对社会开放，这使得一些与该染色体相关的疾病和其他基因的研究因此受益[13-19]。

基因组测序

为了识别作为 22 号染色体测序底物的基因组克隆，科研人员用黏粒、F 黏粒、细菌人工染色体（BAC）和 P1 衍生人工染色体（PAC）构建了大量的该染色体的克隆图谱。使用已知的来自该染色体的序列标签位点（STS）标记的序列，对覆盖超过 20 倍基因组的 BAC 和 PAC 克隆文库进行筛选，或使用 22 号染色体流式分选的 DNA 的黏粒和 F 黏粒文库进行筛选，识别来自部分 22 号染色体的克隆。在限制性内切酶指纹和 STS 含量数据的基础上组装重叠的克隆重叠群，并用已建立的该染色体框架图对其进行相应排列[9]。通过用每个重叠群末端的序列进行染色体步移迭代循环，将上述生成的新重叠群延长并连接。其中两处重叠群（AL049708 和 AL049760）的连接或延长使用了酵母人工染色体（YAC）来进行。已知序列图以存在 10 个缺口的 11 个克隆重叠群覆盖 22q，从包含已知的 22 号染色体着丝粒串联重复的序列延伸至 22q 端粒[20]。

在最终序列中，距着丝粒 234 kb 处发现另一个缺口（见下文），这个缺口很难进行测序。克隆重叠群之间的各个缺口分布于图的两端，在与着丝粒邻近的 4.3 Mb 处及端粒末端的 7.3 Mb 处。这些区域被一个 23 Mb 的中央重叠群分开。我们使用与重

end. These regions are separated by a central contig of 23 Mb. We have concluded that the gaps contain sequences that are unclonable with the available host-vector systems, as we were unable to detect clones containing the sequences in these gaps by screening more than 20 genome equivalents of bacterial clones using sequences adjacent to the contig ends.

The size of the seven gaps in the telomeric region has been estimated by DNA fibre fluorescence *in situ* hybridization (FISH). No gap in this region is judged to be larger than ~150 kb. For three of these gaps, a number of BAC and PAC clones that contain STSs on either side of the gap were shown to be deleted for at least a minimal core region by DNA fibre FISH. As these clones come from multiple donor DNA sources, these results are unlikely to be due to deletion in the DNA used to make the libraries. Furthermore, the same result was observed for the gap at 32,600 kb from the centromeric end of the sequence, when the DNA fibre FISH experiments were performed on DNA from two different lymphoblastoid cell lines. One possible explanation for this observation is that DNA fragments containing the gap sequences are initially cloned in the BAC library but clones that delete these sequences have a significant selective advantage as the library is propagated. As the observed size range of the cloned inserts in the BAC libraries ranges from 100 kb to more than 230 kb (http://bacpac.med.buffalo.edu/), such deletion events are not distinguishable on the basis of size from undeleted BACs. Additional analysis of the distribution of BAC end sequences from dbGSS (http://www.ncbi.nlm.nih.gov/dbGSS/index.html) suggests that the BAC coverage is sparser closer to the gaps and that this analysis did not identify any BACs spanning the gaps. The three remaining clone-map gaps in the proximal region of the long arm are in regions that may contain segments of previously characterized low-copy repeats[21]. These gaps could not be sized by DNA fibre FISH because of the extensive intra- and interchromosomal repeat sequences (see below) but were amenable to long-range restriction mapping. The gap between AP000529 and AP000530 was estimated to be shorter than 150 kb by comparison with a previously established long-range restriction map[22]. The gap closest to the centromere, which is less than 2 kb in size, could not be sequenced despite BAC clone coverage as it was unrepresented in plasmid or M13 libraries, and was intractable to all sequencing strategies applied. Detailed descriptions of several of the clone contigs have been published[21,23,24] or will be published elsewhere.

Each sequencing group took responsibility for completion of adjacent areas of the sequence as illustrated in Fig. 1. (Editorial note: this figure was original included as a fold-out insert. For details, see http://www.nature.com/articles/990031/figures/1) A set of minimally overlapping clones (the "tile path") was chosen from the physical map and sequenced using a combination of a random shotgun assembly, followed by directed sequencing to close gaps and resolve ambiguities ("finishing"). The major problems encountered during completion of the sequence in the directed sequencing phase were CpG islands, tandem repeats and apparent cloning biases. Directed sequencing using oligonucleotide primers, very short insert plasmid libraries, or identification of bridging clones by screening high complexity plasmid or M13 libraries solved these problems.

叠群末端相邻的序列对覆盖超过 20 倍基因组的细菌克隆进行筛选，仍无法检测到含有这些缺口序列的克隆，因此我们认为这些缺口包含用现有宿主–载体系统无法克隆的序列。

通过 DNA 纤维荧光原位杂交（FISH）估算端粒区域七个缺口的大小。经判断，这一区域的缺口均不超过 150 kb 左右。据 DNA 纤维荧光原位杂交显示，对其中三个缺口而言，一些在缺口的任意一侧包含 STS 的 BAC 和 PAC 克隆被删去了至少一处最小核心区。由于这些克隆有多个 DNA 供体来源，因此这些结果不可能是由于用于构建文库的 DNA 存在缺失。此外，当在两种不同成淋巴母细胞细胞系 DNA 上进行 DNA 纤维荧光原位杂交实验时，在序列中距着丝粒末端 32,600 kb 的缺口处也观察到同样的结果。这个现象的一种可能解释是，包含缺口序列的 DNA 片段最初在 BAC 文库中进行了克隆，但当文库扩增时，删除这些序列的克隆有显著的选择性优势。观察到的插入 BAC 文库的克隆片段的大小从 100 kb 到 230 kb 以上不等（http://bacpac.med.buffalo.edu/），因此这些缺失状况无法根据未缺失的 BAC 大小进行区分。对来自 dbGSS（http://www.ncbi.nlm.nih.gov/dbGSS/index.html）的 BAC 末端序列的分布进一步分析表明，接近缺口处 BAC 覆盖稀疏，这一分析也没有发现任何跨过缺口的 BAC。剩下的三个在长臂近端区域的克隆图谱缺口位于可能包含早先表征过的低拷贝重复片段的区域[21]。由于大量染色体内和染色体间重复序列（见下文）的存在，这些缺口的大小无法通过 DNA 纤维荧光原位杂交测定，但可以通过长距离限制性作图得到。通过与先前建立的长距离限制图比较可知，AP000529 和 AP000530 之间的缺口估计小于 150 kb[22]。与着丝粒最近的缺口不到 2 kb，但由于在质粒和 M13 文库中没有表达，因此尽管被 BAC 克隆覆盖却仍无法测序，而且在所有测序策略中均无法处理。几个克隆重叠群的详细介绍已经发表[21,23,24]，或将另行发表。

序列相邻区域的测序由每个测序小组负责完成，如图 1 所示。（编者注：这张图在原文中是一个折叠插入图，细节请见 http://www.nature.com/articles/990031/figures/1）从物理图中选出一套最小重叠克隆（“覆瓦式”）并采用随机鸟枪法组件的组合进行测序，继而为封闭缺口和消除模糊（“精加工”）进行直接测序。在直接测序阶段的序列完成期间遇到的主要问题是 CpG 岛、串联重复序列和明显的克隆偏好。这些问题通过以下方法进行解决：使用寡核苷酸引物进行直接测序、使用极短的插入质粒文库，或通过筛选高复杂性质粒或 M13 文库识别桥接克隆。

Fig. 1. The sequence of human chromosome 22. Coloured boxes depict the annotated features of the sequence of human chromosome 22, with the centromere to the left and the telomere to the right. Coordinates are in kilobases. Vertical yellow blocks indicate the positions of the gaps in the sequence and are proportional in size to the estimated size of each gap. From bottom to top the following features are displayed: positions of interspersed repetitive sequences including tandem repeats categorized by nucleotide repeat unit length (at this resolution *Alu* repeats are not visibly separated in some regions); the positions of the microsatellite markers in the genetic map of Dib *et al*[36]; the tiling path of genomic clones used to determine the sequence labelled by their GenBank/EMBL/DDBJ accession number and coloured according to the source of the sequence; and the annotated gene, pseudogene and CpG island content of the sequence. Transcripts and pseudogenes oriented 5′ to 3′ on the DNA strand from centromere to telomere are designated "+", those on the opposite strand "−". In the transcript rows, the annotated genes are subdivided by colour according to the criteria in the text. Annotated genes with approved gene symbols from the HUGO nomenclature committee are labelled. For details of all the genes with their positions in the reference sequence, see Supplementary Information, Table 1. In the case of the immunoglobulin variable region, the entire locus has been drawn as a single block; in reality, this is a complex of variable chain genes (see ref. 27). At the top is a graphical plot of the repeat density for the common interspersed repeats *Alu* and Line1, and the C+G base frequency across the sequence. Each is calculated as a percentage of the sequence using a sliding 100-kb window moved in 50-kb iterations. Since the production of Fig. 1, six accession codes have been updated. The new codes are AL050347 (for Z73987), AL096754 (for Z68686), AL049749 (for Z82197), Z75892 (for Z75891), AL078611 (for Z79997) and AL023733 (for AL023593).

The completed sequence covers 33.4 Mb of 22q with 11 gaps and has been estimated to be accurate to less than 1 error in 50,000 bases, by internal and external checking exercises[25]. The order and size of each of the contiguous pieces of sequence is detailed in Table 1. The largest contiguous segment stretches over 23 Mb. From our gap-size estimates, we calculate that we have completed 33,464 kb of a total region spanning 34,491 kb and that therefore the sequence is complete to 97% coverage of 22q. The complete sequence and analysis is available on the internet (http://www.sanger.ac.uk/HGP/Chr22 and http://www.genome.ou.edu/Chr22.html).

Table 1. Sequence contigs on chromosome 22

Contig*	Size (kb)	
AP000522–AP000529	234	
gap		**1.9**
AP000530–AP000542	406	
gap		**~150**
AP000543–AC006285	1,394	
gap		**~150**
AC008101–AC007663	1,790	
gap		**~100**
AC007731–AL049708	23,006	
gap		**~50**
AL118498–AL022339	767	
gap†		**~50-100**

图 1. 人类 22 号染色体的序列。彩色方块指示人类 22 号染色体序列的注释特征，其着丝粒在左侧，端粒在右侧。坐标单位为千碱基。垂直黄色区域指示序列中缺口的位置，其大小与每个缺口的估算大小成比例。自下而上显示的特征是：散布重复序列的位置，包括根据核苷酸重复单位长度分类的串联重复（在此分辨率下 *Alu* 重复在某些区域无法通过视觉区分）；迪卜等人的遗传图中微卫星标记物的位置[36]；基因组克隆覆瓦式路径用来决定 GenBank/EMBL/DDBJ 检索编号标记的序列，并根据序列来源用不同颜色显示；以及序列中注释的基因、假基因和 CpG 岛含量。在 DNA 链上，从着丝粒到端粒按 5′ 到 3′ 走向的转录本和假基因被命名为“+”，相应在反义链上的为“−”。在转录物各行中，注释基因根据文中的标准用颜色进行了细分。来自 HUGO 命名委员会核准的基因符号的注释基因进行了标记。所有基因及其在参考序列中的位置的详细情况见补充信息表 1。对于免疫球蛋白可变区，整个基因座已绘制为单一区域；实际上这是可变链基因的复合体（见参考文献 27）。顶部是常见散布重复 *Alu* 和 Line1 的重复密度，以及序列中 C+G 碱基频率的图表。二者都是使用 100 kb 的滑动窗口及 50 kb 的迭代进行计算得到的，并表示为序列的百分比。图 1 制成之后有 6 个检索号发生了更新。新编号为：AL050347（原 Z73987）、AL096754（原 Z68686）、AL049749（原 Z82197）、Z75892（原 Z75891）、AL078611（原 Z79997）和 AL023733（原 AL023593）。

全部的序列覆盖 22q 33.4 Mb，有 11 个缺口，通过内部和外部检查，序列准确度预计达到每 50,000 碱基中错误数小于 1[25]。每一个连续的序列片段的顺序和大小详见表 1。最大的连续片段延伸超过 23 Mb。根据对缺口大小的估算，我们已经完成了总长 34,491 kb 区域中的 33,464 kb，因此，序列对 22q 的覆盖率达到 97%。完整的序列及分析可以在网上获得（http://www.sanger.ac.uk/HGP/Chr22 和 http://www.genome.ou.edu/Chr22.html）。

表 1. 22 号染色体的序列重叠群

重叠群＊	大小（kb）	
AP000522–AP000529	234	
缺口		**1.9**
AP000530–AP000542	406	
缺口		**约 150**
AP000543–AC006285	1,394	
缺口		**约 150**
AC008101–AC007663	1,790	
缺口		**约 100**
AC007731–AL049708	23,006	
缺口		**约 50**
AL118498–AL022339	767	
缺口 †		**约 50～100**

Continued

Contig*	Size (kb)	
Z85994–AL049811	1,528	
gap		**~150**
AL049853–AL096853	2,485	
gap‡		**~50**
AL096843–AL078607	190	
gap†		**~100**
AL078613–AL117328	993	
gap		**~100**
AL080240–AL022328	291	
gap†		**~100**
AL096767–AC002055	380	
Total sequence length	33,464	
Total length of 22q	34,491	

* Contigs are indicated by the first and last sequence in the orientation centromere to telomere, and are named by their GenBank/EMBL/DDBJ accession numbers.
† These gaps are spanned by BAC and/or PAC clones with deletions.
‡ This gap shows a complex duplication of AL096853 in DNA fibre FISH.

Sequence Analysis and Gene Content

Analysis of the genomic sequence of the model organisms has made extensive use of predictive computational analysis to identify genes[26-28]. In human DNA, identification of genes by these methods is more difficult because of extensive splicing, lower density of exons and the high proportion of interspersed repetitive sequences. The accuracy of *ab initio* gene prediction on vertebrate genomic sequence has been difficult to determine because of the lack of sequence that has been completely annotated by experiment. To determine the degree of overprediction made by such algorithms, all genes within a region need to be experimentally identified and annotated, however it is virtually impossible to know when this job is complete. A 1.4-Mb region of human genomic sequence around the BRCA2 locus has been subjected to extensive experimental investigation, and it is believed that the 170 exons identified is close to the total number expressed in the region.

The most recent calibration of *ab initio* methods against this region (R.B.S.K. and T.H., manuscript in preparation) shows that with the best methods[29,30] more than 30% of exon predictions do not overlap any experimental exons, in other words, they are overpredictions. Furthermore, having now applied this analysis to larger amounts of data (more than 15 Mb from the Sanger Annotated Genome Sequence Repository which can

续表

重叠群 *	大小（kb）	
Z85994–AL049811	1,528	
缺口		**约 150**
AL049853–AL096853	2,485	
缺口 ‡		**约 50**
AL096843–AL078607	190	
缺口 †		**约 100**
AL078613–AL117328	993	
缺口		**约 100**
AL080240–AL022328	291	
缺口 †		**约 100**
AL096767–AC002055	380	
序列总长度	33,464	
22q 总长度	34,491	

* 重叠群按照从着丝粒到端粒方向的第一个和最后一个序列进行标记，并用其在 GenBank/EMBL/DDBJ 检索编号命名。
† 被删除的 BAC 和/或 PAC 克隆横跨这些缺口。
‡ 该缺口在 DNA 纤维荧光原位杂交中显示出复杂的 AL096853 重复。

序列分析和基因含量

模式生物的基因组序列分析已经广泛用于确定基因的预测计算分析中[26-28]。人类 DNA 中存在剪接多、外显子密度低和散在重复序列比例高的问题，因此通过这些方法鉴定基因更为困难。由于缺乏完全通过实验注释的序列，脊椎动物基因组中从头预测基因的准确性一直难以确定。为确定这些算法产生的过度预测的程度，一个区域内所有的基因都需要用实验方法鉴定并注释，但是这项任务何时能完成又是几乎不可知的。针对 BRCA2 基因座周围 1.4 Mb 的人类基因组序列区域的深入实验研究认为，鉴定出的 170 个外显子接近该区域中表达的总数。

针对这个区域的从头算法的最新校准（布鲁斯克耶维奇和哈伯德，稿件准备中）表明，使用最佳方法[29,30]时，超过 30%的外显子预测未与任何实验证实的外显子重合，换句话说，它们的预测是过度的。此外，将这一分析应用于更大规模的数据（超过 15 Mb，来自桑格注释基因组序列存储库，可作为 Genesafe 集合（http://www.hgmp.mrc.ac.uk/Genesafe/）的一部分而获得）之后证实，在序列不同区域基因模型预

be obtained as part of the Genesafe collection (http://www.hgmp.mrc.ac.uk/Genesafe/)), it is confirmed that prediction accuracy also varies considerably between different regions of sequence. It was hoped that these calibration efforts would lead to rules for reliable gene prediction based on *ab initio* methods alone, perhaps on the basis of combining several different methods, GC content and so on. However, so far this has not been possible. The same analysis also shows that although ~95% of genes are at least partially predicted by *ab initio* methods, few gene structures are completely correct (none in BRCA2) and more than 20% of experimental exons are not predicted at all. The comparison of *ab initio* predictions and the annotated gene structures (see below) in the chromosome 22 sequence is consistent with this, with 94% of annotated genes at least partially detected by a Genscan gene prediction, but only 20% of annotated genes having all exons predicted exactly. Sixteen per cent of all the exons in annotated genes were not predicted at all, although this is only 10% for internal exons (that is, not 5′ and 3′ ends). As a result, we do not consider that *ab initio* gene prediction software can currently be used directly to reliably annotate genes in human sequence, although it is useful when combined with other evidence (see below), for example, to define splice-site boundaries, and as a starting point for experimental studies.

Fortunately, a vast resource of experimental data on human genes in the form of complementary DNA and protein sequences and expressed sequence tags (ESTs) is available which can be used to identify genes within genomic DNA. Furthermore about 60% of human genes have distinctive CpG island sequences at their 5′ ends[31] which can also be used to identify potential genes. Thus, the approach we have taken to annotating genes in the chromosome 22 sequence relies on a combination of similarity searches against all available DNA and protein databases, as well as a series of *ab initio* predictions. Upon completion of the sequence of each clone in the tile path, the sequence was subjected to extensive computational analysis using a suite of similarity searches and prediction tools. Briefly, the sequences were analysed for repetitive sequence content, and the repeats were masked using RepeatMasker (http://ftp.genome.washington.edu/RM/RepeatMasker.html). Masked sequence was compared to public domain DNA and protein databases by similarity searches using the blast family of programs[32]. Unmasked sequence was analysed for C+G content and used to predict the presence of CpG islands, tandem repeat sequences, tRNA genes and exons. The completed analysis was assembled into contigs and visualized using implementations of ACEDB (http://www.sanger.ac.uk/Software/Acedb/). In addition, the contiguous masked sequence was analysed using gene prediction software[29,30].

Gene features were identified by a combination of human inspection and software procedures. Figure 1 shows the 679 gene sequences annotated across 22q. They were grouped according to the evidence that was used to identify them as follows: genes identical to known human gene or protein sequences, referred to as "known genes" (247); genes homologous, or containing a region of similarity, to gene or protein sequences from human or other species, referred to as "related genes" (150); sequences homologous to only ESTs, referred to as "predicted genes" (148); and sequences homologous to a known gene or protein, but with a disrupted open reading frame, referred to "pseudogenes" (134). (See

测的准确度有很大的差别。人们希望这些校准工作可以形成单纯基于从头算法或基于几种不同方法和 GC 含量等相结合的可靠的基因预测标准。然而迄今为止这还从未实现。同样的分析还表明，虽然大约 95%的基因通过从头算法得到了至少部分的预测，但只有很少数基因的预测结构是完全正确的(在 BRCA2 中完全没有)，而且 20%以上实验验证的外显子根本无法被预测到。22 号染色体序列的从头算法预测和已注释的基因结构的对比(见下文)与此相符，通过 Genscan 基因预测，94%的注释基因至少能部分检测到，但只有 20%注释基因的所有外显子能准确预测。注释基因中所有外显子的 16%完全未被预测，然而这只占内部外显子(即不是 5′ 和 3′ 端)的 10%。因此我们认为，目前基因从头预测软件还不能直接用于对人类序列基因进行可靠的注释，但它在结合其他证据时还是有用的(见下文)，例如确定剪接位点边界和作为实验研究的起点。

幸运的是，大量关于人类基因的互补 DNA 和蛋白质序列及表达序列标签(EST)的实验数据资源可以用于确定基因组 DNA 中的基因。另外，约 60%的人类基因在其 5′ 端有独特的 CpG 岛序列[31]，也可以用来确定可能存在的基因。因此，我们采取的用于注释 22 号染色体序列中基因的方法依赖于综合运用所有类似的可用的 DNA 和蛋白质数据库以及一系列从头预测。一旦完成覆瓦式分析中每个克隆的序列，该序列会用一套相似性的搜索和预测工具进行大量计算分析。简言之，分析序列的重复序列含量，并且用 RepeatMasker 对重复序列进行标记(http://ftp.genome.washington.edu/RM/RepeatMasker.html)。使用 blast 系列程序进行相似性搜索，将标记的序列与公共领域的 DNA 和蛋白质数据库进行比对[32]。分析未标记序列 C+G 的含量，并且用于预测 CpG 岛、串联重复序列、tRNA 基因和外显子的存在。将已完成的分析组装为重叠群，并通过 ACEDB 进行查询实现可视化(http://www.sanger.ac.uk/Software/Acedb/)。此外，用基因预测软件对连续标记的序列进行分析[29,30]。

基因特征通过人为识别检查和软件程序相结合确定。图 1 给出了 22q 中已注释的 679 个基因序列。根据识别它们的证据将其分组如下：与已知的人类基因或蛋白质序列一致的基因命名为“已知基因”(247 个)；与来自人类或其他物种的基因或蛋白质序列同源或含有相似区域的基因命名为“相关基因”(150 个)；只与 EST 同源的序列命名为“预测基因”(148 个)；与已知基因或蛋白质序列同源，但其可读框被中断的序列命名为“假基因”(134 个)。(关于这些基因的详细情况见补充信息表 1。)

Supplementary Information, Table 1, for details of these genes.) The *ab initio* gene prediction program, Genscan, predicted 817 genes (6,684 exons) in the contiguous sequence, of which 325 do not form part of the annotated genes categorized above. Given the calibration of *ab initio* prediction methods discussed above, we estimate that of the order of 100 of these will represent parts of "real" genes for which there is currently no supporting evidence in any sequence database, and that the remainder are likely to be false positives.

The total length of the sequence occupied by the annotated genes, including their introns, is 13.0 Mb (39% of the total sequence). Of this, only 204 kb contain pseudogenes. About 3% of the total sequence is occupied by the exons of these annotated genes. This contrasts sharply with the 41.9% of the sequence that represents tandem and interspersed repeat sequences. There is no significant bias towards genes encoded on one strand at the 5% level ($\chi^2 = 3.83$).

A striking feature of the genes detected is their variety in terms of both identity and structure. There are several gene families that appear to have arisen by tandem duplication. The immunoglobulin λ locus is a well-known example, but there also are other immunoglobulin-related genes on the chromosome outside the immunoglobulin λ region. These include the three genes of the immunoglobulin λ-like (IGLL) family plus a fourth possible member of the family (AC007050.7). There are five clustered immunoglobulin κ variable region pseudogenes in AC006548, and an immunoglobulin variable-related sequence (VpreB3) in AP000348. Much further away from the λ genes is a variable region pseudogene, ~123 kb telomeric of IGLL3 in sequence AL008721 (coordinates ~9,420–9,530 kb from the centromeric end of the sequence), and a cluster of two λ constant region pseudogenes and a variable region pseudogene in sequences AL008723/AL021937 (coordinates ~16,060–16,390 kb from the centromeric end).

Human chromosome 22 also contains other duplicated gene families that encode glutathione *S*-transferases, Ret-finger-like proteins[19], phorbolins or APOBECs, apolipoproteins and β-crystallins. In addition, there are families of genes that are interspersed among other genes and distributed over large chromosomal regions. The γ-glutamyl transferase genes represent a family that appears to have been duplicated in tandem along with other gene families, for instance the BCR-like genes, that span the 22q11 region and together form the well-known LCR22 (low-copy repeat 22) repeats (see below).

The size of individual genes encoded on this chromosome varies over a wide range. The analysis is incomplete as not all 5′ ends have been defined. However, the smallest complete genes are only of the order of 1 kb in length (for example, HMG1L10 is 1.13 kb), whereas the largest single gene (LARGE[15]) stretches over 583 kb. The mean genomic size of the genes is 19.2 kb (median 3.7 kb). Some complete gene structures appear to contain only single exons, whereas the largest number of exons in a gene (PIK4CA) is 54. The mean exon number is 5.4 (median 3). The mean exon size is 266 bp (median 135 bp). The smallest complete exon we have identified is 8 bp in the PITPNB gene. The largest single exon is 7.6 kb in the PKDREJ, which is an intron-less gene with a 6.7-kb open reading frame. In addition,

基因从头预测程序 Genscan 在连续序列中预测了 817 个基因(6,684 个外显子)，其中 325 个不属于上面分类的注释基因。鉴于上文讨论的从头预测方法的校准，我们估计其中 100 个将代表部分目前在任何序列数据库中还没有支持性证据的“真正”基因，而其余则可能是假阳性。

这些注释基因包括其内含子所占的序列总长度为 13.0 Mb(占总序列的 39%)，其中只有 204 kb 含有假基因。这些注释基因的外显子约占总序列的 3%，这与代表串联和散布重复序列的序列所占的比例(41.9%)形成鲜明对比。对于一条链上的编码基因，在 5%的水平上不存在显著性偏差($\chi^2 = 3.83$)。

检测到的基因的一个显着特点是它们在一致性和结构上的多样性。有几个基因家族可能是通过串联重复产生的。免疫球蛋白 λ 位点是众所周知的例子，但在免疫球蛋白 λ 区域外的染色体上也有其他免疫球蛋白相关基因。其中包括免疫球蛋白 λ 样(IGLL)家族的 3 个基因以及该家族的第 4 个可能成员(AC007050.7)。AC006548 有五个成簇的免疫球蛋白 κ 可变区假基因，AP000348 有一个免疫球蛋白可变相关序列(VpreB3)。距离 λ 基因更远的地方，在序列 AL008721 中，IGLL3 端粒侧约 123 kb 处(距序列着丝粒端约 9,420 ~ 9,530 kb 处)有一个可变区假基因，在序列 AL008723/AL021937 中(距着丝粒端约 16,060 ~ 16,390 kb 处)有一个由两个 λ 恒定区假基因和一个可变区假基因形成的簇。

人类 22 号染色体还包含其他重复基因家族，这些基因家族编码谷胱甘肽 *S*–转移酶、Ret–指样蛋白[19]、phorbolin 或 APOBEC、载脂蛋白和 β–晶状体蛋白。此外，有些基因家族散布在其他基因中并分散在大的染色体区域内。γ–谷氨酰转移酶基因代表一个可能与其他基因家族串联重复的家族，如 BCR 样基因，该家族跨越 22q11 区域并共同组成了著名的 LCR22(低拷贝重复 22)重复(见下文)。

这条染色体编码的单个基因大小差别很大。并非所有编码基因的 5′ 端都已确定，所以该分析是不完整的。然而，最小的完整基因长度仅为 1 kb 量级(例如，HMG1L10 是 1.13 kb)，而最大的单基因(LARGE[15])超过 583 kb。基因的平均大小是 19.2 kb(中位数为 3.7 kb)。一些完整的基因结构似乎只包含单一的外显子，而单个基因中外显子数量最多达 54 个(PIK4CA)。平均外显子数目是 5.4 个(中位数为 3 个)。外显子的平均大小为 266 bp(中位数为 135 bp)。我们已鉴定的最小的完整外显子为 8 bp，位于 PITPNB 基因中。最大的单一外显子为 7.6 kb，位于 PKDREJ 中，

two genes occur within the introns of other expressed genes. The 61-kb TIMP3 gene, which is involved in Sorsby fundus macular degeneration, lies within a 268-kb intron of the large SYN3 gene, and the 8.5-kb HCF2 gene lies within a 27.5-b intron of the PIK4CA gene. In each case, the genes within genes are oriented in the opposite transcriptional orientation to the outer gene. We also observe pseudogenes frequently lying within the introns of other functional genes.

Peptide sequences for the 482 annotated full-length and partial genes with an open reading frame of greater than or equal to 50 amino acids were analysed against the protein family (PFAM)[33], Prosite[34] and SWISS-PROT[35] databases. These data were processed and displayed in an implementation of ACEDB. Overall, 240 (50%) predicted proteins had matching domains in the PFAM database encompassing a total of 164 different PFAM domains. Of the residues making up these 482 proteins, 25% were part of a PFAM domain. This compares with PFAM's residue coverage of SWISS-PROT/TrEMBL, which is more than 45% and indicates that the human genome is enriched in new protein sequences. Sixty-two PFAM domains were found to match more than one protein, including ten predicted proteins containing the eukaryotic protein kinase domain (PF00069), nine matching the Src homology domain 3 (PF00018) and eight matching the RhoGAP domain (PF00620). Fourteen predicted proteins contain zinc-finger domains (See Supplementary Information, Table 2, for details of the PFAM domains identified in the predicted proteins).

Nineteen per cent of the coding sequences identified were designated as pseudogenes because they had significant similarity to known genes or proteins but had disrupted protein coding reading frames. Because 82% of the pseudogenes contained single blocks of homology and lacked the characteristic intron-exon structure of the putative parent gene, they probably are processed pseudogenes. Of the remaining spliced pseudogenes, most represent segments of duplicated gene families such as the immunoglobulin κ variable genes, the β-crystallins, CYP2D7 and CYP2D8, and the GGT and BCR genes. The pseudogenes are distributed over the entire sequence, interspersed with and sometimes occurring within the introns of annotated expressed genes. However, there also is a dense cluster of 26 pseudogenes in the 1.5-Mb region immediately adjacent to the centromere; the significance of this cluster is currently unclear.

Given that the sequence of 33.4 Mb of chromosome 22q represents 1.1% of the genome and encodes 679 genes, then, if the distribution of genes on the other chromosomes is similar, the minimum number of genes in the entire human genome would be at least 61,000. Previous work has suggested that chromosome 22 is gene rich[6] by a factor of 1.38 (http://www.ncbi.nlm.nih.gov/genemap/page.cgi?F = GeneDistrib.html), which would reduce this estimate to ~45,000 genes. It is important, however, to recognize that the analysis described here only provides a minimum estimate for the gene content of chromosome 22q, and that further studies will probably reveal additional coding sequences that could not be identified with the current approaches.

该基因是一个有 6.7 kb 可读框的无内含子的基因。此外，有两个基因出现在其他表达基因的内含子中。与索斯比眼底黄斑变性有关的 61 kb 的 TIMP3 基因位于较大的 SYN3 基因的一个 268 kb 的内含子中，而 8.5 kb 的 HCF2 基因位于 PIK4CA 基因的 27.5 b 的内含子中。在这两种情况中，内部基因的转录方向与外部基因相反。我们还观察到，假基因经常位于其他功能性基因的内含子中。

使用蛋白质家族（PFAM）[33]、Prosite[34] 和 SWISS-PROT[35] 数据库，对可读框大于等于 50 个氨基酸的 482 个已注释的全长和部分基因的多肽序列进行了分析。这些数据在执行 ACEDB 任务查询中被处理和显示。总体而言，240 个（50%）预测蛋白质在总数为 164 个不同 PFAM 域的 PFAM 数据库中有匹配结构域。在组成这 482 个蛋白质的残基中，有 25%属于一个 PFAM 结构域。与此相比，PFAM 的残基覆盖 SWISS-PROT/TrEMBL 超过 45%，表明人类基因组在新的蛋白质序列中被富集。有 62 个 PFAM 结构域可匹配一个以上的蛋白质，包括 10 个含有真核蛋白质激酶结构域（PF00069）的测预蛋白质，9 个匹配 Src 同源结构域 3（PF00018）的测预蛋白质和 8 个匹配 RhoGAP 域（PF00620）的测预蛋白质。14 个预测蛋白质含有锌指结构域（在预测蛋白中鉴定 PFAM 域的详细信息见补充信息表 2）。

已鉴定的编码序列中有 19%被认为是假基因，其原因是它们与已知基因或蛋白有显著相似性，但蛋白质编码可读框被破坏。由于 82%的假基因含单一区域同源性，而且缺乏假定的亲本基因特征性的内含子–外显子结构，因此它们可能是被加工过的假基因。在其余的剪接假基因中，大多数代表重复的基因家族片段，如免疫球蛋白 κ 可变基因、β–晶状体蛋白、CYP2D7 和 CYP2D8，以及 GGT 和 BCR 基因。假基因分布在整个序列中，分散存在于已注释的表达基因的内含子之间，有时存在于其内部。然而，还有一簇密集的 26 个假基因位于毗邻着丝粒的 1.5 Mb 区域内；这个基因簇的意义目前还不清楚。

考虑到染色体 22q 上 33.4 Mb 的序列代表人类基因组的 1.1%，并编码 679 个基因，因此，如果基因在其他染色体上的分布与此相似，那么整个人类基因组的最小基因总数可能至少为 61,000 个。之前的工作已经显示，22 号染色体基因丰度较高[6]，系数为 1.38（http://www.ncbi.nlm.nih.gov/genemap/page.cgi?F = GeneDistrib.html），据此，该估值将降为约 45,000 个基因。尽管认识到此处描述的分析只提供 22q 染色体最低基因含量的估值很重要，进一步研究可能会揭示更多不能用现有方法鉴定的编码序列。

Two lines of evidence point to the existence of additional genes that are not detected in this analysis. First, the 553 predicted CpG islands, which typically lie at the true 5′ ends of about 60% of human genes[31], are in excess of 60% of the number of genes identified (60% = 327, excluding pseudogenes); 282 of the genes identified have CpG islands at or close to the 5′ end (within 5-kb upstream of the first exon, or 1-kb downstream). Thus, there could be up to 271 additional genes associated with CpG islands undetected in the sequence. Second, there are 325 putative genes predicted by the *ab initio* gene prediction program, Genscan, that are not in regions already containing annotated transcripts. We estimate (see above) that roughly 100 of these will represent parts of real genes. Identifying additional genes will require further computational and experimental studies. These studies are continuing and entail testing candidate sequences for possible messenger RNA expression, implementing new gene prediction software able to detect the regions around or near CpG islands that currently have no identified transcript, and further analysis of sequences that are conserved between human and mouse. Furthermore, full-length cDNA sequences that accumulate in the sequence databases of human and other species will be used to refine the gene structures.

The Long-range Chromosome Landscape

Critical to the utility of the genomic sequence to genetic studies is the integration of established genetic maps. The positions of the commonly used microsatellite markers from the Genethon genetic map[36] are given in Fig. 1. The correlation of the order of markers between the genetic map and the sequence is good, within the limitations of genetic mapping. Only a single marker (D22S1175) is discrepant between the two data sets, and this lies in a sequence that is repeated twice on the chromosome (AL021937, see below). In the telomeric region, four of the Genethon markers must lie in our sequence gaps, and we were unable to identify clones from all libraries tested for these. Comparison of genetic distance against physical distance for all the microsatellites whose order is maintained between the datasets shows a mean value of 1.87 cM Mb^{-1}. However, the relationship between genetic and physical distance across the chromosome partitions into two types of region, areas of high and low recombination (Fig. 2). The areas of high recombination may represent recombinational hot spots, although we have not yet been able to identify any specific sequence characteristics common to these areas.

The mean G+C content of the sequence is 47.8%. This is significantly higher than the G+C content calculated for the sum of all human genomic sequence determined so far (42%). Although this result was expected from previous indirect measurements of the G+C content of chromosome 22[7,8,37], the distribution is not uniform, but regionally segmented as illustrated in Fig. 1. There are clear fluctuations in the base content, resulting in areas that are relatively G+C rich and others that are relatively G+C poor. On chromosome 22 these regions stretch over several megabases. For example, the 2 Mb of sequence closest to the centromeric end of the sequence is relatively G+C poor, with the G+C content dropping below 40%. Similarly, the area between 16,000 and 18,800 kb from the

有两个系列的证据指出存在着更多在这一分析中没有检测到的基因。首先，CpG 岛通常位于约 60%的人类基因的真正 5′端[31]，而 553 个预测 CpG 岛超过已鉴定基因数量的 60%(不包括假基因，则 60% = 327 个)；有 282 个基因被发现在 5′端或接近 5′端处(第一外显子上游 5 kb 或下游 1 kb 内)有 CpG 岛。因此，在序列中可能还有与 CpG 岛有关的多达 271 个额外基因未检测到。其次，有 325 个由基因从头预测程序 Genscan 预测的假定基因未包含在已注释转录物的区域内。我们估计(见上文)，其中大约 100 个代表部分真正的基因。要鉴定更多的基因需要进一步的计算与实验研究。这些研究仍在继续，并需要检测候选序列，寻找可能存在的信使 RNA 表达，应用新的基因预测软件，以检测目前没有鉴定到转录物的 CpG 岛的周围或附近区域，并进一步分析人类与小鼠之间的保守序列。此外，在人类和其他物种序列数据库中积累的全长 cDNA 序列将用于完善基因结构。

长距离染色体概况

用基因组序列进行遗传研究的关键是整合已有的遗传图谱。Genethon 遗传图谱常用的微卫星标记的位置[36]如图 1。遗传图谱和序列之间标记顺序的相关性良好，并在遗传作图的要求限制之内。只有一个标记(D22S1175)在两个数据集之间有差异，它位于一个在染色体中重复两次的序列内(AL021937，见下文)。在端粒区域，一定有四个 Genethon 标志位于我们的序列缺口中，我们无法从为这些缺口而测试的所有文库中识别出克隆。将在这些数据集之间保持顺序的所有微卫星的遗传距离与物理距离进行对比显示其平均值为 1.87 cM · Mb^{-1}。然而，整个染色体中遗传距离和物理距离之间的关系分成两个区域类型：高重组区和低重组区(见图 2)。高重组区可能代表重组热点，不过我们还未能确定任何这些区域常见的特定序列特征。

序列的平均 G+C 含量为 47.8%。显著高于目前确定的所有人类基因组序列计算得到的 G+C 含量(42%)。虽然从先前间接测量的 22 号染色体的 G+C 含量[7,8,37]可以预测这一结果，但分布并不一致，其区域分割如图 1 所示。碱基含量存在明显波动，导致某些区域的 G+C 含量相对较高，而其他区域的 G+C 含量相对较低。在 22 号染色体上，这些区域可延伸达几 Mb。例如，最接近序列着丝粒末端的 2 Mb 序列的 G+C 含量相对较低，降至 40%以下。同样，距序列着丝粒末端 16,000 到 18,800 kb 的区域 G+C 含量也始终低于 45%。G+C 丰富的区域(如距序列着丝粒

centromeric end of the sequence is consistently below 45% G+C. The G+C rich regions often reach more than 55% G+C (for example, at 20,100–23,400 kb from the centromeric end of the sequence). This fluctuation appears to be consistent with previous observations that vertebrate genomes are segmented into "isochores" of distinct G+C content[38] and is similar to the structure seen in the human major histocompatibility complex (MHC) sequence[39]. Isochores correlate with both genes and chromosome structure. The G+C rich isochores are rich in genes and *Alu* repeats, and are located in the G+C rich chromosomal R-bands, whereas the G+C poor isochores are relatively depleted in genes and *Alu* repeats, and are located in the G-bands[8,37,40]. The G+C poor regions of chromosome 22 are depleted in genes and relatively poor in *Alu* sequences. For example, the region between 16,000 and 18,800 kb from the centromeric end contains just three genes, two of which are greater than 400 kb in length. The G+C poor regions also are depleted in CpG islands, which are clustered in the gene-rich, G+C rich regions. Although it is tempting to correlate the sequence features that we see with the chromosome banding patterns, we believe that high-resolution mapping of the chromosome band boundaries will be required to assign definitively these to genomic sequence.

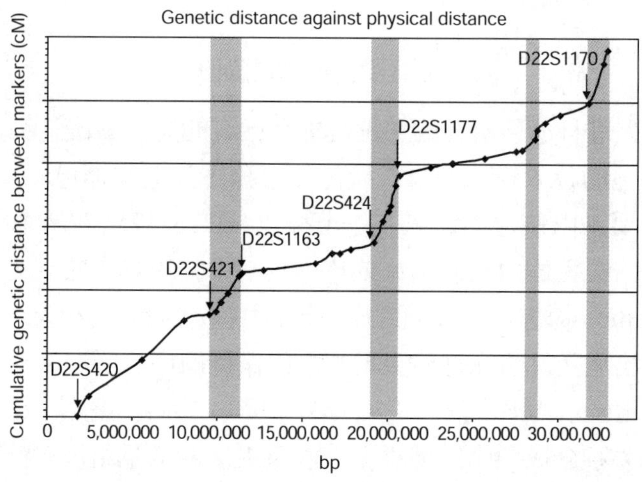

Fig. 2. The relationship between physical and genetic distance. The sex-averaged genetic distances of Dib *et al.*[36] were obtained from ftp://ftp.genethon.fr/pub/Gmap/Nature-1995/ and the cumulative intermarker distances for unambiguously ordered markers (in cM) were plotted against the positions of the microsatellite markers in the genomic sequence. It should be stressed that the y axis does not represent the true genetic distance between distant markers but the sum of the local intermarker distances. The positions of selected genetic markers are labelled. Grey regions are indicative of areas of relatively increased recombination per unit physical distance.

Over 41.9% of the chromosome 22 sequence comprises interspersed and tandem repeat family sequences (Table 2). The density of repeats across the sequence is plotted in Fig. 1. There is variation in the density of *Alu* repeats and some of the regions with low *Alu* density correlate with the G+C poor regions, for example, in the region 16,000–18,800 kb from the centromeric end, and these data support the relationship of isochores with *Alu* distribution. However, in other areas the relationship is less clear. We provide a World-Wide Web interface to the long-range analyses presented here and to further analysis of the

末端 20,100 ~ 23,400 kb 的区域）中，其含量往往超过 55%。这种波动似乎与以前的结果一致，即脊椎动物基因组分割为不同 G+C 含量的“等容线”[38]，与人类主要组织相容性复合物（MHC）序列中的结构相似[39]。等容线与基因和染色体结构有关。G+C 丰富的等容线富含基因和 *Alu* 重复，位于 G+C 丰富的染色体 R 带，而 G+C 稀少的等容线所含的基因和 *Alu* 重复相对较少，且位于 G 带[8,37,40]。22 号染色体 G+C 稀少的区域不含基因且 *Alu* 序列也相对较少。例如，距着丝粒端 16,000 和 18,800 之间的区域只包含三个基因，其中两个基因长度大于 400 kb。G+C 稀少的区域也缺少 CpG 岛，后者集中在基因和 G+C 丰富的区域。虽然将我们看到的序列特征与染色体带型相结合是很吸引人的，但我们认为，要将这些序列特征明确地分配到基因组序列中，需要对染色体带的界限进行高分辨率作图。

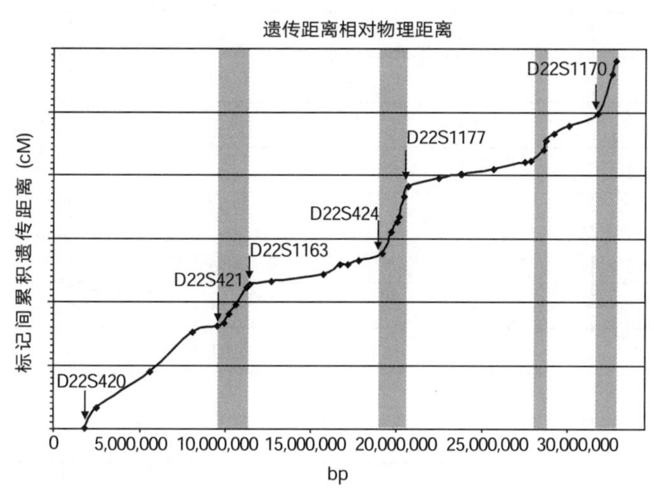

图 2. 物理和遗传距离之间的关系。迪卜等人研究的性别平均遗传距离[36]来自 ftp://ftp.genethon.fr/pub/Gmap/Nature-1995/，将顺序明确的累积的标记间距离（以 cM 为单位）相对基因组序列中微卫星标记的位置进行作图。应当指出，y 轴并不代表远距离标记物间的真正遗传距离，但代表局部标记间的距离总和。选定的遗传标记物的位置已标出。灰色区域指示每单位物理距离相对增加的重组区域。

22 号染色体上超过 41.9%的序列包含散在和串联重复家族序列（见表 2）。序列中的重复密度绘制在图 1 中。*Alu* 重复的密度存在差异，一些低 *Alu* 密度的区域与 G+C 稀少区域（如距着丝粒末端 16,000 ~ 18,800 kb 的区域）有关，这些数据支持等容线与 *Alu* 分布的关系。但是，这种关系在其他区域不太明显。我们为本文进行的长距离分析以及对序列的许多其他重复类型和特征的进一步分析提供一个万维网界

many other repeat types and features of the sequence at http://www.sanger.ac.uk/cgi-bin/cwa/22cwa.pl. The 1-Mb region closest to the centromere contains several interesting repeat sequence features that may be typical of other pericentromeric regions. In addition to the density of pseudogenes described above, there is a large 120-kb block of tandemly repeated satellite sequence (D22Z3) centred 500 kb from the centromeric sequence start (not shown in Fig. 1, but evident from the absence of *Alu* and LINE1 sequences at this point). There is also a cluster of satellite II repeats 80-kb telomeric of the D22Z3 sequences. Isolated alphoid satellite repeats are found closer to the centromeric end of the sequence. Furthermore, this pericentromeric 1 Mb closest to the centromere contains many sequences that are shared with a number of different chromosomes, particularly chromosomes 2 and 14. During map construction, 33 out of 37 STSs designed from sequence that was free of high-copy repeats amplified from more than one chromosome in somatic cell hybrid panel analysis.

Table 2. The interspersed repeat content of human chromosome 22

Repeat type	Total number	Coverage (bp)	Coverage (%)
Alu	20,188	5,621,998	16.80
HERV	255	160,697	0.48
Line1	8,043	3,256,913	9.73
Line2	6,381	1,273,571	3.81
LTR	848	256,412	0.77
MER	3,757	763,390	2.28
MIR	8,426	1,063,419	3.18
MLT	2,483	605,813	1.81
THE	304	93,159	0.28
Other	2,313	625,562	1.87
Dinucleotide	1,775	133,765	0.40
Trinucleotide	166	18,410	0.06
Quadranucleotide	404	47,691	0.14
Pentanucleotide	16	1,612	0.0048
Other tandem	305	102,245	0.31
Total	55,664	14,024,657	41.91

Low-copy Repeats on Chromosome 22

To detect intra- and interchromosomal repeats, we compared the entire sequence of chromosome 22 to itself, and also to all other existing human genomic DNA sequence using Blastn[32] after masking high and medium frequency repeats. The results of the

面：http://www.sanger.ac.uk/cgi-bin/cwa/22cwa.pl。最接近着丝粒的 1 Mb 区域包含几个有趣的可能代表其他着丝粒周边区域的重复序列特征。除了上述假基因密度，有一个大的 120 kb 的串联重复卫星序列（D22Z3）集中在距着丝粒序列起始位置的 500 kb 处（图 1 中未显示，但根据这一位置缺乏 *Alu* 及 LINE1 序列可明显看出）。还有一组卫星 II 重复位于 D22Z3 序列的端粒 80 kb 处。孤立的 alphoid 卫星重复接近序列的着丝粒末端。此外，最接近着丝粒周边的 1 Mb 区域含有许多不同染色体（特别是染色体 2 和 14）共有的序列。在染色体图谱构建过程中，通过无高拷贝重复的序列设计的 37 个 STS 中，有 33 个在体细胞杂交分析中从不止一处染色体发生了扩增。

表 2. 人类 22 号染色体散布的重复情况

重复类型	总数目	覆盖范围 (bp)	覆盖范围 (%)
Alu	20,188	5,621,998	16.80
HERV	255	160,697	0.48
Line1	8,043	3,256,913	9.73
Line2	6,381	1,273,571	3.81
LTR	848	256,412	0.77
MER	3,757	763,390	2.28
MIR	8,426	1,063,419	3.18
MLT	2,483	605,813	1.81
THE	304	93,159	0.28
其他	2,313	625,562	1.87
二核苷酸	1,775	133,765	0.40
三核苷酸	166	18,410	0.06
四核苷酸	404	47,691	0.14
五核苷酸	16	1,612	0.0048
其他串联	305	102,245	0.31
总数	55,664	14,024,657	41.91

22 号染色体上的低拷贝重复

为检测染色体内和染色体间重复，在标记高、中频重复之后，我们用 Blastn 将 22 号染色体的整个序列与其本身进行比对，也与所有其他现有的人类基因组 DNA

intrachromosomal sequence analysis were plotted as a dot matrix (Fig. 3) and reveal a series of interesting features. Locally duplicated gene families lie close to the diagonal axis of the plot. The most striking is the immunoglobulin λ locus that comprises a cluster of 36 potentially functional V-λ gene segments, 56 V-λ pseudogenes, and 27 partial V-λ pseudogenes ("relics"), together with 7 each of the J and C λ segments[24]. Other duplicated gene families that are visible from the dot matrix plot include the clustered genes for glutathione *S*-transferases, β-crystallins, apolipoproteins, phorbolins or APOBECs, the lectins LGALS1 and LGALS2 and the CYP2Ds. A partial inverted duplication of CSF2RB is also observed.

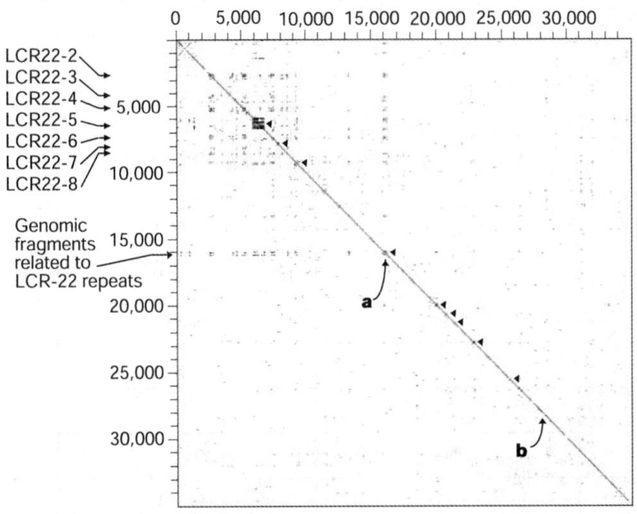

Fig. 3. Intrachromosomal repeats on human chromosome 22. High- and medium-copy repeats and low complexity sequence were masked using RepeatMasker and Dust, and masked sequences were compared using Blastn. The results were filtered to identify regions of more than 50% identity to the query sequence, and were plotted in a 2D matrix with a line proportional to the size of the region of identity. Localized gene family repeats are indicated by arrowheads along the diagonal. From the top, these are the immunoglobulin λ locus, the glutathione *S*-transferase genes, the β-crystallin genes, the Ret-finger-protein-like genes, the apolipoprotein genes, the colony-stimulating factor receptor (CSF2RB) inverted partial duplication, the lectins LGALS1 and LGALS2, the APOBEC genes and the CYP2D genes. Two 60-kb regions of more than 90% homology are labelled "a" (AL008723/AL021937) and "b" (AL031595/AL022339). Seven low-copy repeat regions (LCR22) and a region containing related genomic fragments are indicated at the left margin.

Much more striking are the long-range duplications, which are visible away from the diagonal axis. For example, a 60-kb segment of more than 90% similarity is seen between sequences AL008723/AL021937(at ~16,060–16,390 kb from the centromeric end) and AL031595/AL022339 (at ~27,970–28,110 kb from the centromeric end) separated by almost 12 Mb. The 22q11 region is particularly rich in repeated clusters[41]. Previous work described a low-copy repeat family in 22q11 that might mediate recombination events leading to the chromosomal rearrangements seen in cat eye, velocardiofacial and DiGeorge syndromes[21,42]. The availability of the entire DNA sequence allows detailed dissection of the molecular structure of these low-copy repeats (LCR22s). Edelmann *et al.* described eight

序列进行比对[32]。染色体内序列分析的结果绘制为点阵图（图 3），并显示出一系列有趣的特征。局部重复基因家族位于接近图对角线轴处。最引人注目的是免疫球蛋白 λ 位点，它包含由 36 个潜在功能 V-λ 基因片段、56 个 V-λ 假基因和 27 个部分 V-λ 假基因（“残迹”）形成的一个簇，以及 J 和 C λ 片段形成的 7 个簇[24]。其他重复基因家族在点阵图上都可以看到，包括以下成簇基因：谷胱甘肽 *S*–转移酶、β–晶状体蛋白、载脂蛋白、phorbolin 或 APOBEC、凝集素 LGALS1 和 LGALS2，以及 CYP2D。此外还观察到 CSF2RB 的部分反向重复。

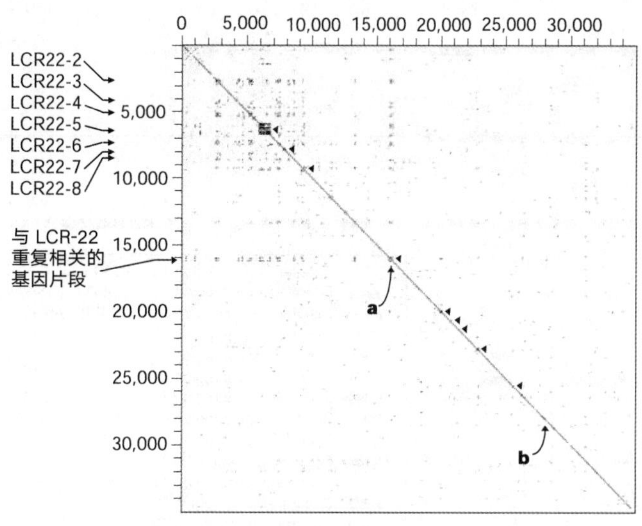

图 3. 人类 22 号染色体上的染色体内重复。用 RepeatMasker 和 Dust 标记高拷贝和中拷贝重复以及低复杂性序列，用 Blastn 比较被标记的序列。结果进行过滤以确定与所查询序列同一性多于 50% 的区域，并用与同一性区域大小成比例的直线绘制在二维矩阵中。沿对角线的箭头指示局部基因家族重复。从上方开始依次是免疫球蛋白 λ 位点、谷胱甘肽 *S*–转移酶基因、β–晶状体蛋白基因、Ret 指蛋白样基因、载脂蛋白基因、集落刺激因子受体（CSF2RB）倒置部分重复、凝集素 LGALS1 和 LGALS2、APOBEC 基因和 CYP2D 基因。同源性大于 90%的两个 60 kb 区域标记为“a”（AL008723/AL021937）和“b”（AL031595/AL022339）。7 个低拷贝重复区域（LCR22）和一个包含相关基因组片段的区域显示在左边空白处。

更引人注目的是远离对角线轴的长距离重复。例如，在相距接近 12 Mb 的 AL008723/AL021937 序列（距着丝粒末端约 16,060 ~ 16,390 kb）和 AL031595/AL022339 序列（距着丝粒末端约 27,970 ~ 28,110 kb）之间可见一个相似性 90%以上的 60 kb 的片段。22q11 区域尤其富含重复簇[41]。以往的工作介绍了 22q11 中有可能介导重组事件的一个低拷贝重复家族，这些重组事件导致猫眼、腭心面和迪格奥尔格综合征中观察到的染色体重排[21,42]。整个 DNA 序列的获得使我们可以详细分析这些低拷贝重复序列（LCR22）的分子结构。埃德尔曼等描述了 8 个 LCR22 区域[21,42]。我们无法

LCR22 regions[21,42]. We were unable to find the LCR22 repeat closest to the centromere, but it may lie in the gap at 700 kb from the centromeric end of the sequence. The other LCR22 regions are distributed over 6.5 Mb of 22q11. Analysis of the sequence shows that each LCR22 contains a set of genes or pseudogenes (Fig. 4). For example, five of the LCR22s contain copies of the γ-glutamyl transferase genes and γ-glutamy-transferase-related genes. There is also evidence that a more distant sequence at ~16,000 kb from the centromeric start of the genomic sequence shares certain sequences with the LCR22 repeats. This similarity involves related genomic fragments including parts of the Ret-finger-protein-like genes, and the IGLC and IGLV genes.

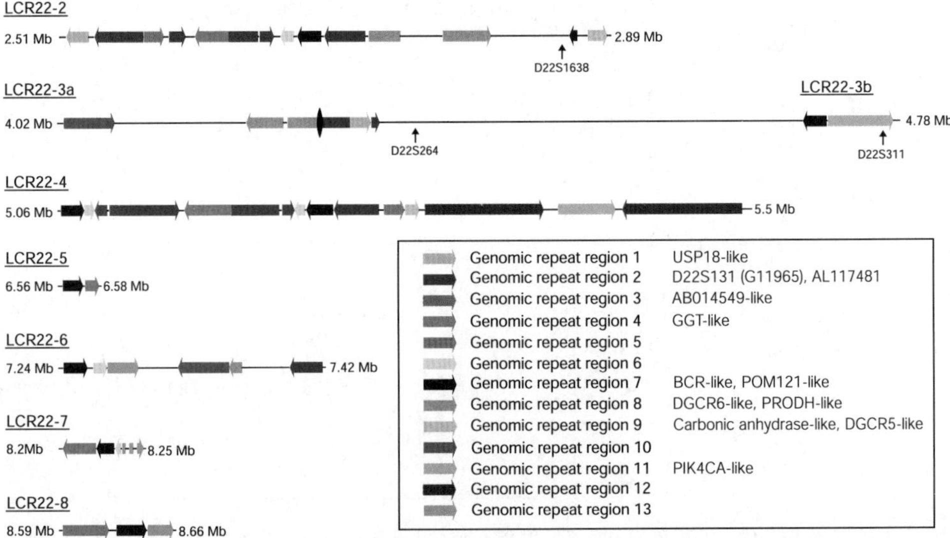

Fig. 4. Sequence composition of the LCR22 repeats. Illustration of the sequence composition of seven LCR22 repeats. The span of each LCR22 region is shown in megabases from the centromere. Coloured arrows indicate the extent of one of the thirteen genomic repeat regions and the orientation of the repeat. The known gene and marker content of these genomic repeat regions is indicated in the key. The black oval indicates the position of the gap in the sequence in LCR22-3.

Regions of Conserved Synteny with the Mouse

The genomic organization of different mammalian species is well known to be conserved[43]. Comparison of genetic and physical maps across species can aid in predicting gene locations in other species, identifying candidate disease genes[13], and revealing various other features relevant to the study of genome organization and evolution. For all the cross-species relationships, that between man and mouse has been most studied. We have examined the relationship of the human chromosome 22 genes to their mouse orthologues.

Of the 160 genes we identified in the human chromosome 22 sequence that have orthologues in mouse, 113 of the murine orthologues have known mouse chromosomal

发现最接近着丝粒的 LCR22 重复，但它可能位于距序列着丝粒末端 700 kb 的缺口中。其他 LCR22 区域分布在 22q11 上超过 6.5 Mb 的范围内。序列分析显示，每个 LCR22 包含一组基因或假基因（图 4）。例如，5 个 LCR22 包含 γ-谷氨酰转移酶基因和 γ-谷氨酰转移酶相关基因的拷贝。还有证据表明，一处更远的，距基因组序列着丝粒起点约 16,000 kb 的序列与 LCR22 重复存在某些共有序列。这种相似性涉及相关的基因组片段，包括部分 Ret 指蛋白样基因以及 IGLC 和 IGLV 基因。

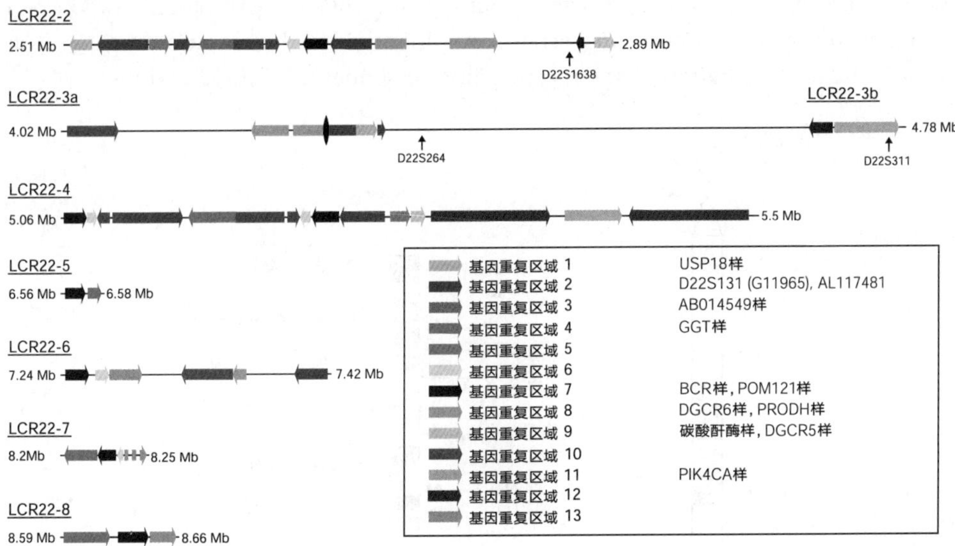

图 4. LCR22 重复的序列组成。7 个 LCR22 重复的序列组成图示。每个 LCR22 区域的跨度从着丝粒开始计算，以 Mb 为单位。彩色箭头表示 13 个基因重复区域中一个的长度和重复方向。这些基因组重复区域的已知基因和标记物含量在图例中注明。黑色椭圆表示序列 LCR22-3 中缺口的位置。

与小鼠具有保守共线性的区域

目前已经清楚地知道，不同哺乳动物物种的基因组的组成是保守的[43]。比较物种间的遗传图谱和物理图谱有助于预测其他物种的基因位置，确定可能的疾病基因[13]，并揭示与基因组的组成和演化等研究相关的各种其他特征。在所有跨物种关系中，人与小鼠之间的关系研究得最多。我们之前已研究过人类 22 号染色体基因与小鼠直系同源基因之间的关系。

我们在人类 22 号染色体序列中鉴定出 160 个基因在小鼠中有直系同源基因，其中 113 个小鼠直系同源基因在小鼠染色体上的位置是已知的（数据可在以下网址获

locations (data available at http://www.sanger.ac.uk/HGP/Chr22/Mouse/). Examination of these mouse chromosomal locations mapped onto the human chromosome 22 sequence confirms the conserved linkage groups corresponding to human chromosome 22 on mouse chromosomes 6, 16, 10, 5, 11, 8 and 15[18,44-46] (Fig. 5). Furthermore, these studies allow placement of the sites of evolutionary rearrangements that have disrupted the conservation of synteny more accurately at the DNA sequence scale. For example, the breakdown of synteny between the mouse 8C1 block and the mouse 15E block occurs between the equivalents of the human HMOX1 and MB genes, which are separated by less than 160 kb that also contains a conspicuous 41-copy 18-nucleotide tandem repeat. A clear prediction from these data is that, for the most part, the unmapped murine orthologues of the human genes lie within these established linkage groups, along with the orthologues of the human genes that currently lack mouse counterparts. Exploitation of the chromosome 22 sequence may hasten the determination of the mouse genomic sequence in these regions.

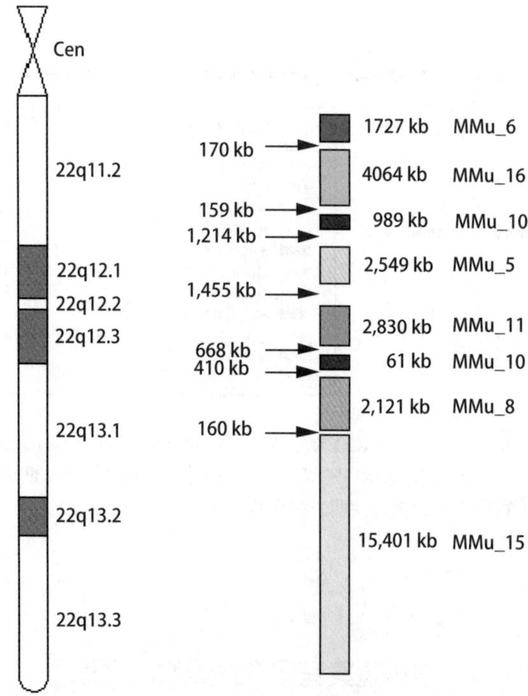

Fig. 5. Regions of conserved synteny between human chromosome 22 and the mouse genome. Regions of mouse chromosomes with conserved synteny to human chromosome 22 are shown as adjacent coloured blocks, determined by the mouse map position of mouse orthologues to human chromosome 22 genes. The size of human chromosome 22 corresponding to each mouse chromosomal region is indicated in kb, as well as the size of the gap between the last orthologue in each conserved block. These data are available at http://www.sanger.ac.uk/Chr22/Mouse.

得：http://www.sanger.ac.uk/HGP/Chr22/Mouse/）。将小鼠染色体中的位置绘制到人类 22 号染色体序列上的检查，确认了小鼠 6、16、10、5、11、8 和 15 号染色体上的与人类 22 号染色体相对应的保守连锁群[18,44-46]（图 5）。此外，这些研究使我们能够在 DNA 序列的尺度上更准确地定位这些已经破坏了共线性的保守序列的演化重排。例如，小鼠 8C1 模块和 15E 模块之间共线性的破坏发生在相应的人 HMOX1 和 MB 基因之间的位置，这两个基因之间的距离不到 160 kb，其中包含一个明显的 41 拷贝的 18 个核苷酸串联重复。根据这些数据可以做出明确的预测：在大多数情况下，未作图的人类基因的小鼠直系同源基因与目前尚无小鼠对应物的人类基因的直系同源基因共同位于这些既定连锁群内。对 22 号染色体序列的开发利用有望加快确定这些区域的小鼠基因组序列。

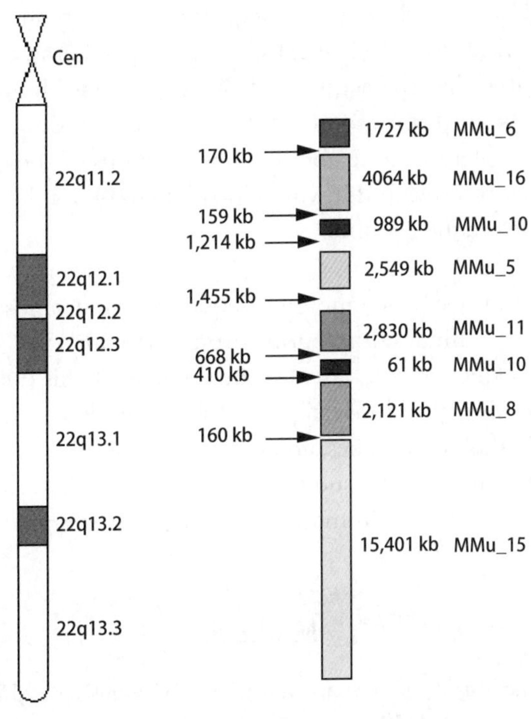

图 5. 人类 22 号染色体与小鼠基因组间具有保守共线性的区域。小鼠染色体上与人类 22 号染色体具有保守共线性的区域用相邻的彩色方块标出，通过小鼠与人类 22 号染色体基因的直系同源基因在小鼠图谱中的位置测得。对应于每个小鼠染色体区域的人类 22 号染色体的大小，以及每个保守模块中最后一个直系同源基因之间的缺口的大小，均以 kb 为单位。这些数据可在以下网站获得：http://www.sanger.ac.uk/Chr22/Mouse。

Conclusions

We have shown that the clone by clone strategy is capable of generating long-range continuity sufficient to establish the operationally complete genomic sequence of a chromosome. In doing so, we have generated the largest contiguous segment of DNA sequence to our knowledge to date. The analysis of the sequence gives a foretaste of the information that will be revealed from the remaining chromosomes.

We were unable to obtain sequence over 11 small gaps using the available cloning systems. It may be possible that additional approaches such as using combinations of cloning systems with small insert sizes and low-copy number could reduce the size of these gaps. Direct cloning of restriction fragments that cross these gaps into small insert plasmid or M13 libraries, or direct sequencing approaches might eventually provide access to all the sequence in the gaps. However, closing these gaps is certain to require considerable time and effort, and might be considered as a specialist activity outside the core genome-sequencing efforts. It also is probable that the sequence features responsible for several of these gaps are unlikely to be specific to chromosome 22. In the best case, similar unclonable sequences might be restricted to the centromeric and telomeric regions of the other chromosomes and areas with large tandem repeats, and it will be possible to obtain large contiguous segments for the bulk of the euchromatic genome.

Over the course of the project, the emerging sequence of chromosome 22 has been made available in advance of its final completion through the internet sites of the consortium groups and the public sequence databases[47]. The benefits of this policy can be seen in both the regular requests received from investigators for materials and information that arise as the result of sequence homology searches, and the publications that have used the data[14-19]. The genome project will continue to pursue this data release policy as we move closer to the anticipated completed sequence of humans, mice and other complex genomes[47,48].

Methods

The methods for construction of clone maps have been previously described[24,49,50] and can also be found at http://www.sanger.ac.uk/HGP/methods/. Details of sequencing methods and software are available at http://www.sanger.ac.uk/HGP/methods/, http://www.genome.ou.edu/proto.html, http://www-alis.tokyo.jst.go.jp/HGS/team_KU/team.html and in the literature[1,24].

(**402**, 489-495; 1999)

I. Dunham, N. Shimizu, B. A. Roe, S. Chissoe *et al.*[†]

[†] A full list of authors and affiliations is given in the original paper.

Received 5 November; accepted 11 November 1999.

结　论

我们已经证实，连续克隆这一方法产生的长距离连续性足以建立一条可操作的完整的染色体基因组序列。在此过程中，我们已经生成了迄今所知最大的连续 DNA 序列片段。对序列的分析预示了其他染色体将要揭示的信息。

使用现有的克隆系统，我们无法获取横跨 11 个小缺口的序列。其他方法，例如将具有小插入尺寸和低拷贝数目的克隆系统结合使用，有望缩减这些缺口的大小。将跨越这些缺口的限制性片段直接克隆到小插入片段质粒或 M13 文库中或直接进行测序，可能最终能够得到缺口中所有的序列。然而，弥合这些缺口必然需要大量的时间和精力，并且可能会被视为核心的基因组测序工作以外的专项活动。也有可能其中一些缺口的序列特征不是 22 号染色体特有的。最理想的情况是，类似的不能克隆的序列会局限于其他染色体的着丝粒和端粒区域，以及存在大串联重复的区域，这样我们就有可能获得大量基因组常染色质大的连续片段。

在项目执行过程中，22 号染色体的序列在最终完成前已通过合作单位的网站和公共序列数据库对外公布[47]。研究者在检索序列同源性后通常要求获取相应的材料和信息，也有一些出版物使用了我们的数据，由此可见该政策确实具有一定的益处[14-19]。在我们获取期望的完整的人类、小鼠和其他复杂基因组的序列的过程中，基因组计划将继续奉行这一数据发布政策[47,48]。

方　法

用于构建克隆图谱的方法此前已描述过[24,49,50]，也可以在 http://www.sanger.ac.uk/HGP/methods/ 找到。测序方法和软件的详细信息可在 http://www.sanger.ac.uk/HGP/methods/，http://www.genome.ou.edu/proto.html 和 http://www-alis.tokyo.jst.go.jp/HGS/team_KU/team.html 以及文献中找到[1,24]。

（李梅 翻译；肖景发 审稿）

References:

1. The Sanger Centre & The Genome Sequencing Centre. Toward a complete human genome sequence. *Genome Res.* **8**, 1097-1108 (1998).

2. Weber, J. L. & Myers, E. W. Human whole-genome shotgun sequencing. *Genome Res.* **7**, 401-409 (1997).

3. Green, P. Against a whole-genome shotgun. *Genome Res.* **7**, 410-417 (1997).

4. Collins, F. S. *et al.* New goals for the U.S. human genome project: 1998-2003. *Science* **282**, 682-689 (1998).

5. Morton, N. E. Parameters of the human genome. *Proc. Natl Acad. Sci. USA* **88**, 7474-7476 (1991).

6. Deloukas, P. *et al.* A physical map of 30,000 human genes. *Science* **282**, 744-746 (1998).

7. Craig, J. M. & Bickmore, W. A. The distribution of CpG islands in mammalian chromosomes. *Nature Genet.* **7**, 376-382 (1994).

8. Saccone, S., Caccio, S., Kusuda, J., Andreozzi, L. & Bernardi, G. Identification of the gene-richest bands in human chromosomes. *Gene* **174**, 85-94 (1996).

9. Collins, J. E. *et al.* A high-density YAC contig map of human chromosome 22. *Nature* **377**, 367-379 (1995).

10. Pulver, A. E. *et al.* Psychotic illness in patients diagnosed with velo-cardio-facial syndrome and their relatives. *J. Nerv. Ment. Dis.* **182**, 476-478 (1994).

11. Gill, M. *et al.* A combined analysis of D22S278 marker alleles in affected sib-pairs: support for a susceptibility locus for schizophrenia at chromosome 22q12. Schizophrenia Collaborative Linkage Group (Chromosome 22). *Am. J. Med. Genet.* **67**, 40-45 (1996).

12. Zu, L., Figueroa, K. P., Grewal, R. & Pulst, S. M. Mapping of a new autosomal dominant spinocerebellar ataxia to chromosome 22. *Am. J. Hum. Genet.* **64**, 594-599 (1999).

13. Southard-Smith, E. M. *et al.* Comparative analyses of the dominant megacolon-SOX10 genomic interval in mouse and human. *Mamm. Genome* **10**, 744-749 (1999).

14. Nishino, I., Spinazzola, A. & Hirano, M. Thymidine phosphorylase gene mutations in MNGIE, a human mitochondrial disorder. *Science* **283**, 689-692 (1999).

15. Peyrard, M. *et al.* The human LARGE gene from 22q12.3-q13.1 is a new, distinct member of the glycosyltransferase gene family. *Proc. natl Acad. Sci. USA* **96**, 598-603 (1999).

16. Kao, H. T. *et al.* A third member of the synapsin gene family. *Proc. Natl Acad. Sci. USA* **95**, 4667-4672 (1998).

17. Mittman, S., Guo, J., Emerick, M. C. & Agnew, W. S. Structure and alternative splicing of the gene encoding alpha1I, a human brain T calcium channel alpha1 subunit. *Neurosci. Lett.* **269**, 121-124 (1999).

18. Seroussi, E. *et al.* TOM1 genes map to human chromosome 22q13.1 and mouse chromosome 8C1 and encode proteins similar to the endosomal proteins HGS and STAM. *Genomics* **57**, 380-388 (1999).

19. Seroussi, E. *et al.* Duplications on human chromosome 22 reveal a novel ret finger protein-like gene family with sense and endogenous antisense transcripts. *Genome Res.* **9**, 803-814 (1999).

20. Ning, Y., Rosenberg, M., Biesecker, L. G. & Ledbetter, D. H. Isolation of the human chromosome 22q telomere and its application to detection of cryptic chromosomal abnormalities. *Hum. Genet.* **97**, 765-769 (1996).

21. Edelmann, L., Pandita, R. K. & Morrow, B. E. Low-copy repeats mediate the common 3-Mb deletion in patients with velo-cardio-facial syndrome. *Am. J. Hum. Genet.* **64**, 1076-1086 (1999).

22. McDermid, H. E. *et al.* Long-range mapping and construction of a YAC contig within the cat eye syndrome critical region. *Genome Res.* **6**, 1149-1159 (1996).

23. Johnson, A. *et al.* A 1.5-Mb contig within the cat eye syndrome critical region at human chromosome 22q11.2. *Genomics* **57**, 306-309 (1999).

24. Kawasaki, K. *et al.* One-megabase sequence analysis of the human immunoglobulin lambda gene locus. *Genome Res.* **7**, 250-261 (1997).

25. Felsenfeld, A., Peterson, J., Schloss, J. & Guyer, M. Assessing the quality of the DNA sequence from the Human Genome Project. *Genome Res.* **9**, 1-4 (1999).

26. Mewes, H. W. *et al.* Overview of the yeast genome. *Nature* **387**, 7-65 (1997).

27. Blattner, F. R. *et al.* The complete genome sequence of Escherichia coli K-12. *Science* **277**, 1453-1474 (1997).

28. The C. elegans Sequencing Consortium. Genome sequence of the nematode C. elegans: a platform for investigating biology. *Science* **282**, 2012-2018 (1998).

29. Solovyev, V. & Salamov, A. The Gene-Finder computer tools for analysis of human and model organisms genome sequences. *Ismb* **5**, 294-302 (1997).

30. Burge, C. & Karlin, S. Prediction of complete gene structures in human genomic DNA. *J. Mol. Biol.* **268**, 78-94 (1997).

31. Cross, S. H. & Bird, A. P. CpG islands and genes. *Curr. Opin. Genet. Dev.* **5**, 309-314 (1995).

32. Altschul, S. F., Gish, W., Miller, W., Myers, E. W. & Lipman, D. J. Basic local alignment search tool. *J. Mol. Biol.* **215**, 403-410 (1990).

33. Bateman, A. *et al.* Pfam 3.1: 1313 multiple alignments and profile HMMs match the majority of proteins. *Nucleic Acids Res.* **27**, 260-262 (1999).

34. Hofmann, K., Bucher, P., Falquet, L. & Bairoch, A. The PROSITE database, its status in 1999. *Nucleic Acis Res.* **27**, 215-219 (1999).

35. Bairoch, A. & Apweiler, R. The SWISS-PROT protein sequence data bank and its supplement TrEMBL in 1999. *Nucleic Acids Res.* **27**, 49-54 (1999).

36. Dib, C. *et al.* A comprehensive genetic map of the human genome based on 5,264 microsatellites. *Nature* **380**, 152-154 (1996).

37. Holmquist, G. P. Chromosome bands, their chromatin flavors, and their functional features. *Am. J. Hum. Genet.* **51**, 17-37 (1992).

38. Bernardi, G. *et al.* The mosaic genome of warm-blooded vertebrates. *Science* **228**, 953-958 (1985).

39. The MHC sequencing consortium. Complete sequence and gene map of a human major histocompatibility complex. *Nature* **401**, 921-923 (1999).

40. Bernardi, G. The isochore organization of the human genome. *Annu. Rev. Genet.* **23**, 637-661 (1989).

41. Collins, J. E., Mungall, A. J., Badcock, K. L., Fay, J. M. & Dunham, I. The organization of the gamma-glutamyl transferase genes and other low copy repeats in human chromosome 22q11. *Genome Res.* **7**, 522-531 (1997).

42. Edelmann, L. *et al.* A common molecular basis for rearrangement disorders on chromosome 22q11. *Hum. Mol. Genet.* **8**, 1157-1167 (1999).

43. Eppig, J. T. & Nadeau, J. H. Comparative maps: the mammalian jigsaw puzzle. *Curr. Opin. Genet. Dev.* **5**, 709-716 (1995).

44. Bucan, M. *et al.* Comparative mapping of 9 human chromosome 22q loci in the laboratory mouse. *Hum. Mol. Genet.* **2**, 1245-1252 (1993).

45. Carver, E. A. & Stubbs, L. Zooming in on the human-mouse comparative map: genome conservation re-examined on a high-resolution scale. *Genome Res.* **7**, 1123-1137 (1997).

46. Puech, A. *et al.* Comparative mapping of the human 22q11 chromosomal region and the orthologous region in mice reveals complex changes in gene organization. *Proc. Natl Acad. Sci.* USA **94**, 14608-14613 (1997).

47. Bentley, D. R. Genomic sequence information should be released immediately and freely in the public domain. *Science* **274**, 533-534 (1996).

48. Guyer, M. Statement on the rapid release of genomic DNA sequence. *Genome Res.* **8**, 413 (1998).

49. Dunham, I., Dewar, K., Kim, U.-J. & Ross, M. T. in *Genome Analysis: A Laboratory Manual Series, Volume 3: Cloning Systems* (eds Birren, B. *et al.*) 1-86 (Cold Spring Harbor Laboratory Press, Cold Spring Harbor, New York, 1999).

50. Asakawa, S. *et al.* Human BAC library: construction and rapid screening. *Gene* **191**, 69-79 (1997).

Supplementary information is available on *Nature*'s World-Wide Web site (http://www.nature.com) or as paper copy from the London editorial office of *Nature*.

Correspondence and requests for materials should be addressed to I.D. (e-mail: id1@sanger.ac.uk).

Molecular Analysis of Neanderthal DNA from the Northern Caucasus

I.V. Ovchinnikov *et al.*

Editor's Note

The only thing better than having a sequence of DNA from an extinct species is having a second one, for comparative purposes. Mitochondrial DNA extracted from the original bones of type specimen Neanderthal 1 in 1997 showed that Neanderthals were not closely related to modern humans. But how secure was this single set of data? Here Igor Ovchinnikov and colleagues describe mitochondrial DNA extracted from 29,000-year-old Neanderthal remains from the northern Caucasus. Although far removed in space and time from the other Neanderthal sequence, the results still showed that the Caucasus DNA grouped more closely with the original Neanderthal finds, further ruling out admixture with modern humans.

The expansion of premodern humans into western and eastern Europe ~40,000 years before the present led to the eventual replacement of the Neanderthals by modern humans ~28,000 years ago[1]. Here we report the second mitochondrial DNA (mtDNA) analysis of a Neanderthal, and the first such analysis on clearly dated Neanderthal remains. The specimen is from one of the eastern-most Neanderthal populations, recovered from Mezmaiskaya Cave in the northern Caucasus[2]. Radiocarbon dating estimated the specimen to be ~29,000 years old and therefore from one of the latest living Neanderthals[3]. The sequence shows 3.48% divergence from the Feldhofer Neanderthal[4]. Phylogenetic analysis places the two Neanderthals from the Caucasus and western Germany together in a clade that is distinct from modern humans, suggesting that their mtDNA types have not contributed to the modern human mtDNA pool. Comparison with modern populations provides no evidence for the multiregional hypothesis of modern human evolution.

THE first successful extraction and sequencing of the mtDNA hypervariable regions (I and II (HVRI & HVRII)) was performed on the Neanderthal type specimen from Feldhofer Cave, the Neander valley, Germany[4,5]. Phylogenetic analysis of the sequence placed the Neanderthal mtDNA outside the mtDNA pool of modern humans. This was regarded as a breakthrough in the study of modern human evolution, providing molecular evidence that Neanderthals did not contribute mtDNA to modern humans. From this sequence the divergence of Neanderthals and modern humans was estimated to have occurred between 317,000 and 741,000 years ago[4,5]. However, these estimates were based on the molecular analysis of a single specimen. The shortage of potentially well preserved Neanderthal material[6] and limited access to Neanderthal remains for destructive analysis have hindered the analysis of additional specimens, but genetic characterization

高加索北部尼安德特人 DNA 的分子分析

奥夫钦尼科夫等

编者按

比从已灭绝的物种得到一个DNA序列更好的事情只有一件，那就是获得第二个序列，以便进行对比。1997年从尼安德特人1的原始骨骼标本中提取的线粒体DNA显示，尼安德特人与现代人的关系并不密切。但这仅有的一组数据的可信度怎样呢？在本文中，伊戈尔·奥夫钦尼科夫及其同事描述了从高加索北部绝对年龄为29,000年的尼安德特人遗骸中提取的线粒体DNA。尽管在时空上与另一个尼安德特人序列相距甚远，但是结果仍表明在高加索发现的该DNA与最初发现的尼安德特人关系更近，并进一步排除了与现代人有基因交流的可能性。

距今约40,000年前，古老型智人扩散至西欧和东欧，并最终导致尼安德特人在约28,000年前被现代人取代[1]。在本文中我们报道了第二例对尼安德特人线粒体DNA(mtDNA)的分子分析，这也是首例对有清晰年代尼安德特人遗骸进行的此类分析。标本来自尼安德特人最东部的种群之一，发现于高加索北部的梅兹迈斯卡亚洞[2]。根据放射性碳同位素年代测定，该标本绝对年龄约为29,000年，因此是生存最晚的尼安德特人之一[3]。该序列显示其与费尔德霍费尔原始的尼安德特人之间存在3.48%的差异[4]。系统发育分析将高加索和德国西部的这两个尼安德特人置于有别于现代人的同一分支，表明他们的mtDNA类型对现代人的mtDNA库没有贡献。与现代人种群进行的比较没有为现代人演化的多地起源说提供证据。

第一次对mtDNA高变区(I和II，即HVRI和HVRII)的成功提取和测序是用德国尼安德谷费尔德霍费尔洞出土的尼安德特人模式标本完成的[4,5]。该序列的系统发育分析将此尼安德特人的mtDNA置于现代人的mtDNA库之外。这曾被认为是现代人类演化研究的一项突破，为证明尼安德特人对现代人的mtDNA没有贡献提供了分子证据。根据这一序列推测尼安德特人和现代人的分化发生在741,000年至317,000年前[4,5]。然而，这些推测是以单一标本的分子分析为基础的。保存完好的尼安德特人原材料的缺乏[6]以及对尼安德特人遗骸进行破坏性分析的局限性阻碍了对更多样本的分析。然而，为了理解尼安德特人的分子多样性及不同尼安德特人种

of additional Neanderthals is essential to understand their molecular diversity and the relationship between different Neanderthal populations, and to assess their relationship to modern humans further.

The Caucasus, which is located on the southeastern boundary between Europe and Asia, is one of the areas through which pre-modern humans and anatomically modern *Homo sapiens* may have entered Europe from the Near East and Africa. Neanderthals invaded the region at an unknown point in time[7,8] and may have occupied the region alongside modern humans from ~40,000 years before the present (B.P.). During the excavation of the Mezmaiskaya Cave[2], which is located in the northern Caucasus (Fig. 1), a fragmentary skeleton of an infant was found that contained a set of morphological characteristics which indicated clear affinities to the Neanderthals of western and central Europe[2]. Mitochondrial DNA analysis was undertaken using one of this Neanderthal's ribs.

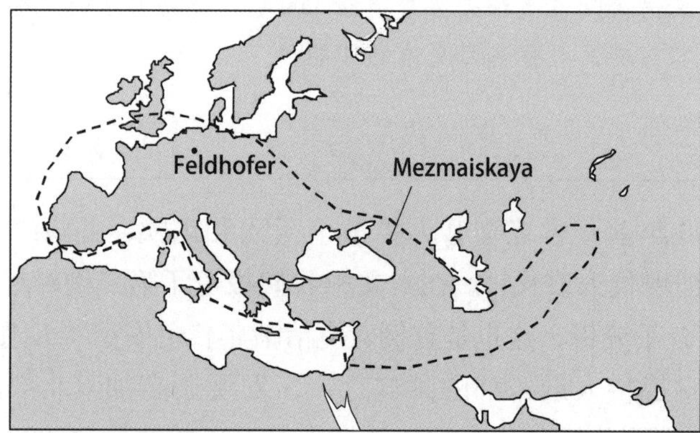

Fig. 1. The area within which Neanderthal remains have been found (dotted line). The Mezmaiskaya Cave is situated 1,310 m above sea level within the northern Caucasus at 44° 10′ N 40° 00′ E on the bank of the Sukhoy Kurdzhips river. The location of the Feldhofer Cave in Germany is also shown.

The preservation of collagen-type debris was used as an indicator of macromolecule preservation in the bone. The amount of collagen-type debris extracted[9] from 130 mg of the Mezmaiskaya Neanderthal rib fragment was 22% of the average level extracted from modern bones, and the extracted collagen contained 41.6% carbon and 14.7% nitrogen. This is within the values recovered from prehistoric samples displaying good preservation[10]. These data suggested that there were low levels of diagenetic modification. The high collagen yield made it possible to date the Neanderthal infant to 29,195 ± 965 (Ua-14512) years B.P. by using a radio-carbon accelerator. This date does not agree with the previously published dates of more than 45,000 (Le-3841) and 40,660 ± 1,600 (Le-3599) years for the Mousterian layers in the Mezmaiskaya Cave[2] with which the skeleton was associated. The most likely reason for this discrepancy is the incorrect identification of the poorly defined layers in this area of the cave. The value obtained from the bone itself rather than from associated material gives the most reliable date for this individual.

Two sections of one rib (90 mg and 123 mg) were used for DNA extraction in two

群之间的关系，并进一步评估他们与现代人的关系，了解更多尼安德特人的遗传特征是至关重要的。

高加索位于欧洲和亚洲的东南交界处，可能是古老型智人和早期现代人从近东和非洲进入欧洲途经的地区之一。尼安德特人在某个未知的时间侵入了这一地区[7,8]，并且在距今(BP)约 40,000 年与现代人共同占据这一地区。梅兹迈斯卡亚洞[2]位于高加索北部(图 1)，在挖掘该洞期间发现了一具破碎的幼儿骨架，其蕴含的一组形态学特征展现了与西欧和中欧尼安德特人明确的亲缘关系[2]。线粒体 DNA 分析正是使用这个尼安德特人的一根肋骨进行的。

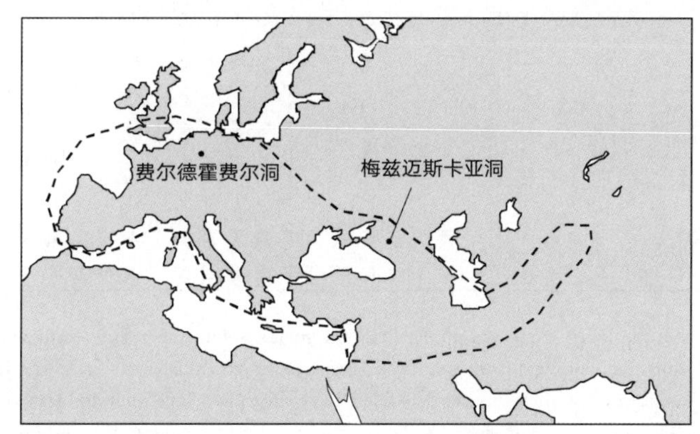

图 1. 发现尼安德特人遗骸的区域(虚线所示)。梅兹迈斯卡亚洞海拔 1,310 米，位于高加索北部苏霍伊库尔吉普斯河畔，北纬 44°10′，东经 40°00′。德国费尔德霍费尔洞的位置也在图中标明。

骨骼中胶原蛋白型片断的保存状况被用作大分子保存状况的指标。从 130 mg 梅兹迈斯卡亚尼安德特人肋骨碎片提取的胶原蛋白型片断[9]为现代骨骼平均提取水平的 22%，提取出的胶原蛋白含碳量为 41.6%，含氮量为 14.7%。这一数值处于保存良好的史前样本的数值范围内[10]。这些数据表明成岩改造作用水平较低。该肋骨碎片的高胶原含量使得我们可以利用放射性碳加速器测定该尼安德特幼儿的年龄为距今 29,195±965(Ua-14512)年。这一年代与之前发表的梅兹迈斯卡亚洞[2]此骨架出土的莫斯特文化层年代为距今 45,000(Le-3841)年和 40,660±1,600(Le-3599)年是不一致的。导致这一差异最可能的原因是对该洞穴中这一区域不明确的骨架出土层位的错误鉴定。从骨骼本身而非间接关联材料得到的数值才能给出该个体最可信的年代信息。

两个独立的实验室分别选取了一段肋骨的两个切片(90 mg 和 123 mg)来提取

independent laboratories. In the Glasgow laboratory, a total of 345 base pairs (bp) of HVRI was determined from two overlapping polymerase chain reaction (PCR) fragments with lengths of 232 and 256 bp. Forty PCR amplification cycles produced sufficient product to enable direct sequencing. Products from independent PCR amplifications were also cloned into a TA vector and sequenced (Fig. 2).

	16037	16078	16086	16093	16107	16108	16111	16112	16118	16129	16139	16148	16154	16156	16169	16182	16183	16189	16209	16223	16230	16234	16243	16244	16250	16256	16258	16261	16262	16263.1	16278	16299	16311	16320	16344	16362	16365	16393
Reference	A	A	T	T	C	C	C	C	G	G	A	C	T	G	C	A	A	T	T	C	A	C	T	G	C	C	A	C	C	-	C	A	T	C	C	T	C	C
Direct 1		.	C	.	.	.	.	.	.	A	T	T	.	A	T	C	C	C	C	T	G	T	.	A	.	A	.											
P1		.	C	.	.	.	.	.	.	A	T	T	.	A	T	C	C	C	C	T	G	T	.	A	.	A	.	.	T	A	T							
P2		.	C	.	.	.	.	.	.	A	T	T	.	A	T	C	C	C	C	T	G	T	.	A	.	A	.	.	T	A	T							
P3		.	C	.	.	.	.	.	A	A	T	T	.	A	T	C	C	C	C	T	G	T	.	A	.	A	.	.	T	A	T							
Direct 2																				T	G	T	.	A	.	A	.	.	T	A	T	G	C	T	T	C	.	.
577.1																				T	G	T	.	A	.	A	.	T	T	A	T	G	C	T	T	C	.	T
557.2																				T	G	T	.	A	T	A	.	.	T	A	T	G	C	T	T	C	.	.
581.2																				T	G	T	.	A	.	A	.	.	T	A	T	G	C	T	T	C	.	.
581.3																				T	G	T	C	A	.	A	.	.	T	A	T	G	C	T	T	C	T	.
Mezmaiskaya		.	C	.	.	.	.	.	.	A	T	T	.	A	T	C	C	C	C	**T**	**G**	**T**	**.**	**A**	**.**	**A**	**.**	**.**	**T**	**A**	**T**	**G**	**C**	**T**	T	C	.	.
Feldhofer	G	G	.	C	T	T	T	T	.	A	T	T	C	.	T	.	C	C	C	T	G	T	.	A	.	A	G	.	T	A	T	G	C	T	.	C	.	.

Fig. 2. Variable sites in the DNA sequences of the PCR fragments obtained by direct sequencing (Direct 1 and 2) and cloned PCR products derived from the Neanderthal from Mezmaiskaya Cave. The human reference sequence[11] and the sequence of the Neanderthal from Feldhofer Cave[4] are included for comparison. A full stop indicates that the sequence is the same as the reference. The sequence that could be duplicated in the Stockholm laboratory is shown in bold within the compiled Mezmaiskaya sequence.

The authenticity of the DNA sequence obtained in the Glasgow laboratory is supported by a number of factors. First, a section of the mtDNA was isolated and sequenced with congruent results in the Stockholm laboratory. Second, the PCR products were generated using Neanderthal-specific primer pairs that, under the amplification conditions, failed to amplify any fragments using modern DNA controls from individuals of different ethnic origins. Third, the retrieval of the sequence was not dependent on the primers used. Fourth, the low level of diagenetic modification indicated that the sample could theoretically contain amplifiable DNA. Last, and most convincingly, the sequence is similar to, and after phylogenetic analysis clusters strongly with, the previously analysed Neanderthal sequence[4].

Comparison of the 345-bp fragment of HVRI with the Anderson reference sequence[11] and the Neanderthal from Feldhofer Cave[4] revealed 22 differences (17 transitions, 4 transversions and 1 insertion) and 12 differences (11 transitions and 1 transversion), respectively. The Feldhofer Neanderthal HVRI contained 27 differences to Anderson reference sequence[11] (over the equivalent 345 bp[4]). The two Neanderthals share 19 substitutions relative to the reference sequence. The cloned PCR products contained all the substitutions that were detected by direct sequencing; six other non-reproducible substitutions occurred in seven different clones. No

DNA。格拉斯哥实验室通过长度分别为 232 bp 和 256 bp 的两个重叠的聚合酶链式反应(PCR)测定了片段总长度为 345 个碱基对(bp)的 HVRI。通过 40 个 PCR 扩增循环产生了足够的产物，可以进行直接测序。同时也将独立 PCR 扩增的产物克隆到 TA 载体中并进行了测序(图 2)。

	1	1	1	1	1	1	1	1	1	1	1	1	1	1	1	1	1	1	1	1	1	1	1	1	1	1	1	1	1	1	1	1	1	1	1	1	1	1
	6	6	6	6	6	6	6	6	6	6	6	6	6	6	6	6	6	6	6	6	6	6	6	6	6	6	6	6	6	6	6	6	6	6	6	6	6	6
	0	0	0	0	1	1	1	1	1	1	1	1	1	1	1	1	1	1	2	2	2	2	2	2	2	2	2	2	2	2	2	2	3	3	3	3	3	3
	3	7	8	9	0	0	1	1	1	2	3	4	5	5	6	8	8	8	0	2	3	3	4	4	5	5	5	6	6	6	7	9	1	2	4	6	6	9
	7	8	6	3	7	8	1	2	8	9	9	8	4	6	9	2	3	9	9	3	0	4	3	4	0	6	8	1	2	3.1	8	9	1	0	4	2	5	3
参考序列	A	A	T	T	C	C	C	C	G	G	A	C	T	G	C	A	A	T	T	C	A	C	T	G	C	C	A	C	C	-	C	A	T	C	C	T	C	C
Direct 1		.	C	.	.	.	.	.	.	A	T	T	.	A	T	C	C	C	C	T	G	T	.	A	.	A	.											
P1		.	C	.	.	.	.	.	.	A	T	T	.	A	T	C	C	C	C	T	G	T	.	A	.	A	.	.	T	A	T							
P2		.	C	.	.	.	.	.	.	A	T	T	.	A	T	C	C	C	C	T	G	T	.	A	.	A	.	.	T	A	T							
P3		.	C	.	.	.	.	.	A	A	T	T	.	A	T	C	C	C	C	T	G	T	.	A	.	A	.	.	T	A	T							
Direct 2																				T	G	T	.	A	.	A	.	.	T	A	T	G	C	T	T	C	.	.
577.1																				T	G	T	.	A	.	A	.	T	T	A	T	G	C	T	T	C	.	T
557.2																				T	G	T	.	A	T	A	.	.	T	A	T	G	C	T	T	C	.	.
581.2																				T	G	T	.	A	.	A	.	.	T	A	T	G	C	T	T	C	.	.
581.3																				T	G	T	C	A	.	A	.	.	T	A	T	G	C	T	T	C	T	.
梅兹迈斯卡亚		.	C	.	.	.	.	.	.	A	T	T	.	A	T	C	C	C	C	**T**	**G**	**T**	**.**	**A**	**.**	**A**	**.**	**.**	**T**	**A**	**T**	**G**	**C**	**T**	T	C	.	.
费尔德霍费尔	G	G	.	C	T	T	T	T	.	A	T	T	C	.	T	.	C	C	C	T	G	T	.	A	.	A	G	.	T	A	T	G	C	T	.	C	.	.

图 2. 通过直接测序(Direct 1 和 2)PCR 片段及梅兹迈斯卡亚洞尼安德特人的克隆 PCR 产物，得到 DNA 序列上的可变位点。现代人参考序列[11]及费尔德霍费尔洞的尼安德特人序列[4]都包含在内，进行比对。圆点表示序列与参考序列相同。在斯德哥尔摩实验室中可被复制的序列在梅兹迈斯卡亚合成序列中用黑体表示。

有多个因素支持格拉斯哥实验室得到的 DNA 序列的可靠性。首先，一段线粒体 DNA 被分离并测序，取得了与斯德哥尔摩实验室一致的结果。第二，PCR 产物是使用针对尼安德特人的特异性引物对生成的，在扩增条件下，这些引物无法扩增来自不同种族个体的现代 DNA 对照片段。第三，序列的获取不取决于所使用的引物。第四，低水平的成岩改造作用表明这一样本理论上含有可扩增的 DNA。最后且最有说服力的是，该序列与以前分析过的尼安德特人序列[4]相似，且系统发育分析显示它们强烈类聚。

该 345 bp 的 HVRI 片段与安德森参考序列[11]和费尔德霍费尔洞的尼安德特人[4]序列的分别对比显示出 22 处碱基差异(17 处转换，4 处颠换及 1 处插入)和 12 处碱基差异(11 处转换及 1 处颠换)。费尔德霍费尔尼安德特人的 HVRI 与安德森参考序列[11]相比有 27 处碱基差异(在相应的 345 bp 中[4])。与参考序列相比，这两个尼安德特人共同拥有 19 处碱基替换。克隆的 PCR 产物包含了直接测序检测到的所有碱基替换；六处其他不可重复的替换出现于七个不同的克隆中。在格拉斯哥实验室中，

modern sequences were found in the Glasgow laboratory either by direct sequencing or by sequencing cloned PCR products. The Stockholm laboratory experienced problems with contamination: most of the cloned PCR products that they analysed contained sequences that are found in the modern human mtDNA pool, with two haplotypes predominant. However, three clones contained DNA that was the same as the sequence determined in Glasgow (two of these contained non-reproducible substitutions).

The preservation of 256-bp DNA fragments in bone that is ~29,000 years old, that has not been preserved in permafrost and that contained sufficient DNA to enable direct DNA sequencing after amplification is unprecedented and may be attributed to specific features of the microenvironment of the limestone cavern[2]. The retrieval of mtDNA showed a positive correlation to the preservation of collagen content and the skeletal morphology.

Phylogenetic analysis using both distance and parsimony optimizations places the two Neanderthal sequences together, in a distinct clade, basal to modern humans. Neighbour-joining analysis supports this separation (Fig. 3a). Parsimony analysis, which makes minimal assumptions about the model of evolution and optimizes the fit between the tree and data, produced similar results (Fig. 3b).

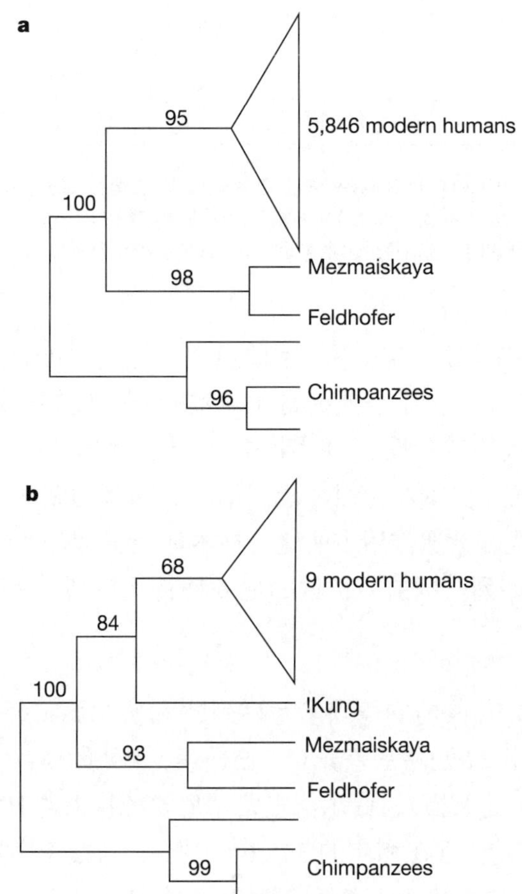

Fig. 3. Phylogenetic relationship of the two Neanderthals and modern humans. **a**, A neighbour-joining

直接测序和克隆 PCR 产物测序都没有发现现代序列。斯德哥尔摩实验室则遇到了污染问题：他们分析的大部分克隆 PCR 产物都含有现代人线粒体 DNA 库中存在的序列，主要为两个单倍型。不过，有三个克隆含有的 DNA 与格拉斯哥实验室确定的序列相同(其中两个含有不可重复的替换)。

年龄约 29,000 年的骨骼中保存了 256 bp 的 DNA 片段，没有保存在永久冻土中却含有足够的 DNA 可供扩增后直接 DNA 测序，这种史无前例的情况或许可以归因于这一石灰岩洞穴中特殊的微环境[2]。线粒体 DNA 的获得表明胶原蛋白成分的保存状况与骨骼形态是正相关的。

使用距离和简约性优化法则进行的系统发育分析将这两个尼安德特人的序列归并到一起，构成一个独立的分支，并位于现代人基底位置。邻接分析支持这一分化(图 3a)。简约性分析得到了类似的结果，该分析对演化模式的假设最少，并会优化系统树与数据之间的拟合(图 3b)。

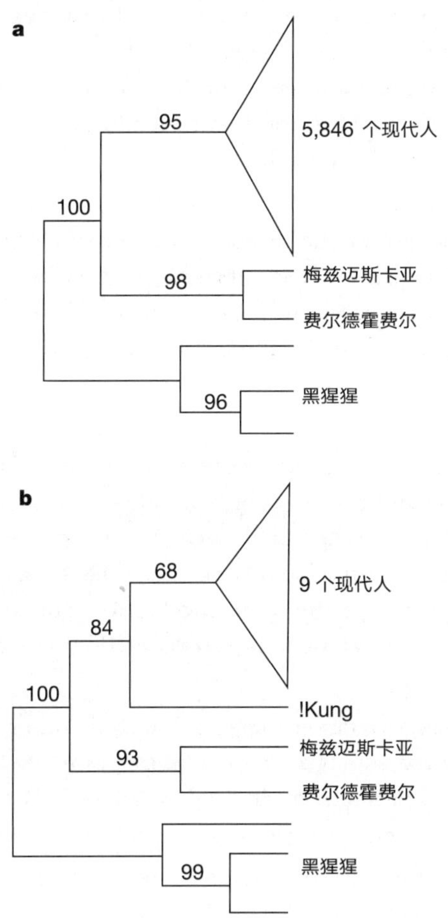

图 3. 两个尼安德特人与现代人的系统发育关系。**a**，使用来自 5,846 个现代人的共 1,897 个单倍型计算

tree computed using a total of 1,897 haplotypes derived from 5,846 modern humans[19]. **b**, A maximum parsimony branch and bound search with the two Neanderthal sequences along with the sequences of one !Kung, three other Africans, three Asians and three Europeans, all randomly selected[19]. This result is congruent with four additional data sets that were analysed. In both analyses, three chimpanzee sequences[13] were used as an outgroup. The numbers in both diagrams refer to the bootstrap frequencies (%) obtained from 1,000 replicates.

The level of pairwise difference found between the two Neanderthals was higher than the average values found in random samples of 300 Caucasoids (5.28 ± 2.24) and Mongoloids (6.27 ± 2.29)—less than 1% of Caucasoid and Mongoloid pairs differ at 12 or more positions—but comparable to a random sample of 300 Africans (8.36 ± 3.2), where 37% of pairs differed at 12 or more positions. When analysing ancient DNA there is the possibility of misincorporating nucleotides in the early stages of PCR, especially when the target DNA is possibly damaged and present in low copy number[12]. As both Neanderthals were analysed in replicate and the results were consistent, however, errors of this type can be discounted.

The Feldhofer and Mezmaiskaya Neanderthals were separated geographically by over 2,500 km. Given that these two individuals contained closely related mtDNA, which is phylogenetically distinct from modern humans, and displays only a moderate level of sequence diversity compared with some primates[13], these data provide further support for the hypothesis of a very low gene flow between the Neanderthals and modern humans. In particular, these data reduce the likelihood that Neanderthals contained enough mtDNA sequence diversity to encompass modern human diversity.

The "out-of-Africa" hypothesis for the origin of modern humans predicts equal distances between the Neanderthal sequences and all modern sequences. We observed this in our analysis—the average pairwise differences between the Neanderthals and 300 randomly selected Africans, Mongoloids and Caucasoids were calculated to be 23.09 ± 2.86, 23.27 ± 4.06 and 25.45 ± 3.27, respectively.

We estimated the age of the most recent common ancestor (MRCA) of the mtDNA of the eastern and western Neanderthals to be 151,000–352,000 years. This coincides with the time of emergence of the Neanderthal lineage in the palaeontological records[14]. The divergence of modern human and Neanderthal mtDNA was estimated to be between 365,000 and 853,000 years. Using the same model, we estimated the age of the earliest modern human divergences in mtDNA to be between 106,000 and 246,000 B.P.

The results obtained from this specimen suggest that some other Neanderthal samples may be amenable to molecular analysis. To obtain a more complete picture of the relationship of Neanderthals to modern humans, additional Neanderthals and early modern humans must be analysed, especially from the regions where they may have co-existed. The excellent preservation of this specimen leads to the potential of analysing the entire Neanderthal mitochondrial genome.

出的邻接树[19]。**b**，两个尼安德特人序列与随机选择的一个 !Kung 人、三个其他非洲人、三个亚洲人和三个欧洲人的序列采用分支限界法搜索到的最简约系统树[19]。这一结果与另外四组数据的分析结果一致。以上两个分析使用三只黑猩猩的序列[13]作为外群。两个图中的数字代表 1,000 次重复得到的自举频率（%）。

两个尼安德特人的配对差异水平高于随机选择的 300 个高加索人种（5.28±2.24）和蒙古人种（6.27±2.29）样本的平均值——不到 1% 的高加索人种和蒙古人种配对组合有 12 个或更多位点存在差异——但是与一个 300 个非洲人（8.36±3.2）的随机样本相当，该非洲样本在 12 个及更多位点有 37% 的配对组合存在差异。在分析古 DNA 时，PCR 初始阶段可能会发生核苷酸被错误插入的情况，特别是当标靶 DNA 可能损坏且拷贝数较低时[12]。因为这两个尼安德特人都进行了重复分析而且结果是一致的，所以该类失误可以被忽略。

费尔德霍费尔尼安德特人和梅兹迈斯卡亚尼安德特人在地理上相距 2,500 多千米。考虑到这两个个体拥有密切相关的线粒体 DNA，且在系统发育上不同于现代人，与某些灵长类相比[13]仅显示出中等水平的序列多样性，这些证据为尼安德特人和现代人之间基因交流很少这一假说提供了进一步的支持。需要特别指出的是，这些证据同时反驳了尼安德特人拥有足够的线粒体 DNA 序列多样性且足以覆盖现代人类多样性这一说法。

关于现代人类起源的“走出非洲”的假说认为尼安德特人序列与所有现代序列间的差异距离相等。在我们的分析中也观察到了这一点——经计算，尼安德特人与 300 个随机选择的非洲人、蒙古人和高加索人间的平均配对差异分别为 23.09±2.86、23.27±4.06 和 25.45±3.27。

我们推测东西方尼安德特人线粒体 DNA 的最近共同祖先（MRCA）的年龄约为距今 151,000 至 352,000 年。这与古生物学记录中尼安德特人谱系出现的时间是一致的[14]。据推测现代人和尼安德特人线粒体 DNA 的分化是在距今 365,000 至 853,000 年。使用同一模型，我们推测最早的现代人类线粒体 DNA 分化发生在距今 106,000 至 246,000 年。

从该标本得到的结果表明，一些其他尼安德特人标本也可以用来进行分子分析。为了得到尼安德特人与现代人类关系更完整的图谱，必须分析其他尼安德特人和早期现代人类，特别是来自他们共同生存地区的标本。该标本极佳的保存状况使得对尼安德特人整个线粒体基因组进行分析成为可能。

Methods

DNA extraction, PCR, cloning and sequencing

The DNA extraction methods used in Glasgow[4] and Stockholm[15] have been described. We took precautions to prevent contamination from modern DNA[4]. The Neanderthal-specific primer pairs, NL16,055 and NH16,262, and NL16,209 and NH16,400 (5'-TGATTTCACGGAGGATGGTGA-3') were used in Glasgow and the primers L16,212 (5'-ATGCTTACAAGCAAGCACA-3') and H16,332 (5'-TTGACTGTAATGTGCTATG-3') were used in Stockholm. The annealing temperatures for the primer pairs NL16,055–NH16,262, NL16,209–NH16,400 and L16,212–H16,332 were 50°C, 60°C and 50°C respectively; 40 cycles were used for the first two pairs, 55 cycles for the third. AmpliTaq Gold (Perkin Elmer Cetus) was used in all PCRs. PCR products were purified using the QIAquick Gel Extraction kit (Qiagen) before direct sequencing using the Dye Terminator sequencing kit (Perkin Elmer) or cloning into the TA vector (Invitrogen) before sequencing with the same kit using the M13 and T7 primers.

Sequence analyses

The neighbour-joining and the maximum parsimony branch and bound trees were both constructed using PAUP* 4.0 (ref. 16). For the neighbour-joining analysis, the Tamura-Nei DNA substitution model[17] was used with a gamma distribution of 0.4 (ref. 18), for all other parameters the defaults provided by PAUP* 4.0 were used. The MRCA was calculated using the described methods and assumptions[5]. PAUP* 4.0 was used to calculate pairwise differences between sequences: the data sets used for this were constructed by randomly selecting appropriate samples from a published data set[19].

(**404**, 490-493; 2000)

Igor V. Ovchinnikov[*†‡], **Anders Götherström**[§], **Galina P. Romanova**[‖], **Vitaliy M. Kharitonov**[¶], **Kerstin Lidén**[§] & **William Goodwin**[*]

[*] Human Identification Centre, University of Glasgow, Glasgow G12 8QQ, Scotland, UK

[†] Institute of Gerontology, Moscow 129226, Russia

[§] Archaeological Research Laboratory, Stockholm University, 106 91 Stockholm, Sweden

[‖] Institute of Archaeology, Moscow 117036, Russia

[¶] Institute and Museum of Anthropology, Moscow State University, Moscow 103009, Russia

[‡] Present address: Department of Medicine, Columbia University, New York, New York 10032 USA

Received 15 November 1999; accepted 31 January 2000.

References:

1. Stringer, C. B. & Mackie, R. *African Exodus: the Origin of Modern Humanity* (Cape, London, 1996).

2. Golovanova, L. V., Hoffecker, J. F., Kharitonov, V. M. & Romanova, G. P. Mezmaiskaya Cave: A Neanderthal occupation in the Northern Caucasus. *Curr. Anthropol.* **40**, 77-86 (1999).

3. Smith, F. H., Trinkaus, E., Pettitt, P. B., Karavanic, I. & Paunovic, M. Direct radiocarbon dates for Vindija G1 and Velika Pecina Late Pleistocene hominid remains.

方　法

DNA 提取、PCR、克隆及测序

格拉斯哥[4]和斯德哥尔摩[15]实验室使用的 DNA 提取方法已经介绍过。我们采取了一些预防措施以防止来自现代 DNA 的污染[4]。格拉斯哥实验室使用的尼安德特人特异性引物对包括 NL16,055 和 NH16,262，以及 NL16,209 和 NH16,400(5′-TGATTTCACGGAGGATGGTGA-3′)；斯德哥尔摩实验室使用的引物对为 L16,212(5′-ATGCTTACAAGCAAGCACA-3′)和 H16,332(5′-TTGACTGTAATGTGCTATG-3′)。引物对 NL16,055–NH16,262、NL16,209–NH16,400 和 L16,212–H16,332 的退火温度分别为 50℃、60℃和 50℃；前两对引物扩增了 40 个循环，第三对扩增了 55 个循环。所有 PCR 均使用 AmpliTaq Gold(Perkin Elmer Cetus) 试剂。用 QIAquick 凝胶回收试剂盒(Qiagen) 对 PCR 产物进行纯化，然后使用 Dye Terminator 测序试剂盒(Perkin Elmer) 进行直接测序，或者使用 M13 和 T7 引物，用相同的试剂盒在测序之前克隆到 TA 载体(Invitrogen) 上。

序列分析

使用 PAUP* 4.0 构建邻接分析和最简约分支界限树(参考文献 16)。邻接分析中使用 Tamura-Nei DNA 替换模型[17]，γ 分布取 0.4(参考文献 18)，其他所有参数使用 PAUP* 4.0 提供的默认值。使用文献中描述的方法和假设计算 MRCA[5]。使用 PAUP* 4.0 计算序列之间的成对差异：进行该计算的数据集是通过从已发表的数据集中随机选择适当样本而构建的[19]。

（刘皓芳 翻译；张颖奇 审稿）

Proc. Natl Acad. Sci. USA **96**, 12281-12286 (1999).

4. Krings, M. *et al.* Neandertal DNA sequence and the origin of modern humans. *Cell* **90**, 19-30 (1997).
5. Krings, M., Geisert, H., Schmitz, R. W., Krainitzki, H. & Pääbo, S. DNA sequence of the mitochondrial hypervariable region II from the Neandertal type specimen. *Proc. Natl Acad. Sci.* USA **96**, 5581-5585 (1999).
6. Cooper, A. *et al.* Neandertal genetics. *Science* **277**, 1021-1023 (1997).
7. Gabunia, L. & Vekua, A. A Plio-Pleistocene hominid from Dmanisi, East Georgia, Caucasus. *Nature* **373**, 509-512 (1995).
8. Kozlowski, J. K. in *Neandertals and Modern Humans in Western Asia* 461-482 (Plenum, New York-London, 1998).
9. Brown, T. A., Nelson, D. E., Vogel, J. S. & Southon, J. R. Improved collagen extraction by modified Longin method. *Radiocarbon* **30**, 171-177 (1988).
10. DeNiro, M. J. Postmortem preservation and alteration of *in vivo* bone collagen isotope ratios in relation to palaeodietary reconstruction. *Nature* **317**, 806-809 (1985).
11. Anderson, S. *et al.* Sequence and organisation of the human mitochondrial genome. *Nature* **290**, 457-474 (1981).
12. Höss, M. *et al.* DNA damage and DNA sequence retrieval from ancient tissue. *Nucleic Acids Res.* **24**, 1304-1307 (1996).
13. Gagneux, P. *et al.* Mitochondrial sequences show diverse evolutionary histories of African hominoids. *Proc. Natl Acad. Sci. USA* **96**, 5077-5082 (1999).
14. Gamble, C. in *Prehistoric Europe* 5-41 (Oxford Univ. Press, Oxford, 1998).
15. Lidén, K., Götherström, A. & Eriksson, E. Diet, gender and rank. *ISKOS* **11**, 158-164 (1997).
16. Swofford, D. L. *PAUP*: Phylogenetic Analysis Using Parsimony (* and Other Methods)* Version 4. (Sinauer Associates, Sunderland, Massachusetts, 1998).
17. Tamura, K. & Nei, M. Estimation of the number of nucleotide substitutions in the control region of mitochondrial DNA in humans and chimpanzees. *J. Mol. Evol.* **10**, 512-526 (1993).
18. Excoffier, L. & Yang, Z. Substitution rate variation among sites in mitochondrial hypervariable region I of humans and chimpanzees. *Mol. Biol. Evol.* **16**, 1357-1368 (1999).
19. Burckhardt, F., von Haeseler, A. & Meyer, S. HvrBase: compilation of mtDNA control region sequences from primates. *Nucleic Acids Res.* **27**, 138-142 (1999).

Acknowledgements. We are indebted to L. V. Golovanova for the excavations in Mezmaiskaya Cave that provided materials for analysis. We thank V. P. Ljubin and P. Vanezis for encouragement and support; B. L. Cohen for numerous discussions; J. L. Harley, O. I. Ovtchinnikova, E. B. Druzina and J. Wakefield for technical help and assistance; R. Page for help with the phylogenetic analysis; and P. Beerli, A. Cooper, M. Cusack, M. Nordborg and M. Ruvolo for useful comments. I.V.O. thanks his host G. Curry. I.V.O. was supported by a Royal Society/NATO Fellowship. We thank the Swedish Royal Academy of Sciences and the Swedish Research Council for Natural Sciences for partial financial support.

Correspondence and requests for material should be addressed to W.G. (e-mail: w.goodwin@formed.gla.ac.uk).

A Refugium for Relicts?

M. Manabe *et al.*

Editor's Note

The Yixian Formation of Liaoning Province in north-east China has yielded one of the finest assemblages of fossils, including feathered dinosaurs and early flowering plants. Estimates of the age of the Yixian varied over a 25-million-year timespan, from the Late Jurassic to Early Cretaceous. Evidence placing the age at 124 million years—the younger, Cretaceous end of the spectrum—made the Yixian look like a refuge for ancient relics: a real-life Jurassic Park. This view was put forward in a News and Views article in *Nature* by Zhexi Luo of the Carnegie Museum of Natural History. But it is countered in this paper by palaeontologist Makoto Manabe and colleagues, who show that Cretaceous forms from Japan also have a Jurassic cast, making the picture more complex than Luo had indicated.

LUO[1] suggests that the vertebrate fauna from the Yixian Formation (Liaoning Province, China) shows that this region of eastern Asia was a refugium, in which several typically Late Jurassic lineages (compsognathid theropod dinosaurs, "rhamphorhynchoid" pterosaurs, primitive mammals) survived into the Early Cretaceous[1] (Fig. 1). Data from slightly older sediments in the Japanese Early Cretaceous, however, suggest that the faunal composition of this region can only be partly explained by the concept of a refugium.

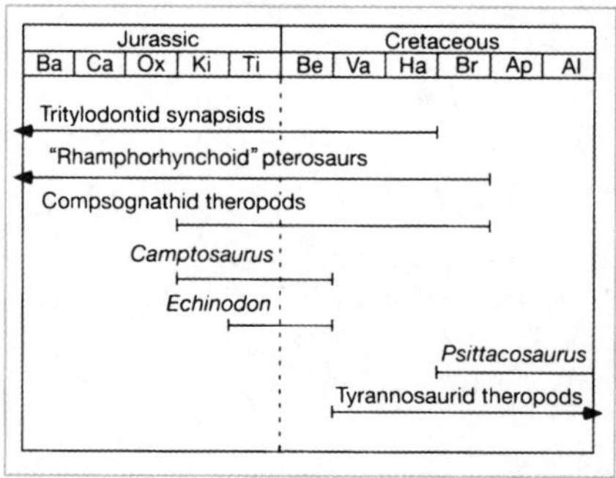

Fig. 1. Stratigraphic ranges of clades that include taxa recovered from the Yixian Formation, China, and the Kuwajima and Itsuki Formations, Japan[1,4,9]. Data on *Camptosaurus* and *Echinodon* are from ref. 13. Arrows, lineage extends beyond the time range shown here; solid bars, first and last occurrences. Al, Albian; Ap, Aptian; Ba, Bathonian; Be, Berriasian; Br, Barremian; Ca, Callovian; Ha, Hauterivian; Ki, Kimmeridgian; Ox, Oxfordian; Ti, Tithonian; Va, Valanginian.

The Kuwajima Formation of Ishikawa Prefecture, central Japan, is yielding an important Early Cretaceous vertebrate fauna. This unit is a lateral equivalent of the Okurodani Formation that outcrops in neighbouring Gifu Prefecture[2]. Stratigraphic, biostratigraphic

残遗种庇护所？

真锅真等

编者按

中国东北部的辽宁省义县组已经产出了最完美的化石群之一，这些化石包括带羽毛的恐龙和早期的被子植物。对义县组的年代估测有着超过2,500万年的差异，从晚侏罗世到早白垩世。有证据表明该地层年龄为1亿2,400万年，对应于更年轻的白垩纪，这使得义县看起来似乎是古代孑遗的避难所：一个真实的侏罗纪公园。这个观点由美国卡内基自然历史博物馆的罗哲西博士发表于《自然》的“新闻与观点”栏目中。但是在本文中，它被古生物学家真锅真和他的同事们反驳了，他们指出日本的白垩纪形成时也有一个“侏罗纪公园”，使得情况远比罗哲西提出的要复杂得多。

罗哲西[1]认为义县组（中国辽宁省）的脊椎动物群说明该亚洲东部地区是一个残遗种庇护所。在这个区域内，晚侏罗世的几个典型谱系（兽脚亚目的美颌龙类、翼龙目的“喙嘴翼龙类”、原始的哺乳类）存活到早白垩世[1]（图1）。但是，日本早白垩世的一些更古老沉积物的数据显示，这个区域的动物群组成不能仅通过残遗种庇护所的概念进行解释。

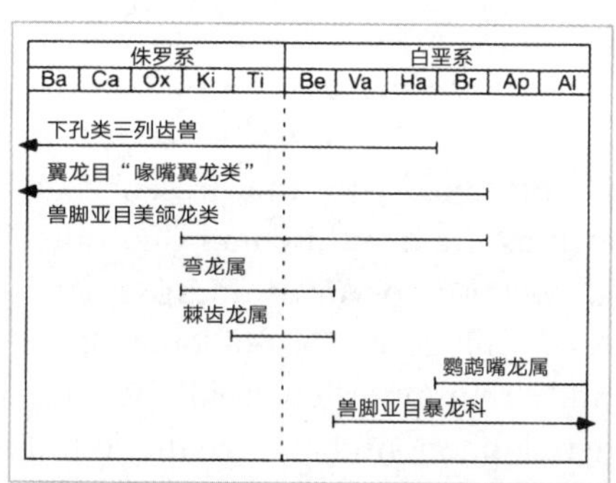

图1. 包括中国义县组和日本桑岛组、五木村组恢复的分类单元在内的分支的地层分布图[1,4,9]。弯龙属和棘齿龙属的数据来自参考文献13。箭头：谱系延伸的时间范围超过图示；实心棒：最先和最后出现。Al：阿尔必阶；Ap：阿普特阶；Ba：巴通阶；Be：贝里阿斯阶；Br：巴列姆阶；Ca：卡洛维阶；Ha：欧特里沃阶；Ki：基默里奇阶；Ox：牛津阶；Ti：提塘阶；Va代表凡兰吟阶。

日本中部石川县桑岛组正在出产一个重要的早白垩世脊椎动物群。桑岛组与邻近的岐阜县大黑谷组层位相同[2]。地层学、生物地层学和放射性测量的数据表明大

and radiometric data show that the Okurodani Formation is basal Cretaceous (Valanginian or Hauterivian) in age[3]. The Kuwajima Formation has yielded more than one hundred isolated teeth of a new genus of tritylodontid synapsid[4]. Before these discoveries, tritylodontids were thought to have become extinct sometime in the Middle or early Late Jurassic, as the youngest-known tritylodontid (*Bienotheroides*) was recovered from late Middle Jurassic deposits. This discovery supports the concept of an East Asian refugium, but other evidence suggests that different factors may have had an equally strong influence on faunal composition.

A theropod dinosaur referable to the unnamed clade Oviraptorosauria + Therizinosauroidea[5] has also been found in the tritylodontid locality. This clade is best known from the Late Cretaceous of mainland Asia, although several taxa referable to this clade are known from the late Early Cretaceous of Liaoning (*Beipiaosaurus*[6] and *Caudipteryx*[7]), and possibly from the Early Jurassic of Yunnan Province, China[8]. The Japanese material, consisting of a single manual ungual (Fig. 2) with a pronounced posterodorsal lip (a feature synapomorphic of this group of theropods[5]), is one of the earliest representatives of this group. The Itsuki Formation of Fukui Prefecture, a lateral equivalent of the Okurodani and Kuwajima Formations[2], has produced an isolated tyrannosaurid tooth, identifiable by its D-shaped cross-section—a synapomorphy of tyrannosaurids[9].

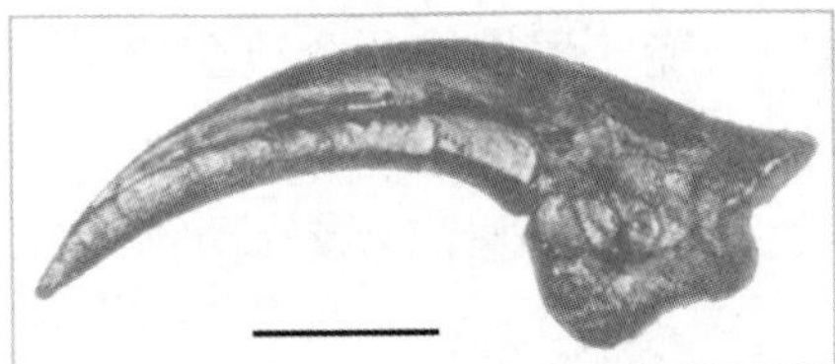

Fig. 2. Manual ungual of a theropod dinosaur from the Kuwajima Formation (Valanginian or Hauterivian) of Shiramine, Ishikawa Prefecture, Japan. Note the prominent lip posterodorsal to the articular surface of the ungual, a synapomorphy of the clade Oviraptorosauria + Therizinosauroidea[5]. Scale bar, 5 mm.

These Japanese discoveries, combined with the presence of late Early Cretaceous taxa in the Yixian Formation (such as the ornithischian dinosaur *Psittacosaurus*[10]), suggest that several dinosaur clades (such as tyrannosaurids and psittacosaurids) may have originated and diversified in eastern Asia while a number of other lineages (tritylodontid synapsids, compsognathid dinosaurs and "rhamphorhynchoid" pterosaurs) persisted in this region. Moreover, the presence of hypsilophodontid and iguanodontid ornithopod dinosaurs in the Japanese Early Cretaceous[11] suggests faunal connections with western Asia and Europe. The historical biogeography of this region appears to be much more complex than was thought previously.

Alternatively, the so-called relict taxa in eastern Asia may indicate that faunal turnover at the Jurassic–Cretaceous boundary was not as marked as has been suggested[12]. The presence of camptosaurid (*Camptosaurus*) and heterodontosaurid (*Echinodon*) ornithopods in European Early Cretaceous faunas[13] indicates faunal similarities to the Late Jurassic

黑谷组在白垩系基部（凡兰吟阶或者欧特里沃阶）[3]。桑岛组发现了一个下孔类的三列齿兽新属[4]的一百多颗独立的牙齿。在这些发现之前，由于最晚的三列齿兽（似卞氏兽）在中侏罗世晚期沉积物中发现，因此人们认为三列齿兽在中侏罗世或晚侏罗世早期的某个时间早已灭绝。尽管桑岛组这个发现支持亚洲东部残遗种庇护所的观点，但是其他证据表明不同的因素可能同样强烈地影响了动物群的组成。

一种未命名的兽脚亚目恐龙也在三列齿兽的产地发现，可能属于窃蛋龙类和镰刀龙类共同组成的一个未命名分支[5]。这个分支大部分属种主要发现自晚白垩世的亚洲大陆，只有少数几个分类单元在中国辽宁省的早白垩世晚期地层（北票龙[6]和尾羽龙[7]）以及可能在中国云南省的早侏罗世地层[8]被发现。日本发现的材料包括一个带有明显后背向的唇突（该兽脚类类群的共同衍征[5]）的单一手爪尖（图 2），这是该类群最早的代表之一。在福井县五木村组（与大黑谷组和桑岛组层位相同[2]）发现了一颗单独的恐龙牙齿。通过 D 形横断面——暴龙的共同衍征[9]可以将其鉴定为暴龙牙齿。

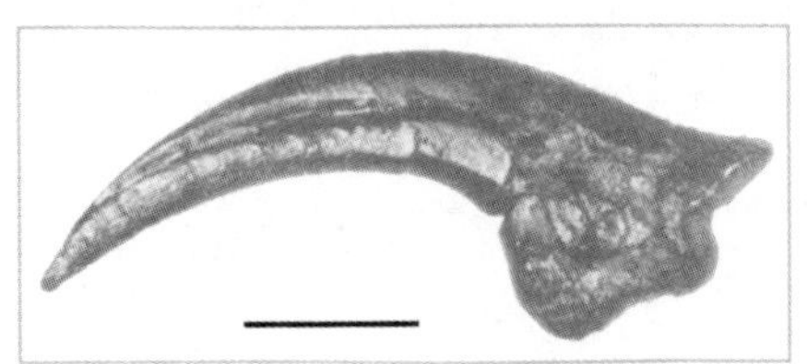

图 2. 日本石川县白峰村桑岛组（凡兰吟阶或欧特里沃阶）发现的兽脚亚目恐龙的手爪尖。注意爪尖关节面后背向的明显的唇突，此为窃蛋龙类 + 镰刀龙类分支[5]的共同衍征。比例尺为 5 毫米。

日本的这些发现，结合义县组早白垩世晚期存在的分类单元（如鸟臀目的鹦鹉嘴龙[10]），表明一些恐龙分支（例如暴龙科和鹦鹉嘴龙科）可能从亚洲东部起源并且发生分化，而许多其他谱系（下孔类的三列齿兽、美颌龙类恐龙和翼龙目“喙嘴翼龙类”）在该地区一直存在。此外，日本早白垩世的棱齿龙类和鸟脚亚目的禽龙类的发现[11]揭示在亚洲西部和欧洲之间存在动物群的联系。该地区的历史生物地理学似乎比之前预想的要复杂得多。

或者说，亚洲东部所谓的残遗分类单元可能表明侏罗纪–白垩纪边界的动物群转化不像人们猜测的那么明显[12]。在欧洲早白垩世动物群中存在的鸟脚亚目恐龙——弯龙类（弯龙属）和异齿龙类（棘齿龙属）[13]表现出与北美洲晚侏罗世莫里森组

Morrison Formation of North America. The presence of "Late Jurassic" taxa in eastern Asia may simply represent another example of this more gradual Jurassic–Cretaceous faunal transition (Fig. 1), although more evidence is needed to distinguish between these alternatives.

(**404**, 953; 2000)

Makoto Manabe[*], **Paul M. Barrett**[†], **Shinji Isaji**[‡]

[*] Department of Geology, National Science Museum, 3-23-1 Hyakunin-cho, Shinjuku-ku, Tokyo 169-0073, Japan

[†] Department of Zoology, University of Oxford, South Parks Road, Oxford OX1 3PS, UK e-mail: paul.barrett@zoo.ox.ac.uk

[‡] Chiba Prefectural Museum of Natural History, Chiba 260-0682, Japan

---

References:

1. Luo, Z. *Nature* **400**, 23-25 (1999).
2. Maeda, S. *J. Collect. Arts Sci. Chiba Univ.* **3**, 369-426 (1961).
3. Evans, S. E. *et al. New Mexico Mus. Nat. Hist. Sci. Bull.* **14**, 183-186 (1998).
4. Setoguchi, T., Matsuoka, H. & Matsuda, M. in *Proc. 7th Annu. Meet. Chinese Soc. Vert. Paleontol.* (eds Wang, Y.-Q. & Deng, T.) 117-124 (China Ocean, Beijing, 1999).
5. Makovicky, P. J. & Sues, H.-D. *Am. Mus. Novitates* **3240**, 1-27 (1998).
6. Xu, X., Tang, Z.-L. & Wang, X.-L. *Nature* **399**, 350-354 (1999).
7. Sereno, P. C. *Science* **284**, 2137-2147 (1999).
8. Zhao, X.-J. & Xu, X. *Nature* **394**, 234-235 (1998).
9. Manabe, M. *J. Paleontol.* **73**, 1176-1178 (1999).
10. Xu, X. & Wang, X.-L. *Vertebrata PalAsiatica* **36**, 147-158 (1998).
11. Manabe, M. & Hasegawa, Y. in *6th Symp. Mesozoic Terrest. Ecosyst. Biotas* (eds Sun, A.-L. & Wang, Y.-Q.) 179 (China Ocean, 1995).
12. Bakker, R. T. *Nature* **274**, 661-663 (1978).
13. Norman, D. B. & Barrett, P. M. *Spec. Pap. Palaeontol.* (in the press).

的动物群的相似性。尽管我们需要更多的证据去区分这两种可能性，但是亚洲东部“晚侏罗世”分类单元的存在可能仅仅代表侏罗纪向白垩纪的动物群逐渐演变的另外一个例证(图1)。

（吕静 张茜楠 翻译；汪筱林 审稿）

Guide to the Draft Human Genome

T. G. Wolfsberg *et al.*

Editor's Note

The release of the first draft of the human genome—the genetic sequence of all human chromosomes—presented the scientific community with a data set that was almost overwhelming in its implications, questions, complications and repercussions. To accompany the paper reporting the results obtained by the International Human Genome Sequencing Consortium, *Nature* here presented an analysis of how the data might be used. As the authors point out, the sequence data joined a considerable existing body of information about the structure and evolution of the human genome, ranging from detailed studies of individual genes and their protein products to linkages between particular inherited phenotypes. The paper provides some pointers for how these disparate types of data might be integrated.

There are a number of ways to investigate the structure, function and evolution of the human genome. These include examining the morphology of normal and abnormal chromosomes, constructing maps of genomic landmarks, following the genetic transmission of phenotypes and DNA sequence variations, and characterizing thousands of individual genes. To this list we can now add the elucidation of the genomic DNA sequence, albeit at "working draft" accuracy. The current challenge is to weave together these disparate types of data to produce the information infrastructure needed to support the next generation of biomedical research. Here we provide an overview of the different sources of information about the human genome and how modern information technology, in particular the internet, allows us to link them together.

THE ultimate goal of the Human Genome Project is to produce a single continuous sequence for each of the 24 human chromosomes and to delineate the positions of all genes. The working draft sequence described by the International Human Genome Sequencing Consortium was constructed by melding together sequence segments derived from over 20,000 large-insert clones[1]. All of the results of this analysis are available on a web site maintained by the University of California at Santa Cruz (http://genome.ucsc.edu). Over the next few years, draft quality sequence will be steadily replaced by more accurate data. The National Center for Biotechnology Information (NCBI) has developed a system for rapidly regenerating the genomic sequence and gene annotation as sequences of the underlying clones are revised (http://www.ncbi.nlm.nih.gov/genome/guide). Undoubtedly, others will apply a variety of approaches to large-scale annotation of genes and other features. One such project is Ensembl, a joint project of the European Bioinformatics Institute (EBI) and the Sanger Centre (http://www.ensembl.org).

人类基因组草图导读

沃尔夫斯伯格等

编者按

人类基因组——全人类染色体遗传序列第一份草图的发布向科学界提供了一个数据集，这个数据集所带来的启发和疑问，以及所具有的复杂性和影响力都是空前的。随着国际人类基因组测序联盟发表文章公布其成果，《自然》杂志在这里刊出了一篇关于如何使用这些数据的分析文章。正如作者所指出的，这些序列数据将大量已知的关于人类基因组结构和进化的信息连接起来，范围从单个基因及其蛋白产物的详细研究到与特定遗传表型之间的关联。本文为如何整合这些不同类型的数据提供了一些指南。

已经有大量的方法被用于研究人类基因组的结构、功能和进化。这些方法包括检测正常和异常染色体的形态，构建基因组标记图谱，追踪表型和DNA序列变异的遗传传递，以及描述数千个单基因的特征。虽然这还处于“工作草图”的准确度阶段，但我们已经可以在此列表中加上基因组DNA序列的解析。我们目前面临的挑战是将这些类型各异的数据编织在一起，进而形成能够支持下一代生物医学研究的信息学基础框架。在这里，我们将会概述这些与人类基因组相关的不同来源的信息，以及如何利用现代信息技术，特别是互联网，将它们联系在一起。

人类基因组计划的最终目标是为人类24条染色体中的每一条都构建一个单独的连续序列，并描述所有基因的位置。国际人类基因组测序联盟所绘制的工作草图序列，是通过把超过20,000个大片断插入克隆的序列合并在一起而构建得到的[1]。所有的分析结果均可在加利福尼亚大学圣克鲁兹分校维护的网站（http://genome.ucsc.edu）上获取。在未来几年间，草图质量序列会不断地被更精确的数据替代。美国国家生物技术信息中心（NCBI）已经开发了一个系统，它能在相应的克隆序列被修订后，迅速更新对应的基因组序列和基因注释（http://www.ncbi.nlm.nih.gov/genome/guide）。毋庸置疑的是其他研究小组将会采用各种各样的方法对基因以及其他一些特征进行大规模注释。其中一个项目是Ensembl，它是由欧洲生物信息研究所（EBI）和桑格中心（http://www.ensembl.org）联合建立的一个合作项目。

Pinpointing New Genes

Recent estimates have placed the number of human genes at 25,000–35,000 (refs 2, 3). More than 10,000 human genes have been catalogued in the Online *Mendelian Inheritance in Man*[4] (OMIM), which documents all inherited human diseases and their causal gene mutations. Integration of the information contained in OMIM with the working draft is facilitated by the fact that it has already been tied to reference messenger RNA sequences through a collaborative effort between OMIM, the Human Gene Nomenclature Committee and the NCBI[5,6]. As a result, the positions of many known genes have been determined by alignment of mRNAs with genomic sequences. For the remaining genes, we must currently resort to computational gene-finding methods (reviewed in refs 7, 8).

When mRNA species align differently to a genomic sequence, this indicates that alternative splicing has taken place. In the current set of full-length reference mRNAs, 11,174 transcripts have been sequenced from 10,742 distinct genes (2.4% of the genes have multiple splicing variants). Alignments of expressed sequence tag (EST) sequences to the working draft sequence, however, suggest that about 60% of human genes have multiple splicing variants, which has important implications for the complexity of human gene expression[1]. By their sheer numbers (currently over 2.5 million) we might expect ESTs to sample a larger fraction of splicing variants than would be the case for more traditional targeted approaches. For example, alignment of the mRNA of the membrane-bound metalloprotease-disintegrin ADAM23 (ref. 9) to the draft genome reveals that the gene consists of at least 23 exons. Of the many ESTs that also align to the *ADAM23* locus, one lacks the exon that encodes the transmembrane domain, which suggests an alternatively spliced, soluble protein. Although this is a biologically plausible conclusion, one should exercise caution when interpreting such results: ESTs are partial single-pass sequences that have been associated with a variety of artefacts, including sequencing errors and improper splicing[10,11].

Finding Relatives

Genes can be found through an implied relationship to something else—for example, being a putative orthologue (related to a gene in another species). To do this, it is useful to search the genomic sequence or, preferably, its mRNA sequences and protein products, using BLAST[12]. As an example, we use the mouse *Lmx1b* gene, which encodes a LIM homeobox protein that is important in pattern development[13]. When we used the protein sequence encoded by *Lmx1b* as a query in a BLAST search against the working draft human genome sequence, the best match was to a region of 9q34, and the positions of the alignments line up with the exons of the human *LMX1B* gene (Fig. 1a).

查明新基因

最近的估计结果认定人类基因的数目大概是介于 25,000 ~ 35,000 之间（参考文献 2 和 3）。已经有超过 10,000 个基因被编入在线“人类孟德尔遗传”（OMIM）数据库[4]中，这个系统记录了所有的人类遗传性疾病以及导致疾病发生的基因突变。通过 OMIM、人类基因命名委员会和 NCBI 的协作，OMIM 中包含的信息已经关联到相关的信使 RNA（mRNA）序列上，这促进了 OMIM 中的信息与工作草图的整合[5,6]。因此，很多已知基因的位置都已经通过 mRNA 与基因组序列的比对确定下来。对于剩下的基因，目前我们必须要求助于各种基于计算的基因寻找方法（参考文献 7 和 8 中的相关综述）。

当 mRNA 种类与基因组序列的比对结果不同时，意味着发生了选择性剪接。在目前的一整套全长参考 mRNA 中，有 11,174 个转录产物从 10,742 个特定基因中（2.4% 的基因有多剪接变异体的情况）测序得到。然而，将表达序列标签（EST）序列与基因组草图序列进行比对，结果却发现大概有 60% 的人类基因具有多剪接变异体，这对于解释人类基因表达的复杂性具有重要意义[1]。仅考虑其数目的话（目前超过了 250 万），我们可以期望 EST 的样本数相对于传统靶向方法具有更多的剪接变异体。举例来讲，将细胞膜结合金属蛋白酶—解整联蛋白 ADAM23（参考文献 9）所对应的 mRNA 与基因组草图进行对比发现该基因至少包含 23 个外显子。在与 *ADAM23* 位点比对的 EST 中，其中一个 EST 缺少编码跨膜区的外显子，这意味着 mRNA 发生了选择性剪接，得到的是可溶性蛋白。尽管这是生物学上合理的结论，但当解释这一类结果时我们仍然需要保持谨慎：EST 是部分单向的序列，可能会与各种各样的人为修饰相关联，包括测序错误以及不适当的剪接[10,11]。

寻找亲缘关系

通过基因与其他一些事物之间暗含的关系寻找基因——例如，推定直系同源基因（与其他物种某个基因相关）。这可以利用 BLAST[12] 搜索基因组序列，或者最好使用 mRNA 序列或者蛋白质产物来寻找基因。以小鼠的 *Lmx1b* 基因为例，这个基因编码的是一个 LIM 同源异形框蛋白，该蛋白在模式发育方面具有重要作用[13]。当我们用 *Lmx1b* 编码的蛋白质序列作为查询序列，通过 BLAST 在人类基因组草图序列数据库中进行检索时，最匹配的区域位于基因组 9p34 区域，其对应序列的位置是人类 *LMX1B* 基因外显子的区域（图 1a）。

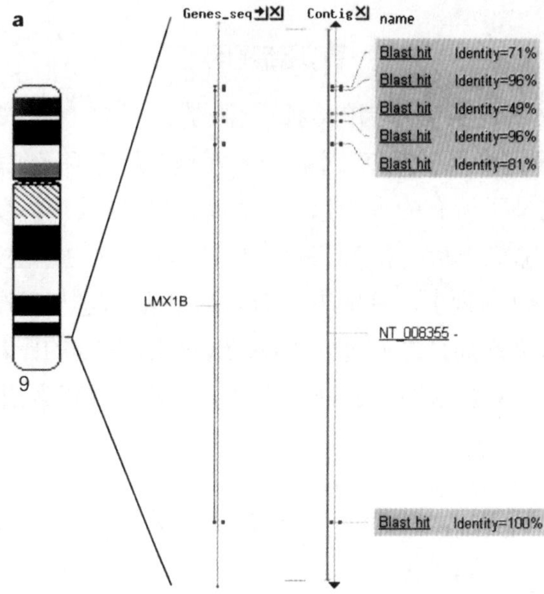

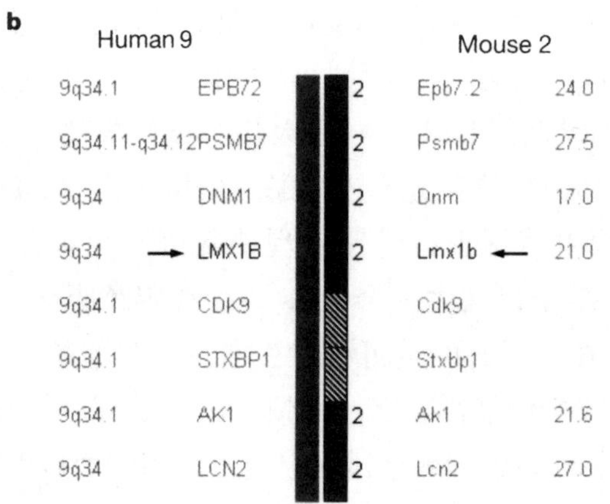

Fig. 1. *Lmx1b* encodes a transcription factor that helps to control the trajectory of motor axons during mammalian limb development. **a**, The results of a search of the six-frame translation of the draft genome sequence on the NCBI site using mouse *Lmx1b* protein NP_034855.1 as a query and using TBLASTN with standard search parameters. The best match was to a region of chromosome 9 that contains the human *LMX1B* gene. **b**, The mouse *Lmx1b* and the human *LMX1B* genes lie within a conserved syntenic block of genes, in mouse on chromosome 2 and in human on chromosome 9. This conservation of gene order supports the theory that *Lmx1b* and *LMX1B* are orthologous.

Additional support for two genes being orthologous comes from the mouse-human homology map. Despite being separated by 200 million years of evolution, mouse and human genes often fall into homologous chromosomal regions that share a conserved gene order (synteny). In fact, the working draft sequence has helped to refine the homology map and provides inferred map positions for many mouse genes[1]. The two homeobox genes fall within a conserved syntenic block between mouse chromosome 2 and human

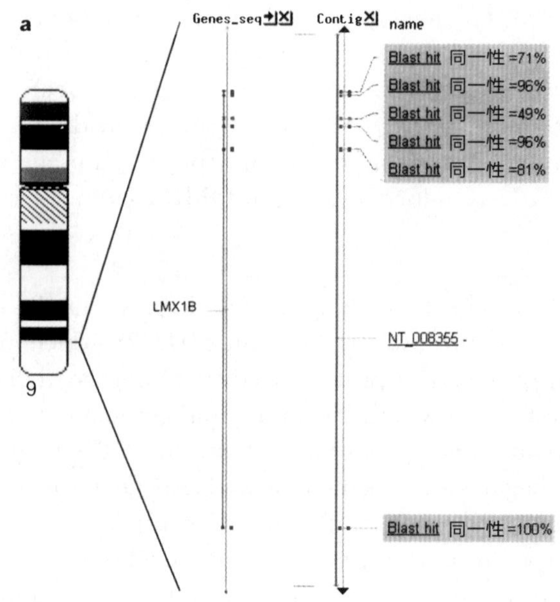

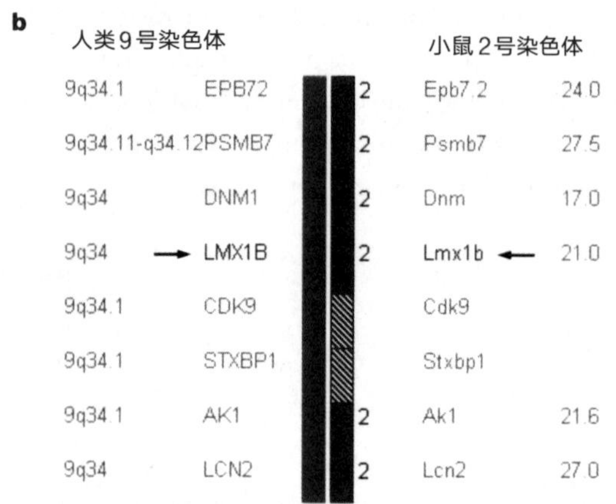

图 1. *Lmx1b* 编码一个转录因子，该转录因子在哺乳动物肢体发育过程中，帮助控制运动神经元轴突的轨迹方向。**a**，在 NCBI 网站上，以小鼠的 Lmx1b 蛋白 NP_034855.1 作为查询序列，使用 TBLASTN 及标准搜索参数，在人类基因组草图序列的六种读码框翻译中进行检索的结果。最佳匹配是 9 号染色体上的一段包含有人类 *LMX1B* 基因的区域。**b**，小鼠的 *Lmx1b* 基因和人类的 *LMX1B* 基因都位于一个保守的线性基因区域中，小鼠的这个基因在 2 号染色体上，而人类的在 9 号染色体上。这种保守的基因排布支持 *Lmx1b* 和 *LMX1B* 是直系同源基因的推测。

进一步支持两个基因是直系同源的证据来自小鼠–人类同源基因图谱。尽管各自独立地进化了 2 亿年，小鼠和人类的基因常常位于具有保守基因序列(共线性)的同源染色体区域内。事实上，人类基因组草图已经帮助修正了小鼠和人类之间的同源基因图谱，为许多小鼠基因的定位提供了推测的图谱位置[1]。这两个同源异形框基因定位于小鼠的 2 号染色体和人类的 9 号染色体上的一个共同的保守线性区域中(图 1b)。除

chromosome 9 (Fig. 1b). Furthermore, the human *LMX1B* gene has been implicated in nail patella syndrome (NPS), an autosomal recessive disorder characterized by limb and kidney defects. A mouse in which *Lmx1b* has been inactivated shows a phenotype that is strikingly similar to NPS[13]. Besides providing additional support for the conclusion of orthology, this connection may provide a useful mouse model for the human disorder. In this way, information from OMIM and mouse mutants can further define human genes.

Another way to find a gene is by looking for paralogues—family members derived by gene duplication. As an example, we used human *ADAM23*, which maps to 2q33 (ref. 14). In a BLAST search against a set of proteins predicted from the draft sequence, aside from matching itself, the best match was to a peptide from chromosome 20. No ADAM family member has previously been mapped to this chromosome. The predicted protein encoded by this gene does not begin with a methionine and appears to be incomplete at its amino terminus when aligned with other family members: this could be due to an erroneous protein prediction or a gap in the draft sequence. Computational analysis can reveal protein family domains and their relationships to three-dimensional protein structures. In this case, the putative ADAM paralogue contains both the zinc metalloprotease and disintegrin motifs characteristic of the ADAM family. Critical amino acids of the metalloprotease domain are conserved in the putative paralogue (Fig. 2b), including a trio of histidine residues in the active site (shaded yellow), which are important in complexing the zinc ion (Fig. 2a). The finding that the sequence of the new predicted ADAM member has an intact active site suggests that the predicted gene is functional, rather than being a pseudogene.

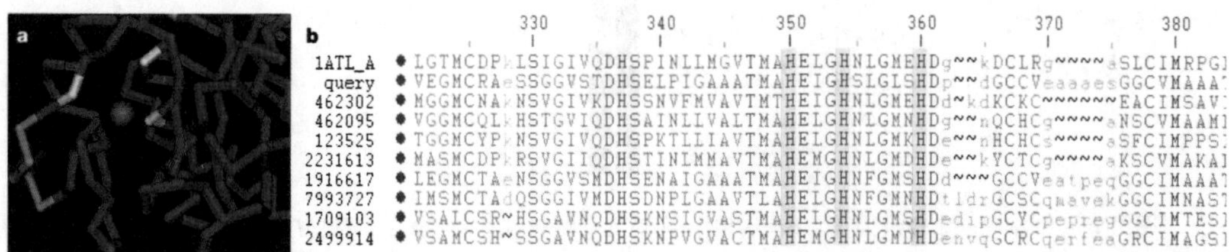

Fig. 2. The ADAM23 protein sequence NP_003803.1 was used in a BLASTP search of the Ensembl confirmed peptides produced from the 5 Sep 2000 version of the working draft sequence. As of October 2000, the best match was to the peptide ENSP00000025626 derived from chromosome 2, the predicted peptide for ADAM23. The second match was to peptide ENSP00000072108, derived from chromosome 20 and falling within GenBank Acc. No. AC055771.2. We used this predicted peptide to search a database of Pfam[24] and SMART[25] protein domains that are aligned with protein structures. This Conserved Domain Database (CDD) search at NCBI resulted in hits to reprolysin and disintegrin domains. **a**, Pfam family 01421, Reprolysin, has a structure associated with it: the zinc-dependent metalloprotease Atrolysin C (PDB: 1ATL). **b**, The query ENSP00000072108 aligns with the 1ATL protein sequence and eight other ADAMs from the Pfam Reprolysin entry. In the structure and alignment, red indicates conserved residues; grey indicates non-aligned sequences. The three histidines of the metalloprotease active site, which complex with the zinc ion, are highlighted in yellow. The structure and alignment were created using the structure viewing program Cn3D.

此之外，人类的 *LMX1B* 基因被认为与指甲髌骨综合征(NPS)的发生相关，它是一种常染色体隐性疾病，其特征是肢体和肾脏的缺陷。*Lmx1b* 基因失活的小鼠会表现出与 NPS 极为相似的症状[13]。除了可以为同源基因的结论提供更多的支持以外，这种关联性也可以为研究人类疾病提供一种有用的小鼠模型。通过这样一种方式，来自 OMIM 和小鼠的突变体的信息可以进一步定义人类的基因。

另外一种寻找基因的方法是寻找旁系同源序列——由基因复制得到的家族成员。以人类的 *ADAM23* 基因为例，它在基因组草图上对应的是 2q33(参考文献 14)。通过 BLAST 检索人类基因组草图预测的所有蛋白质发现，除了与其自身蛋白匹配外，最佳的匹配结果为 20 号染色体上的一个多肽。在此之前还没有任何一个 ADAM 蛋白质家族的成员被定位到这条染色体上。通过与其他家族成员的比对发现，推测的这个基因所编码的蛋白质并不是以甲硫氨酸起始的，它的氨基末端似乎是不完整的：这可能是由于蛋白质预测错误或者是由于在人类基因草图序列上存在缺口所引起的。通过计算分析可以揭示蛋白家族结构域以及它们与蛋白质三维结构之间的关系。在这个例子中，预测得到的 ADAM 旁系同源蛋白同时具有 ADAM 家族特征性的锌金属蛋白酶和解整联蛋白基序。金属蛋白酶结构域的关键氨基酸在推定的旁系同源蛋白中是保守的(图 2b)，这其中包括位于活性部位的组氨酸残基三联体(黄色阴影表示)，它对于络合锌离子至关重要(图 2a)。新预测的 ADAM 家族成员的序列具有一个完整的活性位点，说明这个预测的基因是有功能的，并不是一个假基因。

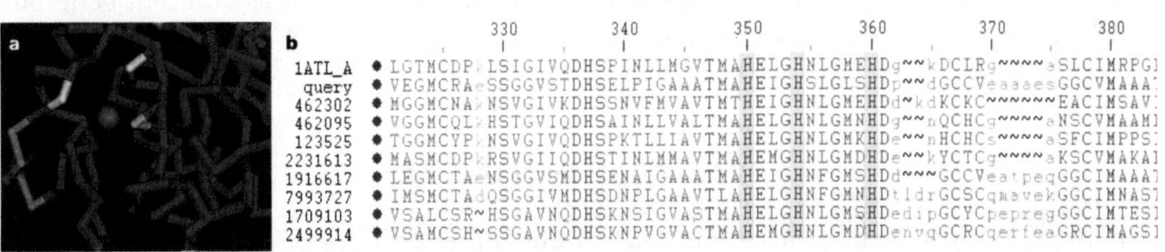

图 2. 在 Ensembl 中用 BLASTP 检索 ADAM23 蛋白序列 NP_003803.1，确认了 2000 年 9 月 5 日版本中人类基因草图序列得到的多肽产物。截至 2000 年 10 月，最好的匹配结果是来自 2 号染色体的多肽 ENSP00000025626，也就是预测的 ADAM23 多肽。匹配结果排在第二位的多肽是 ENSP00000072108，来自 20 号染色体，其在 GenBank 中的编号是 AC055771.2。我们将这个预测的多肽在一个数据库中检索，该数据库包含了与蛋白结构相匹配的 Pfam[24] 和 SMART[25] 蛋白结构域。在 NCBI 上搜索的保守结构域数据库(CDD)，对应的是锌金属蛋白酶和解整联蛋白结构域。**a**，Pfam 家族 01421——锌金属蛋白酶具有一个和它相关的结构：锌依赖的金属蛋白酶 Atrolysin C(PDB 代码：1ATL)。**b**，ENSP00000072108 作为查询序列，与 1ATL 蛋白序列以及 Pfam 锌金属蛋白酶条目下的另外八个 ADAMs 进行比对。在结构图和比对图中，红色代表保守性残基；灰色代表没有比对上的序列。金属蛋白酶活性位点上的三个组氨酸残基，采用黄色突出显示，它们可以与锌离子络合。图中的结构和比对结果采用结构可视化程序 Cn3D 生成。

Searching by Position

It is sometimes desirable to find genes by their position in the genome, rather than by sequence similarity. For example, when genetic or cytogenetic analysis has implicated a particular region in the aetiology of a disease, it is of interest to see what genes lie in the region. A natural way to describe positions in a sequence would be by base coordinates, but this is impractical for the working draft sequence, as the sequence is still being revised. Cytogenetic band nomenclature is more commonly used to describe positions in the genome, and many human diseases are linked to chromosomal deletions, amplifications and translocations[15]. However, to be useful in conjunction with the working draft, these designations must be related to the sequence. Towards this end, a consortium has integrated this information using fluorescence *in situ* hybridization (FISH) to localize BAC clones that also bear sequence tags that can be found in the draft sequence[16]. In addition to providing cytogenetic coordinates as entry points into the genome, they also provide mapped clone reagents that may be useful in further experimental work.

Another way to describe positions in the genome is relative to mapped sequence tagged site (STS) markers. This is particularly useful in positional cloning projects, where candidate regions are usually defined by polymorphic STSs used in genetic linkage analysis. For example, the breast cancer susceptibility locus *BRCA2* was originally localized by fine genetic mapping to a 600-kilobase (kb) interval on chromosome 13 centred around the STS marker D13S171 (ref. 17). STS markers from several genetic and physical maps have been localized in the working draft sequence using a procedure known as electronic PCR[18]. Thus, by simply looking up the position of D13S171, we can see the region around what is now known to be *BRCA2*, together with other features such as adjacent genes and markers, translocation breakpoints, and genetic variations.

Variations on a Theme

A map of DNA sequence variations will aid our understanding of complex diseases and human population dynamics. The most common class of variation is the single nucleotide polymorphism (SNP) and the total number of SNPs in the public database (dbSNP)[19] now exceeds 2.5 million, representing 1.5 million unique SNP loci. Because database entries include flanking sequence surrounding the polymorphic base(s), it is possible to localize variations within the working draft by simple sequence alignment[20].

Histone deacetylase 3 (HDAC3) is a nucleosome-remodelling enzyme that deacetylates the lysine residues of histones, affecting transcriptional repression[21]. Its genomic region contains seven mapped SNPs near the locus, one of which falls within the coding region—a G-to-C substitution that results in the nonsynonymous substitution Arg265Pro in the protein product. The three-dimensional structure of an *Aquifex aeolicus* homologue shows that this residue is at the lip of the active-site pocket[22]. Of the two classes of eukaryotic histone

基于位置的检索

有时候，通过基因在基因组上的位置来寻找基因比通过序列相似性来寻找基因更能获得令人满意的结果。举例来讲，当遗传学或者细胞遗传学分析已经暗示了某一个染色体特定区域和疾病发生密切相关时，搞清楚有什么基因位于这个区域内就显得非常有意义。描述序列位置最自然的一种方法就是采用碱基坐标，但是这种方法用于描述基因组草图序列是行不通的，因为序列还在不断的修正过程中。细胞遗传学条带系统命名法是一种更为普遍的用于描述基因组序列位置的方法，许多人类疾病都与染色体缺失、扩增以及易位相关[15]。然而，为了更好地与草图衔接使用，这些名称必须与序列相关联。为了达到这个目的，某个联盟利用荧光原位杂交技术(FISH)定位了在人类基因组草图序列中可以找到的含有序列标签的BAC克隆，从而整合了这些信息[16]。他们不仅可以为细胞遗传学切入基因组提供相匹配的坐标，还能提供已经比对好的克隆用于后续的实验工作。

另外一种描述基因组序列位置的方法为序列标签位点(STS)标记物定位法。这种方法特别适用于定位克隆的项目，这些候选区域通常采用遗传连锁分析中使用的多态性STS来定义。举例来讲，乳腺癌易感位点*BRCA2*最初就采用精细遗传图谱定位在13号染色体上以STS标记D13S171为中心的大约600 kb区域内(参考文献17)。现在，来自多个遗传和物理图谱上的STS标记已经通过一种叫电子PCR的方法被定位在基因组草图序列上[18]。因而，简单地查询D13S171的位置，我们就可以看到这个区域位于*BRCA2*附近，并且还可以看到其他特征，如临近基因和标记物、易位断点以及遗传变异等。

基因组序列上的变异

DNA序列变异图谱将帮助我们理解复杂疾病和人类种群的动态变化。最常见的一类变异是单核苷酸多态性(SNP)，在公共数据库(dbSNP)[19]中存储的SNP总数目前已经超过了250万个，代表了150万个唯一SNP位点。由于数据库条目包含了多态性碱基的侧翼序列，因此采用简单的序列比对方法就可以确定这些变异在基因组草图上的位置[20]。

组蛋白脱乙酰酶3(HDAC3)是一个核小体重构酶，可以催化组蛋白中的赖氨酸残基去乙酰化，从而影响转录抑制[21]。它所在的基因组区域包含七个确定位置的SNP，其中一个就位于其编码区域内——一个G碱基到C碱基的替换，这会导致蛋白质产物发生非同义突变，即265位的精氨酸突变为脯氨酸。来自超嗜热菌的同源物的三维结构显示，这个残基位于活性位点口袋的边缘[22]。真核细胞组蛋白脱乙酰

deacetylase, one most often has Arg at this position and the other most often has Pro[23], a trait shared with the bacterial members of the histone deacetylase superfamily. This SNP might therefore occur at a functionally interesting site, and also gives pause for speculation: as bacterial members of the superfamily predominantly have a Pro in this position, perhaps Pro is the ancient residue at this site, and not Arg. Note that several high-throughput SNP discovery methods have been used to generate these data and not all SNPs have been rigorously validated.

Conclusions

The draft sequence provides us with the first comprehensive integration of diverse genomic resources. The mapping of ESTs, gene predictions, STSs and SNPs onto the draft sequence can enable identification of alternative splicing, orthologues, paralogues, map positions and coding sequence variations. Users should remember, though, that these genomic resources represent a work-in-progress, and will evolve as the genome is finished and computation methods further refined.

(**409**, 824-826; 2001)

Tyra G. Wolfsberg*, Johanna McEntyre† & Gregory D. Schuler†

* Genome Technology Branch, National Human Genome Research Institute, National Institutes of Health, Bethesda, Maryland 20892, USA

† National Center for Biotechnology Information, National Library of Medicine, National Institutes of Health, Bethesda, Maryland 20894, USA

References:

1. International Human Genome Sequencing Consortium. Initial sequencing and analysis of the human genome. *Nature* **409**, 860-921 (2001).
2. Ewing, B. & Green, P. Analysis of expressed sequence tags indicates 35,000 human genes. *Nature Genet.* **25**, 232-234 (2000).
3. Roest Crollius, H. *et al.* Estimate of human gene number provided by genome-wide analysis using *Tetraodon nigroviridis* DNA sequence. *Nature Genet.* **25**, 235-238 (2000).
4. McKusick, V. A. *Mendelian Inheritance in Man. Catalogs of Human Genes and Genetic Disorders* (Johns Hopkins Univ. Press, Baltimore, 1998).
5. Maglott, D. R., Katz, K. S., Sicotte, H. & Pruitt, K. D. NCBI's LocusLink and RefSeq. *Nucleic Acids Res.* **28**, 126-128 (2000).
6. Pruitt, K. D., Katz, K. S., Sicotte, H. & Maglott, D. R. Introducing RefSeq and LocusLink: curated human genome resources at the NCBI. *Trends Genet.* **16**, 44-47 (2000).
7. Guigo, R., Agarwal, P., Abril, J. F., Burset, M. & Fickett, J. W. An assessment of gene prediction accuracy in large DNA sequences. *Genome Res.* **10**, 1631-1642 (2000).
8. Stormo, G. D. Gene-finding approaches for eukaryotes. *Genome Res.* **10**, 394-397 (2000).
9. Sagane, K., Ohya, Y., Hasegawa, Y. & Tanaka, I. Metalloproteinase-like, disintegrin-like, cysteine-rich proteins MDC2 and MDC3: novel human cellular disintegrins highly expressed in the brain. *Biochem. J.* **334**, 93-98 (1998).
10. Wolfsberg, T. G. & Landsman, D. A comparison of expressed sequence tags (ESTs) to human genomic sequences. *Nucleic Acids Res.* **25**, 1626-1632 (1997).
11. Wolfsberg, T. G. & Landsman, D. in *Bioinformatics: A Practical Guide to the Analysis of Genes and Proteins* (eds Baxevanis, A. D. & Ouellette, B. F. F.) (Wiley-Liss, Inc., New York, 2001).
12. Altschul, S. F. *et al.* Gapped BLAST and PSI-BLAST: a new generation of protein database search programs. *Nucleic Acids Res.* **25**, 3389-3402 (1997).
13. Chen, H. *et al.* Limb and kidney defects in Lmx1b mutant mice suggest an involvement of LMX1B in human nail patella syndrome. *Nature Genet.* **19**, 51-55 (1998).
14. Poindexter, K., Nelson, N., DuBose, R. F., Black, R. A. & Cerretti, D. P. The identification of seven metalloproteinase-disintegrin (ADAM) genes from genomic libraries. *Gene* **237**, 61-70 (1999).
15. Mitelman, F., Mertens, F. & Johansson, B. A breakpoint map of recurrent chromosomal rearrangements in human neoplasia. *Nature Genet.* **15**, 417-474 (1997).
16. The BAC Resource Consortium. Integration of cytogenetic landmarks into the draft sequence of the human genome. *Nature* **409**, 953-958 (2001).

酶分为两类，一类在这个位置上通常是精氨酸，而另一类则与细菌组蛋白脱乙酰酶超家族一样，在这个位置上通常是脯氨酸。因此这个 SNP 位点可能出现在一个有趣的功能性位点上，同时我们可以推测：鉴于细菌组蛋白脱乙酰酶超家族在这个位置上主要是脯氨酸，或许脯氨酸才是这个位置上的古老残基，而非精氨酸。需要注意的是，已经有几种高通量 SNP 发现方法被用来产生这些数据，并且不是所有的 SNP 位点都能被严格验证。

结　论

基因组草图序列使我们得以首次全面整合不同的基因组资源。将 EST、基因预测、STS 和 SNP 定位到草图序列上，可以帮助我们确认可变剪接、直系同源序列、旁系同源序列、比对位置以及编码序列的变异。但是，使用者也要注意，这些基因组资源尚在“加工”之中，会随着基因组测序的最终完成和计算方法的进一步优化而不断地修改。

（刘振明 翻译；陈捷胤 审稿）

17. Wooster, R. *et al.* Localization of a breast cancer susceptibility gene, BRCA2, to chromosome 13q12-13. *Science* **265**, 2088-2090 (1994).
18. Schuler, G. D. Sequence mapping by electronic PCR. *Genome Res.* **7**, 541-550 (1997).
19. Smigielski, E. M., Sirotkin, K., Ward, M. & Sherry, S. T. dbSNP: a database of single nucleotide polymorphisms. *Nucleic Acids Res.* **28**, 352-355 (2000).
20. The International SNP Map Working Group. A map of human genome sequence variation containing 1.42 million single nucleotide polymorphisms. *Nature* **409**, 928-933 (2001).
21. Struhl, K. Histone acetylation and transcriptional regulatory mechanisms. *Genes Dev.* **12**, 599-606 (1998).
22. Finnin, M. S. *et al.* Structures of a histone deacetylase homologue bound to the TSA and SAHA inhibitors. *Nature* **401**, 188-193 (1999).
23. Leipe, D. D. & Landsman, D. Histone deacetylases, acetoin utilization proteins and acetylpolyamine amidohydrolases are members of an ancient protein superfamily. *Nucleic Acids Res.* **25**, 3693-3697 (1997).
24. Bateman, A. *et al.* The Pfam protein families database. *Nucleic Acids Res.* **28**, 263-266 (2000).
25. Schultz, J., Copley, R. R., Doerks, T., Ponting, C. P. & Bork, P. SMART: a web-based tool for the study of genetically mobile domains. *Nucleic Acids Res.* **28**, 231-234 (2000).

Acknowledgements. We thank G. Marth, S. Sherry, D. Landsman, D. Church and D. Lipman for suggestions and review of the manuscript.

Correspondence should be addressed to G.D.S. (e-mail: schuler@ncbi.nlm.nih.gov).

Mining the Draft Human Genome

E. Birney *et al.*

Editor's Note

The first draft of the human genome provided a wealth of information of potential interest to researchers in many fields, ranging from medicine to palaeontology. The sequence was made publicly available, but making productive use of it required some knowledge both of the nature of the information it contained and the tools provided for navigating these data. Here some researchers from the team involved in the genome project supply a quick guide to the possibilities of such data mining.

Now that the draft human genome sequence is available, everyone wants to be able to use it. However, we have perhaps become complacent about our ability to turn new genomes into lists of genes. The higher volume of data associated with a larger genome is accompanied by a much greater increase in complexity. We need to appreciate both the scale of the challenge of vertebrate genome analysis and the limitations of current gene prediction methods and understanding.

IN this issue, accompanying the description of the sequence[1], there are nine data-mining papers that interrogate the genome from distinct biological perspectives. These range from broad topics—cancer[2], addiction[3], gene expression[4], immunology[5] and evolutionary genomics[6]—to the more focused: membrane trafficking[7], cytoskeleton[8], cell cycle[9] and circadian clock[10]. The findings reported by these authors are likely to be indicative of many people's experiences with the draft human genome: frustrating and rewarding in equal measures.

The Current Data Set

The human genome—the first vertebrate genome sequence to be determined—seems likely to be quite representative of what we will find in other vertebrate genomes. It is around 30 times larger than the recently sequenced worm and fly genomes, and 250 times larger than that of yeast, the first eukaryotic genome to be sequenced[11]. Despite its size, it seems likely to have only two or three times as many genes as the fly and worm genomes, with the coding regions of genes accounting for only 3% of the DNA. Repeat sequences form a large proportion of the remaining DNA, around 46%. These repeats may or may not have a function, but they are certainly characteristic of large vertebrate genomes. The rest of the sequence contains promoters, transcriptional regulatory sequences and other

挖掘人类基因组草图

伯尼等

编者按

人类基因组的第一份草图为从医学到古生物学等众多领域的研究者提供了大量有用的信息。序列已经可以公开获取，但是要高效利用这些序列既需要了解这些序列所包含信息的本质，也需要了解浏览这些数据的工具。在这里，一些参与该基因组项目的研究者为挖掘这些数据的可能性提供了快速指南。

既然已经有了人类基因组草图序列，每个人都希望能够使用它。然而，我们可能已经为我们拥有将新基因组变成一个个基因列表的能力而沾沾自喜。基因组越大，包含的数据越多，同时还伴随着复杂程度大大增加。我们需要意识到对脊椎动物基因组进行分析所面临挑战的规模，以及目前在基因预测方法上和认识上的局限性。

在这一期中，除了对序列的描述[1]，还有九篇数据挖掘方面的研究论文从截然不同的生物学视角对基因组进行审视。这些研究的范围从宽泛的主题——癌症[2]、成瘾性[3]、基因表达[4]、免疫学[5]和进化基因组学[6]，到更具体的主题——膜转运[7]、细胞骨架[8]、细胞周期[9]和生物钟[10]。这些作者报道的发现可能代表了许多人类基因组草图研究者的经验：挫折和收获一样多。

目前的数据集

人类基因组——作为第一个被测序的脊椎动物基因组——在我们将要发现的其他脊椎动物基因组中，似乎很具有代表性。它大约是目前已经测序的蠕虫和苍蝇基因组的 30 倍，是第一个测序的真核生物——酵母的基因组的 250 倍[11]。尽管很大，但是它所包含的基因数目看起来只有蠕虫和苍蝇基因组的两到三倍，因为它的基因编码区仅占整个 DNA 长度的 3%。重复序列构成了其余 DNA 的一大部分，大概占到 46%。这些重复序列可能有功能，也可能没有功能，但是它们确实是大型脊椎动物基因组的典型特征。剩下的序列包含启动子、转录调控序列和其他一些尚未知晓

features, as yet unknown.

The International Human Genome Sequencing Consortium has been sequencing the genome in fragments of about 100–200 kilobases (kb). These fragments exist as bacterial artificial chromosome (BAC) clones, which are derived from sequences whose chromosomal location is known. Each newly generated sequence is deposited in the high-throughput genome sequence (HTGS) division of the International Nucleotide Database (GenBank/EMBL/DDBJ) within 24 hours of being assembled and is assigned a unique identifier (its accession number). For the working draft, about 75% of clones are "unfinished": each still consists of about 10–20 unassembled sequence fragments. Sequencing centres are continuously reading new sequence data from these clones until all the gaps are eliminated, at which point the sequence is declared "finished". As the HTGS entries are updated they retain the same accession numbers, but their version numbers increase.

There is a great deal of overlap between BAC clones, so it is typically more convenient to view a cleaned up version of the raw data, in which the sequences of the clones are correctly ordered and overlapped to remove redundancy and create a contiguous DNA sequence for each chromosome. These virtual chromosome sequences change continuously as gaps are closed and fragment ordering is refined.

Finding Genes

With over 30 genomes sequenced, the casual observer could be forgiven for thinking that gene prediction, or annotation, was a problem filed neatly under "solved". Unfortunately this is far from true. The large size of the genome makes finding the genes much more difficult. The protein-coding parts of human genes, called exons, are split into pieces in the genome and these pieces are separated by non-coding sequence called introns. Nearly all of the increase in gene size in human compared with fly or worm is due to the introns becoming much longer (about 50 kb versus 5 kb). The protein-encoding exons, on the other hand, are roughly the same size. This decrease in signal (exon) to noise (intron) ratio in the human genome leads to misprediction by computational gene-finding strategies.

Many methods for predicting genes are based on compositional signals that are found in the DNA sequence. These methods detect characteristics that are expected to be associated with genes, such as splice sites and coding regions, and then piece this information together to determine the complete or partial sequence of a gene. Unfortunately, these *ab initio* methods tend to produce false positives, leading to overestimates of gene numbers, which means that we cannot confidently use them for annotation. They also do not work well with unfinished sequence that has gaps and errors, which may give rise to frameshifts, when the reading frame of the gene is disrupted by the addition or removal of bases.

Thankfully, there is a wealth of data that we can use to produce more reliable gene predictions. Information on expressed sequences (expressed sequence tags (ESTs) and

的特征性序列。

国际人类基因组测序联盟一直在对长约 100 ~ 200 kb 的基因组片段进行测序。这些片段以细菌人工染色体（BAC）克隆的形式存在，这些克隆来自那些染色体位置已知的序列。每一条新测序的序列都会在组装完成后的 24 小时内存入国际核酸序列数据库（GenBank/EMBL/DDBJ）中的高通量测序基因组序列（HTGS）子库，并且被分配一个唯一的标识符（它的编号）。对于整个工作草图而言，还有大概 75% 的克隆处于“未完成”的状态：它们中的每一个大约包含 10 ~ 20 个没有完成组装的序列片段。测序中心正在不断地从这些克隆中读取新的序列数据，直到序列上所有的空缺都被消除为止，直到此时，整个序列的测序工作才能宣告“完成”。在 HTGS 条目更新的过程中，原有编号会保留下来，但是它们的版本号会增加。

细菌人工染色体克隆之间会有大量重叠，因此通常查看原始数据处理后的版本会比较方便，在整洁版中，这些克隆的序列按照正确的顺序进行排列、相互重叠时移除冗余部分，为每一条染色体构造一个连续的 DNA 序列。伴随着序列内空缺的闭合和片段排列顺序的修正，这些实际的染色体序列会不断地改变。

寻找基因

面对 30 多个已经测序的基因组，对旁观者而言，认为“基因的预测或注释领域的问题已经解决”也情有可原。但不幸的是，事实远非如此。庞大的基因组让寻找基因的工作变得更加困难。人类基因中蛋白编码的部分称为外显子，在基因组中被非编码序列的内含子分隔成为片段。与苍蝇和蠕虫基因相比，几乎所有人类基因大小的增加都是由内含子变长造成的（约为 50 kb 比 5 kb）。而另一方面，编码蛋白的外显子，长度则大致相同。人类基因组中的这种信（外显子）噪（内含子）比降低的情况，会导致通过计算寻找基因的策略给出错误的预测。

许多预测基因的方法都是基于在 DNA 序列中找到的组成信号。这些方法首先探测某些期望与基因相关的特征，例如剪接位点和编码区，然后把这些信息拼接在一起，从而确定基因的完整序列或部分序列。然而不幸的是，这些从头预测的方法倾向于给出假阳性的结果，导致基因数目的高估，这就意味着我们不能自信地应用它们来进行基因注释。此外，这些方法也不能很好地处理含有缺口和错误的未完成序列，因为如果增加或者移除的碱基破坏了基因的阅读框，就可能会产生移码。

万幸的是，已经有丰富的数据可以用来进行更为可靠的基因预测。来自人和其他生物的表达序列（表达序列标签（EST）和互补 DNA）和蛋白质信息，为从浩瀚的

complementary DNAs) and proteins from humans and other organisms provide a more accurate resource for resolving gene structures against the vast genomic background. The most effective algorithms integrate gene-prediction methods with similarity comparisons. Such algorithms are integral to software programs such as GeneWise[12], Genomescan[13] and Genie[14], which provide accurate, automatic predictions, whereas BLAST or FASTA programs typically require considerable manual effort to determine the complete structure of a single gene.

The most powerful tool for finding genes may be other vertebrate genomes. Comparing conserved sequence regions between two closely related organisms will enable us to find genes and other important regions in both genomes with no previous knowledge of the gene content of either. The next couple of years should see the sequencing of the mouse, zebrafish and *Tetraodon* genomes. The preliminary sequence of *Tetraodon* has already proved useful in estimating gene numbers[15], and shows much promise for the use of comparative genomics in gene prediction.

Resources Available to the User

There are a number of resources currently available for perusing the human genome. "Human Genome Central"[16] attempts to gather together the most useful web sites (see http://www.ensembl.org/genome/central/ or http://www.ncbi.nlm.nih.gov/genome/central). The best starting point for the uninitiated will be a site such as those of NCBI, Ensembl or the University of California Santa Cruz (UCSC). These sites offer a mixture of genomic viewers and web-searchable datasets, and allow analysis of the human genome sequence without the need to run complex software locally.

For more involved analysis, it might be necessary to download some of the data locally. Useful downloadable sequence-oriented datasets include protein datasets (available from Ensembl and NCBI) and the assembled DNA sequence for regions of the genome, available at UCSC. Other genomic datasets are also available, such as the global physical map from The Genome Sequencing Center in St Louis and the single nucleotide polymorphism (SNP) database from NCBI. Raw sequence data is available from the International Nucleotide Database (GenBank/EMBL/DDBJ), but this data is generally more difficult to handle because it is very fragmentary, can contain contaminating non-human DNA and may include misleading information such as incorrect map assignment.

This loose network of sites will probably coalesce into a more coordinated network of sites offering informative web pages and resources. NCBI, Ensembl and UCSC are developing new, more accessible resources that will become available within the next year.

基因组背景中解析出基因结构提供了更为精确的数据源。最有效的算法是把基因预测方法和相似性比较整合在一起。这类算法已经被整合进一系列的软件程序中，例如 GeneWise[12]、Genomescan[13]以及 Genie[14]，它们可以提供准确的自动化预测。与之相比，使用 BLAST 和 FASTA 程序通常需要相当多的人工工作来确定一个基因的完整结构。

寻找基因最为有效的工具也许就是其他脊椎动物的基因组。比较两个近缘生物的序列保守区可以使我们不必知道两个基因组中任何一个基因的内容就能发现这两个基因组中的基因和其他重要区域。未来几年应该会看到对小鼠、斑马鱼和河鲀鱼基因组的测序。河鲀鱼基因组的初步测序结果在预测基因数目方面的作用已经得到证实[15]，同时显示出比较基因组学在基因预测方面的应用很有前景。

用户可以利用的资源

目前，已经有了一些可用于详尽分析人类基因组的资源。“人类基因组中心”[16]正试图将最有用的网站（参见 http://www.ensembl.org/genome/central/或者 http://www.ncbi.nlm.nih.gov/genome/central）集合在一起。对于缺少经验的人而言，最好的起点就是类似于 NCBI、Ensembl 或者加州大学圣克鲁兹分校（UCSC）这样的一些站点。这些站点都提供了基因组浏览器以及可通过网络搜索的数据库，这样可以允许使用者对人类基因组序列进行分析，而无需在本地运行复杂的软件。

对于更进一步的分析，可能就需要将一些数据下载到本地了。有用的可供下载的定位于序列的数据库，包括蛋白质数据库（可以在 Ensembl 和 NCBI 上获得）以及基因组区域组装好的 DNA 序列，可以在 UCSC 上获得。其他一些基因组数据库也是可以利用的，比如来自圣路易斯的基因组测序中心的完整物理图谱，以及 NCBI 提供的单核苷酸多态性（SNP）数据库。原始的序列数据可以从国际核酸数据库（GenBank/EMBL/DDBJ）获得，但是这类数据通常更难处理，因为它们的片段化非常严重，可能包含受污染的非人类 DNA 序列，还可能包含误导性信息，比如错误的图谱排列。

这些松散的网络站点有可能合并成为一个更为协同的网络，可以提供信息丰富的网页以及资源。NCBI、Ensembl 以及 UCSC 等正在开发新的、更容易使用的资源，这些资源在接下来的一年内将可以供人使用。

How to Use the Resources

There are two main ways to use the human genome sequence. First, we can look for a homologue of a protein that is known from another organism. For example, Clayton *et al.*[10] looked for relatives of the *Drosophila* period clock protein and found the three known relatives and a possible fourth cousin on chromosome 7. Or we can try and find all of the proteins belonging to a particular family—in ref. 4, Tupler *et al.* catalogue all homeobox domains[4]. The easiest way to approach these problems is to use a protein set. This sidesteps the frustration of predicting genes, but makes the researcher reliant on the quality of the predictions being provided. For most of the accompanying reports, a single protein set was the most useful resource provided. For example, Nestler *et al.* searched for G-protein receptor kinases[3] using PSI-BLAST, which searches only protein datasets.

What are the potential pitfalls of the data? Human genes are hard to predict and are often fragmented. If each end of a query protein matches to a different predicted protein, we should suspect that the query sequence may in fact be two parts of a fragmented gene. The two matched human genes should be in the same or adjacent genomic locations. Pollard[8] discovered that fragmentation complicated the analysis of myosin genes. In addition, the unfinished human genomic DNA may contain contamination, particularly from bacteria but also from other sources. Contaminating DNA is routinely removed from finished sequence, but some is still present in unfinished sequence. If the predicted gene matches a bacterial gene more closely than any vertebrate gene then it will almost always be a contaminant. Futreal *et al.*[2] were led up a blind alley for a week before they discovered that cDNA contamination in draft genomic sequences was giving the false impression of multiple p53 proteins in the genome.

During the assembly of unfinished human genomic data it is possible to create artificial duplications, which can result in artefacts in the subsequent analysis. Very similar gene sequences found within the same clone may represent duplicate genes, but could also be the result of an assembly error. This also means that predicted protein sets may contain artificial duplications, leading to overestimation of the number of members in a family.

What does this analysis tell us? For Bock *et al.*[7], the draft genome revealed a list of the molecular players involved in membrane trafficking, providing a platform for experiments that may complete our understanding of this area of biology. In contrast, Murray and Marks[9] found no new cyclin-dependent kinases, indicating that they were all found by traditional experimental techniques. Futreal *et al.* had a similar experience for known cancer genes, but suggest that with new techniques the genome will provide new avenues of cancer research[2].

The interpretation of unfinished draft genomic data may seem like hard work. But it is something to become accustomed to, because we expect future vertebrate genomes to be

如何使用这些资源

使用人类基因组序列有两种主要方式。首先，我们可以寻找与其他生物中已知蛋白质同源的蛋白质。例如，克莱顿等人[10]在人类基因组序列中寻找与果蝇周期性时钟蛋白具有亲缘关系的蛋白，结果在7号染色体上发现了三个已知的同源蛋白以及可能的第四个远亲蛋白。或者，我们可以尝试寻找属于一个特定家族的所有蛋白质——在参考文献4中，图普勒等人整理了所有的同源异型框结构域[4]。解决这些问题最简单的途径是使用一个蛋白质数据库。这样就规避了预测基因的困难，但这使研究者要依赖提供的数据质量。对于大多数研究产生的报告，单一的蛋白质数据库是所提供的最为有效的数据源。举例来讲，内斯特勒等人使用PSI-BLAST程序搜索G蛋白受体激酶[3]，这个过程只搜索了蛋白质数据库。

这些数据潜在的缺陷是什么？人类基因难以预测，并且通常是片段化的。如果所查询蛋白的每一端匹配上了不同的预测蛋白，我们应该怀疑查询序列实际上可能是一个片段化基因的两个部分。两个匹配上的人类基因应当在基因组上处于相同或者邻近的位置。波拉德[8]发现基因的片段化使得对肌球蛋白基因的分析变得复杂。除此之外，没有完成的人类基因组DNA可能会含有污染，特别是来自细菌的污染，当然，也有可能是其他来源的污染。按照规程，污染DNA会从完成的序列中移除，但是仍然会有一些留存在尚未完成的序列中。如果预测的基因与细菌基因的相似性超过了任何脊椎动物的基因，那么几乎可以肯定就是污染物。富特雷亚尔等人[2]被带进了死胡同里，他们花了将近一个星期的时间，才发现基因组草图序列中的cDNA污染造成了人类基因组中含有多个p53蛋白这一错误印象。

在对未完成的人类基因组数据进行组装的过程中，有可能会造成人为的重复，从而导致后续序列分析中的假象。在同一个克隆中找到的非常相似的基因序列可能代表基因复制，但也可能是组装错误的结果。这也意味着预测得到的蛋白质数据库中可能包含人为引入的重复，从而导致高估某个家族中所包含成员的数目。

这个分析能告诉我们什么？对于博克等人而言[7]，人类基因组草图揭示了一系列参与细胞膜转运过程的分子成员，为实验提供了一个平台，可能完善我们对生物学该领域的理解。与之相反，默里和马克斯[9]没有找到新的周期蛋白依赖性激酶，说明它们都已经被传统的实验技术发现了。弗特利尔等人对已知的癌症基因也有相似的经验，但他们的研究说明，随着新技术的使用，基因组将会为癌症的研究提供新的途径[2]。

对尚未完成的基因组草图数据的解析看起来是一项困难的工作。但是它会成为一件平常的事情，因为我们预计未来脊椎动物基因组最先会以草图的形式发布出来。

released initially in draft form. The database providers must develop better ways of viewing the data; and researchers need to be educated in how to use them. That said, there are many undiscovered treasures in the current data set waiting to be found by intuition, hard work and experimental verification. Good luck, and happy hunting!

(**409**, 827-828; 2001)

Ewan Birney[*], **Alex Bateman**[†], **Michele E. Clamp**[†] **& Tim J. Hubbard**[†]

[*] The European Bioinformatics Institute, Wellcome Trust Genome Campus, Hinxton, Cambridge, CB10 1SA, UK

[†] The Sanger Centre, Wellcome Trust Genome Campus, Hinxton, Cambridge CB10 1SA, UK

References:

1. International Human Genome Sequencing Consortium. Initial sequencing and analysis of the human genome. *Nature* **409**, 860-921 (2001).
2. Futreal, A., Wooster, R., Kasprzyk, A., Birney, E. & Stratton, S. Cancer and genomics. *Nature* **409**, 850- 852 (2001).
3. Nestler, E. J. & Landsman, E. Learning about addiction from the genome. *Nature* **409**, 834-835 (2001).
4. Tupler, R., Perini, G. & Green, M. R. Expressing the human genome. *Nature* **409**, 832-833 (2001).
5. Fahrer, A. M., Bazan, J. F., Papathanasiou, P., Nelms, K. A. & Goodnow, C. C. A genomic view of immunology. *Nature* **409**, 836-838 (2001).
6. Li, W-H., Gu, Z., Wang, H. & Nekrutenko, A. Evolutionary analyses of the human genome. *Nature* **409**, 847-849 (2001).
7. Bock, J. B., Matern, H. T., Peden, A. A. & Scheller, R. H. A genomic perspective on membrane compartment organization. *Nature* **409**, 839-841 (2001).
8. Pollard, T. D. Genomics, the cytoskeleton and motility. *Nature* **409**, 842-843 (2001).
9. Murray, A. W. & Marks, D. Can sequencing shed light on cell cycling? *Nature* **409**, 844-846 (2001).
10. Clayton, J. D., Kyriacou, C. P. & Reppert, S. M. Keeping time with the human genome. *Nature* **409**, 829-831 (2001).
11. Goffeau, A. *et al.* The Yeast Genome Directory. *Nature* **387** (suppl.), 1-105 (1997).
12. Birney, E. & Durbin, R. Using GeneWise in the *Drosophila* annotation experiment. *Genome Res.* **10**, 547-548 (2000).
13. Burge *et al. Nature Genet.* (submitted).
14. Reese, M. G., Kulp, D., Tammana, H., Haussler, D. Genie—gene finding in *Drosophila melanogaster*. *Genome Res.* **10**, 529-538 (2000).
15. Crollius, H. R. *et al.* Characterization and repeat analysis of the compact genome of the freshwater pufferfish *Tetraodon nigroviridis*. *Genome Res.* **10**, 939-949 (2000).
16. Genome website set up to help with sequence analysis. *Nature* **406**, 929 (2000).

Correspondence should be addressed to E.B. (e-mail: birney@ebi.ac.uk).

数据库的提供者必须开发更好的方式来查看数据；研究者则需要接受培训以运用这些方法。也就是说，在现有的数据库中，还有许多未知的宝藏等待我们通过感知、努力工作以及实验验证去发现。祝大家好运，并开始快乐的探索旅程！

（刘振明 翻译；解彬彬 审稿）

Initial Sequencing and Analysis of the Human Genome*

International Human Genome Sequencing Consortium

Editor's Note

This paper details the first draft sequence of the human genome. The data, which were published at the same time as Celera Genomics' privately-funded human genome sequence, were the results of an international collaboration between 20 sequencing centres in 6 different countries. This draft sequence, which covers about 94% of the human genome, suggests a presence of some 30,000 to 40,000 protein-coding genes. The final count, revealed with the publication of the complete genome sequence two years later, was down-graded to around 25,000. But the paper remains important because it was the largest extensively sequenced genome of its time, the first vertebrate genome to be extensively sequenced, and uniquely, a first glimpse at the genome of our own species, holding clues to human development, physiology, medicine and evolution.

The human genome holds an extraordinary trove of information about human development, physiology, medicine and evolution. Here we report the results of an international collaboration to produce and make freely available a draft sequence of the human genome. We also present an initial analysis of the data, describing some of the insights that can be gleaned from the sequence.

THE rediscovery of Mendel's laws of heredity in the opening weeks of the 20th century[1-3] sparked a scientific quest to understand the nature and content of genetic information that has propelled biology for the last hundred years. The scientific progress made falls naturally into four main phases, corresponding roughly to the four quarters of the century. The first established the cellular basis of heredity: the chromosomes. The second defined the molecular basis of heredity: the DNA double helix. The third unlocked the informational basis of heredity, with the discovery of the biological mechanism by which cells read the information contained in genes and with the invention of the recombinant DNA technologies of cloning and sequencing by which scientists can do the same.

* This is a shortened version of the original paper. Some, but by no means all, of the omissions have been indicated in the text. Lists of authors and affiliations are given in the original paper. Citations are numbered herein as they are in the original paper; only those references that are cited in this version are listed in the "References" section.

人类基因组的初步测序与分析*

国际人类基因组测序联盟

编者按

本文详细解析了人类基因组的第一个草图序列。本文的数据是6个国家的20个测序中心的国际合作的结果，与塞莱拉基因组公司资助的人类基因组序列同时发表。草图序列覆盖了约94%的人类基因组，提示可能存在30,000～40,000个蛋白质编码基因。随着两年后完整基因组序列的发表，最终蛋白质编码基因计数结果降至25,000个。但是，本文仍旧占据重要地位，因为它是其所处时代最大规模的测序基因组，也是第一个经大规模测序的脊椎动物的基因组，而且非常独特的是，这也是对我们自身物种基因组的第一次探索，获得了关于人类发育、生理学、医药和进化的诸多研究线索。

人类基因组蕴含着与人类发育、生理学、医药和进化相关的海量信息。本文报道了国际合作组织的测序结果，该组织完成了人类基因组草图序列，该信息可以免费获取。我们也对这些数据进行了初步分析，描述了从序列中获得的启示。

20世纪初，孟德尔遗传定律的重新发现[1-3]激发了人们对遗传信息的本质与内涵的探索，推动了生物学近一百年的发展。这些科学进展自然地分成了四个主要的阶段，大体上对应了20世纪的四个二十五年。第一个阶段建立了遗传的细胞学基础：染色体。第二个阶段定义了遗传的分子基础：DNA双螺旋。第三个阶段解密了遗传的信息学基础，发现了细胞读取包含在基因中的遗传信息的生物学机制，并发明了克隆和测序的DNA重组技术，使得科学家可以进行DNA重组。

---

* 这是原文的缩略版。有一些删减在文本中标示出来了，但并非所有的删减都进行了标注。作者及单位名单请参见原文。本文中引用参考文献处的编号与原文一致；文末“References”部分仅保留了本文中有所引用的文献。

The last quarter of a century has been marked by a relentless drive to decipher first genes and then entire genomes, spawning the field of genomics. The fruits of this work already include the genome sequences of 599 viruses and viroids, 205 naturally occurring plasmids, 185 organelles, 31 eubacteria, seven archaea, one fungus, two animals and one plant.

Here we report the results of a collaboration involving 20 groups from the United States, the United Kingdom, Japan, France, Germany and China to produce a draft sequence of the human genome. The draft genome sequence was generated from a physical map covering more than 96% of the euchromatic part of the human genome and, together with additional sequence in public databases, it covers about 94% of the human genome. The sequence was produced over a relatively short period, with coverage rising from about 10% to more than 90% over roughly fifteen months. The sequence data have been made available without restriction and updated daily throughout the project. The task ahead is to produce a finished sequence, by closing all gaps and resolving all ambiguities. Already about one billion bases are in final form and the task of bringing the vast majority of the sequence to this standard is now straightforward and should proceed rapidly.

The sequence of the human genome is of interest in several respects. It is the largest genome to be extensively sequenced so far, being 25 times as large as any previously sequenced genome and eight times as large as the sum of all such genomes. It is the first vertebrate genome to be extensively sequenced. And, uniquely, it is the genome of our own species.

Much work remains to be done to produce a complete finished sequence, but the vast trove of information that has become available through this collaborative effort allows a global perspective on the human genome. Although the details will change as the sequence is finished, many points are already clear.

- The genomic landscape shows marked variation in the distribution of a number of features, including genes, transposable elements, GC content, CpG islands and recombination rate. This gives us important clues about function. For example, the developmentally important HOX gene clusters are the most repeat-poor regions of the human genome, probably reflecting the very complex coordinate regulation of the genes in the clusters.

- There appear to be about 30,000–40,000 protein-coding genes in the human genome—only about twice as many as in worm or fly. However, the genes are more complex, with more alternative splicing generating a larger number of protein products.

- The full set of proteins (the "proteome") encoded by the human genome is more complex than those of invertebrates. This is due in part to the presence of vertebrate-specific protein domains and motifs (an estimated 7% of the total), but more to the fact that vertebrates appear to have arranged pre-existing components into a richer collection of domain architectures.

20 世纪的最后二十五年，科学家们通过不懈努力破译了第一组基因，随后进一步破译了完整的基因组，从而开创了新的研究领域——基因组学。这项工作的研究成果包括了 599 个病毒和类病毒、205 个天然质粒、185 个细胞器、31 株真细菌、7 株古菌、1 株真菌、2 种动物和 1 种植物的基因组序列。

本文报道了来自美国、英国、日本、法国、德国和中国的 20 多个团队的合作研究成果，他们合作完成了人类基因组草图的构建。基因组草图序列由物理图谱产生，该图谱覆盖了基因组常染色质 96% 的序列，加上公共数据库中的其他序列，大约覆盖了人类基因组 94% 的序列。序列的构建所耗费的时间相对较短，序列覆盖度从大约 10% 达到 90% 以上只用了大约 15 个月。序列数据可开放获取，没有任何限制，且在项目进行过程中每天更新。项目下一步的任务是构建基因组完成图，修补所有序列缺口，并解决所有序列歧义。人类基因组中已经有 10 亿个碱基得到了确认，将大部分序列按上述标准落实是项目的首要任务，需要迅速完成。

人类基因组序列从多个方面来看都十分有趣。它是目前被大规模测序的基因组中最大的一个，是之前测序的任意一个基因组的 25 倍，也是之前测序的基因组总和的 8 倍。它是第一个被大规模测序的脊椎动物的基因组。而且，特别的是，它还是我们人类自己的基因组。

为了构建基因组完成图，仍有大量工作需要做，但是在合作者的共同努力下，我们已经获得了海量有价值的信息，可以从全球视角认识人类基因组。尽管当序列全部完成时一些细节会改变，但是很多要点已经理清。

- 基因组全景显示，基因组的大量特征在分布上存在显著差异，包括基因、转座元件、GC 含量、CpG 岛和重组率，这为我们提供了重要的功能线索。例如，对发育非常重要的 HOX 基因簇是基因组上重复序列最贫瘠的区域，这可能反映了基因簇中复杂的基因间协同调控。

- 人类基因组上约有 30,000 ~ 40,000 个蛋白质编码基因，仅约为线虫或果蝇的两倍。然而，人类基因更为复杂，具有更多的可变剪接，可产生更多的蛋白质产物。

- 人类基因组所编码的全部蛋白质（“蛋白质组”）比无脊椎动物更为复杂。这一现象部分是由于脊椎动物特异性蛋白质结构域和基序的出现（约占总量的 7%），但更重要的是，脊椎动物似乎已经将先前存在的组分组装为更丰富的结构域集合体系。

- Hundreds of human genes appear likely to have resulted from horizontal transfer from bacteria at some point in the vertebrate lineage. Dozens of genes appear to have been derived from transposable elements.

- Although about half of the human genome derives from transposable elements, there has been a marked decline in the overall activity of such elements in the hominid lineage. DNA transposons appear to have become completely inactive and long-terminal repeat (LTR) retroposons may also have done so.

- The pericentromeric and subtelomeric regions of chromosomes are filled with large recent segmental duplications of sequence from elsewhere in the genome. Segmental duplication is much more frequent in humans than in yeast, fly or worm.

- Analysis of the organization of Alu elements explains the longstanding mystery of their surprising genomic distribution, and suggests that there may be strong selection in favour of preferential retention of Alu elements in GC-rich regions and that these "selfish" elements may benefit their human hosts.

- The mutation rate is about twice as high in male as in female meiosis, showing that most mutation occurs in males.

- Cytogenetic analysis of the sequenced clones confirms suggestions that large GC-poor regions are strongly correlated with "dark G-bands" in karyotypes.

- Recombination rates tend to be much higher in distal regions (around 20 megabases (Mb)) of chromosomes and on shorter chromosome arms in general, in a pattern that promotes the occurrence of at least one crossover per chromosome arm in each meiosis.

- More than 1.4 million single nucleotide polymorphisms (SNPs) in the human genome have been identified. This collection should allow the initiation of genome-wide linkage disequilibrium mapping of the genes in the human population.

In this paper, we start by presenting background information on the project and describing the generation, assembly and evaluation of the draft genome sequence. We then focus on an initial analysis of the sequence itself: the broad chromosomal landscape; the repeat elements and the rich palaeontological record of evolutionary and biological processes that they provide; the human genes and proteins and their differences and similarities with those of other organisms; and the history of genomic segments. (Comparisons are drawn throughout with the genomes of the budding yeast *Saccharomyces cerevisiae*, the nematode worm *Caenorhabditis elegans*, the fruitfly *Drosophila melanogaster* and the mustard weed *Arabidopsis thaliana*; we refer to these for convenience simply as yeast, worm, fly and mustard weed.) Finally, we discuss applications of the sequence to biology and medicine and describe next steps in the project. A full description of the methods is provided as Supplementary Information on *Nature*'s web site (http://www. nature.com).

- 几百个人类基因似乎来自细菌的横向转移，这可能发生在脊椎动物世系的某个点。很多基因似乎起源于转座元件。

- 尽管人类基因组约半数起源于转座元件，类人猿世系中的这一类元件的整体活性出现了显著的降低。DNA 转座子似乎已完全失活，长末端重复序列(LTR)反转录转座子可能也已失活。

- 染色体的近着丝粒和亚端粒区域被近期发生的大片段的重复序列所填补，这些序列来自基因组的其他区域。在人类基因组中，片段重复的发生概率远高于酵母、果蝇或线虫。

- 对 Alu 元件结构组织的分析解释了长期困扰我们的谜团，即它们惊人的全基因组范围的分布，提示了可能存在强烈的选择，使 Alu 元件偏好性地保留在 GC 富集区。这些"自私"元件可能使其人类宿主收益。

- 男性减数分裂的突变率约为女性的两倍，显示多数突变发生在男性中。

- 对已测序的克隆的细胞遗传学分析确认了一种假设，即大片段的 GC 匮乏区与染色体核型中的"暗 G 带"存在强烈关联。

- 染色体远端(约 20 兆碱基(Mb))的重组率显著升高，在染色体短臂上则为平均值，这种模式促使每条染色体臂在每次减数分裂时至少发生一次交换。

- 在人类基因组中识别出了超过 140 万个单核苷酸多态性(SNP)，这使我们可以在人类种群中建立基因组范围的连锁不平衡图谱。

在本文中，我们从展示人类基因组计划的背景信息开始，描述了基因组草图序列的产生、组装和评估。随后，我们聚焦于序列本身的初步分析：宽泛的染色体全景；重复序列元件，以及丰富的关于进化的古生物学证据以及它们提供的生物学过程；人类基因和蛋白质与其他生物的基因和蛋白质的差异和相似性；基因组片段的历史。(这些比较是通过出芽生殖的酿酒酵母、线虫类的秀丽隐杆线虫、黑腹果蝇、芥草拟南芥的基因组提取的；方便起见，我们将这四种生物简称为酵母、线虫、果蝇和芥草。)最后，我们讨论了这些序列在生物学和医药方面的应用，并描述了人类基因组计划的下一步工作。《自然》杂志网站(http://www.nature.com)上的补充信息提供了关于研究方法的完整描述。

We recognize that it is impossible to provide a comprehensive analysis of this vast dataset, and thus our goal is to illustrate the range of insights that can be gleaned from the human genome and thereby to sketch a research agenda for the future.

Background to the Human Genome Project

The Human Genome Project arose from two key insights that emerged in the early 1980s: that the ability to take global views of genomes could greatly accelerate biomedical research, by allowing researchers to attack problems in a comprehensive and unbiased fashion; and that the creation of such global views would require a communal effort in infrastructure building, unlike anything previously attempted in biomedical research.

The idea of sequencing the entire human genome was first proposed in discussions at scientific meetings organized by the US Department of Energy and others from 1984 to 1986 (refs 21, 22). A committee appointed by the US National Research Council endorsed the concept in its 1988 report[23], but recommended a broader programme, to include: the creation of genetic, physical and sequence maps of the human genome; parallel efforts in key model organisms such as bacteria, yeast, worms, flies and mice; the development of technology in support of these objectives; and research into the ethical, legal and social issues raised by human genome research. The programme was launched in the US as a joint effort of the Department of Energy and the National Institutes of Health. In other countries, the UK Medical Research Council and the Wellcome Trust supported genomic research in Britain; the Centre d'Etude du Polymorphisme Humain and the French Muscular Dystrophy Association launched mapping efforts in France; government agencies, including the Science and Technology Agency and the Ministry of Education, Science, Sports and Culture supported genomic research efforts in Japan; and the European Community helped to launch several international efforts, notably the programme to sequence the yeast genome. By late 1990, the Human Genome Project had been launched, with the creation of genome centres in these countries. Additional participants subsequently joined the effort, notably in Germany and China. In addition, the Human Genome Organization (HUGO) was founded to provide a forum for international coordination of genomic research. Several books[24-26] provide a more comprehensive discussion of the genesis of the Human Genome Project.

Through 1995, work progressed rapidly on two fronts. The first was construction of genetic and physical maps of the human and mouse genomes[27-31], providing key tools for identification of disease genes and anchoring points for genomic sequence. The second was sequencing of the yeast[32] and worm[33] genomes, as well as targeted regions of mammalian genomes[34-37]. These projects showed that large-scale sequencing was feasible and developed the two-phase paradigm for genome sequencing. In the first, "shotgun", phase, the genome is divided into appropriately sized segments and each segment is covered to a high degree of redundancy (typically, eight- to tenfold) through the sequencing of randomly selected subfragments. The second is a "finishing" phase, in which sequence gaps are closed and

我们认识到本文无法提供关于这些海量数据的全面分析，因此我们的目标是，举例说明从人类基因组中所得到的具有普遍意义的新发现，从而建立起未来的研究框架。

人类基因组计划的背景

人类基因组计划起源于20世纪80年代初期的两个重要的远见：通过让研究者以深刻而无偏见的方式挑战难题，全面把握基因组的能力可以极大地促进生物医药的研究；这一全局视野的产生需要全世界在科研的基础设施建设方面共同努力，而不像之前试图进行的任何生物医药研究。

对人类全基因组进行测序的想法是在一次学术会议的讨论中产生的，该会议由美国能源部和其他机构在1984～1986年组织(参考文献21和22)。美国国家研究委员会在其1988年的报告中签署同意了这一概念[23]，但是他们建议将项目的目标拓宽，包括：建立人类基因组的遗传图谱、物理图谱和序列图谱；同时建立关键模式生物如细菌、酵母、线虫、果蝇和小鼠的上述三种图谱；发展支持上述目标的技术；研究因人类基因组研究而产生的伦理、法律和社会议题。该项目由美国能源部和美国国立卫生研究院联合启动。在其他国家，英国医学研究理事会和维康信托基金会支持了英国的基因组研究；法国人类多态性研究中心和肌肉萎缩症协会落实了法国的测序工作；政府机构，包括日本科学振兴机构、文部科学省支持了日本的基因组研究；欧洲共同体发起了若干国际项目，尤其是酵母基因组测序项目。至1990年末，人类基因组计划已经展开，并在这些国家建立了基因组研究中心。随后有更多的国家加入这一项目，主要是德国和中国。而且还成立了国际人类基因组组织(HUGO)，为基因组研究的国际协作提供论坛。一些著作[24-26]提供了关于人类基因组计划产生的更加深入的讨论。

1995年，研究工作在以下两个前沿快速进展。第一，建立了人类和小鼠基因组的遗传和物理图谱[27-31]，提供了识别疾病基因和基因组序列锚定位点的关键工具。第二，测定了酵母[32]和线虫[33]的基因组以及哺乳动物基因组的靶区域[34-37]。这些项目表明了大规模测序是灵活可变的，并建立了基因组测序的两步法范例。在第一步即“鸟枪”阶段，基因组被打断成合适大小的片段，通过随机挑取亚片段测序，每个片段都被覆盖且高度冗余(通常来说，8～10倍)。第二步是“完成”阶段，在这一阶段，通过直接分析，填补序列的缺口，纠正序列歧义。上述结果也显示，完整

remaining ambiguities are resolved through directed analysis. The results also showed that complete genomic sequence provided information about genes, regulatory regions and chromosome structure that was not readily obtainable from cDNA studies alone.

The human genome sequencing effort moved into full-scale production in March 1999. The idea of first producing a draft genome sequence was revived at this time, both because the ability to finish such a sequence was no longer in doubt and because there was great hunger in the scientific community for human sequence data. In addition, some scientists favoured prioritizing the production of a draft genome sequence over regional finished sequence because of concerns about commercial plans to generate proprietary databases of human sequence that might be subject to undesirable restrictions on use[42-44].

The consortium focused on an initial goal of producing, in a first production phase lasting until June 2000, a draft genome sequence covering most of the genome. Such a draft genome sequence, although not completely finished, would rapidly allow investigators to begin to extract most of the information in the human sequence. Experiments showed that sequencing clones covering about 90% of the human genome to a redundancy of about four- to fivefold ("half-shotgun" coverage) would accomplish this[45,46]. The draft genome sequence goal has been achieved, as described below.

The second sequence production phase is now under way. Its aims are to achieve full-shotgun coverage of the existing clones during 2001, to obtain clones to fill the remaining gaps in the physical map, and to produce a finished sequence (apart from regions that cannot be cloned or sequenced with currently available techniques) no later than 2003.

Strategic Issues

Hierarchical shotgun sequencing

Soon after the invention of DNA sequencing methods[47,48], the shotgun sequencing strategy was introduced[49-51]; it has remained the fundamental method for large-scale genome sequencing[52-54] for the past 20 years. The approach has been refined and extended to make it more efficient. For example, improved protocols for fragmenting and cloning DNA allowed construction of shotgun libraries with more uniform representation. The practice of sequencing from both ends of double-stranded clones ("double-barrelled" shotgun sequencing) was introduced by Ansorge and others[37] in 1990, allowing the use of "linking information" between sequence fragments.

Practical difficulties arise because of repeated sequences and cloning bias. The human genome is filled (> 50%) with repeated sequences, including interspersed repeats derived from transposable elements, and long genomic regions that have been duplicated in tandem, palindromic or dispersed fashion. Such features complicate the assembly of a correct and finished genome sequence.

的基因组序列提供了关于基因、调控区和染色体结构的信息，这些信息是无法仅从cDNA研究中获得的。

1999年3月，人类基因组测序经过努力，进入了全序列产生阶段。首先生成一个基因组草图序列的想法在这一时期重新流行起来，这一方面是因为完成这样的序列的能力已经不再是问题，另一方面也是因为科学共同体对人类序列数据有着强大的需求。而且，一些科学家支持优先产生基因组草图序列，而不是局部完成序列。因为一些商业计划企图对人类基因组序列数据库申请专利，这可能会对数据的使用造成不良限制[42-44]。

科学共同体聚焦于产生数据这一首要目标，即第一阶段持续到2000年6月，在此之前使基因组草图序列覆盖基因组的大部分区域。这一基因组草图序列尽管没有彻底完成，但也使得研究者们能够尽快地提取到人类基因组中的大部分信息。实验显示，测序克隆覆盖约90%的人类基因组，约4～5倍冗余（“半鸟枪”覆盖率）[45,46]。如下所示，人类基因组草图序列目标已完成。

第二阶段正在进行中，旨在于2001年获得目前所有克隆的全鸟枪覆盖，并获得能够填补物理图谱缺口的克隆，以便在2003年之前生成基因组序列完成图（除了现有技术无法克隆和测序的基因组区域之外）。

策略问题

层次鸟枪法测序

DNA测序方法[47,48]发明以后，很快就出现了鸟枪法测序[49-51]的策略；在过去的二十多年里，鸟枪法仍然是大规模基因组测序[52-54]的基础方法。鸟枪法也在不断地补充和改进，以便更加高效。例如，DNA片段化和克隆的方法改进后，鸟枪法构建的文库更加均一。1990年，安佐格等人[37]提出了双链克隆的两端测序法（“双管”鸟枪法测序），使测序片段之间“连接信息”的使用成为可能。

由于重复序列和克隆偏好性的存在，实际操作存在不少困难。人类基因组含有大量（>50%）重复序列，包括起源于转座元件的散在重复序列，以及以串联、回文或散在等重复形式存在的长基因组区域。这些情况使一个正确完整的基因序列的组装工作更加复杂。

There are two approaches for sequencing large repeat-rich genomes. The first is a whole-genome shotgun sequencing approach, as has been used for the repeat-poor genomes of viruses, bacteria and flies, using linking information and computational analysis to attempt to avoid misassemblies. The second is the "hierarchical shotgun sequencing" approach (Fig. 2), also referred to as "map-based", "BAC-based" or "clone-by-clone". This approach involves generating and organizing a set of large-insert clones (typically 100–200 kb each) covering the genome and separately performing shotgun sequencing on appropriately chosen clones. Because the sequence information is local, the issue of long-range misassembly is eliminated and the risk of short-range misassembly is reduced.

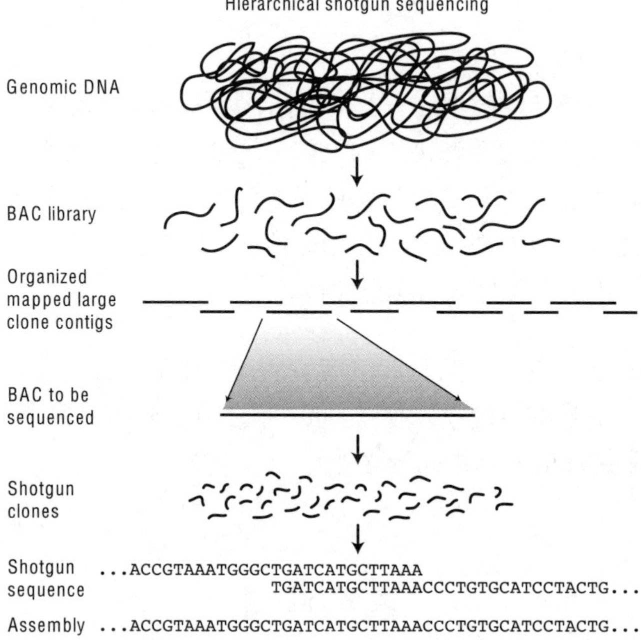

Fig 2. Idealized representation of the hierarchical shotgun sequencing strategy. A library is constructed by fragmenting the target genome and cloning it into a large-fragment cloning vector; here, BAC vectors are shown. The genomic DNA fragments represented in the library are then organized into a physical map and individual BAC clones are selected and sequenced by the random shotgun strategy. Finally, the clone sequences are assembled to reconstruct the sequence of the genome.

A biotechnology company, Celera Genomics, has chosen to incorporate the whole-genome shotgun approach into its own efforts to sequence the human genome. Their plan[60,61] uses a mixed strategy, involving combining some coverage with whole-genome shotgun data generated by the company together with the publicly available hierarchical shotgun data generated by the International Human Genome Sequencing Consortium. If the raw sequence reads from the whole-genome shotgun component are made available, it may be possible to evaluate the extent to which the sequence of the human genome can be assembled without the need for clone-based information. Such analysis may help to refine sequencing strategies for other large genomes.

对富含重复序列的大型基因组进行测序，目前有两种途径。一种是全基因组鸟枪测序法，这种方法用于完成病毒、细菌、果蝇等重复较少的基因组测序，使用片段链接的信息和电脑分析来避免错误组装。第二种是“层次鸟枪法测序”(图 2)，也称为“基于图谱”“基于 BAC”或“连续克隆”法。这种方法包括生成并组装一组能覆盖整个基因组的大片段克隆(通常每个 100 ~ 200 kb)，选择其中合适的克隆进行鸟枪法测序。由于所得的序列信息都是局部的，这样就不存在长程错误组装的问题，也降低了短程错误组装的风险。

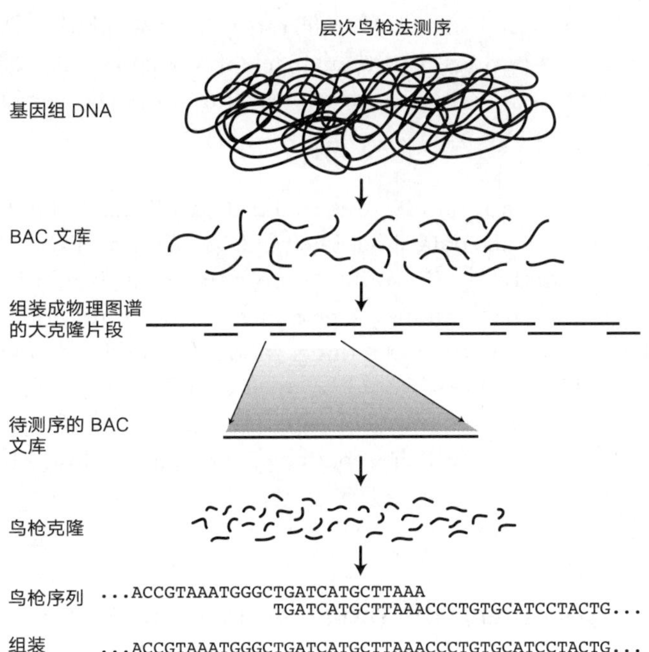

图 2. 理想的层次鸟枪法测序策略。通过目的基因组片段化并克隆到大片段载体上构建文库；此处显示的是 BAC 文库。库中的基因组 DNA 片段组建成物理图谱，挑选出单个的 BAC 克隆利用随机鸟枪法测序。最后，对克隆序列进行组装并重建基因组序列。

一家生物技术公司——塞莱拉基因选择在他们的方法中融入全基因组鸟枪法来完成人类基因组测序。他们计划[60,61]使用混合式策略，将公司全基因组鸟枪法测序得到的数据与国际人类基因组测序联盟公布的层次鸟枪法数据联合。如果可获得全基因组鸟枪法组分的原始数据，就有可能估算出在不依赖克隆信息的前提下，多大程度上可对人类基因组进行组装。这样的分析会帮助修正今后其他大型基因组测序的策略。

Coordination and public data sharing

The Human Genome Project adopted two important principles with regard to human sequencing. The first was that the collaboration would be open to centres from any nation. Although potentially less efficient, in a narrow economic sense, than a centralized approach involving a few large factories, the inclusive approach was strongly favoured because we felt that the human genome sequence is the common heritage of all humanity and the work should transcend national boundaries, and we believed that scientific progress was best assured by a diversity of approaches. The collaboration was coordinated through periodic international meetings (referred to as "Bermuda meetings" after the venue of the first three gatherings) and regular telephone conferences. Work was shared flexibly among the centres, with some groups focusing on particular chromosomes and others contributing in a genome-wide fashion.

The second principle was rapid and unrestricted data release. The centres adopted a policy that all genomic sequence data should be made publicly available without restriction within 24 hours of assembly[79,80]. Pre-publication data releases had been pioneered in mapping projects in the worm[11] and mouse genomes[30,81] and were prominently adopted in the sequencing of the worm, providing a direct model for the human sequencing efforts. We believed that scientific progress would be most rapidly advanced by immediate and free availability of the human genome sequence. The explosion of scientific work based on the publicly available sequence data in both academia and industry has confirmed this judgement.

Generating the Draft Genome Sequence

Generating a draft sequence of the human genome involved three steps: selecting the BAC clones to be sequenced, sequencing them and assembling the individual sequenced clones into an overall draft genome sequence.

The draft genome sequence is a dynamic product, which is regularly updated as additional data accumulate en route to the ultimate goal of a completely finished sequence. The results below are based on the map and sequence data available on 7 October 2000, except as otherwise noted. At the end of this section, we provide a brief update of key data.

Clone selection

The hierarchical shotgun method involves the sequencing of overlapping large-insert clones spanning the genome. For the Human Genome Project, clones were largely chosen from eight large-insert libraries containing BAC or P1-derived artificial chromosome (PAC) clones (refs 82–88). The libraries were made by partial digestion of genomic DNA with restriction

项目协调与公共数据共享

关于人类基因组全序列，人类基因组计划采用了两项重要的原则。一是此项计划的合作面向全世界所有国家的研究中心。尽管从狭隘的经济学角度看，集中于几个大的中心的工作效率会更高，但我们强烈赞成更具包容性的合作，因为我们认为，人类基因组序列是全人类共同的财富，此项工作应跨越国家的界限，而且我们相信，多样化的途径是获得科学进展的最好保证。合作中的协调工作是通过定期的国际会议（百慕大会议，由前三次的会议地点命名）和日常电话会议完成。某些小组集中于特定的染色体，而另一些则着眼于全基因组的范围，各中心间的进展分享非常灵活。

第二点原则是快速的不受限的数据发布。所有的测序中心都遵循一个政策：在基因组序列完成组装的 24 小时内，要将其不受限地公布于众[79,80]。在线虫[11]和小鼠[30,81]基因组的图谱工作中，公开发表前的数据发布已经为人类基因组测序做了很好的榜样。我们确信，伴随人类基因组数据的及时和无偿使用，科学研究将获得迅速的进展。在学术界和产业界，基于公共可获取序列的科学研究成果的激增充分肯定了我们的判断。

生成基因组草图序列

生成人类基因组草图序列包括三步：选择要测序的 BAC 克隆，测序，将测序完成的单个克隆组装为整个基因组草图序列。

基因组草图序列是动态产物，草图序列会因数据的不断积累而经常更新，最终目标是完整的完成序列。除非特别注明，下文的结论都是基于 2000 年 10 月 7 日的图谱与序列数据。在本节的末尾，我们会提供关键数据的简要更新。

克隆选择

层次鸟枪法的测序包括遍布整个基因组的相互重叠的大片段插入克隆。对人类基因组计划来说，克隆大部分选自 8 个大片段插入文库，包括 BAC 或 P1 人工染色体（PAC）克隆（参考文献 82 ~ 88）。文库是由限制性内切酶部分消化基因组 DNA 后

enzymes. Together, they represent around 65-fold coverage (redundant sampling) of the genome. Libraries based on other vectors, such as cosmids, were also used in early stages of the project.

The libraries were prepared from DNA obtained from anonymous human donors in accordance with US Federal Regulations for the Protection of Human Subjects in Research (45CFR46) and following full review by an Institutional Review Board. Briefly the opportunity to donate DNA for this purpose was broadly advertised near the two laboratories engaged in library construction. Volunteers of diverse backgrounds were accepted on a first-come, first-taken basis. Samples were obtained after discussion with a genetic counsellor and written informed consent. The samples were made anonymous as follows: the sampling laboratory stripped all identifiers from the samples, applied random numeric labels, and transferred them to the processing laboratory, which then removed all labels and relabelled the samples. All records of the labelling were destroyed. The processing laboratory chose samples at random from which to prepare DNA and immortalized cell lines.

Because the sequencing project was shared among twenty centres in six countries, it was important to coordinate selection of clones across the centres. Most centres focused on particular chromosomes or, in some cases, larger regions of the genome. We also maintained a clone registry to track selected clones and their progress. In later phases, the global map provided an integrated view of the data from all centres, facilitating the distribution of effort to maximize coverage of the genome. Before performing extensive sequencing on a clone, several centres routinely examined an initial sample of 96 raw sequence reads from each subclone library to evaluate possible overlap with previously sequenced clones.

Sequencing

The selected clones were subjected to shotgun sequencing. Detailed protocols are available on the web sites of many of the individual centres.

The overall sequencing output rose sharply during production (Fig. 4). Following installation of new sequence detectors beginning in June 1999, sequencing capacity and output rose approximately eightfold in eight months to nearly 7 million samples processed per month, with little or no drop in success rate (ratio of useable reads to attempted reads). By June 2000, the centres were producing raw sequence at a rate equivalent to onefold coverage of the entire human genome in less than six weeks. This corresponded to a continuous throughput exceeding 1,000 nucleotides per second, 24 hours per day, seven days per week. This scale-up resulted in a concomitant increase in the sequence available in the public databases (Fig. 4).

构建的。总的来说，样品覆盖 65 倍的基因组(样品冗余)。在计划早期，我们也应用了其他的一些载体，比如黏粒等。

制备文库所使用的 DNA 来自匿名的捐献者，符合美国联邦法律对于保护研究中的人类受试对象的规定(45CFR46)，并且接受机构审查委员会的监督。简单地说就是构建文库的两个实验室在其附近广泛宣传了此次捐献 DNA 的机会。不同背景的志愿者的接收采取先到先取的方式。在与遗传顾问讨论并签署知情同意书后，样品被采集。样品的匿名程序如下：采样实验室去掉所有样品的标签，用随机数字进行编号，将样品移交给处理实验室，后者再次去掉标签并重新标记。所有的标记记录都被销毁。处理实验室会随机选取制备 DNA 的样品，并且将细胞系永生化。

由于测序计划是在 6 个国家的 20 个中心共同完成的，协调各中心的克隆选择是非常重要的。大部分中心的测序集中于特定染色体，另一些则着眼于全基因组范围。我们维护了一个克隆注册表，用以跟踪选出的克隆及后续进展。在后期，全局图谱提供了来自所有研究中心的数据整合视图，这样有助于在各中心间分配工作从而获得基因组的最大覆盖度。在对某个克隆进行详细测序以前，几个中心通常会对最初样品的每一个亚克隆文库的 96 个原始读序进行检验，评估其与已完成测序的克隆出现重叠的可能性。

测序

选出的克隆随后被送去做鸟枪法测序。具体的操作方法可在各中心的网站上获得。

在计划完成过程中，测序结果的输出量在急剧增加(图 4)。1999 年 6 月开始安装新的测序仪后，测序能力和输出量在 8 个月内增长至原来的 8 倍，每月可处理近 700 万样品，而成功率(可用读序与尝试读序的比率)几乎没有下降。到 2000 年 6 月，所有中心获得原始序列的速度相当于在不到 6 周时间内将整个人类基因组覆盖一遍。这相当于每周七天、每天工作 24 小时、每秒测出的核苷酸超过 1,000 个的连续通量。这种扩容也同步增加了公共数据库中的可用序列(图 4)。

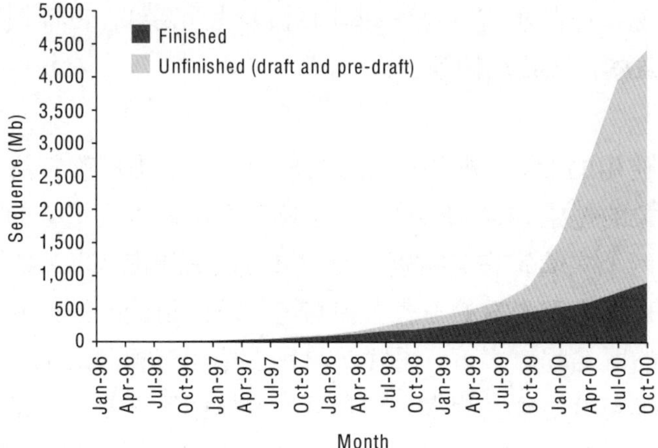

Fig 4. Total amount of human sequence in the High Throughput Genome Sequence (HTGS) division of GenBank. The total is the sum of finished sequence (red) and unfinished (draft plus predraft) sequence (yellow).

Although the main emphasis was on producing a draft genome sequence, the centres also maintained sequence finishing activities during this period, leading to a twofold increase in finished sequence from June 1999 to June 2000 (Fig. 4). The total amount of human sequence in this final form stood at more than 835 Mb on 7 October 2000, or more than 25% of the human genome. This includes the finished sequences of chromosomes 21 and 22 (refs 93, 94). As centres have begun to shift from draft to finished sequencing in the last quarter of 2000, the production of finished sequence has increased to an annualized rate of 1 Gb per year and is continuing to rise.

Assembly of the draft genome sequence

We then set out to assemble the sequences from the individual large-insert clones into an integrated draft sequence of the human genome. The assembly process had to resolve problems arising from the draft nature of much of the sequence, from the variety of clone sources, and from the high fraction of repeated sequences in the human genome. (Editorial note: several details of the assembly procedure have been omitted in this shortened version.)

The result of the assembly process is an integrated draft sequence of the human genome. (Editorial note: details of sequence quality assessment have been omitted.)

The contiguity of the draft genome sequence at each level is an important feature.

Genome coverage

We assessed the nature of the gaps within the draft genome sequence, and attempted to

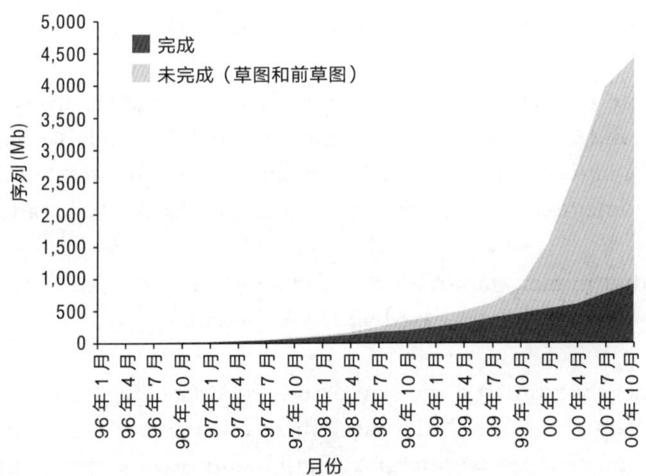

图 4. GenBank 中经高通量基因组测序（HTGS）的人类基因组序列总量。总数为完成序列（红色）和未完成（草图和前草图）序列（黄色）的总量。

尽管工作重点在于生成基因组草图序列，但各中心在此期间也一直在做最终序列的整理，使 1999 年 6 月到 2000 年 6 月之间完成序列的数量增加了一倍（图 4）。2000 年 10 月 7 日，最终形式的人类基因组序列总量超过 835 Mb，大于人类基因组的 25%。这其中包含了测序完成的第 21 和 22 号染色体（参考文献 93 和 94）。2000 年最后一个季度开始，各中心的工作从草图测序转向完成序列，完成序列的产量每年增加 1 Gb，并持续上升。

基因组草图序列的组装

随后，我们着手将单独的大片段插入克隆的序列组装成人类基因组草图序列。装配过程必须解决大部分序列以草图形式存在产生的问题、不同克隆来源产生的问题和人类基因组中高比例的重复序列产生的问题。（编者注：在这篇删减版中组装过程的细节被删减了。）

组装过程的结果是产生一个整合的人类基因组草图序列。（编者注：序列质量的测试细节被删减了。）

基因组草图序列在各层次上的连续性是一个重要特征。

基因组覆盖度

我们对基因组草图序列中间隙的性质进行了评估，并尝试估计人类基因组未表

estimate the fraction of the human genome not represented within the current version.

Our results indicate that about 88% of the human genome is represented in the draft genome sequence and about 94% in the combined publicly available sequence databases. The figure of 88% agrees well with our independent estimates above that about 3%, 5% and 4% of the genome reside in the three types of gap in the draft genome sequence.

We arrived at a total human genome size estimate of around 3,200 Mb, which compares favourably with previous estimates based on DNA content.

We also independently estimated the size of the euchromatic portion of the genome by determining the fraction of the 5,615 random raw sequences that matched the finished portion of the human genome (whose total length is known with greater precision). Twenty-nine per cent of these raw sequences found a match among 835 Mb of nonredundant finished sequence. This leads to an estimate of the euchromatic genome size of 2.9 Gb. This agrees reasonably with a prediction based on the length of the draft genome sequence.

Broad Genomic Landscape

What biological insights can be gleaned from the draft sequence? In this section, we consider very large-scale features of the draft genome sequence: the distribution of GC content, CpG islands and recombination rates, and the repeat content and gene content of the human genome. The draft genome sequence makes it possible to integrate these features and others at scales ranging from individual nucleotides to collections of chromosomes. Unless noted, all analyses were conducted on the assembled draft genome sequence described above.

Figure 9 (Editorial note: since the Figure is oversized and cannot fit here, it is provided on *Nature*'s web site) provides a high-level view of the contents of the draft genome sequence, at a scale of about 3.8 Mb per centimetre. Of course, navigating information spanning nearly ten orders of magnitude requires computational tools to extract the full value. We have created and made freely available various "Genome Browsers". These web-based computer tools allow users to view an annotated display of the draft genome sequence, with the ability to scroll along the chromosomes and zoom in or out to different scales.

Fig. 9. Overview of features of draft human genome. The Figure shows the occurrences of twelve important types of feature across the human genome. Large grey blocks represent centromeres and centromeric heterochromatin (size not precisely to scale). Each of the feature types is depicted in a track, from top to bottom as follows. (1) Chromosome position in Mb. (2) The approximate positions of Giemsa-stained chromosome bands at the 800 band resolution. (3) Level of coverage in the draft genome sequence. Red, areas covered by finished clones; yellow, areas covered by predraft sequence. Regions covered by draft sequenced clones are in orange, with darker shades reflecting increasing shotgun sequence coverage. (4) GC content. Percentage of bases in a 20,000 base window that are C or G. (5) Repeat density. Red line, density of SINE class repeats in a 100,000-base window; blue

示在当前版本中的部分。

我们的结果显示，约 88% 的人类基因组存在于草图序列中，94% 在公开数据库中可获取。我们在前文的独立评估显示，约 3%、5% 和 4% 的基因组存在于基因组草图序列的三种间隙中，88% 的数字与这一结果良好吻合。

我们估计人类基因组约 3,200 Mb，与先前基于 DNA 含量的估计值接近。

我们还通过测算 5,615 个随机原始序列与人类基因组完成序列(精度更高的整体长度)的匹配度，独立估算了基因组中常染色质部分的体量。这些原始序列中的 29% 在 835 Mb 的无冗余完成序列中找到了匹配，由此估计常染色质体量为 2.9 Gb，这与依据基因组草图序列长度所做的预测结果相当。

宏观基因组蓝图

从草图序列中我们可以获得哪些生物学启示呢？在这一部分，我们将考虑那些基因组草图序列体现出的非常大范围上的特征：GC 含量分布，CpG 岛和重组率，人类基因组的重复含量和基因含量。基因组草图序列使得从单核苷酸到染色体集合等各个尺度范围上的特征的整合成为可能。除特殊说明外，以下所有分析都是基于前面所描述的组装后的基因组草图序列。

图 9(编者注：图片过大，此处无法呈现，请在《自然》杂志网站获取)提供了基因组草图序列的高级视图，每厘米约对应 3.8 Mb。当然，浏览的内容跨越近 10 个数量级，需要计算工具来提取出完整的信息。我们已经创建并免费提供各种“基因组浏览器”。这些基于网页的电脑工具使使用者能看到基因组草图序列的注释，也能沿着染色体滚动并放大或缩小到不同的比例。

图 9. 人类基因组草图特征概览。本图展示了人类基因组上 12 种重要的特征序列。大灰色方块代表着丝粒和着丝粒异染色质(大小与实际并非准确相关)。每一种特征类型都标注在基因组上，从上到下依次如下文所述。(1)染色体位置，单位为 Mb。(2)吉姆萨染色法染出的染色体条带位置，分辨率为 800 个条带。(3)基因组草图序列的覆盖程度。红色，该区域被最终克隆覆盖；黄色，该区域被预组装序列覆盖。被草图序列覆盖的区域用橙色表示，深色阴影表示不断增长的鸟枪测序覆盖率。(4)GC 含量。在 20,000 碱基窗口中碱基 C 或 G 的百分比。(5)重复序列密度。红线，SINE 类重复序列在 100,000 碱基窗口的密度；蓝线，LINE 类重复序列在 100,000 碱基窗口的密度。(6)SNP 在 50,000 碱基窗口的密

line, density of LINE class repeats in a 100,000-base window. (6) Density of SNPs in a 50,000-base window. The SNPs were detected by sequencing and alignments of random genomic reads. Some of the heterogeneity in SNP density reflects the methods used for SNP discovery. Rigorous analysis of SNP density requires comparing the number of SNPs identified to the precise number of bases surveyed. (7) Non-coding RNA genes. Brown, functional RNA genes such as tRNAs, snoRNAs and rRNAs; light orange, RNA pseudogenes. (8) CpG islands. Green ticks represent regions of ~200 bases with CpG levels significantly higher than in the genome as a whole, and GC ratios of at least 50%. (9) Exofish ecores. Regions of homology with the pufferfish *T. nigroviridis*[292] are blue. (10) ESTs with at least one intron when aligned against genomic DNA are shown as black tick marks. (11) The starts of genes predicted by Genie or Ensembl are shown as red ticks. The starts of known genes from the RefSeq database[110] are shown in blue. (12) The names of genes that have been uniquely located in the draft genome sequence, characterized and named by the HGM Nomenclature Committee. Known disease genes from the OMIM database are red, other genes blue. This Figure is based on an earlier version of the draft genome sequence than analysed in the text, owing to production constraints. We are aware of various errors in the Figure, including omissions of some known genes and misplacements of others. Some genes are mapped to more than one location, owing to errors in assembly, close paralogues or pseudogenes. Manual review was performed to select the most likely location in these cases and to correct other regions. For updated information, see http://genome.ucsc.edu/ and http://www.ensembl.org/.

In addition to using the Genome Browsers, one can download from these sites the entire draft genome sequence together with the annotations in a computer-readable format. The sequences of the underlying sequenced clones are all available through the public sequence databases. URLs for these and other genome websites are listed in Box 2.

Box 2. Sources of publicly available sequence data and other relevant genomic information

http://genome.ucsc.edu/

University of California at Santa Cruz

Contains the assembly of the draft genome sequence used in this paper and updates

http://genome.wustl.edu/gsc/human/Mapping/

Washington University

Contains links to clone and accession maps of the human genome

http://www.ensembl.org

EBI/Sanger Centre

Allows access to DNA and protein sequences with automatic baseline annotation

http://www.ncbi.nlm.nih.gov/genome/guide/

NCBI

Views of chromosomes and maps and loci with links to other NCBI resources

http://www.ncbi.nlm.nih.gov/genemap99/

Gene map 99: contains data and viewers for radiation hybrid maps of EST-based STSs

度。SNP 是通过读取随机基因组测序和比对检测到的。SNP 密度的异质性反映了 SNP 检测方法的不同。SNP 密度的精确分析需要将识别出的 SNP 数量与所研究的精确的碱基数量做比较。(7) 非编码 RNA 基因。棕色，功能 RNA 基因如 tRNA、snoRNA 和 rRNA；浅橙色，RNA 假基因。(8) CpG 岛。绿色钩代表约 200 个碱基的区域，该区域的 CpG 水平显著高于全基因组，且 GC 含量至少为 50%。(9) 鱼类外显子。与河豚序列[292]同源的区域为蓝色。(10) 与基因组序列比对，带有至少一个内含子的 EST 用黑色钩表示。(11) 通过 Genie 或 Ensembl 预测的基因起始位点用红色钩表示。从 RefSeq 数据库[110]中得到的已知基因的起始位点用蓝色表示。(12) 每一个基因特有的名字已经定位在基因组草图序列中，经 HGM 命名委员会表征并命名。来自 OMIM 数据库的已知疾病基因用红色表示，其他基因用蓝色表示。限于输出限制，本图是基于基因组草图序列的早期版本而不是本文分析的内容绘制的。我们意识到图中有几处错误，包括遗漏了一些已知基因以及搞错了其他一些基因的位置。由于组装、相近的同源序列或假基因方面的错误，一些基因被标注在多个位点。通过人工检查我们挑出了上述情况中最有可能的位置，并将其更正。关于更新的信息，请见 http://genome.ucsc.edu/ 和 http://ensembl.org/。

除了使用基因组浏览器之外，使用者还可以从这些网站下载计算机可读的基因组草图序列和相应的注释。那些基础测序的克隆的序列都可以从公共序列数据库获得。上述网站和其他基因组网站的 URL 列表在框 2 中。

框 2. 可公开获取的序列数据及其他相关基因组信息的来源

http://genome.ucsc.edu/
加州大学圣克鲁兹分校
包含本文及后续更新中使用的基因组草图序列的组装

http://genome.wustl.edu/gsc/human/Mapping/
华盛顿大学
包含人类基因组的克隆和接入图谱的链接

http://www.ensembl.org
EBI/桑格中心
可以访问带有自动基线注释的 DNA 和蛋白质序列

http://www.ncbi.nlm.nih.gov/genome/guide/
NCBI
可浏览染色体、图谱和基因座，并带有其他 NCBI 资源的链接

http://www.ncbi.nlm.nih.gov/genemap99/
Gene map 99：包含基于 EST 的 STS 辐射杂交图谱的数据和阅读器

http://compbio.ornl.gov/channel/index.html

Oak Ridge National Laboratory

Java viewers for human genome data

http://hgrep.ims.u-tokyo.ac.jp/

RIKEN and the University of Tokyo

Gives an overview of the entire human genome structure

http://snp.cshl.org/

The SNP Consortium

Includes a variety of ways to query for SNPs in the human genome

http://www.ncbi.nlm.nih.gov/Omim/

Online *Mendelian Inheritance in Man*

Contain information about human genes and disease

http://www.nhgri.nih.gov/ELSI/and http://www.ornl.gov/hgmis/elsi/elsi.html

NHGRI and DOE

Contains information, links and articles on a wide range of social, ethical and legal issues

Long-range variation in GC content

The existence of GC-rich and GC-poor regions in the human genome was first revealed by experimental studies involving density gradient separation, which indicated substantial variation in average GC content among large fragments. Subsequent studies have indicated that these GC-rich and GC-poor regions may have different biological properties, such as gene density, composition of repeat sequences, correspondence with cytogenetic bands and recombination rate[112-117]. Many of these studies were indirect, owing to the lack of sufficient sequence data.

The draft genome sequence makes it possible to explore the variation in GC content in a direct and global manner. Visual inspection (Fig. 9) confirms that local GC content undergoes substantial long-range excursions from its genome-wide average of 41%.

There are huge regions (> 10 Mb) with GC content far from the average.

Long-range variation in GC content is evident not just from extreme outliers, but throughout the genome.

The correlation between GC content domains and various biological properties is of great interest, and this is likely to be the most fruitful route to understanding the basis of variation

http://compbio.ornl.gov/channel/index.html
橡树岭国家实验室
用于人类基因组数据的 Java 阅读器

http://hgrep.ims.u-tokyo.ac.jp/
RIKEN 及东京大学
给出整个人类基因组结构的总览

http://snp.cshl.org/
SNP 联盟
包含查询人类基因组中 SNP 的多种方式

http://www.ncbi.nlm.nih.gov/Omim/
在线人类孟德尔遗传数据库
包含关于人类基因和疾病的信息

http://www.nhgri.nih.gov/ELSI/ and http://www.ornl.gov/hgmis/elsi/elsi.html
NHGRI 及 DOE
包含大量与社会、伦理和法律问题相关的信息、链接及文章

GC 含量的远距离变化

人类基因组中存在 GC 富集区和 GC 匮乏区，这一特征最初是通过密度梯度离心的实验发现的，揭示了大片段之间平均 GC 含量存在显著变化。随后的研究表明，GC 富集区和 GC 匮乏区可能具有不同的生物学特性，如基因密度、重复序列组成以及细胞遗传学带和重组率的对应关系[112-117]。由于缺乏足够的序列数据，许多研究是间接的。

基因组草图序列的完成使直接而全面地研究 GC 含量变化成为可能。直接观察(图 9)证实，全基因组范围内的平均 GC 含量是 41%，而局部 GC 含量与之相比则有大量的远程偏移。

有很大的区域(> 10 Mb)的 GC 含量远高于平均值。

GC 含量的远程变化是很明显的，不仅来自极端的异常点，而是存在于整个基因组。

区域的 GC 含量与各种生物学性质之间的相关性引起了人们极大的兴趣，这可

in GC content. We confirm the existence of strong correlations with both repeat content and gene density.

CpG islands

A related topic is the distribution of so-called CpG islands across the genome. The dinucleotide CpG is notable because it is greatly under-represented in human DNA, occurring at only about one-fifth of the roughly 4% frequency that would be expected by simply multiplying the typical fraction of Cs and Gs (0.21 × 0.21). The deficit occurs because most CpG dinucleotides are methylated on the cytosine base, and spontaneous deamination of methyl-C residues gives rise to T residues. (Spontaneous deamination of ordinary cytosine residues gives rise to uracil residues that are readily recognized and repaired by the cell.) As a result, methyl-CpG dinucleotides steadily mutate to TpG dinucleotides. However, the genome contains many "CpG islands" in which CpG dinucleotides are not methylated and occur at a frequency closer to that predicted by the local GC content. CpG islands are of particular interest because many are associated with the 5′ ends of genes[122-127].

We searched the draft genome sequence for CpG islands.

The density of CpG islands varies substantially among some of the chromosomes. Most chromosomes have 5–15 islands per Mb, with a mean of 10.5 islands per Mb. However, chromosome Y has an unusually low 2.9 islands per Mb, and chromosomes 16, 17 and 22 have 19–22 islands per Mb. The extreme outlier is chromosome 19, with 43 islands per Mb. Similar trends are seen when considering the percentage of bases contained in CpG islands. The relative density of CpG islands correlates reasonably well with estimates of relative gene density on these chromosomes, based both on previous mapping studies involving ESTs and on the distribution of gene predictions discussed below.

Comparison of genetic and physical distance

The draft genome sequence makes it possible to compare genetic and physical distances and thereby to explore variation in the rate of recombination across the human chromosomes. We focus here on large-scale variation. Finer variation is examined in an accompanying paper[131].

Two striking features emerge from analysis of these data. First, the average recombination rate increases as the length of the chromosome arm decreases. A similar trend has been seen in the yeast genome[132,133], despite the fact that the physical scale is nearly 200 times as small.

The second observation is that the recombination rate tends to be suppressed near the centromeres and higher in the distal portions of most chromosomes, with the increase largely in the terminal 20–35 Mb. The increase is most pronounced in the male meiotic map.

能是了解GC含量变化本质的最有效途径。我们确定重复序列与基因密度存在强相关性。

CpG岛

一个相关的话题是所谓的CpG岛在基因组内的分布。CpG二核苷酸引人注意是因为它在人类DNA中所占比例非常少，如果按CpG出现的理论概率计算，简单将C和G的占比相乘（0.21×0.21）即大约4%，而CpG的出现率仅为4%的五分之一。发生缺失的原因是大多数CpG二核苷酸的胞嘧啶发生甲基化，甲基化的C残基自发脱氨基而使T残基增加（普通胞嘧啶的自发脱氨基使尿嘧啶残基增加，尿嘧啶很容易被细胞识别和修复）。结果，甲基化CpG二核苷酸稳步突变为TpG二核苷酸。然而，基因组含有许多“CpG岛”，岛内CpG二核苷酸没有被甲基化，发生频率与局部GC含量预测频率接近。CpG岛非常有趣，因为许多CpG岛与基因的5′端相关[122-127]。

我们在基因组草图序列中搜寻CpG岛。

CpG岛的密度在不同染色体之间差异很大。大多数染色体每Mb有5～15个岛，平均每Mb有10.5个岛。不过，Y染色体的CpG岛密度异常低，每Mb仅有2.9个，而16号、17号和22号染色体每Mb则有19～22个岛。最为异常的是19号染色体，每Mb约有43个CpG岛。考虑CpG岛所含碱基的百分比时能看到类似的趋势。基于之前包含EST的图谱研究和下面讨论的基因预测的分布，CpG岛的相对密度与这些染色体上估算的相对基因密度之间相关性良好。

遗传距离与物理距离的比较

基因组草图序列的完成，使遗传距离和物理距离的比较成为可能，进而可以探索人类染色体间重组率的变化。此处，我们将重点放在大范围内的变化上，更细微的变化在附带的论文中讨论[131]。

数据分析显示出两个明显特征：一是平均重组率随着染色体臂长的减少而增加。尽管酵母基因组的物理大小约为人类的200分之一，但相似的趋势仍然存在[132,133]。

第二个观察结果是，在大多数染色体上，接近着丝粒的重组率倾向于被抑制，而重组率在远端部分更高，在末端20～35 Mb更是大幅增加。这种重组率的增长在男性减数分裂图谱中更加显著。

Why is recombination higher on smaller chromosome arms? A higher rate would increase the likelihood of at least one crossover during meiosis on each chromosome arm, as is generally observed in human chiasmata counts[135]. Crossovers are believed to be necessary for normal meiotic disjunction of homologous chromosome pairs in eukaryotes.

Mechanistically, the increased rate of recombination on shorter chromosome arms could be explained if, once an initial recombination event occurs, additional nearby events are blocked by positive crossover interference on each arm. Evidence from yeast mutants in which interference is abolished shows that interference plays a key role in distributing a limited number of crossovers among the various chromosome arms in yeast[136]. An alternative possibility is that a checkpoint mechanism scans for and enforces the presence of at least one crossover on each chromosome arm.

Variation in recombination rates along chromosomes and between the sexes is likely to reflect variation in the initiation of meiosis-induced double-strand breaks (DSBs) that initiate recombination. DSBs in yeast have been associated with open chromatin[137,138], rather than with specific DNA sequence motifs. With the availability of the draft genome sequence, it should be possible to explore in an analogous manner whether variation in human recombination rates reflects systematic differences in chromosome accessibility during meiosis.

Repeat Content of the Human Genome

A puzzling observation in the early days of molecular biology was that genome size does not correlate well with organismal complexity. For example, *Homo sapiens* has a genome that is 200 times as large as that of the yeast *S. cerevisiae*, but 200 times as small as that of *Amoeba dubia*[139,140]. This mystery (the C-value paradox) was largely resolved with the recognition that genomes can contain a large quantity of repetitive sequence, far in excess of that devoted to protein-coding genes (reviewed in refs 140, 141).

In the human, coding sequences comprise less than 5% of the genome (see below), whereas repeat sequences account for at least 50% and probably much more. Broadly, the repeats fall into five classes: (1) transposon-derived repeats, often referred to as interspersed repeats; (2) inactive (partially) retroposed copies of cellular genes (including protein-coding genes and small structural RNAs), usually referred to as processed pseudogenes; (3) simple sequence repeats, consisting of direct repetitions of relatively short *k*-mers such as $(A)_n$, $(CA)_n$ or $(CGG)_n$; (4) segmental duplications, consisting of blocks of around 10–300 kb that have been copied from one region of the genome into another region; and (5) blocks of tandemly repeated sequences, such as at centromeres, telomeres, the short arms of acrocentric chromosomes and ribosomal gene clusters. (These regions are intentionally under-represented in the draft genome sequence and are not discussed here.)

Repeats are often described as "junk" and dismissed as uninteresting. However, they actually represent an extraordinary trove of information about biological processes. The repeats

为什么在较小的染色体臂上重组率较高？通常在人类交叉计数中观察到，更高的重组率会增加每条染色体臂在减数分裂过程中至少进行一次交换的可能性[135]。交换被认为是真核生物中同源染色体配对正常减数分裂分离所必需的。

从机制上来看，较短染色体臂上重组率的增加可以这样解释：一旦发生第一次重组事件，每条染色体臂上的交换都会被干扰，进而阻止附近区域重组事件的发生。对干扰作用被破坏的酵母突变体的研究表明，在酵母各染色体臂之间分配有限的交换次数时，干扰作用发挥了关键作用[136]。另一种可能性是检查点机制扫描基因组并促成每个染色体臂上至少存在一个交换。

染色体上以及性别之间的重组率的差异可能反映了减数分裂导致的双链 DNA 断裂（DSB）起始过程的差异，该过程是重组的起始阶段。酵母中的 DSB 与开放的染色质而不是特定的 DNA 序列基序有关[137,138]。如果得到了基因组草图序列，就能用相似的方法探索人类重组率的差异是否反映了减数分裂过程中染色体可接近性的系统性差异。

人类基因组重复序列含量

早期分子生物学领域令人困惑的一个观察结果是基因组大小与生物复杂性的相关性不强。例如，智人的基因组大小是酿酒酵母的 200 倍，但也是变形虫的两百分之一[139,140]。这个谜团（C 值悖论）在很大程度上得到了解决，是由于认识到基因组可以包含大量的重复序列，远远超过了蛋白质编码基因的序列（参考文献 140 和 141 中有综述）。

人类的编码序列在基因组占比小于 5%（见下文），而重复序列至少占 50%，甚至更多。概括地说，重复序列分为五类：(1) 转座子起源的重复，通常称为散在重复；(2) 基因（包括蛋白质编码基因和小的结构 RNA）的退化失活（或部分失活）拷贝，通常称为加工过的假基因；(3) 简单重复序列，由相对比较短的核苷酸如 $(A)_n$、$(CA)_n$ 或 $(CGG)_n$ 正向重复组成；(4) 片段重复，由 10 ~ 300 kb 的大片段组成，这些片段是从基因组的一个区域复制到另一区域形成；(5) 串联重复序列，如着丝粒、端粒、近端着丝粒染色体的短臂和核糖体基因簇等区域。（这些区域在基因组草图序列中代表性不足，本文不讨论。）

重复序列通常被描述为“垃圾”，并且被认为是没有意义的。但实际上它们代表了一个关于生物学过程的特殊信息库。这些重复构成了丰富的古生物学记录，包含

constitute a rich palaeontological record, holding crucial clues about evolutionary events and forces. As passive markers, they provide assays for studying processes of mutation and selection. It is possible to recognize cohorts of repeats "born" at the same time and to follow their fates in different regions of the genome or in different species. As active agents, repeats have reshaped the genome by causing ectopic rearrangements, creating entirely new genes, modifying and reshuffling existing genes, and modulating overall GC content. They also shed light on chromosome structure and dynamics, and provide tools for medical genetic and population genetic studies.

The human is the first repeat-rich genome to be sequenced, and so we investigated what information could be gleaned from this majority component of the human genome. Although some of the general observations about repeats were suggested by previous studies, the draft genome sequence provides the first comprehensive view, allowing some questions to be resolved and new mysteries to emerge.

Transposon-derived repeats

Most human repeat sequence is derived from transposable elements[142,143]. We can currently recognize about 45% of the genome as belonging to this class. Much of the remaining "unique" DNA must also be derived from ancient transposable element copies that have diverged too far to be recognized as such.

In mammals, almost all transposable elements fall into one of four types, of which three transpose through RNA intermediates and one transposes directly as DNA. These are long interspersed elements (LINEs, an ancient and common innovation in eukaryotes), short interspersed elements (SINEs, which are freeloaders on LINEs), LTR retrotransposons and DNA transposons.

Transposable elements employ different strategies to ensure their evolutionary survival. LINEs and SINEs rely almost exclusively on vertical transmission within the host genome[154] (but see refs 148, 155). DNA transposons are more promiscuous, requiring relatively frequent horizontal transfer. LTR retroposons use both strategies, with some being long-term active residents of the human genome (such as members of the ERVL family) and others having only short residence times.

Currently recognized SINEs, LINEs, LTR retroposons and DNA transposon copies comprise 13%, 20%, 8% and 3% of the sequence, respectively. We expect these densities to grow as more repeat families are recognized, among which will be lower copy number LTR elements and DNA transposons, and possibly high copy number ancient (highly diverged) repeats.

Age distribution. The age distribution of the repeats in the human genome provides a rich "fossil record" stretching over several hundred million years. The ancestry and

关于进化事件和进化驱动力的重要线索。作为一种被动标记，它们提供了用于研究突变过程和选择过程的材料。我们可以识别在同一时间“出生”的一群重复序列，追踪它们在基因组的不同区域或不同物种中的命运。作为活跃分子，重复序列通过引起异位重排、创建全新基因、修饰和重组现有基因及调整总体 GC 含量等重塑了基因组。它们还揭示了染色体结构和动力学，并为医学遗传学和群体遗传学研究提供了工具。

人类是第一个经过测序的富含重复序列的基因组，因此我们研究了从人类基因组大多数组分中可以收集到哪些信息。虽然以前的研究也提出了一些关于重复的普遍观察结果，但基因组草图序列第一次提供了序列的全貌，在解决某些问题的同时产生了新的谜团。

转座子起源的重复

大多数人类重复序列来源于转座元件[142,143]。目前，我们可以认定约 45%的基因组属于这一类。剩下的“独特” DNA 中，许多应该也来自远古转座元件的拷贝，这些拷贝已经因分化得太远而无法识别。

在哺乳动物中，几乎所有的转座元件都可归入四个大类，其中三类是由 RNA 介导，一类是 DNA 直接转座完成。这四类分别是：长散在重复序列(LINE，在真核生物中古老而常见的创新序列)，短散在重复序列(SINE，LINE 中的自由插入序列)、LTR 逆转座子和 DNA 转座子。

转座元件采用不同的策略来确保它们的进化生存。LINE 和 SINE 几乎完全依赖宿主基因组内的垂直传递[154](请参见参考文献 148 和 155)。DNA 转座子更加杂乱，需要相对频繁的水平转移。LTR 逆转座子则使用上述两种策略，一些在人类基因组长期活跃驻扎(例如 ERVL 家族的成员)，其他则是短暂停留。

目前公认的 SINE、LINE、LTR 逆转座子和 DNA 转座子的拷贝数分别占序列的 13%、20%、8%和 3%。我们预计上述数字会随着更多重复家族的识别而增加，其中包括较低拷贝数的 LTR 元件和 DNA 转座子，可能还有高拷贝数的远古(高度分化)重复。

年代分布 人类基因组中重复序列的年龄分布提供了绵延数亿年的丰富的“化

approximate age of each fossil can be inferred by exploiting the fact that each copy is derived from, and therefore initially carried the sequence of, a then-active transposon and, being generally under no functional constraint, has accumulated mutations randomly and independently of other copies. We can infer the sequence of the ancestral active elements by clustering the modern derivatives into phylogenetic trees and building a consensus based on the multiple sequence alignment of a cluster of copies. Using available consensus sequences for known repeat subfamilies, we calculated the per cent divergence from the inferred ancestral active transposon for each of three million interspersed repeats in the draft genome sequence.

The percentage of sequence divergence can be converted into an approximate age in millions of years (Myr) on the basis of evolutionary information.

Several facts are apparent from analysis.

First, most interspersed repeats in the human genome predate the eutherian radiation. This is a testament to the extremely slow rate with which nonfunctional sequences are cleared from vertebrate genomes.

Second, LINE and SINE elements have extremely long lives. The monophyletic LINE1 and Alu lineages are at least 150 and 80 Myr old, respectively. In earlier times, the reigning transposons were LINE2 and MIR[148,158]. The SINE MIR was perfectly adapted for reverse transcription by LINE2, as it carried the same 50-base sequence at its 3′ end. When LINE2 became extinct 80–100 Myr ago, it spelled the doom of MIR.

Third, there were two major peaks of DNA transposon activity. The first involved Charlie elements and occurred long before the eutherian radiation; the second involved Tigger elements and occurred after this radiation. Because DNA transposons can produce large-scale chromosome rearrangements[159-162], it is possible that widespread activity could be involved in speciation events.

Fourth, there is no evidence for DNA transposon activity in the past 50 Myr in the human genome. The youngest two DNA transposon families that we can identify in the draft genome sequence (MER75 and MER85) show 6–7% divergence from their respective consensus sequences representing the ancestral element, indicating that they were active before the divergence of humans and new world monkeys. Moreover, these elements were relatively unsuccessful, together contributing just 125 kb to the draft genome sequence.

Finally, LTR retroposons appear to be teetering on the brink of extinction, if they have not already succumbed. For example, the most prolific elements (ERVL and MaLRs) flourished for more than 100 Myr but appear to have died out about 40 Myr ago[163,164].

More generally, the overall activity of all transposons has declined markedly over the past

石记录”。每个“化石”的祖先和大致年龄可以通过探究以下事实来推断：每个拷贝来源于一个当时活跃的转座子并因此在最初携带其序列，同时在没有功能选择压力的情况下独立于其他拷贝之外随机累积突变。我们可以通过将现代衍生物聚类为系统发生树并基于拷贝簇的多重序列比对建立共识，来推断祖先活性元件的序列。根据已知的重复亚家族的共有序列，我们计算了基因组草图序列中三百万个散在重复序列各自与推测的祖先活跃转座子的百分比差异。

依据进化的信息，序列差异的百分比可近似转化为以百万年计（Myr）的时间。

分析后有几点明显的特征：

首先，人类基因组中大多数散在重复序列早于真兽亚纲的扩张，这证明了非功能性序列从脊椎动物基因组中被清除的速率非常缓慢。

其次，LINE 元件和 SINE 元件具有很长的寿命。单源 LINE1 和 Alu 谱系至少有 1.5 亿年和 8,000 万年的历史。早期，居于统治地位的转座子是 LINE2 和 MIR[148,158]。SINE MIR 的 3′ 端携带与 LINE2 相同的 50 个碱基序列，因而完全适用于 LINE2 的逆转录。当 LINE2 在 8,000 万 ~ 1 亿年前灭绝时，MIR 也难逃厄运。

第三，DNA 转座子的活性曾有两次峰值。第一次涉及 Charlie 元件，发生在真兽亚纲扩张之前很久；第二次涉及 Tigger 元件，发生在扩张后。因为 DNA 转座子可以产生大规模的染色体重排[159-162]，所以转座子的广泛活动可能涉及形成物种的事件。

第四，在人类基因组中，过去 5,000 万年里没有 DNA 转座子活性的证据。我们在基因组草图序列中鉴定出的最年轻的两个 DNA 转座子家族（MER75 和 MER85）显示出与各自的代表祖先元件的共有序列有 6% ~ 7%的差异，这表明它们在人类与新世界猴分化之前是有活性的。而且，这些元件没那么成功，合在一起仅占基因组草图序列 125 kb 的长度。

最后，即使 LTR 逆转座子还没有屈服，但似乎已在灭绝的边缘了。例如，最丰富的元件（ERVL 和 MaLRs）兴盛了 1 亿年，但它们已经在 4,000 万年前消失了[163,164]。

更普遍的是，在过去的 3,500 万 ~ 5,000 万年间，所有转座子的整体活性显著下

35–50 Myr, with the possible exception of LINE1. Indeed, apart from an exceptional burst of activity of Alus peaking around 40 Myr ago, there would appear to have been a fairly steady decline in activity in the hominid lineage since the mammalian radiation. The extent of the decline must be even greater than it appears because old repeats are gradually removed by random deletion and because old repeat families are harder to recognize and likely to be under-represented in the repeat databases.

What explains the decline in transposon activity in the lineage leading to humans? There is no similar decline in the mouse genome.

Comparison with other organisms. In contrast to their possible extinction in humans, LTR retroposons are alive and well in the mouse. These evolutionary findings are consistent with the empirical observations that new spontaneous mutations are 30 times more likely to be caused by LINE insertions in mouse than in human (~3% versus 0.1%)[170] and 60 times more likely to be caused by transposable elements in general. It is estimated that around 1 in 600 mutations in human are due to transpositions, whereas 10% of mutations in mouse are due to transpositions (mostly IAP insertions).

The contrast between human and mouse suggests that the explanation for the decline of transposon activity in humans may lie in some fundamental difference between hominids and rodents. Population structure and dynamics would seem to be likely suspects. Rodents tend to have large populations, whereas hominid populations tend to be small and may undergo frequent bottlenecks. Evolutionary forces affected by such factors include inbreeding and genetic drift, which might affect the persistence of active transposable elements[171]. Studies in additional mammalian lineages may shed light on the forces responsible for the differences in the activity of transposable elements[172]. (Editorial note: a detailed discussion of inter-species comparisons has been omitted.)

Distribution by GC content. We next focused on the correlation between the nature of the transposons in a region and its GC content. We calculated the density of each repeat type as a function of the GC content in 50-kb windows. As has been reported[142,173-176], LINE sequences occur at much higher density in AT-rich regions (roughly fourfold enriched), whereas SINEs (MIR, Alu) show the opposite trend (for Alu, up to fivefold lower in AT-rich DNA). LTR retroposons and DNA transposons show a more uniform distribution, dipping only in the most GC-rich regions.

The preference of LINEs for AT-rich DNA seems like a reasonable way for a genomic parasite to accommodate its host, by targeting gene-poor AT-rich DNA and thereby imposing a lower mutational burden.

The contrary behaviour of SINEs, however, is baffling. How do SINEs accumulate in GC-rich DNA, particularly if they depend on the LINE transposition machinery[178]?

降，也许只有 LINE1 是个例外。事实上，除了 4,000 万年前 Alu 的活性爆发并达到顶峰之外，自哺乳动物扩张以来，类人猿世系的转座子活性似乎已经相当稳定地下降了。由于古老的重复序列逐渐地随机删除，同时古老重复序列家族难以识别且可能在数据库中代表不足，实际下降幅度应该更大。

什么能够解释人类谱系中的转座子活性下降呢？在小鼠的基因组中并未观察到同样的现象。

与其他生物的比较 与在人类中可能灭绝的情况相反，LTR 逆转座子在小鼠中存活得很好。这些进化上的发现与经验观察值相吻合，在小鼠中，由 LINE 插入产生新的自发突变的可能性是人类中的 30 倍（约 3% 相对于 0.1%）[170]，由转座元件插入产生的可能性是人类的 60 倍。据估计，人类突变中约六百分之一由转座引起，在小鼠中由转座引起的突变占到 10%（大多数为 IAP 插入）。

人类与小鼠的对比表明，人类转座子活性的下降可能基于类人猿世系和啮齿类动物的某些根本差异。种群结构和种群动态似乎是个疑点。啮齿动物往往拥有大种群，而人类种群倾向于变小，并经常遇到瓶颈。进化动力受近亲繁殖和遗传漂变等因素的影响，可能会影响活跃转座元件的持续[171]。其他哺乳动物谱系的研究也许能揭示转座元件活性差异的原因[172]。（编者注：关于种群间比较的详细讨论被删减了。）

GC 含量的分布 下面我们将重点关注区域内转座子的性质和 GC 含量之间的关系。以 50 kb 为窗口，我们将每类重复序列的密度与 GC 含量关联起来。正如已报道的[142,173-176]，LINE 序列分布在 AT 富集区（大约 4 倍富集），而 SINE（MIR、Alu）呈现相反的趋势（Alu 在富含 AT 的 DNA 区域密度低至五分之一）。逆转录子 LTR 和 DNA 转座子呈现更均匀的分布，仅处于 GC 富集区。

LINE 序列对 AT 富集 DNA 的偏好似乎是适应基因组寄生生活的合理途径，AT 富集 DNA 通常基因分布较少，插入后引起的突变负担更小。

然而，SINE 的相反行为令人困惑。SINE 如何在 GC 富集 DNA 中积累，特别是如果它们依赖于 LINE 转座机制[178]？

We used the draft genome sequence to investigate this mystery by comparing the proclivities of young, adolescent, middle-aged and old Alus. Strikingly, recent Alus show a preference for AT-rich DNA resembling that of LINEs, whereas progressively older Alus show a progressively stronger bias towards GC-rich DNA. These results indicate that the GC bias must result from strong pressure: there is a 13-fold enrichment of Alus in GC-rich DNA within the last 30 Myr, and possibly more recently.

These observations indicate that there may be some force acting particularly on Alus. This could be a higher rate of random loss of Alus in AT-rich DNA, negative selection against Alus in AT-rich DNA or positive selection in favour of Alus in GC-rich DNA. The first two possibilities seem unlikely because AT-rich DNA is gene-poor and tolerates the accumulation of other transposable elements. The third seems more feasible, in that it involves selecting in favour of the minority of Alus in GC-rich regions rather than against the majority that lie in AT-rich regions. But positive selection for Alus in GC-rich regions would imply that they benefit the organism.

Our results may support the controversial idea that SINEs actually earn their keep in the genome. Clearly, much additional work will be needed to prove or disprove the hypothesis that SINEs are genomic symbionts.

Fast living on chromosome Y. The pattern of interspersed repeats can be used to shed light on the unusual evolutionary history of chromosome Y. Our analysis shows that the genetic material on chromosome Y is unusually young, probably owing to a high tolerance for gain of new material by insertion and loss of old material by deletion. Overall, chromosome Y seems to maintain a youthful appearance by rapid turnover.

Interspersed repeats on chromosome Y can also be used to estimate the relative mutation rates, α_m and α_f, in the male and female germlines. Chromosome Y always resides in males, whereas chromosome X resides in females twice as often as in males. The substitution rates, μ_Y and μ_X, on these two chromosomes should thus be in the ratio $\mu_Y:\mu_X = (\alpha_m):(\alpha_m + 2\alpha_f)/3$, provided that one considers equivalent neutral sequences. Page and colleagues[192] obtained an estimate of $\mu_Y:\mu_X = 1.36$, corresponding to $\alpha_m:\alpha_f = 1.7$.

Our estimate is in reasonable agreement with that of Page *et al.*, although it is based on much more total sequence (360 kb on Y, 1.6 Mb on X) and a much longer time period. Various theories have been proposed for the higher mutation rate in the male germline, including the greater number of cell divisions in the formation of sperm than eggs and different repair mechanisms in sperm and eggs. (Editorial note: a detailed discussion of transposon dynamics and simple sequence repeats has been omitted.)

Segmental duplications

A remarkable feature of the human genome is the segmental duplication of portions of

我们使用基因组草图序列，通过比较“青年”“青少年”“中年”和“老年”Alu 序列的倾向性来调查这个谜团。引人注目的是，较年轻的 Alu 展示了一种类似于 LINE 对 AT 富集 DNA 优先选择的倾向性，而逐渐变老的 Alu 对富含 GC 的 DNA 的偏好逐渐增强。这些结果表明，GC 偏好性应该是强大的压力造成的。Alu 序列在 GC 富集 DNA 中约 13 倍富集，这发生在距今 3,000 万年内或更近。

这些观察结果表明，可能会有一些力量在 Alu 上发挥独特的作用。这可能是富含 AT 的 DNA 随机丢失 Alu 的概率较高，AT 富集 DNA 对 Alu 的负选择或者 GC 富集 DNA 对 Alu 的正选择。前两种似乎可能性不大，因为 AT 富集 DNA 缺乏基因，并且容忍其他转座元件的积累。第三种看起来更可行，因为它意味着在 GC 富集区选择偏好少数 Alu，而不是在 AT 富集区排斥多数 Alu。但是 GC 富集区正向选择 Alu 意味着生物体将因此获益。

我们的研究结果也许支持了这样一个有争议的观点，那就是 SINE 在基因组中争得了一席之地。显然，若想证实或否定 SINE 是基因组共生体的假设，需要做更多的工作。

在 Y 染色体上快速生存 散在重复的模式可用于阐明 Y 染色体不寻常的进化历史。我们的分析表明，Y 染色体上的遗传物质异常年轻，可能对于是对插入新物质和删除旧物质具有较高耐受性。总的来说，Y 染色体似乎通过快速更新而保持年轻的外貌。

Y 染色体上的散布重复也可用于估算男性和女性生殖系中的相对突变率 α_m 和 α_f。染色体 Y 总是存在于雄性中，而染色体 X 存在于女性中的频率是男性的两倍。这两种染色体上的替代率 μ_Y 和 μ_X 应符合比值 $\mu_Y:\mu_X=(\alpha_m):(\alpha_m+2\alpha_f)/3$，前提是考虑相等的中性序列。佩奇及其同事[192]估算 $\mu_Y:\mu_X=1.36$，相应的 $\alpha_m:\alpha_f=1.7$。

尽管我们的估计基于更多的全序列（Y 染色体 360 kb，X 染色体 1.6 Mb）和更长的时间周期，但是与佩奇等人的估计相当吻合。对于男性生殖系中的较高突变率，目前已经提出了不同理论，包括精子形成中细胞分裂的数量多于卵子以及精子和卵子中修复机制不同。（编者注：关于转座子动力学和简单序列重复的详细讨论被删除了。）

片段重复

人类基因组的一个显著特征是基因组序列的部分片段重复[215-217]。这种片段重复

genomic sequence[215-217]. Such duplications involve the transfer of 1–200-kb blocks of genomic sequence to one or more locations in the genome. The locations of both donor and recipient regions of the genome are often not tandemly arranged, suggesting mechanisms other than unequal crossing-over for their origin. They are relatively recent, inasmuch as strong sequence identity is seen in both exons and introns (in contrast to regions that are considered to show evidence of ancient duplications, characterized by similarities only in coding regions). Indeed, many such duplications appear to have arisen in very recent evolutionary time, as judged by high sequence identity and by their absence in closely related species.

Segmental duplications can be divided into two categories. First, interchromosomal duplications are defined as segments that are duplicated among nonhomologous chromosomes.

The second category is intrachromosomal duplications, which occur within a particular chromosome or chromosomal arm.

Until now, the identification and characterization of segmental duplications have been based on anecdotal reports. The availability of the entire genomic sequence will make it possible to explore the nature of segmental duplications more systematically.

We performed a global genome-wide analysis to characterize the amount of segmental duplication in the genome.

The finished sequence consists of at least 3.3% segmental duplication. Interchromosomal duplication accounts for about 1.5% and intrachromosomal duplication for about 2%, with some overlap (0.2%) between these categories. We analysed the lengths and divergence of the segmental duplications. The duplications tend to be large (10–50 kb) and highly homologous, especially for the interchromosomal segments. The sequence divergence for the interchromosomal duplications appears to peak between 96.5% and 97.5%. This may indicate that interchromosomal duplications occurred in a punctuated manner. It will be intriguing to investigate whether such genomic upheaval has a role in speciation events.

We compared the entire human draft genome sequence (finished and unfinished) with itself to identify duplications with 90–98% sequence identity. The draft genome sequence contains at least 3.6% segmental duplication. The actual proportion will be significantly higher, because we excluded many true matches with more than 98% sequence identity (at least 1.1% of the finished sequence). Although exact measurement must await a finished sequence, the human genome seems likely to contain about 5% segmental duplication, with most of this sequence in large blocks (> 10 kb). Such a high proportion of large duplications clearly distinguishes the human genome from other sequenced genomes, such as the fly and worm.

涉及 1 ~ 200 kb 的序列转移至基因组一个或多个位置。基因组的供体和受体区域的位置通常不是串联排列的，这提示可能存在与其起源的不等交换不同的其他机制。这类重复是相对较晚出现的，因为不管是外显子还是内含子都能观察到较强的序列一致性(与此相反，那些被认为古老的重复序列仅在编码区域中具有相似性)。事实上，许多这样的重复似乎是在最近的进化年代中出现的，它们不仅序列一致性高，而且不会出现在关系很近的物种之间。

片段重复可以分为两类。第一类，染色体间的重复被定义为非同源染色体间的重复片段。

第二类是染色体内重复，发生在特定的染色体或染色体臂内。

到目前为止，对片段重复的确定和描述基本来自坊间的报道。全基因组序列的可用性使得人们可以更系统地探索片段重复的性质。

我们在全基因组范围内进行分析，识别片段重复的数量。

完成序列包含至少 3.3%的片段重复。染色体间重复约占 1.5%，染色体内重复约占 2%，两类之间有部分重叠(0.2%)。我们分析了片段重复的长度和差异性。重复的片段往往很大(10 ~ 50 kb)并且高度同源，尤其是染色体间重复。染色体间重复的序列差异的峰值似乎在 96.5%和 97.5%之间。这可能表明染色体间重复以间断的形式发生。调查此类基因组的巨变在物种形成中是否发挥作用将是非常有趣的。

我们将人类全基因组草图序列(完成和未完成)与其自身进行比较，用以鉴定具有 90% ~ 98%序列一致性的重复序列。基因组草图序列至少包含 3.6%的片段重复，我们排除了许多序列一致性在 98%以上(至少 1.1%的完成序列)的真实匹配，因此实际比例将更高。精确值必须等到基因组序列完成后才可测量，但人类基因组很可能含有约 5%的片段重复，其中大部分重复序列是大片段的(> 10 kb)。人类基因组内大段重复的比例如此之高是其与其他已测序基因组(如果蝇和线虫)的明显区别。

Gene Content of the Human Genome

Genes (or at least their coding regions) comprise only a tiny fraction of human DNA, but they represent the major biological function of the genome and the main focus of interest by biologists. They are also the most challenging feature to identify in the human genome sequence.

The ultimate goal is to compile a complete list of all human genes and their encoded proteins, to serve as a "periodic table" for biomedical research[243]. But this is a difficult task. In organisms with small genomes, it is straightforward to identify most genes by the presence of long open reading frames (ORFs). In contrast, human genes tend to have small exons (encoding an average of only 50 codons) separated by long introns (some exceeding 10 kb). This creates a signal-to-noise problem, with the result that computer programs for direct gene prediction have only limited accuracy. Instead, computational prediction of human genes must rely largely on the availability of cDNA sequences or on sequence conservation with genes and proteins from other organisms. This approach is adequate for strongly conserved genes (such as histones or ubiquitin), but may be less sensitive to rapidly evolving genes (including many crucial to speciation, sex determination and fertilization).

Noncoding RNAs

Although biologists often speak of a tight coupling between "genes and their encoded protein products", it is important to remember that thousands of human genes produce noncoding RNAs (ncRNAs) as their ultimate product[244]. There are several major classes of ncRNA. (1) Transfer RNAs (tRNAs) are the adapters that translate the triplet nucleic acid code of RNA into the amino-acid sequence of proteins; (2) ribosomal RNAs (rRNAs) are also central to the translational machinery, and recent X-ray crystallography results strongly indicate that peptide bond formation is catalysed by rRNA, not protein[245,246]; (3) small nucleolar RNAs (snoRNAs) are required for rRNA processing and base modification in the nucleolus[247,248]; and (4) small nuclear RNAs (snRNAs) are critical components of spliceosomes, the large ribonucleoprotein (RNP) complexes that splice introns out of pre-mRNAs in the nucleus.

We can identify genomic sequences that are homologous to known ncRNA genes, using BLASTN or, in some cases, more specialized methods.

It is sometimes difficult to tell whether such homologous genes are orthologues, paralogues or closely related pseudogenes (because inactivating mutations are much less obvious than for protein-coding genes). For tRNA, there is sufficiently detailed information about the cloverleaf secondary structure to allow true genes and pseudogenes to be distinguished with high sensitivity. For many other ncRNAs, there is much less structural information and so we employ an operational criterion of high sequence similarity (> 95% sequence identity

人类基因组的基因含量

基因（至少是它们的编码区）只占人类 DNA 的很小一部分，但体现了基因组的主要生物学功能，是生物学家研究的焦点。它们也是人类基因组序列中识别起来最富挑战性的一项特征。

我们的终极目标是编制一张用于生物医学研究的“元素周期表”，罗列所有人类基因及其编码的蛋白质[243]。但是这是一项很困难的工作。对于基因组较小的物种来说，通过长的开放阅读框（ORF）可以直接确定大部分基因。与此相反，人类基因倾向于含有被大的内含子（有些超过 10 kb）分隔开的小的外显子（仅平均编码 50 个密码子）。这就产生了信噪比问题，因此利用电脑程序直接预测基因的准确性很有限。因此，人类基因的电脑预测要极大地依赖于 cDNA 序列或来自其他物种的基因和蛋白质的保守序列。这个方法对高度保守的基因（如组蛋白或泛素）足以胜任，但是对于快速进化的基因（包括很多对物种形成、性别决定和受精非常重要的基因）就没那么敏感了。

非编码 RNA

虽然生物学家经常说“基因与其编码的蛋白质产物”连接紧密，但请务必记得还有数以千计的人类基因产生非编码 RNA（ncRNA）作为其最终产物[244]。ncRNA 主要分以下几大类：(1) 转运 RNA（tRNA），负责将 RNA 的三联体密码翻译为蛋白质的氨基酸序列；(2) 核糖体 RNA（rRNA），是转录机器的核心，最近的 X 射线晶体学结果显示肽键的形成是由 rRNA 而非蛋白质催化[245,246]；(3) 小核仁 RNA（snoRNA），参与核仁中的 rRNA 编辑和碱基修饰[247,248]；(4) 小核 RNA（snRNA），剪切体的关键组分，剪切体是大型核糖核蛋白（RNP）复合物，负责在细胞核中将前 mRNA 的内含子切除。

我们可以利用 BLASTN 或其他更专业的方法识别与已知的 ncRNA 基因同源的基因组序列。

有时，很难讲清这些同源关系是直系同源、旁系同源还是密切相关的假基因（因为与编码蛋白质的基因相比，ncRNA 基因的失活突变更不明显）。对 tRNA 来说，三叶草的二级结构提供了足够多的信息，用来判定基因和假基因的灵敏度很高。但是对于其他 ncRNA 来说，结构信息要少得多，因此我们引入了一条序列高度相似（大于 95% 的一致性和大于 95% 的全长）的操作标准来区别基因和假基因。这些工

and > 95% full length) to distinguish true genes from pseudogenes. These assignments will eventually need to be reconciled with experimental data.

Transfer RNA genes. In the draft genome sequence, we find only 497 human tRNA genes. This appears to include most of the known human tRNA species. The draft genome sequence contains 37 of 38 human tRNA species listed in a tRNA database[253], allowing for up to one mismatch.

The results indicate that the human has fewer tRNA genes than the worm, but more than the fly. This may seem surprising, but tRNA gene number in metazoans is thought to be related not to organismal complexity, but more to idiosyncrasies of the demand for tRNA abundance in certain tissues or stages of embryonic development. For example, the frog *Xenopus laevis*, which must load each oocyte with a remarkable 40 ng of tRNA, has thousands of tRNA genes[254].

The tRNA genes are dispersed throughout the human genome. However, this dispersal is nonrandom. tRNA genes have sometimes been seen in clusters at small scales[262,263] but we can now see striking clustering on a genome-wide scale. More than 25% of the tRNA genes (140) are found in a region of only about 4 Mb on chromosome 6. This small region, only about 0.1% of the genome, contains an almost sufficient set of tRNA genes all by itself. The 140 tRNA genes contain a representative for 36 of the 49 anticodons found in the complete set; and of the 21 isoacceptor types, only tRNAs to decode Asn, Cys, Glu and selenocysteine are missing. Many of these tRNA genes, meanwhile, are clustered elsewhere; 18 of the 30 Cys tRNAs are found in a 0.5-Mb stretch of chromosome 7 and many of the Asn and Glu tRNA genes are loosely clustered on chromosome 1. More than half of the tRNA genes (280 out of 497) reside on either chromosome 1 or chromosome 6. Chromosomes 3, 4, 8, 9, 10, 12, 18, 20, 21 and X appear to have fewer than 10 tRNA genes each; and chromosomes 22 and Y have none at all (each has a single pseudogene). (Editorial note: much of the discussion of RNA genes has been omitted.)

Our observations confirm the striking proliferation of ncRNA-derived pseudogenes. There are hundreds or thousands of sequences in the draft genome sequence related to some of the ncRNA genes. The most prolific pseudogene counts generally come from RNA genes transcribed by RNA polymerase III promoters, including U6, the hY RNAs and SRP-RNA. These ncRNA pseudogenes presumably arise through reverse transcription. The frequency of such events gives insight into how ncRNA genes can evolve into SINE retroposons, such as the tRNA-derived SINEs found in many vertebrates and the SRP-RNA-derived Alu elements found in humans.

Protein-coding genes

(Editorial note: this section has been significantly shortened.) Identifying the protein-coding genes in the human genome is one of the most important applications of the

作最终需要实验数据来协调。

转运 RNA 基因 在基因组草图序列中，我们仅找到了 497 个人类 tRNA 基因。这包含了目前已知的大多数 tRNA 种类。在 tRNA 数据库[253]列举的 38 种人类 tRNA 中，基因组草图中出现了 37 种，允许最多出现一次错配。

研究结果显示，人类的 tRNA 基因少于线虫，但多于果蝇。虽然出人意料，但在多细胞生物中，tRNA 基因的数量被认为与生物体的复杂程度无关，而是与某些组织的对 tRNA 需求或胚胎发育的阶段等特征有关。如非洲爪蟾，它的卵母细胞会携带多达 40 ng 的 tRNA，而它的 tRNA 基因也多达上千种[254]。

tRNA 基因分散在整个人类基因组中。但并非随机分布。曾经也发现过 tRNA 基因小规模聚集成簇[262,263]，但是我们现在看到，tRNA 基因在基因组范围内有非常显著的基因簇。超过 25% 的 tRNA（140 个）位于 6 号染色体的长约 4 Mb 的一个区域内。这个小区域仅约占人类基因组总长的 0.1%，但却包含了几乎一整套必需的 tRNA 基因。这 140 个 tRNA 基因包含 49 个反密码子中典型的 36 个；而 21 个同工型中只缺少天冬酰胺、半胱氨酸、谷氨酸和硒代半胱氨酸的解码 tRNA。这些 tRNA 基因大多成簇存在于其他位置，30 个半胺氨酸 tRNA 中的 18 个在 7 号染色体的 0.5 Mb 区域内，很多天冬酰胺和谷氨酸 tRNA 基因则松散地成簇出现在 1 号染色体上。超过一半的 tRNA 基因（497 个中的 280 个）位于 1 号或 6 号染色体上，第 3、4、8、9、10、12、18、20、21 号染色体和 X 染色体上各自分布不到 10 个 tRNA 基因，22 号和 Y 染色体则完全没有 tRNA 基因（各自含有一个假基因）。（编者注：关于 RNA 基因的大部分讨论被删减了。）

我们的观察证实了起源于 ncRNA 的假基因的显著增殖过程。基因组草图中，成千上万的序列与某些 ncRNA 基因有关。大量的假基因一般来自 RNA 聚合酶Ⅲ启动子转录的 RNA 基因，包括 U6、hY RNA 和 SRP-RNA。这些 ncRNA 假基因很可能起源于逆转录。这种事件发生的频率也让我们有机会了解，ncRNA 基因是怎样进化为 SINE 逆转座子的，就像 tRNA 起源的 SINE 存在于很多脊椎动物基因组中，SRP-RNA 起源的 Alu 元件存在于人类基因组中。

蛋白质编码基因

（编者注：这部分有大量的删减。）在人类基因组中识别出蛋白质编码基因，是

sequence data, but also one of the most difficult challenges.

Towards a complete index of human genes. We focused on creating an initial index of human genes and proteins. This index is quite incomplete, owing to the difficulty of gene identification in human DNA and the imperfect state of the draft genome sequence. Nonetheless, it is valuable for experimental studies and provides important insights into the nature of human genes and proteins.

Gene identification is difficult in human DNA. The signal-to-noise ratio is low: coding sequences comprise only a few per cent of the genome and an average of about 5% of each gene; internal exons are smaller than in worms; and genes appear to have more alternative splicing. The challenge is underscored by the work on human chromosomes 21 and 22. Even with the availability of finished sequence and intensive experimental work, the gene content remains uncertain, with upper and lower estimates differing by as much as 30%. The initial report of the finished sequence of chromosome 22 (ref. 94) identified 247 previously known genes, 298 predicted genes confirmed by sequence homology or ESTs and 325 *ab initio* predictions without additional support. Many of the confirmed predictions represented partial genes. In the past year, 440 additional exons (10%) have been added to existing gene annotations by the chromosome 22 annotation group, although the number of confirmed genes has increased by only 17 and some previously identified gene predictions have been merged[286].

Creating an initial gene index. We set out to create an initial integrated gene index (IGI) and an associated integrated protein index (IPI) for the human genome.

Evaluation of IGI/IPI. We used several approaches to evaluate the sensitivity, specificity and fragmentation of the IGI/IPI set.

Comparison with "new" known genes. One approach was to examine newly discovered genes arising from independent work that were not used in our gene prediction effort. We identified 31 such genes: 22 recent entries to RefSeq and 9 from the Sanger Centre's gene identification program on chromosome X. Of these, 28 were contained in the draft genome sequence and 19 were represented in the IGI/IPI. This suggests that the gene prediction process has a sensitivity of about 68% (19/28) for the detection of novel genes in the draft genome sequence and that the current IGI contains about 61% (19/31) of novel genes in the human genome.

Comparison with RIKEN mouse cDNAs. In a less direct but larger-scale approach, we compared the IGI gene set to a set of mouse cDNAs sequenced by the Genome Exploration Group of the RIKEN Genomic Sciences Center[309]. Around 81% of the genes in the RIKEN mouse set showed sequence similarity to the human genome sequence, whereas 69% showed sequence similarity to the IGI/IPI. This suggests a sensitivity of 85% (69/81).

序列数据最重要的应用之一，也是最艰巨的挑战之一。

生成人类基因的完整索引 我们聚焦于产生一个人类基因与蛋白质的初始索引。由于在人类 DNA 中识别基因的困难，以及基因组草图序列的未完成状态，该索引离完成之日尚早。尽管如此，它对实验性研究仍有价值，并可从中洞察人类基因与蛋白质的本质。

在人类 DNA 中识别基因是很困难的。有信噪比很低的问题：编码序列只构成了基因组很小的一部分，平均只占每个基因的 5%；基因内部的外显子比线虫的还小；基因可能存在更多的可变剪接。对人类 21 和 22 号染色体进行测序所面临的挑战被低估了。即使能够得到最终序列和深入的实验研究，基因含量仍旧是不确定的，估计值上限和下限的差异可达 30%。22 号染色体完成序列的初始报告（参考文献 94）确认了 247 个已知基因、298 个根据序列同源性或表达序列标签 EST 预测的基因以及 325 个没有其他证据支持的从头预测的基因。许多预测基因不完整。在过去的一年里，在 22 号染色体的注释数据组中，尽管已确定基因的数量只增长了 17 个，一些之前识别出来的基因也有所重叠，我们在现有基因注释的基础上又发现了 440 个外显子（10%）[286]。

生成初始基因索引 我们着手对人类基因组生成一个初始的整合基因索引（IGI），和一个相关的整合蛋白质索引（IPI）。

IGI/IPI 的评估 我们用几种方法评估 IPI/IGI 数据组的灵敏性、特异性和片段化。

与“新的”已知基因的比较 一种方法是检查那些新发现的基因，它们来自我们的基因预测中尚未使用的独立工作。我们发现了 31 个这样的基因：有 22 个最近刚加入 RefSeq 数据库，有 9 个来自桑格中心的 X 染色体基因识别项目。在这些基因中，有 28 个包含在基因组草图序列中，有 19 个出现在 IGI/IPI 数据组中。这提示，基因预测过程中，从基因组草图序列中预测新基因的灵敏性约为 68%（19/28），而目前的 IGI 数据组包含了人类基因组中 61%（19/31）的新基因。

与 RIKEN 小鼠 cDNA 比较 通过一个较为间接而规模较大的方法，我们将 IGI 基因数据组与小鼠的 cDNA 数据组进行比较，该小鼠的 cDNA 序列是由日本理化研究所（RIKEN）基因组科学中心的基因组探索团队测序的[309]。在 RIKEN 小鼠数据组中，有 81% 的基因与人类基因组存在序列相似性，而有 69% 的基因与 IGI/IPI 数据组存在序列相似性。这提示了 85%（69/81）的灵敏度。

Chromosomal distribution. Finally, we examined the chromosomal distribution of the IGI gene set. The average density of gene predictions is 11.1 per Mb across the genome, with the extremes being chromosome 19 at 26.8 per Mb and chromosome Y at 6.4 per Mb. It is likely that a significant number of the predictions on chromosome Y are pseudogenes (this chromosome is known to be rich in pseudogenes) and thus that the density for chromosome Y is an overestimate.

Summary. We are clearly still some way from having a complete set of human genes. The current IGI contains significant numbers of partial genes, fragmented and fused genes, pseudogenes and spurious predictions, and it also lacks significant numbers of true genes. This reflects the current state of gene prediction methods in vertebrates even in finished sequence, as well as the additional challenges related to the current state of the draft genome sequence. Nonetheless, the gene predictions provide a valuable starting point for a wide range of biological studies and will be rapidly refined in the coming year.

The analysis above allows us to estimate the number of distinct genes in the IGI, as well as the number of genes in the human genome. The IGI set contains about 15,000 known genes and about 17,000 gene predictions. Assuming that the gene predictions are subject to a rate of overprediction (spurious predictions and pseudogenes) of 20% and a rate of fragmentation of 1.4, the IGI would be estimated to contain about 24,500 actual human genes. Assuming that the gene predictions contain about 60% of previously unknown human genes, the total number of genes in the human genome would be estimated to be about 31,000. This is consistent with most recent estimates based on sampling, which suggest a gene number of 30,000–35,000.

Comparative proteome analysis

Knowledge of the human proteome will provide unprecedented opportunities for studies of human gene function. Often clues will be provided by sequence similarity with proteins of known function in model organisms. Such initial observations must then be followed up by detailed studies to establish the actual function of these molecules in humans.

For example, 35 proteins are known to be involved in the vacuolar protein-sorting machinery in yeast. Human genes encoding homologues can be found in the draft human sequence for 34 of these yeast proteins, but precise relationships are not always clear. In nine cases there appears to be a single clear human orthologue (a gene that arose as a consequence of speciation); in 12 cases there are matches to a family of human paralogues (genes that arose owing to intra-genome duplication); and in 13 cases there are matches to specific protein domains[311-314]. Hundreds of similar stories emerge from the draft sequence, but each merits a detailed interpretation in context. To treat these subjects properly, there will be many following studies, the first of which appear in accompanying papers[315-323].

Genes shared with fly, worm and yeast. IPI.1 contains apparent homologues of 61%

染色体分布　最后，我们检验了 IGI 数据组的染色体分布。基因预测在全基因组范围内的平均密度是每 Mb 分布 11.1 个基因，极限是 19 号染色体上每 Mb 分布 26.8 个基因，Y 染色体上每 Mb 分布 6.4 个基因。可能 Y 染色体上大量预测出来的是假基因（已知这条染色体富含假基因），因此 Y 染色体的密度被高估了。

小结　我们显然还未得到完整的人类基因数据组。目前的 IGI 包含了大量的不完整基因、片段化和融合基因、假基因以及错误的预测；它也同样缺少相当数量的真基因。这反映了在脊椎动物完成序列中基因预测方法的当前状态，以及与基因组草图序列当前状态相关的更多挑战。尽管如此，基因预测还是为大范围的生物学研究提供了宝贵的起点，并将在未来快速优化。

上述分析使我们能够估计 IGI 中不同基因的数量以及人类基因组中的基因数量。IGI 数据组中包含约 15,000 个已知基因和约 17,000 个预测基因。假设赋予基因预测 20% 的高估率（错误预测和假基因）以及 1.4 倍的片段化率，则 IGI 数据组估计应含 24,500 个实际人类基因。假设基因预测包含 60% 未知人类基因，则人类基因组中的基因总数估算为约 31,000 个。这与最近基于采样的估计结果一致，该结果为 30,000 至 35,000 个。

比较蛋白质组学分析

人类蛋白质组学的知识将为研究人类基因功能提供前所未有的重大机遇。这些线索常常由模式生物中已知功能的蛋白质的序列相似性所提供。必须进行详细的研究跟进这些初始的观察，从而发现这些分子在人类中的实际功能。

例如，已知酵母的液泡蛋白分选机制涉及 35 个蛋白质。在人类基因组草图序列中，可以找到 34 个酵母此类蛋白质的人类基因编码的同源物，但它们之间精细的关系还不清楚。有 9 个可能是人类的直系同源物（由于物种形成产生的基因）；有 12 个可能与人类旁系同源物（基因组内的序列倍增产生的基因）家族匹配；有 13 个与特异性的蛋白质结构域匹配[311-314]。在草图序列中产生了几百个相似的例子，但每一个都值得在文中详细解析。为妥善处理这些研究对象，我们需要许多后续研究，最先开始的工作就出现在附随的文章中[315-323]。

与果蝇、线虫和酵母共有的基因　IPI.1 包含了果蝇蛋白质组中 61% 的显著同源

of the fly proteome, 43% of the worm proteome and 46% of the yeast proteome. We next considered the groups of proteins containing likely orthologues and paralogues (genes that arose from intragenome duplication) in human, fly, worm and yeast.

We identified 1,308 groups of proteins, each containing at least one predicted orthologue in each species and many containing additional paralogues. The 1,308 groups contained 3,129 human proteins, 1,445 fly proteins, 1,503 worm proteins and 1,441 yeast proteins. These 1,308 groups represent a conserved core of proteins that are mostly responsible for the basic "housekeeping" functions of the cell, including metabolism, DNA replication and repair, and translation.

Most proteins do not show simple 1-1-1 orthologous relationships across the three animals.

New architectures from old domains. Whereas there appears to be only modest invention at the level of new vertebrate protein domains, there appears to be substantial innovation in the creation of new vertebrate proteins. This innovation is evident at the level of domain architecture, defined as the linear arrangement of domains within a polypeptide. New architectures can be created by shuffling, adding or deleting domains, resulting in new proteins from old parts.

We quantified the number of distinct protein architectures found in yeast, worm, fly and human by using the SMART annotation resource[339]. The human proteome set contained 1.8 times as many protein architectures as worm or fly and 5.8 times as many as yeast. This difference is most prominent in the recent evolution of novel extracellular and transmembrane architectures in the human lineage. Human extracellular proteins show the greatest innovation: the human has 2.3 times as many extracellular architectures as fly and 2.0 times as many as worm.

Conclusion. Five lines of evidence point to an increase in the complexity of the proteome from the single-celled yeast to the multicellular invertebrates and to vertebrates such as the human. Specifically, the human contains greater numbers of genes, domain and protein families, paralogues, multidomain proteins with multiple functions, and domain architectures. According to these measures, the relatively greater complexity of the human proteome is a consequence not simply of its larger size, but also of large-scale protein innovation.

An important question is the extent to which the greater phenotypic complexity of vertebrates can be explained simply by two- or threefold increases in proteome complexity. The real explanation may lie in combinatorial amplification of these modest differences, by mechanisms that include alternative splicing, post-translational modification and cellular regulatory networks. The potential numbers of different proteins and protein–protein interactions are vast, and their actual numbers cannot readily be discerned from the genome sequence. Elucidating such system-level properties presents one of the great challenges for modern biology. (Editorial note: a discussion of genome segmental history and comparison of conserved segments between human and mouse genomes has been omitted.)

物、线虫蛋白质组中 43% 的显著同源物及酵母蛋白质组中 46% 的显著同源物。我们接下来考虑的蛋白质类群包括人类、果蝇、线虫和酵母中可能的直系同源物和旁系同源物（来自基因组内的基因加倍）。

我们识别出 1,308 个蛋白质类群，每个类群至少包含每个物种中一个预测的直系同源物，有许多类群包含更多的旁系同源物。1,308 个类群中包含了 3,129 个人类蛋白质、1,445 个果蝇蛋白质、1,503 个线虫蛋白质和 1,441 个酵母蛋白质。这 1,308 个类群代表了一组保守的核心蛋白质，它们大多数主要负责细胞中最基本的“管家”功能，包括代谢、DNA 复制和修复以及翻译。

在这三种动物中，大多数蛋白质未显示出 1-1-1 的直系同源关系。

来自旧结构域的新型组织结构 尽管看起来在脊椎动物新的蛋白质结构域水平只有少量发现，但在产生脊椎动物新的蛋白质方面有实质性新发现。这一发现在结构域组织结构水平十分显著，即在多肽中结构域的线性排列。新的组织结构可通过变换结构域位置、添加或删除结构域来实现，使旧的组件中产生新的蛋白质。

通过 SMART 注释资源[339]，我们对酵母、线虫、果蝇和人类中找到的不同蛋白质组织结构进行计数。人类蛋白质组包含的蛋白质组织结构数量是线虫或果蝇的 1.8 倍，是酵母的 5.8 倍。在人类谱系新的细胞外和跨膜组织结构的近期演化中，这些差异最为显著。人类细胞外蛋白质中的新发现是最多的：人类细胞外组织结构数量是果蝇的 2.3 倍、线虫的 2.0 倍。

结论 从单细胞的酵母到多细胞的无脊椎动物再到脊椎动物，例如人类，蛋白质组的复杂性逐渐升高，有 5 条证据均指向这一点。具体地说，人类包含更多的基因、结构域和蛋白质家族、旁系同源物、具有多功能的多结构域蛋白质以及结构域组织结构。通过这些衡量可以发现，人类蛋白质组相对高的复杂性不仅仅是由于它的基因组有更大的尺寸，也是由于大规模的蛋白质新组合。

一个重要的问题是，脊椎动物表型的复杂程度可简单地用蛋白质组复杂性的两倍或者三倍的增长来解释。真正的解释可能存在于这些基本差异的组合扩增，可通过包括可变剪接、翻译后修饰和细胞调控网络在内的机制实现。不同的蛋白质和蛋白质–蛋白质相互作用的潜在数量是巨大的，也无法从基因组序列中可靠地辨明实际数量。对这些系统水平的性质的阐释，代表了现代生物学面临的巨大挑战之一。（编者注：对基因组片段历史以及人类和小鼠基因组保守片段的比较被删除了。）

Applications to Medicine and Biology

In most research papers, the authors can only speculate about future applications of the work. Because the genome sequence has been released on a daily basis over the past four years, however, we can already cite many direct applications. We focus on a handful of applications chosen primarily from medical research.

Disease genes

A key application of human genome research has been the ability to find disease genes of unknown biochemical function by positional cloning[388]. This method involves mapping the chromosomal region containing the gene by linkage analysis in affected families and then scouring the region to find the gene itself. Positional cloning is powerful, but it has also been extremely tedious. When the approach was first proposed in the early 1980s[9], a researcher wishing to perform positional cloning had to generate genetic markers to trace inheritance; perform chromosomal walking to obtain genomic DNA covering the region; and analyse a region of around 1 Mb by either direct sequencing or indirect gene identification methods. The first two barriers were eliminated with the development in the mid-1990s of comprehensive genetic and physical maps of the human chromosomes, under the auspices of the Human Genome Project. The remaining barrier, however, has continued to be formidable.

All that is changing with the availability of the human draft genome sequence. The human genomic sequence in public databases allows rapid identification *in silico* of candidate genes, followed by mutation screening of relevant candidates, aided by information on gene structure. For a mendelian disorder, a gene search can now often be carried out in a matter of months with only a modestly sized team.

At least 30 disease genes[55,389-422] have been positionally cloned in research efforts that depended directly on the publicly available genome sequence. As most of the human sequence has only arrived in the past twelve months, it is likely that many similar discoveries are not yet published. In addition, there are many cases in which the genome sequence played a supporting role, such as providing candidate microsatellite markers for finer genetic linkage analysis.

The genome sequence has also helped to reveal the mechanisms leading to some common chromosomal deletion syndromes. In several instances, recurrent deletions have been found to result from homologous recombination and unequal crossing over between large, nearly identical intrachromosomal duplications. Examples include the DiGeorge/velocardiofacial syndrome region on chromosome 22 (ref. 238) and the Williams-Beuren syndrome recurrent deletion on chromosome 7 (ref. 239).

在医学与生物学上的应用

对基因组序列在未来的应用，大多数研究性论文的作者仅能做出预测。然而，由于基因组序列的数据在过去的四年中每天都在增加，现在我们已经可以引用很多直接的应用实例了。我们在这里精选了几个医学上的研究应用。

疾病基因

目前，人类基因组研究一项最关键的应用是通过定位克隆来找到生化功能未知的致病基因[388]。这种方法涉及通过家系的连锁分析将包括该基因的染色体进行定位，再从该区域搜寻基因。定位克隆是有效的，但也极度冗余。当此方法在 20 世纪 80 年代早期被首次提出时[9]，研究者要做定位克隆首先要生成一个遗传标记跟踪该基因；然后通过染色体步移获得覆盖整个区域的基因组 DNA；再通过直接测序或间接鉴定基因的方法来分析 1 Mb 左右的区域。在 20 世纪 90 年代中期，在人类基因组计划的助力下，人类染色体遗传图谱和物理图谱的广泛绘制使前两步障碍已经扫除。即便如此，逾越剩下的障碍仍然非常困难。

当人类基因组草图完成后，一切都为之改观。我们可以利用储存在公共数据库中的人类基因组序列，通过生物信息学方法对目标基因进行快速鉴定，随后还可根据基因结构信息对目标基因的突变进行筛选。对某种孟德尔遗传疾病来说，一个中等规模的团队现在大约只需数月就能完成寻找特定基因的工作。

直接基于已发表的基因组序列，至少 30 个致病基因[55,389-422]已经完成定位克隆。由于大多数的人类基因组序列是在过去 12 个月内进入数据库的，可能还有很多此类研究尚未发表。此外，基因组序列还在很多研究中起到辅助作用，比如为精细的遗传连锁分析提供可用的微卫星标记。

基因组序列还帮助揭示了某些常见染色体缺失综合征的致病机制。在一些疾病中，我们发现频发性缺失是由大段的、几乎完全相同的染色体内部重复序列的同源重组和不等交换导致的。比如 22 号染色体上区域缺失引起的迪格奥尔格/腭心面综合征（参考文献 238）和 7 号染色体上频发性缺失引起的威廉姆斯综合征（参考文献 239）。

Drug targets

Over the past century, the pharmaceutical industry has largely depended upon a limited set of drug targets to develop new therapies. A recent compendium[426,427] lists 483 drug targets as accounting for virtually all drugs on the market. Knowing the complete set of human genes and proteins will greatly expand the search for suitable drug targets. Although only a minority of human genes may be drug targets, it has been predicted that the number will exceed several thousand, and this prospect has led to a massive expansion of genomic research in pharmaceutical research and development. (Editorial note: the original text includes discussion of some specific applications of this new genomic information to drug development.)

Basic biology

Response to certain bitter tastes. Recently, investigators mapped this trait in both humans and mice and then searched the relevant region of the human draft genome sequence for G-protein coupled receptors. These studies led, in quick succession, to the discovery of a new family of such proteins, the demonstration that they are expressed almost exclusively in taste buds, and the experimental confirmation that the receptors in cultured cells respond to specific bitter substances[433-435].

The Next Steps

Considerable progress has been made in human sequencing, but much remains to be done to produce a finished sequence. Even more work will be required to extract the full information contained in the sequence. Many of the key next steps are already underway.

Developing the IGI and IPI

A high priority will be to refine the IGI and IPI to the point where they accurately reflect every gene and every alternatively spliced form. Several steps are needed to reach this ambitious goal.

Finishing the human sequence will assist in this effort, but the experiences gained on chromosomes 21 and 22 show that sequence alone is not enough to allow complete gene identification. One powerful approach is cross-species sequence comparison with related organisms at suitable evolutionary distances. The sequence coverage from the pufferfish *T. nigroviridis* has already proven valuable in identifying potential exons[292]; this work is expected to continue from its current state of onefold coverage to reach at least fivefold coverage later this year. The genome sequence of the laboratory mouse will provide a

药物靶点

在过去的一个世纪里，制药产业极大地依赖有限的药物靶点来寻找新的治疗方法。最近发布的一项纲要总结了目前市场上所有药物的 483 个靶点[426,427]。而对人类的所有基因和蛋白质的揭示将极大地扩展对适当药物靶点的寻找。尽管目前只有少数人类基因可作为药物靶点，但可预见的是，这个数量将超过几千种，而此预期已经带来了药物研发领域中基因组研究的大发展。（编者注：原文包括新的基因信息在药物研发中的特定应用的讨论。）

基础生物学

对某些苦味的反应　最近，研究者们将这一特征定位在人类和小鼠的基因组上，并在人类基因组草图序列的相关区域中寻找 G 蛋白偶联受体。这些连贯的研究导致了这类蛋白质新家族的发现，证明了这些基因只在味蕾中表达，并实验验证了在体外培养的细胞中这些受体对苦味底物有信号反应[433-435]。

后　续

尽管人类基因组序列已经取得了相当的进展，但要获得完整的最终序列，还有很多工作要做。而挖掘序列中包含的所有信息需要更多的工作。后续计划中的很多关键步骤已在进行中。

发展 IGI 和 IPI

高度优先的任务是将 IGI 和 IPI 精细到点，以精确地反映每一个基因和每一个可变剪接形式。这个宏伟的目标还需要几个步骤。

人类基因组序列的完成对这项工作有很大的帮助，但是从 21 号和 22 号染色体上获取的经验表明，仅有序列信息还不足以完成全部基因的识别。与演化距离适当的物种进行跨物种序列比对是一个有力的工具。河豚鱼基因组序列的覆盖率对识别潜在的外显子很有价值[292]，这项工作有望在今年从一倍覆盖率达到至少五倍覆盖率。实验室小鼠的基因组序列将为外显子识别提供相当有力的工具，序列相似性有

particularly powerful tool for exon identification, as sequence similarity is expected to identify 95–97% of the exons, as well as a significant number of regulatory domains[436-438].

Another important step is to obtain a comprehensive collection of full-length human cDNAs, both as sequences and as actual clones. The Mammalian Gene Collection project has been underway for a year[18] and expects to produce 10,000–15,000 human full-length cDNAs over the coming year, which will be available without restrictions on use. The Genome Exploration Group of the RIKEN Genomic Sciences Center is similarly developing a collection of cDNA clones from mouse[309], which is a valuable complement because of the availability of tissues from all developmental time points.

Large-scale identification of regulatory regions

The one-dimensional script of the human genome, shared by essentially all cells in all tissues, contains sufficient information to provide for differentiation of hundreds of different cell types, and the ability to respond to a vast array of internal and external influences. Much of this plasticity results from the carefully orchestrated symphony of transcriptional regulation. Although much has been learned about the *cis*-acting regulatory motifs of some specific genes, the regulatory signals for most genes remain uncharacterized. It will also be of considerable interest to study epigenetic modifications such as cytosine methylation on a genome-wide scale, and to determine their biological consequences[446,447]. Towards this end, a pilot Human Epigenome Project has been launched[448,449].

Sequencing of additional large genomes

More generally, comparative genomics allows biologists to peruse evolution's laboratory notebook—to identify conserved functional features and recognize new innovations in specific lineages. Determination of the genome sequence of many organisms is very desirable. Already, projects are underway to sequence the genomes of the mouse, rat, zebrafish and the pufferfishes *T. nigroviridis* and *Takifugu rubripes*. Plans are also under consideration for sequencing additional primates and other organisms that will help define key developments along the vertebrate and nonvertebrate lineages.

To realize the full promise of comparative genomics, however, it needs to become simple and inexpensive to sequence the genome of any organism. Sequencing costs have dropped 100-fold over the last 10 years, corresponding to a roughly twofold decrease every 18 months. This rate is similar to "Moore's law" concerning improvements in semiconductor manufacture.

望识别 95% ~ 97% 的外显子以及大量的调控结构域[436-438]。

另一个重要的工作是获得人类基因组的全长 cDNA 库，既有序列信息，又有实际使用的克隆。哺乳动物基因库计划已经执行了一年[18]，预计将在明年产生 10,000 ~ 15,000 个人类全长 cDNA，并向公众开放使用。同样，日本理化研究所基因组科学中心的基因组探索团队也用小鼠建立了 cDNA 克隆库[309]，该库是极有价值的补充，因为它含有所有发育时间点的组织。

调控区域的大规模识别

人类基因组的一维文本信息在所有组织中为所有细胞共享，为几百种不同细胞类型的分化提供充足的信息，并对大量内部和外部的影响做出反应。这种可塑性多数来自精细安排的转录调控。尽管已经对一些特定基因的顺式作用调控模体做了很多研究，大多数基因的调控信号仍然未知。人们对研究表观遗传学修饰，例如基因组范围内的胞嘧啶甲基化，以及确定它们的生物学效果也有很大的兴趣[446,447]。一个领航性的人类表观基因组计划已经为实现此目标而发起[448,449]。

更多的大基因组测序

通常来说，比较基因组学使得生物学家能够利用演化的实验室数据识别保守的功能特征以及特定谱系中的新序列。我们需要确定大量物种的基因组序列。已经有一些项目在测定小鼠、大鼠、斑马鱼、河豚鱼和红旗东方鲀的基因组。人们也在酝酿测定更多灵长类动物和其他生物的基因组，这将帮助解释脊椎动物与非脊椎动物的关键发育过程。

要建立完善的比较基因组学，就需要简单而廉价地测定生物体的基因组。在过去的 10 年中，测序费用已经下降为最初的 100 分之一，大约每 18 个月下降一半。这一速率与半导体制造业发展的“摩尔定律”相符。

Completing the catalogue of human variation

The human draft genome sequence has already allowed the identification of more than 1.4 million SNPs, comprising a substantial proportion of all common human variation. This program should be extended to obtain a nearly complete catalogue of common variants and to identify the common ancestral haplotypes present in the population. In principle, these genetic tools should make it possible to perform association studies and linkage disequilibrium studies[376] to identify the genes that confer even relatively modest risk for common diseases. Launching such an intense era of human molecular epidemiology will also require major advances in the cost efficiency of genotyping technology, in the collection of carefully phenotyped patient cohorts and in statistical methods for relating large-scale SNP data to disease phenotype.

From sequence to function

The scientific program outlined above focuses on how the genome sequence can be mined for biological information. In addition, the sequence will serve as a foundation for a broad range of functional genomic tools to help biologists to probe function in a more systematic manner. These will need to include improved techniques and databases for the global analysis of: RNA and protein expression, protein localization, protein–protein interactions and chemical inhibition of pathways.

Concluding Thoughts

The Human Genome Project is but the latest increment in a remarkable scientific program whose origins stretch back a hundred years to the rediscovery of Mendel's laws and whose end is nowhere in sight. In a sense, it provides a capstone for efforts in the past century to discover genetic information and a foundation for efforts in the coming century to understand it.

We find it humbling to gaze upon the human sequence now coming into focus. In principle, the string of genetic bits holds long-sought secrets of human development, physiology and medicine. In practice, our ability to transform such information into understanding remains woefully inadequate.

The scientific work will have profound long-term consequences for medicine, leading to the elucidation of the underlying molecular mechanisms of disease and thereby facilitating the design in many cases of rational diagnostics and therapeutics targeted at those mechanisms. But the science is only part of the challenge. We must also involve society at large in the work ahead. We must set realistic expectations that the most important benefits will not be reaped overnight. Moreover, understanding and wisdom will be required to ensure that

完成人类变异目录

在人类基因组草图中已经识别出超过1,400,000个SNP，它们构成了人类基因组序列变异的大部分。这一项目应扩展以建立基本包括所有常见变异的目录，并识别出种群中常见的祖先单倍型。原则上，这些遗传学工具应当使我们能够开展相关性研究和连锁不平衡研究[376]以辨别出即便是仅有中度风险的常见疾病基因。要开启这样一个人类分子流行病学的时代，也需要基因分型技术的成本效率、精细分型的患者群体的收集以及将大规模SNP数据与疾病分型关联起来的统计学方法等方面的重大进展。

从序列到功能

科学项目勾勒出了以上研究热点，关注如何从基因组数据中挖掘生物学信息。而且，序列数据还是大规模功能基因组学工具的基础，能帮助生物学家以更加系统化的方式验证功能。这些都需要优化技术和数据库，用于如下全局性分析：RNA和蛋白质表达、蛋白质定位、蛋白质–蛋白质相互作用和通路的化学抑制。

结　语

人类基因组计划只是一个举世瞩目的科学研究计划中的最新增长点，它的起源可回溯至一百年前孟德尔遗传定律的重新发现，而它的终点则延伸至未知的未来。从某种意义上讲，它是过去一个世纪以来发现遗传学信息科学研究的顶峰，并为下一个世纪解析这些数据提供了基础。

只关注目前聚焦的人类基因组序列令人感到惭愧。理论上，这些遗传信息蕴含着关于人类发育、生理和医学中我们长期探求的秘密。而实际上，我们将这些信息转化理解的能力严重不足。

科学研究工作蕴含着深刻而长远的医学成果，能够阐释疾病的分子机制，也因此方便根据这些机制设计合理的诊断学和治疗学手段。但是，科学只是挑战的一部分。在开展工作之前，我们也必须关注到全社会。我们必须正确面对现实的期待，不能急于求成。而且，我们需要理解和智慧来确保这些好处能够公平合理地普惠大

these benefits are implemented broadly and equitably. To that end, serious attention must be paid to the many ethical, legal and social implications (ELSI) raised by the accelerated pace of genetic discovery. This paper has focused on the scientific achievements of the human genome sequencing efforts. This is not the place to engage in a lengthy discussion of the ELSI issues, which have also been a major research focus of the Human Genome Project, but these issues are of comparable importance and could appropriately fill a paper of equal length.

Finally, it is has not escaped our notice that the more we learn about the human genome, the more there is to explore.

"We shall not cease from exploration. And the end of all our exploring will be to arrive where we started, and know the place for the first time. "—T. S. Eliot[450]

(**409**, 860-921; 2001)

Received 7 December 2000; accepted 9 January 2001.

References:

1. Correns, C. Untersuchungen u ber die Xenien bei Zea mays. *Berichte der Deutsche Botanische Gesellschaft*, **17**, 410-418 (1899).

2. De Vries, H. Sur la loi de disjonction des hybrides. *Comptes Rendue Hebdemodaires, Acad. Sci. Paris* **130**, 845-847 (1900).

3. von Tschermack, E. Uber Künstliche Kreuzung bei Pisum sativum. *Berichte der Deutsche Botanische Gesellschaft*, **18**, 232-239 (1900).

9. Botstein, D., White, R. L., Skolnick, M. & Davis, R. W. Construction of a genetic linkage map in man using restriction fragment length polymorphisms. *Am. J. Hum. Genet.* **32**, 314-331 (1980).

11. Coulson, A., Sulston, J., Brenner, S. & Karn, J. Toward a physical map of the genome of the nematode *Caenorhabditis elegans*. *Proc. Natl Acad. Sci. USA* **83**, 7821-7825 (1986).

18. Strausberg, R. L., Feingold, E. A., Klausner, R. D. & Collins, F. S. The mammalian gene collection. *Science* **286**, 455-457 (1999).

21. Sinsheimer, R. L. The Santa Cruz Workshop-May 1985. *Genomics* **5**, 954-956 (1989).

22. Palca, J. Human genome-Department of Energy on the map. *Nature* **321**, 371 (1986).

23. National Research Council *Mapping and Sequencing the Human Genome* (National Academy Press, Washington DC, 1988).

24. Bishop, J. E. & Waldholz, M. *Genome* (Simon and Schuster, New York, 1990).

25. Kevles, D. J. & Hood, L. (eds) *The Code of Codes: Scientific and Social Issues in the Human Genome Project* (Harvard Univ. Press, Cambridge, Massachusetts, 1992).

26. Cook-Deegan, R. *The Gene Wars: Science, Politics, and the Human Genome* (W. W. Norton & Co., New York, London, 1994).

27. Donis-Keller, H. *et al.* A genetic linkage map of the human genome. *Cell* **51**, 319-337 (1987).

28. Gyapay, G. *et al.* The 1993-94 Genethon human genetic linkage map. *Nature Genet.* **7**, 246-339 (1994).

29. Hudson, T. J. *et al.* An STS-based map of the human genome. *Science* **270**, 1945-1954 (1995).

30. Dietrich, W. F. *et al.* A comprehensive genetic map of the mouse genome. *Nature* **380**, 149-152 (1996).

31. Nusbaum, C. *et al.* A YAC-based physical map of the mouse genome. *Nature Genet.* **22**, 388-393 (1999).

32. Oliver, S. G. *et al.* The complete DNA sequence of yeast chromosome III. *Nature* **357**, 38-46 (1992).

33. Wilson, R. *et al.* 2.2 Mb of contiguous nucleotide sequence from chromosome III of *C. elegans*. *Nature* **368**, 32-38 (1994).

34. Chen, E. Y. *et al.* The human growth hormone locus: nucleotide sequence, biology, and evolution. *Genomics* **4**, 479-497 (1989).

35. McCombie, W. R. *et al.* Expressed genes, Alu repeats and polymorphisms in cosmids sequenced from chromosome 4p16.3. *Nature Genet.* **1**, 348-353 (1992).

36. Martin-Gallardo, A. *et al.* Automated DNA sequencing and analysis of 106 kilobases from human chromosome 19q13.3. *Nature Genet.* **1**, 34-39 (1992).

37. Edwards, A. *et al.* Automated DNA sequencing of the human HPRT locus. *Genomics* **6**, 593-608 (1990).

42. Marshall, E. A second private genome project. *Science* **281**, 1121 (1998).

43. Marshall, E. NIH to produce a 'working draft' of the genome by 2001. *Science* **281**, 1774-1775 (1998).

44. Pennisi, E. Academic sequencers challenge Celera in a sprint to the finish. *Science* **283**, 1822-1823 (1999).

45. Bouck, J., Miller, W., Gorrell, J. H., Muzny, D. & Gibbs, R. A. Analysis of the quality and utility of random shotgun sequencing at low redundancies. *Genome Res.* **8**, 1074-1084 (1998).

46. Collins, F. S. *et al.* New goals for the U. S. Human Genome Project: 1998-2003. *Science* **282**, 682-689 (1998).

众。为此，我们应当严肃地关注加速发展的遗传学发现导致的伦理、法律和社会问题(ELSI)。本文聚焦于人类基因组测序产生的科学成就。这里并未就 ELSI 问题进行长篇大论，但这些问题已经成为人类基因组研究的主要热点之一，这些问题也同等重要，可以写一篇同样长的论文。

最终我们意识到，对人类基因组了解得越多，就有越多未知需要去探索。

“我们不会停止求索，万般求索终将抵达最初的起点，此时才把这个地方看个透彻。”——艾略特[450]

（王海纳 任奕 翻译；于军 审稿）

47. Sanger, F. & Coulson, A. R. A rapid method for determining sequences in DNA by primed synthesis with DNA polymerase. *J. Mol. Biol.* **94**, 441-448 (1975).

48. Maxam, A. M. & Gilbert, W. A new method for sequencing DNA. *Proc. Natl Acad. Sci. USA* **74**, 560- 564 (1977).

49. Anderson, S. Shotgun DNA sequencing using cloned DNase I-generated fragments. *Nucleic Acids Res.* **9**, 3015-3027 (1981).

50. Gardner, R. C. *et al.* The complete nucleotide sequence of an infectious clone of cauliflower mosaic virus by M13mp7 shotgun sequencing. *Nucleic Acids Res.* **9**, 2871-2888 (1981).

51. Deininger, P. L. Random subcloning of sonicated DNA: application to shotgun DNA sequence analysis. *Anal. Biochem.* **129**, 216-223 (1983).

52. Chissoe, S. L. *et al.* Sequence and analysis of the human ABL gene, the BCR gene, and regions involved in the Philadelphia chromosomal translocation. *Genomics* **27**, 67-82 (1995).

53. Rowen, L., Koop, B. F. & Hood, L. The complete 685-kilobase DNA sequence of the human beta T cell receptor locus. *Science* **272**, 1755-1762 (1996).

54. Koop, B. F. *et al.* Organization, structure, and function of 95 kb of DNA spanning the murine T-cell receptor C alpha/C delta region. *Genomics* **13**, 1209-1230 (1992).

55. Wooster, R. *et al.* Identification of the breast cancer susceptibility gene BRCA2. *Nature* **378**, 789-792 (1995).

60. Venter, J. C. *et al.* Shotgun sequencing of the human genome. *Science* **280**, 1540-1542 (1998).

61. Venter, J. C. *et al.* The sequence of the human genome. *Science* **291**, 1304-1351 (2001).

79. Bentley, D. R. Genomic sequence information should be released immediately and freely in the public domain. *Science* **274**, 533-534 (1996).

80. Guyer, M. Statement on the rapid release of genomic DNA sequence. *Genome Res.* **8**, 413 (1998).

81. Dietrich, W. *et al.* A genetic map of the mouse suitable for typing intraspecific crosses. *Genetics* **131**, 423-447 (1992).

82. Kim, U. J. *et al.* Construction and characterization of a human bacterial artificial chromosome library. *Genomics* **34**, 213-218 (1996).

83. Osoegawa, K. *et al.* Bacterial artificial chromosome libraries for mouse sequencing and functional analysis. *Genome Res.* **10**, 116-128 (2000).

84. Marra, M. A. *et al.* High throughput fingerprint analysis of large-insert clones. *Genome Res.* **7**, 1072- 1084 (1997).

85. Marra, M. *et al.* A map for sequence analysis of the *Arabidopsis thaliana* genome. *Nature Genet.* **22**, 265-270 (1999).

86. The International Human Genome Mapping Consortium. A physical map of the human genome. *Nature* **409**, 934-941 (2001).

87. Zhao, S. *et al.* Human BAC ends quality assessment and sequence analyses. *Genomics* **63**, 321-332 (2000).

88. Mahairas, G. G. *et al.* Sequence-tagged connectors: A sequence approach to mapping and scanning the human genome. *Proc. Natl Acad. Sci. USA* **96**, 9739-9744 (1999).

93. Hattori, M. *et al.* The DNA sequence of human chromosome 21. *Nature* **405**, 311-319 (2000).

94. Dunham, I. *et al.* The DNA sequence of human chromosome 22. *Nature* **402**, 489-495 (1999).

110. Pruit, K. D. & Maglott, D. R. RefSeq and LocusLink: NCBI gene-centered resources. *Nucleic Acids Res.* **29**, 137-140 (2001).

112. Hurst, L. D. & Eyre-Walker, A. Evolutionary genomics: reading the bands. *Bioessays* **22**, 105-107 (2000).

113. Saccone, S. *et al.* Correlations between isochores and chromosomal bands in the human genome. *Proc. Natl Acad. Sci. USA* **90**, 11929-11933 (1993).

114. Zoubak, S., Clay, O. & Bernardi, G. The gene distribution of the human genome. *Gene* **174**, 95-102 (1996).

115. Gardiner, K. Base composition and gene distribution: critical patterns in mammalian genome organization. *Trends Genet.* **12**, 519-524 (1996).

116. Duret, L., Mouchiroud, D. & Gautier, C. Statistical analysis of vertebrate sequences reveals that long genes are scarce in GC-rich isochores. *J. Mol. Evol.* **40**, 308-317 (1995).

117. Saccone, S., De Sario, A., Della Valle, G. & Bernardi, G. The highest gene concentrations in the human genome are in telomeric bands of metaphase chromosomes. *Proc. Natl Acad. Sci. USA* **89**, 4913-4917 (1992).

122. Bird, A., Taggart, M., Frommer, M., Miller, O. J. & Macleod, D. A fraction of the mouse genome that is derived from islands of nonmethylated, CpG-rich DNA. *Cell* **40**, 91-99 (1985).

123. Bird, A. P. CpG islands as gene markers in the vertebrate nucleus. *Trends Genet.* **3**, 342-347 (1987).

124. Chan, M. F., Liang, G. & Jones, P. A. Relationship between transcription and DNA methylation. *Curr. Top. Microbiol. Immunol.* **249**, 75-86 (2000).

125. Holliday, R. & Pugh, J. E. DNA modification mechanisms and gene activity during development. *Science* **187**, 226-232 (1975).

126. Larsen, F., Gundersen, G., Lopez, R. & Prydz, H. CpG islands as gene markers in the human genome. *Genomics* **13**, 1095-1107 (1992).

127. Tazi, J. & Bird, A. Alternative chromatin structure at CpG islands. *Cell* **60**, 909-920 (1990).

131. Yu, A. Comparison of human genetic and sequence-based physical maps. *Nature* **409**, 951-953 (2001).

132. Kaback, D. B., Guacci, V., Barber, D. & Mahon, J. W. Chromosome size-dependent control of meiotic recombination. *Science* **256**, 228-232 (1992).

133. Riles, L. *et al.* Physical maps of the 6 smallest chromosomes of *Saccharomyces cerevisiae* at a resolution of 2.6-kilobase pairs. *Genetics* **134**, 81-150 (1993).

135. Laurie, D. A. & Hulten, M. A. Further studies on bivalent chiasma frequency in human males with normal karyotypes. *Ann. Hum. Genet.* **49**, 189-201 (1985).

136. Roeder, G. S. Meiotic chromosomes: it takes two to tango. *Genes Dev.* **11**, 2600-2621 (1997).

137. Wu, T.-C. & Lichten, M. Meiosis-induced double-strand break sites determined by yeast chromatin structure. *Science* **263**, 515-518 (1994).

138. Gerton, J. L. *et al.* Global mapping of meiotic recombination hotspots and coldspots in the yeast *Saccharomyces cerevisiae. Proc. Natl Acad. Sci. USA* **97**, 11383-11390 (2000).

139. Li, W. -H. *Molecular Evolution* (Sinauer, Sunderland, Massachusetts, 1997).

140. Gregory, T. R. & Hebert, P. D. The modulation of DNA content: proximate causes and ultimate consequences. *Genome Res.* **9**, 317-324 (1999).

141. Hartl, D. L. Molecular melodies in high and low C. *Nature Rev. Genet.* **1**, 145-149 (2000).

142. Smit, A. F. Interspersed repeats and other mementos of transposable elements in mammalian genomes. *Curr. Opin. Genet. Dev.* **9**, 657-663 (1999).

143. Prak, E. L. & Haig, H. K. Jr Mobile elements and the human genome. *Nature Rev. Genet.* **1**, 134-144 (2000).

148. Smit, A. F. The origin of interspersed repeats in the human genome. *Curr. Opin. Genet. Dev.* **6**, 743- 748 (1996).

154. Malik, H. S., Burke, W. D. & Eickbush, T. H. The age and evolution of non-LTR retrotransposable elements. *Mol. Biol. Evol.* **16**, 793-805 (1999).

155. Kordis, D. & Gubensek, F. Bov-B long interspersed repeated DNA (LINE) sequences are present in *Vipera ammodytes* phospholipase A2 genes and in genomes of Viperidae snakes. *Eur. J. Biochem.* **246**, 772-779 (1997).

158. Smit, A. F., Toth, G., Riggs, A. D., & Jurka, J. Ancestral, mammalian-wide subfamilies of LINE-1 repetitive sequences. *J. Mol. Biol.* **246**, 401-417 (1995).

159. Lim, J. K. & Simmons, M. J. Gross chromosome rearrangements mediated by transposable elements in *Drosophila melanogaster. Bioessays* **16**, 269-275 (1994).

160. Caceres, M., Ranz, J. M., Barbadilla, A., Long, M. & Ruiz, A. Generation of a widespread Drosophila inversion by a transposable element. *Science* **285**, 415-418 (1999).

161. Gray, Y. H. It takes two transposons to tango: transposable-element-mediated chromosomal rearrangements. *Trends Genet.* **16**, 461-468 (2000).

162. Zhang, J. & Peterson, T. Genome rearrangements by nonlinear transposons in maize. *Genetics* **153**, 1403-1410 (1999).

163. Smit, A. F. Identification of a new, abundant superfamily of mammalian LTR-transposons. *Nucleic Acids Res.* **21**, 1863-1872 (1993).

164. Cordonnier, A., Casella, J. F. & Heidmann, T. Isolation of novel human endogenous retrovirus-like elements with foamy virus-related pol sequence. *J. Virol.* **69**, 5890-5897 (1995).

170. Kazazian, H. H. Jr & Moran, J. V. The impact of L1 retrotransposons on the human genome. *Nature Genet.* **19**, 19-24 (1998).

171. Malik, H. S. & Eickbush, T. H. NeSL-1, an ancient lineage of site-specific non-LTR retrotransposons from *Caenorhabditis elegans. Genetics* **154**, 193-203 (2000).

172. Casavant, N. C. *et al.* The end of the LINE?: lack of recent L1 activity in a group of South American rodents. *Genetics* **154**, 1809-1817 (2000).

173. Meunier-Rotival, M., Soriano, P., Cuny, G., Strauss, F. & Bernardi, G. Sequence organization and genomic distribution of the major family of interspersed repeats of mouse DNA. *Proc. Natl Acad. Sci. USA* **79**, 355-359 (1982).

174. Soriano, P., Meunier-Rotival, M. & Bernardi, G. The distribution of interspersed repeats is nonuniform and conserved in the mouse and human genomes. *Proc. Natl Acad. Sci. USA* **80**, 1816- 1820 (1983).

175. Goldman, M. A., Holmquist, G. P., Gray, M. C., Caston, L. A. & Nag, A. Replication timing of genes and middle repetitive sequences. *Science* **224**, 686-692 (1984).

176. Manuelidis, L. & Ward, D. C. Chromosomal and nuclear distribution of the *Hind*III 1.9-kb human DNA repeat segment. *Chromosoma* **91**, 28-38 (1984).

178. Jurka, J. Sequence patterns indicate an enzymatic involvement in integration of mammalian retroposons. *Proc. Natl Acad. Sci. USA* **94**, 1872-1877 (1997).

192. Bohossian, H. B., Skaletsky, H. & Page, D. C. Unexpectedly similar rates of nucleotide substitution found in male and female hominids. *Nature* **406**, 622-625 (2000).

215. Ji, Y., Eichler, E. E., Schwartz, S. & Nicholls, R. D. Structure of chromosomal duplicons and their role in mediating human genomic disorders. *Genome Res.* **10**, 597-610 (2000).

216. Eichler, E. E. Masquerading repeats: paralogous pitfalls of the human genome. *Genome Res.* **8**, 758- 762 (1998).

217. Mazzarella, R. & D. Schlessinger, D. Pathological consequences of sequence duplications in the human genome. *Genome Res.* **8**, 1007-1021 (1998).

238. Shaikh, T. H. *et al.* Chromosome 22-specific low copy repeats and the 22q11.2 deletion syndrome: genomic organization and deletion endpoint analysis. *Hum. Mol. Genet.* **9**, 489-501 (2000).

239. Francke, U. Williams-Beuren syndrome: genes and mechanisms. *Hum. Mol. Genet.* **8**, 1947-1954 (1999).

243. Lander, E. S. The new genomics: Global views of biology. *Science* **274**, 536-539 (1996).

244. Eddy, S. R. Noncoding RNA genes. *Curr. Op. Genet. Dev.* **9**, 695-699 (1999).

245. Ban, N., Nissen, P., Hansen, J., Moore, P. B. & Steitz, T. A. The complete atomic structure of the large ribosomal subunit at 2.4 angstrom resolution. *Science* **289**, 905-920 (2000).

246. Nissen, P., Hansen, J., Ban, N., Moore, P. B. & Steitz, T. A. The structural basis of ribosome activity in peptide bond synthesis. *Science* **289**, 920-930 (2000).

247. Weinstein, L. B. & Steitz, J. A. Guided tours: from precursor snoRNA to functional snoRNP. *Curr. Opin. Cell Biol.* **11**, 378-384 (1999).

248. Bachellerie, J.-P. & Cavaille, J. in *Modification and Editing of RNA* (ed. Benne, H. G. a. R.) 255-272 (ASM, Washington DC, 1998).

253. Sprinzl, M., Horn, C., Brown, M., Ioudovitch, A. & Steinberg, S. Compilation of tRNA sequences and sequences of tRNA genes. *Nucleic Acids Res.* **26**, 148-153 (1998).

254. Long, E. O. & Dawid, I. B. Repeated genes in eukaryotes. *Annu. Rev. Biochem.* **49**, 727-764 (1980).

262. Buckland, R. A. A primate transfer-RNA gene cluster and the evolution of human chromosome 1. *Cytogenet. Cell Genet.* **61**, 1-4 (1992).

263 Gonos, E. S. & Goddard, J. P. Human tRNA-Glu genes: their copy number and organization. *FEBS Lett.* **276**, 138-142 (1990).

286. Dunham, I. The gene guessing game. *Yeast* **17**, 218-224 (2000).

292. Roest Crollius, H. *et al.* Estimate of human gene number provided by genome-wide analysis using *Tetraodon nigroviridis* DNA sequence. *Nature Genet.* **25**, 235-238 (2000).

309. The RIKEN Genome Exploration Research Group Phase II Team and the FANTOM Consortium. Functional annotation of a full-length mouse cDNA collection. *Nature* **409**, 685-690 (2001).

311. Janin, J. & Chothia, C. Domains in proteins: definitions, location, and structural principles. *Methods Enzymol.* **115**, 420-430 (1985).

312. Ponting, C. P., Schultz, J., Copley, R. R., Andrade, M. A. & Bork, P. Evolution of domain families. *Adv. Protein Chem.* **54**, 185-244 (2000).

313. Doolittle, R. F. The multiplicity of domains in proteins. *Annu. Rev. Biochem.* **64**, 287-314 (1995).

314. Bateman, A. & Birney, E. Searching databases to find protein domain organization. *Adv. Protein Chem.* **54**, 137-157 (2000).

315. Futreal, P. A. *et al.* Cancer and genomics. *Nature* **409**, 850-852 (2001).

316. Nestler, E. J. & Landsman, D. Learning about addiction from the human draft genome. *Nature* **409**, 834-835 (2001).

317. Tupler, R., Perini, G. & Green, M. R. Expressing the human genome. *Nature* **409**, 832-835 (2001).

318. Fahrer, A. M., Bazan, J. F., Papathanasiou, P., Nelms, K. A. & Goodnow, C. C. A genomic view of immunology. *Nature* **409**, 836-838 (2001).

319. Li, W. -H., Gu, Z., Wang, H. & Nekrutenko, A. Evolutionary analyses of the human genome. *Nature* **409**, 847-849 (2001).

320. Bock, J. B., Matern, H. T., Peden, A. A. & Scheller, R. H. A genomic perspective on membrane compartment organization. *Nature* **409**, 839-841 (2001).

321. Pollard, T. D. Genomics, the cytoskeleton and motility. *Nature* **409**, 842-843 (2001).

322. Murray, A. W. & Marks, D. Can sequencing shed light on cell cycling? *Nature* **409**, 844-846 (2001).

323. Clayton, J. D., Kyriacou, C. P. & Reppert, S. M. Keeping time with the human genome. *Nature* **409**, 829-831 (2001).

339. Schultz, J., Copley, R. R., Doerks, T., Ponting, C. P. & Bork, P. SMART: a web-based tool for the study of genetically mobile domains. *Nucleic Acids Res.* **28**, 231-234 (2000).

376. Lander, E. S. & Schork, N. J. Genetic dissection of complex traits. *Science* **265**, 2037-2048 (1994).

388. Collins, F. S. Positional cloning moves from perditional to traditional. *Nature Genet.* **9**, 347-350 (1995).

389. Nagamine, K. *et al.* Positional cloning of the APECED gene. *Nature Genet.* **17**, 393-398 (1997).

390. Reuber, B. E. *et al.* Mutations in PEX1 are the most common cause of peroxisome biogenesis disorders. *Nature Genet.* **17**, 445-448 (1997).

391. Portsteffen, H. *et al.* Human PEX1 is mutated in complementation group 1 of the peroxisome biogenesis disorders. *Nature Genet.* **17**, 449-452 (1997).

392. Everett, L. A. *et al.* Pendred syndrome is caused by mutations in a putative sulphate transporter gene (PDS). *Nature Genet.* **17**, 411-422 (1997).

393. Coffey, A. J. *et al.* Host response to EBV infection in X-linked lymphoproliferative disease results from mutations in an SH2-domain encoding gene. *Nature Genet.* **20**, 129-135 (1998).

394. Van Laer, L. *et al.* Nonsyndromic hearing impairment is associated with a mutation in DFNA5. *Nature Genet.* **20**, 194-197 (1998).

395. Sakuntabhai, A. *et al.* Mutations in ATP2A2, encoding a Ca2+pump, cause Darier disease. *Nature Genet.* **21**, 271-277 (1999).

396. Gedeon, A. K. *et al.* Identification of the gene (SEDL) causing X-linked spondyloepiphyseal dysplasia tarda. *Nature Genet.* **22**, 400-404 (1999).

397. Hurvitz, J. R. *et al.* Mutations in the CCN gene family member WISP3 cause progressive pseudorheumatoid dysplasia. *Nature Genet.* **23**, 94-98 (1999).

398. Laberge-le Couteulx, S. *et al.* Truncating mutations in CCM1, encoding KRIT1, cause hereditary cavernous angiomas. *Nature Genet.* **23**, 189-193 (1999).

399. Sahoo, T. *et al.* Mutations in the gene encoding KRIT1, a Krev-1/rap1a binding protein, cause cerebral cavernous malformations (CCM1). *Hum. Mol. Genet.* **8**, 2325-2333 (1999).

400. McGuirt, W. T. *et al.* Mutations in COL11A2 cause non-syndromic hearing loss (DFNA13). *Nature Genet.* **23**, 413-419 (1999).

401. Moreira, E. S. *et al.* Limb-girdle muscular dystrophy type 2G is caused by mutations in the gene encoding the sarcomeric protein telethonin. *Nature Genet.* **24**, 163-166 (2000).

402. Ruiz-Perez, V. L. *et al.* Mutations in a new gene in Ellis-van Creveld syndrome and Weyers acrodental dysostosis. *Nature Genet.* **24**, 283-286 (2000).

403. Kaplan, J. M. *et al.* Mutations in ACTN4, encoding alpha-actinin-4, cause familial focal segmental glomerulosclerosis. *Nature Genet.* **24**, 251-256 (2000).

404. Escayg, A. *et al.* Mutations of SCN1A, encoding a neuronal sodium channel, in two families with GEFS+2. *Nature Genet.* **24**, 343-345 (2000).

405. Sacksteder, K. A. *et al.* Identification of the alpha-aminoadipic semialdehyde synthase gene, which is defective in familial hyperlysinemia. *Am. J. Hum. Genet.* **66**, 1736-1743 (2000).

406. Kalaydjieva, L. *et al.* N-myc downstream-regulated gene 1 is mutated in hereditary motor and sensory neuropathy-Lom. *Am. J. Hum. Genet.* **67**, 47-58 (2000).

407. Sundin, O. H. *et al.* Genetic basis of total colourblindness among the Pingelapese islanders. *Nature Genet.* **25**, 289-293 (2000).

408. Kohl, S. *et al.* Mutations in the CNGB3 gene encoding the beta-subunit of the cone photoreceptor cGMP-gated channel are responsible for achromatopsia (ACHM3) linked to chromosome 8q21. *Hum. Mol. Genet.* **9**, 2107-2116 (2000).

409. Avela, K. *et al.* Gene encoding a new RING-B-box-coiled-coil protein is mutated in mulibrey nanism. *Nature Genet.* **25**, 298-301 (2000).

410. Verpy, E. *et al.* A defect in harmonin, a PDZ domain-containing protein expressed in the inner ear sensory hair cells, underlies usher syndrome type 1C. *Nature Genet.* **26**, 51-55 (2000).

411. Bitner-Glindzicz, M. *et al.* A recessive contiguous gene deletion causing infantile hyperinsulinism, enteropathy and deafness identifies the usher type 1C gene. *Nature Genet.* **26**, 56-60 (2000).

412. The May-Hegglin/Fechtner Syndrome Consortium. Mutations in MYH9 result in the May-Hegglin anomaly, and Fechtner and Sebastian syndromes. *Nature Genet.* **26**, 103-105 (2000).

413. Kelley, M. J., Jawien, W., Ortel, T. L. & Korczak, J. F. Mutation of MYH9, encoding non-muscle myosin heavy chain A, in May-Hegglin anomaly. *Nature Genet.* **26**, 106-108 (2000).

414. Kirschner, L. S. *et al.* Mutations of the gene encoding the protein kinase A type I-a regulatory subunit in patients with the Carney complex. *Nature Genet.* **26**, 89-92 (2000).

415. Lalwani, A. K. *et al.* Human nonsyndromic hereditary deafness DFNA17 is due to a mutation in non-muscle myosin MYH9. *Am. J. Hum. Genet.* **67**, 1121-1128 (2000).

416. Matsuura, T. *et al.* Large expansion of the ATTCT pentanucleotide repeat in spinocerebellar ataxia type 10. *Nature Genet.* **26**, 191-194 (2000).

417. Delettre, C. *et al.* Nuclear gene OPA1, encoding a mitochondrial dynamin-related protein, is mutated in dominant optic atrophy. *Nature Genet.* **26**, 207-210 (2000).

418. Pusch, C. M. *et al.* The complete form of X-linked congenital stationary night blindness is caused by mutations in a gene encoding a leucine-rich repeat protein. *Nature Genet.* **26**, 324-327 (2000).

419. The ADHR Consortium. Autosomal dominant hypophosphataemic rickets is associated with mutations in FGF23. *Nature Genet.* **26**, 345-348 (2000).

420. Bomont, P. *et al.* The gene encoding gigaxonin, a new member of the cytoskeletal BTB/kelch repeat family, is mutated in giant axonal neuropathy. *Nature Genet.* **26**, 370-374 (2000).

421. Tullio-Pelet, A. *et al.* Mutant WD-repeat protein in triple-A syndrome. *Nature Genet.* **26**, 332-335 (2000).

422. Nicole, S. *et al.* Perlecan, the major proteoglycan of basement membranes, is altered in patients with Schwartz-Jampel syndrome (chondrodystrophic myotonia). *Nature Genet.* **26**, 480-483 (2000).

426. Drews, J. Research & development. Basic science and pharmaceutical innovation. *Nature Biotechnol.* **17**, 406 (1999).

427. Drews, J. Drug discovery: a historical perspective. *Science* **287**, 1960-1964 (2000).

433. Matsunami, H., Montmayeur, J. P. & Buck, L. B. A family of candidate taste receptors in human and mouse. *Nature* **404**, 601-604 (2000).

434. Adler, E. *et al.* A novel family of mammalian taste receptors. *Cell* **100**, 693-702 (2000).

435. Chandrashekar, J. *et al.* T2Rs function as bitter taste receptors. *Cell* **100**, 703-711 (2000).

436. Hardison, R. C. Conserved non-coding sequences are reliable guides to regulatory elements. *Trends Genet.* **16**, 369-372 (2000).

437. Onyango, P. *et al.* Sequence and comparative analysis of the mouse 1-megabase region orthologous to the human 11p15 imprinted domain. *Genome Res.* **10**, 1697-1710 (2000).

438. Bouck, J. B., Metzker, M. L. & Gibbs, R. A. Shotgun sample sequence comparisons between mouse and human genomes. *Nature Genet.* **25**, 31-33 (2000).

446. Feil, R. & Khosla, S. Genomic imprinting in mammals: an interplay between chromatin and DNA methylation? *Trends Genet.* **15**, 431-434 (1999).

447. Robertson, K. D. & Wolffe, A. P. DNA methylation in health and disease. *Nature Rev. Genet.* **1**, 11-19 (2000).

448. Beck, S., Olek, A. & Walter, J. From genomics to epigenomics: a loftier view of life. *Nature Biotechnol.* **17**, 1144-1144 (1999).

449. Hagmann, M. Mapping a subtext in our genetic book. *Science* **288**, 945-946 (2000).

450. Eliot, T. S. in *T. S. Eliot. Collected Poems 1909-1962* (Harcourt Brace, New York, 1963).

Supplementary Information is available on *Nature*'s World-Wide Web site (http://www.nature.com) or as paper copy from the London editorial office of *Nature*.

Correspondence and requests for materials should be addressed to E. S. Lander (e-mail: lander@genome.wi.mit.edu), R. H. Waterston (e-mail: bwaterst@watson.wustl.edu), J. Sulston (e-mail: jes@sanger.ac.uk) or F. S. Collins (e-mail: fc23a@nih.gov).

DNA sequence databases

GenBank, National Center for Biotechnology Information, National Library of Medicine, National Institutes of Health, Bldg. 38A, 8600 Rockville Pike, Bethesda, Maryland 20894, USA

EMBL, European Bioinformatics Institute, Wellcome Trust Genome Campus, Hinxton, Cambridge CB10 1SD, UK

DNA Data Bank of Japan, Center for Information Biology, National Institute of Genetics, 1111 Yata, Mishima-shi, Shizuoka-ken 411-8540, Japan

A Map of Human Genome Sequence Variation Containing 1.42 Million Single Nucleotide Polymorphisms

The International SNP Map Working Group

Editor's Note

This study reveals the first detailed map of the 1.42 million single nucleotide polymorphisms (SNPs) found across the human genome. The data, amassed by The International SNP Map Working Group, provide interesting first glimpses into the patterns of variation found across the human genome. Although the vast majority of human DNA sequences are the same, single base pair changes in the code (SNPs) can affect our response to disease, drugs and environmental factors such as toxins and viruses. This makes SNPs a valuable tool for biomedical research and drug design. SNPs also change little between generations, making them useful for studies of evolutionary history. This map integrates all of the publicly available SNPs with known genes and other genomic features, and was made possible by and was published with the first draft sequence of the human genome.

We describe a map of 1.42 million single nucleotide polymorphisms (SNPs) distributed throughout the human genome, providing an average density on available sequence of one SNP every 1.9 kilobases. These SNPs were primarily discovered by two projects: The SNP Consortium and the analysis of clone overlaps by the International Human Genome Sequencing Consortium. The map integrates all publicly available SNPs with described genes and other genomic features. We estimate that 60,000 SNPs fall within exon (coding and untranslated regions), and 85% of exons are within 5 kb of the nearest SNP. Nucleotide diversity varies greatly across the genome, in a manner broadly consistent with a standard population genetic model of human history. This high-density SNP map provides a public resource for defining haplotype variation across the genome, and should help to identify biomedically important genes for diagnosis and therapy.

INHERITED differences in DNA sequence contribute to phenotypic variation, influencing an individual's anthropometric characteristics, risk of disease and response to the environment. A central goal of genetics is to pinpoint the DNA variants that contribute most significantly to population variation in each trait. Genome-wide linkage analysis and positional cloning have identified hundreds of genes for human diseases[1] (http://ncbi.nlm.nih.gov/OMIM), but nearly all are rare conditions in which mutation of a single gene is necessary and sufficient to cause disease. For common diseases, genome-wide linkage studies have had limited success, consistent with a more complex genetic architecture.

包含 142 万个单核苷酸多态性的人类基因组序列变异图谱

国际 SNP 图谱研究组

编者按

这项研究首次展示了人类基因组中已发现的 142 万个单核苷酸多态性（SNP）的详细图谱。这份由国际 SNP 图谱研究组积累的数据让人们首次目睹了人类基因组的变异图谱。尽管绝大多数的人类 DNA 序列是一致的，但是序列中单个碱基对的改变（SNP）就能够影响我们对疾病、药物和环境因素（例如毒素和病毒）的反应。这使得 SNP 成为生物医药研究和药物设计行业的有力工具。SNP 在不同世代中变化不大，这对于进化史的研究是很有帮助的。这份图谱整合了所有可公开获取的 SNP 与已知基因及其他基因组特征，依据人类基因组第一个草图序列得以完成，并与后者共同发表。

我们描绘了一张遍及人类基因组的包含 142 万个单核苷酸多态性（SNP）的图谱，这张图谱给我们提供了可用序列中 SNP 的平均密度：每 1,900 个碱基一个 SNP。这些 SNP 最初是在两个项目中发现的：SNP 国际联合会和国际人类基因组测序协会的克隆重叠分析。该图谱整合了所有可公开获取的 SNP 与已知基因和其他基因组特征。我们估计有 60,000 个 SNP 在外显子（编码和非翻译区）中，85% 的外显子距离最近的 SNP 5kb 以内。不同基因组间的核苷酸多样性变化很大，但变化方式大致符合人类历史标准群体遗传学模型。高密度的 SNP 图谱可以为定义整个基因组的单体型变异提供公开资源，帮助识别生物医学相关的重要基因，为诊断与治疗提供帮助。

可遗传的 DNA 序列差异导致了表型上的变异，这影响了个体的人体测量特征、疾病的风险和环境应激反应。遗传学的一个核心目标是精确找到在各个性状中对种群变异贡献最显著的 DNA 变异。全基因组连锁分析和定位克隆已经鉴定出人类疾病的数百个基因[1]（http://ncbi.nlm.nih.gov/OMIM），但单个基因的突变足以引起疾病的情况十分罕见。对于常见疾病，全基因连锁研究的成果有限，这与基因结构的复

If each locus contributes modestly to disease aetiology, more powerful methods will be required.

One promising approach is systematically to explore the limited set of common gene variants for association with disease[2-4]. In the human population most variant sites are rare, but the small number of common polymorphisms explain the bulk of heterozygosity[3] (see also refs 5–11). Moreover, human genetic diversity appears to be limited not only at the level of individual polymorphisms, but also in the specific combinations of alleles (haplotypes) observed at closely linked sites[8,11-14]. As these common variants are responsible for most heterozygosity in the population, it will be important to assess their potential impact on phenotypic trait variation.

If limited haplotype diversity is general, it should be practical to define common haplotypes using a dense set of polymorphic markers, and to evaluate each haplotype for association with disease. Such haplotype-based association studies offer a significant advantage: genomic regions can be tested for association without requiring the discovery of the functional variants. The required density of markers will depend on the complexity of the local haplotype structure, and the distance over which these haplotypes extend, neither of which is yet well defined.

Current estimates (refs 13–17) indicate that a very dense marker map (30,000–1,000,000 variants) would be required to perform haplotype-based association studies. Most human sequence variation is attributable to SNPs, with the rest attributable to insertions or deletions of one or more bases, repeat length polymorphisms and rearrangements. SNPs occur (on average) every 1,000–2,000 bases when two human chromosomes are compared[5,6,9,18-20], and are thus present at sufficient density for comprehensive haplotype analysis. SNPs are binary, and thus well suited to automated, high-throughput genotyping. Finally, in contrast to more mutable markers, such as microsatellites[21], SNPs have a low rate of recurrent mutation, making them stable indicators of human history. We have constructed a SNP map of the human genome with sufficient density to study human haplotype structure, enabling future study of human medical and population genetics.

Identification and Characteristics of SNPs

The map contains all SNPs that were publicly available in November 2000. Over 95% were discovered by The SNP Consortium (TSC) and the public Human Genome Project (HGP). TSC contributed 1,023,950 candidate SNPs (http://snp.cshl.org) identified by shotgun sequencing of genomic fragments drawn from a complete (45% of data) or reduced (55% of data) representation of the human genome[18,22]. Individual contributions were: Whitehead Institute, 589,209 SNPs from 2.57 million (M) passing reads; Sanger Centre, 262,279 SNPs from 1.16M passing reads; Washington University, 172,462 SNPs from 1.69M passing reads. TSC SNPs were discovered using a publicly available panel of 24 ethnically diverse individuals[23]. Reads were aligned to one another and to the available genome

杂性有关。如果每个突变位点的发现对疾病病因学贡献不大，我们就需要更有力的方法。

一个可行的方法是系统地探索出一组数目有限的常规基因变异来研究相关疾病[2-4]。人类群体中大多数变异位点是罕见的，但是少数的常见多态性解释了大量杂合性的存在[3]（也可见于参考文献 5~11）。此外，对人类基因多样性的研究不该受限于个体多态性水平，而且要研究相邻连锁位点观察到的等位基因（单体型）的特异组合[8,11-14]。正因为这些常见的变异是引起种群中大多数杂合性的原因，因此，评价它们在表型性状变异中的潜在影响就十分重要。

如果有限的单体型多态性普遍存在，那么用多态性标记物的密集组合来定义常见的单体型，评估疾病相关的每个单体型，应该会变得切实可行。这些基于单体型的相关研究提供给我们一个重要的优势：无需发现功能型变异，基因组区域可用来测试相关性。标记所需的密度将取决于局部单体型结构的复杂程度，以及这些单体型延伸的范围，这两者均未被很好地界定。

现有的估计（参考文献 13~17）指出，做基于单体型的关联研究需要一个十分密集的标记图谱（30,000~1,000,000 个突变体）。大部分人类序列变异可归因于 SNP，而剩下的则归因于一个或更多碱基的插入或缺失、重复长度多态性和重排。当将两个人类染色体进行比较时，（平均）每 1,000~2,000 个碱基出现一个 SNP[5,6,9,18-20]。对于复杂的单体型分析，SNP 的这种出现频率足够了。SNP 是二等位型的，因此很适合自动化的、高通量基因分型。最后，与更可变的标记如微卫星[21]相比，SNP 有较低的频发突变，这使得它们可以作为人类历史的稳定指标。我们已经绘制了一个具有足够密度的人类基因组 SNP 图谱，用来研究人类单体型结构，从而使未来人类医学和群体遗传学的研究成为可能。

SNP 的鉴定及特性

这张图谱包含 2000 年 11 月公开可获取的所有 SNP。超过 95% 的 SNP 是由 SNP 国际联合会（TSC）和公共人类基因组计划（HGP）发现的。TSC 贡献了 1,023,950 个候选 SNP（http://snp.cshl.org），它们是通过鸟枪测序法测定来自完整的（45% 的数据）或者缩短的（55% 的数据）人类基因组碎片确定的[18,22]。各机构的贡献如下：怀特黑德研究所，从 257 万合格读长中得到 589,209 个 SNP 位点；桑格中心，从 116 万合格读长中得到 262,279 个 SNP 位点；华盛顿大学，从 169 万合格读长中得到 172,462 个 SNP 位点。TSC 的 SNP 位点是从 24 个种族不同个体可公开获取的基因组检测

sequence, followed by detection of single base differences using one of two validated algorithms: Polybayes[24] and the neighbourhood quality standard (NQS[18,22]).

An additional 971,077 candidate SNPs were identified as sequence differences in regions of overlap between large-insert clones (bacterial artificial chromosomes (BACs) or P1-derived artificial chromosomes (PACs)) sequenced by the HGP. Two groups (NCBI/Washington University (556,694 SNPs): G.B., P.Y.K. and S.S.; and The Sanger Centre (630,147 SNPs): J.C.M. and D.R.B.) independently analysed these overlaps using the two detection algorithms. This approach contributes dense clusters of SNPs throughout the genome. The remaining 5% of SNPs were discovered in gene-based studies, either by automated detection of single base differences in clusters of overlapping expressed sequence tags[24-28] or by targeted resequencing efforts (see ftp://ncbi.nlm.nih.gov/snp/human/submit_format/*/*publicat.rep.gz).

It is critical that candidate SNPs have a high likelihood of representing true polymorphisms when examined in population studies. Although many methods and contributors are represented on the map (see above), most SNPs (> 95%) were contributed by two large-scale efforts that uniformly applied automated methods. Random samples of these SNPs have been evaluated by confirmation in the original DNA samples (where possible) to rule out false positives, and in independent population samples to determine allele frequency. The TSC centres and two outside laboratories (Orchid and Cold Spring Harbor Laboratory) successfully genotyped 1,585 TSC SNPs in the 24 DNA samples used for discovery (http://snp.cshl.org); having surveyed all chromosomes in which each SNP could have been identified, any non-polymorphic candidates must represent false positives. In these tests, 1,500 SNPs (95%) were polymorphic, 67 (4%) non-polymorphic (false positives) and 18 (1%) uniformly heterozygous (previously unrecognized repeats). These high validation rates were observed separately for subsets of SNPs discovered by reduced representation shotgun and genomic alignment, and for subsets identified with Polybayes and the NQS. Thus, these algorithms appear to generate few false positive SNPs. The small number (1%) of uniformly "heterozygous" candidate SNPs show that the methods also exclude nearly all low-copy repeats.

The allele frequencies of a set of SNPs have been evaluated[29] in independent populations using pooled resequencing. Samples of TSC ($n = 502$) and overlap SNPs ($n = 774$) were studied in population samples of European, African American and Chinese descent, revealing 82% to be polymorphic in at least one ethnic group at frequencies above the detection threshold of pooled resequencing (~10%). The remaining 18% presumably represent SNPs with a frequency less than 10% in the populations surveyed and false positives. Furthermore, 77% of SNPs had a minor allele frequency of more than 20% in at least one population, and 27% had an allele frequency higher than 20% in all three ethnic groups. TSC and overlap SNPs had similar distributions across the populations, showing that they are comparable in quality and frequency. The high proportion of SNPs with significant population frequency is expected after SNP discovery in two or a few chromosomes, given standard assumptions about human population history[18,29,30].

区域发现的[23]。序列进行拼接并与已有基因组序列进行比对，接着用 Polybayes[24] 和周边质量标准（NQS[18,22]）两个有效算法之一进行单个碱基差异性的检测。

额外的 971,077 个候选 SNP 是在 HGP 测序得到的大片段插入克隆（细菌人工染色体（BAC）或者 P1 衍生人工染色体（PAC））的重叠区中发现的序列差异。两个机构（美国国家生物技术信息中心/华盛顿大学（556,694 个 SNP 位点）：邦尼、郭沛恩和谢里；桑格中心（630,147 个 SNP 位点）：马利金和本特利）用两种检测算法对这些重叠区域进行单独的分析。这种方法有助于形成遍及基因组的密集的 SNP 簇。剩下 5% 的 SNP 则是在基于基因的研究中发现的，要么是通过成簇的重叠表达序列标签中单碱基差异性的自动检测[24-28]，要么是通过靶向重测序（见于 ftp://ncbi.nlm.nih.gov/snp/human/submit_format/*/*publicat.rep.gz）。

至关重要的是，检测群体研究发现，候选 SNP 极有可能代表真正的多态性。尽管图谱来源于许多方法以及贡献者（如上所述），但大多数 SNP（＞95%）都是由运用同样自动化方法的两大机构发现的。这些 SNP 的随机样本都经过了原始 DNA 样本（可获得的）的评估来排除假阳性，并且在独立群体样本中检测等位基因的频率。TSC 中心和两个外部实验室（兰花和冷泉港实验室）从 24 个用于研究的样本中，成功地对 1,585 个 TSC SNP 位点进行了基因分型（http://snp.cshl.org）；对所有鉴定到 SNP 的染色体都进行了调研，任何非多态性候选位点都被认定为是假阳性。在这些测试中，1,500 个 SNP（95%）是多态性的，67 个 SNP（4%）是非多态性的（假阳性），而 18 个 SNP（1%）是一致的杂合子（先前无法识别的重复）。这些高验证率分别是在简化鸟枪和基因组比对发现的 SNP 子集以及 Polybayes 和 NQS 鉴定的子集中观察到的。因此，这些算法几乎很少出现假阳性的 SNP。少量（1%）一致的“杂合子”候选 SNP 表明这些方法也几乎排除了所有的低拷贝重复序列。

采用混合重测序的方法，对一组 SNP 在独立群体中的等位基因频率进行了评估[29]。在欧洲人、非洲裔美国人和中国人的人群样本中，研究了 TSC（n = 502）和重叠 SNP（n = 774）的样本，发现在至少一个族群中，82% 的样本具有多态性，其频率高于混合重测序检测阈值（约 10%）。剩下的 18% 可能代表 SNP 低于 10% 的受访群体以及假阳性。此外，在至少一个群体中，77% 的 SNP 等位基因频率超过 20%，在三个种族群体中，27% 的 SNP 等位基因频率高于 20%。TSC 和重叠 SNP 在各群体间有相似的分布，这表明他们在性质和频率上相似。考虑到关于人类群体历史的标准假设，在两个或少数染色体中发现 SNP 后，预计有高比例的 SNP 伴随显著的群体频率[18,29,30]。

Description of the SNP Map

We mapped the sequence flanking each SNP by alignment to the genomic sequence of large-insert clones in GenBank. These alignments were converted into chromosomal coordinates according to the publicly available genome assemblies of July and September 2000 (http://genome.ucsc.edu). Candidate SNPs were included in the final map only if they mapped to a single location in the genome assembly. Integrated displays of SNPs, genes and other features are available at the ENSEMBL (http://www.ensembl.org), NCBI (National Center for Biotechnology Information; http://www.ncbi.nlm.nih.gov), UCSC (University of California at Santa Cruz; http://genome.ucsc.edu) and TSC (http://snp.cshl.org) websites.

The nonredundant SNP total of 1,433,393 is fewer than the sum of individual submissions (2,067,476) because some SNPs (mainly in regions of BAC overlap) were discovered by more than one effort. Of these, 1,419,190 mapped to unique locations in the 2.7 gigabases (Gb) of assembled human genome sequence, providing an average density of one SNP every 1.91 kb. TSC SNPs, which are more evenly distributed than those from clone overlaps, were found on average every 3.05 kb. SNP density (Table 1) is relatively constant across the autosomes. To characterize the distribution of SNPs, we examined 366,192 SNPs that fell within finished sequence. Most of the genome contains SNPs at high density (Fig. 1): 90% of contiguous 20-kb windows contain one or more SNPs, as do 63% of 5-kb windows and 28% of 1-kb windows. Only 4% of genome sequence falls in gaps between SNPs of > 80 kb, and some of these gaps are covered by SNPs that are discovered but not yet mapped owing to gaps in the genome assembly.

To evaluate the density of SNPs in regions within and surrounding genes, we used the September 2000 release of RefSeq[31]. In total, 14,534 SNPs map to within these 7,000 carefully annotated, non-redundant messenger RNAs, equivalent to about two exonic SNPs per gene (coding and untranslated regions). Extrapolating two exonic SNPs per gene to the approximately 30,000 human genes[32], we estimate there to be 60,000 exonic SNPs in this collection. The density of SNPs in exons (one SNP per 1.08 kb; Table 1) is higher than in the genome as a whole, owing to the contribution of efforts targeted to exonic regions.

We also assessed the distribution of SNPs in the genomic locus surrounding each of the RefSeq mRNAs. We assigned the RefSeq exons to their genomic locations, restricting analysis to the 2,960 RefSeq mRNAs mapping onto finished sequence. As we cannot define the extent of the noncoding (regulatory) regions of each gene, we arbitrarily defined each "gene locus" as extending from 10 kb upstream of the start of the first exon to the end of the last exon. By this definition, 93% of gene loci contain at least one SNP, and 98% are within 5 kb of the nearest SNP; also, 59% of gene loci contained five or more SNPs, and 39% ten or more. Of 24,953 exons, 85% were within 5 kb of the nearest SNP. Thus, most exons should be close enough to at least one SNP for haplotype-based association studies, where the functional variant may be some distance from the SNPs used in the study.

SNP 图谱的描述

我们将每个 SNP 的侧翼序列与 GenBank 中的大片段插入克隆基因组序列进行比对。依照 2000 年 7 月和 9 月发布的可公开获取的基因组组装图(http://genome.ucsc.edu)，将比对转换成染色体坐标。候选 SNP 只有在基因组组装中被比对到单一位置才能包含在最终图谱中。SNP、基因和其他特征的综合信息可在如下网站获得：ENSEMBL(http://www.ensembl.org)、NCBI(美国国家生物技术中心；http://www.ncbi.nlm.nih.gov)、UCSC(加州大学圣克鲁兹分校；http://genome.ucsc.edu)和 TSC(http://snp.cshl.org)。

非冗余的 SNP 总计有 1,433,393 个，少于各机构提交的总数(2,067,476)，这是因为一些 SNP(主要在 BAC 重叠区)被发现的次数多于一次。其中的 1,419,190 个 SNP 在人类基因组 2.7 千兆碱基(Gb)的序列中坐标唯一，平均密度为每 1.91kb 一个 SNP。TSC SNP 比克隆重叠中得到的 SNP 分布更加均匀，其平均密度为每 3.05kb 一个。SNP 密度(表 1)在常染色体间相对恒定。为了描绘 SNP 的分布，我们检查了位于已完成测序区域的 366,192 个 SNP。多数基因组包含高密度 SNP(图 1)：90% 的相邻 20 kb 阅读窗内含有一个或多个 SNP，63% 的 5 kb 阅读窗和 28% 的 1 kb 阅读窗同样如此。只有 4% 的基因组序列位于相距大于 80 kb 的两个 SNP 之间的缺口中，有些缺口被已发现的 SNP 覆盖，但由于基因组组装上的缺陷还没有被绘制出来。

为了评估基因区域内和周围的 SNP 密度，我们采用了 2000 年 9 月发布的参考序列[31]。总体说来，14,534 个 SNP 比对到了 7,000 个精确注释、非冗余的信使 RNA 上，这等同于每个基因约有两个外显子 SNP(编码区和非翻译区)。以此推算到大约 30,000 个人类基因中[32]，我们估计会有 60,000 个外显子 SNP。外显子中 SNP 的密度(每 1.08 kb 一个 SNP；表 1)比全部 SNP 在基因组中的密度要高，这是由把外显子区作为测序靶标导致的。

我们也估算了每个参考序列 mRNA 周围基因座上 SNP 的分布。我们将参考序列外显子分配到它们的基因组位置，将分析限制在已完成序列图谱比对的 2,960 个 mRNA 参考序列上。由于不能界定每个基因非编码(调控)区的大小，我们人为地定义每个“基因座”延伸范围为从第一个外显子起始位置上游 10 kb 到最后一个外显子。通过此定义，93% 的基因座至少包含一个 SNP，而 98% 的基因座在 5 kb 内就有一个 SNP；此外，59% 的基因座包含 5 个或者更多的 SNP，而 39% 拥有 10 个及以上的 SNP。24,953 个外显子中，85% 的外显子在 5 kb 以内有一个 SNP。因此，多数外显子应该至少靠近一个基于单体型关联研究的 SNP，然而功能性变异可能与研究中使用的 SNP 有一定的距离。

Table 1. SNP distribution by chromosome

Chromosome	Length (bp)	All SNPs		TSC SNPs	
		SNPs	kb per SNP	SNPs	kb per SNP
1	214,066,000	129,931	1.65	75,166	2.85
2	222,889,000	103,664	2.15	76,985	2.90
3	186,938,000	93,140	2.01	63,669	2.94
4	169,035,000	84,426	2.00	65,719	2.57
5	170,954,000	117,882	1.45	63,545	2.69
6	165,022,000	96,317	1.71	53,797	3.07
7	149,414,000	71,752	2.08	42,327	3.53
8	125,148,000	57,834	2.16	42,653	2.93
9	107,440,000	62,013	1.73	43,020	2.50
10	127,894,000	61,298	2.09	42,466	3.01
11	129,193,000	84,663	1.53	47,621	2.71
12	125,198,000	59,245	2.11	38,136	3.28
13	93,711,000	53,093	1.77	35,745	2.62
14	89,344,000	44,112	2.03	29,746	3.00
15	73,467,000	37,814	1.94	26,524	2.77
16	74,037,000	38,735	1.91	23,328	3.17
17	73,367,000	34,621	2.12	19,396	3.78
18	73,078,000	45,135	1.62	27,028	2.70
19	56,044,000	25,676	2.18	11,185	5.01
20	63,317,000	29,478	2.15	17,051	3.71
21	33,824,000	20,916	1.62	9,103	3.72
22	33,786,000	28,410	1.19	11,056	3.06
X	131,245,000	34,842	3.77	20,400	6.43
Y	21,753,000	4,193	5.19	1,784	12.19
RefSeq	15,696,674	14,534	1.08		
Totals	2,710,164,000	1,419,190	1.91	887,450	3.05

Length (bp) is from the public Genome Assembly of 5 September 2000. Density of SNPs on each chromosome is influenced by the amount of available genome sequence included in the Genome Assembly, depth of overlap coverage from TSC reads and clone overlaps, and the underlying heterozygosity (Table 2). Data are presented for the entire dataset (All SNPs) and for those from the SNP consortium (TSC SNPs), as the latter are more evenly spaced than those from clone overlaps.

表1. SNP在染色体上的分布

染色体	长度(bp)	所有SNP		TSC SNP	
		SNP	SNP间的距离kb	SNP	SNP间的距离kb
1	214,066,000	129,931	1.65	75,166	2.85
2	222,889,000	103,664	2.15	76,985	2.90
3	186,938,000	93,140	2.01	63,669	2.94
4	169,035,000	84,426	2.00	65,719	2.57
5	170,954,000	117,882	1.45	63,545	2.69
6	165,022,000	96,317	1.71	53,797	3.07
7	149,414,000	71,752	2.08	42,327	3.53
8	125,148,000	57,834	2.16	42,653	2.93
9	107,440,000	62,013	1.73	43,020	2.50
10	127,894,000	61,298	2.09	42,466	3.01
11	129,193,000	84,663	1.53	47,621	2.71
12	125,198,000	59,245	2.11	38,136	3.28
13	93,711,000	53,093	1.77	35,745	2.62
14	89,344,000	44,112	2.03	29,746	3.00
15	73,467,000	37,814	1.94	26,524	2.77
16	74,037,000	38,735	1.91	23,328	3.17
17	73,367,000	34,621	2.12	19,396	3.78
18	73,078,000	45,135	1.62	27,028	2.70
19	56,044,000	25,676	2.18	11,185	5.01
20	63,317,000	29,478	2.15	17,051	3.71
21	33,824,000	20,916	1.62	9,103	3.72
22	33,786,000	28,410	1.19	11,056	3.06
X	131,245,000	34,842	3.77	20,400	6.43
Y	21,753,000	4,193	5.19	1,784	12.19
参考序列	15,696,674	14,534	1.08		
总计	2,710,164,000	1,419,190	1.91	887,450	3.05

长度(bp)来自2000年9月5日公布的基因组组装信息。每条染色体上SNP的密度受基因组组装收录的可用基因组序列的数量、TSC读长和克隆重叠的覆盖度的深度，以及潜在的杂合性影响(表2)。表中数据来自整个数据集(所有SNP)和SNP国际联合会(TSC SNP)，后者比来自克隆重叠的SNP分布更均匀。

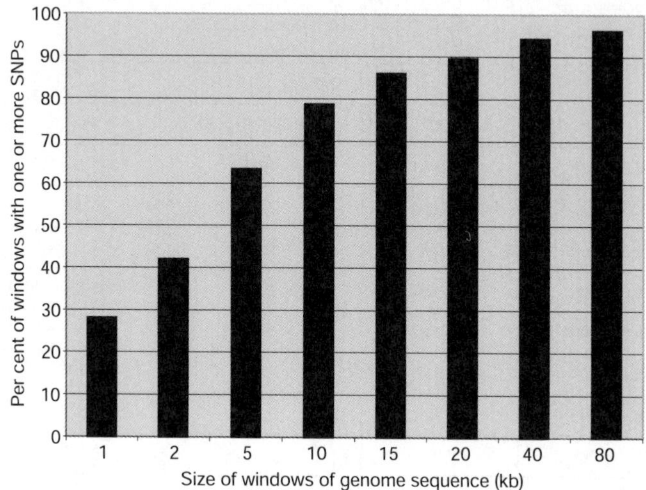

Fig. 1. Distribution of SNP coverage across intervals of finished sequence. Windows of defined size (in chromosome coordinates) were examined for whether they contained one or more SNPs. Analysis was restricted to the 900 Mb of available finished sequence.

The density of SNPs obtained at any given location depends upon the methods of SNP discovery contributing at each position (TSC, BAC overlap or targeted), the availability of genome sequence for SNP discovery and mapping, and the rate of nucleotide diversity. Of these, only nucleotide diversity is a fundamental characteristic of the region and population studied. To chart the landscape of human genome sequence polymorphism, we performed a genome-wide analysis of nucleotide diversity.

Analysis of Nucleotide Diversity

Describing the underlying pattern of nucleotide diversity required a polymorphism survey performed at high density, in a single, defined population sample, and analysed with a uniform set of tools. We reanalysed 4.5M passing sequence reads generated by TSC using genomic alignment using the NQS (see Methods). This set contained 1.2 billion aligned bases and 920,752 heterozygous positions. We measured nucleotide sequence variation using the normalized measure of heterozygosity (π), representing the likelihood that a nucleotide position will be heterozygous when compared across two chromosomes selected randomly from a population. π also estimates the population genetic parameter $\Theta = 4N_e\mu$ in a model in which sites evolve neutrally, with mutation rate μ, in a constant-sized population of effective size N_e. For the human genome, π was 7.51×10^{-4}, or one SNP for every 1,331 bp surveyed in two chromosomes drawn from the NIH diversity panel. This value agrees with smaller surveys of human genome variation[18-20].

We next examined the heterozygosity of individual chromosomes (Table 2). The autosomes were quite similar to one another, with 20 of 22 within 10% of the genome-wide average for autosomes (7.65×10^{-4}). Two had more extreme values: chromosome 21 ($\pi = 5.19 \times 10^{-4}$) and

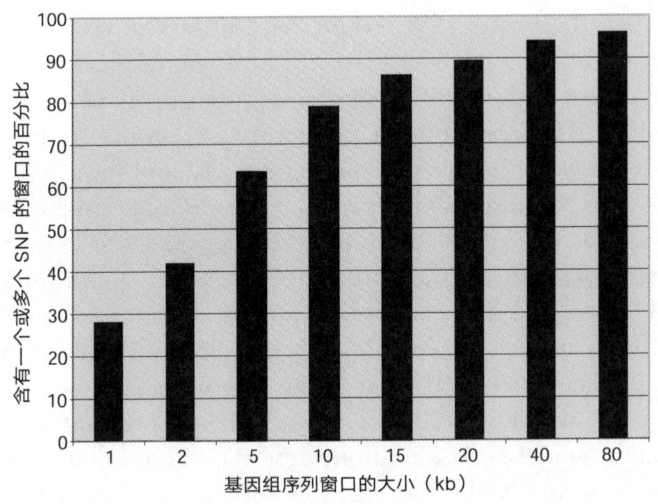

图 1. 已完成序列区间里 SNP 的覆盖分布。检查已定义大小的窗口（在染色体坐标内）是否包含一个或多个 SNP。分析限于 900 Mb 可用的已完成序列。

任何给定位置的 SNP 密度取决于各位点发现 SNP 时所用的方法（TSC、BAC 重叠或者靶向）、基因组序列对于 SNP 发现和定位的可利用性以及核苷酸多样性的比率。这其中，只有核苷酸多样性是区域和群体研究的基本特性。为了绘制人类基因组序列的多态性图谱，我们进行了一个全基因组范围内的核苷酸多样性分析。

核苷酸多样性分析

描述核苷酸多样性的潜在模式需要在高密度、单一的群体样本中进行多态性调查，并且使用一组统一的工具进行分析。我们使用 NQS 方法（见方法）对 TSC 产生的 4.5M 数据进行了重新分析（见方法）。这一组包含了 12 亿对碱基和 920,752 个杂合位点。我们用测量杂合性的标准方法（π）测量核苷酸序列的差异，这代表着当随机从一个群体中选出两条染色体比对时一个核苷酸位置是杂合子的可能性。π 也估算了群体遗传参数 $\Theta = 4N_e\mu$，在模型中其位点进化中立，突变率为 μ，群体有效大小为常量 N_e。对于人类基因组，π 是 7.51×10^{-4}，或者来自 NIH 多样性模型中两个染色体调查得到每 1,331 bp 一个 SNP。这个估值与人类基因组变异的小型调查一致[18-20]。

我们接下来检测了单个染色体的杂合性（表 2）。常染色体彼此十分相似，22 对常染色体中有 20 对在全基因组平均值的 10% 以内（7.65×10^{-4}）。另外两对的值则非常极端：21 号染色体（$\pi = 5.19 \times 10^{-4}$），15 号染色体（$\pi = 8.79 \times 10^{-4}$）。这些观测结

chromosome 15 ($\pi = 8.79 \times 10^{-4}$). Whether these observations are due to statistical fluctuations or methodological issues, or are biologically meaningful, will require investigation. The most striking difference in heterozygosity is the lower diversity of the sex chromosomes. The lower rate of polymorphism on the X chromosome may be explained by both a lower effective population size (N_e) and lower mutation rate (μ) in $\Theta = 4N_e\mu$. Because the X chromosome is hemizygous in males, the effective population size is three-quarters of that of the autosomes. In addition, μ is higher in male than in female meiosis, with $\mu_{male}/\mu_{female} \approx 1.7/1.0$ (ref. 33). As the X chromosome undergoes male meiosis only 1/3 of the time, the overall rate of mutation in the X chromosome is expected to be 91% that of the autosomes ($\mu_X = 1.23/1.35 = 0.91$). Thus, the diversity of the X chromosome is predicted to be 69% that of the autosomes. The observed heterozygosity of the X chromosome was 4.69×10^{-4}, or 61% of the average value of the autosomes. Thus, the population genetic considerations described above could largely explain the lower heterozygosity on the X chromosome. It is possible that strong selection on the X chromosome (owing to hemizygosity in males) or other factors might partially explain this observation.

Table 2. Nucleotide diversity by chromosome

Chromosome	Heterozygous positions	High-quality bp examined	$\pi (\times 10^{-4})$
1	71,483	92,639,616	7.72
2	81,860	111,060,861	7.37
3	61,190	81,359,748	7.52
4	59,922	74,162,156	8.08
5	56,344	77,924,663	7.23
6	53,864	72,380,717	7.44
7	52,010	68,527,550	7.59
8	44,477	57,476,056	7.74
9	41,329	50,834,047	8.13
10	43,040	52,184,561	8.25
11	47,477	56,680,783	8.38
12	38,607	51,160,578	7.55
13	35,250	43,915,606	8.03
14	35,083	47,425,180	7.40
15	27,847	31,682,199	8.79
16	22,994	27,736,356	8.29
17	21,247	27,124,496	7.83
18	24,711	30,357,102	8.14
19	11,499	15,060,544	7.64
20	22,726	31,795,754	7.15
21	26,160	50,367,158	5.19

果无论是由统计学波动或方法论的问题造成的，还是具有生物学意义，都需要进行研究。杂合性方面最显著的差异是性染色体中较低的多样性。X 染色体的低多态性也许可以用 $\Theta = 4N_e\mu$ 公式中较小的有效群体大小 (N_e) 和较低的突变率 (μ) 来解释。因为 X 染色体在男性中是半合子，其有效群体大小是常染色体的四分之三。此外，男性减数分裂比女性减数分裂时的 μ 更高，$\mu_{male}/\mu_{female} \approx 1.7/1.0$（参考文献 33）。由于 X 染色体在男性减数分裂中只占 1/3 的时间，那么 X 染色体的总体突变率应该是常染色体的 91%（$\mu_X = 1.23/1.35 = 0.91$）。因此，X 染色体的多样性预计是常染色体的 69%。观测到的 X 染色体的杂合性为 4.69×10^{-4}，或者说是常染色体平均值的 61%。因此，上述关于群体遗传学的考虑很大程度上可以解释 X 染色体的低杂合性。X 染色体上的强烈选择（由于在男性中是半合子），或者其他因素可能会部分地解释这项观察结果。

表 2. 染色体的核苷酸多样性

染色体	杂合性位置	检查到的高质量 bp	$\pi(\times 10^{-4})$
1	71,483	92,639,616	7.72
2	81,860	111,060,861	7.37
3	61,190	81,359,748	7.52
4	59,922	74,162,156	8.08
5	56,344	77,924,663	7.23
6	53,864	72,380,717	7.44
7	52,010	68,527,550	7.59
8	44,477	57,476,056	7.74
9	41,329	50,834,047	8.13
10	43,040	52,184,561	8.25
11	47,477	56,680,783	8.38
12	38,607	51,160,578	7.55
13	35,250	43,915,606	8.03
14	35,083	47,425,180	7.40
15	27,847	31,682,199	8.79
16	22,994	27,736,356	8.29
17	21,247	27,124,496	7.83
18	24,711	30,357,102	8.14
19	11,499	15,060,544	7.64
20	22,726	31,795,754	7.15
21	26,160	50,367,158	5.19

Continued

Chromosome	Heterozygous positions	High-quality bp examined	$\pi(\times 10^{-4})$
22	17,469	20,478,378	8.53
X	23,818	50,809,568	4.69
Y	348	2,304,916	1.51
Total	920,752	1,225,448,590	7.51

Heterozygosity (π) of each chromosome. The data were filtered to remove repetitive sequences and heterozygosity calculated as described in the methods. Heterozygous positions and high-quality bases examined were counted separately for each pairwise comparison of read to genome, and then summed over each chromosome.

The Y chromosome has the lowest observed heterozygosity of any chromosome. It is divided into two regions: a pseudoautosomal region at either telomeric end that recombines with the X chromosome and is highly heterozygous[34], and the non-recombining Y (NRY). The genome assembly used for this analysis contains only the NRY, which shows very little diversity: 348 SNPs in 2,304,916 bases ($\pi = 1.51 \times 10^{-4}$). These values agree reasonably with previous estimates for NRY[35,36]. The lower diversity of NRY is influenced by a smaller effective population size (20% that of the autosomes), counterbalanced by the higher mutation rate of male meiosis ($\mu_Y = 1.7/1.35 = 1.26 \times$ that of the autosomes). These factors predict that the Y chromosome would have a diversity 31% that of the autosomes, as compared to the observed 20%. Other influences might include selection against deleterious alleles, patterns of male dispersal[35] and a correlation of diversity with recombination rate[19].

To look at diversity on a finer scale, we divided each chromosome into contiguous 200,000-bp bins according to the public Genome Assembly of 5 September 2000. The distribution of heterozygosity among these bins ranges from zero (12 bins, each with zero SNPs over an average of 24,720 bp examined) to 60×10^{-4} (357 SNPs in a bin surveying 58,755 bp). Although 95% of bins display nucleotide diversity values between 2.0×10^{-4} and 15.8×10^{-4}, the pattern is variable (Fig. 2a, b; see also Supplementary Information). One measure of the spread in the data is the coefficient of variation (CV), the ratio of the standard deviation (σ) to the mean (μ) of the heterozygosity π of each individual read. For the observed data, the CV($\sigma_{\text{observed}}/\mu_{\text{observed}}$) was 1.93, considerably larger than would be expected if every base had uniform diversity, corresponding to a Poisson sampling process ($\sigma_{\text{Poisson}}/\mu_{\text{Poisson}} = 1.73$). It was expected that the observed distribution would be much more variable than a Poisson process, because both biochemical and evolutionary forces cause diversity to be nonuniform across the genome. Biological factors may include rates of mutation and recombination at each locus. For example, heterozygosity is correlated with the GC content for each read (Fig. 2c), reflecting, at least in part, the high frequency of CpG to TpG mutations arising from deamination of methylated 5-methylcytosine. Population genetic forces are likely to be even more important: each locus has its own history, with samples at some loci tracing back to a recent common ancestor, and other loci describing more ancient genealogies. The time to the most recent common ancestor at a particular stretch of DNA is variable, and represents the opportunity for sequence divergence; thus, the expected pattern of heterozygosity is more heterogeneous than if every locus shared the same history[37,38].

续表

染色体	杂合性位置	检查到的高质量 bp	$\pi(\times 10^{-4})$
22	17,469	20,478,378	8.53
X	23,818	50,809,568	4.69
Y	348	2,304,916	1.51
总计	920,752	1,225,448,590	7.51

每条染色体的杂合性(π)。数据过滤掉了重复序列，杂合性按照方法里描述的进行计算。杂合位置和高质量碱基的检查是由读长与基因组两两比较分别计算得到的，然后将每条染色体进行合计。

Y 染色体观察到的杂合性比任何一个染色体都低。它分成了两个区：一个是假常染色体区，在与 X 染色体结合的两个端粒末端之一，是高度杂合的[34]；另一个是 Y 染色体非重组区(NRY)。本研究对基因组进行组装的分析只包含 NRY 区，该区域有较低的多样性：2,304,916 个碱基中只有 348 个 SNP 位点($\pi = 1.51 \times 10^{-4}$)。这些值与之前对 NRY 的估计一致[35,36]。NRY 的低多样性受较小的有效群体大小的影响(常染色体的 20%)，这抵消了男性减数分裂时的高突变率($\mu_Y = 1.7/1.35 = 1.26 \times$ 常染色体的 μ)。这些因素预示与观察到的 20% 相比，Y 染色体将有常染色体 31% 的多样性。其他的影响可能包括对有害等位基因的选择，男性的传播模式[35]以及多样性与重组率的相关性[19]。

为了在更精细的尺度上研究多样性，我们根据 2000 年 9 月 5 日公布的基因组组装结果，把每条染色体分成了连续 200,000 bp 的分箱。这些分箱的杂合性分布变化从 0(12 个分箱平均长度是 24,720 bp，每个分箱有 0 个 SNP)到 60×10^{-4}(该分箱长度是 58,755 bp，有 357 个 SNP)。尽管 95% 的分箱呈现的核苷酸多样性值在 2.0×10^{-4} 到 15.8×10^{-4} 之间，但模式是不同的(图 2a 和 2b；也可见于补充信息)。衡量数据分散性的一种方法是差异系数(CV)，即每个个体序列杂合性(π)的标准差(σ)与平均值(μ)之比。对于观察数据，CV 值($\sigma_{观测}/\mu_{观测}$)是 1.93，比如果每个碱基具有统一的多样性所预期的值要大得多，相当于泊松抽样过程($\sigma_{泊松}/\mu_{泊松} = 1.73$)。预期观测到的分布将会比泊松过程变化多得多，因为生化和进化力都会导致整个基因组的多样性不均匀。生物因素可能包括每个基因座的突变和重组率。例如，每个序列的杂合性与其 GC 含量相关(图 2c)，至少在一定程度上能反映由于甲基化 5–甲基胞嘧啶脱氨引起的 CpG 到 TpG 突变高频发生的原因。群体遗传学力量也许更重要：每个基因座有它自己的历史，一些基因座可以追溯至一个最近的共同祖先，而其他的基因座可能有更多的古老家系。在特定的 DNA 分支上追溯到最近共同祖先的时间是不同的，这代表了序列分化的可能性；因此，相比如果每个基因座都具有相同的历史情况，杂合性的期望模式会更多样[37,38]。

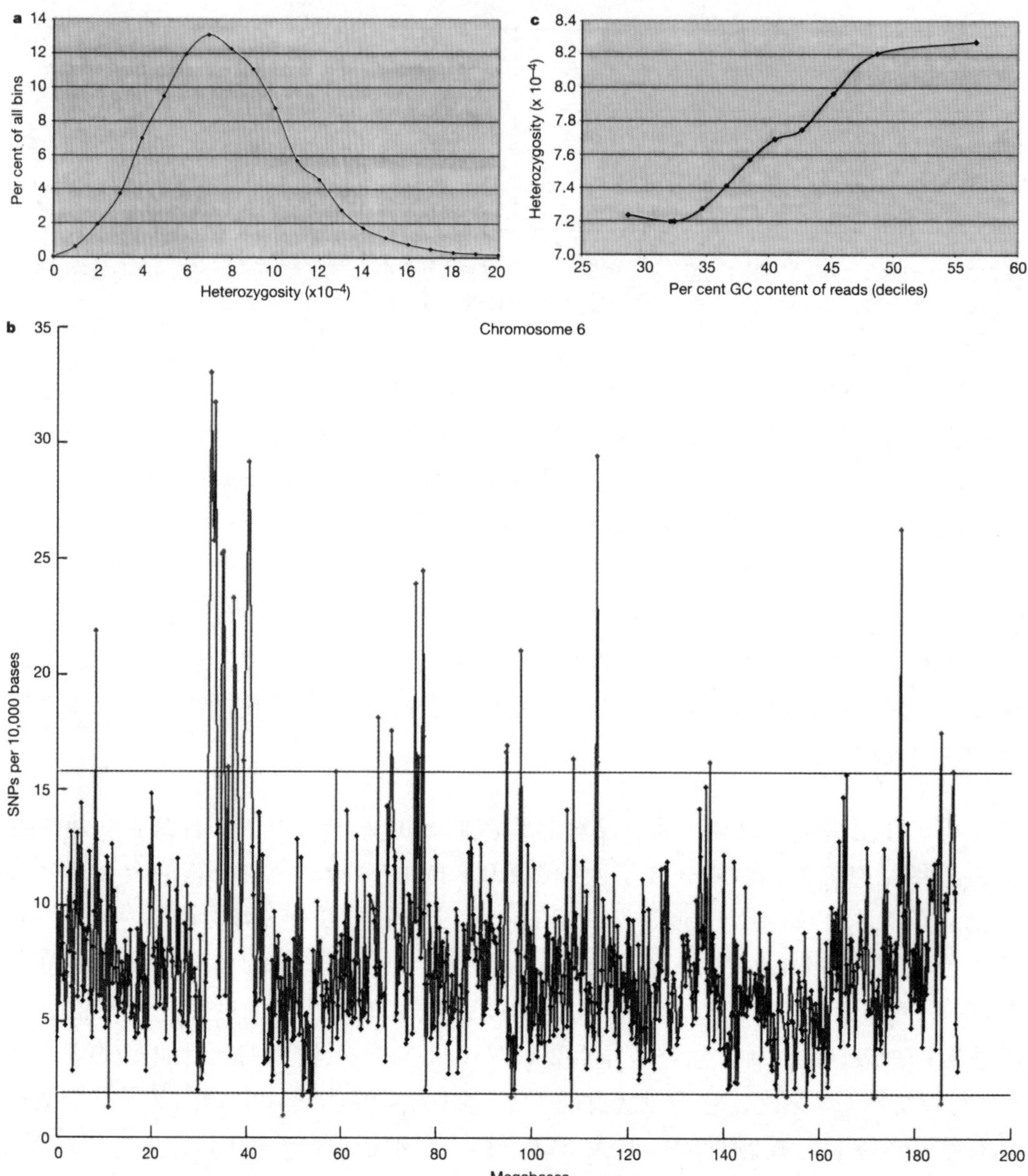

Fig. 2. Distribution of heterozygosity. **a**, The genome was divided into contiguous bins of 200,000 bp based on chromosome coordinates, and the number of high-quality bases examined and heterozygosity calculated for each. A histogram was generated of the distribution of heterozygosity values across all such bins. **b**, Heterozygosity was calculated across contiguous 200,000-bp bins on Chromosome 6. The blue lines represent the values within which 95% of regions fall: 2.0×10^{-4}–15.8×10^{-4}. Red, bins falling outside this range. The extended region of unusually high heterozygosity centred at 34 Mb corresponds to the HLA. **c**, Correlation of nucleotide diversity with GC content of each read (autosomes only). The GC content and heterozygosity of reads from the heterozygosity analysis was calculated after sorting of reads by GC content and separation into 10 bins of equal size. Each bin contains ~150 Mb of aligned, high-quality sequence.

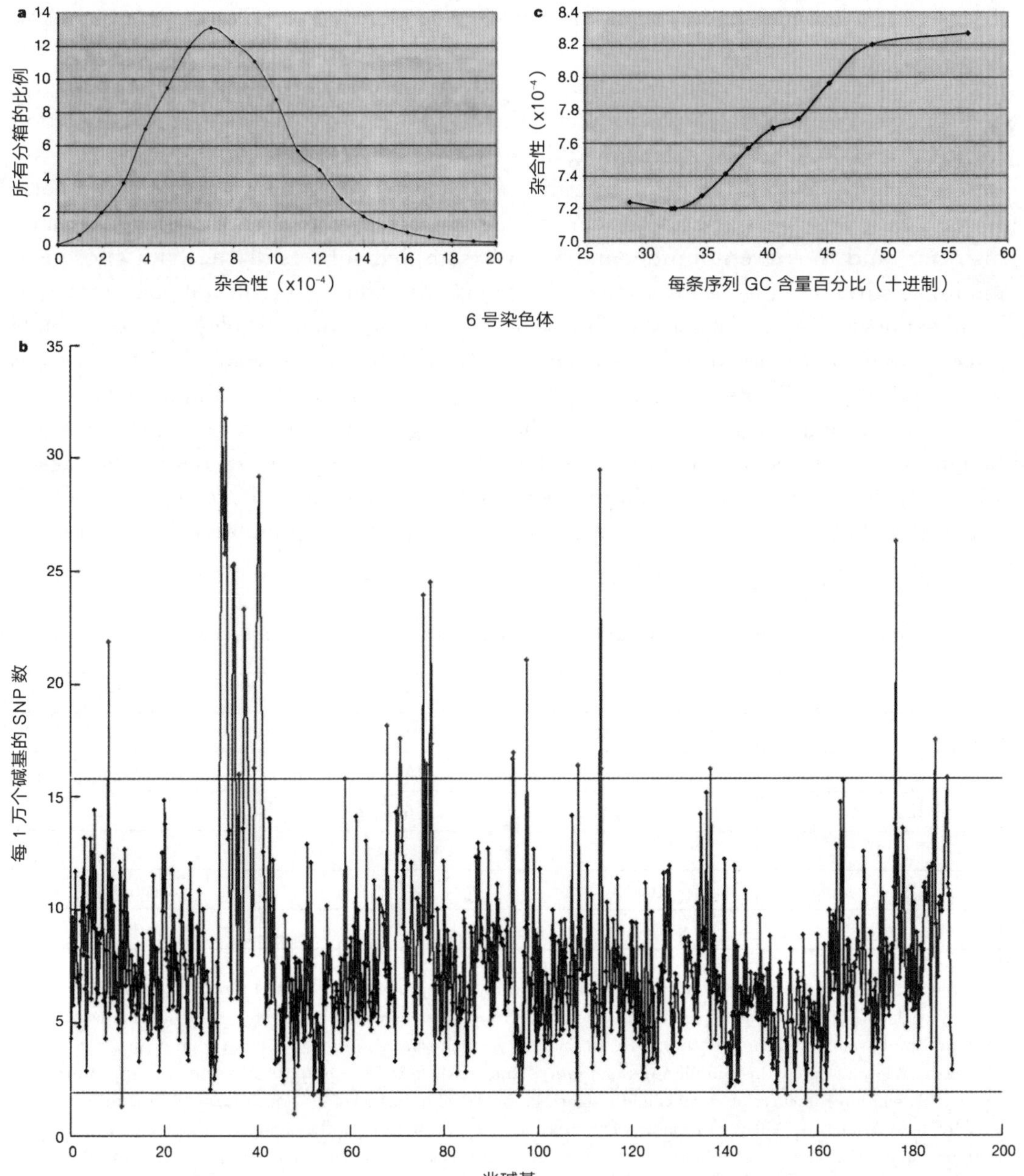

图 2. 杂合性分布。**a**，根据染色体坐标，基因组被分成 200,000 bp 的连续分箱，并计算每个分箱高质量碱基的数目和杂合性。所有这些分箱的杂合性值的分布形成一个直方图。**b**，6 号染色体上所有连续的 200,000 bp 分箱的杂合性。蓝线代表的区间为 2.0×10^{-4}~15.8×10^{-4}，95% 的值落在这之间。红色代表落在该范围外面的分箱。相对于 HLA，显著高杂合性的扩展区集中在 34 Mb。**c**，每条序列（只有常染色体）的核苷酸多样性与 GC 含量的相关性。首先将序列按 GC 含量进行排序，然后平均分成 10 个分箱再进行杂合性分析，包括 GC 含量和序列杂合性计算。每个分箱包含约 150 Mb 比对上的高质量的序列。

To assess whether gene history would account for the observed variation in heterozygosity, we compared the observed CV to that expected under a standard coalescent population genetic model. For each read, we adjusted μ on the basis of its per cent GC and length, and simulated genealogical histories under the assumption of a constant-sized population with $N_e = 10{,}000$. The CV determined under this model ($\sigma_{\text{constant-size}}/\mu_{\text{constant-size}} = 1.96$) is a close match to the observed data. To estimate standard deviations around these estimates of the CV, it was necessary to consider that tightly linked regions may display correlated histories, and thus are nonindependent. We sampled subsets of the data chosen to minimize correlation among reads (see Methods), providing estimates of the mean and standard deviation of CV for the observed and simulated data (Table 3). These results indicate that the observed pattern of genome-wide heterozygosity is broadly consistent with predictions of this standard population genetic model (for comparison, see an analysis of variation in heterozygosity in the mouse genome)[39]. However, much work will be required to assess additional factors that could influence this distribution: biological factors such as variation in mutation and recombination rates, historical forces such as bottlenecks[40,41], expansions or admixture of differentiated populations, evolutionary selection, and methodological artefacts.

Table 3. Coefficients of variation for the observed data and the Poisson and coalescent models

SNPs per read	Observed	Poisson	Coalescent
0	8,796 ± 43	8,256 ± 52	8,767 ± 50
1	2,247 ± 44	3,040 ± 49	2,332 ± 46
2	668 ± 24	617 ± 24	663 ± 26
3	214 ± 14	99 ± 9	200 ± 15
4	102 ± 10	16 ± 4	66 ± 9
σ/μ	1.94 ± 0.02	1.72 ± 0.02	1.96 ± 0.03

Observed distribution of heterozygosity and comparison to expectation under Poisson and coalescent population genetic models. The autosomes were divided into 200,000-bp bins according to chromosome coordinates and one read randomly selected from each bin. This procedure was chosen to minimize the correlation in gene history of nearby regions, under the simplifying assumption that reads 200,000 bp apart and selected from unrelated individuals will have uncorrelated genealogies. Correlation of gene history does not influence the expected mean value of the CV, but does effect its variance. The random selection of reads and generation of expected distributions were repeated 100 times: presented are the mean and standard deviation of the number of reads in which 0, 1, 2, 3, or 4 SNPs were observed or predicted under each scenario. The Poisson model reports the number of such reads expected to display 0–4 SNPs under Poisson sampling of each read with a heterozygosity adjusted for length and GC content (Fig. 2c). Even in this reduced data set, the Poisson model can be rejected at $P < 10^{-99}$. The coalescent simulation[38] assumed a constant-sized population of effective size 10,000 and free recombination among reads. For each read, μ was scaled according to its length and GC content (Fig. 2c). Each sampled read was assigned a coalescent history from a simulated distribution and the number of SNPs predicted. The coefficient of variation of the estimate of heterozygosity is presented, with the mean and standard deviation of the 100 sampling runs shown.

Regions of low diversity were more prevalent on the sex chromosomes. Whereas only 2.5% of 200,000-bp bins across the genome had $\pi < 2.0 \times 10^{-4}$, 15% of bins on the X chromosome[42] and 89% on the Y chromosome (NRY) had these levels of diversity. Regions

为了评估基因历史能否解释观察到的杂合性的变异，我们将观测到的 CV 与标准的联合群体遗传模型下的预期值进行了对比。对每条序列来说，我们依据 GC 的百分比和长度来调整 μ，并在恒定种群大小 $N_e = 10,000$ 的假定条件下模拟系谱历史。这个模式（$\sigma_{恒定大小}/\mu_{恒定大小} = 1.96$）下得到的 CV 与观测数据高度匹配。为了评估这些 CV 估值的标准差，必须要考虑紧密相连区域可能表现出相关联的历史，这些区域是非独立的。为了使关联性降到最小，我们抽取了这些数据的子集样本（见方法），并提供了观测和模拟数据 CV 的平均值和标准差（表 3）。这些结果表明观测到的全基因组杂合性的模式与该标准群体遗传学模型的预测大体一致（为了进行比较，请参阅小鼠基因组杂合性变异的分析）[39]。然而，我们还需要做许多工作来评价可能影响分布的其他因素：生物因素如突变和重组率的变异，历史力比如瓶颈[40,41]，不同群体的扩张或混合，进化选择以及方法学误差。

表 3. 观测数据、泊松模型以及联合模型的差异系数

每个读长的 SNP	观测值	泊松模型	联合模型
0	8,796 ±43	8,256 ± 52	8,767 ± 50
1	2,247 ± 44	3,040 ± 49	2,332 ± 46
2	668 ± 24	617 ± 24	663 ± 26
3	214 ± 14	99 ± 9	200 ± 15
4	102 ± 10	16 ± 4	66 ± 9
σ/μ	1.94 ± 0.02	1.72 ± 0.02	1.96 ± 0.03

观测到的杂合性分布，以及在泊松模型和联合群体遗传模型下的比较结果。依据染色体坐标，常染色体被分成 200,000 bp 的分箱，并随机从每个分箱中选择一条序列。该过程是为了使相邻区域基因座的相关性降到最小，在一种简化思维的假设下，每条序列间隔是 200,000 bp，并且选自不相关的个体，具有不相关的家族系谱。基因史的关联性并不影响 CV 预期的平均值，但会影响它的方差。为了达到期望的分布情况，对序列和世代进行随机选择并重复 100 次：表中呈现的是能观测到的或预测的包含 0、1、2、3 或者 4 个 SNP 位点的序列数量的平均值和标准差。泊松模型报告了这样的序列数量，期望在每次利用泊松抽样抽取序列时，预期能显示 0~4 个 SNP，并且依据长度和 GC 含量对杂合性进行调整（图 2c）。即使在这个简化的数据集中，在 $P < 10^{-99}$ 时，泊松模型仍然可以被拒绝。联合模拟[38]假定了一个具有 10,000 个有效个体的恒定大小的群体，序列之间可以进行自由重组。依据每条序列的长度和 GC 含量来衡量其 μ 值（图 2c）。每个抽样的序列从模拟分布和预测 SNP 的数目中分配一个联合历史。表中呈现的是杂合性估算的差异系数以及 100 次运行的平均值和标准差。

在性染色体上，低多样性区域更普遍。尽管整个基因组中只有 2.5% 的 200,000 bp 分箱的 π 值小于 2.0×10^{-4}，但 X 染色体上 15% 的分箱[42]和 Y 染色体（NRY）上 89%

of low diversity may be explained by the smaller effective population size of the sex chromosomes and the variable underlying distribution of heterozygosity. Strong selection acting on the sex chromosomes in males might also have a role, but this hypothesis requires further testing. Regions of high heterozygosity were also observed. One was found on chromosome 6 (Fig. 2b, centred on 34 Mb), and was confirmed to represent the HLA locus, which has high nucleotide diversity owing to balancing selection[43]. Other regions of varying size were observed on this and other chromosomes (Fig. 2c and Supplementary Information). Some of these highly diverse regions might have also experienced balancing selection, but there are other possible explanations: for example, sampling fluctuations of the coalescent distribution, regions with high rates of mutation and/or recombination, unrecognized duplications in the human genome and sequencing of a rare haplotype by the HGP (to which the TSC reads were compared).

Given the unfinished state of publicly available sequence data and genome assembly, it will be important to reevaluate these estimates as more complete genome sequence becomes available.

Implications for Medical and Population Genetics

We describe a map of publicly available SNPs (as of November 2000), fully integrated with the sequence, physical and genetic maps of the human genome. We anticipate immediate application to studies of human population genetics, candidate-gene studies for disease association, and eventually unbiased, genome-wide association scans. First, the map provides an unprecedented tool for studying the character of human sequence variation. We use these data to describe the first genome-wide view of how human DNA sequence varies in the population, and the public availability of these data should fuel future research into biological and population genetic influences on human genetic diversity.

Second, insights into human evolutionary history will be obtained by using SNPs from the map to characterize haplotype diversity throughout the genome. Human haplotype structure remains largely unexplored, and this map makes it possible to define the extent and variation of haplotype identity, the number and frequencies of common haplotypes, and their distribution among and within existing ethnic groups.

Most practically, where a gene has been implicated in causing disease (by chromosomal position relative to linkage peaks, known biological function or expression pattern), it is desirable exhaustively to survey allelic variation for any association to disease. Using the SNP map, it should be possible to evaluate the extent to which common haplotypes contribute to disease risk. As the speed and efficiency of SNP genotyping increases, such studies will fuel increasingly comprehensive tests of the hypothesis that common variants contribute significantly to the risk of common diseases. To the extent that such studies are successful, they should profoundly affect our understanding of disease, methods of diagnosis, and ultimately the development of new and more effective therapies.

的分箱具有这些多样性水平。这些低多样性的区域可能由性染色体有效种群大小较小以及潜在的杂合性分布不同所致。作用在男性性染色体上的强选择可能也起到一定作用，但这个假说需要进一步的验证。此外还发现了一些高度杂合性的区域。其中一个在 6 号染色体上（图 2b，集中于 34 Mb），并被证实是 HLA 的基因座；由于平衡选择[43]，该区域具有高的核苷酸多样性。在 6 号和其他染色体上观察到大小不一的其他区域（图 2c 及补充信息）。一些高度多样化区域可能经历了平衡选择，但是也有一些其他可能的解释：比如联合模型分布抽样的波动性，带有高突变率和（或）重组率的区域，人类基因组未识别到的重复序列，以及由 HGP 进行的罕见单体型测序（以 TSC 读长作为对比）。

考虑到可公开获取序列数据和基因组组装的未完成状态，当更完整的基因组序列可以获取之时，重新评价这些估算显得尤为重要。

对医学和群体遗传学的影响

我们充分地整合了人类基因组的序列、物理图谱以及遗传图谱，描绘了一张可公开获取的 SNP 图谱（截至 2000 年 11 月）。我们预计该图谱可直接应用于人类群体基因组学研究和疾病相关的候选基因研究，并最终应用于无偏移的全基因组关联扫描。首先，图谱为人类序列差异特性的研究提供了空前的工具。我们用这些数据来描述关于群体中人类 DNA 序列变化的第一个全基因组视图，而这些数据的可公开获取性将促进未来关于生物和群体遗传学对人类遗传多样性的影响的研究。

其次，我们将利用图谱中的 SNP 来洞悉人类进化史，从而描述整个基因组的单体型多样性。人类单体型结构仍然有大部分未被探索，而这张图谱使得对单体型特征的范围和变异，常见单体型的数量和频率，以及它们在现有种群内的分布等的界定变得可能。

最实际的应用就是，当一个基因引起疾病时（通过与连锁峰相关的染色体位置，已知生物学功能或表达模式），可以详尽地调查与疾病有任何相关的等位基因的变异。利用 SNP 图谱，我们可以评估常见单体型促成疾病风险的程度。随着 SNP 基因分型的速度和效率的增加，这样的研究有助于更加全面地检验常见突变显著增加常见疾病风险这一假说。从某种程度上说这些研究是成功的，它们会深刻影响我们对疾病的理解和诊断方法，并最终开发新的更有效的治疗方法。

Methods

SNP identification

Candidate SNPs were identified by detection of high-confidence base differences in aligned sequences. For TSC, sequence reads were filtered to exclude low quality reads and those containing predominantly known repetitive sequence. Sequences were aligned to each other using the reduced representation shotgun (RRS) method, and by genomic alignment (GA) as described[18,22]. For GA of TSC data, reads were compared to available large-insert clones (finished and draft with available PHRAP quality scores) in GenBank. For the analysis of clone overlaps, all available finished and unfinished genomic sequence accessions were aligned. Two methods were used to detect SNPs. The NQS relies upon the sequence trace quality surrounding the SNP base to increase base-calling confidence[18,22]; most data discovered using the NQS was processed using SsahaSNP, an ultrafast, hash-based implementation of the algorithm (Z.N., A. Cox and J.C.M, manuscript in preparation). The second method calculates confidence scores on the basis of a Bayesian analysis of confidence scores[24]. A variety of methods were used to find SNPs in expressed sequence tag (EST) overlaps[24,25,27] and for targeted resequencing; details of the remaining SNPs can be found in the individual dbSNP entries (www.ncbi.nlm.nih.gov/SNP/).

Mapping of SNPs and features

MEGABLAST[44] was used to align TSC SNP flanking sequences to the genomic sequence accessions. A SNP was considered mapped if a high-quality match (99% identity or greater) was found across the available flanking sequence of no less than 270 bp. SNPs that matched more than three accessions with identity > 98% were judged to be possible repetitive regions and set aside. SNP coordinates were generated relative to the OO18 build of the genome assembly (5 September 2000) and the OO15 build (15 July 2000), using the AGP format files provided by D. Haussler (http://genome.ucsc.edu).

The NCBI RefSeq mRNA transcripts[31] were aligned to the Genome Assembly using the NCBI SPIDEY alignment tool. Alignment required > 97% sequence similarity between mRNA and genome sequence; alignments were refined by taking into account the donor/acceptor sites. In cases where CDS annotations were available in the GenBank record, exons of the CDS were aligned within the confines of the mRNA alignment. Regions of known human repeats were annotated directly using RepeatMasker (A. Smit, unpublished).

Nucleotide diversity analysis

To characterize nucleotide diversity, we required a data set in which all data could be analysed both for the number of high-quality bases meeting quality standards for SNP detection, and for

方　法

SNP 的鉴定

利用比对序列中高置信度碱基的差异进行候选 SNP 的鉴定。对 TSC 来说，将序列读长进行过滤处理，从而排除低质量读长以及那些包含显著已知重复序列的读长。利用简化代表性鸟枪法（RRS）和描述的基因组比对（GA）[18,22] 方法进行序列比对。对 TSC 数据的 GA 而言，序列与 GenBank 中可用的长的插入克隆（包含 PHRAP 质量值的已完成和未完成序列）进行比较。为了分析克隆重叠群，所有可用的完成和未完成基因组序列（含有基因库中编号）均被进行比对。两种方法用来对 SNP 进行鉴定。一种是 NQS 方法，该方法依赖围绕在 SNP 碱基周围序列的质量追踪来增加碱基识别的置信度[18,22]；多数数据使用 NQS 方法都是通过使用 SsahaSNP 软件来进行分析的，SsahaSNP 是一种超快的基于哈希算法实现的 SNP 鉴定工具（宁泽民、考克斯和马利金，稿件准备中）。第二种方法在基于置信评分的贝叶斯分析基础上计算置信评分[24]。我们使用多种方法在表达序列标签（EST）中鉴定了 SNP[24,25,27] 并进行靶向测序；剩余 SNP 的信息可以在单独的 dbSNP 条目中找到（www.ncbi.nlm.nih.gov/SNP/）。

SNP 图谱及其特征

利用 MEGABLAST 软件[44] 将 TSC SNP 侧翼序列与基因组含有基因库编号的序列进行比对。对于任何一个 SNP 位点，如果发现在可用侧翼序列中不少于 270 bp 的高质量碱基与库中序列相匹配（99% 的一致性或更高），则认为该 SNP 被比对上。如果一个 SNP 匹配三个以上的基因组序列（98% 相似性），那么这个 SNP 定义为重复区，将被去掉。基于 OO18（2000 年 9 月 5 日）和 OO15（2000 年 7 月 15 日）构建的基因组组装结果，使用豪斯勒提供的 AGP 文件格式（http://genome.ucsc.edu）生成 SNP 坐标。

利用 NCBI SPIDEY 比对工具将 NCBI 参考序列中的 mRNA 转录本[31] 与基因组组装结果进行比对。比对要求 mRNA 与基因组序列间有 > 97% 的序列相似性；考虑供 / 受体位置对结果进行调整。当 CDS 在 GenBank 中有注释信息时，CDS 的外显子区域在信使 RNA 的范围内进行比对。人类基因组中已知的重复片段区域直接使用 RepeatMasker 软件进行注释（斯米特，未发表）。

核苷酸多样性分析

为了描述核苷酸的多样性，我们需要一个数据集，该数据集里所有的数据都满足 SNP 检测的质量标准中高质量碱基的数量以及 SNP 数量。为了确保分析的均一性，我们只分

the number of SNPs. To ensure homogeneity of analysis, we performed a single analysis of 4.5 million high-quality TSC reads from the Sanger Centre, Washington University in St. Louis and the Whitehead Center for Genome Research. The GC content of these reads was 41%, the same as the genome as a whole[32], and the distribution of read GC content across deciles of the genome (sorted by GC content) was within 10% of the expected value for all bins. The read coverage was well distributed: 88% of contiguous 200,000-bp windows contained over 10,000 aligned bases (5%) surveyed for SNPs (see below). Using a single analytic tool (SsahaSNP, an implementation of the NQS; Z.N., A. Cox and J.C.M, in preparation), these reads were aligned to the available genome sequence (finished and draft with quality scores) and the number of high-quality bases (meeting NQS) and SNPs counted. We limited the analysis to SNPs found by genomic alignment so that the cluster depth of each comparison would be exactly two chromosomes. We precisely measured the target size for SNP discovery by counting the number of positions meeting the NQS. This is desirable because alignments contain positions of both high and low quality, but only those meeting the NQS are candidates for SNP discovery. Where a single TSC read aligned to multiple (overlapping) BACs from the HGP, we averaged the number of SNPs and aligned bp for all pairwise alignments of that read; this weighted evenly those reads mapping to a single BAC and those aligning to a region of overlap. Reads representing repeat loci were excluded using validated criteria[18,22]: alignments of reads to genome were excluded if they were less than 99% identical. The genome was then divided into contiguous bins of 200,000 bp (based on chromosome-relative coordinates). Individual reads were filtered for repeats: any that aligned to more than one bin in the genome assembly were rejected. Finally, heterozygous positions and bases meeting the NQS were counted. As a final filter for regions containing a high proportion of repeats, we reject any bin for which more than 10% of the reads mapping to that bin also mapped to another chromosome. Finally, to avoid statistical fluctuation due to inadequate sampling, we examined only the 88% of bins in which at least 10,000 aligned bases met the NQS and thus could be examined for SNPs.

Coalescent modelling was performed by simulation[38], and assumed a constant-sized population of 10,000 individuals and a mutation rate adjusted for each read on the basis of its GC content (Fig. 2c) and length. To assess the standard deviation around this estimate, the simulation was repeated 100 times. For the observed data, calculating a standard deviation around the CV is difficult owing to the correlation of gene history for closely linked sites. In expectation, this correlation should not alter the mean of the observed coefficient of variation, but does influence its variance. To estimate the variance around the CV for the observed data, we selected 100 reduced data sets, each containing one randomly chosen read from each 200,000-bp bin along the autosomes. In using this approach, we assume that these reads, 200,000 bp apart and sampled from unrelated individuals, have independent genealogies. This random sampling procedure was repeated 100 times to estimate the mean and variance of the observed CV.

The data for the heterozygosity analysis, including the coordinates of each bin, the number of bases examined and number of SNPs identified, is available as Supplementary Information.

(**409**, 928-933; 2001)

析了来自桑格中心、圣路易斯华盛顿大学以及怀特黑德基因组研究中心的 450 万个高质量 TSC 序列。这些序列的 GC 含量是 41%，与整体基因组一致[32]，序列 GC 含量跨基因组十分位数的分布（按 GC 含量分类）在所有分箱预期值的 10%以内。序列覆盖分布良好：对连续 200,000 bp 窗口区域进行 SNP 分析发现，有 10,000 多个碱基（5%）分布在 88% 的区域中（见下文）。只利用一种分析软件（SsahaSNP，NQS 的一种实现方法；宁泽民、考克斯和马利金，稿件准备中）将这些序列与可用的基因组序列（含有质量值的完成的或未完成的序列）进行比对，计算高质量碱基（符合 NQS）和 SNP 的数目。我们对由基因组比对发现的 SNP 设置限制条件，使每个比较的重叠群深度都正好是两条染色体。通过计算符合 NQS 的位置数量，可以精确计算 SNP 发现区域的目标大小。因为比对信息中包括高质量和低质量碱基的位置，只有那些符合 NQS 的才能作为候选 SNP 位点，这一点是可取的。当单个 TSC 序列比对到来自 HGP 的多个（重叠）BAC 上时，我们对该序列上 SNP 的数目和比对的长度取平均数，这对那些映射到图谱的单一 BAC 和那些与一个重叠区域匹配的序列进行了加权平均。重复基因座上的序列使用如下确认准则进行排除[18,22]：如果与基因组比对的相似性小于 99%，那么该序列被去掉。然后，基因组被分成连续 200,000 bp 的分箱（基于染色体相对坐标）。对每条序列都进行如下重复序列过滤：如果匹配到基因组中多于一个分箱，序列被去掉。最终，计算符合 NQS 的杂合子的位置和碱基数目。最后一步过滤是针对高比例重复区的筛选，当某一分箱中 10% 以上的序列既比对到该分箱中，同时又比对到其他染色体上，那么该分箱被去掉。最终，为了避免由于采样不足引起的统计波动，我们只选择了其中 88% 的分箱进行 SNP 鉴定，这些分箱中至少有 10,000 个比对的碱基符合 NQS。

我们模拟了联合模型[38]，该模型假定了恒定大小为 10,000 个体的群体，并根据 GC 含量（图 2c）和长度对每个序列的突变率进行了调整。将模拟重复 100 次以评估该估算附近的标准差。对于观测到的数据来说，由于紧密相连位点的基因史具有相关性，根据 CV 计算标准差很困难。在预期中，这种相关性应该不会改变观测到的变异系数的均值，但是会影响其方差。为了依据 CV 对观测数据的方差进行估算，我们选定 100 个简化数据集，每个数据集均包含从每 200,000 bp 的常染色体分箱中随机选择的一个序列。运用这种方法，我们假定这些序列相距 200,000 bp 并且从来自不相关的个体中抽样，有独立的家系图谱。该随机抽样过程重复 100 次，以估算观测到的 CV 的平均值和方差。

杂合性分析的数据在补充信息中，这些信息包括每个分箱的坐标、被检查的碱基数目和被鉴定的 SNP 数目。

（杨晶 翻译；胡松年 审稿）

The International SNP Map Working Group[*]
[*] A full list of authors appears at the end of this paper.

Received 28 November; accepted 27 December 2000.

References:

1. Collins, F. S. Of needles and haystacks: finding human disease genes by positional cloning. *Clin. Res.* **39**, 615-623 (1991).
2. Collins, F. S., Guyer, M. S. & Charkravarti, A. Variations on a theme: cataloging human DNA sequence variation. *Science* **278**, 1580-1581 (1997).
3. Lander, E. S. The new genomics: global views of biology. *Science* **274**, 536-539 (1996).
4. Risch, N. & Merikangas, K. The future of genetic studies of complex human diseases. *Science* **273**, 1516-1517 (1996).
5. Li, W. H. & Sadler, L. A. Low nucleotide diversity in man. *Genetics* **129**, 513-523 (1991).
6. Cargill, M. *et al.* Characterization of single-nucleotide polymorphisms in coding regions of human genes [published erratum appears in *Nature Genet.* **23**, 373 (1999)]. *Nature Genet.* **22**, 231-238(1999).
7. Cambien, F. *et al.* Sequence diversity in 36 candidate genes for cardiovascular disorders. *Am. J. Hum. Genet.* **65**, 183-191 (1999).
8. Fullerton, S. M. *et al.* Apolipoprotein E variation at the sequence haplotype level: implications for the origin and maintenance of a major human polymorphism. *Am. J. Hum. Genet.* **67**, 881-900 (2000).
9. Halushka, M. K. *et al.* Patterns of single-nucleotide polymorphisms in candidate genes for blood-pressure homeostasis. *Nature Genet.* **22**, 239-247 (1999).
10. Nickerson, D. A. *et al.* DNA sequence diversity in a 9.7-kb region of the human lipoprotein lipase gene. *Nature Genet.* **19**, 233-240 (1998).
11. Rieder, M. J., Taylor, S. L., Clark, A. G. & Nickerson, D. A. Sequence variation in the human angiotensin converting enzyme. *Nature Genet.* **22**, 59-62 (1999).
12. Templeton, A. R., Weiss, K. M., Nickerson, D. A., Boerwinkle, E. & Sing, C. F. Cladistic structure within the human lipoprotein lipase gene and its implications for phenotypic association studies. *Genetics* **156**, 1259-1275 (2000).
13. Eaves, I. A. *et al.* The genetically isolated populations of Finland and sardinia may not be a panacea for linkage disequilibrium mapping of common disease genes. *Nature Genet.* **25**, 320-323 (2000).
14. Taillon-Miller, P. *et al.* Juxtaposed regions of extensive and minimal linkage disequilibrium in human Xq25 and Xq28. *Nature Genet.* **25**, 324-328 (2000).
15. Kruglyak, L. Prospects for whole-genome linkage disequilibrium mapping of common disease genes. *Nature Genet.* **22**, 139-144 (1999).
16. Collins, A., Lonjou, C. & Morton, N. E. Genetic epidemiology of single-nucleotide polymorphisms. *Proc. Natl Acad. Sci. USA* **96**, 15173-15177 (1999).
17. Reich, D. E. *et al.* Linkage disequilibrium in the human genome. *Nature* (submitted).
18. Altshuler, D. *et al.* An SNP map of the human genome generated by reduced representation shotgun sequencing. *Nature* **407**, 513-516 (2000).
19. Nachman, M. W., Bauer, V. L., Crowell, S. L. & Aquadro, C. F. DNA variability and recombination rates at X-linked loci in humans. *Genetics* **150**, 1133-1141 (1998).
20. Wang, D. G. *et al.* Large-scale identification, mapping, and genotyping of single-nucleotide polymorphisms in the human genome. *Science* **280**, 1077-1082 (1998).
21. Jorde, L. B. Linkage disequilibrium and the search for complex disease genes. *Genome Res.* **10**, 1435- 1444 (2000).
22. Mullikin, J. C. *et al.* An SNP map of human chromosome 22. *Nature* **407**, 516-520 (2000).
23. Collins, F. S., Brooks, L. D. & Chakravarti, A. A DNA polymorphism discovery resource for research on human genetic variation [published erratum appears in *Genome Res.* **9**, 210 (1999)]. *Genome Res.* **8**, 1229-1231 (1998).
24. Marth, G. T. *et al.* A general approach to single-nucleotide polymorphism discovery. *Nature Genet.* **23**, 452-456 (1999).
25. Buetow, K. H., Edmonson, M. N. & Cassidy, A. B. Reliable identification of large numbers of candidate SNPs from public EST data. *Nature Genet.* **21**, 323-325 (1999).
26. Gu, Z., Hillier, L. & Kwok, P. Y. Single nucleotide polymorphism hunting in cyberspace. *Hum. Mutat.* **12**, 221-225 (1998).
27. Irizarry, K. *et al.* Genome-wide analysis of single-nucleotide polymorphisms in human expressed sequences. *Nature Genet.* **26**, 233-236 (2000).
28. Picoult-Newberg, L. *et al.* Mining SNPs from EST databases. *Genome Res.* **9**, 167-174 (1999).
29. Marth, G. T. *et al.* Single nucleotide polymorphisms in the public database: how useful are they? *Nature Genet.* (submitted).
30. Yang, Z. *et al.* Sampling SNPs. *Nature Genet.* **26**, 13-14 (2000).
31. Pruitt, K. D., Katz, K. S., Sicotte, H. & Maglott, D. R. Introducing RefSeq and LocusLink: curated human genome resources at the NCBI. *Trends Genet.* **16**, 44-47 (2000).
32. International Human Genome Sequencing Consortium. Initial sequencing and analysis of the human genome. *Nature* **409**, 860-921 (2001).
33. Bohossian, H. B., Skaletsky, H. & Page, D. C. Unexpectedly similar rates of nucleotide substitution found in male and female hominids. *Nature* **406**, 622-625 (2000).
34. Cooke, H. J., Brown, W. R. & Rappold, G. A. Hypervariable telomeric sequences from the human sex chromosomes are pseudoautosomal. *Nature* **317**, 687-692 (1985).
35. Shen, P. *et al.* Population genetic implications from sequence variation in four Y chromosome genes. *Proc. Natl Acad. Sci. USA* **97**, 7354-7359 (2000).
36. Underhill, P. A. *et al.* Detection of numerous Y chromosome biallelic polymorphisms by denaturing high-performance liquid chromatography. *Genome Res.* **7**, 996-1005 (1997).
37. Tajima, F. Evolutionary relationship of DNA sequences in finite populations. *Genetics* **105**, 437-460 (1983).
38. Hudson, R. R. in *Oxford Surveys in Evolutionary Biology* (eds Futuyma, D. & Antonovics, J.) 1-44 (Oxford Univ. Press, Oxford, 1991).
39. Lindblad-Toh, K. *et al.* Large-scale discovery and genotyping of single-nucleotide polymorphisms in the mouse. *Nature Genet.* **24**, 381-386 (2000).

40. Kimmel, M. *et al.* Signatures of population expansion in microsatellite repeat data. *Genetics* **148**, 1921-1930 (1998).

41. Reich, D. E. & Goldstein, D. B. Genetic evidence for a Paleolithic human population expansion in Africa [published erratum appears in *Proc. Natl Acad. Sci. USA* **95**, 11026 (1998)]. *Proc. Natl Acad. Sci. USA* **95**, 8119-8123 (1998).

42. Miller, R. D., Taillon-Miller, P. & Kwok, P. Y. Regions of low single-nucleotide polymorphism (SNP) incidence in human and orangutan Xq: deserts and recent coalescences. *Genomics* (in the press).

43. Horton, R. *et al.* Large-scale sequence comparisons reveal unusually high levels of variation in the HLA-DQB1 locus in the class II region of the human MHC. *J. Mol. Biol.* **282**, 71-97 (1998).

44. Zhang, Z., Schwartz, S., Wagner, L. & Miller, W. A greedy algorithm for aligning DNA sequences. *J. Comput. Biol.* **7**, 203-214 (2000).

Supplementary Information is available on *Nature*'s World-Wide Web site (http://www.nature.com) or as paper copy from the London editorial office of *Nature*.

Acknowledgements. The SNP Consortium, the Wellcome Trust and the National Human Genome Research Institute funded SNP discovery and data management at Cold Spring Harbor Laboratories, The Sanger Centre, Washington University in St. Louis, and the Whitehead/MIT Center for Genome Research. Work in P.Y.K.'s laboratory is supported in part by grants from the SNP Consortium and the National Human Genome Research Institute. P.Y.K. thanks Q. Li, M. Minton, R. Donaldson and S. Duan for technical assistance. D.M.A. was supported during a phase of this work under a Postdoctoral Fellowship for Physicians from the Howard Hughes Medical Institute. For full list of contributors to TSC programme, see www.snp.cshl.org.

Correspondence and requests for materials should be addressed to D.A. (e-mail: altshul@genome.wi.mit.edu) or D.B. (e-mail: drb@sanger.ac.uk).

* **The International SNP Map Working Group** (contributing institutions are listed alphabetically).

Cold Spring Harbor Laboratories: Ravi Sachidanandam[1], David Weissman[1], Steven C. Schmidt[1], Jerzy M. Kakol[1] & Lincoln D. Stein[1]

National Center for Biotechnology Information: Gabor Marth[2] & Steve Sherry[2]

The Sanger Centre: James C. Mullikin[3], Beverley J. Mortimore[3], David L. Willey[3], Sarah E. Hunt[3], Charlotte G. Cole[3], Penny C. Coggill[3], Catherine M. Rice[3], Zemin Ning[3], Jane Rogers[3], David R. Bentley[3]

Washington University in St. Louis: Pui-Yan Kwok[4], Elaine R. Mardis[4], Raymond T. Yeh[4], Brian Schultz[4], Lisa Cook[4], Ruth Davenport[4], Michael Dante[4], Lucinda Fulton[4], LaDeana Hillier[4], Robert H. Waterston[4] & John D. McPherson[4]

Whitehead/MIT Center for Genome Research: Brian Gilman[5], Stephen Schaffner[5], William J. Van Etten[5,6], David Reich[5], John Higgins[5], Mark J. Daly[5], Brendan Blumenstiel[5], Jennifer Baldwin[5], Nicole Stange-Thomann[5], Michael C. Zody[5], Lauren Linton[5], Eric S. Lander[5,7] & David Altshuler[5,8]

1, Cold Spring Harbor, New York 11724, USA; 2, Building 38A, 8600 Rockville Pike, Bethesda, Maryland 20894, USA; 3, Wellcome Trust Genome Campus, Hinxton, Cambridge, CB10 1SA, UK; 4, 660 S. Euclid Ave, St. Louis, Missouri 63110, USA; 5, 9 Cambridge Center, Cambridge, Massachusetts 02139, USA; 6, Present address: Blackstone Technology Group, Boston, Massachusetts 02110, USA; 7, Department of Biology, Massachusetts Institute of Technology, Cambridge, Massachusetts 02142, USA; 8, Departments of Genetics and Medicine, Harvard Medical School; Department of Molecular Biology and Diabetes Unit, Massachusetts General Hospital, Boston, Massachusetts 02114, USA.

New Hominin Genus from Eastern Africa Shows Diverse Middle Pliocene Lineages

M. G. Leakey *et al.*

Editor's Note

The story of human evolution is not a linear progress of forms, each passing the baton to its successor. Instead, it is more like a bush, with several coexistent lineages. However, most interpretations of human evolution tend to converge on a single lineage between four and three million years ago, represented by *Australopithecus afarensis*. This model was exploded at the turn of the millennium in this paper by Meave Leakey and colleagues, who describe a 3.5-million-year-old cranium from West Turkana that looked radically different from *A. afarensis*. It had a much flatter and more human-like face, a signal that different lineages ran very deep. The status of this form, *Kenyanthropus platyops*, was soon debated, given the difficulties of reconstructing a skull from many fragments.

Most interpretations of early hominin phylogeny recognize a single early to middle Pliocene ancestral lineage, best represented by *Australopithecus afarensis*, which gave rise to a radiation of taxa in the late Pliocene. Here we report on new fossils discovered west of Lake Turkana, Kenya, which differ markedly from those of contemporary *A. afarensis*, indicating that hominin taxonomic diversity extended back, well into the middle Pliocene. A 3.5 Myr-old cranium, showing a unique combination of derived facial and primitive neurocranial features, is assigned to a new genus of hominin. These findings point to an early diet-driven adaptive radiation, provide new insight on the association of hominin craniodental features, and have implications for our understanding of Plio–Pleistocene hominin phylogeny.

THE eastern African hominin record between 4 and 3 Myr is represented exclusively by a single species, *A. afarensis*, and its possible ancestor, *Australopithecus anamensis*, which are commonly thought to belong to the lineage ancestral to all later hominins[1,2]. This apparent lack of diversity in the middle Pliocene contrasts markedly with the increasingly bushy phylogeny evident in the later hominin fossil record. To study further the time interval between 4 and 3 Myr, fieldwork in 1998 and 1999 focused on sites of this age at Lomekwi in the Nachukui Formation, west of Lake Turkana. New hominin discoveries from Lomekwi, as well as two mandibles and isolated molars recovered previously[3] (Table 1), indicate that multiple species existed between 3.5 and 3.0 Myr. The new finds include a well-preserved temporal bone, two partial maxillae, isolated teeth, and most importantly a largely complete, although distorted, cranium. We assign the latter specimen to a new hominin genus on the basis of its unique combination of primitive and derived features.

东非发现的古人类新属表现出中上新世多样的人类谱系

利基等

编者按

人类演化的故事并不是形式逐个传递的线性变化过程。相反，人类演化是一种灌木式的发展，存在许多并存的谱系。但是许多人试图将人类演化集中到300万～400万年前的一个谱系上去，即南方古猿阿法种。在千禧年之际，米芙·利基及其同事描述了图尔卡纳湖西岸发现的一件350万年前的颅骨，它看上去与南方古猿阿法种形态迥异，这件标本使得这一理论发生了翻天覆地的变革。这件标本有着更扁平且更加像人的脸，这是不同谱系各自深入演化的证据。由于从许多块碎片复原头骨的难度较大，所以肯尼亚扁脸人的地位很快就引起了争论。

大多数关于早期古人类系统发育的解释认为，从早上新世到中上新世，人类只有一个祖先谱系，以南方古猿阿法种最具代表性，它在晚上新世产生了辐射演化。本文中我们报道了在肯尼亚图尔卡纳湖西部发现的新化石，它与同时代的南方古猿阿法种化石显著不同，暗示着古人类的分类学多样性可以上溯到中上新世时期。一具350万年的颅骨显示出衍生的面部特征和原始的脑颅特征镶嵌的独特的综合特征，我们将其划分到了一个古人类的新属中。这些发现指向一种早期的受饮食驱动的适应性辐射，为古人类的颅牙特征的相关性提供了新的视角，对我们理解上新世–更新世时期的古人类的系统发育也具有提示性。

在400万年前到300万年前之间，东非的古人类记录只有一个物种，那就是南方古猿阿法种以及它可能的祖先——南方古猿湖畔种，它们通常被认为是后来所有古人类的祖先[1,2]。中上新世这种明显的多样性缺乏，与后来的古人类化石记录中显著的灌木式系统发育形成了鲜明对比。为了进一步研究400万年前到300万年前这段时间间隔，于1998年和1999年进行的野外考察工作主要集中在图尔卡纳湖西部纳丘库伊组的洛梅奎的这一年代的遗址上。在洛梅奎取得的新的古人类发现以及之前发掘的两个下颌骨和游离的臼齿[3]（表1）暗示着在350万年前到300万年前之间存在着多个物种。这些新发现包括一个保存完好的颞骨、两块上颌骨部分、游离的牙齿和一块虽然变形但是大部分完整的非常重要的颅骨。由于后一标本独特地结合了原始和衍生的特征，所以我们将其划分到了一个古人类新属中。

Table 1. Hominin specimens from the lower Lomekwi and Kataboi Members

KNM-WT	Description	Year	Discoverer	Locality	Measurements (mm)
8556	Mandible fragment: symphysis, right body with RP_3–RM_1, isolated partial RM_2, RM_3, LP_3	1982	N. Mutiwa	LO-5	RP_3, 9.8, 12.4; RP_4, 11.3, 12.6; RM_1, 13.7, 12.9; RM_2, NA, NA; RM_3, (17.5), (14.1); LP_3, 9.8, 12.5
8557	$LM_{1/2}$	1982	N. Mutiwa	LO-4	NA, (11.5)
16003	RM^3	1985	M. Kyeva	LO-5	13.3, 14.6
16006	Left mandible fragment with M_2 fragment and M_3	1985	N. Mutiwa	LO-4E	M_2, NA, NA; M_3, 15.3, 13.1
38332	Partial RM^3 crown	1999	M. Eregae	LO-4E	NA, 14.8
38333	$LM_{1/2}$ crown	1999	M. Eregae	LO-4E	13.1, 12.1
38334	$LM_{1/2}$	1999	M. Eregae	LO-4W	12.1, 11.5
38335	$RM_{1/2}$ crown fragment	1999	M. Eregae	LO-4E	NA
38337	$RM^{1/2}$	1999	R. Moru	LO-4E	11.5, 12.3
38338	Partial $RM^{1/2}$ crown	1999	N. Mutiwa	LO-4E	NA
38339	$LM_{1/2}$ crown	1999	J. Erus	LO-4W	12.8, 12.7
38341	Partial $LM_{2/3}$	1999	G. Ekalale	LO-4E	NA
38342	$LM_{1/2}$ crown	1999	J. Erus	LO-4E	12.8, (11.3)
38343	Right maxilla fragment with I^2 and P^3 roots and partial C; mandible fragment with partial P_4 and M_1 roots	1998	J. Erus	LO-4W	NA
38344	$RM_{1/2}$ crown	1998	M. Eregae	LO-9	12.8, 12.2
38346	Partial $RM^{1/2}$	1998	M. Mutiwa	LO-5	NA
38347	LdM_2 crown	1998	R. Moru	LO-5	11.7, 9.6
38349	$RM_{1/2}$ crown	1998	W. Mangao	LO-5	13.5, 12.6
38350	Left maxilla fragment with P^3 and P^4 roots and partial M^1	1998	B. Onyango	LO-5	LM^1: (10.5), (12.0)
38352	Partial $RM_{1/2}$	1998	W. Mangao	LO-5	NA, 11.5
38355	Partial $RM^{1/2}$ crown	1998	M. Eregae	LO-9	NA
38356	Partial $RM^{1/2}$ crown	1998	M. Eregae & J. Kaatho	LO-9	12.8, NA
38357	$RM_{1/2}$	1998	G. Ekalale	LO-5	12.8, 11.8
38358	Associated RI^2,LM_2 fragment, LM_3 RM^3 fragment, four crown fragments	1998	G. Ekalale	LO-5	RI^2, 7.5, 7.5, 9.1; LM_3, 15.3, 13.2
38359	Associated RM_1,RM_2	1998	M. Eregae	LO-5	RM_1, 12.7, 11.6; RM_2, 13.9, 12.2
38361	Associated (partial) germs of I^1, LI^2, RC, LRP^3, LRP^4	1998	R. Moru	LO-5	I^1, NA, (8.0), (11.5); LI^2, 7.6, > 5.9, 8.3; LP^3, (9.3), (12.0)
38362	Associated partial $LM^{1/2}$,$RM^{1/2}$	1998	R. Moru	LO-5	$RM^{1/2}$, 12.9, 14.3
39949	Partial LP_4	1998	R. Moru	LO-5	NA
39950	RM_3	1998	R. Moru	LO-5	16.0, 14.5
39951	$RM_{1/2}$ fragment	1998	R. Moru	LO-5	NA

表 1. 从下洛梅奎段和卡塔博伊段发掘的古人类标本

KNM-WT	描述	年份	发现者	产地	测量（毫米）
8556	下颌骨碎片：联合部、右下颌体带有 RP_3~RM_1、游离的部分 RM_2、RM_3、LP_3	1982	N. 穆蒂瓦	LO-5	RP_3, 9.8, 12.4; RP_4, 11.3, 12.6; RM_1, 13.7, 12.9; RM_2, NA, NA; RM_3, (17.5), (14.1); LP_3, 9.8, 12.5
8557	$LM_{1/2}$	1982	N. 穆蒂瓦	LO-4	NA, (11.5)
16003	RM^3	1985	基耶瓦	LO-5	13.3, 14.6
16006	带有 M_2 碎片与 M_3 的左下颌骨碎片	1985	N. 穆蒂瓦	LO-4E	M_2, NA, NA; M_3, 15.3, 13.1
38332	部分 RM^3 齿冠	1999	埃雷加埃	LO-4E	NA, 14.8
38333	$LM_{1/2}$ 齿冠	1999	埃雷加埃	LO-4E	13.1, 12.1
38334	$LM_{1/2}$	1999	埃雷加埃	LO-4W	12.1, 11.5
38335	$RM_{1/2}$ 齿冠碎片	1999	埃雷加埃	LO-4E	NA
38337	$RM^{1/2}$	1999	莫鲁	LO-4E	11.5, 12.3
38338	部分 $RM^{1/2}$ 齿冠	1999	N. 穆蒂瓦	LO-4E	NA
38339	$LM_{1/2}$ 齿冠	1999	埃鲁斯	LO-4W	12.8, 12.7
38341	部分 $LM_{2/3}$	1999	埃卡拉莱	LO-4E	NA
38342	$LM_{1/2}$ 齿冠	1999	埃鲁斯	LO-4E	12.8, (11.3)
38343	有 I^2 和 P^3 齿根与部分 C 的右上颌骨；有部分 P_4 与 M_1 齿根的下颌骨碎片	1998	埃鲁斯	LO-4W	NA
38344	$RM_{1/2}$ 齿冠	1998	埃雷加埃	LO-9	12.8, 12.2
38346	部分 $RM^{1/2}$	1998	M. 穆蒂瓦	LO-5	NA
38347	LdM_2 齿冠	1998	莫鲁	LO-5	11.7, 9.6
38349	$RM_{1/2}$ 齿冠	1998	曼加奥	LO-5	13.5, 12.6
38350	有 P^3 和 P^4 齿根和部分 M^1 的左上颌骨碎片	1998	翁扬戈	LO-5	LM^1: (10.5), (12.0)
38352	部分 $RM_{1/2}$	1998	曼加奥	LO-5	NA, 11.5
38355	部分 $RM^{1/2}$ 齿冠	1998	埃雷加埃	LO-9	NA
38356	部分 $RM^{1/2}$ 齿冠	1998	埃雷加埃 & 卡阿索	LO-9	12.8, NA
38357	$RM_{1/2}$	1998	埃卡拉莱	LO-5	12.8, 11.8
38358	齿列 RI^2、LM_2 碎片、LM_3、RM^3 碎片、4 个齿冠碎片	1998	埃卡拉莱	LO-5	RI^2, 7.5, 7.5, 9.1; LM_3, 15.3, 13.2
38359	齿列 RM_1、RM_2	1998	埃雷加埃	LO-5	RM_1, 12.7, 11.6; RM_2, 13.9, 12.2
38361	I^1、LI^2、RC、LRP^3、LRP^4 齿列（部分）牙胚	1998	莫鲁	LO-5	I^1, NA, (8.0), (11.5); LI^2, 7.6, $>$ 5.9, 8.3; LP^3, (9.3), (12.0)
38362	齿列 $LM^{1/2}$ 部分、$RM^{1/2}$	1998	莫鲁	LO-5	$RM^{1/2}$, 12.9, 14.3
39949	部分 LP_4	1998	莫鲁	LO-5	NA
39950	RM_3	1998	莫鲁	LO-5	16.0, 14.5
39951	$LM_{1/2}$ 碎片	1998	莫鲁	LO-5	NA

Continued

KNM-WT	Description	Year	Discoverer	Locality	Measurements (mm)
39952	$LM_{1/2}$	1998	R. Moru	LO-5	NA
39953	$LM_{1/2}$ fragment	1998	R. Moru	LO-5	NA
39954	Two tooth fragments	1998	R. Moru	LO-5	NA
39955	L_C fragment	1998	R. Moru	LO-5	NA
40000	Cranium	1999	J. Erus	LO-6N	RM^2, 11.4, 12.4
40001	Right temporal bone	1998	P. Gathogo	LO-5	NA

Dental measurements taken as in ref. 34. Mesiodistal crown diameter followed by buccolingual or labiolingual diameter, and for incisors and canines, labial crown height. Values in parentheses are estimates. NA, Not available. L or R in the "Description" column indicates the left or right side. $\underline{C}$, upper canine; d, deciduous.

Description of *Kenyanthropus platyops*

Order Primates LINNAEUS 1758
Suborder Anthropoidea MIVART 1864
Superfamily Hominoidea GRAY 1825
Kenyanthropus gen. nov.

Etymology. In recognition of Kenya's contribution to the understanding of human evolution through the many specimens recovered from its fossil sites.

Generic diagnosis. A hominin genus characterized by the following morphology: transverse facial contour flat at a level just below the nasal bones; tall malar region; zygomaticoalveolar crest low and curved; anterior surface of the maxillary zygomatic process positioned over premolars and more vertically orientated than the nasal aperture and nasoalveolar clivus; nasoalveolar clivus long and both transversely and sagittally flat, without marked juga; moderate subnasal prognathism; incisor alveoli parallel with, and only just anterior to, the bicanine line; nasal cavity entrance stepped; palate roof thin and flexed inferiorly anterior to the incisive foramen; upper incisor (I^1 and I^2) roots near equal in size; upper premolars (P^3, P^4) mostly three-rooted; upper first and second molars (M^1 and M^2) small with thick enamel; tympanic element mediolaterally long and lacking a petrous crest; external acoustic porus small. *Kenyanthropus* can be distinguished from *Ardipithecus ramidus* by its buccolingually narrow M^2, thick molar enamel, and a temporal bone with a more cylindrical articular eminence and deeper mandibular fossa. It differs from *A. anamensis*, *A. afarensis*, *A. africanus* and *A. garhi* in the derived morphology of the lower face, particularly the moderate subnasal prognathism, sagittally and transversely flat nasoalveolar clivus, anteriorly positioned maxillary zygomatic process, similarly sized I^1 and I^2 roots, and small M^1 and M^2 crowns. From *A. afarensis* it also differs by a transversely flat midface, a small, external acoustic porus, and the absence of an occipital/marginal venous sinus system, and from *A. africanus* by a tall malar region, a low and curved zygomaticoalveolar crest, a narrow nasal aperture, the absence of anterior facial pillars,

续表

KNM-WT	描述	年份	发现者	产地	测量（毫米）
39952	$LM_{1/2}$	1998	莫鲁	LO-5	NA
39953	$LM_{1/2}$ 碎片	1998	莫鲁	LO-5	NA
39954	2 个牙齿碎片	1998	莫鲁	LO-5	NA
39955	L_C 碎片	1998	莫鲁	LO-5	NA
40000	颅骨	1999	埃鲁斯	LO-6N	RM^2, 11.4, 12.4
40001	右颞骨	1998	加索戈	LO-5	NA

牙齿的尺寸测量方法见参考文献 34。近中远中端齿冠的直径之后是颊舌侧或唇舌侧直径，门齿和犬齿的直径之后是唇侧齿冠高度。圆括号中的数值是估计值。NA，不可用。“描述”栏中的 L 或 R 表示左侧或右侧。$\underline{C}$，上犬齿；d，乳齿。

肯尼亚扁脸人的描述

灵长目 Primates LINNAEUS 1758

类人猿亚目 Anthropoidea MIVART 1864

人超科 Hominoidea GRAY 1825

肯尼亚人（新属）*Kenyanthropus* gen. nov.

词源学　考虑了肯尼亚通过其化石遗址发掘出来的许多标本对于理解人类演化的贡献。

属的鉴别特征　这是一个具有以下形态特征的人属：横向的面部轮廓平坦，恰好与鼻骨下方相平；高的颧骨区；颧骨齿槽脊低而弯曲；上颌骨颧突的前表面位于前臼齿之上，比鼻孔和鼻齿槽斜坡的方向更加垂直；鼻齿槽斜坡长，水平方向和矢状方向都很平，没有明显的隆起；适中的鼻下凸颌；门齿槽与双犬齿线平行并且刚刚位于其前面；鼻腔开口呈梯形；颚顶薄，在门齿孔前面向下弯曲；上门齿（I^1 和 I^2）齿根大小几乎相等；上前臼齿（P^3、P^4）大部分是三齿根的；第一和第二上臼齿（M^1 和 M^2）小，牙釉质厚；鼓室部分中间外侧方向上很长，无岩脊；外耳门小。可以通过肯尼亚人的颊舌侧窄的 M^2、厚的臼齿釉质以及具有更接近圆柱体的关节隆起和更深的下颌窝的颞骨将其与地猿原始种区别开来。在下面部的衍生形态方面，它与南方古猿湖畔种、南方古猿阿法种、南方古猿非洲种和南方古猿惊奇种都不同，尤其是适中的鼻下凸颌、矢状方向和横向都很平的鼻齿槽斜坡、位于前面的上颌骨颧突、相似大小的 I^1 和 I^2 齿根、小的 M^1 和 M^2 齿冠等。其与南方古猿阿法种的不同之处还有：横向扁平的中面部、小的外耳门以及缺少枕窦/边缘窦系统等；与南方古猿非洲种的不同之处包括：高的颧骨区、低而弯曲的颧骨齿槽脊、狭窄的鼻孔、缺少前面柱、管状的长而无脊的鼓室部分以及小的外耳门等。肯尼亚人缺少在傍人埃

a tubular, long and crestless tympanic element, and a small, external acoustic porus. *Kenyanthropus* lacks the suite of derived dental and cranial features found in *Paranthropus aethiopicus*, *P. boisei* and *P. robustus* (Table 2), and the derived cranial features of species indisputably assigned to *Homo* (For example, *H. erectus s.l.* and *H. sapiens*, but not *H. rudolfensis* and *H. habilis*)[4].

Table 2. Derived cranial features of *Paranthropus*, and their character state in *K. platyops* and *H. rudolfensis*

	Paranthropus aethiopicus	*Paranthropus boisei*	*Paranthropus robustus*	*Kenyanthropus platyops*	*Homo rudolfensis*
Upper molar size	Large	Large	Moderate	Small	Moderate
Enamel thickness	Hyperthick	Hyperthick	Hyperthick	Thick	Thick
Palatal thickness	Thick	Thick	Thick	Thin	Thin
Incisor alveoli close to bicanine line*	Present	Present	Present	Present	Present
Nasoalveolar clivus	Gutter	Gutter	Gutter	Flat	Flat
Midline subnasal prognathism	Strong	Moderate	Moderate	Weak	Weak
Upper I^2 root to lateral nasal aperture	Medial	Medial	Medial	Lateral	Lateral
Nasal cavity entrance	Smooth	Smooth	Smooth	Stepped	Stepped
Zygomaticoalveolar crest	Straight, high	Straight, high	Straight, high	Curved, low	Curved, low
Anteriorly positioned zygomatic process of maxilla*	Present	Present	Present	Present	Present
Midface transverse contour	Concave, dished	Concave, dished	Concave, dished	Flat	Flat
Malar region	Wide	Wide	Wide	Tall	Tall
Malar orientation to lateral nasal margin	Aligned	Aligned	Aligned	More vertical	More vertical
Facial hafting, frontal trigone	High, present	High, present	High, present	Low, absent	Low, absent
Postorbital constriction	Marked	Marked	Marked	Moderate	Moderate
Initial supraorbital course of temporal lines	Medial	Medial	Medial	Posteromedial	Posteromedial
Tympanic vertically deep and plate-like	Present	Present	Present	Absent	Absent
Position external acoustic porus	Lateral	Lateral	Lateral	Medial	Medial
Mandibular fossa depth	Shallow	Deep	Deep	Moderate	Moderate
Foramen magnum heart shaped	Present	Present	Absent	Absent	Absent
Occipitomarginal sinus	Unknown	Present	Present	Absent	Absent

Hypodigm of *H. rudolfensis* as in ref. 35. See refs 1, 8, 11, 36–40 for detailed discussions of the features.
* Character states shared by *Paranthropus* and *K. platyops*.

塞俄比亚种、傍人鲍氏种和傍人粗壮种（表 2）中观察到的一系列牙齿和颅骨衍生特征，还缺少那些无可置疑地被划分到人属的物种（例如广义直立人和智人，但不包括鲁道夫人和能人）所具有的颅骨衍生特征[4]。

表 2. 傍人颅骨的衍生特征及其在肯尼亚扁脸人和鲁道夫人中的特征状态

	傍人埃塞俄比亚种	傍人鲍氏种	傍人粗壮种	肯尼亚扁脸人	鲁道夫人
上臼齿尺寸	大	大	中	小	中
牙釉质厚度	超厚	超厚	超厚	厚	厚
颚厚度	厚	厚	厚	薄	薄
门齿槽靠近双犬齿线 *	存在	存在	存在	存在	存在
鼻齿槽斜坡	沟状	沟状	沟状	平坦	平坦
中线鼻下凸颌	强	中	中	弱	弱
上 I^2 齿根相对梨状孔位置	中间	中间	中间	侧向	侧向
鼻腔开口	平滑	平滑	平滑	梯形	梯形
颧骨齿槽脊	直而高	直而高	直而高	弯而低	弯而低
位于前方的上颌骨颧骨突 *	存在	存在	存在	存在	存在
中面部横向轮廓	中凹碟形	中凹碟形	中凹碟形	平坦	平坦
颧骨区	宽	宽	宽	高	高
颧骨相对于鼻侧边缘	齐平	齐平	齐平	更垂直	更垂直
面部，额三角	高，存在	高，存在	高，存在	低，缺失	低，缺失
眶后缩狭	明显	明显	明显	中	中
颞骨线的初始上眶路径	中间	中间	中间	中后	中后
鼓室垂直方向深，呈板状	存在	存在	存在	缺失	缺失
外耳门位置	侧向	侧向	侧向	中间	中间
下颌窝深度	浅	深	深	适中	适中
心形枕骨大孔	存在	存在	缺失	缺失	缺失
枕边缘窦	未知	存在	存在	缺失	缺失

鲁道夫人的种型群见参考文献 35。对特征的详细讨论见参考文献 1、8、11、36 ~ 40。
* 傍人和肯尼亚扁脸人共有的特征状态。

Type species. *Kenyanthropus platyops* sp. nov.

Etymology. From the Greek *platus*, meaning flat, and *opsis*, meaning face; thus referring to the characteristically flat face of this species.

Specific diagnosis. Same as for genus.

Types. The holotype is KNM-WT 40000 (Fig. 1a–d), a largely complete cranium found by J. Erus in August 1999. The paratype is KNM-WT 38350 (Fig. 1e), a partial left maxilla found by B. Onyango in August 1998. The repository is the National Museums of Kenya, Nairobi.

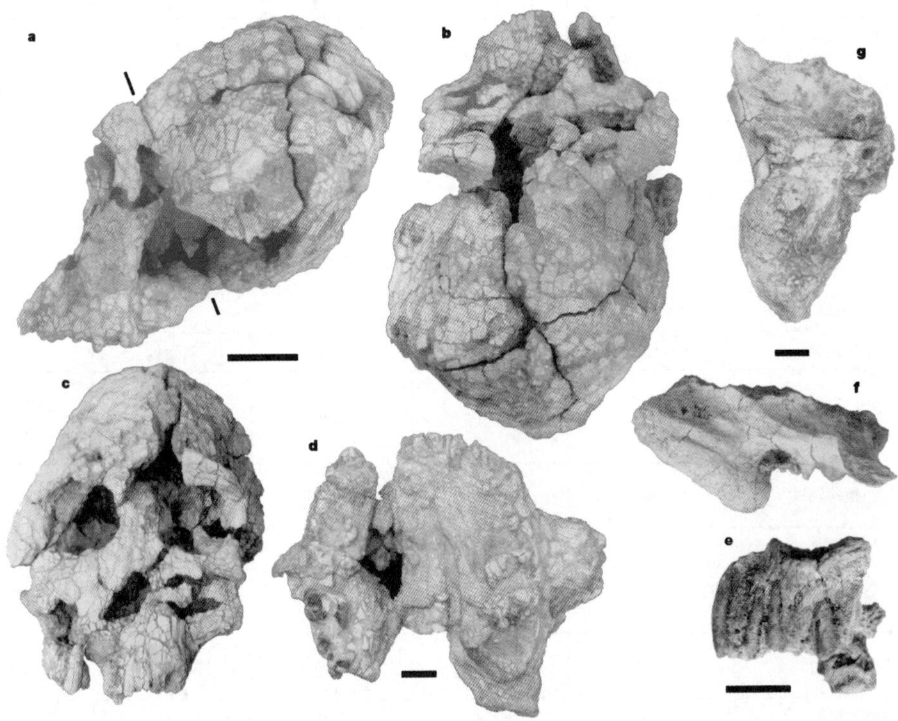

Fig. 1. Holotype KNM-WT 40000. **a**, Left lateral view (markers indicate the plane separating the distorted neurocranium and the well-preserved face). **b**, Superior view. **c**, Anterior view. **d**, Occlusal view of palate. Paratype KNM-WT 38350. **e**, Lateral view. KNM-WT 40001. **f**, Lateral view. **g**, Inferior view. Scale bars: **a**–**c**, 3 cm; **d**–**g**, 1 cm.

Localities. Lomekwi localities are situated in the Lomekwi and Topernawi river drainages in Turkana district, northern Kenya (Fig. 2). The type locality LO-6N is at 03° 54.03′ north latitude, 035° 44.40′ east longitude.

Horizon. The type specimen is from the Kataboi Member, 8 m below the Tulu Bor Tuff and 12 m above the Lokochot Tuff, giving an estimated age of 3.5 Myr. The paratype is from the lower Lomekwi Member, 17 m above the Tulu Bor Tuff, with an estimated age of 3.3 Myr.

模式种 肯尼亚扁脸人新种。

词源 希腊语中 *platus* 意思是扁平的，*opsis* 意思是面部；因此是指该物种特征性的扁平面部。

种的鉴别特征 与属的特征是一样的。

类型 正模标本是 KNM-WT 40000(图 1a ~1d)，是由埃鲁斯于 1999 年 8 月发现的一个大部分完整的颅骨。副模标本是 KNM-WT 38350(图 1e)，是由翁扬戈于 1998 年 8 月发现的部分左上颌骨。它们都被保存在位于内罗毕的肯尼亚国家博物馆中。

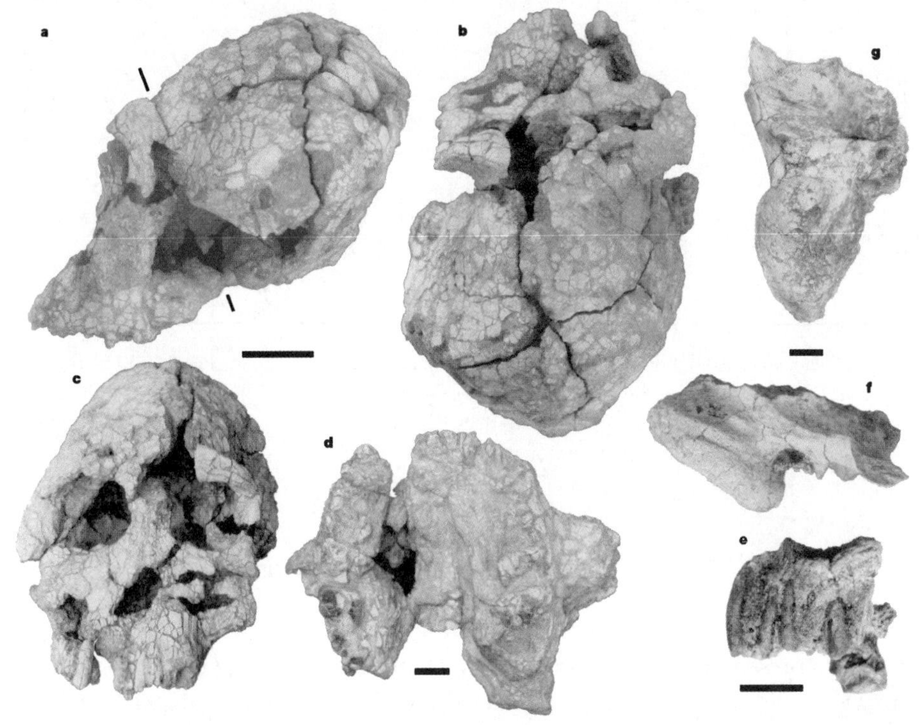

图 1. 正模标本 KNM-WT 40000。**a**，左侧面观(标记表示将变形的脑颅与保存完好的面部分离开来的平面)。**b**，上面观。**c**，前面观。**d**，颚的咬合面观。副模标本 KNM-WT 38350。**e**，侧面观。KNM-WT 40001。**f**，侧面观。**g**，下面观。比例尺：**a**~**c**，3 厘米；**d**~**g**，1 厘米。

产地 洛梅奎遗址位于肯尼亚北部的图尔卡纳地区的洛梅奎和托佩尔纳维河流域(图 2)。LO-6N 典型地点位于北纬 03°54.03′，东经 035°44.40′。

地层 该类型标本是从卡塔博伊段发掘出来的，位于图卢博尔凝灰岩之下 8 米、洛科乔特凝灰岩之上 12 米处，估计其年代为 350 万年。副模标本是在下洛梅奎段发掘出来的，位于图卢博尔凝灰岩之上 17 米处，估计其年代为 330 万年。

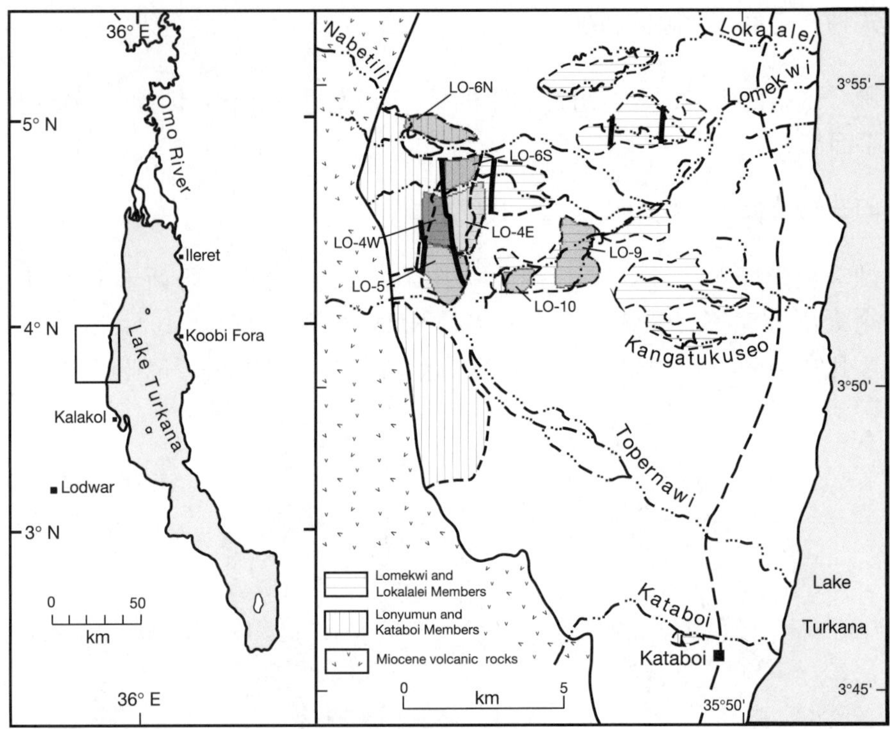

Fig. 2. Map showing localities of fossil collection in upper Lomekwi and simplified geology. The boundary between the Kataboi and Lomekwi Members is the base of the Tulu Bor Tuff, indicated as a dashed line through LO-4E and LO-4W. Faults are shown as thick lines; minor faults are omitted. LO-4E and LO-4W are of different shades to distinguish them from each other.

Cranial Description and Comparisons

The overall size of the KNM-WT 40000 cranium falls within the range of *A. afarensis* and *A. africanus*. It is preserved in two main parts, the neurocranium with the superior and lateral orbital margins, but lacking most of the cranial base; and the face, lacking the premolar and anterior tooth crowns and the right incisor roots. Most of the vault is heavily distorted, both through post-mortem diploic expansion and compression from an inferoposterior direction (Fig. 1a, b). The better preserved facial part shows some lateral skewing of the nasal area, anterior displacement of the right canine, and some expansion of the alveolar and zygomatic processes (Fig. 1c–d), but allows for reliable assessment of its morphology.

Only the right M^2 crown is sufficiently preserved to allow reliable metric dental comparisons. It is particularly small, falling below the known ranges of other early hominin species (Fig. 3a). Likewise, the estimated M^1 crown size of KNM-WT 38350 (Table 1) corresponds to minima for *A. anamensis*, *A. afarensis* and *H. habilis*, and is below the ranges for other African early hominins[5-7]. Molar enamel thickness in both specimens is comparable to that in *A. anamensis* and *A. afarensis*. CT scans show that both P^3 and P^4 of KNM-WT 40000 have a lingual root and two well-separated buccal roots. This morphology, thought

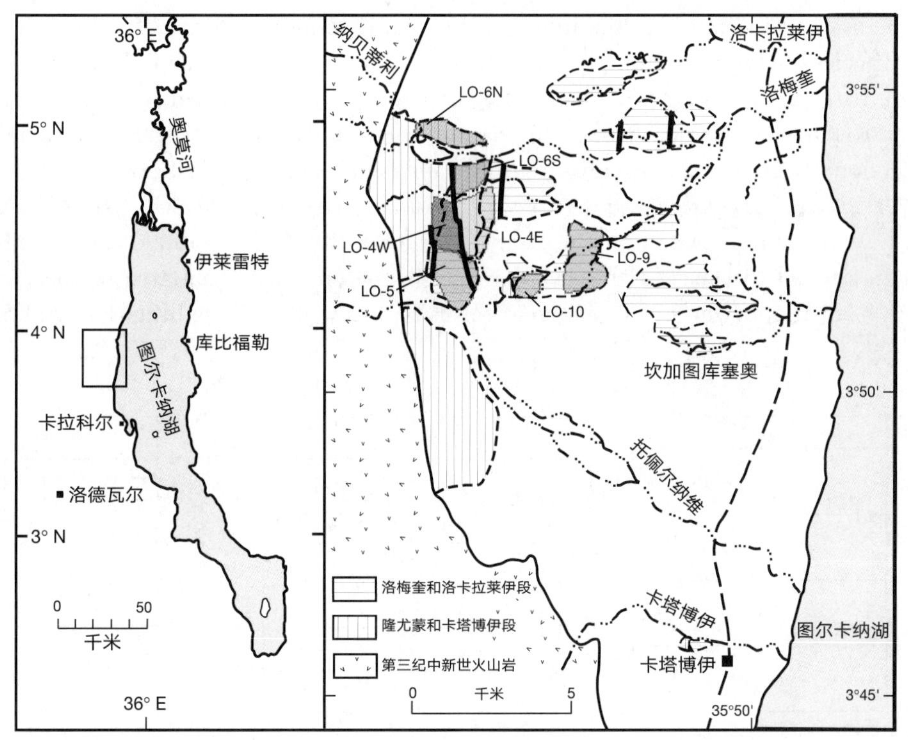

图 2. 该地图表示了在上洛梅奎段搜集到的化石的位置和简化的地质情况。卡塔博伊和洛梅奎段间的界线是图卢博尔凝灰岩的基部，用一条穿过 LO-4E 和 LO-4W 的虚线表示。断层用粗线表示；小断层都省略掉了。LO-4E 和 LO-4W 用不同的明暗度表示以相互区分。

颅骨的描述与比较

KNM-WT 40000 颅骨的总大小处于南方古猿阿法种和南方古猿非洲种的范围内。其保存下来了两个主要部分：脑颅，具有眼窝上缘和侧缘、缺少大部分颅底；面部，缺少前臼齿、前部牙齿齿冠和右门齿根。颅顶大部分由于死后板障扩张和来自下后方向的压缩而严重变形了（图 1a 和 1b）。保存较好的面部区域显示出鼻区有些侧斜，右犬齿向前移位以及齿槽和颧骨突有些扩张（图 1c ~ 1d），但是还可以根据它们可靠地估计其形态。

只有右 M^2 的齿冠保存下来的部分足以用来可靠地比较牙齿的尺寸。它非常小，比其他早期的古人类的已知范围还要小（图 3a）。同样地，KNM-WT 38350 的 M^1 的齿冠的估值（表 1）与南方古猿湖畔种、南方古猿阿法种和能人的最小值一致，而比其他的非洲早期古人类的尺寸范围要小[5-7]。两个标本的臼齿釉质厚度都与南方古猿湖畔种和南方古猿阿法种的相当。CT 扫描表明 KNM-WT 40000 的 P^3 和 P^4 都有一个舌侧的齿根和两个完全分离的颊侧齿根。这种形态被认为是人科动物的祖先种才

to be the ancestral hominoid condition[8], is commonly found in *Paranthropus*, but is variable among species of *Australopithecus*. The P^3 of KNM-WT 38350 has three well-separated roots (Fig. 1e). Its P^4 seems to be two-rooted, but the deeply grooved buccal root may split more apically. Relative to M^2 crown size, the canine roots of KNM-WT 40000 are smaller in cross-section at the alveolar margin than in *Ardipithecus ramidus* and *A. anamensis*, similar in size to *A. afarensis*, *A. africanus* and *H. habilis*, and larger than in *P. boisei*. Exposed surfaces and CT scans demonstrate that the I^1 and I^2 roots in KNM-WT 40000 are straight and similar in size. At the level of the alveolar margin the cross-sectional area of the I^2 root is about 90% of that of the I^1 root, whereas this is typically 50–70% in other known hominid taxa.

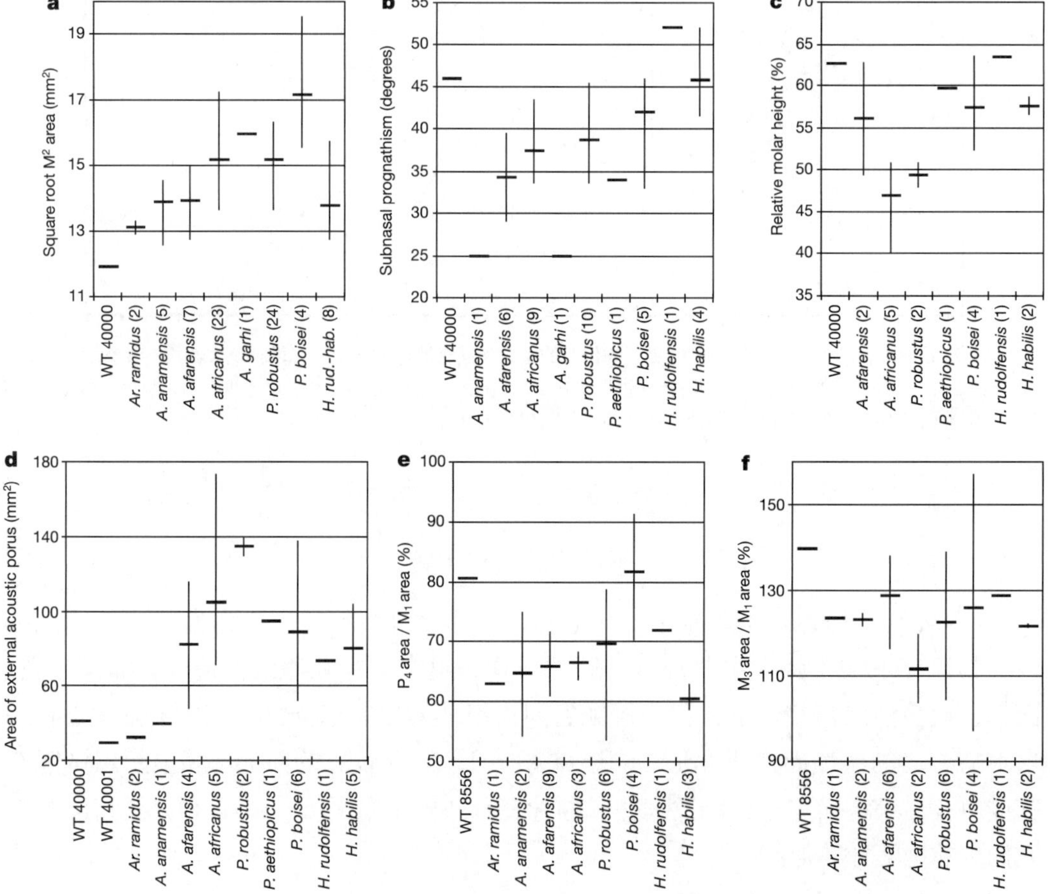

Fig. 3. Mean and range of characters of specified hominins. **a**, Square root of M^2 crown area (buccolingual × mesiolingual diameters). **b**, Angle of subnasal prognathism (nasospinale–prosthion to postcanine alveolar plane). **c**, Malar height[8] relative to orbitoalveolar height (orbitale to alveolar margin aligned with malar surface). **d**, Area of the external acoustic porus (π × long axis × short axis). **e**, Crown area of P_4 relative to that of M_1 × 100. **f**, Crown area of M_3 relative to that of M_1 × 100. All measurements are taken from originals, directly or as given in refs 8, 9, 14, 18, 37, 41–44, except for some South African crania taken from casts. Numbers in parentheses indicate sample size.

具有的特征[8]，通常是在傍人属中发现的，但是在南方古猿的物种中具有较多变异。KNM-WT 38350 的 P^3 具有三个完全分开的齿根（图 1e）。其 P^4 看上去似乎具有两个齿根，但是有较深槽的颊侧齿根可能在靠近根尖部分开。相对于 M^2 的齿冠尺寸而言，KNM-WT 40000 的犬齿根在齿槽边缘的横截面处比地猿原始种和南方古猿湖畔种的要小，而与南方古猿阿法种、南方古猿非洲种和能人的尺寸相近，比傍人鲍氏种的要大。暴露出来的表面和 CT 扫描证实 KNM-WT 40000 的 I^1 和 I^2 的齿根很直，大小相似。在齿槽边缘水平面上，I^2 齿根的横截面面积大约是 I^1 齿根的 90%，而这一数值在其他已知的人科动物分类单元中多为 50% ~ 70%。

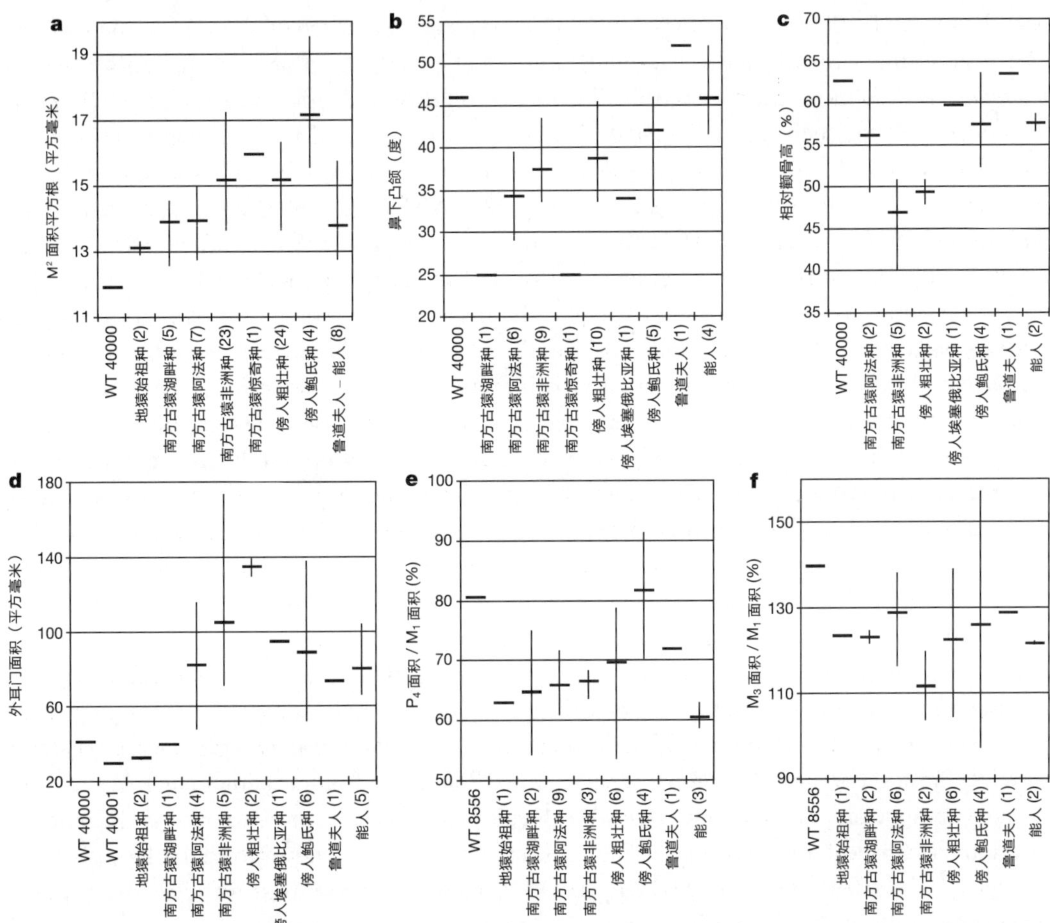

图 3. 特定古人类特征的平均值和范围。**a**，M^2 齿冠面积（颊舌侧 × 近中舌侧直径）的平方根。**b**，鼻下凸颌的角度（鼻棘点–齿槽中点与犬齿后部牙齿牙槽平面之间）。**c**，相对于眶齿槽高度（眶最下点到与颧骨表面在一条直线上的齿槽边缘的距离）的颧骨高度[8]。**d**，外耳门的面积（π × 长轴 × 短轴）。**e**，P_4 的齿冠面积与 M_1 的齿冠面积之比 × 100。**f**，M_3 的齿冠面积与 M_1 的齿冠面积之比 × 100。所有测量值都是直接从原始标本测量得到或者来自参考文献 8、9、14、18、37、41~44 中给出的数据，除了有些南非颅骨的尺寸是根据模型测量的。圆括号中的数字代表标本尺寸。

The incisor alveoli of KNM-WT 40000 are aligned coronally, just anterior to the bicanine line, and the overlying nasoalveolar clivus is flat both sagittally and transversely. There is no canine jugum visible on the preserved left side, reflecting the modest size of the canine root. At 32 mm (chord distance nasospinale to prosthion) the clivus is among the longest of all early hominins. Subnasal prognathism is moderate, expressed by a more vertically orientated clivus than in nearly all specimens of *Australopithecus* and *Paranthropus* (Fig. 3b). The nasal aperture lies in the same coronal plane as the nasoalveolar clivus and there are no anterior facial pillars (Fig. 1a–c). The nasal aperture is small and narrow, in contrast to the large, wide aperture in *A. africanus* and *P. robustus*. The midface of KNM-WT 40000 is dominated by the tall malar region (Fig. 3c) with a low and curved zygomaticoalveolar crest. At a level just below the nasal bones the transverse facial contour is flat (Fig. 1b). In both KNM-WT 40000 and KNM-WT 38350 the anterior surface of the zygomatic process of the maxilla is positioned between P^3 and P^4 (Fig. 1a, d, e), as is commonly seen in *Paranthropus*, but more anteriorly than in most *Australopithecus* specimens[9] or in *H. habilis*. The supraorbital region is *Australopithecus*-like, lacking both a frontal trigon as seen in *Paranthropus*, and a supratoral sulcus as seen in *H. habilis* (but not *H. rudolfensis*). Relative postorbital constriction (frontofacial index) of KNM-WT 40000 is similar to that in *Australopithecus*, *H. rudolfensis* and *H. habilis*, and less than in *P. boisei* (estimated frontofacial index[9] = 70). Its temporal lines converging on the frontal squama have a posteromedial course throughout (Fig. 1b). Around bregma the midline morphology is not well preserved, but the posterior half of the parietals show double, slightly raised temporal lines about 6 mm apart. These contribute posteriorly to indistinct compound temporal/nuchal lines. The original shape of the severely distorted mastoids cannot be reconstructed, but other parts of the left temporal are well preserved. The tubular tympanic lacks a petrous crest and forms a narrow external acoustic meatus with a small aperture. This combination constitutes the primitive hominin morphology, also seen in *Ar. ramidus* and *A. anamensis* (Fig. 3d). The mandibular fossa resembles that of specimens of *A. afarensis* and *A. africanus*. It is moderately deep, and the articular eminence, missing its lateral margin, is cylindrical with a moderately convex sagittal profile. The preserved posterior half of the foramen magnum suggests that it was probably oval in shape, rather than the heart shape seen in *P. boisei* and probably *P. aethiopicus*. Regarding the endocranial aspect, the reasonably well preserved occipital surface lacks any indication of the occipital/marginal venous sinus system characteristic of *A. afarensis*, *P. boisei* and *P. robustus*. Bilateral sulci suggest that the transverse/sigmoid sinus system was well developed. Endocranial capacity is difficult to estimate because of the distorted vault. However, comparing hominin glabella–opisthion arc lengths[8] with that of KNM-WT 40000 (259 mm; an estimate inflated by diploic expansion) suggests a value in the range of *Australopithecus* or *Paranthropus*.

The sex of KNM-WT 40000 is difficult to infer. Interpretation of the canine root size proves inconclusive without a suitable comparative context. The small M^2 crown size could suggest that the specimen is female. However, the close proximity and slightly raised aspect of the temporal lines on the posterior half of the parietals is not seen in known female hominin crania, including the *Paranthropus* specimens KNM-ER 732, KNM-ER 407 and DNH7, and suggests that KNM-WT 40000 could be male.

KNM-WT 40000的门齿齿槽呈冠状排列，刚刚位于双犬齿线的前面，上面的鼻齿槽斜坡在矢状方向和横向上都是平的。在保存下来的左侧没有看到犬齿轭，这反映了犬齿齿根具有最适度的大小。斜坡的长度(从鼻棘点到牙槽中点的弦距离)是32毫米，这是所有早期古人类中最长的之一。鼻下凸颌适中，具有比几乎所有南方古猿和傍人属标本都更垂直的斜坡(图3b)。鼻孔与鼻齿槽斜坡位于同一冠状面上，没有前面柱(图1a~1c)。鼻孔小而窄，与南方古猿非洲种和傍人粗壮种的大而宽的鼻孔形成对比。KNM-WT 40000的中面部主要有一个高高的颧骨区(图3c)，颧骨区有一个低而弯曲的颧骨齿槽脊。在刚刚位于鼻骨之下的水平面上，横向的面部轮廓是平的(图1b)。在KNM-WT 40000和KNM-WT 38350中，上颌骨颧突的前表面都位于P^3和P^4之间(图1a、1d和1e)，这在傍人中是很常见的，但是比大部分南方古猿标本[9]或能人中的位置要更靠前一些。眶上区域与南方古猿相似，既缺少傍人中所见的额三角，也没有能人(并非鲁道夫人)中所见到的圆枕上沟。KNM-WT 40000的相对眶后缩狭(额面指数)与南方古猿、鲁道夫人和能人的都相似，而比傍人鲍氏种的(额面指数估计值[9] = 70)要小。其收敛于额鳞部的颞线有一个贯穿后中的路径(图1b)。前囟周围，中线的形态没有很好地保存下来，但是顶骨的后半部分显示出大约相隔6毫米远的、双重的、稍微抬高的颞线。再向后形成了不明确的复合颞/项线。严重变形的乳突的原始形状无法重建出来了，但是左颞骨的其余部分保存完好。管状鼓室缺少岩脊，形成了一个具有小孔的狭窄的外耳道。这种组合构成了原始的古人类形态，在地猿原始种和南方古猿湖畔种中也见到过(图3d)。其下颌窝与南方古猿阿法种和南方古猿非洲种标本的很像。其深度适中；关节隆起呈圆柱形，侧边缘丢失，有一个凸度适中的矢状面。保存下来的枕骨大孔的后半部分表明，其形状可能是卵圆形，而非傍人鲍氏种中见到(傍人埃塞俄比亚种中也可能见到)的心形。至于颅内方面，保存相当完好的枕骨表面缺少任何南方古猿阿法种、傍人鲍氏种和傍人粗壮种特征性的枕窦/边缘窦系统。双侧沟表明横窦/乙状窦系统很发达。颅容量很难估计，因为颅顶已经变形了。但是，将古人类眉间–枕后点弧长[8]与KNM-WT 40000的(259毫米；由板障扩张获得的估计值)比较，得到的数值处于南方古猿或傍人的范围内。

KNM-WT 40000的性别很难推断。对于犬齿齿根尺寸的说明证明是不确定的，没有适当的可用于比较的方面。小的M^2齿冠尺寸能够表明该标本是雌性。但是，在顶骨后半部分颞线极为接近并且较高，这在已知的雌性古人类颅骨(包括傍人标本KNM-ER 732、KNM-ER 407和DNH7)中并没有见到，这一特点表明KNM-WT 40000可能是雄性。

With incisor alveoli close to the bicanine line and anteriorly positioned zygomatic processes, the face of KNM-WT 40000 resembles the flat, orthognathic-looking faces of both *Paranthropus* and *H. rudolfensis* cranium KNM-ER 1470. However, KNM-WT 40000 lacks most of the derived features that characterize *Paranthropus* (Table 2), and its facial architecture differs from the latter in much the same way as has been described for KNM-ER 1470 (refs 8, 10). Facial flatness in *Paranthropus* results from the forward position of the anteroinferiorly sloping malar region, whose main facial surface approximates the plane of the nasal aperture, but whose orientation contrasts with the more horizontally inclined nasoalveolar gutter[11]. In KNM-WT 40000 and KNM-ER 1470, it is the flat and orthognathic nasoalveolar clivus that aligns with the plane of the nasal aperture, whereas the anteriorly set, tall malar region is more vertically orientated. KNM-WT 40000 lacks the derived short nasal bones and everted lateral nasal margin of KNM-ER 1470, and is less orthognathic in the midfacial region than this specimen; however, on balance this is the hominin face that KNM-WT 40000 most closely resembles.

Additional Material

The right maxilla fragment KNM-WT 38343A preserves three well-separated P^3 roots, and its damaged canine seems low-crowned when compared with *A. afarensis* canines of similar size and degree of wear. The right temporal bone KNM-WT 40001 lacks the squama and petrous apex, but is otherwise well preserved (Fig. 1f, g). It shows a combination of characters not seen in any other hominin specimen. The projecting mastoid process is rounded, with an anteriorly positioned tip. It has a well-developed digastric fossa in the form of a deep, narrow groove that runs posterolaterally from the stylomastoid foramen, fully demarcating the mastoid process from the adjacent nuchal plane. The tympanic element is long, inferosuperiorly shallow and lacks a petrous crest. The external acoustic porus is the smallest of any known hominin temporal bone (Fig. 3d). The articular eminence is as broad mediolaterally (38 mm) as in *P. aethiopicus* and *P. boisei*, and similar to the largest found in *A. afarensis*. Compared with KNM-WT 40000 the eminence is relatively flat sagittally, and the mandibular fossa is shallow.

The partial mandibles KNM-WT 8556 and KNM-WT 16006 have been assigned to *A. afarensis*[3]. However, KNM-WT 8556 shows a more derived morphology than this species by having a flat, more horizontal post-incisive plane, a more superiorly positioned genioglossal pit, a molarized lower fourth premolar (P_4) and a large M_3 (ref. 3). Indeed, relative to its *Australopithecus*-sized M_1 (refs 5, 6, 12, 13), the P_4 and M_3 crowns of KNM-WT 8556 are enlarged to an extent only seen in *P. boisei* (Fig. 3e, f). All unworn molars in the Lomekwi sample are characterized by low occlusal relief and numerous secondary fissures. Most of the lower molars, including the KNM-WT 16006 M_3, have a well-developed protostylid, a feature that is usually absent in *A. afarensis*, but common in *A. africanus*[14]. The two I^2s are lower crowned than in *A. afarensis*, *A. africanus*[14] and *P. robustus*[14]. Inability to distinguish between first and second molars makes meaningful intertaxon comparisons of these elements difficult.

KNM-WT 40000的门齿槽靠近双犬齿线，颧突前置，面部与傍人和鲁道夫人颅骨KNM-ER 1470的扁平而正颌的面部很像。然而，KNM-WT 40000缺少大部分表征傍人的衍生特征(表2)，其面部结构与后者不同，而与KNM-ER 1470描述过的情况几乎一样(参考文献8和10)。傍人的面部扁平的特点是由于向前下方倾斜的颧骨区的位置向前，其主面部表面接近鼻孔平面，但是其方向与更加水平倾斜的鼻齿槽沟形成对比[11]。在KNM-WT 40000和KNM-ER 1470中，扁平而正颌的鼻齿槽斜坡与鼻孔平面走向一致，而位于前面的高的颧骨区更加垂直。KNM-WT 40000缺少KNM-ER 1470所具有的衍生的短小的鼻骨和外翻的侧鼻缘，中面部的正颌性比该标本差；但是，总的说来，这是与KNM-WT 40000最为相像的古人类面部。

其他材料

右上颌骨碎片KNM-WT 38343A保存下来了三个完全分离的P^3齿根，当与南方古猿阿法种的具有相似大小和磨损程度的犬齿比较时，发现其损坏的犬齿似乎是低齿冠的。右颞骨KNM-WT 40001缺少鳞部和岩部尖，但是同样保存状况很好(图1f和1g)。其显示出在其他古人类标本中所没有看到过的特征的组合。突出的乳突呈圆形，有一个位于前面的尖端。它有一个发达的二腹肌窝，呈深而狭窄的沟状，从茎乳孔向后外侧延伸，完全将乳突与相邻的项面分隔开来。鼓室长，上下方向浅，缺少岩脊。外耳门是现在已知的古人类颞骨中最小的(图3d)。关节隆起在中外侧像傍人埃塞俄比亚种和傍人鲍氏种的一样宽阔(38毫米)，与南方古猿阿法种中发现的最大值很相似。与KNM-WT 40000相比，该隆起在矢状方向上相对平一些，下颌窝很浅。

部分下颌骨KNM-WT 8556和KNM-WT 16006都被划分到了南方古猿阿法种中[3]。但是，KNM-WT 8556显示出比该物种更加衍生的形态，具有扁平而更加水平的门齿后平面、位置靠上的颏舌肌窝、臼齿化的下第四前臼齿(P_4)和一个大的M_3(参考文献3)。实际上，相对于其尺寸与南方古猿相当的M_1(参考文献5、6、12和13)，KNM-WT 8556的P_4和M_3的齿冠扩大到了只有在傍人鲍氏种中才见到过的程度(图3e和3f)。洛梅奎标本中所有没磨损的臼齿都具有的特征是低的咬合面脊纹和许多次级沟纹。大部分下臼齿，包括KNM-WT 16006的M_3，都具有发达的原副尖，这是南方古猿阿法种中通常不具备而南方古猿非洲种中很常见的特征[14]。两颗I^2的齿冠都比南方古猿阿法种、南方古猿非洲种[14]和傍人粗壮种[14]的低。无法区分第一和第二臼齿，这使得很难对这些部位在分类单元之间进行有意义的比较。

Taxonomic Discussion

The hominin specimens recovered from the Kataboi and lower Lomekwi Members show a suite of features that distinguishes them from established hominin taxa, including the only contemporaneous eastern African species, *A. afarensis*. Compared with the latter, the morphology of *K. platyops* is more derived facially, and more primitive in its small external acoustic porus and the absence of an occipital/marginal sinus system. These finds not only provide evidence for a taxonomically more diverse middle Pliocene hominin record, but also show that a more orthognathic facial morphology emerged significantly earlier in hominin evolutionary history than previously documented. This early faciodental diversity concerns morphologies that functionally are most closely associated with mastication. It suggests a diet-driven adaptive radiation among hominins in this time interval, which perhaps had its origins considerably earlier. Furthermore, the presence in *K. platyops* of an anteriorly positioned zygomatic process in combination with a small M^1 and M^2 indicates that such characters are more independent than is suggested by developmental and functional models that link such facial morphology in *Paranthropus* with postcanine megadontia[11,15].

At present it is unclear whether the Lomekwi hominin fossils sample multiple species. Apart from the paratype maxilla KNM-WT 38350 with its small molar size and anteriorly positioned zygomatic process, the other specimens cannot be positively associated morphologically with the *K. platyops* holotype. These are therefore not included in the paratype series, and are left unassigned until further evidence emerges. Differences between the tympanic and mandibular fossa morphologies of the KNM-WT 40000 and KNM-WT 40001 temporal bones can perhaps be accommodated within a single species, but their shared primitive characters do not necessarily imply conspecificity. Affiliation of the KNM-WT 8556 mandible with the *K. platyops* types is not contradicted by its molarized P_4, which is consistent with an anteriorly positioned zygomatic process. However, its M_1 is larger than would be inferred from the smaller upper molars of the types, and with a 177 mm^2 crown area it is also larger than any in the combined sample of ten isolated M_1s and M_2s (139–172 mm^2). One isolated $M^{1/2}$ (KNM-WT 38362) is significantly larger than the molars of the *K. platyops* types, whereas another (KNM-WT 38337) is similar in size to the holotype's M^2.

The marked differences of the KNM-WT 40000 cranium from established hominin taxa, both with respect to individual features and their unique combination, fully justify its status as a separate species. It is worth noting that comparisons with *Australopithecus bahrelghazali* cannot be made directly, because this species was named on the basis of the limited evidence provided by an anterior mandible fragment[16]. Specific distinction of *A. bahrelghazali* from *A. afarensis* has yet to be confirmed[17], and Lomekwi specimens differ from *A. bahrelghazali* in symphyseal morphology and incisor crown height.

The generic attribution of KNM-WT 40000 is a more complex issue, in the absence of

分类学讨论

从卡塔博伊和下洛梅奎段发掘到的古人类标本显示出一套与已确定的古人类分类单元不同的特征，与唯一的同时代的东非物种——南方古猿阿法种也不相同。与后者相比，肯尼亚扁脸人的面部形态更具有衍生性，其较小的外耳门和枕窦/边缘窦系统的缺失则更加原始。这些发现不仅为分类学上更加多样的中上新世古人类记录提供了证据，而且也表明，更加正颌的面部形态在古人类演化史上出现的时间比之前文件证明的早得多。这一早期的面牙多样性涉及那些功能上与咀嚼关系最密切的形态。它表明在这一时间段的古人类中存在着饮食驱动的适应性辐射，其起源的时间可能更早。此外，肯尼亚扁脸人位置靠前的颧突与小型的 M^1 和 M^2 组合暗示了这些特征更具有独立性，而非发育和功能学模型所倡导的将傍人中的这种面部形态与较大的犬齿后齿联系起来[11,15]。

现在还不清楚洛梅奎古人类化石标本是否包含多个物种。除了副模标本上颌骨 KNM-WT 38350 及其小型臼齿尺寸和前置的颧突外，不能肯定其余标本在形态上与肯尼亚扁脸人正模标本有关。因此这些不被包括在副模标本系列中，直到出现进一步的证据时，才能对这些标本的归属进行判定。KNM-WT 40000 与 KNM-WT 40001 颞骨的鼓室和下颌窝形态之间的差异或许可以被归结为同一个物种内的差异，但是它们共有的原始特征未必意味着同种性。KNM-WT 8556 下颌骨归属于肯尼亚扁脸人类型，与其臼齿化的 P_4 并不冲突，这与前置的颧突是一致的。然而，其 M_1 比根据该类型较小的上臼齿推断出来的要大，齿冠面积达 177 平方毫米，比十颗游离的 M_1 和 M_2 的所有标本中的任何一个（139~172 平方毫米）都大。一颗游离的 $M^{1/2}$（KNM-WT 38362）比肯尼亚扁脸人类型的臼齿要大得多，但是另一个标本（KNM-WT 38337）则与正模标本的 M^2 的大小相似。

KNM-WT 40000 颅骨与已确定的古人类分类单元在个体特征与其特异的结合方面的明显差异，都充分地说明了将其当成一个单独的物种的合理性。值得说明的是，不能直接与南方古猿羚羊河种进行比较，因为该物种是根据一块前下颌骨碎片提供的有限证据而命名的[16]。南方古猿羚羊河种与南方古猿阿法种的特异差别还没有得到证实[17]，洛梅奎标本与南方古猿羚羊河种在下颌联合部的形态和门齿齿冠高度方面有差异。

KNM-WT 40000 属一级别的划分归类是一个更加复杂的问题，对于该属分类的

consensus over the definition of the genus category[4]. The specimen lacks almost all of the derived features of *Paranthropus* (Table 2), and there are no grounds for assigning it to this genus unless it can be shown to represent a stem species. However, the fact that the facial morphology of KNM-WT 40000 is derived in a markedly different way renders this implausible. As KNM-WT 40000 does not show the derived features associated with *Homo*[4] (excluding *H. rudolfensis* and *H. habilis*) or the strongly primitive morphology of *Ardipithecus*[18], the only other available genus is *Australopithecus*. We agree with the taxonomically conservative, grade-sensitive approach to hominin classification that for the moment accepts *Australopithecus* as a paraphyletic genus in which are clustered stem species sharing a suite of key primitive features, such as a small brain, strong subnasal prognathism, and relatively large postcanine teeth. However, with its derived face and small molar size, KNM-WT 40000 stands apart from species assigned to *Australopithecus* on this basis. All it has in common with such species is its small brain size and a few other primitive characters in the nasal, supraorbital and temporal regions. Therefore, there is no firm basis for linking KNM-WT 40000 specifically with *Australopithecus*, and the inclusion of such a derived but early form could well render this genus polyphyletic. In a classification in which *Australopithecus* also includes the "robust" taxa and perhaps even species traditionally known as "early *Homo*"[4], this genus subsumes several widely divergent craniofacial morphologies. It could thus be argued that the inclusion of KNM-WT 40000 in *Australopithecus* would merely add yet another hominin species with a derived face. This amounts to defining *Australopithecus* by a single criterion, those hominin species not attributable to *Ardipithecus* or *Homo*, which in our view constitutes an undesirable approach to classification. Thus, given that KNM-WT 40000 cannot be grouped sensibly with any of the established hominin genera, and that it shows a unique pattern of facial and dental morphology that probably reflects a distinct dietary adaptive zone, we assign this specimen to the new genus *Kenyanthropus*.

Despite being separated by about 1.5 Myr, KNM-WT 40000 is very similar in its facial architecture to KNM-ER 1470, the lectotype of *H. rudolfensis*. The main differences amount to the more primitive nasal and neurocranial morphology of KNM-WT 40000. This raises the possibility that there is a close phylogenetic relationship between the two taxa, and affects our interpretation of *H. rudolfensis*. The transfer of this species to *Australopithecus* has been recommended[4,19], but *Kenyanthropus* may be a more appropriate genus. The identification of *K. platyops* has a number of additional implications. As a species contemporary with *A. afarensis* that is more primitive in some of its morphology, *K. platyops* weakens the case for *A. afarensis* being the sister taxon of all later hominins, and thus its proposed transfer to *Praeanthropus*[1,20]. Furthermore, the morphology of *K. platyops* raises questions about the polarity of characters used in analyses of hominin phylogeny. An example is the species' small molar size, which, although probably a derived feature, might also imply that the larger postcanine dentition of *A. afarensis* or *A. anamensis* does not represent the primitive hominin condition. Finally, the occurrence of at least one additional hominin species in the middle Pliocene of eastern Africa means that the affiliation of fragmentary specimens can now be reassessed. For example, the attribution of the 3.3 Myr old KNM-ER 2602 cranial fragment to *A. afarensis*[21] has been questioned[8],

定义还没有达成一致[4]。标本缺少几乎所有的傍人的衍生特征（表 2），除非可以表明其代表了主干物种，否则没有依据将其划分到该属。但是，KNM-WT 40000 的面部形态的衍生方式如此迥异，使其不可能代表主干物种。鉴于 KNM-WT 40000 并没有显示出与人属（除了鲁道夫人和能人）相关的衍生特征[4]或者地猿的非常原始的形态[18]，其他唯一一个可考虑的属就是南方古猿了。我们同意分类学上保守的、级别敏感的古人类的分类方式，该方式目前将南方古猿认为是一种并系属，在该属中聚集了共有一套关键的原始特征的主干物种，这里所说的关键的原始特征包括较小的脑、强壮的鼻下凸颌和相对大的犬齿后部牙齿等。然而，KNM-WT 40000 具有衍生的面部和小的臼齿，这些将其与划分为南方古猿的物种分开了。它与该类物种共有的全部特征就是其小的脑和鼻骨、眶上区和颞区的一些其他原始特征。因此，没有确定的依据将 KNM-WT 40000 特异性地与南方古猿联系起来，将这样一个衍生但尚属早期的形式包括进来可以很好地使这个属复杂化。在一种分类中，南方古猿也包括"粗壮"的分类单元，甚至可能包括传统意义上称为"早期人属"的物种[4]，该属包含几种差别很大的颅面形态。因此可以这样说，将 KNM-WT 40000 囊括进南方古猿仅会增加另一个具有衍生面部的古人类物种。这等同于将南方古猿用唯一的标准定义，即那些不能被归入地猿或人属的古人类物种，依我们看来，这形成了一个不理想的分类方式。因此，鉴于 KNM-WT 40000 不能明智地与任何已确定的古人类属聚类到一起，以及鉴于其表明了一种独特的面部和牙齿形态模式，该模式可能反映了一个与众不同的饮食适应区，我们将该标本划分到一个新属中，即肯尼亚人。

尽管分隔约 150 万年，但是 KNM-WT 40000 在面部结构上与鲁道夫人的选模标本 KNM-ER1470 很相似。主要差异包括 KNM-WT 40000 更加原始的鼻骨和脑颅形态。这提出了一种可能性，即这两种分类单元之间存在密切的系统发育关系，并且影响了我们对鲁道夫人的解释。已经有人建议将鲁道夫人归入南方古猿[4,19]，但是肯尼亚人可能是一个更适合的属。对肯尼亚扁脸人的鉴定具有许多额外的含义。作为与某些形态更加原始的南方古猿阿法种同时代的一个物种，肯尼亚扁脸人削弱了南方古猿阿法种作为所有后来的古人类的姐妹分类单元的理由，以及将其归入 *Praeanthropus* 的建议[1,20]。此外，肯尼亚扁脸人的形态引发了关于分析古人类系统发育时使用的特征的极性的问题。一个例子是该物种的小型臼齿尺寸，这尽管可能是一种衍生性状，但是也可能意味着南方古猿阿法种或南方古猿湖畔种的较大犬齿后部牙齿并不代表原始的古人类状态。最后，在东非的中上新世时期至少出现了另外一个古人类物种，这意味着现在可以重新估计碎片标本的归属了。例如，将 330 万年前的 KNM-ER 2602 颅骨碎片划分为南方古猿阿法种[21]已经被质疑了[8]，现在

and evaluating its affinities with *K. platyops* is now timely.

Geological Context and Dating

KNM-WT 40000 was collected near the contact of the Nachukui Formation with Miocene volcanic rocks in the northern tributary of Lomekwi (Nabetili). It is situated 12 m above the Lokochot Tuff, and 8 m below the β-Tulu Bor Tuff (Fig. 4). Along Nabetili, the Lokochot Tuff is pinkish-grey and contains much clay and volcanic detritus. It is overlain by a volcanic pebble conglomerate, followed by a pale brown quartz-rich fine sandstone that includes a burrowed fine-sandstone marker bed 10–15 cm thick. The Lokochot Tuff is replaced by a thick volcanic clast conglomerate in the central part of Lomekwi. The contact between the fine sandstone and the overlying dark mudstone can be traced from Nabetili to the hominin locality. Locally the mudstone contains volcanic pebbles at the base, and it has thin pebble conglomerate lenses in the upper part at the hominin locality, and also contains $CaCO_3$ concretions. The hominin specimen and other vertebrate fossils derive from this mudstone. Overlying the dark mudstone at the hominin site is a brown mudstone (8 m) that directly underlies the β-Tulu Bor Tuff.

New ^{40}Ar–^{39}Ar determinations on alkali feldspars from pumice clasts in the Moiti Tuff and the Topernawi Tuff, stratigraphically beneath the Lokochot Tuff, were instrumental in re-investigating the lower portion of this section. The new results yield a mean age for the Topernawi Tuff of 3.96 ± 0.03 Myr; this is marginally older than the pooled age for the Moiti Tuff of 3.94 ± 0.03 Myr. Previous investigations[22,23] placed the Topernawi Tuff above the Moiti Tuff, mainly on the basis of the K/Ar ages on alkali feldspar from pumice clasts in the Topernawi Tuff (3.78, 3.71, 3.76 and 3.97 Myr, all ± 0.04 Myr)[23]. The older determination (3.97 Myr) was thought to result from contamination by detrital feldspar. South of Topernawi, however, the Topernawi Tuff has now been shown to underlie the Moiti Tuff, and to be in turn underlain by a tephra informally termed the "Nabwal tuff", previously thought to be a Moiti Tuff correlative. The correct sequence is shown in Fig. 4, and the new ^{40}Ar–^{39}Ar age data on the Moiti Tuff and Topernawi Tuff are provided as Supplementary Information.

正是对其与肯尼亚扁脸人的亲缘关系进行评估的时候。

地质学环境与定年

KNM-WT 40000 是在洛梅奎（纳贝蒂利）北部支流的中新世火山岩与纳丘库伊组接壤部位附近搜集到的。它位于洛科乔特凝灰岩之上 12 米，β 图卢博尔凝灰岩之下 8 米处（图 4）。沿着纳贝蒂利，洛科乔特凝灰岩呈略带粉红的灰色，包含大量黏土和火山碎屑。其上覆盖着火山卵石砾岩，紧接着是淡褐色的富含石英的细砂岩，其中包括挖掘出来的 10~15 厘米厚的细砂岩标准层。洛科乔特凝灰岩在洛梅奎中部被一层厚的火山碎屑砾岩取代了。在细砂岩和覆盖在上面的深色泥岩之间的衔接部分可以从纳贝蒂利一直追踪到古人类遗址处。在当地，泥岩在基部包含火山卵石，在古人类遗址的上半部分有薄层的卵石砾岩透镜体，还含有 $CaCO_3$ 结核。古人类标本和其他脊椎动物化石都是从这层泥岩中发现的。覆盖在古人类遗址深色泥岩上面的是一种褐色的泥岩（8 米），位于 β 图卢博尔凝灰岩的紧下面。

根据在地层学上位于洛科乔特凝灰岩下面的莫伊蒂凝灰岩和托佩尔纳维凝灰岩中的浮岩碎屑的碱性长石确定的 $^{40}Ar-^{39}Ar$ 新数据，对于重新调查该剖面的下半部分是有帮助的。这些新结果产生的托佩尔纳维凝灰岩的平均年代是 396 万 ± 3 万年；这比莫伊蒂凝灰岩的并合年代 394 万 ± 3 万年古老的程度有限。之前的调查[22,23]将托佩尔纳维凝灰岩置于莫伊蒂凝灰岩之上，主要的依据是从托佩尔纳维凝灰岩中的浮岩碎屑得到的碱性长石的 K/Ar 年代（378 万、371 万、376 万和 397 万年，都是 ± 4 万年）[23]。确定的年代中较为古老的（397 万年）被认为是长石碎屑的污染造成的。但是，在托佩尔纳维南部，现在表明托佩尔纳维凝灰岩是位于莫伊蒂凝灰岩之下的，而反过来，一种非正式称为“纳布瓦尔凝灰岩”的火山碎屑又位于它的下面，后者之前被认为是莫伊蒂凝灰岩的相关物。正确的地层顺序见图 4 所示，莫伊蒂凝灰岩和托佩尔纳维凝灰岩的新的 $^{40}Ar-^{39}Ar$ 年代数据在补充信息中给出。

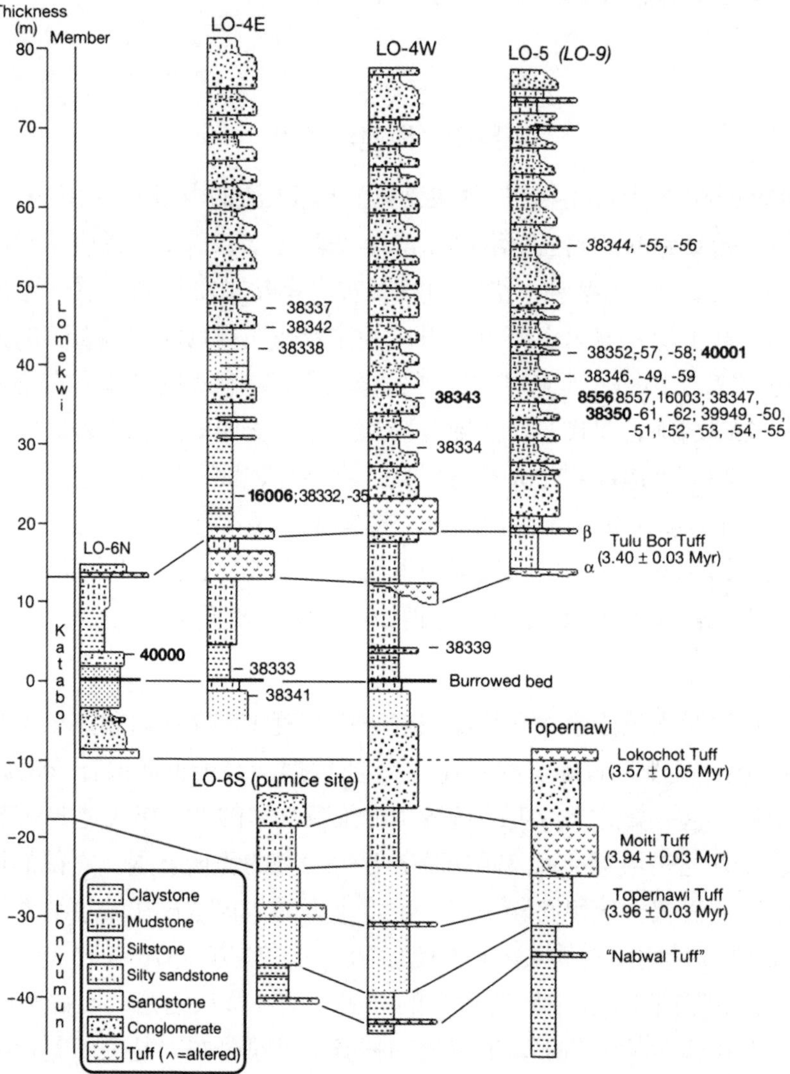

Fig. 4. Stratigraphic sections and placement of hominin specimens at sites in upper part of the Lomekwi drainage, west of Lake Turkana, northern Kenya. Specimen numbers are given without the prefix KNM-WT, and those in bold are discussed in the text. Placement of specimens is relative to the nearest marker bed in the section. Italicized numbers show the relative placement of specimens at LO-9 on section LO-5. The burrowed bed, a useful local marker, is used as stratigraphic datum (0 m). Representative ^{40}Ar–^{39}Ar analytical data on the Moiti Tuff and on the Topernawi Tuff are given as Supplementary Information. The date for the Tulu Bor Tuff is taken as the age of the Sidi Hakoma Tuff at Hadar[25], which is consistent with the age of the Toroto Tuff (3.32 ± 0.03 Myr)[45] that overlies the Tulu Bor at Koobi Fora. The age of the Lokochot Tuff is assigned from its placement at the Gilbert/Gauss Chron boundary[24,46]. The tuff formerly thought to be the Moiti Tuff at Lomekwi[22] has been informally called the "Nabwal tuff"[47]. Ages on the Tulu Bor and the Lokochot Tuffs are consistent with orbitally tuned ages of correlative ash layers in Ocean Drilling Program Core 722A in the Arabian Sea[48].

Linear interpolation between the Lokochot Tuff (3.57 Myr old)[24] and Tulu Bor Tuff (3.40 Myr)[25] yields an age of 3.5 Myr for KNM-WT 40000, and 3.53 Myr for the burrowed bed.

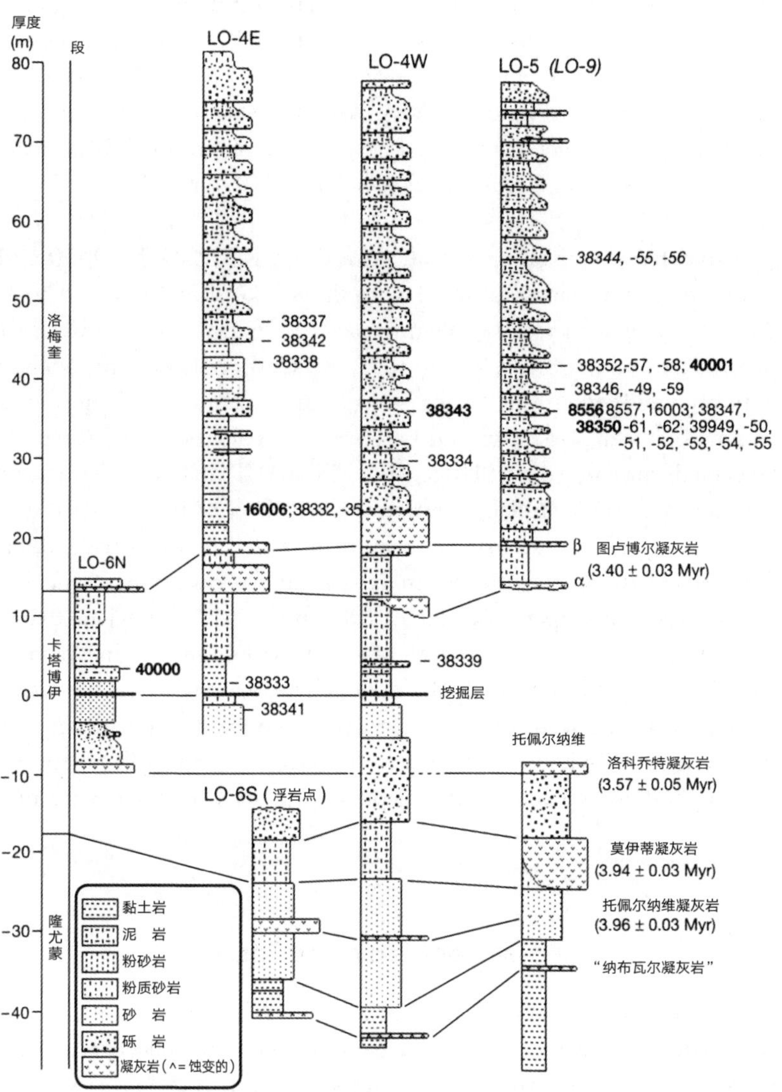

图 4. 肯尼亚北部图尔卡纳湖西侧的洛梅奎流域上游区的地层剖面和发现古人类标本的位置。标本号缺省了前缀 KNM-WT，黑体表示的是在正文中讨论的标本。标本的位置是与剖面中最近的标准层对应的。斜体数字表示 LO-9 处的标本在剖面 LO-5 的相对位置。挖掘层是一个有用的局部标志，被用来当作地层基准面（0 米）。对莫伊蒂凝灰岩和托佩尔纳维凝灰岩进行的代表性的 $^{40}Ar-^{39}Ar$ 分析数据作为补充信息给出。图卢博尔凝灰岩的年代根据哈达尔的西迪哈克玛凝灰岩的年代得来[25]，这与覆盖在库比福勒的图卢博尔之上的托罗托凝灰岩的年代（332 万 ± 3 万年）[45] 一致。洛科乔特凝灰岩的年代根据其在吉尔伯特/高斯年代界线处的位置来判定[24,46]。在洛梅奎处之前认为属于莫伊蒂的凝灰岩[22] 被非正式地称作“纳布瓦尔凝灰岩”[47]。图卢博尔和洛科乔特凝灰岩的年代与阿拉伯海[48] 的海洋钻探计划岩芯 722A 中的相关灰层的轨道调谐年代一致。

洛科乔特凝灰岩（357 万年）[24] 与图卢博尔凝灰岩（340 万年）[25] 之间的线性插入得到 KNM-WT 40000 的年代为 350 万年，挖掘层的年代为 353 万年。紧邻挖掘层之下的

KNM-WT 38341 from immediately below the burrowed bed has an age near 3.53 Myr. KNM-WT 38333 and 38339, from between the burrowed bed and the α-Tulu Bor Tuff lie between 3.4 and 3.5 Myr. Other specimens from LO-4, LO-5, and LO-6 lie 16–24 m above the β-Tulu Bor Tuff, with ages near 3.3 Myr. Assuming linear sedimentation between the Tulu Bor Tuff and the Lokalalei Tuff (2.5 Myr)[23], specimens from LO-9 are around 3.2 Myr. The probable error on these age estimates is less than 0.10 Myr.

Palaeogeographically, the mudstone that contained KNM-WT 40000 at LO-6N was deposited along the northern margin of a shallow lake that extended to Kataboi and beyond[26,27]. Laterally discontinuous volcanic pebble conglomerates within the mudstone record small streams draining from hills to the west. Carbonate concretions at the hominin level are probably pedogenic, and indicate regional conditions with net evaporative loss. Other specimens between the burrowed bed and the Tulu Bor Tuff were also preserved in lake-margin environments, as is the case for KNM-WT 38341 that was collected below the burrowed bed. At LO-5, and in the upper part of LO-4E, strata were laid down by ephemeral streams draining the basin margin, principally the ancestral Topernawi, which deposited gravels in broad, shallow channels, and finer grained materials in interfluves. Specimens preserved in floodplain deposits of the ancestral Omo River that occupied the axial portion of the basin include those at LO-9, those less than 6 m above the Tulu Bor Tuff at LO-4E, and KNM-WT 38338. Thus, there is evidence for hominins occupying floodplains of major rivers, alluvial fans, and lake-margin environments 3.0–3.5 Myr ago. There is reasonable evidence that water sources were available to these hominins in channels of the ephemeral streams, and also possibly as seeps or springs farther out into the basin.

Palaeoecology and Fauna

Faunal assemblages from Lomekwi sites LO-4, LO-5, LO-6 and LO-9 indicate palaeoenvironments that were relatively well watered and well vegetated. The relative proportions of the bovids in the early collections from these sites indicate a mosaic of habitats, but with predominantly woodland and forest-edge species dominating[22]. Comparisons of the Lomekwi faunal assemblages with those from the few known hominin sites of similar age, Laetoli in Tanzania, Hadar in Ethiopia and Bahr el Ghazal in Chad, are of interest in view of the different hominin taxa represented. Hadar and Bahr el Ghazal, like Lomekwi, represent lakeshore or river floodplain palaeoenvironments[28,29], whereas Laetoli was not located near a water source; no aquatic taxa nor terrestrial mammals indicative of swamp or grassy wetlands were recovered[30]. The faunal assemblages of all four sites indicate a mosaic of habitats that seems to have included open grasslands and more wooded or forested environments[22,28,29,31,32]; the assemblages differ primarily in the indication of the nature of the dominant vegetation cover.

Although the mammalian faunal assemblage from Lomekwi is more similar to that from Hadar than to that from Laetoli, some mammalian species represented are different.

KNM-WT 38341 的年代接近 353 万年。从挖掘层与 α 图卢博尔凝灰岩之间发掘到的 KNM-WT 38333 和 38339 的年代介于 340 万年到 350 万年之间。从 LO-4、LO-5 和 LO-6 挖掘出来的其他标本位于 β 图卢博尔凝灰岩之上 16~24 米处，年代接近 330 万年。假定图卢博尔凝灰岩与洛卡拉莱伊凝灰岩（250 万年）[23]之间存在线性沉积作用，那么从 LO-9 挖掘出来的标本的年代就约为 320 万年。这些年代估计值的可能误差小于 10 万年。

从古地理学角度看，在 LO-6N 处包含 KNM-WT 40000 的泥岩是沿着一个浅湖的北部边缘沉积下来的，这个湖一直延伸到卡塔博伊甚至更远处[26,27]。在泥岩中的侧面的不连续的火山卵石砾岩记录下了从山丘流到西部的小溪流。在古人类地层中的碳酸盐结核可能是成土的，暗示着具有净蒸发损失的区域性情况。位于挖掘层与图卢博尔凝灰岩之间的其他标本也在湖缘环境中保存下来了，正如 KNM-WT 38341 的情况一样，该标本是在挖掘层之下的地层中搜集到的。在 LO-5，以及 LO-4E 的上半部分，地层由从盆地边缘流出的季节河（主要是原始的托佩尔纳维）沉积而成，它们将砾石在宽阔而浅的河道中沉积下来，细粒的材料在江河分水区沉积下来。原始的奥莫河占据了盆地的轴向部分，在其河漫滩沉积物中保存的标本包括在 LO-9 的那些标本，在 LO-4E 的图卢博尔凝灰岩之上不到 6 米处的标本，以及 KNM-WT 38338。因此，有证据表明，300 万 ~ 350 万年前，有古人类居住在大河的河漫滩、冲积扇以及湖缘环境中。有可靠证据表明，这些古人类可以得到的水源来自季节河的河道，可能还有深入盆地内部的渗流或泉水。

古生态学与动物群

来自洛梅奎遗址 LO-4、LO-5、LO-6 和 LO-9 的动物群组合暗示当时的古环境是相对水源充足、植物茂盛的。从这些遗址搜集到的早期标本中牛科动物的相对比例暗示着生境的镶嵌性，但林地和森林边缘物种占主导地位[22]。将洛梅奎动物群组合与已知的少数相似年代的古人类遗址（坦桑尼亚的莱托里、埃塞俄比亚的哈达尔和乍得的加扎勒河省）发现的动物群组合相比，从所代表的不同古人类分类单元的视点来看很有趣。哈达尔和加扎勒河省像洛梅奎一样，代表着湖岸或河流河漫滩的古环境[28,29]，而莱托里并不位于水源附近；没有发现暗示沼泽或多草湿地存在的水生分类单元或陆地哺乳动物[30]。所有四处遗址的动物群组合都暗示着生境的镶嵌性，它似乎包括开放的草地和更加多树木或多森林的环境[22,28,29,31,32]；这些动物群组合的主要区别在于优势植被的性质的含义不同。

尽管洛梅奎的哺乳动物群组合与哈达尔的相似性胜过其与莱托里的，但是有些代表性的哺乳动物物种是不同的。在洛梅奎，布鲁姆狮尾狒很常见，并且是主要的

At Lomekwi, *Theropithecus brumpti* is common and is the dominant cercopithecid, as it is elsewhere in the Turkana Basin at this time. This species is generally considered to indicate more forested or closed woodland habitats. In the Hadar Formation, *Theropithecus darti* is the common *Theropithecus* species and is associated with lower occurrences of the water-dependent reduncines and higher occurrences of alcelaphines and/or *Aepyceros*, which indicates drier woodlands and grasslands[33]. Differences in the representation of other common species at the two sites that are less obviously linked to habitat include *Kolpochoerus limnetes*, *Tragelaphus nakuae* and *Aepyceros shungurensis* at Lomekwi, as opposed to *K. afarensis*, *T. kyaloae* and an undescribed species of *Aepyceros* at Hadar (K. Reed, personal communication). The general indication is that the palaeoenvironment at Lomekwi may have been somewhat more vegetated and perhaps wetter than that persisting through much of the Hadar Formation. At both sites more detailed analyses will be essential to further develop an understanding of how subtle temporal changes in the faunal assemblages relate to hominin occurrences.

Note added in proof: If the hominin status of the recently published Lukeino craniodental specimens[49] is confirmed, this would support the suggestion that small molar size is the primitive rather than the derived hominin condition.

(**410**, 433-440; 2001)

Meave G. Leakey*, Fred Spoor†, Frank H. Brown‡, Patrick N. Gathogo‡, Christopher Kiarie*, Louise N. Leakey* & Ian McDougall§

* Division of Palaeontology, National Museums of Kenya, P.O. Box 40658, Nairobi, Kenya

† Department of Anatomy & Developmental Biology, University College London, WC1E 6JJ, UK

‡ Department of Geology & Geophysics, University of Utah, Salt Lake City, Utah 84112, USA

§ Research School of Earth Sciences, The Australian National University, Canberra ACT 0200, Australia

Received 31 January; accepted 16 February 2001.

References:

1. Strait, D. S., Grine, F. E. & Moniz, M. A. A reappraisal of early hominid phylogeny. *J. Hum. Evol.* **32**, 17-82 (1997).

2. Ward, C., Leakey, M. & Walker, A. The new hominid species *Australopithecus anamensis. Evol. Anthrop.* **7**, 197-205 (1999).

3. Brown, B., Brown, F. & Walker, A. New hominids from the Lake Turkana Basin, Kenya. *J. Hum. Evol.* (in the press).

4. Wood, B. A. & Collard, M. C. The human genus. *Science* **284**, 65-71 (1999).

5. Leakey, M. G., Feibel, C. S., McDougall, I. & Walker, A. C. New four-million-year-old hominid species from Kanapoi and Allia Bay, Kenya. *Nature* **376**, 565-571 (1995).

6. Leakey, M. G., Feibel, C. S., McDougall, I., Ward, C. & Walker, A. New specimens and confirmation of an early age for *Australopithecus anamensis. Nature* **393**, 62-66 (1998).

7. Kimbel, W. H., Johanson, D. C. & Rak, Y. Systematic assessment of a maxilla of *Homo* from Hadar, Ethiopia. *Am. J. Phys. Anthrop.* **103**, 235-262 (1997).

8. Wood, B. A. *Koobi Fora Research Project* Vol. 4 (Clarendon Press, Oxford, 1991).

9. Lockwood, C. A. & Tobias, P. V. A large male hominin cranium from Sterkfontein, South Africa, and the status of *Australopithecus africanus. J. Hum. Evol.* **36**, 637-685 (1999).

10. Bilsborough, A. & Wood, B. A. Cranial morphometry of early hominids: facial region. *Am. J. Phys. Anthrop.* **76**, 61-86 (1988).

11. Rak, Y. *The Australopithecine Face* (Academic, New York, 1983).

12. Lockwood, C. A., Kimbel, W. H. & Johanson, D. C. Temporal trends and metric variation in the mandibles and dentition of *Australopithecus afarensis. J. Hum. Evol.* **39**, 23-55 (2000).

13. Moggi-Cecchi, J., Tobias, P. V. & Beynon, A. D. The mixed dentition and associated skull fragments of a juvenile fossil hominid from Sterkfontein, South Africa. *Am. J.*

猴科动物，就像当时在图尔卡纳盆地的其他地方一样。该物种通常被认为暗示着更加森林化或者更加封闭的林地生境。在哈达尔组，达提狮尾狒是常见的狮尾狒属的物种，与较少出现的依赖水的苇羚科和经常出现的狷羚科和（或）高角羚属相关，暗示着更加干旱的林地和草地[33]。在这两处遗址，与生境关联不那么明显的其他常见物种的差别包括洛梅奎的 *Kolpochoerus limnetes*、*Tragelaphus nakuae* 和 *Aepyceros shungurensis*，而哈达尔的 *K. afarensis*、*T. kyaloae* 和一种未描述的高角羚属物种（里得，个人交流）与之刚好形成对照。通常认为，与哈达尔组大部分地区长期存在的古环境相比，洛梅奎的古环境可能植被更加丰富，或许也更加潮湿。在这两处遗址，需要进行更加详细的分析以进一步理解动物群组合在时间上的微妙变化如何与古人类出现相关。

附加说明：如果最近发表的卢凯伊诺颅牙标本[49]的古人类地位得到了证实，那么将支持如下建议，即小型的臼齿尺寸是原始的，而非衍生的古人类特征。

（刘皓芳 翻译；赵凌霞 审稿）

Phys. Anthrop. **106**, 425-465 (1998).

14. Robinson, J. T. The dentition of the Australopithecinae. *Transv. Mus. Mem.* **9**, 1-179 (1956).
15. McCollum, M. A. The robust australopithecine face: a morphogenetic perspective. *Science* **284**, 301-305 (1999).
16. Brunet, M. *et al. Australopithecus bahrelghazali*, une nouvelle espèce d'Hominidé ancien de la région de Koro Toro (Tchad). *C.R. Acad. Sci. Ser. IIa* **322**, 907-913 (1996).
17. White, T. D., Suwa, G., Simpson, S. & Asfaw, B. Jaws and teeth of *A. afarensis* from Maka, Middle Awash, Ethiopia. *Am. J. Phys. Anthrop.* **111**, 45-68 (2000).
18. White, T. D., Suwa, G. & Asfaw, B. *Australopithecus ramidus*, a new species of early hominid from Aramis, Ethiopia. *Nature* **371**, 306-312 (1994).
19. Wood, B. & Collard, M. The changing face of genus *Homo. Evol. Anthrop.* **8**, 195-207 (2000).
20. Harrison, T. in *Species, Species Concepts, and Primate Evolution* (eds Kimbel, W. H. & Martin, L. B.) 345-371 (Plenum, New York, 1993).
21. Kimbel, W. H. Identification of a partial cranium of *Australopithecus afarensis* from the Koobi Fora Formation. *J. Hum. Evol.* **17**, 647-656 (1988).
22. Harris, J. M., Brown, F. & Leakey, M. G. Stratigraphy and paleontology of Pliocene and Pleistocene localities west of Lake Turkana, Kenya. *Cont. Sci. Nat. Hist. Mus. Los Angeles* **399**, 1-128 (1988).
23. Feibel, C. S., Brown, F. H. & McDougall, I. Stratigraphic context of fossil hominids from the Omo Group deposits, northern Turkana Basin, Kenya and Ethiopia. *Am. J. Phys. Anthrop.* **78**, 595-622 (1989).
24. McDougall, I., Brown, F. H., Cerling, T. E. & Hillhouse, J. W. A reappraisal of the geomagnetic polarity time scale to 4 Ma using data from the Turkana Basin, East Africa. *Geophys. Res. Lett.* **19**, 2349-2352 (1992).
25. Walter, R. C. & Aronson, J. L. Age and source of the Sidi Hakoma Tuff, Hadar Formation, Ethiopia. *J. Hum. Evol.* **25**, 229-240 (1993).
26. Brown, F. H. & Feibel, C. S. in *Koobi Fora Research Project*, Vol. 3, *Stratigraphy, artiodactyls and paleoenvironments* (ed. Harris, J. M.) 1-30 (Clarendon, Oxford, 1991).
27. Feibel, C. S., Harris, J. M. & Brown, F. H. in *Koobi Fora Research Project*, Vol. 3, *Stratigraphy, artiodactyls and paleoenvironments* (ed. Harris, J. M.) 321-346 (Clarendon, Oxford, 1991).
28. Johanson, D. C., Taieb, M. & Coppens, Y. Pliocene hominids from the Hadar Formation, Ethiopia (1973-1977): stratigraphic, chronologic and paleoenvironmental contexts, with notes on hominid morphology and systematics. *Am. J. Phys. Anthrop.* **57**, 373-402 (1982).
29. Brunet, M. *et al.* The first australopithecine 2,500 kilometres west of the Rift Valley (Chad). *Nature* **378**, 273-240 (1995).
30. Leakey, M. D. & Harris, J. M. *Laetoli, a Pliocene Site in Northern Tanzania* (Clarendon, Oxford, 1987).
31. Harris, J. M. in *Laetoli, a Pliocene Site in Northern Tanzania* (eds Leakey, M. D. & Harris, J. M.) 524-531 (Clarendon, Oxford, 1987).
32. Kimbel, W. H. *et al.* Late Pliocene *Homo* and Oldowan tools from the Hadar Formation (Kadar Hadar Member), Ethiopia. *J. Hum. Evol.* **31**, 549-561 (1996).
33. Eck, G. G. in *Theropithecus, the Rise and Fall of a Primate Genus* (ed. Jablonski, N. G.) 15-83 (Cambridge Univ. Press, 1993).
34. White, T. D. New fossil hominids from Laetolil, Tanzania. *Am. J. Phys. Anthrop.* **46**, 197-230 (1977).
35. Wood, B. Origin and evolution of the genus *Homo. Nature* **355**, 783-790 (1992).
36. Suwa, G. *et al.* The first skull of *Australopithecus boisei. Nature* **389**, 489-492 (1997).
37. Keyser, A. W. The Drimolen skull: the most complete australopithecine cranium and mandible to date. *S. Afr. J. Sci.* **96**, 189-193 (2000).
38. Kimbel, W. H., White, T. D. & Johanson, D. C. Cranial morphology of *Australopithecus afarensis*: a comparative study based on a composite reconstruction of the adult skull. *Am. J. Phys. Anthrop.* **64**, 337-388 (1984).
39. Walker, A., Leakey, R. E., Harris, J. H. & Brown, F. H. 2.5-Myr *Australopithecus boisei* from west of Lake Turkana, Kenya. *Nature* **322**, 517-522 (1986).
40. Kimbel, W. H., White, T. D. & Johanson, D. C. in *Evolutionary History of the "Robust" Australopithecines* (ed. Grine, F. E.) 259-268 (Aldine de Gruyter, New York, 1988).
41. Asfaw, B. *et al. Australopithecus garhi*: A new species of early hominid from Ethiopia. *Science* **284**, 629-635 (1999).
42. Grine, F. E. & Strait, D. S. New hominid fossils from Member 1 "Hanging Remnant" Swartkrans Formation, South Africa. *J. Hum. Evol.* **26**, 57-75 (1994).
43. Johanson, D. C., White, T. D. & Coppens, Y. Dental remains from the Hadar Formation, Ethiopia: 1974-1977 collections. *Am. J. Phys. Anthrop.* **57**, 545-603 (1982).
44. Tobias, P. V. in *Olduvai Gorge* Vol. 4 (Cambridge Univ. Press, 1991).
45. McDougall, I. K-Ar and ^{40}Ar/^{39}Ar dating of the hominid-bearing Pliocene-Pleistocene sequence at Koobi Fora, Lake Turkana, northern Kenya. *Geol. Soc. Am. Bull.* **96**, 159-175 (1985).
46. Brown, F. H., Shuey, R. T. & Croes, M. K. Magnetostratigraphy of the Shungura and Usno Formations, southwestern Ethiopia: new data and comprehensive reanalysis. *Geophys. J. R. Astron. Soc.* **54**, 519-538 (1978).
47. Haileab, B. *Geochemistry, Geochronology and Tephrostratigraphy of Tephra from the Turkana Basin, Southern Ethiopia and Northern Kenya*. Thesis, Univ. Utah (1995).
48. deMenocal, P. B. & Brown, F. H. in *Hominin Evolution and Climatic Change in Europe* Vol. 1 (eds Agustí, J., Rook, L. & Andrews, P.) 23-54 (Cambridge Univ. Press, 1999).
49. Senut, B. *et al.* First hominid from the Miocene (Lukeino Formation, Kenya). *C.R. Acad. Sci. Paris* **332**,137-144 (2001).

Supplementary information is available on *Nature*'s World-Wide Web site (http://www.nature.com) or as paper copy from the London editorial office of *Nature*.

Acknowledgements. We thank the Government of Kenya for permission to carry out this research and the National Museums of Kenya for logistical support. The National Geographic Society funded the field work and some laboratory studies. Neutron irradiations were facilitated by the Australian Institute of Nuclear Science and Engineering and the Australian Nuclear Science and Technology Organisation. We also thank the Ethiopian Ministry of Information and Culture, the National

Museum of Ethiopia, B. Asfaw, Y. Bayene, C. Howell, D. Johanson, W. Kimbel, G. Suwa and T. White for permission to make comparisons with the early Ethiopian hominins and numerous people including N. Adamali, B. Asfaw, C. Dean, C. Feibel, A. Griffiths, W. Kimbel, R. Kruszynski, K. Kupczik, R. Leakey, D. Lieberman, J. Moore, K. Patel, D. Plummer, K. Reed, B. Sokhi, M. Tighe, T. White, and B. Wood for their help. Caltex (Kenya) provided fuel for the field expeditions, and R. Leakey allowed us the use of his aeroplane. The field expedition members included U. Bwana, S. Crispin, G. Ekalale, M. Eragae, J. Erus, J. Ferraro, J. Kaatho, N. Kaling, P. Kapoko, R. Lorinyok, J. Lorot, S. Hagemann, B. Malika, W. Mangao, S. Muge, P. Mulinge , D. Mutinda, K. Muthyoka, N. Mutiwa, W. Mutiwa, B. Onyango, E. Weston and J. Wynn. A. Ibui, F. Kyalo, F. Kirera, N. Malit, E. Mbua, M. Muungu, J. Ndunda, S. Ngui and A. Mwai provided curatorial assistance. The Leakey Foundation awarded a grant to F.B.

Correspondence and requests for materials should be addressed to M.G.L. (e-mail: meave@swiftkenya.com).

Linkage Disequilibrium in the Human Genome

D. E. Reich *et al.*

Editor's Note

By this time, it was known that single nucleotide polymorphisms (SNPs)—the single-letter DNA changes thought to underlie disease susceptibility and individual variation—"travel" together, with one SNP carrying information about its SNP neighbours. So armed with the recent SNP map, geneticist Eric Lander and colleagues set out to see if the method of linkage disequilibrium (LD), where blocks of SNPs are correlated back to an ancestral chromosome, can be used to map disease-causing genes. This study shows, at least in the northern European population tested, that using LD to map disease-related genes might be easier and more practical than had been thought. The results also shed light on human history, suggesting this particular LD pattern might reflect the relatively recent migration of anatomically modern humans.

With the availability of a dense genome-wide map of single nucleotide polymorphisms (SNPs)[1], a central issue in human genetics is whether it is now possible to use linkage disequilibrium (LD) to map genes that cause disease. LD refers to correlations among neighbouring alleles, reflecting "haplotypes" descended from single, ancestral chromosomes. The size of LD blocks has been the subject of considerable debate. Computer simulations[2] and empirical data[3] have suggested that LD extends only a few kilobases (kb) around common SNPs, whereas other data have suggested that it can extend much further, in some cases greater than 100 kb[4-6]. It has been difficult to obtain a systematic picture of LD because past studies have been based on only a few (1–3) loci and different populations. Here, we report a large-scale experiment using a uniform protocol to examine 19 randomly selected genomic regions. LD in a United States population of north-European descent typically extends 60 kb from common alleles, implying that LD mapping is likely to be practical in this population. By contrast, LD in a Nigerian population extends markedly less far. The results illuminate human history, suggesting that LD in northern Europeans is shaped by a marked demographic event about 27,000–53,000 years ago.

TO characterize LD systematically around genes, each of the 19 regions that we studied was anchored at a "core" SNP in the coding region of a gene. The core SNP was chosen from a database of more than 3,000 coding SNPs that had been identified by screening in a multi-ethnic panel (see Methods), subject to two requirements. First, "finished" genomic sequence was available for 160 kb in at least one direction from the core SNP; second, the frequency of the minor (less common) allele was at least 35% in the multi-ethnic panel.

人类基因组中的连锁不平衡

赖希等

编者按

如今众所周知，单核苷酸多态性(SNP)——被认为影响疾病的易感性和个体差异性的DNA单个碱基变化——是结伴“旅行”的，即一个SNP会携带关于其邻位的信息。因此，结合目前的SNP图谱，遗传学家埃瑞克·兰德及其同事着手研究连锁不平衡(LD，即SNP的板块都是和一个祖先染色体相关联的)的方法能否被用于发现致病基因。本研究显示，至少在北欧人群的测试中，利用LD来寻找致病基因也许比预想的要更简单更实际。该结果还可用于阐释人类历史，提示这种独特的LD模式可能反映出解剖学上现代人类的迁徙。

随着高密度全基因组单核苷酸多态性(SNP)图谱的获得[1]，人类遗传学的一个核心问题就是现在是否有可能利用连锁不平衡(LD)绘制出致病基因图谱。LD指的是相邻等位基因之间的相互联系，表现为来自单一祖先染色体的“单体型”。LD区域的大小一直是争议的焦点。计算机模拟[2]和经验性数据[3]显示LD位于常见SNP周围，大约只延伸数千个碱基(kb)，而其他数据显示其范围更广，在某些情况下可能大于100 kb[4-6]。一直以来，LD的系统性图谱难以获得，因为过去的研究都仅仅基于几个(1 ~ 3)位点以及不同的人群。本文我们报告一个采用了统一规范，针对19个随机选取的基因组区域的大规模实验。结果发现拥有北欧血统的美国人群其LD一般从常见等位位点延伸60 kb，说明在这一群体中进行LD图谱的绘制是可行的。相反，在一个尼日利亚人群中，LD区域的缩短十分明显。这些结果勾画出人类演化历史，提示大约27,000 ~ 53,000年前的某个重大的人口统计学事件导致了北欧人群中LD的形成。

为了系统性地鉴定基因附近的LD，在我们研究的19个区域中，每个区域都在其编码区内锚定了一个“核心”SNP。核心SNP选自一个含有超过3,000个编码SNP的数据库，这些SNP从多民族人群筛查中选定(见方法)，要求满足以下两个条件。其一，从核心SNP起始，至少一个方向上160 kb范围内可获得完整基因组序列；其二，次等位基因位点(频率较低的那个位点)在这个多族群数据库中的频率至少为35%。

We focused on high-frequency SNPs for several reasons. First, they tend to be of high frequency in all populations[7], facilitating cross-population comparisons. Second, LD around common alleles represents a "worst case" scenario: LD around rare alleles is expected to extend further because such alleles are generally young[8] and there has been less historical opportunity for recombination to break down ancestral haplotypes[2]. Third, LD around common alleles can be measured with a modest sample size of 80–100 chromosomes to a precision within 10–20% of the asymptotic limit (see Methods). Last, LD around common alleles will probably be particularly relevant to the search for genes predisposing to common disease[9].

To identify SNPs at various distances from the core SNP, we resequenced subregions of around 2 kb centred at distances 0, 5, 10, 20, 40, 80 and 160 kb in one direction from the core SNP using 44 unrelated individuals from Utah. Altogether, we screened 251,310 bp (see Methods) and found an average heterozygosity of $\pi = 0.00070$, consistent with past studies[1]. A total of 272 "high frequency" polymorphisms were identified (Table 1).

Table 1. Distribution of regions and SNPs within regions

Gene identification*	Chromosome	Local recombination rate (cM Mb^{-1})†	Span of region in physical map (cR)‡	Number of high-frequency polymorphisms at distances (kb) from core SNPs§						
				1	5	10	20	40	80	160
BMP8‖	1	1.4	60.88–61.04	1	1	1	2	2	1	3
ACVR2B‖	3	1.1	37.27–37.43	2	1	2	3	0	3	1
TGFBI	5	1.4	152.16–152.00	1	3	1	0	1	7	0
DDR1‖	6	–	46.046–45.89	1	4	1	4	2	2	0
GTF2H4	6	–	46.059–46.22	3	3	3	6	2	0	0
COL11A2‖	6	–	48.27–48.43	1	0	1	2	3	1	1
LAMB1‖	7	2.3	106.58–106.42	0	0	5	3	3	2	4
WASL	7	0.5	122.99–122.83	2	1	0	4	2	2	1
SLC6A12	12	3.3	3.62–3.78	2	8	2	1	6	0	0
KCNA1	12	3.3	8.50–8.66	1	1	5	2	1	1	0
SLC2A3	12	2.1	16.33–16.49	1	2	1	1	5	0	2
ARHGDIB‖	12	1.2	21.84–22	3	9	1	2	2	1	2
PCI‖	14	4.3	98.41–98.57	0	7	0	0	4	14	1
PRKCBI	16	1.0	32.50–32.66	0	2	3	0	4	3	0
NFI	17	1.0	38.10–38.26	1	0	2	2	5	1	0
SCYA2‖	17	3.0	40.21–40.37	0	5	1	1	3	1	1
PAI2	18	2.7	64.17–64.01	2	0	2	1	5	2	10
IL17R‖	22	5.9	14.48–14.32	1	1	2	1	2	0	4
HCF2‖	22	2.0	17.91–17.75	1	1	2	2	1	0	3

* Abbreviations from LocusLink (www.ncbi.nlm.nih.gov/LocusLink/list.cgi).

我们着眼于高频率的 SNP 有以下几点原因。首先，它们倾向于在所有人群中高频出现[7]，这有利于跨人群的比对。其次，常见等位位点附近的 LD 实际代表了一种“最差”的情况：而罕见位点附近的 LD 能够延伸更长，因为这些等位基因通常出现得比较晚[8]，历史上发生重组而破坏原始单体型的概率也更低一些[2]。第三，常见等位基因位点附近的 LD 可以通过对 80 ~ 100 个染色体这样一个适中的样本数量测量获得，精度在渐进极限的 10% ~ 20% 之间（见方法）。最后，常见等位基因位点附近的 LD 很可能与寻找常见疾病易感基因更为相关[9]。

为了确定距离核心 SNP 不同位置的 SNP，我们选取同一方向上距离核心 SNP 0、5、10、20、40、80 和 160 kb 的位置作为中心，对中心附近 2 kb 的子区域进行重测序，样本来自犹他州 44 个不相关的个体。总体来说，我们对 251,310 个碱基对（见方法）进行了上述筛查，发现平均杂合度 π = 0.00070，与既往的研究一致[1]。总共检测到 272 个“高频”多态性位点（表 1）。

表 1. 区域分布以及区域内的 SNP

基因代码 *	染色体	局部重组率 (cM · Mb^{-1}) †	物理图谱中该区域的范围 (cR) ‡	离核心 SNP 不同距离 (kb) 的高频多态位点数目 §						
				1	5	10	20	40	80	160
BMP8‖	1	1.4	60.88 ~ 61.04	1	1	1	2	2	1	3
ACVR2B‖	3	1.1	37.27 ~ 37.43	2	1	2	3	0	3	1
TGFBI	5	1.4	152.16 ~ 152.00	1	3	1	0	1	7	0
DDR1‖	6	–	46.046 ~ 45.89	1	4	1	4	2	2	0
GTF2H4	6	–	46.059 ~ 46.22	3	3	3	6	2	0	0
COL11A2‖	6	–	48.27 ~ 48.43	1	0	1	2	3	1	1
LAMB1‖	7	2.3	106.58 ~ 106.42	0	0	5	3	3	2	4
WASL	7	0.5	122.99 ~ 122.83	2	1	0	4	2	2	1
SLC6A12	12	3.3	3.62 ~ 3.78	2	8	2	1	6	0	0
KCNA1	12	3.3	8.50 ~ 8.66	1	1	5	2	1	1	0
SLC2A3	12	2.1	16.33 ~ 16.49	1	2	1	1	5	0	2
ARHGDIB‖	12	1.2	21.84 ~ 22	3	9	1	2	2	1	2
PCI‖	14	4.3	98.41 ~ 98.57	0	7	0	0	4	14	1
PRKCBI	16	1.0	32.50 ~ 32.66	0	2	3	0	4	3	0
NFI	17	1.0	38.10 ~ 38.26	1	0	2	2	5	1	0
SCYA2‖	17	3.0	40.21 ~ 40.37	0	5	1	1	3	1	1
PAI2	18	2.7	64.17 ~ 64.01	2	0	2	1	5	2	10
IL17R‖	22	5.9	14.48 ~ 14.32	1	1	2	1	2	0	4
HCF2‖	22	2.0	17.91 ~ 17.75	1	1	2	2	1	0	3

* 缩写来自 LocusLink（www.ncbi.nlm.nih.gov/LocusLink/list.cgi）。

† For three regions the genetic and physical maps were inconsistent and no estimates were made.

‡ Span of region within a radiation hybrid map (http://www.ncbi.nlm.nih.gov/genome/seq/HsHome.shtml): 1 Mb ≈ 1 centirad. Position of the core SNP is listed first.

§ Number of SNPs discovered with at least 15 copies of the minor allele (successfully genotyped in at least 32 individuals) and in Hardy-Weinberg equilibrium using a significance criterion of $P < 0.02$.

‖ The ten regions selected for follow-up genotyping.

We measured LD between two SNPs using the classical statistic D′ (see Methods)[10]. D′ has the same range of values regardless of the frequencies of the SNPs compared[11]. Its sign (positive or negative) depends on the arbitrary choice of the alleles paired at the two loci. We chose the pair of SNPs that caused $D' > 0$ in Utah so that, in comparisons with other populations, the sign of D′ indicates whether the same or opposite allelic association is present. In a large sample, $|D'|$ of 1 indicates complete LD; 0 corresponds to no LD. The degree of LD needed for effective mapping depends on the details of a particular study[2]. A useful measure is the "half-length" of LD (the distance at which the average $|D'|$ drops below 0.5).

Comparing the 19 randomly selected regions, LD has a half-length of about 60 kb (Fig. 1). Significant *P*-values for LD occur in greater than 50% of cases at distances of $\leqslant 80$ kb. LD therefore extends much further than a previous prediction[2], and our data indicate that, in general, blocks of LD are large. Although the average extent is large, there is great variation in LD across the genomic regions (see also ref. 12), which is apparent in the different rates at which LD declines around the core SNP (Fig. 2). For example, $|D'| > 0.5$ for at least 155 kb around the *WASL* gene, but for less than 6 kb around *PCI* (Fig. 2). The variability across different genomic regions within the same population sample provides a context for explaining why past empirical studies, each based on one to three regions[3-5], have produced such different results. Large variations in LD are expected because of stochastic factors, such as different gene histories across loci[13]. Differences in recombination rates among regions can also affect the extent of LD. We observe a significant and important correlation ($P < 0.005$) between LD and the estimated local recombination rate (Fig. 2, inset).

Another feature of the data is that, near the range of distances at which LD drops off, there is often considerable variability in $|D'|$ values at neighbouring SNPs (for example, around *IL17R*, *SCYA2*, *TGFBI* and *BMP8* in Fig. 2). Such a wide scatter of LD, even for markers close to each other, has been noted before[14], and is due to the underlying haplotypic structure of LD. SNPs marking sections of chromosome with short extents of correlation are likely to display much lower $|D'|$ values than SNPs marking long haplotypes. In regions of high haplotype diversity, several SNPs may have to be genotyped to have a good chance of tagging most haplotypes. LD-based gene mapping may therefore require clusters of closely spaced SNPs to have maximal power.

† 其中三个区域的基因图谱和物理图谱不一致，因此没有进行估算。

‡ 放射性杂交图谱中的区域范围（www.ncbi.nlm.nih.gov/genome/seq/HsHome.shtml）：1Mb≈1 厘拉德。核心 SNP 的位置列在前。

§ 在至少 15 个次等位位点拷贝（在至少 32 个个体中成功确定基因型）中发现的 SNP 数量，并且在哈迪–温伯格平衡中显著性标准 $P < 0.02$。

‖ 这十个区域用于后续的基因分型。

我们使用经典的 D′ 值量化两个 SNP 之间的 LD(见方法)[10]。不论相比较的 SNP 出现频率如何，D′ 都具有相同的取值范围[11]。其符号（正或者负）取决于两位点间等位位点对的随机选择。我们选择了在犹他州人群中 D′ > 0 的 SNP 对，这样在和其他人群进行比较时，D′ 的符号表示是否存在相同或者相反的等位相关性。在大样本中，|D′| 为 1 表示完全的 LD，0 表示不存在 LD。有效的图谱绘制所需 LD 的程度需要根据具体研究细节确定[2]。一种可行的度量方式是利用 LD 的“半长”(即平均 |D′| 降至 0.5 以下的距离)。

对比这 19 个随机选取的区域，LD 的半长度大约是 60 kb(图 1)。在距离 ≤ 80 kb 的情况下，超过 50% 的样本 LD 的 P 值具有显著性。因此 LD 延伸的距离比先前的预计要长得多[2]，而且我们的数据显示，总体上 LD 区域都相当大。尽管平均尺度很大，整个基因组区域中 LD 的变化也很大(也可见参考文献 12)，但很明显在核心 SNP 附近 LD 衰减的速率不同(图 2)。比如，WASL 基因附近，|D′| > 0.5 的区域至少有 155 kb, 但是对于 PCI 基因附近则小于 6 kb(图 2)。同一人群中这种不同基因组区域间的变异，可以用来解释为什么过去的经验性研究尽管只是基于 1 到 3 个区域[3-5]，却产生如此多不同的结果。因为存在随机因素，比如不同位点间的基因历史不同[13]，LD 的这种巨大变异是可以预期的。区域间重组率的不同也能影响到 LD 的程度。我们观察到 LD 和预测的局部重组率之间存在显著且重要的相关性($P <$ 0.005)(图 2，右下角插图)。

这些数据的另一个特征是在 LD 急剧下降的区间附近，相邻 SNP 间的 |D′| 值有较大波动(例如，图 2 中 IL17R、SCYA2、TGFBI 和 BMP8 附近)。LD 这种即使在相邻位点也产生大范围波动以前也曾发现过[14]，是由于 LD 潜在的单体型结构引起的。标记染色体区域的 SNP 如果相关性延伸较短，相比于长单体型的 SNP 可能显示出更低的 |D′| 值。在具有高度单体型变化的区域，可能需要多个 SNP 被分型确定才有可能代表大多数的单体型。因此，绘制基于 LD 的基因图谱需要位置关系紧密的 SNP 集合才能达到最大效果。

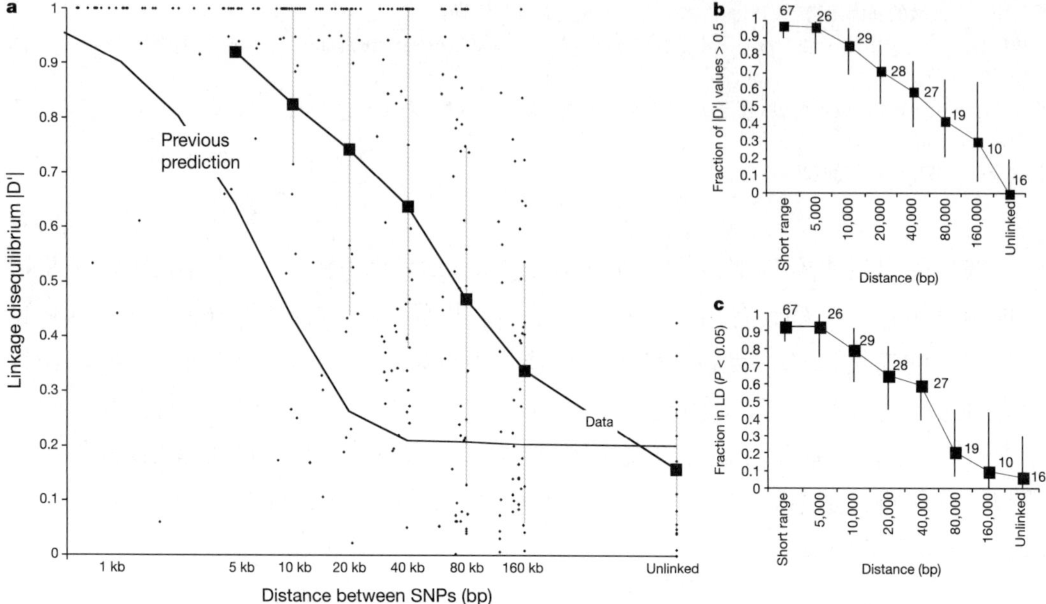

Fig. 1. LD versus physical distance between SNPs. For each distance from the core SNP (Table 1), we chose the SNP with the largest number of copies of the minor allele for comparison to SNPs at other distances. At a given distance, all comparisons are independent. **a**, Average |D′| values for each distance separation ("Data"; dotted lines indicate the 25th and 75th percentiles), compared with a prediction[2] based on simulations (see Methods). |D′| values for shorter physical distances were calculated by looking within contiguously sequenced stretches of DNA containing at least two SNPs, and picking the two with the most minor alleles. Unlinked marker comparisons are obtained by comparing SNPs in the 40-kb bin in each row of Table 1 to those in the next row. **b**, **c**, Fraction of |D′| values greater than 0.5 (**b**) and proportion of significant ($P < 0.05$) associations (**c**) between two SNPs separated by a given distance (as assessed by a likelihood ratio test[10]). Bars indicate 95% central confidence intervals. The number of data points used to make the calculations are shown.

Why does LD extend so far? LD around an allele arises because of selection or population history—a small population size, genetic drift or population mixture—and decays owing to recombination, which breaks down ancestral haplotypes[15]. The extent of LD decreases in proportion to the number of generations since the LD-generating event. The simplest explanation for the observed long-range LD is that the population under study experienced an extreme founder effect or bottleneck: a period when the population was so small that a few ancestral haplotypes gave rise to most of the haplotypes that exist today. Our simulations show that a severe bottleneck (inbreeding coefficient $F \geqslant 0.2$) occurring 800–1,600 generations ago (about 27,000–53,000 years ago assuming 25 years per generation) could have generated the LD observed (Fig. 3). In principle, long-range LD could also be generated by population mixture[16], but the degree of LD is much greater than would arise from the mixing of even extremely differentiated populations. An alternative explanation for the observed long-range LD is that the recombination rates in the regions studied might be markedly less than the genome-wide average. This could happen if recombination occurred primarily in well-separated hotspots and our regions fell between them (Fig. 3). However, under this hypothesis, the regions would be expected to show long-range LD in all populations, and this pattern is not observed (see below).

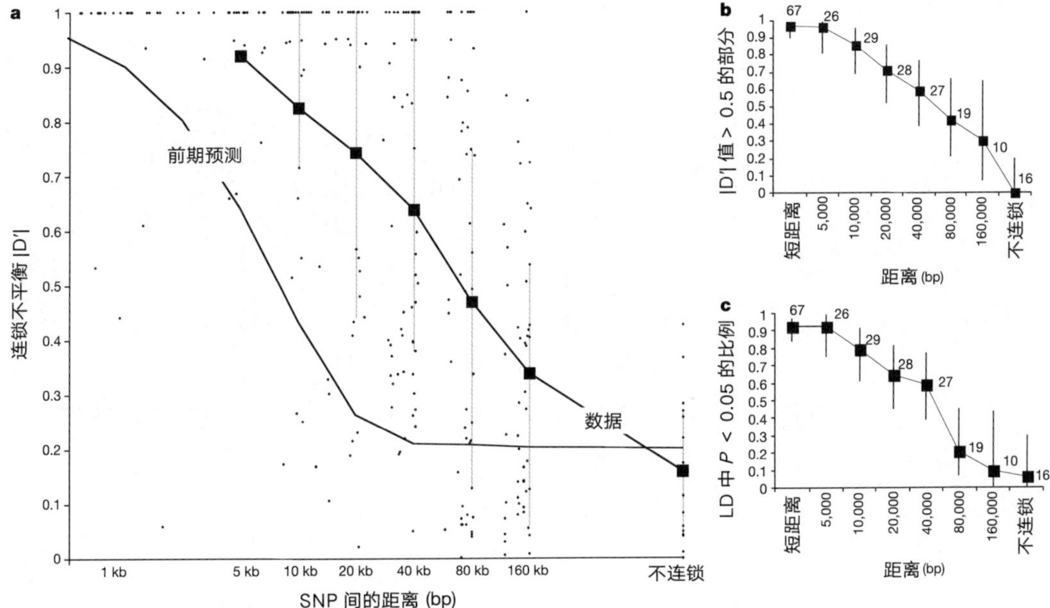

图 1. LD 与 SNP 间的物理距离。对于距离核心 SNP 的不同距离（表 1），我们选择具有次等位基因位点中最大拷贝数目的 SNP 用于与其他距离的 SNP 进行对比。在距离给定的前提下，所有的对比都是独立的。**a**，每个距离间隔的平均 |D'| 值（“数据”；虚线表示 25 和 75 百分位），与模拟推算得出的预期值 [2] 进行比较（见方法）。更短物理距离的 |D'| 值是通过找出含有至少两个 SNP 的连续 DNA 序列，并且挑选最罕见的两个等位位点计算获得。不连锁的标志是通过对照表 1 每行中每 40 kb 分箱的 SNP 与其下一行进行比较获得的。**b** 和 **c**, 在一定距离（通过似然比检验的估算获得 [10]）的两个 SNP 之间，|D'| 值超过 0.5（**b**）和具有显著性关联（$P < 0.05$）（**c**）的部分。条形表示 95% 的置信区间。图中显示了计算所用的数据点的个数。

LD 为什么一直在延伸？等位基因附近 LD 的出现是由于自然选择或者种群历史——一个小规模种群，遗传漂变或者种群混合——由于发生了打破祖先单体型的重组而衰减导致的 [15]。LD 长度的减小与 LD 产生事件之后世代的数目成比例。对较长 LD 的最简单解释就是我们所研究的人群经历了极端的建立者效应或者瓶颈效应：有一段时间种群太小以至于一些祖先单体型衍生出目前存在的大部分单体型。我们的模拟显示大约 800 ~ 1,600 代以前（大约 27,000 ~ 53,000 年前，假设每代 25 年）发生了一次严重的瓶颈效应（近交相关系数 $F \geqslant 0.2$），由此可能产生了我们观察到的 LD（图 3）。原则上说，长 LD 还能够通过种群混合产生 [16]，但 LD 的程度要远远超过极端分化的种群混合而产生的 LD。另一种对长 LD 的解释就是所研究区域的重组率显著低于基因组的平均水平。如果重组主要发生在分离热点区且我们的研究区域刚好落入其中，这种现象可能出现（图 3）。但是，根据这个假设，这些区域应该在所有的人群中都显示长 LD，而事实观察到的并非如此（见下文）。

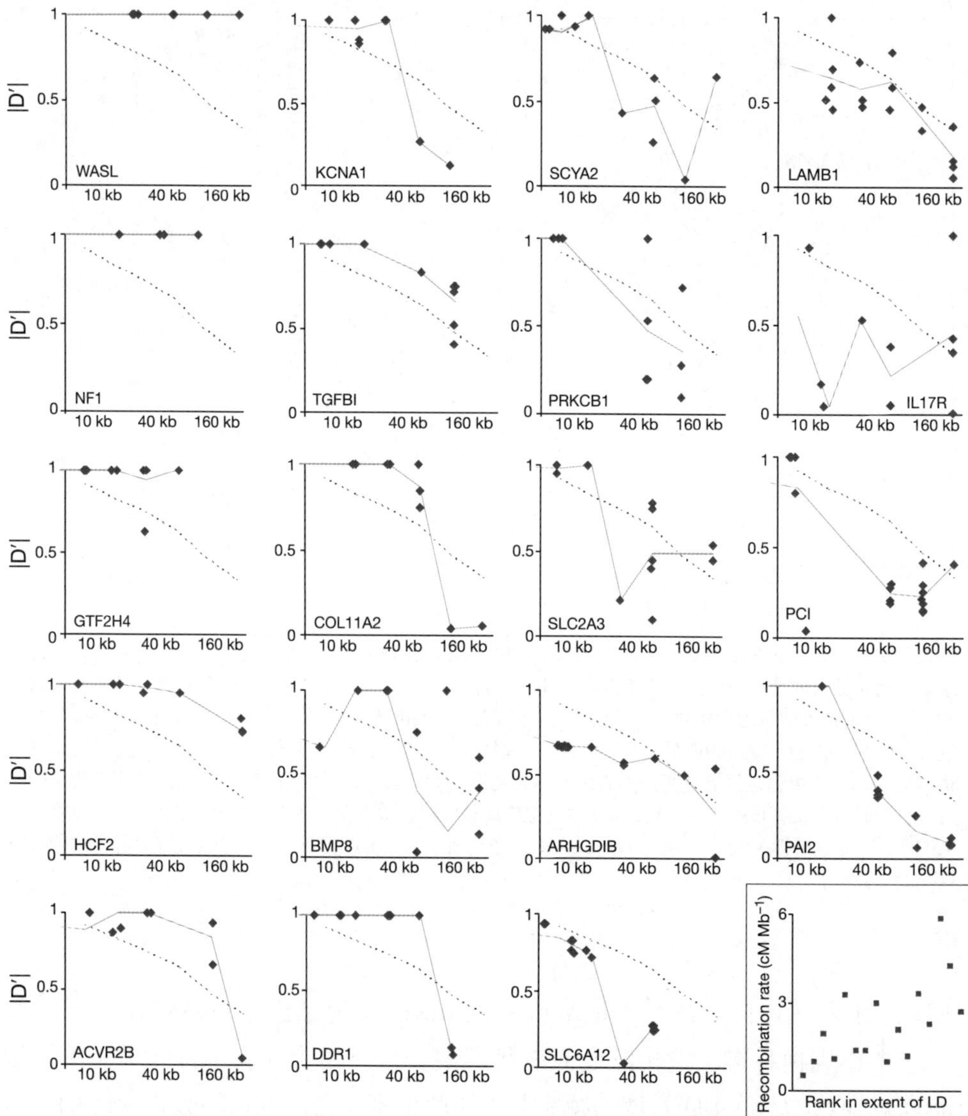

Fig. 2. LD profiles for each genomic region. The chosen SNP is usually the core SNP itself, unless the core SNP could not be readily genotyped or another SNP with more minor alleles had been identified within 1 kb. In both cases, we substituted the closest high-frequency SNP. The chosen SNP is compared with every other high-frequency SNP in the same genomic region. Solid line indicates average $|D'|$ values for each distance for which SNPs were available; for comparison, the dashed line indicates the consensus LD curve from Fig. 1. The extent of LD was calculated by performing a least-squares linear regression to the average $|D'|$ values at each distance from the chosen SNP; more sloped lines indicate less LD. The regions are ordered according to the extent of LD (most extensive LD, top left; least extensive LD, bottom right). Inset (bottom right) shows the rank of each region in terms of LD extent versus the estimated recombination rate per unit of physical distance. For each 160-kb region of interest, we looked for the closest pair of flanking genetic markers from the Marshfield map[34] subject to the condition that they were separated by a non-zero genetic distance on the map. We divided genetic map distance by the physical map distance on the basis of the available draft genome sequence[35]. We analysed the 16 regions for which the genetic and physical map orderings of markers were locally consistent (one-sided Spearman rank correlation, $P < 0.005$).

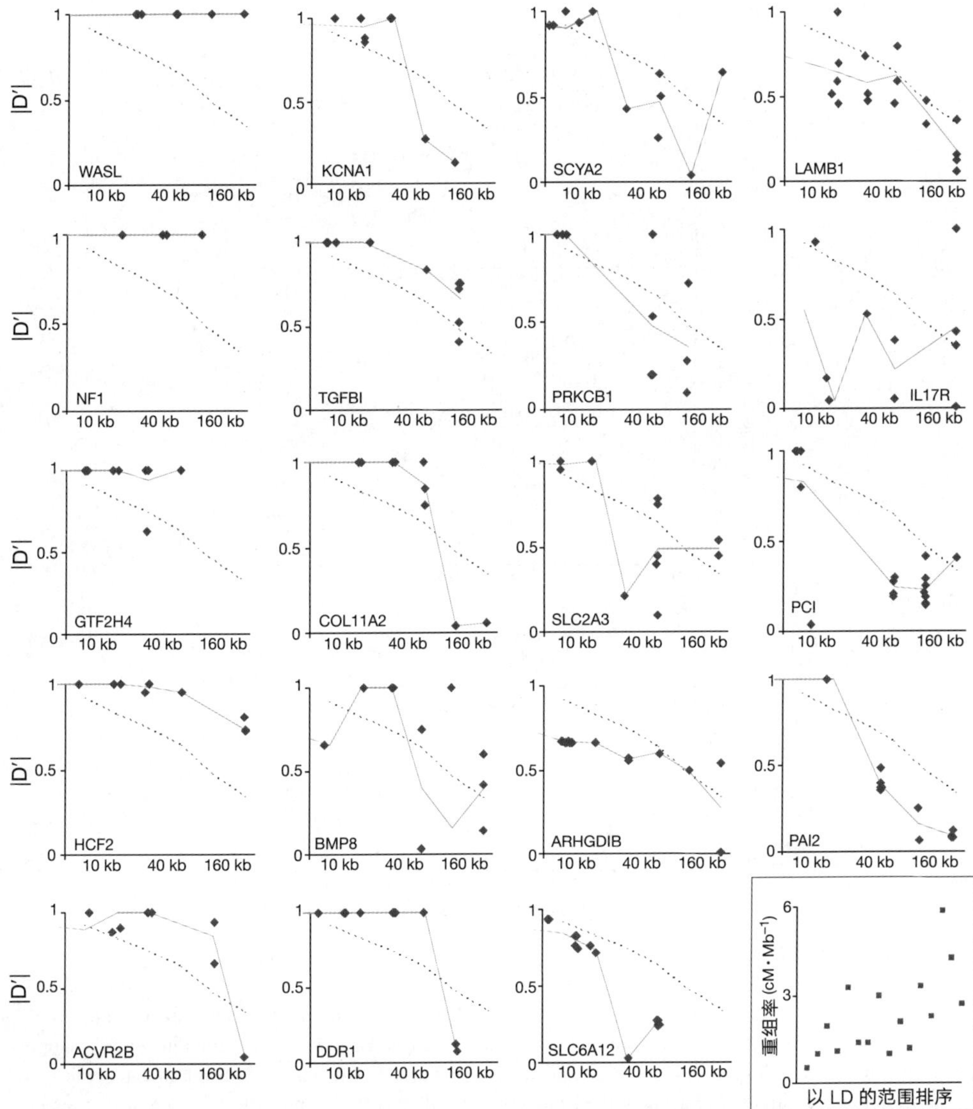

图 2. 每个基因组区域的 LD 概况。选择的 SNP 通常是核心 SNP 本身，除非核心 SNP 不能准确分型或者在 1 kb 范围内发现具有更小的次等位基因频率的 SNP。在这两种情况下，我们都替换成最近的高频率 SNP。所选取的 SNP 与同一个基因组区域的每个其他高频率 SNP 进行对比。实线表示每个距离上可获得 SNP 的平均 |D'| 值；作为对照，虚线表示从图 1 中获得的一致的 LD 曲线。LD 长度的计算是通过对每个距离上选定 SNP 的平均 |D'| 值进行最小二乘线性回归拟合；斜率越大表示 LD 越小。不同区域根据 LD 的长度进行排列（最长的位于左上，最小的位于右下）。插图（右下角）横纵轴分别代表依照 LD 范围大小将各个区域排序而成的各个序列与每一单位物理距离所估算的重组率。对于每个感兴趣的 160 kb 区域，我们根据它们在图谱上被一个非零遗传距离分隔的条件，从马什菲尔德图谱 [34] 中寻找最近的两侧基因标志。根据可获得的基因草图序列 [35]，我们用物理图谱距离来划分基因图谱的距离。我们分析了基因和物理图谱标志顺序局部一致的 16 个区域（单侧斯皮尔曼等级相关性，$P < 0.005$）。

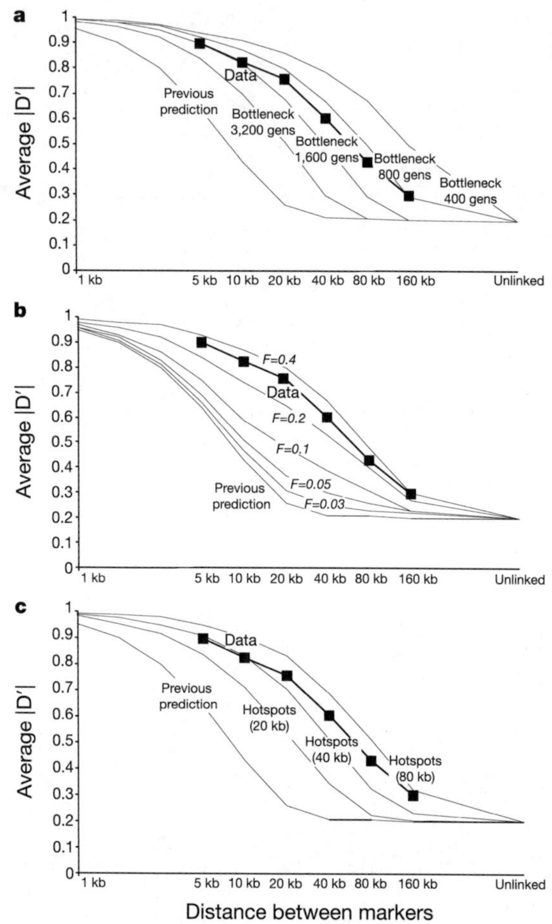

Fig. 3. Effect on LD of assumptions about population history and recombination. **a**, The effect of a population bottleneck instantaneously reducing the population to a constant size of 50 individuals for 40 generations ($F = 0.4$) and occurring 400, 800, 1,600 or 3,200 generations ago. **b**, The effect of bottleneck intensity (F) for a bottleneck that occurred 800 generations ago. **c**, The effect of variation in recombination rate, assuming that all recombination in the genome occurs at hotspots randomly distributed according to a Poisson process with an average density of one every 20, 40 and 80 kb, and with a genome-wide average rate of 1.3 centiMorgans per megabase per generation[35]. Results are compared to a no-hotspot model for the same historical hypothesis. ("Data" refers to the mean |D′| values at each physical distance separation, obtained from Fig. 1a).

To confirm our findings of long-range LD and to investigate the reasons for its occurrence, we next examined a representative subset of SNPs in two additional samples. We first studied another north-European sample (48 southern Swedes) and found LD in a nearly identical pattern to that observed in Utah, both in terms of the overall magnitude of LD and the particular alleles that were associated (indicated by the sign of D′) (Fig. 4). The similarity in LD patterns may be due to the same historical event, which occurred deep in European prehistory before the separation of the ancestors of these two groups. This suggests that the long-range LD pattern is general in northern Europeans[3,17].

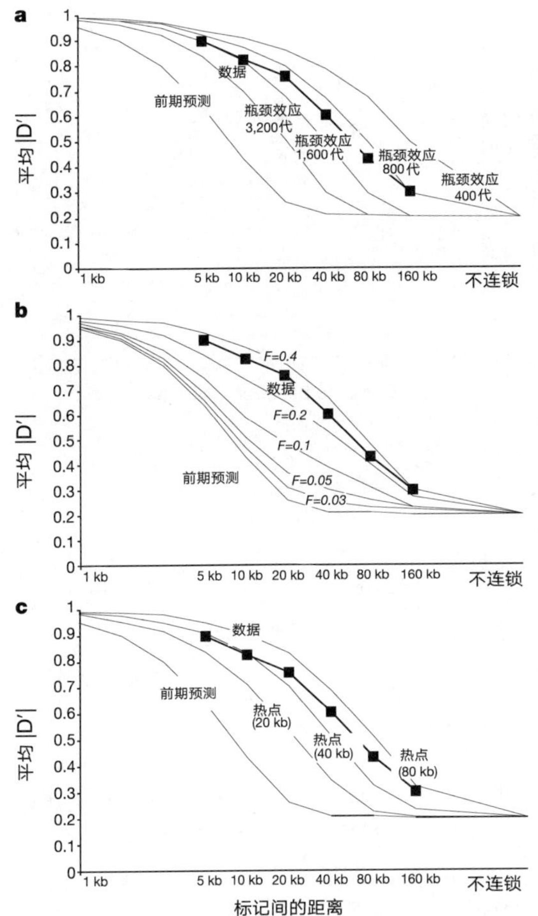

图 3. 关于种群历史和重组的假设对 LD 的影响。**a**，按照 400，800，1,600 或者 3,200 代以前发生的将人群数量即刻减少到 50 个恒定个体并持续 40 代（$F = 0.4$）的人口瓶颈效应影响。**b**，一个发生于 800 代前的瓶颈效应强度（F）的影响。**c**，变异对重组率的影响，假设基因组中所有的重组发生在热点上且服从泊松过程，每 20、40 和 80 kb 的平均密度为 1，并且整个基因组范围内的平均速率是 1.3 厘摩每百万碱基每代 [35]。所有结果均和相同的历史事件假设但没有热点的模型进行比较。（“数据”指的是图 1a 中得到的每个物理距离分隔的 |D′| 平均值）。

为了证实我们发现的长 LD 并且研究其出现的原因，接下来我们在另外两个不同样本中检查了一个有代表性的 SNP 子集。首先我们研究了另一个北欧人群样本（48 名南瑞典人）并且发现无论是总体的 LD 大小还是相关的特定等位基因（由 D′ 结果推测）都和犹他州样本发现的模式几乎一致（图 4）。LD 整体模式的相似性可能是由于群体经历了相同的历史事件。该事件可能发生于史前的欧洲，彼时两个群体的祖先尚未分离。这提示在北欧人群中长 LD 的模式普遍存在 [3,17]。

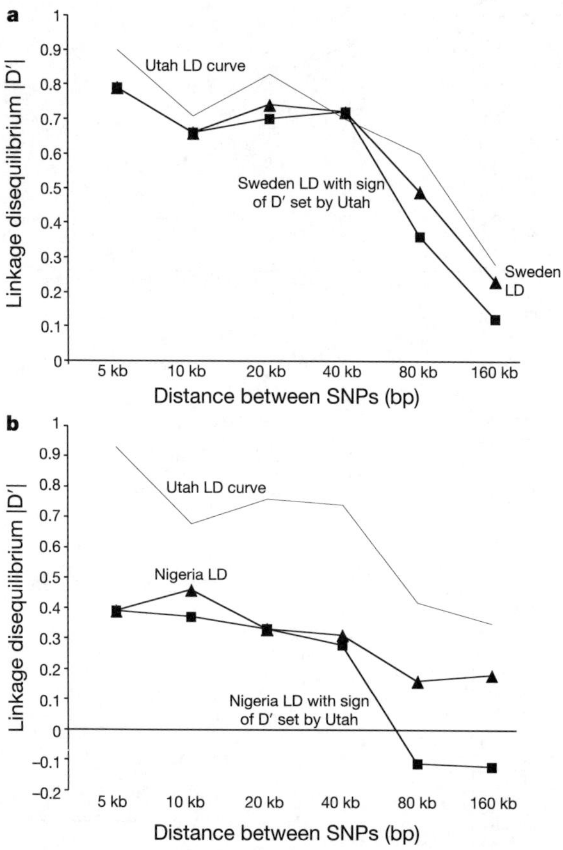

Fig. 4. LD curve for Swedish and Yoruban samples. To minimize ascertainment bias, data are only shown for marker comparisons involving the core SNP. Alleles are paired such that $D' > 0$ in the Utah population. $D' > 0$ in the other populations indicates the same direction of allelic association and $D' < 0$ indicates the opposite association. **a**, In Sweden, average D' is nearly identical to the average $|D'|$ values up to 40-kb distances, and the overall curve has a similar shape to that of the Utah population (thin line in **a** and **b**). **b**, LD extends less far in the Yoruban sample, with most of the long-range LD coming from a single region, *HCF2*. Even at 5 kb, the average values of $|D'|$ and D' diverge substantially. To make the comparisons between populations appropriate, the Utah LD curves are calculated solely on the basis of SNPs that had been successfully genotyped and met the minimum frequency criterion in both populations (Swedish and Yoruban).

We next studied 96 Yorubans (from Nigeria), believed to share common ancestry with northern Europeans about 100,000 years ago[18]. At short distances, the Nigerian and European-derived populations typically show the same allelic combinations (Fig. 4): D' has the same sign and a similar magnitude, indicating a common LD-generating event tracing far back in human history. However, the half-length of LD seems to extend less than 5 kb (Fig. 4) in the Yorubans. Markedly shorter range LD in sub-Saharan Africa has also been observed in several studies of single regions[19,20] (although two other studies did not show a clear trend[21,22]). Our results indicate that the pattern of shorter LD in sub-Saharan African populations may be general.

Notably, LD in the Nigerians is largely a subset of what is seen in the northern Europeans.

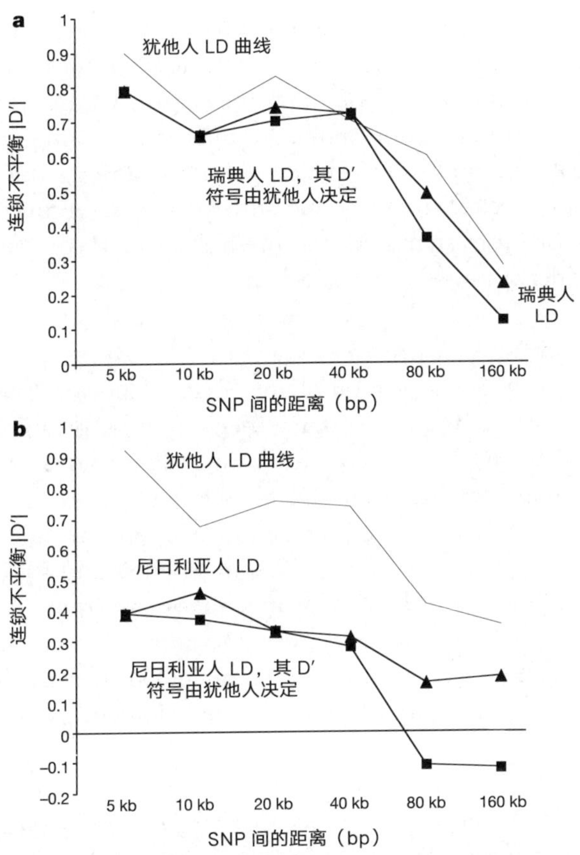

图 4. 瑞典和约鲁巴人的 LD 曲线。为了使确认偏倚最小化，只显示含有核心 SNP 的标志对比数据。等位基因位点进行配对使犹他州人群中 D′ > 0。其他人群中 D′ > 0 说明具有相同方向的等位位点具有相关性，而 D′ < 0 则说明具有相反的关系。**a**，在瑞典人中，平均 D′ 与长达 40 kb 距离的平均 |D′| 值几乎一致，总体曲线的形状也与犹他州人群类似（**a** 和 **b** 中的细线）。**b**，约鲁巴样本中 LD 的长度短一些，其中最长的 LD 来自 *HCF2* 一个单独的区域。即便只有 5 kb，平均 | D′| 值和 D′ 值最终还是发生了分离。为了在两个人群间进行恰当的比较，犹他州人群的 LD 曲线完全基于成功分型并且符合两个人群（瑞典人和约鲁巴人）的最低频率标准的 SNP 进行计算。

随后我们研究了 96 名约鲁巴人（来自尼日利亚），他们被认为在 100,000 年前与北欧人拥有相同的祖先[18]。在短的 LD 中，来自尼日利亚和欧洲的群体都显示出相同的等位基因位点重组（图 4）：D′ 有相同的迹象和类似的大小，说明在人类历史中经历了共同的 LD 产生事件。但是，约鲁巴人中 LD 的半长似乎少于 5 kb（图 4）。数个针对单个区域的研究也发现撒哈拉沙漠以南的非洲人 LD 范围明显缩短[19,20]（尽管另外两个研究没有显示明显的趋势[21,22]）。我们的结果表明撒哈拉沙漠以南非洲人群中更短的 LD 模式可能是普遍存在的。

显然，尼日利亚人的 LD 在很大程度上是观察到的北欧人 LD 的子集。约鲁巴

The Yoruban haplotypes are generally contained within the longer Utah haplotypes, and there is little Yoruban-specific LD (85% of observations of substantial LD ($|D'| > 0.5$) in Yorubans are also substantial and of the same sign in Utah). The vast difference in the extent of LD between populations points to differences owing to population history, probably a bottleneck or founder effect that occurred among the ancestors of north Europeans after the divergence from the ancestors of the Nigerians. The short extent of LD in Nigerians is more consistent with the predictions of a computer simulation study assuming a simple model of population expansion[2].

Could the apparent differences in the extent of LD among populations be due to "ascertainment bias" in the identification of the SNPs? The core SNPs are probably not subject to bias because they were identified in a multi-ethnic population. The neighbouring SNPs were identified in the Utah population and subsequently studied in the other populations, and thus they may be susceptible to ascertainment bias. However, we selected only SNPs with high frequency in Utah and most of these satisfied the high-frequency criterion for use in the other populations (87% in Sweden and 71% in Nigeria). Thus, the inferences about LD are not likely to be much different from what would have been obtained had we used SNPs ascertained in the Yoruban sample. Moreover, the cross-population comparisons (Fig. 4) minimize ascertainment bias because they involve only the core SNP, and because they calculate LD in each population using only the SNPs present in both.

What was the nature of the population event that created the long-range LD? The event could be specific to northern Europe, which was substantially depopulated during the Last Glacial Maximum (30,000–15,000 years ago), and subsequently recolonized by a small number of founders[23,24]. Alternatively, the long-range LD could be due to a severe bottleneck that occurred during the founding of Europe or during the dispersal of anatomically modern humans from Africa[19,20,25,26] (the proposed "Out of Africa" event) as recently as 50,000 years ago. Under the first hypothesis, the strong LD at distances $\geqslant$ 40 kb would be absent in populations not descended from northern Europeans. Under the second hypothesis, the same pattern of long-range LD could be observed in a variety of non-African populations. Regardless of the timing and context of the bottleneck, the severity of the event (in terms of inbreeding) can be assessed from our data. To have a strong effect on LD, a substantial proportion of the modern population would have to be derived from a population that had experienced an event leading to an inbreeding coefficient of at least $F = 0.2$ (Fig. 3). This corresponds to an effective population size (typically less than the true population size[15]) of 50 individuals for 20 generations; 1,000 individuals for 400 generations; or any other combination with the same ratio.

Our results have implications for disease gene mapping, suggesting a possible two-tiered strategy for using LD. The presence of large blocks of LD in northern European populations suggests that genome-wide LD mapping is likely to be possible with existing SNP resources[1]. Although the large blocks should make initial localization easier, they may also limit the resolution of mapping to blocks of DNA in the range of 100 kb[27]. Populations

单体型基本上都包含在更长的犹他州单体型中，几乎没有约鲁巴人独有的LD(85%约鲁巴人常见的LD(| D′| > 0.5)在犹他州人中也很常见，而且具有相同的标志)。不同人群间LD长度的明显差别归因于人群历史的差异，北欧人祖先与尼日利亚人祖先分离之后可能发生了瓶颈效应或者建立者效应。尼日利亚人中的短LD与基于人群扩张简单模型假设的计算机模拟研究预测结果一致[2]。

这种人群中LD的长度有明显差别的现象是否是由于SNP鉴定过程中存在的“确认偏倚”导致的？核心SNP不太可能受偏倚的影响，因为它们是在多种族人群中鉴定出来的。在犹他州人群中鉴定出的相邻SNP，随后在其他人群中进行了研究，因此它们有可能受到确认偏倚的影响。但是，我们在犹他州人群中仅仅选择了高频出现的SNP，其大部分都满足了可以用于其他人群的高频出现的标准(瑞典人群有87%，尼日利亚人群有71%)。因而我们现在推断出的LD与这些在约鲁巴人群样本中确定出的SNP所具有的LD不会有太大的差别。此外，跨人群比较使确认偏倚最小化(图4)，因为这种比较仅仅使用了核心SNP，还因为仅仅使用了共有的SNP来计算每个人群的LD。

那么产生长LD的人群事件的本质是什么？这些事件可能是北欧人特有的，即在末次冰盛期(30,000 ~ 15,000年前)人口的大量减少，随后少量的建立者重新定居[23,24]。另一种可能是，这种长LD是由于在欧洲的建立过程中或是50,000年前解剖学上的现代人从非洲分离[19,20,25,26](所谓“走出非洲”事件)所产生的严重瓶颈效应。根据第一个假设，长度 ≥ 40 kb的大型LD应该不存在于非北欧后代人群中。根据第二个假设，这类长LD应该可以出现于各种非非洲人群中。不考虑这种瓶颈效应的时间和背景，从我们的数据可以推断出该事件的严重程度(以近亲繁殖的程度表示)。如果对LD产生巨大影响，很大部分的现代人类将是来自经历过能导致近交系数至少是 $F = 0.2$ 的严重事件的群体(图3)。这就相当于一个有效群体(通常小于真实的人口[15])有50个个体，经历20代繁衍，或1,000个个体，经历400代繁衍，或是任何相同比例的组合。

我们的结论对疾病的基因图谱绘制具有重要意义，给出了在两个层次上利用LD的可能策略。北欧人群中存在大的LD板块，说明在现有SNP资源下获得基因组水平的图谱是可能实现的[1]。尽管大的LD板块可能会使初始的定位变得容易，但是这同时也限制了将图谱的DNA解析为100 kb大小的分辨率[27]。含有更小LD

with much smaller blocks of LD (for example, Yorubans) may allow fine-structure mapping to identify the specific nucleotide substitution responsible for a phenotype[12]. Our study also has implications for LD as a tool to study population history[19-22]. Simultaneous assessment of LD at multiple regions of the genome provides an approach for studying history with potentially greater sensitivity to certain aspects of history than traditional methods based on properties of a single locus.

Methods

Core SNPs were identified by screening more than 3,000 genes in a multi-ethnic panel of 15 European Americans, 10 African Americans, and 7 East Asians (see ref. 28 for details; a full description of this database will be presented elsewhere). DNA used for sequencing was obtained from the Coriell Cell Repositories. Identification numbers for these Utah samples from the CEPH mapping panel were NA12344, 06995, 06997, 07013, 12335, 06990, 10848, 07038, 06987, 10846, 10847, 07029, 07019, 07048, 06991, 10851, 07349, 07348, 10857, 10852, 10858, 10859, 10854, 10856, 10855, 12386, 12456, 10860, 10861, 10863, 10830, 10831, 10835, 10834, 10837, 10836, 10838, 10839, 10841, 10840, 10842, 10843, 10845 and 10844. We did follow-up genotyping in 48 Swedes (healthy individuals from a case/control study of adult-onset diabetes) and in 96 Yoruban males from Nigeria (healthy individuals from a case/control study of hypertension).

SNPs were discovered by DNA sequencing[28] in the 44 individuals from Utah; we sequenced about 2 kb centred at each distance from the core SNP $\geqslant$ 5 kb, with about 1 kb sequenced around the core SNP itself. When no polymorphism of sufficiently high frequency was found, a nearby subregion of about two further kilobases was resequenced; this occurred in only 18% of the cases. Polymorphisms were identified and genotypes were scored automatically using Polyphred[29] and checked manually by at least two different scorers. SNPs in Hardy-Weinberg disequilibrium or showing evidence of breakdown of LD over short physical distances (< 2.5 kb) were triple-checked. Of the 275 high-frequency SNPs (that is, SNPs with at least 15 observed copies of the minor allele), three were discarded because of a Hardy-Weinberg P value of < 0.02; one of the SNPs used in the analysis had a nominally significant P value ($P < 0.05$). To assess the accuracy of scoring, we rescored 26 randomly chosen high-frequency SNPs; only seven discrepancies were found among 1,144 genotypes. For cases in which follow-up genotyping was done, the discrepancy rate was 47 out of 1,484 (3%) between genotypes obtained by both methods.

Genotyping of SNPs was performed by single-base extension followed by mass spectroscopy (Sequenom)[30], fluorescence polarization (LJL Biosystems)[31] or detection on a sequencing gel (Applied Biosystems) . For the ten regions selected for follow-up genotyping (Table 1), we chose at most one "representative" SNP at each distance from the core SNP (each column in Table 1) according to the criterion that it had the highest number of minor alleles of all SNPs at that distance from the core SNP. For other populations, only those SNPs that, when genotyped, had a minimum number of minor alleles were included in studies of LD. For Yorubans, the cutoff was 25 alleles (76% of SNPs met the criterion); for Swedes, the cutoff was 15 alleles (89%). The fact that most of the SNPs we studied in Utah are also present in high frequency in these other populations indicates that the

板块的人群(比如约鲁巴人)则能够进行精细结构的图谱绘制，便于鉴定出造成某种表型的特异性核苷酸替换[12]。我们的研究同时也显示LD可以作为研究人群历史的工具[19-22]。同时评估基因组中多个区域的LD可以作为研究人群历史的方法，相对于基于单个位点性质的传统方法来说在一些方面具有更高的敏感度。

方 法

核心SNP是通过对一个多民族人群中超过3,000个基因进行筛查确定的，其中包括15个欧裔美国人、10个非裔美国人和7个东亚人(详见参考文献28，此数据库的详细描述将另文表述)。用于测序的DNA来自科里尔细胞库。CEPH图谱标记中这些犹他州人群样本的编号是NA12344、06995、06997、07013、12335、06990、10848、07038、06987、10846、10847、07029、07019、07048、06991、10851、07349、07348、10857、10852、10858、10859、10854、10856、10855、12386、12456、10860、10861、10863、10830、10831、10835、10834、10837、10836、10838、10839、10841、10840、10842、10843、10845和10844。我们对另外48个瑞典人(一项成人糖尿病的病例/对照研究中的健康个体)和96个尼日利亚约鲁巴男性(一项高血压的病例/对照研究中的健康个体)进行了基因分型。

SNP是从44个犹他州人群个体中用DNA测序[28]方法获得的。我们测定了每个距离核心SNP ≥ 5 kb处周边2 kb的序列，同时也对距离核心SNP周边1 kb区域进行了测序。如果没有足够高频率的基因多态位点，则将周边测序范围再扩大2 kb；这种情况出现的比例是18%。多态位点发现以后，就用Polyphred软件[29]自动进行基因型的评分，并且至少两个不同的计分人进行手工复查。哈迪–温伯格不平衡中的SNP或者在短物理距离(< 2.5 kb)显示LD断开证据的SNP需要进行三重检查。在这275个高频SNP中(即SNP含有至少15个次等位基因位点的可见拷贝)，有3个由于哈迪–温伯格P值< 0.02而被舍弃；用于分析的其中一个SNP只是在名义上有显著性P值($P < 0.05$)。为了评分的准确性，我们重新将26个随机选取的高频SNP进行了评分；在1,144个基因型中只发现了7处差异。对于进行基因分型的样本，两种方法获得的基因型差异率是47/1,484(3%)。

SNP的基因分型使用的是单碱基延伸法，随后使用质谱(Sequenom)[30]、荧光偏振(LJL Biosystems)[31]或者测序胶(Applied Biosystems)[32]检测。对于选定进行后续分型的10个区域而言(表1)，我们以距离核心SNP一定距离的所有SNP中出现最多的次等位位点作为标准，选择一个最具代表性的SNP(表1中的每一栏)。对于其他人群，只有那些分型后具有最低数量要求的次等位基因位点才纳入LD的研究。约鲁巴人的阈值是25个等位基因位点(76%的SNP满足标准)；瑞典人的阈值是15个等位位点(89%的SNP满足标准)。我们在犹他州人群中研究的SNP大多数也高频存在于其他人群的事实表明，LD的评估不太可能受

assessment of LD is not likely to be subject to large ascertainment bias.

Heterozygosity[15] (π) was calculated as the average of $2jk/n(n-1)$ for all base pairs screened, with j and k equal to the number of copies of the minor and major alleles, respectively ($n=j+k$). A base was considered screened if it had Phred quality scores[29] of $\geqslant 15$ in $\geqslant 10$ individuals. D′ values between markers with alleles A/a and B/b (allele frequencies, c_A, c_a, c_B and c_b; haplotype frequencies, c_{AB} , c_{Ab}, c_{aB} and c_{ab}) were obtained by dividing $c_{AB}-c_Ac_B$ by its maximum possible value: $\min(c_Ac_b , c_ac_B)$ if $D>0$ and $\min(c_Ac_B, c_ac_b)$ otherwise. An implementation of the expectation maximization algorithm was used to infer haplotype frequencies for pairs of SNPs both for actual and simulated data[33]. A likelihood ratio test was used to assess significance of associations between pairs of SNPs[10].

Computer simulations were based on a model related to that in ref. 2, assuming a population that was constant at an effective size of 10,000 individuals until 5,000 generations ago, when it expanded instantaneously to a size of 100,000,000 (an arbitrary value). This model captures many of the features of more complicated growth, as the effect of population growth on LD is not dependent on the precise details of population growth or the final population size when the growth is moderately fast[2]. Bottlenecks were modelled as described, with the population crashing to a constant size for a fixed number of generations before re-expanding. (The effect of a bottleneck on LD depends primarily on the F-value, the inbreeding coefficient, which is defined as the probability that two alleles randomly picked from the population after the bottleneck derive from the same ancestral allele just before the bottleneck.) Coalescent simulations were used to generate gene genealogies under these models for markers separated by a specified recombination distance (see ref. 13 for a more detailed description of the theory behind these simulations). Simulations were run 2,000 times with sample-size distributions mimicking our data. SNPs were generated by distributing mutations on the simulated gene genealogies at a mutation rate of 6×10^{-5} per generation, under an "infinite alleles"mutation model. The mutation rate was chosen such that the probability of high frequency SNPs in a 2-kb stretch of DNA sequenced in 44 samples (for the model of a simple expansion 5,000 generations ago) was similar to what we observed (about 70%). We also tested mutation rates ten times higher and found that inferences about LD were essentially unchanged.

An extreme hypothesis of population mixture and its effect on LD were assessed in a simulated, mixed population of European Americans and sub-Saharan Africans. For the first simulated mixture, we constructed a mixture of 22 Yorubans and 22 samples from Utah, and used data from the ten core SNPs genotyped in both populations. For the second simulation, we used 26 SNPs from a previous study[7] that had been found to have a minor allele frequency of at least 15% in either African Americans or European Americans; we chose at most one SNP per gene, picking the first listed SNP (in Table A1 of ref. 7) that met our minimum frequency criterion. We found much stronger LD even at 40 kb (56% with $|D'|>0.5$) (Fig. 1) in our actual data than in the simulated, admixed populations. For the 45 possible pairwise comparisons of the ten core SNPs in the simulated mix of Yoruban and Utah samples, no values of $|D'|>0.5$ were observed. For the 325 possible pairwise comparisons from the second study, only 11% showed $|D'|>0.5$. This suggests that admixture probably did not generate the strong signal of LD at long physical distances seen in Utah.

(**411**, 199-204; 2001)

到大量的确认偏倚的影响。

对于所有筛选的碱基对而言，杂合度[15](π)等于$2jk/n(n-1)$的均值，其中j和k分别相当于次等位位点和主等位位点的拷贝数，$n=j+k$。如果一个碱基在≥10个个体中的Phred质量评分[29]≥15，则被认为通过筛选。等位基因A/a和B/b(等位基因频率c_A、c_a、c_B和c_b；单体型频率c_{AB}、c_{Ab}、c_{aB}和c_{ab})间标志的D′值的计算是将$c_{AB}-c_Ac_B$除以其最大可能值：如果D > 0，取c_Ac_b和c_ac_B的较小值，反之则是c_Ac_B和c_ac_b的较小值。最大期望值算法用来推断实际和模拟数据中成对SNP的单体型频率[33]。似然比检验用来评价成对SNP之间相关性的显著性[10]。

计算机模拟是基于参考文献2中的一个相关模型，模型假设一个种群一直保持稳定在有效群体数量为10,000个个体直到5,000代，突然扩张到100,000,000(一个随意假设的值)个个体。在增长速度比较快的情况下，人群增长对LD的影响不取决于其增长的准确细节或最终大小[2]，因此该模型具有更复杂增长模式的许多特点。瓶颈效应的模型如前述进行设置，在重新扩张前人群的数量在固定的代数中坍缩至常数值。(瓶颈效应对LD的影响主要取决于F值，即近交系数，其定义是瓶颈效应后从群体中随机选取的两个等位基因位点来源于瓶颈效应之前同一个祖先位点的可能性。)联合模拟用来生成这些模型设定下的基因谱系以获得特定的重组距离分隔的标志物(有关模拟的详细理论陈述详见参考文献13)。在样本量的分布类似于我们数据的条件下进行了2,000次模拟。SNP是在突变率为每代6×10^{-5}，“无限等位位点”突变模型下，通过将突变分布在模拟的基因谱系中产生。突变率的设定是为了保证44个样本中所测序的2 kb序列中出现高频SNP的可能性(对于5,000代之前简单扩张的模型)与我们所观察到的类似(大约70%)。我们也检验了突变率增高至10倍以后的情况，发现关于LD相关结论没有改变。

对于人群混合的极端假设及其对LD的影响是在模拟的、混合了欧裔美国人和南撒哈拉非洲人中进行评估的。对于第一组模拟的混合，我们构建了22个约鲁巴人和22个犹他州人的混合样本，并且使用两个人群中都进行了分型的10个核心SNP的数据。对于第二个模拟，我们使用先前研究[7]的26个SNP，已发现无论在非裔美国人还是欧裔美国人中，出现次等位基因位点的频率都不小于15%；每个基因根据第一个满足我们最小频率标准的SNP最多选择一个位点(见参考文献7表A1)。我们发现即便是40 kb处，实际数据比模拟的混合人群中的LD程度更强(56%的|D′| > 0.5)(图1)。在模拟混合的约鲁巴人和犹他州人样本中10个核心SNP的45种设对比较中，没有看到|D′| > 0.5的值。在第二个研究的325个可设对比较中，只有11%出现|D′| > 0.5。这说明人群混合可能无法形成像犹他州人群体中所见到的那种在长物理距离上产生LD的强信号。

(毛晨晖 翻译；曾长青 审稿)

David E. Reich[*], **Michele Cargill**[*†], **Stacey Bolk**[*], **James Ireland**[*], **Pardis C. Sabeti**[‡], **Daniel J. Richter**[*], **Thomas Lavery**[*], **Rose Kouyoumjian**[*], **Shelli F. Farhadian**[*], **Ryk Ward**[‡] **& Eric S. Lander**[*§]

[*] Whitehead Institute/MIT Center for Genome Research, Nine Cambridge Center, Cambridge, Massachusetts 02142, USA

[‡] Institute of Biological Anthropology, University of Oxford, Oxford OX2 6QS, UK

[§] Department of Biology, MIT, Cambridge, Massachusetts 02139, USA

[†] Present address: Celera Genomics, 45 West Gude Drive, Rockville, Maryland 20850, USA.

Received 11 December 2000; accepted 13 March 2001.

References:

1. Sachidanandam, R. *et al.* A map of human genome sequence variation containing 1.42 million single nucleotide polymorphisms. *Nature* **409**, 928-933 (2001).
2. Kruglyak, L. Prospects for whole-genome linkage disequilibrium mapping of common disease genes. *Nature Genet.* **22**, 139-144 (1999).
3. Dunning, A. M. *et al.* The extent of linkage disequilibrium in four populations with distinct demographic histories. *Am. J. Hum. Genet.* **67**, 1544-1554 (2000).
4. Abecasis, G. R. *et al.* Extent and distribution of linkage disequilibrium in three genomic regions. *Am. J. Hum. Genet.* **68**, 191-197 (2001).
5. Taillon-Miller, P. *et al.* Juxtaposed regions of extensive and minimal linkage disequilibrium in human Xq25 and Xq28. *Nature Genet.* **25**, 324-328 (2000).
6. Collins, A., Lonjou, C. & Morton, N. E. Genetic epidemiology of single-nucleotide polymorphisms. *Proc. Natl Acad. Sci. USA* **96**, 15173-15177 (1999).
7. Goddard, K. A. B., Hopkins, P. J., Hall, J. M. & Witte, J. S. Linkage disequilibrium and allele-frequency distributions for 114 single-nucleotide polymorphisms in five populations. *Am. J. Hum. Genet.* **66**, 216-234 (2000).
8. Watterson, G. A. & Guess, H. A. Is the most frequent allele the oldest? *Theor. Pop. Biol.* **11**, 141-160 (1977).
9. Lander, E. S. The new genomics: global views of biology. *Science* **274**, 536-539 (1996).
10. Schneider, S., Kueffler, J. M., Roessli, D. & Excoffier, L. Arlequin (ver. 2.0): A software for population genetic data analysis (Genetics and Biometry Laboratory, Univ. Geneva, Switzerland, 2000).
11. Lewontin, R. C. On measures of gametic disequilibrium. *Genetics* **120**, 849-852 (1988).
12. Jorde, L. B. Linkage disequilibrium and the search for complex disease genes. *Genome Res.* **10**, 1435-1444 (2000).
13. Hudson, R. R. in *Oxford Surveys in Evolutionary Biology* (eds Futuyma, D. J. & Antonovics, J.) 1-44 (Oxford Univ. Press, Oxford, 1990).
14. Clark, A. G. *et al.* Haplotype structure and population genetic inferences from nucleotide-sequence variation in human lipoprotein lipase. *Am. J. Hum. Genet.* **63**, 595-612 (1998).
15. Hartl, D. L. & Clark, A. G. *Principles of Population Genetics* (Sinauer, Massachusetts, 1997).
16. Chakraborty, R. & Weiss, K. M. Admixture as a tool for finding linked genes and detecting that difference from allelic association between loci. *Proc. Natl Acad. Sci. USA* **85**, 9119-9123 (1988).
17. Eaves, I. A. *et al.* The genetically isolated populations of Finland and Sardinia may not be a panacea for linkage disequilibrium mapping of common disease genes. *Nature Genet.* **25**, 320-322 (2000).
18. Goldstein, D. B., Ruiz Linares, A., Cavalli-Sforza, L. L. & Feldman, M. W. Genetic absolute dating based on microsatellites and the origin of modern humans. *Proc. Natl Acad. Sci. USA* **92**, 6723-6727 (1995).
19. Tishkoff, S. A. *et al.* Global patterns of linkage disequilibrium at the CD4 locus and modern human origins. *Science* **271**, 1380-1387 (1996).
20. Tishkoff, S. A. *et al.* Short tandem-repeat polymorphism/*Alu* haplotype variation at the *PLAT* locus: Implications for modern human origins. *Am. J. Hum. Genet.* **67**, 901-925 (2000).
21. Kidd, J. R. *et al.* Haplotypes and linkage disequilibrium at the phenylalanine hydroxylase locus, *PAH*, in a global representation of populations. *Am. J. Hum. Genet.* **66**, 1882-1899 (2000).
22. Mateu, E. *et al.* Worldwide genetic analysis of the *CFTR* region. *Am. J. Hum. Genet.* **68**, 103-117 (2001).
23. Housley, R. A., Gamble, C. S., Street, M. & Pettitt, P. Radiocarbon evidence for the Late glacial human recolonisation of northern Europe. *Proc. Prehist. Soc.* **63**, 25-54 (1994).
24. Richards, M. *et al.* Tracing European founder lineages in the Near Eastern mtDNA pool. *Am. J. Hum. Genet.* **67**, 1251-1276 (2000).
25. Reich, D. E. & Goldstein, D. B. Genetic evidence for a Paleolithic human population expansion in Africa. *Proc. Natl Acad. Sci. USA* **95**, 8119-8123 (1998).
26. Ingman, M., Kaessmann, H., Pääbo, S. & Gyllensten, U. Mitochondrial genome variation and the origin of modern humans. *Nature* **408**, 708-713 (2000).
27. Altshuler, D., Daly, M. & Kruglyak, L. Guilt by association. *Nature Genet.* **26**, 135-137 (2000).
28. Cargill, M. *et al.* Characterization of single-nucleotide polymorphisms in coding regions of human genes. *Nature Genet.* **22**, 231-238 (1999).
29. Nickerson, D. B., Tobe, V. O. & Taylor, S. L. PolyPhred: automating the detection and genotyping of single nucleotide substitutions using fluorescence-based sequencing. *Nucleic Acids Res.* **25**, 2745-2751 (1997).
30. Ross, P. Hall, L., Smirnov, I. & Haff, L. High level multiplex genotyping by MALDI-TOF mass spectroscopy. *Nature Biotech.* **16**, 1347-1351 (1998).
31. Chen, X., Levine, L. & Kwok, P. Y. Fluorescence polarization in homogenous nucleic acid analysis. *Genome Res.* **9**, 492-498 (1999).
32. Lindblad-Toh, K. *et al.* Large-scale discovery and genotyping of single-nucleotide polymorphisms in the mouse. *Nature Genet.* **24**, 381-386 (2000).
33. Excoffier, L. & Slatkin, M. Maximum-likelihood estimation of molecular haplotype frequencies in a diploid population. *Mol. Biol. Evol.* **12**, 921-927 (1995).

34. Broman, K. W., Murray, J. C. Sheffield, V. C., White, R. L. & Weber, J. L. Comprehensive human genetic maps: individual and sex-specific variation in recombination. *Am. J. Hum. Genet.* **63**, 861-689 (1998).

35. Lander, E. S. *et al.* Initial sequencing and analysis of the human genome. *Nature* **409**, 860-921.

Acknowledgements. We thank L. Groop for the Swedish samples; R. Cooper and C. Rotimi for the Yoruban samples; and D. Altshuler, M. Daly, D. Goldstein, J. Hirschhorn, C. Lindgren and S. Schaffner for discussions. This work was supported in part by grants from the National Institutes of Health, Affymetrix, Millennium Pharmaceuticals, and Bristol-Myers Squibb Company, and by a National Defense Science and Engineering Fellowship to D.E.R.

Correspondence and requests for materials should be addressed to D.E.R. (e-mail: reich@genome.wi.mit.edu) or E.S.L. (e-mail: lander@genome.wi.mit.edu).

Estimating the Human Health Risk from Possible BSE Infection of the British Sheep Flock

N. M. Ferguson *et al.*

Editor's Note

The neurodegenerative disease bovine spongiform encephalopathy (BSE) was first recognized in British cattle in the 1980s. The government's response was to monitor herds closely and restrict beef sales to ensure that organs known to harbour BSE were not sold. Nevertheless, an outbreak occurred in the late 1980s, leading to the culling of British cattle and a European ban on British beef exports. BSE is believed to trigger a similar, fatal disease in humans called Creutzfeldt–Jakob disease, of which more than 160 people have died following the 1980s BSE outbreak. Here epidemiologist Roy Anderson and colleagues model the likely course of the human epidemic and predict future deaths to be in the range 50–50,000.

Following the controversial failure of a recent study[1] and the small numbers of animals yet screened for infection[2], it remains uncertain whether bovine spongiform encephalopathy (BSE) was transmitted to sheep in the past via feed supplements and whether it is still present. Well grounded mathematical and statistical models are therefore essential to integrate the limited and disparate data, to explore uncertainty, and to define data-collection priorities. We analysed the implications of different scenarios of BSE spread in sheep for relative human exposure levels and variant Creutzfeldt–Jakob disease (vCJD) incidence. Here we show that, if BSE entered the sheep population and a degree of transmission occurred, then ongoing public health risks from ovine BSE are likely to be greater than those from cattle, but that any such risk could be reduced by up to 90% through additional restrictions on sheep products entering the food supply. Extending the analysis to consider absolute risk, we estimate the 95% confidence interval for future vCJD mortality to be 50 to 50,000 human deaths considering exposure to bovine BSE alone, with the upper bound increasing to 150,000 once we include exposure from the worst-case ovine BSE scenario examined.

THE aim of this study was not to evaluate the probability that BSE has entered the sheep flock, but rather, given the pessimistic assumption that infection has occurred, to explore its potential extent and pattern of spread. In this, we used epidemiological parameter estimates from experimental BSE infections of sheep, and, where data are unavailable, assumed (given the observed similarities in BSE and scrapie pathogenesis in sheep) that other aspects of disease epidemiology resemble those of scrapie. Analyses were constrained to be consistent with the failure to detect the BSE agent in a small sample of 180 brains[2] collected between 1996 and 2000 from sheep diagnosed with scrapie (giving

英国羊群可能的BSE感染带来的人类健康风险评估

弗格森等

编者按

神经退行性疾病牛海绵状脑病(BSE)于20世纪80年代首次在英国牛中发现。英国政府的回应是密切监控牧群,限制牛肉销售,以确保患有BSE的牛的器官不被出售。然而,20世纪80年代末爆发了一场疫情,导致英国牲畜遭到扑杀,欧洲禁止英国牛肉出口。据信,BSE在人类中引发了一种类似的致命疾病(称为克雅氏病)。自20世纪80年代BSE爆发以来,已有160多人死于这种疾病。流行病学家罗伊·安德森和他的同事们模拟了人类流行病的可能进程,并预测未来因这种疾病死亡的人数将在50~50,000之间。

随着最近一项有争议的研究的失败[1]加上少数动物已经接受过感染筛查[2],但牛海绵状脑病(BSE)过去是否是通过饲料添加剂传染到羊身上以及现在是否仍然存在都还不能确定。因此,有充分依据的数学和统计学模型对于整合有限的不同数据、探索不确定性以及确定数据收集的重点都是十分必要的。我们分析了BSE在羊群中传播的不同方式与相关的人类暴露水平以及变异型克雅氏病(vCJD)患病率之间的关系,我们发现,如果BSE进入了羊群并且发生了一定程度的传播,那么羊BSE导致的不断增加的公共卫生风险很可能比牛的大,但是通过附加限制羊产品进入食物供应,可以将任何这类风险减少高达90%。将这个分析延伸到绝对风险,在95%置信区间内,我们估计单纯暴露于牛BSE的未来vCJD死亡率是50到50,000人,而一旦暴露于最坏方式传播的羊BSE,则上限增加到150,000人。

本研究的目的不是评估BSE进入羊群的可能性,而是悲观地假设感染已经发生,以此来研究感染传播的潜在程度和方式。在本文中,我们使用了实验羊的BSE感染得出的流行病学参数估计,对于那些无法获得的数据,就假定该病流行病学的这些数据与瘙痒病类似(考虑到羊身上BSE和瘙痒病发病机理的相似性)。此次分析的是从1996年到2000年之间患有瘙痒病的羊中收集的180份脑标本[2],要求在这些脑标本中不能检测到BSE病原体的存在(考虑到在明显患有瘙痒病的绵羊

an upper bound for BSE prevalence within apparently scrapie-affected sheep of 2%; Fig. 1a), and are also broadly consistent with an assessment of historical exposure of the ovine population to meat and bonemeal (MBM)[3].

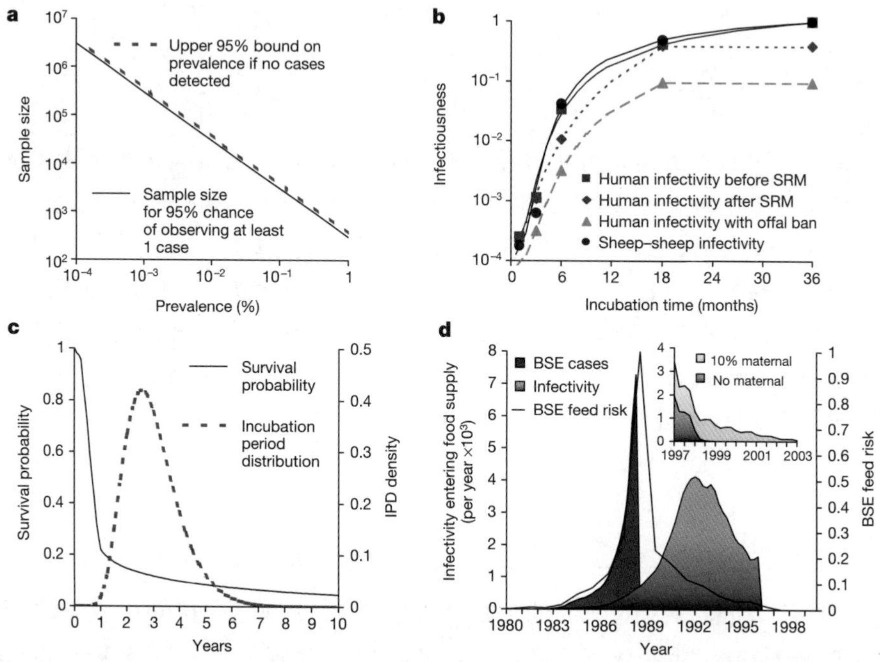

Fig. 1. Epidemiological inputs to transmission model. **a**, Relationship between sample size and detectable prevalence in screening studies. **b**, Infectiousness of sheep as a function of time from infection (see Methods). Sheep and human profiles are separately normalized to give maxima of 1. **c**, Survival probability of sheep as a function of age, and assumed incubation period distribution (IPD) of BSE in sheep. The survivorship function was estimated from annual data from the June census, slaughter, export and disappearance statistics, and data on the seasonality of lamb slaughter. **d**, Before mid-1988, both reported and unreported clinical BSE cases could be used for food (red-shaded curve). After that time, BSE was made notifiable and cases were destroyed, so we assume none entered food. The blue-shaded curve represents the rate of slaughter (per year) of pre-clinical infected cattle weighted by infectiousness relative to disease onset, assuming infectiousness grows at the exponential rate of 4 per year. The over-30-month scheme ended most bovine exposure in 1996, with estimates of residual levels (inset) being dependent on the extent of maternal transmission of BSE. The solid black curve represents the estimated infection hazard to cattle and sheep posed by infectivity in contaminated feed, relative to the maximum level reached in 1988.

Key to our analysis is estimates of the infectivity in animal tissues during disease incubation. Data are limited for BSE in sheep acquired by oral challenge, but using new experimental results and published data from studies of both scrapie[4-7] and sheep BSE[8,9] pathogenesis, we constructed an infectiousness profile. This profile was based on temporal changes in the density of the agent in different tissues, weighted by the proportion of such tissue in the host's body (Fig. 1b). This profile differs from that of BSE in cattle[10,11], with a more rapid rise in overall infectivity early in the incubation period in a wide range of tissues (for example, spleen and lymph nodes). The sheep–human infectiousness profiles (Fig. 1b) adjust for tissue-specific usage in food[12] and the effect of the 1997 specified risk materials (SRM) ban in sheep.

中 BSE 的患病率上限是 2%；图 1a)，并且要求与对这些绵羊群曾经与肉类和骨粉(MBM)接触的历史评估大致一样[3]。

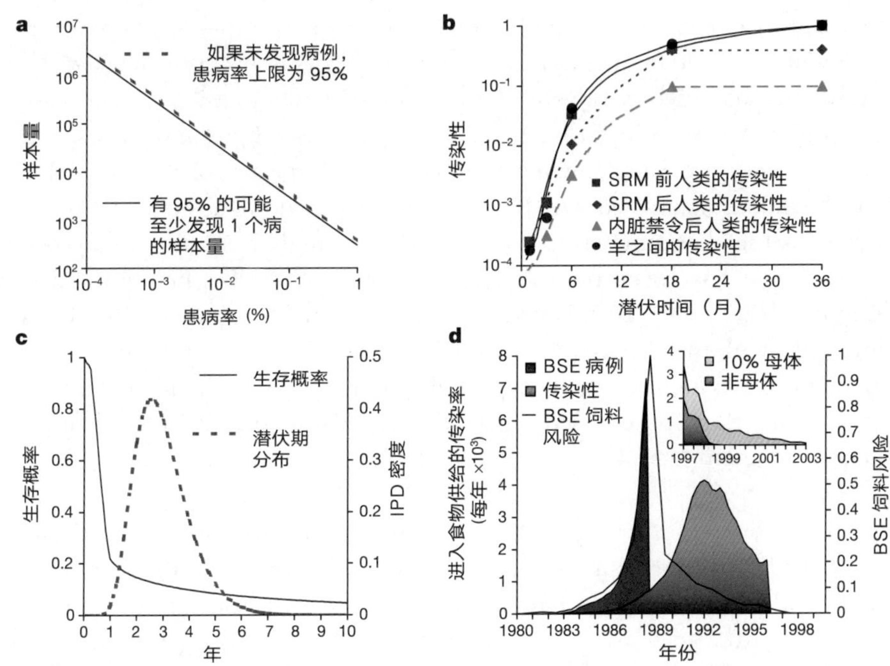

图 1. 传播模型中的流行病学数据。**a**，在筛选研究中样本量和检测到的患病率间的关系。**b**，羊传染性随感染时间的变化(见方法)。羊和人的患病率概况分别归一化到最大值为 1。**c**，羊随着年龄增长的生存概率，以及假设的羊群中 BSE 的潜伏期分布状况(IPD)。生存函数是根据每年 6 月份进行的羊群普查、屠宰、出口和失踪的数据以及羊羔屠宰的季节性数据进行估计的。**d**，1988 年中期之前，报告的和未报告的 BSE 患病牲畜都可以做成食品(红色阴影曲线)。此后，BSE 病例都要上报，并且患畜都被销毁，因此我们假定没有患畜做成食品。蓝色阴影曲线表示临床感染前期牛的屠宰率(每年)，以相对于疾病发病的传染性来表示，假设传染性以每年 4 的速度指数式增长。这个超过 30 个月的方案在 1996 年结束了大部分牛群的暴露，而残存水平的估计(插入图)取决于 BSE 母体传播的程度。黑色的实线表示相对于 1988 年达到的最高水平，在受污染饲料中的传染性对牛羊造成的感染风险的估计。

我们分析的关键是评估疾病潜伏期内动物组织的传染性。通过进食感染的 BSE 羊群的数据有限，但是根据对瘙痒病[4-7]和羊 BSE[8,9]发病机理研究所获得的新的研究结果和公布的数据，我们构建出了一个传染性的概况。这个概况基于不同组织中病原体密度的暂时性改变，以该组织在宿主体内所占的比例来表示(图 1b)。此概况与牛中的 BSE 感染情况不同[10,11]，表现为在不同组织(例如脾和淋巴结)的潜伏期早期，总体传染性增加更快。羊–人传染的概况(图 1b)根据食物中特殊组织的含量[12]以及 1997 年针对羊的高风险食品(SRM)禁令的效果进行调整。

The distribution of the BSE incubation period in sheep is not well characterized, but on the basis of the limited available data, we used an offset gamma distribution with a mean of 3 years (Fig. 1c) and substantial variance (intended to capture variation caused by dose dependency and host genotype) both for sheep infected by feed and those infected horizontally. Pathogenesis and susceptibility are dependent on the genetic background of the host[13-15], and genotype frequencies of the key polymorphisms vary considerably within and between flocks of different breeds[16,17]. Collation of limited available data suggests that roughly one-third of sheep in Great Britain (England, Scotland and Wales) have BSE-susceptible genotypes (see Supplementary Information). Exposure of the sheep flock to the BSE agent via contaminated feed (Fig. 1d) is assumed to mirror (albeit at a much lower level) that of British cattle, as estimated in back-calculation analyses of the BSE epidemic[18,19]. Exposure is likely to have occurred as far back as the early 1980s, before the disease was identified in cattle[18,19]. As for BSE in cattle[20], host survivorship is important given the long incubation period of the disease. Best estimates (see Supplementary Information) are presented in Fig. 1c (giving a mean life expectancy of 1.5 years), although more precise data are urgently required, perhaps based on a sheep equivalent of the British Cattle Tracing System (http://www.bcms.gov.uk).

Capturing the information in Fig. 1 requires a framework that integrates the temporal evolution of BSE pathogenesis in the individual host (incorporating age-dependent susceptibility and exposure) into a mathematical model of transmission within (including seeding and spread) and between flocks. This model builds on previous analyses of within- and between-flock transmission of scrapie[21,22], and consists of a set of nonlinear partial differential equations detailing the transmission dynamics of the agent and the demography of the sheep flock under time-dependent exposure to BSE-contaminated feed. The dynamics of disease transmission within the sheep population are determined by the magnitude of the respective basic reproduction numbers of the agent within a flock (R_0^A) and between flocks (R_0^F). The reproduction numbers define the average number of secondary cases or flocks generated by one primary case or infected flock in a susceptible flock or population of flocks. We considered three representative scenarios: (I) $R_0^A > 1$, $R_0^F < 1$—self-sustaining transmission within a flock but not between flocks; (II) $R_0^A > 1$, $R_0^F > 1$—the worst-case scenario for future spread, with spread within and between flocks inducing an expanding epidemic; and (III) $R_0^A < 1$, $R_0^F < 1$—the best-case scenario, with non-self-sustaining transmission both within and between flocks (see Fig. 2 for precise parameter values).

For each scenario, the level of flock infection due to MBM exposure was adjusted to give BSE prevalence below 2% of scrapie prevalence at present (consistent with the results of ongoing studies screening sheep brains). Judgements of scenario consistency therefore depend on the estimated prevalence of scrapie in British sheep, with such estimates being based on limited data from detailed surveys of specific flocks (using clinical criteria for diagnosis) and a postal survey of farmers intended to characterize national historical patterns of scrapie incidence[23,24]. The uncertainties in interpreting these data (giving estimates of infection prevalence anywhere between 0.1% and 1%, see Supplementary

羊群中BSE潜伏期的分布没有很明显的特征，但是根据有限的可用数据，我们对饲料感染以及水平感染的羊均使用平均数为3年的偏移γ分布（图1c）以及显著的差异（试图发现由剂量依赖性和宿主基因型引起的变化）。发病机制和易感性由宿主的遗传背景决定[13-15]，重要多态性的基因型频率在不同品种的羊群内部和羊群之间的差别相当大[16,17]。对有限可用数据的整理表明，英国（英格兰、苏格兰和威尔士）大约三分之一的羊具有BSE易感基因型（见补充信息）。正如对BSE流行病的反算法分析所估计的那样，这些羊群通过污染的饲料暴露于BSE病原体中（图1d），这与英国中的情况差不多（尽管水平要低得多）[18,19]。暴露很可能在20世纪80年代早期就已经存在，那时尚未在牛群中发现此病[18,19]。对于牛的BSE[20]，由于潜伏期很长，宿主的生存概率就显得很重要。尽管迫切需要更加精确的数据，但图1c已经显示了最佳估计值（见补充信息）（平均预期寿命是1.5年），这可能基于在羊群中使用的英国牛群追踪系统（http://www.bcms.gov.uk）。

获得图1所示的信息需要一个模型，而这个模型能够将个体宿主中BSE发病机制的时间演变（包括年龄依赖的易感性和暴露）整合到在羊群内部（包括播种和传播）和羊群之间传播的数学模型中。该模型基于先前对羊群内部和羊群之间瘙痒病传播的分析[21,22]，并且由一组非线性偏微分方程组成，详细描述了病原体传播的动力学和暴露在受BSE污染饲料下的具有时间依赖性的羊群的种群统计学。羊群中疾病传播的动力学是由群内部（R_0^A）和群之间（R_0^F）病原体各自的基础繁殖数量的大小决定的。繁殖数目定义了以下四种情况的平均数量，或者次要病例，或者由一个主要病例产生的羊群，或者易感群体中受感染的群体，或者种群。我们考虑了三个代表性情况：(I) $R_0^A > 1$，$R_0^F < 1$——群内部是自我维持的传播，而群之间不是；(II) $R_0^A > 1$，$R_0^F > 1$——未来最差的情况，在群内部和群之间传播，导致流行病不断扩大；(III) $R_0^A < 1$，$R_0^F < 1$——最好的情况，群内部和群之间都是非自我维持的传播（精确的参数值见图2）。

对于每一种情况，由于MBM暴露导致的群体感染的水平都调整到使得BSE的患病率低于目前2%的瘙痒病患病率（与正在进行的羊脑筛查的研究结果一致）。因此，对情况一致性的判断就取决于英国羊群中估计的瘙痒病患病率，而这个估计是基于对特定羊群（使用临床诊断标准进行诊断）的详细调查以及对农民的邮政调查所得到的有限数据，目的是描述全国瘙痒病发病的历史模式[23,24]。解释这些数据的不确定性（感染的患病率估计值在0.1%～1%之间，见补充信息）使得大规模（图1a）

Information) make large-scale (Fig. 1a) screening of the national flock for transmissible spongiform encephalopathies (TSEs) a priority. In constructing the scenarios, we assumed a scrapie prevalence of about 0.3%, with scenarios II and III then corresponding to a BSE prevalence of 0.5% that of scrapie, and scenario I to a prevalence of about 2% that of scrapie. Scenario I was thus intended to represent something of a worst case in terms of the numbers of animals infected to date, although we cannot exclude the possibility of even larger epidemics given the limited data currently available.

Figure 2 displays the epidemiological characteristics of these scenarios in terms of within-flock and overall prevalence (Fig. 2a–c), and their implications for human exposure via food (Fig. 2d–f). The estimates of exposure incorporate data on human consumption of ovine material (see Supplementary Information), which indicate that 67% of lambs and 83% of sheep older than 12 months slaughtered for consumption in 1999 were consumed domestically in the UK. Very few live sheep are imported into Great Britain, and most imported lamb meat originates from New Zealand, which has never detected signs of BSE or scrapie infection in either its bovine or ovine populations. Thus the potential risk from BSE-infected sheep arises from home-bred animals.

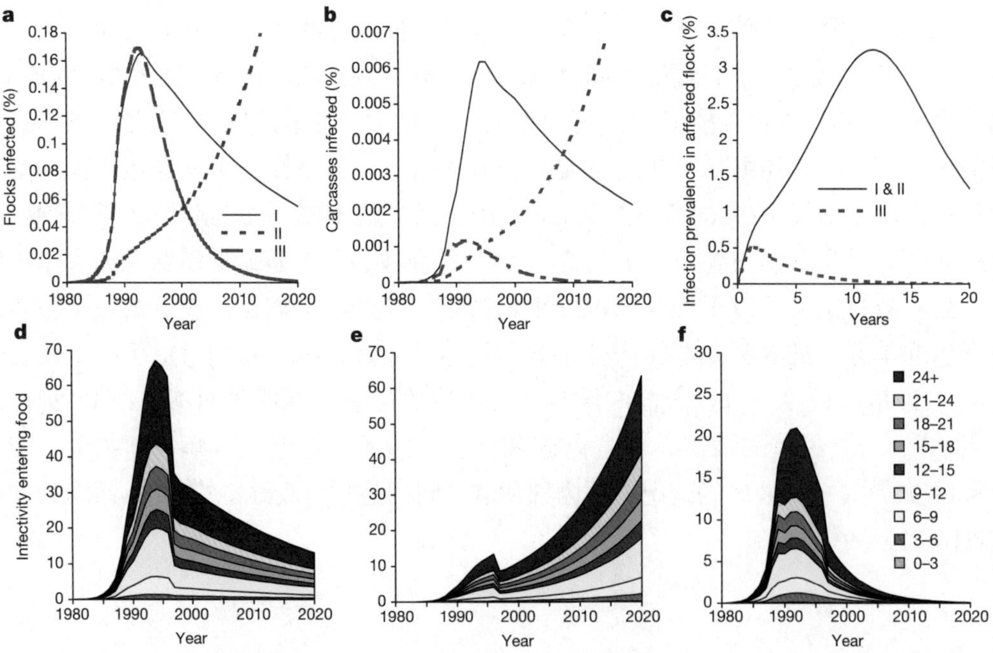

Fig. 2. Epidemiological characteristics of BSE transmission scenarios in sheep. $R_0^A = 2$, $R_0^F = 0.8$ and $\beta_B = 0.2\%$ per year for scenario I; $R_0^A = 2$, $R_0^F = 1.5$ and $\beta_B = 0.025\%$ per year for scenario II; $R_0^A = 0.8$, $R_0^F = 0.5$ and $\beta_B = 0.2\%$ per year for scenario III; where β_B is assumed flock infection incidence rate per unit of feed risk profile shown in Fig. 1c. By comparison, β_B values of 0.2% and 0.025% represent per-animal infection hazards about 50- and 400-fold less than that experienced by cattle. **a**, Proportion of flocks affected through time for scenarios I–III. **b**, Proportion of carcasses entering the food supply infected (at any incubation stage) with BSE. **c**, Prevalence in affected flock as a function of time since initial entry of infection into the flock. **d–f**, Estimated infectivity (in units of maximally infectious carcasses) entering the food supply under scenarios I, II and III, respectively, derived from the infectiousness profiles shown in

筛查全国羊群中的传染性海绵状脑病(TSE)成为优先事项。在构建上述情况时，我们假定瘙痒病的患病率大约是 0.3%，情况 II 和 III 相对应的 BSE 的患病率是瘙痒病的 0.5%，而情况 I 对应的 BSE 患病率是瘙痒病的 2%。因此，情况 I 预期能够代表目前为止感染动物数量的最差状况，尽管我们不能排除由于目前可用的数据有限而发生更大范围流行病的可能性。

图 2 显示了这些情况在群内部和总体患病率方面的流行病学特征(图 2a ~ 2c)，以及它们在人类通过食物暴露时的影响(图 2d ~ 2f)。暴露估计包括人类对绵羊身体组成部分进行消费的数据(见补充信息)，这表明 1999 年屠宰的 67% 的羊羔和 83% 的超过 12 个月龄的绵羊都是在英国国内消费的。很少有活羊进口到英国，大部分进口的羔羊肉来自新西兰，而新西兰从未在羊群或者牛群中检测到 BSE 或者瘙痒病感染的迹象。因此，BSE 感染的羊的潜在风险都来自家养动物。

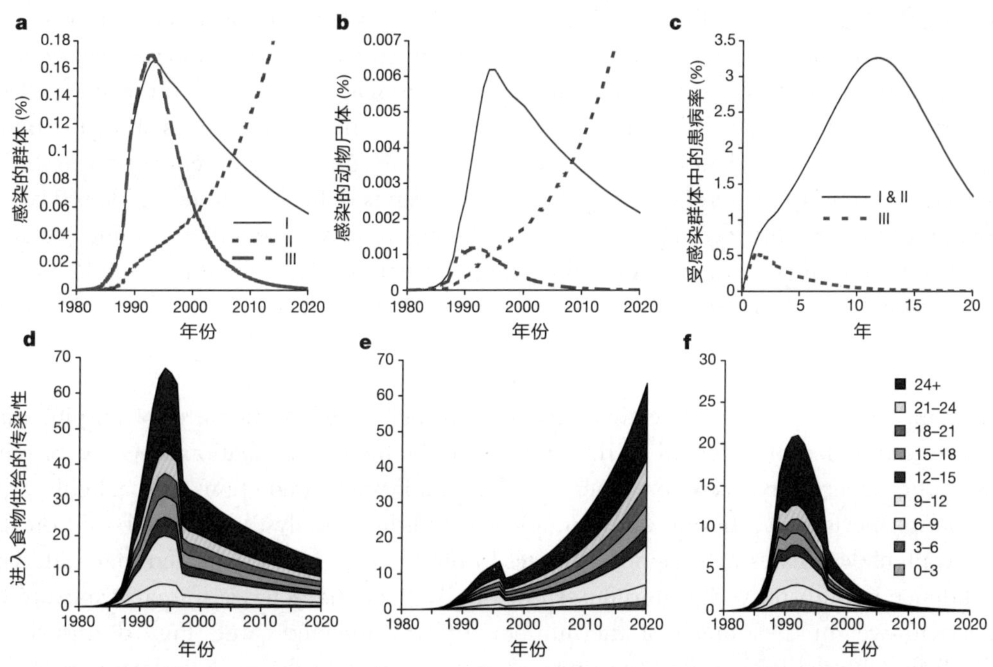

图 2. 羊群中 BSE 传播情况的流行病学特征。情况 I，$R_0^A = 2$，$R_0^F = 0.8$，$\beta_B = 0.2\%$ 每年；情况 II，$R_0^A = 2$，$R_0^F = 1.5$，$\beta_B = 0.025\%$ 每年；情况 III，$R_0^A = 0.8$，$R_0^F = 0.5$，$\beta_B = 0.2\%$ 每年；其中 β_B 是每单位饲料风险分布的假定群体感染发病率，如图 1c 所示。相比之下，β_B 值为 0.2% 和 0.025% 分别表示每个动物感染的风险是牛感染的风险的五十分之一和四百分之一。**a**，三种情况下，随着时间的推移，受感染群体的比例。**b**，感染了 BSE 的动物尸体进入食物供给的比例(包括各种潜伏阶段)。**c**，从感染进入群体开始，受感染群体的患病率随时间的变化的函数。**d~f**，根据图 1a 所示的传染性概况，在 I、II、III 三种情况下进入食物供给的估计传染性(以最大传染性尸体为单位)。改变受 BSE 感染的羊群的不同比例

Fig. 1a. Varying the proportion of flocks affected with BSE (shown in **a**, and determined by β_B) scales the exposure curves in **d–f** proportionately.

The sudden drop in human exposure in the late 1990s (Fig. 2d–f) is due to the SRM ban for sheep (banning the consumption of the skull, tonsils and spinal column of animals older than 12 months and the spleen of all animals). In the worst-case scenario (II), the degree of risk is still rising steeply, whereas for the other scenarios risk is falling, approaching very low levels in the best case (III). Critically, even in the worst-case scenario, the great majority of past exposure of the human population to BSE arose from cattle (Fig. 1d). However, for all three scenarios considered (and therefore assuming BSE did enter the sheep flock), the number of maximally infectious sheep entering the food supply in Great Britain at present is estimated to be greater than the number of maximally infectious cattle[18,19] (Fig. 1d, inset). This comparison equates the risk from one maximally infectious sheep with that of a cow, despite the difference in body mass, an assumption motivated by the evidence of more widespread distribution of infectivity in sheep.

Potential risk-reduction strategies include restrictions on the age of sheep slaughtered for consumption and enhanced tissue-based controls to reduce the amount of infectivity entering the food supply; additional measures based on flock history or ram genotype might also be possible but are not considered here. Combined tissue- and age-based restrictions are estimated to reduce current and future risk by at least 80% for all three sheep BSE scenarios considered (Fig. 3a). This is encouraging, but the precise values shown depend on the accuracy of the data on the development of infectivity in different tissues of BSE-infected sheep. Furthermore, verifying the age of sheep is difficult without an identification scheme, so in the short term it would be feasible only to exploit seasonal birthing patterns and dental indicators to impose approximate age restrictions. How the impact of 12-month age restrictions might be reduced by compensatory changes in slaughter patterns is shown in Fig. 3b.

Translation of the patterns of relative exposure through time into measures of absolute risk requires estimation of potential vCJD mortality in the human population of Great Britain. Given the uncertainty in the parameters determining such predictions (in particular, the incubation period distribution and human susceptibility), analyses must be based on the best available estimates of temporal changes in human exposure to infected material. The confidence bounds on vCJD mortality shown in Table 1 characterize human exposure by infectivity-weighted estimates of the numbers of BSE-infected cattle (Fig. 1d) and sheep (Fig. 2) slaughtered for consumption through time (correcting for early under-reporting of cattle BSE incidence). We modelled the vCJD epidemic solely in the 40% of the population that are methionine homozygous at prion protein (PrP) codon 129, assuming no other genetic variation in susceptibility. Currently no case data exist with which to constrain epidemic scenarios for other genotypes, but if future cases are diagnosed then upper bounds on epidemic size will increase. Compared with our previous estimates[25], upper bounds on epidemic size in the absence of BSE in sheep have reduced (largely as a result of a change in statistical methods), being in the range 50,000–100,000 depending on the assumptions

（如图 **a** 所示，由 β_B 决定）按比例缩放 **d~f** 的暴露曲线。

20 世纪 90 年代末人类暴露的骤降（图 2d ~ 2f）是因为人们发布了针对羊的 SRM 禁令（禁止食用 12 个月龄以上动物的头骨、扁桃体和脊柱以及所有动物的脾脏）。风险程度在最坏的情况下（II）仍然急剧升高，而在其他情况下都下降了，在最好的情况下（III）达到最低。准确地说，即便是在最坏的情况下，人类既往暴露于 BSE 的大部分情况来源于牛（图 1d）。但是，对所有三种情况来说（假设 BSE 确实进入了羊群），目前英国进入食物供给的受传染的羊的最大数目估计超过了感染的牛的数量[18,19]（图 1d，插图）。这个比较将一只传染性最强的羊的风险与一头奶牛的风险等同起来，尽管羊和牛的体重差别很大，但是做这一比较是因为有证据表明羊的传染性分布更为广泛。

可能的降低风险的措施包括限制供消费而屠宰的羊的年龄以及加强基于动物组织的控制，以减少进入食物供给的传染性，基于羊群历史以及公羊基因型的其他措施可能也起作用，但是这里不做考虑。对于羊 BSE 三种情况的考虑，结合基于组织和年龄的限制，预计能够将目前和未来的风险降低至少 80%（图 3a）。这很令人鼓舞，但是显示的精确的数值取决于 BSE 感染羊中不同组织传染性发展的数据的准确性。此外，在没有鉴定方案的情况下确定羊的年龄是很困难的，因此在短期内可行的仅仅是根据季节性生育模式以及牙齿指标来施加近似年龄的限制。图 3b 中显示了 12 个月龄限制的效果是如何被屠宰模式的补偿性改变所降低的。

将相对暴露随时间改变的模式转换成绝对风险值需要估计英国人群中潜在的 vCJD 死亡率。考虑到决定这些预测的参数的不确定性（尤其是潜伏期分布以及人类易感性），我们的分析必须基于人类暴露于感染物质的时间变化的最佳现有估计。表 1 显示的 vCJD 死亡率的置信区间以不同时间下屠宰用于消费的 BSE 感染的牛（图 1d）和羊（图 2）数量的传染力加权估计值（对早期牛 BSE 发病率的漏报进行了更正）特征性地描述了人类暴露的情况。我们仅在 40% 的朊蛋白（PrP）密码子第 129 位是蛋氨酸纯合子的人群中模拟 vCJD 流行病，假设这些人在易感性方面没有其他遗传变异。目前，还没有病例数据来限制其他基因型的流行情况，但是如果将来的病例得到诊断，那么流行范围的上限将会增加。与我们之前的估计相比[25]，在没有疯牛病的情况下，羊疫情的规模上限已经降低（主要是由于统计方法的改变）。根据所作的假设，其范围在 50,000 ~ 100,000 之间。这些数值远远超过了最近发表的估计，它是根据人类先前暴露于 BSE 传染性趋势的粗略描述得出的[26,27]。2001 ~ 2080 年的

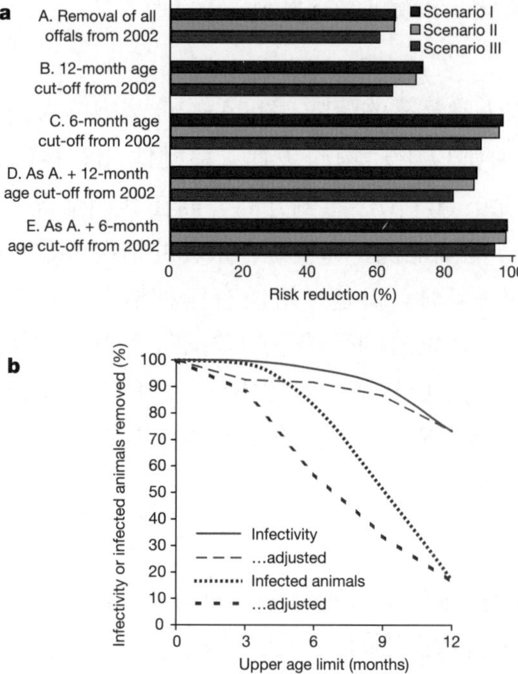

Fig. 3. Impact of risk-reduction measures. **a**, Estimated relative impact of various measures on current exposure to BSE infectivity in sheep food products for scenarios I–III. "Removal of all offals" corresponds to a ban on all internal organs and central nervous system tissue of sheep entering the human food supply, but does not assume removal of lymph nodes. Estimates presented were calculated assuming that age restrictions would not change slaughter patterns. **b**, Estimated impact of imposing age restrictions as a function of the upper age limit imposed for scenario I. Effects on infectivity and total number of infected animals entering the food supply are shown, with dashed curves showing how the impact is reduced if slaughter patterns are adjusted after such a measure, such that all animals currently slaughtered under 12 months old are then slaughtered under the upper age limit imposed.

made. These values are substantially greater than recently published estimates derived assuming a cruder representation of past trends in human exposure to BSE infectivity[26,27]. Best-fit estimates associated with 2001–2080 confidence bounds generally lie in the range 100–1,000, but the fits of the model vary little for fewer than 10,000 deaths. It should be noted (Table 1, E) that large epidemics are still possible even if the mean incubation period is less than 60 years. In the presence of BSE in sheep, the upper bound is substantially increased only if BSE is capable of becoming endemic in the national flock (scenario II).

Table 1. 95% confidence intervals for future vCJD deaths

Cattle only*	2000–2002	2001–2005	2001–2010	2001–2020	2001–2040	2001–2080
A	20–100	40–400	40–1,200	40–5,000	40–20,000	50–50,000
B	20–100	40–400	40–1,200	40–5,000	40–20,000	40–40,000
C	20–100	30–400	40–1,400	40–7,000	40–40,000	40–90,000
D	20–100	30–400	30–1,300	40–7,000	40–40,000	40–100,000
E	20–100	40–350	40–1,100	40–5,000	40–20,000	50–35,000

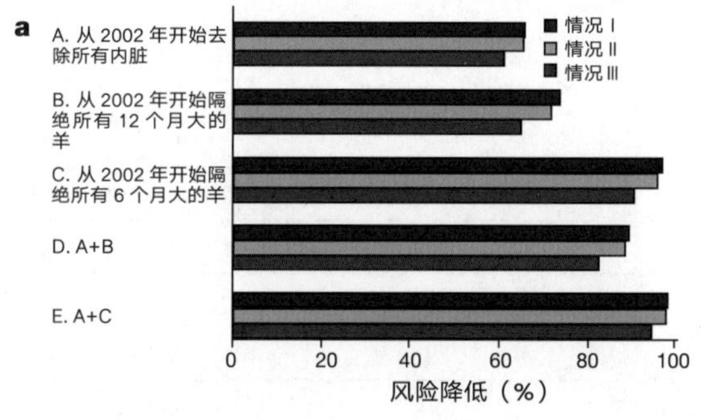

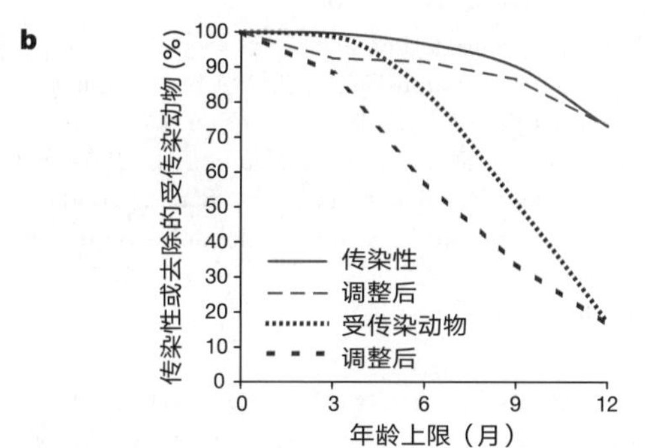

图3. 降低风险的措施的效果。**a**，三种情况下，估计各种措施对于目前羊类食品暴露于BSE传染性的相对影响。“去除所有内脏”表示禁止羊的内脏和中枢神经系统组织进入人类的食物供给，但并不意味着要去除淋巴结。显示出来的估计效果是假设年龄限制不会改变屠宰状况的情况下得到的。**b**，对于情况I，将年龄限制强行作为年龄上限的函数的估计效果。图中显示了传染性的影响以及进入食物供应的受感染动物总数，而虚线显示的是假如屠宰状况根据下面的措施进行调整后这种影响是如何降低的。该措施是将目前在12个月龄以下屠宰的所有动物限定到年龄上限以下屠宰。

置信区间相关的最佳估计在100～1,000之间，但对于少于10,000的死亡数，这个模型的最佳值变化不大。需要注意的是（表1，E），大规模的流行仍然是可能的，即便平均潜伏期小于60年。羊群中BSE存在的情况下，只有BSE能够在全国的羊群中流行时，其上限才会大幅度地增加（情况II）。

表1. 未来vCJD死亡数的95%置信区间

仅仅牛*	2001～2002	2001～2005	2001～2010	2001～2020	2001～2040	2001～2080
A	20～100	40～400	40～1,200	40～5,000	40～20,000	50～50,000
B	20～100	40～400	40～1,200	40～5,000	40～20,000	40～40,000
C	20～100	30～400	40～1,400	40～7,000	40～40,000	40～90,000
D	20～100	30～400	30～1,300	40～7,000	40～40,000	40～100,000
E	20～100	40～350	40～1,100	40～5,000	40～20,000	50～35,000

Continued

Cattle and sheep†	Sheep:cattle infectivity ratio	2001–2020	2001–2040	2001–2080
Scenario I	1:1	40–5,000	40–20,000	40–60,000
	10:1	40–5,000	50–30,000	50–70,000
Scenario II	1:1	20–5,000	20–20,000	20–70,000
	10:1	40–5,000	70–30,000	110–150,000
Scenario III	1:1	40–5,000	40–20,000	40–50,000
	10:1	40–5,000	50–30,000	40–50,000

* Cattle-only calculations assume: A, BSE exposure in humans was proportional to the BSE cases before the mid-1988 in addition to infected animal slaughter rates shown in Fig. 1c; B, as A but excluding exposure to reported BSE cases; C, as A but additionally fitting to at least three vCJD deaths reported before 2001 being infected before 1986, as suggested by analysis of the Queniborough cluster; D, as A but assuming the 1989 specified bovine offal ban reduced human exposure by at least 80%; E, as A but restricting the mean incubation period to be no longer than 60 years.

† Calculations for cattle and sheep assume exposure as A from cattle plus exposure from sheep as in Fig. 2d–f from scenarios I–III, assuming sheep are equally or 10 times as infectious as cattle. Short-term predictions for these scenarios are similar to those excluding sheep.

Although the risk analysis presented here incorporates a wide variety of available information into a single integrated framework, its reliability depends on the quality and volume of data available for parameter estimation. The limited data highlight the need for further studies to measure: (1) scrapie and BSE prevalence in sheep (stratified by age), employing sample sizes sufficiently large to detect low prevalence; (2) sheep survivorship more precisely; (3) BSE infectivity in sheep quantitatively, by stage of incubation, tissue and sheep genotype; (4) age-dependent susceptibility to TSE infection in a variety of species, including sheep and cattle; and (5) historical trends in bovine and ovine tissue consumption. Molecular typing methods giving rapid results[28,29] are clearly valuable for prevalence screening and strain typing (whether for BSE or scrapie), but study design should take account of test sensitivity and therefore consider which tissue should be tested (brain not necessarily being optimal). Given the uncertainty regarding the presence of BSE in the British sheep flock, such large-scale testing is a priority. In the interim, this analysis informs prevalence survey design and policy consideration of the potential benefits of additional risk-reduction measures.

Methods

Additional detail is provided as Supplementary Information.

Modelling transmission of BSE in sheep

Infectiousness of a BSE-infected sheep as a function of time τ since infection (relative to the incubation period T) was estimated by fitting the parametric form $\rho(\frac{\tau}{T}) = \exp\,(a(\frac{\tau}{T})^b/((\frac{\tau}{T})^b+c))$

续表

牛和羊†	羊：牛传染性比例	2001 ~ 2020	2001 ~ 2040	2001 ~ 2080
情况 I	1∶1	40 ~ 5,000	40 ~ 20,000	40 ~ 60,000
	10∶1	40 ~ 5,000	50 ~ 30,000	50 ~ 70,000
情况 II	1∶1	20 ~ 5,000	20 ~ 20,000	20 ~ 70,000
	10∶1	40 ~ 5,000	70 ~ 30,000	110 ~ 150,000
情况 III	1∶1	40 ~ 5,000	40 ~ 20,000	40 ~ 50,000
	10∶1	40 ~ 5,000	40 ~ 20,000	40 ~ 50,000

* 只计算牛的假设：A，除了图 1c 中所示的受感染动物的屠宰率之外，人类中的 BSE 暴露与 1988 年中期以前的 BSE 病例成比例；B，和 A 一样，但是排除了暴露于已经报告过的 BSE 病例；C，和 A 一样，但是额外地加入了至少 3 例 2001 年前报告的并且在 1986 年前感染的 vCJD 死亡病例，正如对 Queniborough 群体的分析所提示的一样；D，和 A 一样，但是假设 1989 年规定的牛内脏禁令将人类的暴露减少了至少 80%；E，和 A 一样，但是限制平均潜伏期不超过 60 年。

† 计算牛和羊：假设来自牛的暴露和 A 一样，来自羊的暴露如图 2d~2f 所示的三种情况。假定羊的传染性与牛相同或是牛的 10 倍。这些情况下的短期估计值和不包括羊的情况类似。

尽管这里所描述的风险分析将很多可用信息整合到了单个的整体框架中，但是其可靠性依赖于用于参数估计的数据的质量和数量。有限的数据强调了进一步测量研究的必要性：(1) 羊的瘙痒病和 BSE 患病率（根据年龄分层），需要足够大的样本量来检测低患病率；(2) 羊的更加精确的生存期；(3) 根据潜伏期的阶段、组织和羊基因型，定量分析羊 BSE 的传染性；(4) 不同物种对于 TSE 感染的年龄依赖的易感性，包括牛和羊；以及 (5) 对牛和羊组织消费的历史趋势。分子分型方法能够快速得出结果 [28,29]，对于患病率筛查和种类分型都非常有价值（无论是 BSE 还是瘙痒病），但是研究设计需要考虑测试的敏感性，因此需要考虑用什么组织进行测试（脑组织并不是最佳选择）。由于英国羊群中是否存在 BSE 仍不确定，这种大规模的测试是需要优先考虑的。在此期间，这个分析为患病率调查的设计以及政策考量提供了依据，说明了额外减少风险措施的潜在好处。

方　法

补充信息中提供了详细内容。

BSE 在羊中传播的模型

通过将参数式 $\rho(\frac{\tau}{T}) = \exp\,(a(\frac{\tau}{T})^b/((\frac{\tau}{T})^b+c))$ 与羊 BSE[8,9] 和瘙痒病 [4-7] 发病机制研究的组织特异性的传染力数据进行组合，估算感染后（相对于潜伏期 T）羊 BSE 传染性随时间 τ

to tissue-specific infectivity data from studies of pathogenesis of ovine BSE[8,9] and scrapie[4-7]. The data were weighted by total tissue mass when estimating sheep-to-sheep infectiousness, and by the proportion of tissue mass entering food when estimating human exposure to infectivity (Fig. 1b). This approach captures the impact on overall infectiousness of between-tissue variation in PrP accumulation during pathogenesis of ovine BSE—namely, that some tissues (for example, lymph nodes) rapidly develop detectable infectivity, the growth of which later slows and/or saturates, whereas others (for example, brain) only exhibit high levels of infectivity close to clinical onset.

Because mass-action models cannot capture the observed clustering of cases[20], we developed a model of TSE transmission in sheep that incorporated three relevant tiers: (1) the individual animal—capturing age-dependent susceptibility and exposure, and pathogenesis of disease; (2) the individual flock—capturing within-flock sheep-to-sheep transmission; and (3) the national population of flocks—capturing between-flock transmission and exposure to contaminated feed.

Assuming homogenous mixing within a flock, the deterministic susceptible-infected model is:

$$\frac{\partial x}{\partial t}+\frac{\partial x}{\partial a}=-(\Lambda(t)\kappa(a)+\mu(a))x \qquad x(t,0)=B \qquad \mu(a)=-\frac{1}{S}\frac{\mathrm{d}S}{\mathrm{d}a}$$

$$\frac{\partial y}{\partial t}+\frac{\partial y}{\partial a}+\frac{\partial y}{\partial \tau}=-(F_A(\tau)+\mu(a))y \qquad F_A(\tau)=\frac{f_A(\tau)}{1-\int_0^{\tau} f_A(\tau')\mathrm{d}\tau'}$$

$$y(t,a,0)=\Lambda(t)\kappa(a)x(t,a)$$

$$\Lambda(t)=\beta_A\int_{\tau=0}^{t}\int_{T=\tau}^{\infty}F_A(T)\,\rho(\tau/T)y(t,a,\tau)\mathrm{d}T\,\mathrm{d}\tau \qquad y(0,a,0)=\frac{\kappa(a)\,Y_0}{\int\kappa(a)\,\mathrm{d}a}$$

where we denote the densities of susceptible and infected animals of age a at time t by $x(t, a)$ and $y(t, a, \tau)$; time from infection by τ; the incubation period distribution by $f_A(\tau)$ (see Fig. 1c); the force of infection by $\Lambda(t)$; the transmission coefficient for horizontal spread by β_A; the fixed birth rate by B; the mortality rate by $\mu(a)$; and age-dependent susceptibility by $\kappa(a)$ (conservatively assumed to be constant for the first 12 months of life and zero thereafter). The number of animals infected at the start of a flock outbreak is Y_0. Assuming that the infectiousness at time t of a flock infected at $t = 0$ to other flocks is proportional to the within-flock force of infection $\Lambda(t)$, we modelled transmission within a population of 100,000 flocks of 400 animals each. Genetic heterogeneity in susceptibility to infection was modelled at the flock level, with 33,000 flocks assumed to have all animals susceptible and the remainder completely resistant. The introduction of infection into a flock from an external source (contaminated feed or other sheep) was modelled as a rare event, initially infecting 1% of animals.

The infection dynamics of flocks were modelled using an SIRS framework, whereby flocks are initially "susceptible" to infection, enter an "infected" state of extended duration, "recover" and

的变化。估算羊之间的传染性时，数据通过组织总质量进行加权，而在估计人类暴露的传染性时则通过组织进入食物供给的比例进行加权（图 1b）。这种方法捕捉到羊 BSE 发病过程中不同组织间 PrP 聚集的不同对总体传染性的影响——也就是说，一些组织（比如淋巴结）很快就可以检测到传染性，其增长随后放缓和（或）达到饱和，而另一些组织（比如脑）仅仅显示出高水平的传染，接近于临床发作。

由于质量作用模型不能捕捉观察到的病例群[20]，我们建立了一个 TSE 在羊中传播的模型，包含了三个相关的层次：(1) 个体动物——得到依赖年龄的易感性和暴露以及疾病的发病机制；(2) 独立的羊群——捕捉群内羊之间的传播；(3) 全国的羊群——捕捉群间传播以及对污染饲料的暴露。

假定群内的混合是均匀的，那么确定的易感染模型是：

$$\frac{\partial x}{\partial t}+\frac{\partial x}{\partial a}=-(\Lambda(t)\kappa(a)+\mu(a))x \qquad x(t,0)=B \quad \mu(a)=-\frac{1}{S}\frac{\mathrm{d}S}{\mathrm{d}a}$$

$$\frac{\partial y}{\partial t}+\frac{\partial y}{\partial a}+\frac{\partial y}{\partial \tau}=-(F_A(\tau)+\mu(a))y \qquad F_{\mathrm{A}}(\tau)=\frac{f_{\mathrm{A}}(\tau)}{1-\int_0^{\tau}f_{\mathrm{A}}(\tau')\mathrm{d}\tau'}$$

$$y(t,a,0)=\Lambda(t)\kappa(a)x(t,a)$$

$$\Lambda(t)=\beta_{\mathrm{A}}\int_{\tau=0}^{t}\int_{T=\tau}^{\infty}F_{\mathrm{A}}(T)\,\rho(\tau/T)y(t,a,\tau)\mathrm{d}T\,\mathrm{d}\tau \qquad y(0,a,0)=\frac{\kappa(a)\,Y_0}{\int\kappa(a)\,\mathrm{d}a}$$

其中，我们用 $x(t, a)$ 和 $y(t, a, \tau)$ 表示年龄为 a 的易感动物和已感染的动物在 t 时刻的密度；感染后的时间是 τ；潜伏期分布是 $f_{\mathrm{A}}(\tau)$(见图 1c)；感染的强度是 $\Lambda(t)$；水平传播系数是 β_{A}；固定生产率是 B；死亡率是 $\mu(a)$；依赖年龄的易感性是 $\kappa(a)$(保守地假定生命的前 12 个月为常数，之后是 0)。羊群疫情暴发开始时感染动物的数量是 Y_0。假设一个羊群在 $t=0$ 时受感染，在 t 时刻对其他羊群的传染性与群内传染性 $\Lambda(t)$ 成正比。我们在含有 100,000 个羊群的群体中建立传播模型，每个羊群中包含 400 只羊。对感染易感性的遗传异质性在羊群水平进行建模，假定 33,000 个羊群中所有的动物都是易感的，而剩余的羊群都是完全耐药的。通过外部来源将感染引入羊群（污染的食物或者其他羊）被建模为小概率事件，最初感染 1% 的动物。

使用 SIRS 框架对羊群的感染动态进行建模，其中羊群最初对感染是“易感状态”，然后进入长时间的“感染”状态，“恢复”，并对再次感染具有耐药性（大约持续 20 年），然后恢复

become resistant to further infection (for a period of 20 years), then revert to the "susceptible" state. The flock-level "recovery" process approximates flock outbreak extinction mechanisms, such as demographic stochasticity (likely to be critical[30] given the low reported prevalence of scrapie), control measures and selection for scrapie-resistant genotypes.

Denoting the number of susceptible flocks at time t by $s(t)$, flocks infected time θ ago by $h(t, \theta)$ (with infectivity $\Lambda(\theta)$ from the within-flock model) and recovered/resistant flocks by $r(t)$, model dynamics are described by:

$$\frac{ds}{dt} = \xi r - \Omega(t)s \qquad \Omega(t) = \beta_F \int_{\theta=0}^{\infty} \Lambda(\theta)h(t, \theta)d\theta + \beta_B \phi(t)$$

$$\frac{\partial h}{\partial t} + \frac{\partial h}{\partial \theta} = -F_F(\theta)h \qquad F_F(\theta) = \frac{f_F(\theta)}{1 - \int_0^{\theta} f_F(\theta')d\theta'}$$

$$\frac{dr}{dt} = \int F_F(\theta)h(t, \theta)d\theta - \xi r \qquad s(0) = N \quad h(t, 0) = \Omega(t)x \quad h(0, \theta) = 0$$

Here, $f_F(\theta)$ is the "incubation period" distribution for flocks (gamma distributed with mean 6 years; see Supplementary Information); $\Omega(t)$ the force of infection for flocks at time t; β_F and β_B the coefficients for between-flock transmission and exposure of flocks to contaminated feed; $\phi(t)$ the relative risk from contaminated feed at time t (estimates from refs 18, 19); and ξ (= 0.05 per year) the rate at which recovered flocks re-enter the susceptible pool. The values of β_A and β_F corresponding to required values of R_0^A and R_0^F were determined numerically.

Exposure of the human population to BSE infectivity in sheep was represented by the number of infected animals slaughtered for food (stratified by age and incubation stage, weighted by infectivity; Fig. 2d–f). The effectiveness of risk-reduction measures was evaluated by examining the distribution of exposure as a function of animal age, and by comparing results obtained from the infectivity profiles corresponding to current SRM controls and a ban on all offal.

It should be noted that this model framework reproduces observed within-flock and overall scrapie incidence patterns well if run to endemicity (results not shown).

Prediction of vCJD incidence

As in earlier work[25], the probability density that an individual develops clinical disease at time t and age a is

$$p(t, a) = S_H(t, a) \int_{t-a}^{t} f(t-u)\, I(u, a-t+u) \exp\left[-\int_0^u I(u', a-t+u')du' \right] du$$

where $S_H(t, a)$ is the probability that someone born at time $t - a$ will survive to age a (derived from

到“易感”状态。羊群水平“恢复”的过程类似于羊群爆发灭绝的机制，比如人口统计学上的随机性（由于报道的瘙痒病的患病率很低，因此这可能很关键[30]）、控制措施和选择耐瘙痒病基因型。

用 $s(t)$ 来表示 t 时刻易感羊群的数量，用 $h(t,\theta)$ 来表示 θ 时刻之前感染的羊群数量（根据羊群内的模型，传染性是 $\Lambda(\theta)$），用 $r(t)$ 来表示恢复的/耐药性的羊群数量，动态模型表示为：

$$\frac{\mathrm{d}s}{\mathrm{d}t}=\xi r-\Omega(t)s \qquad \Omega(t)=\beta_{\mathrm{F}}\int_{\theta=0}^{\infty}\Lambda(\theta)h(t,\theta)\mathrm{d}\theta+\beta_{\mathrm{B}}\phi(t)$$

$$\frac{\partial h}{\partial t}+\frac{\partial h}{\partial \theta}=-F_{\mathrm{F}}(\theta)h \qquad F_{\mathrm{F}}(\theta)=\frac{f_{\mathrm{F}}(\theta)}{1-\int_{0}^{\theta}f_{\mathrm{F}}(\theta')\mathrm{d}\theta'}$$

$$\frac{\mathrm{d}r}{\mathrm{d}t}=\int F_{\mathrm{F}}(\theta)h(t,\theta)\mathrm{d}\theta-\xi r \qquad s(0)=N \quad h(t,0)=\Omega(t)x \quad h(0,\theta)=0$$

这里，$f_{\mathrm{F}}(\theta)$ 是羊群的“潜伏期”分布（平均 6 年的 γ 分布，见补充信息）；$\Omega(t)$ 是 t 时刻的羊群感染强度；β_{F} 和 β_{B} 是羊群间传播和羊群暴露于污染食物的系数；$\phi(t)$ 是 t 时刻污染食物的相对风险（从文献 18 和 19 预估）；ξ（= 每年 0.05）是恢复的羊群重新进入易感群体的速度。与 R_0^{A} 和 R_0^{F} 的需求值对应的 β_{A} 和 β_{F} 的数值用数字来确定。

人群对羊群 BSE 传染性的暴露程度由屠宰后作为食物的感染动物数量来表示（根据年龄和潜伏阶段分层，用传染性加权，图 2d ~ 2f）。降低风险措施的有效性通过以下两种方式评估：检查以动物年龄为函数的暴露分布；将目前 SRM 控制和禁止所有内脏相对应的传染性结果进行比较。

需要注意的是，如果遇到区域性流行，这个模型框架可以很好地复制观察到的羊群内以及总体的瘙痒病发病率模式（结果没有显示）。

vCJD 发病率的预测

如之前的工作所述[25]，个体在 t 时间和 a 年龄发生临床疾病的概率密度是

$$p(t,a)=S_{\mathrm{H}}(t,a)\int_{t-a}^{t}f(t-u)\,I(u,a-t+u)\exp\left[-\int_{0}^{u}I(u',a-t+u')\mathrm{d}u'\right]\mathrm{d}u$$

其中 $S_{\mathrm{H}}(t,a)$ 是某个生于时间 $t-a$ 的个体能够活到 a 年龄的可能性（来自人口普查数据），

census data) and $f(t-u)$ is the incubation period distribution (modified lambda distribution). The infection hazard is given by

$$I(t, a) = \beta g(a) \left(v_c(t) \int \Omega_c(z)\omega_c(z, t)\mathrm{d}z + v_s(t) \int \Omega_s(z)\omega_s(z, t)\mathrm{d}z \right)$$

where β is the transmission coefficient and $g(a)$ an age-dependent susceptibility/exposure function (uniform with gamma-distributed tails). For each infectious species i (C for cattle, S for sheep), $\Omega_i(z)$ is the relative infectiousness of an animal time z from disease onset, $v_i(t)$ is the time-dependent effectiveness of risk-reduction measures (for cattle, this is parameterized as a step reduction in 1989 due to the specified bovine offal ban), and $\omega_i(z, t)$ is the number of animals slaughtered for consumption stratified by time and incubation stage. (Values for sheep were obtained from the above model, and those for cattle used updated back-calculation estimates[18,19], with $\int \Omega_c(z)\omega_c(z, t)\mathrm{d}z$ plotted in Fig. 1d.)

The relative infectiousness of cattle by incubation stage was assumed to increase exponentially from a baseline level to a maximum value before onset of clinical signs[25]. We restricted analysis to the 40% of the population that are methionine homozygous at PrP codon 129, assuming no other genetic variation in susceptibility.

Through numerical solution of the inverse problem (see Supplementary Information or further details), β was calculated as a function of case incidence, allowing incidence of vCJD deaths in any time interval to be treated as a model parameter. This enabled nonlinear optimization techniques to be used to obtain likelihood profiles by fitting the model to the joint age- and time-stratified mortality data to the end of 2000. We obtained 95% confidence bounds from the one-dimensional likelihood profiles. Infection prevalence is poorly constrained by the observed incidence data, and can range from being equal to mortality to 100-fold larger, with little effect on model fit.

(**415**, 420-424; 2002)

N. M. Ferguson, A. C. Ghani, C. A. Donnelly, T. J. Hagenaars & R. M. Anderson
Department of Infectious Disease Epidemiology, Faculty of Medicine, Imperial College of Science, Technology and Medicine, St Mary's Campus, Norfolk Place, London W2 1PG, UK

Received 21 November; accepted 12 December 2001. Published online 9 January 2002, DOI 10.1038/nature709.

References:

1. Frankish, H. Samples blunder renders sheep-BSE study useless. *Lancer* **358**, 1436 (2001).
2. Beckett, M. *House of Commons Hansard Debates, 22 October 2001* 373 (2001) ⟨http://www.parliament.the-stationery-office.co.uk/pa/cm200102/cmhansrd/cm011022/debindx/11022-x.htm⟩ .
3. Det Norske Veritas *Assessment of Exposure to BSE Infectivity in the UK Sheep Flock* Report no. C782506 (The Meat and Livestock Commission, London, 1998).
4. Hadlow, W. J., Kennedy, R. C. & Race, R. E. Natural infection of Suffolk sheep with scrapie virus. *J. Infect. Dis.* **146**, 652-664 (1982).
5. van Keulen, L. J. M., Schreuder, B. E. C., Vromans, M. E. W., Langeveld, J. P. M. & Smits, M. A. Scrapie-associated prion protein in the gastrointestinal tract of sheep with natural scrapie. *J. Comp. Pathol.* **121**, 55-63 (1999).
6. van Keulen, L. J. M., Schreuder, B. E. C., Vromans, M. E. W., Langeveld, J. P. M. & Smits, M. A. Pathogenesis of natural scrapie in sheep. *Arch. Virol. Suppl.* **16**, 52-21 (2000).

$f(t-u)$ 是潜伏期分布（修改后的 λ 分布）。感染风险计算如下

$$I(t, a)=\beta g(a)\left(v_c(t)\int \Omega_c(z)\omega_c(z, t)\mathrm{d}z+v_s(t)\int \Omega_s(z)\omega_s(z, t)\mathrm{d}z\right)$$

其中 β 是传输系数，$g(a)$ 是依赖年龄的易感性/暴露函数（与 γ 分布一致）。对于每个感染性物种 i（C 表示牛，S 表示羊），$\Omega_i(z)$ 是疾病出现后 z 时刻某个动物的相对传染性，$v_i(t)$ 是降低风险措施的时间依赖有效性（对牛来说，由于特定的牛内脏禁令，因此在 1989 年发病率骤减），$\omega_i(z, t)$ 是根据时间和潜伏阶段分层的用于消费的动物屠宰数量。（羊的数据从以上的模型得出，而牛的数据用更新后的反算法估计 [18,19]，$\int\Omega_c(z)\omega_c(z, t)\mathrm{d}z$ 在图 1d 中绘出。）

在潜伏期，牛的相对传染性假定从基线水平指数式地增长到临床症状出现前的最高水平 [25]。我们将分析限制在那些 PrP 密码子第 129 位是蛋氨酸纯合的 40% 的人群中，这些人被认为在易感性方面没有其他遗传变异。

通过对反问题的数值解（详情请参阅补充资料），β 以病例发病率的函数来计算，将任意时间段 vCJD 死亡发生率作为模型参数。这使得非线性优化技术通过将截至 2000 年底的年龄和时间分层的死亡率数据套入到该模型中来获得似然分布。我们从单维似然分布中获得了 95% 置信区间。感染的患病率几乎不受观察到的发病率数据的限制，其范围从与死亡率相等到是死亡率的 100 多倍，对模型的拟合影响不大。

（毛晨晖 翻译；肖景发 审稿）

7. Andreoletti, O. *et al.* Early accumulation of PrP^{Sc} in gut-associated lymphoid and nervous tissues of susceptible sheep from a Romanov flock with natural scrapie. *J. Gen. Virol.* **81**, 3115-3126 (2000).

8. Foster, J. D., Parnham, D. W., Hunter, N. & Bruce, M. Distribution of the prion protein in sheep terminally infected with BSE following experimental oral transmission. *J. Gen. Virol.* **82**, 2319-2326 (2001).

9. Jeffrey, M. *et al.* Oral inoculation of sheep with the agent of bovine spongiform encephalopathy (BSE). 1. Onset and distribution of disease-specific PrP accumulation in brain and viscera. *J. Comp. Pathol.* **124**, 280-289 (2001).

10. Fraser, H., Bruce, M. E., Chree, A., McConnell, I. & Wells, G. A. H. Transmission of bovine spongiform encephalopathy and scrapie to mice. *J. Gen. Virol.* **73**, 1891-1897 (1992).

11. Wells, G. A. H. *et al.* Infectivity in the ileum of cattle challenged orally with bovine spongiform encephalopathy. *Ver. Rec.* **135**, 40-41 (1994).

12. Hart, R. J., Church, P. N., Kempster, A. J. & Matthews, K. R. *Audit of Bovine and Ovine Slaughter and By-products Sector (Ruminant Products Audit)* (Leatherhead Food Research Association, London, 1997).

13. Hunter, N. PrP genetics in sheep and the implications for scrapie and BSE. *Trends Microbiol.* **5**, 331-334 (1997).

14. Hunter, N., Goldmann, W., Marshall, E. & O'Neill, G. Sheep and goats: natural and experimental TSEs and factors influencing incidence of disease. *Arch. Virol. Suppl.* **16**, 181-188 (2000).

15. Jeffrey, M. *et al.* Frequency and tissue distribution of infection-specific PrP in tissues of clinical scrapie and cull sheep obtained from scrapie affected farms in Shetland. *J. Comp. Pathol.* (submitted).

16. Baylis, M., Houston, F., Goldmann, W., Hunter, N. & McLean, A. R. The signature of scrapie: differences in the PrP genotype profile of scrapie-affected and scrapie-free UK sheep flocks. *Proc. R. Soc. Lond.* B **267**, 2029-2035 (2000).

17. Hunter, N. *et al.* Is scrapie solely a genetic disease? *Nature* **386**, 137 (1997).

18. Anderson, R. M. *et al.* Transmission dynamics and epidemiology of BSE in British cattle. *Nature* **382**, 779-788 (1996).

19. Ferguson, N. M., Donnelly, C. A., Woolhouse, M. E. J. & Anderson, R. M. The epidemiology of BSE in cattle herds in Great Britain II: Model construction and analysis of transmission dynamics. *Phil. Trans. R. Soc. Lond.* B **352**, 803-838 (1997).

20. Donnelly, C. A. & Ferguson, N. M. *Statistical Aspects of BSE and vCJD—Models for Epidemics* (Chapman & Hall and CRC, London, 2000).

21. Woolhouse, M. E. J. *et al.* Population dynamics of scrapie in a sheep flock. *Phil. Trans. R. Soc. Lond. B* **354**, 751-756 (1999).

22. Gravenor, M. B., Cox, D. R., Hoinville, L. J., Hoek, A. & McLean, A. R. The flock-to-flock force of infection for scrapie in Britain. *Proc. R. Soc. Lond. B* **268**, 587-592 (2001).

23. Hoinville, L. J. A review of the epidemiology of scrapie in sheep. *Rev. Sci. Techn. Office Int. Epizoories* **15**, 827-852 (1996).

24. Hoinville, L. J., Hoek, A., Gravenor, M. B. & McLean, A. R. Descriptive epidemiology of scrapie in Great Britain: results of a postal survey. *Vet. Rec.* **146**, 455-461 (2000).

25. Ghani, A. C., Ferguson, N. M., Donnelly, C. A. & Anderson, R. M. Predicted vCJD mortality in Great Britain. *Nature* **406**, 583-584 (2000).

26. d'Aignaux, J. N. H., Cousens, S. N. & Smith, P. G. Predictability of the UK variant Creutzfeldt–Jakob disease epidemic. *Science* **294**, 1729-1731; published online 25 October 2001 (10.1126/ science.1064748).

27. Valleron, A.-J., Boelle, P.-Y., Will, R. & Cesbron, J.-Y. Estimation of epidemic size and incubation time based on age characteristics of vCJD in the United Kingdom. *Science* **294**, 1726-1728(2001).

28. Hill, A. F. *et al.* Molecular screening of sheep for bovine spongiform encephalopathy. *Neurosci. Lett.* **255**, 159-162 (1998).

29. Maissen, M., Roeckl, F., Glatzel, M., Goldmann, W. & Aguzzi, A. Plasminogen binds to disease-associated prion protein of multiple species. *Lancer* **357**, 2026-2028 (2001).

30. Hagenaars, T. J., Ferguson, N. M., Donnelly, C. A. & Anderson, R. M. Persistence patterns of scrapie in a sheep flock. *Epidemiol. Infect.* **127**, 157-167 (2001).

Supplementary Information accompanies the paper on *Nature*'s website (http://www.nature.com).

Acknowledgements. This work was funded by the Food Standards Agency. N.M.F. and A.C.G. acknowledge funding from The Royal Society. C.A.D., T.J.H. and R.M.A. acknowledge funding from the Wellcome Trust. We are very grateful to M. Bruce, S. Bellworthy and M. Jeffrey for access to pre-publication data on infectivity. We also thank L. Hoinville, J. Wilesmith and R. Will for access to data. We thank N. Hunter, L. Green, J. Anderson, A. James, A. Bromley, H. Mason, P. Comer and members of the Spongiform Encephalopathy Advisory Committee for discussions.

Competing interests statement. The authors declare that they have no competing financial interests.

Correspondence and requests for materials should be addressed to N.M.F. (e-mail: neil.ferguson@ic.ac.uk).

Detecting Recent Positive Selection in the Human Genome from Haplotype Structure

P. C. Sabeti *et al.*

Editor's Note

In this paper, geneticist Eric Lander and colleagues present a method for detecting the genetic imprint of natural selection. The technique focuses on haplotypes, sets of closely linked genetic markers present on one chromosome that tend to be inherited together. Knowing that variants of two key genes confer malarial resistance in some individuals, the team looked for evidence that a particular haplotype associated with the resistance variant arose faster in certain populations than would have happened by chance. This "signature" of natural selection was found in both the *G6PD* and CD40 ligand genes, and the team suggest that the test could be used to scan the entire human genome for evidence of recent positive selection.

The ability to detect recent natural selection in the human population would have profound implications for the study of human history and for medicine. Here, we introduce a framework for detecting the genetic imprint of recent positive selection by analysing long-range haplotypes in human populations. We first identify haplotypes at a locus of interest (core haplotypes). We then assess the age of each core haplotype by the decay of its association to alleles at various distances from the locus, as measured by extended haplotype homozygosity (EHH). Core haplotypes that have unusually high EHH and a high population frequency indicate the presence of a mutation that rose to prominence in the human gene pool faster than expected under neutral evolution. We applied this approach to investigate selection at two genes carrying common variants implicated in resistance to malaria: *G6PD*[1] and CD40 ligand[2]. At both loci, the core haplotypes carrying the proposed protective mutation stand out and show significant evidence of selection. More generally, the method could be used to scan the entire genome for evidence of recent positive selection.

THE recent history of the human population is characterized by great environmental change and emergent selective agents[3]. The domestication of plants and animals at the start of the Neolithic, roughly 10,000 years ago, yielded an increase in human population density. Humans were confronted with the spread of new infectious diseases, new food sources and new cultural environments. The last 10,000 years have thus been some of the most interesting times in human biological history, and may be when many important genetic adaptations and disease resistances arose.

We sought to design a powerful approach for detecting recent selection. Our method relies

从单倍型结构检测人类基因组近期的正选择

萨贝提等

编者按

在这篇文章中，遗传学家埃里克·兰德和他的同事们提出了一种检测自然选择的遗传印记的方法。这项技术关注的是单倍型，即存在于一条染色体上的一系列紧密相连的遗传标记，这些标记往往是一起遗传的。研究小组了解到两个关键基因的变异型在某些个体中产生了疟疾抗药性，因此他们寻找证据来证明在某些人群中，与抗药性变异相关的一种特定单倍型出现的速度比偶然出现的速度更快。在 *G6PD* 和 CD40 配体基因中都发现了这种自然选择的"特征"，研究小组认为，该测试可以用来扫描整个人类基因组，以寻找近期正选择的证据。

检测人群中最近发生的自然选择对于研究人类历史和医学具有深远的意义。这里我们介绍一个通过分析人群中长范围的单倍型来检测最近正选择的基因印记的框架。首先我们在感兴趣的基因座识别单倍型（核心单倍型）。然后我们通过其与该基因座不同距离的等位基因联系的衰减评估每个核心单倍型的年龄，即通过扩展单倍型纯合性（EHH）进行测定。具有异常高的 EHH 和高人群频率的核心单倍型表明，在自然中性进化下，一种突变在人类基因库中上升的速度快于预期。我们应用这种方法研究了两个携带涉及抗疟疾能力的寻常变异型的基因上的选择，*G6PD*[1] 和 CD40 配体 [2]。在这两个基因座上，携带假设的保护性突变的核心单倍型凸现出来并显示出重要的选择证据。一般来说，该方法可以用于扫描整个基因组，以寻找近期正选择的证据。

人类近代史的特点是巨大的环境变化和涌现的选择性媒介 [3]。大约 10,000 年前新石器时代开始时，人类培育植物和驯化动物导致人群密度的增长。人类面临着新的传染病、新的食物来源和新的文化环境的传播。因此，过去的 10,000 年成为了人类生物学史上最有意义的一段时间，也可能是许多重要的基因适应和疾病抵抗力出现的时间。

我们试图设计一种强大的方法来检测最近的选择。我们的方法主要基于等位基

on the relationship between an allele's frequency and the extent of linkage disequilibrium (LD) surrounding it. (LD often refers to association between two alleles. Here, we use it to measure the association between a single allele at one locus with multiple loci at various distances.) Under neutral evolution, new variants require a long time to reach high frequency in the population, and LD around the variants will decay substantially during this period owing to recombination[4,5]. As a result, common alleles will typically be old and will have only short-range LD. Rare alleles may be either young or old and thus may have long- or short-range LD. The key characteristic of positive selection, however, is that it causes an unusually rapid rise in allele frequency, occurring over a short enough time that recombination does not substantially break down the haplotype on which the selected mutation occurs. A signature of positive natural selection is thus an allele having unusually long-range LD given its population frequency. The decay of LD, and therefore the relative scale of "short"- and "long"-range LD, is dependent on local recombination rates. A general test for selection on the basis of these principles must therefore control for local variation in recombination rates.

We developed an experimental design to detect positive selection at a locus using the breakdown of LD as a clock for estimating the ages of alleles. We began by genotyping a collection of single nucleotide polymorphisms (SNPs) in a small "core region" to identify the "core haplotypes". We selected SNPs of sufficient density, so that recombination between them would be extremely rare and the core haplotypes could be explained in terms of a single gene genealogy (Supplementary Fig. 1). Zones of very low historical recombination were identified by looking for clusters of SNPs where Hudson's R_M was 0 and $|D'|$ was one[6,7] (see Supplementary Fig. 1).

We then added increasingly distant SNPs to study the decay of LD from each core haplotype. To visualize this process, we generated haplotype bifurcation diagrams that branch to reflect the creation of new, extended haplotypes by historical recombination proximal and distal to the core region. We measured LD at a distance x from the core region by calculating the extended haplotype homozygosity (EHH). EHH is defined as the probability that two randomly chosen chromosomes carrying the core haplotype of interest are identical by descent (as assayed by homozygosity at all SNPs[8]) for the entire interval from the core region to the point x. EHH thus detects the transmission of an extended haplotype without recombination. Our test for positive selection involves finding a core haplotype with a combination of high frequency and high EHH, as compared with other core haplotypes at the locus. An attractive aspect of this approach is that the various core haplotypes at a locus serve as internal controls for one another, adjusting for any unevenness in the local recombination rate.

We applied our approach to two genes that have been implicated in resistance to the malaria parasite *Plasmodium falciparum*. Glucose-6-phosphate dehydrogenase (*G6PD*) is a classical example of a gene where variants can confer malaria resistance[9]. Evidence over the past 40 years has shown that the common variant *G6PD*-202A confers partial protection against malaria, with a case-control study estimating a reduction in disease risk

因的频率与其附近连锁不平衡(LD)程度之间的关系。(LD 通常指两个等位基因间的联系。这里我们用它来测量一个基因座的单个等位基因与不同距离处的多个基因座等位基因之间的联系。)在中性进化下，新的变异型需要一段很长的时间才能达到人群中较高的频率，而且在此期间，变异型附近的 LD 会由于重组大幅度地衰减[4,5]。结果，常见等位基因通常是老的并只有短范围的 LD。很少的等位基因可能是年轻的也可能是老的，因此可能具有长或者短范围的 LD。但是正选择的关键特征是它能引起等位基因频率在足够短的时间内异常快速的增长，以至于重组并未能充分打破这些选择突变发生处的单倍型。因此一个正自然选择的标志就是一个等位基因在其人类频率下具有异常长范围的 LD。LD 的衰减以及因此相对的短和长范围 LD 规模，取决于局部的重组率。根据这个原理设计的选择试验必须控制重组率的局部变异。

我们开发了一种实验设计，利用 LD 的分解作为时钟来估算等位基因的年龄，以检测一个基因座的正选择。我们首先在一个小的“核心区域”对一组单核苷酸多态性(SNP)进行基因分型，以识别“核心单倍型”。我们选取具有足够密度的 SNP，这样它们之间的重组会非常少，而且核心单倍型可以用单个基因的谱系来表示(见补充图 1)。通过寻找那些赫德森 $R_M = 0$ 且 $|D'| = 1$ 的 SNP 簇来找出历史重组率非常低的区域[6,7](见补充图 1)。

然后我们加入逐渐增加距离的 SNP 研究每个核心单倍型 LD 的衰减。为了可视化这个过程，我们设计了单倍型分支图，用分支的方式反映核心区域近端和远端通过历史重组产生的新的扩展单倍型。我们通过计算扩展单倍型纯合性(EHH)来测量距离核心区域 x 的 LD。EHH 可以被定义为从核心区域到距离 x 的整个区间内，随机选择的两条携带感兴趣的核心单倍型的染色体在下降过程(通过所有 SNP 的纯合性来测定[8])中的相同概率。因此，EHH 能够检测不含有重组的扩展单倍型的传播。我们对正选择的检测包含了找到一个与该基因座的其他核心单倍型相比具有高频率和高 EHH 组合的核心单倍型。这种方法令人满意的方面在于一个基因座的多个核心单倍型可以作为相互之间的内部对照，对任何局部重组率上的不均衡进行调节。

我们将方法用于与疟疾寄生虫恶性疟原虫的抗药性有关的两个基因上。葡萄糖–6–磷酸脱氢酶(*G6PD*)是一个基因变异型能够产生疟疾抗性的经典例子[9]。过去 40 年的证据表明常见变异型 *G6PD*-202A 对疟疾具有部分抗性，其中一项病例对照研究估计可将疾病风险降低约 50%(文献 1)。CD40 配体基因(*TNFSF5*)编码一种对

of about 50% (ref. 1). The CD40 ligand gene (*TNFSF5*) encodes a protein with a critical role in immune response to infectious agents. One case-control study suggested that a common variant in the promoter region, *TNFSF5*-726C, is associated with a similar degree of protection against malaria[2].

We first studied *G6PD* (Fig. 1). We defined a core region of 15 kilobases (kb) at *G6PD* and genotyped 11 SNPs in 3 African and 2 non-African populations. The SNPs defined 9 core haplotypes (Table 1a) (denoted *G6PD*-CH1 to 9, for core haplotypes 1 to 9). The *G6PD*-202A allele, which has been associated with protection from malaria, was carried on only one core haplotype, *G6PD*-CH8. Notably, *G6PD*-CH8 is common in Africa (18%), where malaria is endemic, but is absent outside of Africa. For carrying out our test for selection, we focused on the three African populations, which did not differ significantly with respect to core haplotype frequencies (by Fisher's exact test[10]) and hence were pooled for the main analysis. (Analyses were also performed separately for each population, yielding qualitatively similar results; see below.)

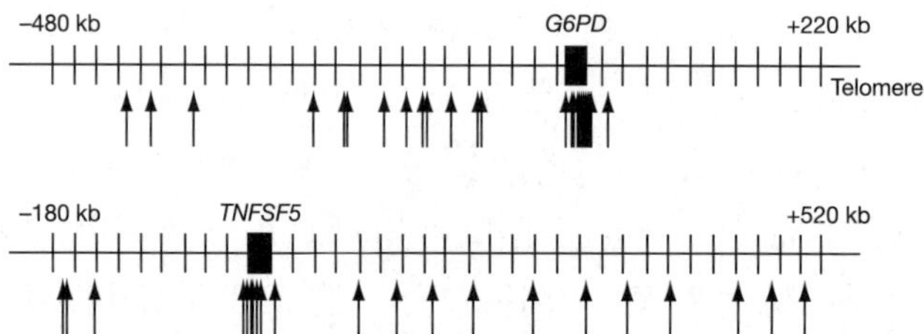

Fig. 1. Experimental design of core and long-range SNPs for *G6PD* and *TNFSF5*. The core region is highlighted by a cluster of densely spaced SNPs (arrows) at the gene. Additional, widely separated flanking SNPs, used to examine the decay of LD from each core haplotype, are also shown. Markers distal to *G6PD* were within repetitive subtelomeric sequence and could not be genotyped.

Table 1. Core haplotype frequencies in six populations

(**a**) *G6PD*	Core SNP alleles (kb)											Core haplotype frequencies in six populations						
Core haplotype	−10	−2	−2	−1	0	1	1	3	3	4	4	Total	Beni	Yoruba	Shona	African American	European American	Asian
1	C	G	C	G	G	A	C	C	G	C	C	0.13 (54)	0.15 (9)	0.21 (18)	0.13 (11)	0.13 (12)	0.03 (2)	0.07 (2)
2	–	–	–	–	–	–	–	–	–	–	T	0.16 (67)	0 (0)	0 (0)	0 (0)	0.07 (6)	0.57 (37)	0.80 (24)
3	–	–	–	–	–	–	–	–	–	T	–	0.25 (102)	0.23 (14)	0.24 (21)	0.45 (37)	0.19 (17)	0.17 (11)	0.07 (2)
4	–	–	–	–	–	–	–	T	–	–	–	0.01 (5)	0.02 (1)	0 (0)	0.04 (3)	0.01 (1)	0 (0)	0 (0)
5	–	–	–	–	–	–	–	T	A	–	–	0.10 (41)	0.22 (13)	0.11 (10)	0.06 (5)	0.14 (13)	0 (0)	0 (0)
6	–	–	–	–	–	G	–	–	–	–	–	0.12 (48)	0.17 (10)	0.10 (9)	0.11 (9)	0.22 (20)	0 (0)	0 (0)
7	–	–	T	A	–	G	G	–	–	–	–	0.04 (17)	0.05 (3)	0.07 (6)	0.06 (5)	0.03 (3)	0 (0)	0 (0)
8	–	A*	T	A	A	G	G	–	–	–	–	0.13 (53)	0.17 (10)	0.23 (20)	0.13 (11)	0.13 (12)	0 (0)	0 (0)
9	T	–	–	–	–	–	–	–	–	–	T	0.07 (28)	0 (0)	0.03 (3)	0.02 (2)	0.07 (6)	0.23 (15)	0.07 (2)
											N	415	60	87	83	90	65	30

感染性病原体的免疫应答起关键作用的蛋白。一个病例对照研究表明启动子区域的一个常见变异型 *TNFSF5*-726C 对疟疾具有类似程度的抗性[2]。

我们首先研究了 *G6PD*(图 1)。在 *G6PD* 上定义了一个 15 千碱基(kb)的核心区域，并在 3 个非洲人群和 2 个非非洲人群中对 11 个 SNP 进行了基因分型。这些 SNP 定义了 9 个核心单倍型(表 1a)(分别表示为 *G6PD*-CH1 到 *G6PD*-CH9，指代核心单倍型 1 到 9)。只有一个核心单倍型(*G6PD*-CH8)携带了 *G6PD*-202A 等位基因，该等位基因与疟疾抗病性相关。显然，*G6PD*-CH8 在疟疾流行的非洲很常见(18%)，而在非洲以外地区不存在。为了实施选择测试，我们将重点放在三个非洲人群上，他们在核心单倍型频率上没有显著差异(通过 Fisher 精确测试[10])，因此被集中在一起进行主要分析。(对每个人群也分别进行了分析，得到了定性上相似的结果；见下文)。

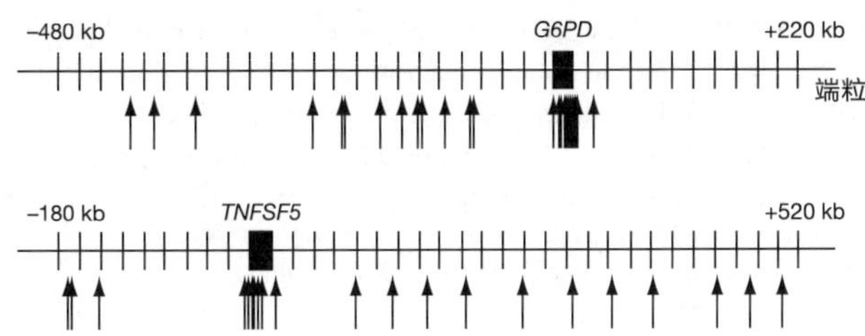

图 1. *G6PD* 和 *TNFSF5* 的核心 SNP 和长范围 SNP 的实验设计。基因的核心区域被一群密集排列的 SNP(箭头)突出显示。此外，图中还显示了高度分离的侧翼 SNP，用于检验每个核心单倍型的 LD 衰减。*G6PD* 远端的标记位于重复的亚端粒序列中，不能进行基因分型。

表 1. 六个群体中的核心单倍型频率

(a) *G6PD* 核心单倍型	核心 SNP 等位基因(kb)											六个群体中的核心单倍型频率						
	−10	−2	−2	−1	0	1	1	3	3	4	4	总计	贝尼人	约鲁巴人	绍纳人	非裔美国人	欧裔美国人	亚洲人
1	C	G	C	G	G	A	C	C	G	C	C	0.13(54)	0.15(9)	0.21(18)	0.13(11)	0.13(12)	0.03(2)	0.07(2)
2	–	–	–	–	–	–	–	–	–	–	T	0.16(67)	0(0)	0(0)	0(0)	0.07(6)	0.57(37)	0.80(24)
3	–	–	–	–	–	–	–	–	–	T	–	0.25(102)	0.23(14)	0.24(21)	0.45(37)	0.19(17)	0.17(11)	0.07(2)
4	–	–	–	–	–	–	–	T	–	–	–	0.01(5)	0.02(1)	0(0)	0.04(3)	0.01(1)	0(0)	0(0)
5	–	–	–	–	–	–	–	T	A	–	–	0.10(41)	0.22(13)	0.11(10)	0.06(5)	0.14(13)	0(0)	0(0)
6	–	–	–	–	–	G	–	–	–	–	–	0.12(48)	0.17(10)	0.10(9)	0.11(9)	0.22(20)	0(0)	0(0)
7	–	–	T	A	–	G	G	–	–	–	–	0.04(17)	0.05(3)	0.07(6)	0.06(5)	0.03(3)	0(0)	0(0)
8	–	A*	T	A	A	G	G	–	–	–	–	0.13(53)	0.17(10)	0.23(20)	0.13(11)	0.13(12)	0(0)	0(0)
9	T	–	–	–	–	–	–	–	–	–	T	0.07(28)	0(0)	0.03(3)	0.02(2)	0.07(6)	0.23(15)	0.07(2)
											N	415	60	87	83	90	65	30

Continued

(b) *TNFSF5* Core haplotype	Core SNP alleles (kb)					Core haplotype frequencies in six populations						
	−6	0	1	3	4	Total	Beni	Yoruba	Shona	African American	European American	Asian
1	T	T	T	T	G	0.03 (12)	0.06 (4)	0.03 (3)	0.02 (2)	0.04 (3)	0 (0)	0 (0)
2	C	–	–	–	–	0.50 (200)	0.32 (20)	0.38 (33)	0.46 (38)	0.40 (32)	0.78 (49)	1.00 (28)
3	C	–	–	C	–	0.08 (32)	0.10 (6)	0.12 (10)	0.06 (5)	0.14 (11)	0 (0)	0 (0)
4	–	C*	–	–	–	0.25 (100)	0.38 (24)	0.38 (33)	0.26 (21)	0.27 (22)	0 (0)	0 (0)
5	–	–	C	–	–	0.12 (47)	0.14 (9)	0.07 (6)	0.18 (15)	0.14 (11)	0.10 (6)	0 (0)
6	–	–	C	–	A	0.02 (10)	0 (0)	0 (0)	0.01 (1)	0.01 (1)	0.13 (8)	0 (0)
7	–	C*	C	–	A	0 (1)	0 (0)	0.01 (1)	0 (0)	0 (0)	0 (0)	0 (0)
8	C	–	C	–	–	0 (1)	0 (0)	0 (0)	0 (0)	0.01 (1)	0 (0)	0 (0)
					N	403	63	86	82	81	63	28

Observed core haplotypes at *G6PD* and *TNFSF5* in six populations of African, European and Asian descent. Relative distances of core SNP alleles from the putative malaria resistance mutations are given in kb. Frequencies for haplotypes (and numbers of observations) are given for all populations. There are no apparent recombinants among the *G6PD* core haplotypes, and R_M is 0. There are 2 recombinant haplotypes among the 403 *TNFSF5* chromosomes, and R_M would also be 0 if the 2 haplotypes appearing only once were removed from the analysis[6].
*Both proposed mutations associated with malaria resistance (*G6PD*-202A and *TNFSF5*-726C) are observed only in Africans and occur on *G6PD*-CH8 and on *TNFSF5*-CH4 and *TNFSF5*-CH7, respectively.

G6PD-CH8 demonstrates clear long-range LD (as seen by the predominance of one thick branch in the haplotype bifurcation diagram (Fig. 2a)) and has correspondingly high EHH. The EHH is 0.38 at the largest distance tested; that is, 413 kb (Fig. 2c). For each core haplotype we calculated the relative EHH; specifically, the factor by which EHH decays on the tested core haplotype compared with the decay of EHH on all other core haplotypes combined (Methods).

To test formally for selection, for each core haplotype, we compared the allele frequency to the relative EHH at various distances (Fig. 2e shows the comparison at 413 kb proximal to *G6PD*). *G6PD*-CH8 has a much higher relative EHH than other haplotypes of comparable frequency, but is this statistically significant? To obtain a sense of how unusual our observation is, we simulated haplotypes using a coalescent process (see Methods and Fig. 2e)[11]. The deviation from the simulation results is highly significant and becomes progressively more marked with increasing distance (Fig. 2g) (*P*-values at 413 kb proximal are: constant-sized population, $P < 0.0008$; expansion, $P < 0.0006$; bottleneck, $P < 0.0008$; population structure, $P < 0.0008$; see Methods and Supplementary Table 2 for details of the demographic models we considered). The frequency and LD properties of *G6PD*-CH8 are incompatible with what is expected under a model of neutral evolution for a wide range of demographies. Furthermore, when the three African populations comprising the pooled sample were considered separately, a signal of selection was identified independently in each population (Yoruba, $P < 0.0012$; Beni, $P < 0.0440$; and Shona, $P < 0.0030$, based on

续表

(b) TNFSF5核心单倍型	核心 SNP 等位基因(kb)					六个群体中的核心单倍型频率						
	−6	0	1	3	4	总计	贝尼人	约鲁巴人	绍纳人	非裔美国人	欧裔美国人	亚洲人
1	T	T	T	T	G	0.03(12)	0.06(4)	0.03(3)	0.02(2)	0.04(3)	0(0)	0(0)
2	C	–	–	–	–	0.50(200)	0.32(20)	0.38(33)	0.46(38)	0.40(32)	0.78(49)	1.00(28)
3	C	–	–	C	–	0.08(32)	0.10(6)	0.12(10)	0.06(5)	0.14(11)	0(0)	0(0)
4	–	C*	–	–	–	0.25(100)	0.38(24)	0.38(33)	0.26(21)	0.27(22)	0(0)	0(0)
5	–	–	C	–	–	0.12(47)	0.14(9)	0.07(6)	0.18(15)	0.14(11)	0.10(6)	0(0)
6	–	–	C	–	A	0.02(10)	0(0)	0(0)	0.01(1)	0.01(1)	0.13(8)	0(0)
7	–	C*	C	–	A	0(1)	0(0)	0.01(1)	0(0)	0(0)	0(0)	0(0)
8	C	–	C	–	–	0(1)	0(0)	0(0)	0(0)	0.01(1)	0(0)	0(0)
					N	403	63	86	82	81	63	28

在非洲、欧洲和亚洲后代的 6 个群体中观察到 *G6PD* 和 *TNFSF5* 的核心单倍型。核心 SNP 等位基因与假定的疟疾抗性突变之间的相对距离用 kb 表示。所有群体的单倍型频率(以及观察次数)也显示出来。*G6PD* 核心单倍型中没有明显的重组，R_M 是 0。在 403 个 *TNFSF5* 染色体中有两个重组的单倍型，但是如果将仅出现一次的这两个单倍型从分析中去除，R_M 仍然是 0[6]。

* 两种与疟疾抗性相关的假定突变(*G6PD*-202A 和 *TNFSF5*-726C)仅在非洲人中观察到，分别出现在 *G6PD*-CH8、*TNFSF5*-CH4 和 *TNFSF5*-CH7 中。

G6PD-CH8 具有清晰的长范围 LD(在单倍型分支图中显示为明显的一个粗分支优势型(图 2a))和相应的高 EHH。EHH 在最大的检测距离下是 0.38，也就是 413 kb (图 2c)。对每一个核心单倍型，我们计算了相对的 EHH。具体来说，就是将检测的核心单倍型中 EHH 衰减的因素与所有其他核心单倍型组合起来的 EHH 衰减进行了对比(见方法)。

为了正式地检测选择，对每一个核心单倍型在不同距离下的等位基因频率与相对 EHH 进行了比较(图 2e 显示了 *G6PD* 近端 413 kb 的对比)。*G6PD*-CH8 与其他具有类似频率的单倍型相比有更高的相对 EHH，但其是否有统计学显著意义？为了了解我们的发现是多么得不寻常，我们用联合的过程模拟了单倍型(见方法和图 2e)[11]。模拟结果得出的偏离是高度显著的，并且随着距离的增加变得越来越显著(图 2g)(近端 413 kb 的 P 值：数量稳定的人群 $P < 0.0008$；扩张的人群 $P < 0.0006$；瓶颈效应 $P < 0.0008$；人群结构 $P < 0.0008$；我们使用的人口统计学模型的详细信息见方法和补充表 2)。*G6PD*-CH8 的频率和 LD 特性与广泛人口分布中的中性进化模型的预期不一致。此外，当把组成融合样本的三个非洲群体分开考虑时，每个群体中各自识别出了一个选择信号(约鲁巴人 $P < 0.0012$；贝尼人 $P < 0.0440$；绍纳人 $P < 0.0030$，基于数量稳定的群体的模型)，证明选择信号不是将三个群体样本融合

simulation of a constant-sized population), demonstrating that the signal of selection is not an artefact of pooling the three population samples.

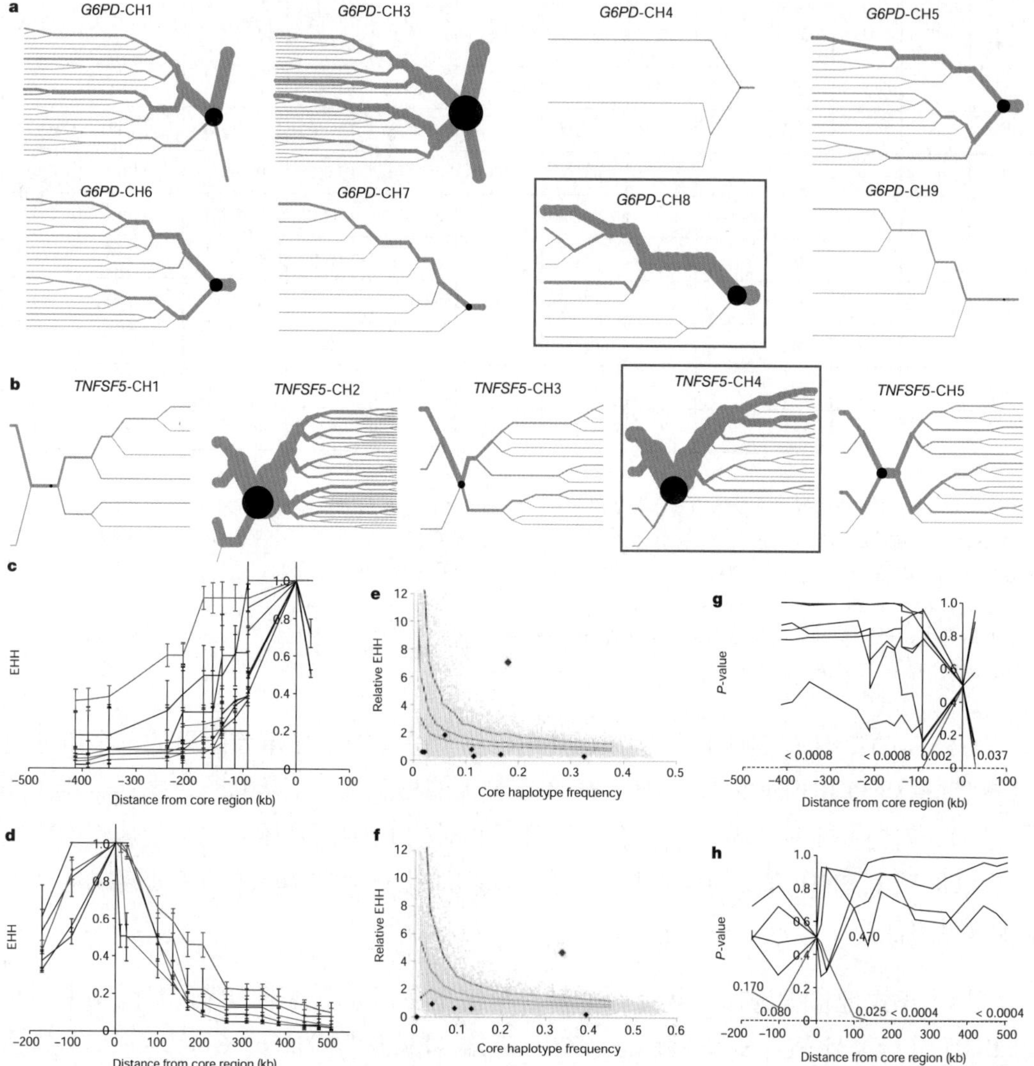

Fig. 2. Core haplotype frequency and relative EHH of *G6PD* and *TNFSF5*. **a**, **b**, Haplotype bifurcation diagrams (see Methods) for each core haplotype at *G6PD* (**a**) and *TNFSF5* (**b**) in pooled African populations demonstrate that *G6PD*-CH8 and *TNFSF5*-CH4 (boxed or labelled in red) have long-range homozygosity that is unusual given their frequency. **c**, **d**, The EHH at varying distances from the core region on each core haplotype at *G6PD* (**c**) and *TNFSF5* (**d**) demonstrates that *G6PD*-CH8 and *TNFSF5*-CH4 have persistent, high EHH values. **e**, **f**, At the most distant SNP from *G6PD* (**e**) and *TNFSF5* (**f**) core regions, the relative EHH plotted against the core haplotype frequency is presented and compared with the distribution of simulated core haplotypes (on the basis of simulation of 5,000 data sets; represented by grey dots and given with 95th, 75th and 50th percentiles). The observed non-selected core haplotypes in our data are represented by black diamonds. **g**, **h**, We calculated the statistical significance of the departure of the observed data from the simulated distribution at each distance from the core. *G6PD*-CH8 (**g**) and *TNFSF5*-CH4 (**h**) demonstrate increasing deviation from a model of neutral drift at further distances from the core region in both directions.

在一起的产物。

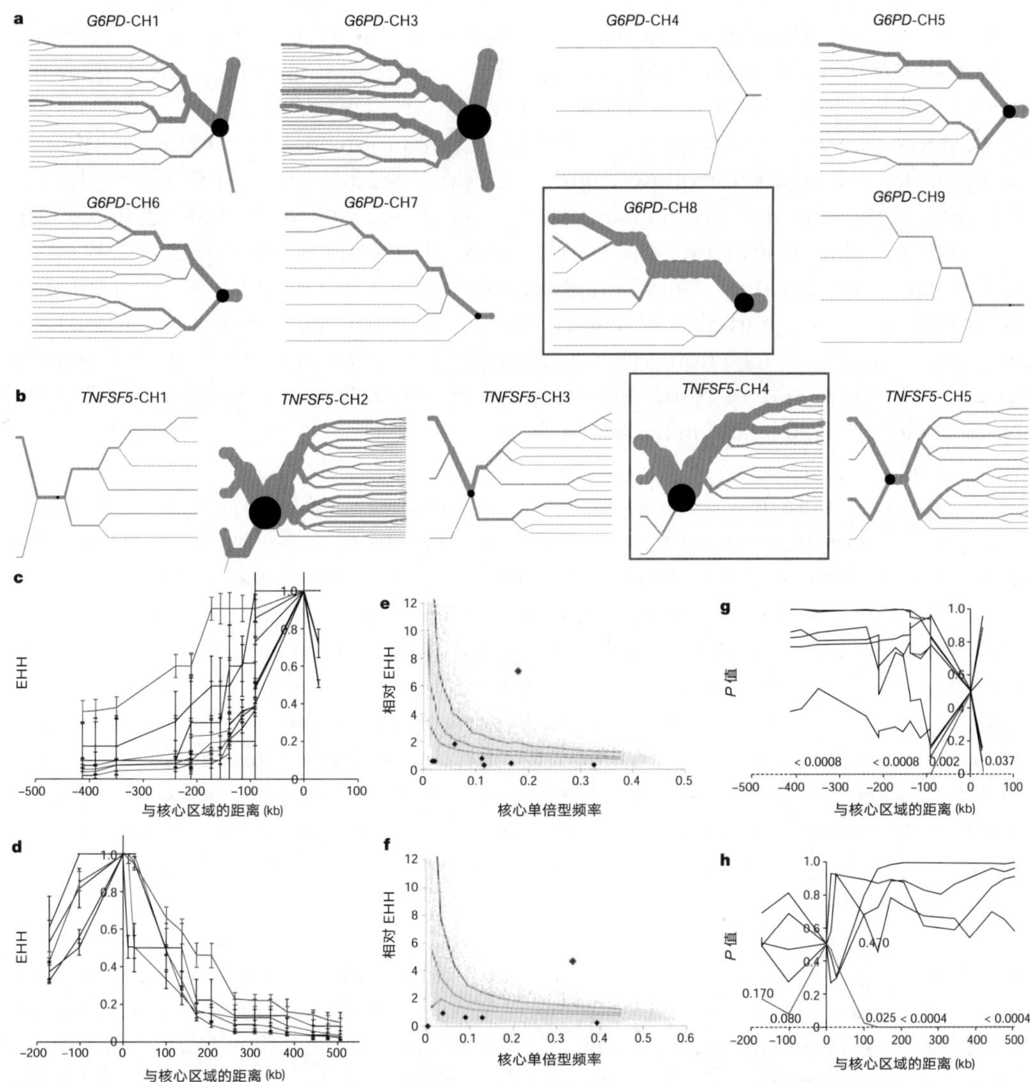

图 2. *G6PD* 和 *TNFSF5* 的核心单倍型频率和相对 EHH。**a**、**b**，在总体非洲人群中，*G6PD*(**a**) 和 *TNFSF5*(**b**) 的每个核心单倍型的分支图（见方法）表明，*G6PD*-CH8 和 *TNFSF5*-CH4（红色框）具有长范围的纯和性，这在其频率下是不寻常的。**c**、**d**，*G6PD*(**c**) 和 *TNFSF5*(**d**) 的每个核心单倍型中核心区域不同距离的 EHH 显示了 *G6PD*-CH8 和 *TNFSF5*-CH4 具有持续高水平的 EHH 值。**e**、**f**，*G6PD*(**e**) 和 *TNFSF5*(**f**) 核心区域最远处的 SNP，其相对 EHH 与核心单倍型的频率进行作图，并与模拟的核心单倍型分布图进行对比（根据 5,000 个数据点的模拟；以灰点表示，并给出第 95、75 和 50 个百分位数）。我们的数据中观察到的非选择的核心单倍型用黑色的菱形表示。**g**、**h**，我们观察到了在离核心单倍型每一距离处模拟分布的数据，并计算了这些数据偏差的显著性差异。*G6PD*-CH8(**g**) 和 *TNFSF5*-CH4(**h**) 显示出在距离核心区域更远的两个方向上，偏离中性漂移模型的程度更大。

We next applied our approach to the CD40 ligand gene (Fig. 1). We defined a core region of 10 kb and genotyped 5 SNPs. The SNPs defined seven core haplotypes (Table 1b). The *TNFSF5*-726C allele, which has been associated with protection from malaria, was present on *TNFSF5*-CH4, which is common in Africa (34%), but is absent outside of Africa. *TNFSF5*-CH4 demonstrates high LD as seen in the haplotype bifurcation diagrams (Fig. 2b) and has high EHH at long distances (Fig. 2d). *TNFSF5*-CH4 is a clear outlier when compared with other haplotypes (Fig. 2f), and its frequency and LD properties are incompatible with neutral evolution under multiple demographic models (*P*-values at 506 kb distal are: constant-sized population, $P < 0.0012$; expansion, $P < 0.0008$; bottleneck, $P < 0.0012$; population structure, $P < 0.0008$; see Methods and Supplementary Table 2 for details). Again, the *P*-value is increasingly significant at further distances from the core region both proximally and distally (Fig. 2h). When each African population was analysed separately, a signal of selection was significant (Yoruba, $P < 0.0008$; Beni, $P < 0.0023$; Shona, $P < 0.0242$). These results thus provide independent evidence supporting the proposed role of CD40 ligand in malaria resistance[2].

We tested our conclusion of positive selection by performing a similar analysis on 17 randomly chosen control regions across the human genome in the same African populations. We only used data from each control if it was closely matched to our data in terms of the number of chromosomes studied and the homozygosity at the core haplotype and at long distances from the core (Fig. 3). *G6PD*-CH8 and *TNFSF5*-CH4 clearly stand out from the other loci, showing that the *P*-values determined by simulation are also supported by direct, empirical comparison. In measuring *P*-values for the controls where there is no prior hypothesis of selection, a Bonferonni correction for multiple-hypothesis testing was applied. Notably, one core haplotype, from the monocyte chemotactic protein 1 region, shows frequency and LD properties similar to *G6PD* and *TNFSF5*, although this nominally significant result may be simply a false-positive owing to the large number of hypotheses examined.

We used a linkage-disequilibrium-based technique[12] to estimate dates of origin of the two resistance variants. The estimates were about 2,500 years for *G6PD* and about 6,500 years for *TNFSF5* (see Supplementary Information for details). The date for *G6PD* is consistent with a recent independent age estimate for *G6PD*-202A based primarily on microsatellite data[13].

Finally, we explored whether positive selection could have been detected with traditional tests (Supplementary Table 3)[14]. We performed Tajima's *D*-test[15], Fu and Li's *D*-test[16], Fay and Wu's *H*-test[17], the Ka/Ks test[18], the McDonald and Kreitman test[19], and the Hudson–Kreitman–Aguadè (HKA) test[20]. None showed significant deviation from neutral evolution for either *G6PD* or *TNFSF5*, consistent with their low power to detect recent selection.

接下来我们将方法应用到 CD40 配体基因中（图 1）。我们定义了一个 10 kb 的核心区域，对 5 个 SNP 进行基因分型。这些 SNP 确定了 7 个核心单倍型（表 1b）。与疟疾抗性相关的 *TNFSF5*-726C 等位基因存在于 *TNFSF5*-CH4 上，后者在非洲很常见（34%），但是在非洲以外地区不存在。*TNFSF5*-CH4 在单倍型分支图中显示高 LD（图 2b），在长距离时显示高 EHH（图 2d）。*TNFSF5*-CH4 与其他单倍型相比是一个明确的离群值（图 2f），其频率和 LD 特性在多种人口统计学模型下都与中性进化相矛盾（506 kb 远端的 P 值是：数量稳定的人群 $P < 0.0012$；扩张的人群 $P < 0.0008$；瓶颈效应 $P < 0.0012$；群体结构 $P < 0.0008$；细节见方法及附表 2）。同样，P 值显著性随着距核心区域的距离增加而增加，无论是近端还是远端（图 2h）。当分别分析非洲群体时，选择的信号都有显著性意义（约鲁巴人 $P < 0.0008$；贝尼人 $P < 0.0023$；绍纳人 $P < 0.0242$）。因此这些结果为支持 CD40 配体在疟疾抗性中的作用提供了独立证据[2]。

我们通过在相同的非洲人群中对人类基因组中的 17 个随机选取的对照区域进行类似的分析，验证了我们关于正选择的结论。只有当对照区域在染色体数目、核心单倍型所在位置以及距离核心区较远处的纯合性方面与我们的数据密切匹配时，我们才使用其数据（图 3）。*G6PD*-CH8 和 *TNFSF5*-CH4 与其他基因座明显不同，表明通过模拟得出的 P 值也得到了直接的、实证比较分析的支持。在没有关于选择的优先假设的情况下，我们测量了对照 P 值，并采用邦费罗尼校正方法进行多假设检验。显然，来自单核细胞趋化蛋白 1 区域的一个核心单倍型显示了类似于 *G6PD* 和 *TNFSF5* 的频率和 LD 特征，然而这一表面上显著的结果可能仅仅是由于大量的假设被检验，而呈现的假阳性。

我们使用一种基于连锁不平衡的技术[12]估计这两个抗性变异体的起源日期。*G6PD* 估计大约 2,500 年前，*TNFSF5* 大约 6,500 年前（详情见补充信息）。*G6PD* 的起源日期与最近主要基于微卫星数据预估的 *G6PD*-202A 的独立年龄相一致[13]。

最后，我们探讨了正选择是否能够用传统的方法检测到（补充表 3）[14]。我们进行了 Tajima 的 D 检验[15]，Fu 和 Li 的 D 检验[16]，Fay 和 Wu 的 H 检验[17]，Ka/Ks 检验[18]，McDonald 和 Kreitman 检验[19]以及 Hudson-Kreitman-Aguadè（HKA）检验[20]。无论是 *G6PD* 还是 *TNFSF5*，均未表现出明显的中性进化偏差，这与它们不能检测到最近的选择相一致。

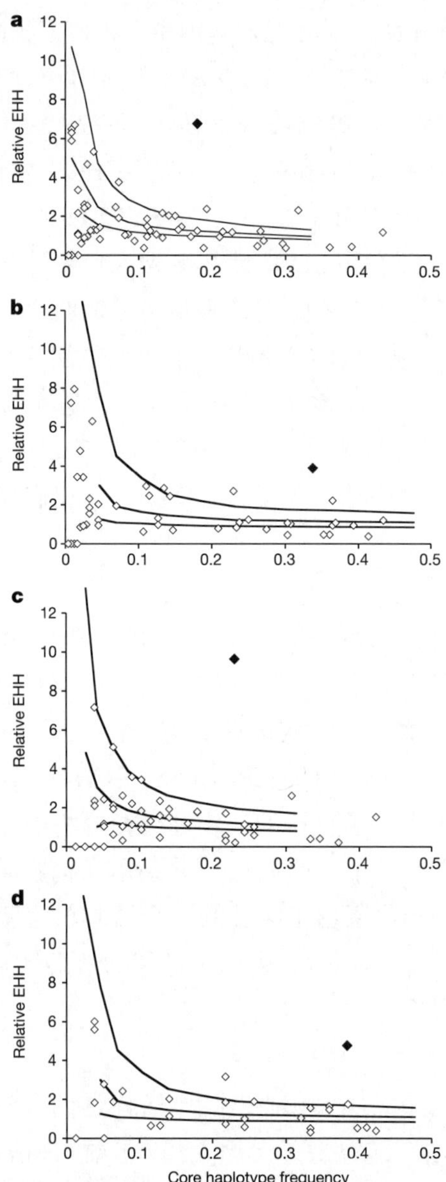

Fig. 3. Control regions: core haplotype frequency against relative EHH. To provide an empirical, non-simulation-based evaluation of the signal of selection, we compared the frequency and relative EHHs for *G6PD* (**a**) and *TNFSF5* (**b**) with patterns observed in randomly chosen genes in the genome (see Methods). **c**, **d**, We performed the entire analysis again on *G6PD* (**c**) and *TNFSF5* (**d**) for a subset of 78 Yoruban haplotypes (using family trios where phase could be determined). We were able to match 30 to 87 core haplotypes (indicated by outlined diamonds) from the control regions to our data. The 95th, 75th and 50th percentiles for simulated data are also shown. *G6PD*-CH8 and *TNFSF5*-CH4 (indicated by black diamonds) clearly stand out from the pattern seen at other loci in the genome, suggesting a true signal of selection.

Our approach, which we refer to as the long-range haplotype (LRH) test, provides a way to detect recent positive selection by analysing haplotype structure in random individuals from a population. How far back in human history can one detect positive selection?

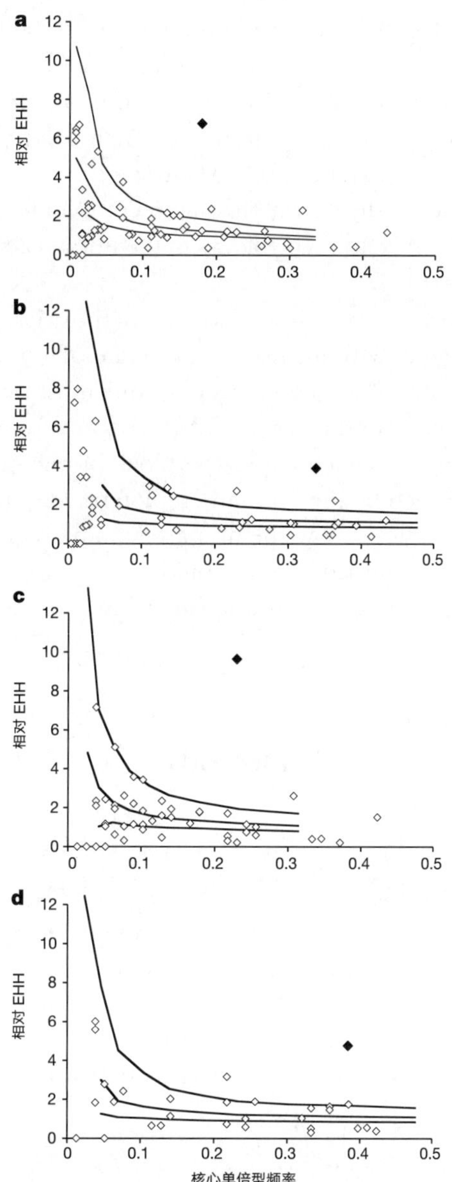

图 3. 对照区域：核心单倍型频率与相对 EHH。为了提供实验性、非模拟性的选择信号评估，我们将 *G6PD* (**a**) 和 *TNFSF5* (**b**) 的频率和相对 EHH 与基因组中随机选择的基因中观察到的模式进行了比较（见方法）。**c**、**d**，我们再次对 78 个约鲁巴人的单倍型子集的 *G6PD*(**c**) 和 *TNFSF5*(**d**) 进行了完整的分析（使用能够确定阶段的三体家系）。我们能够将对照区域的 30 到 87 个核心单倍型（以空心的菱形表示）与我们的数据匹配起来。图中也显示了模拟数据的第 95、75 和 50 个百分位数。*G6PD*-CH8 和 *TNFSF5*-CH4（以黑色菱形表示）与基因组中其他基因座所表现的形式完全不同，表明存在真正的选择信号。

我们称为长范围单倍型（LRH）检测的方法，提供了一种通过分析群体中随机个体的单倍型结构来检测近期正选择的方法。在人类历史上向前追溯多久才能检测到正选择呢？发生在不到 400 代之前的选择性事件（假设 25 年一代的话就是 10,000

Selective events occurring less than 400 generations ago (10,000 years assuming 25 years per generation) should leave a clear imprint at distances of over 0.25 centiMorgans (cM). The signal of such long-range LD should be distinguishable from the background extent of LD for common haplotypes in the genome[21], which are typically tens of thousands of generations old[4] and hence extend 0.02 cM or less. Over many tens of thousands of years, the signal of selection will become lost as recombination whittles the long-range haplotypes to the typical size of haplotype blocks in the human genome[21].

The LRH test can be used to search for evidence of positive selection by testing each common haplotype in a gene, without prior knowledge of a specific variant or selective advantage. Once the signature of selection is found, one must then decipher its cause. The LRH test could be applied to scan the entire human genome for evidence of recent positive selection simply by applying it to each haplotype block in reference data sets from human populations, as will be collected by the Human Haplotype Map project[21]. In this fashion, it should be possible to shed light on how the human genome was shaped by recent changes in culture and environment. The LRH test should also be useful for studying selection in other organisms, including domestic animals and parasites such as the malaria parasite *P. falciparum*[22].

Methods

Human subjects

DNA samples from 252 males from Africa were used in the study: 92 Yoruba and 73 Beni from Nigeria, and 87 Shona from Zimbabwe. Additional DNA samples from 29 Yoruban trios (father–mother–child clusters) were genotyped at the 17 control regions. The Yoruba and Shona males were healthy individuals obtained as part of the International Collaborative Study of Hypertension in Blacks. The Beni samples were from civil servants in Benin City. Samples from four non-African populations and four primates were also used (Supplementary Information).

SNP genotyping

We genotyped 49 SNPs distributed around *G6PD* and 37 SNPs distributed around *TNFSF5* using mass spectrometry (Sequenom)[23]. The SNPs were identified by our own resequencing and through previous discovery efforts[2,24,25]. For *G6PD*, we focused on genotyping SNPs proximal to the gene, as SNPs in the repetitive subtelomeric sequence distal to *G6PD* could not be genotyped. A total of 25 SNPs around *G6PD* and 21 SNPs around *TNFSF5* were successfully genotyped and used in analysis (see Supplementary Information for details).

年）应该在超过 0.25 厘摩（cM）的距离上留下明显的印记。这种长范围的 LD 信号应该能够和基因组中普通单倍型的背景长度的 LD 区别开来[21]，因为后者通常有几万代的历史[4]，因此延伸的长度不超过 0.02 cM。经过数万年的时间，随着重组的发生，长范围的单倍型就会变成人类基因组中典型大小的单倍型区域[21]，选择的信号会逐渐丢失。

LRH 检测可以通过分析基因中的每个常见单倍型来寻找正选择的证据，而无须提前知道特定的变异型或者选择优势。一旦发现了正选择的特征，就必须进一步研究其原因。LRH 检测可用于扫描整个人类基因组，仅仅通过将其用于人类群体中参考数据集内的每个单倍型区域，来寻找近期正选择的证据。这些数据信息将收集在人类单倍型图谱计划中[21]。通过这种方式，我们有可能弄清楚人类基因组是如何被近期的文化和环境变化塑造的。LRH 检测对研究其他生物的选择也有效，包括家畜和寄生虫（比如恶性疟原虫[22]）。

方　法

人群对象

研究中使用了 252 名非洲男性的 DNA 样本：来自尼日利亚的 92 名约鲁巴人、73 名贝尼人，来自津巴布韦的 87 名绍纳人。在 17 个对照区对来源于 29 个约鲁巴三体家系（父亲–母亲–孩子集群）的额外 DNA 样本进行了基因分型。约鲁巴人和绍纳人是健康的男性，是黑人高血压国际合作研究项目的对象。贝尼人样本来自贝宁城的公务员。还使用了 4 个非非洲人群和 4 个灵长类动物的样本（补充资料）。

SNP 基因分型

我们使用质谱法（Sequenom）[23] 对 *G6PD* 附近的 49 个 SNP 以及 *TNFSF5* 附近的 37 个 SNP 进行了基因分型。这些 SNP 是通过我们自己的重新测序以及先前的发现相结合鉴定出来的[2,24,25]。对于 *G6PD*，我们主要着眼于对基因近端的 SNP 进行基因分型，因为其远端的重复性亚端粒序列中的 SNP 无法基因分型。总共有 *G6PD* 附近的 25 个 SNP 和 *TNFSF5* 附近的 21 个 SNP 成功基因分型并用于分析（详细内容见补充资料）。

Haplotype bifurcation diagrams

To visualize the breakdown of LD on core haplotypes, we created bifurcation diagrams using MATLAB. The root of each diagram is a core haplotype, identified by a black circle. The diagram is bi-directional, portraying both proximal and distal LD. Moving in one direction, each marker is an opportunity for a node; the diagram either divides or not based on whether both or only one allele is present. Thus the breakdown of LD on the core haplotype background is portrayed at progressively longer distances. The thickness of the lines corresponds to the number of samples with the indicated long-distance haplotype.

Extended haplotype homozygosity and relative EHH

EHH at a distance x from the core region is defined as the probability that two randomly chosen chromosomes carrying a tested core haplotype are homozygous at all SNPs[8] for the entire interval from the core region to the distance x. EHH is on a scale of 0 (no homozygosity, all extended haplotypes are different) to 1 (complete homozygosity, all extended haplotypes are the same). Relative EHH is the ratio of the EHH on the tested core haplotype compared with the EHH of the grouped set of core haplotypes at the region not including the core haplotype tested. Relative EHH is therefore on a scale of 0 to infinity.

Coalescent simulations

We used a computer program by Hudson that simulates gene history with recombination[11]. The program was modified to generate data such as we collected. We simulated a long region of DNA (1.3 cM), with one end defined as the "core". We progressively added SNPs at the core until they matched our data (for the *G6PD* or *TNFSF5* core) to within ± 12.5% in terms of the homozygosity. To mimic the SNP selection strategy used by The SNP consortium[25], which was the source of most of the SNPs in our study, we only included simulated SNPs in our analysis if different alleles were observed at two randomly chosen chromosomes from the sample. At longer distances, we added additional SNPs, only choosing SNPs for analysis that matched our data in terms of frequency (within a ± 12.5% window) and also broke down EHH to the same extent as was observed in our data (within ± 12.5%).

We repeated the simulations for 5,000 data sets (each producing typically 6–8 core haplotypes) to generate many data points with which to compare our data. *P*-values were obtained by first binning the simulated data by core haplotype frequency into 30 bins of equal size, each containing about 1,000 data points. We then ranked an observed core haplotype's relative EHH compared with that of all simulated data points within the bin containing haplotypes of the same frequency—the rank determines the *P*-value. For simulations of additional demographic histories, we considered two models of expansion, an extreme bottleneck, and a highly structured population. For expansions, we simulated a population that was constant at size 10,000 until 200 or 5,000 generations ago, when it

单倍型分支图

为了可视化核心单倍型中 LD 的分解，我们使用 MATLAB 创建了分支图。每个图标的根部是核心单倍型，以黑圈表示。图表是双向的，表示近端和远端的 LD。向一个方向移动的话，每个标记都是一个形成节点的机会；图标分支与否取决于存在的等位基因是一个还是两个。因此核心单倍型背景中的 LD 的分解就表示为进行性变长的距离。线条的厚度对应于含有表示的长距离单倍型的样本的数量。

扩展单倍型纯合性和相对 EHH

距离核心区域 x 的 EHH 被定义为从核心区域到距离 x 的整个区间内，两个随机选择的携带被测的核心单倍型的染色体在所有的 SNP 中是纯合的可能性 [8]。EHH 的范围在 0(没有纯合性，所有的扩展单倍型都是不同的)到 1(完全纯合性，所有扩展单倍型都是一样的)之间。相对 EHH 是被测的核心单倍型的 EHH 与不含有被测单倍型的区域内部的核心单倍型集群的 EHH 的比值，因此相对 EHH 的范围是从 0 到无穷大。

联合模拟

我们使用赫德森的计算机程序，通过重组模拟基因历史 [11]。该程序经过修改以生成数据，比如我们之前收集的数据。我们模拟了一个长 DNA 区域(1.3 cM)，其一端定义为“核心”。我们逐渐增加核心部位的 SNP，直到它们匹配我们的数据(*G6PD* 或者 *TNFSF5* 核心)使得纯合性偏差在±12.5% 之内。为了模拟 SNP 联盟使用的 SNP 选择策略 [25](我们研究中大部分 SNP 来源于此)，我们只在样本中随机选择的两个染色体上观察到不同的等位基因时，才将模拟的 SNP 用于分析。在更远的距离处，我们增加额外的 SNP，仅仅选择那些在频率上与我们的数据匹配(±12.5% 之内)，而且 EHH 的分解程度与我们数据中观察到的一致(±12.5% 之内)的 SNP 用于分析。

我们对 5,000 个数据集合进行重复模拟(每个通常能够产生 6 ~ 8 个核心单倍型)以产生多个数据点，用于和我们的数据进行对比。首先将模拟的数据按照核心单倍型频率分类成 30 个大小相等的集合，每个集合约含有 1,000 个数据点，计算 *P* 值。然后我们对观察到的核心单倍型的相对 EHH 与含有相同频率单倍型的集合内的所有模拟数据点的 EHH 进行排序，这个排列就确定了 *P* 值。为了模拟更多的人口统计学历史，我们考虑两种扩张模型，一个极端的瓶颈效应和一个高度结构化的人群。对于扩张，我们模拟了一个人群，在 200 或者 5,000

expanded suddenly by a factor of 1,000. For an extreme bottleneck, we simulated a population that was constant at size 10,000 except for a brief bottleneck (inbreeding coefficient 0.18) that occurred 800 generations ago. (An inbreeding coefficient[26] of 0.18 is generated by dropping the population size to 800 chromosomes for 160 generations.) For a structured population, we simulated two equal-sized populations of size $N/2$ that exchanged migrants throughout history with a probability of $1/8N$ per generation per chromosome.

Results of the simulations remained qualitatively similar when we explored additional demographies and when we varied the stringency of matching to SNP allele frequencies and to haplotype homozygosities. A comprehensive, simulation-based exploration of the LRH test will be presented elsewhere, along with explorations of the statistical power of the test and computer code for implementing the LRH test on other data sets, including those with missing or unphased data (D.E.R., manuscript in preparation).

Control regions

To obtain control data for comparison to the *G6PD* and *TNFSF5* haplotypes, we genotyped the same population samples in 17 randomly chosen autosomal genes (*ACVR2B*, *TGFB1*, *DDR1*, *GTF2H4*, *COL11A2*, *LAMB1*, *WASL*, *SLC6A12*, *KCNA1*, *ARGHDIB*, *PCI*, *PRKCB1*, *NF1*, *SCYA2*, *PAI2*, *IL17R* and *HCF2*) selected previously as part of a genome-wide survey of linkage disequilibrium[26]. For our analyses we randomly picked chromosomes to match the numbers sampled for *G6PD* and *TNFSF5*, and we only included those control regions that we could match to our data in terms of homozygosity at the core and homozygosity at long distance (±25% stringency of matching). After the filtering process, we evaluated *G6PD* at 240.3 kb proximal to the gene and *TNFSF5* at 343.9 kb distal to the gene, because we could not make enough comparisons to control regions at the further distances. Seven genes matched to *G6PD* and seven genes matched to *TNFSF5*. We repeated the analysis using 78 Yoruban chromosomes for which phase information was known experimentally (because of genotyping in trios) and for which phase for the most part did not have to be inferred computationally. Six genes matched to *G6PD* and six genes matched to *TNFSF5* (see Supplementary Information for details).

(**419**, 832-837; 2002)

Pardis C. Sabeti[*†#], David E. Reich[*], John M. Higgins[*] Haninah Z. P. Levine[*], Daniel J. Richter[*], Stephen F. Schaffner[*], Stacey B. Gabriel[*], Jill V. Platko[*], Nick J. Patterson[*], Gavin J. McDonald[*], Hans C. Ackerman[‡], Sarah J. Campbell[‡], David Altshuler[*§], Richard Cooper[‖], Dominic Kwiatkowski[‡], Ryk Ward[†] & Eric S. Lander[*¶]

[*] Whitehead Institute/MIT Center for Genome Research, Nine Cambridge Center, Cambridge, Massachusetts 02142, USA

[†] Institute of Biological Anthropology, University of Oxford, Oxford, OX2 6QS, UK

[‡] Wellcome Trust Centre for Human Genetics, Roosevelt Drive, Oxford, OX3 7BN, UK

[§] Departments of Genetics and Medicine, Harvard Medical School, Department of Molecular Biology and Diabetes Unit, Massachusetts General Hospital, Boston, Massachusetts 02114, USA

[‖] Department of Preventive Medicine and Epidemiology, Loyola University Medical School, Maywood, Illinois 60143, USA

[¶] Department of Biology, MIT, Cambridge, Massachusetts 02139, USA

[#] Harvard Medical School, Boston, Massachusetts 02115, USA

代之前规模一直稳定在 10,000，然后它突然扩张了 1,000 倍。对于极端的瓶颈效应，我们模拟了一个大小稳定在 10,000 的人群，除了在 800 代前出现了一次短暂的瓶颈效应（近交系数 0.18）。（近交系数 [26]0.18 的产生是在 160 代内将人群的大小减少至 800 条染色体。）对于结构化的人群，我们模拟了两个大小均是 *N*/2 的人群，其在整个历史中以每代每个染色体 1/8*N* 的概率交换迁移者。

当我们探索其他人口统计学时，当我们改变与 SNP 等位基因频率和单倍型纯合性匹配的严格程度时，模拟的结果在定性上保持相似。我们还将对 LRH 测试进行全面的、基于模拟的探索，并探讨该测试的统计能力和在其他数据集（包括那些丢失或者非阶段数据的数据集）上实现 LRH 测试的计算机代码（赖希，稿件准备中）。

对照区域

为了获得对照数据，用来和 *G6PD* 以及 *TNFSF5* 单倍型进行比较，我们对同一个人群样本中的 17 个随机选择的常染色体基因进行基因分型（*ACVR2B*，*TGFB1*，*DDR1*，*GTF2H4*，*COL11A2*，*LAMB1*，*WASL*，*SLC6A12*，*KCNA1*，*ARGHDIB*，*PCI*，*PRKCB1*，*NF1*，*SCYA2*，*PAI2*，*IL17R* 和 *HCF2*），之前选择它们作为基因组水平连锁不平衡研究的一部分 [26]。对于我们的分析，我们随机选取了匹配 *G6PD* 以及 *TNFSF5* 样本数目的染色体，而且我们仅仅使用那些能够在核心的纯合性和长距离的纯合性与我们的数据匹配的对照区域（匹配的严格度是 ±25%）。经过筛选，我们对 *G6PD* 近端 240.3 kb 处和 *TNFSF5* 远端 343.9 kb 处进行了评估，因为我们无法对更远距离的对照区域进行比较。发现分别有 7 个基因对应于 *G6PD* 和 *TNFSF5*。我们用 78 名约鲁巴人的染色体进行了重复分析，它们的阶段信息已经通过实验得到（因为在三体家系中进行了基因分型），而且大部分的阶段信息不需要通过计算机推断。各有 6 个基因分别对应于 *G6PD* 和 *TNFSF5*（详细内容见补充资料）。

（毛晨晖 翻译；胡松年 审稿）

Received 7 June; accepted 19 September 2002; doi:10.1038/nature01140.
Published online 9 October 2002.

References:

1. Ruwende, C. & Hill, A. Glucose-6-phosphate dehydrogenase deficiency and malaria. *J. Mol. Med.* **76**, 581-588 (1998).
2. Sabeti, P. *et al.* CD40L association with protection from severe malaria. *Genes Immun.* **3**, 286-291 (2002).
3. Cavalli-Sforza, L. L., Menozzi, P. & Piazza, A. *The History and Geography of Human Genes* (Princeton Univ. Press, Princeton, 1994).
4. Kimura, M. *The Neutral Theory of Molecular Evolution* (Cambridge Univ. Press, Cambridge/New York, 1983).
5. Stephens, J. C. *et al.* Dating the origin of the CCR5-Delta32 AIDS-resistance allele by the coalescence of haplotypes. *Am. J. Hum. Genet.* **62**, 1507-1515 (1998).
6. Hudson, R. R. & Kaplan, N. L. Statistical properties of the number of recombination events in the history of a sample of DNA sequences. *Genetics* **111**, 147-164 (1985).
7. Lewontin, R. The interaction of selection and linkage. I. General considerations; heterotic models. *Genetics* **49**, 49-67 (1964).
8. Nei, M. *Molecular Evolutionary Genetics* Eqn. 8.4 (Columbia Univ. Press, New York, 1987).
9. Luzatto, L., Mehta, A. & Vulliamy, T. *The Metabolic & Molecular Bases of Inherited Disease* 4517-4553 (McGraw-Hill, New York, 2001).
10. Raymond, M. & Rousset, F. An exact test for population differentiation. *Evolution* **49**, 1280-1283 (1995).
11. Hudson, R. R. Properties of a neutral allele model with intragenic recombination. *Theor. Popul. Biol.* **23**, 183-201 (1983).
12. Reich, D. E. & Goldstein, D. B. *Microsatellites: Evolution and Applications* 128-138 (Oxford Univ. Press, Oxford/New York, 1999).
13. Tishkoff, S. A. *et al.* Haplotype diversity and linkage disequilibrium at human G6PD: recent origin of alleles that confer malarial resistance. *Science* **293**, 455-462 (2001).
14. Rozas, J. & Rozas, R. DnaSP version 3: an integrated program for molecular population genetics and molecular evolution analysis. *Bioinformatics* **15**, 174-175 (1999).
15. Tajima, F. Statistical method for testing the neutral mutation hypothesis by DNA polymorphism. *Genetics* **123**, 585-595 (1989).
16. Fu, Y. X. & Li, W. H. Statistical tests of neutrality of mutations. *Genetics* **133**, 693-709 (1993).
17. Fay, J. C. & Wu, C. I. Hitchhiking under positive Darwinian selection. *Genetics* **155**, 1405-1413 (2000).
18. Hughes, A. L. & Nei, M. Pattern of nucleotide substitution at major histocompatibility complex class I loci reveals overdominant selection. *Nature* **335**, 167-170 (1988).
19. McDonald, J. H. & Kreitman, M. Adaptive protein evolution at the Adh locus in *Drosophila. Nature* **351**, 652-654 (1991).
20. Hudson, R. R., Kreitman, M. & Aguade, M. A test of neutral molecular evolution based on nucleotide data. *Genetics* **116**, 153-159 (1987).
21. Gabriel, S. B. *et al.* The structure of haplotype blocks in the human genome. *Science* **23**, 2225-2229 (2002).
22. Wootton, J. C. *et al.* Genetic diversity and chloroquine selective sweeps in *Plasmodium falciparum. Nature* **418**, 320-323 (2002).
23. Tang, K. *et al.* Chip-based genotyping by mass spectrometry. *Proc. Natl Acad. Sci. USA* **96**, 10016-10020 (1999).
24. Vulliamy, T. J. *et al.* Linkage disequilibrium of polymorphic sites in the G6PD gene in African populations and the origin of G6PD A. *Gene Geogr.* **5**, 13-21 (1991).
25. Sachidanandam, R. *et al.* A map of human genome sequence variation containing 1.42 million single nucleotide polymorphisms. *Nature* **409**, 928-933 (2001).
26. Reich, D. E. *et al.* Linkage disequilibrium in the human genome. *Nature* **411**, 199-204 (2001).

Supplementary Information accompanies the paper on *Nature*'s website (http://www.nature.com/nature).

Acknowledgements. We thank B. Blumenstiel, M. DeFelice, A. Lochner, J. Moore, H. Nguyen and J. Roy for assistance in genotyping the 17 control regions. We also thank L. Gaffney, S. Radhakrishna, T. DiCesare and T. Lavery for graphics and technical support, B. Ferrell for the Beni samples, and A. Adeyemo and C. Rotimi for helping to collect the Yoruba and Shona samples. Finally, we thank M. Daly, E. Cosman, B. Gray, V. Koduri, T. Herrington and L. Peterson for comments on the manuscript. P.C.S. was supported by grants from the Rhodes Trust, the Harvard Office of Enrichment, and by a Soros Fellowship. This work was supported by grants from the National Institute of Health.

Competing interests statement. The authors declare that they have no competing financial interests.

Correspondence and requests for materials should be addressed to E.S.L. (e-mail: lander@genome.wi.mit.edu).

Initial Sequencing and Comparative Analysis of the Mouse Genome*

Mouse Genome Sequencing Consortium

Editor's Note

Here, a high-quality draft sequence of the mouse genome is made freely available for the first time, thanks to the work of an international research consortium. At the time, the data provided an invaluable reference point for geneticists struggling to understand the draft human genome which had been published a year earlier. But it continues to shed light on human biology and disease today, as the mouse remains an invaluable and stalwart model in biomedical research. Comparisons here between the DNA of mouse and man reveal immense similarities, confirming the credentials of this tiny rodent to help fathom the workings of the human genome. But the study also highlights some 300 mouse-specific genes, with the biggest disparities linked to smell, immunity and detoxification.

The sequence of the mouse genome is a key informational tool for understanding the contents of the human genome and a key experimental tool for biomedical research. Here, we report the results of an international collaboration to produce a high-quality draft sequence of the mouse genome. We also present an initial comparative analysis of the mouse and human genomes, describing some of the insights that can be gleaned from the two sequences. We discuss topics including the analysis of the evolutionary forces shaping the size, structure and sequence of the genomes; the conservation of large-scale synteny across most of the genomes; the much lower extent of sequence orthology covering less than half of the genomes; the proportions of the genomes under selection; the number of protein-coding genes; the expansion of gene families related to reproduction and immunity; the evolution of proteins; and the identification of intraspecies polymorphism.

WITH the complete sequence of the human genome nearly in hand[1,2], the next challenge is to extract the extraordinary trove of information encoded within its roughly 3 billion nucleotides. This information includes the blueprints for all RNAs and proteins, the regulatory elements that ensure proper expression of all genes, the structural elements that govern chromosome function, and the records of our evolutionary history.

* This is a shortened version of the original paper that appeared in *Nature*. In particular, it omits much of the detailed quantitative data in the original figures and tables, and the full list of authors and affiliations has been removed. The original text is freely available in the *Nature* online archive. Citations are numbered herein as they are in the original paper; only those references that are cited in this version are listed in the "References" section.

小鼠基因组的初步测序和对比分析*

小鼠基因组测序协作组

编者按

感谢国际协作组的工作，本文首次免费提供一个高质量的小鼠基因组序列草图。当时，这些数据为遗传学家努力理解一年前发表的人类基因组草图提供了宝贵的参考点。但是现在，它仍然继续为揭示人类生物学和疾病提供帮助。因为在生物医学研究中，小鼠始终是一个宝贵而有力的模型。小鼠和人类DNA之间的比较揭示了两者巨大的相似性，证实了这种小型啮齿动物能帮助了解人类基因组的运行。但是，该研究也强调了大约300个小鼠特异性基因，其中最大的不同与嗅觉、免疫和解毒作用有关。

小鼠基因组序列是理解人类基因组内容的关键信息工具，也是生物医学研究的关键实验工具。在此，我们报告国际合作的成果——一个高质量的小鼠基因组序列草图。我们还对小鼠和人类基因组进行了初步对比分析，描述了从这两个序列中获得的一些启示。我们讨论的主题包括对影响基因组大小、结构和序列的进化力的分析；存在于大部分基因组中的同源染色单体(同线性)的大规模保留；比之低得多只覆盖不到一半基因组的序列直系同源；受选择基因组的比例；蛋白质编码基因的数量；与生殖和免疫相关的基因家族的扩增；蛋白质的进化；物种内多态性的确定。

随着即将掌握人类基因组的完整序列[1,2]，下一个挑战是提取大约30亿个核苷酸中编码的非常巨大的信息宝藏。这些信息包括所有RNA和蛋白质的蓝图、确保所有基因正确表达的调控元件、控制染色体功能的结构元件以及我们进化史的记录。这些特征当中，一些可以很容易在人类序列中识别，但很多特征是微妙而且难辨别

* 这是《自然》杂志中原文的缩略版。此版本在于省略了原始图表中的大部分详细量化数据并去除了所有作者及单位名称。原文可在《自然》杂志在线免费获取。参考文献的编码与原文一致，文末“References”仅保留了此版本中有所引用的文献。

Some of these features can be recognized easily in the human sequence, but many are subtle and difficult to discern. One of the most powerful general approaches for unlocking the secrets of the human genome is comparative genomics, and one of the most powerful starting points for comparison is the laboratory mouse, *Mus musculus*.

Metaphorically, comparative genomics allows one to read evolution's laboratory notebook. In the roughly 75 million years since the divergence of the human and mouse lineages, the process of evolution has altered their genome sequences and caused them to diverge by nearly one substitution for every two nucleotides (see below) as well as by deletion and insertion. The divergence rate is low enough that one can still align orthologous sequences, but high enough so that one can recognize many functionally important elements by their greater degree of conservation. Studies of small genomic regions have demonstrated the power of such cross-species conservation to identify putative genes or regulatory elements[3-12]. Genome-wide analysis of sequence conservation holds the prospect of systematically revealing such information for all genes. Genome-wide comparisons among organisms can also highlight key differences in the forces shaping their genomes, including differences in mutational and selective pressures[13,14].

Literally, comparative genomics allows one to link laboratory notebooks of clinical and basic researchers. With knowledge of both genomes, biomedical studies of human genes can be complemented by experimental manipulations of corresponding mouse genes to accelerate functional understanding. In this respect, the mouse is unsurpassed as a model system for probing mammalian biology and human disease[15,16]. Its unique advantages include a century of genetic studies, scores of inbred strains, hundreds of spontaneous mutations, practical techniques for random mutagenesis, and, importantly, directed engineering of the genome through transgenic, knockout and knockin techniques[17-22].

For these and other reasons, the Human Genome Project (HGP) recognized from its outset that the sequencing of the human genome needed to be followed as rapidly as possible by the sequencing of the mouse genome. In early 2001, the International Human Genome Sequencing Consortium reported a draft sequence covering about 90% of the euchromatic human genome, with about 35% in finished form[1]. Since then, progress towards a complete human sequence has proceeded swiftly, with approximately 98% of the genome now available in draft form and about 95% in finished form.

Here, we report the results of an international collaboration involving centres in the United States and the United Kingdom to produce a high-quality draft sequence of the mouse genome and a broad scientific network to analyse the data. The draft sequence was generated by assembling about sevenfold sequence coverage from female mice of the C57BL/6J strain (referred to below as B6). The assembly contains about 96% of the sequence of the euchromatic genome (excluding chromosome Y) in sequence contigs linked together into large units, usually larger than 50 megabases (Mb).

With the availability of a draft sequence of the mouse genome, we have undertaken an

的。解开人类基因组秘密最有力的通用方法之一是比较基因组学，而进行比较最有力的出发点就是实验用小鼠，拉丁文名称为 *Mus musculus*。

打个比方，比较基因组学允许人们阅读进化的实验笔记。在人类和小鼠的谱系分开后的大约 7,500 万年中，进化历程改变了它们的基因组序列，通过几乎每两个核苷酸就替换一个核苷酸（见下文）以及核苷酸的缺失和插入产生分化。分歧率足够低的部分可以比对直系同源序列，但分歧率较高的部分可以通过它们更高的保守性来识别许多功能上重要的元件。对基因组小块区域的研究已经证实了这种跨物种的保守性在识别推测基因或调控元件方面的能力 [3-12]。全基因组范围的序列保守性分析有望系统地揭示所有基因的这类信息。物种之间的全基因组范围比较还可以使得包括突变和选择压力差异在内的形成基因组的关键差异作用更加突出地显示出来 [13,14]。

从字面上看，比较基因组学可以让人们将临床和基础研究人员的实验笔记联系起来。有了这两个基因组的知识，人类基因的生物医学研究可以通过实验操作相应的小鼠基因来补充，加速对于基因功能的理解。在这方面，小鼠是探索哺乳动物生物学和人类疾病无与伦比的模型系统 [15,16]。它的独特优势包括百年的遗传学研究、大量的近交系、成百个自发突变、随机诱变的实用实验技术，以及尤其重要的通过转基因、基因敲除和敲入技术进行的基因组定向操作 [17-22]。

基于这些和其他原因，人类基因组计划（HGP）从一开始就意识到人类基因组的测序需要尽可能快地接着进行小鼠基因组的测序。2001 年初，国际人类基因组测序协作组报告了一份覆盖了 90% 人类常染色质的基因组序列草图，其中 35% 是已经完成了的形式 [1]。从那时起，向着完整人类序列的发展进展迅速，目前大约 98% 的基因组以草图形式提供，约 95% 的基因组以完成图形式提供。

在此，我们报告由美国和英国的多个中心参加的国际合作所产生的一个高质量小鼠基因组序列草图和一个数据分析的广泛科学网络。这个序列草图是由 C57BL/6J 雌性小鼠（以下称为 B6），大约 7 倍的序列覆盖度组装而成。组装结果包括 96% 的常染色质基因组序列（不包括 Y 染色体），由测序序列片段重叠拼接在一起形成大的单元，通常大于 50 兆碱基（Mb）。

利用小鼠基因组的序列草图，我们进行了初步的对比分析，以检测人类和小鼠

initial comparative analysis to examine the similarities and differences between the human and mouse genomes. Some of the important points are listed below.

- The mouse genome is about 14% smaller than the human genome (2.5 Gb compared with 2.9 Gb). The difference probably reflects a higher rate of deletion in the mouse lineage.

- Over 90% of the mouse and human genomes can be partitioned into corresponding regions of conserved synteny, reflecting segments in which the gene order in the most recent common ancestor has been conserved in both species.

- At the nucleotide level, approximately 40% of the human genome can be aligned to the mouse genome. These sequences seem to represent most of the orthologous sequences that remain in both lineages from the common ancestor, with the rest likely to have been deleted in one or both genomes.

- The neutral substitution rate has been roughly half a nucleotide substitution per site since the divergence of the species, with about twice as many of these substitutions having occurred in the mouse compared with the human lineage.

- By comparing the extent of genome-wide sequence conservation to the neutral rate, the proportion of small (50–100 bp) segments in the mammalian genome that is under (purifying) selection can be estimated to be about 5%. This proportion is much higher than can be explained by protein-coding sequences alone, implying that the genome contains many additional features (such as untranslated regions, regulatory elements, non-protein-coding genes, and chromosomal structural elements) under selection for biological function.

- The mammalian genome is evolving in a non-uniform manner, with various measures of divergence showing substantial variation across the genome.

- The mouse and human genomes each seem to contain about 30,000 protein-coding genes. These refined estimates have been derived from both new evidence-based analyses that produce larger and more complete sets of gene predictions, and new *de novo* gene predictions that do not rely on previous evidence of transcription or homology. The proportion of mouse genes with a single identifiable orthologue in the human genome seems to be approximately 80%. The proportion of mouse genes without any homologue currently detectable in the human genome (and vice versa) seems to be less than 1%.

- Dozens of local gene family expansions have occurred in the mouse lineage. Most of these seem to involve genes related to reproduction, immunity and olfaction, suggesting that these physiological systems have been the focus of extensive lineage-specific innovation in rodents.

- Mouse–human sequence comparisons allow an estimate of the rate of protein evolution

基因组之间的相似性和差异。下面列出了一些要点。

- 小鼠基因组大概比人类基因组小约 14%(2.5 Gb 比 2.9 Gb)。这种差异可能反映了小鼠谱系中较高的缺失率。

- 超过 90% 的小鼠和人类基因组可以被划分成相应的保守同线性区域，反映了两个物种所保存的最近共同祖先区段上的基因顺序。

- 在核苷酸水平上，大约 40% 的人类基因组可以比对到小鼠基因组上。这些序列似乎代表了来自共同祖先的两个谱系中的大部分直系同源序列，余下的序列可能在一个或两个基因组中被删除了。

- 自物种分化以来，每个位点的中性替代率大约是核苷酸替代率的一半，与人类谱系相比较，在小鼠中这些替代出现的次数大概是人类的两倍。

- 通过对全基因组序列的保守程度和中性率进行比较，估计哺乳动物基因组中处于(纯化)选择的小片段(50 ~ 100 bp)的比例大约为 5%。这个比例远远高于仅用蛋白质编码序列所能解释的，这意味着基因组很多额外性质(如非翻译区、调节元件、非蛋白质编码基因和染色体结构元件)也受到了生物学功能的选择。

- 哺乳动物基因组正在以不均匀的方式进化，对于分化的不同度量结果显示了基因组中的巨大变化。

- 小鼠和人类的基因组似乎都包含大约 30,000 个蛋白质编码基因。这些精准的估计分别来自基于新证据获得的更大、更完整的基因预测集和不依赖于先前转录或同源性证据的基因重新预测。具有单一可识别直系同源基因的小鼠基因在人类基因组中的比例大约为 80%。没有在人类基因组中检测到任何同源性的小鼠基因的比例似乎小于 1%，反之亦然。

- 小鼠谱系中已经发生了数十个局部基因家族的扩张。大部分看来与生殖、免疫和嗅觉相关，提示这些生理系统是啮齿动物的谱系特异性更新的焦点。

- 小鼠–人类序列的比较可以预测哺乳动物蛋白质的进化速率。某些与生殖、宿

in mammals. Certain classes of secreted proteins implicated in reproduction, host defence and immune response seem to be under positive selection, which drives rapid evolution.

- Despite marked differences in the activity of transposable elements between mouse and human, similar types of repeat sequences have accumulated in the corresponding genomic regions in both species. The correlation is stronger than can be explained simply by local (G+C) content and points to additional factors influencing how the genome is moulded by transposons.

- By additional sequencing in other mouse strains, we have identified about 80,000 single nucleotide polymorphisms (SNPs). The distribution of SNPs reveals that genetic variation among mouse strains occurs in large blocks, mostly reflecting contributions of the two subspecies *Mus musculus domesticus* and *Mus musculus musculus* to current laboratory strains.

The mouse genome sequence is freely available in public databases (GenBank accession number CAAA01000000) and is accessible through various genome browsers (http://www.ensembl.org/Mus_musculus/, http://genome.ucsc.edu/ and http://www.ncbi.nlm.nih.gov/genome/guide/mouse/).

In this paper, we begin with information about the generation, assembly and evaluation of the draft genome sequence, the conservation of synteny between the mouse and human genomes, and the landscape of the mouse genome. We then explore the repeat sequences, genes and proteome of the mouse, emphasizing comparisons with the human. This is followed by evolutionary analysis of selection and mutation in the mouse and human lineages, as well as polymorphism among current mouse strains. A full and detailed description of the methods underlying these studies is provided as Supplementary Information. In many respects, the current paper is a companion to the recent paper on the human genome sequence[1]. Extensive background information about many of the topics discussed below is provided there.

Background to the Mouse Genome Sequencing Project

Origins of the mouse

The precise origin of the mouse and human lineages has been the subject of recent debate. Palaeontological evidence has long indicated a great radiation of placental (eutherian) mammals about 65 million years ago (Myr) that filled the ecological space left by the extinction of the dinosaurs, and that gave rise to most of the eutherian orders[23]. Molecular phylogenetic analyses indicate earlier divergence times of many of the mammalian clades. Some of these studies have suggested a very early date for the divergence of mouse from other mammals (100–130 Myr[23-25]) but these estimates partially originate from the fast molecular clock in rodents (see below).Recent molecular studies that are less sensitive to the differences in evolutionary rates have suggested that the eutherian mammalian radiation

主防御和免疫应答相关的分泌蛋白质似乎是处于正选择，这些推动了蛋白质快速进化。

• 尽管小鼠和人类的转座元件的活性有明显差异，但是两个物种的相应基因组区域中积累了相似类型的重复序列。这种相关性要比简单通过局部(G+C)含量来解释更强，并且指向影响基因组如何被转座子塑造的其他因素。

• 通过对其他品系小鼠的额外测序，我们已经鉴定出了大概 80,000 个单核苷酸多态性(SNP)。SNP 的分布表明小鼠品系间的遗传变异发生在大的区块内，主要反映了两个小鼠亚种：*Mus musculus domesticus* 和 *Mus musculus musculus* 对当前实验室品系的贡献。

小鼠基因组序列可在公共数据库(GenBank 编号 CAAA01000000)中免费获得，并可通过多种基因组浏览器(http://www.ensembl.org/Mus_musculus/，http://genome.ucsc.edu/ 和 http://www.ncbi.nlm.nih.gov/genome/guide/mouse/)访问。

在本文中，我们从基因组序列草图的产生、组装和评估，小鼠和人类基因组之间的同线性保守性，以及小鼠基因组的概貌信息入手。然后，我们探索小鼠的重复序列、基因和蛋白质组，尤其是与人类的比较。接着是对小鼠和人类谱系的选择和突变以及目前小鼠品系中的多态性的进化分析。补充材料对这些研究的方法进行了完整而详细的描述。在很多方面，这篇论文与最近一篇关于人类基因组序列的文章[1]的相配套。那里提供了关于下面讨论的许多主题的充分的背景信息。

小鼠基因组测序项目的背景

小鼠的起源

小鼠和人类谱系的确切起源是最近的争论话题。古生物学的证据长期表明，大约 6,500 万年前胎盘哺乳类动物(真兽亚纲动物)的大量扩张，填充了恐龙灭绝留下的生态空间，真兽亚纲大多数目的动物由此产生[23]。分子系统发育分析表明，许多哺乳动物分支的分化时间较早。其中一些研究表明，小鼠从其他哺乳动物中分化出来的时间非常早(1 亿 ~ 3 亿年前[23-25])，但是这些估计部分来自啮齿动物的快速分子钟(见下文)。最近对进化率差异不那么敏感的分子研究表明真兽亚纲动物的扩

took place throughout the Late Cretaceous period (65–100 Myr), but that rodents and primates actually represent relatively late-branching lineages[26,27]. In the analyses below, we use a divergence time for the human and mouse lineages of 75 Myr for the purpose of calculating evolutionary rates, although it is possible that the actual time may be as recent as 65 Myr.

Origins of the mouse genetics

With the rediscovery of Mendel's laws of inheritance in 1900, pioneers of the new science of genetics (such as Cuenot, Castle and Little) were quick to recognize that the discontinuous variation of bred ("fancy") mice was analogous to that of Mendel's peas, and they set out to test the new theories of inheritance in mice. Mating programmes were soon established to create inbred strains, resulting in many of the modern, well-known strains (including C57BL/6J)[30].

Genetic mapping in the mouse began with Haldane's report[31] in 1915 of linkage between the pink-eye dilution and albino loci on the linkage group that was eventually assigned to mouse chromosome 7, just 2 years after the first report of genetic linkage in *Drosophila*. The genetic map grew slowly over the next 50 years as new loci and linkage groups were added—chromosome 7 grew to three loci by 1935 and eight by 1954. The accumulation of serological and enzyme polymorphisms from the 1960s to the early 1980s began to fill out the genome, with the map of chromosome 7 harbouring 45 loci by 1982 (refs 29, 31).

The real explosion, however, came with the development of recombinant DNA technology and the advent of DNA-sequence-based polymorphisms. Initially, this involved the detection of restriction-fragment length polymorphisms (RFLPs)[32]; later, the emphasis shifted to the use of simple sequence length polymorphisms (SSLPs; also called microsatellites), which could be assayed easily by polymerase chain reaction (PCR)[33-36] and readily revealed polymorphisms between inbred laboratory strains.

Origins of mouse genomics

When the Human Genome Project (HGP) was launched in 1990, it included the mouse as one of its five central model organisms, and targeted the creation of genetic, physical and eventually sequence maps of the mouse genome.

By 1996, a dense genetic map with nearly 6,600 highly polymorphic SSLP markers ordered in a common cross had been developed[34], providing the standard tool for mouse genetics. Subsequent efforts filled out the map to over 12,000 polymorphic markers, although not all of these loci have been positioned precisely relative to one another. With these and other loci, Haldane's original two-marker linkage group on chromosome 7 had now swelled to about 2,250 loci.

张发生在整个白垩纪晚期(6,500 万年 ~ 1 亿年前)，但啮齿类动物和灵长类动物实际代表了相对较晚的分支谱系 [26,27]。在下面的分析中，我们用 7,500 万年的人类和小鼠的谱系分歧时间来计算进化速率，尽管实际时间可能近至 6,500 万年前。

小鼠遗传学的起源

1900 年，随着孟德尔遗传定律的重新发现，遗传的新科学先驱们(如屈埃诺、卡斯尔和利特尔)很快意识到培育出来的精致小鼠的不连续变异类似于孟德尔的豌豆，于是他们开始测试小鼠的新遗传理论。很快建立了交配方案培育近交系，从而产生了很多现代众所周知的品系(包括 C57BL/6J) [30]。

小鼠的遗传作图始于 1915 年霍尔丹 [31] 关于连锁群上粉色眼稀释和白化基因座之间连锁的报告，该连锁群最终被定位于小鼠 7 号染色体，距首次报道果蝇的遗传连锁仅仅两年。随着新的基因座和连锁群的增加，遗传图谱在随后的 50 年中慢慢发展起来——7 号染色体到 1935 年增长到 3 个基因座，到 1954 年增长到 8 个基因座。从 20 世纪 60 年代到 80 年代早期，血清学和酶学多态性的积累开始充实基因组，到 1982 年，7 号染色体上包含了 45 个基因座(见参考文献 29、31)。

然而，真正的爆发来自重组 DNA 技术的发展和基于 DNA 序列的多态性的出现。最初，这只涉及限制性内切酶片段长度的多态性(RFLP)检测 [32]；后来，重点转移到了简单序列长度多态性(SSLP，也称作微卫星)的应用上，这种多态性可以很容易地通过聚合酶链反应(PCR)进行检测 [33-36]，并且容易发现实验室近交系之间的多态性。

小鼠基因组学的起源

1990 年人类基因组计划(HGP)启动时就把小鼠作为五个中心模式生物之一，将构建遗传、物理和最终小鼠基因组序列图谱列为目标。

到 1996 年，已绘制出一个密度包括了常见杂交中顺序排列的近 6,600 个高多态性 SSLP 标记的遗传图谱 [34]，为小鼠遗传学提供了标准工具。随后的工作将该图谱上补充为超过 12,000 个多态性标记，尽管并不是所有的基因座都精确地一一相对定位。利用这些和其他基因座，霍尔丹当初在 7 号染色体上的双标记连锁群现在已经扩大到含有约 2,250 个标记。

Physical maps of the mouse genome also proceeded apace, using sequence-tagged sites (STS) together with radiation-hybrid panels[37,38] and yeast artificial chromosome (YAC) libraries to construct dense landmark maps[39]. Together, the genetic and physical maps provide thousands of anchor points that can be used to tie clones or DNA sequences to specific locations in the mouse genome.

With these resources, it became straightforward (but not always easy) to perform positional cloning of classic single-gene mutations for visible, behavioural, immunological and other phenotypes. Many of these mutations provide important models of human disease, sometimes recapitulating human phenotypes with uncanny accuracy. It also became possible for the first time to begin dissecting polygenic traits by genetic mapping of quantitative trait loci (QTL) for such traits.

Continuing advances fuelled a growing desire for a complete sequence of the mouse genome. The development of improved random mutagenesis protocols led to the establishment of large-scale screens to identify interesting new mutants, increasing the need for more rapid positional cloning strategies. QTL mapping experiments succeeded in localizing more than 1,000 loci affecting physiological traits, creating demand for efficient techniques capable of trawling through large genomic regions to find the underlying genes. Furthermore, the ability to perform directed mutagenesis of the mouse germ line through homologous recombination made it possible to manipulate any gene given its DNA sequence, placing an increasing premium on sequence information. In all of these cases, it was clear that genome sequence information could markedly accelerate progress.

Origin of the Mouse Genome Sequencing Consortium

With the sequencing of the human genome well underway by 1999, a concerted effort to sequence the entire mouse genome was organized by a Mouse Genome Sequencing Consortium (MGSC). The MGSC originally consisted of three large sequencing centres—the Whitehead/Massachusetts Institute of Technology (MIT) Center for Genome Research, the Washington University Genome Sequencing Center, and the Wellcome Trust Sanger Institute—together with an international database, Ensembl, a joint project between the European Bioinformatics Institute and the Sanger Institute.

In addition to the genome-wide efforts of the MGSC, other publicly funded groups have been contributing to the sequencing of the mouse genome in specific regions of biological interest. Together, the MGSC and these programmes have so far yielded clone-based draft sequence consisting of 1,859 Mb (74%, although there is redundancy) and finished sequence of 477 Mb (19%) of the mouse genome. Furthermore, Mural and colleagues[45] recently reported a draft sequence of mouse chromosome 16 containing 87 Mb (3.5%).

To analyse the data reported here, the MGSC was expanded to include the other publicly funded sequencing groups and a Mouse Genome Analysis Group consisting of scientists

通过联合使用序列标签位点(STS)、辐射杂交组合[37,38]和酵母人工染色体(YAC)文库来建立高密度标记物图谱[39]，小鼠基因组的物理图谱进展也很快。遗传和物理图谱提供了数千个锚定点，可以将克隆或DNA序列与小鼠基因组中的特定位置联系起来。

有了这些资源，对于可见的、行为的、免疫的以及其他表型的经典单基因突变进行定位克隆变得很直接(但并非总是容易的)。这些突变中的很多突变为人类疾病提供了重要模型，有时以不可思议的精准性描述人类表型。而且，通过进行数量性状基因座(QTL)的遗传作图也使首次开始解析多基因性状成为可能。

持续的进步激发了人们对小鼠基因组完整序列不断增长的渴望。改良的随机诱变方案的发展促成对于感兴趣的新突变的大规模筛选，增加了对更快的定位克隆策略的需求。数量性状基因座绘图实验成功地定位了超过1,000个影响生理性状的基因座，产生了对搜寻大片基因组区域定位基因的高效性技术的需求。而且通过同源重组技术对小鼠胚系进行定向诱变使得人们可以操控给定DNA序列的任何基因，从而增加了对序列信息的重视。所有这些例子中显然基因组序列信息最能显著加速进展。

小鼠基因组测序协作组的起源

1999年，随着人类基因组测序工作的顺利进行，小鼠基因组测序协作组(MGSC)组织协调了对小鼠整个基因组的测序。MGSC最初由三个大型测序中心——怀特黑德研究所/麻省理工学院(MIT)基因组研究中心、华盛顿大学基因组测序中心和威康信托桑格研究所，以及由欧洲生物信息研究所和桑格研究所合作建立的国际数据库Ensembl组成。

除了MGSC在全基因组范围的努力，其他公共资助的团体也一直在生物学感兴趣的特定区域为小鼠基因组测序做贡献。目前为止，MGSC和这些项目一起产生了基于克隆的小鼠基因组的1,859 Mb序列草图(74%，尽管存在冗余)和477 Mb(19%)的完成序列。此外，穆拉及其同事们[45]最近报道了一个包含87 Mb(3.5%)的小鼠16号染色体的序列草图。

为了分析这里报告的数据，MGSC扩大到包括其他公共项目资助的测序小组和

from 27 institutions in 6 countries.

Generating the Draft Genome Sequence

Sequencing strategy

Sanger and co-workers developed the strategy of random shotgun sequencing in the early 1980s, and it has remained the mainstay of genome sequencing over the ensuing two decades. The approach involves producing random sequence "reads", generating a preliminary assembly on the basis of sequence overlaps, and then performing directed sequencing to obtain a "finished" sequence with gaps closed and ambiguities resolved[46]. Ansorge and colleagues[47] extended the technique by the use of "paired-end sequencing", in which sequencing is performed from both ends of a cloned insert to obtain linking information, which is then used in sequence assembly. More recently, Myers and co-workers[48], and others, have developed efficient algorithms for exploiting such linking information.

The ultimate aim of the MGSC is to produce a finished, richly annotated sequence of the mouse genome to serve as a permanent reference for mammalian biology. In addition, we wished to produce a draft sequence as rapidly as possible to aid in the interpretation of the human genome sequence and to provide a useful intermediate resource to the research community. Accordingly, we adopted a hybrid strategy for sequencing the mouse genome. The strategy has four components: (1) production of a BAC-based physical map of the mouse genome by fingerprinting and sequencing the ends of clones of a BAC library[44]; (2) whole-genome shotgun (WGS) sequencing to approximately sevenfold coverage and assembly to generate an initial draft genome sequence; (3) hierarchical shotgun sequencing[46] of BAC clones covering the mouse genome combined with the WGS data to create a hybrid WGS-BAC assembly; and (4) production of a finished sequence by using the BAC clones as a template for directed finishing. This mixed strategy was designed to exploit the simpler organizational aspects of WGS assemblies in the initial phase, while still culminating in the complete high-quality sequence afforded by clone-based maps.

We chose to sequence DNA from a single mouse strain, rather than from a mixture of strains[45], to generate a solid reference foundation, reasoning that polymorphic variation in other strains could be added subsequently (see below). After extensive consultation with the scientific community[52], the B6 strain was selected because of its principal role in mouse genetics, including its well-characterized phenotype and role as the background strain on which many important mutations arose. We elected to sequence a female mouse to obtain equal coverage of chromosome X and autosomes. Chromosome Y was thus omitted, but this chromosome is highly repetitive (the human chromosome Y has multiple duplicated regions exceeding 100 kb in size with 99.9% sequence identity[53]) and seemed an unwise target for the WGS approach. Instead, mouse chromosome Y is being sequenced by a purely clone-based (hierarchical shotgun) approach.

由 6 个国家 27 个科研机构的科学家组成的小鼠基因组分析小组。

生成基因组草图

测序策略

20 世纪 80 年代早期，桑格和同事们研发了随机的鸟枪法测序策略，在随后的 20 年里，它依然是基因组测序的主流手段。该方法包括产生随机序列"读取片段"，基于序列重叠产生初步组装，然后进行定向测序，以获得填补了空缺和解决了歧义的"完成"序列[46]。安索奇及其同事们[47]通过进行"双末端测序"扩展了该技术，从克隆插入片段的两端进行测序，获得连接信息，然后用于序列组装。最近，迈尔斯和同事们[48]以及其他人已经开发了利用这种连接信息的有效算法。

MGSC 的最终目标是产生一个完整的、注释丰富的小鼠基因组序列，作为哺乳动物生物学的永久参考。此外，我们希望尽可能快地产生一个序列草图，以帮助解释人类基因组序列，并为学界提供一个有用的中间资源。因此，我们采用混合策略对小鼠基因组进行测序。这个策略包括四个部分：(1) 通过指纹法和对 BAC 文库克隆末端的测序产生基于 BAC 的小鼠基因组物理图谱[44]；(2) 全基因组鸟枪 (WGS) 测序到大约 7 重覆盖范围，组装产生最初的基因组序列草图；(3) 覆盖小鼠基因组的 BAC 克隆的分级鸟枪测序[46]与 WGS 数据结合，以构建 WGS-BAC 组装；以及 (4) 利用 BAC 克隆作为模板进行定向整理，生产完成序列。这种混合策略的设计主要是在初始阶段先解析较为简单的 WGS 组装的组织构架，最终通过克隆图谱获得完整的高质量序列。

我们选择一个单一的小鼠品系进行 DNA 测序，而不是通过品系的混合[45]产生坚实的参考基础，理由是随后可以不断添加其他品系的多态性变化 (见下文)。经过科学界的广泛磋商[52]，B6 品系因其在小鼠遗传中的重要作用而被选中。B6 品系有良好的特征表型，是一个背景品系，许多重要的突变都发生在该品系上。我们选择对雌性小鼠进行测序，以获得 X 染色体和常染色体的相等覆盖率。因此染色体 Y 被省略了，但是这条染色体具有高度重复性 (人类 Y 染色体具有多个重复区域，这些区域超过 100 kb，具有 99.9% 的序列一致性[53])，似乎并不是 WGS 方法的明智目标。取而代之的是，将通过纯克隆的方法 (分级鸟枪) 对小鼠 Y 染色体进行测序。

Sequencing and assembly

The genome assembly was based on a total of 41.4 million sequence reads derived from both ends of inserts (paired-end reads) of various clone types prepared from B6 female DNA. The inserts ranged in size from 2 to 200 kb. The three large MGSC sequencing centres generated 40.4 million reads, and 0.6 million reads were generated at the University of Utah. In addition, we used 0.4 million reads from both ends of BAC inserts reported by The Institute for Genome Research[54].

A total of 33.6 million reads passed extensive checks for quality and source, of which 29.7 million were paired; that is, derived from opposite ends of the same clone. The assembled reads represent approximately 7.7-fold sequence coverage of the euchromatic mouse genome (6.5-fold coverage in bases with a Phred quality score of >20)[55]. Together, the clone inserts provide roughly 47-fold physical coverage of the genome.

The sequence reads, together with the pairing information, were used as input for two recently developed sequence-assembly programs, Arachne[56,57] and Phusion[58]. No mapping information and no clone-based sequences were used in the WGS assembly, with the exception of a few reads ($<0.1\%$ of the total) derived from a handful of BACs, which were used as internal controls. The assembly programs were tested and compared on intermediate data sets over the course of the project and were thereby refined. The programs produced comparable outputs in the final assembly. The assembly generated by Arachne was chosen as the draft sequence described here because it yielded greater short-range and long-range continuity with comparable accuracy.

The assembly contains 224,713 sequence contigs, which are connected by at least two read-pair links into supercontigs (or scaffolds). There are a total of 7,418 supercontigs at least 2 kb in length, plus a further 37,125 smaller supercontigs representing $<1\%$ of the assembly. The contigs have an N50 length of 24.8 kb, whereas the supercontigs have an N50 length that is approximately 700-fold larger at 16.9 Mb (N50 length is the size x such that 50% of the assembly is in units of length at least x). In fact, most of the genome lies in supercontigs that are extremely large: the 200 largest supercontigs span more than 98% of the assembled sequence, of which 3% is within sequence gaps.

Anchoring to chromosomes

We assigned as many supercontigs as possible to chromosomal locations in the proper order and orientation. Supercontigs were localized largely by sequence alignments with the extensively validated mouse genetic map[34], with some additional localization provided by the mouse radiation-hybrid map[37] and the BAC map[44]. We found no evidence of incorrect global joins within the supercontigs (that is, multiple markers supporting two discordant locations within the genome), and thus were able to place them directly. Altogether, we

测序和组装

基因组组装基于共计 4,140 万个序列读取片段而成，均来自 B6 雌鼠 DNA 制备的各种克隆类型的双末端插入片段（成对末端序列读取片段）。插入片段大小为 2 ~ 200 kb。三个大型 MGSC 测序中心产生了 4,040 万个序列读取片段，犹他大学产生了 60 万个序列读取片段。此外，我们使用了基因组研究所报告的 BAC 插入片段双末端的 40 万个序列读取片段 [54]。

总计 3,360 万个序列读取片段通过了质量和来源的充分检查，其中 2,970 万个是配对的；也就是说，来自同一个克隆的两端。组装了的序列读取片段代表了小鼠常染色体基因组约 7.7 重序列覆盖率（Phred 质量分数大于 20 的碱基覆盖率为 6.5 重）[55]。克隆插入片段总共提供了大约 47 倍的基因组物理覆盖。

序列读取片段以及配对信息被最近开发的两个序列组装程序（Arachne[56,57] 和 Phusion[58]）用来作为输入信息进行组装。除了从少量作为内参的 BAC 中得到一些序列读取片段（小于总数的 0.1%），WGS 组装中不使用图谱信息和基于克隆的序列。在项目过程中，对组装程序进行了中间数据集的测试和比较，从而进行了改进。这些程序在最终的组装中产生可比较的序列输出。Arachne 生成的组装被选为本文所描述的序列草图，因为它产生了更好的短区段和长区段的连续性，具有相当的精确度。

这个组装包括 224,713 个序列重叠群，它们通过至少两个配对序列读取片段连接成超级重叠群（或者支架）。总共有 7,418 个超级重叠群至少 2 kb 长，另外还有 37,125 个代表了小于 1% 组装的较小超级重叠群。重叠群的 N50 长度为 24.8 kb，而超级重叠群的 N50 长度大概是其 700 倍达到 16.9 Mb（如果 N50 长度是 x，则 50% 组装件的长度至少为 x）。事实上，大部分基因组都包括在非常大的超级重叠群中：200 个最大的超级重叠群跨越了 98% 以上的组装序列，其中还有 3% 处于测序空缺区域内。

染色体锚定

我们将尽可能多的超级重叠群按照适当的顺序和方向配到染色体位置上。超级重叠群主要通过与充分验证了的小鼠遗传图谱 [34] 进行比对而定位，一些额外的定位由小鼠辐射杂交图谱 [37] 和 BAC 图谱 [44] 提供。我们没有发现超级重叠群中有不正确的总体连接的证据（即两个不一致位置受到基因组中多个标记的支持），因此可以直接将它们放入染色体中。我们总共放置了 377 个超级重叠群，包括所有长度超过

placed 377 supercontigs, including all supercontigs > 500 kb in length.

Once much of the sequence was anchored, it was possible to exploit additional read-pair and physical mapping information to obtain greater continuity. For example, some adjacent supercontigs were connected by BAC-end (or other) links, satisfying appropriate length and orientation constraints, including single links. Furthermore, some adjacent extended supercontigs were connected by means of fingerprint contigs in the BAC-based physical map. These additional links were used to join sequence into ultracontigs. In the end, a total of 88 ultracontigs with an N50 length of 50.6 Mb (exclusive of gaps) contained 95.7% of the assembled sequence (Fig. 1).

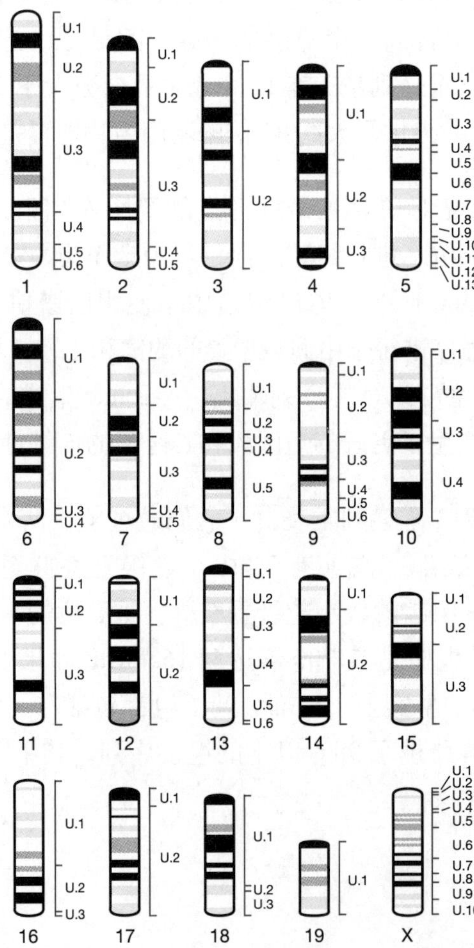

Fig. 1. The mouse genome in 88 sequence-based ultracontigs. The position and extent of the 88 ultracontigs of the MGSCv3 assembly are shown adjacent to ideograms of the mouse chromosomes. All mouse chromosomes are acrocentric, with the centromeric end at the top of each chromosome. The supercontigs of the sequence assembly were anchored to the mouse chromosomes using the MIT genetic map. Neighbouring supercontigs were linked together into ultracontigs using information from single BAC links and the fingerprint and radiation-hybrid maps, resulting in 88 ultracontigs containing 95% of the bases in the euchromatic genome.

500 kb 的超级重叠群。

一旦锚定了大部分序列，就可以利用额外的配对序列读取片段和物理图谱信息来获得更大的连接。例如，一些相邻的超级重叠群通过 BAC 的末端（或者其他）连接相接，如果这些连接包括单连接满足长度和方向约束。此外，在以 BAC 为基础的物理图谱中，一些连接是通过指纹图谱重叠群连接一些相邻延伸的超级重叠群。这些额外的连接用于将序列连成特大超级重叠群。最终，共有 88 个特大超级重叠群，N50 长度为 50.6 Mb（不包括空缺），包括了 95.7% 的组装序列（图 1）。

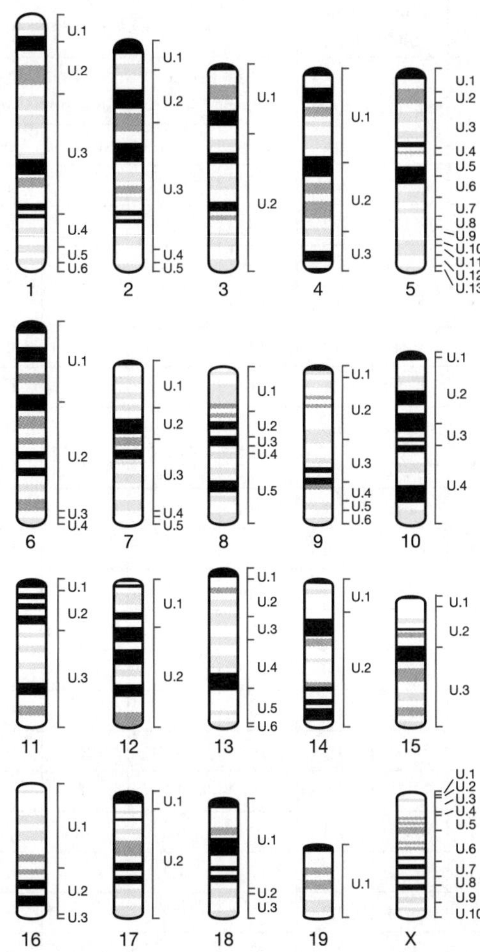

图 1. 88 个基于序列的特大超级重叠群中的小鼠基因组。MGSCv3 组装的 88 个特大超级重叠群的位置和范围显示在小鼠染色体核型模式图相邻的位置。小鼠所有染色体都是近端着丝粒的，着丝粒位于每条染色体的顶端。利用 MIT 基因图谱，将序列组装的超级重叠群锚定在小鼠染色体上。利用来自单个 BAC 连接、指纹图谱和辐射杂交图谱的信息，将相邻的超级重叠群连在一起形成特大超级重叠群，产生 88 个包括常染色体基因组中 95% 碱基的特大超级重叠群。

Of the 187 Mb of finished mouse sequence, 96% was contained in the anchored assembly. This finished sequence, however, is not a completely random cross-section of the genome (it has been cloned as BACs, finished, and in some cases selected on the basis of its gene content). Of 11,452 cDNA sequences from the curated RefSeq collection, 99.3% of the cDNAs could be aligned to the genome sequence (see Supplementary Information). These alignments contained 96.4% of the cDNA bases. Together, this indicates that the draft genome sequence includes approximately 96% of the euchromatic portion of the mouse genome, with about 95% anchored.

Table 3. Mouse chromosome size estimates

Chromosome	Actual bases in sequence (Mb)	Ultracontigs (Mb)		Gaps within supercontigs		Gaps between supercontigs					Total estimated size (Mb) ‡
						Captured by additional read pairs		Captured by fingerprint contigs*		Uncaptured†	
		Number	N50 size	Number	Mb	Number	Mb	Number	Mb	Number	
All	2,372	88	52.7	176,094	104.5	252	14.0	37	2.30	68	2,493
1	183	6	52.7	13,178	7.8	16	1.1	1	0.32	5	192
2	169	5	111.1	12,141	6.5	4	0.1	1	0.20	4	176
3	149	2	108.9	10,630	6.8	17	0.7	3	0.16	1	157
4	140	3	83.1	10,745	6.3	14	0.4	3	0.26	2	147
5	137	13	17.8	11,288	6.7	11	0.5	3	0.11	12	144
6	138	4	91.4	10,021	6.6	19	1.1	2	0.26	3	146
7	122	5	45.1	9,484	5.7	55	3.4	4	0.12	4	131
8	119	5	35.0	9,186	6.1	7	0.2	2	0.12	4	125
9	116	6	26.8	8,479	4.5	6	0.6	1	0.06	5	121
10	121	4	50.4	9,490	5.4	9	0.6	0	0	3	127
11	115	3	80.4	8,681	4.3	2	0.0	1	0.05	2	119
12	105	2	77.4	7,577	4.0	27	1.2	2	0.00	1	110
13	107	6	28.0	7,910	4.7	13	0.8	4	0.19	5	113
14	107	2	93.6	7,605	4.0	10	0.5	2	0.12	1	112
15	96	3	65.3	7,025	4.3	2	0.1	0	0	2	100
16	91	3	62.3	6,695	4.4	1	0.0	0	0	2	95
17	85	2	80.8	6,584	3.7	17	1.2	4	0.19	1	90
18	84	3	73.5	6,192	3.2	2	0.0	0	0	2	87
19	55	1	57.7	3,934	2.4	7	0.6	2	0.12	0	58
X	134	10	19.9	9,249	7.0	13	0.8	2	0.00	9	142

* These gaps had fingerprint contigs spanning them. The size for 18 out of 37 were estimated using conserved synteny to determine the size of the region in the human genome. The remaining gaps were arbitrarily given the average size of the assessed gaps (59 kb), adjusted to reflect the 16% difference in genome size.

† Uncaptured gaps were estimated by mouse–human synteny to have a total size of 5 Mb. However, because some of these gaps are due to repetitive expansions in mouse (absent in human), the actual total for the uncaptured gaps

在 187 Mb 已完成的小鼠序列中，96% 包含在锚定到染色体的组装中。然而，这个已完成的序列不是一个完全随机的基因组横截面（它已经被克隆成 BAC，完成了测序，并且在一些情况下是根据其基因含量被选择的）。在 RefSeq 数据库收集的 11,452 个 cDNA 序列中，99.3% 的 cDNA 可以和基因组序列比对（见补充信息）。这些比对包含 96.4% 的 cDNA 碱基。这表明基因组序列草图包括小鼠基因组大约 96% 的常染色体部分，其中约 95% 被锚定在染色体上。

表 3. 小鼠染色体长度估算

染色体	序列中实际碱基（Mb）	特大超级重叠群（Mb）		超级重叠群内的测序空缺		超级重叠群之间的测序空缺					总估计长度（Mb）‡
						由其他读取片段对捕获		由指纹图谱重叠群捕获 *		未捕获 †	
		数量	N50 大小	数量	Mb	数量	Mb	数量	Mb	数量	
合计	2,372	88	52.7	176,094	104.5	252	14.0	37	2.30	68	2,493
1	183	6	52.7	13,178	7.8	16	1.1	1	0.32	5	192
2	169	5	111.1	12,141	6.5	4	0.1	1	0.20	4	176
3	149	2	108.9	10,630	6.8	17	0.7	3	0.16	1	157
4	140	3	83.1	10,745	6.3	14	0.4	3	0.26	2	147
5	137	13	17.8	11,288	6.7	11	0.5	3	0.11	12	144
6	138	4	91.4	10,021	6.6	19	1.1	2	0.26	3	146
7	122	5	45.1	9,484	5.7	55	3.4	4	0.12	4	131
8	119	5	35.0	9,186	6.1	7	0.2	2	0.12	4	125
9	116	6	26.8	8,479	4.5	6	0.6	1	0.06	5	121
10	121	4	50.4	9,490	5.4	9	0.6	0	0	3	127
11	115	3	80.4	8,681	4.3	2	0.0	1	0.05	2	119
12	105	2	77.4	7,577	4.0	27	1.2	2	0.00	1	110
13	107	6	28.0	7,910	4.7	13	0.8	4	0.19	5	113
14	107	2	93.6	7,605	4.0	10	0.5	2	0.12	1	112
15	96	3	65.3	7,025	4.3	2	0.1	0	0	2	100
16	91	3	62.3	6,695	4.4	1	0.0	0	0	2	95
17	85	2	80.8	6,584	3.7	17	1.2	4	0.19	1	90
18	84	3	73.5	6,192	3.2	2	0.0	0	0	2	87
19	55	1	57.7	3,934	2.4	7	0.6	2	0.12	0	58
X	134	10	19.9	9,249	7.0	13	0.8	2	0.00	9	142

* 这些测序空缺含有横跨的指纹图谱重叠群。利用人类基因组中该区域长度进行保守同线性估计，得到了 37 个中的 18 个的大小。剩余的空缺且用被估算空缺的平均值（59 kb）给出，由此调节反映了基因组的 16% 差异。

† 通过小鼠–人类同线性估计未捕获的测序空缺共有 5 Mb。然而，因为这些空缺中的一部分是由于小鼠中的重复扩张（人类中缺失），所以未捕获空缺的实际总数可能会高得多。例如，1 号染色体（Sp-100rs 区域）上一个大的未捕获空缺就大约有 6 Mb（见正文）。

is probably substantially higher. For example, one large uncaptured gap on chromosome 1 (the Sp-100rs region) is roughly 6 Mb (see text).

‡ Omitting centromeres and telomeres. These would add, on average, approximately 8 Mb per chromosome, or about 160 Mb to the genome. Also omitting uncaptured gaps between supercontigs.

On the basis of the estimated sizes of the ultracontigs and gaps between them, the total length of the euchromatic mouse genome was estimated to be about 2.5 Gb (see Supplementary Information), or about 14% smaller than that of the euchromatic human genome (about 2.9 Gb) (Table 3).

Quality assessment at intermediate scale

Although no evidence of large-scale misassembly was found when anchoring the assembly onto the mouse chromosomes, we examined the assembly for smaller errors.

To assess the accuracy at an intermediate scale, we compared the positions of well-studied markers on the mouse genetic map and in the genome assembly (see Supplementary Information). Out of 2,605 genetic markers that were unambiguously mapped to the sequence assembly (BLAST match using 10^{-100} or better as an *E*-value to a single location) we found 1.8% in which the chromosomal assignment in the genetic map conflicted with that in the sequence. This is well within the known range of erroneous assignments within the genetic map[34].

We also found 19 instances (0.7%) of conflicts in local marker order between the genetic map and sequence assembly. A conflict was defined as any instance that would require changing more than a single genotype in the data underlying the genetic map to resolve. We studied ten cases by re-mapping the genetic markers, and eight were found to be due to errors in the genetic map. On the basis of this analysis, we estimate that chromosomal misassignment and local misordering affects $< 0.3\%$ of the assembled sequence.

At the single nucleotide level in the assembly, the observed discrepancy rates varied in a manner consistent with the quality scores assigned to the bases in the WGS assembly (see Supplementary Information). Overall, 96% of nucleotides in the assembly have Arachne quality scores $\geqslant 40$, corresponding to a predicted error rate of 1 per 10,000 bases. Such bases had an observed discrepancy rate against finished sequence of 0.005%, or 5 errors per 100,000 bases.

Collapse of duplicated regions

The human genome contains many large duplicated regions, estimated to comprise roughly 5% of the genome[59], with nearly identical sequence. If such regions are also common in the mouse genome, they might collapse into a single copy in the WGS assembly. Such artefactual collapse could be detected as regions with unusually high read coverage, compared with the average depth of 7.4-fold in long assembled contigs. We searched for

‡ 去除着丝粒和端粒。这些将使每个染色体平均增加大约 8 Mb，或者基因组增加大约 160 Mb。也除去了超级重叠群之间的未捕获空缺。

以特大超级重叠群和它们之间的空缺的预估尺寸为基础，小鼠常染色体基因组的总长度估计约为 2.5 Gb(见补充信息)，或者比人类常染色体基因组(约 2.9 Gb)小了约 14%(表 3)。

中等规模的质量评估

虽然在将组装锚定到小鼠染色体上时没有大规模错误组装的证据，我们还是检查了组装以发现较小的错误。

为了在中等规模上评估准确性，我们比较了小鼠遗传图谱和基因组组装中研究透彻的标记位置(见补充信息)。在明确定位到序列组装的 2,605 个遗传标记中(用 10^{-100} 或更好值作为 E 值与单个位置进行 BLAST 匹配)，我们发现遗传图谱中有 1.8% 的染色体分配和序列不符。这完全在遗传图谱已知的错误分配范围内[34]。

我们还发现局部标记的顺序在遗传图谱和序列组装之间有 19 例(0.7%)不符。矛盾的定义是需要在遗传图谱基础数据中进行多于一个基因型的更改才能解决不符的任何情况。我们通过重新放置映射遗传标记研究了 10 个例子，发现其中 8 个是遗传图谱的错误。在此分析基础上，我们估计染色体分配错误和局部顺序错误影响小于 0.3% 的组装序列。

在组装的单核苷酸水平上，观测到的差异率的变化一定程度上与 WGS 组装中的匹配碱基的质量分数相一致(参见补充信息)。总体来说，该组装中 96% 的核苷酸 Arachne 质量分数 ≥ 40，相当于预测误差率为每 10,000 个碱基有 1 个错误。这类错误碱基在完成序列上所观测到的差异率为 0.005%，或者每 100,000 个碱基有 5 个错误。

复制区域的分解

人类基因组包含很多大规模的复制区域，估计约占基因组的 5%[59]，它们具有几乎相同的序列。如果这些区域在小鼠的基因组中也很常见，在 WGS 组装中它们可能会分解成单个拷贝。与平均深度 7.4 倍的长组装重叠群相比，这种人工折叠分解会被检测为具有异常高的读取片段覆盖度的区域。我们搜索了长度大于 20 kb 且其

contigs that were > 20 kb in size and contained > 10 kb of sequence in which the read coverage was at least twofold higher than the average. Such regions comprised only a tiny fraction (< 0.0001) of the total assembly, of which only half had been anchored to a chromosome. None of these windows had coverage exceeding the average by more than threefold. This may indicate that the mouse genome contains fewer large regions of near-exact duplication than the human. Alternatively, regions of near-exact duplication may have been systematically excluded by the WGS assembly programme. This issue is better addressed through hierarchical shotgun than WGS sequencing and will be examined more carefully in the course of producing a finished mouse genome sequence.

Unplaced reads and large tandem repeats

We expected that highly repetitive regions of the genome would not be assembled or would not be anchored on the chromosomes. Indeed, 5.9 million of the 33.6 million passing reads were not part of anchored sequence, with 88% of these not assembled into sequence contigs and 12% assembled into small contigs but not chromosomally localized.

Evaluation of WGS assembly strategy

The WGS assembly described here involved only random reads, without any additional map-based information. By many criteria, the assembly is of very high quality. The N50 supercontig size of 16.9 Mb far exceeds that achieved by any previous WGS assembly, and the agreement with genome-wide maps is excellent. The assembly quality may be due to several factors, including the use of high-quality libraries, the variety of insert lengths in multiple libraries, the improved assembly algorithms, and the inbred nature of the mouse strain (in contrast to the polymorphisms in the human genome sequences). Another contributing factor may be that the mouse differs from the human in having less recent segmental duplication to confound assembly.

Notwithstanding the high quality of the draft genome sequence, we are mindful that it contains many gaps, small misassemblies and nucleotide errors. It is likely that these could not all be resolved by further WGS sequencing, therefore directed sequencing will be needed to produce a finished sequence. The results also suggest that WGS sequencing may suffice for large genomes for which only draft sequence is required, provided that they contain minimal amounts of sequence associated with recent segmental duplications or large, recent interspersed repeat elements.

Adding finished sequence

As a final step, we enhanced the WGS sequence assembly by substituting available finished BAC-derived sequence from the B6 strain. In total, we replaced 3,528 draft sequence

中大于 10 kb 序列的读取覆盖度至少是平均值两倍的重叠群。这些区域只包括总组装的小部分（小于 0.0001），其中只有一半锚定在染色体上。这些窗口的覆盖度都没有超过平均值三倍以上。这可能表明小鼠基因组含有比人类更少的近精准复制的大区域。另一种可能是，WGS 组装程序可能已经系统地排除了近精确复制的区域。与 WGS 测序相比，分级鸟枪法能更好解决这个问题，因此将在构建小鼠基因组完成图序列的过程中更仔细地检查这个问题。

未入选的序列读取片段和大串联重复

我们预计基因组中高度重复区域不会被组装或者不会被锚定到染色体上。确实，3,360 万个合格的序列读取中有 590 万个不是被锚定序列的一部分，其中的 88% 不会被组装成序列重叠群，12% 组装成小的重叠群，但是没定位在染色体上。

WGS 组装策略的评估

这里描述的 WGS 组装只涉及随机的序列读取片段，没有任何附加的基于图谱本身的信息。按照很多标准，这种组装是高质量的。超级重叠群的 N50 长度为 16.9 Mb，远远超过之前任何 WGS 组装的结果，并且与全基因组图谱有极好的一致性。组装质量可能归因于几个因素，包括高质量文库的使用、多个文库中插入片段长度的多样性、改进的组装算法以及小鼠品系的近交特性（与人类基因组序列的多态性相比较）。另外一个因素可能是小鼠与人类的不同之处，它具有较少的混淆组装的近期片段复制。

尽管基因组序列草图的质量很高，我们注意到了它包含很多测序空缺、小的组装错误和核苷酸错误。这些问题很可能无法通过进一步的 WGS 测序来解决，因此需要定向测序来产生一个完成图序列。这些结果也提示，WGS 测序对于只需要序列草图的大基因组可能就足够了，前提是近期片段复制或者大量近期分散开的重复元件相关序列最少。

添加完成序列

作为最后一步，我们通过替换来自 B6 品系可用的 BAC 完成序列来增强 WGS 序列组装。总共用组装时能够提供的 210 个完成 BAC 的 48.2 Mb 完成序列替换了

contigs with 48.2 Mb of finished sequence from 210 finished BACs available at the time of the assembly. The resulting draft genome sequence, MGSCv3, was submitted to the public databases and is freely available in electronic form through various sources (see below).

As the MGSC produces additional BAC assemblies and finished sequence, we plan to continue to revise and release enhanced versions of the genome sequence *en route* to a completely finished sequence[66], thereby providing a permanent foundation for biomedical research in the twenty-first century.

Conservation of Synteny between Mouse and Human Genomes

With the draft sequence in hand, we began our analysis by investigating the strong conservation of synteny between the mouse and human genomes. Beyond providing insight into evolutionary events that have moulded the chromosomes, this analysis facilitates further comparisons between the genomes.

Starting from a common ancestral genome approximately 75 Myr, the mouse and human genomes have each been shuffled by chromosomal rearrangements. The rate of these changes, however, is low enough that local gene order remains largely intact. It is thus possible to recognize syntenic (literally "same thread") regions in the two species that have descended relatively intact from the common ancestor.

The earliest indication that genes reside in similar relative positions in different mammalian species traces to the observation that the albino and pink-eye dilution mutants are genetically closely linked in both mouse and rat[67,68]. Significant experimental evidence came from genetic studies of somatic cells[69]. A recent gene-based synteny map[37] used more than 3,600 orthologous loci to define about 200 regions of conserved synteny. However, it is recognized that such maps might still miss regions owing to insufficient marker density.

With a robust draft sequence of the mouse genome and > 90% finished sequence of the human genome in hand, it is possible to undertake a more comprehensive analysis of conserved synteny. Rather than simply relying on known human–mouse gene pairs, we identified a much larger set of orthologous landmarks as follows. We performed sequence comparisons of the entire mouse and human genome sequences using the PatternHunter program[71] to identify regions having a similarity score exceeding a high threshold (> 40, corresponding to a minimum of a 40-base perfect match, with penalties for mismatches and gaps), with the additional property that each sequence is the other's unique match above this threshold. Such regions probably reflect orthologous sequence pairs, derived from the same ancestral sequence.

About 558,000 orthologous landmarks were identified; in the mouse assembly, these sequences have a mean spacing of about 4.4 kb and an N50 length of about 500 bp. The landmarks had a total length of roughly 188 Mb, comprising about 7.5% of the mouse

3,528 个序列草图重叠群。最终的基因组序列草图 MGSCv3 被提交给各公共数据库，并且通过各种资源以电子形式免费提供（见下文）。

随着 MGSC 产生更多的 BAC 组装和完成序列，我们计划继续修订和发布基因组序列的改进版本，使其成为完整的完成序列 [66]，从而为二十一世纪的生物医学研究提供永久的基础。

小鼠和人类基因组同线性的保守性

随着序列草图在手，我们开始分析小鼠和人类基因组之间同线性的强保守性。这项研究不仅可以深入了解染色体构造中的进化事件，还可以促进不同物种基因组之间的进一步比较。

从大约 7,500 万年前的共同的祖先基因组开始，小鼠和人类的基因组都被染色体重排彻底改变了。然而，这个改变率相当低，以至于局部基因序列仍然基本保持完整。因此我们才有可能识别两个物种从共同祖先那里相对完整地传下来的同线性区域（字面意思是“同一条线”）。

最早的迹象表明，在不同哺乳动物物种中，基因位于相似的相对位置，这可追溯到对小鼠和大鼠中白化和粉色眼稀释突变都在遗传学上紧密连锁的观察 [67,68]。重要的实验证据来自体细胞遗传学研究 [69]。最近发布的一项基于基因的同线性图谱 [37] 使用超过 3,600 个直系同源基因座，定义了约 200 个保守的同线性区域。然而，大家还是意识到由于标记密度不足，这些图谱仍然可能遗漏一些同线性区域。

利用稳定的小鼠基因组序列草图和人类基因组已完成的 90% 的序列，可以进行更全面的保守同线性分析。这项工作并不是简单地依赖于已知的人类–小鼠基因对，而是确定了如下所示的更多数量的直系同源标记物。我们使用 PatternHunter 软件 [71] 对整个小鼠和人类基因组序列进行对比，识别出相似分值较高（大于 40 分，即至少有 40 碱基完全匹配，错配和空缺会罚分）的序列区域，此外每个序列必须是另一序列高于阈值的唯一匹配。这些区域可能反映了来自同一个祖先序列的直系同源序列对。

我们识别出约 558,000 个直系同源标记；在小鼠基因组组装中，这些序列的平均间隔约为 4.4 kb，N50 长度约为 500 bp。这些标记物总长度大约为 188 Mb，约占小鼠基因组的 7.5%。应当强调的是，这些标记物仅代表了小鼠和人类基因组中可以

genome. It should be emphasized that the landmarks represent only a small subset of the sequences, consisting of those that can be aligned with the highest similarity between the mouse and human genomes. (Indeed, below we show that about 40% of the human genome can be aligned confidently with the mouse genome.)

The locations of the landmarks in the two genomes were then compared to identify regions of conserved synteny. We define a syntenic segment to be a maximal region in which a series of landmarks occur in the same order on a single chromosome in both species. A syntenic block in turn is one or more syntenic segments that are all adjacent on the same chromosome in human and on the same chromosome in mouse, but which may otherwise be shuffled with respect to order and orientation. To avoid small artefactual syntenic segments owing to imperfections in the two draft genome sequences, we only considered regions above 300 kb and ignored occasional isolated interruptions in conserved order (see Supplementary Information). Thus, some small syntenic segments have probably been omitted—this issue will be addressed best when finished sequences of the two genomes are completed.

Marked conservation of landmark order was found across most of the two genomes (Fig. 2). Each genome could be parsed into a total of 342 conserved syntenic segments. On average, each landmark resides in a segment containing 1,600 other landmarks. The segments vary greatly in length, from 303 kb to 64.9 Mb, with a mean of 6.9 Mb and an N50 length of 16.1 Mb. In total, about 90.2% of the human genome and 93.3% of the mouse genome unambiguously reside within conserved syntenic segments. The segments can be aggregated into a total of 217 conserved syntenic blocks, with an N50 length of 23.2 Mb.

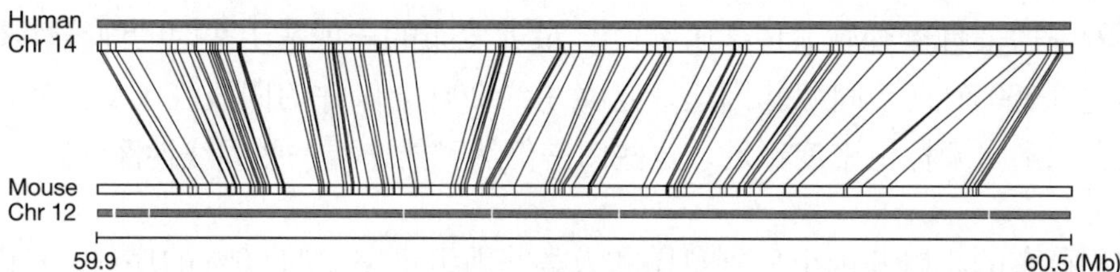

Fig. 2. Conservation of synteny between human and mouse. We detected 558,000 highly conserved, reciprocally unique landmarks within the mouse and human genomes, which can be joined into conserved syntenic segments and blocks (defined in text). A typical 510-kb segment of mouse chromosome 12 that shares common ancestry with a 600-kb section of human chromosome 14 is shown. Blue lines connect the reciprocal unique matches in the two genomes. The cyan bars represent sequence coverage in each of the two genomes for the regions. In general, the landmarks in the mouse genome are more closely spaced, reflecting the 14% smaller overall genome size.

The nature and extent of conservation of synteny differs substantially among chromosomes (Fig. 3). In accordance with expectation, the X chromosomes are represented as single, reciprocal syntenic blocks[72]. Human chromosome 20 corresponds entirely to a portion of mouse chromosome 2, with nearly perfect conservation of order along almost the entire length, disrupted only by a small central segment (Fig. 4a, d). Human chromosome 17

用于比对的高度一致序列的一个小子集。(的确，下文中我们表明大约 40% 人类基因组肯定能够与小鼠的基因组比对。)

随后在两个基因组中比较直系同源标记物的位置以识别保守的同线性区域。我们将同线性片段定义为在两个基因组中的单条染色体按相同顺序出现的一系列直系同源标记的最长区域。一个同线性区域首先是一个或多个在人类基因组和小鼠基因组中一条染色体上的毗邻同线性区段，但是在两个基因组上的排列方向和顺序可以打乱而不同。为了避免由两个基因组序列草图不完善而导致的人为的小同线性片段，我们只考虑大于 300 kb 的区域，并忽略保守排列顺序中偶然出现的孤立的中断部分(见补充信息)。因此，一些小的同线性片段可能会被遗漏了，但当两个基因组完整序列的测序完成时，这个问题会得到彻底解决。

我们在两个基因组的大部分序列中发现了标记顺序的显著保守性(图 2)。每个基因组可以被分解为共计 342 个保守的同线性片段。平均每个标记位于包含其他 1,600 个标记的片段中。这些片段长度差异很大，从 303 kb 到 64.9 Mb，平均 6.9 Mb，N50 长度为 16.1 Mb。总体而言，大约 90.2% 的人类基因组和 93.3% 的小鼠基因组明确置于保守的同线性片段中。这些片段可以组成共计 217 个保守的同线区，N50 长度为 23.2 Mb。

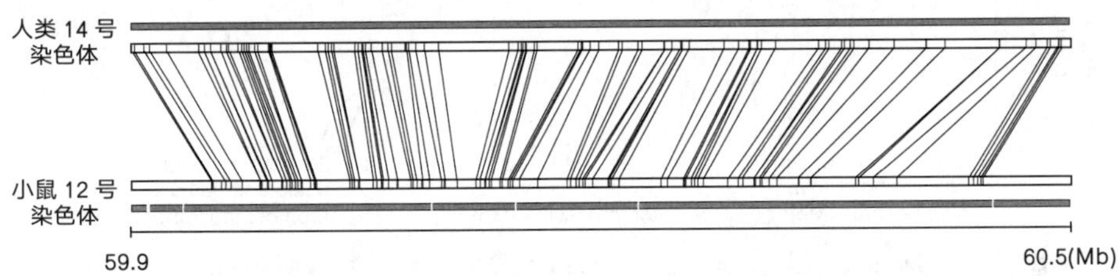

图 2. 人类和小鼠之间保守的同线性。我们在小鼠和人类基因组中检测到 558,000 个高度保守且相互独立的标记，这些标记可以连成保守的同线性片段和区块(定义见正文)。这里显示小鼠 12 号染色体一个 510 kb 与人类 14 号染色体的一个 600 kb 片段具有共同祖先的典型片段。深蓝线连接了两个基因组中相互独立匹配的片段。浅蓝色条带代表该区域在每个基因组的序列覆盖度。一般而言，小鼠基因组中的标记间隔更紧密，反映出其总基因组比人类减少了 14%。

不同染色体之间同线性的保守性质和程度有很大差异(图 3)。与预期一致，X 染色体表现为单一、相互同线性区块[72]。整条人类 20 号染色体和小鼠 2 号染色体的一部分完全对应，保守性的顺序在整个长度上近乎完美，仅被一个小的中心片段打断(图 4a、4d)。整条人类 17 号染色体与小鼠 11 号染色体的一部分完全对应，但

corresponds entirely to a portion of mouse chromosome 11, but extensive rearrangements have divided it into at least 16 segments (Fig. 4b, e). Other chromosomes, however, show evidence of much more extensive interchromosomal rearrangement than these cases (Fig. 4c, f).

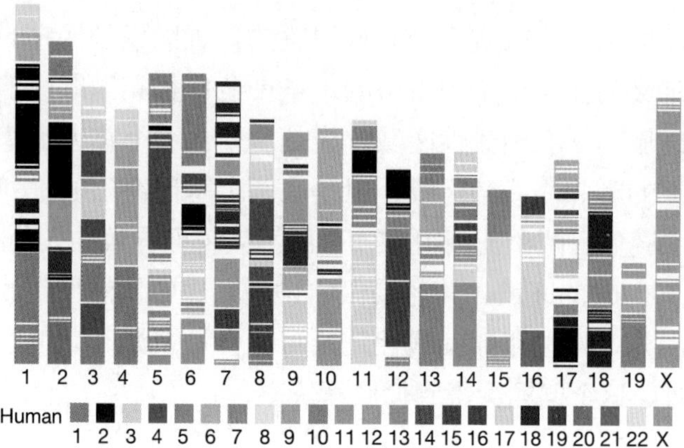

Fig. 3. Segments and blocks > 300 kb in size with conserved synteny in human are superimposed on the mouse genome. Each colour corresponds to a particular human chromosome. The 342 segments are separated from each other by thin, white lines within the 217 blocks of consistent colour.

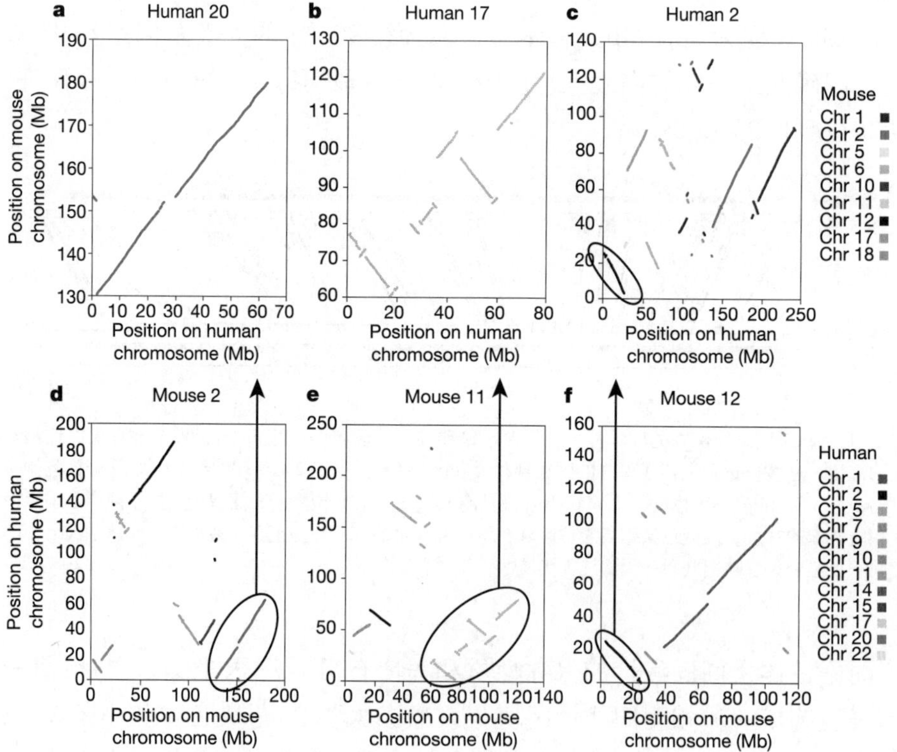

Fig. 4. Dot plots of conserved syntenic segments in three human and three mouse chromosomes. For each of three human (**a–c**) and mouse (**d–f**) chromosomes, the positions of orthologous landmarks are plotted along the *x* axis and the corresponding position of the landmark on chromosomes in the other genome

是大量的重组将其分成了至少16个片段(图4b、4e)。对其他染色体来说，证据则显示比这些例子更为剧烈的染色体之间的重组(图4c、4f)。

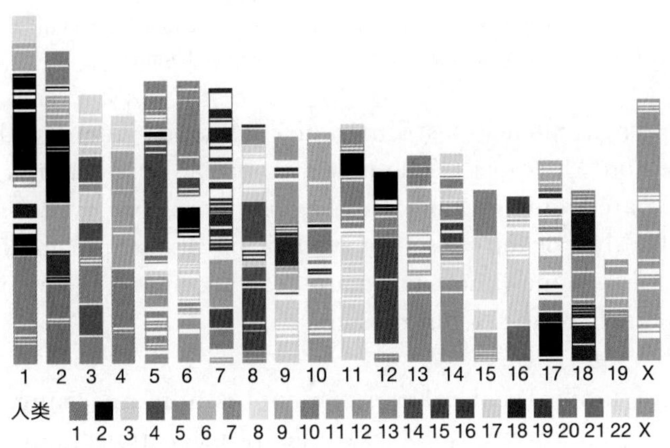

图3. 人类基因组中大于300 kb的保守同线性片段和区块叠加在小鼠的基因组上。每种颜色对应一个特定的人类染色体。在217个具有同一颜色的同线性区块中，共有342个同线性片段通过细白线彼此分开。

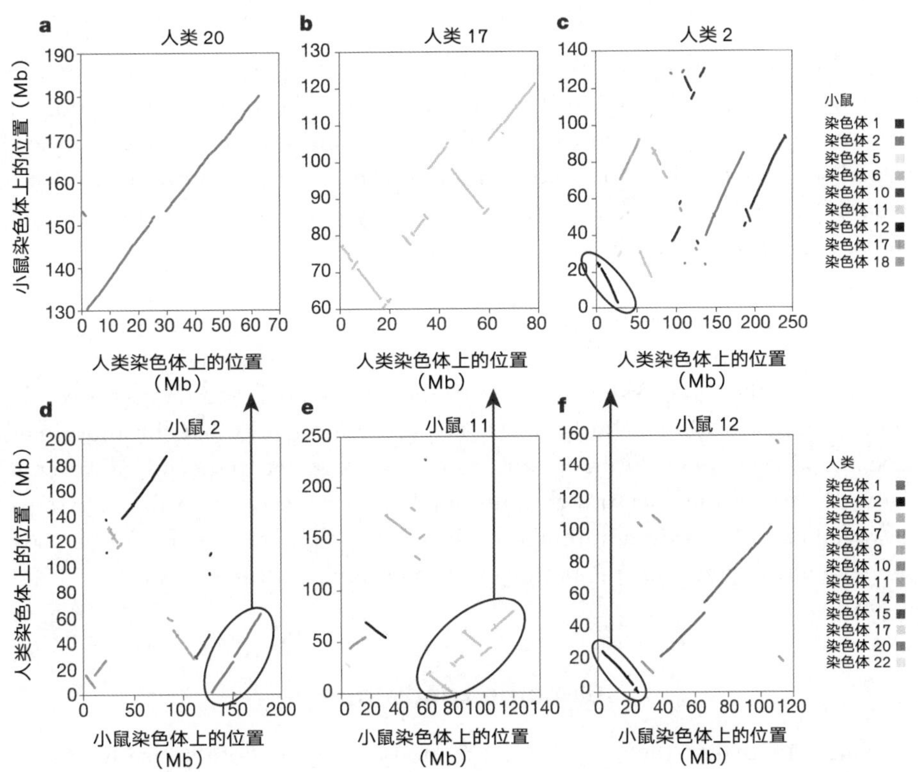

图4. 人类三个染色体和小鼠三个染色体中保守同线性片段的点图。x轴为每条人类(**a**~**c**)和小鼠(**d**~**f**)染色体上直系同源标记的位置，y轴为该标记在另一个基因组染色体上的对应位置。不同颜色区分位于同一条染色体上的多个同源标记对应的另一个物种的不同染色体。保守同线性的一个典型例子是

is plotted on the *y* axis. Different chromosomes in the corresponding genome are differentiated with distinct colours. In a remarkable example of conserved synteny, human chromosome 20 (**a**) consists of just three segments from mouse chromosome 2 (**d**), with only one small segment altered in order. Human chromosome 17 (**b**) also shares segments with only one mouse chromosome (11) (**e**), but the 16 segments are extensively rearranged. However, most of the mouse and human chromosomes consist of multiple segments from multiple chromosomes, as shown for human chromosome 2 (**c**) and mouse chromosome 12 (**f**). Circled areas and arrows denote matching segments in mouse and human.

We compared the new sequence-based map of conserved synteny with the most recent previous map based on 3,600 loci[30]. The new map reveals many more conserved syntenic segments (342 compared with 202) but only slightly more conserved syntenic blocks (217 compared with 170). Most of the conserved syntenic blocks had previously been recognized and are consistent with the new map, but many rearrangements of segments within blocks had been missed (notably on the X chromosome).

The occurrence of many local rearrangements is not surprising. Compared with interchromosomal rearrangements (for example, translocations), paracentric inversions (that is, those within a single chromosome and not including the centromere) carry a lower selective disadvantage in terms of the frequency of aneuploidy among offspring. These are also seen at a higher frequency in genera such as *Drosophila*, in which extensive cytogenetic comparisons have been carried out[73,74].

The block and segment sizes are broadly consistent with the random breakage model of genome evolution[75]. At this gross level, there is no evidence of extensive selection for gene order across the genome. Selection in specific regions, however, is by no means excluded, and indeed seems probable (for example, for the major histocompatibility complex). Moreover, the analysis does not exclude the possibility that chromosomal breaks may tend to occur with higher frequency in some locations.

With a map of conserved syntenic segments between the human and mouse genomes, it is possible to calculate the minimal number of rearrangements needed to "transform" one genome into the other[70,76,77]. When applied to the 342 syntenic segments above, the most parsimonious path has 295 rearrangements. The analysis suggests that chromosomal breaks may have a tendency to reoccur in certain regions. With only two species, however, it is not yet possible to recover the ancestral chromosomal order or reconstruct the precise pathway of rearrangements. As more mammalian species are sequenced, it should be possible to draw such inferences and study the nature of chromosome rearrangement.

Genome Landscape

We next sought to analyse the contents of the mouse genome, both in its own right and in comparison with corresponding regions of the human genome.

All of the mouse genome information is accessible in electronic form through various

人类 20 号染色体（**a**）仅由来自小鼠 2 号染色体（**d**）上的三个同源片段组成，其中只有一个很小的片段的顺序不一致。人类 17 号染色体（**b**）也仅与小鼠一条染色体（11）（**e**）同源片段共享，不过 16 个片段都被彻底重组了。然而，小鼠和人类大多数染色体由来自多个染色体的多个同源片段组成，如人类 2 号染色体（**c**）和小鼠 12 号染色体（**f**）所示。圆圈圈住的区域和箭头表示小鼠和人类相匹配的片段。

我们比较了全新的基于序列保守同线性图谱和最近发布的基于 3,600 个基因座的前期图谱 [30]。新图谱展示的保守同线性片段数量增加了很多（342 个比 202 个），但保守同线性区域数量只有少量增加（217 个比 170 个）。这表明大多数保守同线性区先前已经被识别了，并且与新图谱一致，但是位于区域内的很多重组片段被错过了（特别是 X 染色体）。

发生许多局部重组并不令人意外。就子代间非整倍体频率而言，臂内倒位（即单个染色体内且不包括着丝粒的倒位）相比于染色体间重组（如易位）选择劣势更低。这种局部重组在果蝇等广泛用于细胞遗传学比较的物种中的频率也很高 [73,74]。

同源区域和片段的长度与基因进化的随机断裂模型大体一致 [75]。总体而言，没有证据表明在基因组中存在广泛的对基因排布顺序的选择。然而，在特殊区域绝不排除这种选择，而且看起来确实可能存在（例如，对主要组织相容性复合体的选择）。此外，分析并不排除染色体断裂可能倾向于在某些位置以更高频率发生的可能性。

利用人类和小鼠基因组之间的保守同线性片段图谱，可以计算将一个基因组“转化”为另一个基因组所需要的最小重组数目 [70,76,77]。应用于上述的 342 个同线性片段，最简捷的途径有 295 个重组。分析表明，染色体断裂可能倾向于在某个区域重复出现。然而仅凭两个物种不太可能复原祖先染色体上的排列顺序或重建精确的染色体重组过程。随着更多哺乳动物物种的基因组测序完成，理论上可以得出这样的推论并研究染色体重组的特性。

基因组概貌

下一步，我们试图分析小鼠基因组的内容，包括其本身以及与人类基因组的相应区域的比较。

所有小鼠基因组信息可以通过各种浏览器以电子形式访问：Ensembl（http://

browsers: Ensembl (http://www.ensembl.org), the University of California at Santa Cruz (http://genome.ucsc.edu) and the National Center for Biotechnology Information (http://www.ncbi.nlm.nih.gov). These browsers allow users to scroll along the chromosomes and zoom in or out to any scale, as well as to display information at any desired level of detail. The mouse genome information has also been integrated into existing human genome browsers at these same organizations. In this section, we compare general properties of the mouse and human genomes.

Genome expansion and contraction

The projected total length of the euchromatic portion of the mouse genome (2.5 Gb) is about 14% smaller than that of the human genome (2.9 Gb). To investigate the source of this difference, we examined the relative size of intervals between consecutive orthologous landmarks in the human and mouse genomes. The mouse/human ratio has a mean at 0.91 for autosomes, but varies widely, with the mouse interval being larger than the human in 38% of cases. Chromosome X, by contrast, shows no net relative expansion or contraction, with a mouse/human ratio of 1.03. What accounts for the smaller size of the mouse genome? We address this question below in the sections on repeat sequences and on genome evolution.

(G+C) content

The overall distribution of local (G+C) content is significantly different between the mouse and human genomes. Such differences have been noted in biochemical studies[78-81] and in comparative analyses of fourfold degenerate sites in codons of mouse and human genes[82-85], but the availability of nearly complete genome sequences provides the first detailed picture of the phenomenon.

The mouse has a slightly higher overall (G+C) content than the human (42% compared with 41%), but the distribution is tighter. When local (G+C) content is measured in 20-kb windows across the genome, the human genome has about 1.4% of the windows with (G+C) content > 56% and 1.3% with (G+C) content < 33%. Such extreme deviations are virtually absent in the mouse genome. The contrast is even seen at the level of entire chromosomes. The human has extreme outliers with respect to (G+C) content (the most extreme being chromosome 19), whereas the mouse chromosomes tend to be far more uniform.

There is a strong positive correlation in local (G+C) content between orthologous regions in the mouse and human genomes, but with the mouse regions showing a clear tendency to be less extreme in (G+C) content than the human regions. This tendency is not uniform, with the most extreme differences seen at the tails of the distribution.

In mammalian genomes, there is a positive correlation between gene density and (G+C)

www.ensembl.org)、加州大学圣克鲁斯分校(http://genome.ucsc.edu)和美国国家生物技术信息中心(http://www.ncbi.nlm.nih.gov)。这些浏览器允许用户沿着染色体滚动查看，放大或缩小到任何比例，并且以任何期望的细节水平来显示信息。这些组织机构在现有的人类基因组浏览器中也整合了小鼠基因组。在本节中，我们比较了小鼠和人类基因组的一般特性。

基因组扩增和收缩

小鼠基因组的常染色体部分(2.5 Gb)的预计总长度比人类基因组(2.9 Gb)小约14%。为了研究这种差异的来源，我们检测了人类和小鼠基因组中连续直系同源标记之间的间隔的相对大小。常染色体的小鼠/人的比率平均值为0.91，但是变化范围很大，38%的情况是小鼠的间隔大于人类。相比之下，X染色体没有表现出对应的净扩张或收缩，小鼠/人的比率为1.03。小鼠基因组较小的原因是什么？我们在下面关于重复序列和基因组进化部分中讨论了这个问题。

(G+C)含量

小鼠和人类基因组之间的局部(G+C)含量的总体分布有显著差异。在以往生化研究[78-81]以及小鼠与人类基因组密码子的四倍简并位点的比较分析[82-85]中已经注意到了这些差异，但是近乎完整的可用基因组序列为这种现象提供了第一个详细的图像。

小鼠总体(G+C)含量略高于人类(42%比41%)，但是分布更紧密。在基因组中以20 kb窗口测量局部(G+C)含量时，人类基因组具有(G+C)含量大于56%的窗口约为1.4%,(G+C)含量小于33%的约1.3%。这种极端偏差在小鼠基因组中几乎不存在。这种差别甚至在染色体水平上都能看到。人类有极端的(G+C)含量的离群值(最极端的是19号染色体)，而小鼠的染色体则趋向于更加均匀。

在小鼠和人类基因组的直系同源区域之间，局部(G+C)含量有很强的正相关性，但小鼠区域(G+C)含量的趋势明显不如人类区域那样极端。这种趋势并不均匀，最大差异分布在尾部。

在哺乳动物基因组中，基因密度和(G+C)含量呈正相关[81,86-89]。鉴于人类和小

content[81,86-89]. Given the differences in (G+C) content between human and mouse, we compared the distribution of genes—using the sets of orthologous mouse and human genes described below—with respect to (G+C) content for both genomes. The density of genes differed markedly when expressed in terms of absolute (G+C) content, but was nearly identical when expressed in terms of percentiles of (G+C) content. For example, both species have 75–80% of genes residing in the (G+C)-richest half of their genome. Mouse and human thus show similar degrees of homogeneity in the distribution of genes, despite the overall differences in (G+C) content. Notably, the mouse shows similar extremes of gene density despite being less extreme in (G+C) content.

What accounts for the differences in (G+C) content between mouse and human? Does it reflect altered selection for (G+C) content[90,91], altered mutational or repair processes[92-94], or possibly both? Data from additional species will probably be needed to address these issues. Any explanation will need to account for various mysterious phenomena. For example, although overall (G+C) content in mouse is slightly higher than in human (42% compared with 41%), the (G+C) content of chromosome X is slightly lower (39.0% compared with 39.4%). The effect is even more pronounced if one excludes lineage-specific repeats (see below), thereby focusing primarily on shared DNA. In that case, mouse autosomes have an overall (G+C) content that is 1.5% higher than human autosomes (41.2% compared with 39.7%) whereas mouse chromosome X has a (G+C) content that is 1% lower than human chromosome X (37.8% compared with 36.8%).

CpG islands

In mammalian genomes, the palindromic dinucleotide CpG is usually methylated on the cytosine residue. Methyl-CpG is mutated by deamination to TpG, leading to approximately fivefold under-representation of CpG across the human[1,95] and mouse genomes. In some regions of the genome that have been implicated in gene regulation, CpG dinucleotides are not methylated and thus are not subject to deamination and mutation. Such regions, termed CpG islands, are usually a few hundred nucleotides in length, have high (G+C) content and above average representation of CpG dinucleotides.

We applied a computer program that attempts to recognize CpG islands on the basis of (G+C) and CpG content of arbitrary lengths of sequence[96,97] to the non-repetitive portions of human and mouse genome sequences. The mouse genome contains fewer CpG islands than the human genome (about 15,500 compared with 27,000), which is qualitatively consistent with previous reports[98]. The absolute number of islands identified depends on the precise definition of a CpG island used, but the ratio between the two species remains fairly constant.

The reason for the smaller number of predicted CpG islands in mouse may relate simply to the smaller fraction of the genome with extremely high (G+C) content[99] and its effect on the computer algorithm. Approximately 10,000 of the predicted CpG islands in each species show significant sequence conservation with CpG islands in the orthologous intervals in

鼠之间(G+C)含量的差异，我们用下面描述的小鼠和人类的同源基因集合来对比两种基因组的(G+C)含量来比较基因的分布。两物种间基因密度就绝对(G+C)含量而言，差异显著，但是就(G+C)含量百分数而言，几乎相同。例如，两个物种都有75%~80%的基因位于它们基因组中(G+C)含量最高的一半的区域。因此，尽管在(G+C)含量上存在总体差异，但小鼠和人类在基因分布上显示出相似程度的同质性。值得注意的是，尽管在(G+C)含量上不那么极端，但在基因密度上，小鼠表现出了相似的极端性。

小鼠和人类(G+C)含量差异的原因是什么？它是否反映了对(G+C)含量的不同选择[90,91]，不同的突变或修复过程[92-94]，或者可能两者都有？可能需要来自其他物种的数据来解决这些问题。各种神秘现象都需要解释。例如，尽管小鼠总(G+C)含量略高于人类(42%比41%)，但是X染色体的(G+C)含量略低于人类(39.0%比39.4%)。如果排除了特定谱系的重复序列(见下文)，这一现象更加明显，因此主要关注共享DNA。在这种情况下，小鼠常染色体总(G+C)含量比人类常染色体高1.5%(41.2%比39.7%)，而小鼠X染色体的总(G+C)含量比人类X染色体低1%(37.8%比36.8%)。

CpG岛

在哺乳动物的基因组中，回文二核苷酸CpG通常在胞嘧啶残基上甲基化。甲基化的CpG通过脱氨基作用突变为TpG，导致人类和小鼠基因组中CpG的含量大概降低为原来的1/5[1,95]。在一些与基因调控有关的基因组区域，CpG二核苷酸不甲基化，因此不受脱氨基和突变的影响。这些区域被称为CpG岛，通常是几百个核苷酸的长度，高(G+C)含量和高于平均数量的CpG二核苷酸。

我们用一个计算机程序根据(G+C)含量和任意长度序列CpG的含量[96,97]识别人类和小鼠基因组序列中非重复部分的CpG岛。小鼠基因组包含的CpG岛比人类基因组略少(大约15,500比27,000)，这与之前的报道[98]在数量级上是一致的。确定的CpG岛的绝对数量依赖于所使用的CpG岛的精确界定，但是两个物种之间的比例保持相当一致。

小鼠中预测的CpG岛数量较少的原因可能与基因组中仅有较少区域具有极高(G+C)含量[99]及这一现象对计算机算法的影响有关。每个物种中约有10,000个预测的CpG岛显示出与其他物种的直系同源区间中的CpG岛存在显著的序列保守性，

the other species, falling within the orthologous landmarks described above. Perhaps these represent functional CpG islands, a proposition that can now be tested experimentally[84].

Repeats

The single most prevalent feature of mammalian genomes is their repetitive sequences, most of which are interspersed repeats representing "fossils" of transposable elements. Transposable elements are a principal force in reshaping the genome, and their fossils thus provide powerful reporters for measuring evolutionary forces acting on the genome. A recent paper on the human genome sequence[1] provided extensive background on mammalian transposons, describing their biology and illustrating many applications to evolutionary studies. Here, we will focus primarily on comparisons between the repeat content of the mouse and human genomes.

Mouse has accumulated more new repetitive sequence than human

Approximately 46% of the human genome can be recognized currently as interspersed repeats resulting from insertions of transposable elements that were active in the last 150–200 million years. The total fraction of the human genome derived from transposons may be considerably larger, but it is not possible to recognize fossils older than a certain age because of the high degree of sequence divergence. Because only 37.5% of the mouse genome is recognized as transposon-derived, it is tempting to conclude that the smaller size of the mouse genome is due to lower transposon activity since the divergence of the human and mouse lineages. Closer analysis, however, shows that this is not the case. As we discuss below, transposition has been more active in the mouse lineage. The apparent deficit of transposon-derived sequence in the mouse genome is mostly due to a higher nucleotide substitution rate, which makes it difficult to recognize ancient repeat sequences.

Lineage-specific versus ancestral repeats

Interspersed repeats can be divided into lineage-specific repeats (defined as those introduced by transposition after the divergence of mouse and human) and ancestral repeats (defined as those already present in a common ancestor). Such a division highlights the fact that transposable elements have been more active in the mouse lineage than in the human lineage. Approximately 32.4% of the mouse genome (about 818 Mb) but only 24.4% of the human genome (about 695 Mb) consists of lineage-specific repeats. Contrary to initial appearances, transposon insertions have added at least 120 Mb more transposon-derived sequence to the mouse genome than to the human genome since their divergence. This observation is consistent with the previous report that the rate of transposition in the human genome has fallen markedly over the past 40 million years[1,100].

属于之前描述的直系同源性标记。或许这些区域代表了功能性的 CpG 岛，这个假设可以通过实验进行验证 [84]。

重复序列

哺乳动物基因组中最普遍的特征是它们的重复序列，其中大多数是离散的重复序列，代表着转座元件的“化石”。转座元件是重塑基因组的主要力量，因此它们的化石为测量作用于基因组的进化力提供了有力的报告。最近一篇关于人类基因组序列的论文 [1] 给出了哺乳动物转座子的大量背景，描述了它们的生物学特性并说明了在进化研究中的许多应用。这里，我们将主要关注于小鼠和人类基因组中重复内容之间的比较。

小鼠比人类积累了更多的新的重复序列

目前，大约 46% 的人类基因组是由过去 1.5 亿至 2 亿年中活跃的转座子插入引起的离散重复序列。来自转座子的人类基因组的总比例可能相当大，但是由于序列的高度分化，不可能识别比某个时代更久远的化石。因为只有 37.5% 的小鼠基因组被认为来自转座子，很容易得出结论，小鼠基因组更小是由于人类和小鼠谱系分化后，转座子活性降低。然而，更仔细的分析显示情况并非如此。正如我们下面讨论的，转座现象在小鼠谱系中更为活跃。小鼠基因组中转座衍生序列的明显缺失主要是由于更高的核苷酸替换率，以至于很难识别古老的重复序列。

谱系特异性与祖源重复

离散的重复可以分为谱系特异性重复（定义为小鼠和人类分化后通过转座引入的那些重复）和祖源重复（定义为存在于共同祖先中的重复）。这样的划分凸显了一个事实，即转座元件在小鼠谱系中比在人类谱系中更为活跃。约 32.4% 的小鼠基因（大约 818 Mb）但只有 24.4% 的人类基因（大约 695 Mb）由谱系特异性重复组成。与最初的表现相反，自从小鼠和人类的祖先分开之后，在小鼠基因组的转座子插入中，其转座子衍生序列比人类基因组增加了至少 120 Mb。这个观察与之前报道的人类基因组的转座率在过去的 4,000 万年中显著下降一致 [1,100]。

The overall lower interspersed repeat density in mouse is the result of an apparent lack of ancestral repeats: they comprise only 5% of the mouse genome compared with 22% of the human genome. The ancestral repeats recognizable in mouse tend to be those of more recent origin, that is, those that originated closest to the mouse–human divergence. This difference may be due partly to a higher deletion rate of non-functional DNA in the mouse lineage, so that more of the older interspersed repeats have been lost. However, the deficit largely reflects a much higher neutral substitution rate in the mouse lineage than in the human lineage, rendering many older ancestral repeats undetectable with available computer programs.

Higher substitution rate in mouse lineage

The hypothesis that the neutral substitution rate is higher in mouse than in human was suggested as early as 1969 (refs 101–103). The idea has continued to be challenged on the basis that the apparent differences may be due to inaccuracies in mammalian phylogenies[104,105]. The explanation, however, remains unclear, with some attributing it to generation time[101,106] and others pointing to a closer correlation with body size[107,108].

Ancestral repeats provide a powerful measure of neutral substitution rates, on the basis of comparing thousands of current copies to the inferred consensus sequence of the ancestral element. The large copy number and ubiquitous distribution of ancestral repeats overcome issues of local variation in substitution rates (see below). Most notably, differences in divergence levels are not affected by phylogenetic assumptions, as the time spent by an ancestral repeat family in either lineage is necessarily identical.

The median divergence levels of 18 subfamilies of interspersed repeats that were active shortly before the human–rodent speciation indicates an approximately twofold higher average substitution rate in the mouse lineage than in the human lineage, corresponding closely to an early estimate by Wu and Li[109]. In human, the least-diverged ancestral repeats have about 16% mismatch to their consensus sequences, which corresponds to approximately 0.17 substitutions per site. In contrast, mouse repeats have diverged by at least 26–27% or about 0.34 substitutions per site, which is about twofold higher than in the human lineage. The total number of substitutions in the two lineages can be estimated at 0.51. Below, we obtain an estimate of a combined rate of 0.46–0.47 substitutions per site, on the basis of an analysis that counts only substitutions since the divergence of the species.

Assuming a speciation time of 75 Myr, the average substitution rates would have been 2.2×10^{-9} and 4.5×10^{-9} in the human and mouse lineages, respectively. This is in accord with previous estimates of neutral substitution rates in these organisms. (Reports of highly similar substitution rates in human and mouse lineages relied on a much earlier divergence time of rodents from other mammals[104].)

Comparison of ancestral repeats to their consensus sequence also allows an estimate of the

小鼠总的离散重复密度较低归因于祖源重复序列的明显缺乏：它们仅占小鼠基因组的 5%，而在人类基因组中占了 22%。在小鼠基因组中可识别的祖源重复序列倾向于那些更为近期起源的，也就是起源于最接近小鼠和人类分化的那些。这种差异存在的部分可能原因是小鼠谱系中非功能性 DNA 的删除率较高，因此丢失了更多的古老的离散重复序列。但是，这种缺陷很大程度上反映了小鼠谱系中的中性替代率明显高于人类谱系，使得许多更古老的祖源重复无法用可用的计算机程序检测到。

小鼠谱系中的高替代率

早在 1969 年就提出了小鼠的中性替代率高于人类的假说（参考文献 101 ~ 103）。这一观点持续受到挑战，因为这种明显的差异可能是由哺乳动物系统发育不准确造成的[104,105]。然而，这一解释还是不够清楚，一些人将其归因于世代时间[101,106]，而另一些人则指出与身体尺寸关系更紧密[107,108]。

祖源重复序列为中性替代率提供了一个强有力的度量方法，该方法基于将数千份当前拷贝与所推断的祖源元件的共有序列进行比较。祖源重复的大量拷贝数和广泛分布克服了替代率局部差异的问题（见下文）。更值得注意的是，分化水平的差异不受系统发育假设的影响，因为祖源重复家族在任何一个谱系中花费的时间是必然相同的。

在人类–啮齿动物物种形成之前不久活跃的 18 个离散重复序列亚家族的中位分化水平表明，小鼠谱系中的平均替代率大约是人类谱系的两倍，这与 Wu 和 Li[109] 的早期估计非常接近。在人类中，分化最小的祖源重复序列与其共有序列有大约 16% 的错配，相当于每个位点大约 0.17 个替换。相比之下，小鼠的重复序列有至少 26% ~ 27% 的差异或是每个位点大约 0.34 个替换，大约是人类谱系的两倍。这两个谱系的替换总数估计为 0.51。下面，我们根据仅计算物种分化以来产生的替换的分析，得出了每个位点 0.46 ~ 0.47 个替换的综合估计值。

假设物种形成时间为 7,500 万年，人类和小鼠谱系的平均替换率分别为 2.2×10^{-9} 和 4.5×10^{-9}。这与之前对这些生物体的中性替代率的估计是一致的。（关于人类和小鼠谱系中高度相似的替代率的报道取决于啮齿动物与其他哺乳动物更早的分化时间[104]。）

将祖源重复序列和它们的一致性序列进行比较，还可以估计小片段插入（小于

rate of occurrence of small (< 50 bp) insertions and deletions (indels). Both species show a net loss of nucleotides (with deleted bases outnumbering inserted bases by at least 2–3-fold), but the overall loss owing to small indels in ancestral repeats is at least twofold higher in mouse than in human. This may contribute a small amount (1–2%) to the difference in genome size noted above.

It should be noted that the roughly twofold higher substitution rate in mouse represents an average rate since the time of divergence, including an initial period when the two lineages had comparable rates. Comparison with more recent relatives (mouse–rat and human–gibbon, each about 20–25 Myr) indicate that the current substitution rate per year in mouse is probably much higher, perhaps about fivefold higher (see Supplementary Information). Also, note that these estimates refer to substitution rate per year, rather than per generation. Because the human generation time is much longer than that of the mouse (by at least 20-fold), the substitution rate is greater in human than mouse when measured per generation.

Higher substitution rate obscures old repeats

We measured the impact of the higher substitution rate in mouse on the ability to detect ancestral repeats in the mouse genome. By computer simulation, the ability of the RepeatMasker[100] program to detect repeats was found to fall off rapidly for divergence levels above about 37%. If we simulate the events in the mouse lineage by adjusting the ancestral repeats in the human genome for the higher substitution levels that would have occurred in the mouse genome, the proportion of the genome that would still be recognizable as ancestral repeats falls to only 6%. This is in close agreement with the proportion actually observed for the mouse. Thus, the current analysis of repeated sequences allows us to see further back into human history (roughly 150–200 Myr) than into mouse history (roughly 100–120 Myr).

A higher rate of interspersed repeat insertion does not explain the larger size of the human genome. Below, we suggest that the explanation lies in a higher rate of large deletions in the mouse lineage.

Comparison of mouse and human repeats

All mammals have essentially the same four classes of transposable elements: (1) the autonomous long interspersed nucleotide element (LINE)-like elements; (2) the LINE-dependent, short RNA-derived short interspersed nucleotide elements (SINEs); (3) retrovirus-like elements with long terminal repeats (LTRs); and (4) DNA transposons. The first three classes procreate by reverse transcription of an RNA intermediate (retroposition), whereas DNA transposons move by a cut-and-paste mechanism of DNA sequence (see refs 1, 100 for further information about these classes).

50 bp）和缺失的发生率。两个物种都有核苷酸的净损失（缺失的碱基数量至少是插入碱基的 2 ~ 3 倍），但是祖源重复中小片段插入缺失导致的整体损失在小鼠中至少是在人类中的两倍。这可能是导致上述基因组大小差异的一小部分原因（1% ~ 2%）。

应该注意的是，小鼠中大约 2 倍高的替代率代表了自分化时间以来的平均比率，包括两个谱系的替代率相当的初始时期。与最近的近源物种（小鼠–大鼠及人类–猩猩，每个大约 2,000 ~ 2,500 万年）的比较表明，小鼠每年的替代率可能要高得多，大约是 5 倍（见补充信息）。此外，注意这些估计指的是每年的替代率，而不是每代的替代率。因为人类的世代时间要比小鼠长很多（至少 20 倍），所以当测量每个世代时，人类的替代率比小鼠高很多。

较高的替代率掩盖了旧的重复

我们测量了小鼠高替代率对检测小鼠基因组祖源重复序列能力的影响。通过计算机模拟，发现当差异水平超过 37% 时，RepeatMasker[100] 程序检测重复的能力迅速下降。如果我们模拟小鼠谱系中的事件时利用人类基因组中的祖源重复来校正小鼠基因组中的高替代水平，那么仍然可识别的祖源重复在基因组中的比例下降到仅有 6%。这与实际观察到的小鼠比例非常一致。因此，对重复序列进行的当前分析，可以让我们追溯人类历史（大约 1.5 亿 ~ 2 亿年）到比小鼠历史更远的年代（大约 1 亿 ~ 1.2 亿年）。

比较高的散布的重复插入率并不能解释为何人类的基因组比较大。通过下面的分析，我们认为原因在于小鼠谱系中较高的缺失率。

小鼠和人类重复序列的比较

所有哺乳动物基本上都有四类转座元件：(1) 自主的长散在核苷酸元件（LINE）一类元件；(2) 依赖于 LINE、由短 RNA 衍生的短散在核苷酸元件（SINE）；(3) 具有长末端重复（LTR）的逆转录病毒样元件；(4) DNA 转座子。前三类通过 RNA 中间体介导的逆转录（逆转录转座）产生，而 DNA 转座通过 DNA 的剪切–粘贴机制进行（更多信息见参考文献 1、100）。

A comparison of these repeat classes in the mouse and human genomes can be enlightening. On the one hand, differences between the two species reveal the dynamic nature of transposable elements; on the other hand, similarities in the location of lineage-specific elements point to common biological factors that govern insertion and retention of interspersed repeats.

Differences between mouse and human

The most notable difference is in the changing rate of transposition over time: the rate has remained fairly constant in mouse, but markedly increased to a peak at about 40 Myr in human, and then plummeted. This phenomenon was noted in our initial analysis of the human genome; the availability of the mouse genome sequence now confirms and sharpens the observation. Beyond this overall tendency, there are specific differences in each of the four repeat classes.

The first class that we discuss is LINEs. Copies of LINE1 (L1) form the single largest fraction of interspersed repeat sequence in both human and mouse. No other LINE seems to have been active in either lineage. The extant L1 elements in both species derive from a common ancestor by means of a series of subfamilies defined primarily by the rapidly evolving 3′ non-coding sequences[110]. The L1 5′-untranslated regions (UTRs) in both lineages have been even more variable, occasionally through acquisition of entirely new sequences[111]. Indeed, the three active subfamilies in mouse, which are otherwise > 97% identical, have unrelated or highly diverged 5′ ends[112-114]. L1 seems to have remained highly active in mouse, whereas it has declined in the human lineage. Goodier and co-workers[113] estimated that the mouse genome contains at least 3,000 potentially active elements (full-length with two intact open reading frames (ORFs)). The current draft sequence of the mouse genome contains only 400 young, full-length elements; of these only 12 have two intact ORFs. This is probably a reflection of the WGS shotgun approach used to assemble the genome. Indeed, most of the young elements in the draft genome sequence are incomplete owing to internal sequence gaps, reflecting the difficulty that WGS assembly has with highly similar repeat sequences. This is a notable limitation of the draft sequence.

The second repeat class is SINEs. Whereas only a single SINE (Alu) was active in the human lineage, the mouse lineage has been exposed to four distinct SINEs (B1, B2, ID, B4). Each is thought to rely on L1 for retroposition, although none share sequence similarity, as is the rule for other LINE–SINE pairs[115,116]. The mouse B1 and human Alu SINEs are unique among known SINEs in being derived from 7SL RNA; they probably have a common origin[117]. The mouse B2 is typical among SINEs in having a transfer RNA-derived promoter region. Recent ID elements seem to be derived from a neuronally expressed RNA gene called *BC1*, which may itself have been recruited from an earlier SINE. This subfamily is minor in mouse, with 2–4,000 copies, but has expanded rapidly in rat where it has produced more than 130,000 copies since the mouse–rat speciation[118]. Both B2 and ID closely resemble Ala-tRNA, but seem to have independent origins. The B4 family resembles

对小鼠和人类基因组中这些重复类型进行比较可以获得启发。一方面，两个物种之间的差异揭示了转座元件的动态性质；另一方面，谱系特殊元件位置上的相似性表明控制离散重复的插入和保留具有共同的生物学因素。

小鼠和人类之间的差异

最显著的区别在于转座率随着时间发生的变化：小鼠的转座率保持相当稳定，但人类在大约 4,000 万年的时候，转座率明显增加并达到峰值，然后下降。我们在对人类基因组进行初步分析时注意到了这个现象，小鼠基因组序列的获得则证实并印证了这一观察。除了这个整体趋势外，四个重复类型中的每一个都有特定的差异。

我们首先讨论第一类 LINE。在人类和小鼠中，LINE1(L1)的拷贝构成了离散重复序列中的最大部分。在这两个谱系中似乎都没有其他具有活性的 LINE。两个物种中存在的 L1 元件是从一个共同祖先，通过 3′非编码序列的快速进化形成的一系列亚家族衍生而来[110]。两个谱系中 L1 的 5′非翻译区变异性很大，偶尔会获得全新的序列[111]。事实上，小鼠中三个活跃的亚家族在其他方面的相似性大于 97%，但却具有不相关或者高度分化的 5′端[112-114]。L1 似乎在小鼠中保持高度活性，而在人类谱系中却已经下降了。古迪尔和同事们[113]估计，小鼠基因组中至少包含 3,000 个潜在活性元件(全长可达两个完整的开放阅读框)。目前小鼠基因组序列草图只含有 400 个年轻的全长元件，其中只有 12 个含有两个完整的开放阅读框。这可能是用 WGS 鸟枪法组装基因组的一种现象。事实上，由于内部序列空缺，基因组序列草图中大多数年轻元件是不完整的，这反映了 WGS 在组装高度相似重复序列时的难度。这是该序列草图的一个显著不足。

第二类重复是 SINE。虽然人类谱系中只有一个 SINE(Alu)有活性，但是在小鼠谱系中已经显露了四个不同的 SINE(B1、B2、ID、B4)。尽管它们没有共享序列相似性，每个都被认为依赖于 L1 进行逆转录转座，这也是其他 LINE-SINE 配对的规则[115,116]。小鼠 B1 和人类 Alu 的 SINE 在已知的 SINE 中比较特殊，均衍生于 7SL RNA，所以它们可能有共同起源[117]。小鼠的 B2 则是 SINE 中比较典型的，具有转运 RNA 衍生的启动子区域。近期出现的 ID 元件似乎来自神经组织表达的称为 *BC1* 的 RNA 基因，它本身可能是被一个早期 SINE 募集的。这个亚家族在小鼠体内是次要的，有 2 ~ 4,000 个拷贝数，但是在大鼠中迅速扩张，自小鼠–大鼠物种形成以来已经产生超过 130,000 个拷贝数[118]。B2 和 ID 都与 Ala-tRNA(丙氨酸–转

a fusion between B1 and ID[119,120]. We found that 25% of the 75,000 identified ID elements were located within 50 bp of a B1 element of similar orientation, suggesting that perhaps most older ID elements are mislabelled or truncated B4 SINEs.

The third repeat class is LTR elements. All interspersed LTR-containing elements in mammals are derivatives of the vertebrate-specific retrovirus clade of retrotransposons. The earliest infectious retroviruses probably originated from endogenous retroviral-like (ERV) elements that acquired mechanisms for horizontal transmission[121], whereas many current endogenous retroviral elements have probably arisen from infection by retroviruses.

Endogenous retroviruses fall into three classes (I–III), which show a markedly dissimilar evolutionary history in human and mouse. Notably, ERVs are nearly extinct in human whereas all three classes have active members in mouse.

Class III accounts for 80% of recognized LTR element copies predating the human–mouse speciation. This class includes the non-autonomous MaLRs: with 388,000 recognizable copies in mouse, it is the single most successful LTR element. It is still active in mouse (represented by MERVL and the MT and ORR1 MaLRs), but died out some 50 Myr in human[122].

Copies of class II elements are tenfold denser in mouse than in human. In contrast, class I element copies are fourfold more common in the human than the mouse genome (although it is possible that some have not yet been recognized in mouse). In mouse, this class includes active ERVs, such as the murine leukaemia virus, MuRRS, MuRVY and VL30 (several of which have caused insertional mutations in mouse)—no similar activity is known to exist in human.

The fourth repeat class is the DNA transposons. Although most transposable elements have been more active in mouse than human, DNA transposons show the reverse pattern. Only four lineage-specific DNA transposon families could be identified in mouse (the mariner element MMAR1, and the hAT elements URR1, RMER30 and RChar1), compared with 14 in the primate lineage.

For evolutionary survival, DNA transposons are thought to depend on frequent horizontal transfer to new host genomes by means of vectors such as viruses and other intracellular parasites[116,125]. The mammalian immune system probably forms a large obstacle to the successful invasion of DNA transposons. Perhaps the rodent germ line has been harder to infiltrate by horizontal transfer than the primate genome. Alternatively, it is possible that highly diverged families active in early rodent evolution have not been detected yet. Notably, most copies in the human genome were deposited early in primate evolution.

Some of the above differences in the nature of interspersed repeats in human and mouse could reflect systematic factors in mouse and human biology, whereas others may represent

运 RNA）相似，但似乎是独立起源。B4 家族类似于 B1 和 ID 之间的融合 [119,120]。我们发现，在 75,000 个已经识别的 ID 元件中，25% 定位于具有相似方向的 B1 元件的 50 bp 中，这表明可能大多古老的 ID 元件是错误标记或者截短的 B4 SINE。

第三类重复是 LTR 元件。哺乳动物中所有离散的含 LTR 元件都是脊椎动物特有的逆转录病毒分支的逆转录转座子衍生物。最早的感染性逆转录病毒可能起源于获得了水平传播机制的内源性逆转录病毒（ERV）样元件 [121]，而现在很多内源性逆转录病毒元件可能来自逆转录病毒的感染。

内源性逆转录病毒分为三类（I ~ III），显示了人类和小鼠进化史明显不同。值得注意的是，ERV 在人类中几乎灭绝，而这三类在小鼠中都有活跃的成员。

在人类–小鼠物种形成之前，第 III 类元件占可识别的 LTR 元件拷贝数的 80%。这类包括非自主的 MaLR：小鼠中包括 388,000 个可识别的拷贝数，它是唯一最成功的 LTR 元件。仍然活跃在小鼠身上（以 MERVL、MT 和 ORR1 MaLR 为代表），但在人类中已经死亡了约 5,000 万年 [122]。

第 II 类元件的拷贝数在小鼠中的密度是人类中的十倍。相比之下，第 I 类元件拷贝数在人类基因组中是在小鼠基因组中的四倍多（尽管在小鼠中可能还有一些没有被识别）。在小鼠中，这类包括活跃的 ERV（如小鼠白血病病毒）、MuRRS、MuRVY 和 VL30（其中一些已经导致小鼠插入突变）——在人类中已经没有类似的活性。

第四类重复是 DNA 转座子。尽管大多数转座元件在小鼠体内比在人类中更活跃，但 DNA 转座子显示出相反的模式。在小鼠中，只鉴定出四个谱系特异性 DNA 转座家族（mariner 元件 MMAR1，hAT 元件 URR1、RMER30 和 RChar1），而灵长类谱系中有 14 个。

为了进化生存，DNA 转座子被认为依赖于通过病毒和其他细胞内的寄生虫等载体频繁水平转移到新的宿主基因组 [116,125]。哺乳动物的免疫系统可能对 DNA 转座子成功入侵形成巨大的屏障。或许啮齿动物的生殖种系比灵长类基因组更难被水平转移渗透。也可能在啮齿动物进化中活跃的高分化家族可能还没有被检测到。值得注意的是，人类基因组中 DNA 转座子的大多拷贝在灵长类进化早期就被保存下来了。

上述人类和小鼠离散重复的性质差异可以反映小鼠和人类生物学中的系统因素，

random fluctuations. Deeper understanding of the biology of transposable elements and detailed knowledge of interspersed repeat populations in other mammals should clarify these issues.

Similar repeats accumulate in orthologous locations

One of the most notable features about repeat elements is the contrast in the genomic distribution of LINEs and SINEs. Whereas LINEs are strongly biased towards (A+T)-rich regions, SINEs are strongly biased towards (G+C)-rich regions. The contrast is all the more notable because both elements are inserted into the genome through the action of the same endonuclease[126,127].

Such preferences were studied in detail in the initial analysis of the human genome[1], and essentially equivalent preferences are seen in the mouse genome. With the availability of two mammalian genomes, however, it is possible to extend this analysis to explore whether (A+T) and (G+C) content are truly causative factors or merely reflections of an underlying biological process.

Towards that end, we studied the insertion of lineage-specific repeat elements in orthologous segments in the human and mouse genomes. Each insertion represents a new, independent event occurring in one lineage, and thus any correlation between the two species reflects underlying proclivity to insert or retain repeats in particular regions. Visual inspection reveals a strong correlation in the sites of lineage-specific repeats of the various classes. Lineage-specific repeats also correlate with other genomic features, as discussed in the section on genome evolution.

The correlation of local lineage-specific SINE density is extremely strong. Moreover, local SINE density in one species is better predicted by SINE density in the other species than it is by local (G+C) content. The local density of each distinct rodent-specific type of SINE is a strong predictor of Alu density at the orthologous locus in human, although the Alu equivalent B1 SINEs show the strongest correlation ($r^2 = 0.784$).

We interpret these results to mean that SINE density is influenced by genomic features that are correlated with (G+C) content but that are distinct from (G+C) content *per se*. The fact that (G+C) content alone does not determine SINE density is consistent with the observation that some (G+C)-rich regions of the human genome are not Alu rich[128,129].

Simple sequence repeats

Mammalian genomes are scattered with simple sequence repeats (SSRs), consisting of short perfect or near-perfect tandem repeats that presumably arise through slippage during DNA replication. SSRs have had a particularly important role as genetic markers

而其他可能代表了随机波动。对转座元件生物机制更深入的了解和对其他哺乳动物的离散重复序列的详细了解应该可以澄清这些问题。

类似重复在直系同源位置的积累

重复元件的一个最显著的特征是 LINE 和 SINE 的基因组分布差异。LINE 强烈倾向于(A+T)富集区，而 SINE 强烈倾向于(G+C)富集区。这种差异如此显著是因为两种元件都是通过相同的内切核酸酶的作用插入到基因组中[126,127]。

在人类基因组的最初分析中就对这种倾向性进行了详细研究[1]，在小鼠基因组中也发现了类似的倾向。但是，随着两个哺乳动物基因组的获得，就可以扩展这一分析以探索(A+T)和(G+C)含量是真正的影响因素或者仅仅是对于潜在生物学过程的一种反映。

为此，我们研究了在人类和小鼠基因组的直系同源片段中插入的谱系特异性重复元件。每个插入代表了一个新的、独立的事件发生在一个谱系中，因此两个物种之间的任何相关性反映了在特定区域插入或保留重复序列的潜在倾向。目测显示，不同类型的谱系特异的重复位点有很强的相关性。谱系特异的重复也与其他基因组特征相关，如在基因组进化一节中讨论所述。

局部谱系特异的 SINE 密度的相关性非常强。此外，与根据局部(G+C)含量进行预测相比，一个物种的局部 SINE 密度可以通过另一个物种的 SINE 密度进行更好的预测。尽管 Alu 与 B1 SINE 表现出最强的关联性($r^2 = 0.784$)，每种啮齿动物特定类型的 SINE 局部密度都是人类直系同源性基因座上的 Alu 密度的强力预测因子。

我们将这些结果解释为 SINE 密度受到与(G+C)含量相关但与(G+C)含量本身不同的基因组特征的影响。(G+C)含量不单独决定 SINE 密度，这一事实与人类基因组中某些(G+C)富集的区域 Alu 并不富集的观察结果一致[128,129]。

简单序列重复

哺乳动物基因组散在分布着简单重复序列(SSR)，可能由 DNA 复制过程中的滑脱引起的短的完全或近似完全的串联重复序列组成。SSR 在小鼠和人类的连锁研

in linkage studies in both mouse and human, because their lengths tend to be polymorphic in populations and can be readily assayed by PCR. It is possible that such SSRs, arising as they do through replication errors, would be largely equivalent between mouse and human; however, there are impressive differences between the two species[135].

Overall, mouse has 2.25–3.25-fold more short SSRs (1–5 bp unit) than human; the precise ratio depends on the percentage identity required in defining a tandem repeat. The mouse seems to represent an exception among mammals on the basis of comparison with the small amount of genomic sequence available from dog (4 Mb) and pig (5 Mb), both of which show proportions closer to human[136] (E. Green, unpublished data).

The analysis can be refined, however, by excluding transposable elements that contain SSRs at their 3′ ends. When these sources are eliminated, the contrast between mouse and human grows to roughly fourfold.

The reason for the greater density of SSRs in mouse is unknown. Analysis of the distribution of SSRs across chromosomes also reveals an interesting feature common to both organisms. In both human and mouse, there is a nearly twofold increase in density of SSRs near the distal ends of chromosome arms. Because mouse chromosomes are acrocentric, they show the effect only at one end. The increased density of SSRs in telomeric regions may reflect the tendency towards higher recombination rates in subtelomeric regions[1].

Mouse Genes

Genes comprise only a small portion of the mammalian genome, but they are understandably the focus of greatest interest. One of the most notable findings of the initial sequencing and analysis of the human genome[1] was that the number of protein-coding genes was only in the range of 30,000–40,000, far less than the widely cited textbook figure of 100,000, but in accord with more recent, rigorous estimates[55,139-141]. The lower gene count was based on the observed and predicted gene counts, statistically adjusted for systematic under- and overcounting.

Our goal here is to produce an improved catalogue of mammalian protein-coding genes and to revisit the gene count. Genome analysis has been enhanced by a number of recent developments. These include burgeoning mammalian EST and cDNA collections, knowledge of the genomes and proteomes of a growing number of organisms, increasingly complete coverage of the mouse and human genomes in high-quality sequence assemblies, and the ability to use *de novo* gene prediction methodologies that exploit information from two mammalian genomes to avoid potential biases inherent in using known transcripts or homology to known genes.

We focus here on protein-coding genes, because the ability to recognize new RNA genes remains rudimentary. As used below, the terms "gene catalogue" and "gene count" refer to

究中作为遗传标记物具有特别重要的作用，因为它们的长度在群体中倾向于多态形式存在，可以轻易地通过 PCR 进行检测。这类 SSR 可能在人类和小鼠之间大致相同，都是由于复制错误而产生的，尽管其序列在两个物种之间存在着显著的差异[135]。

总体而言，小鼠的短 SSR（1 ~ 5 bp 单位）是人类的 2.25 ~ 3.25 倍；精确比率取决于定义一个串联重复所需要的百分比的一致性。根据与狗（4 Mb）和猪（5 Mb）这两个比例都接近于人类的物种[136]的可获得的少量基因组序列的比较，小鼠似乎代表了哺乳动物中的例外。（格林，未发表数据）

然而，通过排除 3′端包含 SSR 的转座元件可以优化分析。当这些来源被消除后，小鼠和人类之间的差异将增加为大约四倍。

小鼠体内的 SSR 密度较大的原因未知。对 SSR 跨染色体分布的分析也揭示了两种生物体共有的一个有趣特征。在人类和小鼠中，靠近染色体臂末端的 SSR 密度几乎增加了两倍。因为小鼠的染色体是近端着丝点的，所以它们仅在一端显示出效果。端粒区的 SSR 密度增加可能反映了亚端粒区高重组率的趋势[1]。

小鼠基因

虽然基因只占哺乳动物基因组的一小部分，但它们是最受关注的焦点。人类基因组初期测序分析[1]最显著的发现之一是，编码蛋白质的基因数量只在 30,000 ~ 40,000 范围内，远低于被广泛引用的教科书上的数字 100,000，但是符合近期更为严谨的估计[55,139-141]。较低的基因计数是基于对于观察的和预测的基因计数进行系统的不足和过度计数的统计校正。

我们的目标是建立一个改进的哺乳动物蛋白质编码基因目录和重新审视基因计数。最近的很多进展增强了基因组分析能力。这包括迅速发展的哺乳动物 EST 和 cDNA 收集，对于越来越多的生物的基因组和蛋白质组的知识，对小鼠和人类基因组的覆盖度越来越完整的高质量序列组装，以及使用新型基因预测方法从两个哺乳动物基因组获取信息，从而避免因使用已知转录本或已知同源基因时固有的潜在偏差的能力。

我们在此关注蛋白质编码基因，因为识别新的 RNA 基因的能力仍然处于初级阶段。如下文所用，术语“基因目录”和“基因计数”仅涉及蛋白质编码基因。我们在

protein-coding genes only. We briefly discuss RNA genes at the end of the section.

Evidence-based gene prediction

We constructed catalogues of human and mouse gene predictions on the basis of available experimental evidence. The main computational tool was the Ensembl gene prediction pipeline[142] augmented with the Genie gene prediction pipeline[143]. Briefly, the Ensembl system uses three tiers of input. First, known protein-coding cDNAs are mapped onto the genome. Second, additional protein-coding genes are predicted on the basis of similarity to proteins in any organism using the GeneWise program[144]. Third, *de novo* gene predictions from the GENSCAN program[145] that are supported by experimental evidence (such as ESTs) are considered. These three strands of evidence are reconciled into a single gene catalogue by using heuristics to merge overlapping predictions, detect pseudogenes and discard misassemblies. These results are then augmented by using conservative predictions from the Genie system, which predicts gene structures in the genomic regions delimited by paired 5′ and 3′ ESTs on the basis of cDNA and EST information from the region.

The computational pipeline produces predicted transcripts, which may represent fragmentary products or alternative products of a gene. They may also represent pseudogenes, which can be difficult in some cases to distinguish from real genes. The predicted transcripts are then aggregated into predicted genes on the basis of sequence overlaps. The computational pipeline remains imperfect and the predictions are tentative.

Initial and current human gene catalogue

The initial human gene catalogue[1] contained about 45,000 predicted transcripts, which were aggregated into about 32,000 predicted genes containing a total of approximately 170,000 distinct exons (Table 10). Many of the predicted transcripts clearly represented only gene fragments, because the overall set contained considerably fewer exons per gene (mean 4.3, median 3) than known full-length human genes (mean 10.2, median 8).

This initial gene catalogue was used to estimate the number of human protein-coding genes, on the basis of estimates of the fragmentation rate, false positive rate and false negative rate for true human genes. Such corrections were particularly important, because a typical human gene was represented in the predictions by about half of its coding sequence or was significantly fragmented. The analysis suggested that the roughly 32,000 predicted genes represented about 24,500 actual human genes (on the basis of fragmentation and false positive rates) out of the best-estimate total of approximately 31,000 human protein-coding genes on the basis of estimated false negatives[1]. We suggested a range of 30,000–40,000 to allow for additional genes.

本节末尾简要讨论了 RNA 基因。

基于证据的基因预测

我们在现有实验证据的基础上构建了人类和小鼠基因预测目录。主要的计算工具 Ensembl 基因预测流程 [142]，并用 Genie 基因预测流程 [143] 加强预测结果。简而言之，Ensembl 系统使用了三层输入。首先，已知编码蛋白质的 cDNA 标记到基因组上。第二，利用 GeneWise 程序 [144]，根据在任何生物体中蛋白质相似性预测剩下的蛋白质编码基因。第三，基于 GENSCAN 程序 [145] 中的由实验证据（如 EST）支持的新基因预测。通过使用试探法合并重叠预测，检测假基因和去掉错误组装，这三组证据被整合成单个基因目录。然后，这些结果通过 Genie 系统的保守预测得到进一步证实。Genie 是基于 cDNA 和 EST 信息，由配对的 5′和 3′ EST 界定基因组区域的基因结构预测系统。

计算流程产生预测的转录产物，这些可能代表了一个基因的片段产物或可变产物。它们也可能代表假基因，在某些情况下很难将其与真正的基因区分开。然后预测的转录产物根据序列重叠汇集到预测的基因中。目前计算流程还不完善，因此预测的产物也是试验性的。

初始和目前的人类基因目录

最初的人类基因目录 [1] 包含大概 45,000 个预测的转录本，被比对到大约 32,000 个预测基因上，其中包括总计约 170,000 个不同的外显子（表 10）。大多数预测的转录本明显只代表基因片段，因为整个序列中每个基因的外显子数目（平均值为 4.3，中位数为 3）比在已知的人类全长基因组上的数目（平均值为 10.2，中位数为 8）少很多。

基于对于碎片化比率、假阳性率和假阴性率的评估，这个初始基因目录被用来估算人类蛋白质编码基因的数量。这种校正非常重要，因为一个典型的人类基因在预测中只代表了其编码序列的一半，或者呈现明显的片段化。分析表明，约 32,000 个预测基因代表了基于假阴性估计的大约 31,000 个人类蛋白编码基因的最佳估计总数 [1] 中的 24,500 个实际的人类基因（基于碎片化和假阳性率）。我们建议使用 30,000 ~ 40,000 的范围以允许增加其他基因。

Table 10. Gene count in human and mouse genomes

Genome feature	Human		Mouse	
	Initial	Current	Initial*	Extended†
	(Feb. 2001)	(Sept. 2002)	(this paper)	(this paper)
Predicted transcripts	44,860	27,048	28,097	29,201
Predicted genes	31,778	22,808	22,444	22,011
Known cDNAs	14,882	17,152	13,591	12,226
New predictions	16,896	5,656	8,853	9,785
Mean exons/transcript‡	4.2 (3)	8.7 (6)	8.2 (6)	8.4 (6)
Total predicted exons	170,211	198,889	191,290	213,562

* Without RIKEN cDNA set.
† With RIKEN cDNA set.
‡ Median values are in parentheses.

Since the initial paper[1], the human gene catalogue has been refined as sequence becomes more complete and methods are revised. The current catalogue (Ensembl build 29) contains 27,049 predicted transcripts aggregated into 22,808 predicted genes containing about 199,000 distinct exons (Table 10). The predicted transcripts are larger, with the mean number of exons roughly doubling (to 8.7), and the catalogue has increased in completeness, with the total number of exons increasing by nearly 20%. We return below to the issue of estimating the mammalian gene count.

Mouse gene catalogue

We sought to create a mouse gene catalogue using the same methodology as that used for the human gene catalogue (Table 10). An initial catalogue was created by using the same evidence set as for the human analysis, including cDNAs and proteins from various organisms. This set included a previously published collection of mouse cDNAs produced at the RIKEN Genome Center[41].

We also created an extended mouse gene catalogue by including a much larger set of about 32,000 mouse cDNAs with significant ORFs that were sequenced by RIKEN (see ref. 150). These additional mouse cDNAs improved the catalogue by increasing the average transcript length through the addition of exons (raising the total from about 191,000 to about 213,000, including many from untranslated regions) and by joining fragmented transcripts. The set contributed roughly 1,200 new predicted genes. The total number of predicted genes did not change significantly, however, because the increase was offset by a decrease due to mergers of predicted genes. These mouse cDNAs have not yet been used to extend the human gene catalogue. Accordingly, comparisons of the mouse and human gene catalogues below use the initial mouse gene catalogue.

表 10. 人类和小鼠基因组中的基因计数

基因组特性	人类		小鼠	
	初始	现在	初始 *	扩展 †
	(2001 年 2 月)	(2002 年 9 月)	(本文)	(本文)
预测的转录本	44,860	27,048	28,097	29,201
预测的基因	31,778	22,808	22,444	22,011
已知 cDNA	14,882	17,152	13,591	12,226
新预测	16,896	5,656	8,853	9,785
平均外显子数值 / 转录本 ‡	4.2(3)	8.7(6)	8.2(6)	8.4(6)
预测的外显子合计	170,211	198,889	191,290	213,562

* 不包括 RIKEN 的 cDNA 集。
† 包括 RIKEN 的 cDNA 集。
‡ 括号中为中位数。

从最初的文献 [1] 开始，人类基因目录随着序列的完整和方法的修订而不断完善。目前的目录(Ensembl 版本 29)包含 27,049 个预测转录本，比对到 22,808 个预测基因上，其中包含大约 199,000 个不同的外显子(表 10)。预测的转录本更大，外显子的平均数大约翻了一番(至 8.7)，目录的完整性增加，外显子的总数增加了近 20%。下面我们重新回到估计哺乳动物基因计数问题。

小鼠基因目录

我们试图利用与人类基因目录相同的方式创建一个小鼠基因目录(表 10)。最初的目录是使用与人类分析相同的证据集合创建的，包括不同生物体的 cDNA 和蛋白质。这一集合包括 RIKEN(日本理化学研究所)基因组中心之前发表的小鼠 cDNA 数据集 [41]。

我们还创建了一个扩展的小鼠基因目录，包括一个由 RIKEN 测序且具有重要开放阅读框的更大的约 32,000 个小鼠 cDNA 的集合(见参考文献 150)。这些额外的小鼠 cDNA 通过增加平均转录本长度，即通过增加外显子(总数从约 191,000 增加到约 213,000，包括许多来自非翻译区)和通过连接片段化的转录产物来改进目录。该集合贡献了大约 1,200 个新的预测基因。然而，预测基因的总数没有明显变化，因为增加量被预测基因的合并造成的减少相抵消了。这些小鼠 cDNA 尚未用来扩展人类基因目录。因此，下面的小鼠和人类基因目录的比较使用的是初始小鼠基因目录。

The extended mouse gene catalogue contains 29,201 predicted transcripts, corresponding to 22,011 predicted genes that contain about 213,500 distinct exons. These include 12,226 transcripts corresponding to cDNAs in the public databases, with 7,481 of these in the well-curated RefSeq collection[151]. There are 9,785 predicted transcripts that do not correspond to known cDNAs, but these are built on the basis of similarity to known proteins.

The new mouse and human gene catalogues contain many new genes not previously identified in either genome. It should be emphasized that the human and mouse gene catalogues, although increasingly complete, remain imperfect. Both genome sequences are still incomplete. Some authentic genes are missing, fragmented or otherwise incorrectly described, and some predicted genes are pseudogenes or are otherwise spurious. We describe below further analysis of these challenges.

Pseudogenes

An important issue in annotating mammalian genomes is distinguishing real genes from pseudogenes, that is, inactive gene copies. Processed pseudogenes arise through retrotransposition of spliced or partially spliced mRNA into the genome; they are often recognized by the loss of some or all introns relative to other copies of the gene. Unprocessed pseudogenes arise from duplication of genomic regions or from the degeneration of an extant gene that has been released from selection. They sometimes contain all exons, but often have suffered deletions and rearrangements that may make it difficult to recognize their precise parentage. Over time, pseudogenes of either class tend to accumulate mutations that clearly reveal them to be inactive, such as multiple frameshifts or stop codons. More generally, they acquire a larger ratio of non-synonymous to synonymous substitutions than functional genes. These features can sometimes be used to recognize pseudogenes, although relatively recent pseudogenes may escape such filters.

To assess the impact of pseudogenes on gene prediction, we focused on two classes of gene predictions: (1) those that lack a corresponding gene prediction in the region of conserved synteny in the human genome (2,705); and (2) those that are members of apparent local gene clusters and that lack a reciprocal best match in the human genome (5,143). A random sample of 100 such predicted genes was selected, and the predictions were manually reviewed. We estimate that about 76% of the first class and about 30% of the second class correspond to pseudogenes. Overall, this would correspond to roughly 4,000 of the predicted genes in mouse.

Comparison of mouse and human gene sets

We then sought to assess the extent of correspondence between the mouse and human gene sets. Approximately 99% of mouse genes have a homologue in the human genome. For 96% the homologue lies within a similar conserved syntenic interval in the human genome.

扩展的小鼠基因目录包括 29,201 个预测转录本，对应的 22,011 个预测基因包含约 213,500 个不同的外显子。这些包括对应于 cDNA 公共数据库中的 12,226 个转录本，其中 7,481 个在 RefSeq 集合 [151] 中有很好的注释。有 9,785 个预测转录本与已知的 cDNA 不对应，但这些转录本是建立在已知蛋白质相似性的基础上。

新的小鼠和人类基因目录包括很多新基因，这些新基因以前没有在任何一个基因组中被识别出来。应该强调的是，人类和小鼠基因目录虽然日趋完整，但仍然不完善。两个基因组序列仍然不完整。一些真实的基因缺失，碎片化或以其他方式被错误描述，一些预测基因是假基因或其他形式的假象。下面我们将进一步叙述对这些挑战的分析。

假基因

哺乳动物基因组注释的一个重要问题是区分真基因和假基因，即非活性基因拷贝。经过加工的假基因是通过剪接或部分剪接 mRNA 再逆转录到基因组中而产生的；相比于基因的其他拷贝，假基因经常通过部分或全部内含子的丢失而被识别。未经过加工的假基因产生于基因组区域的重复或是自然选择所释放的某个现存基因的退化。它们有时包含所有的外显子，但是经常遭受缺失和重排，这可能使得很难辨别它们的精确亲缘关系。随着时间的推移，任何一类的假基因都倾向于积累突变，明确显示它们是非活性的，例如存在多个移码或者终止密码子。一般而言，它们比功能基因获得更大比例的非同义或同义替代。这些特征有的时候会被用来识别假基因，尽管相对较新的假基因可能会逃过这种筛选。

为了评估假基因对基因预测的影响，我们关注两类基因预测：(1) 与人类基因组的保守同线性区域内缺乏相应基因预测的 (2,705)；(2) 在人类基因组中缺乏相互最佳匹配但有明显局部基因簇的成员 (5,143)。随机抽取 100 个这样的预测基因样本，并对预测进行人工检查。我们估计大约 76% 的第一类和大约 30% 的第二类与假基因相对应。总体而言，这相当于小鼠中大约有 4,000 个预测基因。

小鼠和人类基因集合的比较

然后，我们试图评估小鼠和人类基因集合之间的对应程度。大约 99% 的小鼠基因在人类基因组中具有同源性。96% 的同源基因位于人类基因组中相似的保守同线性间隔内。对于 80% 的小鼠基因，人类基因组中的最佳匹配反过来在保守同线性间

For 80% of mouse genes, the best match in the human genome in turn has its best match against that same mouse gene in the conserved syntenic interval. These latter cases probably represent genes that have descended from the same common ancestral gene, termed here 1:1 orthologues.

Comprehensive identification of all orthologous gene relationships, however, is challenging. If a single ancestral gene gives rise to a gene family subsequent to the divergence of the species, the family members in each species are all orthologous to the corresponding gene or genes in the other species. Accordingly, orthology need not be a 1:1 relationship and can sometimes be difficult to discern from paralogy (see protein section below concerning lineage-specific gene family expansion).

There was no homologous predicted gene in human for less than 1% (118) of the predicted genes in mouse. Genes that seem to be mouse-specific may correspond to human genes that are still missing owing to the incompleteness of the available human genome sequence. Alternatively, there may be true human homologues present in the available sequence, but the genes could be evolving rapidly in one or both lineages and thus be difficult to recognize.

Mammalian gene count

To re-estimate the number of mammalian protein-coding genes, we studied the extent to which exons in the new set of mouse cDNAs sequenced by RIKEN[132] were already represented in the set of exons contained in our initial mouse gene catalogue, which did not use this set as evidence in gene prediction.

Our estimates suggest that the mammalian gene count may fall at the lower end of (or perhaps below) our previous prediction of 30,000–40,000 based on the human draft sequence[1]. Although small, single-exon genes may add further to the count, the total seems unlikely to greatly exceed 30,000. This lower estimate for the mammalian gene number is consistent with other recent extrapolations[141].

RNA genes

The genome also encodes many RNAs that do not encode proteins, including abundant RNAs involved in mRNA processing and translation (such as ribosomal RNAs and tRNAs), and more recently discovered RNAs involved in the regulation of gene expression and other functions (such as micro RNAs)[165,166]. There are probably many new RNAs not yet discovered, but their computational identification has been difficult because they contain few hallmarks. Genomic comparisons have the potential to significantly increase the power of such predictions by using conservation to reveal relatively weak signals, such as those arising from RNA secondary structure[167]. We illustrate this by showing how comparative

隔中与相同的小鼠基因也具有最佳匹配。后面这些例子可能代表来自同一共同祖先的基因，这里称为 1∶1 直系同源性基因。

然而，全面鉴定所有直系同源基因关系具有挑战性。如果单个祖先基因在物种分化后产生了基因家族，则每个物种的家族成员都与相应基因或其他物种中的基因是直系同源的。因此，直系同源不一定是 1∶1 关系，有时很难从种内同源中辨别出来(参考下文关于谱系特异性基因家族扩增的蛋白质章节)。

小鼠中，少于 1%(118)的预测的基因与人类中预测的基因没有同源性。看起来是小鼠特有的基因如果在人类基因中还是缺失状态可能缘于现有的人类基因组序列仍然不完整。也有可能是在现有的序列中存在真正的人源基因，但是基因可能在一个或两个谱系中快速进化，因此很难识别。

哺乳动物基因计数

为了重新估计哺乳动物蛋白编码基因，我们研究了 RIKEN 测序 [132] 的小鼠 cDNA 新集合中的外显子在我们最初的小鼠基因目录中的外显子集合中的富集程度，而且没有使用这一组数据作为基因预测的证据。

我们的估计表明，哺乳动物基因计数可能会处于(或可能低于)我们先前基于人类序列草图 [1] 预测的 30,000 ~ 40,000 的下限值。虽然小的单外显子基因可能进一步增加计数，但总数似乎不可能大大超过 30,000。对于哺乳动物基因数的这种较低估计与最近的其他推断相符 [141]。

RNA 基因

基因组还编码许多不编码蛋白质的 RNA，包括参与 mRNA 加工和翻译的大量 RNA(如核糖体 RNA 和 tRNA)，以及最近发现的参与基因表达调控和其他功能的 RNA(例如微 RNA) [165,166]。可能还有很多新 RNA 没有被发现，但由于其特征较少导致它们的计算识别很困难。通过基因组之间的比较，使用保守性来揭示相对较弱的信号(例如由 RNA 二级结构产生的信号 [167])能显著增加此类预测的潜在能力。我们通过展示比较基因组学如何提高一个即便是非常了解的基因家族(tRNA 基因)的识

genomics can improve the recognition of even an extremely well understood gene family, the tRNA genes.

In our initial analysis of the human genome[1], the program tRNAscan-SE[168] predicted 518 tRNA genes and 118 pseudogenes. A small number (about 25 of the total) were filtered out by the RepeatMasker program as being fossils of the MIR transposon, a long-dead SINE element that was derived from a tRNA[169,170].

To improve discrimination of functional tRNA genes, we exploited comparative genomic analysis of mouse and human. True functional tRNA genes would be expected to be highly conserved. Indeed, the 498 putative mouse tRNA genes differ on average by less than 5% (four differences in about 75 bp) from their nearest human match, and nearly half are identical. In contrast, non-genic tRNA-related sequences (those labelled as pseudogenes by tRNAscan-SE or as SINEs by RepeatMasker) differ by an average of 38% and none is within 5% divergence. Notably, the 19 suspect predictions that violate the wobble rules show an average of 26% divergence from their nearest human homologue, and none is within 5% divergence.

On the basis of these observations, we identified the set of tRNA genes having cross-species homologues with < 5% sequence divergence. The set contained 335 tRNA genes in mouse and 345 in human. In both cases, the set represents all 46 expected anti-codons and exactly satisfies the expected wobble rules. The sets probably more closely represent the true complement of functional tRNA genes.

Although the excluded putative genes (163 in mouse and 167 in human) may include some true genes, it seems likely that our earlier estimate of approximately 500 tRNA genes in human is an overestimate. The actual count in mouse and human is probably closer to 350.

Mouse Proteome

Eukaryotic protein invention appears to have occurred largely through two important mechanisms. The first is the combination of protein domains into new architectures. (Domains are compact structures serving as evolutionarily conserved functional building blocks that are often assembled in various arrangements (architectures) in different proteins[174].) The second is lineage-specific expansions of gene families that often accompany the emergence of lineage-specific functions and physiologies[175] (for example, expansions of the vertebrate immunoglobulin superfamily reflecting the invention of the immune system[1], receptor-like kinases in *A. thaliana* associated with plant-specific self-incompatibility and disease-resistance functions[49], and the trypsin-like serine protease homologues in *D. melanogaster* associated with dorsal–ventral patterning and innate immune response[176,177]).

The availability of the human and mouse genome sequences provides an opportunity to

别度来说明这一点。

在我们对人类基因组最初的分析中[1]，tRNAscan-SE[168]程序预测了518个tRNA基因和118个假基因。少量（一共25个）被RepeatMasker程序识别为MIR转座子化石（来源于tRNA的早已死亡的SINE元件[169,170]）而被过滤。

为了提高对功能性tRNA基因的识别，我们进行了小鼠和人类的比较基因组分析。真正的功能性tRNA基因预期是高度保守的。事实上，498个假定的小鼠tRNA基因与最接近的人类匹配基因的平均差异度小于5%（75 bp中大约有4个差异），近一半是完全相同的。相比之下，非tRNA基因相关序列（那些被tRNAscan-SE标记为假基因或者被RepeatMasker标记为SINE的序列）平均差异度为38%，没有一个在5%的差异范围内。值得注意的是，违反了摆动法则的19个可疑预测显示，与它们最接近的人类同系物的差异度平均为26%，没有一个在5%的差异范围内。

在此观察基础上，我们确定了具有小于5%序列差异的跨物种同系物的tRNA基因集合。这个集合包括335个小鼠tRNA基因和345个人类tRNA基因。这两个集合代表了所有46个预期的反密码子，并且完全满足预期的摆动法则。这些集合可能更接近功能性tRNA基因的真实补充。

尽管被排除掉的假定基因（小鼠163个，人类167个）中可能包括一些真正的基因，但我们早期以为人类大约有500个tRNA基因似乎是高估了。小鼠和人类的实际计数可能更接近350个。

小鼠蛋白质学

真核蛋白的出现看来主要通过两个重要机制。第一种是将蛋白质结构域重组到新结构里。（结构域是一种紧凑的结构，作为进化上保守的功能构建块，经常以不同的排列（结构）组装到不同的蛋白质中[174]。）第二种是伴随着谱系特异功能和生理现象出现的基因家族的谱系特异性扩张[175]（例如，脊椎动物免疫球蛋白超家族的扩增反映了免疫系统的出现[1]，拟南芥的受体样激酶与植物特异性自交不亲和性和抗病性功能相关[49]，黑腹果蝇中的胰蛋白酶样丝氨酸蛋白酶同系物与背腹图案和先天免疫反应相关[176,177]）。

人类和小鼠这两个密切相关的基因组序列为探索蛋白质进化课题提供了可能，

explore issues of protein evolution that are best addressed through the study of more closely related genomes. The great similarity of the two proteomes allows extensive comparison of orthologous proteins (those that descended by speciation from a single gene in the common ancestor rather than by intragenome duplication), permitting an assessment of the evolutionary pressures exerted on different classes of proteins. The differences between the mouse and human proteomes, primarily in gene family expansions, might reveal how physiological, anatomical and behavioural differences are reflected at the genome level.

Overall proteome comparison

We compared the largest transcript for each gene in the mouse gene catalogue to the National Center for Biotechnology Information (NCBI) database ("nr" set; ftp://ftp.ncbi.nih.gov/blast/db/nr.z) using the BLASTP program[178]. Mouse proteins predicted to be homologues ($E < 10^{-4}$) of other proteins were classified into one of six taxonomic groupings: (1) rodent-specific; (2) mammalian-specific; (3) chordate-specific; (4) metazoan-specific; (5) eukaryote-specific; and (6) other (Fig. 17). The results were similar to those from an analysis of human proteins[1].

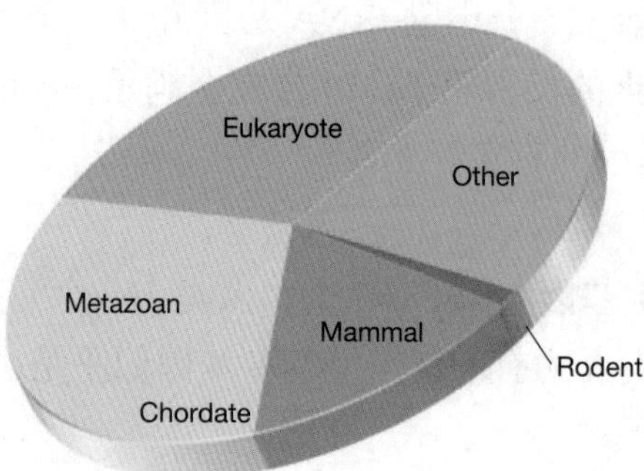

Fig. 17. Taxonomic breakdown of homologues of mouse proteins according to taxonomic range. Note that only a small fraction of genes are possibly rodent-specific (< 1%) as compared with those shared with other mammals (14%, not rodent-specific); shared with chordates (6%, not mammalian-specific); shared with metazoans (27%, not chordate-specific); shared with eukaryotes (29%, not metazoan-specific); and shared with prokaryotes and other organisms (23%, not eukaryotic-specific).

As expected, most of the protein or domain families have similar sizes in human and mouse. However, 12 of the 50 most populous InterPro[179] families in mouse show significant differences in numbers between the two proteomes, most notably high mobility group HMG1/2 box and ubiquitin domains. On close analysis, the differences for six of these families can be accounted for by differential expansion of endogenous retroviral sequences in the genomes. We return below to the issue of expansion of gene families.

这些问题通过研究更密切相关的基因组得到了最好解决。这两个蛋白质组的巨大相似性允许同源蛋白质(那些由共同祖先的单个基因形成而不是基因组内重复形成的蛋白质)进行广泛比较，从而可以评估施加在不同类型蛋白质上的进化压力。小鼠和人类蛋白质组之间的差异，主要是基因家族扩展，可能揭示了生理、解剖和行为的差异如何在基因组水平上反映出来。

总蛋白质组比较

我们使用 BLASTP 程序[178]将小鼠基因目录中每个基因的最大转录本与美国生物技术信息中心(NCBI)数据库("nr"集合；ftp://ftp.ncbi.nih.gov/blast/db/nr.z)进行了比较。根据与其他物种蛋白质的同源性($E < 10^{-4}$)比较，小鼠蛋白质的同系性被分类为六个类群中的一个：(1)啮齿动物特异性；(2)哺乳动物特异性；(3)脊索动物特异性；(4)后生动物特异性；(5)真核生物特异性；(6)其他(图 17)。这些结果与人类蛋白质分析结果相似[1]。

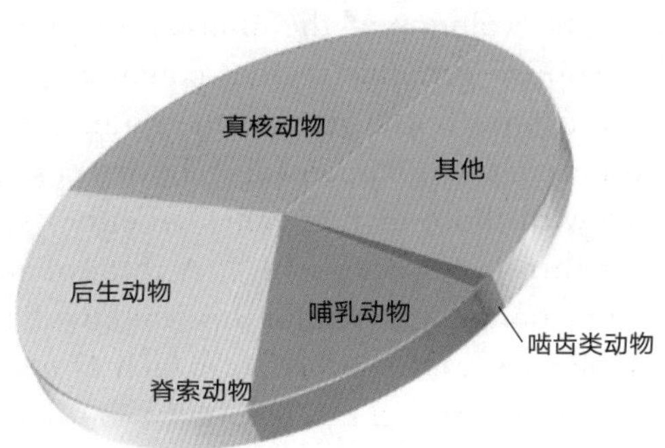

图 17. 按照范围对小鼠蛋白质进行同系物分类。注意，与其他哺乳动物(14%，非啮齿动物特有)共享的基因相比，只有小部分基因可能是啮齿动物特有的(小于 1%)；与脊索动物共享(6%，非哺乳动物特有)；与后生动物共享(27%，非脊索动物特有)；与真核生物共享(29%，非后生动物特有)；与原核生物和其他生物体共享(23%，非真核生物特有)。

正如所料，大多数蛋白质或是其结构域家族在人类和小鼠中的大小相似。然而，小鼠中最常见的 InterPro 家族[179]的 50 个蛋白质中有 12 个在两个蛋白质组之间的数量上存在显著差异，最显著的是高迁移率一组中 HMG1/2 盒和泛素化结构域。仔细分析，这些家族中的六个的差异可以通过基因组中内源性逆转录病毒序列的差异扩张来解释。下面我们回到基因家族的扩张这一问题。

Evolution of gene families in mouse

As noted above, 80% of mouse proteins seem to have strict 1:1 orthologues in the human genome. Many of the remainder belong to gene families that have undergone differential expansion in at least one of the two genomes, resulting in the lack of a strict 1:1 relationship. Such gene family changes represent an insight into aspects of physiology that have emerged since the last common ancestor.

A well-documented example of family expansion is the olfactory receptor gene family, which represents a branch of the larger G-protein-coupled receptor superfamily tree[193,194]. Duplication of olfactory receptor genes seems to have occurred frequently in both rodent and primate lineages, and differences in number and sequence have been seen as distinguishing the degrees and repertoires of odorant detection between mice and humans. Moreover, an estimated 20% of the mouse olfactory receptor homologues[194] and a higher percentage of human homologues[195,196] are pseudogenes, indicating that there is a dynamic interplay between gene birth and gene death in the recent evolution of this family. The importance of these genes in reproductive behaviour is evident from defects in pheromone responses that result from deletion of the VR1 vomeronasal olfactory receptor gene cluster[197].

To explore systematically recent evolution of the mouse proteome, we searched for mouse-specific gene clusters. We identified genomic regions containing four or more homologous mouse genes that descended from a single gene in the human–mouse common ancestor; these represent local expansions in the mouse lineage. To detect such clusters, we compared all transcripts of each gene with those of five genes on either side (using the BLAST-2-Sequences program with a threshold of $E < 10^{-4}$). A total of 4,563 mouse genes were found to have at least one such homologue within this window. A total of 147 such clusters containing at least four homologues was identified, of which 47 contained multiple olfactory receptor genes, which have been studied elsewhere[193,199] and are not discussed further here. For the remaining 100 clusters, we then constructed dendrograms to examine the evolutionary relationship among the mouse proteins and their human homologues. This allowed us to identify those clusters containing mouse genes that are descendants of a single ancestral gene or for which multiple gene deletions had occurred in the human lineage.

In total, 25 such mouse-specific clusters were identified. In most cases (16), the mouse-specific cluster corresponds to only a single gene in the human genome. Among these 25 clusters, two major functional themes emerge: 14 contain genes involved in rodent reproduction and 5 contain genes involved in host defence and immunity. Each of the 14 "reproduction" clusters contains at least one gene whose expression is modulated by androgens, is involved in the biosynthesis or metabolism of hormones, has an established role in the placenta, gonads or spermatozoa, or has documented roles in mate selection, including pheromone olfaction. The fact that so many of the 25 clusters are related to reproduction is unlikely to be coincidental. Many of the most pronounced physiological differences between rodents and primates relate to reproduction, including substantial

小鼠基因家族的进化

如上所述，80% 的小鼠蛋白质在人类基因组中似乎有严格的 1:1 的直系同源物。其余的大多属于基因家族，它们在两个基因组中的至少一个发生了差异性扩增，导致缺乏严格的 1:1 关系。这种基因家族变化代表了自上一个共同祖先生理学方面出现变化的现象。

一个有文献记载的例子是嗅觉受体基因家族的扩张，它隶属于 G 蛋白偶联受体超家族树的一个分支 [193,194]。嗅觉受体基因的重复似乎在啮齿动物和灵长类动物谱系中经常出现，并且其数量和序列的差异被视为小鼠和人类对于气味的检测程度和组成的区分。此外，估计 20% 的小鼠嗅觉受体同源物 [194] 和更高比例的人类同源物 [195,196] 是假基因，这表明在该家族最近的进化过程中，基因出现和基因死亡之间存在着动态的相互作用。这些基因在生殖行为中的重要性从 VR1 鼻骨的嗅觉受体基因簇的缺失所导致的信息素反应缺陷中显而易见 [197]。

为了系统地探索小鼠蛋白质组的最新进化，我们寻找了小鼠特异性基因簇。我们确定了包含四个或者更多同源小鼠基因的基因组区域，这些同源小鼠基因是人类–小鼠共同祖先中一个单基因的后代；这些基因代表了小鼠谱系中的局部扩增。为了检测这些基因簇，我们将每个基因的所有转录产物与人类和小鼠任意一边的五个基因的转录本进行了比较(使用 BLAST2 序列程序，阈值 $E < 10^{-4}$)。发现总共有 4,563 个小鼠基因在这个窗口中至少有一个这样的同源基因。共计鉴定出 147 个这样的簇，包含至少 4 个同源性基因，其中 47 个包括多种嗅觉受体基因，这些基因已经在别处进行了研究 [193,199]，在此不再做进一步讨论。对于剩下的 100 个簇，我们构建了树状图来研究小鼠蛋白质和它们的人类同源物之间的进化关系。这使我们能够辨别那些小鼠基因簇，它们在小鼠中是单个祖先基因的后代，在人类谱系中则发生了多个基因的删除。

总共鉴定出 25 个这样的小鼠特异性基因簇。在大多数情况下(16)，小鼠特异性簇只对应人类基因组中的单个基因。在这 25 个簇中，出现了两个主要的功能主题：14 个包括啮齿动物生殖相关基因，5 个包括宿主防御和免疫相关基因。14 个“生殖”簇中的每一个都包含至少一个其表达受雄激素调节的基因，参与荷尔蒙的生物合成或代谢，在胎盘、性腺或者精子中具有确定的作用，或在配偶选择中有确实的作用，包括信息素嗅觉。25 个簇中这么多与生殖相关的事实不太可能是巧合。啮齿动物和灵长类动物之间许多显著的生理差异与生殖相关，包括胎盘结构、产仔数、

variations in placental structures, litter sizes, oestrous cycles and gestation periods. It seems likely that reproductive traits have been responsible for some of the most powerful evolutionary pressures on the mouse genome, and that the demand for innovation has been met through gene family expansions. Examination of the human genome in this way may similarly reveal gene clusters that reflect particular aspects of human reproduction.

The five mouse clusters that encode genes involved in immunity suggest that another major evolutionary force is acting on host defence genes. The five clusters include the major histocompatibility complex (MHC) class Ib genes, two clusters of antimicrobial β-defensins, a cluster of WAP domain antimicrobial proteins and a cluster of type A ribonucleases. Ribonuclease A genes appear to have been under strong positive selection, possibly due to their significant role in host-defence mechanisms[224].

The two major themes—reproduction and immunity—may not be entirely unrelated; that is, the MHC class Ib genes have roles in both pregnancy and immunity. MHC genotype is also known from ethological studies to influence mate selection, although the molecular mechanisms underlying this effect remain unknown. Within the MHC complex, the class I genes are the most divergent, having arisen after the rodent–human divergence[227].

Genome Evolution: Selection

Investigation of the two principal forces that shape the evolution of the mouse and human genomes—mutation and selection—requires looking beyond coarse-scale identification of regions of conserved synteny and purely codon-based analysis of orthologues, to fine-scale alignment of the two genomes at the nucleotide level.

The substantial sequence divergence between the mouse and human genomes is still low enough that orthologous sequences undergoing neutral drift remain conserved enough for them to be aligned reliably. The challenge then is to use such alignments to tease apart the effects of neutral drift, which can teach us about underlying mutational processes, and selection, which can inform us about functionally important elements. It should be emphasized that sequence similarity alone does not imply functional constraint.

In this section, we use whole-genome alignments to explore the extent of sequence conservation in neutral sites (such as ancestral repeat sequences), known functional elements (such as coding regions) and the genome as a whole. By comparing these, we are able to estimate the proportion of regions of the mammalian genome under evolutionary selection (about 5%), which far exceeds the amount attributable to protein-coding sequences. In the next section, we then use the neutral sites to study how mutational forces vary across the genome.

发情周期和妊娠期的实质性变化。生殖相关性状似乎对应了小鼠基因组上受到的一些最强大的进化压力，并且通过基因家族扩张满足了创新的需求。以这种方式对人类基因组进行检查，可以类似地揭示反映人类生殖的特定方面的基因簇。

编码免疫基因的五个小鼠簇显示作用于宿主防御基因的另一个主要的进化力量。这五个簇包括主要组织相容性复合物（MHC）Ib 类基因，两组抗微生物的 β-防御素，一组 WAP 结构域抗微生物蛋白和一组 A 型核糖核酸酶。核糖核酸酶 A 基因似乎一直处于强正选择中，可能是由于它们在宿主防御机制中的重要作用[224]。

生殖和免疫这两个主题可能并非完全没有关系；也就是说，MHC 中 Ib 类基因在妊娠和免疫中都有作用。从行为学研究中也知道 MHC 基因型会影响配偶选择，尽管这种影响的分子机制仍然不清楚。在 MHC 复合体中，I 类基因是最为分化的，出现在啮齿动物-人类分化之后[227]。

基因组进化：选择

研究塑造小鼠和人类基因组进化的两个主要力量——突变和选择——需要超越粗略识别保守同线性区域和完全基于密码子的直系同源性分析，达到在核苷酸水平对两个基因组进行精准比对。

小鼠和人类基因组之间的实质性序列差异仍然很小，使得经过中性漂变的同源序列仍足够保守到能够可靠地比对到一起。接下来的挑战就是使用这种比对梳理中性漂变和选择的影响，中性漂变可以告诉我们潜在的突变过程，选择可以告诉我们功能性重要元件。应该强调的是序列相似性本身并不意味着功能约束。

在本节中，我们使用全基因组比对来探索中性位点（例如祖源重复序列）、已知功能元件（例如编码区）和整个基因组的序列保守程度。通过比较这些，我们能够估计哺乳动物基因组受到进化选择的区域的比例（大约 5%），远远超过蛋白质编码序列的数量。在下一节，我们用中性位点来研究突变力如何在基因组中变化。

Fine-scale alignment of genomes

We began by creating a catalogue of sequence alignments between the mouse and human genomes. The alignments were produced by the BLASTZ[328] program by comparing all non-repeat sequences across the genome to identify all high-scoring matches, then, using these as seeds, we extended the alignments into the surrounding regions, including into repeat sequences. Regions that could be aligned clearly at the nucleotide level totalled about 1.1 Gb, corresponding to roughly 40% of the human genome.

Proportion of genome under selection

To investigate the fraction of a mammalian genome under evolutionary selection for biological function, we estimated the proportion of the genome that is better conserved than would be expected given the underlying neutral rate of substitution. We compared the overall distribution S_{genome} of conservation scores for the genome to the neutral distribution $S_{neutral}$ of conservation scores for ancestral repeats using a genome-wide set of 14.3 million non-overlapping 50-bp (human) windows, each containing at least 45 bp (mean 48.67 bp) of aligned sequence. The genome-wide score distribution for these windows has a prominent tail extending to the right, reflecting a substantial excess of windows with high conservation scores relative to the neutral rate. The excess can be estimated by decomposing the genome-wide distribution S_{genome} as a mixture of two components: $S_{neutral}$ and $S_{selected}$ (reflecting windows under selection).

The mixture coefficients indicate that at least 20.8% of the windows are under selection, with the remainder consistent with neutral substitution. Because about 25.2% of all human bases are contained in the windows, this suggests that at least 5.25% (25.2% of 20.8%) of the 50-base windows in the human genome is under selection. Repeating the analysis on more stringently filtered alignments (with non-syntenic and non-reciprocal best matches removed) requiring different numbers of aligned bases per window and with 100-bp windows, yields similar estimates, ranging mostly from 4.8% to about 6.1% of windows under selection (D. Haussler, unpublished data), as does using an alternative score function that considers flanking base context effects and uses a gap penalty[330].

The analysis thus suggests that about 5% of small segments (50 bp) in the human genome are under evolutionary selection for biological functions common to human and mouse. This corresponds to regions totalling about 140 Mb of human genomic DNA, although not all of the nucleotides in these windows are under selection. In addition, some bases outside these windows are likely to be under selection. In a loose sense, these regions might be regarded as containing the "functional" conserved subset of the mammalian genome. Of course, it should be noted that nonconserved sequence may have important roles, for example, as a passive spacer or providing a function specific to one lineage. Notably, protein-coding regions of genes can account for only a fraction of the genome under selection. From

基因组的精细比对

我们首先创建了小鼠和人类基因组之间的序列比对目录。通过 BLASTZ[328] 程序完成比对，该程序通过比较基因组中所有非重复序列来确定所有高得分匹配，然后用这些序列作为种子，我们扩展这些对比到周围区域，包括重复序列在内。在核苷酸水平上可以清楚比对的区域大约为 1.1 Gb，对应于人类基因组的大约 40%。

处于选择的基因组比例

为了研究生物功能进化选择下哺乳动物基因组的比例，我们估计这部分基因组比预期的潜在中性替代更保守。我们使用全基因组的 1,430 万个非重叠的 50 bp（人类）窗口，将基因组保守分数 S_{genome} 的总分布和祖源重复序列的保守分数 $S_{neutral}$ 的中性分布进行了比较，每个窗口包含至少 45 bp（平均 48.67 bp）的比对序列。这些窗口的全基因组得分分布具有向右延伸的突出尾部，反映出相对于中性率而言，具有高保守分数的窗口明显超量。超量值可通过将全基因组分布 S_{genome} 分解为两个成分的混合物——$S_{neutral}$ 和 $S_{selected}$（反映经历选择的窗口）进行估计。

混合系数表明，至少 20.8% 的窗口处于选择中，其余的窗口与中性替换一致。因为人类所有碱基的大约 25.2% 都包含在窗口中，这表明人类基因组中 50 个碱基窗口中至少有 5.25%（20.8% 中的 25.2%）处于选择中。以更严格筛选比对条件重复上述分析（去除了非同线性和单向的最佳匹配），每个窗口需要数量不同的比对碱基和 100 bp 的窗口，得到了类似的估计值，经历选择作用的窗口多数在 4.8% 到 6.1% 范围内（豪斯勒，未发表数据），如同使用一个考虑碱基两端关系效应和使用空缺惩罚的替代评分函数 [330]。

因此，分析表明人类基因组中大约 5% 的小片段（50 bp）正处于人类和小鼠共同具有的生物学功能的选择性进化中。这相当于人类基因组 DNA 总量中的大约 140 Mb 的区域，尽管并非所有处于此窗口的核苷酸都处在选择中。除此之外，这些窗口之外的一些碱基可能也在经历选择。非严格意义上讲，这些区域可能被认为包含哺乳动物基因组的“功能性”保守亚群。当然，应该注意的是，非保守序列可能具有重要作用，例如作为被动间隔子或者在某一特定谱系发挥功能。值得注意的是，基因的蛋白编码区域只占被选择基因组的小部分。从我们对基因的数量和性质的

our analysis of the number and properties of genes, coding regions comprise only about 1.5% of the human genome and account for less than half of the segments under selection.

What accounts for the remainder of the genome under selection? About 1% of the genome is contained in untranslated regions of protein-coding genes, and some of this sequence is under some functional constraint. Another main class of interest are those sequences that control gene expression; if a typical gene contains a few such regulatory sequences, there may be tens to hundreds of thousands of such elements. In addition, conserved sequences probably encode non-protein-coding RNAs (which remain difficult to discern) and chromosomal structural elements. Furthermore, some of the conserved fraction may correspond to sequences that were under selection for some period of time but are no longer functional; these could include recent pseudogenes. Characterization of the conserved sequences should be a high priority for genomics in the years ahead.

The analysis above allows us to infer the proportion of the genome under selection by decomposing the curve S_{genome} into curves $S_{neutral}$ and $S_{selected}$. Importantly, it does not definitively assign an individual conserved sequence as being neutral or selected. One can calculate, for a sequence with conservation score S, the probability $P_{selected}(S)$ that the window of sequence belongs to the selected subset. The probability exceeds 83% for sequences with $S > 3$ and 93% for $S > 4$, but is only 52% for $S = 2$. In other words, some functionally important sequence cannot be separated cleanly from the tail of the distribution of neutral conservation.

Genome Evolution: Mutation

Genome-wide alignments also allow us to investigate how the patterns of neutral substitution, deletion and insertion vary across the genome, providing an insight on the underlying mutational processes.

Substitution rate varies across the genome

Significant variation in the level of sequence conservation has been reported in several small-scale studies of human and mouse genomic regions[10,248-254] and in several larger-scale studies of coding sequences[255-260]. It has not been clear in all cases whether the variation reflects differences in neutral substitution rates or in selection. The human–mouse genome alignments allow us to address the variation more comprehensively and to test for co-variation with the rates of other processes, such as insertions of transposable elements[255] and meiotic recombination[258].

We used the collection of aligned ancestral repeats and aligned fourfold degenerate sites to calculate the apparent neutral substitution rate for about 2,500 overlapping 5-Mb windows across the human genome. To accurately follow fluctuations while accounting for

分析来看，编码区域仅大约占人类基因组的 1.5%，可解释少于一半的处于选择中的片段。

是什么原因导致基因组的剩余部分处于选择中？大约 1% 的基因组是在蛋白质编码基因的非翻译区，其中一些序列受到某些功能限制。另一类引起关注的主要类别是那些控制基因表达的序列；如果一个典型的基因包括一些这样的调节序列，则可能会有数以万计的这种元件的存在。此外，保守序列可能编码非蛋白质的 RNA（这些仍然难以识别）和染色体结构元件。进而，部分这些保守区域可能与某段时间内被选择但不再具有功能的序列相符；这些可能包括最近出现的假基因。未来几年中，解析保守序列应该是基因组学研究的高度优先事项。

上述分析允许我们通过将 S_{genome} 曲线分解为 $S_{neutral}$ 曲线和 $S_{selected}$ 曲线来推断处于选择的基因组的比例。重要的是，它没有明确地将单个保守序列指定为中性或被选择。对于具有保守分数 S 的序列，可以计算出属于选择作用子集的序列窗口的概率 $P_{selected}(S)$。$S > 3$ 的序列概率超过 83%，$S > 4$ 的序列概率超过 93%，但 $S = 2$ 的序列只占 52%。换句话说，一些重要功能序列不能从中性保守分布的尾部被清楚地分离出来。

基因组进化：突变

全基因组的比对还使得我们能够研究中性替代、缺失和插入在基因组中如何发生和变化，从而提供了对潜在突变过程的解析。

替代率在基因组中的变化

在人类和小鼠基因组区域的几个小规模研究 [10,248-254] 和编码序列的几个大规模研究 [255-260] 中，已经报道了序列保守性上的显著变化。尚不清楚在所有情况中，这个变化是否反映了中性替代率或选择上的差异。人类–小鼠基因组比对使我们更全面地研究变异，检测与其他过程比率的共突变情况，例如转座元件的插入 [255] 和减数分裂的重组 [258]。

我们通过收集比对的祖源重复序列和比对的四倍简并位点来计算人类基因组中大约 2,500 个重叠 5 Mb 窗口的中性替代率。为了在考虑碱基组成的区域变化的同时能准确地跟踪波动，通过仅使用窗口中的祖源重复位点（平均大约 280,000 个 / 窗

regional changes in base composition, the regional nucleotide substitution rate in ancestral repeat sites, t_{AR}, was calculated separately for each 5-Mb window by maximum likelihood estimation of the parameters of the REV model using only the ancestral repeat sites in the window (average of about 280,000 sites per window). The regional nucleotide substitution rate in fourfold degenerate sites, t_{4D}, was calculated similarly from an average of about 3,700 fourfold degenerate sites per window. Windows with fewer than 800 ancestral repeats or fourfold degenerate sites were discarded.

The mean and standard deviations across the windows were $t_{AR} = 0.467 \pm 0.022$ and $t_{4D} = 0.447 \pm 0.067$ substitutions per site. The standard deviation is much larger (over tenfold and threefold, respectively) than would be expected from sampling variance. These data clearly indicate substantial regional fluctuation.

What properties of chromosomal DNA could account for the variation in substitution rate? One possible explanation is local (G+C) content, but previous studies disagree on whether it correlates strongly with divergence[92,255,262,263]. We find that t_{AR} and t_{4D} vary with local (G+C) content, although the dependence is nonlinear[262,264] and is better fitted by regression with a quadratic curve[263]. In other words, the substitution rate seems to be higher in regions of extremely high or low (G+C) content, with the sign of the correlation differing in regions with high versus low (G+C) content. This pattern persists if CpG substitutions are removed from the analysis (data not shown).

All three forces that alter the genome (nucleotide substitution, deletion and insertion) thus vary substantially across the genome. Moreover, they are significantly correlated and tend to co-vary along chromosomes. Notably, these three measures of interspecies divergence are also correlated with recent substitutions in the human genome, as measured by the density of SNPs identified by the SNP Consortium[265].

Possible explanations for variation

What explains the correlation among these many measures of genome divergence? It seems unlikely that direct selection would account for variation and co-variation at such large scales (about 5 Mb) and involving abundant neutral sites taken from ancestral transposon relics. Selection against deleterious mutations can remove linked polymorphisms[270,271], but it is not clear that such effects or related effects[272] could extend to such large scales or to interspecies divergence over such large time periods[273].

It seems more probable that these features reflect local variation in underlying mutation rate, caused by differences in DNA metabolism or chromosome physiology. The causative factors may include recombination-associated mutagenesis[258,266], transcription-associated mutagenesis[274], transposon-associated deletion and genomic rearrangement[275-278], and replication timing[279,280]. Nuclear location may also be involved, including proximity to matrix attachment sites, heterochromatin, nuclear membrane, and origins of replication.

口）对 REV 模型的参数进行最大似然估计，分别计算每 5 Mb 窗口中祖源重复位点 t_{AR} 的区域核苷酸替代率。四倍简并位点 t_{4D} 的区域核苷酸替代率类似地由每个窗口大约 3,700 个四倍简并位点的平均值计算得到。祖源重复或者四倍简并位点少于 800 个的窗口被放弃。

窗口的平均值和标准偏差为每个位点 $t_{AR} = 0.467 \pm 0.022$，$t_{4D} = 0.447 \pm 0.067$ 个碱基替换。标准偏差比抽样方差的估计值大很多（分别超过 10 倍和 3 倍）。这些数据明确显示出巨大的区域波动。

染色体 DNA 的哪些特性可以解释替代率的变化呢？一种可能的解释是局部（G+C）含量，但之前的研究在它是否与这种变化有很强的相关性上结果并不一致 [92,255,262,263]。我们发现，t_{AR} 和 t_{4D} 随着局部（G+C）含量的变化而变化，尽管这种依赖性是非线性的 [262,264]，用二次曲线回归更合适 [263]。换句话说，在（G+C）含量极高或极低的区域，替代率似乎更高，在（G+C）含量高和低的区域，具有不同的相关性。如果从分析中删除 CpG 位点上的碱基替换，这种模式仍然存在（数据未显示）。

因此，这三种改变基因组的力量（核苷酸替换、删除和插入）在基因组中都有很大的差异。此外，它们之间明显相关而且倾向于沿着染色体共同变异。值得注意的是，这三种变异的种间差异的测量也与人类基因组中的最近替换具有相关性，这是由 SNP 协作组进行的 SNP 密度测量所确定的 [265]。

变化的可能解释

用什么解释基因组分化的各种测量指标之间的关联性？直接选择似乎不太能解释如此大规模（大约 5 Mb）的变异和共变异，而且这些变异还涉及大量从祖源转座子遗物中获取的大量中性位点。对有害突变的选择可以消除连锁多态性 [270,271]，但是尚不清楚这种效应或者相关效应 [272] 能否扩展到如此大的规模或在如此长的时间内 [273] 扩展到种间分化。

这些特征似乎更可能反映了 DNA 代谢或染色体生理学上的差异所引起的潜在突变率的局部变化。诱发因素可能包括重组相关的突变 [258,266]，转录相关的突变 [274]，转座相关缺失和基因组重排 [275-278] 和复制时间 [279,280]。也可能涉及核定位，包括接近基质附着点、异染色质、核膜和复制起始点。

It is clear that the mammalian genome is evolving under the influence of non-uniform local forces. It remains an important challenge to unravel the mechanistic basis and evolutionary consequences of such variation.

Genetic Variation among Strains

Implications for the laboratory mouse

The promise of genomics is the ability to connect phenotypes with genotypes for a wide variety of traits and to use the resulting molecular insights to develop new approaches for the cure and prevention of disease. The laboratory mouse occupies a central place in this vision, both as a prototype for all mammalian biology and as a well-characterized organism for modelling human disease states[15,16,123]. In this section, we briefly discuss ways in which the mouse genome sequence will accelerate biomedical progress in the future. Because the sequence has been made available in public databases in advance of publication, examples for many of the predictions can already be cited.

Positional cloning of genes for mendelian phenotypes

More than 1,000 spontaneously arising and radiation-induced mouse mutants causing heritable mendelian phenotypes are catalogued in the Mouse Genome Informatics (MGI) database (http://www.informatics.jax.org). Largely through positional cloning, the molecular defect is now known for about 200 of these mutants. The availability of an annotated mouse genome sequence now provides the most efficient tool yet in the gene hunter's toolkit. One can move directly from genetic mapping to identification of candidate genes, and the experimental process is reduced to PCR amplification and sequencing of exons and other conserved elements in the candidate interval. With this streamlined protocol, it is anticipated that many decades-old mouse mutants will be understood precisely at the DNA level in the near future. An example of how the draft genome sequence has already been successfully used is the recent identification of the mouse mutation "chocolate" in the melanosome protein Rab38 (ref. 284).

Identification of quantitative trait loci

The availability of more than 50 commonly used laboratory inbred strains of mice, each with its own phenotype for multiple continuously variable traits, has provided an important opportunity to map QTLs that underlie heritable phenotypic variation. A systematic initiative is currently underway[285] to define parameters such as body weight, behavioural patterns, and disease susceptibility among a standard set of inbred lines, and to make these data freely available to the scientific community in the Mouse Phenome Database (www.jax.org/phenome). Appropriate crosses between such lines, followed by genotyping, will enable

很明显，哺乳动物基因组是在非均匀局部力量的影响下进化的。揭示这种变异的机制基础和进化后果仍然是一个很重要的挑战。

品系间遗传变异

对实验小鼠的影响

基因组学的前景是将各种性状的表型和基因型联系起来，并利用由此产生的分子发现来开发治疗和预防疾病的新方案。实验小鼠在这里占据了中心位置，既是作为所有哺乳动物生物学的原型，也是模拟人类疾病状态的具有良好特征的生物体[15,16,123]。在本节中，我们简单讨论小鼠基因组序列在未来将如何加速生物医学进程。因为该序列已经在出版前公布在公共数据库中，所以可以引用许多预测的示例。

孟德尔表型基因的定位克隆

在小鼠基因组信息（MGI）数据库（http://www.informatics.jax.org）中收录了超过1,000 个自发产生和辐射诱导产生的小鼠突变体，这些突变体引发可遗传的孟德尔表型。主要通过定位克隆，目前已知大约 200 个这些突变体的分子缺陷。小鼠基因组序列的注释信息提供了基因猎人工具箱中最有效的工具。研究人员可以直接从基因定位转移到候选基因的鉴定，实验过程简化为对候选区间外显子和其他保守元件进行 PCR 扩增和测序。通过这种简化的方案，预计在不久的将来，将在 DNA 水平上精确地解释许多几十年前的小鼠突变体。成功使用基因组序列草图的一个例子就是最近在黑素体蛋白 Rab38 中鉴定出了小鼠突变体“chocolate”（参考文献 284）。

数量性状基因座的鉴定

常用的实验室近交系小鼠超过 50 种，每种小鼠具有多种连续变化的性状作为自身表型，这为定位可遗传表型的变异的数量性状基因座（QTL）提供了重要机会。目前正在进行一项系统性举措[285]，以确定标准近交系中例如体重、行为模式和疾病易感性等参数，并在小鼠表观数据库（www.jax.org/phenome）中将这些数据免费提供给科学界。这些品系之间适当杂交，然后进行基因分型，能够进行 QTL 的定位，然

the mapping of QTLs, which can then be subjected to positional cloning. The degree of difficulty is substantially greater for a QTL cloning project than for a mendelian disorder, however, as the responsible intervals are usually much larger, the boundaries more difficult to delineate precisely, and the causative variant often much more subtle[286]. For these reasons, only a handful of the approximately 1,000 mapped QTLs have been identified at the molecular level. The availability of the mouse sequence should greatly improve the chances for future success.

Creation of knockout and knockin mice

The wide application of homologous recombination in embryonic stem cells has provided a remarkable abundance of "custom" mice with specifically engineered loss- or gain-of-function mutations in specific genes of biological or medical interest. Yet this remains a time-consuming process. The design of recombinant DNA constructs for injection has often been delayed by incomplete knowledge of gene structure, requiring tedious restriction mapping or sequencing, and occasionally giving rise to unsatisfying outcomes due to incorrect information. The availability of the mouse genome sequence will both speed the design of such constructs and reduce the likelihood of unfortunate choices. Furthermore, the long-range continuity of the sequence should facilitate the generation of models of contiguous gene-deletion syndromes.

Creation of transgenic animals

For many transgenic experiments, it is important to maintain copy-dependent, tissue-specific expression of the transgene. This is most readily accomplished through BAC transgenesis. The availability of a deep, end-sequenced BAC library from the B6 strain mapped to the genome sequence now makes it straightforward to obtain a desired gene in a BAC for such experiments; end-sequenced BAC libraries from other strains should be available in the future. BACs also provide the ability to make mutant alleles with relative ease, by taking advantage of powerful genetic engineering techniques for custom mutagenesis in the *Escherichia coli* host.

Applications to cancer

The mouse genome sequence also has powerful applications to the molecular characterization of the somatic mutations that result in neoplasia. High-density SNP mapping to identify loss of heterozygosity[288,289], combined with comparative genomic hybridization using cDNA or BAC arrays[290,291], can be used to identify chromosomal segments showing loss or gain of copy number in particular tumour types. The combination of such approaches with expression arrays that include all mouse genes should further enhance the ability to pinpoint the molecular lesions that result in carcinogenesis. Full sequencing of all the exons and

后进行定位克隆。然而，QTL 克隆项目的难度远远大于孟德尔疾病，因为相关的区间通常要大得多，边界更难精确描绘，而致病突变通常也更加微妙[286]。由于这些原因，在大约 1,000 个定位的 QTL 中，只有少数在分子水平上得到鉴定。而小鼠序列的产生在未来应该大大提高成功的机会。

构建敲除和敲入小鼠

同源重组技术在胚胎干细胞中的广泛应用提供了大量“定制”小鼠，它们携带了专门设计的生物学或医学感兴趣的特定基因功能丧失或获得的突变。然而，这仍然是一个耗时的过程。用于注射的重组 DNA 构建体的设计常常因为对基因结构的不完全了解而延迟，需要繁琐的限制性作图或测序，并且偶尔由于不正确的信息而导致令人不满意的结果。小鼠基因组序列的出现将加速这种结构设计并降低产生不幸选择的可能性。此外，序列的长程连续性有助于构建连续基因缺失综合征的模型。

构建转基因动物

对于很多转基因实验来说，保持转基因的拷贝依赖性、组织特异性表达是很重要的。这是最容易通过 BAC 转基因实现的。来自 B6 品系基因组序列的深度末端测序的 BAC 库的应用，使得现在在此类实验中可以直接获得 BAC 中所需的基因；其他品系的末端测序 BAC 库在将来也将可用。通过利用强大的基因工程技术在大肠杆菌宿主中进行定制诱变，BAC 还能够相对容易地产生突变等位基因。

应用于癌症

小鼠基因组序列在研究导致瘤形成的体细胞突变的分子特征上也具有强大的应用。用于鉴定杂合性缺失的高密度 SNP[288,289] 比对、结合使用 cDNA 或 BAC 阵列的基因组杂交进行比较[290,291] 都可以用于鉴定在特定肿瘤类型中拷贝数缺失或增加的染色体区段。这些方法与包含所有小鼠基因的表达阵列的结合使用，将进一步精确定位导致癌症产生的分子损伤。目前可以对已知肿瘤抑制基因、癌基因和其他候

regulatory regions of known tumour suppressors, oncogenes, and other candidate genes can now be contemplated, as has been initiated in a few centres for human tumours[292].

Making better mouse models

Not all mouse models replicate the human phenotype in the expected way. The availability of the full human and mouse sequences provides an opportunity to anticipate these differences, and perhaps to compensate for them. In some instances, it may turn out that the murine mutation did not reside in the true orthologue of the human disease gene. Alternatively, in a circumstance where the human genome contains only a single gene family member, but the mouse genome contains a paralogue as well as the orthologue, one can anticipate that knockout of the orthologue alone may give a much milder phenotype (or none at all). Such was the case, for instance, with the occulocerebrorenal syndrome described by Lowe and colleagues[296]. Creating double knockout mice may then provide a closer match to the human disease phenotype.

Understanding gene regulation

Of the approximately 5% of windows of the mammalian genome that are under selection, most do not appear to code for protein. Much of this sequence is probably involved in the regulation of gene expression. It should be possible to pinpoint these regulatory elements more precisely with the availability of additional related genomes. However, mouse is likely to provide the most powerful experimental platform for generating and testing hypotheses about their function. An example is the recent demonstration, based on mouse–human sequence alignment followed by knockout manipulation, of several long-range locus control regions that affect expression of the Il4/Il13/Il5 cluster[4].

Conclusion

The mouse provides a unique lens through which we can view ourselves. As the leading mammalian system for genetic research over the past century, it has provided a model for human physiology and disease, leading to major discoveries in such fields as immunology and metabolism. With the availability of the mouse genome sequence, it now provides a model and informs the study of our genome as well.

Comparative genome analysis is perhaps the most powerful tool for understanding biological function. Its power lies in the fact that evolution's crucible is a far more sensitive instrument than any other available to modern experimental science: a functional alteration that diminishes a mammal's fitness by one part in 10^4 is undetectable at the laboratory bench, but is lethal from the standpoint of evolution.

选基因的所有外显子和调控区域进行测序，类似工作在几个人类肿瘤中心已经开始进行[292]。

制作更好的小鼠模型

并非所有的小鼠模型都以预期的方式复制人类表型。完整的人类和小鼠序列提供了一个机会去预测这些差异，并且进行弥补。在某些状况下，可能会发现，小鼠的突变不在人类疾病基因的真正的直系同源物中。或者，在一种情况下，人类基因组仅包含单个基因家族，但是小鼠基因组包含了旁系同源物和直系同源物，人们可以预测单独的直系同源物的敲除可能会产生更温和的表型(或者没有表型)。例如，洛及其同事们[296]描述的闭塞性肾综合征就是这种情况。通过建立双敲除小鼠可以获得与人类疾病表型更接近的模型。

了解基因调控

哺乳动物基因组中大约 5% 的窗口序列处于选择，大多数的区域似乎不编码蛋白质。大部分这样的序列可能参与基因表达的调控。通过更多相关基因组可以更精准地确定这些调控元件。然而，小鼠很可能提供最强大的实验平台用于产生和测试关于这些元件功能的假设。最近的一个例子是，基于小鼠–人类序列比对后进行的基因敲除证明了影响 Il4/Il13/Il5 基因簇表达的几个长程调控区域[4]。

结　论

小鼠提供了一个独特的镜头，通过它我们可以看到自己。在过去一个世纪里，作为主要的哺乳动物遗传研究系统，它为人类生理和疾病提供了一个模型，从而在免疫学和新陈代谢领域有了重大发现。随着小鼠基因组序列的获得，它正在为我们的基因组研究提供模型和信息。

比较基因组分析可能是了解生物学功能最有力的工具。它的作用在于，进化的熔炉是一种比现代实验科学中任何其他仪器都敏感的仪器：一个会降低哺乳动物身体健康度万分之一的功能性改变是在实验台上检测不到，但是从进化角度来看却是致命的。

Comparative analysis of genomes should thus make it possible to discern, by virtue of evolutionary conservation, biological features that would otherwise escape our notice. In this way, it will play a crucial role in our understanding of the human genome and thereby help lay the foundation for biomedicine in the twenty-first century.

The initial sequence of the mouse genome reported here is merely a first step in this intellectual programme. The sequencing of many additional mammalian and other vertebrate genomes will be needed to extract the full information hidden within our chromosomes. Moreover, as we begin to understand the common elements shared among species, it may also become possible to approach the even harder challenge of identifying and understanding the functional differences that make each species unique.

Methods

Production of sequence reads

Paired-end reads from libraries with different insert sizes were produced as previously described[1] using 384-well trays to ensure linkages.

Availability of sequence and assembly data

Unprocessed sequence reads are available from the NCBI trace archive (ftp://ftp.ncbi.nih.gov/pub/TraceDB/mus_musculus/). Raw assembly data (before removal of contaminants, anchoring to chromosomes, and addition of finished sequence) are available from the Whitehead Institute for Biomedical Research (WIBR) (ftp://wolfram.wi.mit.edu/pub/mouse_contigs/Mar10_02/). The released assembly MGSCv3 is available from Ensembl (http://www.ensembl.org/Mus_musculus/), NCBI (ftp://ftp.ncbi.nih.gov/genomes/M_musculus/MGSCv3_Release1/), UCSC (http://genome.ucsc.edu/downloads.html) and WIBR (ftp://wolfram.wi.mit.edu/pub/mouse_contigs/MGSC_V3/). (See Supplementary Information for detailed Methods.)

(**420**, 520-562; 2002)

Received 18 September; accepted 31 October 2002.

References:

1. International Human Genome Sequencing Consortium Initial sequencing and analysis of the human genome. *Nature* **409**, 860-921 (2001).

2. Venter, J. C. *et al.* The sequence of the human genome. *Science* **291**, 1304-1351 (2001).

3. O'Brien, S. J. *et al.* The promise of comparative genomics in mammals. *Science* **286**, 458-462, 479-481 (1999).

4. Loots, G. G. *et al.* Identification of a coordinate regulator of interleukins 4, 13, and 5 by cross-species sequence comparisons. *Science* **288**, 136-140 (2000).

5. Pennacchio, L. A. & Rubin, E. M. Genomic strategies to identify mammalian regulatory sequences. *Nature Rev. Genet.* **2**, 100-109 (2001).

6. Oeltjen, J. C. *et al.* Large-scale comparative sequence analysis of the human and murine Bruton's tyrosine kinase loci reveals conserved regulatory domains. *Genome Res.* **7**, 315-329 (1997).

7. Ellsworth, R. E. *et al.* Comparative genomic sequence analysis of the human and mouse cystic fibrosis transmembrane conductance regulator genes. *Proc. Natl*

因此，基因组的比较分析应该能够通过进化保守性分辨出我们不会注意到的生物学特征。通过这种方式，它将对我们理解人类基因组发挥重要作用，从而为二十一世纪的生物医学奠定基础。

这里报道的小鼠基因组的初始序列仅仅是这个充满智慧的项目的第一步。我们还需要对许多其他哺乳动物和其他脊椎动物进行测序，以提取隐藏在我们染色体中的全部信息。此外，当我们开始了解物种之间共享的相同元件时，也可能接近识别和理解每个物种所拥有的独特功能及其彼此差异。

方 法

序列读数的产生

如前所述[1]，利用 384 孔板获得不同片段长度的文库并生成配对末端读取片段，以确保序列的连锁性。

序列和组装数据的可用性

未处理的序列读取片段可以从 NCBI 存档中获得（ftp://ftp.ncbi.nih.gov/pub/TraceDB/mus_musculus/）。原始组装数据（在去掉污染物、锚定染色体和添加完成序列之前）可以从怀特黑德生物医学研究所（WIBR）获得（ftp://wolfram.wi.mit.edu/pub/mouse_contigs/Mar10_02/）。程序集 MGSCv3 可以从 Ensembl（http://www.ensembl.org/Mus_musculus/）、NCBI（ftp://ftp.ncbi.nih.gov/genomes/M_musculus/MGSCv3_Release1/）、UCSC（http://genome.ucsc.edu/downloads.html）和 WIBR（ftp://wolfram.wi.mit.edu/pub/mouse_contigs/MGSC_V3/）获得。（具体方法见补充信息。）

（张瑶楠 翻译；曾长青 审稿）

Acad. Sci. USA **97**, 1172-1177 (2000).

8. Mallon, A. M. *et al.* Comparative genome sequence analysis of the Bpa/Str region in mouse and man. *Genome Res.* **10**, 758-775 (2000).
9. Dehal, P. *et al.* Human chromosome 19 and related regions in mouse: conservative and lineage-specific evolution. *Science* **293**, 104-111 (2001).
10. DeSilva, U. *et al.* Generation and comparative analysis of approximately 3.3 Mb of mouse genomic sequence orthologous to the region of human chromosome 7q11.23 implicated in Williams syndrome. *Genome Res.* **12**, 3-15 (2002).
11. Toyoda, A. *et al.* Comparative genomic sequence analysis of the human chromosome 21 down syndrome critical region. *Genome Res.* **12**, 1323-1332 (2002).
12. Ansari-Lari, M. A. *et al.* Comparative sequence analysis of a gene-rich cluster at human chromosome 12p13 and its syntenic region in mouse chromosome 6. *Genome Res.* **8**, 29-40 (1998).
13. Lercher, M. J., Williams, E. J. & Hurst, L. D. Local similarity in evolutionary rates extends over whole chromosomes in human-rodent and mouse-rat comparisons: implications for understanding the mechanistic basis of the male mutation bias. *Mol. Biol. Evol.* **18**, 2032-2039 (2001).
14. Makalowski, W. & Boguski, M. S. Evolutionary parameters of the transcribed mammalian genome: an analysis of 2,820 orthologous rodent and human sequences. *Proc. Natl Acad. Sci. USA* **95**, 9407-9412 (1998).
15. Rossant, J. & McKerlie, C. Mouse-based phenogenomics for modelling human disease. *Trends Mol. Med.* **7**, 502-507 (2001).
16. Paigen, K. A miracle enough: the power of mice. *Nature Med.* **1**, 215-220 (1995).
17. Hogan, B., Beddington, R., Costantini, F. & Lacy, E. *Manipulating the Mouse Embryo: A Laboratory Manual* (Cold Spring Harbor Laboratory Press, Woodbury, New York, 1994).
18. Joyner, A. L. *Gene Targeting: A Practical Approach* (Oxford Univ. Press, New York, 1999).
19. Copeland, N. G., Jenkins, N. A. & Court, D. L. Recombineering: a powerful new tool for mouse functional genomics. *Nature Rev. Genet.* **2**, 769-779 (2001).
20. Yu, Y. & Bradley, A. Engineering chromosomal rearrangements in mice. *Nature Rev. Genet.* **2**, 780-790 (2001).
21. Bucan, M. & Abel, T. The mouse: genetics meets behaviour. *Nature Rev. Genet.* **3**, 114-123 (2002).
22. Silver, L. M. *Mouse Genetics: Concepts and Practice* (Oxford Univ. Press, New York, 1995).
23. Bromham, L., Phillips, M. J. & Penny, D. Growing up with dinosaurs: molecular dates and the mammalian radiation. *Trends Ecol. Evol.* **14**, 113-118 (1999).
24. Nei, M., Xu, P. & Glazko, G. Estimation of divergence times from multiprotein sequences for a few mammalian species and several distantly related organisms. *Proc. Natl Acad. Sci. USA* **98**, 2497-2502 (2001).
25. Kumar, S. & Hedges, S. B. A molecular timescale for vertebrate evolution. *Nature* **392**, 917-920 (1998).
26. Madsen, O. *et al.* Parallel adaptive radiations in two major clades of placental mammals. *Nature* **409**, 610-614 (2001).
27. Murphy, W. J. *et al.* Molecular phylogenetics and the origins of placental mammals. *Nature* **409**, 614-618 (2001).
29. Morse, H. *The Mouse in Biomedical Research* (eds Foster, H. L., Small, J. D. & Fox, J. G.) 1-16 (Academic, New York, 1981).
30. Morse, H. C. *Origins of Inbred Mice* (ed. Morse, H. C.) 1-21 (Academic, New York, 1978).
31. Haldane, J. B. S., Sprunt, A. D. & Haldane, N. M. Reduplication in mice. *J. Genet.* **5**, 133-135 (1915).
32. Botstein, D., White, R. L., Skolnick, M. & Davis, R. W. Construction of a genetic linkage map in man using restriction fragment length polymorphisms. *Am. J. Hum. Genet.* **32**, 314-331 (1980).
33. Dietrich, W. *et al. Genetic Maps* (ed. O'Brien, S.) 4.110-4.142, (1992).
34. Dietrich, W. F. *et al.* A comprehensive genetic map of the mouse genome. *Nature* **380**, 149-152 (1996).
35. Love, J. M., Knight, A. M., McAleer, M. A. & Todd, J. A. Towards construction of a high resolution map of the mouse genome using PCR-analysed microsatellites. *Nucleic Acids Res.* **18**, 4123-4130 (1990).
36. Weber, J. L. & May, P. E. Abundant class of human DNA polymorphisms which can be typed using the polymerase chain reaction. *Am. J. Hum. Genet.* **44**, 388-396 (1989).
37. Hudson, T. J. *et al.* A radiation hybrid map of mouse genes. *Nature Genet.* **29**, 201-205 (2001).
38. Van Etten, W. J. *et al.* Radiation hybrid map of the mouse genome. *Nature Genet.* **22**, 384-387 (1999).
39. Nusbaum, C. *et al.* A YAC-based physical map of the mouse genome. *Nature Genet.* **22**, 388-393 (1999).
41. Kawai, J. *et al.* Functional annotation of a full-length mouse cDNA collection. *Nature* **409**, 685-690 (2001).
44. Gregory, S. G. *et al.* A physical map of the mouse genome. *Nature* **418**, 743-750 (2002).
45. Mural, R. J. *et al.* A comparison of whole-genome shotgun-derived mouse chromosome 16 and the human genome. *Science* **296**, 1661-1671 (2002).
46. Green, E. D. Strategies for the systematic sequencing of complex genomes. *Nature Rev. Genet.* **2**, 573-583 (2001).
47. Edwards, A. *et al.* Automated DNA sequencing of the human HPRT locus. *Genomics* **6**, 593-608 (1990).
48. Huson, D. H. *et al.* Design of a compartmentalized shotgun assembler for the human genome. *Bioinformatics* **17**, S132-S139 (2001).
49. Analysis of the genome sequence of the flowering plant *Arabidopsis thaliana*. *Nature* **408**, 796-815 (2000).
52. Battey, J., Jordan, E., Cox, D. & Dove, W. An action plan for mouse genomics. *Nature Genet.* **21**, 73-75 (1999).
53. Kuroda-Kawaguchi, T. *et al.* The AZFc region of the Y chromosome features massive palindromes and uniform recurrent deletions in infertile men. *Nature Genet.* **29**, 279-286 (2001).
54. Zhao, S. *et al.* Mouse BAC ends quality assessment and sequence analyses. *Genome Res.* **11**, 1736-1745 (2001).
55. Ewing, B. & Green, P. Analysis of expressed sequence tags indicates 35,000 human genes. *Nature Genet.* **25**, 232-234 (2000).
56. Batzoglou, S. *et al.* ARACHNE: a whole-genome shotgun assembler. *Genome Res.* **12**, 177-189 (2002).

57. Jaffe, D. B. *et al.* Whole-genome sequence assembly for mammalian genomes: Arachne **2**. *Genome Res.* (in the press).

58. Mullikin, J. & Ning, Z. The Phusion Assembler. *Genome Res.* (in the press).

59. Bailey, J. A. *et al.* Recent segmental duplications in the human genome. *Science* **297**, 1003-1007 (2002).

66. Mouse Genome Sequencing Consortium Progress in sequencing the mouse genome. *Genesis* **31**, 137-141 (2001).

67. Clark, F. H. Inheritance and linkage relations of mutant characteristics in the deermouse. *Contrib. Lab. Vert. Biol.* 7, 1-11 (1938).

68. Castle, W. W. Observations of the occurrence of linkage in rats and mice. *Car. Inst. Wash. Pub.* **288**, 29-36 (1919).

69. Lalley, P. A., Minna, J. D. & Francke, U. Conservation of autosomal gene synteny groups in mouse and man. *Nature* **274**, 160-163 (1978).

70. Nadeau, J. H. & Taylor, B. A. Lengths of chromosomal segments conserved since divergence of man and mouse. *Proc. Natl Acad. Sci. USA* **81**, 814-818 (1984).

71. Ma, B., Tromp, J. & Li, M. PatternHunter: faster and more sensitive homology search. *Bioinformatics* **18**, 440-445 (2002).

72. Ohno, S. *Sex Chromosomes and Sex-Linked Genes* (Springer, Berlin, 1996).

73. Sturtevant, A. H. & Beadle, G. W. The relations of inversions in the X chromosome of *Drosophila melanogaster* to crossing over and disjunction. *Genetics* **21**, 554-604 (1936).

74. Ranz, J. M., Casals, F. & Ruiz, A. How malleable is the eukaryotic genome? Extreme rate of chromosomal rearrangement in the genus *Drosophila*. *Genome Res.* **11**, 230-239 (2001).

75. Nadeau, J. H. & Sankoff, D. The lengths of undiscovered conserved segments in comparative maps. *Mamm. Genome* **9**, 491-495 (1998).

76. Ferretti, V., Nadeau, J. H. & Sankoff, D. *Combinatorial Pattern Matching, 7th Annual Symposium* (eds Hirschberg, D. & Myers, G.) 159-167 (Springer, Berlin, 1996).

77. Bourque, G. & Pevzner, P. A. Genome-scale evolution: reconstructing gene orders in the ancestral species. *Genome Res.* **12**, 26-36 (2002).

78. Thiery, J. P., Macaya, G. & Bernardi, G. An analysis of eukaryotic genomes by density gradient centrifugation. *J. Mol. Biol.* **108**, 219-235 (1976).

79. Salinas, J., Zerial, M., Filipski, J. & Bernardi, G. Gene distribution and nucleotide sequence organization in the mouse genome. *Eur. J. Biochem.* **160**, 469-478 (1986).

80. Sabeur, G., Macaya, G., Kadi, F. & Bernardi, G. The isochore patterns of mammalian genomes and their phylogenetic implications. *J. Mol. Evol.* **37**, 93-108 (1993).

81. Zerial, M., Salinas, J., Filipski, J. & Bernardi, G. Gene distribution and nucleotide sequence organization in the human genome. *Eur. J. Biochem.* **160**, 479-485 (1986).

82. Mouchiroud, D., Fichant, G. & Bernardi, G. Compositional compartmentalization and gene composition in the genome of vertebrates. *J. Mol. Evol.* **26**, 198-204 (1987).

83. Mouchiroud, D., Gautier, C. & Bernardi, G. The compositional distribution of coding sequences and DNA molecules in humans and murids. *J. Mol. Evol.* **27**, 311-320 (1988).

84. Mouchiroud, D. & Gautier, C. Codon usage changes and sequence dissimilarity between human and rat. *J. Mol. Evol.* **31**, 81-91 (1990).

85. Robinson, M., Gautier, C. & Mouchiroud, D. Evolution of isochores in rodents. *Mol. Biol. Evol.* **14**, 823-828 (1997).

86. Bernardi, G. *et al.* The mosaic genome of warm-blooded vertebrates. *Science* **228**, 953-958 (1985).

87. Mouchiroud, D. *et al.* The distribution of genes in the human genome. *Gene* **100**, 181-187 (1991).

88. Zoubak, S., Clay, O. & Bernardi, G. The gene distribution of the human genome. *Gene* **174**, 95-102 (1996).

89. Saccone, S., Pavlicek, A., Federico, C., Paces, J. & Bernard, G. Genes, isochores and bands in human chromosomes 21 and 22. *Chromosome Res.* **9**, 533-539 (2001).

90. Bernardi, G. Compositional constraints and genome evolution. *J. Mol. Evol.* **24**, 1-11 (1986).

91. Bernardi, G., Mouchiroud, D. & Gautier, C. Compositional patterns in vertebrate genomes: conservation and change in evolution. *J. Mol. Evol.* **28**, 7-18 (1988).

92. Wolfe, K. H., Sharp, P. M. & Li, W. H. Mutation rates differ among regions of the mammalian genome. *Nature* **337**, 283-285 (1989).

93. Sueoka, N. Directional mutation pressure and neutral molecular evolution. *Proc. Natl Acad. Sci. USA* **85**, 2653-2657 (1988).

94. Sueoka, N. On the genetic basis of variation and heterogeneity of DNA base composition. *Proc. Natl Acad. Sci. USA* **48**, 582-592 (1962).

95. Bird, A. P. DNA methylation and the frequency of CpG in animal DNA. *Nucleic Acids Res.* **8**, 1499-1504 (1980).

96. Larsen, F., Gundersen, G., Lopez, R. & Prydz, H. CpG islands as gene markers in the human genome. *Genomics* **13**, 1095-1107 (1992).

97. Gardiner-Garden, M. & Frommer, M. CpG islands in vertebrate genomes. *J. Mol. Biol.* **196**, 261-282 (1987).

98. Antequera, F. & Bird, A. Number of CpG islands and genes in human and mouse. *Proc. Natl Acad. Sci. USA* **90**, 11995-11999 (1993).

99. Adams, R. L. & Eason, R. Increased G+C content of DNA stabilizes methyl CpG dinucleotides. *Nucleic Acids Res.* **12**, 5869-5877 (1984).

100. Smit, A. F. Interspersed repeats and other mementos of transposable elements in mammalian genomes. *Curr. Opin. Genet. Dev.* **9**, 657-663 (1999).

101. Laird, C. D., McConaughy, B. L. & McCarthy, B. J. Rate of fixation of nucleotide substitutions in evolution. *Nature* **224**, 149-154 (1969).

102. Kohne, D. E. Evolution of higher-organism DNA. *Q. Rev. Biophys.* **3**, 327-375 (1970).

103. Goodman, M., Barnabas, J., Matsuda, G. & Moore, G. W. Molecular evolution in the descent of man. *Nature* **233**, 604-613 (1971).

104. Kumar, S. & Subramanian, S. Mutation rates in mammalian genomes. *Proc. Natl Acad. Sci. USA* **99**, 803-808 (2002).

105. Easteal, S., Collet, C. & Betty, D. *The Mammalian Molecular Clock* (Landes, Austin, Texas, 1995).

106. Li, W. H., Ellsworth, D. L., Krushkal, J., Chang, B. H. & Hewett-Emmett, D. Rates of nucleotide substitution in primates and rodents and the generation-time effect hypothesis. *Mol. Phylogenet. Evol.* **5**, 182-187 (1996).

107. Martin, A. P. & Palumbi, S. R. Body size, metabolic rate, generation time, and the molecular clock. *Proc. Natl Acad. Sci. USA* **90**, 4087-4091 (1993).

108. Bromham, L. Molecular clocks in reptiles: life history influences rate of molecular evolution. *Mol. Biol. Evol.* **19**, 302-309 (2002).

109. Wu, C. I. & Li, W. H. Evidence for higher rates of nucleotide substitution in rodents than in man. *Proc. Natl Acad. Sci. USA* **82**, 1741-1745 (1985).

110. Smit, A. F., Toth, G., Riggs, A. D. & Jurka, J. Ancestral, mammalian-wide subfamilies of LINE-1 repetitive sequences. *J. Mol. Biol.* **246**, 401-417 (1995).

111. Adey, N. B. *et al.* Rodent L1 evolution has been driven by a single dominant lineage that has repeatedly acquired new transcriptional regulatory sequences. *Mol. Biol. Evol.* **11**, 778-789 (1994).

112. Mears, M. L. & Hutchison, C. A. III The evolution of modern lineages of mouse L1 elements. *J. Mol. Evol.* **52**, 51-62 (2001).

113. Goodier, J. L., Ostertag, E. M., Du, K. & Kazazian, H. H. Jr A novel active L1 retrotransposon subfamily in the mouse. *Genome Res.* **11**, 1677-1685 (2001).

114. Hardies, S. C. *et al.* LINE-1 (L1) lineages in the mouse. *Mol. Biol. Evol.* **17**, 616-628 (2000).

115. Ohshima, K., Hamada, M., Terai, Y. & Okada, N. The 3′ ends of tRNA-derived short interspersed repetitive elements are derived from the 3′ ends of long interspersed repetitive elements. *Mol. Cell Biol.* **16**, 3756-3764 (1996).

116. Smit, A. F. The origin of interspersed repeats in the human genome. *Curr. Opin. Genet. Dev.* **6**, 743-748 (1996).

117. Quentin, Y. A master sequence related to a free left Alu monomer (FLAM) at the origin of the B1 family in rodent genomes. *Nucleic Acids Res.* **22**, 2222-2227 (1994).

118. Kim, J. & Deininger, P. L. Recent amplification of rat ID sequences. *J. Mol. Biol.* **261**, 322-327 (1996).

119. Lee, I. Y. *et al.* Complete genomic sequence and analysis of the prion protein gene region from three mammalian species. *Genome Res.* **8**, 1022-1037 (1998).

120. Serdobova, I. M. & Kramerov, D. A. Short retroposons of the B2 superfamily: evolution and application for the study of rodent phylogeny. *J. Mol. Evol.* **46**, 202-214 (1998).

121. Coffin, J. M., Hughes, S. H. & Varmus, H. E. (eds) *Retroviruses* (Cold Spring Harbor Laboratory Press, Cold Spring Harbor, New York, 1997).

122. Smit, A. F. Identification of a new, abundant superfamily of mammalian LTR-transposons. *Nucleic Acids Res.* **21**, 1863-1872 (1993).

123. Hamilton, B. A. & Frankel, W. N. Of mice and genome sequence. *Cell* **107**, 13-16 (2001).

125. Kidwell, M. G. Horizontal transfer. *Curr. Opin. Genet. Dev.* **2**, 868-873 (1992).

126. Feng, Q., Moran, J. V., Kazazian, H. H. Jr & Boeke, J. D. Human L1 retrotransposon encodes a conserved endonuclease required for retrotransposition. *Cell* **87**, 905-916 (1996).

127. Jurka, J. Sequence patterns indicate an enzymatic involvement in integration of mammalian retroposons. *Proc. Natl Acad. Sci. USA* **94**, 1872-1877 (1997).

128. Bernardi, G. The isochore organization of the human genome. *Annu. Rev. Genet.* **23**, 637-661 (1989).

129. Holmquist, G. P. Chromosome bands, their chromatin flavors, and their functional features. *Am. J. Hum. Genet.* **51**, 17-37 (1992).

132. Lyon, M. F. X-chromosome inactivation: a repeat hypothesis. *Cytogenet. Cell Genet.* **80**, 133-137 (1998).

135. Beckman, J. S. & Weber, J. L. Survey of human and rat microsatellites. *Genomics* **12**, 627-631 (1992).

136. Toth, G., Gaspari, Z. & Jurka, J. Microsatellites in different eukaryotic genomes: survey and analysis. *Genome Res.* **10**, 967-981 (2000).

139. Dunham, I. *et al.* The DNA sequence of human chromosome 22. *Nature* **402**, 489-495 (1999).

140. Hattori, M. *et al.* The DNA sequence of human chromosome 21. *Nature* **405**, 311-319 (2000).

141. Roest Crollius, H. *et al.* Estimate of human gene number provided by genome-wide analysis using *Tetraodon nigroviridis* DNA sequence. *Nature Genet.* **25**, 235-238 (2000).

142. Hubbard, T. *et al.* The Ensembl genome database project. *Nucleic Acids Res.* **30**, 38-41 (2002).

143. Kulp, D., Haussler, D., Reese, M. G. & Eeckman, F. H. Integrating database homology in a probabilistic gene structure model. *Pac. Symp. Biocomput.* 232-244 (1997).

144. Birney, E. & Durbin, R. Using GeneWise in the *Drosophila* annotation experiment. *Genome Res.* **10**, 547-548 (2000).

145. Burge, C. & Karlin, S. Prediction of complete gene structures in human genomic DNA. *J. Mol. Biol.* **268**, 78-94 (1997).

150. The FANTOM Consortium and the RIKEN Genome Exploration Research Group Phase I & II Team. Analysis of the mouse transcriptome based on functional annotation of 60,770 full-length cDNAs. *Nature* **420**, 563-573 (2002).

151. Pruitt, K. D. & Maglott, D. R. RefSeq and LocusLink: NCBI gene-centered resources. *Nucleic Acids Res.* **29**, 137-140 (2001).

165. Eddy, S. R. Non-coding RNA genes and the modern RNA world. *Nature Rev. Genet.* **2**, 919-929 (2001).

166. Storz, G. An expanding universe of noncoding RNAs. *Science* **296**, 1260-1263 (2002).

167. Eddy, S. R. Computational genomics of noncoding RNA genes. *Cell* **109**, 137-140 (2002).

168. Lowe, T. M. & Eddy, S. R. tRNAscan-SE: a program for improved detection of transfer RNA genes in genomic sequence. *Nucleic Acids Res.* **25**, 955-964 (1997).

169. Daniels, G. R. & Deininger, P. L. Repeat sequence families derived from mammalian tRNA genes. *Nature* **317**, 819-822 (1985).

170. Lawrence, C., McDonnell, D. & Ramsey, W. Analysis of repetitive sequence elements containing tRNA-like sequences. *Nucleic Acids Res.* **13**, 4239-4252 (1985).

174. Ponting, C. P. & Russell, R. R. The natural history of protein domains. *Annu. Rev. Biophys. Biomol. Struct.* **31**, 45-71 (2002).

175. Lespinet, O., Wolf, Y. I., Koonin, E. V. & Aravind, L. The role of lineage-specific gene family expansion in the evolution of eukaryotes. *Genome Res.* **12**, 1048-1059 (2002).

176. Ponting, C. P., Mott, R., Bork, P. & Copley, R. R. Novel protein domains and repeats in *Drosophila melanogaster*: insights into structure, function, and evolution. *Genome Res.* **11**, 1996-2008 (2001).

177. Rubin, G. M. *et al.* Comparative genomics of the eukaryotes. *Science* **287**, 2204-2215 (2000).

178. Altschul, S. F. *et al.* Gapped BLAST and PSI-BLAST: a new generation of protein database search programs. *Nucleic Acids Res.* **25**, 3389-3402 (1997).

179. Zdobnov, E. M. & Apweiler, R. InterProScan—an integration platform for the signature-recognition methods in InterPro. *Bioinformatics* **17**, 847-848 (2001).

193. Young, J. M. *et al.* Different evolutionary processes shaped the mouse and human olfactory receptor gene families. *Hum. Mol. Genet.* **11**, 535-546 (2002).

194. Zhang, X. & Firestein, S. The olfactory receptor gene superfamily of the mouse. *Nature Neurosci.* **5**, 124-133 (2002).

195. Glusman, G., Yanai, I., Rubin, I. & Lancet, D. The complete human olfactory subgenome. *Genome Res.* **11**, 685-702 (2001).

196. Rouquier, S. *et al.* Distribution of olfactory receptor genes in the human genome. *Nature Genet.* **18**, 243-250 (1998).

197. Del Punta, K. *et al.* Deficient pheromone responses in mice lacking a cluster of vomeronasal receptor genes. *Nature* **419**, 70-74 (2002).

199. Lane, R. P. *et al.* Genomic analysis of orthologous mouse and human olfactory receptor loci. *Proc. Natl Acad. Sci. USA* **98**, 7390-7395 (2001).

224. Zhang, J., Dyer, K. D. & Rosenberg, H. F. Evolution of the rodent eosinophil-associated RNase gene family by rapid gene sorting and positive selection. *Proc. Natl Acad. Sci. USA* **97**, 4701-4706 (2000).

227. Yeager, M. & Hughes, A. L. Evolution of the mammalian MHC: natural selection, recombination, and convergent evolution. *Immunol. Rev.* **167**, 45-58 (1999).

248. Koop, B. F. Human and rodent DNA sequence comparisons: a mosaic model of genomic evolution. *Trends Genet.* **11**, 367-371 (1995).

249. DeBry, R. W. & Seldin, M. F. Human/mouse homology relationships. *Genomics* **33**, 337-351 (1996).

250. Gottgens, B. *et al.* Long-range comparison of human and mouse SCL loci: localized regions of sensitivity to restriction endonucleases correspond precisely with peaks of conserved noncoding sequences. *Genome Res.* **11**, 87-97 (2001).

251. Shiraishi, T. *et al.* Sequence conservation at human and mouse orthologous common fragile regions, FRA3B/FHIT and Fra14A2/Fhit. *Proc. Natl Acad. Sci. USA* **98**, 5722-5727 (2001).

252. Wilson, M. D. *et al.* Comparative analysis of the gene-dense ACHE/TFR2 region on human chromosome 7q22 with the orthologous region on mouse chromosome 5. *Nucleic Acids Res.* **29**, 1352-1365 (2001).

253. Hardison, R. C. Conserved noncoding sequences are reliable guides to regulatory elements. *Trends Genet.* **16**, 369-372 (2000).

254. Chiaromonte, F. *et al.* Association between divergence and interspersed repeats in mammalian noncoding genomic DNA. *Proc. Natl Acad. Sci. USA* **98**, 14503-14508 (2001).

255. Matassi, G., Sharp, P. M. & Gautier, C. Chromosomal location effects on gene sequence evolution in mammals. *Curr. Biol.* **9**, 786-791 (1999).

256. Williams, E. J. & Hurst, L. D. The proteins of linked genes evolve at similar rates. *Nature* **407**, 900-903 (2000).

257. Chen, F. C., Vallender, E. J., Wang, H., Tzeng, C. S. & Li, W. H. Genomic divergence between human and chimpanzee estimated from large-scale alignments of genomic sequences. *J. Hered.* **92**, 481-489 (2001).

258. Lercher, M. J. & Hurst, L. D. Human SNP variability and mutation rate are higher in regions of high recombination. *Trends Genet.* **18**, 337-340 (2002).

259. Castresana, J. Genes on human chromosome 19 show extreme divergence from the mouse orthologs and a high GC content. *Nucleic Acids Res.* **30**, 1751-1756 (2002).

260. Smith, N. G. C., Webster, M. & Ellegren, H. Deterministic mutation rate variation in the human genome. *Genome Res.* **12**, 1350-1356 (2002).

262. Bernardi, G. The human genome: organization and evolutionary history. *Ann. Rev. Genet.* **23**, 637-661 (1995).

263. Hurst, L. D. & Willliams, E. J. B. Covarication of GC content and the silent site substitution rate in rodents: implications for methodology and for the evolution of isochores. *Gene* **261**, 107-114 (2000).

264. Bernardi, G. Misunderstandings about isochores. Part 1. *Gene* **276**, 3-13 (2001).

265. The SNP Consortium An SNP map of the human genome generated by reduced representation shotgun sequencing. *Nature* **407**, 513-516 (2000).

266. Perry, J. & Ashworth, A. Evolutionary rate of a gene affected by chromosomal position. *Curr. Biol.* **9**, 987-989 (1999).

270. Charlesworth, B. The effect of background selection against deleterious mutations on weakly selected, linked variants. *Genet. Res.* **63**, 213-227 (1994).

271. Hudson, R. R. & Kaplan, N. L. Deleterious background selection with recombination. *Genetics* **141**, 1605-1617 (1995).

272. Maynard Smith, J. & Haigh, J. The hitch-hiking effect of a favourable gene. *Genet. Res.* **23**, 23-35 (1974).

273. Birky, C. W. & Walsh, J. B. Effects of linkage on rates of molecular evolution. *Proc. Natl Acad. Sci. USA* **85**, 6414-6418 (1988).

274. Francino, M. P. & Ochman, H. Strand asymmetries in DNA evolution. *Trends Genet.* **13**, 240-245 (1997).

275. Gilbert, N., Lutz-Prigge, S. & Moran, J. Genomic deletions created upon LINE-1 retrotransposition. *Cell* **110**, 315-325 (2002).

276. Symer, D. *et al.* Human l1 retrotransposition is associated with genetic instability *in vivo*. *Cell* **110**, 327-338 (2002).

277. Moran, J. *et al.* High frequency retrotransposition in cultured mammalian cells. *Cell* **87**, 917-927 (1996).

278. Hughes, J. F. & Coffin, J. M. Evidence for genomic rearrangements mediated by human endogenous retroviruses during primate evolution. *Nature Genet.* **29**, 487-489 (2001).

279. Wolfe, K. H. Mammalian DNA replication: mutation biases and the mutation rate. *J. Theor. Biol.* **149**, 441-451 (1991).

280. Gu, X. & Li, W. H. A model for the correlation of mutation rate with GC content and the origin of GC-rich isochores. *J. Mol. Evol.* **38**, 468-475 (1994).

284. Loftus, S. K. *et al.* Mutation of melanosome protein RAB38 in chocolate mice. *Proc. Natl Acad. Sci. USA* **99**, 4471-4476 (2002).

285. Paigen, K. & Eppig, J. T. A mouse phenome project. *Mamm. Genome* **11**, 715-717 (2000).

286. Doerge, R. W. Mapping and analysis of quantitative trait loci in experimental populations. *Nature Rev. Genet.* **3**, 43-52 (2002).

288. Mei, R. *et al.* Genome-wide detection of allelic imbalance using human SNPs and high-density DNA arrays. *Genome Res.* **10**, 1126-1137 (2000).

289. Lindblad-Toh, K. *et al.* Loss-of-heterozygosity analysis of small-cell lung carcinomas using single-nucleotide polymorphism arrays. *Nature Biotechnol.* **18**, 1001-1005 (2000).

290. Heiskanen, M. *et al.* CGH, cDNA and tissue microarray analyses implicate FGFR2 amplification in a small subset of breast tumors. *Anal. Cell Pathol.* **22**, 229-234 (2001).

291. Cai, W. W. *et al.* Genome-wide detection of chromosomal imbalances in tumors using BAC microarrays. *Nature Biotechnol.* **20**, 393-396 (2002).

292. Davies, H. *et al.* Mutations of the *BRAF* gene in human cancer. *Nature* **417**, 949-954 (2002).

296. Janne, P. A. *et al.* Functional overlap between murine Inpp5b and Ocrl1 may explain why deficiency of the murine ortholog for OCRL1 does not cause Lowe syndrome in mice. *J. Clin. Invest.* **101**, 2042-2053 (1998).

328. Schwartz, S. *et al.* Human-mouse alignments with Blastz. *Genome Res.* (in the press).

330. Roskin, K. M. Score Functions for Assessing Conservation in Locally Aligned Regions of DNA from Two Species. UCSC Tech Report UCSC-CRL-02-30, School of Engineering, Univ. California (2002).

Supplementary Information accompanies the paper on *Nature*'s website (http://www. nature. com/nature).

Acknowledgements. We thank J. Takahashi and M. Johnston for comments on the manuscript; the Mouse Liaison Group for strategic advice; L. Gaffney, D. Leja and K.-S. Toh for graphical help; B. Graham and G. Roberts for administrative work on sequencing of individual mouse BACs; and P. Kassos and M. McMurtry for secretarial assistance. We thank D. Hill and L. Corbani of the Mouse Genome Informatics Group for their contributions to the GO analysis for mouse and human, and the members of the Bork group at EMBL for discussions. Funding was provided by the National Institutes of Health (National Human Genome Research Institute, National Cancer Institute, National Institute of Dental and Craniofacial Research, National Institute of Diabetes and Digestive and Kidney Diseases, National Institute of General Medical Sciences, National Eye Institute, National Institute of Environmental Health Sciences, National Institute of Aging, National Institute of Arthritis and Musculoskeletal and Skin Diseases, National Institute on Deafness and Other Communication Disorders, National Institute of Mental Health, National Institute on Drug Abuse, National Center for Research Resources, the National Heart Lung and Blood Institute and The Fogarty International Center); the Wellcome Trust; the Howard Hughes Medical Institute; the United States Department of Energy; the National Science Foundation; the Medical Research Council; NSERC; BMBF (German Ministry for Research and Education); the European Molecular Biology Laboratory; Plan Nacional de I+D and Instituto Carlos III; Swiss National Science Foundation, NCCR Frontiers in Genetics, the Swiss Cancer League and the "Childcare" and "J. Lejeune" Foundations; and the Ministry of Education, Culture, Sports, Science and Technology of Japan. The initial threefold sequence coverage was partly supported by the Mouse Sequencing Consortium (GlaxoSmithKline, Merck and Affymetrix) through the Foundation for the National Institutes of Health. We acknowledge A. Holden for coordinating the Mouse Sequencing Consortium. We thank the Sanger Institute systems group for maintenance and provision of the computer resource. The MGSC also used Hewlett-Packard Company's BioCluster, a configuration of 27 HP AlphaServer ES40 systems with 100 CPUs and 1 terabyte of storage. The BioCluster is housed in Hewlett-Packard's IQ Solutions Center, and was accessed remotely. The computing resource greatly accelerated the analysis.

Competing interests statement. The authors declare that they have no competing financial interests.

Correspondence and requests for materials should be addressed to R.H.W. (e-mail: waterston@gs.washington.edu), K.L.T. (e-mail: kersli@genome.wi.mit.edu) or E.S.L. (e-mail: lander@genome.wi.mit.edu).

Authors' contributions. The following authors contributed to project leadership: R. H. Waterston, K. Lindblad-Toh, E. Birney, J. Rogers, M. R. Brent, F. S. Collins, R. Guigó, R. C. Hardison, D. Haussler, D. B. Jaffe, W. J. Kent, W. Miller, C. P. Ponting, A. Smit, M. C. Zody and E. S. Lander.

The Principle of Gating Charge Movement in a Voltage-dependent K^+ Channel

Y. Jiang *et al.*

Editor's Note

Proteins in cell membranes that act as gates for alkali-metal ions, opening and shutting in response to changes in the electrochemical potential (voltage) across the membrane, play key roles in important biological processes of higher organisms, such as muscle action and nerve signal transmission. But the mechanism by which these "ion channels" operate was not fully known before these papers from biochemist Roderick MacKinnon and colleagues. This understanding demanded a detailed picture of the molecular structure of the proteins, which was hard to obtain because membrane proteins seldom crystallize well for X-ray diffraction studies. Nonetheless, having succeeded in obtaining a detailed crystal structure, reported in a companion paper to this one, the researchers here describe the changes in molecular shape that control potassium-ion movement through the central pore.

The steep dependence of channel opening on membrane voltage allows voltage-dependent K^+ channels to turn on almost like a switch. Opening is driven by the movement of gating charges that originate from arginine residues on helical S4 segments of the protein. Each S4 segment forms half of a "voltage-sensor paddle" on the channel's outer perimeter. Here we show that the voltage-sensor paddles are positioned inside the membrane, near the intracellular surface, when the channel is closed, and that the paddles move a large distance across the membrane from inside to outside when the channel opens. KvAP channels were reconstituted into planar lipid membranes and studied using monoclonal Fab fragments, a voltage-sensor toxin, and avidin binding to tethered biotin. Our findings lead us to conclude that the voltage-sensor paddles operate somewhat like hydrophobic cations attached to levers, enabling the membrane electric field to open and close the pore.

VOLTAGE-DEPENDENT K^+ channel opening follows a very steep function of membrane voltage[1]. To allow channels to switch to the open state, gating charges—charged amino acids on the channel protein—move within the membrane electric field to open the pore[1-3]. The crystal structure of KvAP, a voltage-dependent K^+ channel, suggests how these gating charge movements might occur[4]. Four arginine residues are located on a predominantly hydrophobic helix–turn–helix structure called the voltage-sensor paddle. One paddle on each subunit is present at the outer perimeter of the channel. By moving across the membrane near the protein–lipid interface, the paddles could carry the arginine residues through the electric field, coupling pore opening to membrane voltage. We test this

电压依赖性 K^+通道的门控电荷运动的原理

姜有星等

编者按

那些位于细胞膜中的、对碱金属离子起着门控作用、随着膜两侧电化学势(电压)的变化而开放和关闭的蛋白质在高等生物的重要生物过程中起着关键作用，比如肌肉动作和神经信号传导。但是，在生物化学家罗德里克·麦金农及其同事们发表相关的文章之前，这些“离子通道”运转的机制并不完全为人们所知。对这种机制的了解需要有对蛋白质的分子结构的详细描述，然而这一点很难做到，因为膜蛋白很难得到足够好的晶体进行X射线衍射分析。尽管如此，研究者们还是成功地获得了其详细的晶体结构并在本文的姊妹篇中进行了报道，在本文中他们则描述了这些蛋白分子形状的变化，这些变化控制了钾离子穿过中央小孔的运动。

通道开放对膜电压的严格依赖使电压依赖性 K^+ 通道像开关一样打开。其开放受门控电荷运动驱动，门控电荷来自这个蛋白的 S4 螺旋片段上的精氨酸残基。每个 S4 片段在通道的外围形成半个“电压感受桨”。在这里，我们证明当通道关闭时，电压感受桨位于膜内，接近细胞内表面，当通道打开时桨从内向外跨膜移动很大的距离。将 KvAP 通道重建到平面脂质膜中，并用单克隆的 Fab (抗原表位) 片段、电压传感器毒素和连接了生物素的抗生物素蛋白对 KvAP 通道进行研究。通过研究结果我们可以得出这样的结论：电压感受桨的运转有点像疏水性阳离子依附到杠杆上，使得膜电场可打开和关闭孔道。

电压依赖性 K^+ 通道开放严格受控于膜电压 [1]。为了让通道切换到开放状态，门控电荷——通道蛋白上的带电氨基酸——在膜电场内运动以打开孔道 [1-3]。KvAP 的晶体结构(一种电压依赖性 K^+ 通道)，可能表明了这些门控电荷运动是如何发生的 [4]。四个精氨酸残基位于一个显著疏水的被称为电压感受桨的螺旋-转角-螺旋结构上。每个亚基上有一个桨位于通道外围。通过跨膜运动接近蛋白-膜脂质界面，桨可携带精氨酸残基通过电场，将孔道开放与膜电压耦联。我们通过测定通道关闭和打开时膜内电压感受桨的位置来验证该门控电荷运动假说。

hypothesis for the movement of gating charges by estimating the positions of the voltage-sensor paddles inside the membrane when the channel is closed and opened.

Using Fabs and a Toxin to Detect Paddle Motions

Two monoclonal Fabs, 6E1 and 33H1, were used to crystallize and determine the structures of the full-length KvAP channel and the isolated voltage sensor, respectively[4]. Both Fabs were found attached to the same epitope on the tip of the voltage-sensor paddle between S3b and S4. We used these same Fabs to examine the position of the voltage-sensor paddles when the KvAP channel functions in lipid membranes (Fig. 1a), and to assess whether they change their position when the channel gates open in response to membrane depolarization. Figure 1b shows that both the 6E1 and 33H1 Fabs inhibit channel function when applied to the external solution. By contrast, neither Fab affected channel function from the internal solution.

Inhibition by external Fabs requires membrane depolarization (Fig. 1c). When the 33H1 Fab is added to the external chamber while the channel is held closed at −100 mV for 10 min (Fig. 1c, interval between the black and red data points), no inhibition is observed. The slightly larger current of the first red data point (20-min point) reflects recovery from a small amount of steady-state inactivation of channels occurring during the control pulse period (0–10-min data points)[5]. The important point, however, is that inhibition of current is detectable only on the second pulse following the addition of the Fab, as if the channel has first to open in order for the Fabs to bind. Inhibition progresses as the membrane is repeatedly depolarized. We also observe gradual recovery from inhibition if, once the Fabs are bound, the membrane is held at a negative voltage for a prolonged period (30–40-min interval), implying that negative membrane voltages destabilize the interactions between channels and Fabs, causing the Fabs to dissociate. The same properties of inhibition are observed for Fab 6E1. Based on these results, we conclude that the voltage-sensor paddles must remain inaccessible as long as the channel is held closed by the negative membrane voltage, and that the entire epitope (two helical turns of S3b and one turn of S4) becomes exposed to the external side in response to depolarization.

Why do the Fabs inhibit the channel? If they bind to the voltage-sensor paddles from the external side when the membrane is depolarized, why do the Fabs not simply hold the channel permanently open? Inhibition can be explained by the fact that the KvAP channel, like most voltage-dependent K^+ channels, inactivates[5]. That is, its pore stops conducting ions spontaneously during prolonged depolarizations of the membrane, even though the voltage sensors remain in their open conformation. Inactivation could also occur when the Fabs bind and hold the voltage sensors in their open conformation, thus explaining inhibition.

使用 Fabs 和毒素检测桨的运动

两个单克隆的 Fabs 片段——6E1 和 33H1，分别用于结晶并确定全长 KvAP 通道和独立的电压传感器的结构[4]。我们发现两个 Fabs 都附在位于 S3b 和 S4 之间的电压感受桨末端的相同抗原表位。当 KvAP 通道在脂质膜内起作用时，我们用相同的 Fabs 检查电压感受桨的位置（图 1a），并评估当膜去极化使通道开放时它们是否改变自己的位置。图 1b 显示，当 Fabs 片段 6E1 和 33H1 被加入外侧溶液中时都能抑制通道功能。相反，当被加入内侧溶液中时两个 Fabs 片段都不影响通道功能。

外侧 Fabs 抑制需要膜去极化（图 1c）。当 33H1 Fab 加到外侧，同时通道在 –100 mV 保持关闭 10 min 时（图 1c，黑色和红色的数据点之间的间隔），没有观察到抑制。第一个红色数据点所显示的略大的电流（20 min 点）反映了在控制脉冲周期内发生的从小量通道稳态失活的恢复（0 ~ 10 min 数据点）[5]。然而，重点是电流抑制只在加入 Fab 后的第二个脉冲检测到，好像通道必须首先打开才能使得 Fabs 进行结合。当膜反复去极化时抑制会继续。若是一旦 Fabs 被结合，且使膜维持更长时间的负电压（30 ~ 40 min 间隔），我们还可以观察到从抑制中逐渐恢复的过程，这意味着负的膜电压使通道和 Fabs 间相互作用不稳定，造成 Fabs 脱离。在 Fab 6E1 的实验中也观察到了相同的抑制特性。基于这些结果，我们得出这样的结论：只要通道通过负膜电压保持关闭，电压感受桨必定是难以接近的，只有去极化时整个抗原表位（S3b 的两个螺旋转角和 S4 的一个转角）才会相应地暴露在外侧。

为什么 Fabs 会抑制通道？当膜去极化时，如果它们从外侧结合到电压感受桨，那为什么 Fabs 不能简单地维持通道永久开放？抑制可以解释为，KvAP 通道会像大多数电压依赖性 K^+ 通道一样失活[5]。即当膜去极化时间延长，即使电压传感器保持其开放构象，其孔道也会自发停止导通离子。当 Fabs 结合并保持电压传感器开放的构象中，失活也会发生，从而解释了抑制现象。

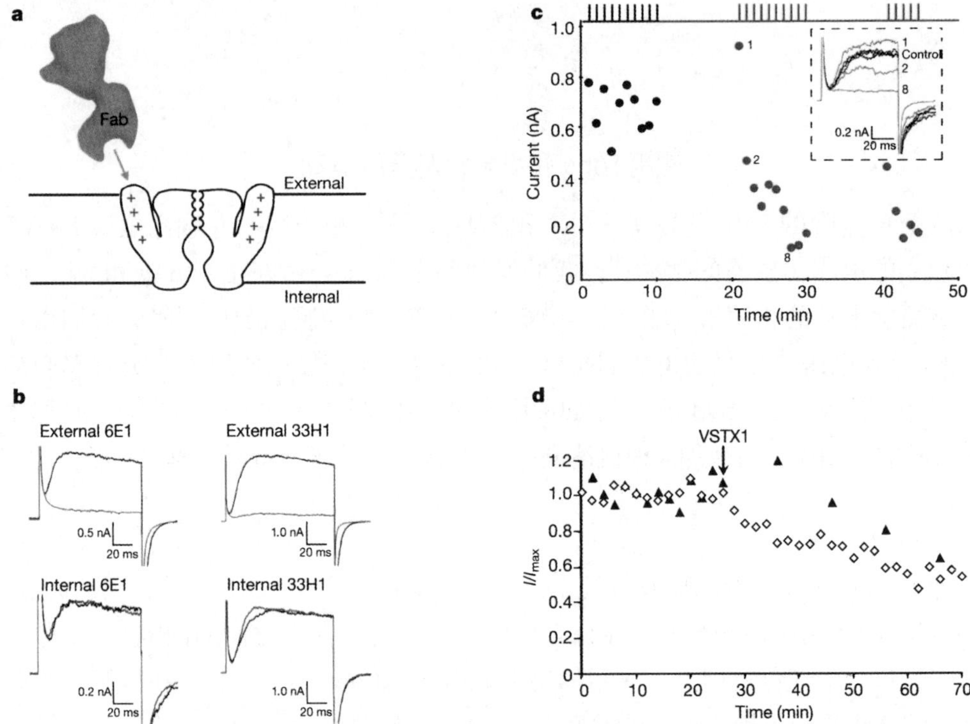

Fig. 1. Inhibition of KvAP channels by Fabs and a tarantula venom toxin that bind to the voltage-sensor paddles. **a**, Experimental strategy: Fab (green) is added to the external or internal side of a planar lipid membrane to determine whether the epitope is exposed. **b**, Fabs used in crystallization (6E1 and 33H1) inhibit from the external (red traces) but not the internal (blue traces) side of the membrane. Currents before (black traces) and after the addition of about 500 nM Fab (red and blue traces) were elicited by membrane depolarization to 100 mV from a holding voltage of −100 mV. **c**, Fab 33H1 binds to the voltage-sensor paddle only when the membrane is depolarized. Current elicited by depolarization from −100 mV to 100 mV at times indicated by the stimulus trace (above) in the absence (black symbols) or presence of 500 nM Fab (red symbols) is presented as a function of time. Selected current traces corresponding to the numbered symbols are shown (inset). **d**, VSTX1 binds from the external side only when the membrane is depolarized. Currents, normalized to the average control value, were elicited by a 100-ms depolarization to 100 mV every 120 s (diamonds) or every 600 s (triangles). VSTX1 (30 nM) was added to the external solution at the point indicated by the arrow.

It is interesting that tarantula venom toxins also bind to residues on S3 and S4 (ref. 6), which we now know to be on the voltage-sensor paddles[4]. On addition of an approximately half-inhibitory concentration of the tarantula venom toxin VSTX1 to the external side of KvAP, we observe that the rate of inhibition is faster if the membrane is depolarized at a higher frequency (Fig. 1d). Therefore, the toxins, like the Fabs, require membrane depolarization.

We conclude from experiments with Fab fragments and a tarantula venom toxin that the voltage-sensor paddles are exposed to the extracellular solution during membrane depolarization (positive inside) but not during hyperpolarization (negative inside). In view of the KvAP crystal structure[4], these results imply that the voltage-sensor paddles can move a large distance across the lipid membrane. How deep inside the membrane do the paddles sit

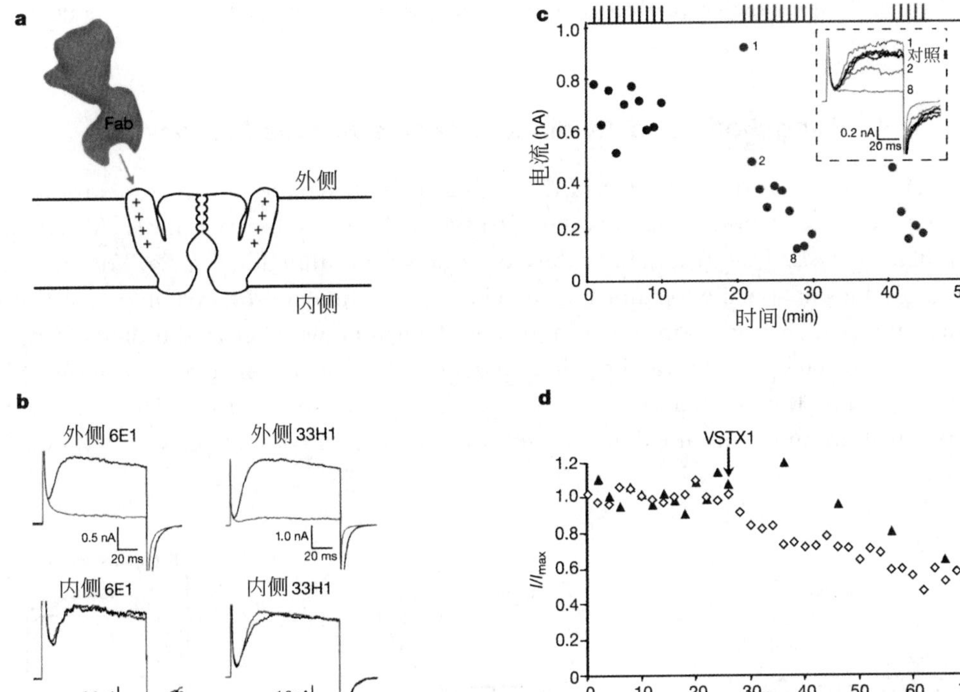

图 1. 结合到电压感受桨的 Fabs 和狼蛛毒液毒素对 KvAP 通道的抑制。**a**，实验策略：将 Fab(绿色)加入平面脂质膜的外侧或内侧，以确定抗原表位是否暴露出来。**b**，结晶所使用的 Fabs(6E1 和 33H1)从膜的外侧(红色轨迹)抑制而非内侧(蓝色轨迹)。加入约 500 nM Fab 之前(黑色轨迹)和之后(红色和蓝色轨迹)的电流是由膜从 −100 mV 钳制电压去极化到 100 mV 时产生的。**c**，只有当膜去极化时，Fab 33H1 才结合到电压感受桨。从 −100 mV 到 100 mV 去极化时产生的电流，在不存在(黑色符号)或存在(红色符号)500 nM Fab 时表示为时间的函数。图中对应于数字符号的电流值，来自选定的电流轨迹(插图中)。**d**，只有当膜去极化时，VSTX1 才从外侧结合。100 ms 去极化至 100 mV，将每 120 s(菱形)或每 600 s(三角形)时间点产生的电流归一化到平均值。在箭头所指示的时间点，将 VSTX1(30 nM)加到外侧溶液。

有趣的是，狼蛛毒液毒素也结合 S3 和 S4 的残基(参考文献 6)，现在我们知道它们位于电压感受桨上[4]。另外加入大约一半抑制浓度的狼蛛毒液毒素 VSTX1 到 KvAP 外侧，我们看到，如果膜的去极化频率越高，则抑制速率也越快(见图 1d)。因此，毒素像 Fabs 一样需要膜去极化。

我们从 Fab 片段和狼蛛毒液毒素实验可得出结论，电压感受桨在膜去极化时(内部为正)，而不是超极化时(内部为负)暴露在细胞外的溶液中。鉴于 KvAP 晶体结构[4]，这些结果意味着，电压感受桨可以跨越脂质膜移动一段较大的距离。当膜

when the membrane is hyperpolarized, and how far do they move when the channel opens?

Using Biotin and Avidin to Measure Paddle Motions

Finkelstein and co-workers used avidin binding to biotinylated colicin to examine the movements of its components across the lipid bilayer[7-9]. We subjected the KvAP channel to a similar analysis. The idea behind the experiments is outlined in Fig. 2a. We introduce cysteine residues at specific locations in the channel, biotinylate the cysteine, reconstitute channels into planar lipid membranes, and then determine whether avidin binds from the internal or external side, and whether binding depends on membrane depolarization. Wild-type KvAP channels contain a single cysteine on the carboxy-terminus, which was mutated to serine (without affecting function) to work with a channel without background cysteine residues.

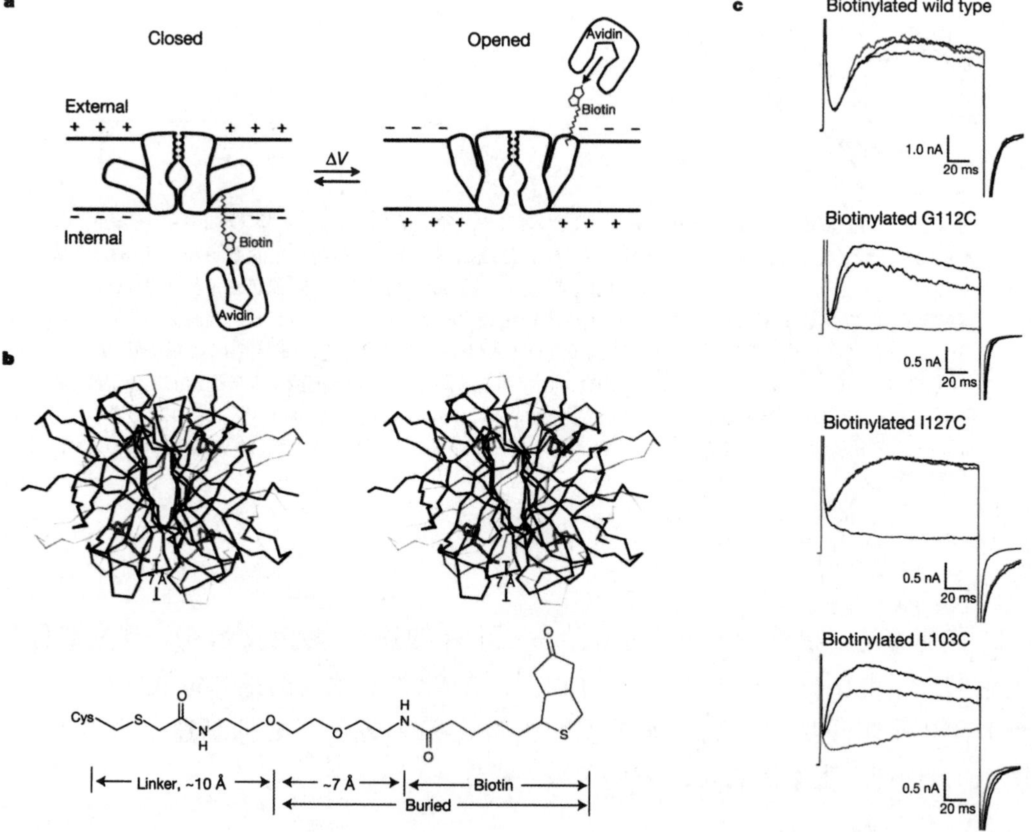

Fig. 2. Using avidin and tethered biotin as a molecular ruler to measure positions of the voltage-sensor paddles. **a**, Experimental strategy: KvAP channels with biotin tethered to a site-directed cysteine can be "grabbed" by avidin in solution to affect channel function. **b**, Stereo view of an avidin tetramer (Cα trace, Protein Data Bank code 1AVD) with biotin (red) in its binding pockets. Chemical structure of biotin and its PEO-iodoacetyl linker with buried (inside avidin) and exposed segments indicated. **c**, Representative traces

超极化时，桨在膜内多深的地方？通道打开时它们会移动多远？

利用生物素和抗生物素蛋白测量桨运动

芬克尔斯坦和他的同事们用结合大肠杆菌生物素的抗生物素蛋白来检测其组成部分跨越脂质双层的运动[7-9]。我们对 KvAP 通道做了类似的分析。图 2a 概述了实验构想。我们在通道的特殊位置引入半胱氨酸残基，用生物素标记这些半胱氨酸，将通道重建到平面脂质膜，然后确定抗生物素蛋白是从内侧还是从外侧结合，以及结合是否取决于膜的去极化。野生型 KvAP 通道蛋白在羧基末端包含一个半胱氨酸，将它突变成丝氨酸(不影响功能)，与无本底半胱氨酸残基背景的通道一起进行实验。

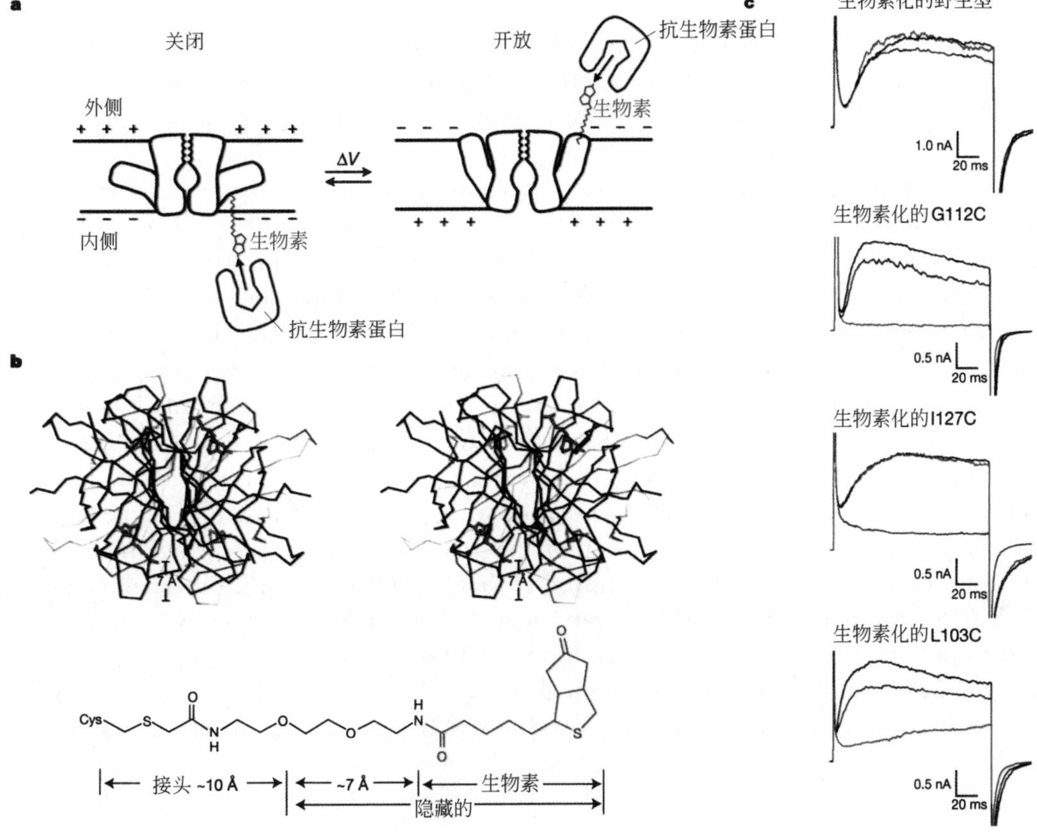

图 2. 用结合了生物素的抗生物素蛋白作为分子尺来测量电压感受桨的位置。**a**，实验策略：生物素定点结合到 KvAP 通道的半胱氨酸上，就可以被溶液内抗生物素蛋白"捕获"进而影响通道功能。**b**，抗生物素蛋白四聚体(α 碳原子示踪，蛋白质数据库索引号 1AVD)与结合在其口袋处的生物素(红色)的立体构象。生物素及其 PEO–碘乙酰基连接接头的化学结构，其中隐藏的(即在抗生物素蛋白内的)和暴

showing the effects of avidin on wild-type and mutant biotinylated KvAP channels. Currents were elicited with depolarizing steps in the absence (black traces) or presence of internal (blue traces) or external (red traces) avidin.

The crystal structure of avidin and the chemical structure of biotin attached to its linker are shown in Fig. 2b. Biotin binds within a deep cleft inside the core of avidin, a rigid protein molecule[10]. The atom on the biotin molecule to which the linker is attached is 7 Å beneath the surface of avidin (Fig. 2b). Therefore, when avidin binds to a biotinylated cysteine on the channel, the distance from the cysteine α-carbon to the surface of avidin is 10 Å; this is the pertinent linker length from the α-carbon to avidin. Since avidin is large (the tetramer is a 57-kDa protein), it cannot fit into clefts on the channel and therefore cannot penetrate the surface. The important point is that, for avidin to bind to a tethered biotin, the cysteine α-carbon has to be within 10 Å of the bulk aqueous solution on either side of the membrane. Applying this restraint, we used linked biotin and avidin as a molecular ruler to measure positions of the voltage-sensor paddles.

Detection of avidin binding to biotinylated channels depends on the demonstration of a functional effect when avidin "grabs" tethered biotin. Control experiments and examples are illustrated in Fig. 2c. Our convention for representing data will be black, red and blue traces corresponding to control (no avidin), external and internal avidin, respectively. Biotinylated wild-type channels are not affected by external avidin and show a small reduction in current when avidin is applied to the internal solution. On the basis of protein gel assays, we conclude that the wild-type channels do not contain detectable biotin on them after the biotinylation procedure (not shown). Three examples of biotinylated cysteine mutant channels are shown. The G112C mutant is inhibited completely by external but not internal avidin; the small reduction by internal avidin is similar to the wild-type control. The I127C mutant is inhibited completely by internal but not by external avidin. The L103C mutant shows an outcome different from complete inhibition: external avidin reduces the current but does not abolish it, and changes the kinetics of current activation. Another outcome that is observed at certain positions is partial inhibition by avidin without changing the gating kinetics (Fig. 3). These cases probably reflect incomplete biotinylation of buried cysteine residues. We excluded the possibility that "no effect" (that is, internal avidin on G112C or external avidin on I127C) represents "silent" avidin binding by adding avidin first to the "no effect" side of the membrane and then to the opposite side (not shown). Because avidin binding to biotin is essentially irreversible, silent binding to one side should protect from binding to the other, but this was never observed. Thus, "no effect" signifies no binding.

露的部分在图中标示出来了。**c**，代表性电流轨迹显示了抗生物素蛋白对野生型和突变体生物素标记的 KvAP 通道的影响。在没有（黑色轨迹）或存在内侧（蓝色轨迹）或外侧（红色轨迹）抗生物素蛋白的情况下，去极化可以引起电流。

抗生物素蛋白的晶体结构和结合到其接头的生物素的化学结构如图 2b 所示。生物素结合在一个刚性蛋白分子——抗生物素蛋白——核心处的一个深深的裂缝内[10]。生物素分子上与接头结合的原子在距抗生物素蛋白表面 7 Å 处（图 2b）。因此，当抗生物素蛋白结合到通道上的生物素化的半胱氨酸时，从半胱氨酸 α 碳到抗生物素蛋白表面的距离就是 10 Å；这是从 α 碳到抗生物素蛋白的合适连接长度。因为抗生物素蛋白很大（该四聚体是一个 57 kDa 的蛋白），它不能进入通道上的裂缝，因此不能穿越表面。重要的一点是，为使抗生物素蛋白结合到与通道绑定的生物素上，半胱氨酸的 α 碳必须位于膜任意一侧距大量水溶液 10 Å 范围内。运用这个尺度，我们将连接的生物素和抗生物素蛋白作为分子尺来测量电压感受桨的位置。

检测结合到生物素标记通道的抗生物素蛋白取决于当抗生物素蛋白“捕获”绑定的生物素时的功能影响的论证。对照组和实验组如图 2c 所示。我们给出的代表性数据中黑色、红色和蓝色的轨迹分别为对照（无抗生物素）、外侧和内侧加入抗生物素蛋白。生物素标记的野生型通道不受外部抗生物素蛋白影响，并且当抗生物素加入内侧溶液后电流轻微减弱。在蛋白质凝胶电泳分析基础上，我们得出这样的结论：野生型通道在生物素化处理后不含有可检测的生物素（未显示）。这里显示了三个生物素化半胱氨酸突变通道实例。G112C 突变体被外侧的抗生物素蛋白完全抑制，而内侧抗生物素蛋白无此作用；内侧抗生物素蛋白引起的轻微减弱类似于野生型对照。I127C 突变体被内侧的抗生物素蛋白完全抑制，而外侧抗生物素蛋白无此作用。L103C 突变体显示了与完全抑制不同的结果：外侧抗生物素蛋白降低电流，但并不完全消除电流，它还改变了电流激活的动力学。另一项结果是观察到的某些位置被抗生物素蛋白部分抑制而不改变门控动力学（见图 3）。这些例子可能反映了包埋的半胱氨酸残基不能被生物素完全标记。我们通过先将抗生物素蛋白加到膜“没有影响”的一侧然后加到另一侧，排除了“没有影响”（内侧抗生物素蛋白对 G112C 或外侧抗生物素蛋白对 I127C）代表“沉默”抗生物素蛋白结合的可能性（未显示）。由于抗生物素蛋白结合生物素基本上是不可逆的，沉默结合到一侧时应不再结合到另一侧，事实上也从未被观察到，因此，“没有影响”意味着没有结合。

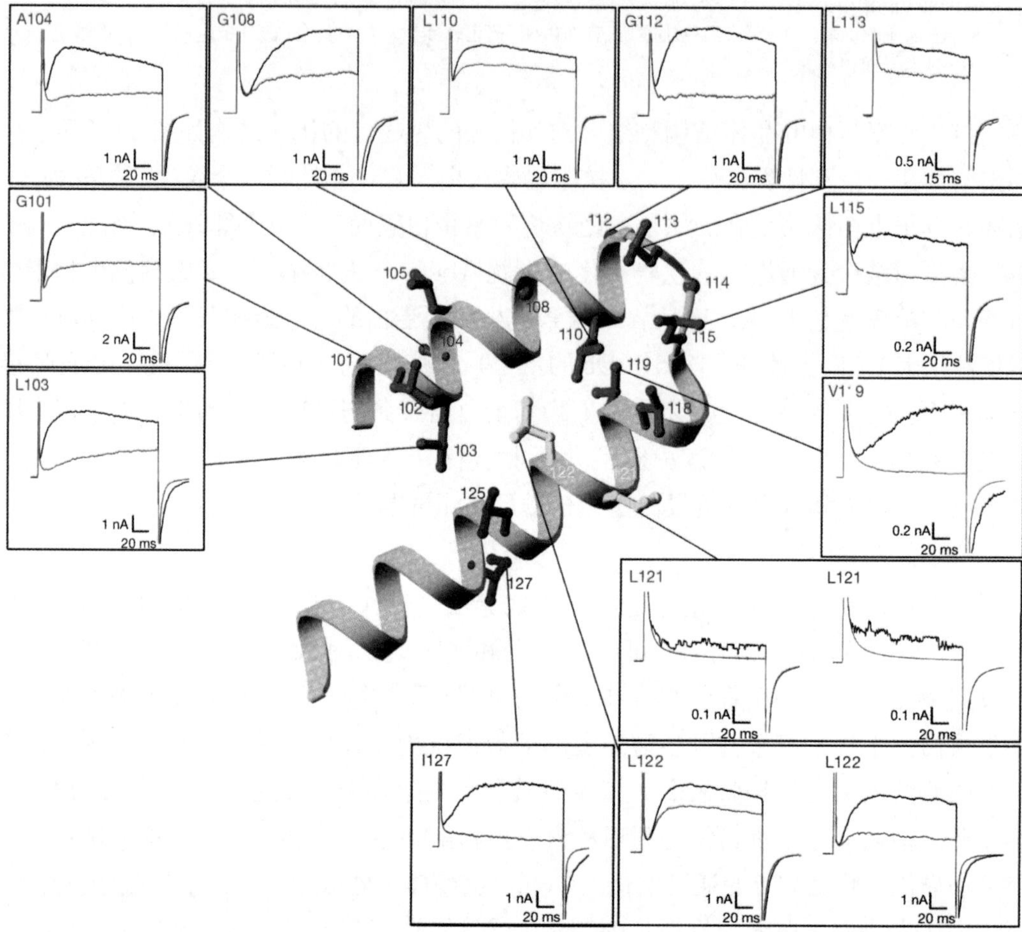

Fig. 3. Accessibility of voltage-sensor paddle residues to the internal and external sides of the membrane. Membrane was depolarized every 120 s in the absence (control) or presence of avidin on the internal and external side of each biotinylated mutant. Red side chains on the voltage-sensor paddle and selected traces indicate inhibition by avidin from the external side only; blue side chains and selected traces indicate inhibition by avidin from the internal side only; yellow side chains and blue and red traces indicate inhibition by avidin from both sides. Black traces show control currents before avidin addition and coloured traces show currents after adding avidin. Each trace is the average of 5–10 measured traces, with the exception of the single traces for position 121.

Data from a scan of the voltage-sensor paddles from amino-acid position 101 on S3b to position 127 on S4 are shown in Fig. 3. Positions were mutated individually to cysteine, biotinylated and studied in planar lipid membranes. Certain biotinylated mutants showed altered gating even before avidin addition (for example, valine 119), or appeared to be less abundant in the membrane (for example, leucine 121). But in all cases, channels could be held closed at negative membrane voltages (typically −100 mV) and opened by membrane depolarizations (typically 100 mV for 200 ms) every 120 s. To compensate for mutations that shifted the voltage-activation curve, we sometimes held the membrane as negative as −140 mV, or used opening depolarizations as positive as 200 mV. Avidin bound to the

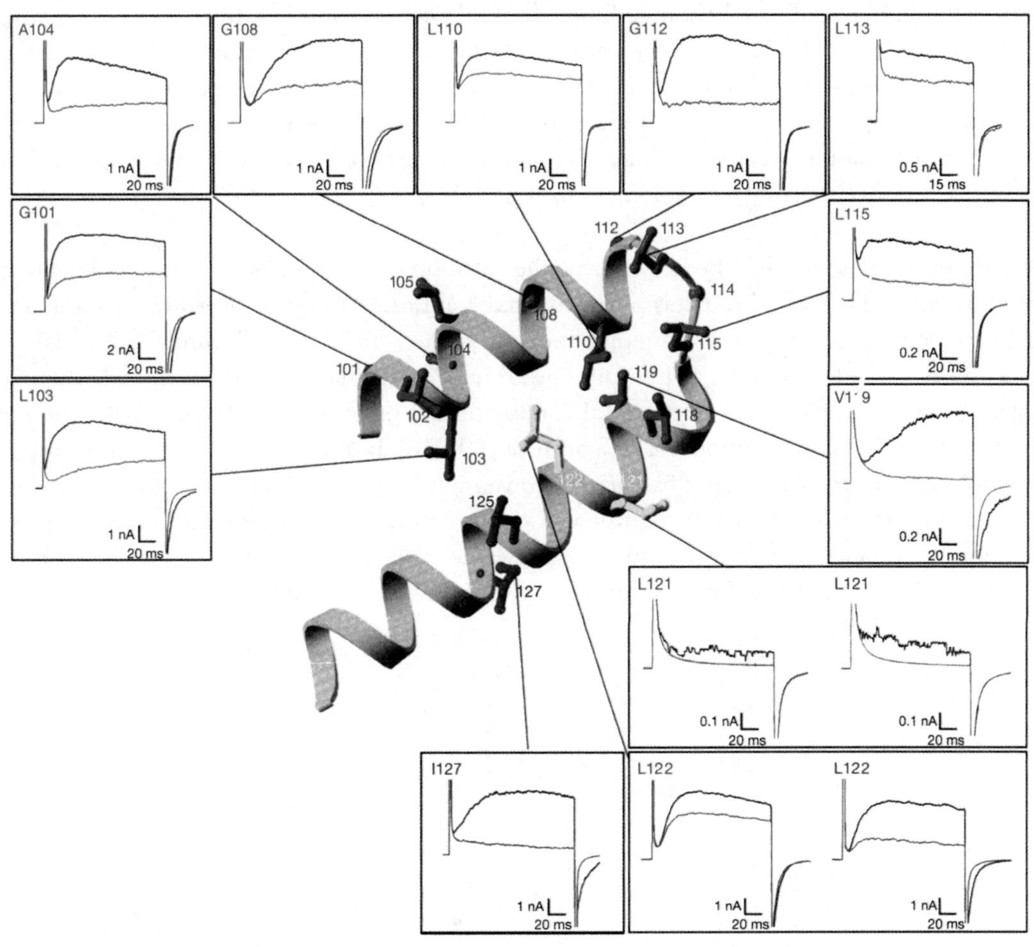

图 3. 电压感受桨残基与膜内侧和外侧的可结合性。在每个生物素标记的突变体内侧和外侧没有（对照）或存在抗生物素蛋白的情况下，每 120 s 膜被去极化。电压感受桨上的红色侧链和对应的电流轨迹表明抗生物素蛋白仅从外侧抑制，蓝色侧链和对应的电流轨迹表明抗生物素蛋白仅从内侧抑制，黄色侧链及蓝色和红色的轨迹表明抗生物素蛋白从两侧抑制。黑色轨迹表明抗生物素蛋白加入前的对照电流，彩色轨迹表明加入抗生物素蛋白后的电流。每个轨迹是 5 ~ 10 个测量轨迹的平均值，除了位点 121 是单一的轨迹外。

图 3 显示了对电压感受桨从 S3b 上 101 位点到 S4 上 127 位点的氨基酸进行扫描得到的数据。对突变为半胱氨酸的位点，用生物素标记后在平面脂质膜内对其进行了研究。某些生物素标记的突变体甚至在加入抗生物素蛋白之前就显示了门控的改变（例如，缬氨酸 119），或在膜中的含量减少（例如，亮氨酸 121）。但在所有情况下，通道可以在负膜电压时维持关闭（通常为 −100 mV），并通过每 120 s 一次的膜去极化（通常为 100 mV，200 ms）而开放。为了抵消突变体电压激活曲线的移动，我们有时将膜电位维持在 −140 mV，或使用更高的去极化电位（200 mV）使其开放。

tethered biotin and affected channel function at all positions studied. In many cases, avidin caused complete inhibition, and in some cases partial inhibition with altered kinetics. Partial inhibition at certain sites (for example, leucine 110 and leucine 122) can be explained on the basis of incomplete biotinylation because the side chain is buried within the "core" of the voltage-sensor paddle between S3b and S4. This finding is consistent with the proposal that S3b and S4 move together as a voltage-sensor paddle unit.

We are most interested in whether a particular position on the voltage-sensor paddle allows binding to avidin from the external or internal solution. Amino acids on the paddle are colour-coded according to the membrane side from which avidin bound: red, outside; blue, inside; yellow, both. For avidin binding from the external side, we examined whether membrane depolarization is required, by studying the effect of depolarization frequency on the rate of channel inhibition by avidin (Fig. 4). Inhibition from the external solution required depolarization: higher frequencies gave higher rates of inhibition. In other words, the voltage-sensor paddles can be protected from external avidin binding by keeping the membrane at negative voltages, and the paddles are exposed to external avidin at positive voltages. This was the case for all external positions (Fig. 3, red and yellow).

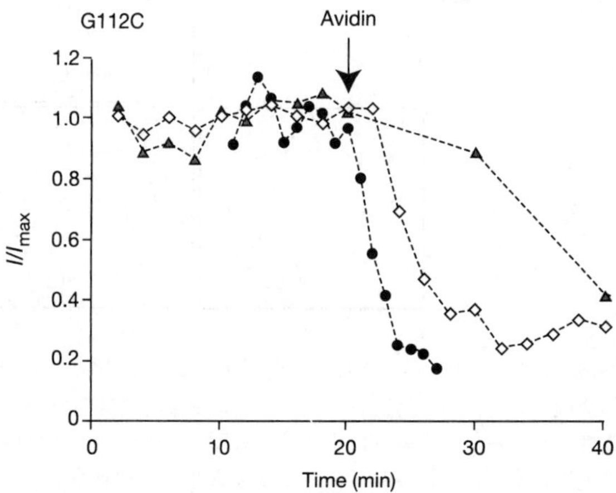

Fig. 4. Exposure of the voltage-sensor paddle to the external solution occurs only when the membrane is depolarized. For the biotinylated G112C mutant, normalized (to the average control) currents elicited by depolarization to 100 mV every 60 s (circles), 120 s (diamonds) and 600 s (triangles) are shown before and after the addition of avidin to the external side.

However, protection from inhibition by external avidin, Fabs and a voltage-sensor toxin, by holding the membrane at negative voltages, is incomplete. In particular, if the wait at negative voltages is long enough, inhibition occurs but at a low rate (Fig. 4). This finding is explained on the basis of thermal fluctuations of the voltage sensors. Probably four voltage-sensor paddles have to move to open the pore[1,11-13]. But individual paddle movements must occasionally occur even at negative voltages. These sensor movements underlying incomplete protection are consistent with gating currents preceding pore opening[2], and longer delays before pore opening when the membrane is depolarized from more negative

抗生物素蛋白结合到绑定的生物素上，在被研究的所有位点上影响通道功能。在许多情况下，抗生物素蛋白引起完全抑制，而在另一些情况下，通过改变分子动力学而引起部分抑制作用。在某些位点上（例如，110 位亮氨酸和 122 位亮氨酸）部分抑制可以在不完全生物素标记的基础上进行解释，因为其侧链被包埋在位于 S3b 和 S4 之间的电压感受桨的“核心”内。这一发现与认为 S3b 和 S4 作为电压感受桨单元一起移动的观点一致。

我们最感兴趣的是抗生物素蛋白是否从外侧或内侧溶液中结合到电压感受桨上某一特定位置。根据抗生物素蛋白从膜的哪一侧结合，对桨上的氨基酸进行了颜色标记：红色，外侧；蓝色，内侧；黄色，两侧。对于抗生物素蛋白从外侧结合，我们通过研究去极化频率对抗生物素蛋白抑制通道的程度的影响，观察膜的去极化是否必需（图 4）。从外侧溶液抑制需要去极化：频率越高抑制率也越高。换言之，通过保持膜的负电压可以保护电压感受桨免受外侧抗生物素蛋白结合，而正电压时桨暴露于外侧的抗生物素蛋白中。这是所有外侧位置的情况（图 3，红色和黄色）。

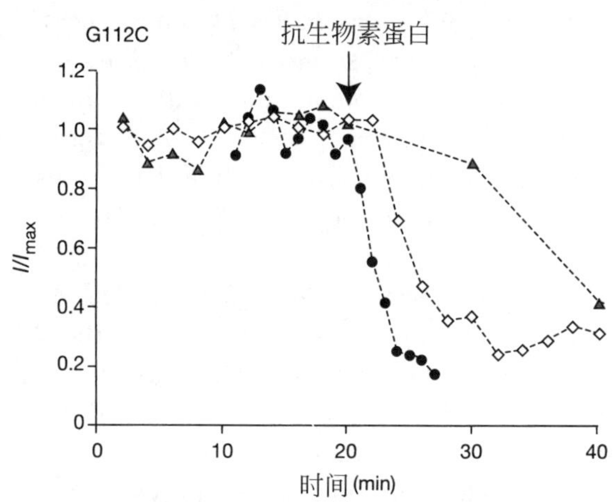

图 4. 只有当膜去极化时，才会发生电压感受桨暴露在外侧溶液中的情况。对于生物素标记的 G112C 突变体，图中显示了抗生物素蛋白加入外侧前后，由每 60 s（圆圈）、120 s（菱形）、600 s（三角形）去极化到 100 mV 所引起的归一化（至平均值）电流。

然而，通过维持膜负电压，并不能完全避免外侧抗生物素、Fabs 和电压传感器毒素造成的抑制。特别是，如果在负电压保留的时间足够长，抑制可发生但发生率低（图 4）。这个发现可用电压传感器的热波动来解释。大概有 4 个电压感受桨必须移动以打开孔道 [1,11-13]。但是，即使在负电压下也必然会偶尔发生个别的桨运动。这些不完全保护下的传感器运动与先于孔道开放的门控电流一致 [2]，且当膜从更负的

holding voltages, a phenomenon known as the Cole–Moore effect[14].

Biotin molecules tethered at two positions were captured by avidin from both sides of the membrane (121 and 122; Fig. 3, yellow). At these positions, inhibition was complete from either side alone, because subsequent addition of avidin to the opposite side caused no further inhibition. Thus, the dual accessibility cannot be ascribed to two structurally distinct populations of channels. We conclude that positions 121 and 122 actually drag biotin and its linker all the way across the lipid membrane from the solution on one side to that on the other when the channel gates. This finding is very important, because it indicates that the voltage-sensor paddles must move a large distance through the membrane, and that they must move through a lipid environment where a bulky chemical structure such as biotin and its linker (Fig. 2b) would be unimpeded. Biotin and its linker could not be dragged through the core of a protein, as would be required by conventional models, which invoke an S4 helix buried within the protein.

Discussion

Positional constraints on the voltage-sensor paddles are summarized in Fig. 5a, b. Horizontal solid lines show the external and internal surfaces of the cell membrane (~35 Å thick) and dashed lines show the 10 Å limit from the membrane surface set by the linker length (Fig. 2b). If the α-carbon of a cysteine residue comes within 10 Å of the membrane surface, then avidin can bind to the attached biotin, otherwise the biotin is inaccessible. Avidin is too large to enter crevices on the channel, so it cannot penetrate below the membrane surface.

At negative membrane voltages when the channel is closed, no positions are accessible to the external side; all residues on the voltage-sensor paddles in their channel-closed position must lie deeper in the membrane than the 10-Å limit below the external surface (Fig. 5a). At negative voltages, the blue and yellow residues on S4 bind from inside, and therefore must come within 10 Å of the internal solution. The red residues, including all of S3b, the tip of the paddle and the first helical turn into S4, are protected from both sides at negative voltages and must therefore lie further than 10 Å from both surfaces; that is, between the two dashed lines. This pattern of accessibility in the closed (negative voltage) conformation constrains the voltage-sensor paddles to lie near the internal surface of the membrane with S3b above S4, as shown (Fig. 5a), similar to the orientation in the crystal structure[4].

At positive membrane voltage when the channel is opened, the entire S3b helix, the tip of the paddle and the first two-and-a-half helical turns of S4 become accessible to external avidin and must therefore be within 10 Å of the external solution (Fig. 5b). The next helical turn on S4 (positions 125 and 127, blue residues) remains more distant than 10 Å from the external surface. The Fabs inform us that, in the opened conformation, two helical turns of S3b and one turn of S4 must actually protrude clear into the external solution, otherwise the epitope would not be exposed (Fig. 1b). The Fabs and pattern of avidin accessibility in

保持电压去极化时孔道开放前的延迟更长，这一现象被称为科尔–摩尔效应[14]。

抗生物素蛋白从膜两侧捕获标记在两个位置的生物素分子（121 位和 122 位；图 3，黄色）。在这些位置，来自任一侧的抑制都是完全的，因为随后在对侧加入抗生物素蛋白并没有引起进一步的抑制。因此，双重可结合性不能归因于两种在结构上截然不同的通道类型。我们断定当通道处于门控状态时，位于 121 位和 122 位的氨基酸残基实际上拖动生物素和其接头一路跨越脂质膜，从一侧溶液到达另一侧溶液。这一发现很重要，因为它说明电压感受桨必须移动很长一段距离穿过膜，必须移动通过脂质环境，在该脂质环境里大的化学结构如生物素及其接头（图 2b）不受阻碍。生物素及接头无法通过蛋白质的核心，正如传统模型所述，该过程要借助于包埋在蛋白质内的 S4 螺旋。

讨　论

图 5a 和 b 总结了电压感受桨上的位置约束。水平实线显示了细胞膜的外表面和内表面（内外表面间的厚度约 35 Å），虚线显示了由接头长度决定的距离膜表面的 10 Å 距离（图 2b）。如果半胱氨酸残基 α 碳到达距膜表面 10 Å 的范围内，那么抗生物素蛋白可以结合到与它连接的生物素上，否则无法接近生物素。抗生物素蛋白太大不能进入通道裂缝，所以不能穿透膜表面。

当膜电压为负，通道关闭时，没有可连接到外侧的位点；电压感受桨上所有残基在通道处于关闭位置时均处于膜外表面下 10 Å 以下的深处（图 5a）。负电压时，S4 上蓝色和黄色残基可从内侧结合，因此，必须到达距细胞内液 10 Å 以内。红色的残基，包括所有 S3b，桨的尖端和 S4 的第一个螺旋，在负电压时它们的两侧都受到保护，因此必须位于离两个表面远于 10 Å 的地方，即两虚线之间。在关闭（负电压）构象中，S3b 在上、S4 在下形成的电压感受桨处于接近膜内表面附近，如图所示（图 5a），类似于晶体结构中的取向[4]。

当膜电压为正，通道开放时，整个 S3b 螺旋、桨的末端和 S4 最初的两个半螺旋转角变得易于接触到外侧抗生物素蛋白，因此必然在距外侧溶液 10 Å 内（图 5b）。S4 上紧接的一个螺旋转角（125 位和 127 位，蓝色残基）与外表面保持着超过 10 Å 的距离。结合位点告诉我们，在开放构象中，S3b 的两个螺旋转角和 S4 的一个转角需从膜内突出进入外侧溶液，否则抗原表位将不会暴露出来（图 1b）。在开放（正电压）构象时，抗原表位和抗生物素蛋白的可结合模式迫使电压感受桨以一个更加垂

the opened (positive voltage) conformation constrain the voltage-sensor paddles to be near the external membrane surface with a more vertical orientation, as shown (Fig. 5b).

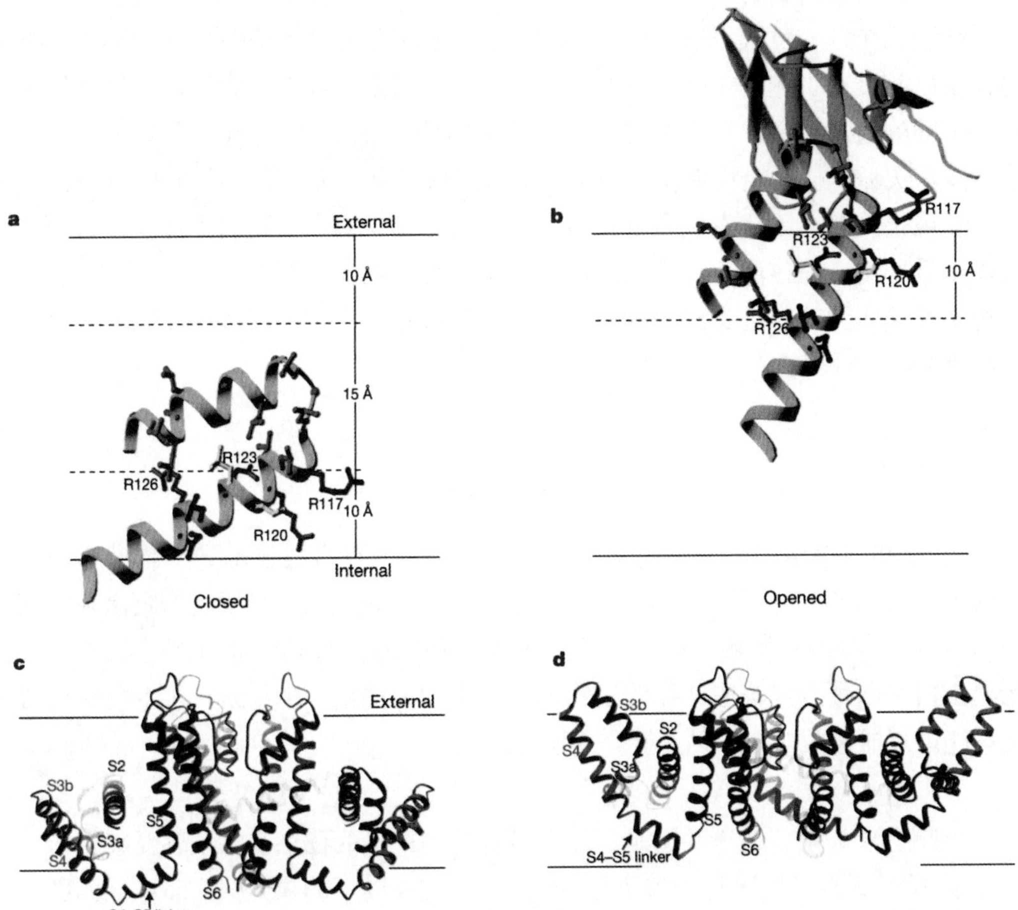

Fig. 5. Positions within the membrane of the voltage-sensor paddles during closed and opened conformations, and a hypothesis for coupling to pore opening. **a**, **b**, Closed (**a**) and opened (**b**) positions of the paddles derived from the tethered biotin–avidin measurements, and structural and functional measurements with Fabs. A voltage-sensor paddle is shown as a cyan ribbon with side chains colour-coded as in Fig. 3. Grey side chains show four arginine residues on the paddle, and the green ribbon (**b**) shows part of a bound Fab from the crystal structures. Solid horizontal lines show the external and internal membrane surfaces, and dashed lines indicate the 10-Å distance from the surfaces set by biotin and its linker. **c**, The closed KvAP structure is based on the paddle depth and orientation in **a** (red), and adjusting the S5 and S6 helices of KvAP to the positions in KcsA, a closed K^+ channel. **d**, The opened KvAP structure is based on the paddle depth and orientation in **b** (red), and the pore of KvAP.

In moving from their closed to their opened position, the voltage-sensor paddles' centre of mass translates approximately 20 Å (assuming a membrane thickness of 35 Å) through the membrane from inside to out, and the paddles tilt from a somewhat horizontal to a more vertical orientation. Arginines 117, 120, 123 and 126, the first four arginines on S4 (Fig. 1a), are distributed along the paddles. In the closed conformation, arginines 126,

直的取向靠近膜的外表面，如图所示(图 5b)。

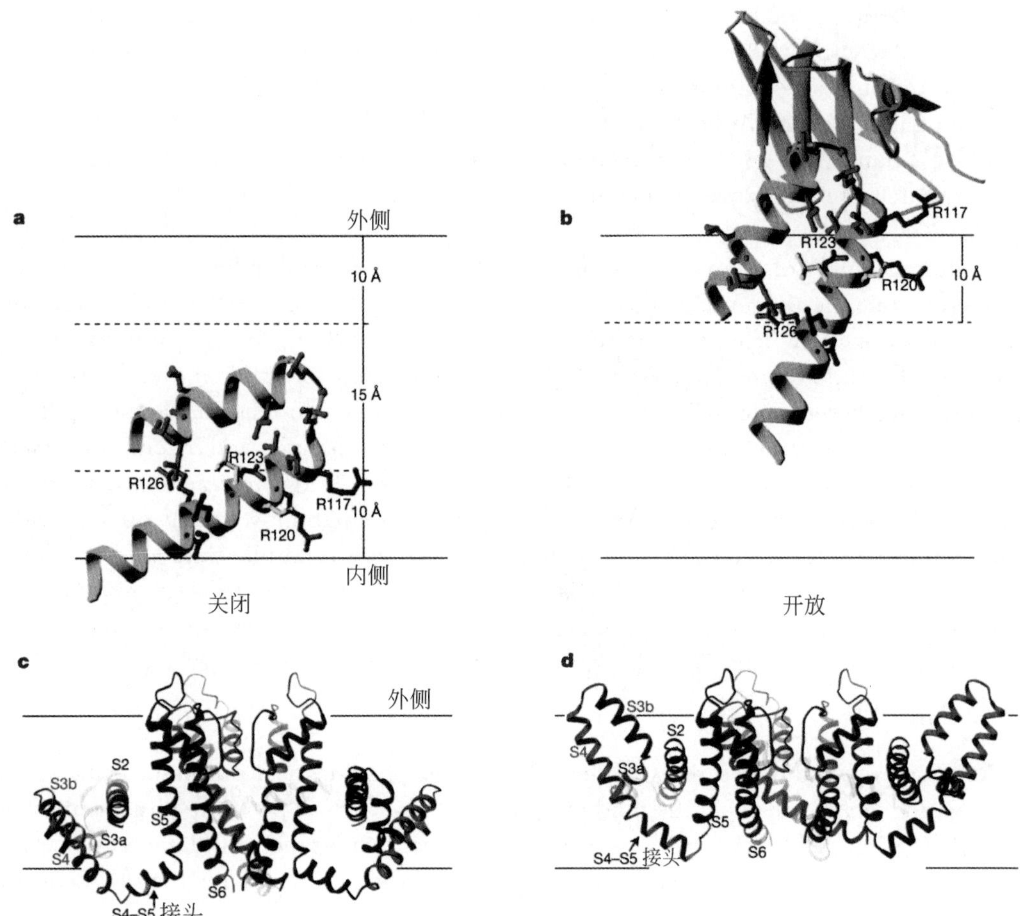

图 5. 关闭和开放构象期间电压感受桨在膜内的位置，以及与孔道开放结合的假设。**a** 和 **b** 分别表示关闭和开放状态时桨的位置，是基于生物素–抗生物素的连接以及借助 Fabs 结合进行的结构和功能测定。如图 3 中显示电压感受桨为一个带有彩色编码侧链的蓝绿色带。灰色侧链显示桨上的 4 个精氨酸残基，(**b**) 中的绿色带为晶体中一个结合的 Fab 的局部结构。水平实线显示外侧和内侧膜表面，虚线表明由生物素及其接头决定的距膜表面 10 Å 的距离。**c**，关闭的 KvAP 结构是基于 **a**(红色) 中桨的深度和位置，并调整 KvAP 的 S5 和 S6 螺旋到 KcsA 中位置，使通道关闭。**d**，开放的 KvAP 结构基于 **b**(红色) 中桨的深度和方向，以及 KvAP 孔道。

在它们从关闭位置移动到开放位置的过程中，电压感受桨的质量中心从内到外穿过膜移动大约 20 Å(假设膜厚度为 35 Å)，而桨从相对水平向更加垂直的方向倾斜。S4 上位于 117 位、120 位、123 位和 126 位的前 4 个精氨酸(图 1a) 沿桨分布。在关闭的构象中，126 位、123 位，也许还有 120 位的精氨酸可将带正电荷的胍基延

123 and perhaps 120 can probably extend their positively charged guanidinium group to the internal lipid head group layer, whereas arginine 117 is near the internal side but still within the membrane. In the opened conformation, arginines 117, 120 and probably 123 can extend to the external solution or lipid head group layer, whereas 126 is near the external side but still within the membrane. The near-complete transfer of these four arginine residues across the membrane (in each of the four subunits) is compatible with the total gating charge in the *Shaker* K^+ channel of 12–14 electrons (3.0–3.5 electrons per subunit)[15-17], and with the demonstration that each of these four arginine residues carries approximately one electron charge unit[16,17].

A positional aspect of the voltage-sensor paddles not constrained by these experiments is whether they lie tangential to the channel's outer surface or whether they point in a radial direction away from it. The flexible S3 loops and S4–S5 linkers probably do not constrain the paddles much. However, we have two reasons for thinking that the paddles are positioned tangentially, which is the way they are positioned in Fig. 5c. First, the paddles are oriented tangentially in the crystal structure of the full channel; and second, the crystal structure of the isolated voltage sensor shows interesting salt-bridge interactions between S4 and S2, and between S3a and S2 (ref. 4). A tangential orientation would favour salt-bridge interactions between arginine residues on the voltage-sensor paddles and acidic residues on the S2 and S3a helices. Studies by Papazian and co-workers on the *Shaker* K^+ channel indeed suggest that salt-bridge pairs are important, and that they might exchange as the paddles move between their closed and opened positions on the outer perimeter of the channel[18-20].

To a first approximation, we describe the voltage-sensor paddles as hydrophobic cations that carry gating charges through the lipid bilayer. Ionic interactions between S4 arginines and S2 and S3 acidic residues probably assist the gating charge movement, and the presence of a polar, sometimes acidic loop between S3b and S4 in certain voltage-dependent K^+ channels (for example, the *Shaker* K^+ channel[21]) raises interesting questions about the structure of the lipid–water interface above the paddles when they are in their closed channel position. However, there is no escaping the basic finding that the paddles are located at the protein–lipid interface, and move while contacting the lipid membrane. Given that the paddles move through a lipid environment, it is interesting to ask why the basic residues on the voltage-sensor paddles are nearly always arginine and not lysine? One reason is that arginine ($pK_a \approx 12.5$ in water) will nearly always move through the membrane with a protonated, charged side chain, whereas lysine ($pK_a \approx 10.5$ in water) will sometimes be unprotonated. Another reason may be that arginine can easily participate in multiple hydrogen bonds (perhaps with acidic side chains) in a spatially directed way. Yet a third possible reason is that it might be energetically less costly to transfer from water to lipid the diffuse positive charge on a guanidinium group (arginine) than the more focused charge on an amino group (lysine). Studies of the transfer of hydrophobic peptides from water to octanol by White and co-workers[22] provide evidence for this idea by showing that the charged lysine side chain is energetically more costly in this assay than the charged arginine by about 1.0 kcal mol^{-1}. Given that four voltage-sensor paddles must move through the membrane to open the pore,

伸至内侧的脂质头部基团，而 117 位精氨酸接近内侧但仍然在膜内。在开放的构象中，117 位、120 位，或许还有 123 位的精氨酸可以延伸至外侧溶液或脂质头部基团层，而 126 位精氨酸接近外侧但仍然在膜内。这 4 个精氨酸残基跨膜接近完全的转移（在 4 个亚基中的每一个内）与 12 ~ 14 个电子的 *Shaker* K^+ 通道的总门控电荷（每个亚基 3.0 ~ 3.5 个电子）相一致 [15-17]，并证明了每组 4 个精氨酸残基搬运了约 1 个电荷单位 [16,17]。

不受这些实验限制的电压感受桨要么位于通道外表面的切面上，要么指向一个远离通道外表面的径向方向。柔性的 S3 无规卷曲和 S4–S5 接头可能不太限制桨的位置。但是，我们有两个理由认为桨是切向定位的，它们的定位方式见图 5c。首先，在通道的完整晶体结构中桨是切向定位的；其次，独立的电压感受器的晶体结构显示了 S2 与 S4 之间和 S3a 与 S2 之间存在有趣的盐桥相互作用（参考文献 4）。切线方向将有利于电压感受桨上的精氨酸残基和 S2 与 S3a 螺旋上的酸性残基之间的盐桥相互作用。帕帕济安及其同事们对 *Shaker* K^+ 通道的研究表明盐桥对确实具有重要意义，当桨在关闭和开放位置间运动时，它们可能在通道外口周围发生交换 [18-20]。

我们已经描述了电压感受器桨作为疏水阳离子可携带门控电荷通过脂质双层。S4 精氨酸与 S2 和 S3 上酸性残基之间的离子相互作用可能协助门控电荷的运动并产生极性，有时候，在某些电压依赖性 K^+ 通道（例如，*Shaker* K^+ 通道 [21]）中的 S3b 和 S4 之间的酸性环带来当通道在关闭位置时桨上方脂–水界面结构的有趣问题。然而，无法回避的基本结果是，即桨位于蛋白质–脂质界面，且随着脂质膜移动。已知桨是在脂质环境中移动，有趣的是，为什么电压感受桨上的碱性残基几乎总是精氨酸而不是赖氨酸？一个原因是，精氨酸（在水中 $pK_a \approx 12.5$）几乎总是带有一个质子化的、带电的侧链进行跨膜运动，而赖氨酸（在水中 $pK_a \approx 10.5$）有时是非质子化的。另外一个原因可能是精氨酸易于在空间定向上形成多个氢键（也许是与酸性侧链）。第三个可能的原因是，从水向脂质转移在胍基（精氨酸）上分散的正电荷比转移氨基（赖氨酸）上更集中的电荷耗费较少的能量。怀特及其同事 [22] 对疏水多肽从水向辛醇转移的研究为这一想法提供了证据，他们的研究显示带电赖氨酸侧链在这一实验中比带电精氨酸耗费的能量大约多 1.0 $kcal \cdot mol^{-1}$。鉴于 4 个电压感受桨必须通过膜内运动才能打开孔道，我们推测从进化的角度讲更偏向于精氨酸而非赖氨酸。在这方面，有趣的是，最近发现机械敏感性通道 MscS 的结构在其跨膜区段有两个碱性氨基酸，

we might expect there to be a strong evolutionary bias in favour of arginine over lysine. In this regard, it is interesting that the recent structure of MscS, a mechanosensitive channel, has two basic amino acids in its transmembrane segments that are arginine[23]. MscS is not gated by voltage *per se*, but its mechanical force-induced opening can be modulated by voltage[24], and the candidate residues that are proposed to underlie voltage modulation are arginine, not lysine[23].

On the basis of the KvAP crystal structure[4], the deduced positions of its voltage-sensor paddles in the functioning channel (Fig. 5a, b), and previous studies of opened and closed K^+ channels[25,26], we propose a model for how membrane voltage gates the pore (Fig. 5c, d). This is a working model to envision how the electromechanical coupling process might occur, and it will need to be revised as more data are obtained. In the closed conformation (Fig. 5c), the positively charged voltage-sensor paddles (red) are near the intracellular membrane surface, held there by the large electric field (mean value $> 10^7$ V m^{-1}) imposed by the negative resting membrane voltage. In this conformation, the S5 and S6 (outer and inner) helices are arranged as they are in KcsA, a closed K^+ channel[25]; favourable packing interactions between the inner helices have been proposed to help to stabilize the closed conformation[27]. In response to depolarization, the voltage-sensor paddles move across the membrane to their external position, which exerts a force on the S4–S5 linker, pulling the S5 helices away from the pore axis (Fig. 5d). The crystal structure of KvAP shows that the S5 helices form a cuff outside the S6 helices, and suggests that when the S5 cuff is expanded, the S6 helices follow, opening the pore[4].

Although previous mutational studies of voltage-dependent gating have been interpreted in the context of the conventional models, many of the data are consistent with the structural model presented here, and indeed help to constrain it. For example, second-site suppressor mutations in the *Shaker* K^+ channel suggest that certain salt bridges probably break and reform as the voltage-sensor paddles move[20]. In addition to four S4 arginine residues, a component of the gating charge in *Shaker* was shown to come from an S2 acidic residue (glutamate 293, corresponding to aspartate 72 in KvAP)[16], implying that S2 might change its position in the membrane as the voltage-sensor paddles move; the loose attachment of S2 to the pore in the KvAP crystal structure certainly would allow this to happen. Disulphide cross-linking of S4 to the turret loop between S5 and the pore helix[28], and of two S4 segments within a tetramer to each other[29] (an observation that was very difficult to understand in the context of conventional models), are in agreement with the mobile voltage-sensor paddles in Fig. 5c, d. The accessibility of thiol-reactive compounds to cysteine residues introduced into the *Shaker* K^+ channel S4 (refs 30–32), and the ability of histidine residues to shuttle protons across the membrane[33], are in reasonable agreement with our data on voltage-sensor paddle movements. However, our structural and mechanistic interpretation, summarized in Fig. 5, differs fundamentally from past models of voltage-dependent gating. This new picture is based on the elucidation of the voltage-sensor paddle structure, its flexible attachments and disposition relative to the pore, and the paddles' positions in the membrane when the channel is closed and opened.

均是精氨酸[23]。MscS 不受电压本身门控，而它由机械外力诱导的开放可被电压调节[24]，并且认为产生电压调节的候补残基是精氨酸而不是赖氨酸[23]。

基于 KvAP 晶体结构[4]、推导的电压感受桨在功能通道中的位置（图 5a 和 b）和以往对开放和关闭 K^+ 通道的研究[25,26]，我们提出了一个膜电压如何门控孔道的模型（图 5c 和 d）。即一个如何产生电–机械耦联的工作模式的假想模型，该模型在获得更多的数据时需要做修正。在关闭构象下（图 5c），带正电荷的电压感受桨（红色）接近细胞内膜表面，在这附近由于负的静息膜电位而存在较大电场（平均值＞$10^7\,V\cdot m^{-1}$）。在此构象中，S5 和 S6（外和内）螺旋的排列与它们在 KcsA（一个关闭的 K^+ 通道）中一致[25]；推测内螺旋间的相互作用有利于稳定关闭构象[27]。当发生去极化反应时，电压感受桨跨膜移动到其位于膜外的位置上，也对 S4–S5 连接施加力量，拉动 S5 螺旋远离孔轴（图 5d）。KvAP 晶体结构表明，S5 螺旋在 S6 螺旋外形成一个袖口，并推测当 S5 袖口扩大时，S6 螺旋跟随其后，打开孔道[4]。

虽然已进行的电压依赖性门控的突变研究已经解释了本文介绍的模型，许多数据与这里呈现的结构模型是一致的，并且确实有助于修正它。例如，*Shaker* K^+ 通道中第二位点抑制突变表明，当电压感受桨运动时某些盐桥可能断裂或重组[20]。除了 4 个 S4 精氨酸残基以外，*Shaker* 中门控电荷的一个组成被证明来自 S2 上的一个酸性残基（293 位谷氨酸，对应于 KvAP 的 72 位天门冬氨酸）[16]，这意味着当电压感受桨移动时 S2 可能改变其在膜中的位置；在 KvAP 晶体结构中 S2 松散连接到孔上将允许这种情况发生。二硫键将 S4 交联到 S5 和孔区螺旋之间的塔形无规卷曲上[28]，并将四聚体内的 2 个 S4 片段相互交联[29]（这种现象在传统模型背景下很难理解），这种二硫交联与图 5c 和 d 中可移动的电压感受桨一致。巯基化合物与 *Shaker* K^+ 通道 S4 的半胱氨酸残基的可结合性（参考文献 30 ~ 32），以及组氨酸残基跨膜穿梭运送质子的能力[33]，与我们关于电压感受桨移动的数据相当一致。然而，我们关于结构和机械运动的解释（归纳在图 5 中）本质上不同于过去的电压依赖性门控模型。这幅新图是以电压感受桨的结构阐明、它与孔道的柔性连接及与孔道功能的关系，以及当通道关闭和开放时桨在膜中的位置这三点为前提的。

Conclusion

Here and in an accompanying paper[4], we have shown that (1) the gating charges are carried on voltage-sensor paddles, which are helix–turn–helix structures attached to the channel through flexible S3 loops and S4–S5 linkers; (2) the paddles are located at the channel's outer perimeter and move within the lipid membrane; (3) the S2 helices lie beside the pore and contain acidic amino acids that could help to stabilize positive charges on the paddles as they move across the membrane; (4) the total displacement of the voltage-sensor paddles is approximately 20 Å perpendicular to the membrane; and (5) the large displacement of the paddles could open the pore by pulling on the S4–S5 linker. We conclude that the voltage sensor operates by an extraordinarily simple principle based on hydrophobic cations attached to levers, which enables the membrane electric field to perform mechanical work to open and close the ion-conduction pore.

Methods

Biotinylation

All biotinylation studies were carried out using a KvAP channel in which the single endogenous cysteine was mutated to serine (C247S). This mutant showed no detectable electrophysiological differences when compared to wild-type KvAP. Single cysteine mutations were then added to the voltage-sensor paddle (positions 101 to 127) using the QuickChange method (Stratagene) and confirmed by sequencing the entire gene. Mutant channels were expressed and purified by the same protocol as wild-type KvAP channels[5], except that before gel filtration, mutant KvAP channels were incubated with 10 mM DTT for 1 h. Immediately after gel filtration, mutant channels (at 0.5–1.0 mg ml^{-1}) were incubated with 500 μM PEO-iodoacetylbiotin (Pierce) for 2–3 h at room temperature, and then either reconstituted into lipid vesicles for electrophysiological analysis, or purified away from the excess biotin reagent on a desalting column, complexed with avidin (Pierce) and run on an SDS gel to assess the extent of biotinylation.

Electrophysiology

Fabs (6E1 and 33H1) and VSTX1 were purified as described in the companion paper[4] and used in electrophysiological assays. Electrophysiological studies of wild-type and biotinylated KvAP channels were carried out as described[5]. To measure inhibition, Fabs (~500 nM), VSTX1 (30 nM) or avidin (40 μg ml^{-1}) were added to wild-type and/or cysteine-mutant, biotinylated channels reconstituted into lipid membranes, and studied using various voltage protocols.

(**423**, 42-48; 2003)

结 论

在本文及一篇相关论文[4]中，我们已经表明：(1)在电压感受桨上携带有门控电荷，它们通过灵活的S3无规卷曲和S4–S5接头连接到通道上的螺旋–转角–螺旋结构；(2)桨位于通道的外围并在脂质膜内移动；(3)S2螺旋位于孔道旁边并含有酸性氨基酸，当桨跨膜移动时，这些酸性氨基酸可能有助于稳定桨上的正电荷；(4)电压感受桨垂直于膜的总位移约20 Å；(5)桨的较大位移可以通过拉动S4–S5接头打开孔道。我们总结如下：电压传感器的操作是基于一个非常简单的将疏水阳离子连接到杠杆上的原理，使得膜电场转为机械操作来打开和关闭离子导通孔。

方 法

生物素化

所有的生物素化研究都是在KvAP通道上进行的，在KvAP通道中单个内源性半胱氨酸被突变为丝氨酸(C247S)。与野生型KvAP相比，这个突变体没有可检测到的电生理方面的差异。使用QuickChange法(Stratagene公司)，将电压感受桨上的单个半胱氨酸(101位至127位)突变，并通过全基因测序验证突变。采用与野生型KvAP通道相同的方法表达和纯化了突变通道[5]，与野生型不同的是，在凝胶过滤前，突变的KvAP通道需先与10 mM DTT(二硫苏糖醇)温育1 h。凝胶过滤后，突变的通道(0.5～1.0 mg · ml^{-1})立即与500 μM PEO–碘代乙酰生物素(皮尔斯公司)在室温下温育2～3 h，然后将突变通道重建到脂质膜上进行电生理分析，或者用脱盐柱纯化以除去过量的生物素试剂，与抗生物素蛋白(皮尔斯公司)结合并通过SDS凝胶电泳评估生物素化的程度。

电生理学

如本文的相关论文中[4]所描述，Fabs(6E1和33H1)以及VSTX1被纯化并用于电生理实验。野生型和生物素化KvAP通道的电生理研究按描述的方法进行[5]。为检测抑制情况，Fabs(约500 nM)，VSTX1(30 nM)或抗生物素蛋白(40 μg · ml^{-1})被加到野生型和(或)半胱氨酸突变体的生物素标记的通道上，通道重建在脂质膜中，并利用各种电压刺激方式对它们进行研究。

(李梅 翻译；王晓良 审稿)

Youxing Jiang*, Vanessa Ruta, Jiayun Chen, Alice Lee & Roderick MacKinnon
Howard Hughes Medical Institute, Laboratory of Molecular Neurobiology and Biophysics, Rockefeller University, 1230 York Avenue, New York, New York 10021, USA
* Present address: University of Texas Southwestern Medical Center, Department of Physiology, 5323 Harry Hines Blvd, Dallas, Texas 75390-9040, USA

Received 19 February; accepted 11 March 2003; doi: 10.1038/nature01581.

References:

1. Sigworth, F. J. Voltage gating of ion channels. *Q. Rev. Biophys.* **27**, 1-40 (1994).
2. Armstrong, C. M. & Bezanilla, F. Charge movement associated with the opening and closing of the activation gates of the Na^+ channels. *J. Gen. Physiol.* **63**, 533-552 (1974).
3. Bezanilla, F. The voltage sensor in voltage-dependent ion channels. *Physiol. Rev.* **80**, 555-592 (2000).
4. Jiang, Y. *et al.* X-ray structure of a voltage-dependent K^+ channel. *Nature* **423**, 33-41 (2003).
5. Ruta, V., Jiang, Y., Lee, A., Chen, J. & MacKinnon, R. Functional analysis of an archeabacterial voltage-dependent K^+ channel. *Nature* **422**, 180-185; advance online publication, 2 March 2003 (doi:10.1038/nature01473).
6. Swartz, K. J. & MacKinnon, R. Mapping the receptor site for hanatoxin, a gating modifier of voltage-dependent K^+ channels. *Neuron* **18**, 675-682 (1997).
7. Slatin, S. L., Qiu, X. Q., Jakes, K. S. & Finkelstein, A. Identification of a translocated protein segment in a voltage-dependent channel. *Nature* **371**, 158-161 (1994).
8. Qiu, X. Q., Jakes, K. S., Finkelstein, A. & Slatin, S. L. Site-specific biotinylation of colicin Ia. A probe for protein conformation in the membrane. *J. Biol. Chem.* **269**, 7483-7488 (1994).
9. Qiu, X. Q., Jakes, K. S., Kienker, P. K., Finkelstein, A. & Slatin, S. L. Major transmembrane movement associated with colicin Ia channel gating. *J. Gen. Physiol.* **107**, 313-328 (1996).
10. Pugliese, L., Coda, A., Malcovati, M. & Bolognesi, M. Three-dimensional structure of the tetragonal crystal form of egg-white avidin in its functional complex with biotin at 2.7 Å resolution. *J. Mol. Biol.* **231**, 698-710 (1993).
11. Zagotta, W. N., Hoshi, T., Dittman, J. & Aldrich, R. W. Shaker potassium channel gating. II. Transitions in the activation pathway. *J. Gen. Physiol.* **103**, 279-319 (1994).
12. Zagotta, W. N., Hoshi, T. & Aldrich, R. W. Shaker potassium channel gating. III. Evaluation of kinetic models for activation. *J. Gen. Physiol.* **103**, 321-362 (1994).
13. Schoppa, N. E. & Sigworth, F. J. Activation of Shaker potassium channels. III. An activation gating model for wild-type and V2 mutant channels. *J. Gen. Physiol.* **111**, 313-342 (1998).
14. Cole, K. S. & Moore, J. W. Potassium ion current in the squid giant axon: dynamic characteristic. *Biophys. J.* **1**, 1-14 (1960).
15. Schoppa, N. E., McCormack, K., Tanouye, M. A. & Sigworth, F. J. The size of gating charge in wild-type and mutant Shaker potassium channels. *Science* **255**, 1712-1715 (1992).
16. Seoh, S. A., Sigg, D., Papazian, D. M. & Bezanilla, F. Voltage-sensing residues in the S2 and S4 segments of the Shaker K^+ channel. *Neuron* **16**, 1159-1167 (1996).
17. Aggarwal, S. K. & Mackinnon, R. Contribution of the S4 segment to gating charge in the Shaker K^+ channel. *Neuron* **16**, 1169-1177 (1996).
18. Papazian, D. M. *et al.* Electrostatic interactions of S4 voltage sensor in Shaker K^+ channel. *Neuron* **14**, 1293-1301 (1995).
19. Tiwari-Woodruff, S. K., Lin, M. A., Schulteis, C. T. & Papazian, D. M. Voltage-dependent structural interactions in the Shaker K^+ channel. *J. Gen. Physiol.* **115**, 123-138 (2000).
20. Papazian, D. M., Silverman, W. R., Lin, M. C., Tiwari-Woodruff, S. K. & Tang, C. Y. Structural organization of the voltage sensor in voltage-dependent potassium channels. *Novartis Found. Symp.* **245**, 178-190 (2002).
21. Gonzalez, C., Rosenman, E., Bezanilla, F., Alvarez, O. & Latorre, R. Periodic perturbations in Shaker K^+ channel gating kinetics by deletions in the S3-S4 linker. *Proc. Natl Acad. Sci. USA* **98**, 9617-9623 (2001).
22. Wimley, W. C., Creamer, T. P. & White, S. H. Solvation energies of amino acid side chains and backbone in a family of host-guest pentapeptides. *Biochemistry* **35**, 5109-5124 (1996).
23. Bass, R. B., Strop, P., Barclay, M. & Rees, D. C. Crystal structure of *Escherichia coli* MscS, a voltage-modulated and mechanosensitive channel. *Science* **298**, 1582-1587 (2002).
24. Martinac, B., Buechner, M., Delcour, A. H., Adler, J. & Kung, C. Pressure-sensitive ion channel in *Escherichia coli*. *Proc. Natl Acad. Sci. USA* **84**, 2297-2301 (1987).
25. Doyle, D. A. *et al.* The structure of the potassium channel: molecular basis of K^+ conduction and selectivity. *Science* **280**, 69-77 (1998).
26. Jiang, Y. *et al.* The open pore conformation of potassium channels. *Nature* **417**, 523-526 (2002).
27. Yifrach, O. & MacKinnon, R. Energetics of pore opening in a voltage-gated K^+ channel. *Cell* **111**, 231-239 (2002).
28. Laine, M. *et al.* Structural interactions between voltage sensor and pore in the Shaker K^+ channels. *Biophys. J.* **82**, 231a (2002).
29. Aziz, Q. H., Partridge, C. J., Munsey, T. S. & Sivaprasadarao, A. Depolarization induces intersubunit cross-linking in a S4 cysteine mutant of the Shaker potassium channel. *J. Biol. Chem.* **277**, 42719-42725 (2002).
30. Larsson, H. P., Baker, O. S., Dhillon, D. S. & Isacoff, E. Y. Transmembrane movement of the Shaker K^+ channel S4. *Neuron* **16**, 387-397 (1996).

31. Yusaf, S. P., Wray, D. & Sivaprasadarao, A. Measurement of the movement of the S4 segment during the activation of a voltage-gated potassium channel. *Pflugers Arch.* **433**, 91-97 (1996).

32. Baker, O. S., Larsson, H. P., Mannuzzu, L. M. & Isacoff, E. Y. Three transmembrane conformations and sequence-dependent displacement of the S4 domain in Shaker K^+ channel gating. *Neuron* **20**, 1283-1294 (1998).

33. Starace, D. M., Stefani, E. & Bezanilla, F. Voltage-dependent proton transport by the voltage sensor of the Shaker K^+ channel. *Neuron* **19**, 1319-1327 (1997).

Acknowledgements. We thank D. Gadsby and O. Andersen for helpful discussions and advice on the manuscript. This work was supported in part by a grant from the National Institutes of Health (NIH) to R.M. V.R. is supported by a National Science Foundation Graduate Student Research Fellowship, and R.M. is an Investigator in the Howard Hughes Medical Institute.

Competing interests statement. The authors declare that they have no competing financial interests.

Correspondence and requests for materials should be addressed to R.M. (mackinn@rockvax.rockefeller.edu).

The International HapMap Project

The International HapMap Consortium*

Editor's Note

This paper describes the International HapMap Project, a large international collaborative effort to identify and catalog the common patterns of human DNA sequence variation. By making the information freely available, the project aimed to help researchers identify genes related to health, disease and responses to drugs and environmental factors. Here, they describe the strategy and key components of the project, including the genotyping of over a million sequence variants, their frequencies and degrees of association, in DNA samples from populations rooted in parts of Africa, Asia and Europe. The Project, officially launched in 2002, successfully published the completed Haplotype Map in 2005, then a more detailed second-generation map in 2007.

The goal of the International HapMap Project is to determine the common patterns of DNA sequence variation in the human genome and to make this information freely available in the public domain. An international consortium is developing a map of these patterns across the genome by determining the genotypes of one million or more sequence variants, their frequencies and the degree of association between them, in DNA samples from populations with ancestry from parts of Africa, Asia and Europe. The HapMap will allow the discovery of sequence variants that affect common disease, will facilitate development of diagnostic tools, and will enhance our ability to choose targets for therapeutic intervention.

COMMON diseases such as cardiovascular disease, cancer, obesity, diabetes, psychiatric illnesses and inflammatory diseases are caused by combinations of multiple genetic and environmental factors[1]. Discovering these genetic factors will provide fundamental new insights into the pathogenesis, diagnosis and treatment of human disease. Searches for causative variants in chromosome regions identified by linkage analysis have been highly successful for many rare single-gene disorders. By contrast, linkage studies have been much less successful in locating genetic variants that affect common complex diseases, as each variant individually contributes only modestly to disease risk[2,3]. A complementary approach to identifying these specific genetic risk factors is to search for an association between a specific variant and a disease, by comparing a group of affected individuals with a group of unaffected controls[4]. In the absence of strong natural selection, there is likely to be a broad spectrum of frequency of such variants, many of which are likely to be common in the population. A number of association studies, focused on candidate genes, regions of linkage

---

* A full list of participants and affiliations is given in the original paper.

国际人类基因组单体型图计划

国际人类基因组单体型图协作组 *

编者按

本文描述了国际人类基因组单体型图计划，这是一个大型的国际合作项目，旨在识别和编目人类 DNA 序列变异的常见模式。该计划通过免费提供这些信息，帮助研究人员识别与健康、疾病以及对药物和环境因素产生响应的相关基因。本文中，他们描述了该计划的策略和关键组成部分，包括从部分非洲、亚洲和欧洲地区提供的 DNA 样本中对 100 多万个序列变异进行基因分型以及它们的频率和关联程度。该计划于 2002 年正式启动，2005 年成功发布了完整的单体型图谱，2007 年又发布了更详细的第二代图谱。

国际人类基因组单体型图计划的目标是确定人类基因组中 DNA 序列变异的共同模式，并将这些资料免费向公众开放。一个国际协作组将通过来自部分非洲、亚洲和欧洲族群的 DNA 样本，确定 100 万或更多序列变异的基因型以及它们的频率和它们之间的相关程度，从而绘制整个基因组的模式图谱。HapMap 将支持我们发现影响常见疾病的序列变异，促进疾病诊断工具的发展，并且提高我们选择治疗干预靶标的能力。

常见的疾病如心血管疾病、癌症、肥胖、糖尿病，精神疾病和炎症性疾病，是由多种基因和环境因素的组合共同造成的 [1]。发现这些遗传因素将为研究人类疾病的发病机理、诊断和治疗提供重要的新见解。对许多罕见的单基因疾病来说，通过连锁分析的方法鉴定染色体区域中的致病变异已经非常成功了。相比之下，通过连锁分析寻找影响常见复杂疾病的遗传变异却不怎么成功，因为每个单独的变异只会增加轻微的疾病风险 [2,3]。通过比较一组病例和正常对照来寻找具体某个遗传变异和某一种疾病间的关联，是确定这些具体遗传风险因素的一个补充方法 [4]。缺乏强大的自然选择时，这种变异的频率范围可能是广泛的，其中许多变异可能是群体中常见的。经过一些侧重于候选基因、与疾病连锁的区域或更大范围的关联研究，人

---

* 协作组的参与者及所属单位的名单请参见原文。

to a disease or more large-scale surveys, have already led to the discovery of genetic risk factors for common diseases. Examples include type 1 diabetes (human leukocyte antigen (HLA[5]), insulin[6] and *CTLA4* (ref. 7)), Alzheimer's disease (*APOE*)[8], deep vein thrombosis (factor V)[9], inflammatory bowel disease (*NOD2* (refs 10, 11) and also 5q31 (ref. 12)), hypertriglyceridaemia (*APOAV*)[13], type 2 diabetes (*PPARG*)[14,15], schizophrenia (neuregulin 1)[16], asthma (*ADAM33*)[17], stroke (*PDE4D*)[18] and myocardial infarction (*LTA*)[19].

One approach to doing association studies involves testing each putative causal variant for correlation with the disease (the "direct" approach)[2]. To search the entire genome for disease associations would entail the substantial expense of whole-genome sequencing of numerous patient samples to identify the candidate variants[3]. At present, this approach is limited to sequencing the functional parts of candidate genes (selected on the basis of a previous functional or genetic hypothesis) for potential disease-associated candidate variants. An alternative approach (the "indirect" approach) has been proposed[20], whereby a set of sequence variants in the genome could serve as genetic markers to detect association between a particular genomic region and the disease, whether or not the markers themselves had functional effects. The search for the causative variants could then be limited to the regions showing association with the disease.

Two insights from human population genetics suggest that the indirect approach is able to capture most human sequence variation, with greater efficiency than the direct approach. First, ~90% of sequence variation among individuals is due to common variants[21]. Second, most of these originally arose from single historical mutation events, and are therefore associated with nearby variants that were present on the ancestral chromosome on which the mutation occurred. These associations make the indirect approach feasible to study variants in candidate genes, chromosome regions or across the whole genome. Prior knowledge of putative functional variants is not required. Instead, the approach uses information from a relatively small set of variants that capture most of the common patterns of variation in the genome, so that any region or gene can be tested for association with a particular disease, with a high likelihood that such an association will be detectable if it exists.

The aim of the International HapMap Project is to determine the common patterns of DNA sequence variation in the human genome, by characterizing sequence variants, their frequencies, and correlations between them, in DNA samples from populations with ancestry from parts of Africa, Asia and Europe. The project will thus provide tools that will allow the indirect association approach to be applied readily to any functional candidate gene in the genome, to any region suggested by family-based linkage analysis, or ultimately to the whole genome for scans for disease risk factors.

Common variants responsible for disease risk will be most readily approached by this strategy, but not all predisposing variants are common. However, it should be noted that even a relatively uncommon disease-associated variant can potentially be discovered using this approach. Reflecting its historical origins, the uncommon variant will be travelling on a chromosome that carries a characteristic pattern of nearby sequence variants. In a group

们已经发现了一些常见疾病的遗传风险因素。例如 1 型糖尿病（人类白细胞抗原(HLA[5])，胰岛素[6]和 *CTLA4*（参考文献 7）)，阿尔茨海默病（*APOE*）[8]，深静脉血栓症（V 因子）[9]，炎症性肠病（*NOD2*（参考文献 10 和 11）和 5q31（参考文献 12）)，高甘油三酯血症（*APOAV*）[13]，2 型糖尿病（*PPARG*）[14,15]，精神分裂症（神经调节蛋白 1）[16]，哮喘（*ADAM33*）[17]，中风（*PDE4D*）[18]和心肌梗死（*LTA*）[19]。

进行关联研究的一个方法是测试每个潜在致病变异与疾病的相关性（“直接”方法）[2]。寻找整个基因组中的疾病相关变异，则需要对众多患者样本开展全基因组测序来确定候选变异，这需要投入大量费用[3]。目前，这种做法仅限于对候选基因的功能部分（基于前期的功能或遗传假说进行选择）进行测序以寻找潜在的疾病相关候选变异。人们已提出了另一种方法（“间接”方法）[20]，利用基因组内一些序列变异作为遗传标签，来检测特定基因组区域和疾病之间的关联，无论这些标签本身是否有功能效应。这样一来，就可以限制在与疾病相关的区域寻找致病变异了。

来自人类群体遗传学的两个观点表明，间接方法能捕获绝大多数的人类序列变异，比直接方法的效率更高。首先，个体间约 90%序列变异是由常见变异组成的[21]。其次，这些变异中大多数最初产生于单一的历史突变事件，因此，与出现在突变发生的祖先染色体上的邻近变异有关联。这些关联使间接方法易于研究候选基因、染色体区域或整个基因组内的变异。这不需要与假定功能变异相关的先验知识。相反，这个方法用相对较小的一组变异信息来涵盖基因组中绝大多数变异的常见模式，因此可以检测任何基因组区域或基因与特定疾病的关联性，如果存在此种关联，被检测到的可能性很高。

国际人类基因组单体型图计划的目的是通过描述部分非洲、亚洲和欧洲族群的 DNA 样本序列变异、序列变异的频率和它们之间的相关性等特征，确定人类基因组中 DNA 序列变异的常见模式。因此该计划将提供工具，使间接关联方法适用于基因组中任何功能候选基因、家族型连锁分析提示的任何区域或最终应用于寻找疾病危险因素的全基因组扫描。

与疾病风险关联的常见变异通过这一策略很容易找到，但并非所有致病变异都是常见的。但是，应该指出的是，即使是相对罕见的疾病，相关变异也可能通过这种方法被发现。罕见的变异将在携带邻近序列变异特有模式的染色体上移动，反映

of people affected by a disease, the rare variant will be enriched in frequency compared with its frequency in a group of unaffected controls. This observation, for example, was of considerable assistance in the identification of the genes responsible for cystic fibrosis[22] and diastrophic dysplasia[23], after linkage had pointed to the general chromosomal region.

Below we provide a brief description of human sequence variation, and then describe the strategy and key components of the project. These include the choice of samples and populations for study, the process of community engagement or public consultation, selection of single-nucleotide polymorphisms (SNPs), genotyping, data release and analysis.

Human DNA Sequence Variation

Any two copies of the human genome differ from one another by approximately 0.1% of nucleotide sites (that is, one variant per 1,000 bases on average)[24-27]. The most common type of variant, a SNP, is a difference between chromosomes in the base present at a particular site in the DNA sequence (Fig. 1a). For example, some chromosomes in a population may have a C at that site (the "C allele"), whereas others have a T (the "T allele"). It has been estimated that, in the world's human population, about 10 million sites (that is, one variant per 300 bases on average) vary such that both alleles are observed at a frequency of $\geqslant 1\%$, and that these 10 million common SNPs constitute 90% of the variation in the population[21,28]. The remaining 10% is due to a vast array of variants that are each rare in the population. The presence of particular SNP alleles in an individual is determined by testing ("genotyping") a genomic DNA sample.

Nearly every variable site results from a single historical mutational event as the mutation rate is very low (of the order of 10^{-8} per site per generation) relative to the number of generations since the most recent common ancestor of any two humans (of the order of 10^4 generations). For this reason, each new allele is initially associated with the other alleles that happened to be present on the particular chromosomal background on which it arose. The specific set of alleles observed on a single chromosome, or part of a chromosome, is called a haplotype (Fig. 1b). New haplotypes are formed by additional mutations, or by recombination when the maternal and paternal chromosomes exchange corresponding segments of DNA, resulting in a chromosome that is a mosaic of the two parental haplotypes[29].

The coinheritance of SNP alleles on these haplotypes leads to associations between these alleles in the population (known as linkage disequilibrium, LD). Because the likelihood of recombination between two SNPs increases with the distance between them, on average such associations between SNPs decline with distance. Many empirical studies have shown highly significant levels of LD, and often strong associations between nearby SNPs, in the human genome[30-34]. These strong associations mean that in many chromosome regions there are only a few haplotypes, and these account for most of the variation among people in those regions[31,35,36].

其历史起源。相比于不受影响的对照组，罕见变异将会显著富集于受某种疾病影响的人群中。在连锁分析已经显示大致染色体易感区域之后，这个发现对鉴定如引起囊性纤维症[22]和畸形发育不良[23]的基因具有相当大的帮助。

下面我们将简要说明人类序列变异，然后描述这个项目的策略和关键组成部分。其中包括选择研究的样本和群体，社会参与或公共咨询的过程，单核苷酸多态性(SNP)的选择，基因分型，数据的公布和分析。

人类 DNA 序列变异

任何两份人类基因组具有大约 0.1%的核苷酸位点差异(即，平均每 1,000 个碱基有一个变异)[24-27]。最常见的变异类型——单核苷酸多态性(SNP)，是染色体间 DNA 序列特定位点的碱基差异(图 1a)。例如，在某一群体中，有些染色体在某个位点可能有一个 C(“C 等位基因”)，而其他染色体有一个 T(“T 等位基因”)。据估计，在世界人群中，约 1,000 万位点(即，平均每 300 个碱基有一个变异)在两个等位基因的变异频率 ≥ 1%，并且这 1,000 万常见的 SNP 贡献了种群变异的 90%[21,28]。其余 10%是群体中各种各样的罕见变异。个体中存在特定 SNP 等位基因是通过检测(“基因分型”)一个基因组 DNA 样本确定的。

相对于任何两个人最近的共同祖先之后的世代数(大约 10^4 代)，几乎每个来自单一历史突变事件的可变异位点的突变率都是非常低的(每代每个位点大约 10^{-8})。出于这个原因，每一个新的等位基因最初都是与它所在的特定染色体背景上的恰好存在的其他等位基因相联系。在单一染色体上或染色体的一部分上观察到的一套具体的等位基因，称为一个单体型(图 1b)。或者额外突变或者母本和父本染色体交换相应 DNA 片段的重组，形成了新的单体型，产生了双方父母单体型嵌合的染色体[29]。

这些单体型上的 SNP 等位基因的共同继承导致群体中这些等位基因间的相互关联(称为连锁不平衡，LD)。因为两个 SNP 间重组的可能性随它们之间距离的增加而增加，因此平均来说，SNP 之间的这种关联会随距离的增加而减少。许多实证研究表明在人类基因组中存在高显著性水平的 LD，而且相邻的 SNP 通常具有显著关联[30-34]。这些紧密关联意味着在许多染色体区域中只有少数单体型，而它们代表了

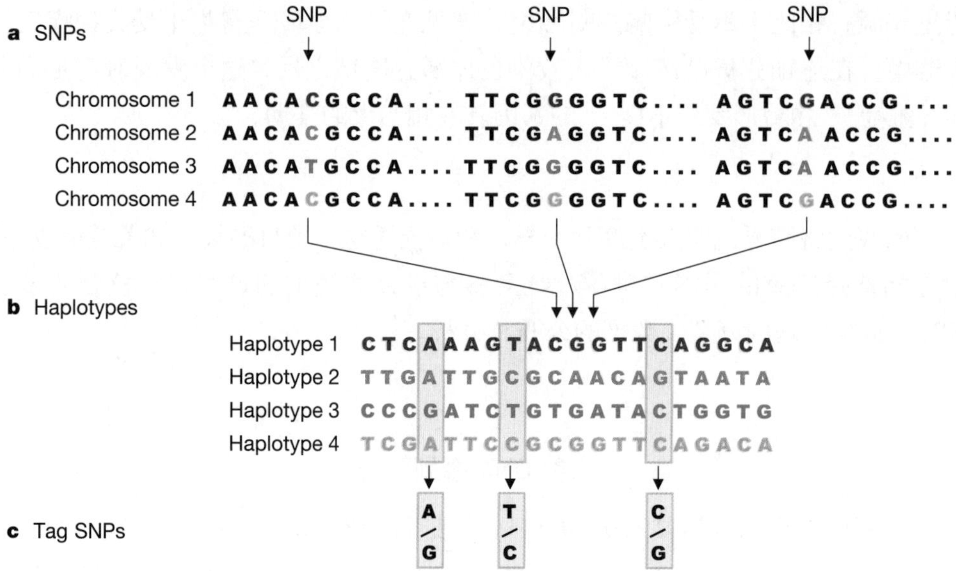

Fig. 1. SNPs, haplotypes and tag SNPs. **a**, SNPs. Shown is a short stretch of DNA from four versions of the same chromosome region in different people. Most of the DNA sequence is identical in these chromosomes, but three bases are shown where variation occurs. Each SNP has two possible alleles; the first SNP in panel **a** has the alleles C and T. **b**, Haplotypes. A haplotype is made up of a particular combination of alleles at nearby SNPs. Shown here are the observed genotypes for 20 SNPs that extend across 6,000 bases of DNA. Only the variable bases are shown, including the three SNPs that are shown in panel **a**. For this region, most of the chromosomes in a population survey turn out to have haplotypes 1–4. **c**, Tag SNPs. Genotyping just the three tag SNPs out of the 20 SNPs is sufficient to identify these four haplotypes uniquely. For instance, if a particular chromosome has the pattern A–T–C at these three tag SNPs, this pattern matches the pattern determined for haplotype 1. Note that many chromosomes carry the common haplotypes in the population.

The strong associations between SNPs in a region have a practical value: genotyping only a few, carefully chosen SNPs in the region will provide enough information to predict much of the information about the remainder of the common SNPs in that region. As a result, only a few of these "tag" SNPs are required to identify each of the common haplotypes in a region[35,37-39] (Fig. 1c).

As the extent of association between nearby markers varies dramatically across the genome[30-32,34,35,40], it is not efficient to use SNPs selected at random or evenly spaced in the genome sequence. Instead, the patterns of association must be empirically determined for efficient selection of tag SNPs. On the basis of empirical studies, it has been estimated that most of the information about genetic variation represented by the 10 million common SNPs in the population could be provided by genotyping 200,000 to 1,000,000 tag SNPs across the genome[31,36,38,39]. Thus, a substantial reduction in the amount of genotyping can be obtained with little loss of information, by using knowledge of the LD present in the genome.

For common SNPs, which tend to be older than rare SNPs, the patterns of LD largely reflect historical recombination and demographic events[41]. Some recombination events occur

这些区域人群间的大多数变异[31,35,36]。

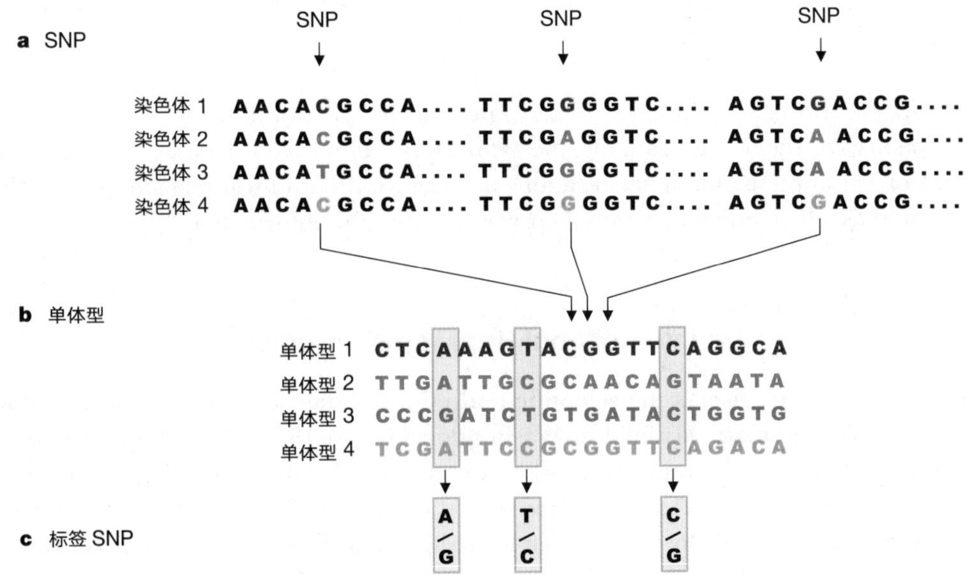

图 1. SNP，单体型和标签 SNP。**a**，SNP。所示的是不同人的同一染色体区域四个版本的一小段 DNA。在这些染色体中大多数 DNA 序列是相同的，但显示了三个发生变异的碱基。每个 SNP 有两个可能的等位基因；第一个 SNP 在 **a** 图中有等位基因 C 和 T。**b**，单体型。单体型是由邻近 SNP 的等位基因的特定组合构成的。此处显示的是观察到的跨越 6,000 个 DNA 碱基的 20 个 SNP 的基因型。只显示变异的碱基，包括 **a** 图中显示的 3 个 SNP。对此区域，调查群体中大多数染色体有单体型 1 ~ 4。**c**，标签 SNP。只对 20 个 SNP 中的三个标签 SNP 进行基因分型就足以确定这四个独特的单体型。例如，如果某一特定染色体在这三个标签 SNP 上有 A–T–C 模式，这种模式匹配由单体型 1 确定的模式。请注意，群体中许多染色体携带共同的单体型。

一个区域内 SNP 之间的紧密关联有着实际的意义：只对该区域内少数、精心挑选的 SNP 进行基因分型将提供足够的信息来预测那个区域剩余的常见 SNP 的大量信息。因此，只需要少量这种“标签”SNP，就可以确定一个区域内的每个常见单体型[35,37-39]（图 1c）。

因为基因组内邻近标记间的关联程度变化显著[30-32,34,35,40]，用随机选择或基因组序列间均匀分隔的 SNP 都不是很有效。相反，相互关联的模式必须凭实验确定，以选择更为有效的标签 SNP。在实验研究的基础上，我们估计群体中 1,000 万常见 SNP 所代表的大部分遗传变异信息，可以通过对基因组内的 200,000 至 1,000,000 标签 SNP 进行基因分型来提供[31,36,38,39]。因此，运用基因组中 LD 的知识，基因分型的数量大幅度减少而得到的信息几乎没有损失。

repeatedly at "hotspots"[30,42]. The result of these processes is that current chromosomes are mosaics of ancestral chromosome regions[29]. This explains the observations that haplotypes and patterns of LD are shared by apparently unrelated chromosomes within a population and generally among populations[43].

These observations are the conceptual and empirical foundation for developing a haplotype map of the human genome, the "HapMap". This map will describe the common patterns of variation, including associations between SNPs, and will include the tag SNPs selected to most efficiently and comprehensively capture this information.

The International HapMap Consortium

An initial meeting to discuss the scientific and ethical issues associated with developing a human haplotype map was held in Washington DC on 18–19 July 2001 (http://www.genome.gov/10001665). Groups were organized to consider the ethical issues, to develop the scientific plan and to choose the populations to include. The International HapMap Project (http://www.hapmap.org/) was then formally initiated with a meeting in Washington DC on 27–29 October 2002 (http://www.genome.gov/10005336). The participating groups and funding sources are listed in Table 1.

Table 1. Groups participating in the International HapMap Project

Country	Research Group	Institution	Role	Per cent genome	Chromosomes	Funding agency
Japan	Yusuke Nakamura	RIKEN, Univ. of Tokyo	Genotyping: Third Wave	25.1%	5, 11, 14, 15, 16, 17, 19	Japanese MEXT
	Ichiro Matsuda	Health Science Univ. of Hokkaido, Eubios Ethics Inst., Shinshu Univ.	Public consultation, Samples			
United Kingdom	David Bentley	Sanger Inst.	Genotyping: Illumina	24.0%	1, 6, 10, 13, 20	Wellcome Trust
	Peter Donnelly	Univ. of Oxford	Analysis			TSC, US NIH
	Lon Cardon	Univ. of Oxford	Analysis			Wellcome Trust, TSC, US NIH
Canada	Thomas Hudson	McGill Univ. and Génome Québec Innovation Centre	Genotyping: Illumina	10.0%	2, 4p	Genome Canada, Génome Québec

常见的 SNP 往往比罕见的 SNP 更古老，其 LD 的模式基本上可以反映历史重组和人口事件[41]。一些重组事件在“热点区”反复发生[30,42]。这些过程的结果是：现在的染色体是祖先染色体区域的嵌合体[29]。这解释了单体型和 LD 的模式被某一群体内显然不相关的染色体或群体之间共享这一观察结果[43]。

这些观察是开发人类基因组单体型图“HapMap”的概念和经验基础。这个图谱将介绍常见的变异模式，包括 SNP 间的关联，未来还将包括用于最有效和全面捕获这些遗传信息的标签 SNP。

国际 HapMap 协作组

最初的讨论关于开发人类单体型图的科学和伦理问题的会议是 2001 年 7 月 18 日至 19 日在华盛顿举行的（http://www.genome.gov/10001665）。协作组被组织起来考虑道德问题，制定科学计划，并选择包括的人群。之后国际 HapMap 计划（http://www.hapmap.org/）于 2002 年 10 月 27 日至 29 日在华盛顿会议上正式启动（http://www.genome.gov/10005336）。参加的团体和资金来源列于表 1。

表 1. 参与国际人类基因组单体型图计划的团队

<table>
<tr><th>国家</th><th>研究团队</th><th>机构</th><th>任务</th><th>基因组百分数</th><th>染色体</th><th>资助机构</th></tr>
<tr><td rowspan="2">日本</td><td>中村佑辅</td><td>日本理化学研究所，东京大学</td><td>基因型分型：Third Wave 公司</td><td>25.1%</td><td>5, 11, 14, 15, 16, 17, 19</td><td rowspan="2">日本文部科学省</td></tr>
<tr><td>松田一郎</td><td>北海道医疗大学，生物伦理研究所，信州大学</td><td>公众咨询，样本</td><td></td><td></td></tr>
<tr><td rowspan="3">英国</td><td>戴维·本特利</td><td>桑格尔研究所</td><td>基因型分型：Illumina 公司</td><td>24.0%</td><td>1, 6, 10, 13, 20</td><td>维康基金会</td></tr>
<tr><td>彼得·唐纳利</td><td>牛津大学</td><td>分析</td><td></td><td></td><td>SNP 协作组，美国国立卫生研究院</td></tr>
<tr><td>朗·卡登</td><td>牛津大学</td><td>分析</td><td></td><td></td><td>维康基金会，SNP 协作组，美国国立卫生研究院</td></tr>
<tr><td>加拿大</td><td>托马斯·哈德森</td><td>麦吉尔大学和魁北克基因组创新中心</td><td>基因型分型：Illumina 公司</td><td>10.0%</td><td>2, 4p</td><td>加拿大基因组，魁北克基因组</td></tr>
</table>

Continued

<table>
<tr><th>Country</th><th colspan="2">Research Group</th><th>Institution</th><th>Role</th><th colspan="2">Per cent genome</th><th>Chromosomes</th><th>Funding agency</th></tr>
<tr><td rowspan="5">China</td><td rowspan="3">Huanming Yang
The Chinese HapMap Consortium</td><td>Changqing Zeng</td><td>Beijing Genomics Inst., Chinese National Human Genome Center at Beijing</td><td>Genotyping: PerkinElmer and Sequenom</td><td>4.8%</td><td rowspan="3">10%</td><td rowspan="3">3, 8p, 21</td><td rowspan="3">Chinese MOST, Chinese Academy of Sciences, Natural Science Foundation of China, Hong Kong Innovation and Technology Commission, University Grants Committee of Hong Kong</td></tr>
<tr><td>Wei Huang</td><td>Chinese National Human Genome Centre at Shanghai, Inst. of Biomedical Sciences (Taiwan)</td><td>Genotyping: Illumina</td><td>3.2%</td></tr>
<tr><td>Lap-Chee Tsui</td><td>Univ. of Hong Kong, Hong Kong Univ. of Sci. & Tech., Chinese Univ. of Hong Kong</td><td>Genotyping: Sequenom</td><td>2.0%</td></tr>
<tr><td colspan="2">Houcan Zhang</td><td>Beijing Normal Univ.</td><td>Community engagement</td><td colspan="2" rowspan="2"></td><td rowspan="2"></td><td rowspan="2">Chinese MOST</td></tr>
<tr><td colspan="2">Changqing Zeng</td><td>Beijing Genomics Inst.</td><td>Samples</td></tr>
<tr><td rowspan="7">United State</td><td colspan="2">Mark Chee</td><td>Illumina</td><td>Genotyping: Illumina</td><td>15.5%</td><td rowspan="5">30.9%</td><td>8q, 9, 18q, 22, X</td><td rowspan="6">US NIH</td></tr>
<tr><td colspan="2" rowspan="2">David Altshuler</td><td rowspan="2">Whitehead Inst.</td><td>Genotyping: Sequenom and Illumina</td><td>9.1%</td><td rowspan="2">4q, 7q, 18p, Y</td></tr>
<tr><td>Analysis</td><td></td></tr>
<tr><td colspan="2">Richard Gibbs</td><td>Baylor College of Medicine, ParAllele</td><td>Genotyping: ParAllele</td><td>4.4%</td><td>12</td></tr>
<tr><td colspan="2">Pui-Yan Kwok</td><td>UCSF, Washington Univ.</td><td>Genotyping: PerkinElmer</td><td>1.9%</td><td>7p</td></tr>
<tr><td colspan="2">Aravinda Chakravarti</td><td>Johns Hopkins Univ.</td><td>Analysis</td><td colspan="2"></td><td></td></tr>
<tr><td colspan="2">Mark Leppert</td><td>Univ. of Utah</td><td>Community engagement Samples</td><td colspan="2"></td><td></td><td>W. M. Keck Foundation, Delores Dore Eccles Foundation, US NIH</td></tr>
<tr><td>Nigeria</td><td colspan="2">Charles Rotimi</td><td>Howard Univ., Univ. of Ibadan</td><td>Community engagement Samples</td><td colspan="2"></td><td></td><td>US NIH</td></tr>
<tr><td></td><td colspan="2">Lincoln Stein</td><td>Cold Spring Harbor Lab., New York</td><td>Data coordination centre</td><td colspan="2"></td><td></td><td>TSC</td></tr>
</table>

DNA Samples and Populations

Human populations are the products of numerous social, historical and demographic processes. As a result, no populations are typical, special or sharply bounded[44,45]. As

续表

<table>
<tr><th>国家</th><th colspan="2">研究团队</th><th>机构</th><th>任务</th><th colspan="2">基因组百分数</th><th>染色体</th><th>资助机构</th></tr>
<tr><td rowspan="5">中国</td><td rowspan="3">杨焕明
中国人类基因组单体型图协作组</td><td>曾长青</td><td>北京基因组研究所，国家人类基因组北方研究中心</td><td>基因型分型：PerkinElmer 公司和 Sequenom 公司</td><td>4.8%</td><td rowspan="3">10%</td><td rowspan="3">3, 8p, 21</td><td rowspan="3">中国科技部，中国科学院，国家自然科学基金，香港创新科技署，香港大学教育资助委员会</td></tr>
<tr><td>黄薇</td><td>国家人类基因组南方研究中心，生物医学科学研究所（台湾）</td><td>基因型分型：Illumina 公司</td><td>3.2%</td></tr>
<tr><td>徐立之</td><td>香港大学，香港科技大学，香港中文大学</td><td>基因型分型：Sequenom 公司</td><td>2.0%</td></tr>
<tr><td colspan="2">张厚粲</td><td>北京师范大学</td><td>社会参与</td><td colspan="2" rowspan="2"></td><td rowspan="2"></td><td rowspan="2">中国科技部</td></tr>
<tr><td colspan="2">曾长青</td><td>北京基因组研究所</td><td>样本</td></tr>
<tr><td rowspan="7">美国</td><td colspan="2">陈马克（音译）</td><td>Illumina 公司</td><td>基因型分型：Illumina 公司</td><td>15.5%</td><td rowspan="5">30.9%</td><td>8q, 9, 18q, 22, X</td><td rowspan="6">美国国立卫生研究院</td></tr>
<tr><td colspan="2" rowspan="2">戴维·阿特舒勒</td><td rowspan="2">怀特黑德研究所</td><td>基因型分型：Sequenom 公司和 Illumina 公司</td><td>9.1%</td><td rowspan="2">4q, 7q, 18p, Y</td></tr>
<tr><td>分析</td><td></td></tr>
<tr><td colspan="2">理查德·吉布斯</td><td>贝勒医学院，ParAllele 公司</td><td>基因型分型：ParAllele 公司</td><td>4.4%</td><td>12</td></tr>
<tr><td colspan="2">郭沛恩</td><td>加州大学旧金山分校，华盛顿大学</td><td>基因型分型：PerkinElmer 公司</td><td>1.9%</td><td>7p</td></tr>
<tr><td colspan="2">阿拉温达·查克拉瓦蒂</td><td>约翰斯·霍普金斯大学</td><td>分析</td><td colspan="2"></td><td></td></tr>
<tr><td colspan="2">马克·莱珀特</td><td>犹他大学</td><td>社会参与，样本</td><td colspan="2"></td><td></td><td>凯克基金会，德洛丽丝·多尔·埃克尔斯基金会，美国国立卫生研究院</td></tr>
<tr><td>尼日利亚</td><td colspan="2">查尔斯·罗蒂米</td><td>霍华德大学，伊巴丹大学</td><td>社会参与，样本</td><td colspan="2"></td><td></td><td>美国国立卫生研究院</td></tr>
<tr><td></td><td colspan="2">林肯·斯坦</td><td>组约冷泉港实验室</td><td>数据协调中心</td><td colspan="2"></td><td></td><td>SNP 协作组</td></tr>
</table>

DNA 样本和群体

人类群体是许多社会、历史和人口学过程的产物。因此，没有哪个人群是典型的、特殊的或有明显界限的[44,45]，因为最常见的变异模式可以在任一人群中找到[46]，

most common patterns of variation can be found in any population[46], no one population is essential for inclusion in the HapMap. Nonetheless, we decided to include several populations from different ancestral geographic locations to ensure that the HapMap would include most of the common variation and some of the less common variation in different populations, and to allow examination of various hypotheses about patterns of LD.

Studies of allele frequency distributions suggest that ancestral geography is a reasonable basis for sampling human populations[44,47,48]. Pilot studies using samples from the Yoruba, Japanese, Chinese and individuals with ancestry from Northern and Western Europe have shown substantial similarity in their haplotype patterns, although the frequencies of haplotypes often differ[31,44]. Given these scientific findings, coupled with consideration of ethical, social and cultural issues, these populations were approached for inclusion in the HapMap through a process of community engagement or consultation (see Box 1).

Box 1. Community engagement, public consultation and individual consent

As no personally identifiable information will be linked to the samples, the risk that an individual will be harmed by a breach of privacy, or by discrimination based on studies that use the HapMap, is minimal. However, because tag SNPs for future disease studies will be chosen on the basis of haplotype frequencies in the populations included in the HapMap, the data will be identified as coming from one of the four populations involved, and it will be possible to make comparisons between the populations. As a result, the use of population identifiers may create risks of discrimination or stigmatization, as might occur if a higher frequency of a disease-associated variant were to be found in a group and this information were then overgeneralized to all or most of its members[64]. It is possible that there are other culturally specific risks that may not be evident to outsiders[65]. To identify and address these group risks, a process of community engagement, or public consultation, was undertaken to confer with members of the populations being approached for sample donation about the implications of their participation in the project[66,67]. The goal was to give people in the localities where donors were recruited the opportunity to have input into the informed consent and sample collection processes, and into such issues as how the populations from which the samples were collected would be named. Community engagement is not a perfect process, but it is an effort to involve potential donors in a more extended consideration of the implications of a research project before being asked to take part in it[68]. Community engagement and individual informed consent were conducted under the auspices of local governments and ethics committees, taking into account local ethical standards and international ethical guidelines. As in any cross-cultural endeavour, the form and outcome of the processes varied from one population to another. A Community Advisory Group is being set up for each community to serve as a continuing liaison with the sample repository, to ensure that future uses of the samples are consistent with the uses described in the informed consent documents. A more detailed article discussing ethical, social and cultural issues relevant to the project, and describing the processes used to engage donor populations in identifying and evaluating these issues, is in preparation.

没有哪个人群是必须列入 HapMap 的。尽管如此，我们依旧决定将几个位于不同祖先地理位置的群体列入其中，以确保 HapMap 可以包括大多数常见变异和不同群体中的一些不太常见的变异，并允许检查各种关于 LD 模式的假说。

对等位基因频率分布的研究表明，祖先地理是选取人群样本的一个合理的依据[44,47,48]。对约鲁巴、日本和中国样本以及对北欧和西欧后裔个体的先导研究表明，虽然单体型的频率常常不同，但其单体型模式有很大相似性[31,44]。鉴于这些科学成果，加上对道德、社会和文化问题的考虑，通过社会参与或咨询后，这些群体被列入了 HapMap 中（见框 1）。

框 1. 社会参与、公众咨询及个人同意

由于个人身份信息不会与样本相联系，侵犯个人隐私或基于利用 HapMap 进行研究而受到的歧视的风险非常小。然而，因为未来用于疾病研究的标记 SNP 将在 HapMap 群体单体型频率的基础上选择，这些数据将被认为是来自四个群体中的某一个，而且将有可能在群体间进行比较。因此，使用群体标识可能产生歧视或侮辱的风险，如果在一个组中发现某种疾病有关的变异频率较高，然后泛化过渡到所有或大部分成员中，就会产生这种风险[64]。还有可能存在其他对外不明显的文化特异的风险[65]。为查明并解决这些风险，将通过社会参与或公共咨询来与样本捐助群体的成员协商参与此计划的意义[66,67]。该计划的目标是让居住在招募捐助者地方的人们有机会加入知情同意和样本收集过程，并参与这些问题的讨论，如收集样本的群体将如何被命名。社会参与不是一个完美的过程，但它努力让潜在的捐助者在被要求参加这个计划前更广泛会考虑该计划研究的意义[68]。社会参与和个人知情同意是在地方政府和道德委员会的帮助下实施的，并考虑了当地的道德标准和国际道德准则。正如在任何跨越文化中的努力一样，这个过程的形式和结果在不同群体间是变化的。正在建立的社区咨询团体充当了使每个社区与样本资源库保持持续联系的角色，确保今后样本的用途与知情同意书中所描述的用途一致。目前正在编写一篇更详细的文章，该文章讨论了有关此项目的伦理、社会和文化问题，并说明招募捐助群体参与确定和评价这些问题的过程。

The HapMap developed with samples from these four large populations will include a substantial amount of the genetic variation found in all populations throughout the world. The goal of the HapMap is medical, and the common patterns of variation identified by the project will be useful to identify genes that contribute to disease and drug response in many other populations. Samples from several other populations are being collected for studies that will examine how similar their haplotype patterns are to those in the HapMap. If the patterns found are very different, samples from some of these populations may be genotyped on a large scale to make the HapMap more applicable to them. Further follow-up studies in other populations, small and large, are likely to be undertaken by scientists in many nations for common disease gene discovery.

The project will study a total of 270 DNA samples: 90 samples (see Supplementary Information, part 1) from a US Utah population with Northern and Western European ancestry (samples collected in 1980 by the Centre d'Etude du Polymorphisme Humain (CEPH)[49] and used for other human genetic maps, 30 trios of two parents and an adult child), and new samples collected from 90 Yoruba people in Ibadan, Nigeria (30 trios), 45 unrelated Japanese in Tokyo, Japan, and 45 unrelated Han Chinese in Beijing, China. All donors gave specific consent for their inclusion in the project. Population membership was determined in ways appropriate for each culture: for the Yoruba by asking the donor whether all four grandparents were Yoruba, for the Han Chinese by asking the donor whether at least three of four grandparents were Han Chinese, and for the Japanese by self-identification. The CEPH samples are available from the non-profit Coriell Institute of Medical Research (http://locus.umdnj.edu/nigms/); cell lines and DNA from the new samples will be available from Coriell in early 2004 for future studies with research protocols approved by appropriate ethics committees. It is anticipated that other researchers will genotype additional SNPs in these samples in the future, and that these data will continuously improve the HapMap.

These samples will have population and sex identifiers without information that could link them to individual donors. As the goal of the project is solely to identify patterns of genetic variation, no medical or other phenotypic information will be included. About 50% more samples were collected than will be used, so that inclusion of a sample from any particular donor cannot be known.

Samples of 45 unrelated individuals should be sufficient to find 99% of haplotypes with a frequency of 5% or greater in a population. Studies of LD can use random individual samples, trios or larger pedigrees; each design has advantages (ease of sampling) and disadvantages (decreasing efficiency with increasing numbers of related individuals). Analysis of existing data and computer simulations suggested that unrelated individuals and trios have considerable power for estimating local LD patterns. The trios will provide useful information on the accuracy of the genotyping platforms being used for the project.

用这四大群体的样本开发的 HapMap 将包括世界各地所有人群中发现的大量的遗传变异。HapMap 计划的目标是医疗，通过此计划确定的共同变异模式可用于鉴定在其他许多人群中对疾病和药物响应的基因。其他几个群体的样本正在收集以用于检测它们的单体型模式与 HapMap 中的有多大相似性。如果发现模式有很大的不同，将对它们中的某些群体样本进行更大规模的基因分型，以使 HapMap 更适用于它们。许多国家的科学家为了发现常见的疾病基因，可能会对其他小型和大型群体进行后续研究。

这个项目将总共研究 270 份 DNA 样本：90 份（见补充资料，第 1 部分）来自祖先是北欧和西欧的美国犹他州居民的样本（样本是由人类多态性研究中心（CEPH）于 1980 年采集的[49]，并用于其他人类的遗传图谱，30 个由父母双方和一个成年孩子组成的三体家系），以及从以下来源收集到的新的样本：90 个尼日利亚伊巴丹的约鲁巴人（30 个三体家系）、45 个互不相关的来自日本东京的日本人和 45 个互不相关的来自中国北京的汉族人。所有捐助者都同意加入该计划，通过与各自背景相符的方式确定群体成员资格：对约鲁巴人，询问捐助者他们的四个祖父母是否都是约鲁巴人；对汉族人，询问捐助者他们的四个祖父母是否至少有三个是汉族人；对日本人只是询问祖先是否来自日本。人类多态性研究中心的样本来自非营利机构科里尔医学研究所（http://locus.umdnj.edu/nigms/）。2004 年，研究协议经相应的伦理委员会批准后，科里尔研究所将为进一步的研究提供新样本的细胞系和 DNA。预计，今后其他研究人员将对这些样本中更多的 SNP 进行基因型分析，这些数据将不断改善 HapMap。

这些样本有群体和性别标识而无捐助者个人标识。正如这一计划的目的仅仅是确定遗传变异的模式，不包含医疗或其他表型信息。收集的样本超过 50% 未被采用，因此并不能知道列入的样本来自哪个捐助者。

从 45 个无关的个体样本中应当足以找到群体中 99%的、频率大于 5% 的单体型。LD 研究可以使用随机个体样本、三体家系或较大的家系；每个设计都有优势（方便抽样）和缺点（随着亲缘个体数目的增加而效率降低）。分析现有的数据和计算机模拟表明，无血缘关系的个体和三体家系对估计 LD 模式有相当强的能力。三体家系将对用于此计划的基因分型平台的准确性提供有益的信息。

Choice of SNPs

A high density of SNPs is needed to describe adequately the genetic variation across the entire genome. When the project started, the average density of markers in the public database dbSNP (http://www.ncbi.nlm.nih.gov/SNP/)[50] was approximately one every kilobase (2.8 million SNPs) but, given their variable distribution, many regions had a lower density of SNPs.

Further SNPs were obtained by random shotgun sequencing from whole-genome and whole-chromosome (flow-sorted) libraries[51], using methods developed for the initial human SNP map[52], and also by collaboration with Perlegen Sciences[36] and through the purchase of sequence traces from Applied Biosystems[53] for SNP detection (see Supplementary Information, part 2). One useful result of this search for more SNPs is the confirmation of SNPs found previously. SNPs for which each allele has been seen independently in two or more samples ("double-hit" SNPs) have a higher average minor allele frequency than do "single-hit" SNPs[28]. This leads to substantial savings in assay development. On 4 November 2003, the number of SNPs (with a unique genomic position) in dbSNP (build 118) was 5.7 million, and the number of double-hit SNPs was over 2 million. By February of 2004, 6.8 million SNPs (with a unique genomic position) are expected to be in dbSNP and available for the project, including 2.7 million double-hit SNPs.

As the extent of LD and haplotypes varies by 100-fold across the genome[30-32,34,35], a hierarchical genotyping strategy has been adopted. In an initial round of genotyping, the project aims to genotype successfully 600,000 SNPs spaced at approximately 5-kilobase intervals and each with a minor allele frequency of at least 5%, in the 270 DNA samples. Priority is being given to previously validated SNPs, double-hit SNPs and SNPs causing amino-acid changes (as these may alter protein function). When these genotyping data are produced (by mid-2004; see below for details of data release), they will be analysed for associations between neighbouring SNPs. Additional SNPs will then be genotyped in the same DNA samples at a higher density only in regions where the associations are weak. Further rounds of analysis and genotyping will be carried out as required. It is expected that more than one million SNPs will be genotyped overall. This hierarchical strategy will permit regions of the genome with the least LD to be characterized at densities of up to one SNP per kilobase, maximizing the characterization of regions with associations only over short distances.

Genotyping

Each genotyping centre is responsible for genotyping all the samples for all the selected SNPs on the chromosome regions allocated (Table 1). Among the centres, a total of five high-throughput genotyping technologies are being used, which will provide an opportunity to compare their accuracy, success rate, throughput and cost. Access to several platforms is

SNP 的 选 择

充分描述全基因组的遗传变异需要高密度的 SNP。当这个计划开始时，公共数据库 dbSNP(http://www.ncbi.nlm.nih.gov/SNP/)[50] 中的平均标记密度为大约每千碱基中有一个 SNP(280 万个 SNP)，但鉴于其分布不同，许多区域的 SNP 密度比较低。

使用最初为人类 SNP 图谱开发的方法[52]，通过对全基因组和整个染色体(流式分离)文库[51] 随机鸟枪法测序得到更多的 SNP，通过与 Perlegen Sciences[36] 合作并从 Applied Biosystems 购买序列示踪剂[53] 来检测 SNP(见补充资料，第 2 部分)也可得到更多的 SNP。通过上述方法搜索更多 SNP 得到的一个有用的结果是对以前发现的 SNP 进行确认。当 SNP 的每个等位基因在两个或多个样本中分别被发现时(“两次发现”SNP)，这些 SNP 比“单次发现”的 SNP 具有更高的平均最小等位基因型频率[28]，这在研发分析中节省了大笔开支。2003 年 11 月 4 日，dbSNP(118 版本)中的 SNP(每个具有独特的基因组位置)的数目是 570 万，而“两次发现”SNP 已经超过 200 万。到 2004 年 2 月，预计 dbSNP 有 680 万 SNP(每个具有独特的基因组位置)可用于此计划，其中包括 270 万“两次发现”SNP。

因为 LD 和单体型在基因组内有着 100 倍的变化范围[30-32,34,35]，因此已采取分级基因分型策略。在最初一轮的基因分型中，该计划的目标是在 270 个 DNA 样本中，对间隔为大约 5 kb 的 600,000 个 SNP 成功进行基因分型，并且每一个 SNP 的最小等位基因型频率至少达到 5%。优先考虑先前确定的 SNP、“两次发现”SNP 及造成氨基酸变化的 SNP(因为这些可能会改变蛋白质的功能)。当得到这些基因分型数据后(截至 2004 年中期；发布的详细数据见下文)，将分析相邻 SNP 间的关联性。其余的 SNP 将在相同的 DNA 样本中以高密度方式对相关性很弱的区域进行基因分型，这需要进一步进行几轮分析和基因分型。预计总共进行基因分型的 SNP 超过 100 万。这一分级战略将允许对 LD 密度最小的基因组区域以高达每千碱基一个 SNP 的密度基因分型，最大限度地确定仅有短距离联系的区域。

基 因 分 型

每个基因分型中心负责对所有的样本中分配的染色体区域内全部选定的 SNP 进行基因分型(表 1)。在这些中心内，一共使用了 5 个高通量基因分型技术，用来比较其准确性、成功率、通量和成本。可以使用几个平台是这个计划的优势，因为一

an advantage for the project, as a SNP assay that fails on one platform may be developed successfully using another method in order to fill a gap in the HapMap. All platforms will be evaluated using a common set of performance criteria to ensure that the quality of data produced for the project meets a uniformly high standard.

Genotype quality is being assessed in three ways. First, at the beginning of the project, all centres were assigned the same randomly selected set of 1,500 SNPs for assay development and genotyping in the 90 CEPH DNA samples being used for the project. Genotyping centres produced data that were on average more than 99.2% complete and more than 99.5% accurate (as compared to the consensus of at least two other platforms). Second, every genotyping experiment includes samples for internal quality checks, with each 96-well plate containing duplicates of five different samples, and one blank. In addition, the data from trios provide a check for consistent mendelian inheritance of SNP alleles. For all the populations, the data from the unrelated samples provide a check that the SNPs are in Hardy–Weinberg equilibrium (a test of genetic mating patterns). Although a small proportion of SNPs may fail these checks for biological reasons, they more typically fail if a genotyping platform makes consistent errors, such as undercalling heterozygotes. Third, a sample of SNP genotypes deposited by each centre will be selected at random and re-genotyped by other centres. These stringent third-party evaluations of quality will ensure the completeness and reliability of the data produced by the project.

Data Release

The project is committed to rapid and complete data release, and to ensuring that project data remain freely available in the public domain at no cost to users. The project follows the data-release principles of a "community resource project" (http://www.wellcome.ac.uk/en/1/awtpubrepdat.html).

All data on new SNPs, assay conditions, and allele and genotype frequencies will be released rapidly into the public domain on the internet at the HapMap Data Coordination Center (DCC) (http://www.hapmap.org/) and deposited in dbSNP. Individual genotype and haplotype data initially will be made available at the DCC under a short-term "click-wrap" licence agreement. This strategy has been adopted to ensure that data from the project cannot be incorporated into any restrictive patents, and will thus remain freely available in the long term. The only condition for data access is that users must agree not to restrict use of the data by others and to share the data only with others who have agreed to the same condition. When haplotypes are defined in a region, then the individual genotypes, haplotypes and tag SNPs in that region will be publicly released to dbSNP, where there are no licensing conditions. Project participants have agreed that their own laboratories will access the data through the DCC and under the click-wrap licence, ensuring that all scientists have equal access to the data for research.

The consortium believes that SNP, genotype and haplotype data in the absence of specific

个 SNP 分析在一个平台失败，可以用其他方法成功地开展以填补 HapMap 的缺口。所有平台将使用一套共同的性能标准评价，以确保为此计划产生的数据质量符合一致的高标准。

基因分型质量有三种方法进行评估。首先，在计划开始时，所有的中心分配同一套随机挑选的 1,500 个 SNP 进行实验开发和对用于该计划的 90 个 CEPH DNA 样本进行基因分型。分型中心产生数据的完整性和准确性平均分别超过 99.2%和 99.5%(相对于至少有两个其他平台的一致性)。其次，每个基因分型实验包括样本的内部质量检查，每 96 孔板含有五个不同样本的重复和一个空白对照。此外，三体家系数据检查 SNP 等位基因符合孟德尔遗传。对于所有群体，无血缘关系的样本数据检查 SNP 的哈迪–温伯格平衡(一个遗传交配模式测试)。虽然因为生物学原因，一小部分 SNP 在这些检查中可能会失败，但如果一个基因分型平台产生一致的错误，它们可能会出现更典型的失败，例如复杂杂合子。第三，每个中心存放的 SNP 基因型的样本将随机抽样并由其他中心重新基因分型。这些严格的第三方质量评价将确保这个计划产生的数据的完整性和可靠性。

数据公布

该计划致力于迅速和完全发布数据，并确保此计划的数据免费向公众开放。该计划遵循“公众资源计划”的数据发布原则(http://www.wellcome.ac.uk/en/1/awtpubrepdat.html)。

关于新 SNP、实验条件、等位基因和基因型频率的所有数据将迅速公布在位于 HapMap 数据协调中心(DCC)(http://www.hapmap.org/)的公共领域并储存在 dbSNP 中。个体基因型和单体型数据最初将在短期“点击许可”协议下由 DCC 提供。采用这一策略以确保该计划的数据不被纳入任何限制性专利中，因此将继续长期免费提供。数据存取唯一的条件是，用户必须同意不限制其他人使用数据并且仅与其他已同意相同条件的用户共享。当单体型在一个区域被确定后，同一区域的单个基因型、单体型和标签 SNP 将公开发布到 dbSNP 上，没有任何附加条款。该计划的参与者已同意自己的实验室也将在“点击许可”协议下通过 DCC 获得数据，以确保所有科学家有平等机会获得数据进行研究。

协作组认为，没有具体效用的 SNP、基因型和单体型数据，并不适于申请专

utility do not constitute appropriately patentable inventions. Specific utility would involve, for example, finding an association of a SNP or haplotype with a medically important phenotype such as a disease risk or drug response. The project does not include any phenotype association studies. However, the data-release policy does not block users from filing for appropriate intellectual property on such associations, as long as any ensuing patent is not used to prevent others' access to the HapMap data.

Data Analysis

The project will apply existing and new methods for analysis and display of the data. LD between pairs of markers will be calculated using standard measures such as D′ (ref. 54), r^2 (refs 55, 56) and others. Various methods are being evaluated to define regions of high LD and haplotypes along chromosomes. Existing methods include "sliding window" LD profiles[57,58], LD unit maps[59], haplotype blocks[31,35] and estimates of meiotic recombination rates along chromosomes[35,60-62]. After analysis of the LD in the first phase of the project, regions in which there is little or no LD will be identified and ranked for further SNP selection and genotyping. Methods to select optimal collections of tag SNPs will be developed and evaluated (see above). The project will thus provide views of the data and tag SNPs that will be useful to the research community. As all data and analysis methods will be made available, other researchers will also be able to analyse the data and improve the analysis methods.

To assist optimization of SNP selection and analysis of LD and haplotypes, a pilot study is underway to produce a dense set of genotypes across large genomic regions. Ten 500-kilobase regions of the genome (see Supplementary Information, part 3) will be sequenced in 48 unrelated HapMap DNA samples (16 CEPH (currently being sequenced), 16 Yoruba, 8 Japanese and 8 Han Chinese). All SNPs identified, as well any additional SNPs in the public databases, will be genotyped in all of the 270 HapMap DNA samples, and the genotype data will be released following the guidelines described above. This study will provide dense genotype data for developing methods for SNP selection and for assessing the completeness of the information extracted, to guide the later stages of genotyping.

When the HapMap is used to examine large genomic regions, the problem of multiple comparisons will arise from testing tens to hundreds of thousands of SNPs and haplotypes for disease associations. This will lead to difficulty in separating true from false-positive results. Thus, new statistical methods, replication studies and functional analyses of variants will be important to confirm the findings and identify the functionally important SNPs.

Conclusion

The goal of the International HapMap Project is to develop a research tool that will help investigators across the globe to discover the genetic factors that contribute to susceptibility

利。具体的效用应涉及如找到一个 SNP 或单体型与医学重要的表型如疾病风险或药物反应的关联。该计划不包括任何表型关联研究。但是，数据发布政策并不阻止用户对这些关联申请适当的知识产权，只要任何随后的专利不被用于防止他人使用 HapMap 数据。

数据分析

该计划将采用现有的和新的方法来分析和显示数据。成对标记间的 LD 将使用如 D'(参考文献 54) 和 r^2(参考文献 55 和 56) 等标准的措施计算。各种方法正在接受评估以确定高 LD 区域和沿染色体的各个单体型。现有的方法包括"滑动窗口"LD 描绘 [57,58]，LD 单位图 [59]，单体型区块 [31,35] 和沿染色体估计减数分裂重组率 [35,60-62]。经过该计划第一阶段 LD 分析，只有很少或没有 LD 的区域将被确定和分级用于进一步 SNP 选择和基因分型。选择最佳的收集标签 SNP 的方法将被制定和评估（见上文）。因此，该计划将提供对数据和标签 SNP 的看法，这些看法将对研究团体有帮助。所有数据和分析方法将被公开，其他研究人员也可以分析数据，提高分析方法。

为帮助优化 SNP 选择和 LD、单体型分析，我们正在开展一个先导研究对一些大片段的基因组区域进行致密的基因分型。基因组的 10 个 500-kb 区域（见补充资料，第 3 部分）将在 48 个无血缘关系的 HapMap DNA 样本（16 个 CEPH（目前正在测序）、16 个约鲁巴人、8 个日本人和 8 个中国汉族人）中测序。所有确定的 SNP 以及公共数据库中任何更多的 SNP，将在所有的 270 份 HapMap DNA 样本中进行基因分型，并且基因型数据都将按照上文所述的准则发布。这项研究将为开发 SNP 选择方法和评估提取数据的完整性提供密集的基因型数据，以对后期基因分型进行指导。

用 HapMap 来研究大的基因组区域，检测成千上万 SNP 和单体型与疾病的关联时将产生多重比较的问题。这将导致难以从假阳性结果中分离出真正阳性的结果。因此，新的统计方法、验证研究和功能分析对确认发现和鉴定具有重要功能的 SNP 很重要。

结　论

国际 HapMap 计划的目标是建立一个研究工具，这将有助于在全球范围内调查发现有助于探索疾病的遗传易感因素、对抗疾病的保护方法和药物反应情况。

to disease, to protection against illness and to drug response. The HapMap will provide an important shortcut to carry out candidate-gene, linkage-based and genome-wide association studies, transforming an unfeasible strategy into a practical one. In its scope and potential consequences, the International HapMap Project has much in common with the Human Genome Project, which sequenced the human genome[63]. Both projects have been scientifically ambitious and technologically demanding, have involved intense international collaboration, have been dedicated to the rapid release of data into the public domain, and promise to have profound implications for our understanding of human biology and human health. Whereas the sequencing project covered the entire genome, including the 99.9% of the genome where we are all the same, the HapMap will characterize the common patterns within the 0.1% where we differ from each other.

For the full potential of the HapMap to be realized, several things must occur. The technology for genotyping must become more cost efficient, and the analysis methods must be improved. Pilot studies with other populations must be completed to confirm that the HapMap is generally applicable, with consideration given to expanding the HapMap if needed so that all major world populations can derive the greatest benefit. To use the tools created by the HapMap, later projects must establish carefully phenotyped sets of affected and unaffected individuals for many common diseases in a way that preserves confidentiality but retains detailed clinical and environmental exposure data. Longitudinal cohort studies of hundreds of thousands of individuals will also be invaluable for assessing the genetic and environmental contributions to disease.

Careful and sustained attention must also be paid to the ethical issues that will be raised by the HapMap and the studies that will use it. By consulting members of donor populations about the consent process and the implications of population-specific findings before sample collection, the project has helped to advance the ethical standard for international population genetics research. Future population genetics projects will continue to refine this approach. It will be an ongoing challenge to avoid misinterpretations or misuses of results from studies that use the HapMap. Researchers using the HapMap should present their findings in ways that avoid stigmatizing groups, conveying an impression of genetic determinism, or attaching incorrect levels of biological significance to largely social constructs such as race.

The HapMap holds much promise as a powerful new tool for discovery—to enhance our understanding of the hereditary factors involved in health and disease. Realizing its full benefits will involve the close partnership of basic science researchers, population geneticists, epidemiologists, clinicians, social scientists, ethicists and the public.

(**426**, 789-796; 2003)

doi:10.1038/nature02168.

HapMap 将提供一个实现候选基因、连锁和全基因组关联研究的捷径，将一个不可行战略变为实际。在适用范围和潜在后果方面，国际 HapMap 计划与为人类基因组测序[63]的人类基因组计划在很多方面有相同之处。这两个项目都有远大的科学雄心并对技术要求很高，涉及广泛的国际合作，一直致力于向公众领域迅速发布数据，承诺将对人类生物学和人类健康的理解产生深刻的影响。测序项目涵盖了整个基因组，包括人类 99.9%的相同基因组，而 HapMap 计划将确定人类互不相同的 0.1%的常见模式。

为充分发挥 HapMap 计划的潜力，有几件事情必须考虑。基因分型技术必须更具成本效益，分析方法必须加以改进。关于其他群体的先导研究必须完成以确认 HapMap 计划是普遍适用的，如果需要的话，HapMap 计划可以考虑扩大以使全球所有主要的群体能够获得最大的利益。如果要使用 HapMap 项目所创造的工具，后来的研究必须仔细建立许多受常见疾病影响和不受影响的个体的表型集合。这些集合除了详细的临床及环境暴露数据，其他都处于机密状态。纵向研究几十万个个体对评估疾病的遗传和环境因素将有重大意义。

由 HapMap 和使用 HapMap 进行的研究引发的道德问题需得到认真持续的关注。通过在样本采集前与捐助群体成员就是否同意这个过程以及种群特异性发现的意义进行商议，该项目帮助推动了国际人口遗传学研究的道德标准，未来群体遗传学项目将继续完善这一做法。这将是一项持续的挑战，以避免误解或误用 HapMap 的研究结果。研究人员利用 HapMap 呈现其研究结果时应避免污蔑某一种群，或传达遗传决定论的思想，或将生物意义与大的社会建构如种族联系在一起。

HapMap 项目很有希望成为强大的新搜索工具，以提高我们对涉及健康和疾病的遗传因素的理解。充分发挥它的作用将涉及基础科学的研究人员、群体遗传学家、流行病学家、临床医生、社会科学家、伦理学家和公众的紧密合作。

（李梅 翻译；常江 审稿）

References:

1. King, R. A., Rotter, J. I. & Motulsky, A. G. *The Genetic Basis of Common Diseases* Vol. 20 (eds Motulsky, A. G., Harper, P. S., Scriver, C. & Bobrow, M.) (Oxford Univ. Press, Oxford, 1992).
2. Risch, N. J. Searching for genetic determinants in the new millennium. *Nature* **405**, 847-856 (2000).
3. Botstein, D. & Risch, N. Discovering genotypes underlying human phenotypes: past successes for mendelian disease, future approaches for complex disease. *Nature Genet.* **33** (Suppl.), 228-237 (2003).
4. Risch, N. & Merikangas, K. The future of genetic studies of complex human diseases. *Science* **273**, 1516-1517 (1996).
5. Dorman, J. S., LaPorte, R. E., Stone, R. A. & Trucco, M. Worldwide differences in the incidence of type I diabetes are associated with amino acid variation at position 57 of the HLA-DQ β chain. *Proc. Natl Acad. Sci. USA* **87**, 7370-7374 (1990).
6. Bell, G. I., Horita, S. & Karam, J. H. A polymorphic locus near the human insulin gene is associated with insulin-dependent diabetes mellitus. *Diabetes* **33**, 176-183 (1984).
7. Nisticò, L. *et al.* The *CTLA-4* gene region of chromosome 2q33 is linked to, and associated with, type 1 diabetes. *Hum. Mol. Genet.* **5**, 1075-1080 (1996).
8. Strittmatter, W. J. & Roses, A. D. Apolipoprotein E and Alzheimer's disease. *Annu. Rev. Neurosci.* 19, 53-77 (1996).
9. Dahlbäck, B. Resistance to activated protein C caused by the factor V $R^{506}Q$ mutation is a common risk factor for venous thrombosis. *Thromb. Haemost.* **78**, 483-488 (1997).
10. Hugot, J. P. *et al.* Association of NOD2 leucine-rich repeat variants with susceptibility to Crohn's disease. *Nature* **411**, 599-603 (2001).
11. Ogura, Y. *et al.* A frameshift mutation in NOD2 associated with susceptibility to Crohn's disease. *Nature* **411**, 603-606 (2001).
12. Rioux, J. D. *et al.* Genetic variation in the 5q31 cytokine gene cluster confers susceptibility to Crohn disease. *Nature Genet.* **29**, 223-228 (2001).
13. Pennacchio, L. A. *et al.* An apolipoprotein influencing triglycerides in humans and mice revealed by comparative sequencing. *Science* **294**, 169-173 (2001).
14. Deeb, S. S. *et al.* A Pro12Ala substitution in PPARγ2 associated with decreased receptor activity, lower body mass index and improved insulin sensitivity. *Nature Genet.* **20**, 284-287 (1998).
15. Altshuler, D. *et al.* The common PPARγ Pro12Ala polymorphism is associated with decreased risk of type 2 diabetes. *Nature Genet.* **26**, 76-80 (2000).
16. Stefansson, H. *et al.* Neuregulin 1 and susceptibility to schizophrenia. *Am. J. Hum. Genet.* **71**, 877-892 (2002).
17. Van Eerdewegh, P. *et al.* Association of the *ADAM33* gene with asthma and bronchial hyperresponsiveness. *Nature* **418**, 426-430 (2002).
18. Gretarsdottir, S. *et al.* The gene encoding phosphodiesterase 4D confers risk of ischemic stroke. *Nature Genet.* **35**, 131-138 (2003).
19. Ozaki, K. *et al.* Functional SNPs in the lymphotoxin-α gene that are associated with susceptibility to myocardial infarction. *Nature Genet.* **32**, 650-654 (2002).
20. Collins, F. S., Guyer, M. S. & Chakravarti, A. Variations on a theme: cataloging human DNA sequence variation. *Science* **278**, 1580-1581 (1997).
21. Kruglyak, L. & Nickerson, D. A. Variation is the spice of life. *Nature Genet.* **27**, 234-236 (2001).
22. Kerem, B. *et al.* Identification of the cystic fibrosis gene: genetic analysis. *Science* **245**, 1073-1080 (1989).
23. Hästbacka, J. *et al.* Linkage disequilibrium mapping in isolated founder populations: diastrophic dysplasia in Finland. *Nature Genet.* **2**, 204-211 (1992).
24. Li, W. H. & Sadler, L. A. Low nucleotide diversity in man. *Genetics* **129**, 513-523 (1991).
25. Wang, D. G. *et al.* Large-scale identification, mapping, and genotyping of single-nucleotide polymorphisms in the human genome. *Science* **280**, 1077-1082 (1998).
26. Cargill, M. *et al.* Characterization of single-nucleotide polymorphisms in coding regions of human genes. *Nature Genet.* **22**, 231-238 (1999).
27. Halushka, M. K. *et al.* Patterns of single-nucleotide polymorphisms in candidate genes for blood-pressure homeostasis. *Nature Genet.* **22**, 239-247 (1999).
28. Reich, D. E., Gabriel, S. B. & Altshuler, D. Quality and completeness of SNP databases. *Nature Genet.* **33**, 457-458 (2003).
29. Pääbo, S. The mosaic that is our genome. *Nature* **421**, 409-412 (2003).
30. Jeffreys, A. J., Kauppi, L. & Neumann, R. Intensely punctate meiotic recombination in the class II region of the major histocompatibility complex. *Nature Genet.* **29**, 217-222 (2001).
31. Gabriel, S. B. *et al.* The structure of haplotype blocks in the human genome. *Science* **296**, 2225-2229 (2002).
32. Reich, D. E. *et al.* Linkage disequilibrium in the human genome. *Nature* **411**, 199-204 (2001).
33. Abecasis, G. R. *et al.* Extent and distribution of linkage disequilibrium in three genomic regions. *Am. J. Hum. Genet.* **68**, 191-197 (2001).
34. Dawson, E. *et al.* A first-generation linkage disequilibrium map of human chromosome 22. *Nature* **418**, 544-548 (2002).
35. Daly, M. J., Rioux, J. D., Schaffner, S. F., Hudson, T. J. & Lander, E. S. High-resolution haplotype structure in the human genome. *Nature Genet.* **29**, 229-232 (2001).
36. Patil, N. *et al.* Blocks of limited haplotype diversity revealed by high-resolution scanning of human chromosome 21. *Science* **294**, 1719-1723 (2001).
37. Johnson, G. C. L. *et al.* Haplotype tagging for the identification of common disease genes. *Nature Genet.* **29**, 233-237 (2001).
38. Carlson, C. S. *et al.* Additional SNPs and linkage-disequilibrium analyses are necessary for whole-genome association studies in humans. *Nature Genet.* **33**, 518-521 (2003).
39. Goldstein, D. B., Ahmadi, K. R., Weale, M. E. & Wood, N. W. Genome scans and candidate gene approaches in the study of common diseases and variable drug responses. *Trends Genet.* **19**, 615-622 (2003).
40. Taillon-Miller, P. *et al.* Juxtaposed regions of extensive and minimal linkage disequilibrium in human Xq25 and Xq28. *Nature Genet.* **25**, 324-328 (2000).
41. Chakravarti, A. Population genetics—making sense out of sequence. *Nature Genet.* **21**, 56-60 (1999).
42. Chakravarti, A. *et al.* Nonuniform recombination within the human β-globin gene cluster. *Am. J. Hum. Genet.* **36**, 1239-1258 (1984).

43. Tishkoff, S. A. *et al.* Global patterns of linkage disequilibrium at the CD4 locus and modern human origins. *Science* **271**, 1380-1387 (1996).

44. Cavalli-Sforza, L. L., Menozzi, P. & Piazza, A. *The History and Geography of Human Genes* (Princeton Univ. Press, Princeton, 1994).

45. Foster, M. W. & Sharp, R. R. Race, ethnicity, and genomics: social classifications as proxies of biological heterogeneity. *Genome Res.* **12**, 844-850 (2002).

46. Barbujani, G., Magagni, A., Minch, E. & Cavalli-Sforza, L. L. An apportionment of human DNA diversity. *Proc. Natl Acad. Sci. USA* **94**, 4516-4519 (1997).

47. Rosenberg, N. A. *et al.* Genetic structure of human populations. *Science* **298**, 2381-2385 (2002).

48. Jorde, L. B. *et al.* Microsatellite diversity and the demographic history of modern humans. *Proc. Natl Acad. Sci. USA* **94**, 3100-3103 (1997).

49. Dausset, J. *et al.* Centre d'Etude du Polymorphisme Humain (CEPH): collaborative genetic mapping of the human genome. *Genomics* **6**, 575-577 (1990).

50. Sherry, S. T. *et al.* dbSNP: the NCBI database of genetic variation. *Nucleic Acids Res.* **29**, 308-311 (2001).

51. Ning, Z., Cox, A. J. & Mullikin, J. C. SSAHA: a fast search method for large DNA databases. *Genome Res.* **11**, 1725-1729 (2001).

52. The International SNP Working Group. A map of human genome sequence variation containing 1.42 million single nucleotide polymorphisms. *Nature* **409**, 928-933 (2001).

53. Venter, J. C. *et al.* The sequence of the human genome. *Science* **291**, 1304-1351 (2001).

54. Lewontin, R. C. The interaction of selection and linkage. I. General considerations: heterotic models. *Genetics* **49**, 49-67 (1964).

55. Hill, W. G. & Robertson, A. Linkage disequilibrium in finite populations. *Theor. Appl. Genet.* **38**, 226-231 (1968).

56. Ohta, T. & Kimura, M. Linkage disequilibrium due to random genetic drift. *Genet. Res.* **13**, 47-55 (1969).

57. Dawson, K. J. The decay of linkage disequilibrium under random union of gametes: how to calculate Bennett's principal components. *Theor. Popul. Biol.* **58**, 1-20 (2000).

58. Langley, C. H. & Crow, J. F. The direction of linkage disequilibrium. *Genetics* **78**, 937-941 (1974).

59. Maniatis, N. *et al.* The first linkage disequilibrium (LD) maps: delineation of hot and cold blocks by diplotype analysis. *Proc. Natl Acad. Sci. USA* **99**, 2228-2233 (2002).

60. Hudson, R. R. Estimating the recombination parameter of a finite population model without selection. *Genet. Res.* **50**, 245-250 (1987).

61. Fearnhead, P. & Donnelly, P. Estimating recombination rates from population genetic data. *Genetics* **159**, 1299-1318 (2001).

62. McVean, G., Awadalla, P. & Fearnhead, P. A coalescent-based method for detecting and estimating recombination from gene sequences. *Genetics* **160**, 1231-1241 (2002).

63. International Human Genome Sequencing Consortium. Initial sequencing and analysis of the human genome. *Nature* **409**, 860-921 (2001).

64. Clayton, E. W. The complex relationship of genetics, groups, and health: what it means for public health. *J. Law Med. Ethics* **30**, 290-297 (2002).

65. Foster, M. W. & Sharp, R. R. Genetic research and culturally specific risks: one size does not fit all. *Trends Genet.* **16**, 93-95 (2000).

66. Sharp, R. R. & Foster, M. W. Involving study populations in the review of genetic research. *J. Law Med. Ethics* **28**, 41-51 (2000).

67. Marshall, P. A. & Rotimi, C. Ethical challenges in community-based research. *Am. J. Med. Sci.* **322**, 241-245 (2001).

68. Juengst, E. T. Commentary: what "community review" can and cannot do. *J. Law Med. Ethics* **28**, 52-54 (2000).

Supplementary Information accompanies the paper on www.nature.com/nature.

Acknowledgements. We thank many people who contributed to this project: J. Beck, C. Beiswanger, D. Coppock, J. Mintzer and L. Toji at the Coriell Institute for Medical Research for transforming the samples, distributing the DNA and cell lines, and storing the samples for use in future research; J. Greenberg and R. Anderson of the NIH National Institute of General Medical Sciences (NIGMS) for providing funding and support for cell-line transformation and storage in the NIGMS Human Genetic Cell Repository at the Coriell Institute; K. Wakui at Shinshu University for assistance in transforming the Japanese cell lines; N. Carter and D. Willey at the Wellcome Trust Sanger Institute for flow sorting the chromosomes and for library construction, respectively; M. Deschesnes and B. Godard for assistance at the University of Montréal; C. Darmond-Zwaig, J. Olivier and S. Roumy at McGill University and Génome Québec Innovation Centre; C. Allred, B. Gillman, E. Kloss and M. Rieder for help in implementing data flow protocols; S. Olson for work on the website explanations; S. Adeniyi-Jones, D. Burgess, W. Burke, T. Citrin, A. Clark, D. Cowhig, P. Epps, K. Hofman, A. Holt, E. Juengst, B. Keats, J. Levin, R. Myers, A. Obuoforibo, F. Romero, C. Tamura and A. Williamson for providing advice on the project to NIH; A. Peck and J. Witonsky of the National Human Genome Research Institute (NHGRI) for help with project management; E. DeHaut-Combs and S. Saylor of NHGRI for staff support; M. Gray for organizing phone calls and meetings; the people of Tokyo, Japan, the Yoruba people of Ibadan, Nigeria, and the community at Beijing Normal University, who participated in public consultations and community engagements; and the people in these communities who were generous in donating their blood samples. This work was supported in part by Genome Canada, Génome Québec, the Chinese Ministry of Science and Technology, the Chinese Academy of Sciences, the Natural Science Foundation of China, the Hong Kong Innovation and Technology Commission, the University Grants Committee of Hong Kong, the Japanese Ministry of Education, Culture, Sports, Science and Technology, the Wellcome Trust, the SNP Consortium, the US National Institutes of Health (FIC, NCI, NCRR, NEI, NHGRI, NIA, NIAAA, NIAID, NIAMS, NIBIB, NIDA, NIDCD, NIDCR, NIDDK, NIEHS, NIGMS, NIMH, NINDS, OD), the W.M. Keck Foundation and the Delores Dore Eccles Foundation.

Competing interests statement. The authors declare that they have no competing financial interests.

Correspondence and requests for materials should be addressed to D.B. (drb@sanger.ac.uk) or M.F. (fost1848@msmailhub.oulan.ou.edu).

Mice Cloned from Olfactory Sensory Neurons

K. Eggan *et al.*

Editor's Note

By 2004, various mammals including sheep and mice had been cloned by injecting donor DNA into enucleated eggs. But in this paper, American cell biologist Rudolf Jaenisch and colleagues use donor DNA taken from mature olfactory neurons to generate fertile mouse clones by the process of nuclear transfer. The study demonstrates that DNA taken from a specialised cell that has stopped dividing can re-enter the cell cycle and be reprogrammed to an embryonic-like state after nuclear transfer. The cloned mice possessed the full range of organized, odorant receptor genes that were indistinguishable from those of normal mice, suggesting that selection of a particular receptor gene in a mature olfactory neuron does not cause irreversible changes in DNA.

Cloning by nuclear transplantation has been successfully carried out in various mammals, including mice. Until now mice have not been cloned from post-mitotic cells such as neurons. Here, we have generated fertile mouse clones derived by transferring the nuclei of post-mitotic, olfactory sensory neurons into oocytes. These results indicate that the genome of a post-mitotic, terminally differentiated neuron can re-enter the cell cycle and be reprogrammed to a state of totipotency after nuclear transfer. Moreover, the pattern of odorant receptor gene expression and the organization of odorant receptor genes in cloned mice was indistinguishable from wild-type animals, indicating that irreversible changes to the DNA of olfactory neurons do not accompany receptor gene choice.

THE chromosome is a dynamic structure that undergoes complex changes that underlie the development and differentiated function of an organism. Alterations in chromosome structure include variation in the complement of regulatory proteins, covalent modifications in chromatin proteins or DNA, and in rare instances, DNA rearrangements[1]. The extent to which such chromosomal changes are reversible can be discerned by cloning experiments involving nuclear transfer. Thus far, cloning experiments using nuclei from post-mitotic cells that have irreversibly exited the cell cycle as part of their programme of differentiation have not generated viable embryos or mice[2,3]. These observations have led to the suggestion that post-mitotic cells might be refractory to epigenetic reprogramming or alternatively might have acquired changes in their DNA that could limit their developmental potential[4,5].

Directed DNA rearrangements are rarely observed as part of a normal differentiation programme and, in vertebrates, they have been described only for the generation of the diverse repertoire of antibodies and T-cell receptors[6-8]. Several observations have suggested

从嗅觉神经元克隆小鼠

伊根等

编者按

截至 2004 年，包括绵羊和小鼠在内的多种哺乳动物已通过将供体 DNA 注射到去核卵细胞中被成功克隆。但在本文中，美国细胞生物学家鲁道夫·耶尼施及其同事使用从成熟嗅觉神经元中提取的供体 DNA 通过核移植获得可育的克隆小鼠。该研究表明，从已经停止分裂的特化细胞中提取的 DNA 可以重新进入细胞周期，并在核移植后重编程为胚胎样状态。克隆小鼠具有与正常小鼠无法区分的全系列有组织的气味受体基因，这表明在成熟的嗅觉神经元中选择特定的受体基因不会引起 DNA 的不可逆变化。

通过细胞核移植进行的克隆已成功地在包括小鼠在内的多种哺乳动物中进行。到目前为止，还没有报道称可以利用有丝分裂后的细胞（如神经元）克隆小鼠。在这里，我们通过将有丝分裂后的嗅觉神经元的细胞核转移到卵母细胞中，成功获得可育克隆小鼠。这些结果表明，有丝分裂后终末分化的神经元基因组可以重新进入细胞周期，并在核移植后重编程产生全能性。此外，克隆小鼠的气味受体基因表达模式和气味受体基因的组构与野生型动物没有显著区别，这表明嗅觉神经元 DNA 的不可逆变化不随受体基因的选择同时出现。

染色体是一种动态结构，经历复杂的变化，是生物发育和分化功能的基础。染色体结构的变化包括调节蛋白的结合，染色质蛋白或 DNA 中的共价修饰，以及在极少数情况下发生的 DNA 重排[1]。染色体变化的可逆程度可以通过涉及核移植的克隆实验确定。有丝分裂后的细胞不可逆地退出细胞周期，这是其分化程序的一部分。迄今为止，使用这种细胞的细胞核进行克隆实验，尚未获得存活的胚胎或小鼠[2,3]。这些观察结果表明，有丝分裂后的细胞可能难以进行表观遗传重编程，或者后天获得的 DNA 变化可能会限制其发育潜力[4,5]。

定向 DNA 重排很少作为正常分化过程的一部分被观察到，而在脊椎动物中，它们被描述成只发生在产生抗体和 T 细胞受体多样性过程中[6-8]。一些观察结果表

that post-mitotic neurons may also use irreversible DNA alterations to generate diversity. First, a population of neurons undergoes apoptosis in mice bearing mutations in the double-strand-break DNA-repair enzymes, which are also required for DNA rearrangements in lymphocytes[9,10]. Second, cortical neurons exhibit a higher incidence of aneuploidy than other cell types, although the functional significance of these changes is unknown[11,12]. These observations and the inability to clone mice from neuronal nuclei have led to models in which the DNA of post-mitotic neurons might undergo rearrangements to supply additional genetic diversity that may enhance neural function[4,5,13-15].

One particularly clear example of neuronal diversity is provided by the olfactory sensory epithelium. In the mouse, each of the 2,000,000 cells in the olfactory epithelium expresses only one of about 1,500 odorant receptor genes, such that the functional identity of a neuron is defined by the nature of the receptor it expresses[16]. Thus, the sensory epithelium consists of at least 1,500 neuronal types. The pattern of receptor expression is apparently random within one of four zones in the epithelium, suggesting that the choice of receptor gene may be stochastic[17,18]. One mechanism to permit the stochastic choice of a single receptor could involve DNA rearrangements[19,20].

Here we report the generation of fertile mouse clones by transferring the nuclei of post-mitotic olfactory neurons into enucleated oocytes. Thus, a post-mitotic nucleus can re-enter the cell cycle and be reprogrammed to totipotency. The DNA of mice derived from sensory neurons reveals no evidence for rearrangements of the expressed olfactory receptor gene. The pattern of receptor expression in these mice was indistinguishable from that of wild-type animals, indicating that irreversible changes in DNA do not accompany olfactory receptor gene choice.

Genetic Marking of Olfactory Sensory Neurons

Less than 1% of nuclear transfers result in the production of an embryonic stem (ES) cell line or live animal[1]. It is therefore difficult to identify with certainty the origin of the donor nucleus that contributed to the generation of a particular cloned animal[2]. This problem is particularly apparent in the olfactory epithelium, which contains mature sensory neurons intermingled with stem cells, neural progenitors and support cells. Therefore, we generated mice in which the endogenous olfactory marker protein (OMP) promoter drives simultaneous expression of OMP and Cre recombinase by inserting an internal ribosome entry site (IRES)-Cre cassette 3′ of the OMP stop codon. OMP is expressed only in mature olfactory sensory neurons (OSNs). When OMP–IRES-Cre mice are crossed to a reporter mouse strain (Z/EG), Cre expression catalyses the excision of a transcriptional stop sequence and results in green fluorescent protein (GFP) expression solely in mature OSNs (Fig. 1a and Methods). Transfer of marked OSN nuclei into oocytes should generate embryos in which every cell expresses GFP, as the Z/EG reporter uses the ubiquitous actin promoter.

明，有丝分裂后的神经元也可以利用不可逆的DNA变化来产生多样性。首先，一群神经元在带有双链断裂DNA修复酶突变的小鼠中经历细胞凋亡，该酶是淋巴细胞中DNA重排所必需的[9,10]。其次，尽管这些变化的功能意义尚不清楚，但皮质神经元表现出比其他细胞类型更高的非整倍性发生率[11,12]。这些观察结果和无法从神经元核中克隆小鼠的事实，促使新的模式被提出：有丝分裂后的神经元的DNA可经历重排以提供额外的遗传多样性来增强神经功能[4,5,13-15]。

嗅觉上皮提供了一个特别清楚的神经元多样性的例子。在小鼠嗅觉上皮中约有2,000,000个细胞，每一个细胞仅表达大约1,500个气味受体基因中的一种，因而神经元的功能由所表达受体的性质决定[16]。因此，感觉上皮由至少1,500种神经元类型组成。在上皮的四个区域中的某个区域内，受体表达的模式显然是随机的，这表明受体基因的选择可能是随机的[17,18]。一种允许随机选择单一受体的机制可能涉及DNA重排[19,20]。

在这里，我们报道通过将有丝分裂后的嗅觉神经元的细胞核移植到去核卵母细胞中产生的可育克隆小鼠。因此，有丝分裂后的细胞核可以重新进入细胞周期并被重编程和获得全能性。源自感觉神经元的小鼠DNA并未显示出其表达的嗅觉受体基因重排的证据。这些小鼠中受体表达的模式与野生型动物无法区分，表明DNA的不可逆变化不伴随嗅觉受体基因选择。

嗅觉神经元的遗传标记

只有不到1%的细胞核移植可以产生胚胎干细胞系或活体动物[1]。因此很难确定克隆动物来源于哪个供体核[2]。这个问题在嗅上皮中特别明显，嗅上皮含有成熟的感觉神经元，它们与干细胞、神经祖细胞和支持细胞混合在一起。因此，我们建立了一种小鼠模型，其中在内源性嗅觉标记蛋白（OMP）终止密码子的3′后端插入了内部核糖体进入位点（IRES）-Cre盒。这样OMP启动子同时驱动OMP和Cre重组酶的表达。OMP仅在成熟的嗅觉神经元（OSN）中表达。当OMP–IRES-Cre小鼠与报告小鼠品系（Z/EG）杂交时，Cre的表达催化了转录终止序列的切除，并导致仅在成熟OSN中表达绿色荧光蛋白（GFP）（图1a和方法）。将标记的OSN核转移到卵母细胞应该可以产生胚胎，并且因为Z/EG报告基因使用普遍存在的肌动蛋白启动子，胚胎中每个细胞都表达GFP。

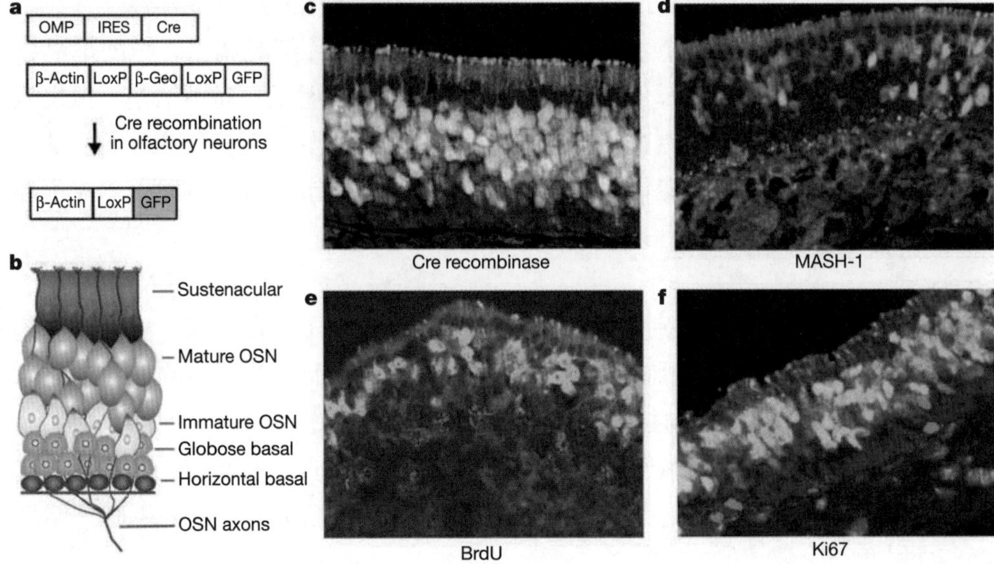

Fig. 1. Genetically marking post-mitotic OSNs. **a**, Donor animals carry one OMP–IRES-Cre allele and one copy of the Z/EG Cre reporter transgene. Cre expression in OSNs catalyses excision of the β-geo/stop cassette, resulting in selective GFP expression and genetic marking of mature OSNs. Thus, GFP^+ neurons could be selectively chosen for nuclear transfer. **b**, Schematic diagram of the approximate laminar distribution of cell types in the mouse olfactory epithelium. **c–f**, Three-colour immunofluorescence staining. Blue, nuclear marker TOTO-3; green, GFP protein fluorescence (**c**, **d**) or antibody to GFP protein (**e**, **f**); red, antibodies to Cre recombinase (**c**), progenitor-specific marker MASH-1 (**d**), markers of dividing cells, BrdU (**e**) and Ki67 (**f**).

In mice bearing the OMP–IRES-Cre allele and the Z/EG reporter gene, GFP expression was observed only in the most mature OSNs. Sections through the entire olfactory epithelium of several adult mice revealed GFP expression in the regions that contain mature OSNs (Fig. 1b–f). Cell counts revealed no GFP^+ cells in either the basal cell layers, where immature progenitors and stem cells reside, or in the apical support cell layer (0 of 835 GFP^+ cells)[21,22]. Double immunostaining for GFP and Cre protein revealed that all GFP^+ cells with visible nuclei also expressed Cre recombinase (599 of 599) (Fig. 1c). In addition, none of the $MASH\text{-}1^+$ basal cell precursors of OSNs expressed GFP (0 of 350 $MASH\text{-}1^+$ cells) (Fig. 1d). These results showed that GFP expression was restricted to mature sensory neurons.

We confirmed that the GFP^+ mature OSNs were post-mitotic by injecting mice with 5-bromodeoxyuridine (BrdU) and staining the olfactory epithelium with anti-BrdU and GFP antibodies (Fig. 1e). BrdU incorporation was observed in the more basal GFP-negative layers as well as in rare support cells in the apical layers. Examination of more than 3,000 GFP^+ cells never revealed a GFP^+ $BrdU^+$ cell. In a separate experiment, labelling with an antibody specific to Ki67, a protein restricted to the nuclei of dividing cells, revealed no staining coincident with GFP[23,24] (Fig. 1f). These data show that GFP^+ cells in the olfactory epithelium of donor animals are mature post-mitotic OSNs.

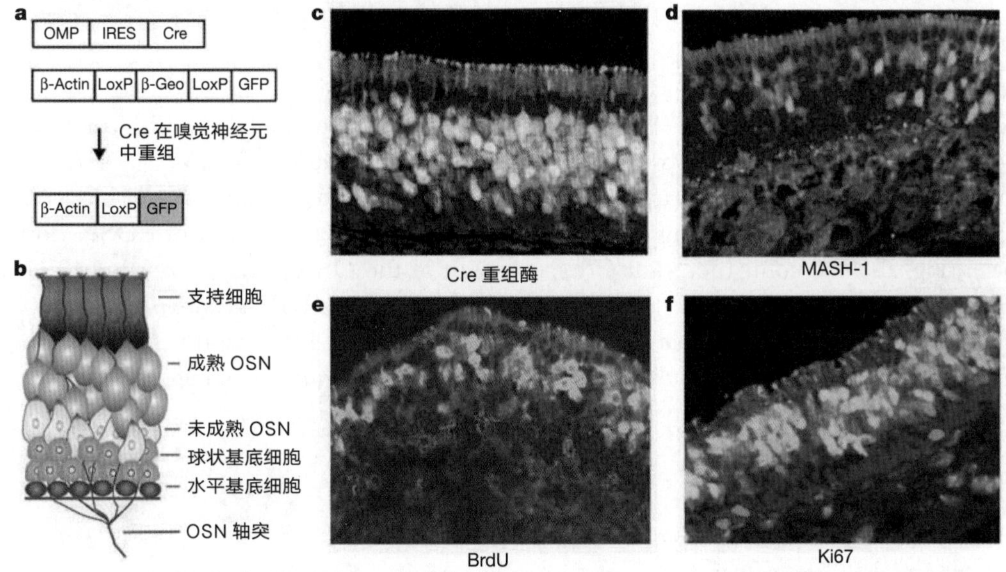

图 1. 有丝分裂后 OSN 的遗传标记。**a**，供体动物携带一个 OMP–IRES-Cre 等位基因和一个 Z/EG Cre 报道转基因拷贝。OSN 中 Cre 的表达催化 β-geo/终止盒的切除，导致选择性 GFP 表达和成熟 OSN 的遗传标记。因此，可以选择性地使用 GFP⁺ 神经元进行核转移。**b**，小鼠嗅上皮各细胞类型近似层状分布的示意图。**c** ~ **f**，三色免疫荧光染色。蓝色，核标记物 TOTO-3；绿色，GFP 蛋白荧光（**c**、**d**）或 GFP 蛋白抗体（**e**、**f**）；红色，Cre 重组酶的抗体（**c**），祖细胞特异性标记物 MASH-1（**d**），分裂细胞标记物 BrdU（**e**）和 Ki67（**f**）。

在携带 OMP–IRES-Cre 等位基因和 Z/EG 报告基因的小鼠中，仅在最成熟的 OSN 中观察到 GFP 表达。数只成年小鼠的整个嗅上皮切片显示，GFP 在含有成熟 OSN 的区域中表达（图 1b ~ 1f）。细胞计数显示，无论是存在未成熟祖细胞和干细胞的基底细胞层，抑或顶端支持细胞层中均没有 GFP^+ 细胞（835 个细胞中 GFP^+ 细胞有 0 个）[21,22]。对 GFP 和 Cre 蛋白的双重免疫染色显示，所有具有可见核的 GFP^+ 细胞均表达 Cre 重组酶（599/599）（图 1c）。此外，OSN 的 MASH-1^+ 基底细胞前体均未表达 GFP（350 个 MASH-1^+ 细胞中 GFP^+ 细胞有 0 个）（图 1d）。这些结果表明 GFP 表达仅限于成熟的感觉神经元。

我们通过向小鼠注射 5–溴脱氧尿苷（BrdU）并用抗 BrdU 抗体和 GFP 抗体染色嗅上皮，来证实 GFP^+ 的成熟 OSN 是经过有丝分裂的（图 1e）。在较多的 GFP 阴性基底细胞和较少的顶层支撑细胞中也观察到 BrdU 整合。检查 3,000 多个 GFP^+ 细胞，并未发现一个 GFP^+ $BrdU^+$ 细胞。在另一个实验中，用仅使分裂细胞核显色的 Ki67 特异性抗体标记细胞，显示没有 GFP^+ $Ki67^+$ 细胞（图 1f）[23,24]。这些数据显示供体动物嗅上皮中的 GFP^+ 细胞是成熟的有丝分裂后的 OSN。

Cloning Mice from Neuronal Nuclei

We asked whether the nucleus of a post-mitotic OSN could re-enter the cell cycle and direct preimplantation development. We dissociated the olfactory epithelium and picked GFP-expressing OSNs for nuclear transfer into enucleated oocytes (Fig. 2a)[25-29]. Of the 352 embryos generated, 48 (14%) developed to the blastocyst stage (Table 1). All blastocysts expressed GFP, demonstrating that they were derived from the mature OSN donor nuclei (Fig. 2b). We confirmed that Cre expression in the OSNs caused constitutive GFP expression after nuclear transfer by generating blastocysts with GFP-negative cells from the donor mice. As expected, none of these blastocysts expressed GFP, demonstrating that Cre is not aberrantly expressed as a result of nuclear transfer or subsequent cloning procedures.

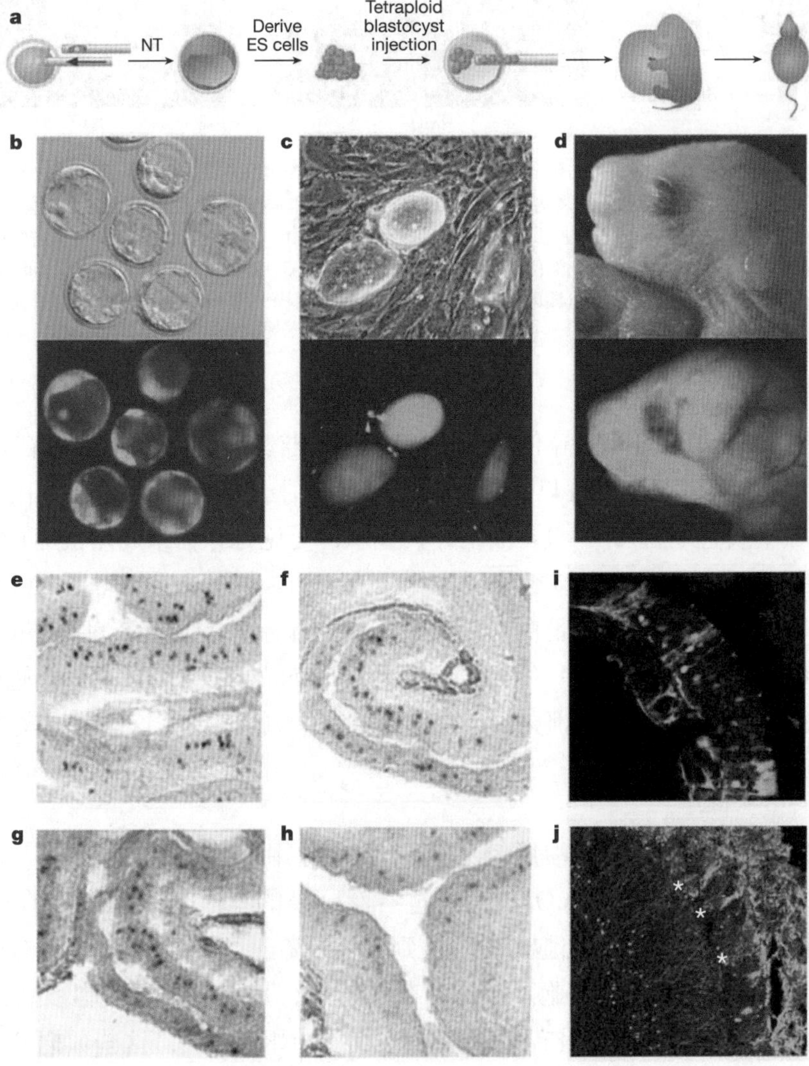

Fig. 2. Mice derived from OSN nuclei. **a**, Strategy to generate cloned ES cell lines and OSN-derived mice by tetraploid complementation with cloned ES cells. **b**, **c**, Bright-field and fluorescence images of nuclear

通过神经元细胞的核移植克隆小鼠

我们探究有丝分裂后 OSN 的细胞核是否可以重新进入细胞周期并指导胚胎植入前发育。我们裂解了嗅上皮组织，并选择了表达 GFP 的 OSN，将其核移植到去核卵母细胞中（图 2a）[25-29]。在产生的 352 个胚胎中，48 个（14%）发育到囊胚期（表 1）。所有囊胚均表达 GFP，证明它们来自成熟的 OSN 供体细胞核（图 2b）。我们通过对来自供体的 GFP 阴性细胞进行核移植产生囊胚验证了组成型 GFP 表达是由 OSN 中的 Cre 表达所致。正如预期的那样，这些囊胚中没有一个表达 GFP，这表明 Cre 不会因核移植或随后的克隆程序而异常表达。

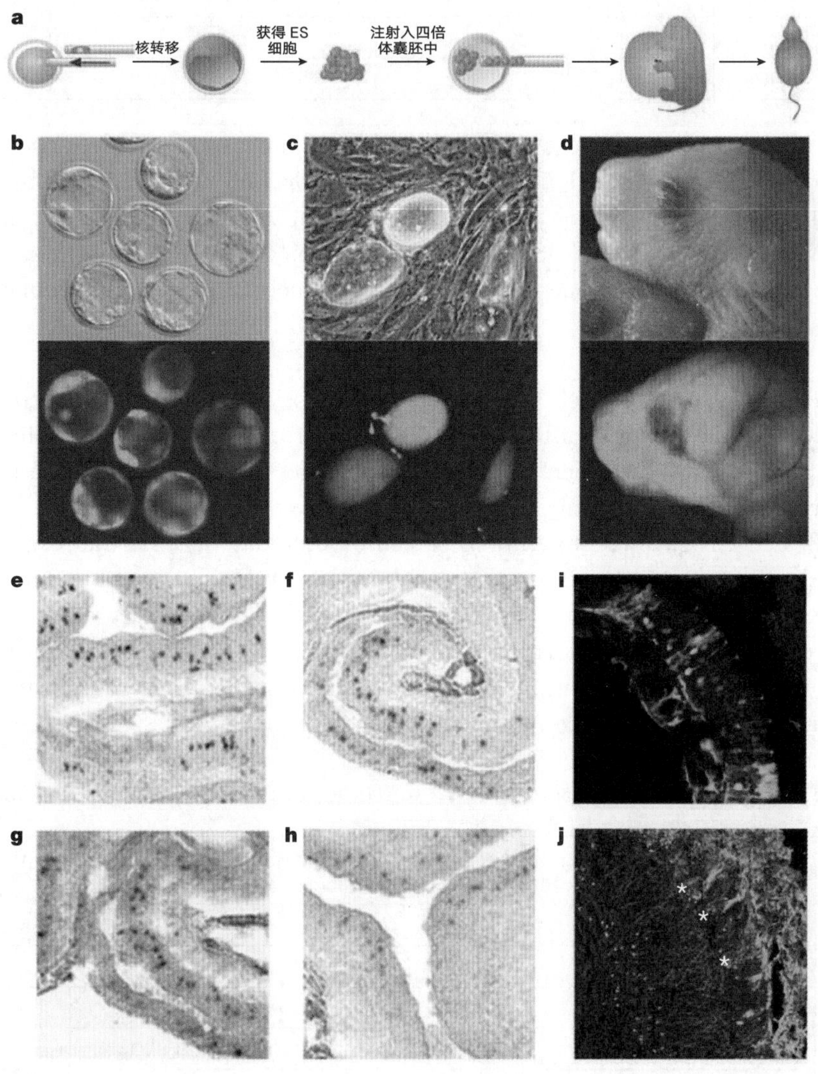

图 2. 源自 OSN 细胞核的小鼠。**a**，获得克隆 ES 细胞系和通过四倍体补偿法获得 OSN 衍生小鼠的方法。**b**、**c**，由 OSN 细胞核产生的核移植囊胚（**b**）和 OSN3 衍生 ES 细胞（**c**）的明场和荧光图像。**d**，通过

transfer blastocysts (**b**) and OSN3 ES cells (**c**) produced from OSN nuclei. **d**, A P0.5 mouse produced by tetraploid complementation with OSN3 ES cells (top) and a P0.5 C57/B6 control (bottom). **e–h**, *In situ* hybridization on the olfactory epithelium of mice wholly derived from OSN3 ES cells with probes for odorant receptors P2 (**e**), M50 (**f**), I7 (**g**) and M71 (**h**). **i, j**, Contribution of GFP^+ OSN2 ES cells to the olfactory epithelium (**i**) and olfactory bulb (**j**) of chimaeras generated by injecting OSN2 ES cells into diploid blastocysts. Several representative glomeruli are demarcated with asterisks.

Table 1. Production of ES cell lines and mice by nuclear transfer

Donor cells	Oocytes surviving (% injected)	Oocytes activated (% surviving)	Two-cell embryos to oviduct (% oocytes activated)	Blastocysts (% oocytes activated)	ES cell lines (% oocytes activated)	Alive at term (% into oviduct)
OSNs	508 (88)	352 (69)	—	48 (14)	3 (1)	—
P2 OSNs	261 (95)	141 (54)	—	18 (13)	2 (1)	—
P2SN1 ES	50 (95)	38 (76)	14 (37)	—	—	1 (7)
P2SN2 ES	47 (90)	37 (79)	18 (49)	—	—	1 (6)

The 48 GFP^+ blastocysts were used to generate ES cell lines and three gave rise to colonies that resembled ES cells (OSN1–3). All three lines expressed GFP, and Southern blotting confirmed the predicted genomic rearrangement at the Z/EG locus (Fig. 2c and data not shown). When the OSN2 and OSN3 ES cells were injected into diploid blastocysts, they contributed extensively to all tissues of the resulting chimaeras, including the germ line (Fig. 2i, j and data not shown).

We injected these cloned ES cells into tetraploid blastocysts to perform the most stringent test of their developmental potency (Table 2). Tetraploid embryo complementation generates an "ES-fetus" in which all embryonic lineages are derived from the injected ES cells, whereas the extraembryonic lineages develop from the host blastocyst[29,30]. Both OSN2 and OSN3 ES cells gave rise to embryos of embryonic day (E) 19.5 that expressed GFP ubiquitously (Table 2 and Fig. 2d). Mice derived from OSN3 survived to adulthood, were fertile and were overtly normal. Histological examination of serial sections through their brains revealed no obvious abnormalities (data not shown). These results indicate that the genome of a post-mitotic, terminally differentiated neuron can be reprogrammed after nuclear transfer and direct the development of all embryonic lineages.

Table 2. Mice produced from sensory neurons by tetraploid embryo complementation

ES cell line	Receptor expressed	Tetraploid blastocysts injected	Mice alive at term (% injected)	Breathing normally (% injected)	Cross-fostered	Survival to maturity (% fostered)
OSN1	?	137	0	0	0	0
OSN2	?	160	1 (1)	0	0	0
OSN3	?	273	49 (18)	29 (11)	21	17 (81)
P2SN1	P2	89	16 (18)	16 (100)	13	12 (92)
P2SN2	P2	151	31 (21)	28 (90)	28	22 (78)

OSN3 ES 细胞利用四倍体补偿法产生的 P0.5 小鼠（上图）和 P0.5 C57/B6 对照小鼠（下图）。**e**～**h**，用针对气味受体 P2（**e**）、M50（**f**）、I7（**g**）和 M71（**h**）的探针，在完全来自 OSN3 衍生 ES 细胞的小鼠嗅上皮进行原位杂交的结果。**i**、**j**，GFP$^+$ OSN2 衍生 ES 细胞对通过将其注射到二倍体囊胚中获得的嵌合体小鼠嗅上皮（**i**）和嗅球（**j**）的贡献。几个代表性的嗅小球用星号标出。

表 1. 通过核移植获得 ES 细胞系和小鼠的情况

供体细胞	存活卵母细胞数（% 占总注射数百分比）	激活卵母细胞数（% 占总存活数百分比）	进入输卵管的双细胞胚胎数（% 占总激活卵母细胞数百分比）	囊胚数（% 占总激活卵母细胞数百分比）	ES 细胞系数（% 占总数活卵母细胞数百分比）	实验过程中存活（% 占总进入输卵管胚胎数百分比）
OSNs	508 (88)	352 (69)	—	48 (14)	3 (1)	—
P2 OSNs	261 (95)	141 (54)	—	18 (13)	2 (1)	—
P2SN1 ES	50 (95)	38 (76)	14 (37)	—	—	1 (7)
P2SN2 ES	47 (90)	37 (79)	18 (49)	—	—	1 (6)

48 个 GFP$^+$ 囊胚用于获得 ES 细胞系，其中 3 个产生类似 ES 细胞的集落（OSN1～OSN3）。所有三个品系都表达 GFP，Southern 印迹证实了之前预测的在 Z/EG 基因座处发生的基因组重排（图 2c，未显示数据）。当将 OSN2 和 OSN3 ES 细胞注射到二倍体囊胚中时，它们对所得嵌合体的所有组织（包括生殖细胞系）产生了广泛嵌合（图 2i、2j，未显示数据）。

我们将这些克隆的 ES 细胞注射到四倍体囊胚中，以对其发育潜力进行最严格的测试（表 2）。四倍体胚胎补偿法获得“ES 胎儿”，其中所有胚胎细胞谱系来自注射的 ES 细胞，而胚外细胞谱系来自宿主囊胚 [29,30]。OSN2 和 OSN3 的 ES 细胞均获得可长至 19.5 天的胚胎，其细胞普遍表达 GFP（表 2 和图 2d）。源自 OSN3 的小鼠存活至成年，可育并且明显正常。大脑连续切片的组织学检查显示没有明显的异常（数据未显示）。这些结果表明，有丝分裂后终末分化的神经元的基因组可以在核移植后重编程，并指导所有胚胎细胞谱系的发育。

表 2. 感觉神经元通过四倍胚胎互补法获得小鼠的情况

ES 细胞系	受体表达	四倍体囊胚注射数	实验过程中小鼠存活数（% 占总注射数百分比）	正常呼吸的小鼠数（% 占总注射数百分比）	交叉抚育小鼠数	存活至性成熟小鼠数（% 占交叉抚育百分比）
OSN1	?	137	0	0	0	0
OSN2	?	160	1 (1)	0	0	0
OSN3	?	273	49 (18)	29 (11)	21	17 (81)
P2SN1	P2	89	16 (18)	16 (100)	13	12 (92)
P2SN2	P2	151	31 (21)	28 (90)	28	22 (78)

Odorant Receptor Expression in Cloned Mice

We examined the patterns of receptor gene expression in the olfactory epithelium of the cloned mice generated by tetraploid complementation. If irreversible genetic rearrangements are required for receptor choice we might expect an altered profile of receptor expression in animals cloned from an OSN. In the simplest model, a rearrangement involving one receptor gene might persist in all neurons such that all neurons will express the same receptor. This scenario was observed in mice derived from a B-cell nucleus, in which all B cells expressed the same immunoglobulin gene[28]. We therefore asked whether sensory neurons from cloned mice express a single receptor or a repertoire of receptor genes. We performed *in situ* hybridization on the olfactory epithelium of mice derived from OSN3 ES cells using probes specific for seven different odorant receptors (Fig. 2e–h and data not shown). The pattern of expression of all seven receptors was indistinguishable from control mice, excluding a simple model of gene rearrangement that would result in the expression of a single receptor gene in all OSNs.

As a neuron expresses a receptor from only one of the two alleles, it remained possible that a rearranged receptor gene expressed in the donor nucleus would be expressed in only half of the sensory neurons. We therefore analysed the repertoire of receptors expressed by polymerase chain reaction with reverse transcription (RT–PCR) to determine whether a single receptor transcript was enriched in the epithelium of OSN-derived mice. RNA from the olfactory epithelium of cloned animals was used in RT–PCR reactions with degenerate primers that recognize conserved motifs present in the majority of odorant receptors. Forty-four PCR products were cloned and restriction digest analysis indicated that they encoded 38 different receptors. Sequence analysis of 20 of these PCR clones revealed that they encoded 20 different receptors, which were located in seven clusters on six different chromosomes. This suggests that a single odorant receptor did not predominate in the olfactory epithelium of mice derived from OSN nuclei.

One additional assay for the diversity of odorant receptor expression is based on the observation that neurons expressing a given receptor, although randomly distributed within a zone of the epithelium, converge on two spatially invariant loci or glomeruli in the olfactory bulb[18,31]. If half of the OSNs from cloned mice expressed the same receptor, then their axonal projections should preferentially innervate a small set of glomeruli. Analysis of the olfactory bulb of chimaeric mice produced with OSN2 ES cells revealed that the OSN2-derived GFP^+ cells innervated most, if not all glomeruli, with no glomerulus receiving predominant input (Fig. 2i, j). These results show that the sensory neurons of mice cloned from an OSN that had expressed a single receptor can express a large repertoire of odorant receptor genes.

克隆小鼠中的气味受体表达

我们检查了由四倍体补偿法获得的克隆小鼠的嗅上皮中受体基因表达的模式。如果受体选择需要不可逆的基因重排，我们可能会从 OSN 克隆的动物中发现受体表达的改变。在最简单的模型中，涉及一个受体基因的重排可能在所有神经元中持续存在，使得所有神经元都表达相同的受体。在源自 B 细胞核的小鼠中观察到这种情况；在这些小鼠中，所有 B 细胞表达相同的免疫球蛋白基因 [28]。因此，我们想知道克隆小鼠的感觉神经元是表达单一受体还是表达受体基因库。我们使用对七种不同气味受体特异的探针对源自 OSN3 的 ES 细胞的小鼠的嗅上皮进行原位杂交（图 2e ~ 2h，未显示数据）。所有七种受体的表达模式与对照小鼠无法区分，而基因重排的简单模型将导致所有 OSN 中表达单一受体基因，因此排除了这种简单模型。

由于神经元仅从两个等位基因中选择一个表达受体，因此在供体细胞核中表达的重排受体基因仍然可能仅在一半感觉神经元中表达。因此，我们通过逆转录聚合酶链反应（RT–PCR）分析了表达的受体库，以确定单个受体转录物是否富集在 OSN 衍生小鼠的上皮细胞中。来自克隆动物嗅上皮的 RNA 用于 RT–PCR 反应，其中简并引物识别大多数气味受体中存在的保守基序。克隆得到 44 种 PCR 产物，限制性消化分析表明它们编码了 38 种不同的受体。对这些 PCR 克隆中的 20 个进行序列分析，发现它们编码了 20 种不同的受体，这些受体位于 6 个不同染色体上的 7 个簇中。这表明源自 OSN 核的小鼠的嗅上皮，并非只表达单一气味受体。

气味受体表达多样性的另一种测定是基于以下观察：表达给定受体的神经元虽然随机分布在上皮特定区域内，但会富集在嗅球中两个空间位置不变的基因座或嗅小球上 [18,31]。如果来自克隆小鼠的 OSN 中有一半表达相同的受体，那么它们的轴突投射应该优先支配一小组嗅小球。对 OSN2 的 ES 细胞衍生的嵌合小鼠的嗅球分析显示，OSN2 衍生的 GFP^+ 细胞支配绝大多数嗅小球，没有嗅小球接受（单一受体对应的）优先支配（图 2i、2j）。这些结果表明，由表达单一受体的 OSN 克隆获得的小鼠，其感觉神经元可以表达大量的气味受体基因。

Mice Cloned from Neurons Expressing the P2 Receptor

The previous experiments suggest that irreversible rearrangements are not required for receptor gene expression. Because these experiments did not allow the prospective identification of the receptor gene expressed by the donor OSN nucleus, we were unable to examine its pattern of expression or its DNA sequence in the cloned mice. Neurons that express the P2 odorant receptor were marked by introducing an IRES directing the translation of GFP into the 3′ untranslated region of the P2 gene (P2–IRES-GFP)[32]. These mice were crossed with strains carrying both the OMP–IRES-Cre alteration and a reporter allele in which the *Rosa26* promoter is separated from a weak GFP gene by a LoxP-flanked transcriptional terminator[33] (Fig. 3a). Mice carrying the P2–IRES-GFP, OMP–IRES-Cre and *Rosa26*–LoxP-Stop-LoxP–weak GFP reporter genes exhibited intense GFP fluorescence in the approximately 0.1% of the OSNs that expressed the P2–IRES-GFP allele, whereas all other neurons appeared dark (Fig. 3b). *Rosa26*-driven GFP was present in mature OSNs, but the GFP signal was too weak to be detected by direct fluorescence. No green cells were observed in animals that contained only the OMP–IRES-Cre and *Rosa26*–LoxP-Stop-LoxP–weak GFP genes (Fig. 3c). Evidence for Cre-mediated excision of the transcription terminator and the resulting expression of weak GFP in mature OSNs was detected by amplifying the signal with GFP antibodies (data not shown).

Fluorescent cells expressing the P2–IRES-GFP allele were picked and their nuclei were injected into enucleated oocytes (Fig. 3d). Eighteen blastocysts developed from 141 reconstructed oocytes generating two ES cell lines: P2SN1 and P2SN2 (Table 1). PCR analysis revealed Cre-mediated excision at the *Rosa26* locus in both cell lines, indicating their origin from cloned nuclei of post-mitotic OSNs (Supplementary Fig. 1). Tetraploid embryos injected with P2SN1 and P2SN2 cells resulted in the birth of multiple viable pups that survived to adulthood (Table 2). These mice exhibited no gross anatomic or behavioural abnormalities and were fertile. PCR analyses of non-neuronal tissues revealed recombination at the *Rosa26* reporter allele, indicating that the cloned mice were derived from mature OMP^+ OSNs (data not shown). In addition, we detected the weak *Rosa26*-driven GFP expression by *in situ* hybridization in both the olfactory epithelium and non-neuronal tissues, demonstrating genetic activation of the reporter (Fig. 3g, h and data not shown).

The expression pattern of P2–IRES-GFP in clones was indistinguishable from that of control donor animals. Approximately 0.1% of the neurons expressed GFP at a high level within zone II of the epithelium (Fig. 3e). The GFP-expressing neurons projected axons to one medial and one lateral glomerulus in the olfactory bulb (Fig. 3f). Moreover, the P2–IRES-GFP allele was expressed at a frequency approximately equal to that of the unmodified P2 allele. *In situ* hybridization was performed on sections through the olfactory epithelium with RNA probes specific to GFP- and P2-coding sequences (Fig. 3g–j). The number of GFP-expressing cells in one section was roughly 50% of the number of P2-expressing cells found in neighbouring sections ($45 \pm 14\%$, s.d.), revealing no preference

由表达 P2 受体的神经元克隆小鼠

先前的实验表明，受体基因表达并不需要不可逆的重排。因为这些实验不允许对供体 OSN 细胞核表达的受体基因进行前瞻性鉴定，所以我们无法检查克隆小鼠中的受体表达模式或其 DNA 序列。表达 P2 气味受体的神经元通过在 P2 基因的 3′ 非翻译区插入 IRES 来标记，该 IRES 可以指导 GFP 翻译（P2–IRES-GFP）[32]。将这些小鼠与同时携带 OMP–IRES-Cre 变异并报告等位基因的小鼠杂交，该报告基因中 *Rosa26* 启动子和弱 GFP 基因通过 LoxP 中间的转录终止子分开 [33]（图 3a）。携带 P2–IRES-GFP，OMP–IRES-Cre 和 *Rosa26*–LoxP-Stop-LoxP– 弱 GFP 报告基因的小鼠在表达 P2–IRES-GFP 等位基因的 OSN 中大约有 0.1%表现出强烈的 GFP 荧光，而其他所有神经元都很暗不发荧光（图 3b）。*Rosa26* 驱动的 GFP 存在于成熟的 OSN 中，但 GFP 信号太弱而不能通过直接荧光检测到。在仅含有 OMP–IRES-Cre 和 *Rosa26*–LoxP-Stop-LoxP– 弱 GFP 基因的动物中未观察到绿色细胞（图 3c）。用 GFP 抗体扩增信号来检测 Cre 介导的转录终止子切除，及其导致的成熟 OSN 中弱 GFP 表达（数据未显示）。

挑选表达 P2–IRES-GFP 等位基因的荧光细胞，并将它们的细胞核注射到去核卵母细胞中（图 3d）。从 141 个重建的卵母细胞发育出 18 个囊胚，产生了两个 ES 细胞系：P2SN1 和 P2SN2（表 1）。PCR 分析显示两种细胞系中均出现 Cre 介导的 *Rosa26* 基因座处的切除，表明它们来自有丝分裂后 OSN 的克隆核（补充图 1）。注射 P2SN1 和 P2SN2 细胞到四倍体囊胚获得存活至成年的多个活幼崽（表 2）。这些小鼠没有表现出严重的解剖学异常或行为异常，且可育。非神经元组织的 PCR 分析揭示了 *Rosa26* 报告等位基因的重组，表明克隆的小鼠源自成熟的 OMP^+ OSN（数据未显示）。此外，我们通过嗅上皮和非神经组织中的原位杂交检测到弱 *Rosa26* 驱动的 GFP 表达，证明了报告基因的遗传激活（图 3g、3h，未显示数据）。

克隆中 P2–IRES-GFP 的表达模式与对照供体动物的表达模式不可区分。大约 0.1%的神经元在上皮区 II 内以高水平表达 GFP（图 3e）。表达 GFP 的神经元将轴突投射到嗅球中的一个内侧嗅小球和一个旁侧嗅小球（图 3f）。此外，P2–IRES-GFP 等位基因的表达频率大约等于未修饰的 P2 等位基因的表达频率。使用对 GFP 和 P2 编码序列特异的 RNA 探针对嗅上皮切片进行原位杂交（图 3g ~ 3j）。在一个切片中表达 GFP 的细胞数量大约是在相邻切片中发现的表达 P2 的细胞数量的 50%（45% ± 14%，标准差），表明 P2–IRES-GFP 等位基因并不优先表达。此外，用对另

for the expression of the P2–IRES-GFP allele. Furthermore, *in situ* hybridization with probes specific to six additional olfactory receptors revealed similar expression patterns in control and cloned mice (Fig. 3k, l and data not shown).

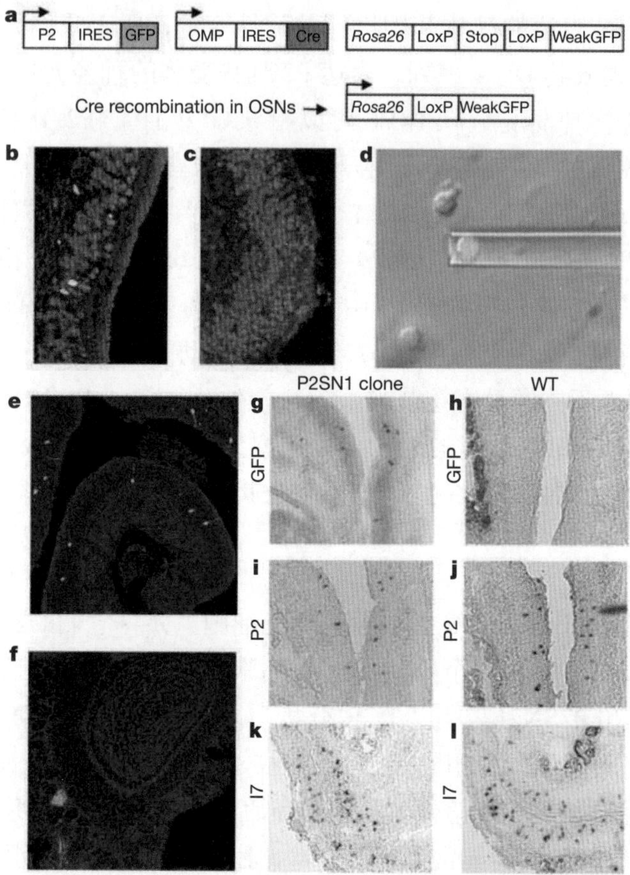

Fig. 3. Mice produced from OSNs expressing the P2 odorant receptor. **a**, Strategy to label P2 sensory neurons with GFP and genetically mark mature OSNs. **b**, **c**, Immunofluorescence staining with nuclear marker TOTO-3 (blue), GFP fluorescence (green) and antibodies to Cre recombinase (red) of the olfactory epithelium of donor animals heterozygous for the P2–IRES-GFP, OMP–IRES-Cre and *Rosa26*–LoxP-Stop-LoxP–weak GFP alleles (**b**) or of control animals heterozygous for the OMP–IRES-Cre and *Rosa26*–LoxP-Stop-LoxP–weak GFP alleles (**c**). **d**, Bright-field and fluorescence merged image of picking a P2–IRES-GFP$^+$ neuron from the dissociated olfactory epithelium of a donor for nuclear transfer. **e**, **f**, GFP fluorescence (green) and TOTO-3$^+$ nuclei (blue) visualized in sections of the olfactory epithelium (**e**) and olfactory bulb (**f**) of an animal generated by tetraploid complementation with P2SN1 ES cells. **g**–**l**, *In situ* hybridization on olfactory epithelium sections derived entirely from P2SN2 ES cells (**g**, **i**, **k**) and a wild-type control (**h**, **j**, **l**) with probes for GFP (**g**, **h**), P2 receptor (**i**, **j**) and I7 receptor (**k**, **l**).

We also performed Southern blotting, PCR and genome sequencing to examine the organization of the P2–IRES-GFP allele in the chromosome of cloned mice in an effort to detect potential DNA rearrangement events. If choice of a single receptor gene involved gene conversion into a single active locus, cells expressing the P2–IRES-GFP allele might contain a second copy of this allele at the active locus. Southern blotting to distinguish

外六种气味受体特异的探针进行的原位杂交揭示了对照和克隆小鼠中相似的受体表达模式(图 3k、3l，未显示数据)。

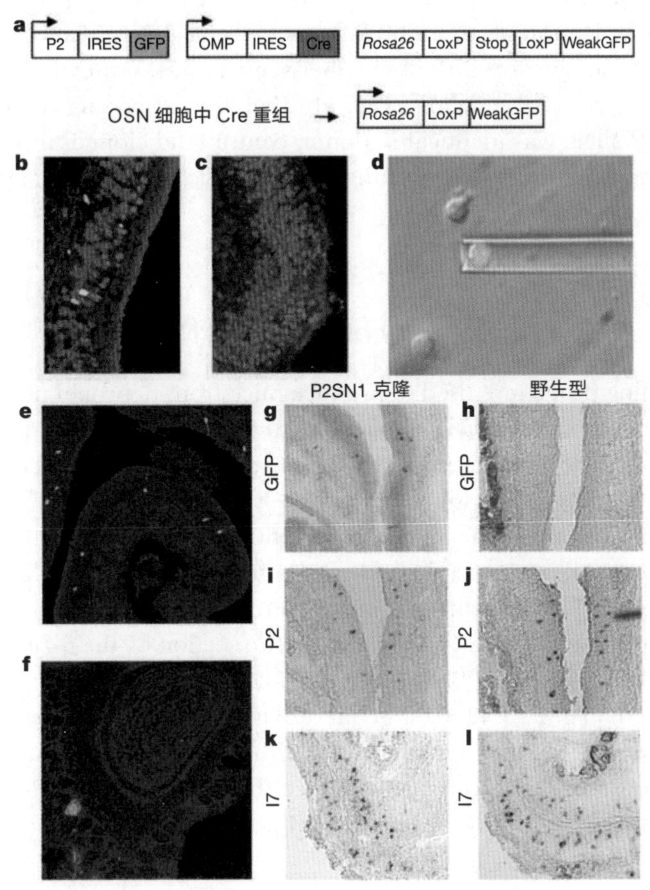

图 3. 由表达 P2 气味受体的 OSN 获得的小鼠。**a**，用 GFP 标记 P2 感觉神经元和遗传标记成熟 OSN 的方法。**b**、**c**，用核标记物 TOTO-3(蓝色)、GFP 荧光(绿色)和 Cre 重组酶抗体(红色)对嗅上皮进行免疫荧光染色。嗅上皮分别来自携带 P2–IRES-GFP、OMP–IRES-Cre 和 *Rosa26*–LoxP-Stop-LoxP– 弱 GFP 等位基因的杂合供体动物(**b**)和携带 OMP–IRES-Cre 和 *Rosa26*–LoxP-Stop-LoxP– 弱 GFP 等位基因的杂合对照动物(**c**)。**d**，从供体裂解的嗅上皮细胞中挑选 P2–IRES-GFP$^+$ 神经元进行核移植的明场和荧光合并图像。**e**、**f**，GFP 荧光(绿色)和 TOTO-3$^+$ 细胞核(蓝色)在通过 P2SN1 ES 细胞与四倍体囊胚补偿得到的动物嗅上皮(**e**)和嗅球(**f**)切片中可视化。**g**～**l**，对完全来源于 P2SN2 ES 细胞的嗅上皮切片(**g**、**i**、**k**)和野生型对照切片(**h**、**j**、**l**)进行原位杂交，探针为 GFP(**g**、**h**)、P2 受体(**i**、**j**)和 I7 受体(**k**、**l**)。

我们还进行了 Southern 印迹、PCR 和基因组测序，以检查克隆小鼠染色体中 P2–IRES-GFP 等位基因的组构，来检测潜在的 DNA 重排事件。如果单个受体基因的选择涉及基因转变成单个活性基因座，则表达 P2–IRES-GFP 等位基因的细胞可能在活性基因座处含有该等位基因的第二个拷贝。用于区分修饰的克隆小鼠 P2–IRES-

the modified P2–IRES-GFP allele from the unmodified endogenous P2 allele in DNA from control and cloned mice revealed only two bands of equal intensity, suggesting that gene conversion into an active locus does not accompany olfactory receptor gene choice (Supplementary Fig. S2). In addition, Southern blot analyses using multiple DNA probes to examine the organization of about 60 kilobases (kb) of DNA 5′ and 3′ of the P2 coding sequence failed to detect any differences between donor, control and cloned animals (Supplementary Fig. S3). Sequencing of 10 kb 3′ of the P2 coding sequence showed that the P2–IRES-GFP allele was identical in donor, control and cloned mice (data not shown). These results suggest that irreversible changes in DNA do not accompany choice of the P2 odorant receptor gene.

Totipotency of Neuronal Nuclei

The two-step cloning procedure used to produce mice from neuronal nuclei generates mice in which the neuronally derived ES cells give rise to all embryonic tissues, whereas cells from the tetraploid host blastocyst contribute to the embryonic trophectoderm[30]. Thus, neither this work nor the cloning of lymphocytes via an ES cell intermediate[28] demonstrated the totipotency of a nucleus from a terminally differentiated cell[34]. To demonstrate totipotency of mature OSN nuclei, we transplanted nuclei from P2SN1 and P2SN2 ES cells into enucleated oocytes[29]. The resulting embryos were cultured for 24 h and transferred to pseudopregnant recipients (Table 1). Upon caesarean section of the recipients, we recovered full-term pups from both the P2SN1 and P2SN2 cell lines. These pups had enlarged placentas (P2SN1, 0.35 g; P2SN2, 0.40 g) but displayed no overt anatomical or behavioural abnormalities, were fertile and survived to adulthood, consistent with previous cloning experiments[29]. These observations demonstrate that nuclei of terminally differentiated olfactory neurons can be reprogrammed to totipotency, directing development of both embryonic and extraembryonic lineages.

Discussion

We have asked whether the nucleus of a post-mitotic olfactory sensory neuron can re-enter the cell cycle and undergo reprogramming to direct development of a mouse. ES cell lines were generated from OSN nuclei at frequencies similar to those obtained with differentiated lymphoid cells that can be induced to proliferate under physiological conditions[2]. Thus the mechanisms that lead to the cell-cycle exit and irreversible mitotic arrest that accompany neural differentiation do not result from irreversible epigenetic or genetic events that would interfere with nuclear totipotency.

The differentiation of neurons requires that neural progenitors exit the cell cycle before a restriction point late in G1. This decision is governed by a complex balance of cell-cycle regulators and proneural genes that drive cells into G0 and prevent them from progressing beyond the restriction point[35]. Although most cells that enter G0 can re-enter the cycle on

GFP 等位基因与来自对照小鼠 DNA 中未修饰的内源性 P2 等位基因的 Southern 印迹仅显示两条相等强度的条带，表明基因转变为活性基因座不伴随气味受体基因的选择(补充图 S2)。此外，使用多个 DNA 探针对 P2 编码序列 5′和 3′端约 60 千碱基(kb)的 DNA 进行 Southern 印迹分析，未能检测到供体、对照和克隆动物之间的任何差异(补充图 S3)。对 P2 编码序列 3′端的 10 kb 进行测序，显示 P2–IRES-GFP 等位基因在供体、对照和克隆小鼠中是相同的(数据未显示)。这些结果表明 DNA 的不可逆变化不伴随 P2 气味受体基因的选择。

神经元细胞核的全能性

使用两步克隆法从神经元细胞核获得的小鼠中，神经元衍生的 ES 细胞发育为所有胚胎组织，而来自四倍体宿主囊胚的细胞发育成滋养外胚层[30]。因此，这一工作和通过 ES 细胞中间体克隆淋巴细胞[28]都不能证明来自终末分化细胞的细胞核具有全能性[34]。为了证明成熟 OSN 细胞核的全能性，我们将 P2SN1 和 P2SN2 衍生 ES 细胞的细胞核移植到去核卵母细胞中[29]。将得到的胚胎培养 24 小时并转移到假孕雌鼠中(表 1)。在对雌鼠进行剖宫产后，我们获得了来源于 P2SN1 和 P2SN2 细胞系的足月幼崽。这些幼仔具有大于常态的胎盘(P2SN1，0.35 g；P2SN2，0.40 g)，但没有显示出明显的解剖学异常或行为异常，具有生育能力并存活至成年期，与先前的克隆实验结果一致[29]。这些观察结果表明，终末分化的嗅觉神经元的细胞核可以重编程恢复全能性，指导胚胎和胚胎外细胞系的发育。

讨　论

我们提出疑问，有丝分裂后嗅觉神经元的细胞核是否可以重新进入细胞周期并进行重编程以指导小鼠的发育。ES 细胞系由 OSN 细胞核产生，该过程的频率与用分化的淋巴细胞获得 ES 细胞系的频率相似，后者可在生理条件下可诱导增殖[2]。因此，伴随着神经分化，导致细胞周期退出和不可逆有丝分裂停滞的机制不是由不可逆的表观遗传或遗传事件引起的，尽管这些事件会干扰核的全能性。

神经元的分化需要神经祖细胞在 G1 晚期的限制点之前退出细胞周期。这一行为受细胞周期调节因子和原神经基因的复杂平衡调控，原神经基因使细胞进入 G0 并阻止它们越过限制点[35]。虽然大多数进入 G0 的细胞可以在适当的刺激下重新进

appropriate stimulation, neurons normally undergo an irreversible mitotic arrest. Whatever mechanisms keep neurons in a post-mitotic state, our experiments demonstrate that they can be overcome in the environment of the egg.

The nervous system contains a diverse array of neural cell types and this diversity is reflected by distinct patterns of gene expression in different neurons. The regulation of gene expression by DNA rearrangements is rare but this mechanism has nonetheless been suggested to explain the diversity inherent in complex nervous systems[36]. DNA recombination events provide *Saccharomyces cerevisiae*, trypanosomes and lymphocytes with a mechanism to stochastically express one member of a set of genes that modulate cellular interactions with the environment. One attractive feature shared by gene rearrangements in trypanosomes and lymphocytes is that gene choice by recombination is a random event. Cells that undergo correct or successful rearrangements are then afforded a selective survival advantage. The stochastic rearrangement of one gene from a gene family and subsequent selection could also provide a mechanism to generate the vast diversity of neuronal cell types. In this manner, neurons with subtly different genotypes would exhibit the array of neuronal phenotypes required for a functioning nervous system.

The olfactory sensory epithelium provides a clear example of neuronal diversity, and it has been suggested that this diversity is generated by stochastic DNA rearrangement events[19,20]. However, efforts to examine the DNA of OSNs expressing a given receptor have been seriously hampered by the inability to obtain homogeneous populations of neurons or clonal cell lines in which each cell expresses the same receptor. We have addressed this problem by generating cloned ES cell lines and mice derived from the nuclei of olfactory sensory neurons expressing the P2 receptor. Analyses of the sequence and organization of the DNA surrounding the expressed P2 allele from cloned ES cells and mice revealed no evidence for either gene conversion or local transpositions at the P2 locus. These results concur with fluorescence *in situ* hybridization studies showing that gene conversion into an active locus is an unlikely mechanism for odorant receptor expression[37]. In addition, the pattern of receptor gene expression in the sensory epithelium of cloned mice was wild type; multiple odorant receptor genes are expressed without preference for the P2 allele expressed in the donor nucleus.

These data demonstrate that the mechanism responsible for the choice of a single odorant receptor gene does not involve irreversible changes in DNA. More dynamic, reversible recombination events might accompany odorant receptor gene choice, but this is unlikely given the current data. Our results, in combination with experiments showing that the expression of various odorant receptor transgenes is influenced by proximal sequence elements, are consistent with epigenetic models of odorant receptor choice[37-42]. In a broader context, the generation of fertile cloned mice that are anatomically and behaviourally indistinguishable from wild-type mice indicates that olfactory sensory neurons do not undergo other irreversible DNA rearrangements that would interfere with either the development or function of the nervous system during adult life.

入细胞周期，但神经元通常会发生不可逆的有丝分裂停滞。无论使神经元保持在有丝分裂后的状态的机制是什么，我们的实验证明神经元可以在卵细胞的环境中重新进入细胞周期。

神经系统包含多种神经细胞类型，这种多样性通过不同神经元中不同的基因表达模式反映出来。通过 DNA 重排调节基因表达是罕见的，但这种机制仍被建议用于解释复杂神经系统固有的多样性 [36]。DNA 重组能够使酿酒酵母、锥虫和淋巴细胞随机表达调节细胞与环境相互作用的一组基因中的一个。锥虫和淋巴细胞中基因重排共有一个明显特征，即通过重组导致的基因选择是随机事件。之后，经历正确或成功重排的细胞具有选择性存活优势。来自基因家族的一个基因的随机重排和随后的选择也可以提供产生大量多样的神经元细胞类型的机制。以这种方式，具有微妙不同基因型的神经元将表现出功能性神经系统所需的多种神经元表型。

嗅觉上皮提供了一个神经元多样性的明显例子，并且已经表明这种多样性是由随机 DNA 重排事件造成的 [19,20]。然而，由于无法获得其中每个细胞表达相同受体的均质神经元群或克隆细胞系，所以一直难以对表达给定受体的 OSN 的 DNA 进行检测。我们通过用表达 P2 受体的嗅觉神经元细胞核获得克隆的 ES 细胞系和小鼠来解决这个问题。对来自克隆的 ES 细胞和小鼠的表达的 P2 等位基因周围的 DNA 的序列和组构的分析显示，在 P2 基因座处没有发生基因转变或局部转座。这些结果与荧光原位杂交研究一致，表明基因转变为活性基因座不太可能是气味受体表达的机制 [37]。此外，克隆小鼠感觉上皮细胞中受体基因表达的模式为野生型；表达多种气味受体基因而并不优先表达在供体细胞核中表达的 P2 等位基因。

这些数据表明，负责选择单一气味受体基因的机制不涉及 DNA 的不可逆变化。更活跃的、可逆的重组事件可能伴随气味受体基因选择，但鉴于目前的数据，这也不太可能。我们的结果与表明各种气味受体转基因的表达受近端序列元件影响的实验相结合，与气味受体选择的表观遗传模型一致 [37-42]。在更广泛的背景下，能够产生在解剖学上和行为上与野生型小鼠无显著区别的可育克隆小鼠表明，嗅觉神经元不经历其他干扰神经系统发育或者成年期功能的不可逆的 DNA 重排。

Methods

Preparation of donor neurons

Olfactory epithelia were dissected in L15 medium (Gibco) at 4 °C, then chopped into small pieces and incubated with type IV collagenase at 1 μg ml^{-1} at 37 °C for 15 min with occasional vigorous shaking. The collagenase digestion was stopped by addition of DMEM plus 10% FBS. Cells were spun down and resuspended in trituration medium (PBS plus 30% glucose plus 10% FBS plus penicillin/streptomycin) and triturated with several widths of pipette tips to produce single cell suspensions. Cells were pelleted and resuspended in L15/10% FBS before picking.

ES cell lines and tetraploid embryo complementation

Production of ES cell lines from nuclear transfer embryos was exactly as described[28], and generation of mice by tetraploid embryo complementation was exactly as described[29].

Generation of cloned embryos and mice

Production of cloned embryos by nuclear transfer was essentially as described[25,29] except that only GFP^+ cells from the two donor populations, as identified by epifluorescence, were picked and used for nuclear transfer.

In situ hybridization and immunohistochemistry

Animals were killed by approved methods and tissues of interest were either fresh frozen in OCT or fixed for 2–12 h in 4% PFA/1 × PBS, then washed and equilibrated in 30% sucrose before freezing. Frozen sections (from 20–30 μm) were placed on slides and standard immunohistochemistry and digoxigenin-labelled probe *in situ* hybridization protocols were used[32]. We used the following antibodies: rabbit anti-GFP antibody (Molecular Probes), rabbit anti-Cre recombinase antibody (Novagen), goat anti-Ki67 antibody (Santa Cruz Biotechnology), rat anti-BrdU antibody (abcam) and rabbit anti-MASH-1 antibody (a gift of J. Johnson).

BrdU labelling and visualization

Mice were injected intraperitoneally with a 5 mg ml^{-1} solution of BrdU in PBS every 2 h for 12 h. Each mouse received BrdU at 100 μg g^{-1} body weight for each injection. Two hours after the last injection mice were killed and tissues were fixed for 2 h in 4% PFA and PBS, washed and sucrose protected, and frozen. To permit simultaneous visualization of BrdU and GFP, sections were first stained with anti-GFP antibody then fixed in 4% PFA/PBS for 15 min, then washed, treated with

方　法

供体神经元的制备

将嗅上皮于 4℃ L15 培养基（Gibco）中裂解，然后切成小块并与 1μg/ml 的 IV 型胶原酶在 37℃下孵育 15 分钟，偶尔剧烈摇动。通过加入含有 10% FBS 的 DMEM 终止胶原酶消化。将细胞离心并重悬于研磨培养基（PBS 加 30%葡萄糖加 10% FBS 加青霉素/链霉素）中，并用不同口径的移液管尖端研磨以获得单细胞悬浮液。收集前将细胞沉淀并重新悬浮于 L15/10% FBS 中。

ES 细胞系和四倍体胚胎补偿法

来自核移植胚胎的 ES 细胞系的获得完全如之前文献描述 [28]，且通过四倍体胚胎补偿获得小鼠的步骤也完全如之前文献描述 [29]。

克隆胚胎和小鼠的获得

通过核移植产生克隆胚胎基本如文献描述 [25,29]，唯一不同点是仅有来自两个供体群的 GFP^+ 细胞通过荧光挑选出来，并用于核移植。

原位杂交和免疫组织化学法

通过经批准的方法处死动物，目标组织放于 OCT 中尽快冷冻，或于 4% PFA/1×PBS 中固定 2～12 小时，然后洗涤并在冷冻前于 30%蔗糖中平衡。将冷冻切片（20～30 μm 厚）置于载玻片上，并进行标准免疫组织染色和用地高辛标记的探针进行原位杂交 [32]。我们使用以下抗体：兔抗 GFP 抗体（Molecular Probes），兔抗 Cre 重组酶抗体（Novagen），山羊抗 Ki67 抗体（Santa Cruz Biotechnology），大鼠抗 BrdU 抗体（abcam）和兔抗 MASH-1 抗体（来自 J.Johnson 的礼物）。

BrdU 标记和可视化

每 2 小时向小鼠腹膜内注射 5 mg/ml 的 BrdU PBS 溶液，持续进行 12 小时。每次注射时，每只小鼠接受 100 μg/g 体重的 BrdU。在最后一次注射后 2 小时，处死小鼠并将组织在 4% PFA 和 PBS 中固定 2 小时，洗涤并用蔗糖保护，之后冷冻。为了同时观察 BrdU 和 GFP，首先用抗 GFP 抗体对切片染色，然后在 4% PFA/PBS 中固定 15 分钟，之后洗涤，用

4 M HCl/0.1% Triton X-100 for 10 min, washed and stained with rat anti-BrdU followed by the appropriate secondary antibodies.

Gene targeting and generation of donor mouse lines

The OMP–IRES-Cre mouse line was generated as described previously except that the Cre recombinase gene sequence was inserted in place of the tTA sequence[32]. An IRES directing the translation of Cre recombinase was introduced into the 3′ untranslated region of the OMP locus by homologous recombination in ES cells. Transgenic mice were generated and crossed with a strain bearing the Z/EG reporter transgene. The *Rosa26*–LoxP-Stop-LoxP–GFP line was generated as described previously except that the GFP sequence replaced the CFP and YFP sequences in the *Rosa26* cassette[33].

Southern blotting, PCR and RT–PCR

Southern blots and PCR screening of ES cells and tails used standard methods[43]. RT–PCR was performed on RNA isolated from mouse olfactory turbinates that had been frozen in Trizol (Invitrogen), and then treated as per the manufacturer's protocol. RNA was reverse transcribed using the Superscript II kit (Invitrogen) and complementary DNA was subjected to PCR using standard conditions.

(**428**, 44-49; 2004)

Kevin Eggan[1*†], Kristin Baldwin[2*], Michael Tackett[1], Joseph Osborne[2†], Joseph Gogos[2], Andrew Chess[1], Richard Axel[2] & Rudolf Jaenisch[1]

[1] Whitehead Institute for Biomedical Research and Department of Biology, Massachusetts Institute of Technology, 9 Cambridge Center, Cambridge, Massachusetts 02142, USA

[2] Department of Biochemistry and Molecular Biophysics, Howard Hughes Medical Institute, College of Physicians and Surgeons, Columbia University, 701 West 168th Street, New York, New York 10032, USA

[*] These authors contributed equally to this work

[†] Present addresses: Department of Physiology and Cellular Biophysics, College of Physicians and Surgeons, Columbia University, 630 West 168th Street, New York, New York 10032, USA (J.O.); Department of Molecular and Cellular Biology, Harvard University, 7 Divinity Avenue, Cambridge, Massachusetts 02138, USA (K.E.)

Received 13 November 2003; accepted 14 January 2004; doi:10.1038/nature02375.
Published online 15 February 2004.

References:

1. Rideout, W. M., Eggan, K. & Jaenisch, R. Nuclear cloning and epigenetic reprogramming of the genome. *Science* **293**, 1093-1098 (2001).

2. Hochedlinger, K. & Jaenisch, R. Nuclear transplantation: lessons from frogs and mice. *Curr. Opin. Cell Biol.* **14**, 741-748 (2002).

3. Gurdon, J. B. & Byrne, J. A. The first half-century of nuclear transplantation. *Proc. Natl Acad. Sci. USA* **100**, 8048-8052 (2003).

4. Osada, T., Kusakabe, H., Akutsu, H., Yagi, T. & Yanagimachi, R. Adult murine neurons: their chromatin and chromosome changes and failure to support embryonic development as revealed by nuclear transfer. *Cytogenet. Genome Res.* **97**, 7-12 (2002).

5. Yamazaki, Y. *et al.* Assessment of the developmental totipotency of neural cells in the cerebral cortex of mouse embryo by nuclear transfer. *Proc. Natl Acad. Sci. USA* **98**, 14022-14026 (2001).

4 M HCl/0.1% Triton X-100 处理 10 分钟，再次洗涤并用大鼠抗 BrdU 染色，然后用适当的二抗染色。

基因靶向和供体小鼠系的获得

除了插入 Cre 重组酶基因序列以代替 tTA 序列[32]，OMP–IRES-Cre 小鼠系的获得方法如前所述。通过 ES 细胞中的同源重组，将指导 Cre 重组酶翻译的 IRES 引入 OMP 基因座的 3′ 非翻译区。获得转基因小鼠，并与携带 Z/EG 报告转基因的小鼠杂交。获得 *Rosa26*–LoxP-Stop-LoxP–GFP 系小鼠，除用 GFP 序列替换 *Rosa26* 盒中的 CFP 和 YFP 序列外其余步骤均已在文献中描述[33]。

Southern 印迹、PCR 和 RT–PCR

对 ES 细胞和小鼠尾部进行的 Southern 印迹和 PCR 筛选均使用标准方法[43]。对从 Trizol (Invitrogen) 中冷冻的小鼠嗅鼻甲中分离的 RNA 进行 RT–PCR，然后按照试剂制造商的方案进行处理。使用 Superscript II 试剂盒 (Invitrogen) 逆转录 RNA，并使用标准条件对互补 DNA 进行 PCR。

（任奕 翻译；王宇 审稿）

6. Chien, Y. H., Gascoigne, N. R., Kavaler, J., Lee, N. E. & Davis, M. M. Somatic recombination in a murine T-cell receptor gene. *Nature* **309**, 322-326 (1984).

7. Hozumi, N. & Tonegawa, S. Evidence for somatic rearrangement of immunoglobulin genes coding for variable and constant regions. *Proc. Natl Acad. Sci. USA* **73**, 3628-3632 (1976).

8. Brack, C., Hirama, M., Lenhard-Schuller, R. & Tonegawa, S. A complete immunoglobulin gene is created by somatic recombination. *Cell* **15**, 1-14 (1978).

9. Gao, Y. *et al.* A critical role for DNA end-joining proteins in both lymphogenesis and neurogenesis. *Cell* **95**, 891-902 (1998).

10. Frank, K. M. *et al.* Late embryonic lethality and impaired V(D)J recombination in mice lacking DNA ligase IV. *Nature* **396**, 173-177 (1998).

11. Rehen, S. K. *et al.* Chromosomal variation in neurons of the developing and adult mammalian nervous system. *Proc. Natl Acad. Sci. USA* **98**, 13361-13366 (2001).

12. Kaushal, D. *et al.* Alteration of gene expression by chromosome loss in the postnatal mouse brain. *J. Neurosci.* **23**, 5599-5606 (2003).

13. Chun, J. & Schatz, D. G. Rearranging views on neurogenesis: neuronal death in the absence of DNA end-joining proteins. *Neuron* **22**, 7-10 (1999).

14. Chun, J. Selected comparison of immune and nervous system development. *Adv. Immunol.* **77**, 297-322 (2001).

15. Yagi, T. Diversity of the cadherin-related neuronal receptor/protocadherin family and possible DNA rearrangement in the brain. *Genes Cells* **8**, 1-8 (2003).

16. Zhang, X. & Firestein, S. The olfactory receptor gene superfamily of the mouse. *Nature Neurosci.* **5**, 124-133 (2002).

17. Ressler, K. J., Sullivan, S. L. & Buck, L. B. A zonal organization of odorant receptor gene expression in the olfactory epithelium. *Cell* **73**, 597-609 (1993).

18. Vassar, R., Ngai, J. & Axel, R. Spatial segregation of odorant receptor expression in the mammalian olfactory epithelium. *Cell* **74**, 309-318 (1993).

19. Buck, L. & Axel, R. A novel multigene family may encode odorant receptors: a molecular basis for odor recognition. *Cell* **65**, 175-187 (1991).

20. Kratz, E., Dugas, J. C. & Ngai, J. Odorant receptor gene regulation: implications from genomic organization. *Trends Genet.* **18**, 29-34 (2002).

21. Goldstein, B. J. & Schwob, J. E. Analysis of the globose basal cell compartment in rat olfactory epithelium using GBC-1, a new monoclonal antibody against globose basal cells. *J. Neurosci.* **16**, 4005-4016 (1996).

22. Holbrook, E. H., Szumowski, K. E. & Schwob, J. E. An immunochemical, ultrastructural, and developmental characterization of the horizontal basal cells of rat olfactory epithelium. *J. Comp. Neurol.* **363**, 129-146 (1995).

23. Ohta, Y. & Ichimura, K. Proliferation markers, proliferating cell nuclear antigen, Ki67, 5-bromo-2′-deoxyuridine, and cyclin D1 in mouse olfactory epithelium. *Ann. Otol. Rhinol. Laryngol.* **109**, 1046-1048 (2000).

24. Gerdes, J. *et al.* Cell cycle analysis of a cell proliferation-associated human nuclear antigen defined by the monoclonal antibody Ki-67. *J. Immunol.* **133**, 1710-1715 (1984).

25. Wakayama, T., Perry, A. C., Zuccotti, M., Johnson, K. R. & Yanagimachi, R. Full-term development of mice from enucleated oocytes injected with cumulus cell nuclei. *Nature* **394**, 369-374 (1998).

26. Wakayama, T., Rodriguez, I., Perry, A. C., Yanagimachi, R. & Mombaerts, P. Mice cloned from embryonic stem cells. *Proc. Natl Acad. Sci. USA* **96**, 14984-14989 (1999).

27. Rideout, W. M. *et al.* Generation of mice from wild-type and targeted ES cells by nuclear cloning. *Nature Genet.* **24**, 109-110 (2000).

28. Hochedlinger, K. & Jaenisch, R. Monoclonal mice generated by nuclear transfer from mature B and T donor cells. *Nature* **415**, 1035-1038 (2002).

29. Eggan, K. *et al.* Hybrid vigor, fetal overgrowth, and viability of mice derived by nuclear cloning and tetraploid embryo complementation. *Proc. Natl Acad. Sci. USA* **98**, 6209-6214 (2001).

30. Nagy, A. *et al.* Embryonic stem cells alone are able to support fetal development in the mouse. *Development* **110**, 815-821 (1990).

31. Wang, F., Nemes, A., Mendelsohn, M. & Axel, R. Odorant receptors govern the formation of a precise topographic map. *Cell* **93**, 47-60 (1998).

32. Gogos, J. A., Osborne, J., Nemes, A., Mendelsohn, M. & Axel, R. Genetic ablation and restoration of the olfactory topographic map. *Cell* **103**, 609-620 (2000).

33. Srinivas, S. *et al.* Cre reporter strains produced by targeted insertion of EYFP and ECFP into the ROSA26 locus. *BMC Dev. Biol.* **1**, 4 (2001).

34. Rossant, J. A monoclonal mouse? *Nature* **415**, 967-969 (2002).

35. Ohnuma, S. & Harris, W. A. Neurogenesis and the cell cycle. *Neuron* **40**, 199-208 (2003).

36. Edelman, G. M. *Neural Darwinism* (Basic Books, New York, NY, 1987).

37. Ishii, T. *et al.* Monoallelic expression of the odourant receptor gene and axonal projection of olfactory sensory neurones. *Genes Cells* **6**, 71-78 (2001).

38. Serizawa, S. *et al.* Negative feedback regulation ensures the one receptor-one olfactory neuron rule in mouse. *Science* **302**, 2088-2094 (2003).

39. Serizawa, S. *et al.* Mutually exclusive expression of odorant receptor transgenes. *Nature Neurosci.* **3**, 687-693 (2000).

40. Ebrahimi, F. A., Edmondson, J., Rothstein, R. & Chess, A. YAC transgene-mediated olfactory receptor gene choice. *Dev. Dyn.* **217**, 225-231 (2000).

41. Qasba, P. & Reed, R. R. Tissue and zonal-specific expression of an olfactory receptor transgene. *J. Neurosci.* **18**, 227-236 (1998).

42. Vassalli, A., Rothman, A., Feinstein, P., Zapotocky, M. & Mombaerts, P. Minigenes impart odorant receptor-specific axon guidance in the olfactory bulb. *Neuron* **35**, 681-696 (2002).

43. Sambrook, J., Fritsch, E. F. & Maniatis, T. *Molecular Cloning: a Laboratory Manual* 2nd edn (Cold Spring Harbor Laboratory Press, Cold Spring Harbor, NY, 1989).

Supplementary Information accompanies the paper on www.nature.com/nature.

Acknowledgements. We thank L. Moring, A. Nemes, M. Mendelsohn, J. Loring, J. Dausman, A. Meissner and K. Hochedlinger for assistance during the course of these experiments; T. Cutforth, J. de Nooij, T. Jessell and J. Johnson for sharing reagents necessary for our experiments; and members of the Jaenisch, Axel and Chess laboratories for discussion and assistance during the course of these experiments, especially F. A. Ebrahimi, M. Rios, W. M. R. Rideout, L. Jackson-Grusby and B. Shykind. This research was sponsored by NIH grants to R.J., R.A., A.C. and M.T.. K.B. is an Associate and R.A. is an Investigator of the Howard Hughes Medical

Institute. K.E. is a Junior Fellow in the Harvard Society of Fellows.

Competing interests statement. The authors declare that they have no competing financial interests.

Correspondence and requests for materials should be addressed to R.J. (jaenisch@wi.mit.edu) or A.C. (achess@wi.mit.edu).

Genetic Evidence Supports Demic Diffusion of Han Culture

B. Wen *et al.*

Editor's Note

Long-established disciplines situated between the natural sciences and the humanities, such as linguistics, anthropology and archaeology, have been transformed in the past several decades by the availability of genetic data for large numbers of people. The distribution of genes in a population encodes a historical record of cultural movements, interactions and relationships, which can be used to test theories about the dissemination of culture, language and traditions. In this paper by geneticist Li Jin and coworkers, genetic profiles of more than 1,000 individuals in China are used to examine alternative theories of how the Han culture came to be dominant throughout China. One possibility is that the culture was spread by migration (demic diffusion); another is that it happened by cultural exchange without mass movement of people. The Han people of the north are genetically distinct from those of the south, but the results show that southern Han are more closely related to northern Han than they are to southern ethnic minorities, especially through paternal descent—supporting the demic model, driven mostly by migration of northern males. This migration happened largely in three waves within the first and early second millennium, during the Western Jin, Tang and Song dynasties.

The spread of culture and language in human populations is explained by two alternative models: the demic diffusion model, which involves mass movement of people; and the cultural diffusion model, which refers to cultural impact between populations and involves limited genetic exchange between them[1]. The mechanism of the peopling of Europe has long been debated, a key issue being whether the diffusion of agriculture and language from the Near East was concomitant with a large movement of farmers[1-3]. Here we show, by systematically analysing Y-chromosome and mitochondrial DNA variation in Han populations, that the pattern of the southward expansion of Han culture is consistent with the demic diffusion model, and that males played a larger role than females in this expansion. The Han people, who all share the same culture and language, exceed 1.16 billion (2000 census), and are by far the largest ethnic group in the world. The expansion process of Han culture is thus of great interest to researchers in many fields.

ACCORDING to the historical records, the Hans were descended from the ancient Huaxia tribes of northern China, and the Han culture (that is, the language and its associated cultures) expanded into southern China—the region originally inhabited by the southern natives, including those speaking Daic, Austro-Asiatic and Hmong-

遗传学证据支持汉文化的扩散源于人口扩张

文波等

编者按

在过去的数十年间，科学家获得了大量的人群遗传数据，也因此改变了处在自然科学与人文科学之间的一批学科，诸如语言学、人类学和考古学。一个群体中基因的分布，记录着文化变革、互动和源流关系的历史，可以用来验证关于文化、语言和传统传播的各种假说。遗传学家金力及合作者在这篇论文中，利用中国一千多个个体的遗传学数据，验证了汉文化如何在中国成为主流的两种假说。一种可能性是移民传播了文化(人口扩张说)；也有可能只是文化交流而没有大规模的人口移动。从遗传学看，北方的汉族人群与南方的显然有差异，但是结果却显示：南方汉族人群更接近北方汉族人群，而不是接近南方少数民族，特别是在父系血缘中。这支持了人口扩张说，主要是北方男性移民传播了文化。这种迁移大量发生于公元后第一个千年，以及第二个千年早期，有西晋、唐代和宋代三次高潮。

语言和文化在人群间的扩散有两种不同的模式：一种是人口扩张、人群迁徙模式；另一种是文化传播模式，人群之间有文化传播，而基因交流却很有限[1]。同一语系的欧洲人群的形成机制争议颇多，争论的焦点在于来自近东的农业文明和语言的扩散是否伴随着大量的农业人口的迁移[1-3]。本文中，通过对汉族群体的 Y 染色体和线粒体 DNA 多态性系统地进行分析，我们发现汉文化向南扩散的格局符合人口扩张模式，而且在扩张过程中男性占主导地位。汉族有着共同的文化和语言，人口超过了十一亿六千万(2000 年人口统计)，无疑是全世界人口最多的民族。因此汉文化的扩散过程广受各领域研究者的关注。

史载汉族源于古代中国北方的华夏部族，在过去的两千多年间，汉文化(汉语和相关的文化传统)扩散到了中国南方，而中国南方原住民族则是说侗台语、南亚

Mien languages—in the past two millennia[4,5]. Studies on classical genetic markers and microsatellites show that the Han people, like East Asians, are divided into two genetically differentiated groups, northern Han and southern Han[6,8], separated approximately by the Yangtze river[9]. Differences between these groups in terms of dialect and customs have also been noted[10]. Such observations seem to support a mechanism involving primarily cultural diffusion and assimilation (the cultural diffusion model) in Han expansion towards the south. However, the substantial sharing of Y-chromosome and mitochondrial lineages between the two groups[11,12] and the historical records describing the expansion of Han people[5] contradict the cultural diffusion model hypothesis of Han expansion. In this study, we aim to examine the alternative hypothesis; that is, that substantial population movements occurred during the expansion of Han culture (the demic diffusion model).

To test this hypothesis, we compared the genetic profiles of southern Hans with their two parental population groups: northern Hans and southern natives, which include the samples of Daic, Hmong-Mien and Austro-Asiatic speaking populations currently residing in China, and in some cases its neighbouring countries. Genetic variation in both the non-recombining region of the Y chromosome (NRY) and mitochondrial DNA (mtDNA)[13-16] were surveyed in 28 Han populations from most of the provinces in China (see Fig. 1 and Supplementary Table 1 for details).

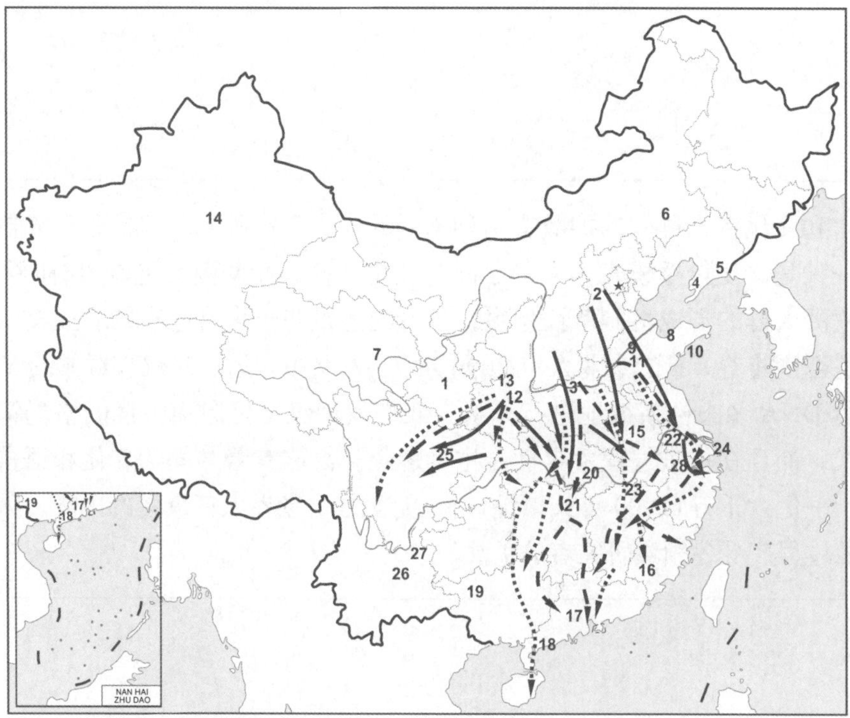

Fig. 1. Geographic distribution of sampled populations. Shown are the three waves of north-to-south migrations according to historical record. The identifications of populations are given in Supplementary Table 1. Populations 1–14 are northern Hans, and 15–28 are southern Hans. The solid, dashed and dotted arrows refer to the first, second and third waves of migrations, respectively. The first wave involving 0.9

语和苗瑶语的人群[4,5]。经典遗传标记和微卫星位点研究显示，汉族和其他东亚人群一样都可以以长江为界[9]分为两个遗传亚群：北方汉族和南方汉族[6,8]。两个亚群之间的方言和习俗差异也很显著[10]。这些现象看似支持文化传播模式，即汉族向南扩张主要是文化传播和同化的结果。然而，两个亚群之间有着许多共同的Y染色体和线粒体类型[11,12]，历史记载的汉族移民史[5]也与汉族的文化传播模式假说相矛盾。本研究对这两种假说进行了检验，证实汉文化的扩散中的确发生了大规模的人群迁徙（人口扩张模式）。

为了验证这些假说，我们把南方汉族的遗传结构与两个亲本群体做比较，其一是北方汉族，其二是南方原住民族，即现居于中国境内和若干邻国的侗台语、苗瑶语和南亚语群体。我们分析了中国汉族28个群体的Y染色体非重组区（NRY）和线粒体DNA（mtDNA）遗传多态[13-16]，这些样本覆盖了中国绝大部分的省份（详见图1和补充信息表1）。

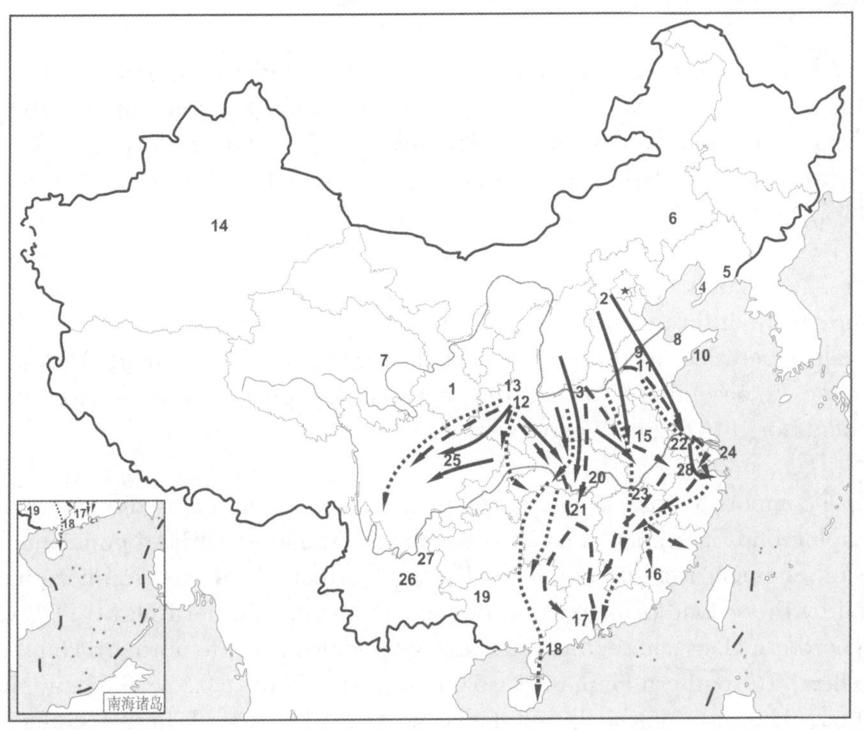

图1. 调查群体的地理分布。图中标出了历史记载中自北而南的三次迁徙浪潮。各群体的详细信息见补充信息表1。群体1～14是北方汉族，15～28是南方汉族。实线、短划线和虚线箭头依次表示三次迁徙浪潮。第一次发生于西晋时期（公元265～316年），迁徙人口约90万（大约是当时南方人口的

million (approximately one-sixth of the southern population at that time) occurred during the Western Jin Dynasty (AD 265–316); the second migration, more extensive than the first, took place during the Tang Dynasty (AD 618–907); and the third wave, including, ~5 million immigrants, occurred during the Southern Song Dynasty (AD 1127–1279).

On the paternal side, southern Hans and northern Hans share similar frequencies of Y-chromosome haplogroups (Supplementary Table 2), which are characterized by two haplogroups carrying the M122-C mutations (O3-M122 and O3e-M134) that are prevalent in almost all Han populations studied (mean and range: 53.8%, 37–71%; 54.2%, 35–74%, for northern and southern Hans, respectively). Haplogroups carrying M119-C (O1* and O1b) and/or M95-T (O2a* and O2a1) (following the nomenclature of the Y Chromosome Consortium) which are prevalent in southern natives, are more frequent in southern Hans (19%, 3–42%) than in northern Hans (5%, 1–10%). In addition, haplogroups O1b-M110, O2a1-M88 and O3d-M7, which are prevalent in southern natives[17], were only observed in some southern Hans (4% on average), but not in northern Hans. Therefore, the contribution of southern natives in southern Hans is limited, if we assume that the frequency distribution of Y lineages in southern natives represents that before the expansion of Han culture that started 2,000 yr ago[5]. The results of analysis of molecular variance (AMOVA) further indicate that northern Hans and southern Hans are not significantly different in their Y haplogroups ($F_{ST} = 0.006$, $P > 0.05$), demonstrating that southern Hans bear a high resemblance to northern Hans in their male lineages.

On the maternal side, however, the mtDNA haplogroup distribution showed substantial differentiation between northern Hans and southern Hans (Supplementary Table 3). The overall frequencies of the northern East Asian-dominating haplogroups (A, C, D, G, M8a, Y and Z) are much higher in northern Hans (55%, 49–64%) than are those in southern Hans (36%, 19–52%). In contrast, the frequency of the haplogroups that are dominant lineages (B, F, R9a, R9b and N9a) in southern natives[12,14,18] is much higher in southern (55%, 36–72%) than it is in northern Hans (33%, 18–42%). Northern and southern Hans are significantly different in their mtDNA lineages ($F_{ST} = 0.006$, $P < 10^{-5}$). Although the F_{ST} values between northern and southern Hans are similar for mtDNA and the Y chromosome, F_{ST} accounts for 56% of the total among-population variation for mtDNA but only accounts for 18% for the Y chromosome.

A principal component analysis is consistent with the observation based on the distribution of the haplogroups in Han populations. For the NRY, almost all Han populations cluster together in the upper right-hand part of Fig. 2a. Northern Hans and southern natives are separated by the second principal component (PC2) and southern Hans' PC2 values lie between northern Hans and southern natives but are much closer to northern Hans (northern Han, 0.58 ± 0.01; southern Han, 0.46 ± 0.03; southern native, -0.32 ± 0.05), implying that the southern Hans are paternally similar to northern Hans, with limited influence from southern natives. In contrast, for mtDNA, northern Hans and southern natives are distinctly separated by PC2 (Fig. 2b), and southern Hans are located between them but are closer to southern natives (northern Han, 0.56 ± 0.02; southern Han, 0.09 ± 0.06; southern native,

六分之一）；第二次发生于唐代（公元618～907年），规模比第一次大得多；第三次发生于南宋（公元1127～1279年），迁徙人口近500万。

父系方面，南方汉族与北方汉族的Y染色体单倍群频率分布非常相近（见补充信息表2），尤其是具有M122-C突变的单倍群（O3-M122和O3e-M134）普遍存在于我们研究的汉族群体中（北方汉族在37%～71%之间，平均53.8%；南方汉族在35%～74%之间，平均54.2%）。南方原住民族中普遍出现的单倍群M119-C(O1*和O1b)和（或）M95-T(O2a*和O2a1)（遵循Y染色体委员会的命名法）在南方汉族中的频率（3%～42%，平均19%）高于北方汉族（1%～10%，平均5%）。而且，南方原住民族中普遍存在的单倍群O1b-M110、O2a1-M88和O3d-M7[17]，在南方汉族中低频存在（平均4%），而北方汉族中却没观察到。如果我们假定在起始于两千多年前的汉文化扩散[5]之前南方原住民族的Y类型频率与现在基本一致的话，南方汉族中南方原住民族的成分应该是不多的。分子方差分析（AMOVA）进一步显示北方汉族和南方汉族的Y染色体单倍群频率分布没有显著差异（$F_{ST}=0.006$，$P>0.05$），说明南方汉族在父系上与北方汉族非常相似。

母系方面，北方汉族与南方汉族的mtDNA单倍群分布非常不同（补充信息表3）。东亚北部的主要单倍群（A，C，D，G，M8a，Y，Z）在北方汉族中的频率（49%～64%，平均55%）比在南方汉族中（19%～52%，平均36%）高得多。另一方面，南方原住民族的主要单倍群（B，F，R9a，R9b，N9a）[12,14,18]在南方汉族中的频率（36%～72%，平均55%）要比在北方汉族中（18%～42%，平均33%）高得多。mtDNA类型的分布在南北汉族之间有极显著差异（$F_{ST}=0.006$，$P<10^{-5}$）。虽然南北汉族之间mtDNA和Y染色体的F_{ST}值相近，但mtDNA的南北差异F_{ST}值占群体间总方差的56%，而Y染色体仅仅占18%。

用汉族群体的单倍群频率数据所做的主成分分析与以上结果相一致。对NRY的分析发现，几乎所有的汉族群体都聚在图2a的右上方。北方汉族和南方原住民族在第2主成分上分离，南方汉族的第2主成分值处于北方汉族和南方原住民族之间，但是更接近于北方汉族（北方汉族0.58 ± 0.01；南方汉族0.46 ± 0.03；南方原住民族-0.32 ± 0.05），这表明南方汉族在父系上与北方汉族相近，受到南方原住民族的影响很小。就mtDNA而言，北方汉族和南方原住民族仍然被第2主成分分开（图2b），南方汉族也在两者之间但稍微接近南方原住民族（北方汉族0.56 ± 0.02；南方汉族0.09 ± 0.06；南方原住民族-0.23 ± 0.04），表明南方汉族的女性基因库比男性

-0.23 ± 0.04), indicating a much more substantial admixture in southern Hans' female gene pool than in its male counterpart.

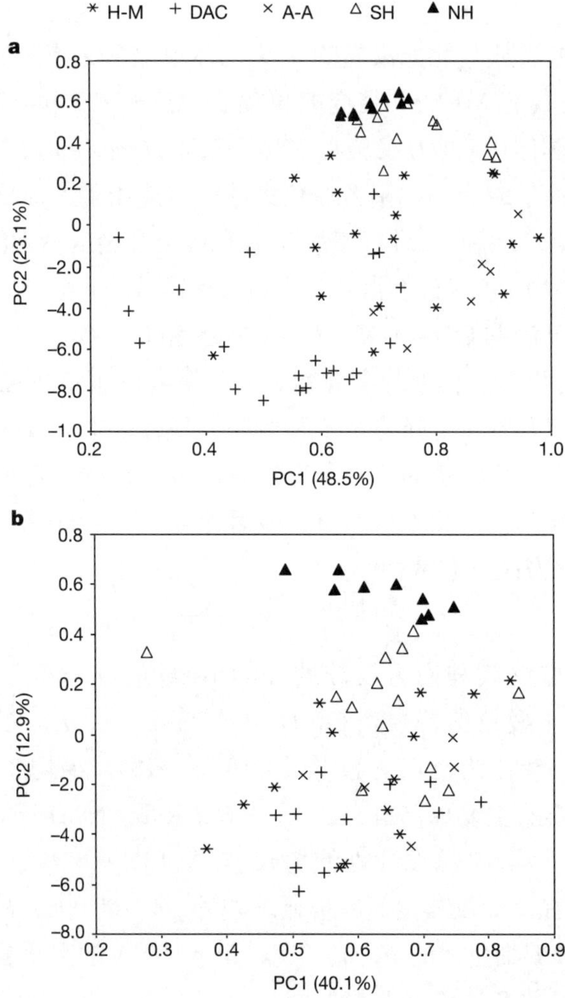

Fig. 2. Principal component plot. **a**, **b**, Plots are of Y-chromosome (**a**) and mtDNA (**b**) haplogroup frequency. Population groups: H-M, Hmong-Mien; DAC, Daic; A-A, Austro-Asiatic; SH, southern Han; NH, northern Han.

The relative contribution of the two parental populations (northern Hans and southern natives) in southern Hans was estimated by two different statistics[19,20], which are less biased than other statistics for single-locus data[21] (Table 1). The estimations of the admixture coefficient (M, proportion of northern Han contribution) from the two methods are highly consistent (for the Y chromosome, $r = 0.922$, $P < 0.01$; for mtDNA, $r = 0.970$, $P < 0.01$). For the Y chromosome, all southern Hans showed a high proportion of northern Han contribution (M_{BE}: 0.82 ± 0.14, range from 0.54 to 1; M_{RH}: 0.82 ± 0.12, range from 0.61 to 0.97) (see refs 20 and 19 for definitions of M_{BE} and M_{RH}, respectively) indicating that males from the northern Hans are the primary contributor to the gene pool of the southern

基因库有更多的混合成分。

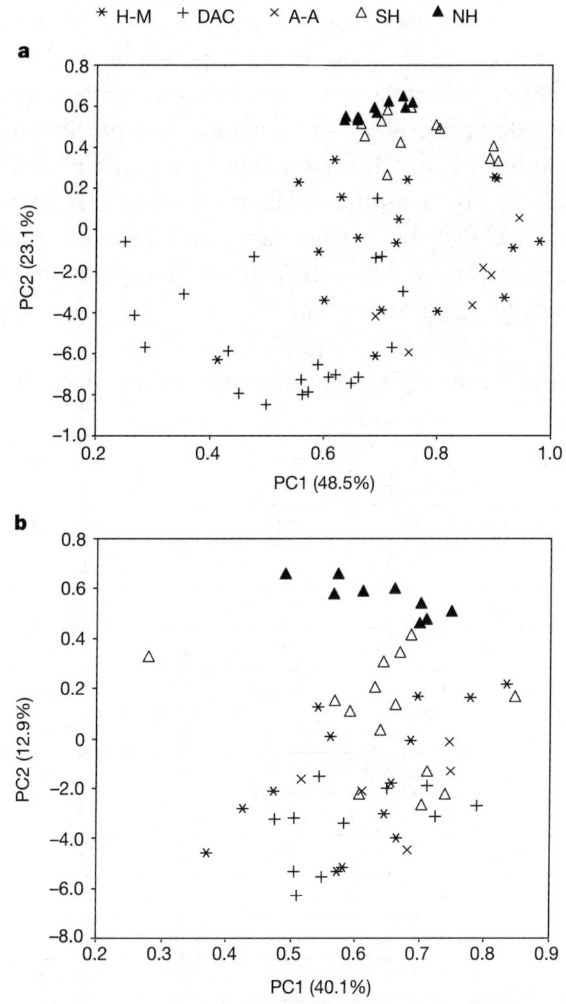

图 2. 主成分散点图。**a** 为 Y 染色体单倍群散点图，**b** 为 mtDNA 单倍群散点图。群体标记：▲北方汉族，△南方汉族，+ 侗台语民族，× 南亚语民族，* 苗瑶语民族。

我们进一步用两种不同的统计方法[19,20]来估计两个亲本(北方汉族和南方原住民)对南方汉族基因库的相对贡献(表 1)，这两个统计量用于单位点分析时比其他的方法更为准确[21]。两种方法得到的混合系数估计值(M，北方汉族的贡献比例)高度一致(Y 染色体，$r = 0.922$，$P < 0.01$；mtDNA，$r = 0.970$，$P < 0.01$)。就 Y 染色体而言，所有的南方汉族都包含很高比例的北方汉族混合比率(M_{BE}: 0.82 ± 0.14，范围 0.54 ~ 1；M_{RH}: 0.82 ± 0.12，范围 0.61 ~ 0.97)(M_{BE} 和 M_{RH} 的定义分别见参考文献 20 和 19)，这表明南方汉族男性基因库的主要贡献成分来自北方汉族。相反，南

Hans. In contrast, northern Hans and southern natives contributed almost equally to the southern Hans' mtDNA gene pool (M_{BE}: 0.56±0.24 [0.15, 0.95]; M_{RH}: 0.50±0.26 [0.07, 0.91]). The contribution of northern Hans to southern Hans is significantly higher in the paternal lineage than in the maternal lineage collectively (*t*-test, $P<0.01$) or individually (11 out of 13 populations for M_{BE}, and 13 out of 13 populations for M_{RH}: $P<0.01$, assuming a null binomial distribution with equal male and female contributions), indicating a strong sex-biased population admixture in southern Hans. The proportions of northern Han contribution (*M*) in southern Hans showed a clinal geographic pattern, which decreases from north to south. The *M*s in southern Hans are positively correlated with latitude ($r^2 = 0.569$, $P<0.01$) for mtDNA, but are not significant for the Y chromosome ($r^2 = 0.072$, $P>0.05$), because the difference of *M*s in the paternal lineage among southern Hans is too small to create a statistically significant trend.

Table 1. Northern Han admixture proportion in southern Hans

Population	Y Chromosome		mtDNA	
	M_{BE} (± s.e.m)	M_{RH}	M_{BE} (± s.e.m)	M_{RH}
Anhui	0.868±0.119	0.929	0.816±0.214	0.755
Fujian	1	0.966	0.341±0.206	0.248
Guangdong1	0.677±0.121	0.669	0.149±0.181	0.068
Guangdong2	ND	ND	0.298±0.247	0.312
Guangxi	0.543±0.174	0.608	0.451±0.263	0.249
Hubei	0.981±0.122	0.949	0.946±0.261	0.907
Hunan	0.732±0.219	0.657	0.565±0.297	0.490
Jiangsu	0.789±0.078	0.821	0.811±0.177	0.786
Jiangxi	0.804±0.113	0.829	0.374±0.343	0.424
Shanghai	0.819±0.087	0.902	0.845±0.179	0.833
Sichuan	0.750±0.118	0.713	0.509±0.166	0.498
Yunnan1	1	0.915	0.376±0.221	0.245
Yunnan2	0.935±0.088	0.924	0.733±0.192	0.645
Zhejiang	0.751±0.084	0.763	0.631±0.180	0.540
Average	0.819	0.819	0.560	0.500

M_{BE} and M_{RH} refer to the statistics described in refs 20 and 19, respectively. The standard error of M_{BE} was obtained by bootstrap with 1,000 replications. The proportions of contribution from northern Hans were estimated using northern Hans and southern natives as the parental populations of the southern Hans. It was assumed that the allele frequency in the southern natives remained unchanged before and after the admixture, which started about 2,000 yr ago, and the genetic exchange between northern Hans and southern natives has been limited. In fact, the gene flow from northern Hans to southern natives has been larger than that from southern natives to northern Hans; therefore, the level of admixture presented in this table is underestimated and is without proper adjustment. The demic expansion of Han would have been more pronounced than was observed in this study.

We provide two lines of evidence supporting the demic diffusion hypothesis for the expansion of Han culture. First, almost all Han populations bear a high resemblance in

方汉族的 mtDNA 基因库中北方汉族和南方原住民族的贡献比例几乎相等(M_{BE}：0.56±0.24 [0.15，0.95]；M_{RH}：0.50±0.26 [0.07，0.91])。总体上北方汉族对南方汉族的遗传贡献父系比母系高得多(t 检验，$P<0.01$)；各群体分别看也是这样：绝大部分南方汉族群体中北方汉族的贡献在父系上大于母系(M_{BE}，11/13，M_{RH}，13/13，$P<0.01$，零假设为男女的贡献相等为二项分布)，这表明南方汉族的群体混合过程有很强的性别偏向。南方汉族中北方汉族贡献的比例(M)呈现出由北向南递减的梯度地理格局。南方汉族 mtDNA 的 M 值与纬度正相关($r^2=0.569$，$P<0.01$)，但 Y 染色体的相关性不显著($r^2=0.072$，$P>0.05$)，因为南方汉族父系的 M 值差异太小，不足以导致统计上的显著性。

表 1. 南方汉族中的北方汉族混合比例

群体	Y 染色体		线粒体 DNA	
	M_{BE}(±s.e.m)	M_{RH}	M_{BE}(±s.e.m)	M_{RH}
安徽	0.868±0.119	0.929	0.816±0.214	0.755
福建	1	0.966	0.341±0.206	0.248
广东 1	0.677±0.121	0.669	0.149±0.181	0.068
广东 2	ND	ND	0.298±0.247	0.312
广西	0.543±0.174	0.608	0.451±0.263	0.249
湖北	0.981±0.122	0.949	0.946±0.261	0.907
湖南	0.732±0.219	0.657	0.565±0.297	0.490
江苏	0.789±0.078	0.821	0.811±0.177	0.786
江西	0.804±0.113	0.829	0.374±0.343	0.424
上海	0.819±0.087	0.902	0.845±0.179	0.833
四川	0.750±0.118	0.713	0.509±0.166	0.498
云南 1	1	0.915	0.376±0.221	0.245
云南 2	0.935±0.088	0.924	0.733±0.192	0.645
浙江	0.751±0.084	0.763	0.631±0.180	0.540
平均	0.819	0.819	0.560	0.500

M_{BE} 和 M_{RH} 分别为参考文献 20 和 19 所描述的统计量。M_{BE} 的标准误差通过 1,000 次自展获得。把南方原住民族和北方汉族作为南方汉族的亲本群体估计北方汉族的遗传贡献比例，假定 2,000 多年前开始的混合过程前后南方原住民族的等位基因频率基本不变，并且北方汉族和南方原始民族之间的遗传交流不多。实际上，从北方汉族到南方原住民族的基因流动比反向的流动大得多，所以表中的估计值在没有适当调整前是低估的。因而汉族实际的人口扩张程度应该大于本项研究得出的数值。

综上所述，我们提出了两项证据支持汉文化扩散的人口扩张假说。首先，几乎所有的汉族群体的 Y 染色体单倍群分布都极为相似，Y 染色体主成分分析也把几

Y-chromosome haplogroup distribution, and the result of principal component analysis indicated that almost all Han populations form a tight cluster in their Y chromosome. Second, the estimated contribution of northern Hans to southern Hans is substantial in both paternal and maternal lineages and a geographic cline exists for mtDNA. It is noteworthy that the expansion process was dominated by males, as is shown by a greater contribution to the Y-chromosome than the mtDNA from northern Hans to southern Hans. A sex-biased admixture pattern was also observed in Tibeto-Burman-speaking populations[22].

According to the historical records, there were continuous southward movements of Han people due to warfare and famine in the north, as illustrated by three waves of large-scale migrations (Fig. 1). Aside from these three waves, other smaller southward migrations also occurred during almost all periods in the past two millennia. Our genetic observation is thus in line with the historical accounts. The massive movement of the northern immigrants led to a change in genetic makeup in southern China, and resulted in the demographic expansion of Han people as well as their culture. Except for these massive population movements, gene flow between northern Hans, southern Hans and southern natives also contributed to the admixture which shaped the genetic profile of the extant populations.

Methods

Samples

Blood samples of 871 unrelated anonymous individuals from 17 Han populations were collected across China. Genomic DNA was extracted by the phenol-chloroform method. By integrating the additional data obtained from the literatures on the Y chromosome and on mtDNA variation, the final sample sizes for analysis expanded to 1,289 individuals (23 Han populations) for the Y chromosome and 1,119 individuals (23 Han populations) for mtDNA. These samples encompass most of the provinces in China (Fig. 1 and Supplementary Table 1).

Genetic markers

Thirteen bi-allelic Y-chromosome markers, YAP, M15, M130, M89, M9, M122, M134, M119, M110, M95, M88, M45 and M120 were typed by polymerase chain reaction-restriction-fragment length polymorphism methods[11]. These markers are highly informative in East Asians[23] and define 13 haplogroups following the Y Chromosome Consortium nomenclature[24].

The HVS-1 of mtDNA and eight coding region variations, 9-bp deletion, 10397 *Alu*I, 5176 *Alu*I, 4831 *Hha*I, 13259 *Hinc*II, 663 *Hae*III, 12406 *Hpa*I and 9820 *Hinf*I were sequenced and genotyped as in our previous report[22]. Both the HVS-1 motif and the coding region variations were used to infer haplogroups following the phylogeny of East Asian mtDNAs[18].

乎所有的汉族群体都集合成一个紧密的聚类。其二，北方汉族对南方汉族的遗传贡献无论父系方面还是母系方面都是可观的，在线粒体 DNA 分布上也存在地理梯度。北方汉族对南方汉族的遗传贡献在父系（Y 染色体）上远大于母系（mtDNA），表明这一扩张过程中汉族男性处于主导地位。性别偏向的混合格局也同样存在于藏缅语人群中[22]。

据历史记载，受北方战乱和饥荒的影响，汉人不断南迁，图 1 中画出了三次大规模移民的浪潮。在两千多年间，除了这三次大潮，各个时期几乎都有小规模的南迁。所以，我们的遗传研究也与历史记载相吻合。大量的北方移民改变了中国南方的遗传构成，而汉族人口扩张的同时也带动了汉文化的扩散。除了大规模的人群迁徙，北方汉族、南方汉族和南方原住民族之间的基因交流造成的族群混合也在很大程度上改变了中国人群的遗传结构。

方　法

样本

采集中国各地的 17 个汉族群体 871 个随机不相关个体的血样。用酚–氯仿法抽提基因组 DNA。结合文献报道的 Y 染色体和 mtDNA 多态性数据，总共分析的样本量是：Y 染色体 23 个群体 1,289 人，mtDNA23 个群体 1,119 人。这些样本涉及了中国的大部分省份（图 1 和补充信息表 1）。

遗传标记

通过聚合酶链式反应–限制性片段长度多态性方法[11]分型 Y 染色体上的 13 个双等位标记：YAP，M15，M130，M89，M9，M122，M134，M119，M110，M95，M88，M45，M120。根据 Y 染色体委员会的命名系统[24]，这些标记构成 13 个单倍群，在东亚人群中具有较高的信息量[23]。

mtDNA 上，对高变 1 区进行测序，对编码区 8 个多态位点（9-bp 缺失，10397 *Alu*I，5176 *Alu*I，4831 *Hha*I，13259 *Hinc*II，663 *Hae*III，12406 *Hpa*I，9820 *Hinf* I）做了分型，有关方法已有报道[22]。根据东亚 mtDNA 系统树[18]，用高变 1 区突变结构和编码区多态性构建单倍群。

Data analysis

Population relationship was investigated by principal component analysis, which was conducted using mtDNA and Y-chromosome haplogroup frequencies and SPSS10.0 software (SPSS Inc.). The genetic difference between northern and southern Hans was tested by AMOVA[25], using ARLEQUIN software[26]. ADMIX 2.0 (ref. 27) and LEADMIX[21] software were used to estimate the level of admixture of the northern Hans and southern natives in the southern Han populations, using two different statistics[19,20]. The selection of parental populations is critical for appropriate estimation of admixture proportion[28,29] and we were careful to minimize bias by using large data sets across East Asia. In this analysis, the average haplogroup frequencies (for Y-chromosome or mtDNA markers, respectively) of northern Hans (arithmetic mean of 10 northern Hans) were taken for the northern parental population. The frequency of southern natives was estimated by the average of three groups including Austro-Asiatic (NRY, 6 populations; mtDNA, 5 populations), Daic (NRY, 22 populations; mtDNA, 11 populations) and Hmong-Mien (NRY, 18 populations; mtDNA, 14 populations). The geographic pattern of Han populations was revealed by the linear regression analysis of admixture proportion against the latitudes of samples[1,3].

(**431,** 302-305, 2004)

Bo Wen[1,2], Hui Li[1], Daru Lu[1], Xiufeng Song[1], Feng Zhang[1], Yungang He[1], Feng Li[1], Yang Gao[1], Xianyun Mao[1], Liang Zhang[1], Ji Qian[1], Jingze Tan[1], Jianzhong Jin[1], Wei Huang[2], Ranjan Deka[3], Bing Su[1,3,4], Ranajit Chakraborty[3] & Li Jin[1,3]

[1] State Key Laboratory of Genetic Engineering and Center for Anthropological Studies, School of Life Sciences and Morgan-Tan International Center for Life Sciences, Fudan University, Shanghai 200433, China

[2] Chinese National Human Genome Center, Shanghai 201203, China

[3] Center for Genome Information, Department of Environmental Health, University of Cincinnati, Cincinnati, Ohio 45267, USA

[4] Key Laboratory of Cellular and Molecular Evolution, Kunming Institute of Zoology, the Chinese Academy of Sciences, Kunming 650223, China

Received 28 April; accepted 20 July, 2004; doi: 10.1038/nature02878.

References:

1. Cavalli-Sforza, L. L., Menozzi, P. & Piazza, A. *The History and Geography of Human Genes* (Princeton Univ. Press, Princeton, 1994).
2. Sokal, R., Oden, N. L. & Wilson, C. Genetic evidence for the spread of agriculture in Europe by demic diffusion. *Nature* **351**, 143-145 (1991).
3. Chikhi, L. *et al.* Y genetic data support the Neolithic demic diffusion model. *Proc. Natl Acad. Sci. USA* **99**, 11008-11013 (2002).
4. Fei, X. T. *The Pattern of Diversity in Unity of the Chinese Nation* (Central Univ. for Nationalities Press, Beijing, 1999).
5. Ge, J. X., Wu, S. D. & Chao, S. J. *Zhongguo yimin shi (The Migration History of China)* (Fujian People's Publishing House, Fuzhou, China, 1997).
6. Zhao, T. M. & Lee, T. D. Gm and Km allotypes in 74 Chinese populations: a hypothesis of the origin of the Chinese nation. *Hum. Genet.* **83**, 101-110 (1989).
7. Du, R. F., Xiao, C. J. & Cavalli-Sforza, L. L. Genetic distances calculated on gene frequencies of 38 loci. *Sci. China* **40**, 613 (1997).
8. Chu, J. Y. *et al.* Genetic relationship of populations in China. *Proc. Natl Acad. Sci. USA* **95**, 11763-11768 (1998).
9. Xiao, C. J. *et al.* Principal component analysis of gene frequencies of Chinese populations. *Sci. China* **43**, 472-481 (2000).
10. Xu, Y. T. A brief study on the origin of Han nationality. *J. Centr. Univ. Natl* **30**, 59-64 (2003).
11. Su, B. *et al.* Y chromosome haplotypes reveal prehistorical migrations to the Himalayas. *Hum. Genet.* **107**, 582-590 (2000).
12. Yao, Y. G. *et al.* Phylogeographic differentiation of mitochondrial DNA in Han Chinese. *Am. J. Hum. Genet.* **70**, 635-651 (2002).
13. Cavalli-Sforza, L. L. & Feldman, M. W. The application of molecular genetic approaches to the study of human evolution. *Nature Genet.* **33**, 266-275 (2003).
14. Wallace, D. C., Brown, M. D. & Lott, M. T. Nucleotide mitochondrial DNA variation in human evolution and disease. *Gene* **238**, 211-230 (1999).
15. Underhill, P. A. *et al.* Y chromosome sequence variation and the history of human populations. *Nature Genet.* **26**, 358-361 (2000).

数据分析

根据 mtDNA 和 Y 染色体单倍群频率，用 SPSS10.0 软件（SPSS 公司）作主成分分析，研究群体间关系。南北汉族的遗传差异用 ARLEQUIN 软件[26] 做 AMOVA 检验[25]。南方汉族中北方汉族和南方原住民族的混合比例估计用两种不同的统计方法[19,20]：ADMIX 2.0[27] 和 LEADMIX[21] 软件。亲本群体的选择对混合比例的适当估计很重要[28,29]，我们通过扩大东亚的参考数据来减小偏差。分析中，10 个北方汉族群体的各单倍群频率（Y 染色体和 mtNDA 标记分别分析）的算术平均作为北方亲本群体。南方原住民族的频率是三个族群的平均：侗台语群（NRY，22 群体；mtNDA，11 群体），南亚语群（NRY，6 群体；mtNDA，5 群体），苗瑶语群（NRY，18 群体；mtNDA，14 群体）。通过样本的混合比例与纬度[1,3] 的线性回归分析揭示汉族群体的地理格局。

（李辉 翻译；徐文堪 审稿）

16. Jobling, M. A. & Tyler-Smith, C. The human Y chromosome: an evolutionary marker comes of age. *Nature Rev. Genet.* **4**, 598-612 (2003).

17. Su, B. *et al.* Y-chromosome evidence for a northward migration of modern humans into eastern Asia during the last ice age. *Am. J. Hum. Genet.* **65**, 1718-1724 (1999).

18. Kivisild, T. *et al.* The emerging limbs and twigs of the East Asian mtDNA tree. *Mol. Biol. Evol.* **19**, 1737-1751 (2002).

19. Roberts, D. F. & Hiorns, R. W. Methods of analysis of the genetic composition of a hybrid population. *Hum. Biol.* **37**, 38-43 (1965).

20. Bertorelle, G. & Excoffier, L. Inferring admixture proportions from molecular data. *Mol. Biol. Evol.* **15**, 1298-1311 (1998).

21. Wang, J. Maximum-likelihood estimation of admixture proportions from genetic data. *Genetics* **164**, 747-765 (2003).

22. Wen, B. *et al.* Analyses of genetic structure of Tibeto-Burman populations revealed a gender-biased admixture in southern Tibeto-Burmans. *Am. J. Hum. Genet.* **74**, 856-865 (2004).

23. Jin, L. & Su, B. Natives or immigrants: modern human origin in East Asia. *Nature Rev. Genet.* **1**, 126-133 (2000).

24. The Y Chromosome Consortium, A nomenclature system for the tree of human Y-chromosomal binary haplogroups. *Genome Res.* **12**, 339-348 (2002).

25. Excoffier, L., Smouse, P. E. & Quattro, J. M. Analysis of molecular variance inferred from metric distances among DNA haplotypes: application to human mitochondrial DNA restriction data. *Genetics* **131**, 479-491 (1992).

26. Schneider, S., *et al.* Arlequin: Ver. 2.000. A software for population genetic analysis. (Genetics and Biometry Laboratory, Univ. of Geneva, Geneva, 2000).

27. Dupanloup, I. & Bertorelle, G. Inferring admixture proportions from molecular data: extension to any number of parental populations. *Mol. Biol. Evol.* **18**, 672-675 (2001).

28. Chakraborty, R. Gene admixture in human populations: Models and predictions. *Yb. Phys. Anthropol.* **29**, 1-43 (1986).

29. Sans, M. *et al.* Unequal contributions of male and female gene pools from parental populations in the African descendants of the city of Melo, Uruguay. *Am. J. Phys. Anthropol.* **118**, 33-44 (2002).

Supplementary Information accompanies the paper on www.nature.com/nature.

Acknowledgements. We thank all of the donors for making this work possible. The data collection was supported by NSFC and STCSM to Fudan and a NSF grant to L.J. L.J., R.D. and R.C. are supported by NIH.

Competing interests statement. The authors declare that they have no competing financial interests.

Correspondence and requests for materials should be addressed to L.J. (lijin@fudan.edu.cn or li.jin@uc.edu). The mtDNA HVS-1 sequences of 711 individuals from 15 Han populations were submitted to GenBank with accession numbers AY594701–AY595411.

A New Small-bodied Hominin from the Late Pleistocene of Flores, Indonesia

P. Brown *et al.*

Editor's Note

When archaeologist Michael Morwood set out to trace how early modern humans got to Australia, the last thing he and his colleagues expected to unearth in Liang Bua cave on Flores was a creature out of folk-tale. But that's what they discovered. "The Hobbit": a human-like creature that stood barely a metre tall, had a very small brain and a body reminiscent of hominids that lived in Africa, three or more million years ago—but which survived until at least 18,000 years ago. *Homo floresiensis* challenged conventional wisdom about the adaptability of the human form, as well as the assumption that *Homo sapiens* reigned alone on Earth. The sensation created by this strangest of fossil humans has yet to fade.

Currently, it is widely accepted that only one hominin genus, *Homo*, was present in Pleistocene Asia, represented by two species, *Homo erectus* and *Homo sapiens*. Both species are characterized by greater brain size, increased body height and smaller teeth relative to Pliocene *Australopithecus* in Africa. Here we report the discovery, from the Late Pleistocene of Flores, Indonesia, of an adult hominin with stature and endocranial volume approximating 1 m and 380 cm^3, respectively—equal to the smallest-known australopithecines. The combination of primitive and derived features assigns this hominin to a new species, *Homo floresiensis*. The most likely explanation for its existence on Flores is long-term isolation, with subsequent endemic dwarfing, of an ancestral *H. erectus* population. Importantly, *H. floresiensis* shows that the genus *Homo* is morphologically more varied and flexible in its adaptive responses than previously thought.

THE LB1 skeleton was recovered in September 2003 during archaeological excavation at Liang Bua, Flores[1]. Most of the skeletal elements for LB1 were found in a small area, approximately 500 cm^2, with parts of the skeleton still articulated and the tibiae flexed under the femora. Orientation of the skeleton in relation to site stratigraphy suggests that the body had moved slightly down slope before being covered with sediment. The skeleton is extremely fragile and not fossilized or covered with calcium carbonate. Recovered elements include a fairly complete cranium and mandible, right leg and left innominate. Bones of the left leg, hands and feet are less complete, while the vertebral column, sacrum, scapulae, clavicles and ribs are only represented by fragments. The position of the skeleton suggests that the arms are still in the wall of the excavation, and may be recovered in the future. Tooth eruption, epiphyseal union and tooth wear indicate an adult, and pelvic

印度尼西亚弗洛勒斯晚更新世－身材矮小人族新成员

布朗等

编者按

当考古学家迈克尔·莫伍德开始研究早期现代人是如何到达澳大利亚的时候，他和他的同事们最不希望在弗洛勒斯的利昂布阿洞穴发现的是一种民间传说中的生物，但这就是他们的发现。“霍比特人”：这是一种类似人类的生物，身高只有一米，大脑非常小，身体让人想起300万年前或更早以前生活在非洲的原始人，但这种原始人至少存活到18,000年前。弗洛勒斯人挑战了关于人类形态适应性的传统观点，也挑战了智人独自统治地球的假设。这种由最奇怪的人类化石产生的轰动还没有消退。

目前，人们普遍认为，更新世人族在亚洲仅存在一个属，即人属，以直立人和智人两个物种为代表。相对于非洲上新世的南方古猿属，这两个种的特征是脑容量更大，身高更高，牙齿更小。在此我们报道，在印度尼西亚弗洛勒斯岛晚更新世发现的一个人族成员成年个体化石，其身高和颅腔容量分别约为1米和380立方厘米，相当于已知最小的南方古猿。其原始与衍生特征的组合，将这一人族成员定为一个新的物种——弗洛勒斯人。对其为何存在于弗洛勒斯岛上，最可能的解释是一个祖先直立人种群的长期隔离，并在随后出现了地方性的矮化。重要的是，弗洛勒斯人显示出人属在形态上的适应性反应比以前所想象的更多样化，也更灵活。

LB1骨架是2003年9月在弗洛勒斯岛的利昂布阿进行考古挖掘期间出土的[1]。LB1的大部分骨骼部位都是在大约500平方厘米的一个很小的区域内被发现的，骨架不少部位仍然互相关节，胫骨弯曲在股骨下方。骨架方位与遗址地层的相对关系表明躯体在被沉积物掩埋之前，沿坡轻微向下移动了一些。骨骼非常脆弱，没有石化，也未被碳酸钙覆盖。出土的部位包括相当完整的颅骨与下颌骨，右腿和左髋骨。左腿、手和脚的骨头的完整性稍差，而脊柱、荐椎、两侧肩胛骨、两侧锁骨和肋骨仅有一些碎片保存。骨架的姿态表明其手臂还埋在发掘探方的侧壁里，将来可能会被发现。牙齿的萌发状况、骨骺的愈合程度及牙齿的磨损程度表明这是一个成年个

anatomy strongly supports the skeleton being that of a female. On the basis of its unique combination of primitive and derived features we assign this skeleton to a new species, *Homo floresiensis*.

Description of *Homo floresiensis*

Order Primates Linnaeus, 1758

Suborder Anthropoidea Mivart, 1864

Superfamily Hominoidea Gray, 1825

Family Hominidae Gray, 1825

Tribe Hominini Gray, 1825

Genus *Homo* Linnaeus, 1758

Homo floresiensis sp. nov.

Etymology. Recognizing that this species has only been identified on the island of Flores, and a prolonged period of isolation may have resulted in the evolution of an island endemic form.

Holotype. LB1 partial adult skeleton excavated in September 2003. Recovered skeletal elements include the cranium and mandible, femora, tibiae, fibulae and patellae, partial pelvis, incomplete hands and feet, and fragments of vertebrae, sacrum, ribs, scapulae and clavicles. The repository is the Centre for Archaeology, Jakarta, Indonesia.

Referred material. LB2 isolated left mandibular P_3. The repository is the Centre for Archaeology, Jakarta, Indonesia.

Localities. Liang Bua is a limestone cave on Flores, in eastern Indonesia. The cave is located 14 km north of Ruteng, the provincial capital of Manggarai Province, at an altitude of 500 m above sea level and 25 km from the north coast. It occurs at the base of a limestone hill, on the southern edge of the Wae Racang river valley. The type locality is at 08° 31′ 50.4″ south latitude 120° 26′ 36.9″ east longitude.

Horizon. The type specimen LB1 was found at a depth of 5.9 m in Sector VII of the excavation at Liang Bua. It is associated with calibrated accelerator mass spectrometry (AMS) dates of approximately 18 kyr and bracketed by luminescence dates of 35 ± 4 kyr and 14 ± 2 kyr. The referred isolated left P_3 (LB2) was recovered just below a discomformity

体，骨盆的解剖学结构特征有力地支持这是一个女性个体的骨架。根据其原始与衍生特征的独特组合，我们将这具骨架定为一新种——弗洛勒斯人。

弗洛勒斯人的描述

灵长目　Primates Linnaeus，1758

类人猿亚目　Anthropoidea Mivart，1864

人猿超科　Hominoidea Gray，1825

人科　Hominidae Gray，1825

人族　Hominini Gray，1825

人属　*Homo* Linnaeus，1758

弗洛勒斯人(新种) *Homo floresiensis* sp. nov.

词源　考虑到该种仅发现于弗洛勒斯岛，并且可能由于长期的隔离才造成了一个岛屿特有类型的演化。

正型标本　2003 年 9 月发掘出土的不完整成年个体骨架 LB1。出土的骨骼部位包括颅骨与下颌骨，两侧股骨，胫骨，腓骨和髌骨，不完整骨盆，不完全的双手和双脚，以及椎体、荐椎、肋骨、两侧肩胛骨和两侧锁骨的碎片。存放在印度尼西亚雅加达考古中心。

归入材料　LB2 游离左下第 3 前臼齿。存放在印度尼西亚雅加达考古中心。

产地　利昂布阿是印度尼西亚东部弗洛勒斯岛上的一个石灰岩洞穴。该洞穴位于芒加莱省会城市鲁滕以北 14 千米处，其海拔 500 米，距北海岸 25 千米。它位于威拉肯河谷南缘的石灰岩山丘底部。模式产地位于南纬 08° 31′ 50.4″ 东经 120° 26′ 36.9″。

层位　模式标本 LB1 发现于利昂布阿发掘 VII 区 5.9 米深处。与其直接相关的校准加速器质谱法(AMS)测定年代大约为 18,000 年，释光法把年代限定在 35,000 ± 4,000 年与 14,000 ± 2,000 年之间。归入材料中的游离左下第 3 前臼齿(LB2)发现于 IV 区 4.7 米处不整合面正下方，其年代限定范围在流石 U 系测定年代

at 4.7 m in Sector IV, and bracketed by a U-series date of 37.7 ± 0.2 kyr on flowstone, and 20 cm above an electron-spin resonance (ESR)/U-series date of 74^{+14}_{-12} kyr on a *Stegodon* molar.

Diagnosis. Small-bodied bipedal hominin with endocranial volume and stature (body height) similar to, or smaller than, *Australopithecus afarensis*. Lacks masticatory adaptations present in *Australopithecus* and *Paranthropus*, with substantially reduced facial height and prognathism, smaller postcanine teeth, and posteriorly orientated infraorbital region. Cranial base flexed. Prominent maxillary canine juga form prominent pillars, laterally separated from nasal aperture. Petrous pyramid smooth, tubular and with low relief, styloid process absent, and without vaginal crest. Superior cranial vault bone thicker than *Australopithecus* and similar to *H. erectus* and *H. sapiens*. Supraorbital torus arches over each orbit and does not form a flat bar as in Javan *H. erectus*. Mandibular P_3 with relatively large occlusal surface area, with prominent protoconid and broad talonid, and either bifurcated roots or a mesiodistally compressed Tomes root. Mandibular P_4 also with Tomes root. First and second molar teeth of similar size. Mandibular coronoid process higher than condyle, and the ramus has a posterior orientation. Mandibular symphysis without chin and with a posterior inclination of the symphysial axis. Posteriorly inclined alveolar planum with superior and inferior transverse tori. Ilium with marked lateral flare. Femur neck long relative to head diameter, the shaft circular and without pilaster, and there is a high bicondylar angle. Long axis of tibia curved and the midshaft has an oval cross-section.

Description and Comparison of the Cranial and Postcranial Elements

Apart from the right zygomatic arch, the cranium is free of substantial distortion (Figs 1 and 2). Unfortunately, the bregmatic region, right frontal, supraorbital, nasal and subnasal regions were damaged when the skeleton was discovered. To repair post-mortem pressure cracks, and stabilize the vault, the calvarium was dismantled and cleaned endocranially before reconstruction. With the exception of the squamous suture, most of the cranial vault sutures are difficult to locate and this problem persists in computed tomography (CT) scans. As a result it is not possible to locate most of the standard craniometric landmarks with great precision.

37,700±200 年以及 20 厘米之上的一个剑齿象臼齿电子自旋共振(ESR)/U 系测定年代 $74,000^{+14,000}_{-12,000}$ 年之间。

特征 身材矮小两足行走的人族成员，其颅腔容积和身材(身高)相当于或小于南方古猿阿法种。缺少南方古猿与傍人所表现出的咀嚼适应性变化，面部高度与凸颌程度明显降低，犬后齿较小，眶下区朝后。颅底角度收缩。突出的上犬齿隆突呈突出柱状，在侧面与鼻孔分离。岩锥平滑、呈管状且凸出程度低，茎突缺失，且没有茎突鞘脊。上部颅顶骨厚于南方古猿而与直立人及智人相当。眶上圆枕在每一个眼眶上呈拱形，且不像爪哇直立人那样呈水平条状。下第 3 前臼齿有相对较大的咀嚼面表面积，下原尖突出，下跟座宽阔，而且牙根或者完全分叉或者为前后压扁的托姆斯式齿根。下第 4 前臼齿也具有托姆斯式齿根。第一和第二臼齿大小相当。颌骨冠状突高于髁突，且上升支向后倾。下颌联合部没有颏，联合部轴线向后倾斜。齿槽面向后倾斜，并且有上下圆枕。髂骨具明显侧向展开。股骨颈部相对于股骨头直径相对显长，骨干截面呈圆形且没有粗线嵴发育，双髁角角度大。胫骨长轴弯曲且骨干中部的横截面为椭圆形。

颅骨与颅后骨骼部位的描述与对比

除了右侧颧弓，颅骨基本没有变形(图 1、图 2)。不幸的是，骨架发现时，其前囟区、右侧额骨、眶上、鼻骨及鼻下区域被损坏。为了修复死后遭受压力产生的裂缝、稳固穹隆，在重建前拆开了颅顶穹隆并从颅内进行了清理。除鳞骨骨缝以外，大部分颅骨穹隆骨缝难以定位，这一问题在计算机断层扫描(CT)中也存在。因此大多数标准颅骨测量标志点都不可能高精度定位。

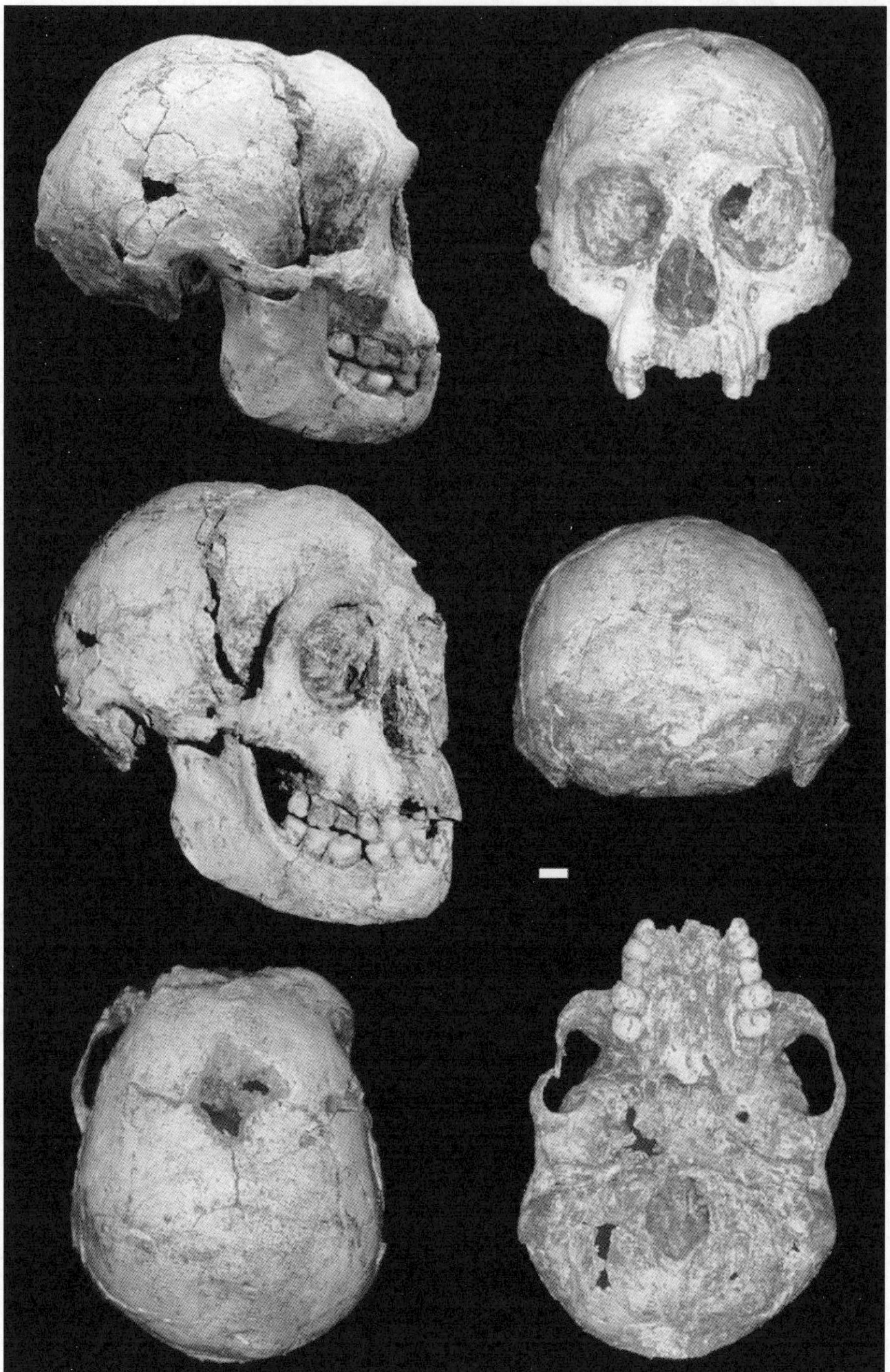

Fig. 1. The LB1 cranium and mandible in lateral and three-quarter views, and cranium in frontal, posterior, superior and inferior views. Scale bar, 1 cm.

The LB1 cranial vault is long and low. In comparison with adult *H. erectus* (including

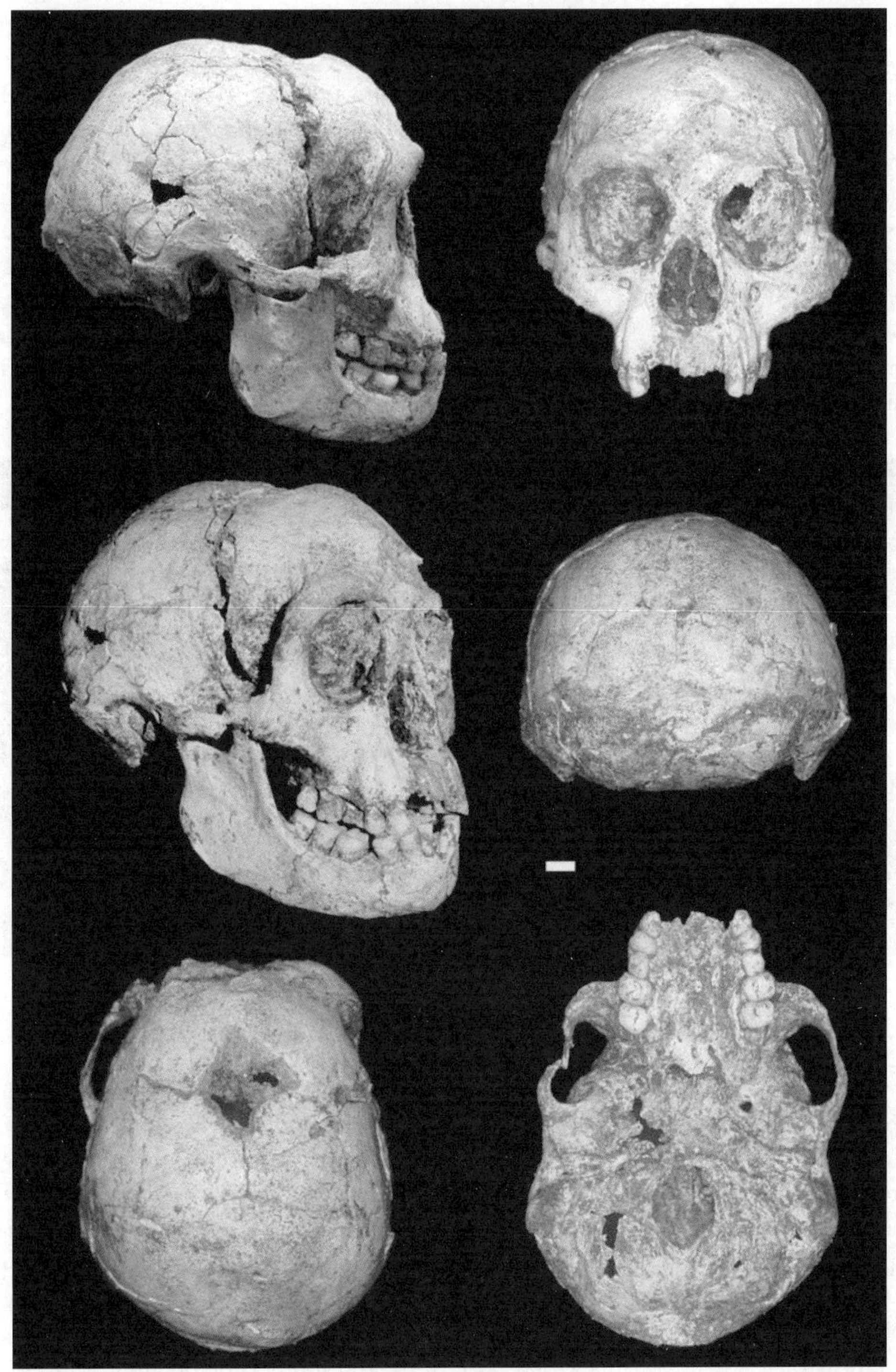

图 1. LB1 颅骨与下颌骨的侧视图与四分之三视图。颅骨的正视、后视、顶视及底视图。比例尺，1 厘米。

LB1 颅骨穹隆长而低。与成年直立人（包括被归属为匠人与格鲁吉亚人的标本）

specimens referred to as *Homo ergaster* and *Homo georgicus*) and *H. sapiens* the calvarium of LB1 is extremely small. Indices of cranial shape closely follow the pattern in *H. erectus* (Supplementary Table 1). For instance, maximum cranial breadth is in the inflated supramastoid region, and the vault is broad relative to its height. In posterior view the parietal contour is similar to *H. erectus* but with reduced cranial height[2,3]. Internal examination of the neurocranium, directly and with CT scan data, indicates that the brain of LB1 had a flattened platycephalic shape, with greatest breadth across the temporal lobes and reduced parietal lobe development compared with *H. sapiens*. The cranial base angle (basion–sella–foramen caecum) of 130° is relatively flexed in comparison with both *H. sapiens* (mean 137°–138° (refs 4, 5)) and Indonesian *H. erectus* (Sambungmacan 4 141° (ref. 6)). Other small-brained hominins, for instance STS 5 *Australopithecus africanus*, have the primitive less-flexed condition.

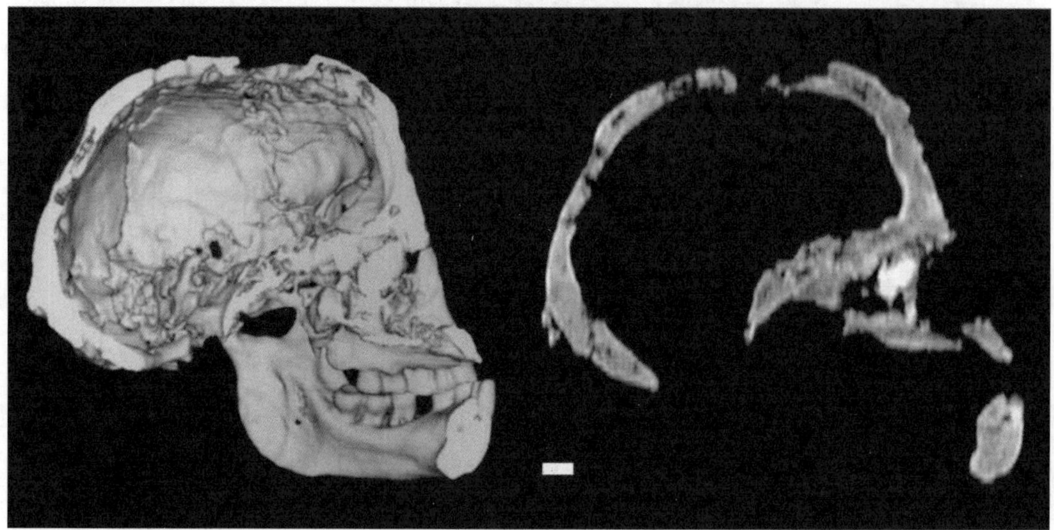

Fig. 2. Rendered three-dimensional and individual midsagittal CT section views of the LB1 cranium and mandible. Scale bar, 1 cm.

The endocranial volume, measured with mustard seed, is 380 cm^3, well below the previously accepted range for the genus *Homo*[7] and equal to the minimum estimates for *Australopithecus*[8]. The endocranial volume, relative to an indicator of body height (maximum femur length 280 mm), is outside the recorded hominin normal range (Fig. 3). Medially, laterally and basally, the cranial vault bone is thick and lies within the range of *H. erectus* and *H. sapiens*[9,10] (Supplementary Table 1 and Fig. 2). Reconstruction of the cranial vault, and CT scans, indicated that for most of the cranial vault the relative thickness of the tabular bone and diploë are similar to the normal range in *H. erectus* and *H. sapiens*. In common with *H. erectus* the vault in LB1 is relatively thickened posteriorly and in areas of pneumatization in the lateral cranial base. Thickened vault bone in LB1, relative to that in *Australopithecus* and early *Homo*[2], results in a substantially reduced endocranial volume in comparison to Plio-Pleistocene hominins with similar external vault dimensions.

及智人相比，LB1 的颅骨穹隆部极小。颅骨形状指数非常符合直立人的模式（补充表 1）。例如，颅骨宽度最大在膨胀的乳突上方区域，穹隆相对于其高度更显宽阔。在后视图中，顶骨轮廓与直立人相似，但是颅骨高度减小[2,3]。对脑颅内部的直接观察与 CT 扫描数据观察表明 LB1 的大脑具有变平了的扁头形状，最大宽度横过颞叶；与智人相比，顶叶发育减弱。颅底角（颅底点–蝶鞍点–孔盲）为 130°，它与智人（平均为 137° ~ 138°（参考文献 4、5））及印度尼西亚直立人（萨姆伯戈默卡 4 141°（参考文献 6））相比，相对屈曲。其他脑容量小的人族成员，例如，STS 5 南方古猿非洲种，拥有屈曲程度较低的原始状态。

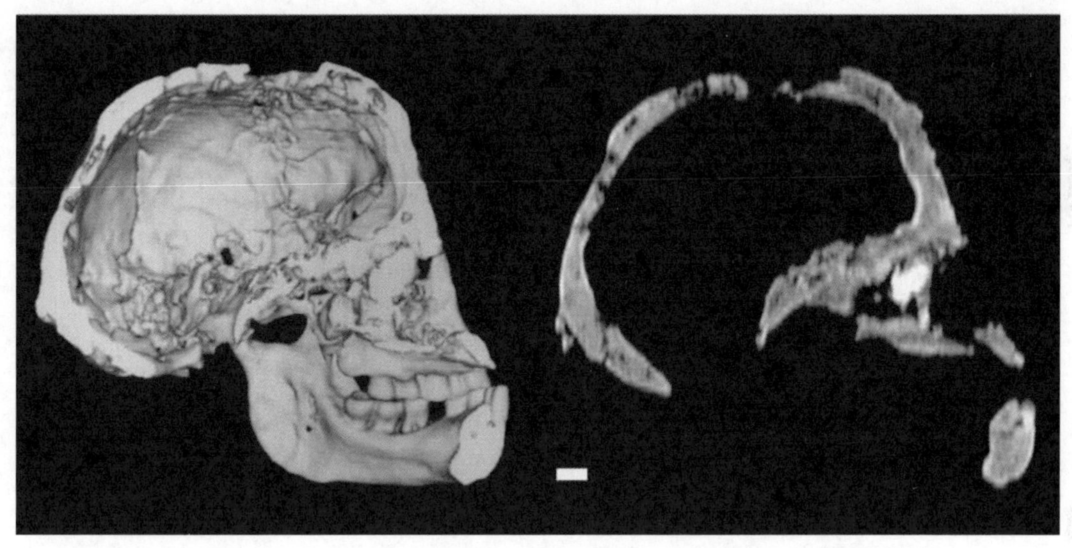

图 2. LB1 颅骨与下颌骨渲染三维视图与单个正中矢状面 CT 截面视图。比例尺，1 厘米。

用芥菜籽来测量的颅腔容积是 380 立方厘米，远低于以前公认的人属[7]的范围，而与南方古猿[8]的最小估计值相当。相对于身高指标（最大股骨长度 280 毫米），颅腔容积位于已知人族成员的正常范围之外（图 3）。颅骨穹隆部骨骼在中间、侧面及底部都厚且在直立人与智人[9,10]的范围之内（补充表 1 和图 2）。颅骨穹隆的重建及 CT 扫描表明，对于颅骨穹隆的大部分来讲，板状骨与板障骨的相对厚度与直立人及智人的正常范围相当。与直立人相同，在 LB1 穹隆后部与颅底侧面气腔区域相对变厚。相对于南方古猿和早期人属成员[2]，LB1 增厚的穹隆骨骼，导致与外部穹隆尺寸相似的上新世－更新世人族成员相比，颅腔容积大幅减少。

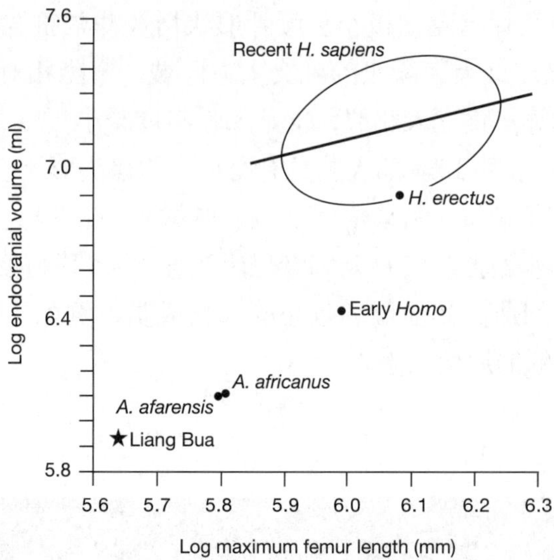

Fig. 3. Relationship between endocranial volume and femur length in LB1, *A. afarensis*, *A. africanus*, early *Homo* sp., *H. erectus* and modern *H. sapiens*. Modern human data, with least squares regression line and 95% confidence ellipse, from a global sample of 155 individuals collected by P.B. Details of the hominin samples are in the Supplementary Information.

The occipital of LB1 is strongly flexed, with an occipital curvature angle of 101° (Supplementary Information), and the length of the nuchal plane dominates over the occipital segment. The occipital torus forms a low extended mound, the occipital protuberance is not particularly prominent compared with Indonesian *H. erectus* and there is a shallow supratoral sulcus. The endinion is positioned 12 mm inferior to the inion, which is within the range of *H. erectus* and *Australopithecus*[10]. Compared with *Australopithecus* and early *Homo*[2] the foramen magnum is narrow (21 mm) relative to its length (28 mm), and mastoid processes are thickened mediolaterally and are relatively deep (20.5 mm). In common with Asian, and some African, *H. erectus* a deep fissure separates the mastoid process from the petrous crest of the tympanic[10,11]. Bilaterally there is a recess between the tympanic plate and the entoglenoid pyramid. These two traits are not seen in modern humans, and show varied levels of development in Asian and African *H. erectus* and Pliocene hominins[10]. The depth and breadth of the glenoid fossae and angulation of the articular eminence are within the range of variation in *H. sapiens*. The inferior surface of the petrous pyramid has numerous similarities with Zhoukoudian *H. erectus*[12], with a smooth tubular external surface as in chimpanzees, and a constricted foramen lacerum. Styloid processes and vaginal crests are not present.

The temporal lines approach to within 33 mm of the coronal suture and have a marked posterior extension. There are no raised angular tori as is common in *H. erectus*[10] and some terminal Pleistocene Australians, and no evidence of parietal keeling. Posteriorly there is some asymmetrical obelionic flattening and CT scans indicate that the parietals reduce in thickness in this slightly depressed area (Fig. 2). A principal component analysis (PCA) of five cranial vault measurements separates LB1, STS5 (*A. africanus*) and KNM-ER 1813

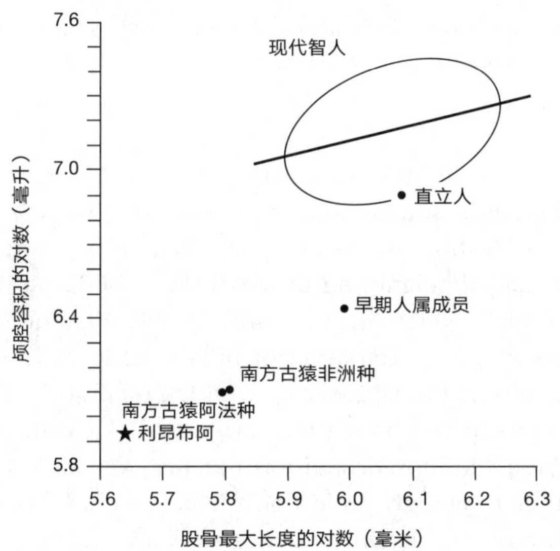

图 3. 在 LB1、南方古猿阿法种、南方古猿非洲种、早期人属未定种、直立人及现代智人中，颅腔容积与股骨长度之间的关系。现代人类的数据带有最小二乘回归线与 95% 置信椭圆，来自布朗收集的全球 155 个个体样本。人族成员样本的详细情况见补充信息。

LB1 的枕骨强烈向下屈曲，枕骨弯曲角 101°(补充信息)，而且项平面长度占了枕节的大部分。枕骨圆枕形成了一个较低的延伸小丘，与印度尼西亚直立人相比，枕骨隆突不是特别突出，并且有一浅上圆枕沟。枕内点位置低于枕外隆凸点 12 毫米，在直立人与南方古猿的范围之内 [10]。与南方古猿及早期人属成员 [2] 相比，枕骨大孔相对于其长度(28 毫米)显狭窄(21 毫米)，乳突由中线向侧面增厚，相对较深(20.5 毫米)。与亚洲和某些非洲直立人相同，有一条很深的裂缝将乳突与鼓室的岩骨脊分开 [10,11]。在鼓板与内颞颌关节锥之间有一两侧凹陷。这两个特征在现代人中是看不到的，显示出亚洲及非洲直立人和上新世人族成员的多样的发育水平 [10]。关节窝的深度和宽度以及关节隆起的角度在智人的变异范围之内。岩锥的下表面与周口店直立人 [12] 有许多相似之处，拥有黑猩猩那样光滑管状的外表面，以及收缩的破裂孔。无茎突与茎突鞘脊突。

颞线与冠状缝相交的位置，距离冠状缝起点在 33 毫米之内，且颞线有明显的向后延伸。不具有直立人 [10] 与一些更新世末尾的澳大利亚人中常见的升高的角圆枕，也没有顶骨矢状隆起的证据。在后方矢状缝的顶骨孔之间区域有一不对称变平，CT 扫描表明顶骨在这个稍微凹陷的区域厚度减少(图 2)。5 个颅骨穹隆测量值的主成分分析(PCA)，将 LB1、STS5(南方古猿非洲种)与 KNM-ER 1813(早期人属成员)

(early *Homo*) from other hominin calvaria in size and shape. Shape, particularly height and breadth relationships, placed LB1 closest to ER-3883, ER-3733 and Sangiran 2 *H. erectus* (Supplementary Fig. 1).

The face of LB1 lacks most of the masticatory adaptations evident in *Australopithecus* and its overall morphology is similar to members of the genus *Homo*[2,3]. In comparison with *Australopithecus*, tooth dimensions and the alveolar segment of the maxillae are greatly reduced, as are facial height and prognathism. The facial skeleton is dominated by pronounced canine juga, which form prominent pillars lateral to the nasal aperture. However, these are distinct from the anterior pillars adjacent to the nasal aperture in *A. africanus*[2,3]. The infraorbital fossae are deep with large infraorbital foramina, the orbits have a particularly arched superior border and a volume of 15.5 cm^3 (ref. 13). On the better preserved right-hand side, the supraorbital torus arches over the orbit and does not form a straight bar, with bulbous laterally projecting trigones, as in Indonesian *H. erectus*[11]. The preserved section of the right torus only extends medially slightly past mid-orbit, and the morphology of the glabella region and medial torus is unknown. In facial view the zygo-maxillary region is medially deep relative to facial height, and the inferior border of the malars are angled at 55° relative to the coronal plane. In lateral view the infraorbital region is orientated posteriorly as in other members of the genus *Homo*, rather than the more vertical orientation in *A. africanus*[2,3]. The root of the maxillary zygomatic process is centred above the first molar, and the incisive canal is relatively large and has an anterior location, contrasting with African and Javan *H. erectus*. In lateral view, curvature of the frontal squama is more similar to African early *Homo* and Dmanisi *H. ergaster*[3,14] than it is to the Javan hominins. The frontal squama is separated from the supraorbital torus by a supraorbital sulcus. In the middle third of the frontal there is a slight sagittal keel, extending into the remains of a low, broad prebregmatic eminence. On the midfrontal squama there is a circular healed lesion, probably the remains of a depressed fracture, which is about 15 mm across.

The mandible is complete, apart from some damage to the right condyle (Fig. 4) and combines features present in a variety of Pliocene and Pleistocene hominins. Post-mortem breaks through the corpus at the right P_3 and M_2, and the left canine have resulted in some lateral distortion of the right ramus. There is a strong Curve of Spee. The ramus root inserts on the corpus above the lateral prominence, and in lateral aspect obscures the distal M_3. The ramus is broadest inferiorly, slopes slightly posteriorly and is thickened medio-laterally, and the coronoid process is higher than the condyle. The right condyle has a maximum breadth of 18 mm. There is a narrow and shallow extramolar sulcus and moderate lateral prominence. The anterior portion of the corpus is rounded and bulbous and without a chin. In the posterior symphyseal region the alveolar planum inclines postero-inferiorly, there is a moderate superior torus, deep and broad digastric fossa, and the inferior transverse torus is low and rounded rather than shelf-like (Fig. 4). There is a strong posterior angulation of the symphyseal axis, and the overall morphology of the symphysis is very similar to LH4 *A. afarensis* and unlike Zhoukoudian and Sangiran *H. erectus*. There are bilaterally double mental foramina, with the posterior foramina smaller and located more inferiorly. Double

在大小与形状上与其他人族成员颅顶分开。形状，特别是高度和宽度之间的关系，把 LB1 放到最靠近 ER-3883、ER-3733 及桑吉兰 2 直立人的位置（补充图 1）。

LB1 面部缺少绝大多数南方古猿中明显的与咀嚼相关的适应特征，且其总体形态与人属成员相似[2,3]。与南方古猿对比，其牙齿大小与上颌骨的齿槽部分都明显减小，面部高度和凸颌程度也是同样的情况。犬齿隆凸突出，在面部骨骼上非常显著，在鼻孔侧面形成突出圆柱。然而，这与南方古猿非洲种中邻近鼻孔的前部圆柱截然不同[2,3]。眶下窝深，眶下孔大，眼窝具有一个特别拱曲的上缘，其容积为 15.5 立方厘米（参考文献 13）。在保存较好的右侧，眶上圆枕在眼眶上成拱形，并不是形成平直条状，具有向侧面突出的球状三角区，正如印度尼西亚直立人中那样[11]。保存的圆枕右侧仅向中间延伸稍微超过眼眶中部，眉间区与中间圆枕形态未知。在正视图中，颧骨上颌骨区域，相对于面高，至面部中线的深度显大，颧骨下缘相对于冠状面所成角度为 55°。侧视图中，眶下区域朝向后方，如人属其他成员一样，而不像南方古猿非洲种那样朝向更加垂直[2,3]。上颌骨颧突根部以第一臼齿上方为中心，门齿管相当大且位置靠前，与非洲和爪哇直立人不同。在侧面图中，额骨鳞部的弯曲程度较爪哇人族成员与非洲的早期人属与德马尼西匠人[3,14]更接近。一眶上沟把额骨鳞部与眶上圆枕分开。在额骨中部三分之一，有一轻微矢状隆起，延伸成为低而宽阔的前囟前部隆起的残留。在中额骨鳞部，有一圆形已愈合的损伤，可能是凹陷骨折的痕迹，大约 15 毫米宽。

下颌骨完整，仅右侧髁突部分损坏（图 4），且复合了多种上新世与更新世人族成员的特征。右下第 3 前臼齿和第 2 臼齿以及左犬齿处的死后破裂贯穿了下颌体，使得右侧上升枝有些侧向变形。下颌齿列表现出强烈的施佩曲线。上升枝根部在侧面下颌突之上嵌入下颌体，从侧面遮挡下第 3 臼齿远中部。上升枝下部最宽，向后稍有倾斜，中侧向变厚，冠状突高于髁突。右侧髁突最大宽度 18 毫米。外臼齿沟窄而浅，下颌突中等程度发育。下颌体前部浑圆肥硕且没有颏。在下颌联合后部区域，齿槽面向后向下倾斜，上圆枕中等程度发育，二腹肌窝深且宽阔，下圆枕低而圆，不呈搁架状（图 4）。下颌联合轴线有一个强烈向后的角度，下颌联合的总体形态与 LH4 南方古猿阿法种很相似，而与周口店及桑吉兰直立人不相像。在两侧都有两个颏孔，后一个孔较小，位置更靠下。双颏孔在印度尼西亚直立人[15]中普遍存在。而

mental foramina are common in Indonesian *H. erectus*[15]. While the mandibular dental arch is narrow anteriorly, and long relative to its breadth, the axis of P_3–M_3 is laterally convex rather than straight (Fig. 4).

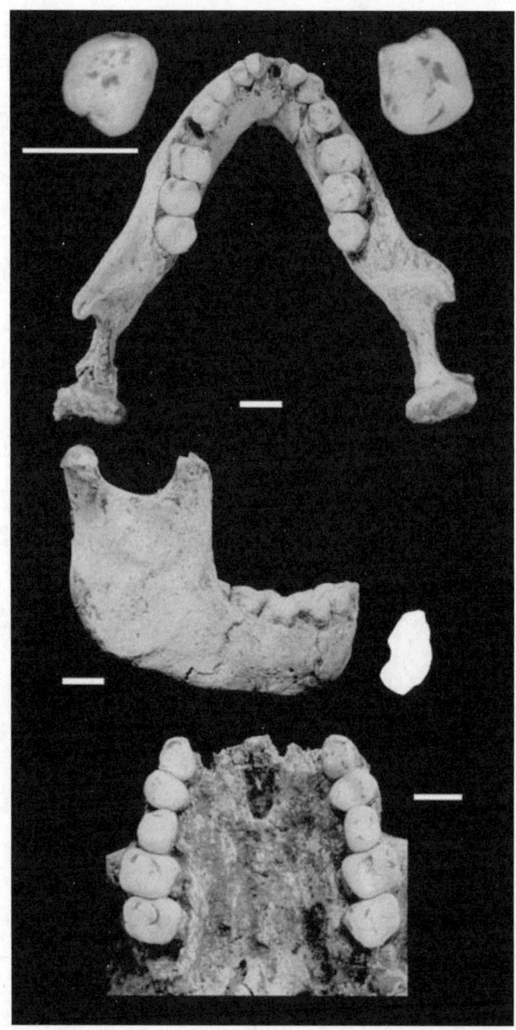

Fig. 4. Right lateral and occlusal views of the LB1 mandible, sagittal profile of the symphysis, occlusal view of the mandibular dentition and occlusal views of the mandibular premolars. Scale bars, 1 cm.

The right P_4 is absent and the alveolus completely fused, the left P_4 was lost after death, and CT scans indicate that the maxillary right M^3 was congenitally absent. The relatively small and conical alveolus for the missing left M^3 suggests that it had a much smaller crown than M^1 and M^2. Size, spacing and angulation of the maxillary incisor alveoli, and absence of a mesial facet on the canines suggest that incisor I^2 was much smaller than I^1, and there may have been a diastema. Occlusal wear has removed details of cusp and fissure morphology from most of the maxillary and mandibular teeth. The canines have worn down to a relatively flat surface and there would have been an edge-to-edge bite anteriorly.

下颌骨齿弓前面窄，相对于宽度显得长，下第 3 前臼齿至第 3 臼齿轴线向侧面凸出而不是直的(图 4)。

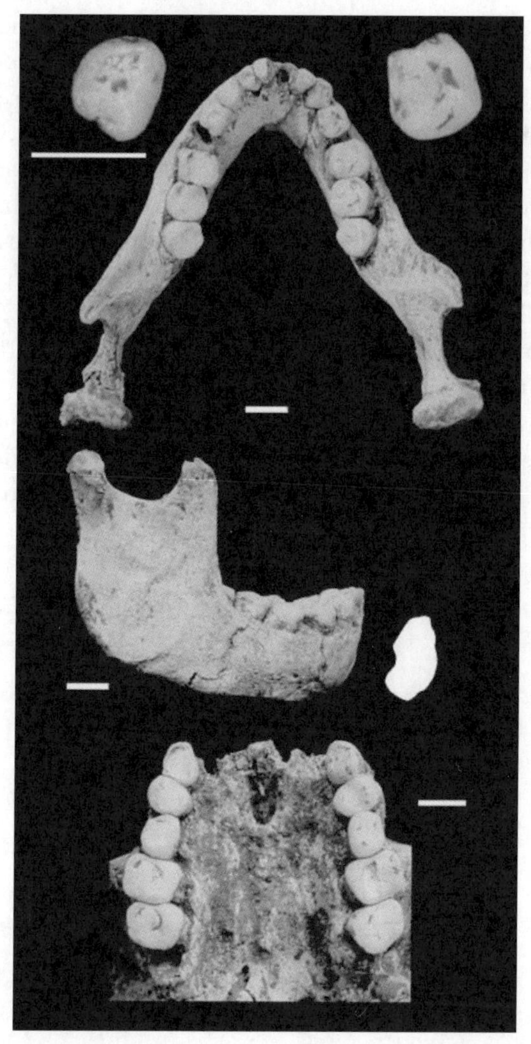

图 4. LB1 下颌骨的右侧视图与咀嚼面视图、联合部矢状面轮廓、下齿列咀嚼面视图及下前臼齿咀嚼面视图。比例尺，1 厘米。

右下第 4 前臼齿缺失且齿槽完全合并，左下第 4 前臼齿为死后丢失，CT 扫描显示右上第 3 臼齿先天缺失。丢失的左上第 3 臼齿齿槽相对小且呈圆锥形，这表明其齿冠要远小于上第 1 臼齿和上第 2 臼齿。上门齿齿槽的大小、间距和角度，以及犬齿近中接触面缺失表明，上第 2 门牙比上第 1 门牙小得多，并且可能存在齿隙。上下颌牙齿咀嚼面的磨耗已经把大部分上下颌牙齿上的齿尖及裂隙形态的细节都磨掉了。犬齿已经磨耗到一个相当平坦的平面，在牙齿前部可能曾有边缘对边缘的咬合。

Interproximal wear is pronounced and in combination with the loss of crown height means that mesio-distal crown dimensions convey little phylogenetic information. With the exception of P_3 the size and morphology of the mandibular teeth follow the pattern in *H. erectus* and *H. sapiens* (Fig. 5, Supplementary Table 2). There is not a great deal of difference between the size of the molar teeth in each quadrant, and the size sequence for both mandibular and maxillary teeth is $M1 \geqslant M2 > M3$. Using the megadontial quotient as a measure of relative tooth size[16], and substituting P_3 crown area for the missing P_4s, LB1 is megadont (1.8) relative to *H. sapiens* (0.9) and *H. ergaster* (0.9), but not *H. habilis* (1.9) (ref. 8) (Supplementary Information). The P_3s have a relatively great occlusal surface area (molariform) and when unworn had a prominent protoconid and broad talonid. Both P_3s have bifurcated roots and the alveolus for the left P_4 indicates a mesiodistally compressed, broad Tomes' root. A larger, less worn, isolated left P_3 from the deposit (LB2) has a more triangular occlusal outline, and a Tomes' root (Supplementary Fig. 2). Mandibular P_3s and P_4s with similar crown and root morphology have been recorded for *Australopithecus* and early *Homo*[17,18], and some Indonesian *H. erectus* mandibular premolars also have bifurcated or Tomes' roots[15]. Unusually, both maxillary P^4s are rotated parallel to the tooth row, a trait that seems to be unrecorded in any other hominin. Maxillary canines and P^3s have long roots and very prominent juga. The P^3 juga are emphasized by the rotation of the adjacent P^4 roots.

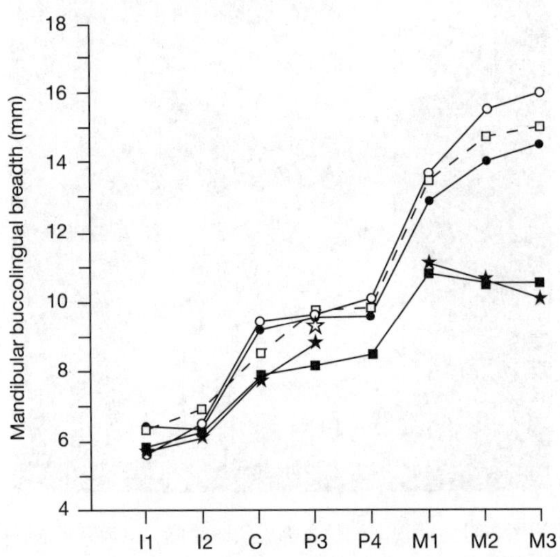

Fig. 5. Mean buccolingual tooth crown breadths for mandibular teeth in *A. afarensis* (filled circles), *A. africanus* (open circles), early *Homo* sp. (open squares), modern *H. sapiens* (filled squares), LB1 (filled stars) and LB2 (open stars). There are no mandibular P_4s preserved for LB1. Data for *Australopithecus* and early *Homo* are from ref. 49. Modern human data from a global sample of 1,199 individuals collected by P.B.

The pelvic girdle is represented by a right innominate, with damage to the iliac crest and pubic region, and fragments of the sacrum and left innominate. The right innominate, which is undistorted, has a broad greater sciatic notch suggesting that LB1 is a female (Fig. 6). In common with all bipedal hominins, the iliac blade is relatively short and wide[19]; however, the ischial spine is not particularly pronounced. Compared with modern humans the LB1

齿间磨耗显著，加上齿冠高度的磨损，意味着齿冠近远中尺寸所含有的系统发育信息很少。除了下第 3 前臼齿，下颌牙齿的大小和形态遵循直立人与智人的模式（图 5 和补充表 2）。在每个四分之一齿枝上，臼齿的大小没有很大差别，上下颌臼齿的大小次序是 M1 ≥ M2 > M3。如果以牙齿巨大系数作为牙齿相对尺寸的量度 [16]，用下第 3 前臼齿齿冠面积替代失去的下第 4 前臼齿，相对于智人（0.9）与匠人（0.9），LB1 属于牙齿巨大（1.8），但相对于能人（1.9）则不是如此（参考文献 8）（补充信息）。下第 3 前臼齿拥有相对大的咀嚼面表面积（臼齿形），未磨耗时下原尖突出，下跟座宽阔。两侧下第 3 前臼齿均拥有二分叉齿根，左下第 4 前臼齿的齿槽孔表明其拥有近远中向压扁且宽阔的托姆斯式齿根。1 枚来自沉积物的更大且磨耗程度较低的左下第 3 前臼齿（LB2）拥有更趋于三角形的咀嚼面轮廓和托姆斯式齿根（补充图 2）。下颌第 3 和第 4 前臼齿具有相似的齿根和齿冠形态，这在南方古猿与早期人属中已有记载 [17,18]，而且一些印度尼西亚直立人下颌前臼齿也有二分叉或托姆斯式齿根 [15]。不同寻常的是，两侧的上颌上第 4 前臼齿均旋转并平行于齿列，这个特征似乎在任何其他人族成员中都没有记载过。上颌犬齿与上第 3 前臼齿牙根长且隆凸十分突出。上第 3 前臼齿隆凸由于相邻上第 4 前臼齿齿根的旋转而显加强。

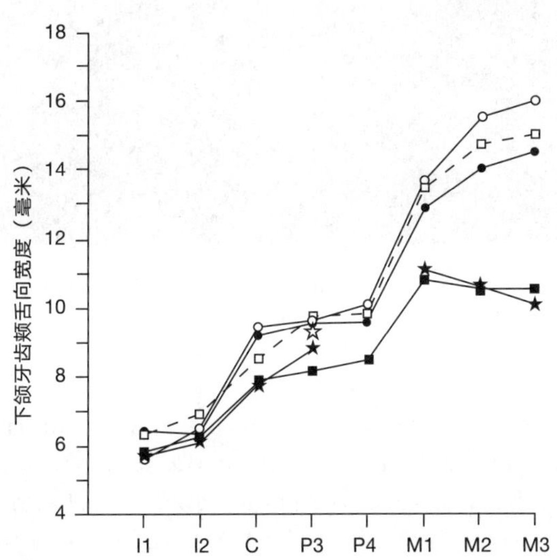

图 5. 南方古猿阿法种（实心圆）、南方古猿非洲种（空心圆）、早期人属（空心正方形）、现代智人（实心正方形）、LB1（实心五角形）及 LB2（空心五角形）的下颌牙齿齿冠颊舌向的平均宽度。LB1 的两侧下第 4 前臼齿未保存。南方古猿与早期人属数据来自参考文献 49。现代人类数据来自布朗收集的一个包括 1,199 个个体的全球样本。

腰带部分保存了髂嵴与耻骨区域破损的右侧髋骨和骶骨及左侧髋骨碎片。右侧髋骨，未变形，坐骨大切迹宽阔，表明 LB1 为一女性（图 6）。和所有两足直立行走的人族成员一样，髂骨刃相对短而宽 [19]；不过，坐骨棘不是特别明显。与现代人相

ilium has marked lateral flare, and the blade would have projected more laterally from the body, relative to the plane of the acetabulum. The left acetabulum is of circular shape, and has a maximum width of 36 mm.

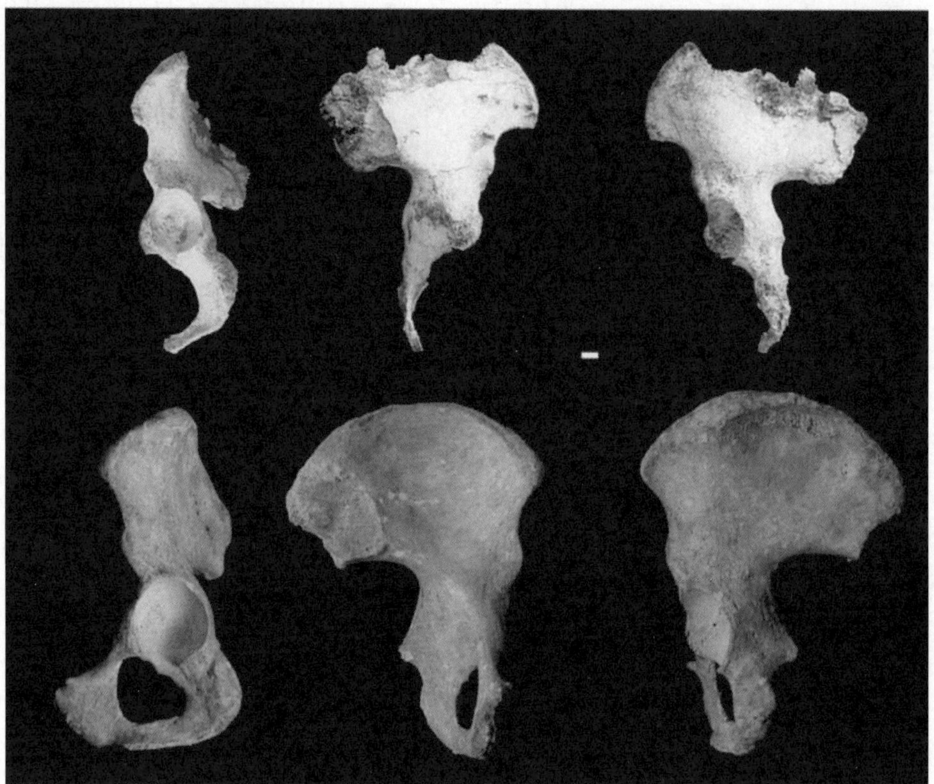

Fig. 6. Comparison of the left innominate from LB1 with a modern adult female *H. sapiens*. Lateral (external), and medial and lateral views of maximum iliac breadth. The pubic region of LB1 is not preserved and the iliac crest is incomplete. Scale bar, 1 cm.

Apart from damage to the lateral condyle and distal shaft, the right femur is complete and undistorted (Fig. 7). The overall anatomy of the femur is most consistent with the broad range of variation in *H. sapiens*, with some departures that may be the result of the allometric effects of very small body size. The femur shaft is relatively straight, and areas of muscle attachment, including the linea aspera, are not well developed. In contrast with some examples of Asian and African *H. erectus*, the femora do not have reduced medullary canals[20]. On the proximal end, the lesser trochanter is extremely prominent and the strong development of the intertrochanteric crest is similar to *H. sapiens* rather than the flattened intertrochanteric area in *Australopithecus* and *H. erectus* (KNM-ER 1481A, KNM-WT 15000). The biomechanical neck length is 55.5 mm and the neck is long relative to the femoral head diameter (31.5 mm), as is common to both *Australopithecus* and early *Homo*[19]. The neck–head junction is 31.5 mm long, with a shaft–neck angle of 130°, and the femur neck is compressed anteroposteriorly (Fig. 7). Several indices of femoral size and shape, for example the relationship between femoral head size and midshaft circumference (66 mm), and femur

比，LB1 髂骨侧面显著变宽，这表明，相对于髋臼平面，髂骨刃会向身体侧面凸出更多。左侧髋臼呈圆形，最大宽 36 毫米。

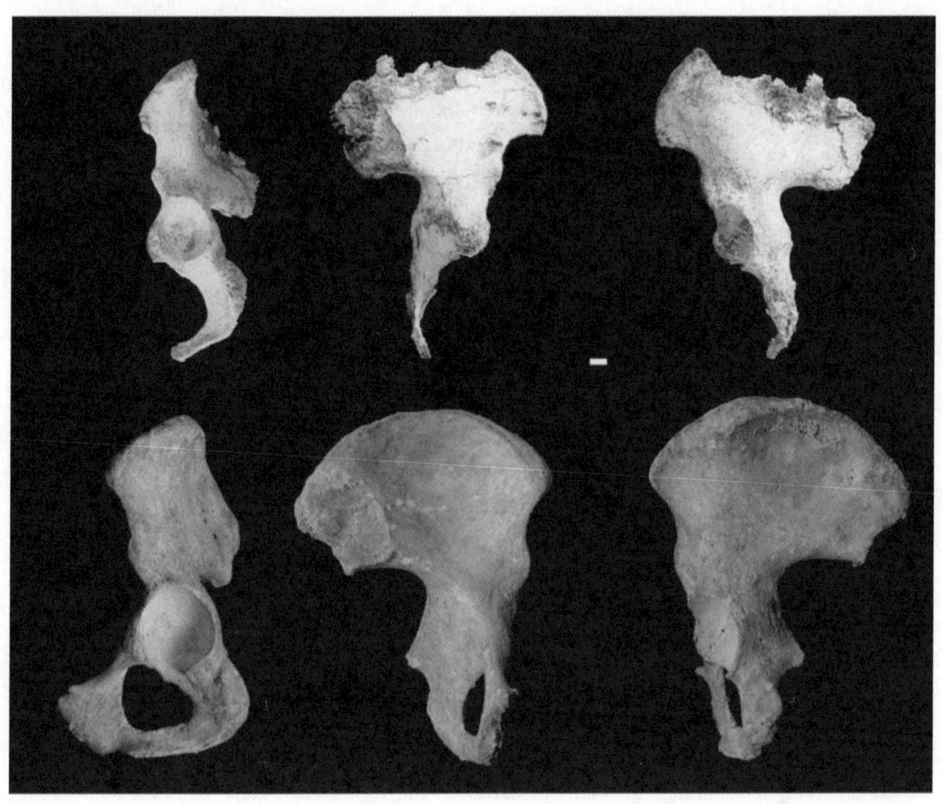

图 6. LB1 左侧髋骨与一现代女性智人的对比。髂骨最大宽度侧（外）视图、内侧视图及侧视图。LB1 的耻骨区域未保存，且髂嵴不完整。比例尺，1 厘米。

除了外侧髁突和远端骨干有破损，右侧股骨完整且未变形（图 7）。股骨的整体解剖学特征与智人宽泛变异范围是最为一致的，一些偏差可能是由体型矮小的异速生长效应导致。股骨干相对较直，包括股骨粗线在内的肌肉附着区不很发育。与亚洲及非洲直立人的一些例子不同，其两侧股骨没有减弱的髓腔 [20]。在近中末端，小转子极其突出，转子间嵴的强烈发育类似于智人，而与南方古猿属和直立人（KNM-ER 1481A，KNM-WT 15000）的平坦的转子间区不同。股骨颈的生物力学长度为 55.5 毫米，且相对于股骨头直径（31.5 毫米）显长，这在南方古猿属及早期人属中也普遍存在 [19]。股骨颈–股骨头联合处长 31.5 毫米，股骨干–股骨颈成 130° 角，股骨颈前后压缩（图 7）。股骨大小与形状的几个指标，例如股骨头大小和股骨干中部周长（66 毫米）间及股骨长度和转子下方骨干大小间的关系 [21]，都落入黑猩猩与南方古猿的变异范围之内。股骨干没有粗线嵴，其横截面呈圆形，骨干中部横截面面积

length and sub-trochanteric shaft size[21], fall within the chimpanzee and australopithecine range of variation. The femur shaft does not have a pilaster, is circular in cross-section, and has cross-sectional areas of 370 mm^2 at the midshaft and 359 mm^2 at the midneck. It is therefore slightly more robust than the best-preserved small-bodied hominin femur of similar length (AL288-1; ref. 21). Distally there is a relatively high bicondylar angle of 14°, which overlaps with that found in *Australopithecus*[22].

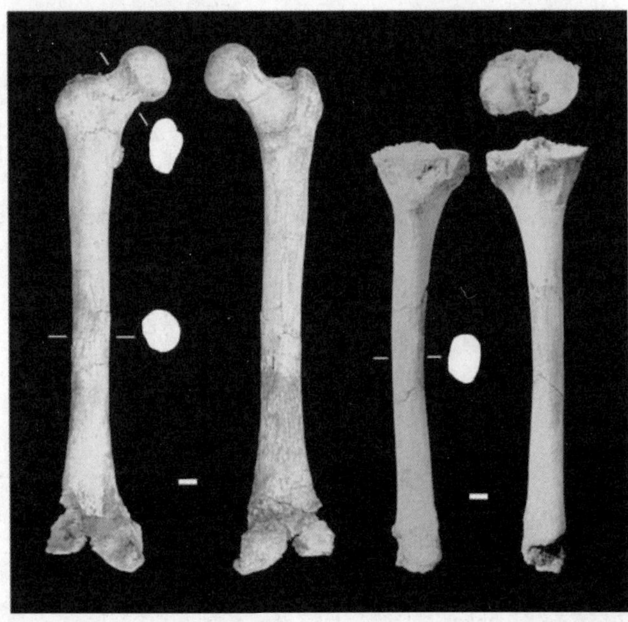

Fig. 7. Anterior and posterior views of the LB1 right femur and tibia, with cross-sections of the femur neck and midshaft, and tibia midshaft. The anterior surfaces of the medial and lateral condyles of the femur are not preserved. With the exception of the medial malleolus, the tibia is complete and undistorted. Scale bar, 1 cm.

The right tibia is complete apart from the tip of the medial malleolus (Fig. 7). Its most distinctive feature, apart from its small size (estimated maximum length 235 mm, bicondylar breadth 51.5 mm) and the slight curvature in the long axis, is a shaft that is oval in cross-section (midshaft 347 mm^2), without a sharp anterior border, and relatively thickened medio-laterally in the distal half. The relationship between the midshaft circumference and the length of the tibia is in the chimpanzee range of variation and distinct from *Homo*[21].

Additional evidence of a small-bodied adult hominin is provided by an unassociated left radius shaft, without the articular ends, from an older section of the deposit (74–95 kyr). The estimated maximum length of this radius when complete is approximately 210 mm. Although the arms of LB1 have not been recovered, the dimensions of this radius are compatible with a hominin of LB1 proportions.

Although there is considerable interspecific variation, stature has been shown to have phylogenetic and adaptive significance among hominins[23]. Broadly speaking, *Australopithecus*

为 370 平方毫米，股骨颈中部横截面面积为 359 平方毫米。因此，它比保存最完好且具有相似长度的身材矮小的人族成员的股骨略显更加粗壮(AL288-1；参考文献 21)。在远端双髁角相对显大，为 14°，与南方古猿属重叠[22]。

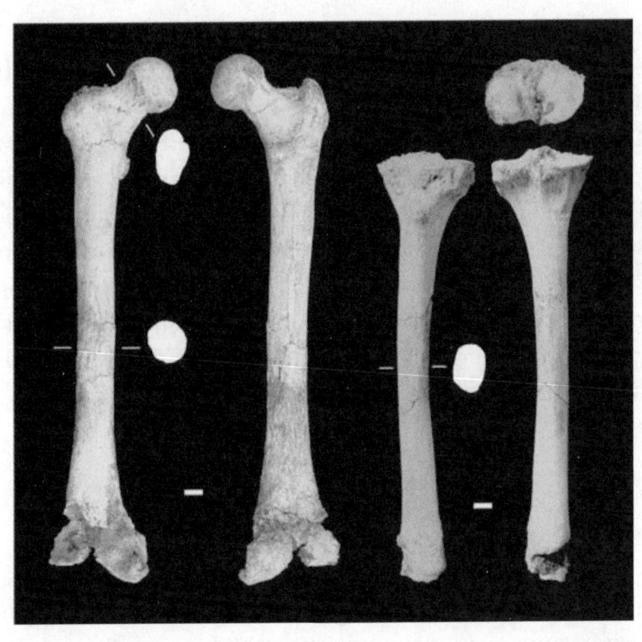

图 7. LB1 右侧股骨和胫骨的前视图与后视图，及股骨颈、股骨干与胫骨干的横截面。股骨的中髁突与侧髁突前方未保存。除了内踝，胫骨完整且未变形。比例尺，1 厘米。

除了内踝末端破损，右胫骨完整(图 7)。除了其尺寸较小(估计最大长为 235 毫米，双髁宽为 51.5 毫米)以及长轴微弯外，它最与众不同的特征是其骨干的横截面呈椭圆形(骨干中部横截面面积为 347 平方毫米)，无明显的前嵴，远端半部的内–侧向加厚。骨干中部周长和胫骨长度之间的关系在黑猩猩的变异范围之内，而与人属截然不同[21]。

另外一根采自更老堆积物(74,000 ~ 95,000 年)且与之不相关的缺失关节部位的左侧桡骨骨干，为身材矮小的成年人族成员提供了附加的证据。该桡骨完整状态下推测最大长度约为 210 毫米。虽然 LB1 的双臂还没有被发现，但该桡骨的尺寸与 LB1 这一人族成员的比例一致。

虽然种间变异相当大，但身高已经在人族之中被证明具有系统发育和适应方面

and the earliest members of the genus *Homo* are shorter than *H. erectus* and more recent hominins[8]. The maximum femur length of LB1 (280 mm) is just below the smallest recorded for *A. afarensis* (AL-288-1, 281 mm[24]) and equal to the smallest estimate for the OH 62 *H. habilis* femur (280–404 mm)[21]. Applying stature estimation formulae developed from human pygmies[25] gives a stature estimate of 106 cm for LB1 (Supplementary Information). This is likely to be an overestimation owing to LB1's relatively small cranial height.

A stature estimate for LB1 of 106 cm gives a body mass of 16 to 28.7 kg, and a femur cross-sectional area of 525 mm^2 gives a mass of 36 kg (Supplementary Information). The brain mass for LB1, calculated from its volume[26], is 433.2 g; this gives an encephalization quotient (EQ)[27] range of 2.5–4.6, which compares with 5.8–8.1 for *H. sapiens*, 3.3–4.4 for *H. erectus/ergaster* and 3.6–4.3 for *H. habilis*, and overlaps with the australopithecine range of variation[28,29]. If LB1 shared the lean and relatively narrow body shape typical of Old World tropical modern humans then the smallest body weight estimate, based on Jamaican school children data[19], is probably most appropriate. This would support the higher EQ estimate and place LB1 within the *Homo* range of variation. Although neurological organization is at least as important as EQ in determining behavioural complexity, these data are consistent with *H. floresiensis* being the Pleistocene toolmaker at Liang Bua.

Origins and Evolution

The LB1 skeleton was recovered from Flores, an island of 14,000 km^2 east of the Wallace Line, in Indonesia. It combines extremely small stature and an endocranial volume in the early australopithecine range, with a unique mosaic of primitive and derived traits in the cranium, mandible and postcranial skeleton. Both its geographic location and comparatively recent date suggest models that differ to those for more expected geological contexts, such as Pliocene eastern Africa. Among modern humans, populations of extremely small average stature were historically found in predominantly rainforest habitat in the equatorial zone of Africa, Asia and Melanesia[30,31]. Explanations for the small body size of these people generally focus on the thermoregulatory advantages for life in a hot and humid forest, either through evaporative cooling[32] or reduced rates of internal heat production[30]. For African pygmies, smaller body size is the result of reduced levels of insulin-like growth factor 1 (IGF-1) throughout the growth period[33], or reduced receptivity to IGF-1 (ref. 34). Although adult stature is reduced, cranio-facial proportions remain within the range of adjacent larger-bodied populations, as does brain size[35,36]. The combination of small stature and brain size in LB1 is not consistent with IGF-related postnatal growth retardation. Similarly, neither pituitary dwarfism, nor primordial microcephalic dwarfism (PMD) in modern humans replicates the skeletal features present in LB1 (refs 37–40).

Other mechanisms must have been responsible for the small body size of these hominins, with insular dwarfing being the strongest candidate. Although small body size was an attribute of Pliocene australopithecines, the facial and dental characteristics of LB1 link it with larger-bodied Pleistocene *Homo*. In this instance, body size is not a direct expression

的意义[23]。宽泛地讲，南方古猿属与人属最早成员矮于直立人与更晚的人族成员[8]。LB1 的股骨最大长度（280 毫米）刚好低于南方古猿阿法种（AL-288-1，281 毫米[24]）的最小记录值，而等于 OH 62 能人股骨长度（280 ~ 404 毫米）的最小估计值[21]。应用基于俾格米人[25]开发的身高估算公式，估算出 LB1 身高为 106 厘米（补充信息）。由于 LB1 的颅骨高度相对较小，这一身高可能是过高估值。

根据 LB1 106 厘米的身高估值，得出其体重为 16 至 28.7 千克，而根据股骨横截面面积为 525 平方毫米，得出体重为 36 千克（补充信息）。根据 LB1 的脑容量来计算[26]，其大脑重量为 433.2 克，从而得到脑系数（EQ）[27]的范围为 2.5 ~ 4.6，而智人为 5.8 ~ 8.1，直立人/匠人为 3.3 ~ 4.4，能人为 3.6 ~ 4.3，与南方古猿的变异范围重叠[28,29]。如果 LB1 享有旧世界热带现代人典型的清瘦和相对狭窄的体型，那么，基于牙买加入学儿童的数据[19]，体重的最小估值或许是最适当的。这会支持较高的 EQ 估值，而把 LB1 置于人属变异范围之内。虽然在决定行为的复杂性上，神经结构至少也与 EQ 一样重要，但这些数据与利昂布阿的弗洛勒斯人是更新世工具制造者是一致的。

起源与演化

LB1 骨架发现于印度尼西亚华莱士线以东方圆 14,000 平方千米的弗洛勒斯岛。它兼具落入早期南方古猿范围的极其矮小的身材和颅内容积，及在颅骨、下颌骨和颅后骨骼中独特的原始和衍生特征的镶嵌。其地理位置和相当晚的年代都表明了与更加期望的地质背景（如上新世非洲东部）不同的模型。在现代人中，平均身高极其矮小的种群在历史时期主要发现于非洲赤道地区、亚洲及美拉尼西亚的热带雨林栖息地[30,31]。对这些人体型小的解释，一般聚焦于在炎热潮湿的森林里生活的热调节优势，无论是通过蒸发降温[32]还是降低体内热量产生的速度[30]。对非洲俾格米人而言，体型较小是由于在整个生长期类胰岛素生长因子 1（IGF-1）分泌水平的降低[33]，或对 IGF-1 的接受能力的降低（参考文献 34）。虽然成年身高降低，但像大脑大小一样，其颅–面比例仍然保持在邻近大体型种群的范围之内[35,36]。LB1 中身材矮小与脑容量小兼备，与 IGF 相关的后天发育延缓不一致。同样，现代人中无论是垂体性侏儒症还是先天性小头侏儒症（PMD）都不能再现 LB1 中出现的骨骼特征（参考文献 37 ~ 40）。

一定是其他机制造成了这些人族成员的身材矮小，而岛屿侏儒化是最佳候选。虽然体型小是上新世南方古猿的特性，但 LB1 的面部与牙齿的特征将其与体型较

of phylogeny. The location of these small hominins on Flores makes it far more likely that they are the end product of a long period of evolution on a comparatively small island, where environmental conditions placed small body size at a selective advantage. Insular dwarfing, in response to the specific ecological conditions that are found on some small islands, is well documented for animals larger than a rabbit[41,42]. Explanations of the island rule have primarily focused on resource availability, reduced levels of interspecific competition within relatively impoverished faunal communities and absence of predators. It has been argued that, in the absence of agriculture, tropical rainforests offer a very limited supply of calories for hominins[43]. Under these conditions selection should favour the reduced energy requirements of smaller individuals. Although the details of the Pleistocene palaeoenvironment on Flores are still being documented, it is clear that until the arrival of Mesolithic humans the faunal suit was relatively impoverished, and the only large predators were the Komodo dragon and another larger varanid. Dwarfing in LB1 may have been the end product of selection for small body size in a low calorific environment, either after isolation on Flores, or another insular environment in southeastern Asia.

Anatomical and physiological changes associated with insular dwarfing can be extensive, with dramatic modification of sensory systems and brain size[44], and certainly exceed what might be predicted by the allometric effects of body size reduction alone. Evidence of insular dwarfing in extinct lineages, or the evolution of island endemic forms, is most often provided by the fossil record. Whereas there is archaeological evidence of hominins being on Flores by approximately 840 kyr[45], there is no associated hominin skeletal material, and the currently limited evidence from Liang Bua is restricted to the Late Pleistocene. The first hominin immigrants may have had a similar body size to *H. erectus* and early *Homo*[21,46], with subsequent dwarfing; or, an unknown small-bodied and small-brained hominin may have arrived on Flores from the Sunda Shelf.

Discussion

When considered as a whole, the cranial and postcranial skeleton of LB1 combines a mosaic of primitive, unique and derived features not recorded for any other hominin. Although LB1 has the small endocranial volume and stature evident in early australopithecines, it does not have the great postcanine tooth size, deep and prognathic facial skeleton, and masticatory adaptations common to members of this genus[2,47]. Instead, the facial and dental proportions, postcranial anatomy consistent with human-like obligate bipedalism[48], and a masticatory apparatus most similar in relative size and function to modern humans[48] all support assignment to the genus *Homo*—as does the inferred phylogenetic history, which includes endemic dwarfing of *H. erectus*. For these reasons, we argue that LB1 is best placed in this genus and have named it accordingly.

On a related point, the survival of *H. floresiensis* into the Late Pleistocene shows that the genus *Homo* is morphologically more varied and flexible in its adaptive responses than is generally recognized. It is possible that the evolutionary history of *H. floresiensis* is unique,

大的更新世人属联系起来。在这种情况下，体型大小不是系统发育的直接表现。弗洛勒斯岛上这些矮小人族成员的发现位置，使其更有可能是在一个相对较小的岛屿上经过长期演化的最终产物，那里的环境条件使得小体型更具有选择优势。岛屿侏儒化，作为对一些小岛上特殊生态条件的响应，在比兔子大的动物中有很详细的记载[41,42]。对岛屿规则的解释主要聚焦于资源的可获得性、相对贫困的动物群落中种间竞争的减少以及捕食者的消失。有人提出，在农业出现前，热带雨林提供给人族成员的卡路里十分有限[43]。在这样的条件下，选择应该偏向于能量需求降低的较矮小个体。虽然弗洛勒斯的更新世古环境资料正处于记载中，但很明显，在中石器时代人类到来前，整个动物群还比较单薄，唯一的大型捕食者是科莫多巨蜥和另一种体型较大的巨蜥。无论是在弗洛勒斯岛上被隔离后，还是在东南亚的另一个孤立环境中，LB1 的侏儒化可能是在低卡路里环境下选择偏向小体型的最终产物。

与岛屿侏儒化相关的解剖学和生理学变化非常之多，伴随对感觉系统和大脑尺寸的改变[44]，并且肯定超过了仅由身体体型减小的异速生长效应所能预测出的程度。已经绝灭的谱系中岛屿侏儒化，或岛屿特有类型的演化的证据，最常由化石记录提供。虽然有考古证据表明，人族成员在弗洛勒斯岛上大约有 840,000 年了[45]，但并没有相关的人族成员的骨骼材料，而现在来自利昂布阿的有限证据仅限于晚更新世。第一批人族成员移民可能有类似于直立人和早期人属成员的身体大小[21,46]，但随后矮化；或者，一种未知的身材矮小、脑容量小的人族成员从巽他大陆架来到了弗洛勒斯。

讨　论

当作为一个整体来看时，LB1 颅骨和颅后骨骼兼具原始和独特的衍生特征的镶嵌，这些衍生特征在其他任何人族成员中都没有记载过。虽然 LB1 像早期南方古猿那样颅内容积及身高小，但其没有硕大的犬后齿尺寸、深而前突的面部骨骼以及这一属的成员所共有的咀嚼适应特征[2,47]。相反，面部和牙齿的比例、颅后骨骼的解剖特征与似人的习惯性两足行走一致[48]，而且其咀嚼器官与现代人在相对大小与功能上最相似[48]，这一切都支持将其归入人属，包括直立人的地方性矮化在内的系统发育史同样支持将其归入人属。基于这些原因，我们认为 LB1 最适合放在这个属，并因此而给她命名。

关于另外相关的一点，弗洛勒斯人能幸存下来延续到晚更新世，显示了人属在其适应性响应中，形态上比人们一般认为的更加多样化和灵活。弗洛勒斯人的演化历史可能是独一无二的，但我们认为更有可能的情况是，随着人属走出非洲，该属

but we consider it more likely that, following the dispersal of *Homo* out of Africa, there arose much greater variation in the morphological attributes of this genus than has hitherto been documented. We anticipate further discoveries of highly endemic, hominin species in locations similarly affected by long-term genetic isolation, including other Wallacean islands.

(**431**, 1055-1061; 2004)

P. Brown[1], T. Sutikna[2], M. J. Morwood[1], R. P. Soejono[2], Jatmiko[2], E. Wayhu Saptomo[2] & Rokus Awe Due[2]

[1] Archaeology & Palaeoanthropology, School of Human & Environmental Studies, University of New England, Armidale, New South Wales 2351, Australia

[2] Indonesian Centre for Archaeology, Jl. Raya Condet Pejaten No. 4, Jakarta 12001, Indonesia

Received 3 March; accepted 8 September 2004.

References:

1. Morwood, M. J. *et al.* Archaeology and age of a new hominin from Flores in eastern Indonesia. *Nature* doi:10.1038/nature02956 **431**, 1087-1091 (2004).

2. Wood, B. A. *Koobi Fora Research Project, Vol. 4: Hominid Cranial Remains* (Clarendon, Oxford, 1991).

3. Vekua, A. K. *et al.* A new skull of early *Homo* from Dmanisi, Georgia. *Science* **297**, 85-89 (2002).

4. Spoor, C. F. Basicranial architecture and relative brain size of STS 5 (*Australopithecus africanus*) and other Plio-Pleistocene hominids. *S. Afr. J. Sci.* **93**, 182-186 (1997).

5. Lieberman, D., Ross, C. F. & Ravosa, M. J. The primate cranial base: ontogeny, function, and integration. *Yearb. Phys. Anthropol.* **43**, 117-169 (2000).

6. Baba, H. *et al. Homo erectus* calvarium from the Pleistocene of Java. *Science* **299**, 1384-1388 (2003).

7. Tobias, P. V. *The Skulls, Endocasts and Teeth of* Homo habilis (Cambridge Univ. Press, Cambridge, 1991).

8. McHenry, H. M. & Coffing, K. E. *Australopithecus* to *Homo*: Transformations of body and mind. *Annu. Rev. Anthropol.* **29**, 125-166 (2000).

9. Brown, P. Vault thickness in Asian *Homo erectus* and modern *Homo sapiens*. *Courier Forschungs-Institut Senckenberg* **171**, 33-46 (1994).

10. Bräuer, G. & Mbua, E. *Homo erectus* features used in cladistics and their variability in Asian and African hominids. *J. Hum. Evol.* **22**, 79-108 (1992).

11. Santa Luca, A. P. *The Ngandong Fossil Hominids* (Department of Anthropology Yale Univ., New Haven, 1980).

12. Weidenreich, F. The skull of *Sinanthropus pekinensis*: a comparative study of a primitive hominid skull. *Palaeontol. Sin.* **D10**, 1-485 (1943).

13. Brown, P. & Maeda, T. Post-Pleistocene diachronic change in East Asian facial skeletons: the size, shape and volume of the orbits. *Anthropol. Sci.* **112**, 29-40 (2004).

14. Gabunia, L. K. *et al.* Earliest Pleistocene hominid cranial remains from Dmanisi, Republic of Georgia: taxonomy, geological setting, and age. *Science* **288**, 1019-1025 (2000).

15. Kaifu, Y. *et al.* Taxonomic affinities and evolutionary history of the Early Pleistocene hominids of Java: dento-gnathic evidence. *Am. J. Phys. Anthropol.* (in the press).

16. McHenry, H. M. in *Evolutionary History of the "Robust" Australopithecines* (ed. Grine, F. E.) 133-148 (Aldine de Gruyter, New York, 1988).

17. Wood, B. A. & Uytterschaut, H. Analysis of the dental morphology of the Plio-Pleistocene hominids. III. Mandibular premolar crowns. *J. Anat.* **154**, 121-156 (1987).

18. Wood, B. A., Abbott, S. A. & Uytterschaut, H. Analysis of the dental morphology of Plio-Pleistocene hominids. IV. Mandibular postcanine root morphology. *J. Anat.* **156**, 107-139 (1988).

19. Aiello, A. & Dean, C. *An Introduction to Human Evolutionary Anatomy* (Academic, London, 1990).

20. Kennedy, G. E. Some aspects of femoral morphology in *Homo erectus*. *J. Hum. Evol.* **12**, 587-616 (1983).

21. Haeusler, M. & McHenry, H. M. Body proportions of *Homo habilis* reviewed. *J. Hum. Evol.* **46**, 433-465 (2004).

22. Stern, J. T. J. & Susman, R. L. The locomotor anatomy of *Australopithecus afarensis*. *Am. J. Phys. Anthropol.* **60**, 279-317 (1983).

23. Ruff, C. B. Morphological adaptation to climate in modern and fossil hominids. *Yearb. Phys. Anthropol.* **37**, 65-107 (1994).

24. Jungers, W. L. Lucy's limbs: skeletal allometry and locomotion in *Australopithecus afarensis*. *Nature* **297**, 676-678 (1982).

25. Jungers, W. L. Lucy's length: stature reconstruction in *Australopithecus afarensis* (A.L.288–1) with implications for other small-bodied hominids. *Am. J. Phys. Anthropol.* **76**, 227-231 (1988).

26. Count, E. W. Brain and body weight in man: their antecendants in growth and evolution. *Ann. NY Acad. Sci.* **46**, 993-1101 (1947).

27. Martin, R. D. Relative brain size and basal metabolic rate in terrestrial vertebrates. *Nature* **293**, 57-60 (1981).

28. Jerison, H. J. *Evolution of the Brain and Intelligence* (Academic, New York, 1973).

29. McHenry, H. M. in *The Primate Fossil Record* (ed. Hartwig, C. H.) 401-406 (Cambridge Univ. Press, Cambridge, 2002).

30. Cavalli-Sforza, L. L. (ed.) *African Pygmies* (Academic, Orlando, 1986).

31. Shea, B. T. & Bailey, R. C. Allometry and adaptation of body proportions and stature in African Pygmies. *Am. J. Phys. Anthropol.* **100**, 311-340 (1996).

32. Roberts, D. F. *Climate and Human Variability* (Cummings Publishing Co., Menlo Park, 1978).

的形态特性产生了比截至目前已记载的更多的变异。我们预期会有更多高度特化的人族成员物种在同样受到长期遗传隔离的地方被发现，包括其他华莱士区的岛屿。

（田晓阳 翻译；张颖奇 审稿）

33. Merimee, T. J., Zapf, J., Hewlett, B. & Cavalli-Sforza, L. L. Insulin-like growth factors in pygmies. *N.* Engl. *J. Med.* **15**, 906-911 (1987).

34. Geffner, M. E., Bersch, N., Bailey, R. C. & Golde, D. W. Insulin-like growth factor I resistance in immortalized T cell lines from African Efe Pygmies. *J. Clin. Endocrinol. Metab.* **80**, 3732-3738 (1995).

35. Hiernaux, J. *The People of Africa* (Charles Scribner's Sons, New York, 1974).

36. Beals, K. L., Smith, C. L. & Dodd, S. M. Brain size, cranial morphology, climate and time machines. *Current Anthropology* **25**, 301-330 (1984).

37. Rimoin, D. L., Merimee, T. J. & McKusick, V. A. Growth-hormone deficiency in man: an isolated, recessively inherited defect. *Science* **152**, 1635-1637 (1966).

38. Jaffe, H. L. *Metabolic, Degenerative and Inflammatory Disease of Bones and Joints* (Lea and Febiger, Philadelphia, 1972).

39. Seckel, H. P. G. *Bird-Headed Dwarfs* (Karger, Basel, 1960).

40. Jeffery, N. & Berkovitz, B. K. B. Morphometric appraisal of the skull of Caroline Crachami, the Sicilian "Dwarf" 1815?–1824: A contribution to the study of primordial microcephalic dwarfism. *Am. J. Med. Genet.* **11**, 260-270 (2002).

41. Sondaar, P. Y. in *Major Patterns in Vertebrate Evolution* (eds Hecht, M. K., Goody, P. C. & Hecht, B. M.) 671-707 (Plenum, New York, 1977).

42. Lomolino, M. V. Body size of *mammals* on islands: The island rule re-examined. *Am. Nat.* **125**, 310-316 (1985).

43. Bailey, R. C. & Headland, T. The tropical rainforest: Is it a productive habitat for human foragers? *Hum. Ecol.* **19**, 261-285 (1991).

44. Köhler, M. & Moyà-Solà, S. Reduction of brain and sense organs in the fossil insular bovid *Myotragus. Brain Behav. Evol.* **63**, 125-140 (2004).

45. Morwood, M. J., O'Sullivan, P. B., Aziz, F. & Raza, A. Fission-track ages of stone tools and fossils on the east Indonesian island of Flores. *Nature* **392**, 173-176 (1998).

46. Walker, A. C. & Leakey, R. (eds) *The Nariokotome* Homo erectus *skeleton* (Harvard Univ. Press, Cambridge, 1993).

47. Rak, Y. The *Australopithecine Face* (Academic, New York, 1983).

48. Wood, B. A. & Collard, M. The human genus. *Science* **284**, 65-71 (1999).

49. Johanson, D. C. & White, T. D. A systematic assessment of early African Hominids. *Science* **202**, 321-330 (1979).

Supplementary Information accompanies the paper on www.nature.com/nature.

Acknowledgements. We would like to thank F. Spoor and L. Aiello for data and discussion. Comments by F. Spoor and D. Lieberman greatly improved aspects of the original manuscript. Conversation with S. Collier, C. Groves, T. White and P. Grave helped clarify some issues. CTscans were produced by CT-Scan KSU, Medical Diagnostic Nusantara, Jakarta. S. Wasisto completed complex section drawings and assisted with the excavation of Sector VII. The 2003 excavations at Liang Bua, undertaken under Indonesian Centre for Archaeology Permit Number 1178/SB/PUS/ BD/24.VI/2003, were funded by a Discovery Grant to M.J.M. from the Australian Research Council. UNE Faculty of Arts, and M. Macklin, helped fund the manufacture of stereolithographic models of LB1.

Authors contributions. P.B. reconstructed the LB1 cranium and was responsible for researching and writing this article, with M.J.M. T.S. directed many aspects of the Liang Bua excavations, including the recovery of the hominin skeleton. M.J.M. and R.P.S. are Principal Investigators and Institutional Counterparts in the ARC project, as well as Co-Directors of the Liang Bua excavations. E.W.S. and Jatmiko assisted T.S., and had prime responsibility for the work in Sector VII. R.A.D. did all of the initial faunal identifications at Liang Bua, including hominin material, and helped clean and conserve it.

Competing interests statement. The authors declare that they have no competing financial interests.

Correspondence and requests for materials should be addressed to P.B. (pbrown3@pobox.une.edu.au).

Stratigraphic Placement and Age of Modern Humans from Kibish, Ethiopia

I. McDougall *et al.*

Editor's Note

Where did our own species come from, and when? Fossils discovered in 1967 in the Kibish Formation in southern Ethiopia yielded hominid skull remains identifiable as anatomically modern humans, but their antiquity had been much debated. Geochronologists McDougall and colleagues sought to settle the matter with accurate argon-argon dates on volcanic layers in the Formation, together with work on the history of the Omo river valley compared with that of the Nile. The final date—195,000 years old, plus or minus 5,000—made the Omo Kibish hominids the oldest examples of *Homo sapiens* from anywhere in the world. Ethiopia was the birthplace of humankind, as it had been for so many other species in the wider human family.

In 1967 the Kibish Formation in southern Ethiopia yielded hominid cranial remains identified as early anatomically modern humans, assigned to *Homo sapiens*[1-4]. However, the provenance and age of the fossils have been much debated[5,6]. Here we confirm that the Omo I and Omo II hominid fossils are from similar stratigraphic levels in Member I of the Kibish Formation, despite the view that Omo I is more modern in appearance than Omo II[1-3]. $^{40}Ar/^{39}Ar$ ages on feldspar crystals from pumice clasts within a tuff in Member I below the hominid levels place an older limit of 198 ± 14 kyr (weighted mean age 196 ± 2 kyr) on the hominids. A younger age limit of 104 ± 7 kyr is provided by feldspars from pumice clasts in a Member III tuff. Geological evidence indicates rapid deposition of each member of the Kibish Formation. Isotopic ages on the Kibish Formation correspond to ages of Mediterranean sapropels, which reflect increased flow of the Nile River, and necessarily increased flow of the Omo River. Thus the $^{40}Ar/^{39}Ar$ age measurements, together with the sapropel correlations, indicate that the hominid fossils have an age close to the older limit. Our preferred estimate of the age of the Kibish hominids is 195 ± 5 kyr, making them the earliest well-dated anatomically modern humans yet described.

THE principal outcrops of the Kibish Formation are along the Omo River where it skirts the Nkalabong Range (Fig. 1), with the highest outcrops close in elevation to that of the watershed between the Omo and the Nile rivers[7,8]. Former hydrographic links are apparent from Nilotic fauna in the Turkana Basin sequence[8-10].

来自埃塞俄比亚基比什的现代人的地层位置和年代

麦克杜格尔等

编者按

我们自己的物种从何而来，何时而来？1967 年，科学家在埃塞俄比亚南部的基比什组中发现了古人类头骨化石，这些化石被认为是解剖学意义上的早期现代人，被归入智人这一类群。但这些化石的出土地点和年代一直备受争议。地质年表学家麦克杜格尔及其同事们试图用地层中火山层的氩精确年代，以及奥莫河谷与尼罗河的历史对比来解决这个问题。最终的年代——19.5 万 ± 0.5 万年——使得奥莫基比什古人类成为目前发现的世界上最古老的智人。埃塞俄比亚是人类的诞生地，同人类大家庭中许多其他物种一样。

1967 年，在埃塞俄比亚南部的基比什组挖掘出来了古人类头盖骨遗迹，该头盖骨被鉴定为解剖学上的早期现代人，被划分到了智人[1-4]。然而，这些化石的出土地点和年代却饱受争议[5,6]。尽管有观点认为奥莫 I 号化石的形态比奥莫 II 号更加接近现代[1-3]，但是我们确证了奥莫 I 号和奥莫 II 号古人类化石是来自于基比什组 I 段的相似层位。人们采集了古人类化石出土层位下方的 I 段的凝灰岩内的浮石碎屑，从中得到的长石晶体确定了 $^{40}Ar/^{39}Ar$ 年代，将古人类化石的年代下限确定在 198 ± 14 kyr(kyr 为千年)(加权均值年代为 196 ± 2 kyr)。III 段凝灰岩内的浮石碎屑中的长石则提供了古人类化石的年代上限，为 104 ± 7 kyr。地质学证据表明基比什组的每一段都经历了快速的沉积作用。基比什组的同位素年龄与地中海腐泥的年代一致，这反映了尼罗河流量的增加，以及由此引起的奥莫河流量的必然增加。因此 $^{40}Ar/^{39}Ar$ 方法得出的年代值，以及它与地中海腐泥形成时间的相关关系，共同暗示了古人类化石的年代接近于上述测年结果的下限。我们更倾向于估计基比什古人类的年代约为 195 ± 5 kyr，这样他们就成为迄今描述过的最早的可以清楚追溯到其年代的解剖学上的现代人。

基比什组的主要露头地是沿着奥莫河一带，那里位于纳卡拉邦山脉的边缘(图 1)，其中位置最高的露头，在海拔上接近奥莫河与尼罗河的分水岭[7,8]。在图尔卡纳盆地的沉积序列里，出现了尼罗河动物群，显然，从前这两条河流有十分接近的水文关系[8-10]。

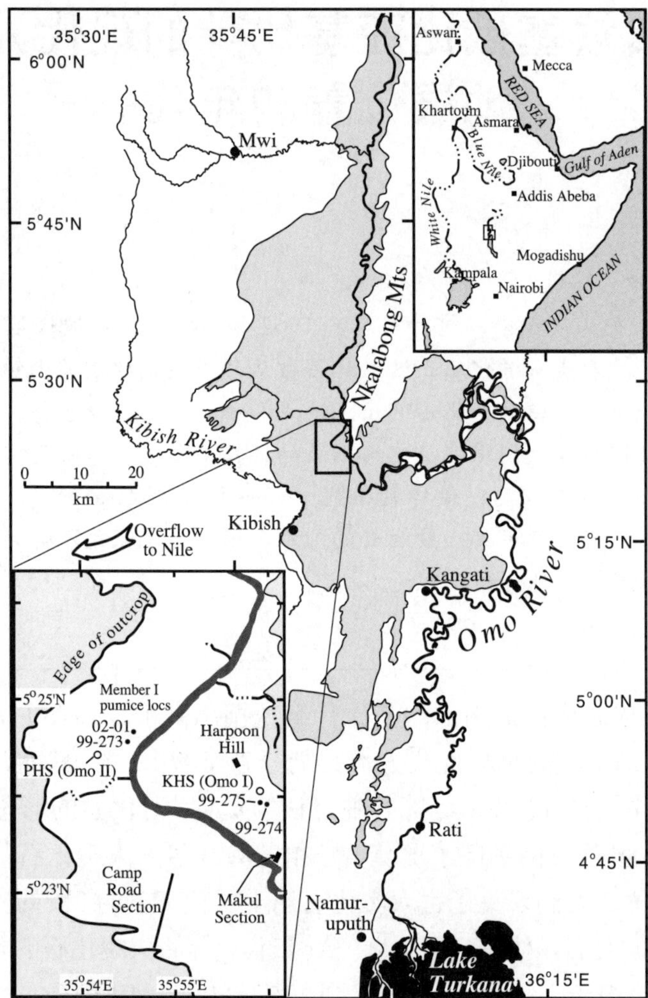

Fig. 1. Map showing the distribution of the Kibish Formation (shaded) in the lower Omo Valley, southern Ethiopia, after Davidson[29]. Inset on lower left, locations of Omo I, Omo II, measured sections, and dated samples.

The Kibish Formation (about 100 m thick) consists of flat-lying, tectonically undisturbed, unconsolidated sediments deposited mainly in deltaic environments over brief periods. It comprises the youngest exposed sedimentary sequence in the Omo Basin, and lies disconformably upon the Nkalabong Formation[11,12] or on the underlying Mursi Formation[12]. Strata are composed principally of claystone and siltstone, with subordinate fine sandstone, conglomerate and tuffs (Fig. 2).

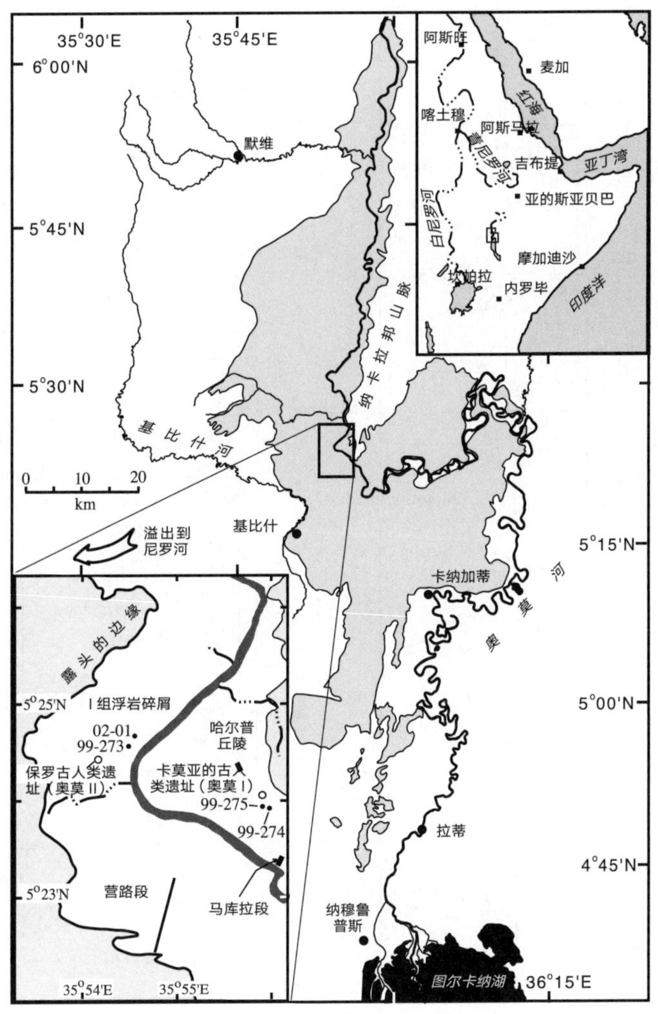

图 1. 埃塞俄比亚南部下奥莫河谷的基比什组（阴影部分）的分布地图（戴维森指定的 [29]）。左下方的插图表示的是奥莫 I 号和奥莫 II 号的位置、测量的剖面以及确定年代的样本。

基比什组（约 100 米厚）主要由短时间内堆积在三角洲环境的平整且未受到干扰的、松散的沉积物组成。其包含奥莫盆地最晚近的暴露沉积序列，以不整合的方式位于纳卡拉邦组之上 [11,12] 或在下方的穆尔斯组上 [12]。地层主要由黏土岩和粉砂岩构成，附着有细砂岩、砾岩和凝灰岩（图 2）。

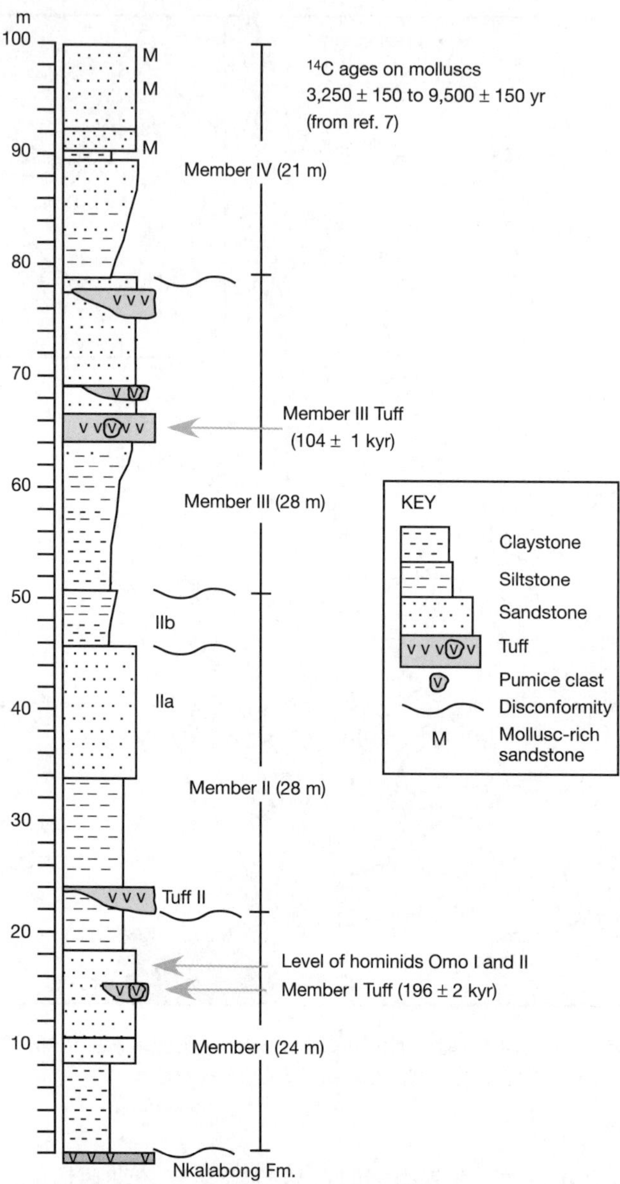

Fig. 2. Composite stratigraphy of the Kibish Formation. Member I was measured near the type section at Makul; Member II was measured near Harpoon Hill, and Members III and IV were measured along Camp Road (see Fig. 1 for locations).

Butzer *et al.*[13] and Butzer[14] divided the Kibish Formation into Members I to IV on the basis of disconformities with up to 30 m relief (Fig. 2). The members record discrete times of deposition when the northern margin of Lake Turkana and the Omo delta lay about 100 km north of their current positions. As the higher lake levels reflect significantly higher precipitation in the region, such periods should be recognizable at least regionally, and perhaps globally.

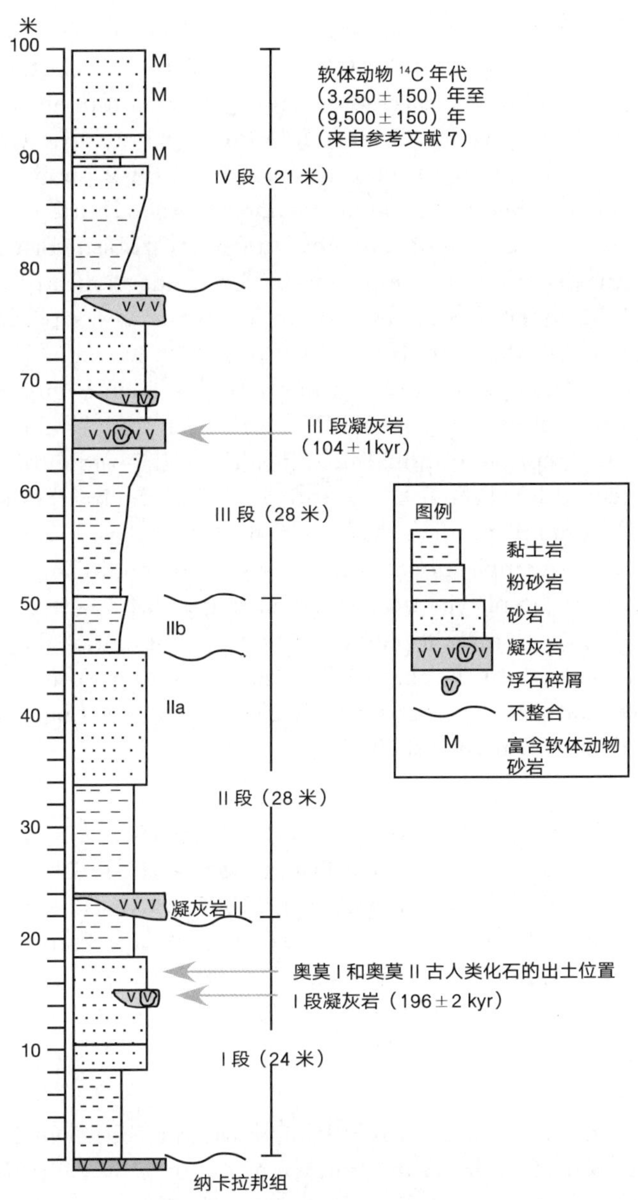

图 2. 基比什组的复合地层学。经测量，I 段位于马库拉的典型剖面附近，II 段位于哈尔普丘陵附近，III 段和 IV 段沿着营路测量的（具体位置见图 1）。

布策等人[13]和布策[14]根据基比什组高达 30 米的地层不整合现象，将其划分成了 I 到 IV 段（图 2）。在那时，图尔卡纳湖的北部边界和奥莫三角洲，在它们现在位置的北边 100 千米，基比什组的四个段精确记录了那时沉积过程的不连续时间。由于湖泊水位的升高反映了该地区降水的显著增加，所以至少在区域上，甚至在全球范围内，都应该可以辨别出这些时期。

Member I (26 m thick; Fig. 2) was deposited disconformably on the Nkalabong Formation in a deltaic environment. Small (less than 30 mm) rounded pumice clasts occur in an impure tuff in Member I, 7 m below the base of Member II. This tuff lies near, but probably slightly below, the levels from which Omo I and Omo II were derived. Glass shards from the tuff are very similar in composition to the glass from three pumice clasts enclosed in the tuff (Supplementary Table 1), indicating a common origin. Further, this composition is distinct from the glass composition of any other tuff in the Kibish Formation. Member II, about 28 m thick, was deposited on a topographic surface developed on Member I with at least 19 m of relief. Member II contains two discrete sequences, separated by an internal disconformity, designated Members IIa and IIb on Fig. 2. Member II was incised by as much as 25 m before the deposition of Member III[11,14], which begins with thin siltstone and claystone beds that drape topography. These beds, averaging about 3 cm thick, fine upward internally, and represent annual flooding. Thus, deposition of the lower part of Member III may record less than 1 kyr. A tuff 3.5–12 m thick, 18 m above the base of Member III, locally contains pumice clasts with alkali feldspar phenocrysts. Glass of the pumices normally differs compositionally from the glass of the tuff (Supplementary Table 1). However, one pumice sample yielded two contrasting sets of analyses: one corresponding to the tuff, the other to the dominant pumices. The youngest unit of the Kibish Formation, Member IV, was deposited on the underlying sediments after up to 30 m of dissection. Member IV comprises at most 21 m of strata deposited between about 9.5 and 3.3 kyr ago, on the basis of ^{14}C dating of mollusc shell[7,14,15].

Omo I was found at Kamoya's Hominid Site (KHS; 5° 24.15′ N, 35° 55.81′ E), which was identified from contemporary photographs, from evidence of the 1967 excavations and by additional hominid bone material conjoining the 1967 finds[16]. The hominid fossils were recovered from a siltstone 2.4 m below the base of Member II[14]. Thus, Omo I derives from near the top of Member I. Butzer *et al.*[13] reported a $^{230}Th/^{234}U$ date of 130 ± 5 kyr on *Etheria* from essentially the same level as the hominid, but they and others have questioned the reliability of the age[5,6].

Omo II was found on the surface at PHS (Paul's Hominid Site), which Butzer[14] mapped about 2.6 km northwest of KHS. A map drawn at the time by Paul Abell (discoverer of Omo II), together with contemporary photographs of the site supplied by K.W. Butzer, have positively identified PHS at 5° 24.55′ N, 35° 54.07′ E, about 3.3 km west by north of KHS (Fig. 1). Although PHS was mislocated on the published map, Butzer's[14] stratigraphic description at the site is correct. The base of Member II lies about 3 m above the approximate level from which Omo II was recovered. The basal tuff of Member II is of unique composition, and correlates from KHS to PHS (Supplementary Table 1).

Ages on alkali feldspars separated from pumice clasts from tuffs in Members I and III reflect the time of their eruption, and provide maximum and minimum ages for the hominid fossils, respectively. $^{40}Ar/^{39}Ar$ results are summarized in Table 1, with details listed in Supplementary Tables 2–4.

I段(26米厚；图2)是在三角洲环境中不整合地沉积在纳卡拉邦组上的。小型(不足30毫米)的圆形浮岩出现于I段的掺杂的凝灰岩中，位于II段基底之下7米处。这种凝灰岩位于奥莫I号和奥莫II号的层位附近，但可能略低于该水平。凝灰岩中的玻璃碎片与凝灰岩中附着的三种浮岩碎屑的玻璃在构成上非常相似(附表1)，这暗示着它们具有共同的来源。此外，这种成分与基比什组的其他凝灰岩中的玻璃成分都不一样。II段约有28米厚，沉积在一个I段上发展起来的表面上，具有至少19米的起伏地势。II段内部有一个不整合面，将这段地层分成了两部分，图2中将它们指定为IIa和IIb段。III段的沉积过程是从覆盖地形的薄的粉砂岩和黏土岩床开始的，但在沉积开始之前，II段被侵入了25米之多[11,14]。这些平均约3厘米厚的河床，内部向上部位的颗粒都很精细，代表了每年一次的洪灾。因此，III段的下半部分的沉积岩，记录的时代可能少于1,000年。一层3.5米到12米厚的凝灰岩，位于III段的基底之上18米处，局部含有浮岩碎屑和碱性长石斑晶。浮岩的玻璃质通常与凝灰岩玻璃质在成分上有所不同(附表1)，但是，对一例浮岩样本的分析，产生了两个截然不同的结果，一套对应凝灰岩，另一套则对应于占优势的浮岩。基比什组中最年轻的单元——IV段是在高达30米的不连续层之后沉积在下面的沉积物上的。根据软体动物外壳进行的^{14}C年代测定发现，IV段沉积形成的时代为9,500年到3,300年前，地层厚度约为21米[7,14,15]。

奥莫I号是在卡莫亚古人类遗址(KHS；北纬5°24.15′，东经35°55.81′)发现的，可以根据1967年挖掘的证据结合1967年发现的其他古人类化石以及同时期的照片上鉴别出来[16]。古人类化石从II段的基底之下的2.4米处的粉砂岩处发掘出来[14]。因此，奥莫I号出土位置接近I段的顶部。布策等[13]使用^{230}Th/^{234}U测年法，测定了一例与古人类化石来自同一层位的艾特利亚人的年代为130±5 kyr，但是他们和其他人都对这一年代的可靠性表示质疑[5,6]。

奥莫II号发现于保罗古人类遗址(PHS)的地表，根据布策[14]的标记，这一地点位于KHS西北约2.6千米处。根据保罗·埃布尔(奥莫II号的发现者)当时绘制的地图与布策提供的同一时期的照片，最终确定PHS位于KHS北部约3.3千米处，地理坐标为北纬5°24.55′东经35°54.07′(图1)。尽管在已发表的地图上将PHS的位置搞错了，但是布策[14]对该遗址进行的地层学描述是正确的。II段的基底在奥莫II号出土位置之上约3米处。II段的基底凝灰岩具有独特的成分，从KHS到PHS都与其具有相关性(附表1)。

从I段和III段的凝灰岩中的浮岩碎屑分离得到的碱性长石的年代反映了它们爆发的时间，并且分别提供了古人类化石的最大和最小年代。^{40}Ar/^{39}Ar的结果在表1中进行了总结，详细情况在附表2～4中列出。

Table 1. Summary of $^{40}Ar/^{39}Ar$ alkali feldspar laser fusion ages from pumice clasts in tuffs of the Kibish area, Turkana Basin, Ethiopia

Sample no.	Tuff	Locality	Irradiation	*n*	*n* used	Simple mean age (kyr)	Weighted mean age (kyr)	Isochron age (kyr)	MSWD	$(^{40}Ar/^{39}Ar)_i$
Kibish Formation, Member III										
99-275A	Member III	0.4 km SSE of KHS	ANU58/L10	10	10	105.4 ± 5.0	106.0 ± 1.6	108.0 ± 3.1	1.03	292.2 ± 4.1
99-275C	Member III	0.4 km SSE of KHS	ANU58/L1	13	11	107.5 ± 7.1	105.4 ± 1.8	101.1 ± 8.4	2.04	303.6 ± 13.8
99-274A	Member III	0.5 km SE of KHS	ANU58/L6	13	13	98.1 ± 5.3	98.2 ± 1.1	97.4 ± 2.2	2.47	298.9 ± 6.8
99-274B	Member III	0.5 km SE of KHS	ANU58/L7	14	13	105.0 ± 8.3	109.6 ± 1.3	110.2 ± 4.4	3.27	292.9 ± 16.2
Kibish Formation, Member I, near Omo II site, just west of Omo River, Nakaa'kire, 5° 24.6′N, 35° 54.5′E										
99-273A	Member I	Nakaa'kire	ANU58/L3	9	9	319.8 ± 18.6	315.3 ± 3.6	318.2 ± 16.6	3.55	287.3 ± 41.5
99-273B	Member I	Nakaa'kire	ANU58/L4	8	7	220.8 ± 13.3	210.9 ± 2.1	209.6 ± 2.0	0.55	301.4 ± 3.1
02-01A	Member I	Nakaa'kire	ANU98/L9	16	16	204.2 ± 14.9	205.3 ± 2.2	194.1 ± 5.4	2.66	315.8 ± 7.3
02-01B	Member I	Nakaa'kire	ANU98/L10	15	14	194.7 ± 10.8	191.5 ± 1.9	184.9 ± 3.8	2.13	304.0 ± 3.5
02-01C	Member I	Nakaa'kire	ANU98/L3	14	14	195.6 ± 15.0	192.7 ± 2.3	192.5 ± 6.7	3.85	295.7 ± 5.8

^{40}K decay constant $\lambda = 5.543 \times 10^{-10}$ yr^{-1}. Fluence monitor: Fish Canyon Tuff sanidine 92-176 of reference age 28.1 Myr. Samples 99-275A and 99-275C (laboratory sample numbers at ANU) = KIB99-47 (field sample number); 99-274A and 99-274B = KIB99-41; 99-273A and 99-273B = KIB99-19. Results with errors are means ± s.d. MSWD, mean square of weighted deviates.

The five pumice clasts measured from Member I yielded three different ages (Table 1), with at least the oldest age interpreted as evidence for reworking of that particular pumice clast. Nine analyses on 99-273A gave a mean age of 320 ± 19 kyr. Seven analyses on 99-273B yielded a mean age of 221 ± 13 kyr after rejecting one outlier (269 ± 18 kyr). Single feldspar crystals from 02-01B yielded a mean age of 195 ± 11 kyr ($n = 14$), forming a concordant data set after one rejection (333 ± 6 kyr). Multiple crystals ($n \leqslant 3$) were analysed in 12 of 16 measurements for 02-01A, and in 8 of 14 analyses for 02-01C ($n \leqslant 4$). No ages were rejected on 02-01A, whose mean age is 204 ± 15 kyr, or on 02-01C, whose mean age is 196 ± 15 kyr. It is probable that these latter three pumice clasts are products of the same volcanic eruption, as shown by their concordant ages and similar mean K/Ca ratios. When combined, all 44 analyses provide an arithmetic mean age of 198 ± 14 kyr and a weighted mean age of 195.8 ± 1.6 kyr. The mean age calculated for pumice 99-273B is statistically older than the ages determined on the three pumice clasts 02-01, so that 99-273B might also be a reworked pumice clast. Clearly, the age of deposition of a tuff must be younger than the youngest igneous component found within it. Thus, this tuffaceous level of Member I of the Kibish Formation was deposited after 196 kyr ago, on the basis of the mean age determined on the three 02-01 pumice clasts.

表 1. 对埃塞俄比亚图尔卡纳盆地基比什地区的凝灰岩中的浮岩碎屑进行的 $^{40}Ar/^{39}Ar$ 碱性长石激光聚变测年概况

样品编号	凝灰岩	产地	放射	取样量	实际使用数目	简单平均年代(kyr)	加权平均年代(kyr)	等龄线法年代(kyr)	加权均方差	$(^{40}Ar/^{39}Ar)_i$
基比什组，III 段										
99-275A	III 段	KHS 东南偏南 0.4 千米处	ANU58/L10	10	10	105.4±5.0	106.0±1.6	108.0±3.1	1.03	292.2±4.1
99-275C	III 段	KHS 东南偏南 0.4 千米处	ANU58/L1	13	11	107.5±7.1	105.4±1.8	101.1±8.4	2.04	303.6±13.8
99-274A	III 段	KHS 东南 0.5 千米处	ANU58/L6	13	13	98.1±5.3	98.2±1.1	97.4±2.2	2.47	298.9±6.8
99-274B	III 段	KHS 东南 0.5 千米处	ANU58/L7	14	13	105.0±8.3	109.6±1.3	110.2±4.4	3.27	292.9±16.2
基比什组，I 段，靠近奥莫 II 遗址，就在奥莫河以西，Nakaa'kire，5°24.6′N，35°54.5′E										
99-273A	I 段	Nakaa'kire	ANU58/L3	9	9	319.8±18.6	315.3±3.6	318.2±16.6	3.55	287.3±41.5
99-273B	I 段	Nakaa'kire	ANU58/L4	8	7	220.8±13.3	210.9±2.1	209.6±2.0	0.55	301.4±3.1
02-01A	I 段	Nakaa'kire	ANU98/L9	16	16	204.2±14.9	205.3±2.2	194.1±5.4	2.66	315.8±7.3
02-01B	I 段	Nakaa'kire	ANU98/L10	15	14	194.7±10.8	191.5±1.9	184.9±3.8	2.13	304.0±3.5
02-01C	I 段	Nakaa'kire	ANU98/L3	14	14	195.6±15.0	192.7±2.3	192.5±6.7	3.85	295.7±5.8

^{40}K 衰变常数 $\lambda = 5.543 \times 10^{-10}\ yr^{-1}$。流量监控器：参考年代为 2,810 万年的菲什峡谷凝灰岩透长石 92-176。样本 99-275A 和 99-275C(ANU 的实验室样本编号) = KIB99-47(野外样本编号)；99-274A 和 99-274B = KIB99-41；99-273A 和 99-273B = KIB99-19。结果与误差用平均值 ± 标准差表示。

对 I 段得到的五份浮岩碎屑的测年产生了三个不同的结果(表 1)，至少最古老的年代被解读为这一例浮岩碎屑曾经被搬运的证据。对 99-273A 进行的九项分析给出的平均年代为 320±19 kyr。对 99-273B 进行的七项分析在排除一个离群值(269±18 kyr)之后，测得的平均年代为 221±13 kyr。来自 02-01B 的唯一长石晶体测得的平均年代为 195±11 kyr($n = 14$)，在剔除一个离群值(333±6 kyr)之后，形成了一套整合的数据集。对 02-01A 的 16 个测量值中的 12 个进行了多晶体($n \leqslant 3$)分析，对 02-01C($n \leqslant 4$)的 14 个测量值中的 8 个进行了分析。得到的 02-01A 的所有年代数值中没有离群值，其平均年代为 204±15 kyr，02-01C 的也没有离群值，其平均年代为 196±15 kyr。有可能后面这三种浮岩碎屑是同一次火山喷发的产物，正如它们的一致年代和相似的平均 K/Ca 比例反映的一样。综合分析，所有 44 个结果提供的算术平均年代是 198±14 kyr，加权平均年代是 195.8±1.6 kyr。统计学上看，浮岩 99-273B 计算出来的平均年代比三种 02-01 浮岩碎屑确定的年代更古老，所以 99-273B 可能也是一种二次搬运的浮岩碎屑。显然，凝灰岩层的年代肯定比其中最年轻的火成组分更年轻。因此，根据这三种 02-01 浮岩碎屑确定的平均年代，基比什组 I 段的凝灰岩层是在 196 kyr 之后沉积下来的。

Feldspars from four pumice clasts from the Member III tuff were analysed. Two were *in situ* (99-275A, C), and two (99-274A, B) had weathered out of the same unit about 150 m farther east (Fig. 1). Ages on 99-275A range from 98.0 ± 7.7 kyr to 114.6 ± 4.3 kyr. Five single-crystal measurements have an arithmetic mean age of 105.5 ± 7.0 kyr, identical to five measurements on groups of two or three crystals (mean age 105.4 ± 2.7 kyr), indicating a homogeneous population. The overall mean age is 105.4 ± 5.0 kyr, with a weighted mean age of 106.0 ± 1.6 kyr. Eleven analyses on sample 99-275C yield a mean age of 107.5 ± 7.1 kyr and a weighted mean of 105.4 ± 1.8 kyr, after rejection of two outliers (142.1 ± 3.3 and 128.8 ± 8.3 kyr). Concordant results from 99-274A on seven single crystals and six pairs of crystals gave an overall arithmetic mean age of 98.1 ± 5.3 kyr. The companion clast 99-274B gave a mean age of 105.0 ± 8.3 kyr, after elimination of one outlier (Supplementary Table 4). Again these results reflect a single age population, as shown by the similar mean ages of the individual pumices, and the overlapping average K/Ca ratios. Combining all these results (except outliers), the overall arithmetic mean age is 103.7 ± 7.4 kyr ($n = 47$) and the weighted mean age is 103.7 ± 0.9 kyr. Thus, the depositional age must be equal to or younger than 104 kyr, providing evidence that Members I and II of the Kibish Formation are older than 104 kyr.

Each of the members of the Kibish Formation was deposited during intervals when Lake Turkana was at a much higher level than at present, and Member II has an internal disconformity. The upper part of Member I was being deposited at or after 196 kyr ago, and the upper part of Member III was being deposited at or after 104 kyr ago; ^{14}C ages on Member IV correspond to deposition between 9.5 and 3.3 kyr ago. These ages are remarkably similar to ages of Mediterranean sapropels S7, S4 and S1. Sapropels S1–S7 have the following estimated ages: 195 kyr (S7), 172 kyr (S6), 124 kyr (S5), 102 kyr (S4), 81 kyr (S3), 55 kyr (S2) and 8 kyr (S1)[17]. In many cases sapropels are related to a greatly increased flow of the Nile River into the Mediterranean Sea as a consequence of intensification of the African monsoon[18], recorded in more negative $\delta^{18}O$ in planktonic foraminifera. As the Omo River shares a divide with the Blue Nile and with tributaries of the White Nile, the Nile and the Omo must be affected similarly. As noted, deposition of each of the members of the Kibish Formation was probably very rapid. Thus, the close correspondence between the ages of Member I (196 kyr) and of sapropel S7 (195 kyr), of Member III (104 kyr) and of sapropel S4 (102 kyr), and of Member IV (3.3–9.5 kyr) and of sapropel S1 (8 kyr) is probably causally related. Sapropel S2 (55 kyr) is absent from or poorly represented in many Mediterranean sedimentary cores and has a very small $\delta^{18}O$ residual; thus, it is not surprising that no deposits of this age have been identified in the Kibish Formation. Sapropel S6, deposited during a European glacial period[19], might also be absent from the Kibish Formation, because it too has a small $\delta^{18}O$ anomaly. The two parts of Member II may be accommodated by sapropels S5 and S6, or they may correspond to the two phases identified in sapropel S5 (119–124 kyr), which are separated by 700–900 yr (ref. 20). This link between sapropel formation in the Mediterranean and very high levels of Lake Turkana is a particularly notable finding. In contrast, S3 is very well represented in many Mediterranean sedimentary cores and is therefore expected to be recorded in the Kibish Formation, but has not been recognized. Given the large expanse of the plain in the

对来自III段凝灰岩的四例浮岩碎屑中的长石进行了分析。两例是在原位(99-275A，C)采集的，另外两例(99-274A，B)也来自同一地层单元，但它们被风化搬运到了东部约150米远的地方(图1)。99-275A的年代从98.0±7.7 kyr到114.6±4.3 kyr不等。五个单晶体测量结果的算术平均值是105.5±7.0 kyr，与两三种晶体一组的五个结果是一致的(均值为105.4±2.7 kyr)，显示它们是同一群体。总平均年代为105.4±5.0 kyr，加权平均年代为106.0±1.6 kyr。对样本99-275C进行的十一项分析在剔除掉两个离群值(142.1±3.3和128.8±8.3 kyr)之后产生的平均年代为107.5±7.1 kyr，加权均值为105.4±1.8 kyr。对99-274A进行的七种单晶体和六对晶体分析得到的一致结果给出的总算术平均值是98.1±5.3 kyr。在剔除掉一个离群值后，伴随碎屑99-274B给出的平均年代为105.0±8.3 kyr(附表4)。这些结果再次反映了这些样本来源于同一次沉积，另外，单个浮岩样本有相似的平均年代，它们的K/Ca比值和滑动平均K/Ca比所表明的一样。将所有这些结果(除了离群值)综合起来，总算术平均年代为103.7±7.4 kyr($n=47$)，加权平均年代为103.7±0.9 kyr。因此，沉积年代必须等于或小于104 kyr，这为基比什组的I段和II段的年代要早于104 kyr提供了证据。

基比什组的每个地层段都是在图尔卡纳湖处于比现在更高的水位上的间隔期间沉积下来的，II段地层具有内部不整合性。I段地层的上半部分是在196 kyr前或晚些时候沉积下来的，III段的上半部分是在104 kyr前或晚些时候沉积下来的；IV段的^{14}C年代分析表明其对应的沉积作用发生于距今9.5 kyr到3.3 kyr前。这些年代与地中海腐泥S7、S4和S1阶段年代非常相似。腐泥S1～S7的估计年代如下：195 kyr(S7)、172 kyr(S6)、124 kyr(S5)、102 kyr(S4)、81 kyr(S3)、55 kyr(S2)和8 kyr(S1)[17]。浮游有孔虫中δ^{18}O的负值显示非洲季风的加剧，尼罗河流入地中海的水量大幅增加，很多情况下这与腐泥形成有关[18]。由于奥莫河与青尼罗河以及白尼罗河支流共有分水岭，所以尼罗河和奥莫河也必然会受到相似的影响。如前所述，基比什组的各成员层的沉积作用可能非常迅速。因此，I段地层(196 kyr)和腐泥S7(195 kyr)的年代之间、III段地层(104 kyr)和腐泥S4(102 kyr)的年代之间、IV段地层(3.3～9.5 kyr)和腐泥S1(8 kyr)的年代之间可能是有因果关系的。腐泥S2层(55 kyr)在许多地中海沉积物岩芯中都不存在或者含量很少，δ^{18}O残值也很小；因此，在基比什组没有测定出该年代的堆积物也不足为奇。腐泥S6是在欧洲冰期[19]堆积下来的，可能在基比什组也不存在，因为它也有一个小的δ^{18}O异常。II段地层存在两个不同的部分，二者可能对应腐泥层S5和S6，或者它们可能对应于腐泥S5层(119～124 kyr)所代表的两个时期，即被700～900年分开的两个时期(参考文献20)。地中海的腐泥地层和图尔卡纳湖的高水位之间的这种联系是一项非常值得注意的发现。相反，很多地中海沉积物岩芯中都能发现S3腐泥层，因此也希望基比什组

region underlain by the Kibish Formation, it is quite possible that deposits correlative with sapropel S3 are present but are not exposed in the immediate Kibish region that we have studied.

Our palaeontological and stratigraphic studies support the original report[14] that Omo I and Omo II are derived from comparable stratigraphic levels within Member I of the Kibish Formation despite their morphological differences[1-3,21,22]. Morphological diversity among fossil hominids from the Middle and Late Pleistocene of Africa is of major importance in understanding the tempo and mode of modern human origins[21,23].

$^{40}Ar/^{39}Ar$ dating of feldspars from tuffs in Member I and Member III of the Kibish Formation shows that its hominid fossils are younger than 195.8 ± 1.6 kyr and older than 103.7 ± 0.9 kyr. Direct isotopic dating of volcanic eruptions recorded in the Kibish Formation does not enable us to place narrower limits on the age. However, the suggested correlations of Member IIa and Member IIb with either the two identified phases of sapropel S5 or sapropels S6 and S5, respectively, indicate that deposition of Member I of the Kibish Formation occurred earlier than about 125 kyr ago or earlier than 172 kyr ago. The geological evidence for rapid deposition of Member I and the remarkably close correspondence of the isotopic ages on the youngest pumice clasts in the tuff of Member I at 196 kyr with the estimated age of sapropel S7 is regarded as strongly supporting the view that Member I was deposited close to 196 ± 2 kyr ago. On this basis we suggest that hominid fossils Omo I and Omo II are relatively securely dated to 195 ± 5 kyr old, somewhat older than the age of between 154 and 160 kyr assigned to the hominid fossils from Herto, Ethiopia[24], making Omo I and Omo II the oldest anatomically modern human fossils yet recovered.

Methods

Age measurements

Alkali feldspar crystals were separated from pumice clasts and cleaned ultrasonically in 7% HF for 5–10 min to remove adhering volcanic glass and surface alteration. Although separations were performed on the coarsest crystals present, in several cases crystals were less than 1 mm, with masses less than 0.8 mg. In such cases several crystals were used for each analysis (see Supplementary Tables 2–4).

Samples were irradiated in facilities X33 or X34 of the High Flux Australian Reactor (Lucas Heights, Sydney, Australia) for 6 h as described in ref. 25. Cadmium shielding 0.2 mm thick was used to reduce the $(^{40}Ar/^{39}Ar)_K$ correction factor, which was measured by analysis of zero-aged synthetic potassium silicate glass co-irradiated with the unknowns. This correction is particularly important because of the young age of the samples, so interpolated values for each sample are given in the Supplementary Tables 2–4. The fluence monitor employed was sanidine 92-176, separated from the Fish Canyon Tuff, with a reference age of 28.1 Myr (ref. 26).

能够有相应的记录，但尚未得到确认。鉴于该地区位于基比什组之下的平原如此之大，与腐泥 S3 相关的堆积物非常可能存在，只是在我们研究过的基比什地区周围没有暴露出来而已。

尽管奥莫 I 号和奥莫 II 号古人类化石存在形态学差异 [1-3,21,22]，但我们的古生物和地层学研究支持原先的报道 [14]，即奥莫 I 号和奥莫 II 号化石都来自基比什组的 I 段地层，其出土层位比较接近。来自非洲中、晚更新世时期的古人类化石之间存在的形态学多样性，对于理解现代人起源的进度和模式具有非常重要的意义 [21,23]。

对基比什组的 I 段和 III 段地层中的凝灰岩里的长石进行 $^{40}Ar/^{39}Ar$ 年代测定，表明古人类化石的年代区间为 195.8 ± 1.6 kyr 至 103.7 ± 0.9 kyr。对基比什组所记录的火山爆发的同位素直接测年并不能帮助我们将年代范围缩小。然而，IIa 组和 IIb 组分别与地中海腐泥 S5 或腐泥 S6 和 S5 两个已知时期的相关性表明，基比什组的 I 段地层的堆积作用发生的年代早于约 125 kyr 前，或者早于 172 kyr 前。地质学证据表明，I 段存在着快速堆积作用，另外，I 段地层的凝灰岩中的最晚的浮岩碎屑的同位素年代（196 kyr）与腐泥 S7 的年代的非常密切的对应性强烈支持了如下观点，即 I 段是在接近 196 ± 2 kyr 前堆积起来的。以此为基础，我们认为奥莫 I 和奥莫 II 的古人类化石比较肯定的年代是 195 ± 5 kyr，比埃塞俄比亚赫托 [24] 的古人类（154 kyr 至 160 kyr）的年代早一些，因此奥莫 I 和奥莫 II 也成为迄今挖掘出来的年代最古老的解剖学上的现代人化石。

方　法

年代测定

碱性长石晶体是从浮岩碎屑中分离出来的，然后用 7% 的氢氟酸超声波洗涤 5 至 10 分钟以去除黏附的火山玻璃和表面蚀变。尽管是在目前最粗糙的晶体上进行分离操作的，但是有些情况下，晶体还不足 1 毫米，质量也不到 0.8 毫克。这些情况下，每个分析都要使用几种晶体（见附表 2 ~ 4）。

样本使用澳大利亚高通量反应器（卢卡斯高地，悉尼，澳大利亚）的 X33 或 X34 仪器照射 6 小时，具体方法见参考文献 25 中的描述。用 0.2 毫米厚的镉屏蔽减小 $(^{40}Ar/^{39}Ar)_K$ 校正因子，该因子通过分析与未知因素共同照射的年代为零的合成硅酸钾玻璃来进行测定。由于样本的年代较近，所以这一修正就格外重要，附表 2~4 中给出了按照此法得到的每个样本的内插值。采用的流量监控器是透长石 92-176，是从菲什峡谷凝灰岩上分离下来的，其参考年代为 2,810 万年（参考文献 26）。

After irradiation, feldspar crystals were loaded into wells in a copper sample tray, installed in the vacuum system and baked overnight. Samples were fused with a focused argon-ion laser beam with up to 10 W of power. After purification of the gases released during fusion, the argon was analysed isotopically in a VG3600 mass spectrometer, using a Daly collector. The overall sensitivity of the system was about 2.5×10^{-17} mol mV^{-1}. Mass discrimination was monitored through regular measurements of atmospheric argon. The irradiation parameter, *J*, for each unknown was derived by interpolation from the measurements made on the fluence monitor crystals, with at least five analyses per level; the precision generally was in the range 0.3–0.75%, standard deviation of the population. Calcium correction factors[27] used in all calculations were $(^{36}Ar/^{37}Ar)_{Ca} = 3.49 \times 10^{-4}$ and $(^{39}Ar/^{37}Ar)_{Ca} = 7.86 \times 10^{-4}$.

Data handling

In calculating the arithmetic mean age for each pumice clast, any result more than two standard deviations from the initial mean of each sample was rejected iteratively until no further outliers were identified. No more than two measurements were rejected in any group of analyses, and no results were rejected for about half the groups. The error quoted in Table 1 is the standard deviation of the population, but because uncertainties on individual ages are variable, a weighted mean age and error is also given, weighting each age by the inverse of the variance. Differences in these mean ages are small (Table 1), but the error of the weighted mean age is usually much lower. In addition, data from each pumice clast, after exclusion of outliers, were plotted in an isotope correlation diagram ($^{36}Ar/^{40}Ar$ versus $^{39}Ar/^{40}Ar$), using the York[28] procedure. The derived ages are quite close to the mean ages (Table 1), and the calculated trapped argon composition generally has the atmospheric argon ratio of 295.5, within uncertainty.

(**433**, 733-736; 2005)

Ian McDougall[1], Francis H. Brown[2] & John G. Fleagle[3]
[1] Research School of Earth Sciences, Australian National University, Canberra, ACT 0200, Australia
[2] Department of Geology and Geophysics, University of Utah, Salt Lake City, Utah 84112, USA
[3] Department of Anatomical Science, Stony Brook University, Stony Brook, New York 11794, USA

Received 22 September; accepted 8 December 2004; doi:10.1038/nature03258.

References:

1. Day, M. H. Omo human skeletal remains. *Nature* **222**, 1135-1138 (1969).
2. Day, M. H. & Stringer, C. B. in *Congrès International de Paléontologie Humaine I, Nice* Vol. 2, 814-846 (Colloque International du CNRS, 1982).
3. Day, M. H. & Stringer, C. B. Les restes crâniens d'Omo-Kibish et leur classification à l'intérieur du genre *Homo*. *Anthropologie* **95**, 573-594 (1991).
4. Day, M. H., Twist, M. H. C. & Ward, S. Les vestiges post-crâniens d'Omo I (Kibish). *Anthropologie* **95**, 595-610 (1991).
5. Howell, F. C. in *Evolution of African Mammals* (eds Maglio, V. J. & Cooke, H. B. S.) 154-248 (Harvard Univ. Press, Cambridge, Massachusetts, 1978).
6. Smith, F. H., Falsetti, A. B. & Donnelly, S. M. Modern human origins. *Yb Phys. Anthropol.* **32**, 35-68 (1989).
7. Butzer, K. W., Isaac, G. Ll., Richardson, J. L. & Washbourn-Kamau, C. Radiocarbon dating of East African lake levels. *Science* **175**, 1069-1076 (1972).
8. Fuchs, V. E. The geological history of the Lake Rudolf Basin, Kenya Colony. *Phil. Trans. R. Soc. Lond. B* **229**, 219-274 (1939).
9. Arambourg, C. *Mission Scientifique de l'Omo 1932–1933. Geologie–Anthropologie–Paleontologie* (Muséum National d'Histoire Naturelle, Paris, 1935-1947).
10. Butzer, K. W. The Lower Omo Basin: Geology, fauna and hominids of Plio-Pleistocene formations. *Naturwissenschaften* **58**, 7-16 (1971).

照射之后，将长石晶体加样到一个铜质样品托盘的加样孔中，安装到真空系统后烘烤过夜。使用聚焦氩离子激光束以高达 10 W 的功率使样本熔化。对熔化期间释放出来的气体进行纯化后，使用戴利收集器在 VG3600 质谱仪中对氩进行同位素分析。该系统的综合灵敏度约为 2.5×10^{-17} mol · mV^{-1}。质量甄别通过大气中的氩的常规含量进行监控。每个未知的照射参数 J 都是通过在流量监控晶体上进行的测量得到的插入值派生出来的，每个水平至少进行五次分析；通常的精确性范围是 0.3% ~ 0.75%，群体的标准差。在所有计算中使用的钙修正因子[27]都是 $(^{36}Ar/^{37}Ar)_{Ca}=3.49\times10^{-4}$ 和 $(^{39}Ar/^{37}Ar)_{Ca}=7.86\times10^{-4}$。

数据处理

在计算每种浮岩碎屑的算术平均年代时，每个样本的最初均值中存在两个以上标准差的任何结果都被反复检查剔除掉了，直到不再出现离群值为止。任何一组分析中，都没有剔除掉两个以上的测量值，而且约有一半的组中都没有剔除掉任何结果。表 1 中引用的误差是群体的标准差，但是由于个体年代的不确定性是不同的，所以给出了加权平均年代和误差，即对变异反向的每个年代进行了加权。这些平均年代的差异很小（表 1），但是加权平均年代的误差通常更低。此外，每种浮岩碎屑的数据在剔除掉离群值之后，都使用 York[28] 程序在同位素相关性表格（$^{36}Ar/^{40}Ar$ 对 $^{39}Ar/^{40}Ar$）中绘图。派生的年代与平均年代非常接近（表 1），而计算出来的捕获氩成分一般具有大气中的氩比值——295.5，属误差之内。

（刘皓芳 翻译；潘雷 审稿）

11. Butzer, K. W. in *Earliest Man and Environments in the Lake Rudolf Basin* (eds Coppens, Y., Howell, F. C., Isaac, G. Ll. & Leakey, R. E. F.) 12-23 (Univ. of Chicago Press, Chicago, 1976).

12. Butzer, K. W. & Thurber, D. L. Some late Cenozoic sedimentary formations of the Lower Omo Basin. *Nature* **222**, 1138-1143 (1969).

13. Butzer, K. W., Brown, F. H. & Thurber, D. L. Horizontal sediments of the lower Omo Valley: the Kibish Formation. *Quaternaria* **11**, 15-29 (1969).

14. Butzer, K. W. Geological interpretation of two Pleistocene hominid sites in the Lower Omo Basin. *Nature* **222**, 1133-1135 (1969).

15. Owen, R. B., Barthelme, J. W., Renaut, R. W. & Vincens, A. Palaeolimnology and archaeology of Holocene deposits north-east of Lake Turkana, Kenya. *Nature* **298**, 523-529 (1982).

16. Fleagle, J. *et al.* The Omo I partial skeleton from the Kibish Formation. *Am. J. Phys. Anthropol. Suppl.* **36**, 95 (2003).

17. Lourens, L. J. *et al.* Evaluation of the Plio-Pleistocene astronomical timescale. *Paleoceanography* **11**, 391-413 (1996).

18. Rossignol-Strick, M., Nesteroff, W., Olive, P. & Vergnaud-Grazzini, C. After the deluge: Mediterranean stagnation and sapropel formation. *Nature* **295**, 05-110 (1982).

19. Rossignol-Strick, M. & Paterne, M. A synthetic pollen record of the eastern Mediterranean sapropels of the last 1 Ma: implications for the time-scale and formation of sapropels. *Mar. Geol.* **153**, 221-237 (1999).

20. Rohling, E. J. *et al.* African monsoon variability during the previous interglacial maximum. *Earth Planet. Sci. Lett.* **202**, 61-75 (2002).

21. Haile-Selassie, Y., Asfaw, B. & White, T. D. Hominid cranial remains from Upper Pleistocene deposits at Aduma, Middle Awash, Ethiopia. *Am. J. Phys. Anthropol.* **123**, 1-10 (2004).

22. Rightmire, G. P. in *The Origins of Modern Humans: A World Survey of the Fossil Evidence* (eds Smith, F. H. & Spencer, F.) 295-325 (Alan R. Liss, New York, 1984).

23. Howell, F. C. in *Origins of Anatomically Modern Humans* (eds Nitecki, M. H. & Nitecki, D. V.) 253-319 (Plenum, New York, 1994).

24. Clark, J. D. *et al.* Stratigraphic, chronological and behavioural contexts of Pleistocene *Homo sapiens* from Middle Awash, Ethiopia. *Nature* **423**, 747-752 (2003).

25. McDougall, I. & Feibel, C. S. Numerical age control for the Miocene–Pliocene succession at Lothagam, a hominoid-bearing sequence in the northern Kenya Rift. *J. Geol. Soc. Lond.* **156**, 731-745 (1999).

26. Spell, T. L. & McDougall, I. Characterization and calibration of $^{40}Ar/^{39}Ar$ dating standards. *Chem. Geol.* **198**, 189-211 (2003).

27. Spell, T. L., McDougall, I. & Doulgeris, A. P. The Cerro Toledo Rhyolite, Jemez Volcanic Field, New Mexico: $^{40}Ar/^{39}Ar$ geochronology of eruptions between two caldera-forming events. *Bull. Geol. Soc. Am.* **108**, 1549-1566 (1996).

28. York, D. Least squares fitting of a straight line with correlated errors. *Earth Planet. Sci. Lett.* **5**, 320-324 (1969).

29. Davidson, A. *The Omo River Project* (Bulletin 2, Ethiopian Institute of Geological Surveys, Addis Ababa, 1983).

Supplementary Information accompanies the paper on www.nature.com/nature.

Acknowledgements. We thank J. Mya, R. Maier and X. Zhang for technical support for the geochronology; participants in the Kibish expeditions between 1999 and 2003, including Z. Assefa, J. Shea, S. Yirga, J. Trapani and especially C. Feibel, B. Passey and C. Fuller for their geological contributions; and R. Leakey, K. Butzer and especially P. Abell for providing us with information and documents about the 1967 expedition to the Kibish area. We thank the Government of Ethiopia, the Ministry of Youth, Sports and Culture, the Authority for Research and Conservation of Cultural Heritage, and the National Museum of Ethiopia for permission to study the Kibish Formation. Support was provided by the National Science Foundation, the Leakey Foundation, the National Geographic Society and the Australian National University. Neutron irradiations were facilitated by the Australian Institute of Nuclear Science and Engineering and the Australian Nuclear Science and Technology Organization.

Competing interests statement. The authors declare that they have no competing financial interests.

Correspondence and requests for materials should be addressed to I.McD (ian.mcdougall@anu.edu.au).

The DNA Sequence of the Human X Chromosome

M. T. Ross *et al.*

Editor's Note

The enigmatic X chromosome reveals its genetic secrets in this paper detailing its DNA sequence. Female mammals carry two X chromosomes, whist males carry one X and one Y. Geneticist Mark Ross and colleagues determined the sequence of over 99% of the gene-containing region of the human X chromosome. The data reveal how X and Y chromosomes evolved from a pair of regular chromosomes around 300 million years ago. It also suggests that nearly 10% of the chromosome's 1,098 genes belong to a group that is upregulated in testicular and other cancers, and links a particular type of repetitive sequence with X-inactivation—the process that silences one of the two copies of the X chromosome in females to avoid a double dose of X chromosome genes.

The human X chromosome has a unique biology that was shaped by its evolution as the sex chromosome shared by males and females. We have determined 99.3% of the euchromatic sequence of the X chromosome. Our analysis illustrates the autosomal origin of the mammalian sex chromosomes, the stepwise process that led to the progressive loss of recombination between X and Y, and the extent of subsequent degradation of the Y chromosome. LINE1 repeat elements cover one-third of the X chromosome, with a distribution that is consistent with their proposed role as way stations in the process of X-chromosome inactivation. We found 1,098 genes in the sequence, of which 99 encode proteins expressed in testis and in various tumour types. A disproportionately high number of mendelian diseases are documented for the X chromosome. Of this number, 168 have been explained by mutations in 113 X-linked genes, which in many cases were characterized with the aid of the DNA sequence.

THE X chromosome has many features that are unique in the human genome. Females inherit an X chromosome from each parent, but males inherit a single, maternal X chromosome. Gene expression on one of the female X chromosomes is silenced early in development by the process of X-chromosome inactivation (XCI), and this chromosome remains inactive in somatic tissues thereafter. In the female germ line, the inactive chromosome is reactivated and undergoes meiotic recombination with the second X chromosome. The male X chromosome fails to recombine along virtually its entire length during meiosis: instead, recombination is restricted to short regions at the tips of the X chromosome arms that recombine with equivalent segments on the Y chromosome.

人类X染色体的DNA序列

罗斯等

编者按

这篇论文通过详细介绍神秘的X染色体的DNA序列，揭示了它的遗传秘密。雌性哺乳动物有两条X染色体，雄性哺乳动物有一条X染色体和一条Y染色体。遗传学家马克·罗斯和他的同事们确定了人类X染色体中99%以上基因区域的序列，这些数据揭示了大约3亿年前X和Y染色体是如何从一对普通染色体进化而来的。它也表明染色体的1,098个基因的近10%属于一个群体，该群体调节睾丸和其他癌症；并且将一个特定类型的重复序列与X染色体失活联系起来，这个过程使得女性当中一个X染色体的两个副本之一沉默，来避免X染色体基因剂量加倍。

人类X染色体有一种独特的生物学特性，这种特性是由其进化形成的，即男性和女性共有的性染色体。我们已经确定了X染色体上99.3%的常染色质序列。我们的分析表明了哺乳动物性染色体起源于常染色体，X和Y之间重组逐渐丢失的过程，以及Y染色体随后的退化程度。LINE1重复元件占X染色体的三分之一，其分布与它们在X染色体失活过程中作为中转站的作用一致。我们在这个序列中发现了1,098个基因，其中有99个编码在睾丸和各种肿瘤类型中表达的蛋白质。X染色体记录了大量不成比例的孟德尔病。在这一数字中，有168个可以用113个X连锁基因的突变解释，这些突变在许多情况下是借助DNA序列来表征的。

X染色体上有许多人类基因组中独特的特征。女性从父母双方各继承一条X染色体，但男性只继承一条母亲的X染色体。其中一条女性X染色体的基因表达在发育早期被X染色体失活(XCI)过程所抑制，此后该染色体在身体组织中仍旧保持失活。在女性生殖系中，失活的染色体被重新激活，经过减数分裂后与第二条X染色体进行重组。在减数分裂过程中，男性的X染色体不能沿着其整个长度进行重组，相反，重组仅限于与Y染色体上的等值段进行重组的X染色体臂端的短区域。这些区域内的基因在性染色体间共享，因此它们的行为被称为“假常染色体”。X染色体

Genes inside these regions are shared between the sex chromosomes, and their behaviour is therefore described as "pseudoautosomal". Genes outside these regions of the X chromosome are strictly X-linked, and the vast majority are present in a single copy in the male genome.

The unique properties of the X chromosome are a consequence of the evolution of sex chromosomes in mammals. The sex chromosomes have evolved from a pair of autosomes within the last 300 million years (Myr)[1]. In the process, the original, functional elements have been conserved on the X chromosome, but the Y chromosome has lost almost all traces of the ancestral autosome, including the genes that were once shared with the X chromosome. The hemizygosity of males for almost all X chromosome genes exposes recessive phenotypes, thus accounting for the large number of diseases that have been associated with the X chromosome[2]. The characteristic pattern of X-linked inheritance (affected males and no male-to-male transmission) was recognized by the eighteenth century for some cases of haemophilia, and gave impetus in the 1980s to the earliest successes in positional cloning—of the genes for chronic granulomatous disease[3] and Duchenne muscular dystrophy[4]. For females, the major consequence of the loss of genes from the Y chromosome is XCI, which equalizes the dosage of X-linked gene products between the sexes.

The biological consequences of sex chromosome evolution account for the intense interest in the human X chromosome in recent decades. However, evolutionary processes are likely to have shaped the behaviour and structure of the X chromosome in many other ways, influencing features such as repeat content, mutation rate, gene content and haplotype structure. The availability of the finished sequence of the human X chromosome, described here, now allows us to explore its evolution and unique properties at a new level.

The X Chromosome Sequence

We constructed a map of the X chromosome using predominantly P1-artificial chromosome (PAC) and bacterial artificial chromosome (BAC) clones (Supplementary Table 1), which were assembled into contigs using restriction-enzyme fingerprinting and integrated with earlier maps using sequence-tagged site (STS) content analysis[5]. Gaps were closed by targeted screening of clone libraries in bacteria or yeast, and by assessing BAC and fosmid end-sequence data for evidence of spanning clones. Fourteen euchromatic gaps remain intractable, despite using libraries with a combined 80-fold chromosome coverage. Five of these gaps are within the 2.7 megabase (Mb) pseudoautosomal region at the tip of the chromosome short arm (PAR1). This is reminiscent of the situation in other human sub-telomeric regions[6], and might reflect cloning difficulties in an area with a high content of (G+C) nucleotides and minisatellite repeats.

We selected 1,832 clones from the map for shotgun sequencing and directed finishing using established procedures[7]. Finished sequences were estimated to be more than 99.99% accurate by independent assessment[8]. The sequence of the X chromosome has been

上这些区域外的基因是严格的 X 连锁的，并且绝大多数都存在于男性基因组的单拷贝中。

X 染色体独特的特性是哺乳动物性染色体进化的结果。性染色体在过去的 3 亿年内从一对常染色体进化而来[1]。在这个过程中，原有的功能元件保留在 X 染色体上，而 Y 染色体已经失去了几乎所有祖先常染色体的痕迹，包括曾经与 X 染色体共享的基因。男性的半杂合性使几乎所有 X 染色体上的基因暴露隐性表型，从而产生大量 X 染色体连锁的疾病[2]。X 连锁遗传的特征模式（影响男性，没有男性之间的传递）因十八世纪血友病的一些案例而被识别出，并推动了 20 世纪 80 年代定位克隆的初步成功——对慢性肉芽肿病[3]和进行性假肥大性肌营养不良[4]的克隆。对女性来说，Y 染色体基因丢失的主要后果是 XCI，它可以使 X 连锁基因产物在两性之间的剂量均衡。

性染色体进化的生物学影响解释了近几十年人们为什么对人类 X 染色体有如此强烈的兴趣。然而，进化过程可能以许多其他方式塑造了 X 染色体的行为和结构，影响如重复序列的含量、突变率、基因含量和单体型结构等特征。这里所描述的已经完成的人类 X 染色体序列让我们在一个新的水平上探索其进化和特性。

X 染色体序列

我们主要通过限制性内切酶指纹法将 P1 人工染色体（PAC）和细菌人工染色体（BAC）克隆（补充表 1）组装成重叠群，并使用序列标签位点（STS）含量分析与早先图谱整合，构建了一个 X 染色体图谱[5]。通过有针对性的筛选细菌或酵母菌克隆文库，并通过评估 BAC 和 F 黏粒末端序列数据来寻找跨域克隆的证据，以封闭缺口。即使使用了染色体覆盖率达 80 倍的文库，仍有 14 个常染色质缺口无法解决。其中五个缺口在染色体短臂末端 2.7 Mb 的假常染色体区域（PAR1）。这使人想起在人类其他亚端粒区域的情况[6]，并可能反映在（G+C）核苷酸和小卫星重复含量高的区域的克隆难度。

我们从图谱中选择 1,832 个克隆进行鸟枪法测序并用已建立的程序进行定向指导[7]。通过独立评价，完成的序列准确率超过 99.99%[8]。该 X 染色体的序列是

assembled from the individual clone sequences and comprises 16 contigs. These extend into the telomeric (TTAGGG)*n* repeat arrays at the ends of the chromosome arms, and include both pseudoautosomal regions (PARs). The data were frozen for the analyses described below, at which point we had determined 150,396,262 base pairs (bp) of sequence (Supplementary Table 2). Subsequently, we obtained a further 609,664 bp of sequence. The 14 euchromatic gaps are estimated to have a combined size of less than 1 Mb (see Methods and Supplementary Table 2), and the sequence therefore covers at least 99.3% of the X chromosome euchromatin. There is also a single heterochromatic gap corresponding to the polymorphic 3.0 ($\pm$0.4) Mb array[9] of alpha satellite DNA at the centromere. On this basis, we conclude that the X chromosome is approximately 155 Mb in length.

The coverage and quality of the finished sequence have been assessed using independent data. All markers from the deCODE genetic map[10] are found in the sequence and the concordance of marker orders is excellent with only one discrepancy. DXS6807 is the most distal Xp marker on the deCODE map (4.39 cM), but in the sequence this marker is proximal to three others with genetic locations of 9–11 cM on the deCODE map. Out of 788 X chromosomal RefSeq[11] messenger RNAs that were assessed, 783 were found completely in the sequence, and parts of four others are also present (T. Furey, personal communication). The missing segments of *GTPBP6*, *CRLF2*, *DHRSX* and *FGF16* lie within gaps 1, 4, 5 and 10, respectively, and the *GAGE3* gene is in gap 7 (Supplementary Table 2). The sequence assembly was assessed using fosmid end-sequence pairs that match the X chromosome sequence. The orientation and separation of end-pairs of more than 17,000 fosmids were consistent with the sequence assembly. In two cases, sequences had been misassembled owing to long and highly similar repeats. There were six instances of large deletions in sequenced clones, which were resolved by determining fosmid sequences through the deleted regions. Finally, there were two cases of apparent length variation between the reference sequence and the DNA used for the fosmid library.

Features of the X Chromosome Sequence

The annotated sequence of the X chromosome is presented in Supplementary Fig. 1, and updates are contained in the Vertebrate Genome Annotation (VEGA) database (http://vega.sanger.ac.uk/Homo_sapiens/). The distribution of a number of sequence features on the chromosome is shown in Fig. 1 (Editorial note: this figure was originally included as a fold-out insert. For details, see *Nature*'s web site). Analysis of the sequence reveals a gene-poor chromosome that is highly enriched in interspersed repeats and has a low (G+C) content (39%) compared with the genome average (41%).

Fig. 1. Features of the X chromosome sequence. **a**, X chromosome ideogram according to Francke[65]. **b**, Evolutionary domains of the X chromosome: the X-added region (XAR), the X-conserved region (XCR; dotted region in proximal Xp does not appear to be part of the XCR), the pseudoautosomal region PAR1, and evolutionary strata S5–S1. **c**, Sequence scale in intervals of 1 Mb. Note that correlation between cytogenetic band positions and physical distance is imprecise, owing to varying levels of condensation of different Giemsa bands. **d**, (G+C) content of 100-kb sequence windows. **e**, Number of genes in 1-Mb

由单个克隆序列组装而成，包括 16 个重叠群。它们延伸到染色体臂末端的端粒(TTAGGG)*n* 重复序列，并包括两个假常染色体区域(PAR)。这些数据被冻结用于下述分析，在此，我们已确定 150,396,262 bp 的序列(补充表 2)。随后，我们进一步获得了长度为 609,664 bp 的序列。估计 14 个常染色质缺口合并大小小于 1 Mb(见方法和补充表 2)，因此，该序列至少覆盖 99.3%的 X 染色体常染色质。还有一个异染色质缺口对应着丝粒上 α 卫星 DNA 的多态性 3.0(±0.4) Mb 阵列[9]。在此基础上，我们得出结论：X 染色体长度大约 155 Mb。

我们用独立的数据对已经完成序列的覆盖度和质量进行了评估，在序列中发现了 deCODE 遗传图谱[10]的所有标记，并且标记顺序一致性很高，只有一个除外。DXS6807 是 deCODE 图上最远端 Xp 标记(4.39 cM)，但在序列中这个标记与其他三个在 deCODE 图上遗传位置距离 9 ~ 11 cM 的标记最接近。我们评估了 788 个 X 染色体的信使 RNA 的参考序列[11]，其中 783 个在序列中被完全发现，其他 4 个也部分出现在序列中(富里，个人交流)。*GTPBP6*、*CRLF2*、*DHRSX* 和 *FGF16* 丢失的片段分别位于缺口 1、4、5 和 10，*GAGE3* 基因在缺口 7 中(补充表 2)。序列组装使用匹配 X 染色体序列的 F 黏粒末端序列对进行评估。超过 1.7 万 F 黏粒的末端对方向和间距与序列组装一致。序列长度长且高度相似的重复序列导致有两例组装错误。在测序的克隆中有六个大的缺失实例，这些实例已通过确定缺失区域的 F 黏粒序列来解决。最后，有两例在参考序列和用于 F 黏粒文库的 DNA 之间有明显的长度差异。

X 染色体序列的特征

X 染色体的注解序列见补充图 1，在脊椎动物基因组注释(VEGA)数据库(http://vega.sanger.ac.uk/Homo_sapiens/)里进行更新。此染色体上若干序列特征的分布如图 1(编者注：在原文中，本图是一个折叠插入图，更多细节请看《自然》官网)所示。序列分析揭示了一个基因贫乏的染色体，其高度富集散在重复序列，与基因组的平均水平(41%)相比(G+C)含量(39%)较低。

图 1. X 染色体序列特征。**a**，根据弗兰克的 X 染色体表意图[65]。**b**，X 染色体进化域：X 增加区(XAR)，X 保守区(XCR；Xp 近端虚线区域不属于 XCR)，假常染色体区域 PAR1 和进化层 S5 ~ S1。**c**，序列刻度间隔为 1 Mb。请注意，由于不同吉姆萨带缩合程度不同，细胞遗传学带位置和物理距离间的相关性并不精确，**d**，100 kb 序列窗口的(G+C)含量。**e**，1 Mb 序列窗口中的基因数量(不包括假基因)。**f** ~ **k**，100 kb 窗口的序列被散在重复序列部分覆盖：L1 重复(**f**)，L1 重复亚族 L1M(**g**)、L1 重复

sequence windows (pseudogenes not included). **f–k**, Fractional coverage of 100-kb sequence windows by interspersed repeats: L1 repeats (**f**), L1M subfamilies of L1 repeats (**g**), L1P subfamilies of L1 repeats (**h**), *Alu* repeats (**i**), L2 repeats (**j**), MIR repeats (**k**). Vertical grey lines in **d–k** represent gaps in the euchromatic sequence of the chromosome. Grey bar centred at approximately 60 Mb shows the position of the centromere. **l**, A selection of landmark genes on the chromosome. *OPN* refers to the three opsin genes in the reference sequence, which are organized as follows: cen-*OPN1LW-OPN1MW-OPN1MW*-tel. **m**, Genes that escape from X-chromosome inactivation as previously identified[48]. **n**, Cancer-testis antigen genes, belonging to the *MAGE* (light green), *GAGE* (dark green), *SSX* (magenta), *SPANX* (orange) or other (grey) CT gene families. For the genes in **l–n**, arrows indicate the direction of transcription.

Genes

Based on a manual assessment of all publicly available human expressed sequences and genes from other organisms, we have annotated 1,098 genes (7.1 genes per Mb) across four different categories (see Methods): known genes (699), novel coding sequences (132), novel transcripts (166), and putative transcripts (101). We have also identified 700 pseudogenes in the sequence (4.6 pseudogenes per Mb), of which 644 are classified as processed and 56 as non-processed. The gene density (excluding pseudogenes) on the X chromosome is among the lowest for the chromosomes that have been annotated to date. This might simply reflect a low gene density on the ancestral autosomes. Alternatively, selection may have favoured transposition of particular classes of gene from the X chromosome to the autosomes during mammalian evolution. These could include developmental genes for which the protein products are required in double dose in males (or in females after XCI has occurred), or genes for which mutation in male somatic tissues is lethal.

Physical characteristics of the genes and pseudogenes are summarized in Supplementary Table 3. Exons of the 1,098 genes account for only 1.7% of the X chromosome sequence. On the basis of the lengths of these gene loci, 33% of the chromosome is transcribed. This is considerably below the recent estimates for chromosomes 6 (ref. 12), 9 (ref. 6), 10 (ref. 13) and 13 (ref. 14), to which the equivalent gene annotation procedure was applied (Supplementary Table 4), and is a reflection not just of low gene density on chromosome X but also of low gene length. For example, mean gene length is 49 kilobases (kb) on chromosome X compared with 57 kb on chromosome 13. Nevertheless, the X chromosome contains the largest known gene in the human genome, the dystrophin (*DMD*) locus in Xp21.1, which spans 2,220,223 bp. Consistent with its low gene density, the frequency of predicted CpG islands on the X chromosome is only 5.25 per Mb, which is exactly half of the estimated genome average[7]. There is an association with a CpG island for 49% of the known genes, the category for which the most complete gene structures are expected in the current annotation.

We identified evolutionarily conserved regions (ECRs) by comparing the X chromosome sequence to the genomes of mouse, rat, zebrafish and the pufferfishes *Tetraodon nigroviridis* and *Fugu rubripes* (Supplementary Table 5). There are 4,493 ECRs that are conserved between the X chromosome and all of the other species. Of these, 4,393 overlap with

亚族 L1P(**h**)、*Alu* 重复(**i**)、L2 重复(**j**)、MIR 重复(**k**)。**d**～**k** 中的垂直灰线代表染色体常染色质序列中的缺口。集中在大约 60 Mb 处的灰色条块代表着丝粒的位置。**l**，染色体上标志性基因的选择。*OPN* 是指参考序列中的三个视蛋白基因，排列如下：着丝粒 -*OPN1LW*-*OPN1MW*-*OPN1MW*- 端粒。**m**，如前所述，从 X 染色体失活中逃逸的基因[48]。**n**，癌/睾丸抗原基因，属于 *MAGE*(浅绿色)、*GAGE*(深绿色)、*SSX*(品红色)、*SPANX*(橙色)或其他(灰色)CT 基因家族。对于 **l**～**n** 中的基因，箭头指示转录的方向。

基　因

在人工评估所有已公布的人类表达序列和其他生物基因的基础上，我们已注释了四个不同类别的 1,098 个基因(每 Mb 有 7.1 个基因)(见方法)：已知基因(699)、新的编码序列(132)、新转录本(166)、假定转录本(101)。我们还在序列中确定了 700 个假基因(每 Mb 有 4.6 个假基因)，其中 644 个为已加工的基因，56 个为未加工的基因。X 染色体上的基因密度(不包括假基因)是迄今为止已注释的染色体中最低的。这可能只是反映了祖先染色体上的低基因密度。或者，在哺乳动物进化中选择可能有利于从 X 染色体向常染色转移某些类别的基因。这些基因可能包括男性(或 XCI 发生后的女性)需要双倍剂量的蛋白质产物发育基因，或男性体细胞组织中突变致死的基因。

补充表 3 总结了这些基因和假基因的物理特性。1,098 个基因的外显子只占 X 染色体序列的 1.7%。根据这些基因座的长度，有 33%的染色体被转录。这大大低于最近采用等效的基因注释程序(补充表 4)对染色体 6(参考文献 12)、9(参考文献 6)、10(参考文献 13)和 13(参考文献 14)的估计，不仅是 X 染色体上低基因密度的反映，同时也是低基因长度的反映。例如，X 染色体上平均基因长度为 49 千碱基(kb)，相比之下 13 号染色体上的为 57 kb。然而，X 染色体包含人类基因组中已知最大的基因——在 Xp21.1 上的肌萎缩蛋白基因座 *DMD*，它跨越 2,220,223 个碱基对。与其低基因密度相符的是，预测 X 染色体上每 Mb 只有 5.25 个 CpG 岛，这正好是基因组平均估计值的一半[7]。在已知基因中有 49% 与 CpG 岛相关，预计在目前的注释中这类基因结构最完整。

我们通过将 X 染色体序列与小鼠、大鼠、斑马鱼、黑青斑河鲀和红鳍东方鲀的基因组序列进行比较，确定了进化上的保守区域(ECR)(补充表 5)。在 X 染色体和所有其他物种间有 4,493 ECR 是保守的。其中，4,393 个与 4,373 个已注释的外显子

4,373 annotated exons. The remaining 100 ECRs are most likely to be unannotated exons, although some could be highly conserved control or structural elements. From these data we conclude that we have annotated at least 97.8% of the protein-coding exons on the X chromosome ([4,373/(4,373 + 100)] × 100).

Non-coding RNA Genes

The gene set described above includes non-coding RNA (ncRNA) genes only when there is supporting evidence of expression from complementary DNA or expressed-sequence-tag (EST) sources. Using a complementary approach, we analysed the X chromosome sequence using the Rfam[15] database of structural RNA alignments, and predicted 173 ncRNA genes and/or pseudogenes (Supplementary Fig. 1 and Supplementary Table 6). These are physically separate from the genes described in the preceding section and are not included in the total gene count, owing to the difficulty in discriminating between genes and pseudogenes for these ncRNA predictions. Using tRNAscan-SE[16], we predicted only two transfer RNA genes on the X chromosome (Supplementary Table 6), out of the several hundred predicted in the human genome[7]. Thirteen microRNAs from the microRNA registry[17] have also been mapped onto the sequence (Supplementary Table 7).

The most prominent of the ncRNA genes on the X chromosome is *XIST* (X (inactive)-specific transcript)[18], which is critical for XCI. The *XIST* locus spans 32,103 bp in Xq13, and its untranslated transcript coats and transcriptionally silences one X chromosome in *cis*. The RefSeq[11] transcript of *XIST* is an RNA of 19,275 bases, which includes the largest exon on the chromosome (exon 1: 11,372 bp). There is also evidence for shorter *XIST* transcripts generated by alternative splicing, particularly in the 3′ region of the gene[19]. In the mouse, *Tsix* is antisense to *Xist*[20], and its transcript (or the process of its transcription) is believed to repress the accumulation of *Xist* RNA. There is evidence for transcription antisense to *XIST* in human[21,22], but we have been unable to annotate the human *TSIX* gene as there are no corresponding expressed sequences in the public databases, and because there is a lack of primary sequence conservation between the human and mouse regions. In the human sequence, two other ncRNA genes are annotated in the 400 kb region distal to *XIST*, which are orthologues of the mouse genes described previously as *Jpx* and *Ftx* (ref. 23). In the mouse, *Xist*, *Jpx* and *Ftx* are located within a smaller area of approximately 200 kb[23].

The Cancer-testis Antigen Genes

On assessing the predicted proteome of the X chromosome for Pfam[24] domains, our most prominent finding was the presence of the MAGE domain (IPR002190) in 32 genes (Supplementary Table 8). In comparison, only four other MAGE genes are reported in the rest of the genome: *MAGEF1* on chromosome 3, and *MAGEL2*, *NDN* and *NDNL2* on chromosome 15. The *MAGE* gene products are members of the cancer-testis (CT) antigen group, which are characterized by their expression in a number of cancer types, while their

重叠。其余 100 个 ECR 最可能是未注释的外显子，尽管有一些可能是高度保守的控制元件或结构元件。从这些数据中我们得出结论：我们至少已经注释了 X 染色体 97.8%的编码蛋白质的外显子（[4,373/(4,373+100)]×100）。

非编码 RNA 基因

上文所述基因集合只有存在互补 DNA 表达或表达序列标签（EST）表达的支持性证据时才包含非编码 RNA（ncRNA）基因。利用互补性方法，我们用 Rfam[15] 数据库的结构 RNA 比对分析了 X 染色体序列，并预测了 173 个 ncRNA 基因和（或）假基因（补充图 1 和补充表 6）。由于这些 ncRNA 预测难以区分基因和假基因，因此它们与前面章节所描述的基因有实质区别，不包括在基因总数中。使用 tRNAscan-SE[16]，我们预测 X 染色体上只有两个转移 RNA 基因（补充表 6），而在人类基因组中预测有几百个 [7]。13 个来自微 RNA 注册表的微 RNA[17] 也绘制在了序列中（补充表 7）。

X 染色体上最明显的 ncRNA 基因是 *XIST*（X（失活）特异转录本）[18]，它对 X 染色体失活至关重要。*XIST* 基因座在 Xq13 中跨越 32,103 bp，其未翻译的转录本覆盖并转录沉默一条 X 染色体。*XIST* 的 RefSeq[11] 转录本是一个包含 19,275 个碱基的 RNA，它包括此染色体上最大的外显子（外显子 1：11,372 bp）。还有证据表明较短的 *XIST* 转录本通过可变剪接产生，特别是在基因 3′ 区 [19]。在小鼠中，*Tsix* 与 *Xist*[20] 是反义的，其转录本（或转录过程）被认为抑制 *Xist* RNA 的积累。有证据证明转录本与人类 *XIST* 是反义的 [21,22]，由于在公共数据库中没有相应的表达序列，且人类和小鼠序列区域缺乏基本的保守性，因此我们无法对人类 *TSIX* 基因进行注释。在人类序列中，其他两个注释的 ncRNA 基因在 *XIST* 远端 400 kb 区域，它们是之前描述的小鼠 *Jpx* 和 *Ftx* 基因的直系同源基因（参考文献 23）。在小鼠中，*Xist*、*Jpx* 和 *Ftx* 都位于一个大约 200 kb 的小区域内 [23]。

癌/睾丸抗原基因

在评估预测 X 染色体蛋白质组的 Pfam[24] 结构域时，我们最突出的发现是在 32 个基因中存在 MAGE 抗原结构域（IPR002190）（补充表 8）。相比之下，在剩余基因组中只报道了另外四个 MAGE 基因：3 号染色体上的 *MAGEF1* 和 15 号染色体上的 *MAGEL2*、*NDN* 和 *NDNL2*。*MAGE* 基因产物是癌/睾丸（CT）抗原组的成员，其特点是在大量癌症类型中表达，而它们在正常组织中的表达仅在或主要是在睾丸中。这

expression in normal tissues is solely or predominantly in testis. This expression profile has led to the suggestion that the CT antigens are potential targets for tumour immunotherapy. A recent report listed 84 CT antigen genes for the human genome[25]. The X chromosome gene set we describe above contains 99 CT antigen genes and includes novel members of the *MAGE, GAGE, SSX, LAGE, CSAGE* and *NXF* families (Supplementary Table 9). Assessment of the most recent RefSeq[11] information shows that this set does not include two known *MAGE* genes (*MAGEA5* and *MAGEA7*) and seven *GAGE* genes (*GAGE3–7, 7B* and *8*), which are expected to lie in gaps 14 and 7, respectively (Supplementary Table 2). Furthermore, gaps 6 and 9 are also within regions of CT antigen gene duplication. Therefore, we predict that approximately 10% of the genes on the X chromosome are of the CT antigen type.

Conclusive data on the normal functions of the CT antigens, or their involvement in disease conditions, are very limited. However, the remarkable enrichment for CT antigen genes on the X chromosome relative to the rest of the genome might be indicative of a male advantage associated with these genes. Recessive alleles that are beneficial to males are expected to become fixed more rapidly on the X chromosome than on an autosome[26]. If these alleles are detrimental to females, their expression could become restricted to male tissues as they rise to fixation. Both the concentration of the CT antigen genes on the X chromosome and their expression profiles are consistent with this model of male benefit. The CT antigen genes on the X chromosome are also notable for the expansion of various gene families by duplication. This degree of duplication is perhaps an indication of selection in males for increased copy number. In this context, it is of interest that the *MAGE* family has independently expanded on the X chromosome in both the human and mouse lineages[27].

Repetitive Sequences

Interspersed repeats account for 56% of the euchromatic X chromosome sequence, compared with a genome average of 45% (Supplementary Table 10). Within this, the *Alu* family of short interspersed nuclear elements (SINEs) is below average, in keeping with the gene-poor nature of the chromosome. Conversely, long terminal repeat (LTR) retroposon coverage is above average; but the most remarkable enrichment is for long interspersed nuclear elements (LINEs) of the L1 family, which account for 29% of the X chromosome sequence compared to a genome average of only 17%. The possible significance of this enrichment for XCI is discussed later.

Applying the criterion of at least 90% sequence identity over at least 5 kb (ref. 28), we estimate that intrachromosomal segmental duplications account for 2.59% of the X chromosome (Supplementary Table 11 and Supplementary Fig. 2). In contrast, interchromosomal segmental duplications indicated by sequence matches to the autosomes account for a very small fraction (0.24%) of the X chromosome (Supplementary Table 12). Six gaps in the X chromosome map are either flanked by or contained within

种基因表达谱表明 CT 抗原是肿瘤免疫治疗的潜在靶标。最近的一份报告列出了人类基因组中 84 个 CT 抗原基因[25]。上述 X 染色体基因集包含 99 个 CT 抗原基因，包括 *MAGE*、*GAGE*、*SSX*、*LAGE*、*CSAGE* 和 *NXF* 家族的新成员（补充表 9）。对最新 RefSeq[11] 信息的评估表明，这个集合不包括两个已知 *MAGE* 基因（*MAGEA5* 和 *MAGEA7*）和 7 个 *GAGE* 基因（*GAGE3* ~ *7*，*7B* 和 *8*），预计分别位于缺口 14 和 7（补充表 2）。此外，缺口 6 和 9 也在 CT 抗原基因复制区域内。因此，我们预测，X 染色体上大约 10%的基因是 CT 抗原类型。

关于 CT 抗原的正常功能，或它们在疾病中的作用的结论性数据非常有限。然而，相比于基因组的其他部分，X 染色体上的 CT 抗原基因显著富集可能表明男性优势与这些基因相关。预计对男性有益的隐性等位基因在 X 染色体上比在常染色体上固定得更快[26]。如果这些等位基因对女性有害，当它们上升到固定状态时表达可能会局限于男性组织。CT 抗原基因在 X 染色体上的浓度和它们的基因表达谱都符合男性受益模式。X 染色体上的 CT 抗原基因也通过复制显著扩大了各种基因家族。复制的程度可能是男性选择增加拷贝数量的标志。在此背景下，有意思的是，在人类和小鼠的谱系中，*MAGE* 家族都独立地扩展到了 X 染色体上[27]。

重复序列

散在重复序列占 X 染色体常染色质序列的 56%，而基因组的平均水平为 45%（补充表 10）。在这个范围内，*Alu* 家族短散在重复序列（SINE）低于平均水平，这与染色体的基因贫乏性质一致。相反，长末端重复序列（LTR）逆转座子覆盖率高于平均水平；L1 家族长散在重复序列（LINE）很多，占 X 染色体序列的 29%，而基因组平均水平只有 17%。稍后将讨论富含 XCI 可能的意义。

应用在至少 5 kb 上至少 90%以上序列同源这一标准（参考文献 28），我们估计，染色体内片段性重复占 X 染色体的 2.59%（补充表 11 和补充图 2）。与此相反，染色体间片段重复的序列与常染色体的匹配只占 X 染色体很小一部分（0.24%）（补充表 12）。X 染色体图上六个缺口或在染色体内重复片段的两侧或包含在其中（见补充表

intrachromosomally-duplicated segments (gaps 2, 3, 6, 7, 9 and 14 in Supplementary Table 2), which might produce instability of clones or otherwise confound mapping progress. The intrachromosomal duplicates are striking in their proximity. Apart from the two segments containing *SSX* gene copies, which are separated by 4.5 Mb, only six of 229 matches are separated by more than 1 Mb. Among these duplications are well-described cases that are associated with genomic disorders[29]. In Xp22.32, deletions of the steroid sulphatase (*STS*) gene, causing X-linked ichthyosis (Online Mendelian Inheritance in Man (OMIM)[2] entry number 308100), result from recombination between flanking duplications that contain copies of the *VCX* gene. Also, some instances of Hunter syndrome (OMIM 309900), red-green colour blindness (OMIM 303800), Emery-Dreifuss muscular dystrophy (OMIM 310300), incontinentia pigmenti (OMIM 308300) and haemophilia A (OMIM 306700) result from rearrangements involving duplicated sequences in Xq28. In haemophilia A, mutations are frequently the result of inversions between a sequence in intron 22 of the *F8* gene and one of two more distally located copies. A novel finding from our analysis of the X chromosome reference sequence is that the two distal copies are in opposite orientations. Therefore, a large deletion involving *F8* and several more distal genes could be an alternative to the inversion rearrangement. A deletion consistent with this prediction has been reported in a family in which carrier females are affected by a high spontaneous-abortion rate in pregnancy[30].

The X Chromosome Centromere

The X chromosome sequence extends from both arms into centromeric, higher-order repeat sequences, which are known to be associated functionally with the X centromere[31-33]. The most proximal 494 kb and 360 kb of the Xp and Xq sequences, respectively, consist of extensive regions of satellite DNA, adjacent to euchromatin of the chromosome arms that is exceptionally high in L1 content (Fig. 2). The satellite region on Xp contains small amounts of other satellite families[31], whereas that on Xq consists entirely of alpha satellite. Similar to all other human chromosome arms that have been examined[33,34], these transition regions consist of monomeric alpha satellite that is not associated with centromere function. Both the Xp and Xq contigs reported here, though, extend more proximally and reach into highly homogeneous, higher-order repeat alpha satellite (DXZ1). Critically, the Xp and Xq contig copies of the DXZ1 repeat are themselves 98–100% identical in sequence, and are oriented in the same direction along the chromosome (Fig. 2). On this basis, the two contigs reach the "end" of each chromosome arm and thus also reach the centromeric locus from either side. This represents a logical endpoint for efforts to complete the sequence of chromosome arms in the human genome, and the first demonstration of this endpoint is provided by the X chromosome sequence.

2 中的缺口 2、3、6、7、9 和 14)，这可能会导致克隆的不稳定或混淆绘图过程。染色体内的重复惊人的接近。除了两个含有 *SSX* 基因拷贝的片段被 4.5 Mb 分开，229 个匹配中只有 6 个匹配之间的距离大于 1 Mb。在这些重复中有与基因组无序相联系的并且充分描述的病例 [29]。在 Xp22.32 中，类固醇硫酸酯酶(*STS*)基因缺失导致了 X 连锁鱼鳞病(在线人类孟德尔遗传数据库(OMIM) [2] 索引号 308100)，这是由包含 *VCX* 基因拷贝的侧翼重复间的重组造成的。还有一些列子：亨特综合征(OMIM 309900)，红绿色盲(OMIM 303800)，埃德二氏肌营养不良症(OMIM 310300)，色素失调症(OMIM 308300)和血友病 A(OMIM 306700)都是 Xq28 重复序列重排的结果。在血友病 A 中，突变通常是在 *F8* 基因内含子 22 中的序列和两个以上远端拷贝中的一个之间倒位的结果。从对 X 染色体参考序列的分析中我们有了一个新发现：这两个远端的拷贝方向相反。因此，涉及 *F8* 和多个远端基因的大片段缺失可能是反向重排的替代方法。符合这一预测的缺失已在一个妊娠期有高自然流产率的女性携带者家族中报道过 [30]。

X 染色体着丝粒

X 染色体序列从两个臂向着丝粒延伸，分布了许多功能上与 X 着丝粒有联系的高度有序的重复序列 [31-33]。Xp 序列近端的 494 kb 和 Xq 序列近端的 360 kb，由广泛的卫星 DNA 区域组成，毗邻 L1 含量非常高的由常染色质组成的染色体臂(见图 2)。Xp 上卫星区域包含少量其他卫星家族 [31]，而 Xq 上的卫星区域包含全部的 α 卫星。与所有其他已检测的人类染色体臂相似 [33,34]，这些过渡区域包括与着丝粒功能无关的单体 α 卫星。但这里报道的 Xp 和 Xq 重叠群以更接近的方式延伸并成了高度均匀的、更高阶重复的 α 卫星(DXZ1)。更为重要的是，DXZ1 重复的 Xp 和 Xq 重叠序列本身序列一致性达 98% ~ 100%，并且沿染色体的方向相同(见图 2)。在此基础上，这两个重叠群到达各自染色体臂的“末端”，从而也从两侧到达着丝粒基因座。这代表了努力完成人类基因组染色体臂序列的一个合乎逻辑的终点，而这个终点是由 X 染色体序列证实的。

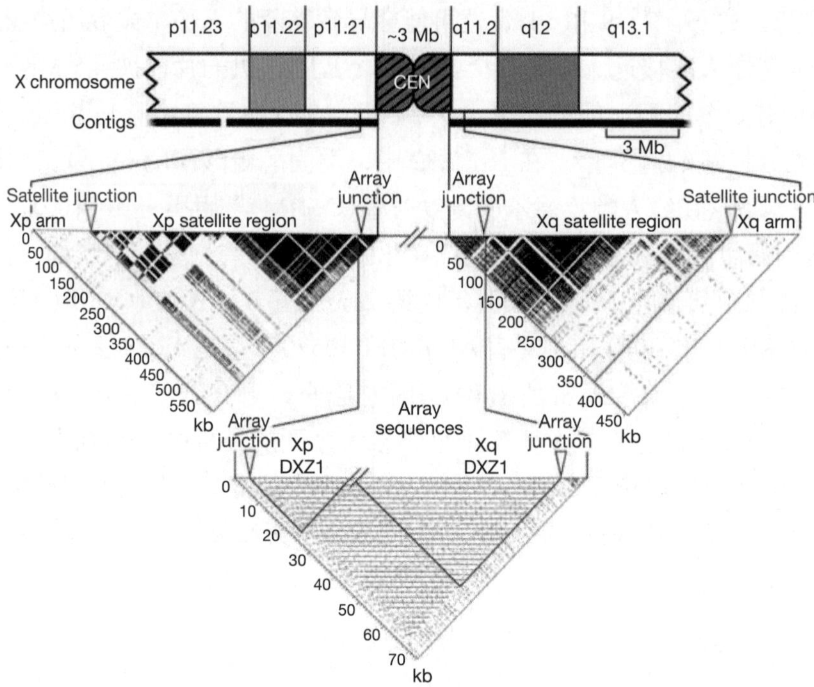

Fig. 2. Xp and Xq pericentromeric contigs extend into the X-chromosome-specific higher-order alpha satellite, DXZ1. The pericentromeric region of the X chromosome is shown as a truncated ideogram. Self-self alignments of proximal sequences from each arm are illustrated by dotter plots below the ideogram. On each plot, the junction between the arm sequence and the arm-specific satellite region is marked by a red arrow, and the junction between the arm-specific satellite region and the X-chromosome-specific alpha satellite array (DXZ1) is marked with a blue arrow. Approximately 594 kb of sequence were analysed from Xp, including ~21 kb of DXZ1 sequence. The ~454 kb of sequence analysed from Xq included ~44 kb of DXZ1 sequence. In each case, ~100 kb of arm sequence were included. The highly repetitive structure of pericentromeric satellites is in stark contrast to the near absence of repetitive structure in the arm sequences, despite an unusually high density of LINE repeats in these regions. Gaps in the dark satellite regions occur where interspersed elements (LINEs, SINEs and LTRs) interrupt the satellite sequences. In the Array Sequences dotter plot, the most proximal ~21 kb of the Xp sequence is joined to the most proximal ~44 kb of the Xq sequence. The periodic nature of the centromeric, higher-order alpha satellite array is evident. Black horizontal lines on the plot reveal near identity of sequences spaced at ~2 kb intervals. This DXZ1 sample represents ~65 kb of the 3 (±0.4) Mb alpha satellite array. The regions outlined in blue are self-self alignments ("Xp DXZ1" and "Xq DXZ1"), and the remaining rectangular region of the plot is an alignment of Xp versus Xq DXZ1, which reveals the close relationship between DXZ1 sequences from each arm.

Single-nucleotide Polymorphisms

A total of 153,146 candidate single-nucleotide polymorphisms (SNPs) have been mapped onto the X chromosome sequence and are displayed in the VEGA database. These include 901 SNPs that result in non-synonymous changes in protein-coding regions, and are therefore candidate functional protein variants. The heterozygosity level on the X chromosome is known to be well below that of the autosomes, and this difference can be explained partly or entirely by population genetic factors[35]. Included in the mapped

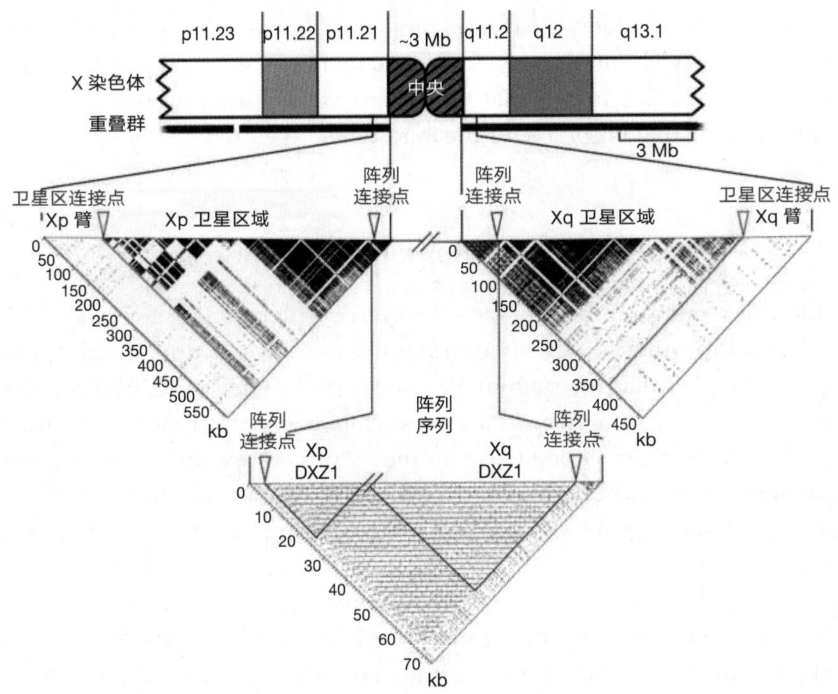

图 2. Xp 和 Xq 着丝粒周边重叠群延伸到 X 染色体特异高阶 α 卫星 DXZ1。X 染色体着丝粒周边区域显示为一个截断表意图。每个臂近端序列自身比对通过表意图下方的带点小块说明。每个小块上，臂序列和臂特异卫星区域间的连接点用红色箭头标记，臂特异卫星区域和 X 染色体特异 α 卫星阵列（DXZ1）的连接点用蓝色箭头标记。从 Xp 中分析了大约 594 kb 的序列，其中包括 ~21 kb 的 DXZ1 序列。从 Xq 中分析了 ~454 kb 的序列，包括 ~44 kb 的 DXZ1 序列。在每一种情况下，都包括 ~100 kb 的臂序列。着丝粒周边卫星序列的高度重复结构与臂序列中几乎没有重复结构形成鲜明对照，尽管这些区域 LINE 重复的密度非常高。黑色卫星区域的缺口发生在散在元件（LINE、SINE 和 LTR）中断卫星序列的地方。在阵列的序列带点小块中，Xp 序列最近端 ~21 kb 与 Xq 序列最近端 ~44 kb 相连接。着丝粒恒定，高阶 α 卫星阵列的周期性是显而易见的。小块上的黑色水平线显示间隔为 ~2 kb 的序列几乎是一致的。这个 DXZ1 样本代表 3（±0.4）Mb 阿尔法卫星阵列的 ~65 kb。蓝色轮廓区域是自身比对（“Xp DXZ1”和“Xq DXZ1”），小块的其余矩形区域是 Xp 与 Xq 的 DXZ1 的比对，它揭示了每个臂的 DXZ1 序列间密切的关系。

单核苷酸多态性

共有 153,146 个候选单核苷酸多态性（SNP）绘制到 X 染色体序列并显示在 VEGA 数据库中。其中包括导致蛋白质编码区域非同义改变的 901 个 SNP，因此，是候选功能蛋白变异。X 染色体上的杂合水平远低于常染色体，这种差异可以部分或者全部由群体遗传因素解释[35]。被绘制到图谱里的 62,334 个 SNP 是通过将流式

SNPs are 62,334 that were identified by alignment of flow-sorted X chromosome shotgun sequence reads to the X chromosome reference sequence. Using comparable sequence data for chromosome 20, we calculated that the heterozygosity level on the X chromosome is approximately 57% of that observed for the autosome.

Evolution of the Human X Chromosome

Males of the three mammalian groups—Eutheria ("placental" mammals), Metatheria (marsupials) and Prototheria (egg-laying mammals)—have X and Y sex chromosomes. Ohno proposed in 1967 that the mammalian sex chromosomes evolved from an autosome pair following their recruitment into a chromosomal system for sex determination[1]. A barrier to recombination developed between these "proto" sex chromosomes, isolating the sex-determining regions and eventually spreading throughout the two homologues. In the absence of recombination, the accumulation of mutation events subsequently led to the degeneration of the Y chromosome. The sex chromosomes of birds are not homologous to those of the mammals. The sex chromosome system of birds evolved independently during the last 300 Myr, giving rise to homogametic (ZZ) male birds and heterogametic (ZW) female birds, in contrast to the mammalian system of XY males and XX females.

The autosomal origin of the mammalian sex chromosomes is vividly illustrated by alignment of the human X and chicken whole genome sequences (Fig. 3a). Orthologues of some human X chromosome genes were previously mapped to chicken chromosomes 1q13-q21 and 4p11-p14 (ref. 36). Using genomic sequence alignment, we identified approximately 30 regions of homology that together cover most of human Xq and are confined to a single section of approximately 20 Mb at the end of chicken chromosome 4p (Fig. 3a). In contrast, most of the short arm (Xp11.3–pter), including the pseudoautosomal region PAR1, matches a single block of chicken chromosome 1q. No clear picture emerges regarding the origin of the remainder of the short arm (Xcen–p11.3). We were unable to detect large regions of conserved synteny using sequence alignment, and genes from this region have orthologues on several chicken autosomes, including chromosomes 12, 1 and 4 (ref. 37). This region is also characterized by the expansion of several families of CT antigen genes (Fig. 1), which have no readily detectable orthologues in chicken. The present analysis supports the notion of a mammalian "X-conserved region" (XCR)[38], which includes the long arm and is descended from the proto-X chromosome. It also supports a separate, large addition ("X-added region" or XAR[38]) to the established X chromosome by translocation from a second autosome, which occurred in the eutherian mammals before their radiation (~105 Myr ago). In contrast to earlier hypotheses, however, it appears that much of the proximal short arm (Xcen–p11.3) should no longer be considered part of an XCR.

X 染色体鸟枪序列数据与 X 染色体参考序列比对后确定的。比较 20 号染色体的序列数据后，我们计算出 X 染色体的杂合水平大约是常染色体的 57%。

人类 X 染色体的进化

三个哺乳动物群体——真兽次亚纲（有“胎盘”的哺乳动物），后兽次亚纲（有袋的哺乳动物）和原兽亚纲（产卵哺乳动物），它们的雄性体内含有 X 和 Y 性染色体。1967 年大野提出哺乳动物性染色体是从一对常染色体进化而来的，它们进入染色体系统之后用来决定性别[1]。这些“原始”性染色体之间形成了重组障碍，分离出决定性别的区域并最终在这两个同源物中传播。如果没有重组，随后的基因突变事件积累导致 Y 染色体退化。鸟的性染色体与哺乳动物的性染色体没有同源性。鸟类性染色体在过去的 3 亿年独立进化，从而形成同型配子（ZZ）的雄鸟和异型配子（ZW）的雌鸟，与哺乳动物系统的 XY 雄性和 XX 雌性相反。

人类 X 染色体和鸡全基因组序列的比对，生动地说明哺乳动物性染色体起源于常染色体（图 3a）。人类的一些 X 染色体基因直系同源物以前被绘制到鸡染色体 1q13-q21 和 4p11-p14 中（参考文献 36）。利用基因组序列比对，我们确定了大约 30 个覆盖了人类 Xq 大部分的同源区域，这些区域都限制在鸡染色体 4p 末端大约 20 Mb 的单一区域内（图 3a）。相比之下，短臂（Xp11.3–pter）的大部分区域，包括假常染色体区域 PARl，与鸡染色体 1q 的单一区块相匹配。关于短臂（Xcen–p11.3）剩余部分的起源没有清楚的图片展示。我们用序列比对无法检测到大的同线性保守区域，并且这些区域的基因在鸡的一些常染色体上有直系同源物，这些常染色体包括 12 号、1 号和 4 号染色体（参考文献 37）。这一区域还表现出若干 CT 抗原基因家族（图 1）扩展的特点，而这在鸡中并没有检测到同源物。这个分析支持哺乳动物“X 保守区”（XCR）的概念[38]，X 保守区包括长臂，它是从原始 X 染色体留下来的。它还支持从第二个常染色体易位到已有的 X 染色体上从而形成的独立的、大型增加片段（“X 补充区域”或 XAR[38]）的概念，这发生在真兽类哺乳动物适应辐射前（~ 1 亿 500 万年前）。但是，与早期假设相反，好像大部分的近端短臂（Xcen–p11.3）不应该被认为是 XCR 的一部分。

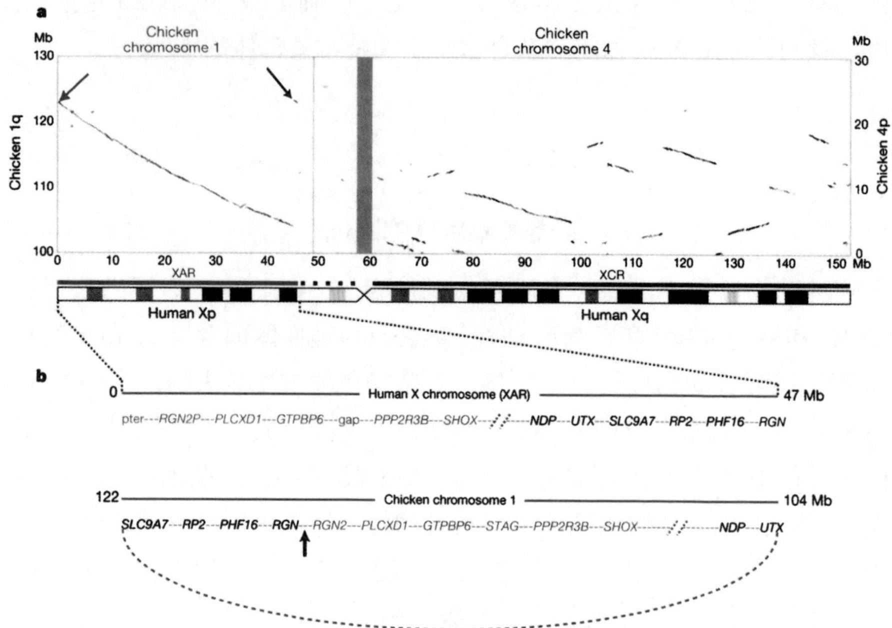

Fig. 3. Homologies between the human X chromosome and chicken autosomes. **a**, Plot of BLASTZ sequence alignments between the X chromosome and chicken chromosomes 1 (red) and 4 (blue). Grey bar centred at approximately 60 Mb shows the position of the X centromere. Only the relevant section of each chicken chromosome is shown (see Mb scale at left for chromosome 1 and at right for chromosome 4). A schematic interpretation of the homologies shows the XAR and XCR as red and blue bars, respectively (see Fig. 1). Homologies at the ends of the XAR are indicated with arrows and are expanded in **b**. **b**, (Top) Genes at the ends of the human XAR. Genes from distal Xp (magenta arrow in **a**) are in magenta and genes from Xp11.3 (black arrow in **a**) are shown in black. (Bottom) Arrangement of the orthologous genes on chicken chromosome 1. A hypothetical ring chromosome, with the equivalent gene order to that observed in the chicken, is indicated by the curved, dotted red line. Recombination between one end of the established X chromosome and the ring chromosome at the arrowed position could, in a single step, have added the XAR and created the gene order observed on the human X chromosome.

The precise location of genes that demarcate the XAR suggests a possible mechanism for the addition. The annotated genes at the extreme ends of the 47 Mb XAR are *PLCXD1* (cU136G2.1 in Supplementary Fig. 1) near Xpter, and *RGN* in Xp11.3. We also found an unprocessed *RGN* pseudogene (*RGN2P*) at Xpter, distal to *PLCXD1*. The orthologues for these three loci are adjacent on chicken chromosome 1, in the order (tel)–*RGN*–*RGN2*–*PLCXD1*–(cen) (Fig. 3b). The generation of these two different gene orders from a common ancestral sequence would require a minimum of two rearrangements as well as the translocation that added the XAR. A more parsimonious model suggested by these data, however, is that the XAR was acquired by recombination between the X chromosome and a ring chromosome in which the ancestral *PLCXD1*, *RGN* and *RGN2* sequences were neighbours (Fig. 3b).

In order to examine more recent patterns of evolution, we compared the human X chromosome with other mammalian sequences. We saw nine major blocks of sequence homology between

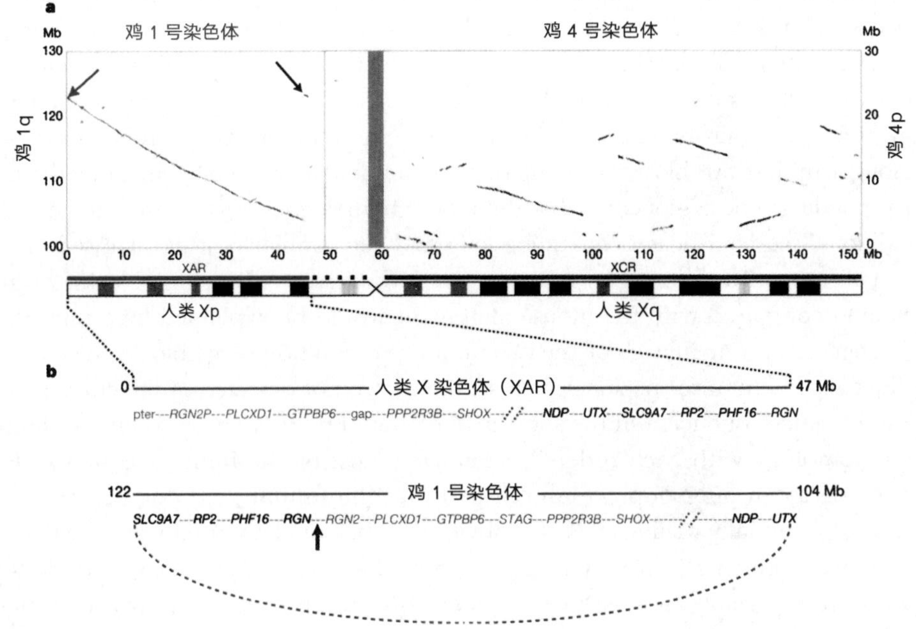

图 3. 人类 X 染色体和鸡常染色体的同源性。**a**，X 染色体和鸡 1 号（红色）和 4 号（蓝色）染色体间 BLASTZ 序列比对。集中在约 60 Mb 的灰色条块显示的是 X 染色体着丝粒的位置。鸡的每条染色体只显示有关的片段（1 号染色体看左边的 Mb 比例尺，4 号染色体看右边的 Mb 比例尺）。一个解释同源性的示意图分别显示了红色条 XAR 和蓝色条 XCR（见图 1）。XAR 末端的同源性用箭头指示，并在 **b** 中扩大显示。**b**，（上）人 XAR 末端的基因。Xp 远端的基因（**a** 中品红色箭头）是品红色，Xp11.3 的基因（**a** 中黑色箭头）显示为黑色。（下）鸡的 1 号染色体上直系同源基因的排列。由弯曲的红色虚线表示的一个假设的环状染色体，其基因顺序与在鸡染色体中观察到的一致。已建立的 X 染色体一端与箭头所指位置的环状染色体之间的重组可以一步完成 XAR 的添加，重组还能创建在人类 X 染色体上观察到的基因顺序。

划分 XAR 的基因的确切位置表明序列增加的可能机制。在 47 Mb XAR 末端的注释基因是接近 Xpter 的 *PLCXD1*（cU136G2.1 见补充图 1）以及在 Xp11.3 中的 *RGN*。我们还在 *PLCXD1* 远端的 Xpter 上发现一个未处理的 *RGN* 假基因（*RGN2P*）。这三个基因座的直系同源物在鸡的 1 号染色体上相邻，顺序为（端粒）–*RGN*–*RGN2*–*PLCXD1*–（着丝粒）（图 3b）。从一个共同祖先的序列产生这两个不同的基因顺序最少需要两个重排，以及增加 XAR 的易位。然而，这些数据暗示了一个更为简洁的模型，即 XAR 是通过 X 染色体与祖先 *PLCXD1*、*RGN* 和 *RGN2* 序列相邻的环状染色体重组获得的（图 3b）。

为了研究更近期的进化模式，我们将人类 X 染色体与其他哺乳动物序列进行了比对。我们发现人类和小鼠 X 染色体之间有 9 个主要的序列同源性区块，人类与

human and mouse X chromosomes, and eleven between human and rat (Fig. 4). The homology blocks occupy almost the entirety of each X chromosome, confirming the remarkable degree of conserved synteny of this chromosome within the eutherian mammalian lineage. This is consistent with Ohno's law, which predicts that the establishment of a dosage compensation mechanism had a stabilizing effect on the gene content of the mammalian X chromosome[1]. On the long arm, just two blocks of homology account for the entire alignment of the human and corresponding mouse sequences, but the mouse homologous regions are punctuated with three additional segments, each containing long and very similar repeats (arrowed in Fig. 4). Alignment of human Xq with the rat sequence reveals four discrete homology blocks; the greater fragmentation compared with the mouse alignment would be explained by a minimum of two rearrangements, one in each of the two mouse–human homology blocks, specifically on the rat lineage. The mouse-specific repeat segments are not detected in the current version of the rat genome sequence. On the short arm of the human X chromosome, seven major blocks of homology with each rodent account for most of the human sequence (Fig. 4). Using the dog as an outgroup, we established that the human and dog X chromosome sequences are essentially collinear (K. Lindblad-Toh, personal communication). Therefore, all of the rearrangements indicated in Fig. 4 occurred in the rodent lineage, and the human X chromosome appears to have been remarkably stable in its organization since the radiation of eutherian mammals. This is consistent with the recent prediction, derived from a comparison of human, rodent and chicken chromosomes, that the human X chromosome is identical to the putative ancestral (eutherian) mammalian X chromosome[39].

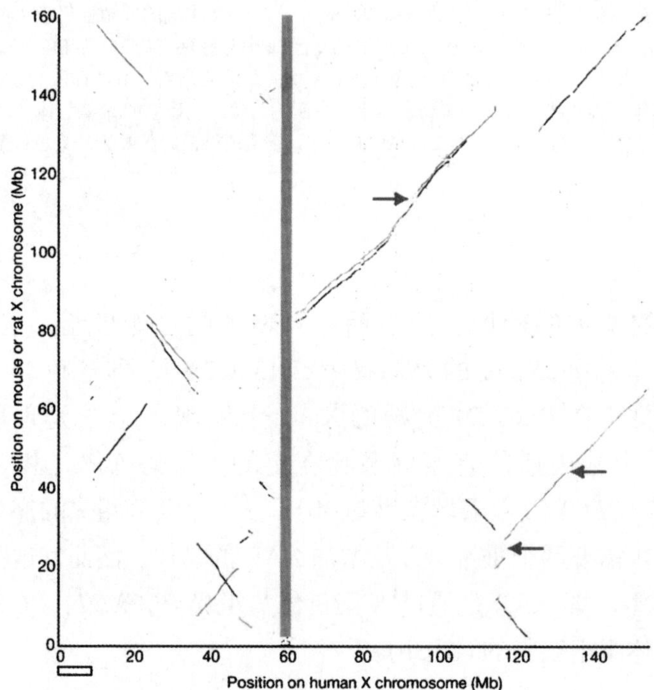

Fig. 4. Conservation of the X chromosome in eutherian mammals. Plot of BLASTZ sequence alignments between the human X chromosome and the mouse (red) and rat (blue) X chromosomes. The rodent

大鼠之间有 11 个(图 4)。这些同源区块几乎占据整个 X 染色体，证实了这条染色体在真兽类哺乳动物谱系内具有显著的保守同线性。这与大野的定律一致，该定律预测剂量补偿机制的建立对哺乳动物 X 染色体的基因含量有稳定作用[1]。在染色体长臂上，人类和小鼠相应序列整体比对后只有两个同源性区块，但小鼠同源区域被另外三个额外的片段隔开，每个片段包含长的、非常相似的重复序列(图 4 中箭头所示)。人类 Xq 与大鼠序列比对揭示四个分离的同源区块；与小鼠比对相比，将通过至少两个重排来解释更大的片段化，即在两个鼠-人同源性区块中分别有一个重排，特别是在大鼠谱系中。小鼠特异性重复片段在目前的大鼠基因组序列中未检测到。人类 X 染色体短臂上与每个啮齿动物同源的 7 个主要区块占了人类序列的大部分(图 4)。使用狗作为外类群，我们确定人和狗 X 染色体的 DNA 序列本质上是共线性的(林德布拉德-托，个人交流)。因此，图 4 所显示的所有重排发生在啮齿动物谱系中，从真兽类哺乳动物适应辐射以来，人类 X 染色体在其系统内似乎非常稳定。这与最近比较人类、啮齿动物和鸡染色体得出的预测一致，即人类 X 染色体与假定哺乳动物祖先(真兽类)X 染色体一致[39]。

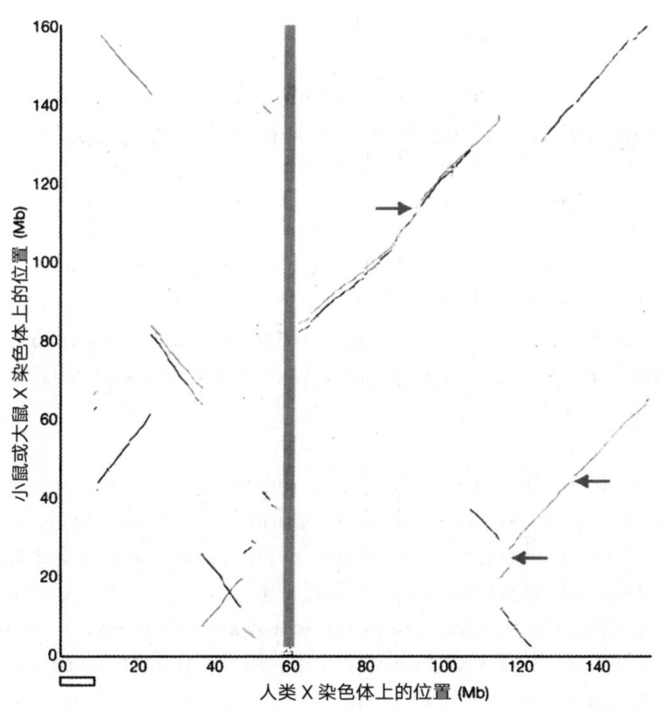

图 4. 在真兽类哺乳动物中 X 染色体的保守性。人类 X 染色体与小鼠(红色)和大鼠(蓝色)X 染色体之间 BLASTZ 序列比对图。啮齿类动物的染色体着丝粒是向下的。箭头指示的区域在小鼠序列中是长度

chromosomes are oriented with their centromeres pointing downwards. Regions indicated with arrows are long, highly similar repeats in the mouse sequence that are absent from the human and rat sequences. These repeats were apparently collapsed in an earlier analysed version of the mouse sequence, which also had a large inversion with respect to the mouse assembly used here (NCBI32)[66]. The NCBI32 assembly has a gap from 0–3 Mb, which explains the absence of homology to the human X sequence in this part of the plot. The open horizontal bar shows the terminal section of human Xp, which is not conserved on the rodent X chromosomes.

The most notable difference we found between the human and rodent X chromosomes is the existence of 9 Mb of sequence at the tip of the human short arm (including human PAR1) that is apparently missing from the rodent X chromosomes (Fig. 4). There are 34 known and novel protein-coding genes in this segment of the human X chromosome (Supplementary Fig. 1), enabling us to investigate how this difference arose. A comprehensive database search of the rodent genome sequences revealed convincing orthologues for only thirteen of these genes in rat and five in mouse. Most of the rat orthologues are located in two groups on chromosome 12, and the only genes for which X-linked orthologues could be found in both rodents were *PRKX* and *STS*. In contrast, we found 24 of these 34 genes on chicken chromosome 1, and the order of these genes is perfectly conserved between the two genomes. Therefore, we conclude that this large terminal segment was present in the XAR and was subsequently removed from the X chromosome in a common murid ancestor of mouse and rat. The relative paucity of rodent ECRs in this segment of the X chromosome sequence (Supplementary Fig. 1) suggests that much of the region may be absent altogether from the genomes of *Mus musculus* and *Rattus norvegicus*.

Comparison of the Human X and Y Chromosomes

The evolutionary process has eradicated most traces of the ancestral relationship between the human X and Y chromosomes. At the cytogenetic level, the Y chromosome has a large and variably sized heterochromatic block and is considerably smaller than the X chromosome, and the euchromatic part of the X chromosome is six times longer than that of Y. Few genes on human chromosome X have an active counterpart on the Y chromosome, and the majority of these are contained in regions where XY homology is of relatively recent origin.

A detailed comparison of the human X and Y chromosome sequences reveals the extent of Y chromosome decay in non-recombining regions. All of the large homologous blocks visible in Fig. 5 (and represented schematically in Fig. 6) are descended from material that was added to the established sex chromosomes. The tip of the short arm of the X and Y chromosomes comprises the 2.7 Mb pseudoautosomal region PAR1. Homology between the X and Y chromosomes in PAR1 is maintained by an obligatory recombination in male meiosis; gene loci in this region are present in two copies in both males and females and are not subject to dosage compensation by XCI. At the tip of the long arm of X and Y is a second pseudoautosomal region, the 330 kb PAR2, which was created by duplication

很长且高度相似的重复序列，在人类和大鼠序列中缺失。这些重复序列在较早分析版本的小鼠序列中显然是塌陷的，该版本的小鼠序列与此处使用的小鼠序列（NCBI32）[66] 相比也有较大的不同。NCBI32 序列有一个 0～3 Mb 的缺口，这就解释了在图的这一部分中缺少与人类 X 序列的同源性。空心的横杠显示人类 Xp 末端部分，它在啮齿动物 X 染色体上并不保守。

我们在人类和啮齿动物的 X 染色体间发现的最显著的差异是人类染色体短臂末端（包括人类 PAR1）存在的 9 Mb 序列在啮齿动物 X 染色体中明显缺失（图 4）。在人类 X 染色体的这一段序列中，有 34 个已知的、新的蛋白编码基因（补充图 1），使我们能够研究这种差异是如何产生的。对啮齿动物基因组序列数据库进行全面搜索显示，大鼠中只有 13 个基因有高可信度直系同源物，而小鼠只有 5 个。大多数大鼠直系同源物位于 12 号染色体的两个组上，在两种啮齿动物中发现的 X 连锁同源基因只有 *PRKX* 和 *STS*。相比之下，我们在鸡的 1 号染色体上发现了这 34 个基因中的 24 个，并且这些基因的顺序在两个基因组间非常保守。因此，我们得出这样的结论：这个大型末端序列曾出现在 XAR 中，后来从小鼠和大鼠共同的鼠科祖先的 X 染色体中被移除。啮齿动物 ECR 在 X 染色体序列的这一段相对缺失（补充图 1）表明，该区域的大部分序列可能在小家鼠和褐家鼠的基因组中共同缺失。

比较人类 X 和 Y 染色体

进化过程已消除人类 X 染色体和 Y 染色体之间祖先关系的大部分痕迹。在细胞遗传学水平上，Y 染色体有一个大的大小可变的异染色质块，但这个异染色质块比 X 染色体小得多，X 染色体常染色质部分是 Y 的 6 倍。人类 X 染色体上很少有基因在 Y 染色体上有活跃的对应基因，其中大多数包含在 XY 同源性起源较晚的区域。

对人类 X 染色体和 Y 染色体序列的详细比较揭示了 Y 染色体在非重组区域衰减的程度。在图 5 中可看到的所有大型同源区块（并在图 6 中示意出）都来自增加到既定的性染色体的物质。X 染色体和 Y 染色体短臂末端包含 2.7 Mb 的假常染色体区域 PAR1。PAR1 中的 X 染色体和 Y 染色体之间的同源性是通过雄性减数分裂中必要的重组来维持的；这个区域的基因座在男性和女性中都有两个拷贝，不受 XCI 的剂量补偿影响。在 X 染色体和 Y 染色体长臂末端是第二个假常染色体区域，即 330 kb 的 PAR2，它是自人类和黑猩猩谱系分化以来，通过复制从 X 到 Y 的物质

of material from X to Y since the divergence of human and chimpanzee lineages[40]. Some genes in PAR2 are subject to XCI, presumably reflecting their status on the X chromosome before the duplication event. Outside the PARs, homologies between the X and Y chromosomes are in non-recombining regions, predominantly in other parts of the XAR, together with a large "X-transposed region" (XTR)[41] in Xq21.3 and Yp11.2–p11.3 (see below). It is thought that the XAR originally formed a large pseudoautosomal region with an equivalent YAR, which is now largely eroded. At a gross level, the homology between the XAR and YAR is continuous for 6 Mb proximal to the pseudoautosomal boundary on the X (PABX), but is considerably more fragmented on the Y chromosome (Figs 5b and 6). Beyond this, the remaining 38.5 Mb of the XAR detects few other remnants of the YAR. Homologies are mostly in small islands around genes with functional orthologues on both sex chromosomes (for example, *AMELX*/*AMELY*, *ZFX*/*ZFY*, see Table 1).

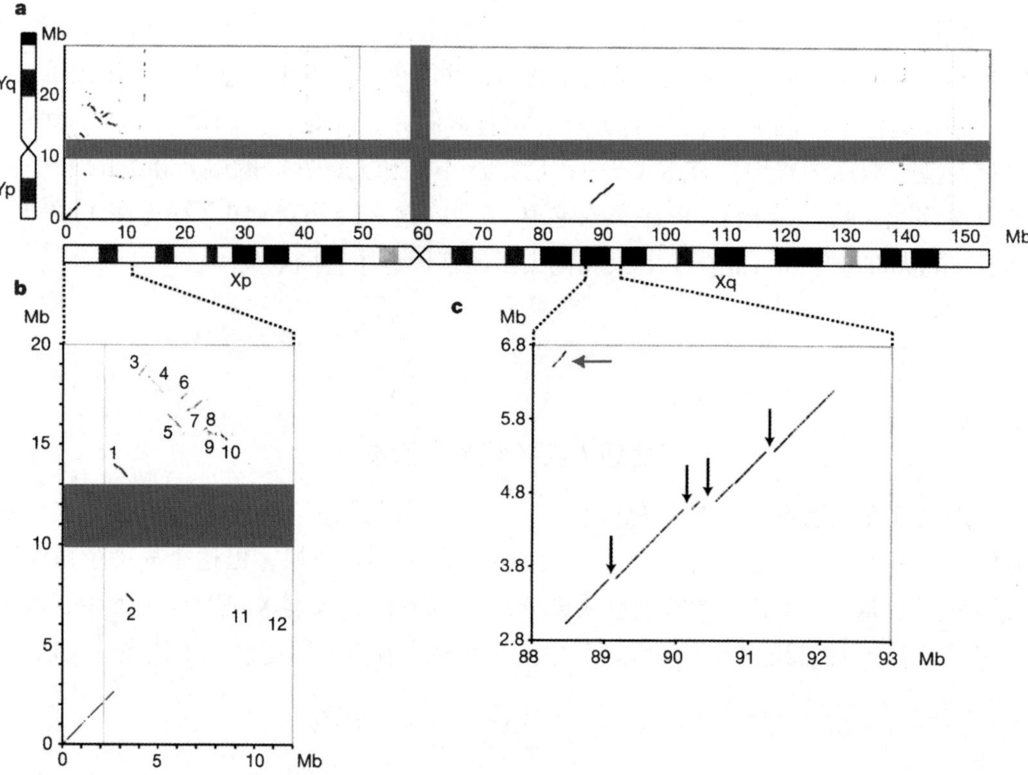

Fig. 5. Limited homology between the human sex chromosomes illustrates the extent of Y chromosome erosion in non-recombining regions. **a**, BLASTN alignments (length ⩾ 80 bp, sequence identity ⩾ 70%) between the finished sequences of the X and Y chromosomes. The centromere positions are represented by grey bars. The analysed Y chromosome sequence ends at the large, heterochromatic segment on Yq, which is indicated by the black bar on the truncated Y chromosome ideogram. **b**, Major blocks of homology remaining between the XAR and the YAR. Expansion of the BLASTN plot from 0–12 Mb on the X chromosome and 0–20 Mb on the Y chromosome. On the X chromosome, the major homologies lie in the terminal 8.5 Mb of Xp: PAR1 (magenta line) and numbered blocks 1–10. Lesser homologies 11 and 12 contain the *TBL1X*/*TBL1Y* and *AMELX*/*AMELY* genes, respectively. **c**, The XTR region in detail (88–93 Mb on X and 2.8–6.8 Mb on Y). Black arrows show large segments deleted from the Y chromosome

产生的[40]。PAR2 中有些基因受 XCI 影响，这很可能反映了它们在复制事件之前在 X 染色体上的状态。PAR 以外，X 染色体和 Y 染色体之间的同源性在非重组区域，主要位于 XAR 的其他区域，以及在 Xq21.3 和 Yp11.2–p11.3 中的大型“X 转置区”(XTR)[41](见下文)。我们认为，XAR 最初形成与 YAR 等同的大型假常染色体区域，现在 YAR 在很大程度上消退了。总的来说，XAR 和 YAR 之间的同源性有 X 假常染色体边界近端(PABX)的连续 6 Mb，但在 Y 染色体上是相当多的分散片段(图 5b 和 6)。除此之外，XAR 其余的 38.5 Mb 检测到其他几个残余的 YAR。同源性大多在两条性染色体上有功能性直系同源物的基因周围(例如，*AMELX*/*AMELY*、*ZFX*/*ZFY*，见表 1)。

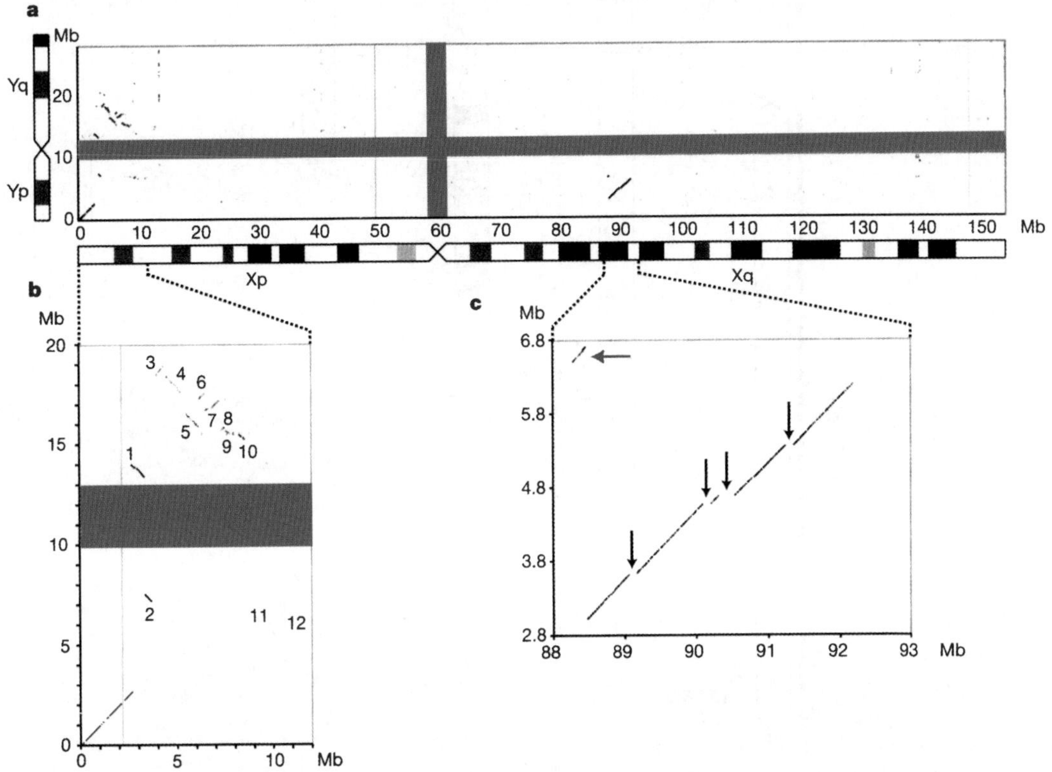

图 5. 人类性染色体之间有限的同源性说明 Y 染色体在非重组区域消退的程度。**a**，X 染色体和 Y 染色体完成的序列之间 BLASTN 比对(长度 ≥ 80 bp，序列一致性 ≥ 70%)。灰色条框指示着丝粒的位置。所分析的 Y 染色体序列末端在 Yq 上大的异染色质片段处，通过图上截断 Y 染色体的黑色条块表示。**b**，XAR 和 YAR 之间主要同源性区块。X 染色体上从 0 ~ 12 Mb，Y 染色体从 0 ~ 20 Mb 的 BLASTN 扩展图。在 X 染色体上，主要的同源性位于 8.5 Mb 的 Xp 末端：PAR1(品红色线)和编号区块 1 ~ 10。较小的同源性片段 11 和 12 分别包含 *TBL1X*/*TBL1Y* 和 *AMELX*/*AMELY* 基因。**c**，XTR 区域详细信息(在 X 染色体上 88 ~ 93 Mb 和在 Y 染色体上 2.8 ~ 6.8 Mb)。黑色箭头显示从 XTR 的 Y 染色体拷贝中删除的大片段。品红色箭头表示通过 Y 染色体臂内倒位后从 XTR 剩余部分分离出的短片段。人类群体中

copy of the XTR. The magenta arrow indicates the short segment that is separated from the rest of the XTR by a paracentric inversion on the Y chromosome. An independent inversion polymorphism on Yp in human populations encompasses this small segment. The position and orientation of the segment shows that the Y chromosome reference sequence is of the less common, derived Y chromosome.

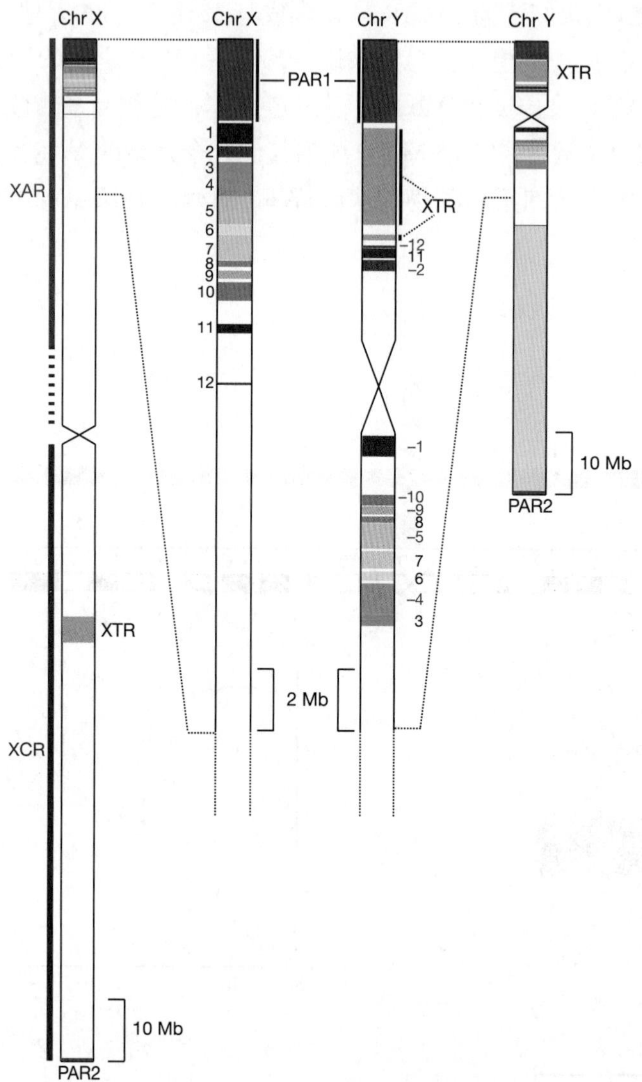

Fig. 6. Schematic representation of major homologies between the human sex chromosomes. The entire X and Y chromosomes are shown using the same scale on the left and right sides of the figure, respectively. The major heterochromatic region on Yq is indicated by the pale grey box proximal to PAR2. Expanded sections of X and Y are shown in the centre of the figure. Homologies coloured in the figure are either part of the XAR (PAR1 and blocks 1–12), or were duplicated from the X chromosome to the Y chromosome since the divergence of human and chimpanzee lineages (XTR and PAR2). The numbering of XAR-YAR blocks follows that in Fig. 5b. Blocks inverted on the Y chromosome relative to the X chromosome are assigned red, negative numbers.

Yp 上的一个独立的倒位多态现象包括这个小片段。这个片段的位置和方向表明，Y 染色体参考序列是不太常见的衍生 Y 染色体。

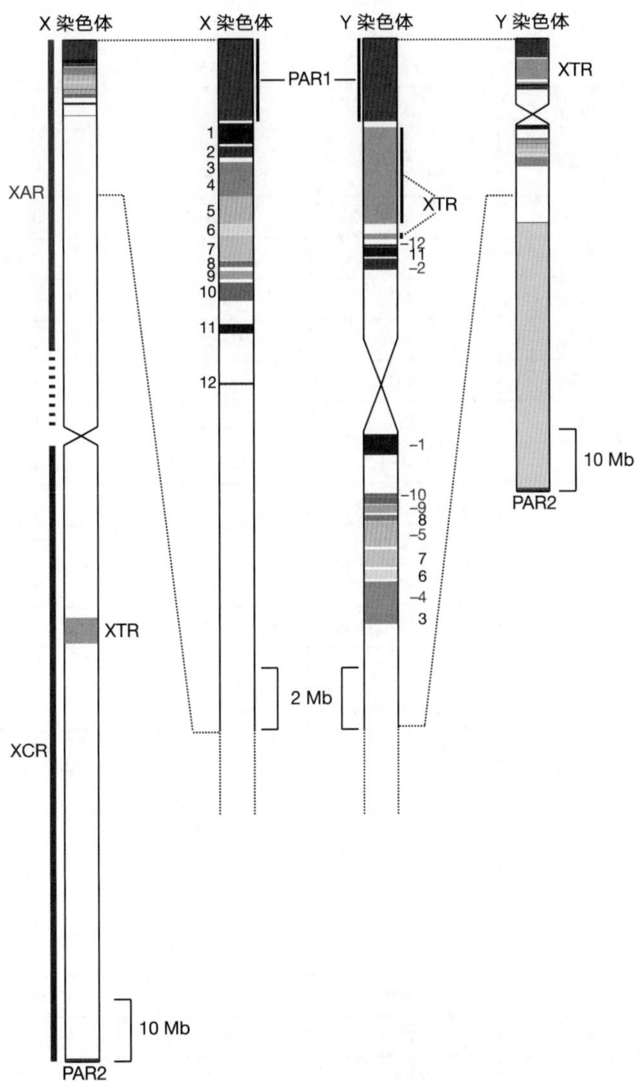

图 6. 人类性染色体之间主要的同源性示意图。整个 X 染色体和 Y 染色体分别以相同的比例显示在图左侧和右侧。Yq 上主要的异染色质区域用接近 PAR2 的浅灰色框指示。图中央显示了 X 和 Y 的展开部分。图中着色的同源物或者是 XAR 的一部分（PAR1 和 1 ~ 12 区块），或者是自人类和黑猩猩谱系分化以来（XTR 和 PAR2）从 X 染色体到 Y 染色体的物质所复制的。XAR-YAR 的编号按照图 5b 所示。Y 染色体相对 X 染色体反转的区块用红色负数标记。

Table 1. Homologous genes on the human X and Y chromosomes

Region	Distance from Xpter (Mb)	X gene*	Y gene	Distance from Ypter (Mb)†	XY homology block‡
Pseudoautosomal region PAR1 (XAR)	0.15	cU136G2.1 (*PLCXD1*)	cU136G2.1 (*PLCXD1*)	0.15	PAR1
	0.17	cU136G2.2 (*GTPBP6*)	cU136G2.2 (*GTPBP6*)	0.17	PAR1
	0.25	cM56G10.2§	cM56G10.2§	0.25	PAR1
	0.29	cM56G10.1 (*PPP2R3B*)	cM56G10.1 (*PPP2R3B*)	0.29	PAR1
	0.57	*SHOX*	*SHOX*	0.57	PAR1
	0.92	bA309M23.1§	bA309M23.1§	0.92	PAR1
	1.31	*CRLF2*	*CRLF2*	1.31	PAR1
	1.38	*CSF2RA*	*CSF2RA*	1.38	PAR1
	1.52	*IL3RA*	*IL3RA*	1.52	PAR1
	1.55	*SLC25A6*	*SLC25A6*	1.55	PAR1
	1.56	bA261P4.5§	bA261P4.5§	1.56	PAR1
	1.57	bA261P4.6 (*CXYorf2*)	bA261P4.6 (*CXYorf2*)	1.57	PAR1
	1.59	*ASMTL*	*ASMTL*	1.59	PAR1
	1.66	bA261P4.4 (*P2RY8*)	bA261P4.4 (*P2RY8*)	1.66	PAR1
	1.76	*DXYS155E* (*CXYorf3*)	*DXYS155E* (*CXYorf3*)	1.76	PAR1
	1.79	*ASMT*	*ASMT*	1.79	PAR1
	1.79	bB297E16.3§	bB297E16.3§	1.79	PAR1
	1.91	bB297E16.4§	bB297E16.4§	1.91	PAR1
	1.93	bB297E16.5§	bB297E16.5§	1.93	PAR1
	2.37	*DHRSX*	*DHRSX*	2.37	PAR1
	2.41	*ALTE* (*ZBED1*)	*ALTE* (*ZBED1*)	2.41	PAR1
	2.54	Em:AC097314.2§	Em:AC097314.2§	2.54	PAR1
	2.53	Em:AC097314.3§	Em:AC097314.3§	2.53	PAR1
	2.63	*CD99*	*CD99*	2.63	PAR1
X-added region (XAR)	3.57	*PRKX*	*PRKY*	7.23	2
	5.81	*NLGN4X*	*NLGN4Y*	15.23	5
	6.31	Em:AC108684.1 (*VCX3A*)	*VCY*, *VCY1B*	14.54, 14.6	6
	7.62	*VCX*	*VCY*, *VCY1B*	14.54, 14.6	9
	7.95	Em:AC097626.1 (*VCX2*)	*VCY*, *VCY1B*	14.54, 14.6	10
	8.24	Em:AC006062.2 (*VCX3B*)	*VCY*, *VCY1B*	14.54, 14.6	10
	9.37	*TBL1X*	*TBL1Y*	6.97	11
	11.07	*AMELX*	*AMELY*	6.78	12
	12.75	*TMSB4X*	*TMSB4Y*	14.25	

表 1. 人类 X 和 Y 染色体上的同源基因

区域	与 Xpter 的距离 (Mb)	X 基因 *	Y 基因	与 Ypter 的距离 (Mb)†	XY 同源性区块 ‡
假常染色体区 PAR1 (XAR)	0.15	cU136G2.1 (*PLCXD1*)	cU136G2.1 (*PLCXD1*)	0.15	PAR1
	0.17	cU136G2.2 (*GTPBP6*)	cU136G2.2 (*GTPBP6*)	0.17	PAR1
	0.25	cM56G10.2§	cM56G10.2§	0.25	PAR1
	0.29	cM56G10.1 (*PPP2R3B*)	cM56G10.1 (*PPP2R3B*)	0.29	PAR1
	0.57	*SHOX*	*SHOX*	0.57	PAR1
	0.92	bA309M23.1§	bA309M23.1§	0.92	PAR1
	1.31	*CRLF2*	*CRLF2*	1.31	PAR1
	1.38	*CSF2RA*	*CSF2RA*	1.38	PAR1
	1.52	*IL3RA*	*IL3RA*	1.52	PAR1
	1.55	*SLC25A6*	*SLC25A6*	1.55	PAR1
	1.56	bA261P4.5§	bA261P4.5§	1.56	PAR1
	1.57	bA261P4.6 (*CXYorf2*)	bA261P4.6 (*CXYorf2*)	1.57	PAR1
	1.59	*ASMTL*	*ASMTL*	1.59	PAR1
	1.66	bA261P4.4 (*P2RY8*)	bA261P4.4 (*P2RY8*)	1.66	PAR1
	1.76	*DXYS155E* (*CXYorf3*)	*DXYS155E* (*CXYorf3*)	1.76	PAR1
	1.79	*ASMT*	*ASMT*	1.79	PAR1
	1.79	bB297E16.3§	bB297E16.3§	1.79	PAR1
	1.91	bB297E16.4§	bB297E16.4§	1.91	PAR1
	1.93	bB297E16.5§	bB297E16.5§	1.93	PAR1
	2.37	*DHRSX*	*DHRSX*	2.37	PAR1
	2.41	*ALTE* (*ZBED1*)	*ALTE* (*ZBED1*)	2.41	PAR1
	2.54	Em:AC097314.2§	Em:AC097314.2§	2.54	PAR1
	2.53	Em:AC097314.3§	Em:AC097314.3§	2.53	PAR1
	2.63	*CD99*	*CD99*	2.63	PAR1
X 增加区 (XAR)	3.57	*PRKX*	*PRKY*	7.23	2
	5.81	*NLGN4X*	*NLGN4Y*	15.23	5
	6.31	Em:AC108684.1 (*VCX3A*)	*VCY*, *VCY1B*	14.54, 14.6	6
	7.62	*VCX*	*VCY*, *VCY1B*	14.54, 14.6	9
	7.95	Em:AC097626.1 (*VCX2*)	*VCY*, *VCY1B*	14.54, 14.6	10
	8.24	Em:AC006062.2 (*VCX3B*)	*VCY*, *VCY1B*	14.54, 14.6	10
	9.37	*TBL1X*	*TBL1Y*	6.97	11
	11.07	*AMELX*	*AMELY*	6.78	12
	12.75	*TMSB4X*	*TMSB4Y*	14.25	

Continued

Region	Distance from Xpter (Mb)	X gene*	Y gene	Distance from Ypter (Mb)†	XY homology block‡
X-added region (XAR)	16.59	*CXorf15*	*CYorf15A, CYorf15B*	20.13, 20.15	
	19.91	*EIF1AX*	*EIF1AY*	21.08	
	23.96	*ZFX*	*ZFY*	2.87	
	40.78	*USP9X*	*USP9Y*	13.33	
	40.96	*DDX3X*	*DDX3Y*	13.46	
	44.61	*UTX*	*UTY*	13.91	
X-conserved region (XCR)	53.00	dJ290F12.2 (*TSPYL2*)	*TSPY* (~35)	9.50	
	53.12	*SMCX*	*SMCY*	20.27	
	71.27	*RPS4X*	*RPS4Y1, RPS4Y2*	2.77, 21.27	
X-transposed region (XTR)	88.50	bB348B13.2§	n/a	2.96	XTR
	88.99	*TGIF2LX*	*TGIF2LY*	3.49	XTR
	91.26	*PCDH11X*	*PCDH11Y*	5.28	XTR
X-conserved region (XCR)	135.68	*RNMX* (*RBMX*)	*RBMY* (6)	22.02, 22.04, 22.37, 22.41, 22.66, 22.85	
	139.31	*SOX3*	*SRY*	2.70	
	148.38	Em:AC016940.3 (*HSFX2*)§	*HSFY1, HSFY2*	19.3, 19.12	
	148.56	Em:AC016939.4 (*HSFX1*)§	*HSFY1, HSFY2*	19.3, 19.12	
Pseudoautosomal region PAR2	154.57	*SPRY3*	*SPRY3*	57.44	PAR2
	154.71	*SYBL1*	*SYBL1*	57.58	PAR2
	154.81	*IL9R*	*IL9R*	57.67	PAR2
	154.81	Em:AJ271736.5§	Em:AJ271736.5§	57.69	PAR2
	154.82	Em:AJ271736.6 (*FAM39A*)§	Em:AJ271736.6 (*FAM39A*)§	57.69	PAR2

Pseudogenes are not included in the table.

* Gene names as shown in Supplementary Fig. 1. HUGO name is in parentheses when the two names differ. Em, EMBL entry.

† Distances refer to Y chromosome sequence assembly NCBI35. Where multiple Y chromosome orthologues exist, the locations of all copies are shown on the Y chromosome. The exception is TSPY, which has ~35 copies in an array centred at approximately 9.5 Mb on the Y chromosome[41].

‡ Major homology blocks as shown in Figs 5 and 6.

§ Novel cases of X genes with Y homologues assigned to these categories.

The XTR arose by duplication of material from X to Y since the divergence of the human and chimpanzee lineages[42]. The duplicated region spans 3.91 Mb on X, but the corresponding region is only 3.38 Mb on the Y chromosome (Fig. 5c). We have aligned the entire X and Y copies of this region. Excluding insertions and deletions, sequence identity between the copies is 98.78%. We estimate that the transposition event occurred

续表

区域	与 Xpter 的距离 (Mb)	X 基因 *	Y 基因	与 Ypter 的距离 (Mb)†	XY 同源性区块 ‡
X 增加区 (XAR)	16.59	*CXorf15*	*CYorf15A*, *CYorf15B*	20.13, 20.15	
	19.91	*EIF1AX*	*EIF1AY*	21.08	
	23.96	*ZFX*	*ZFY*	2.87	
	40.78	*USP9X*	*USP9Y*	13.33	
	40.96	*DDX3X*	*DDX3Y*	13.46	
	44.61	*UTX*	*UTY*	13.91	
X 保守区 (XCR)	53.00	dJ290F12.2 (*TSPYL2*)	*TSPY* (~ 35)	9.50	
	53.12	*SMCX*	*SMCY*	20.27	
	71.27	*RPS4X*	*RPS4Y1*, *RPS4Y2*	2.77, 21.27	
X 转置区 (XTR)	88.50	bB348B13.2§	n/a	2.96	XTR
	88.99	*TGIF2LX*	*TGIF2LY*	3.49	XTR
	91.26	*PCDH11X*	*PCDH11Y*	5.28	XTR
X 保守区 (XCR)	135.68	*RNMX* (*RBMX*)	*RBMY* (6)	22.02, 22.04, 22.37, 22.41, 22.66, 22.85	
	139.31	*SOX3*	*SRY*	2.70	
	148.38	Em:AC016940.3 (*HSFX2*)§	*HSFY1*, *HSFY2*	19.3, 19.12	
	148.56	Em:AC016939.4 (*HSFX1*)§	*HSFY1*, *HSFY2*	19.3, 19.12	
假常染色体区 PAR2	154.57	*SPRY3*	*SPRY3*	57.44	PAR2
	154.71	*SYBL1*	*SYBL1*	57.58	PAR2
	154.81	*IL9R*	*IL9R*	57.67	PAR2
	154.81	Em:AJ271736.5§	Em:AJ271736.5§	57.69	PAR2
	154.82	Em:AJ271736.6 (*FAM39A*)§	Em:AJ271736.6 (*FAM39A*)§	57.69	PAR2

假基因不列入表中。

* 基因名称显示在补充图 1 中。当两个名字不同时，HUGO 用括号括起来。Em，EMBL 索取号。

† 距离参考 Y 染色体序列组装 NCBI35。存在多个 Y 染色体直系同源物，所有拷贝的位置显示在 Y 染色体上。唯一的例外是 TSPY，它在 Y 染色体上有 ~35 个拷贝，位于以 9.5 Mb 为中心的阵列中 [41]。

‡ 主要同源区块如图 5 和图 6 所示。

§ 与 Y 染色体同源的 X 基因的新实例纳入这些类别。

自人类和黑猩猩谱系分化以来，XTR 通过从 X 染色体到 Y 染色体的物质复制产生 [42]。复制区域横跨 X 染色体上的 3.91 Mb，但相应的区域在 Y 染色体上只有 3.38 Mb(图 5c)。我们比对了这一区域整个 X 和 Y 拷贝。不包括插入和缺失，拷贝之间序列一致性达到 98.78%。我们估计，转座事件发生在大约 470 万年前(补充讨

approximately 4.7 Myr ago (Supplementary Discussion 1), which is close to the suggested date of the speciation event that led to humans and chimpanzees, assumed here to be 6 Myr ago. The sequence alignment demonstrates the substantial changes to the XTR on the Y chromosome since the transposition. An inversion is known to have separated a 200-kb section from the rest of the XTR[43] (Fig. 5c). Also, the main block of homology is 540 kb shorter on Y than X, owing in particular to the absence of four large regions from the Y chromosome (Fig. 5c). The detection of these sequences at the expected positions on the chimpanzee X chromosome confirms that they were deleted from the Y chromosome after the transposition.

We found that only 54 of the 1,098 genes annotated on the X chromosome have functional homologues on the Y chromosome (Table 1). We obtained direct evidence for 24 genes in PAR1. Twenty-three of them are annotated (Supplementary Fig. 1), and the location of the 5′ end of *CRLF2* indicates that the rest of this gene is in gap 4 of the human X sequence (see the VEGA database). On the basis of the excellent conservation of synteny between human PAR1 and the chicken sequence, we infer that a stromal antigen gene (orthologue of chicken Ensembl gene ENSGALG00000016716) lies in gap 1 (see Fig. 3b). As the annotated putative transcript cM56G10.2 might represent the 3′ end of this gene, we conclude that PAR1 contains at least 24 genes. Together with the five annotated genes in PAR2, 29 genes lie entirely within the recombining regions of the sex chromosomes. Additionally, the *XG* locus spans the boundary between PAR1 and X-specific DNA, but has been disrupted by rearrangement on the Y chromosome.

Outside the XY-recombining regions of the X chromosome, we observed 25 genes that have functional homologues on the Y chromosome (Table 1). Fifteen of these are within the XAR, and a further three genes are shared by the X and Y copies of the XTR. The seven other XY gene pairs are believed to have descended from the proto-sex chromosomes. Only five cases have been described previously[44,45]: the X chromosome genes are *SOX3*, *SMCX*, *RPS4X*, *RBMX* and *TSPYL2*, which are located on the long arm and proximal short arm (Table 1). The two additional cases we report here involve heat-shock transcription factor genes, designated *HSFX1* and *HSFX2*. They are assigned to the category of XCR genes on the basis of a high degree of divergence from their Y chromosome homologues and their location distal to *SOX3* within the XCR. *HSFX1* and *HSFX2* lie within the separate copies of a palindromic repeat in Xq28 and are identical to each other. By analogy, their Y chromosome homologues (*HSFY1* and *HSFY2*) lie within the arms of a Y chromosome palindrome, the similarity of which is thought to be maintained by gene conversion[41].

On the basis of this and previously published information[41], we can conclude that approximately 15 protein-coding genes on the Y chromosome have no detectable X chromosome homologue.

论 1），这与人类和黑猩猩物种分化的日期接近，分化时间假设是 600 万年前。序列比对显示，自转座以来，Y 染色体上的 XTR 发生了实质性的变化。已知的倒位将一个 200 kb 的片段从 XTR 其余的部分分离出来 [43]（图 5c）。此外，Y 染色体上的主要同源性区块比 X 染色体的短 540 kb，尤其是因为 Y 染色体缺失了四个大型区域（图 5c）。对黑猩猩 X 染色体上预期位置序列的检测证实，它们是在转座后从 Y 染色体上删除的。

我们发现，X 染色体注释的 1,098 个基因中只有 54 个在 Y 染色体上有功能同源基因（表 1）。我们在 PAR1 中获得了 24 个基因的直接证据。其中 23 个是有注释的（补充图 1），*CRLF2* 的 5′ 端的位置表明该基因的其余部分在人类 X 染色体序列缺口 4 中（见 VEGA 数据库）。根据人类 PAR1 和鸡序列间良好的同线性保守关系，我们推断，一个基质抗原基因（鸡 Ensembl 基因 ENSGALG00000016716 的直系同源物）位于缺口 1 中（见图 3b）。因为注释的假定转录本 cM56G10.2 可能代表这个基因的 3′ 末端，我们推断 PAR1 至少包含 24 个基因。连同 PAR2 中 5 个注释基因，29 个基因完全在性染色体重组区域内部。此外，*XG* 基因座跨越 PAR1 和 X 特异性 DNA 边界，但被 Y 染色体上的重排打乱。

在 X 染色体上 XY 重组区域之外，我们观察到 25 个基因在 Y 染色体上有功能同源基因（表 1）。其中 15 个基因在 XAR 内，另外三个基因由 XTR 的 X 和 Y 拷贝共享，其他 7 个 XY 基因对被认为是原始性染色体的后代。这其中只有 5 个实例是之前报道过的 [44,45]：在 X 染色体上的基因是 *SOX3*、*SMCX*、*RPS4X*、*RBMX* 和 *TSPYL2*，它们位于长臂和近端短臂上（表 1）。这里我们报道另外两个涉及热激转录因子基因的实例，这两个基因分别为 *HSFX1* 和 *HSFX2*。根据它们的 Y 染色体同源物与 XCR 远端 *SOX3* 位置的高度分化，我们将其归为 XCR 基因。*HSFX1* 和 *HSFX2* 位于 Xq28 一个回文重复的单独拷贝中并且完全一样。通过类推，其 Y 染色体同源物（*HSFY1* 和 *HSFY2*）位于 Y 染色体回文臂内，其相似性被认为是通过基因转变来维持的 [41]。

基于此信息以及以前公布的信息 [41]，我们可以得出结论，Y 染色体上大约 15 个蛋白编码基因没有可检测的 X 染色体同源物。

The Progressive Loss of XY Recombination

The barrier to recombination between the proto-X and Y chromosomes initially encompassed the sex-determining locus on the Y (*SRY*) and possibly other loci affecting male fitness. It is proposed that rearrangement of the Y chromosome led to the development of this barrier. Thereafter, successive rearrangements that encompassed parts of the pseudoautosomal region resulted in segments of Y-linked DNA that could no longer recombine and consequently degenerated over time. Evidence for the role of Y-specific (as opposed to X-specific) rearrangement in this phenomenon is most clearly illustrated by our analysis of the XAR, which shows very little rearrangement between human and avian lineages (Fig. 3a).

In a previous study[46], four broad physical and evolutionary regions were defined on the X chromosome. The X chromosome genes within a given region all showed a similar level of divergence from their Y chromosome counterparts. However, between regions, levels of divergence were very different, presumably reflecting the stepwise loss of recombination between the X and Y chromosomes.

The physical order of the four regions on the X chromosome was seen to parallel their evolutionary ages, and therefore the chromosome was described as having four "evolutionary strata"[46]. In general, gene pairs were found to be less divergent moving through the strata from Xqter to Xpter. The first two strata (S1 and S2) encompass the long arm and proximal short arm, respectively, and were defined by the genes that survive from the proto-sex chromosomes. Gene pairs were found to be increasingly similar moving through strata 3 and 4, which occupy the proximal and distal sections of the XAR, respectively.

We re-evaluated XY homology in S4 and S3 using finished, genomic sequences from the two chromosomes. For S4 in particular, substantial blocks of homology exist between the chromosomes (blocks 1–10 in Fig. 5b and Fig. 6). Aligning the X and Y chromosome sequences across this region, we observed a bipartite organization, with markedly greater XY identity in the distal 1.0 Mb compared with the proximal 4.5 Mb (Fig. 7a). On this basis, the distal portion containing the *GYG2*, *ARSD*, *ARSE*, *ARSF*, *ADLICAN* and *PRKX* genes can be redefined as a new, fifth stratum, S5 (Figs 1 and 7a). A most parsimonious series of inversions, from the current arrangement of homologous blocks on X to that on Y, is consistent with the proposed strata (Fig. 7b). These data refine the picture of loss of XY recombination during evolution, which occurred by migration of the PABX in a stepwise manner distally through the XAR. The available evidence now suggests that there have been at least four PABX positions within the XAR, which are at the S2/S3, S3/S4 and S4/S5 boundaries (~47 Mb, ~8.5 Mb and ~4 Mb from Xpter, respectively), and at the current position (2.7 Mb from Xpter). We estimate that the two most recent PABX movements, which created first S4 and then S5, occurred 38–44 Myr ago and 29–32 Myr ago, respectively (Supplementary Discussion 2).

XY 重组逐步丧失

原始 X 染色体和 Y 染色体之间重组的障碍最初包括 Y 染色体上的性别决定基因(*SRY*)和其他可能影响男性适合度的基因座。有人提出 Y 染色体的重排导致了这一障碍的发展。之后，包括部分假常染色体区域的连续重排导致 Y 连锁的 DNA 片段不能再重组，因此随着时间的推移这些片段逐渐退化。我们对 XAR 的分析清楚地说明了 Y 染色体特异性(与 X 特异性相反)重排在这一现象中所起的作用。XAR 显示人类和鸟类谱系之间几乎没有重排(图 3a)。

在先前的研究中[46]，在 X 染色体上定义了四个广泛的物理和进化区域。给定区域内 X 染色体基因都表现出与 Y 染色体对应物相似程度的分化。然而，在不同的区域之间，分化程度有很大的不同，这大概反映了 X 和 Y 染色体间重组的逐步丧失。

可以看出 X 染色体上四个区域的物理顺序与它们的进化年龄平行，因此，染色体被描述为有四个“进化阶段”[46]。一般情况下，在从 Xqter 到 Xpter 的基因阶段中，基因对的差异很小。前两个阶段(S1 和 S2)分别包含长臂和近端短臂，这两个阶段通过原始性染色体存留下来的基因确定。位于 XAR 近端和远端的基因对在第 3 阶段和第 4 阶段越来越相似。

我们使用来自两条染色体的完整基因组序列重新评估了 S4 和 S3 中的 XY 同源性。尤其是 S4，大量同源区块存在于染色体之间(在图 5b 和图 6 中的区块 1 ~ 10)。比对这一区域的 X 染色体和 Y 染色体序列，我们观察到了一个由两部分构成的组织。与近端 4.5 Mb 相比，远端 1.0 Mb 的 XY 特性更显著(图 7a)。在此基础上，我们将包含 *GYG2*、*ARSD*、*ARSE*、*ARSF*、*ADLICAN* 和 *PRKX* 基因的远端部分重新定义为一个新的第五阶段——S5(图 1 和 7a)。从目前的 X 染色体上同源区块的排列，到 Y 染色体上同源区块的排列这一系列最简约的倒位与之前提出的阶段一致(图 7b)。这些数据完善了进化过程中 XY 重组缺失的图像，这种缺失是由 PABX 在远端通过 XAR 的逐步迁移造成的。现有的证据表明，在 XAR 内至少有四个 PABX 位置，分别位于 S2/S3、S3/S4 和 S4/S5 边界(分别位于 Xpter 的 ~47 Mb，~8.5 Mb 和 ~4 Mb)和当前位置(位于 Xpter 的 2.7 Mb)。我们估计最近的两次 PABX 运动，先产生了 S4，之后产生了 S5，分别发生在 3,800 万 ~ 4,400 万年前和 2,900 万 ~ 3,200 万年前(补充讨论 2)。

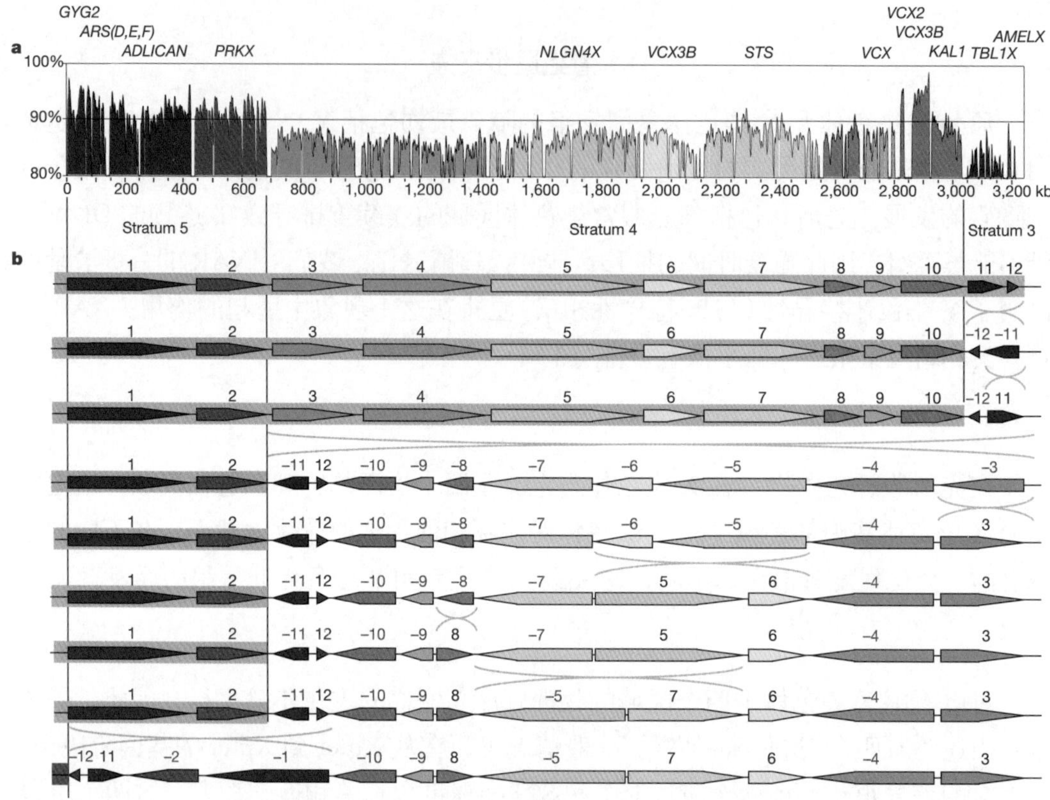

Fig. 7. Evidence for a fifth evolutionary stratum on the X chromosome. **a**, Sequence identity between the X and Y homology blocks 1–12 (see Figs 5b and 6) plotted in 5-kb windows. The scale shows the total amount of sequence aligned, excluding insertions and deletions (see Methods). A 10-kb spacer is placed between each consecutive block of homology. Segments of the plot are coloured according to the system used in Figs 6 and 7b. On the basis of this plot, a new evolutionary stratum S5 is defined, which includes homology blocks 1 and 2. **b**, A most parsimonious series of inversion events from the arrangement of homology blocks 1–12 on the X chromosome (top) to the Y chromosome (bottom), calculated using GRIMM[64]. The grey boxes show the suggested extents of former pseudoautosomal regions within the distal part of the XAR, and the magenta box (bottom row) shows the position of the current pseudoautosomal region. This inversion sequence provides independent support for the proposed pseudoautosomal boundary movements and evolutionary strata. It was previously suggested that *AMELX* (in block 12) is in S4 (ref. 46), or possibly at the boundary between S3 and S4 (ref. 67). However, the more distal location of block 11, which contains *TBL1X* (an S3 gene[46]), is not consistent with these suggestions. The two regions of increased sequence identity within block 10 contain the *VCX2* and *VCX3B* genes on the X chromosome and the *VCY1B* and *VCY* genes on the Y chromosome. This gene family might have arisen *de novo* in the simian lineage[68], which could account for the unusual characteristics of this part of the alignment.

In addition to the varied degree of XY sequence identity within S3, S4, S5 and PAR1, we found marked differences in their sequence composition, which were presumably also caused by the loss of recombination in each region during evolution. Specifically, we observed that L1, L2 and mammalian interspersed repeat (MIR) coverage decrease with each more distal stratum and PAR1 (Table 2 and Fig. 1), but (G+C) levels and *Alu* repeat content increase abruptly at the boundary between S4 and S5 (Table 2 and Fig. 8); variations in the incidence of different *Alu* subfamilies (Y, S and J) also contribute to the

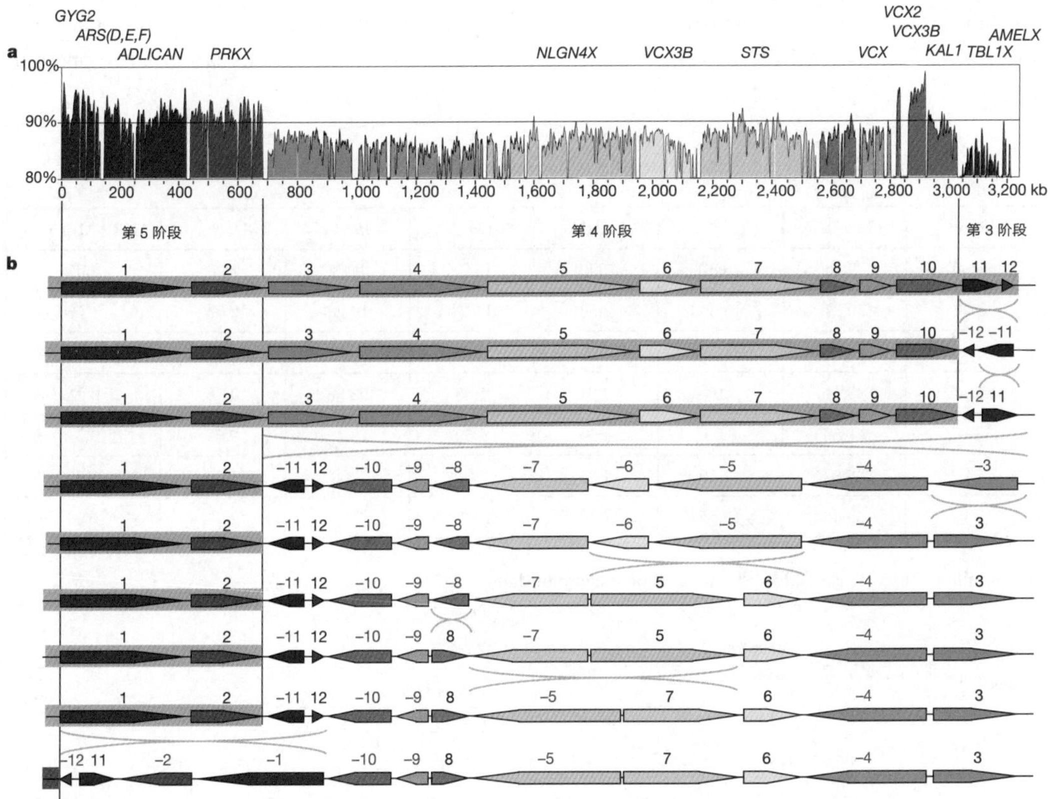

图 7. X 染色体上第五个进化阶段的证据。**a**，在 5 kb 窗口中绘制的 X 染色体和 Y 染色体同源性区块 1～12 的序列一致性（见图 5b 和 6）。该比例显示序列比对的总量，但不包括插入和缺失（见方法）。在每个同源性连续区块之间有 10 kb 的间隔。根据图 6 和 7b 中用的系统对每个片段着色。在此划分的基础上，我们确定了一个新的进化阶段 S5，包括同源块 1 和 2。**b**，用 GRIMM 计算，X 染色体（上）到 Y 染色体（下）同源区块 1～12 排列的一系列最简约的倒位事件 [64]。灰色框显示 XAR 远端部分内之前的假常染色体区域的建议范围，品红色方块（下排）表明目前的假常染色体区域的位置。倒位序列为假常染色体边界运动和进化阶段的观点提供了独立的支持。早前有人提出，*AMELX*（在区块 12）位于 S4 中（参考文献 46），也可能位于 S3 和 S4 间的边界（参考文献 67）。然而，包含 *TBL1X*（一个 S3 基因 [46]）的区块 11 的较远端的位置与这些观点不符。区块 10 内序列一致性增加的两个区域包含 X 染色体上的 *VCX2* 和 *VCX3B* 基因以及 Y 染色体上的 *VCY1B* 和 *VCY* 基因。这一基因家族可能是在类人猿谱系中新产生的 [68]，这可以解释这部分排列的不寻常特征。

除了 S3、S4、S5 和 PAR1 内 XY 序列同源程度不同外，我们还发现它们的序列组成存在显著差异，这也可能是进化时每个区域内重组缺失造成的。具体来说，我们观察到 L1、L2 和哺乳动物散在重复（MIR）覆盖度随每个更远端阶段和 PAR1 减少（表 2 和图 1），但（G+C）水平和 *Alu* 重复含量在 S4 和 S5 边界突然增加（表 2 和图 8）；不同 *Alu* 亚家族（Y、S 和 J）发病率的变化促成了每个阶段和 PAR1 的不同

distinct character of each stratum and PAR1 (Supplementary Table 13). The compositional differences between S4 and S5 provide additional support for the subdivision of the original stratum 4 (Fig. 8).

Table 2. Sequence characteristics of evolutionary domains of the X chromosome

Region	(G+C) (%)	L1 (%)	L1P (%)	L1M (%)	*Alu* (%)	L2 (%)	MIR+MIR3 (%)
X chromosome	39.46	28.87	13.39	15.21	8.23	2.98	2.07
XAR	39.87	17.89	6.60	11.23	10.28	2.63	1.76
XCR	39.28	33.50	16.38	16.97	7.28	3.12	2.19
PAR1	48.11	6.97	2.64	4.38	28.88	0.24	0.21
S5	42.86	8.89	4.36	4.59	18.72	0.66	0.31
S4	38.87	11.10	4.31	6.80	8.60	1.59	0.95
S3	39.46	19.55	7.14	12.34	9.24	2.94	1.98

See Supplementary Table 13 for additional repeat element data.

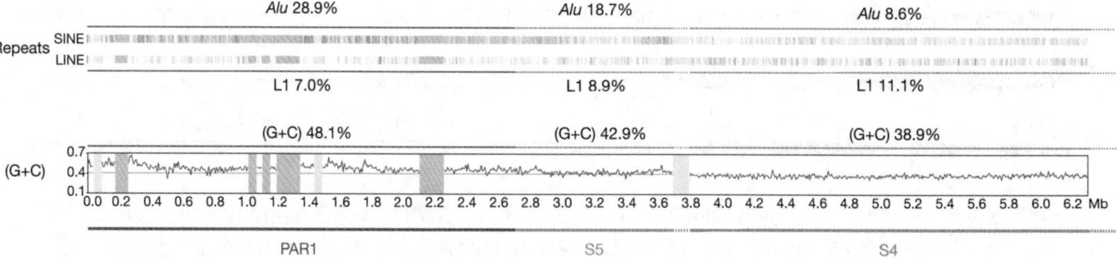

Fig. 8. Sequence compositional changes in the distal evolutionary strata of the X chromosome. Shown are the positions of SINE and LINE repeats and (G+C) content within PAR1, S5 and the distal half of S4. The percentage of *Alu*, L1 and (G+C) are shown for each region (including the whole of S4). There is an abrupt increase in *Alu* repeat levels and (G+C) content from S4 to S5. The five euchromatic gaps in PAR1 are shown as light brown bars. Pale blue bars represent clones for which the sequences were unfinished at the time of the sequence assembly.

X-chromosome Inactivation

XCI in mammals achieves dosage compensation between males and females for X-linked gene products. Inactivation of one X chromosome occurs early in female development and is initiated from the X-inactivation centre (XIC). The *XIST* transcript is expressed initially on both X chromosomes, but later the transcript from the chromosome that is destined for inactivation becomes more stable than the other. Finally, the transcript is expressed only from the inactive X chromosome (X_i). Coating with the *XIST* transcript is the earliest of many chromatin modifications on X_i.

XCI was first proposed based partly on the study of X: autosome translocations in female mice[47]. Studies of derivative chromosomes containing inactivated X chromosome segments

特性（补充表 13）。S4 和 S5 之间的成分差异为原始阶段 4 的细分提供了额外的支持（图 8）。

表 2. X 染色体进化区域的序列特征

区域	(G+C) (%)	L1 (%)	L1P (%)	L1M (%)	*Alu* (%)	L2 (%)	MIR+MIR3 (%)
X 染色体	39.46	28.87	13.39	15.21	8.23	2.98	2.07
XAR	39.87	17.89	6.60	11.23	10.28	2.63	1.76
XCR	39.28	33.50	16.38	16.97	7.28	3.12	2.19
PAR1	48.11	6.97	2.64	4.38	28.88	0.24	0.21
S5	42.86	8.89	4.36	4.59	18.72	0.66	0.31
S4	38.87	11.10	4.31	6.80	8.60	1.59	0.95
S3	39.46	19.55	7.14	12.34	9.24	2.94	1.98

其他重复元件数据见补充表 13。

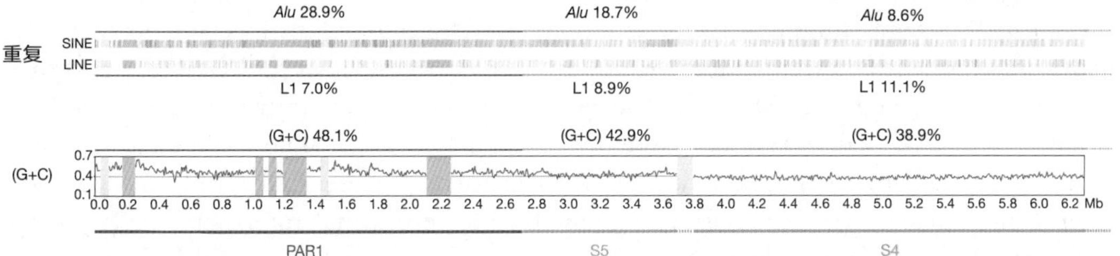

图 8. 在 X 染色体进化最远阶段序列成分变化。图中所示的是 SINE 和 LINE 重复的位置以及 PAR1、S5 和 S4 远端一半的（G+C）含量。每个区域显示了 *Alu*、L1 和（G+C）的百分比（包括整个 S4）。*Alu* 重复水平和（G+C）含量从 S4 到 S5 突然增加。PAR1 中五个常染色质缺口显示为浅褐色方框。淡蓝色方框代表序列组装时未完成序列的克隆。

X 染色体失活

在哺乳动物中，XCI 实现了雄性与雌性之间 X 连锁基因产物的剂量补偿效应。一条 X 染色体失活发生在雌性发育早期，而且是从 X 失活中心（XIC）开始。*XIST* 转录本最初在两条 X 染色体上都有表达，但后来在注定要失活的染色体上的转录本变得比另一条更稳定。最后，转录本只在失活的 X 染色体（X_i）上表达。在 X_i 上的许多染色质修饰中，包覆 *XIST* 转录本是最早的。

XCI 的首次提出部分是基于对 X 染色体的研究：雌性老鼠中常染色体易位 [47]。我们通过对含有失活 X 染色体片段的衍生染色体进行研究后得出结论，失活可以跨

later concluded that the inactivation could spread across the translocation boundary to the autosomal segment, but that inactivation of this segment was incomplete. More recently, it has become clear that more than 15% of the genes on the human X chromosome, including many without functional equivalents on the Y, escape from XCI, as presented in detail elsewhere[48]. The majority of the genes that escape XCI lie within the distal regions of the XAR (Fig. 1): all genes studied in PAR1, S5 and S4 were found to escape from XCI, but there is a lower proportion of escapees in S3, and very few examples in the XCR[48]. This observation correlates with our picture of X chromosome evolution: XCI follows Y chromosome attrition[49], which is less advanced in the distal strata of the XAR.

Inefficient inactivation of the autosomal segment in X_i: autosome translocations led to the proposal that "way stations" on the X chromosome boost the spread of XCI. According to this model, way stations are present throughout the genome but are enriched on the X chromosome, particularly in the region of the XIC[50]. Lyon suggested that L1 elements are good candidates for acting as way stations on account of their enrichment on the mammalian X chromosome[51]. We observe a distribution of L1 elements on the chromosome that is consistent with both the way station and the Lyon hypotheses (Fig. 1 and Table 2). The coverage of L1 repeats is very high in the XCR, especially around the XIC. As noted previously[52], this enrichment in L1 levels is accounted for particularly by elements that were active more recently in mammalian evolution[53] (L1P in Fig. 1). In the XAR, L1 coverage is close to autosome levels, whereas L1 levels are particularly low in the distal evolutionary strata of the XAR, where genes consistently escape inactivation. The *XIST* locus itself lies in a 60 kb region that is virtually devoid of L1 elements, whereas L1 levels are extremely high in the adjacent regions. Based on their distributions, other interspersed repeats are not strong candidates for way stations. For example, although L2 and MIR elements are reduced in S4, S5 and particularly PAR1 relative to the rest of the chromosome, their overall levels on the X chromosome are not enriched relative to the autosomes but are slightly reduced. Furthermore, L2 and MIR levels are low in the region distal to the XIC. These characteristics do not preclude an involvement in XCI, but are not consistent with a role as way stations.

The possible causal relationship of L1 elements to the spread of XCI remains a subject of debate. Some studies have reported significant associations between L1 coverage and inactivation[52], and others have refuted this[54]. Our observations on regional differences in composition emphasize that such studies should compare active and inactivated genes (or domains) from the same evolutionary stratum, in order to avoid correlations that are unrelated to XCI.

Medical Genetics and the X Chromosome Sequence

The X chromosome holds a unique place in the history of medical genetics. Ascertainment of X-linked diseases is enhanced by the relative ease of recognizing this mode of inheritance. More important, however, is the fact that a disproportionately large number of

越易位边界扩散到常染色体片段中，但该片段失活尚不完全。最近，人们已经清楚地看到：人类 X 染色体上超过 15%的基因，包括许多在 Y 染色体上没有功能等同物的基因，没有发生 XCI，这在其他地方也有详细说明 [48]。大多数没发生 XCI 的基因位于 XAR 远端区域（图 1）：在 PAR1、S5 和 S4 中研究的所有基因都没发生 XCI，但 S3 中没发生 XCI 的比例很低，在 XCR 中实例也很少 [48]。这一观察结果与我们的 X 染色体进化情况有关：XCI 发生在 Y 染色体削弱之后 [49]，而 Y 染色体的削弱在 XAR 较远的阶段几乎不发生。

X_i 中常染色体片段低效失活：常染色体易位导致提出 X 染色体上的"中转站"促进 XCI 的扩散这一观点。根据这一模型，中转站出现在整个基因组中，但在 X 染色体上富集，特别是在 XIC 区域 [50]。莱昂认为由于 L1 元件在哺乳动物 X 染色体上富集，所以它是作为中转站的理想选择 [51]。我们观察到 L1 元件在染色体上的分布符合中转站和莱昂假设（图 1 和表 2）。在 XCR 中，L1 重复的覆盖率非常高，特别是 XIC 周围。正如之前所述 [52]，最近在哺乳动物进化中活跃的元件解释了 L1 水平的富集 [53]（图 1 中的 L1P）。在 XAR 中，L1 覆盖率接近常染色体水平，而 L1 水平在 XAR 远端进化阶段特别低，在该阶段基因始终没有发生 XCI。*XIST* 基因座本身位于一个实际上缺少 L1 元件的 60 kb 区域，而在邻近区域 L1 水平极其高。根据它们的分布，其他散在重复中没有强有力的中转站候选。例如，虽然 L2 和 MIR 元件在 S4、S5，尤其是 PAR1 中相对于染色体的其余部分有所减少，但它们在 X 染色体上的总体水平相对于常染色体并没有富集，而是略有降低。此外，L2 和 MIR 水平在 XIC 远端较低。这些特征不排除参与 XCI，但不符合中转站的功能。

L1 元件与 XCI 扩展可能的因果关系仍然是一个有争议的话题。一些研究报告了 L1 的覆盖率和失活之间的显著联系 [52]，而另一些研究则反驳了这一说法 [54]。我们经过对区域组成差异的观察后，强调了这种研究应比较同一进化阶段的活性和失活基因（或域），以避免与 XCI 无关的相关性。

医学遗传学和 X 染色体序列

X 染色体在医学遗传学历史上占有独特的地位。识别这种遗传模式相对容易，于是增强了对 X 连锁疾病的确定。然而更重要的是，实际上大量的疾病条件都与 X 染色体有关，因为任何在 Y 染色体上没有活性对应物的基因，其隐性突变的表型结

disease conditions have been associated with the X chromosome because the phenotypic consequence of a recessive mutation is revealed directly in males for any gene that has no active counterpart on the Y chromosome. Thus, although the X chromosome contains only 4% of all human genes, almost 10% of diseases with a mendelian pattern of inheritance have been assigned to the X chromosome (307 out of 3,199; information obtained from OMIM[2]). These two aspects of the medical genetics of the X chromosome have greatly stimulated progress in the positional cloning of many genes associated with human disease. To date, the molecular basis for 168 X-linked phenotypes has been determined, and the X chromosome sequence has aided this process for 43 of them, by providing positional candidate genes or a reference sequence for comparison to patient samples (Supplementary Table 14).

Identifying genes involved in rare conditions yields important biological insights. For example, discovery of mutations in the *SH2D1A* gene[55] (involved in X-linked lymphoproliferative disease (XLP, OMIM 308240)) led to identification of a new mediator of signal transduction between T and NK cells, and a novel family of proteins involved in the regulation of the immune response. Mental retardation is one of the most common problems in clinical genetics, and affects significantly more males than females. To date, 16 genes from the X chromosome have been associated with cases of non-syndromic X-linked mental retardation (NS-XLMR), in which mental retardation is the only phenotypic feature. These genes encode a range of protein types, and some are also involved in syndromic forms of mental retardation. For example, the *ARX* gene encodes an aristaless-related homeobox transcription factor and is linked to NS-XLMR cases, as well as to syndromic mental retardation associated with epilepsy (infantile spasm syndrome, ISSX, OMIM 308350) or with dystonic hand movements (Partington syndrome, PRTS, OMIM 309510)[56]. The *MECP2* gene, which encodes a methyl-CpG-binding protein, was initially linked to cases of Rett syndrome in girls[57] (RTT, OMIM 312750) but was later also seen to be mutated in males or females with NS-XLMR[58]. The molecular defect has been determined in only a minority of families affected by NS-XLMR, which has led to speculation that there could be as many as 100 genes on the X chromosome that are associated with NS-XLMR[59]. Discovering the genes for these and other rare, monogenic disorders is of critical value in extending our understanding of fundamental new processes in human biology, and the annotated X chromosome will further facilitate this process.

Concluding Remarks

The completion of the X chromosome sequencing project is an essential component of the goal of obtaining a high-quality, annotated human genome sequence for use in studies of gene function, sequence variation, disease and evolution. It also means that for the first time, we now have the finished sex chromosome sequences of an organism. The study of these sequences gives a greater insight into mammalian sex chromosome evolution and its consequences. As these analyses are extended to other genomes, we will gain a greater appreciation of the different evolutionary forces that shape sex chromosome and autosome

果直接在男性中显示。因此，尽管 X 染色体只含有 4%的人类基因，但几乎 10%的孟德尔遗传模式的疾病与 X 染色体有关(3,199 种疾病中有 307 种；信息从 OMIM[2] 获得)。X 染色体的医学遗传学的这两个方面极大地促进了与人类疾病相关的许多基因定位克隆的发展。迄今为止，我们已经确定了 168 个 X 连锁表型的分子基础，通过为患者样本的比较提供位置候选基因或参考序列，X 染色体序列为其中 43 个的分子基础的确定提供了帮助(补充表 14)。

识别罕见情况下涉及的基因可以获得重要的生物学见解。例如，*SH2D1A* 基因突变的发现 [55](涉及 X 连锁淋巴组织增生性疾病(XLP，OMIM 308240))使得我们在 T 细胞和 NK 细胞之间鉴定出一个新的信号转导调节物，以及一个涉及免疫反应调节的新的蛋白家族。精神发育迟缓是临床遗传学最常见问题之一，并且它对男性的影响比对女性的大。迄今为止，已有 16 个来自 X 染色体的基因与非综合征 X 连锁智力低下(NS-XLMR)病例相关，其中智力低下是唯一的表型特征。这些基因编码一系列蛋白类型，其中一些基因还与精神发育迟缓的综合征形式有关。例如，*ARX* 基因编码一个无芒样同源框转录因子，并与 NS-XLMR 病例以及与癫痫(婴儿痉挛综合征，ISSX，OMIM 308350)或手部运动异常(帕廷顿综合征，PRTS，OMIM 309510)有关的精神发育迟缓的综合征相关 [56]。编码甲基化 CpG 结合蛋白的 *MECP2* 基因，最初与女孩中的雷特综合征的病例有关 [57](RTT，OMIM 312750)，但后来也被发现在患有 NS-XLMR 的男性或女性中发生突变 [58]。这种分子缺陷只在少数受 NS-XLMR 影响的家族中被发现，人们推测 X 染色体上可能有多达 100 个基因与 NS-XLMR 相关 [59]。发现上述基因和其他罕见的单基因疾病对于扩展我们对人类生物学新的基本过程的理解具有重要价值，而标注的 X 染色体将进一步促进这一过程。

结 束 语

X 染色体测序项目的完成是获得用于基因功能、序列变异、疾病和进化研究的高质量、带注释的人类基因组序列的重要组成部分。这也意味着，我们第一次拥有了一个有机体的已完成的性染色体序列。对这些序列的研究使我们对哺乳动物性染色体进化及其结果有了更深入的了解。当这些分析扩大到其他基因组时，我们将对形成性染色体和常染色体的不同进化力量有更深入的了解。非常重要的是研究突变过程速率的差异，并考虑不同寻常的雄性重组模式对这些过程的影响。显然，这种

evolution. It will be important to study differences in the rates of mutational processes, and to consider the influence of the unusual pattern of male recombination on these processes. Clearly, this analysis should not be restricted to a consideration of mammalian sex chromosomes, and it will be of great interest to make comparisons with non-mammalian systems that arose independently in evolution.

Methods

The approach used to establish a bacterial clone map of the X chromosome has been previously described[5]. 13,264 clones were identified using 4,363 STS markers derived from published genetic or physical maps, from shotgun sequencing of flow-sorted X chromosomes, or from end-sequences of clones at contig ends. Clones were assembled into contigs using restriction-enzyme fingerprinting, and were integrated with the Washington University Genome Sequencing Center whole genome BAC map[60] in order to identify additional clones. Nine euchromatic gaps were measured using fluorescent *in situ* hybridization of clones to extended DNA fibres, and a tenth gap was estimated on the basis of end-sequence data from spanning, unstable BAC clones (Supplementary Table 2). On the basis of pulsed-field gel electrophoresis experiments, we expect the sizes of the other four euchromatic gaps to have a combined size of less than 400 kb.

Finished sequences of individual clones were determined using procedures described in ref. 7. For the analyses described above, the sequence was frozen in March 2004, at which point 150,396,262 bp of sequence had been determined from a minimal tiling path of 1,832 clones (1,616 sequence accessions). This sequence is available at http://www.sanger.ac.uk/HGP/ChrX/, and its annotation is represented in Supplementary Fig. 1. Updates to the sequence and annotation can be obtained from the VEGA database.

Manual annotation of gene structures has been described elsewhere[14], and used guidelines agreed at the human annotation workshop (HAWK; http://www.sanger.ac.uk/HGP/havana/hawk.shtml). Genes were assigned to one of four groups: (1) known genes that are identical to human cDNAs or protein sequences and have a RefSeq RNA (and RefSeq protein, if the gene encodes a protein); (2) novel coding sequences, which have an open reading frame (ORF) and are identical to spliced ESTs, or have similarity to other genes/proteins (any species); (3) novel transcripts, which are similar to novel coding sequences, except that no ORF can be determined with confidence; and (4) putative transcripts, which are identical to splicing human ESTs but have no ORF. Gene symbols were approved by the HUGO Gene Nomenclature Committee wherever possible. Predicted protein translations were analysed for Pfam domains using InterProScan (http://www.ebi.ac.uk/InterProScan/). CpG islands were predicted using the program GpG (G. Micklem, personal communication).

Interspersed repeats were identified and classified using RepeatMasker (http://repeatmasker.genome.washington.edu). In order to search for segmental duplications, WU-BLASTN (http://blast.wustl.edu) was used to align the current X chromosome sequence to itself or to the NCBI34 autosome assemblies. Duplicated blocks at least 5 kb in length were defined as described in ref. 28.

分析不应局限于对哺乳动物性染色体的考虑，将哺乳动物系统与在进化中独立产生的非哺乳动物系统进行比较将是非常有趣的。

方　法

建立 X 染色体细菌克隆图谱的方法以前已描述过[5]。13,264 个克隆通过 4,363 个 STS 标记被鉴定出来，这些 STS 标记来自已发表的遗传或物理图谱、流式 X 染色体鸟枪法测序、重叠群末端克隆的末端序列。用限制酶指纹法将克隆组装到重叠群，并与华盛顿大学基因组测序中心全基因组 BAC 图谱[60]进行整合，用以确定其他的克隆。使用荧光原位杂交克隆来测量 9 条常染色体缺口以扩展 DNA 纤维，基于不稳定 BAC 克隆末端序列数据，我们对第十个缺口进行了评估（补充表 2）。根据脉冲场凝胶电泳实验，我们预计其他四个常染色体缺口的合并大小小于 400 kb。

使用参考文献 7 中描述的步骤确定单个克隆的完成序列。在上述的分析中，该序列在 2004 年 3 月被冻结，那时从 1,832 个克隆的最小叠瓦式中确定了 150,396,262 bp 的序列（1,616 个序列登记入册）。该序列可在 http://www.sanger.ac.uk/HGP/ChrX/获得，其注解如补充图 1 所示。更新的序列和注解可从 VEGA 数据库获取。

人工注释基因结构在其他地方也有描述[14]，并且使用的准则经过了人类注解研讨会的同意（HAWK；http://www.sanger.ac.uk/HGP/havana/hawk.shtml）。基因被分为四组：（1）与人类 cDNA 或蛋白质序列一致且具有 RNA 参考序列（以及蛋白质参考序列，如果该基因编码了一种蛋白质）的已知基因；（2）有开放阅读框且与剪切表达序列标签一致或与其他基因/蛋白（任何物种）相似的新编码序列；（3）与新编码序列一致的新转录本，但排除了确定没有开放阅读框的情况；（4）与剪切人类表达序列标签一致但没有开放阅读框的假定转录本。基因符号尽可能由 HUGO 基因命名委员会批准。用 InterProScan 分析 Pfam 域的预测蛋白翻译（http://www.ebi.ac.uk/InterProScan/）。使用 CpG 程序分析 CpG 岛（米克勒姆，个人交流）。

使用 RepeatMasker（http://repeatmasker.genome.washington.edu）对散在重复序列进行识别和分类。为了寻找片段重复，我们使用 WU-BLASTN（http://blast.wustl.edu）将当前的 X 染色体序列与自身或与 NCBI34 常染色体组进行比对。如参考文献 28 所述，确定的重复区块长度至少有 5 kb。

SNPs (dbSNP release 119) were mapped onto the X chromosome sequence using first SSAHA[61] and then Cross-match (http://www.phrap.org/phredphrapconsed.html).

Comparative analysis

The genome assemblies used for comparative analyses were: *Gallus gallus* WASHUC1 (Washington University Genome Sequencing Center, http://www.genome.wustl.edu/projects/chicken), *Rattus norvegicus* RGSC3.1 (Rat Genome Sequencing Consortium http://www.hgsc.bcm.tmc.edu/projects/rat/), *Mus musculus* NCBI32 (Mouse Genome Sequencing Consortium, http://www.ncbi.nlm.nih.gov/genome/seq/NCBIContigInfo.html), *Danio rerio* version 3 (Sanger Institute, http://www.sanger.ac.uk/Projects/D_rerio), *T. nigroviridis* version 6 (Genoscope and the Broad Institute, http://www.genoscope.cns.fr/externe/tetraodon/Ressource.html), and *F. rubripes* version 2 (International Fugu Genome Consortium, http://www.fugu-sg.org/project/info.html). ECRs between the X chromosome and the rodent and fish genomes were obtained as described elsewhere[13]. In order to visualize regions of conserved synteny, the X chromosome sequence was aligned to the chicken and rodent genome sequences using BLASTZ (with default parameters), and matches were plotted by chromosome position. Matches to the rodent genomes were filtered to include only those with a sequence identity of at least 70% to the human sequence. The Ensembl database (http://www.ensembl.org) was used to search for orthologous gene pairs between the X chromosome and the other three genomes.

Genomic sequence homologies between the X and Y chromosomes were identified by aligning the two finished chromosome sequences using WU-BLASTN, and then filtering the alignments to include only those of at least 70% sequence identity and 80 bp length. In order to calculate the sequence identity between large, XY-homologous regions, a global alignment of unmasked sequence was generated using LAGAN[62]. Gapped regions, which result from insertions or deletions, were removed from the alignment, and then the nucleotide sequence identity was calculated for the remainder. Sequence identity plots were produced by parsing the LAGAN output into VISTA[63]. GRIMM[64] was used to calculate a most parsimonious series of inversions that would account for differences in homology block order and orientation between the X and Y chromosomes. Homologous protein-coding gene pairs between the X and Y chromosomes were identified by TBLASTN searching with the coding sequences of annotated coding genes on the Y chromosome against the X chromosome genomic sequence.

(**434**, 325-337; 2005)

Mark T. Ross[1], Darren V. Grafham[1], Alison J. Coffey[1], Steven Scherer[2], Kirsten McLay[1], Donna Muzny[2], Matthias Platzer[3], Gareth R. Howell[1], Christine Burrows[1], Christine P. Bird[1], Adam Frankish[1], Frances L. Lovell[1], Kevin L. Howe[1], Jennifer L. Ashurst[1], Robert S. Fulton[4], Ralf Sudbrak[5,6], Gaiping Wen[3], Matthew C. Jones[1], Matthew E. Hurles[1], T. Daniel Andrews[1], Carol E. Scott[1], Stephen Searle[1], Juliane Ramser[7], Adam Whittaker[1], Rebecca Deadman[1], Nigel P. Carter[1], Sarah E. Hunt[1], Rui Chen[2], Andrew Cree[2], Preethi Gunaratne[2], Paul Havlak[2], Anne Hodgson[2], Michael L. Metzker[2], Stephen Richards[2], Graham Scott[2], David Steffen[2], Erica Sodergren[2], David A. Wheeler[2], Kim C. Worley[2], Rachael Ainscough[1], Kerrie D. Ambrose[1], M. Ali Ansari-Lari[2], Swaroop Aradhya[2], Robert I. S. Ashwell[1], Anne K. Babbage[1], Claire L. Bagguley[1], Andrea Ballabio[2], Ruby Banerjee[1], Gary E. Barker[1], Karen F. Barlow[1], Ian P. Barrett[1], Karen N. Bates[1], David M. Beare[1],

首先使用 SSAHA[61] 将 SNP(dbSNP 发布版 119)映射到 X 染色体序列上，然后进行交叉匹配(http://www.phrap.org/phredphrapconsed.html)。

对比分析

用于比较分析的基因组组合是：原鸡 WASHUC1(华盛顿大学基因组测序中心，http://www.genome.wustl.edu/project/chicken)，褐家鼠 RGSC3.1(大鼠基因组测序协作组 http://www.hgsc.bcm.tmc.edu/projects/rat/)，小家鼠 NCBI32(小鼠基因组测序协作组，http://www.ncbi.nlm.nih.gov/genome/seq/NCBIContigInfo.html)，斑马鱼第 3 版(桑格研究所，http://www.sanger.ac.uk/Project/D_rerio)，黑青斑河鲀第 6 版(Genoscope 和布罗德研究所，http://www.genoscope.cns.fr/externe/tetraodon/Ressource.html)和红鳍东方鲀第 2 版(国际河豚基因协作组 http://www.fugu-sg.org/project/info.html)。X 染色体与啮齿动物和鱼类基因组之间的 ECR 如其他地方所述[13]。为了可视化保守的同源区域，X 染色体序列使用 BLASTZ(默认参数)与鸡和啮齿类动物基因组序列进行比对，并通过染色体位置绘制匹配。与啮齿动物基因组匹配的基因经过筛选，只包括那些至少与人类序列一致性达 70%的基因。使用 Ensembl 数据库(http://www.ensembl.org)来寻找 X 染色体和其他三个基因组间直系同源基因对。

通过用 WU-BLASTN 对两个已完成的染色体序列进行比对，确定了 X 染色体和 Y 染色体间的基因组序列同源性，然后对比对结果进行筛选，只包含序列一致性不低于 70%、长度不小于 80 bp 的染色体序列。为了计算大的 XY 同源性区域间的序列一致性，使用 LAGAN[62] 生成了未掩蔽全序列比对。将插入或删除导致的缺口区域从比对中移除，然后计算剩余序列的核苷酸序列一致性。通过将 LAGAN 输出解析为 VISTA[63] 来生成序列标识图。用 GRIMM[64] 计算出一系列最简单的逆序序列，这些逆序序列可以解释 X 染色体和 Y 染色体在同源块顺序和方向上的差异。利用带注释的编码序列进行 TBLASTN 搜索，确定 X 染色体和 Y 染色体之间的同源蛋白编码基因对。

(李梅 翻译；胡松年 审稿)

Helen Beasley[1], Oliver Beasley[1], Alfred Beck[5], Graeme Bethel[1], Karin Blechschmidt[3], Nicola Brady[1], Sarah Bray-Allen[1], Anne M. Bridgeman[1], Andrew J. Brown[1], Mary J. Brown[2], David Bonnin[2], Elspeth A. Bruford[8], Christian Buhay[2], Paula Burch[2], Deborah Burford[1], Joanne Burgess[1], Wayne Burrill[1], John Burton[1], Jackie M. Bye[1], Carol Carder[1], Laura Carrel[9], Joseph Chako[2], Joanne C. Chapman[1], Dean Chavez[2], Ellson Chen[10], Guan Chen[2], Yuan Chen[11], Zhijian Chen[2], Craig Chinault[2], Alfredo Ciccodicola[12], Sue Y. Clark[1], Graham Clarke[1], Chris M. Clee[1], Sheila Clegg[1], Kerstin Clerc-Blankenburg[2], Karen Clifford[1], Vicky Cobley[1], Charlotte G. Cole[1], Jen S. Conquer[1], Nicole Corby[1], Richard E. Connor[1], Robert David[2], Joy Davies[1], Clay Davis[2], John Davis[1], Oliver Delgado[2], Denise DeShazo[2], Pawandeep Dhami[1], Yan Ding[2], Huyen Dinh[2], Steve Dodsworth[1], Heather Draper[2], Shannon Dugan-Rocha[2], Andrew Dunham[1], Matthew Dunn[1], K. James Durbin[2], Ireena Dutta[1], Tamsin Eades[1], Matthew Ellwood[1], Alexandra Emery-Cohen[2], Helen Errington[1], Kathryn L. Evans[13], Louisa Faulkner[1], Fiona Francis[14], John Frankland[1], Audrey E. Fraser[1], Petra Galgoczy[3], James Gilbert[1], Rachel Gill[2], Gernot Glöckner[3], Simon G. Gregory[1], Susan Gribble[1], Coline Griffiths[1], Russell Grocock[1], Yanghong Gu[2], Rhian Gwilliam[1], Cerissa Hamilton[2], Elizabeth A. Hart[1], Alicia Hawes[2], Paul D. Heath[1], Katja Heitmann[5], Steffen Hennig[5], Judith Hernandez[2], Bernd Hinzmann[3], Sarah Ho[1], Michael Hoffs[1], Phillip J. Howden[1], Elizabeth J. Huckle[1], Jennifer Hume[2], Paul J. Hunt[1], Adrienne R. Hunt[1], Judith Isherwood[1], Leni Jacob[2], David Johnson[1], Sally Jones[2], Pieter J. de Jong[15], Shirin S. Joseph[1], Stephen Keenan[1], Susan Kelly[2], Joanne K. Kershaw[1], Ziad Khan[2], Petra Kioschis[16], Sven Klages[5], Andrew J. Knights[1], Anna Kosiura[5], Christie Kovar-Smith[2], Gavin K. Laird[1], Cordelia Langford[1], Stephanie Lawlor[1], Margaret Leversha[1], Lora Lewis[2], Wen Liu[2], Christine Lloyd[1], David M. Lloyd[1], Hermela Loulseged[2], Jane E. Loveland[1], Jamieson D. Lovell[1], Ryan Lozado[2], Jing Lu[2], Rachael Lyne[1], Jie Ma[2], Manjula Maheshwari[2], Lucy H. Matthews[1], Jennifer McDowall[1], Stuart McLaren[1], Amanda McMurray[1], Patrick Meidl[1], Thomas Meitinger[17], Sarah Milne[1], George Miner[2], Shailesh L. Mistry[1], Margaret Morgan[2], Sidney Morris[2], Ines Mü ller[5,18], James C. Mullikin[19], Ngoc Nguyen[2], Gabriele Nordsiek[3], Gerald Nyakatura[3], Christopher N. O'Dell[1], Geoffery Okwuonu[2], Sophie Palmer[1], Richard Pandian[1], David Parker[2], Julia Parrish[2], Shiran Pasternak[2], Dina Patel[1], Alex V. Pearce[1], Danita M. Pearson[1], Sarah E. Pelan[1], Lesette Perez[2], Keith M. Porter[1], Yvonne Ramsey[1], Kathrin Reichwald[3], Susan Rhodes[1], Kerry A. Ridler[1], David Schlessinger[20], Mary G. Schueler[19], Harminder K. Sehra[1], Charles Shaw-Smith[1], Hua Shen[2], Elizabeth M. Sheridan[1], Ratna Shownkeen[1], Carl D. Skuce[1], Michelle L. Smith[1], Elizabeth C. Sotheran[1], Helen E. Steingruber[1], Charles A. Steward[1], Roy Storey[1], R. Mark Swann[1], David Swarbreck[1], Paul E. Tabor[2], Stefan Taudien[3], Tineace Taylor[2], Brian Teague[2], Karen Thomas[1], Andrea Thorpe[1], Kirsten Timms[2], Alan Tracey[1], Steve Trevanion[1], Anthony C. Tromans[1], Michele d'Urso[12], Daniel Verduzco[2], Donna Villasana[2], Lenee Waldron[2], Melanie Wall[1], Qiaoyan Wang[2], James Warren[2], Georgina L. Warry[1], Xuehong Wei[2], Anthony West[1], Siobhan L. Whitehead[1], Mathew N. Whiteley[1], Jane E. Wilkinson[1], David L. Willey[1], Gabrielle Williams[2], Leanne Williams[1], Angela Williamson[2], Helen Williamson[1], Laurens Wilming[1], Rebecca L. Woodmansey[1], Paul W. Wray[1], Jennifer Yen[2], Jingkun Zhang[2], Jianling Zhou[2], Huda Zoghbi[2], Sara Zorilla[2], David Buck[1], Richard Reinhardt[5], Annemarie Poustka[16], André Rosenthal[3], Hans Lehrach[5], Alfons Meindl[7], Patrick J. Minx[4], LaDeana W. Hillier[4], Huntington F. Willard[21], Richard K. Wilson[4], Robert H. Waterston[4], Catherine M. Rice[1], Mark Vaudin[1], Alan Coulson[1], David L. Nelson[2], George Weinstock[2], John E. Sulston[1], Richard Durbin[1], Tim Hubbard[1], Richard A. Gibbs[2], Stephan Beck[1], Jane Rogers[1] & David R. Bentley[1]

[1] The Wellcome Trust Sanger Institute, Wellcome Trust Genome Campus, Hinxton, Cambridge CB10 1SA, UK

[2] Baylor College of Medicine Human Genome Sequencing Center, Department of Molecular and Human Genetics, One Baylor Plaza, Houston, Texas 77030, USA

[3] Genomanalyse, Institut für Molekulare Biotechnologie, Beutenbergstr. 11, 07745 Jena, Germany

[4] Washington University Genome Sequencing Center, Box 8501, 4444 Forest Park Avenue, St. Louis, Missouri 63108, USA

[5] Max Planck Institute for Molecular Genetics, Ihnestrasse 73, 14195 Berlin, Germany

[6] Institute for Clinical Molecular Biology, Christian-Albrechts-University, 24105 Kiel, Germany

[7] Medizinische Genetik, Ludwig-Maximilian-Universität, Goethestr. 29, 80336 München, Germany

[8] HUGO Gene Nomenclature Committee, The Galton Laboratory, Department of Biology, University College London, Wolfson House, 4 Stephenson Way, London NW1 2HE, UK

[9] Department of Biochemistry and Molecular Biology, Pennsylvania State College of Medicine, Hershey, Pennsylvania 17033, USA

[10] Advanced Center for Genetic Technology, PE-Applied Biosystems, Foster City, California 94404, USA

[11] European Bioinformatics Institute, Wellcome Trust Genome Campus, Hinxton, Cambridge CB10 1SD, UK

[12] Institute of Genetics and Biophysics, Adriano Buzzati-Traverso, Via Marconi 12, 80100 Naples, Italy

[13] Medical Genetics Section, University of Edinburgh, Western General Hospital, Edinburgh EH4 2XU, UK

[14] Laboratoire de Génétique et de Physiopathologie des Retards Mentaux, Institut Cochin. Inserm U567, Université Paris V., 24 rue du Faubourg Saint Jacques, 75014 Paris, France

[15] BACPAC Resources, Children's Hospital Oakland Research Institute, 747 52nd Street, Oakland, California 94609, USA

[16] Molekulare Genomanalyse, Deutsches Krebsforschungszentrum, Im Neuenheimer Feld 580, 69120 Heidelberg, Germany
[17] Institute of Human Genetics, GSF National Research Center for Environment and Health, Ingolstädter Landstr. 1, 85764 Neuherberg, Germany
[18] RZPD Resource Center for Genome Research, 14059 Berlin, Germany
[19] National Human Genome Research Institute, National Institutes of Health, Bethesda, Maryland 20892, USA
[20] Laboratory of Genetics, National Institute on Aging, 333 Cassell Drive, Baltimore, Maryland 21224, USA
[21] Institute for Genome Sciences & Policy, Duke University, Durham, North Carolina 27708, USA

Received 1 February; accepted 7 February 2005; doi:10.1038/nature03440.

References:

1. Ohno, S. *Sex Chromosomes and Sex-linked Genes* (Springer, Berlin, 1967).
2. McKusick-Nathans Institute for Genetic Medicine, Johns Hopkins University and National Center for Biotechnology Information, National Library of Medicine. *OMIM: Online Mendelian Inheritance in Man*. 〈http://www.ncbi.nlm.nih.gov/omim/〉 (2000).
3. Royer-Pokora, B. *et al.* Cloning the gene for an inherited human disorder—chronic granulomatous disease—on the basis of its chromosomal location. *Nature* **322**, 32-38 (1986).
4. Monaco, A. P. *et al.* Isolation of candidate cDNAs for portions of the Duchenne muscular dystrophy gene. *Nature* **323**, 646-650 (1986).
5. Bentley, D. R. *et al.* The physical maps for sequencing human chromosomes 1, 6, 9, 10, 13, 20 and X. *Nature* **409**, 942-943 (2001).
6. Humphray, S. J. *et al.* DNA sequence and analysis of human chromosome 9. *Nature* **429**, 369-374 (2004).
7. International Human Genome Sequencing Consortium. Initial sequencing and analysis of the human genome. *Nature* **409**, 860-921 (2001).
8. Schmutz, J. *et al.* Quality assessment of the human genome sequence. *Nature* **429**, 365-368 (2004).
9. Mahtani, M. M. & Willard, H. F. Physical and genetic mapping of the human X chromosome centromere: repression of recombination. *Genome Res.* **8**, 100-110 (1998).
10. Kong, A. *et al.* A high-resolution recombination map of the human genome. *Nature Genet.* **31**, 241-247 (2002).
11. Pruitt, K. D., Katz, K. S., Sicotte, H. & Maglott, D. R. Introducing RefSeq and LocusLink: curated human genome resources at the NCBI. *Trends Genet.* **16**, 44-47 (2000).
12. Mungall, A. J. *et al.* The DNA sequence and analysis of human chromosome 6. *Nature* **425**, 805-811 (2003).
13. Deloukas, P. *et al.* The DNA sequence and comparative analysis of human chromosome **10**. *Nature* 429, 375-381 (2004).
14. Dunham, A. *et al.* The DNA sequence and analysis of human chromosome 13. *Nature* **428**, 522-528 (2004).
15. Griffiths-Jones, S., Bateman, A., Marshall, M., Khanna, A. & Eddy, S. R. Rfam: an RNA family database. *Nucleic Acids Res.* **31**, 439-441 (2003).
16. Lowe, T. M. & Eddy, S. R. tRNAscan-SE: a program for improved detection of transfer RNA genes in genomic sequence. *Nucleic Acids Res.* **25**, 955-964 (1997).
17. Griffiths-Jones, S. The microRNA Registry. *Nucleic Acids Res.* **32** (Database issue), D109-111 (2004).
18. Brown, C. J. *et al.* A gene from the region of the human X inactivation centre is expressed exclusively from the inactive X chromosome. *Nature* 349, 38-44 (1991).
19. Brown, C. J. *et al.* The human XIST gene: analysis of a 17 kb inactive X-specific RNA that contains conserved repeats and is highly localized within the nucleus. *Cell* **71**, 527-542 (1992).
20. Lee, J. T., Davidow, L. S. & Warshawsky, D. Tsix, a gene antisense to Xist at the X-inactivation centre. *Nature Genet.* **21**, 400-404 (1999).
21. Migeon, B. R., Chowdhury, A. K., Dunston, J. A. & McIntosh, I. Identification of TSIX, encoding an RNA antisense to human XIST, reveals differences from its murine counterpart: implications for X inactivation. *Am. J. Hum. Genet.* **69**, 951-960 (2001).
22. Chow, J. C., Hall, L. L., Clemson, C. M., Lawrence, J. B. & Brown, C. J. Characterization of expression at the human XIST locus in somatic, embryonal carcinoma, and transgenic cell lines. *Genomics* **82**, 309-322 (2003).
23. Chureau, C. *et al.* Comparative sequence analysis of the X-inactivation center region in mouse, human, and bovine. *Genome Res.* **12**, 894-908 (2002).
24. Bateman, A. *et al.* The Pfam protein families database. *Nucleic Acids Res.* **32** (Database issue), D138-141 (2004).
25. Scanlan, M. J., Simpson, A. J. & Old, L. J. The cancer/testis genes: review, standardization, and commentary. *Cancer Immun.* [online] **4**, 1 (2004).
26. Hurst, L. D. Evolutionary genomics: Sex and the X. *Nature* **411**, 149-150 (2001).
27. Chomez, P. *et al.* An overview of the MAGE gene family with the identification of all human members of the family. *Cancer Res.* **61**, 5544-5551 (2001).
28. Cheung, J. *et al.* Genome-wide detection of segmental duplications and potential assembly errors in the human genome sequence. *Genome Biol.* **4**, R25 (2003).
29. Stankiewicz, P. & Lupski, J. R. Genome architecture, rearrangements and genomic disorders. *Trends Genet.* **18**, 74-82 (2002).
30. Pegoraro, E. *et al.* Familial skewed X inactivation: a molecular trait associated with high spontaneous- abortion rate maps to Xq28. *Am. J. Hum. Genet.* **61**, 160-170 (1997).
31. Schueler, M. G., Higgins, A. W., Rudd, M. K., Gustashaw, K. & Willard, H. F. Genomic and genetic definition of a functional human centromere. *Science* **294**, 109-115 (2001).
32. Spence, J. M. *et al.* Co-localization of centromere activity, proteins and topoisomerase II within a subdomain of the major human X alpha-satellite array. *EMBO J.* **21**, 5269-5280 (2002).
33. Rudd, M. K. & Willard, H. F. Analysis of the centromeric regions of the human genome assembly. *Trends Genet.* **20**, 529-533 (2004).

34. She, X. *et al.* The structure and evolution of centromeric transition regions within the human genome. *Nature* **430**, 857-864 (2004).
35. The International SNP Map Working Group. A map of human genome sequence variation containing 1.42 million single nucleotide polymorphisms. *Nature* **409**, 928-933 (2001).
36. Schmid, M. *et al.* First report on chicken genes and chromosomes 2000. *Cytogenet. Cell Genet.* **90**, 169-218 (2000).
37. Kohn, M., Kehrer-Sawatzki, H., Vogel, W., Graves, J. A. & Hameister, H. Wide genome comparisons reveal the origins of the human X chromosome. *Trends Genet.* **20**, 598-603 (2004).
38. Graves, J. A. The origin and function of the mammalian Y chromosome and Y-borne genes—an evolving understanding. *Bioessays* **17**, 311-320 (1995).
39. Hillier, L. W. *et al.* Sequence and comparative analysis of the chicken genome provide unique perspectives on vertebrate evolution. *Nature* **432**, 695-716 (2004).
40. Freije, D., Helms, C., Watson, M. S. & Donis-Keller, H. Identification of a second pseudoautosomal region near the Xq and Yq telomeres. *Science* **258**, 1784-1787 (1992).
41. Skaletsky, H. *et al.* The male-specific region of the human Y chromosome is a mosaic of discrete sequence classes. *Nature* **423**, 825-837 (2003).
42. Page, D. C., Harper, M. E., Love, J. & Botstein, D. Occurrence of a transposition from the X-chromosome long arm to the Y-chromosome short arm during human evolution. *Nature* **311**, 119-123 (1984).
43. Sargent, C. A. *et al.* The sequence organization of Yp/proximal Xq homologous regions of the human sex chromosomes is highly conserved. *Genomics* **32**, 200-209 (1996).
44. Toder, R., Wakefield, M. J. & Graves, J. A. The minimal mammalian Y chromosome - the marsupial Y as a model system. *Cytogenet. Cell Genet.* **91**, 285-292 (2000).
45. Delbridge, M. L. *et al.* TSPY, the candidate gonadoblastoma gene on the human Y chromosome, has a widely expressed homologue on the X - implications for Y chromosome evolution. *Chromosome Res.* **12**, 345-356 (2004).
46. Lahn, B. T. & Page, D. C. Four evolutionary strata on the human X chromosome. *Science* **286**, 964-967 (1999).
47. Lyon, M. F. Gene action in the X-chromosome of the mouse (*Mus musculus* L.). *Nature* **190**, 372-373 (1961).
48. Carrel, L. & Willard, H. F. X-inactivation profile reveals extensive variability in X-linked gene expression in females. *Nature* doi:10.1038/nature03479 (this issue).
49. Jegalian, K. & Page, D. C. A proposed path by which genes common to mammalian X and Y chromosomes evolve to become X inactivated. *Nature* **394**, 776-780 (1998).
50. Gartler, S. M. & Riggs, A. D. Mammalian X-chromosome inactivation. *Annu. Rev. Genet.* **17**, 155-190 (1983).
51. Lyon, M. F. X-chromosome inactivation: a repeat hypothesis. *Cytogenet. Cell Genet.* **80**, 133-137 (1998).
52. Bailey, J. A., Carrel, L., Chakravarti, A. & Eichler, E. E. Molecular evidence for a relationship between LINE-1 elements and X chromosome inactivation: the Lyon repeat hypothesis. *Proc. Natl Acad. Sci. USA* **97**, 6634-6639 (2000).
53. Smit, A. F., Toth, G., Riggs, A. D. & Jurka, J. Ancestral, mammalian-wide subfamilies of LINE-1 repetitive sequences. *J. Mol. Biol.* **246**, 401-417 (1995).
54. Ke, X. & Collins, A. CpG islands in human X-inactivation. *Ann. Hum. Genet.* **67**, 242-249 (2003).
55. Coffey, A. J. *et al.* Host response to EBV infection in X-linked lymphoproliferative disease results from mutations in an SH2-domain encoding gene. *Nature Genet.* **20**, 129-135 (1998).
56. Stromme, P. *et al.* Mutations in the human ortholog of *Aristaless* cause X-linked mental retardation and epilepsy. *Nature Genet.* **30**, 441-445 (2002).
57. Amir, R. E. *et al.* Rett syndrome is caused by mutations in X-linked MECP2, encoding methyl-CpG-binding protein 2. *Nature Genet.* **23**, 185-188 (1999).
58. Orrico, A. *et al.* MECP2 mutation in male patients with non-specific X-linked mental retardation.*FEBS Lett.* **481**, 285-288 (2000).
59. Ropers, H. H. *et al.* Nonsyndromic X-linked mental retardation: where are the missing mutations?*Trends Genet.* **19**, 316-320 (2003).
60. The International Human Genome Mapping Consortium. A physical map of the human genome. *Nature* **409**, 934-941 (2001).
61. Ning, Z., Cox, A. J. & Mullikin, J. C. SSAHA: a fast search method for large DNA databases. *Genome Res.* **11**, 1725-1729 (2001).
62. Brudno, M. *et al.* LAGAN and Multi-LAGAN: efficient tools for large-scale multiple alignment of genomic DNA. *Genome Res.* **13**, 721-731 (2003).
63. Frazer, K. A., Pachter, L., Poliakov, A., Rubin, E. M. & Dubchak, I. VISTA: computational tools for comparative genomics. *Nucleic Acids Res.* **32**, W273–W279 (2004).
64. Bourque, G. & Pevzner, P. A. Genome-scale evolution: reconstructing gene orders in the ancestral species. *Genome Res.* **12**, 26-36 (2002).
65. Francke, U. Digitized and differentially shaded human chromosome ideograms for genomic applications. *Cytogenet. Cell Genet.* **65**, 206-218 (1994).
66. Gibbs, R. A. *et al.* Genome sequence of the Brown Norway rat yields insights into mammalian evolution. *Nature* **428**, 493-521 (2004).
67. Iwase, M. *et al.* The amelogenin loci span an ancient pseudoautosomal boundary in diverse mammalian species. *Proc. Natl Acad. Sci. USA* **100**, 5258-5263 (2003).
68. Lahn, B. T. & Page, D. C. A human sex-chromosomal gene family expressed in male germ cells and encoding variably charged proteins. *Hum. Mol. Genet.* **9**, 311-319 (2000).

Supplementary Information accompanies the paper on www.nature.com/nature.

Acknowledgements. We thank the Washington University Genome Sequencing Center for access to chicken and chimpanzee genome sequence data before publication; the Broad Institute for access to dog and chimpanzee genome sequence data before publication, for *T. nigroviridis* genome data, and for fosmid end-sequence data and clones; members of the Sanger Institute zebrafish genome project; the mouse, rat and *F. rubripes* sequencing consortia; Genoscope for *T. nigroviridis* genome data; the Ensembl, UCSC, EMBL and GenBank database groups; G. Schuler for information on sequence overlaps; T. Furey for information on RefSeq RNA coverage; D. Jaffe for data on fosmid end-sequence matches; D. Vetrie, E. Kendall, D. Stephan, J. Trent, A. P. Monaco, J. Chelly, D. Thiselton, A. Hardcastle, G. Rappold and the Resource Centre of the German Human Genome Project (RZPD) for the provision of clones and mapping data; the HUGO Gene Nomenclature

Committee (S. Povey (chair), M. W. Wright, M. J. Lush, R. C. Lovering, V. K. Khodiyar, H. M. Wain and C. C. Talbot Jr) for assigning official gene symbols; C. Rees for assistance with the manuscript; C. Tyler-Smith for critical reading of the manuscript; and the Wellcome Trust, the NHGRI, and the Ministry of Education and Research (Germany) for financial support.

Competing interests statement. The authors declare that they have no competing financial interests.

Correspondence and requests for materials should be addressed to M.T.R. (mtr@sanger.ac.uk). All DNA sequences reported in this study have been deposited in the EMBL or GenBank databases, and accession numbers are given in Supplementary Fig. 1.

First Fossil Chimpanzee

S. McBrearty and N. G. Jablonski

Editor's Note

Fossil hominins are rare, but fossils of chimpanzees—our closest relatives—were entirely absent until this report of the remains of chimpanzee teeth from Kenya. The teeth are around half a million years old, and were buried alongside fossils attributable to our own genus, *Homo*. Most hominin fossils in Africa are found in the Rift Valley, whereas chimps are today confined to forested west and central Africa. But this discovery suggested that hominins and chimps once shared their environment. It still leaves open the question why chimpanzee fossils are so scarce, and why the fossil record of our other relation, the gorilla, is completely blank.

There are thousands of fossils of hominins, but no fossil chimpanzee has yet been reported. The chimpanzee (*Pan*) is the closest living relative to humans[1]. Chimpanzee populations today are confined to wooded West and central Africa, whereas most hominin fossil sites occur in the semi-arid East African Rift Valley. This situation has fuelled speculation regarding causes for the divergence of the human and chimpanzee lineages five to eight million years ago. Some investigators have invoked a shift from wooded to savannah vegetation in East Africa, driven by climate change, to explain the apparent separation between chimpanzee and human ancestral populations and the origin of the unique hominin locomotor adaptation, bipedalism[2-5]. The Rift Valley itself functions as an obstacle to chimpanzee occupation in some scenarios[6]. Here we report the first fossil chimpanzee. These fossils, from the Kapthurin Formation, Kenya, show that representatives of *Pan* were present in the East African Rift Valley during the Middle Pleistocene, where they were contemporary with an extinct species of *Homo*. Habitats suitable for both hominins and chimpanzees were clearly present there during this period, and the Rift Valley did not present an impenetrable barrier to chimpanzee occupation.

THE Kapthurin Formation forms the Middle Pleistocene portion of the Tugen Hills sequence west of Lake Baringo (Figs 1 and 2). It consists of a package of fluvial, lacustrine and volcanic sediments ~125 m thick, exposed over ~150 km^2 (refs 7–9) that contains numerous palaeontological and archaeological sites[9-11]. It is divided into five members informally designated K1–K5 (ref. 7), and the sequence is well calibrated by $^{40}Ar/^{39}Ar$ dating[12].

第一个黑猩猩化石

麦克布里雅蒂，查邦斯基

编者按

古人类化石是很罕见的，但是黑猩猩的化石——我们的近亲——在这篇来自肯尼亚的黑猩猩牙齿化石的报道之前是完全没有的。这些牙齿大约有50万年的历史，和我们人属的化石一起被埋葬。非洲的大部分古人类化石都是在东非大裂谷发现的，而今天的黑猩猩只能生活在非洲西部和中部的森林里。但本文发现表明，古人类和黑猩猩曾经共享它们的生存环境。为什么黑猩猩化石如此稀少？为什么我们的另一个亲戚大猩猩的化石记录是完全空白的？这些问题仍然悬而未决。

我们迄今已发现数以千计的古人类化石，但还没有黑猩猩化石被发现的报道。黑猩猩（黑猩猩属）是现存的与人类亲缘关系最近的动物[1]。今天的黑猩猩种群局限于树木繁茂的非洲西部和中部，而大多数古人类化石遗址都出现在半干旱的东非大裂谷。这种情形更加刺激了人们关于500万到800万年前人类和黑猩猩世系分叉脱离的有关原因的推测。一些研究者提出，在东非，气候变化引起的从森林到热带稀树草原植被的变化，导致黑猩猩与人类祖先种群之间的明显分离，以及古人类独特的运动方式适应——直立行走[2-5]。在某些情况下，东非大裂谷本身对黑猩猩的生存居住起到了障碍的作用[6]。在此我们报道第一个黑猩猩化石。来自肯尼亚卡普图林组的这些化石，显示黑猩猩属在中更新世期间就在东非大裂谷出现了，此时当地还存在着一种灭绝了的人属成员。很显然，这期间的生活环境适于黑猩猩与古人类生存，而且对于黑猩猩的生存居住，东非大裂谷并非不可逾越的障碍。

卡普图林组构成巴林戈湖西部图根丘陵层序的中更新世部分（图1、图2）。它由一套～125米厚的河流、湖泊和火山沉积构成，出露约150平方千米（参考文献7～9），其中有许多古生物和考古遗址[9-11]。它被非正式地划分为K1～K5（参考文献7）5层，且层序由$^{40}Ar/^{39}Ar$测年标度校准[12]。

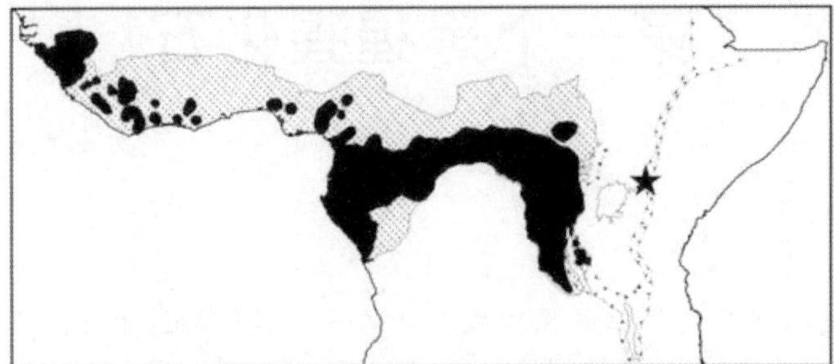

Fig. 1. Map showing current (solid black) and historical (stippled) ranges of *Pan* in equatorial Africa relative to major features of the eastern and western Rift Valleys. The Kapthurin Formation, Kenya, in the Eastern Rift Valley is marked by a star.

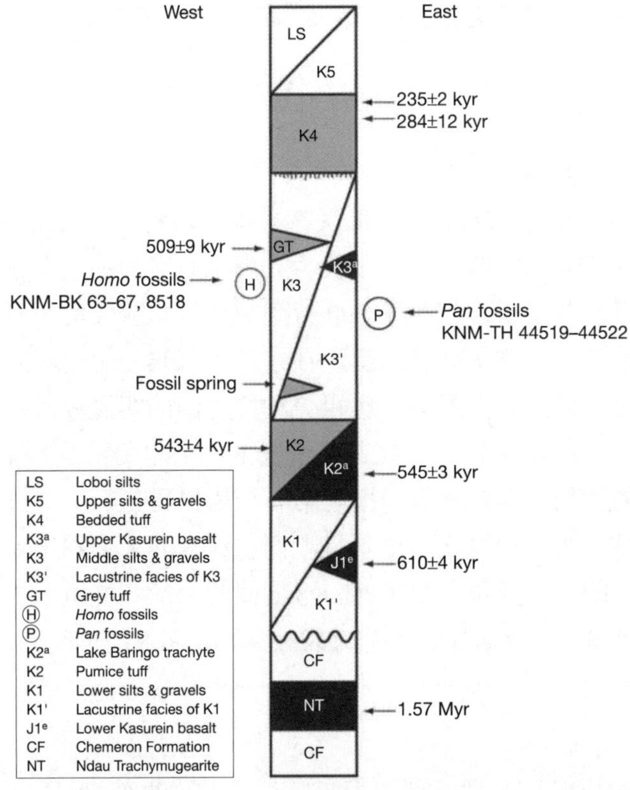

Fig. 2. Idealized stratigraphic column of the Kapthurin Formation, Kenya.

Hominin fossils attributed to *Homo erectus* or *Homo rhodesiensis* have been found in the fluvial sediments of K3 (refs 11, 13, 14). The new chimpanzee fossils were discovered at Locality (Loc.) 99 in K3′, the lacustrine facies of the same geological member. Loc. 99 consists of ~80 m^2 of exposures at an outcrop ~1 km northeast of site GnJh-19 where hominin mandible KNM-BK (Kenya National Museum-Baringo Kapthurin) 8518 was found[14]. Two chimpanzee fossils, KNM-TH (Kenya National Museum-Tugen Hills) 45519 and KNM-TH 45520, were found in surface context within an area of ~12 m^2 within Loc.

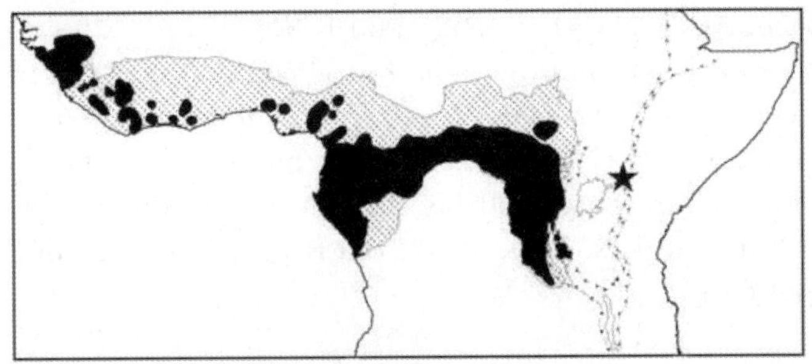

图 1. 与东非大裂谷西部与东部的主要特征相比，黑猩猩属在非洲赤道地区当前范围（实心黑色区）和历史范围（画点区），东非大裂谷肯尼亚卡普图林组用星形标出。

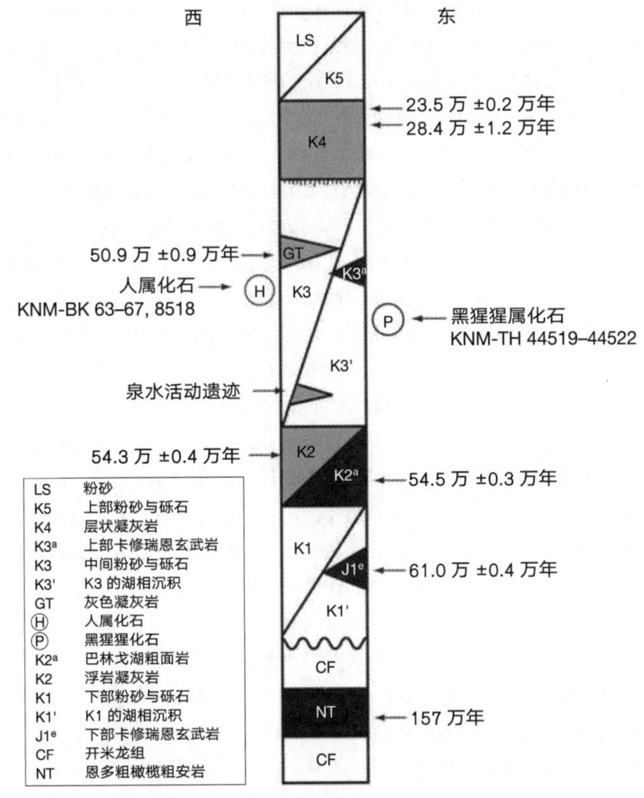

图 2. 肯尼亚卡普图林组的理想地层柱状图

属于直立人或罗得西亚人的化石已经在 K3 的河流沉积中发现（参考文献 11、13、14）。新的黑猩猩化石是在 99 号地点 K3 层中的湖相沉积（编为 K3′）中发现的。在 GnJh-19 遗址东北约 1 千米的断面上，99 号点有约 80 平方米的出露，在那里古人类下颌骨 KNM-BK（肯尼亚国家博物馆–巴林戈卡普图林）8518 被发现 [14]。两个黑猩猩化石，KNM-TH（肯尼亚国家博物馆–图根丘陵）45519 和 KNM-TH 45520，

99; additional specimens (KNM-TH 45521 and KNM-TH 45522) were recovered from sieved superficial sediments within the same restricted area. The age of the chimpanzee fossils is constrained by $^{40}Ar/^{39}Ar$ dates of 545 ± 3 kyr (thousand years) on underlying K2 and 284 ± 12 kyr on overlying K4 (ref. 12). Because they are derived from a position low in this stratigraphic interval, they are probably closer to the maximum age of 545 kyr. *Homo* fossils KNM-BK 63-67 and KNM-BK 8518 from K3 are bracketed by $^{40}Ar/^{39}Ar$ dates of 543 ± 4 kyr and 509 ± 9 kyr[12] (Fig. 2).

K3′ sediments are exposed in an outcrop of ~ 1 km^2 in the eastern portion of the Kapthurin Formation. They consist of black and red zeolitized clays interbedded with sands and heavily altered volcanics. Sedimentary and geochemical features of the clays indicate that they were laid down in a shallow body of water that alternated between fresh and intensely saline-alkaline, probably as a response to changes in outflow geometry controlled by local volcanism[15]. Additional intermittent sources of fresh water are suggested by localized ephemeral stream channel features and the remains of an extensive fossil spring. Loc. 99 has produced fragmentary fossils representing suids, bovids, rodents, cercopithecoid primates and catfish. Eight additional faunal collecting areas in K3′ have also produced elephants, hippopotami, carnivores, crocodiles, turtles, gastropods and additional micromammals. Many K3′ taxa, notably hippopotami (*Hippopotamus*), crocodiles, catfish (*Clarias*), gastropods and turtles, reflect local aquatic conditions. The bulk of K3′ non-aquatic fauna, including a colobine monkey, the elephant, the bovids *Kobus*, *Tragelaphus* and specimens probably belonging to *Syncerus*, and the suids *Potamochoerus porcus* (bushpig) and the extinct *Kolpochoerus majus*[16], are consistent with a closed environment. The presence of the cane rat (*Thryonomys*) indicates localized patches of moist, marshy conditions.

Remains of *Homo* (KNM-BK 63-67 and KNM-BK 8518) were recovered at sites GnJh-01 and GnJh-19 by previous workers[11,13,14] from K3 fluvial sediments to the west that represent a system of braided streams, some of which seem to have debouched into the lake. Fluvial K3 deposits and lacustrine K3′ deposits are interstratified, indicating a shoreline that shifted in position in response to alterations in lake levels. The similarity in the array of fossils encountered in K3 and K3′ sediments suggests that Middle Pleistocene *Pan* and *Homo* lived, or at least died, in broadly similar environmental settings. Taken together, the evidence suggests a locally wooded habitat on the shore of an alternately fresh and saline-alkaline lake, fluctuating lake levels, ephemeral nearshore fluvial channels, a nearby freshwater spring, and a semi-arid climatic regime. These conditions are not unlike those found near the shore of Lake Baringo today, although dense human populations have eliminated much of the woodland that formerly supported chimpanzees and the faunal community of which they were a part.

The chimpanzee specimens comprise a minimum of three teeth, probably from the same individual. Two of these are right and left upper central permanent incisors (I^1; KNM-TH 45519 and KNM-TH 45521, respectively). They exhibit broad, spatulate and moderately worn crowns, with thin dental enamel (Fig. 3). The lingual tubercle is large and flanked at the base by deep mesial and distal foveae, characteristic of *Pan*. This feature imparts great

在 99 号点的地表约 12 平方米范围内发现。另外的标本（KNM-TH 45521 和 KNM-TH 45522）在相同的区域内，通过筛选表层堆积物获得。黑猩猩化石的年龄由 $^{40}Ar/^{39}Ar$ 测年限定在下部 K2 和上部 K4 层位之间，对应年代（54.5 万 ± 0.3 万年～28.4 万 ± 1.2 万年）（参考文献 12）。因为它们源于该地层间隔的低位，所以其年龄比较接近于最大年龄值 54.5 万年。人属化石 KNM-BK 63-67 和 KNM-BK 8518 由 $^{40}Ar/^{39}Ar$ 测年框定在 54.3 万 ± 4.3 万年和 50.9 万 ± 0.9 万年之间 [12]（图 2）。

在卡普图林组东部约 1 平方千米的露头里出露 K3′ 沉积。它们由黑色与红色沸石化黏土构成，并与砂及强变质火山岩互层。黏土的沉积学与地球化学特征显示它们是在淡水与强盐碱水之间交替变化的浅水体中沉积下来的。这种交替可能是作为被当地火山作用控制的流出物质几何学的一种响应 [15]。另外，局部的短暂溪流通道痕迹和泉水活动的相关遗迹提示有额外的间歇性淡水来源。99 号点已经出土的化石碎片分别代表猪科、牛科、啮齿动物、猕猴科灵长类和鲶鱼。在 K3′ 另外 8 个动物化石采集区也出土了象、河马、食肉动物、鳄鱼、海龟、腹足动物及另外的小型哺乳动物。K3′ 的许多动物类，特别是河马、鳄鱼、鲶鱼、腹足动物及海龟，反映了当地的水生环境。K3′ 的大部分非水生动物，包括非洲产疣猴、大象、牛羚、薮羚及可能属于非洲野牛属的标本 [16]，以及猪科的河猪（丛林猪）和灭绝了的阴野猪，是与封闭环境一致的。藤鼠的出现显示出局部有潮湿沼泽的小环境。

人属化石（KNM-BK 63-67 与 KNM-BK 8518）是前人 [11,13,14] 在 GnJh-01 与 GnJh-19 遗址，从 K3 河流沉积到代表辫流系统（其中某些辫流河似乎已经流到了湖里）的西部获得。河流相 K3 沉积与湖相 K3′ 沉积是互层的，这显示水滨线随着湖面变化而产生推移。K3 与 K3′ 沉积中，化石的排列类似，提示出在中更新世的黑猩猩属与人属，是生活在或者至少是死在大致相似的环境里。综上所述，证据提示出这样一个局部树木繁茂的湖岸生活环境：湖泊淡水与盐碱交替变化，湖面上下涨落，季节性近岸河道，靠近淡水泉，气候具有半干旱特征。这些环境条件与今天在靠近巴林戈湖的水岸发现的情况没有什么不同，尽管浓密的人口毁灭了以前黑猩猩赖以为生的很多林地以及它们所属的动物群落。

黑猩猩标本至少有 3 颗牙齿，可能都来自同一个个体的。其中两颗分别是左、右上颌中门齿（I^1，编号 KNM-TH 45519 和 KNM-TH 45521）。其齿冠宽阔，竹片状，磨损中等，釉质薄（图 3）。舌侧结节大，两侧基底连有近远中凹，这是黑猩猩属的

thickness to the labiolingual profiles of the teeth, and clearly distinguishes them from known hominins. The mesial and distal marginal ridges are well formed. The distal corners of the incisal edges are slightly chipped and the labial enamel surfaces exhibit pre-mortem wear as well as slight post-mortem surface weathering. The roots have closed apices and are straight, conical and relatively short. The incisal edges and lingual tubercles exhibit dentinal exposure resulting from wear. Measurements of the specimens, with comparisons to those of extant species of *Pan*, are provided in Table 1. The upper incisors are nearly identical to those of modern *Pan* in all aspects of morphology except their shorter root length. The sub-parallel mesial and distal margins of the incisors bestow a quadrate, rather than triangular, outline to the crowns, a feature that among living chimpanzees is considered to be more common among living *P. troglodytes* than *P. paniscus*[17]. The enamel and cementum coverings are in good condition and the perikymata on the labial surfaces of the crowns and the periradicular striae on the lingual surfaces of the roots can be easily seen. Several of the perikymata near the cervices of the teeth are faintly incised, indicating mild enamel hypoplasia having occurred at about the age of 5 years[18]. The well-matched mesial interproximal wear facets of the Kapthurin Formation *Pan* incisors (KNM-TH 45519 and KNM-TH 45521), the comparable degree of wear on their incisal edges, and the continuity of the enamel hypoplasia on their crowns and the incremental markings on their roots suggest that the two teeth are antimeres.

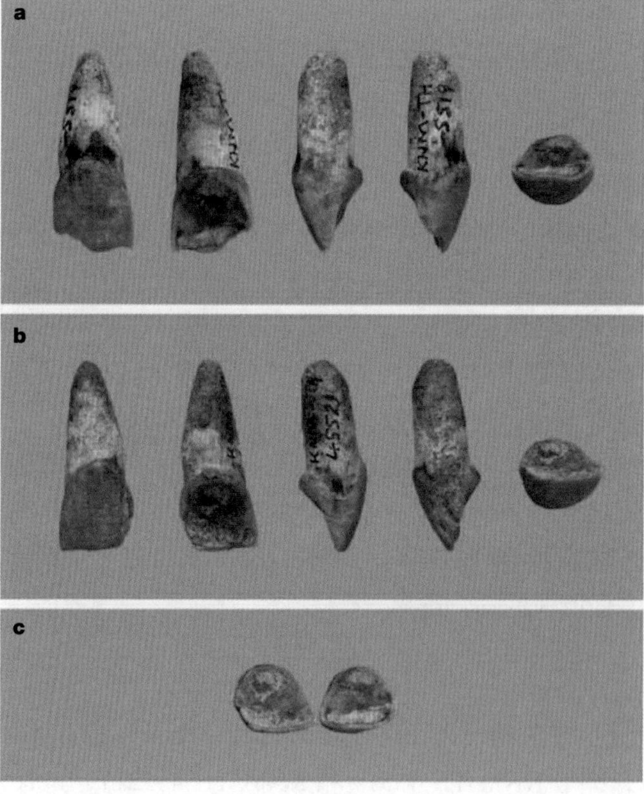

特征。较大的舌侧结节增加了齿冠的唇舌径，并使它们明显有别于已知的古人类。近远中边缘脊发育明显。门齿切缘的近中角有轻微缺口，齿冠唇侧表面有生前磨损和死后轻微风化痕迹。牙根直，尖端封闭，呈圆锥形，较短。门齿切缘和舌侧结节因为磨损而有齿质暴露。化石黑猩猩与其他现生黑猩猩属的测量数据对比列在了表1中。除了其牙根较短外，化石黑猩猩上中门齿的各方面形态特征与现生黑猩猩属几乎是一样的。门齿的近似平行的近远中边缘使牙冠排成的轮廓呈方形而不是三角形，这个特征在现生的黑猩猩属黑猩猩种中要比黑猩猩属倭黑猩猩种中更普遍[17]。釉质和牙骨质状态良好，牙冠唇侧外面釉面横纹和齿根舌侧表面的根周带清晰可见。部分靠近齿颈线的釉面横纹有轻微蚀刻，显示在大约5岁时牙釉质出现轻微发育不全[18]。黑猩猩属的门齿（KNM-TH 45519 和 KNM-TH 45521）相匹配的近中邻接面、程度相近的切缘磨耗、齿冠上对应的釉质发育不良、齿根上对应的生长线都提示这两颗牙齿源自同一个体的两侧。

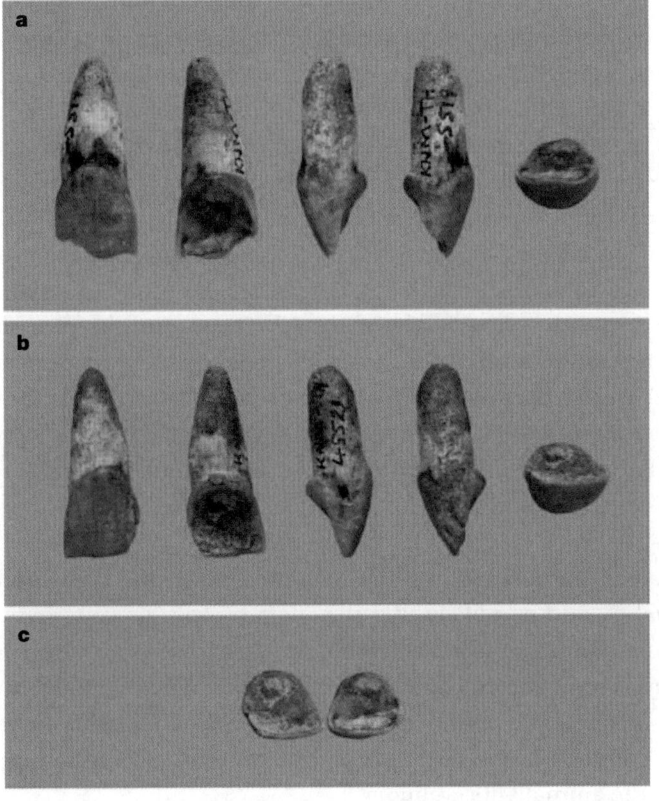

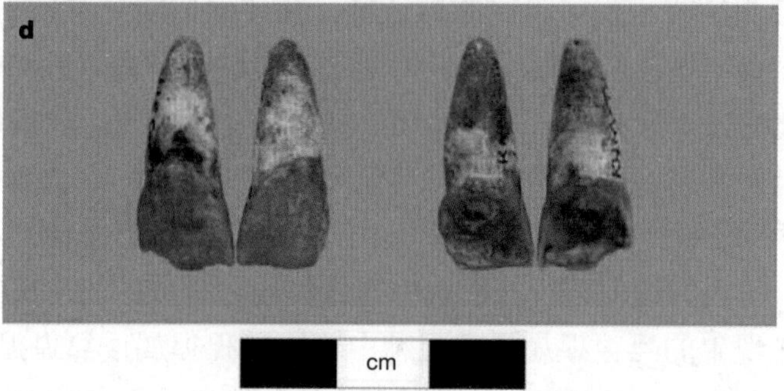

Fig. 3. Central upper incisors of *Pan* from the Kapthurin Formation, Kenya. **a**, KNM-TH 45519. From left to right: labial, lingual, mesial, distal and incisal views. **b**, KNM-TH 45521. Images are in the same sequence as for the previous specimen. **c**, Enlargement of the incisal edge of KNM-TH 45519 (left) and KNM-TH 45521 (right), showing the extreme thinness of the enamel characteristic of modern chimpanzees. **d**, Labial and lingual views of KNM-TH 45519 and KNM-TH 45521.

The third tooth is a lightly worn crown of a left upper permanent molar (KNM-TH 45520) (Fig. 4). It can be problematic to distinguish first from second upper molars in *Pan*, but we identify KNM-TH 45520 as an M^1, judging from the relatively large size of its hypocone, as this cusp is known to decrease in size from M^1 to M^3 (ref. 19). The Kapthurin Formation M^1 is an extremely low molar crown that has lost most of the enamel on its mesial and lingual faces due to breakage after fossilization. The enamel surfaces are pockmarked as a result of chemical and physical weathering. The paracone and metacone are of approximately equal heights and are separated by a sharply incised buccal groove. The hypocone is lower than either of the buccal cusps, but is relatively large and well defined. A shallow trigon basin is delimited by a weak and obliquely oriented postprotocrista (crista obliqua). A deep but short distal fovea lies between the postprotocrista and the low distal marginal ridge. Despite marring of the enamel surface, perikymata are visible on the buccal and distal faces of the paracone, but there is no evidence of enamel hypoplasia. The relative thinness of the enamel can be discerned on the broken mesial and lingual faces of the tooth. The extremely low height of the M^1 crown and the pronounced thinness of the enamel distinguish the tooth from those of known fossil or modern hominins. Among living chimpanzees, the presence of a well-expressed hypocone is more common in *P. troglodytes* than in *P. paniscus*[20]. A fourth tooth (KNM-TH 45522), the crown and proximal roots of a tooth that may be plausibly identified as an aberrant right upper third molar (M^3), will be described elsewhere and is not further discussed here.

The state of wear on the incisors and the M^1 conforms to the known sequence of dental emergence in *Pan*[19,21], and it is likely that they come from the same individual. If they do represent the same animal, its age at death can be estimated at approximately 7–8 years based on standards derived from captive animals[22] and known dental maturation schedules for mandibular molars[23]. The presence of linear enamel hypoplasia on the incisors, but not on the molars, is common in modern apes and seems to be related to nutritional stress that is experienced by the animal after weaning[24].

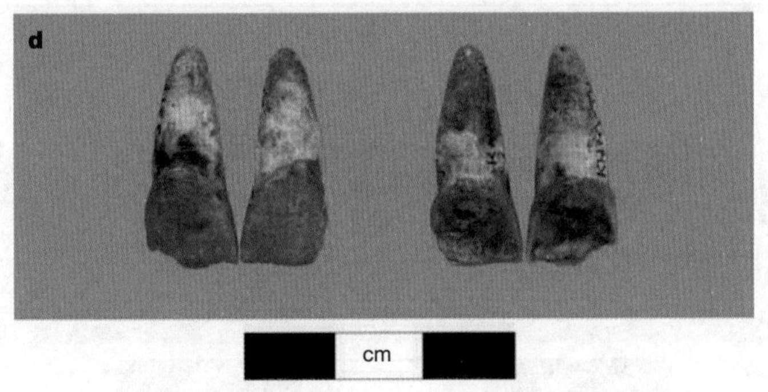

图 3. 来自肯尼亚卡普图林组黑猩猩属的上中门齿。**a**，KNM-TH 45519。从左至右：唇侧，舌侧，近中，远中，切面（咬合面）。**b**，KNM-TH 45521。该牙齿的图片排列顺序与 KNM-TH 45519 一致。**c**，KNM-TH 45519（左）与 KNM-TH 45521（右）放大的切缘，显示出现代黑猩猩的牙釉质非常薄的特征。**d**，KNM-TH 45519 与 KNM-TH 45521 的唇侧和舌侧。

第三颗牙齿是左侧上颌臼齿（KNM-TH 45520）（图 4）轻度磨损的齿冠。要把黑猩猩属上颌第一臼齿和第二臼齿分开是成问题的，但是我们把 KNM-TH 45520 视为上颌第一臼齿，这主要是因为这颗牙齿的次尖较大，而次尖从上颌第一臼齿到上颌第三臼齿是不断减小的（参考 19）。卡普图林组上颌第一臼齿具有较低的齿冠，其近中和舌侧的大部分釉质因石化后遭受的破坏而丧失。釉质表面因为受到物理与化学风化而有麻子坑。前尖和后尖高度接近，被一条较深的颊侧沟分开。次尖比前尖和后尖都低，但尺寸仍然较大、轮廓清晰。发育较弱的斜脊在近中侧定义了一个较浅的三角座凹。斜脊和较低的远中边缘脊之间有一个深但短的远中凹。尽管釉质表面被损伤，但前尖颊侧和远中侧仍保留较明显的釉面横纹，无釉质发育不良。从破损的牙齿近中面和舌侧面可以看出其釉质较薄。较低的齿冠和较薄的釉质使得化石黑猩猩这颗上颌第一臼齿与已知化石及现代人类区分开。在现生黑猩猩中，发育较好的次尖在黑猩猩属黑猩猩种上比在黑猩猩属倭黑猩猩种上更普遍[20]。第四颗牙齿（KNM-TH 45522），其牙冠和近齿冠的齿根部分似乎可以被认定为是一颗奇怪的上颌第三臼齿。这颗牙齿会在其他地方描述，在此不做更进一步的讨论。

门齿与上颌第一臼齿的磨耗级别符合黑猩猩属牙齿萌出顺序[19,21]，且它们很可能来自同一个个体。如果它们确实是代表同一个个体，那么根据从圈养黑猩猩[22]身上得来的标准和已知的下颌骨臼齿的牙齿成熟时间表，可以估计这个化石黑猩猩的死亡年龄约为 7～8 岁[23]。门齿上出现线状牙釉质发育不良，但臼齿上没有出现，这在现代猿类上很常见，而且似乎是与动物断奶后所经历的营养压力有关[24]。

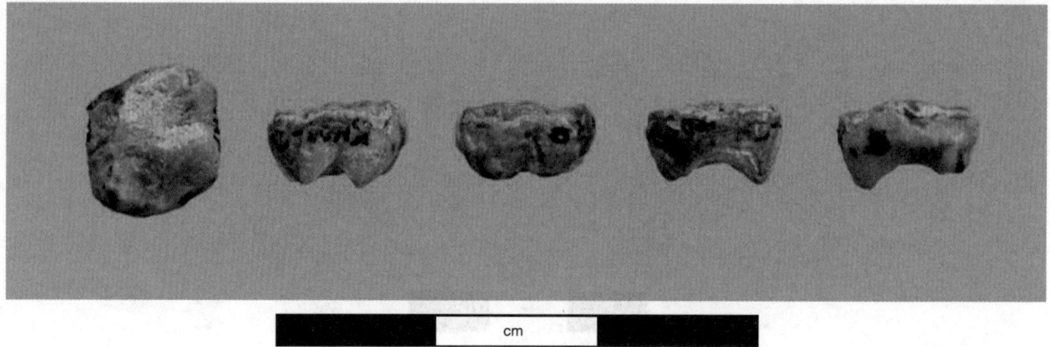

Fig. 4. Upper left first molar (KNM-TH 45520). From left to right: occlusal, labial, lingual, mesial and distal views. Note the thinness of the enamel on the broken mesial face of the paracone in the mesial view.

The morphology of the Kapthurin Formation teeth, especially the pronounced lingual tubercle on the incisors, the thickness of the bases of the incisors, the lowness of the molar crown, and the thinness of the enamel on all the teeth clearly supports their attribution to *Pan* rather than *Homo*. Specific diagnosis of isolated teeth within *Pan*, however, must be approached with caution, and for this reason we assign the Kapthurin Formation specimens to *Pan* sp. indet. Nonmetric characters that have been suggested as diagnostic criteria for *P. troglodytes*, such as a more quadrilateral outline shape to the upper central incisor crowns[17] and a better expressed hypocone on the maxillary molars[19,20], seem to suggest more similarity for the Kapthurin Formation fossils to *P. troglodytes* than to *P. paniscus*, but these features are variably expressed among the living species and subspecies of *Pan*[19,25]. Although mean tooth size is known to be significantly smaller in *P. paniscus* than in *P. troglodytes*[17,25,26], size ranges overlap (Table 1). Furthermore, apart from the present specimens, we lack a fossil record for the Pliocene and Pleistocene from which to assess past variability within the genus, and it is feasible that the Kapthurin Formation fossils represent members of an extinct lineage within the genus *Pan*.

Table 1. Dimensions of the Kapthurin Formation fossil chimpanzee teeth

Sample	Tooth	Mesiodistal dimension (mm)	Mesiodistal range (mm)	Buccolingual dimension (mm)	Buccolingual range (mm)
KNM-TH 45519	Right I^1	10.46	—	9.12	—
KNM-TH 45521	Left I^1	10.50	—	9.33	—
P. troglodytes (male)	I^1	12.6 (n = 14)	10.5–13.5	10.1 (n = 15)	9.0–11.3
P. troglodytes (female)	I^1	11.9 (n = 51)	10.0–13.4	9.6 (n = 50)	8.3–11.7
P. paniscus (male)	I^1	10.3 (n = 15)	8.9–11.9	7.9 (n = 15)	7.2–9.2
P. paniscus (female)	I^1	10.4 (n = 20)	9.0–11.5	7.6 (n = 21)	6.8–8.5
KNM-TH 45520	Left M^1	9.7 (estimate)	—	Damage prevents measurement	—
P. troglodytes (male)	M^1	10.3 (n = 19)	9.3–11.2	11.7 (n = 19)	10.7–13.2
P. troglodytes (female)	M^1	10.1 (n = 51)	9.0–11.9	10.9 (n = 50)	7.0–12.8

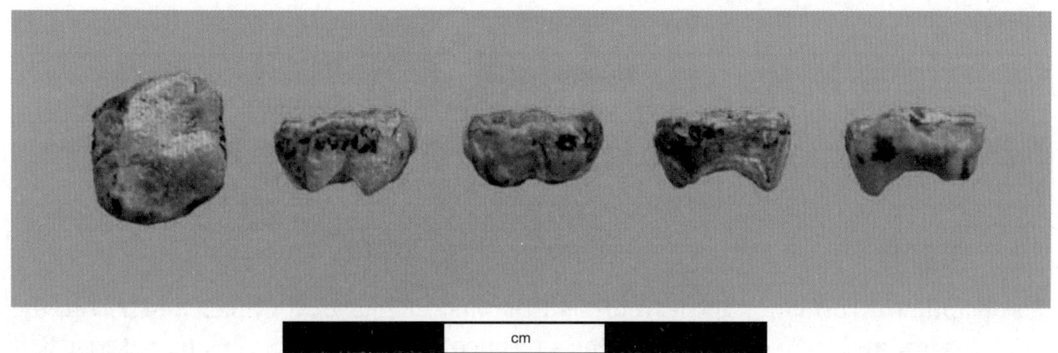

图 4. 左上第一臼齿（KNM-TH 45520）。从左至右：咬合面、颊侧、舌侧、近中侧、远中侧。注意在近中侧视图中前尖破裂的近中面上较薄的釉质。

卡普图林组牙齿的形态，尤其是门齿明显的舌侧结节，较厚的齿冠底部，低的臼齿齿冠，所有牙齿较薄的釉质，都明显地支持它们是属于黑猩猩属而不是人属成员。但是对黑猩猩属单颗牙齿的种一水平上的判断，必须要谨慎，因此我们将卡普图林组标本定为黑猩猩未定种。已经被提出来的非测量特征，比如上中门齿齿冠[17]的轮廓形状要更像四边形，上颌臼齿的次尖发育明显[19,20]，这似乎都提示卡普图林组化石更类似于黑猩猩属黑猩猩种，而与黑猩猩属倭黑猩猩种不同[19,25]。然而，这些特征在黑猩猩属的现生种和亚种上变异较大。虽然已经知道黑猩猩属倭黑猩猩种牙齿的平均尺寸要明显小于黑猩猩属黑猩猩种的[17,25,26]，但仍有重叠（表 1）。此外，除了本文报道的标本外，我们还缺乏用来评估该属过去变异程度的上新世和更新世的化石材料。卡普图林组化石可能代表了黑猩猩属一个已灭绝的谱系成员。

表 1. 卡普图林组黑猩猩牙齿化石的尺寸

标　本	牙齿	近远中径（毫米）	近远中径范围（毫米）	颊舌径（毫米）	颊舌径范围（毫米）
KNM-TH 45519	右 I^1	10.46	–	9.12	–
KNM-TH 45521	左 I^1	10.50	–	9.33	–
黑猩猩属黑猩猩种（雄性）	I^1	12.6 (n = 14)	10.5 ~ 13.5	10.1 (n = 15)	9.0 ~ 11.3
黑猩猩属黑猩猩种（雌性）	I^1	11.9 (n = 51)	10.0 ~ 13.4	9.6 (n = 50)	8.3 ~ 11.7
黑猩猩属倭黑猩猩种（雄性）	I^1	10.3 (n = 15)	8.9 ~ 11.9	7.9 (n = 15)	7.2 ~ 9.2
黑猩猩属倭黑猩猩种（雌性）	I^1	10.4 (n = 20)	9.0 ~ 11. 5	7.6 (n = 21)	6.8 ~ 8.5
KNM-TH 45520	左 M^1	9.7（估计）	–	因损坏而无法进行测量	–
黑猩猩属黑猩猩种（雄性）	M^1	10.3 (n = 19)	9.3 ~ 11.2	11.7 (n = 19)	10.7 ~ 13.2
黑猩猩属黑猩猩种（雌性）	M^1	10.1 (n = 51)	9.0 ~ 11.9	10.9 (n = 50)	7.0 ~ 12.8

Continued

Sample	Tooth	Mesiodistal dimension (mm)	Mesiodistal range (mm)	Buccolingual dimension (mm)	Buccolingual range (mm)
P. paniscus (male)	M^1	8.5 (n = 7)	7.9–9.4	9.5 (n = 6)	9.2–10.4
P. paniscus (female)	M^1	8.3 (n = 6)	7.6–8.8	9.7 (n = 6)	9.3–10.4

Comparative dimensions are given for modern *P. troglodytes* and *P. paniscus* from ref. 19.

The Kapthurin Formation fossils represent the first unequivocal evidence of *Pan* in the fossil record, and they demonstrate the presence of chimpanzees in the eastern Rift Valley of Kenya, ~600 km east of the limit of their current range (Fig. 1). The Rift Valley clearly did not pose a physiographical or ecological barrier to chimpanzee occupation. Chimpanzee habitat is now highly fragmented, in part by human activities, but in historic times chimpanzees ranged over a wide belt of equatorial Africa from southern Senegal to western Uganda and Tanzania (Fig. 1). Although much of this region is rainforest, chimpanzees currently also occupy dry forest, woodland and dry savannah, particularly near the eastern edge of their range[27-29]. The modern Baringo region ecosystem is a mosaic of semi-arid *Acacia* bushland and riverine woodland, with a significant substratum of perennial and annual grasses[30]. The Tugen Hills palaeosol carbon isotope record indicates that the woodland and grassland components of the vegetation have been present there from 16 Myr[30]. Representatives of both *Homo* and *Pan* are present in the same stratigraphic interval of the Kapthurin Formation at sites only ~1 km apart, and faunal data suggest that they occupied broadly similar environments in the Middle Pleistocene. This evidence shows that in the past chimpanzees occupied regions in which the only hominoid inhabitants were thought to have been members of the human lineage. Now that chimpanzees are known to form a component of the Middle Pleistocene fauna in the Rift Valley, it is quite possible that they remain to be recognized in other portions of the fossil record there, and that chimpanzees and hominins have been sympatric since the time of their divergence.

(**437**, 105-108; 2005)

Sally McBrearty[1] & Nina G. Jablonski[2]

[1] Department of Anthropology, University of Connecticut, Box U-2176, Storrs, Connecticut 06269, USA

[2] Department of Anthropology, California Academy of Sciences, 875 Howard Street, San Francisco, California 94103, USA

Received 31 January; accepted 4 July 2005.

References:

1. Ruvolo, M. E. Molecular phylogeny of the hominoids: inferences from multiple independent DNA sequence data sets. *Mol. Biol. Evol.* **14**, 248-265 (1997).
2. Darwin, C. *The Descent of Man and Selection in Relation to Sex* (John Murray, London, 1871).
3. Washburn, S. L. in *Changing Perspectives on Man* (ed. Rothblatt, B.) 193-201 (Univ. Chicago Press, Chicago, 1968).
4. Kortlandt, A. *New Perspectives on Ape and Human Evolution* (Univ. Amsterdam, Amsterdam, 1972).
5. Pilbeam, D. & Young, N. Hominoid evolution: synthesizing disparate data. *C. R. Palevol.* **3**, 305-321 (2004).
6. Coppens, Y. East side story: the origin of mankind. *Sci. Am.* **270**, 88-95 (1994).
7. Martyn, J. *The Geologic History of the Country Between Lake Baringo and the Kerio River, Baringo District, Kenya* (PhD dissertation, Univ. London, 1969).
8. Tallon, P. in *Geological Background to Fossil Man* (ed. Bishop, W. W.) 361-373 (Scottish Academic Press, Edinburgh, 1978).

续表

标　本	牙齿	近远中径（毫米）	近远中径范围（毫米）	颊舌径（毫米）	颊舌径范围（毫米）
黑猩猩属倭黑猩猩种（雄性）	M^1	8.5 (n = 7)	7.9 ~ 9.4	9.5 (n = 6)	9.2 ~ 10.4
黑猩猩属倭黑猩猩种（雌性）	M^1	8.3 (n = 6)	7.6 ~ 8.8	9.7 (n = 6)	9.3 ~ 10.4

参考文献 19 给出了现生黑猩猩属黑猩猩种和黑猩猩属倭黑猩猩种的比较尺寸规格。

卡普图林组标本是第一个明确的黑猩猩属的化石证据，它们证明了在肯尼亚东非大裂谷东部，也就是在它们目前范围界限以东约 600 千米的地方，曾经生活着黑猩猩（图 1）。东非大裂谷在地形上、生态上对黑猩猩的生活栖息无疑都不是一个障碍。黑猩猩的栖息地由于人类活动而变得很零散，但在历史上，黑猩猩分布在非洲从塞内加尔南部到乌干达西部及坦桑尼亚这一广阔赤道地带（图 1）。虽然这个区域大部分为雨林，但黑猩猩现也生活在干燥森林、林地及干燥热带稀树草原，特别是靠近它们分布范围的东部边缘 [27-29]。现代巴林戈区的生态系统，由半干旱金合欢矮灌丛地与有着多年生和一年生草的显著林下层的河边林地嵌合而成 [30]。图根丘陵古土壤碳同位素记录显示，该地区植被的林地与草地成分已经存在了 1,600 万年 [30]。典型的人属和黑猩猩属成员出现在卡普图林组同一地层范围，遗址仅相距大约 1 千米，动物群种类提示它们在中更新世所处环境相似。该证据显示在过去黑猩猩占据的区域中人科居民仅有人类世系成员。现在认为黑猩猩构成了东非大裂谷中更新世动物群的一个组成部分，很有可能在化石记录的其他部分里，它们可能会被发现；而且黑猩猩和古人类自从分离后，分布区一直是重叠的。

（田晓阳 翻译；邢松 审稿）

9. McBrearty, S., Bishop, L. C. & Kingston, J. Variability in traces of Middle Pleistocene hominid behaviour in the Kapthurin Formation, Baringo, Kenya. *J. Hum. Evol.* **30**, 563-580 (1996).
10. McBrearty, S. in *Late Cenozoic Environments and Hominid Evolution: a Tribute to Bill Bishop* (eds Andrews, P. & Banham, P.) 143-156 (Geological Society, London, 1999).
11. McBrearty, S. & Brooks, A. The revolution that wasn't: a new interpretation of the origin of modern human behaviour. *J. Hum. Evol.* **39**, 453-563 (2000).
12. Deino, A. & McBrearty, S. $^{40}Ar/^{39}Ar$ chronology for the Kapthurin Formation, Baringo, Kenya. *J. Hum. Evol.* **42**, 185-210 (2002).
13. Leakey, M., Tobias, P. V., Martyn, J. E. & Leakey, R. E. F. An Acheulian industry with prepared core technique and the discovery of a contemporary hominid at Lake Baringo, Kenya. *Proc. Prehist. Soc.* **35**, 48-76 (1969).
14. Wood, B. A. & Van Noten, F. L. Preliminary observations on the BK 8518 mandible from Baringo, Kenya. *Am. J. Phys. Anthropol.* **69**, 117-127 (1986).
15. Renaut, R. W., Tiercelin, J.-J. & Owen, B. in *Lake Basins Through Space and Time* (eds Gierlowski-Kordesch, E. H. & Kelts, K. R.) 561-568 (Am. Assoc. Petrol. Geol., Tulsa, Oklahoma, 2000).
16. Bishop, L. C., Hill, A. P. & Kingston, J. in *Late Cenozoic Environments and Hominid Evolution: a Tribute to Bill Bishop* (eds Andrews, P. & Banham, P.) 99-112 (Geological Society, London, 1999).
17. Johanson, D. C. Some metric aspects of the permanent and deciduous dentition of the pygmy chimpanzee (*Pan paniscus*). *Am. J. Phys. Anthropol.* **41**, 39-48 (1974).
18. Dean, M. C. & Reid, D. J. Perikymata spacing and distribution on hominid anterior teeth. *Am. J. Phys. Anthropol.* **116**, 209-215 (2001).
19. Swindler, D. R. *Primate Dentition: An Introduction to the Teeth of Non-Human Primates* (CUP, Cambridge, 2002).
20. Kinzey, W. G. in *The Pygmy Chimpanzee* (ed. Susman, R. L.) 65-88 (Plenum, New York, 1984).
21. Smith, B. H., Crummett, T. L. & Brandt, K. L. Ages of eruption of primate teeth: a compendium for aging individuals and comparing life histories. *Yearb. Phys. Anthropol.* **37**, 177-232 (1994).
22. Kuykendall, K. L., Mahoney, C. J. & Conroy, G. C. Probit and survival analysis of tooth emergence ages in a mixed-longitudinal sample of chimpanzees (*Pan troglodytes*). *Am. J. Phys. Anthropol.* **89**, 379-399 (1992).
23. Anemone, R. L., Watts, E. S. & Swindler, D. R. Dental development of known-age chimpanzees, *Pan troglodytes* (Primates, Pongidae). *Am. J. Phys. Anthropol.* **86**, 229-241 (1991).
24. Skinner, M. F. & Hopwood, D. Hypothesis for the causes and periodicity of repetitive linear enamel hypoplasia in large, wild African (*Pan troglodytes* and *Gorilla gorilla*) and Asian (*Pongo pygmaeus*) apes. *Am. J. Phys. Anthropol.* **123**, 216-235 (2004).
25. Uchida, A. *Craniodental Variation Among the Great Apes* (Harvard Univ. Peabody Mus., Cambridge, Massachusetts, 1996).
26. Johanson, D. C. *An Odontological Study of the Chimpanzee with Some Implications for Hominoid Evolution* (PhD dissertation, Univ. Chicago, 1974).
27. Kormos, R., Boesch, C., Bakarr, M. I. & Butynski, T. M. *West African Chimpanzees: Status Survey and Conservation Action Plan* (IUCN Publication Unit, Cambridge, 2003).
28. McGrew, W. C., Baldwin, P. J. & Tutin, C. E. G. Chimpanzees in a hot, dry and open habitat: Mt. Assirik, Senegal. *J. Hum. Evol.* **10**, 227-244 (1981).
29. McGrew, W. C., Marchant, L. F. & Nishida, T. *Great Ape Societies* (CUP, Cambridge, 1996).
30. Kingston, J. D., Marino, B. & Hill, A. P. Isotopic evidence for Neogene hominid palaeoenvironments in the Kenya Rift Valley. *Science* **264**, 955-959 (1994).

Acknowledgements. We wish to thank B. Kimeu, N. Kanyenze and M. Macharwas, who found the chimpanzee fossils reported here. Research in the Kapthurin Formation is carried out with the support of an NSF grant to S.M., and under a research permit from the Government of the Republic of Kenya and a permit to excavate from the Minister for Home Affairs and National Heritage of the Republic of Kenya. Both of these are issued to A. Hill and the Baringo Paleontological Research Project, an expedition conducted jointly with the National Museums of Kenya. We also thank personnel of the Departments of Palaeontology, Ornithology and Mammalogy of the National Museums of Kenya, Nairobi; A. Zihlman; and Y. Hailie-Selassie, L. Jellema and M. Ryan for curation and access to specimens. We express gratitude to A. Hill for his comments on the manuscript. We also thank G. Chaplin for drafting Fig. 1, B. Warren for preparing Figs 3 and 4, and A. Bothell for help with submission of the figures. We are grateful to J. Kelley, J. Kingston, M. Leakey, R. Leakey, C. Tryon, A. Walker and S. Ward for discussions. We thank G. Suwa for his remarks.

Author Information. Reprints and permissions information is available at npg.nature.com/reprintsandpermissions. The authors declare no competing financial interests.

Correspondence and requests for materials should be addressed to S.M. (mcbrearty@uconn.edu).

Molecular Insights into Human Brain Evolution

R. S. Hill and C. A. Walsh

Editor's Note

The knowledge of the human genome is expected to have important consequences throughout biology—not least, to offer insights into how humans diverged from our ancestral species at the genetic level. This paper by two neuroscientists at Harvard Medical School uses information from the human genome sequence to make inferences about the evolution of the human brain. It argues that there has been significant genetic change relatively recently—that is, since the emergence of *Homo sapiens*—particularly in a gene called *FOXP2* associated with articulation and speech in humans. *FOXP2* is often now popularly called a "language gene", although that designation is far too simplistic.

Rapidly advancing knowledge of genome structure and sequence enables new means for the analysis of specific DNA changes associated with the differences between the human brain and that of other mammals. Recent studies implicate evolutionary changes in messenger RNA and protein expression levels, as well as DNA changes that alter amino acid sequences. We can anticipate having a systematic catalogue of DNA changes in the lineage leading to humans, but an ongoing challenge will be relating these changes to the anatomical and functional differences between our brain and that of our ancient and more recent ancestors.

SANTIAGO Ramon y Cajal, widely regarded as the founder of modern neuroscience, recognized as early as the turn of the twentieth century that the human brain was not just larger than that of our ancestors, but it differed in its circuitry as well. Over the course of the last century these differences have been extensively studied at a histological level, although specifying the exact changes that distinguish the human brain has been elusive.

> "The opinion generally accepted at that time that the differences between the brain of [non-human] mammals (cat, dog, monkey, etc) and that of man are only quantitative, seemed to me unlikely and even a little offensive to the human dignity... My investigations showed that the functional superiority of the human brain is intimately bound up with the prodigious abundance and unusual wealth of forms of the so-called neurons with short axon." (Ref. 1, translated by J. DeFelipe).

Comparative Differences in Brain Structure

Understanding the genetic changes that distinguish our brain from that of our ancestors starts with defining the key structural and functional differences between the human brain and that of other primates. Our brain is roughly three times the size of the chimpanzee

从分子角度洞察人类大脑的演化

希尔，沃尔什

编者按

人类基因组的知识预计将在整个生物学领域产生重要影响，尤其是，它将为我们了解在基因层面上人类如何与我们的祖先物种分化提供见解。哈佛医学院的两位神经科学家利用人类基因组序列的信息，对人类大脑的演化做出了推断。该研究认为，自智人出现以来，人类最近发生了重大的基因变化，特别是在一种被称为*FOXP2*的基因中，这种基因与人类的发音和说话有关。*FOXP2*现在通常被称为“语言基因”，尽管这个称谓过于简单了。

基因组结构和序列相关知识的迅速发展使得我们能够用新的手段分析与人类大脑和其他哺乳动物大脑的差异相关的特定DNA改变。最近的研究将演化改变归因于信使RNA和蛋白质表达水平，以及会改变氨基酸序列的DNA变化。我们可以预期将来在导致现代人产生的谱系中会有DNA变化的系统编目，而我们的大脑和我们远古的以及更近的祖先的大脑在解剖和功能上存在差异，我们正在面临的一项挑战就是如何将这些DNA的变化与这些差异联系起来。

圣地亚哥·拉蒙·卡哈尔，被认为是现代神经科学的创始人，早在十九和二十世纪之交就认识到人类的大脑不仅仅是比我们祖先的大，而且脑回路也不同。虽然在上个世纪这些差异已经在组织学层面上得到了广泛的研究，但是确定能够区别人脑的确切的改变仍然是难乎其难。

“在那个年代人们普遍接受的观点是，［非人类］哺乳动物（猫、狗、猴子等）与人类大脑之间的不同仅仅是数量上的差异，在我看来这不太可能，甚至感觉是对人类尊严的冒犯……我的研究表明人类大脑的功能优越性，是与数量惊人、种类非同寻常的具有短轴突的神经元紧密相关的。”（参考文献1，德费利佩译）

大脑结构的比较解剖差异

要理解将我们的大脑同祖先的大脑区分开来的遗传改变，需要首先确定人类的大脑与其他灵长类动物的大脑之间关键结构和功能上的不同。在七百万年至八百万

brain, our nearest living relative, from which we diverged 7–8 million years ago, and about twice the size of pre-human hominids living as recently as 2.5 million years ago[2]. The increased size particularly affects the cerebral cortex, the largest brain structure and seat of most higher cognitive functions. The cortex is a multi-layered sheet that is smooth in rodents, but folded in mammals with larger cortices (Fig. 1), allowing more cortex to squeeze into the limited volume of the head.

The enlarged cortex of great apes reflects a longer period of neuronal formation during pre-natal development, so that each dividing progenitor cell undergoes more cell cycles before stopping cell division[3]. Cortical progenitors undergo 11 rounds of cell division in mice[4], at least 28 in the macaque[3], and probably far more in human. In addition to making a larger cortex, the longer period of neurogenesis adds novel neurons to the cortex, so that the cortical circuit diagram differs between primates and other mammals (Fig. 1). Upper cortical layers, generated late in neurogenesis, are over-represented in the primate cerebral cortex, especially in humans[5]. Additionally, special cell types, such as spindle cells (specialized, deep-layer neurons[6]), are unique to primates. The upper-layer neurons that are so unusually common in great apes represent either locally projecting neurons—the "neurons with short axon"of Cajal—or neurons that connect the cortex to itself, but do not project out of the cortex (Fig. 1).

The cerebral cortex shows remarkable local specialization, reflected as functionally distinct cortical "areas" that are essentially a map of the behaviours and capabilities most essential to each species. For example, whereas rodents show relatively larger areas that respond to odours and sensation from the whiskers, they have small areas subserving their limited vision. In contrast, primates are highly visual, with more than a dozen distinct functional areas analysing various features of a visual scene. Recent work has compared functionally homologous visual regions between humans and macaques, suggesting that some areas are quite similar, whereas other visual areas have been either added or greatly modified during the course of evolution[7]. Primates also have particularly large areas of the frontal lobes anterior to the motor cortex (prefrontal cortex), whereas prefrontal cortex is tiny in non-primates. Prefrontal areas regulate many social behaviours and are preferentially enlarged in great apes. Although it has long been thought that prefrontal cortex is especially enlarged in humans, recent work suggests that other great apes may have equivalent proportions of prefrontal cortex[8].

The human cerebral cortex also shows functional asymmetries, with most of us being right handed and having language function preferentially localized in the left hemisphere. Chimpanzees do not show such strong asymmetry in handedness[9], although their brains show some asymmetries in frontal and temporal lobes (which correspond to language areas in humans)[10]. Recent evidence suggests that the left–right asymmetries of the human cerebral cortex are accompanied by asymmetric gene expression during early fetal development[11], although it is not known whether asymmetries of gene expression are seen in non-human primates. There is some evidence from fossil skulls for cortical asymmetry in human predecessors as well[12].

年前我们与关系最近的现生亲戚黑猩猩发生分化，我们大脑的大小是其3倍左右，而相比于近至250万年前的现代人之前的古人类，我们的大脑是其大脑的2倍[2]。大脑尺寸的增加尤其会影响大脑皮层，后者是大脑中最大的结构，也是大部分高级认知功能的所在地。大脑皮层在啮齿动物中是光滑的多层薄片，但在具有较大面积大脑皮层的哺乳动物中是折叠的(图1)，这样可以将更多的皮层塞入有限的头部空间。

大猿类大脑皮层的增大反映了出生前发育过程中一个较长的神经元形成时期，因此每个将要分裂的祖细胞在细胞分裂停止之前都要经历更多的细胞周期[3]。大脑皮层祖细胞在小鼠中要经历11轮细胞分裂[4]，在猕猴中至少28轮[3]，而在人类中可能更多。除了形成一个更大的大脑皮层之外，更长的神经发生期还会在大脑皮层上增加新的神经元，因此灵长类动物和其他哺乳动物的大脑皮层回路图是不同的(图1)。在神经发生晚期生成的上部皮层层次，在灵长类尤其是人类的大脑皮层中占有优势的比例[5]。除此之外，特殊的细胞类型，如梭形细胞(一类特化的深层神经元[6])，是灵长类独有的。在大猿中极其常见的上层神经元，要么是局部延伸的神经元，即卡哈尔的"短轴突神经元"，要么是将自己连接到皮层但并不延伸到皮层之外的神经元(图1)。

大脑皮层有显著的区域特化，反映为功能截然不同的皮层"区域"，本质上就是对于每个物种最为根本的行为和能力的映射。例如，尽管啮齿类对气味和来自胡须的感觉做出反应的区域相对较大，但服务于其有限视力的区域却很小。相比之下，灵长类更加依赖视觉，有十几处不同的功能区域分析视觉场景中各种不同的特征。最近的研究比较了人和猕猴功能上同源的视觉区域，结果表明，尽管有些区域十分相似，但其他的视觉区域在演化过程中要么被添加要么被大幅度修改[7]。在运动皮层(前额叶皮层)前方灵长类也具有特别大的额叶区域，而在非灵长类中前额叶皮层很小。前额叶区调控着许多社会行为，在大猿当中该区域得到了优先增大。尽管长期以来人们一直认为，人的前额叶皮层尤其增大，但是最近的研究表明，其他的大猿可能具有相同比例的前额叶皮层[8]。

人类大脑皮层也表现出功能的不对称性，就像我们大部分人都是右利手，而语言功能被优先置于左半球。在使用右手或者左手的习惯上，黑猩猩在利手方面并没有表现出强烈的不对称性[9]，尽管它们的大脑在额叶和颞叶(与人类的语言区域相对应)上也表现出一些不对称性[10]。新的证据表明，人类大脑皮层的左右不对称性与胎儿早期发育过程中基因的不对称表达有关[11]，尽管在非人灵长类中还不知道是否也存在基因的不对称表达。也有来自化石头骨的证据表明在人类先祖中也存在皮层不对称性[12]。

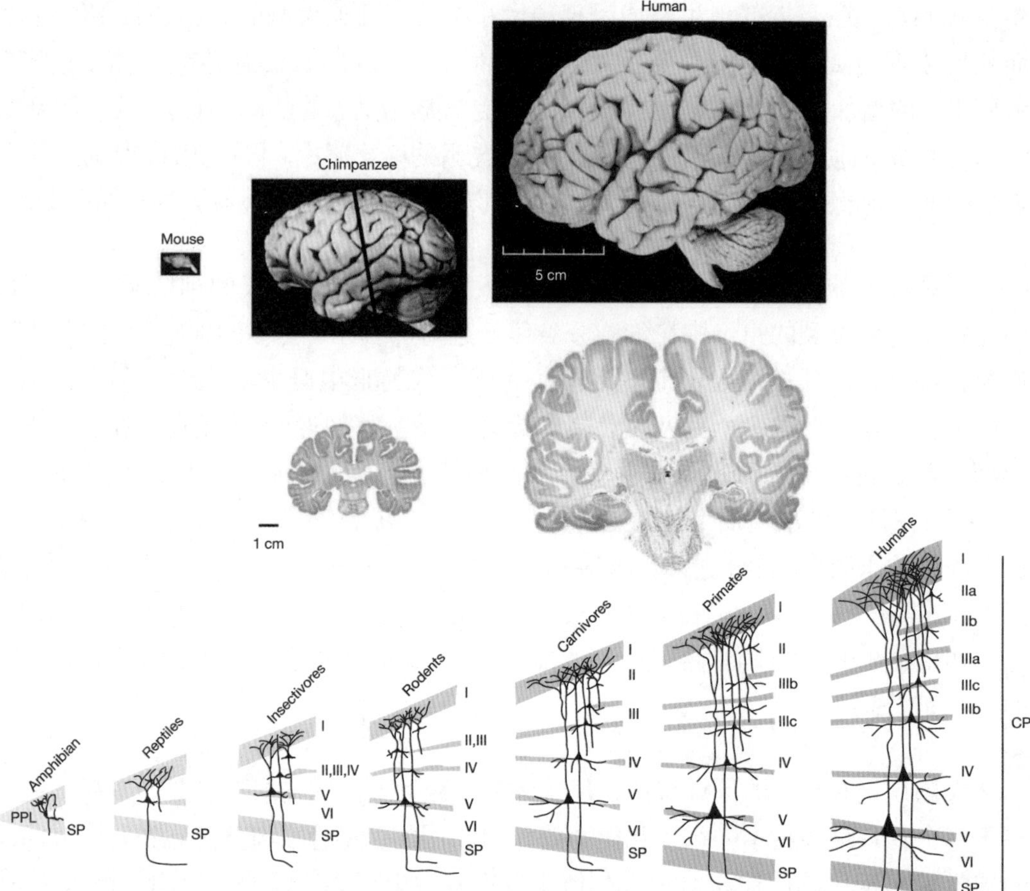

Fig. 1. Differences in cerebral cortical size are associated with differences in the cerebral cortex circuit diagram. The top panel shows side views of the brain of a rodent (mouse), a chimpanzee and a human to show relative sizes. The middle panel shows a cross-section of a human and chimpanzee brain, with the cellular composition of the cortex illustrated in the bottom panel (adapted from ref. 5). The cerebral cortex derives from two developmental cell populations: the primordial plexiform layer (PPL) and the cortical plate (CP). The primordial plexiform layer seems to be homologous to simple cortical structures in Amphibia and Reptilia, and appears first temporally during mammalian brain development. The cortical plate develops as a second population that splits the primordial plexiform layer into two layers (layer I at the top and the subplate (SP) at the bottom; numbering follows the scheme of ref. 31). Cortical-plate-derived cortical layers are added developmentally from deeper first (VI, V) to more superficial (III, II) last. Cortical-plate-derived cortical layers are progressively elaborated in mammals with larger brains (for example, insectivores have a single layer II/III/IV that is progressively subdivided into II, III, IV, then IIa, IIb, and so on), so that humans have a larger proportion of these late-derived neurons, which project locally or elsewhere within the cortex. Images from the top and middle panels are from the Comparative Brain Atlas (http://www.brainmuseum.org).

Evolutionary Mechanisms

What sorts of genetic changes underlie diverse brain shape and size? Approaches to this question have come increasingly into focus, although the answers themselves await further

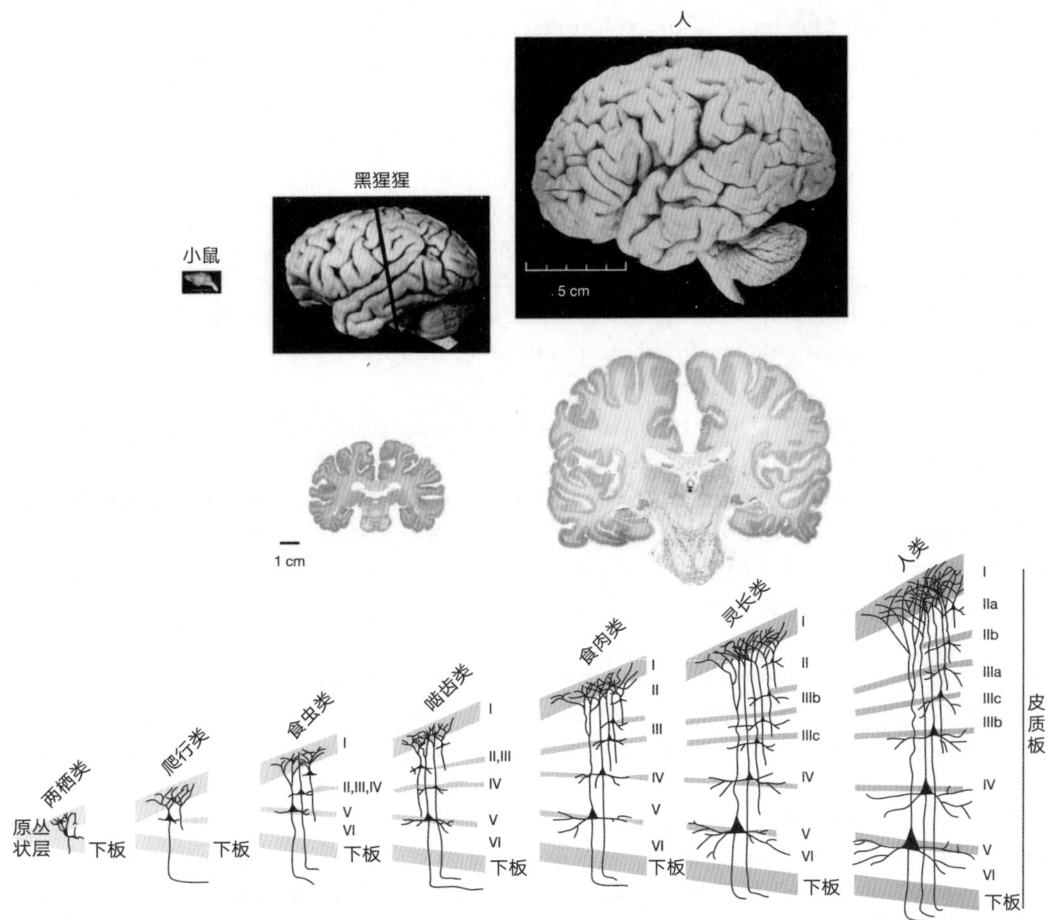

图 1. 大脑皮层大小的不同与大脑皮层回路图的不同相关。上面一排为啮齿动物（小鼠）、黑猩猩以及人类大脑的侧视图，显示相对大小。中间一排为人和黑猩猩大脑的截面，其皮层细胞组成在下面一排图中示意说明（修改自参考文献 5）。大脑皮层源自两种发育细胞群体：原丛状层（PPL）和皮质板（CP）。原丛状层似乎与两栖纲和爬行纲中简单的皮层结构同源，并且在哺乳动物大脑发育过程中首先出现。皮质板作为第二个细胞群体发育，将原丛状层分成两层（在顶部的 I 层和在底部的下板（SP）；编号遵循参考文献 31 的方案）。源自皮质板的皮层层次随着发育从较深层的最初（VI、V）到较浅层的最后（III、II）被逐渐添加。源自皮质板的皮层层次在具有较大大脑的哺乳动物中被逐步精心打造（例如，食虫类有一个单层的 II/III/IV，随着演化过程会被逐步细分成 II、III、IV，然后再分成 IIa、IIb 等），因此人类拥有很大比例的这些晚期衍生的神经元，它们在皮层中局部延伸或向别处延伸。上面和中间一排的图片引自大脑对比图集（http://www.brainmuseum.org）。

演化机制

什么类型的基因改变决定了大脑形态和大小的不同呢？尽管答案本身仍然需要进一步的研究，但是解决问题的途径已经引起越来越多的关注。演化蜕变的三

work. Three major mechanisms of evolutionary changes include: (1) addition or subtraction of entire genes to or from the genome; (2) alterations in levels or patterns of gene expression; and (3) alterations in the coding sequence of genes. Recent evidence suggests roles for all of these mechanisms.

The recent completion of sequencing the chimpanzee genome emphasizes the highly similar composition of the human and chimpanzee genomes[13]. There is evidence for inactivation of genes, especially many olfactory receptor genes, by their conversion into pseudogenes[14]. However, there is currently little evidence to suggest that the addition of novel genes is a major mechanism in human brain evolution[13].

Recent studies suggest that human brain evolution is associated with changes in gene expression specifically within the brain as opposed to other tissues such as liver. A few studies suggest more-accelerated gene expression changes in the brain along the human lineage compared with the chimpanzee lineage[15]. Although the studies differ in design and principal conclusions, they share support for an increase in expression level in a subset of brain-expressed genes in the lineage leading to humans[16,17].

There is also accumulating evidence that some neural genes underwent important changes in their coding sequence over the course of recent brain evolution, although the proportion of neural genes that were targets of positive selection is still in debate. Genes strongly influenced by natural selection can be identified by comparing DNA changes that occur in different, closely related species, for example in different primate species. Synonymous DNA substitutions do not alter the amino acid sequence because they occur at degenerate sites in the codon (such as a CGT to CGG change, as both codons encode arginine). Because synonymous changes do not alter the biochemical properties of the encoded protein, they are usually evolutionarily neutral. In contrast, non-synonymous DNA changes alter the amino acid sequence. The vast majority of non-synonymous DNA changes represent disabling mutations that cause disease, hence decreasing the fitness of the organism, and so most non-synonymous DNA changes are subject to negative, or purifying, selection. In contrast, on rare occasions non-synonymous DNA changes might make the protein work slightly better, hence increasing the fitness of the organism and becoming subject to positive selection (that is, advantageous changes propagated to future generations). A ratio of non-synonymous (K_A) to synonymous (K_S) changes $\ll 1$ is typical of most proteins where change is detrimental[18]; rare proteins show $K_A/K_S > 1$, which can indicate positive selection.

In order to test whether genes expressed in the brain were frequent targets of positive selection in primates, one study[19] analysed 200 brain-expressed genes, comparing them to 200 widely expressed genes. They compared K_A/K_S ratios between rats and mice and between humans and macaque monkeys. They concluded that genes involved in brain development or function had a higher tendency to be under positive selection between macaques and humans than between mice and rats. In contrast, systematic surveys of K_A/K_S ratios across much larger numbers of genes between chimpanzees and humans failed to show that neural genes, as a group, have higher K_A/K_S ratios than genes expressed

个主要机制如下：(1) 从基因组中增加或者减除一个完整的基因；(2) 基因表达水平或者模式上的改变；(3) 基因编码序列的改变。新的证据表明以上所有这些机制都在起作用。

最近黑猩猩的基因组测序工作完成进一步表明人的基因组与黑猩猩基因组组成高度相似[13]。其中也存在基因失效转变成假基因的证据，尤其是许多嗅觉受体基因[14]。然而，目前还没有证据表明新基因的增加是人类大脑演化的主要机制[13]。

最近的研究表明人类大脑演化尤其与大脑内的基因表达变化有关，而与肝脏这样的其他的组织截然相反。一些研究发现与黑猩猩演化谱系相比，在人的演化谱系中大脑中的基因表达变化更加快速[15]。尽管这些研究在目的和主要结论上有所不同，但是他们都认为，在通往现代人的谱系中，大脑表达基因的一个子集的表达水平有所增加[16,17]。

越来越多的证据表明，一些神经基因在大脑最近的演化过程中编码序列发生了重大的变化，尽管作为正选择目标的神经基因所占的比例仍有争议。通过比较不同的或者紧密相关的物种中的 DNA 变化，就可以识别受自然选择强烈影响的基因，例如在不同灵长类物种之间。同义 DNA 替换并不会改变氨基酸序列，因为它们出现在密码子的简并位点(例如，CGT 变成 CGG，但两个密码子都编码精氨酸)。因为同义改变并不改变被编码蛋白质的生物化学性质，所以它们在演化上通常也是中性的。与之相反，非同义 DNA 变化会使氨基酸序列发生变化。绝大多数的非同义 DNA 变化都是失效突变，会引发疾病，从而降低生物的适应性，因此大部分非同义 DNA 变化往往会遭到负面选择或肃清。相比之下，在极少数情况下，非同义 DNA 变化可能会使蛋白质起到稍好的作用，从而提高生物的适应性，受到正选择(也就是说，这种有利的改变会传播给将来的世代)。对于大多数蛋白质来讲，非同义变化(K_A)与同义变化(K_S)的比值通常远小于 1，这些变化是有害的[18]；很少蛋白质显示 K_A/K_S 大于 1，这表明是正选择。

为了检验灵长类大脑中表达的基因是否是正选择的目标，一项研究[19]分析了 200 个大脑表达基因，并将它们同 200 个广泛表达的基因进行了比对。他们比较了大鼠和小鼠之间以及人和猕猴之间的 K_A/K_S 比率。他们得出结论，大脑发育或功能所涉及的基因，在猕猴和人之间比在小鼠和大鼠之间有更高的倾向受到正选择。相比之下，黑猩猩和人之间的更大数量基因的 K_A/K_S 比值的系统考察，并没有显示在这两个物种之间，作为一组的神经基因比大脑之外表达的基因有更高的 K_A/Ks 比值[20,21]。对 50 个 K_A/K_S 比值最高的基因进行分析后发现，它们中只有出奇的极少数

outside of the brain between these two species[20,21]. Analysis of the top 50 genes with the highest K_A/K_S ratios showed surprisingly few with known essential roles in the brain[20]. Analysis of the chimpanzee genome confirms that neural genes, as a group, have much lower average K_A/K_S ratios than genes expressed outside of the brain[13]. However, the more recent study suggested that a substantial fraction of the genes with the highest K_A/K_S ratios had roles in brain development or function[13]. These studies are most easily reconciled by suggesting that a small subset of neural genes may be targets for positive selection (see below), whereas neural genes as a whole are subject to intense negative selection due to the severe disadvantages conferred by mutations that disrupt brain function.

Correlation of Genetic Evolution with Human Brain Function

Whereas genome-wide analyses systematically highlight targets of positive genetic selection in the human lineage, there has been great interest in a subset of human genes that show positive evolutionary selection, and for which correlations between evolutionary patterns and gene function in humans are possible. For example, mutant alleles of *FOXP2* cause a severe disorder of articulation and speech in humans, yet subtle differences in *FOXP2* sequence between humans and non-humans show evidence of positive evolutionary selection by K_A/K_S ratio. Its involvement in speech production suggests that changes in *FOXP2* may have been important in the evolution of language[22,23]. Furthermore, analysis of *FOXP2*'s DNA sequence in diverse human populations suggests that the gene shows unusually low sequence diversity—that is, many human populations share a common ancestral sequence at the *FOXP2* locus. This evidence for a "selective sweep" (explained in detail in several recent reviews[2,24]) within humans suggests that evolutionary selection on this gene may have occurred very recently in human evolution; that is, after the appearance of *Homo sapiens*.

Two genes that cause microcephaly (small cerebral cortex) also show strong evidence for positive evolutionary selection. Microcephaly reduces the human brain to 50% or less of its normal mass; that is, to about the size of the brain of chimpanzees or our pre-human ancestors. Whereas marked mutations in abnormal spindle microcephaly (encoded by the *ASPM* locus) and microcephalin (encoded by the *MCPH1* locus) cause microcephaly, both genes show strong evidence that subtler sequence changes were subject to positive selection in the lineage leading to humans (manifested by a high K_A/K_S ratio)[25-29]. Although the precise functions of the two genes are unknown, both are highly expressed in dividing neural precursor cells in the cerebral cortex, and available evidence suggests roles in cell proliferation. Notably, just as neurons in the upper layers of the cerebral cortex (Fig. 1) are added last during development, and are most highly elaborated in humans and great apes, these upper-layer neurons are preferentially lost in many cases of microcephaly, supporting a requirement for microcephaly genes in the formation of the upper cortical layers.

AHI1, which is essential for axon pathfinding from the cortex to the spinal cord (and hence for normal coordination and gait), is another gene that causes a neurological disease

在大脑中有已知的重要功能[20]。对黑猩猩基因组的分析也确认了一点，神经基因，作为一组，相比在大脑外表达的基因有更低的 K_A/K_S 比值[13]。然而，更新的研究工作表明，有相当一部分具有最高 K_A/K_S 比值的基因在大脑发育和功能中发挥了作用[13]。这些研究最容易这样折中：神经基因的一个非常小的子集或许会是正选择的目标（见下文），而由于这些破坏大脑功能的基因突变所带来的严重的劣势，神经基因，作为一个整体，会遭受很强的负选择。

基因演化与人类大脑功能的相关性

由于基因组范围的分析系统地标明了人类谱系中基因正选择的目标，人类基因中显示正演化选择的一个子集引起了极大的关注，并且对于这些基因来讲，探讨人类中演化模式和基因功能的相关性是可能的。例如，在人类中 *FOXP2* 的突变等位基因会导致严重的语言和表达障碍，但是人类和非人类 *FOXP2* 基因序列的细微差异通过 K_A/K_S 比值显示该基因受到正演化选择。其参与语言产生的这个现象表明，*FOXP2* 中的变化在语言演化中起着很重要的作用[22,23]。况且，不同现代人群体的 *FOXP2* 的 DNA 序列表明该基因显示出非常低的序列多样性，也就是说，许多现代人群体在 *FOXP2* 位点上共有相同的祖先序列。这一人类内部“选择性清除”（在最近的几篇综述[2,24]中对此进行了详细解释）的证据表明，针对这一基因的演化选择在人类进化过程中可能出现得非常晚，也就是说，在智人出现之后。

导致小头畸形（大脑皮层小）的两个基因也显示出正演化选择强有力的证据。小头畸形会使人的大脑减小到正常体积的 50% 或更小，也就是说减小到大约相当于黑猩猩或现代人之前的祖先的脑的大小。尽管异常纺锤体小头畸形蛋白（由 *ASPM* 位点编码）和小头畸形素（由 *MCPH1* 位点编码）的显著突变会导致小头畸形，但是两个基因都显示出强有力的证据：较细微的序列变化在通往现代人的演化谱系中会受到正选择（表现为 K_A/K_S 比值很高）[25-29]。尽管两个基因的确切功能尚不清楚，但二者在要分裂的大脑皮层的神经先驱细胞中被高度表达，现有证据表明它们在细胞增殖中发挥了作用。值得注意的是，正如大脑皮层上层的神经元（图 1）是在发育的过程的最后被添加，并且在现代人和大猿中被最为精心打造一样，在很多小头畸形病例中，这些上层神经元会被专门丢弃，这证明了导致小头畸形的基因参与了皮层上部层次的形成。

AHI1 对于轴突从皮层到脊髓寻路是至关重要的基因（因此对正常的协调和步态也至关重要），该基因的突变也会导致神经系统疾病，但该基因在灵长类之间较细微

when mutated, but for which subtler changes between primate species suggest positive evolutionary selection in the lineage leading to humans[30]. Patients with *AHI1* mutations not only show mental retardation, but can also show symptoms characteristic of autism, such as antisocial behaviour. This raises the intriguing possibility that evolutionary differences in *AHI1* may relate not only to human patterns of gait, but potentially species-specific social behaviour.

The linkage of studies of gene function in humans with evolutionary analysis is just beginning, and is limited mainly by the rate at which the essential functional roles of genes in the human brain are elucidated. As a population, humans show many mutant alleles for every gene that has been extensively studied, so that the human population is likely to represent, to a first approximation, saturation mutagenesis, such that for each gene in the genome there is a human carrying a mutated allele for that gene. Many neurological diseases affect the very processes that define us evolutionarily as human: intelligence (mental retardation), social organization (autism and attention deficit disorder) and higher-order language (dyslexia). As the genes for these uniquely human disorders are characterized, they may give us new insight into our recent evolutionary history.

(**437**, 64-67; 2005)

Robert Sean Hill[1] & Christopher A. Walsh[1]

[1] Division of Neurogenetics and Howard Hughes Medical Institute, Beth Israel Deaconess Medical Center, and Department of Neurology, Harvard Medical School, Room 266, New Research Building, 77 Avenue Louis Pasteur, Boston, Massachusetts 02115, USA

References:

1. Ramon y Cajal, S. *Recuerdos de mi Vida* Vol. 2 *Historia de mi Labour Científica* 345-346 (Moya, Madrid, 1917).

2. Carroll, S. B. Genetics and the making of *Homo sapiens*. *Nature* **422**, 849-857 (2003).

3. Kornack, D. R. & Rakic, P. Changes in cell-cycle kinetics during the development and evolution of primate neocortex. *Proc. Natl Acad. Sci. USA* **95**, 1242-1246 (1998).

4. Takahashi, T., Nowakowski, R. S. & Caviness, V. S. Jr. The cell cycle of the pseudostratified ventricular epithelium of the embryonic murine cerebral wall. *J. Neurosci.* **15**, 6046-6057 (1995).

5. Marin-Padilla, M. Ontogenesis of the pyramidal cell of the mammalian neocortex and developmental cytoarchitectonics: a unifying theory. *J. Comp. Neurol.* **321**, 223-240 (1992).

6. Allman, J., Hakeem, A. & Watson, K. Two phylogenetic specializations in the human brain. *Neuroscientist* **8**, 335-346 (2002).

7. Orban, G. A., Van Essen, D. & Vanduffel, W. Comparative mapping of higher visual areas in monkeys and humans. *Trends Cogn. Sci.* **8**, 315-324 (2004).

8. Semendeferi, K., Lu, A., Schenker, N. & Damasio, H. Humans and great apes share a large frontal cortex. *Nature Neurosci.* **5**, 272-276 (2002).

9. Hopkins, W. D. & Cantalupo, C. Handedness in chimpanzees (*Pan troglodytes*) is associated with asymmetries of the primary motor cortex but not with homologous language areas. *Behav. Neurosci.* **118**, 1176-1183 (2004).

10. Cantalupo, C. & Hopkins, W. D. Asymmetric Broca's area in great apes. *Nature* **414**, 505 (2001).

11. Sun, T. *et al.* Early asymmetry of gene transcription in embryonic human left and right cerebral cortex. *Science* **308**, 1794-1798 (2005).

12. Broadfield, D. C. *et al.* Endocast of Sambungmacan 3 (Sm 3): a new *Homo erectus* from Indonesia. *Anat. Rec.* **262**, 369-379 (2001).

13. The Chimpanzee Sequencing and Analysis Consortium. Initial sequence of the chimpanzee genome and comparison with the human genome. *Nature* doi:10.1038/nature04072 (this issue).

14. Gilad, Y., Man, O. & Glusman, G. A comparison of the human and chimpanzee olfactory receptor gene repertoires. *Genome Res.* **15**, 224-230 (2005).

15. Enard, W. *et al.* Intra- and interspecific variation in primate gene expression patterns. *Science* **296**, 340-343 (2002).

16. Caceres, M. *et al.* Elevated gene expression levels distinguish human from non-human primate brains. *Proc. Natl Acad. Sci. USA* **100**, 13030-13035 (2003).

17. Uddin, M. *et al.* Sister grouping of chimpanzees and humans as revealed by genome-wide phylogenetic analysis of brain gene expression profiles. *Proc. Natl Acad. Sci. USA* **101**, 2957-2962 (2004).

的变化表明，在通向现代人的人类演化谱系中它是受到正选择的[30]。*AHI1* 基因突变的病人不仅表现为智力低下，而且还可能表现出自闭症的典型症状，比如反社会行为。这就产生了一种令人不解的可能性，即在 *AHI1* 基因上的演化差异可能不仅与人类的步态模式有关，而且还可能与物种特定的社会行为有关。

人类基因功能的研究与演化分析之间的结合只是刚刚开始，并且主要受限于阐明这些基因在人类大脑中的根本功能的研究速度。作为一个种群，人类被广泛研究的每一个基因都显示出许多突变等位基因，因此人类种群很可能大致上代表了饱和突变的发生，使得对于基因组中的每个基因，都有一个人携带着这个基因的突变等位基因。许多神经系统疾病会影响这个在演化意义上定义我们为人类的过程：智慧（智力低下），社会组织（自闭症和注意力缺陷症），高级语言（阅读障碍）。随着这些人类特有的疾病的基因的特征都得到详细的描绘，它们可能会让我们对人类最近的演化历史有一个全新的了解。

（吕静 翻译；张颖奇 审稿）

18. Goldman, N. & Yang, Z. A codon-based model of nucleotide substitution for protein-coding DNA sequences. *Mol. Biol. Evol.* **11**, 725-736 (1994).

19. Dorus, S. *et al.* Accelerated evolution of nervous system genes in the origin of *Homo sapiens*. *Cell* **119**, 1027-1040 (2004).

20. Nielsen, R. *et al.* A scan for positively selected genes in the genomes of humans and chimpanzees. *PLoS Biol.* **3**, e170 (2005).

21. Clark, A. G. *et al.* Inferring nonneutral evolution from human-chimp-mouse orthologous gene trios. *Science* **302**, 1960-1963 (2003).

22. Lai, C. S., Fisher, S. E., Hurst, J. A., Vargha-Khadem, F. & Monaco, A. P. A forkhead-domain gene is mutated in a severe speech and language disorder. *Nature* **413**, 519-523 (2001).

23. Enard, W. *et al.* Molecular evolution of *FOXP2*, a gene involved in speech and language. *Nature* **418**, 869-872 (2002).

24. Gilbert, S. L., Dobyns, W. B. & Lahn, B. T. Genetic links between brain development and brain evolution. *Nature Rev. Genet.* **6**, 581-590 (2005).

25. Bond, J. *et al.* *ASPM* is a major determinant of cerebral cortical size. *Nature Genet.* **32**, 316-320 (2002).

26. Evans, P. D., Anderson, J. R., Vallender, E. J., Choi, S. S. & Lahn, B. T. Reconstructing the evolutionary history of microcephalin, a gene controlling human brain size. *Hum. Mol. Genet.* **13**, 1139-1145 (2004).

27. Evans, P. D. *et al.* Adaptive evolution of *ASPM*, a major determinant of cerebral cortical size in humans. *Hum. Mol. Genet.* **13**, 489-494 (2004).

28. Kouprina, N. *et al.* Accelerated evolution of the *ASPM* gene controlling brain size begins prior to human brain expansion. *PLoS Biol.* **2**, E126 (2004).

29. Zhang, J. Evolution of the human *ASPM* gene, a major determinant of brain size. *Genetics* **165**, 2063-2070 (2003).

30. Ferland, R. J. *et al.* Abnormal cerebellar development and axonal decussation due to mutations in *AHI1* in Joubert syndrome. *Nature Genet.* **36**, 1008-1013 (2004).

31. Marin-Padilla, M. Dual origin of the mammalian neocortex and evolution of the cortical plate. *Anat. Embryol.* **152**, 109-126 (1978).

Acknowledgements. This work was supported by grants from the NINDS and Cure Autism Now. We thank M. Ruvolo and D. Reich for comments on an earlier version of this manuscript, and J. DeFilipe for the translation of the Cajal quotation. Owing to space limitations we were unable to cite directly some of the relevant work in this field. C.A.W. is an Investigator of the Howard Hughes Medical Institute.

Author Information. Reprints and permissions information is available at npg.nature.com/reprintsandpermissions. The authors declare no competing financial interests. Correspondence and requests for materials should be addressed to C.A.W. (cwalsh@bidmc.harvard.edu).

Characterization of the 1918 Influenza Virus Polymerase Genes

J. K. Taubenberger *et al.*

Editor's Note

The 1918 "Spanish" influenza pandemic killed more people—about 50 million—than the First World War. There are fears that some recent flu strains could be similarly lethal. Here US virologist Jeffrey Taubenberger and colleagues analyse virus samples extracted from the lung tissue of one victim. They find that the lethal virus was not the result of an animal and human strain merging their DNA, but instead a bird strain that evolved to infect humans. The 1918 virus has some similarities to H5N1 avian flu, which can kill humans but cannot spread between them, and another avian strain, H7N7, which has also killed people, helping researchers identify what separates the relatively mild flu strains from the killers.

The influenza A viral heterotrimeric polymerase complex (PA, PB1, PB2) is known to be involved in many aspects of viral replication and to interact with host factors[1], thereby having a role in host specificity[2,3]. The polymerase protein sequences from the 1918 human influenza virus differ from avian consensus sequences at only a small number of amino acids, consistent with the hypothesis that they were derived from an avian source shortly before the pandemic. However, when compared to avian sequences, the nucleotide sequences of the 1918 polymerase genes have more synonymous differences than expected, suggesting evolutionary distance from known avian strains. Here we present sequence and phylogenetic analyses of the complete genome of the 1918 influenza virus[4-8], and propose that the 1918 virus was not a reassortant virus (like those of the 1957 and 1968 pandemics[9,10]), but more likely an entirely avian-like virus that adapted to humans. These data support prior phylogenetic studies suggesting that the 1918 virus was derived from an avian source[11]. A total of ten amino acid changes in the polymerase proteins consistently differentiate the 1918 and subsequent human influenza virus sequences from avian virus sequences. Notably, a number of the same changes have been found in recently circulating, highly pathogenic H5N1 viruses that have caused illness and death in humans and are feared to be the precursors of a new influenza pandemic. The sequence changes identified here may be important in the adaptation of influenza viruses to humans.

INFLUENZA A viruses cause annual outbreaks in humans and domestic animals. Periodically, new strains emerge in humans that cause global pandemics. The severe "Spanish" influenza pandemic of 1918–1919 infected hundreds of millions, and resulted in the death of approximately 50 million people[12]. We have previously used phylogenetic

1918 年流感病毒聚合酶基因的特点

陶本伯格等

编者按

1918 年的“西班牙”流感大流行造成的死亡人数超过第一次世界大战，约为 5,000 万。人们担心最近的一些流感病毒株可能具有同样的致命性。在这里，美国病毒学家杰弗里·陶本伯格及其同事分析了从一名受害者的肺组织中提取的病毒样本。他们发现这种致命的病毒不是动物和人类病毒株整合 DNA 的结果，而是由禽流感病毒株进化而来感染人类的。1918 年的流感病毒与 H5N1 禽流感病毒（它们可以感染并致人死亡但不能在人与人之间传播）以及另一种禽类病毒 H7N7（同样可以致人死亡）有一些相似之处，这有助于研究人员确定相对温和的流感病毒株与高致病性病毒株的区别。

已知甲型流感病毒异三聚体聚合酶复合体（PA、PB1、PB2）在病毒复制的多个方面发挥作用，并和宿主因子发生作用 [1]，因此在宿主特异性中发挥作用 [2,3]。1918 年人流感病毒的聚合酶蛋白序列只有少数氨基酸与禽流感病毒共有序列不同，从而支持了它们在大流行前不久由禽流感病毒产生的假说。但是，与禽流感序列相比较，1918 年病毒的聚合酶基因的核苷酸序列同义性差异大于预期，提示其与已知的禽流感毒株存在进化距离。这里我们给出了 1918 年流感病毒整个基因组序列并进行了系统发育分析 [4-8]，提出 1918 年病毒不是一种基因重组病毒（如 1957 年和 1968 年大流行的病毒 [9,10]），而更可能是能够感染人类的完全类似禽流感（病毒）的病毒。这些数据支持了先期系统发育研究结果，并表明 1918 年病毒来源于禽流感 [11]。聚合酶蛋白中总共 10 个氨基酸的突变将 1918 年病毒以及随后的人类流感病毒序列同禽流感病毒序列区别开来。显而易见地，许多相同的突变在最近流行的高致病性 H5N1 病毒中被找到，该病毒已经造成了人类的疾病和死亡，并让人担心这是新的流感大流行的先兆。这里鉴定的序列突变可能对于流感病毒适应人类的过程非常重要。

甲型流感病毒每年都在人类和家畜中引起疫情暴发。新的毒株周期性地在人类中出现并引起全球大流行。1918 ~ 1919 年严重的“西班牙”流感大流行波及了数以亿计的人口，并导致了大约五千万人的死亡 [12]。我们之前已经用系统发育分析来

analyses to help understand the origin of the pandemic virus[8,11]; functional studies to understand the pathogenicity of the 1918 virus are underway[6,13-17]. Recent data have shown that viral constructs bearing the 1918 haemagglutinin gene are pathogenic in a mouse model, but the genetic basis of this observation has not yet been mapped[6,13-17]. The overall goals of this project have been to understand the origin and unusual virulence of the 1918 influenza virus.

The influenza virus A polymerase functions as a heterotrimer formed by the PB2, PB1 and PA proteins (see ref. 1 for a review). An additional small open reading frame has recently been identified, coding for a peptide (PB1-F2) that is thought to play a role in virus-induced cell death[18]. It is not yet clear how the polymerase complex must change to adapt to a new host[3]. A single amino acid change in PB2, E627K, was shown (1) to be important for mammalian adaptation[2,3], (2) to distinguish highly pathogenic avian influenza (HPAI) H5N1 viruses in mice[19], and (3) to be present in the single fatal human infection during the HPAI H7N7 outbreak in the Netherlands in 2003 (ref. 20), and in some recent H5N1 isolates from humans in Vietnam and Thailand and wild birds in China[21-23].

The open reading frame sequences of segment 1 (PB2), segment 2 (PB1) and segment 3 (PA) of A/Brevig Mission/1/1918, and theoretical translations of the four identified reading frames, are shown in Supplementary Fig. 1a–c. The 1918 PB2 protein contained five changes from the avian consensus sequence (Table 1). Of these, A199S is in the area mapped as the PB1 binding site, and the L475M change is in a nuclear localization signal[24-26]. Three other changes at residues 567, 627 and 702 occur at sites that are not in known functional domains.

Table 1. Amino acid residues distinguishing human and avian influenza polymerases

Gene	Residue no.	Avian	1918	Human H1N1	Human H2N2	Human H3N2	Classical swine	Equine
PB2	199	A	S	S	S	S	S	A
PB2	475	L	M	M	M	M	M	L
PB2	567	D	N	N	N	N	D	D
PB2	627	E	K	K	K	K	K	E
PB2	702	K	R	R*	R	R	R	K
PB1	375	N/S/T†	S	S	S	S	S	S
PA	55	D	N	N	N	N	N	N
PA	100	V	A	A	A	A	V	A
PA	382	E	D	D	D	D	D	E
PA	552	T	S	S	S	S	S	T

* All human H1N1 PB2 sequences have an Arg residue at position 702, except that two out of three A/PR/8/34 sequences have a Lys residue.

† The majority of avian sequences have an Asn residue at position 375 of PB1, 18% have a Ser residue, 13% a Thr residue.

帮助了解这次大流行病毒的来源[8,11]；了解 1918 年流感病毒致病性的功能研究正在进行中[6,13-17]。最近的数据显示带有 1918 年血凝素基因的重组病毒在小鼠模型中具有致病性，但是这个现象的基因基础尚未明确[6,13-17]。这个项目的总体目标就是了解 1918 年流感病毒的起源和超常毒力。

甲型流感病毒聚合酶以 PB2、PB1 和 PA 蛋白异三聚体的形式发挥功能（见文献 1 综述）。最近发现了一个小的开放阅读框，它编码一种被认为在病毒诱导的细胞死亡中发挥作用的肽（PB1-F2）[18]。目前还不清楚聚合酶复合体必须如何改变以适应新的宿主[3]。PB2 中的单个氨基酸突变——E627K 被发现：(1) 对哺乳动物的适应性非常重要[2,3]；(2) 可用于鉴别小鼠中的高致病性禽流感病毒 H5N1[19]；(3) 出现在 2003 年荷兰高致病性禽流感 H7N7 暴发期间的单例人致死感染（文献 20），以及最近从越南和泰国的人类及中国的野生鸟类中分离出的一些 H5N1 中[21-23]。

A/布雷维格教区/1/1918，PB2、PB1 和 PA 的开放阅读框序列以及四个已识别的阅读框的理论蛋白翻译详见补充材料图 1a ~ 1c。1918 年 PB2 蛋白含有五个相对于禽流感病毒共有序列的突变（表 1）。其中，A199S 位于 PB1 结合位点区域，L475M 突变位于核定位信号中[24-26]。567、627 和 702 三个突变都位于目前已知的功能区域之外。

表 1. 区分人类和禽类流感聚合酶的氨基酸残基

基因	残基号	禽类	1918	人 H1N1	人 H2N2	人 H3N2	猪	马
PB2	199	A	S	S	S	S	S	A
PB2	475	L	M	M	M	M	M	L
PB2	567	D	N	N	N	N	D	D
PB2	627	E	K	K	K	K	K	E
PB2	702	K	R	R*	R	R	R	K
PB1	375	N/S/T†	S	S	S	S	S	S
PA	55	D	N	N	N	N	N	N
PA	100	V	A	A	A	A	V	A
PA	382	E	D	D	D	D	D	E
PA	552	T	S	S	S	S	S	T

* 除了三个 A/PR/8/34 序列中的两个含有赖氨酸残基，所有的人类 H1N1 PB2 序列在位点 702 含有精氨酸残基。
† 大部分禽类序列在 PB1 的位点 375 含有天冬酰胺残基，18% 含有丝氨酸残基，13% 含有苏氨酸残基。

The 1918 PB1 protein differed from the avian consensus by seven residues (one of which is shown in Table 1; see also Supplementary Fig. 2). Of these, K54R is in the overlapping binding domains for complementary (c)RNA and viral (v)RNA. Changes at residues 375, 383 and 473 all occur in between the four conserved polymerase motifs in the cRNA binding domain[27], and changes at residues 576, 645 and 654 occur in the vRNA binding domain[28].

Seven changes were noted in the 1918 PA protein compared with the avian consensus (four of which are shown in Table 1, the other three being C241Y, K312R and I322V). The C241Y change occurs in a nuclear localization signal, but the other six changes (at residues 55, 100, 312, 322, 382 and 552) occur at sites outside of known functional domains[24-26].

Representative phylogenetic analyses of the three polymerase genes are shown in Figs 1–3. The 1918 human pandemic viral polymerase genes were compared to representative avian influenza genes with regards to transition/transversion (T_i/T_v) ratio, synonymous/non-synonymous (S/N) ratio, and the numbers of differences at fourfold degenerate sites (defined in ref. 11). T_i/T_v ratios for most comparisons using the 1918 viral genes and representative sequences of either North American or Eurasian avian genes yielded values between 2 and 4. This range was similar to that observed for comparisons of various avian genes with one another, except for the PB1 gene. For PB1, comparisons of the 1918 viral gene with avian virus PB1 genes was always close to 2, whereas comparisons of various avian genes with one another were in the range of 6–10. There were fewer transversions in comparisons between avian PB1 genes than in comparisons between avian PB1 and 1918 human virus PB1, probably reflecting that transversions more often lead to non-synonymous changes.

S/N ratios for most comparisons using the 1918 viral genes and representative sequences of either North American or Eurasian avian genes usually yielded values in the range of 7–16 for both the PA and PB2 genes, as is the case for most avian versus avian PA and PB2 gene comparisons. Like the T_i/T_v ratios, the S/N ratios were somewhat higher with the PB1 gene (most of the comparisons yielded ratios in the range of 16–25), owing to a smaller number of non-synonymous changes in comparisons of avian PB1 genes with one another. These findings may reflect a more conservative evolution of PB1 in birds.

A subset of synonymous differences occurs at sites that are fourfold degenerate (that is, where a substitution with any base does not result in an amino acid replacement). As these sites are not subject to selective pressure at the protein level, base substitutions at many fourfold degenerate sites may accumulate rapidly. If influenza virus genes have been evolving in birds for long enough to reach evolutionary stasis, as is suggested by the high S/N ratios described above, one would predict that at many of the sites where fourfold degeneracy is possible, all four bases would be present in the avian clade unless the constraints of RNA secondary structure limit the accumulation of synonymous changes. In fact, when avian sequences from geographically distinct lineages (North American versus European) were compared, the per cent difference at fourfold degenerate sites yielded values in the 27–38% range. In contrast, calculating the per cent difference at fourfold degenerate

在蛋白序列上，1918 病毒株的 PB1 蛋白与禽流感株有七个氨基酸残基的差异(其中一个在表 1 中显示，也可见于补充材料图 2)。其中，K54R 位于互补(c)RNA 和病毒(v)RNA 的重叠结合域。375、383 和 473 处的突变都位于 cRNA 结合域的四个保守的聚合酶基序之间 [27]，而 576、645 和 654 处的突变位于 vRNA 结合域 [28]。

1918 病毒株 PA 蛋白与禽流感株在蛋白序列上的差别一共有七处(其中四个在表 1 中显示，其他三个是 C241Y、K312R 和 I322V)。C241Y 突变发生在核定位信号中，但是其他六个突变(55、100、312、322、382 和 552)均在已知的功能区域之外 [24-26]。

三个聚合酶基因的进化分析呈现在图 1 ~ 图 3 中。1918 年人类大流行病毒聚合酶基因与代表性的禽流感病毒基因进行了转换/颠换(T_i/T_v)比、同义/非同义(S/N)比和四倍简并位点(在文献 11 中定义)差别数目的对比。大部分 1918 年病毒基因和北美或者欧亚禽流感病毒基因的代表性序列进行对比获得的 T_i/T_v 数值都在 2 和 4 之间。该范围与不同禽类携带禽流感病毒基因之间对比所得到的结果类似，除了 PB1 基因。对 PB1 基因来说，1918 年病毒基因与禽流感基因的对比始终接近 2，而不同禽类携带禽流感基因之间的对比都在 6 ~ 10 的范围内。禽流感 PB1 基因之间对比得到的颠换数少于禽流感和 1918 年人流感病毒 PB1 基因之间对比得出的颠换数，很可能反映了颠换更常导致非同义突变。

对 PA 和 PB2 基因来说，大部分用 1918 年病毒基因和北美或者欧亚任一禽类流感病毒基因的代表性序列进行对比获得的 S/N 比都在 7 ~ 16 的范围内，大部分禽流感病毒之间对比的结果范围也是一样。就像 T_i/T_v 比一样，PB1 基因的 S/N 比稍微高一些(大部分对比得出的数值在 16 ~ 25 范围内)，因为相对于其他来说，禽流感病毒 PB1 基因的非同义突变数量少一些。这些发现可能反映了禽流感病毒 PB1 进化过程中的保守性更高。

一部分同义突变发生在四倍简并位点(即任何碱基的置换都不会造成氨基酸的替换)。由于这些位点不会在蛋白质水平面临选择压力，许多四倍简并位点的碱基置换可能会迅速地累积。如果流感病毒基因如上面所描述的高 S/N 比所揭示的那样，已经在鸟类中存在足够长的时间并达到进化停滞，我们能够预计在可能出现四倍简并的许多位点，所有四种碱基都可能出现在禽类分支上，除非 RNA 二级结构的限制约束了这种同义突变的积累。事实上，当对比地理上不同谱系(北美和欧亚)的禽类来源流感病毒序列时，四倍简并位点的百分比差别都在 27% ~ 38% 之间。相反的，对比 1918 年病毒的 PA、PB1、PB2 基因序列和禽流感病毒序列时，计算出四倍简

sites in comparisons of the 1918 viral PA, PB1 and PB2 gene sequences with avian sequences yielded consistently higher values (range 41–51%) for all three genes. As with the other 1918 genes[11], this suggests that the donor source of the 1918 virus was in evolutionary isolation from those avian influenza viruses currently represented in the databases.

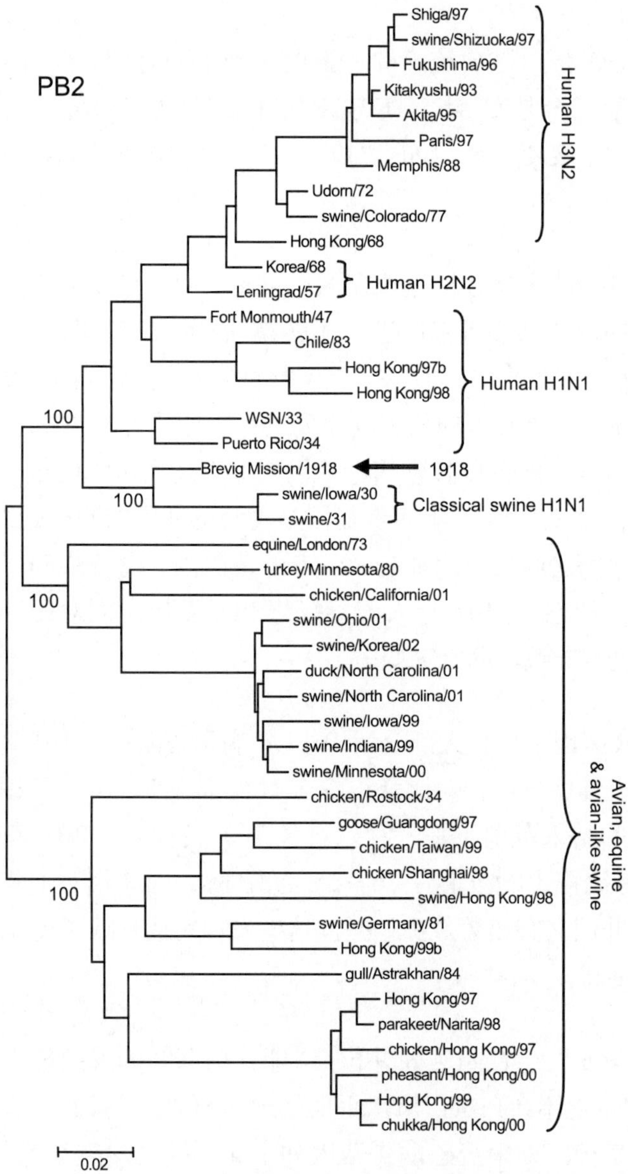

Fig. 1. Phylogenetic tree of the PB2 gene. Sequences were aligned and analysed for phylogenetic relationships using the NJ algorithm, with the proportion of sequence differences as the distance measure. Bootstrap values (100 replications) for key nodes are shown (for clarity, identical and nearly identical sequences have been removed from the trees). Major clades are identified with large brackets. The arrow identifies the position of the 1918 PB2 gene sequence. A distance bar is shown below the tree. Influenza strain abbreviations used in the analyses are listed in Supplementary Table 1.

并位点的百分比差别在三种基因中都处于高水平（41%～51%）。和其他的 1918 年病毒基因一样 [11]，这提示 1918 年病毒的供体来源与目前数据库中显示的禽流感病毒在进化上是隔离的。

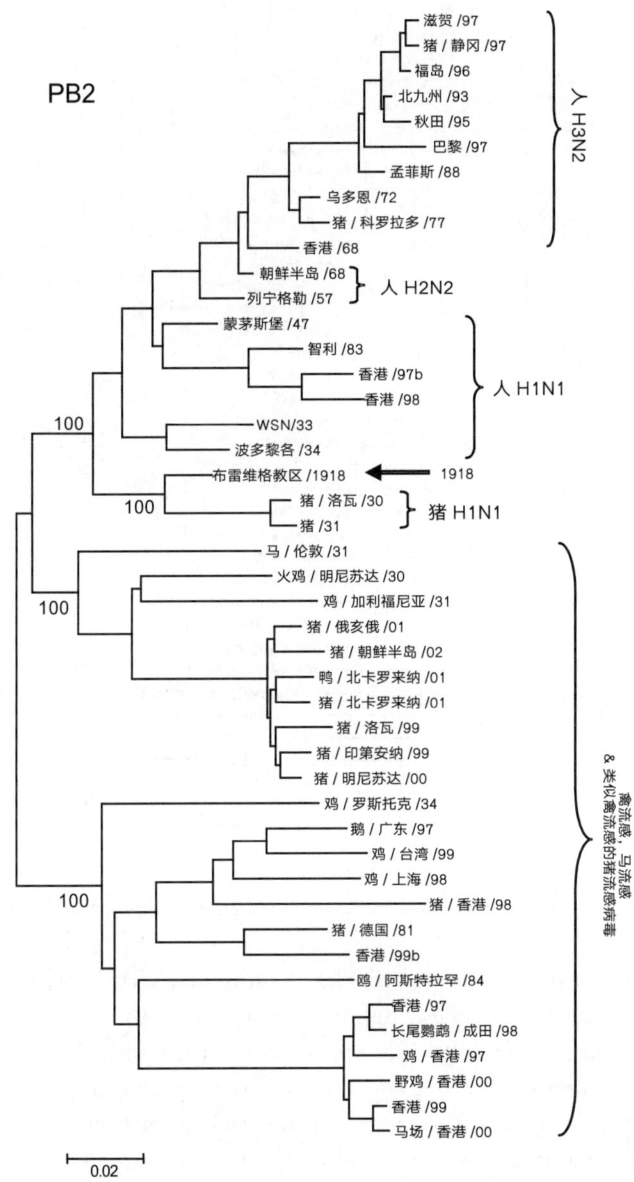

图 1. PB2 基因的系统发育树。用 NJ 算法排比和分析序列的分子进化关系，其距离的度量是序列差别的比例。图上标注了关键节点的自展值（100 个重复）（为了清晰起见，相同的和几乎相同的序列从树中移除）。主要分支用大括号标记出来。箭头指出了 1918 年病毒 PB2 基因序列的位置。距离的比例尺在树下面标出。分析中使用的流感病毒株系的缩写见补充信息表 1。

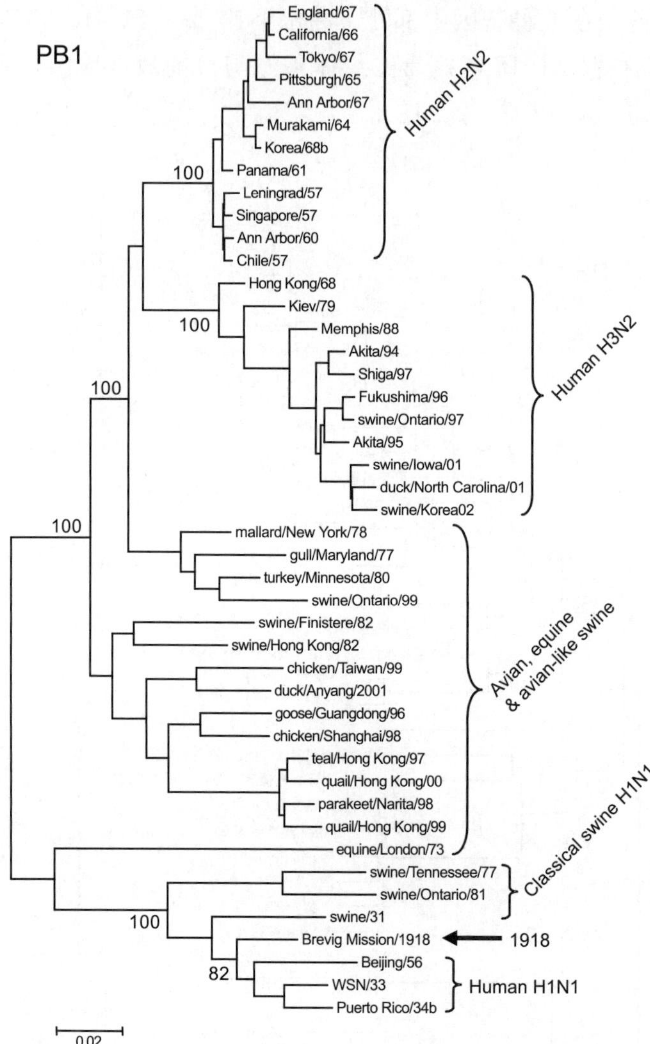

Fig. 2. Phylogenetic tree of the PB1 gene. Sequences were aligned and analysed as detailed in the legend to Fig. 1.

Emphasizing the avian-like nature of the 1918 influenza virus polymerase proteins, out of 19 total amino acid changes from the avian consensus, there are only 10 amino acid positions (out of 2,232 total codons) that consistently distinguish the 1918 and subsequent human polymerase proteins PB2, PB1 and PA from their avian influenza counterparts (these are defined as changes from avian sequences in the 1918 virus that are maintained without change in subsequent human viruses) (Table 1). It is likely that these changes have an important role in human adaptation. Seven of these ten changes were previously noted in an alignment between avian and human influenza polymerases[3]. What follows is a comparison between the 1918 virus changes and recent H5N1 isolates, in order to evaluate possible examples of parallel evolution in the adaptation of avian influenza viruses to humans.

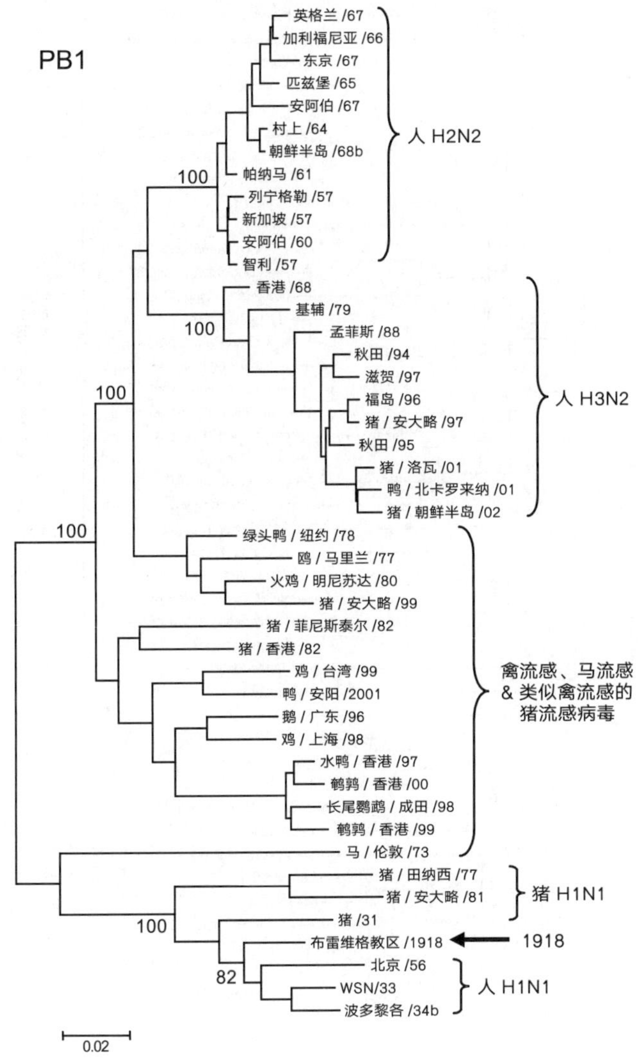

图 2. PB1 基因的系统发育树。如图 1 图注中详细所述进行序列的排比和分析。

在和禽流感病毒共有序列不同的 19 个氨基酸中，只有 10 个氨基酸位置（总共 2,232 个密码子）始终如一地在 1918 年及之后的人类流感病毒聚合酶蛋白 PB2、PB1 和 PA 中与它们的禽流感病毒对应部分保持差别（这些被定义为 1918 年病毒中禽类序列发生的突变，其在后续人类病毒中一直保持稳定）（表 1），这也突出了 1918 年流感病毒聚合酶蛋白的类似禽流感病毒的性质。很有可能这些改变在适应人类的过程中具有重要作用。在之前比对禽流感和人流感聚合酶时就发现了这 10 个突变中的 7 个 [3]。随后就是比较 1918 年病毒变化和最近的 H5N1 病毒分离株，目的是评估禽流感病毒在适应人类的过程中发生平行进化的可能实例。

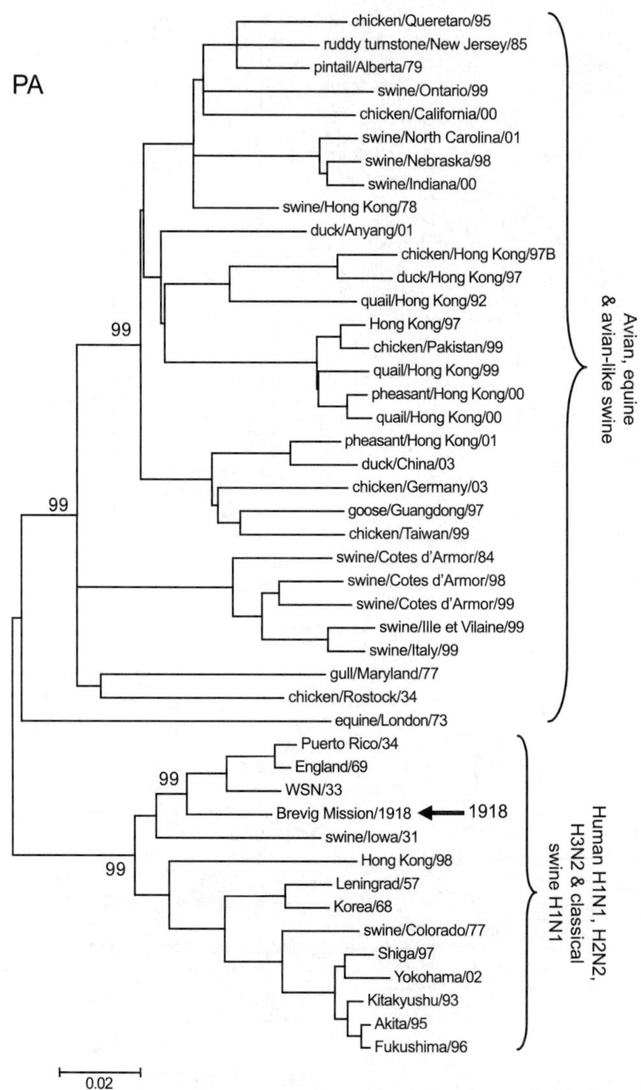

Fig. 3. Phylogenetic tree of the PA gene. Sequences were aligned and analysed as detailed in the legend to Fig. 1.

In the PB2 protein, five changes distinguish the human isolates from avian sequences (Table 1). Out of 253 available PB2 sequences from human H1N1, H2N2 and H3N2 isolates, these five changes are almost completely preserved, with the exception that two recent H3N2 isolates have the avian Lys residue at position 702. Only a small number of avian influenza isolates show any of these five changes, and it is intriguing that almost all of these isolates are from HPAI H5N1 or H7N7 viruses, or from the H9N2 lineage that infected a small number of humans in China in the late 1990s (ref. 29). Only 5 out of 282 available avian PB2 sequences have a Ser residue at position 199, four of these being 1997 H5N1 isolates from Hong Kong. The A199S change was also found in 5 out of 18 H5N1 strains isolated from humans (all five were from the 1997 Hong Kong outbreak). Of the avian viruses, 36 out of 336 have an Arg residue at position 702, 30 of which are H9N2 isolates from China

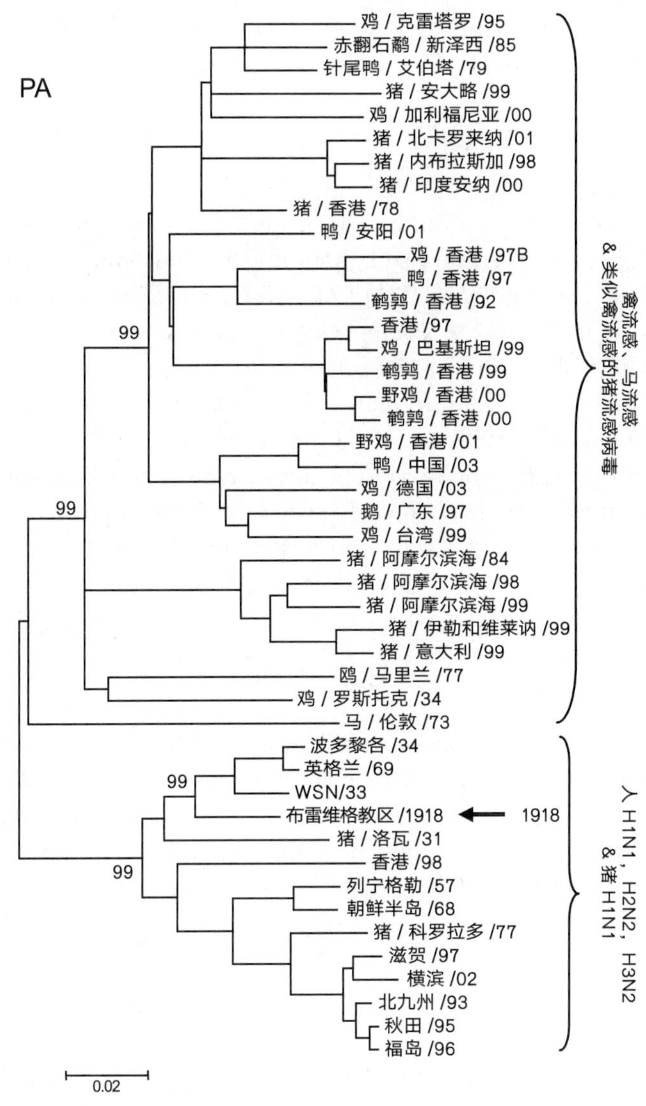

图 3. PA 基因的系统发育树。如图 1 图注中详细所述，进行序列的排比和分析。

在 PB2 蛋白中，人流感病毒分离株和禽病毒序列有 5 处差别(表 1)。在从人类 H1N1、H2N2 和 H3N2 病毒分离株获得的 253 个 PB2 序列中，这 5 处突变几乎完全保留，除了最近分离的两株 H3N2 在位点 702 具有禽流感的赖氨酸残基。只有少量的禽流感病毒分离株含有 5 种突变中的任意一种，而且有趣的是几乎所有含有这些突变的毒株都是从高致病性禽流感 H5N1 或者 H7N7 病毒中分离出来的，或者是从 20 世纪 90 年代晚期感染中国少部分人的 H9N2 病毒株中分离的(文献 29)。获得的 282 个禽流感 PB2 序列中只有 5 个在位点 199 是丝氨酸残基，其中 4 个来自香港的 1997 年 H5N1 病毒株。从人类中分离的 18 个 H5N1 株系中的 5 个发现了 A199S 突变(所有 5 个都来自 1997 年香港流感疫情)。336 个禽流感病毒中，只有 36 个在

around 1996–2000, and 5 are H5N1 isolates from Hong Kong in 1997 and 2001. Out of 18 available 1997 H5N1 strains isolated from humans, three have the K702R change.

Perhaps most interestingly, the 1918 virus and subsequent human isolates have a Lys residue at position 627. This residue has been implicated in host adaptation[2,3], and has previously been shown to be crucial for high pathogenicity in mice infected with the 1997 H5N1 virus[19]. Of the avian isolates, 19 out of 345 have a Lys residue at position 627, 18 of which are HPAI H5N1 or H7N7 avian influenza viruses. Sixteen of these were recently characterized H5N1 isolates from a die-off of wild waterfowl around Qinghai Lake in western China in 2005 (ref. 21). In human H5N1 isolates, 11 out of 37 have the E627K change: A/Hong Kong/483/1997 and A/Hong Kong/485/1997, four out of six isolates from Vietnam in 2004 (ref. 22), and two out of three isolates from Thailand in 2004 (ref. 23). The E627K mutation was seen in six out of seven H5N1 isolates from Thai tigers in 2004, and was also present in the H7N7 virus responsible for the single human fatality during the HPAI H7N7 outbreak in the Netherlands in 2003 (ref. 20). It was not noted in the contemporaneous chicken isolates.

At position 475, only one out of 355 avian isolates has a Met residue (an H5N1 HPAI virus from 2004). Similarly, only one out of 345 avian viruses has an Asn residue at position 567. None of the human H5N1 isolates has the L475M or the D567N changes. None of the available H5N1 or H7N7 sequences has more than one of the proposed human-adaptive PB2 changes determined for the 1918 virus.

The PA protein shows a similar pattern: four residues consistently differ between 1918 and subsequent human isolates and the avian consensus sequence (Table 1). Three other changes (C241Y, K312R and I322V) distinguish 1918, H1N1 and H2N2 human isolates, but most H3N2 isolates have the avian amino acid at these positions. Of 295 available sequences from human H1N1, H2N2 and H3N2 isolates, all have Asn at position 55 (except A/WSN/33), Ala at position 100 and Ser at position 552. Only 5 out of 295 human isolates have the avian Glu residue at position 382. Notably, these five isolates make up a minor clade of recent H3N2 isolates that have a number of unusual changes from typical human H3N2 viruses[30]. When avian influenza sequences are analysed, none (out of 209 sequences) has Asn at position 55 or Ser at position 552. Only 8 out of 209 avian PA protein sequences show the V100A change: six recent H6N2 isolates from chickens in California, and two HPAI 2002 H5N1 duck isolates from China. Of the 209 avian sequences, five have an Asp residue at position 382, including two HPAI H5N2 isolates from chickens in Mexico in 1994.

位点 702 是精氨酸残基，其中的 30 个是 1996 年 ~ 2000 年左右来自中国的 H9N2 分离株，5 个则是来自 1997 年和 2000 年香港的 H5N1 分离株。从人类中获得的 18 个 1997 年 H5N1 病毒株中，三个含有 K702R 突变。

可能最有趣的是，1918 年病毒和之后的人类分离株都在位点 627 含有一个赖氨酸残基。该残基与宿主的适应性相关 [2,3]，并且在之前的研究中已经被证实该位点对于 1997 年 H5N1 病毒感染小鼠时的高致病性非常关键 [19]。在 345 个禽流感分离株中，19 个在位点 627 含有赖氨酸残基，其中 18 个都是高致病性禽流感——H5N1 或者 H7N7 禽流感病毒。其中的 16 个是最近从 2005 年中国西部青海湖周围死去的野生水禽中分离出来的 H5N1 病毒株（文献 21）。在 37 个人类 H5N1 分离株中，11 个具有 E627K 突变：A/香港/483/1997 和 A/香港/485/1997，2004 年从越南分离出的 6 个病毒株中的 4 个（文献 22），以及 2004 年从泰国分离出的 3 个病毒株中的 2 个（文献 23）。该 E627K 突变在 2004 年从泰国老虎身上分离的 7 个 H5N1 毒株的 6 个中发现，并且存在于引起 2003 年荷兰高致病性禽流感 H7N7 大暴发时唯一人类死亡病例的 H7N7 病毒中（文献 20）。在同时期的鸡分离株中未发现该突变。

在位点 475，355 个禽类分离株中只有 1 个含有甲硫氨酸残基（2004 年分离的一个 H5N1 高致病性禽流感病毒）。类似的，345 个禽流感病毒株中只有 1 个在位点 567 含有天冬酰胺残基。人类 H5N1 分离株无 L475M 或者 D567N 突变。在获得的 H5N1 或者 H7N7 序列中，没有一个含有超过一种像 1918 年病毒那样的人类适应性 PB2 突变。

PA 蛋白表现出类似的模式：1918 年大流行及之后的人类分离株和禽类分离株相应序列之间有固定的 4 个残基的差别（表 1）。三个其他的改变（C241Y、K312R、I322V）将 1918 年、H1N1 和 H2N2 人类分离株区别开来，但是大部分 H3N2 分离株在这些位点都有禽流感的氨基酸。在 295 个从人类 H1N1、H2N2 和 H3N2 分离株中获得的序列中，全都在位点 55 含有天冬酰胺残基（除了 A/WSN/33），在位点 100 含有丙氨酸残基，位点 552 含有丝氨酸残基。295 个人类分离株中只有 5 个在位点 382 含有禽流感的谷氨酸残基。值得注意的是，这 5 个分离株形成了近期 H3N2 株的一个小的分支，与经典的人类 H3N2 毒株相比有一些不寻常的改变 [30]。当分析禽流感病毒序列时，没有一个序列（总共 209 个序列）在位点 55 含有天冬酰胺残基或者在位点 552 含有丝氨酸残基。209 个禽类 PA 蛋白序列中只有 8 个含有 V100A 突变：6 个来自近期加利福尼亚的鸡 H6N2 分离株，2 个来自中国 2002 年高致病性禽流感 H5N1 鸭分离株。在这 209 个禽类序列中，5 个在位点 382 含有天冬氨酸残基，包括 2 个从 1994 年墨西哥鸡中分离出的高致病性禽流感 H5N2 病毒。

The PB1 gene segment was replaced by reassortment in both the 1957 and 1968 pandemics[9]. We compared the PB1 protein from the 1918 human virus with those of the avian-derived PB1 segments from the 1957 and 1968 pandemics. Human H1N1, H2N2 and H3N2 viruses derived from the 1918, 1957 and 1968 pandemics, respectively, each possessed a uniquely derived avian-like PB1 gene segment, and so we sought to identify any parallel changes that might shed light on human adaptation. The three human pandemic PB1 proteins differ from the avian consensus by only 4–7 residues each (Supplementary Fig. 2). Only one of these changes is shared among the pandemic isolates: an N375S change. This change to a serine residue is also found in swine and equine influenza A isolates. With few exceptions, all human influenza PB1 proteins have Ser at this site. Of 230 human influenza sequences, only two H1N1 isolates (A/FM/47 and A/Beijing/1956) and the "minor clade" H3N2 isolates described above have the avian Asn residue[30]. In contrast, although this residue is maintained in almost all mammalian isolates, it is variable among avian PB1 proteins. Of 293 avian isolates, 66% have the consensus Asn residue at position 375, 18% have a Ser residue and 12% have a Thr residue.

The data presented here highlight the marked conservation of the PB1 protein in avian influenza viruses. PB1 functions as an RNA-dependent RNA polymerase, and so it is reasonable to hypothesize that its enzymatic function is optimal in this conserved form. In humans, the PB1 proteins experience linear change over time. Indeed, PB1 in humans acquires ~0.4 amino acid changes per year. As there is such strong antigenic selection on human viruses, it is possible that although the observed changes in PB1 are selectively beneficial with respect to antigenicity, they are mildly deleterious to enzyme function. Such complex fitness trade-offs are thought to be commonplace in RNA virus evolution. Supporting this hypothesis, a recent study examining combinations of avian and human influenza polymerases showed that the most efficient influenza transcriptional activity *in vitro* was seen with an avian-derived PB1, even if the PB2, PA and NP proteins were from a human virus[3]. Acquiring an avian PB1 by reassortment might provide a replicative advantage to the new virus, possibly explaining why both of the last two pandemics and the 1918 influenza virus all had very avian-like PB1 proteins.

Both the 1957 and 1968 pandemic influenza viruses were avian/human reassortants in which 2–3 avian gene segments were reassorted with the then-circulating, human-adapted virus[9,10]. Unlike the 1957 and 1968 pandemics, however, the 1918 virus was most likely not a human/avian reassortant virus, but rather an avian-like virus that adapted to humans *in toto*[8,11]. On the basis of amino acid replacement rates in human influenza virus polymerase genes, it is possible that these segments were circulating in human influenza viruses as early as 1900. However, proof that the 1918 virus did not retain gene segments from the previously circulating human influenza A strain would require discovery of a sample of the pre-1918 virus from archival material. The donor source, although avian-like at the protein level, may have come from a subset of avian influenza viruses not currently represented in the sequence databases and may have been in evolutionary isolation.

在 1957 年和 1968 年的大流行中，PB1 基因片段都发生了重组 [9]。我们比较了 1918 年人类病毒的 PB1 蛋白与 1957 年和 1968 年大流行时禽类来源的 PB1 片段。分别来自 1918 年、1957 年和 1968 年大流行的人类 H1N1、H2N2 和 H3N2 病毒，每个都含有独特起源的类似禽流感的 PB1 基因片段，因此我们试图找到任何可能有助于人类适应性的平行突变。这三个人类大流行的 PB1 蛋白与禽类共有序列都仅有 4～7 个残基的差别(补充信息图 2)。只有一个突变在三个大流行分离株中都存在：N375S 突变。在猪和马的甲型流感分离株中也发现了这种到丝氨酸残基的突变。除了少数几个例外，所有的人类流感 PB1 蛋白在这个位点都是丝氨酸。在 230 个人类流感病毒序列中，只有两个 H1N1 分离株(A/FM/47 和 A/北京/1956)以及上面所述的 H3N2 分离株微小分支含有禽类天冬酰胺残基 [30]。相反，尽管该残基在几乎所有的哺乳动物分离株中都存在，其在禽类 PB1 蛋白中是可变的。在 293 个禽类分离株中，66% 在位点 375 含有相同的天冬酰胺残基，18% 含有丝氨酸残基，12% 含有苏氨酸残基。

这里给出的数据凸显出了禽流感病毒 PB1 蛋白的显著保守性。PB1 作为一种 RNA 依赖的 RNA 聚合酶发挥功能，因此有理由假设在这种保守的形式下其酶功能是最佳的。在感染的人类中，PB1 蛋白随时间经历了线性的改变。实际上，PB1 在人群中每年获得大约 0.4 个氨基酸的改变。由于人类病毒存在很强的抗原性选择，尽管 PB1 的改变在抗原性方面具有选择优势，但它们在一定程度上降低了酶功能。这种复杂的适应性权衡被认为在 RNA 病毒的进化中非常普遍。最近一个研究支持了这个假设，它研究了禽流感和人流感聚合酶的组合，结果显示即使 PB2、PA 和 NP 蛋白都来源于人流感，含有禽类来源的 PB1 时，病毒的体外转录活性最有效率 [3]。通过重组获得禽流感 PB1 蛋白可能给新病毒提供复制优势，这可能解释了为什么后两次大流行和 1918 年的流感病毒全都含有类似禽流感的 PB1 蛋白。

1957 年和 1968 年大流行的流感病毒都是禽类/人类的重组病毒，其中 2～3 个禽流感基因片段都与那时流行的适应人类的病毒发生了重组 [9,10]。但是，不像 1957 年和 1968 年大流行，1918 年的病毒很可能不是禽类/人类的重组病毒，而是完全适应人类的类似禽流感的病毒 [8,11]。根据人类流感病毒聚合酶基因氨基酸替换的速度，有可能这些片段早在 1900 年就在人类流感病毒中传播。但是，要证明 1918 年病毒是否保留先前传播的人类甲型流感病毒的基因片段，需要在档案材料中发现 1918 年之前的病毒样本。其来源尽管在蛋白质水平类似于禽流感，但可能是目前序列数据库中未显示的禽流感病毒亚类，并且已经出现了进化隔离。

The fact that amino acid changes identified in the 1918 analysis are also seen in HPAI strains of H5N1 and H7N7 avian viruses that have caused fatalities in humans is intriguing, and suggests that these changes may facilitate virus replication in human cells and increase pathogenicity. It is possible that the high pathogencity of the 1918 virus was related to its emergence as a human-adapted avian influenza virus. These changes may reflect a process of parallel evolution as avian influenza A viruses mutate in response to adaptational pressures, and suggest that the genetic basis of avian influenza virus adaptation to humans can be mapped.

Methods

RNA isolation, amplification and sequencing. RNA was isolated from frozen 1918 human lung tissue using Trizol (Invitrogen) according to the manufacturer's instructions. Each fragment was reverse transcribed, amplified, and sequenced at least twice. Reverse transcription polymerase chain reaction (RT–PCR), isolation of products and sequencing have been previously described[4]. Lists of primers and primer sequences are available upon request. Replicate RT–PCR reactions from independently produced RNA preparations gave identical sequence results. The 2,280-nucleotide complete coding sequence of PB2 was amplified in 33 overlapping fragments. The 2,274-nucleotide coding sequence of PB1 was amplified in 33 overlapping fragments. The 2,151-nucleotide coding sequence of PA was amplified in 32 overlapping fragments. The PCR products ranged in size from 77–138 bp.

Phylogenetic analyses. Phylogenetic analyses of the three polymerase genes were done using standard methods. We generated trees using the neighbour-joining (NJ) algorithm, with proportion of differences as the distance measure using MEGA version 2.1. Character evolution was analysed with the MacClade program after a parsimony analysis using PAUP version 4.0 beta, using ACTRAN as the optimization method. Trees were also generated using maximum-likelihood with midpoint rooting. All algorithms generated comparable trees, with major clades representing human, classical swine and avian-like viruses (NJ trees shown in Figs 1–3; complete data set available upon request). Polymerase segment sequences used in this analysis were obtained from GenBank and the Influenza Sequence Databank (ISD). (See Supplementary Table 1 for a list of sequences used.) For the PB2 gene, 83 sequences were used, all of which were full length. For the PB1 gene, 91 sequences were used, three of which were not full length. For the PA gene, 105 sequences were used, six of which were not full length.

(**437**, 889-893; 2005)

Jeffery K. Taubenberger[1], Ann H. Reid[1]†, Raina M. Lourens[1]†, Ruixue Wang[1], Guozhong Jin[1] & Thomas G. Fanning[1]

[1] Department of Molecular Pathology, Armed Forces Institute of Pathology, Rockville, Maryland 20850, USA

† Present addresses: Board on Life Sciences, The National Academies, 6th Floor, 500 Fifth Street N.W., Washington DC 20001, USA (A.H.R.); University of Iowa, Roy J. and Lucille A. Carver College of Medicine, 200 CMAB, Iowa City, Iowa 52242, USA (R.M.L.)

Received 30 June; accepted 19 September 2005.

1918 年病毒分析中得到的氨基酸改变也在导致人类死亡的 H5N1 和 H7N7 高致病性禽流感株中发现，这非常有意思；同时这也表明上述改变可能促进病毒在人细胞中复制并增加致病性。有可能 1918 年病毒的高致病性与其适应人类的禽流感病毒本质有关。这些改变可能反映了平行进化的过程，即甲型流感病毒（禽流感）应对适应压力做出的突变。它也表明我们可以绘制出禽流感病毒适应人类的遗传基础。

方　法

RNA 分离、扩增和测序　根据生产商的说明，用 Trizol 试剂（Invitrogen）从冰冻的 1918 年人肺组织中提取病毒 RNA。每个片段都进行至少两次逆转录、扩增和测序。逆转录聚合酶链式反应（RT–PCR）、产物的分离和测序之前已经描述过[4]。引物和引物序列的列表可以向我们索取。从单独制备的 RNA 进行重复的 RT–PCR 反应得到了相同的序列产物。PB2 的 2,280 个核苷酸的完整编码序列在 33 个重叠的片段中扩增出来。PB1 的 2,274 个核苷酸的编码序列在 33 个重叠的片段中扩增出来。PA 的 2,151 个核苷酸的编码序列在 32 个重叠的片段中扩增出来。PCR 产物的大小在 77 ~ 138 bp 之间。

系统发育分析　三个聚合酶基因的系统发育分析用标准的方法进行。我们用邻接（NJ）算法绘制进化树，使用 MEGA 2.1 版将差异的比例作为距离度量。使用 ACTRAN 作为优化方法，在使用 PAUP 4.0 beta 版进行简易分析之后用 MacClade 程序分析特征进化。进化树也可用带有中点根的最大似然产生。所有的算法产生了可以比较的树，其主要的分支代表人类病毒、猪病毒和类似禽类的病毒（NJ 树在图 1 ~ 图 3 中显示；详细的数据可以向我们索取）。本分析中使用的聚合酶基因片段序列从 GenBank 和流感序列数据库获得。（使用的序列列表详见补充信息表 1）对于 PB2 基因，使用了 83 个序列，所有都是全长的。对于 PB1 基因，使用了 91 个序列，其中三个不是全长的。对于 PA 基因，使用了 105 个序列，其中六个不是全长的。

（毛晨晖 翻译；陈继征 审稿）

References:

1. Fodor, E. & Brownlee, G. G. in *Influenza* (ed. Potter, C. W.) 1-29 (Elsevier, Amsterdam, 2002).

2. Subbarao, E. K., London, W. & Murphy, B. R. A single amino acid in the PB2 gene of influenza A virus is a determinant of host range. *J. Virol.* **67**, 1761-1764 (1993).

3. Naffakh, N., Massin, P., Escriou, N., Crescenzo-Chaigne, B. & van der Werf, S. Genetic analysis of the compatibility between polymerase proteins from human and avian strains of influenza A viruses. *J. Gen. Virol.* **81**, 1283-1291 (2000).

4. Reid, A. H., Fanning, T. G., Hultin, J. V. & Taubenberger, J. K. Origin and evolution of the 1918 "Spanish" influenza virus hemagglutinin gene. *Proc. Natl Acad. Sci. USA* **96**, 1651-1656 (1999).

5. Reid, A. H., Fanning, T. G., Janczewski, T. A. & Taubenberger, J. K. Characterization of the 1918 "Spanish" influenza virus neuraminidase gene. *Proc. Natl Acad. Sci. USA* **97**, 6785-6790 (2000).

6. Basler, C. F. *et al.* Sequence of the 1918 pandemic influenza virus nonstructural gene (NS) segment and characterization of recombinant viruses bearing the 1918 NS genes. *Proc. Natl Acad. Sci. USA* **98**, 2746-2751 (2001).

7. Reid, A. H., Fanning, T. G., Janczewski, T. A., McCall, S. & Taubenberger, J. K. Characterization of the 1918 "Spanish" influenza virus matrix gene segment. *J. Virol.* **76**, 10717-10723 (2002).

8. Reid, A. H., Fanning, T. G., Janczewski, T. A., Lourens, R. & Taubenberger, J. K. Novel origin of the 1918 pandemic influenza virus nucleoprotein gene segment. *J. Virol.* **78**, 12462-12470 (2004).

9. Kawaoka, Y., Krauss, S. & Webster, R. G. Avian-to-human transmission of the PB1 gene of influenza A viruses in the 1957 and 1968 pandemics. *J. Virol.* **63**, 4603-4608 (1989).

10. Scholtissek, C., Rohde, W., Von Hoyningen, V. & Rott, R. On the origin of the human influenza virus subtypes H2N2 and H3N2. *Virology* **87**, 13-20 (1978).

11. Reid, A. H., Taubenberger, J. K. & Fanning, T. G. Evidence of an absence: the genetic origins of the 1918 pandemic influenza virus. *Nature Rev. Microbiol.* **2**, 909-914 (2004).

12. Johnson, N. P. & Mueller, J. Updating the accounts: global mortality of the 1918–1920 "Spanish" influenza pandemic. *Bull. Hist. Med.* **76**, 105-115 (2002).

13. Geiss, G. K. *et al.* Cellular transcriptional profiling in influenza A virus-infected lung epithelial cells: the role of the nonstructural NS1 protein in the evasion of the host innate defense and its potential contribution to pandemic influenza. *Proc. Natl Acad. Sci. USA* **99**, 10736-10741 (2002).

14. Tumpey, T. M. *et al.* Existing antivirals are effective against influenza viruses with genes from the 1918 pandemic virus. *Proc. Natl Acad. Sci. USA* **99**, 13849-13854 (2002).

15. Tumpey, T. M. *et al.* Pathogenicity and immunogenicity of influenza viruses with genes from the 1918 pandemic virus. *Proc. Natl Acad. Sci. USA* **101**, 3166-3171 (2004).

16. Kash, J. C. *et al.* The global host immune response: contribution of HA and NA genes from the 1918 Spanish influenza to viral pathogenesis. *J. Virol.* **78**, 9499-9511 (2004).

17. Kobasa, D. *et al.* Enhanced virulence of influenza A viruses with the haemagglutinin of the 1918 pandemic virus. *Nature* **431**, 703-707 (2004).

18. Chen, W. *et al.* A novel influenza A virus mitochondrial protein that induces cell death. *Nature Med.* **7**, 1306-1312 (2001).

19. Shinya, K. *et al.* PB2 amino acid at position 627 affects replicative efficiency, but not cell tropism, of Hong Kong H5N1 influenza A viruses in mice. *Virology* **320**, 258-266 (2004).

20. Fouchier, R. A. *et al.* Avian influenza A virus (H7N7) associated with human conjunctivitis and a fatal case of acute respiratory distress syndrome. *Proc. Natl Acad. Sci. USA* **101**, 1356-1361 (2004).

21. Chen, H. *et al.* Avian flu: H5N1 virus outbreak in migratory waterfowl. *Nature* **436**, 191-192 (2005).

22. Li, K. S. *et al.* Genesis of a highly pathogenic and potentially pandemic H5N1 influenza virus in eastern Asia. *Nature* **430**, 209-213 (2004).

23. Puthavathana, P. *et al.* Molecular characterization of the complete genome of human influenza H5N1 virus isolates from Thailand. *J. Gen. Virol.* **86**, 423-433 (2005).

24. Toyoda, T., Adyshev, D. M., Kobayashi, M., Iwata, A. & Ishihama, A. Molecular assembly of the influenza virus RNA polymerase: determination of the subunit-subunit contact sites. *J. Gen. Virol.* **77**, 2149-2157 (1996).

25. Masunaga, K., Mizumoto, K., Kato, H., Ishihama, A. & Toyoda, T. Molecular mapping of influenza virus RNA polymerase by site-specific antibodies. *Virology* **256**, 130-141 (1999).

26. Ohtsu, Y., Honda, Y., Sakata, Y., Kato, H. & Toyoda, T. Fine mapping of the subunit binding sites of influenza virus RNA polymerase. *Microbiol. Immunol.* **46**, 167-175 (2002).

27. Biswas, S. K. & Nayak, D. P. Mutational analysis of the conserved motifs of influenza A virus polymerase basic protein 1. *J. Virol.* **68**, 1819-1826 (1994).

28. Gonzalez, S. & Ortin, J. Distinct regions of influenza virus PB1 polymerase subunit recognize vRNA and cRNA templates. *EMBO J.* **18**, 3767-3775 (1999).

29. Guo, Y. J. *et al.* Characterization of the pathogenicity of members of the newly established H9N2 influenza virus lineages in Asia. *Virology* **267**, 279-288 (2000).

30. Holmes, E. C. *et al.* Whole-genome analysis of human influenza A virus reveals multiple persistent lineages and reassortment among recent H3N2 viruses. *PLoS Biol.* **3**, e300 (2005).

Supplementary Information is linked to the online version of the paper at www.nature.com/nature.

Acknowledgements. The research described in this report was done using stringent safety precautions to protect the laboratory workers, the environment and the public from this virus. The intention of this research is to provide the basis for understanding how influenza pandemic strains form and to help ascertain the risk of future influenza pandemics. This study was partially supported by a grant to J.K.T. from the National Institutes of Health, and by intramural funds from the Armed Forces Institute of Pathology. The opinions contained herein are the private views of the authors and are not to be construed as official or as reflecting the views of the US Department of the Army or the US Department of Defense.

Author Contributions. J.K.T. planned the project, and A.H.R., R.M.L., R.W. and G.J. generated the sequence data. J.K.T., A.H.R. and T.G.F. performed data analysis. J.K.T. wrote the manuscript.

Author Information. Coding sequences of the PB2, PB1 and PA genes have been deposited in GenBank under accession numbers DQ208309, DQ208310 and DQ208311, respectively. Reprints and permissions information is available at npg.nature.com/reprintsandpermissions. The authors declare no competing financial interests. Correspondence and requests for materials should be addressed to J.K.T. (taubenberger@afip.osd.mil).

A Haplotype Map of the Human Genome

The International HapMap Consortium*

Editor's Note

Three years after its official launch, the International HapMap Consortium revealed its completed haplotype map of the human genome. The map reveals the most common genetic differences found across the entire genome for 269 humans from four different populations. The research, led by geneticists David Altshuler and Peter Donnelly, groups single nucleotide polymorphisms (SNPs)—the single-letter differences in the DNA between individuals—into haplotypes, combinations of SNPs that have travelled together over evolutionary time. The goal is to understand the complex genetic changes underlying common diseases such as cardiovascular disease and cancer. Two years later the map was "upgraded" to a second-generation version of over 3 million SNPs.

Inherited genetic variation has a critical but as yet largely uncharacterized role in human disease. Here we report a public database of common variation in the human genome: more than one million single nucleotide polymorphisms (SNPs) for which accurate and complete genotypes have been obtained in 269 DNA samples from four populations, including ten 500-kilobase regions in which essentially all information about common DNA variation has been extracted. These data document the generality of recombination hotspots, a block-like structure of linkage disequilibrium and low haplotype diversity, leading to substantial correlations of SNPs with many of their neighbours. We show how the HapMap resource can guide the design and analysis of genetic association studies, shed light on structural variation and recombination, and identify loci that may have been subject to natural selection during human evolution.

DESPITE the ever-accelerating pace of biomedical research, the root causes of common human diseases remain largely unknown, preventative measures are generally inadequate, and available treatments are seldom curative. Family history is one of the strongest risk factors for nearly all diseases—including cardiovascular disease, cancer, diabetes, autoimmunity, psychiatric illnesses and many others—providing the tantalizing but elusive clue that inherited genetic variation has an important role in the pathogenesis of disease. Identifying the causal genes and variants would represent an important step in the path towards improved prevention, diagnosis and treatment of disease.

More than a thousand genes for rare, highly heritable "mendelian" disorders have been

---

* The full list of authors and affiliations has been removed. The original text is available in the *Nature* online archive.

人类基因组的单体型图谱

国际单体型图谱协作组 *

编者按

计划正式启动三年后，国际HapMap协作组发布了完整的人类基因组单体型图谱。这份图谱揭示了来自四个不同种群的269个人类个体在整个基因组范围内的最常见的遗传差异。由遗传学家戴维·阿特舒勒和彼得·唐纳利领导的这项研究，将单核苷酸多态性(SNP)——个体间DNA上单个字母的差异——集合成单体型，即随着时间共同演化的SNP组合。项目的目标是理解常见疾病如心血管疾病和癌症等背后的复杂的遗传变化。两年后这份图谱"升级"为包含超过300万个SNP的第二代版本。

遗传变异在人类疾病中有着重要作用，但这些作用大部分还未被阐明。在本文中我们报道了一个公开的人类基因组中常见变异的数据库，它包含超过100万个单核苷酸多态性(SNP)。这些位点的准确而全面的基因型是通过来自四个种群的269份DNA样本获得的，我们也在10个500 kb的区域中提取了基本上所有的常见DNA变异的信息。这些数据证明了重组热点、连锁不平衡的区块结构和低单体型密度的普遍性，进而证实了SNP与很多邻近SNP的大量关联。我们展示了HapMap资源如何指导遗传学关联研究的设计和分析，为结构变异和重组提供线索，以及鉴定在人类演化过程中可能经历过自然选择的基因座。

尽管生物医学研究的步伐不断加速，但常见人类疾病的根本原因仍然大部分未知，预防措施普遍不足，现有的治疗手段鲜有疗效。对于几乎所有的疾病——包括心血管疾病、癌症、糖尿病、自身免疫病、精神疾病和很多其他的疾病，家族史是最强的风险因素之一，提供了诱人又难以捉摸的线索，即遗传变异在疾病的发病机理中有着重要的作用。在通往更好地预防、诊断和治疗疾病的道路上，鉴定导致疾病的基因和变异将代表着重要的一步。

对于罕见的、高度遗传的"孟德尔"疾病，已有超过1,000个基因被鉴定出来。

* 本书略去了作者和单位名单。原文可从《自然》在线数据库中获得。

identified, in which variation in a single gene is both necessary and sufficient to cause disease. Common disorders, in contrast, have proven much more challenging to study, as they are thought to be due to the combined effect of many different susceptibility DNA variants interacting with environmental factors.

Studies of common diseases have fallen into two broad categories: family-based linkage studies across the entire genome, and population-based association studies of individual candidate genes. Although there have been notable successes, progress has been slow due to the inherent limitations of the methods; linkage analysis has low power except when a single locus explains a substantial fraction of disease, and association studies of one or a few candidate genes examine only a small fraction of the "universe" of sequence variation in each patient.

A comprehensive search for genetic influences on disease would involve examining all genetic differences in a large number of affected individuals and controls. It may eventually become possible to accomplish this by complete genome resequencing. In the meantime, it is increasingly practical to systematically test common genetic variants for their role in disease; such variants explain much of the genetic diversity in our species, a consequence of the historically small size and shared ancestry of the human population.

Recent experience bears out the hypothesis that common variants have an important role in disease, with a partial list of validated examples including *HLA* (autoimmunity and infection)[1], *APOE4* (Alzheimer's disease, lipids)[2], Factor V^{Leiden} (deep vein thrombosis)[3], *PPARG* (encoding PPARγ; type 2 diabetes)[4,5], *KCNJ11* (type 2 diabetes)[6], *PTPN22* (rheumatoid arthritis and type 1 diabetes)[7,8], insulin (type 1 diabetes)[9], *CTLA4* (autoimmune thyroid disease, type 1 diabetes)[10], *NOD2* (inflammatory bowel disease)[11,12], complement factor H (age-related macular degeneration)[13-15] and *RET* (Hirschsprung disease)[16,17], among many others.

Systematic studies of common genetic variants are facilitated by the fact that individuals who carry a particular SNP allele at one site often predictably carry specific alleles at other nearby variant sites. This correlation is known as linkage disequilibrium (LD); a particular combination of alleles along a chromosome is termed a haplotype.

LD exists because of the shared ancestry of contemporary chromosomes. When a new causal variant arises through mutation—whether a single nucleotide change, insertion/deletion, or structural alteration—it is initially tethered to a unique chromosome on which it occurred, marked by a distinct combination of genetic variants. Recombination and mutation subsequently act to erode this association, but do so slowly (each occurring at an average rate of about 10^{-8} per base pair (bp) per generation) as compared to the number of generations (typically 10^4 to 10^5) since the mutational event.

The correlations between causal mutations and the haplotypes on which they arose have long served as a tool for human genetic research: first finding association to a haplotype, and then subsequently identifying the causal mutation(s) that it carries. This was pioneered in studies of the *HLA* region, extended to identify causal genes for mendelian diseases

在这些疾病中，单个基因的变异对于导致疾病是必要且充分的。相比而言，常见疾病的研究已被证明更具挑战性，因为这些疾病被认为是源于很多不同的易感 DNA 变异与环境因素相互作用的组合效应。

常见疾病的研究分为两大类：基于家系的全基因组连锁研究和基于自然人群的对单个候选基因的关联研究。尽管这些研究已经有令人瞩目的成功，但是由于方法的内在限制，进展一直比较缓慢；连锁分析的效力通常很低，除非单个基因座能解释疾病的很大一部分，而一个或几个候选基因的关联研究则仅仅是检查了一个病人体内序列变异“宇宙”的很小一部分。

全面寻找遗传对疾病的影响将涉及在大量的病例和对照中研究所有的遗传差异。通过完整的基因组重测序，有可能最终实现这一点。同时，系统检测常见的遗传变异在疾病中的作用变得越来越可行；人类种群在历史上有着较小的规模和共同的祖先，因此这样的变异可以解释我们自身物种中很大一部分的遗传多样性。

最新的实践证明了常见变异在疾病中有着重要的作用这个假说，部分得到验证的例子包括 *HLA*(自身免疫病和感染)[1]、*APOE4*(阿尔茨海默病，脂质)[2]，V^{Leiden} 因子(深部静脉血栓形成)[3]、*PPARG*(编码 PPARγ；2 型糖尿病)[4,5]、*KCNJ11*(2 型糖尿病)[6]、*PTPN22*(类风湿性关节炎和 1 型糖尿病)[7,8]、胰岛素(1 型糖尿病)[9]、*CTLA4*(自身免疫性甲状腺病，1 型糖尿病)[10]、*NOD2*(炎性肠病)[11,12]、补体因子 H(老年性黄斑变性)[13-15] 和 *RET*(先天性巨结肠)[16,17]，此外还有很多。

某些个体如果在一个位点携带一个特定的 SNP 等位基因，则常常可预见地在邻近的其他变异位点也携带特定的等位基因，这个事实促进了对常见遗传变异的系统研究。这种关联被称为连锁不平衡(LD)；等位基因在一条染色体上特定的组合被称为单体型。

LD 存在是由于现在的染色体有着共同的祖先。当一种新的导致疾病的变异通过突变——不管是单核苷酸改变、插入/缺失或者结构变异——而产生时，它最初会局限于它所发生的染色体，以独特的遗传变异组合为特征。重组和突变随后削弱这种关联，但是相对于变异发生后的代数(通常 10^4 到 10^5)，这种削弱是很缓慢的(分别以每代每碱基对(bp)大约 10^{-8} 的平均速率发生)。

导致疾病的变异与其所在的单体型的相关性早已被用作人类遗传研究的工具：首先发现与一个单体型的关联，然后鉴定该单体型含有的导致疾病的突变。这种方法最早应用在 *HLA* 区域的研究，后来拓展到孟德尔疾病(如囊性纤维化[18] 和

(for example, cystic fibrosis[18] and diastrophic dysplasia[19]), and most recently for complex disorders such as age-related macular degeneration[13-15].

Early information documented the existence of LD in the human genome[20,21]; however, these studies were limited (for technical reasons) to a small number of regions with incomplete data, and general patterns were challenging to discern. With the sequencing of the human genome and development of high-throughput genomic methods, it became clear that the human genome generally displays more LD[22] than under simple population genetic models[23], and that LD is more varied across regions, and more segmentally structured[24-30], than had previously been supposed. These observations indicated that LD-based methods would generally have great value (because nearby SNPs were typically correlated with many of their neighbours), and also that LD relationships would need to be empirically determined across the genome by studying polymorphisms at high density in population samples.

The International HapMap Project was launched in October 2002 to create a public, genome-wide database of common human sequence variation, providing information needed as a guide to genetic studies of clinical phenotypes[31]. The project had become practical by the confluence of the following: (1) the availability of the human genome sequence; (2) databases of common SNPs (subsequently enriched by this project) from which genotyping assays could be designed; (3) insights into human LD; (4) development of inexpensive, accurate technologies for high-throughput SNP genotyping; (5) web-based tools for storing and sharing data; and (6) frameworks to address associated ethical and cultural issues[32]. The project follows the data release principles of an international community resource project (http://www.wellcome.ac.uk/doc_WTD003208.html), sharing information rapidly and without restriction on its use.

The HapMap data were generated with the primary aim of guiding the design and analysis of medical genetic studies. In addition, the advent of genome-wide variation resources such as the HapMap opens a new era in population genetics, offering an unprecedented opportunity to investigate the evolutionary forces that have shaped variation in natural populations.

The Phase I HapMap

Phase I of the HapMap Project set as a goal genotyping at least one common SNP every 5 kilobases (kb) across the genome in each of 269 DNA samples. For the sake of practicality, and motivated by the allele frequency distribution of variants in the human genome, a minor allele frequency (MAF) of 0.05 or greater was targeted for study. (For simplicity, in this paper we will use the term "common" to mean a SNP with MAF $\geqslant 0.05$.) The project has a Phase II, which is attempting genotyping of an additional 4.6 million SNPs in each of the HapMap samples.

To compare the genome-wide resource to a more complete database of common variation—one in which all common SNPs and many rarer ones have been discovered and tested—a representative collection of ten regions, each 500 kb in length, was selected

畸形性骨发育不良[19])的致病基因的鉴定，最近则应用于复杂疾病，如老年性黄斑变性[13-15]。

早期的信息证实了人类基因组中LD的存在[20,21]；但是这些研究(由于技术原因)局限于数据不完整的少量区域，难以了解其一般的模式。随着人类基因组的测序和高通量基因组方法的发展，我们已经清楚，人类基因组一般比简单种群遗传模型[23]有着更多的LD[22]，此外，与之前的预估相比，LD在不同区域间的变化更大，也呈现更加分节段的结构[24-30]。这些现象说明，基于LD的方法总体来说将会有巨大的价值(因为邻近的SNP通常与很多附近的SNP关联)，也说明需要通过实验研究种群样本中高密度的多态性来确定LD关系。

国际HapMap计划在2002年10月启动，目的是创建一个公开的、全基因组范围的人类常见序列变异数据库，为指导临床表型的遗传学研究提供所需要的信息[31]。这个计划由于以下内容的汇集而已经变得可行：(1)人类基因组序列的获得；(2)设计基因型分型所需的常见SNP数据库的完善(该数据库被此计划进一步丰富)；(3)对人类LD的了解；(4)廉价而准确的高通量SNP基因型分型技术的发展；(5)基于网络的存储和分享数据的工具；以及(6)解决相关的伦理和文化问题的框架[32]。本计划遵循国际共享资源项目的数据释放原则(http://www.wellcome.ac.uk/doc_WTD003208.html)，迅速分享信息，不限制使用。

HapMap数据的主要目的是指导医学遗传学研究的设计和分析。另外，像HapMap这样的全基因组的变异资源将开启种群遗传学的新纪元，为研究在自然种群中塑造了变异的进化动力提供一个前所未有的机会。

第I阶段的HapMap

HapMap计划的第I阶段目标是，对于269个DNA样本的每一个，均能在全基因组范围内的每5,000个碱基中至少对一个常见SNP进行基因分型。为了实用，同时根据人类基因组中变异的等位基因频率分布，我们选择研究最小等位基因频率(MAF)大于或等于0.05的SNP。(为了简便，本文中我们使用“常见的”表示MAF ≥ 0.05的SNP)本计划的第II阶段将在各个HapMap样本中对另外460万个SNP进行基因分型。

为了比较全基因组范围的资源与更完整的常见变异数据库——该数据库中所有的常见SNP和很多相对罕见的SNP已经被发掘和检验，我们从ENCODE(DNA元

from the ENCODE (Encyclopedia of DNA Elements) Project[33]. Each 500-kb region was sequenced in 48 individuals, and all SNPs in these regions (discovered or in dbSNP) were genotyped in the complete set of 269 DNA samples.

The specific samples examined are: (1) 90 individuals (30 parent–offspring trios) from the Yoruba in Ibadan, Nigeria (abbreviation YRI); (2) 90 individuals (30 trios) in Utah, USA, from the Centre d'Etude du Polymorphisme Humain collection (abbreviation CEU); (3) 45 Han Chinese in Beijing, China (abbreviation CHB); (4) 44 Japanese in Tokyo, Japan (abbreviation JPT).

Because none of the samples was collected to be representative of a larger population such as "Yoruba", "Northern and Western European", "Han Chinese", or "Japanese" (let alone of all populations from "Africa", "Europe", or "Asia"), we recommend using a specific local identifier (for example, "Yoruba in Ibadan, Nigeria") to describe the samples initially. Because the CHB and JPT allele frequencies are generally very similar, some analyses below combine these data sets. When doing so, we refer to three "analysis panels" (YRI, CEU, CHB+JPT) to avoid confusing this analytical approach with the concept of a "population".

Important details about the design of the HapMap Project are presented in the Methods, including: (1) organization of the project; (2) selection of DNA samples for study; (3) increasing the number and annotation of SNPs in the public SNP map (dbSNP) from 2.6 million to 9.2 million (Fig. 1); (4) targeted sequencing of the ten ENCODE regions, including evaluations of false-positive and false-negative rates; (5) genotyping for the genome-wide map; (6) intense efforts that monitored and established the high quality of the data; and (7) data coordination and distribution through the project Data Coordination Center (DCC) (http://www.hapmap.org).

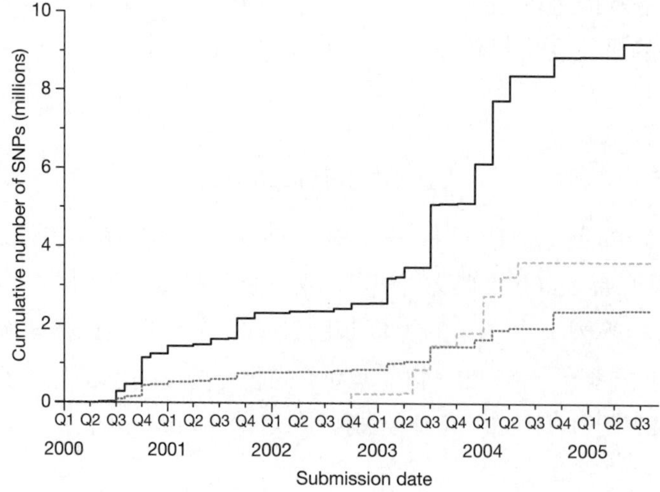

Fig. 1. Number of SNPs in dbSNP over time. The cumulative number of non-redundant SNPs (each mapped to a single location in the genome) is shown as a solid line, as well as the number of SNPs validated by genotyping (dotted line) and double-hit status (dashed line). Years are divided into quarters (Q1–Q4).

件百科全书）计划[33]中选择了10个具有代表性的区域，每个区域的长度是500 kb。先在48个个体中对这10个500 kb的区域进行测序，然后在全部269个DNA样本中对这些区域中所有的SNP(新发现的或已收录在dbSNP数据库的）进行基因分型。

检测的样本包括：(1)尼日利亚伊巴丹市的90个(30个亲本–子孙三联家系）约鲁巴人（简写为YRI)；(2)人类多态性研究中心收集的美国犹他州的90个(30个三联家系）个体（简写为CEU)；(3)中国北京市的45个汉族中国人（简写为CHB)；(4)日本东京市的44个日本人（简写为JPT)。

由于收集的这些样本并不是用来代表更大的种群如“约鲁巴人”“北欧和西欧人”“汉族中国人”或者“日本人”（更不代表“非洲”“欧洲”或者“亚洲”的所有种群），一开始我们建议使用特定的局部地区标识（例如，“尼日利亚伊巴丹市的约鲁巴人”）来描述样本。由于CHB和JPT等位基因频率大体上非常相似，下面的一些分析将其合并到一起。这样操作时，我们使用三个“分析小组”(YRI、CEU、CHB+JPT)的说法来避免将这种分析方法与“种群”的概念混淆。

关于HapMap计划设计的重要细节在方法部分中进行了描述，包括：(1)计划的组织；(2)用于研究的DNA样本的选择；(3)公共SNP图谱（dbSNP数据库）中SNP的数量和注释的增长（从260万到920万）(图1)；(4)对10个ENCODE区域的靶向测序，包括假阳性率和假阴性率的评估；(5)全基因组图谱的基因分型；(6)为建立和监控数据的高质量所付出的巨大努力；(7)通过本计划的数据协调中心进行的数据协调和分配（http://www.hapmap.org)。

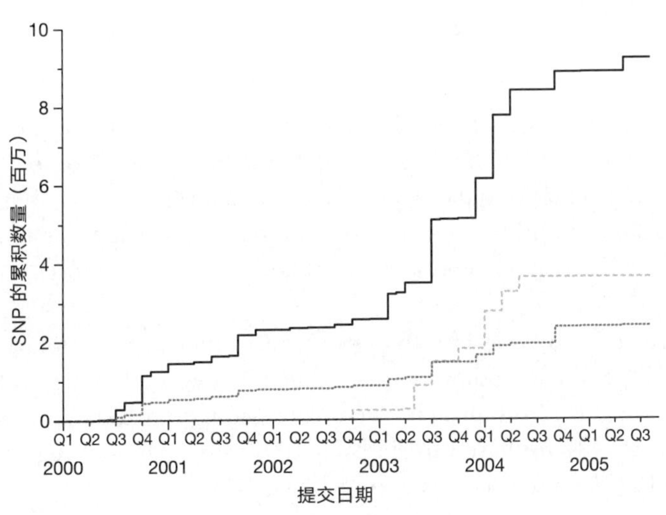

图1. dbSNP数据库中SNP数量随时间的变化。实线表示非冗余SNP(各个SNP比对到基因组的单一位置）的累积数量，虚线表示被基因分型验证的SNP数量，短划线表示“双打击”状态。每年分为四个季度(Q1 ~ Q4)。

Description of the data. The Phase I HapMap contains 1,007,329 SNPs that passed a set of quality control (QC) filters (see Methods) in each of the three analysis panels, and are polymorphic across the 269 samples. SNP genotyping was distributed across centres by chromosomal region, with several technologies employed (Table 1). Each centre followed the same standard rules for SNP selection, quality control and data release; all SNPs were genotyped in the full set of 269 samples. Some centres genotyped more SNPs than required by the rules.

Table 1. Genotyping centres

Centre	Chromosomes	Technology
RIKEN	5, 11, 14, 15, 16, 17, 19	Third Wave Invader
Wellcome Trust Sanger Institute	1, 6, 10, 13, 20	Illumina BeadArray
McGill University and Génome Québec Innovation Centre	2, 4p	Illumina BeadArray
Chinese HapMap Consortium*	3, 8p, 21	Sequenom MassExtend, Illumina BeadArray
Illumina	8q, 9, 18q, 22, X	Illumina BeadArray
Broad Institute of Harvard and MIT	4q, 7q, 18p, Y, mtDNA	Sequenom MassExtend, Illumina BeadArray
Baylor College of Medicine with ParAllele BioScience	12	ParAllele MIP
University of California, San Francisco, with Washington University in St Louis	7p	PerkinElmer AcycloPrime-FP
Perlegen Sciences	5 Mb (ENCODE) on 2, 4, 7, 8, 9, 12, 18 in CEU	High-density oligonucleotide array

* The Chinese HapMap Consortium consists of the Beijing Genomics Institute, the Chinese National Human Genome Center at Beijing, the University of Hong Kong, the Hong Kong University of Science and Technology, the Chinese University of Hong Kong, and the Chinese National Human Genome Center at Shanghai.

Extensive, blinded quality assessment (QA) exercises documented that these data are highly accurate (99.7%) and complete (99.3%, see also Supplementary Table 1). All genotyping centres produced high-quality data (accuracy more than 99% in the blind QA exercises, Supplementary Tables 2 and 3), and missing data were not biased against heterozygotes. The Supplementary Information contains the full details of these efforts.

Although SNP selection was generally agnostic to functional annotation, 11,500 non-synonymous cSNPs (SNPs in coding regions of genes where the different SNP alleles code for different amino acids in the protein) were successfully typed in Phase I. (An effort was made to prioritize cSNPs in Phase I in choosing SNPs for each 5-kb region; all known non-synonymous cSNPs were attempted as part of Phase II.)

数据的描述 HapMap 第 I 阶段包含 1,007,329 个在三个分析小组中均通过了一套质控(QC)过滤(详见方法)，并且在 269 个样本中具有多态性的 SNP。SNP 基因分型任务按照染色体区域分配到各个中心，几种技术也部署到各个中心(表 1)。各个中心遵循相同的标准规则进行 SNP 挑选、质控和数据释放；所有的 SNP 都在整套 269 个样本中进行了基因分型。一些中心进行基因分型的 SNP 数量超出了规则所要求的。

表 1. 基因分型中心

中心	染色体	技术
日本理化研究所	5、11、14、15、16、17、19	Third Wave Invader
维康信托基金会桑格研究院	1、6、10、13、20	Illumina BeadArray
麦吉尔大学和基因组魁北克创新中心	2、4p	Illumina BeadArray
中国 HapMap 协作组 *	3、8p、21	Sequenom MassExtend, Illumina BeadArray
Illumina 公司	8q、9、18q、22、X	Illumina BeadArray
哈佛大学和麻省理工学院布罗德研究院	4q、7q、18p、Y、线粒体 DNA	Sequenom MassExtend, Illumina BeadArray
贝勒医学院和 ParAllele BioScience 公司	12	ParAllele MIP
加州大学旧金山分校和圣路易斯华盛顿大学	7p	PerkinElmer AcycloPrime-FP
Perlegen Sciences 公司	CEU 中 2、4、7、8、9、12、18 号染色体的 5 Mb 区域(ENCODE)	高密度寡核苷酸芯片

* 中国 HapMap 协作组包括北京基因组研究所、中国国家人类基因组北方研究中心、香港大学、香港科技大学、香港中文大学和中国国家人类基因组南方研究中心。

广泛的盲法质量评价(QA)发现这些数据是高度准确(99.7%)和完整的(99.3%，还可详见补充信息表 1)。所有的基因分型中心均产生高质量的数据(在盲法 QA 中准确度超过 99%，补充信息表 2 和 3)，并且缺失的数据没有相对杂合子的偏倚。补充信息包含这些工作的所有细节。

尽管选择 SNP 时通常并不知晓功能注释，但是在第 I 阶段，有 11,500 个非同义 cSNP(即位于基因编码区，且不同的等位基因编码蛋白质的不同氨基酸的 SNP)被成功分型。(对于各个 5 kb 区域，选择 SNP 时优先选择第 I 阶段中的 cSNP；所有已知的非同义 cSNP 作为第 II 阶段的一部分。)

Table 2. ENCODE project regions and genotyping

Region name	Chromosome band	Genomic interval (NCBI) (base numbers)†	Gene density (%)‡	Conservation score (%)§	Pedigree-based recombination rate (cM Mb^{-1})‖	Population-based recombination rate (cM Mb^{-1})¶	G+C content#	Available SNPs			Successfully genotyped SNPs††	Sequencing centre/genotyping centre(s)‡‡
								dbSNP☆	Sequence**	Total		
ENr112	2p16.3	51,633,239–52,133,238	0	3.8	0.8	0.9	0.35	1,570	1,762	3,332	2,275	Broad/McGill-GQIC
ENr131	2q37.1	234,778,639–235,278,638	4.6	1.3	2.2	2.5	0.43	1,736	1,259	2,995	1,910	Broad/McGill-GQIC
ENr113	4q26	118,705,475–119,205,474	0	3.9	0.6	0.9	0.35	1,444	2,053	3,497	2,201	Broad/Broad
ENm010	7p15.2	26,699,793–27,199,792	5.0	22.0	0.9	0.9	0.44	1,220	1,795	3,015	1,271	Baylor/UCSF-WU, Broad
ENm013*	7q21.13	89,395,718–89,895,717	5.5	4.4	0.4	0.5	0.38	1,394	1,917	3,311	1,807	Broad/Broad
ENm014*	7q31.33	126,135,436–126,632,577	2.9	11.2	0.4	0.9	0.39	1,320	1,664	2,984	1,966	Broad/Broad
ENr321	8q24.11	118,769,628–119,269,627	3.2	11.4	0.6	1.1	0.41	1,430	1,508	2,938	1,758	Baylor/Illumina
ENr232	9q34.11	127,061,347–127,561,346	5.9	8.3	2.7	2.6	0.52	1,444	1,523	2,967	1,324	Baylor/Illumina
ENr123	12q12	38,626,477–39,126,476	3.1	1.7	0.3	0.8	0.36	1,877	1,379	3,256	1,792	Baylor/Baylor
ENr213	18q12.1	23,717,221–24,217,220	0.9	7.4	1.2	0.9	0.37	1,330	1,459	2,789	1,640	Baylor/Illumina
Total	–	–	–	–	–	–	–	14,765	16,319	31,084	17,944	–

McGill-GQIC, McGill University and Génome Québec Innovation Centre.

* These regions were truncated to 500 kb for resequencing.

† Sequence build 34 coordinates.

‡ Gene density is defined as the percentage of bases covered either by Ensembl genes or human mRNA best BLAT alignments in the UCSC Genome Browser database.

§ Non-exonic conservation with mouse sequence was measured by taking 125 base non-overlapping sub-windows inside the 500,000 base windows. Sub-windows with less than 75% of their bases in a mouse alignment were discarded. Of the remaining sub-windows, those with at least 80% base identity were used to calculate the conservation

表 2. ENCODE 计划区域和基因分型

区域名称	染色体区带	基因组间距（NCBI）（碱基数量）†	基因密度（%）‡	保守性数值（%）§	基于家系的重组率（cM · Mb^{-1}）‖	基于种群的重组率（cM · Mb^{-1}）¶	G+C 含量 #	已有的 SNP			成功基因分型的 SNP††	测序中心/基因分型中心 ‡‡
								dbSNP☆	测序 **	总计		
ENr112	2p16.3	51,633,239 ~ 52,133,238	0	3.8	0.8	0.9	0.35	1,570	1,762	3,332	2,275	布罗德/McGill-GQIC
ENr131	2q37.1	234,778,639 ~ 235,278,638	4.6	1.3	2.2	2.5	0.43	1,736	1,259	2,995	1,910	布罗德/McGill-GQIC
ENr113	4q26	118,705,475 ~ 119,205,474	0	3.9	0.6	0.9	0.35	1,444	2,053	3,497	2,201	布罗德/布罗德
ENm010	7p15.2	26,699,793 ~ 27,199,792	5	22.0	0.9	0.9	0.44	1,220	1,795	3,015	1,271	贝勒/UCSC-WU，布罗德
ENm013*	7q21.13	89,395,718 ~ 89,895,717	5.5	4.4	0.4	0.5	0.38	1,394	1,917	3,311	1,807	布罗德/布罗德
ENm014*	7q31.33	126,135,436 ~ 126,632,577	2.9	11.2	0.4	0.9	0.39	1,320	1,664	2,984	1,966	布罗德/布罗德
ENr321	8q24.11	118,769,628 ~ 119,269,627	3.2	11.4	0.6	1.1	0.41	1,430	1,508	2,938	1,758	贝勒/Illumina
ENr232	9q34.11	127,061,347 ~ 127,561,346	5.9	8.3	2.7	2.6	0.52	1,444	1,523	2,967	1,324	贝勒/Illumina
ENr123	12q12	38,626,477 ~ 39,126,476	3.1	1.7	0.3	0.8	0.36	1,877	1,379	3,256	1,792	贝勒/贝勒
ENr213	18q12.1	23,717,221 ~ 24,217,220	0.9	7.4	1.2	0.9	0.37	1,330	1,459	2,789	1,640	贝勒/Illumina
合计	–	–	–	–	–	–	–	14,765	16,319	31,084	17,944	–

McGill-GQIC，麦吉尔大学和基因组魁北克创新中心。

* 这些区域被削减到 500 kb 用于重测序。

† 34 版本序列坐标。

‡ 基因密度定义为被 Ensembl 基因或 UCSC 基因组浏览器数据库中人类 mRNA 最佳 BLAT 比对所覆盖的碱基的百分比。

§ 与小鼠序列的非外显子保守性是通过在 500,000 碱基窗口内取 125 碱基非重叠亚窗口来测定的。若一个亚窗口中小于 75% 的碱基比对到小鼠中，则该亚窗口被弃掉。剩下的亚窗口中，含有至少 80% 相同碱基的亚窗口用于计算保守性数值。小鼠比对中下列区域被弃掉：Ensembl 基因、所有的 GenBank mRNA Blastz 比对、FGenesh++ 基

score. The mouse alignments in regions corresponding to the following were discarded: Ensembl genes, all GenBank mRNA Blastz alignments, FGenesh++ gene predictions, Twinscan gene predictions, spliced EST alignments, and repeats.

‖ The pedigree-based sex-averaged recombination map is from deCODE Genetics[48].

¶ Recombination rate based on estimates from LDhat[46].

G+C content calculated from the sequence of the stated coordinates from sequence build 34.

☆ SNPs in dbSNP build 121 at the time the ENCODE resequencing began and SNPs added to dbSNP in builds 122–125 independent of the resequencing.

** New SNPs discovered through the resequencing reported here (not found by other means in builds 122–125).

†† SNPs successfully genotyped in all analysis panels (YRI, CEU, CHB+JPT).

‡‡ Perlegen genotyped a subset of SNPs in the CEU samples.

因预测、Twinscan 基因预测、剪接的 EST 比对以及重复。

‖ 基于家系的性别平均的重组图谱来自 deCODE Genetics[48]。

¶ 基于 LDhat[46] 估计的重组率。

G+C 含量根据来自序列版本 34 的相应序列计算。

☆ 在 ENCODE 重测序开始时 dbSNP 版本 121 中的 SNP，以及加入 dbSNP 的版本 122 ~ 125 中不依赖重测序结果的 SNP。

** 通过本文报道的重测序而新发现的 SNP(在版本 122 ~ 125 中通过其他方法未发现)。

†† 在所有分析小组(YRI、CEU、CHB+JPT)中被成功基因分型的 SNP。

‡‡ Perlegen 公司对 CEU 样本中一部分 SNP 进行了基因分型。

Across the ten ENCODE regions (Table 2), the density of SNPs was approximately tenfold higher as compared to the genome-wide map: 17,944 SNPs across the 5 megabases (Mb) (one per 279 bp).

More than 1.3 million SNP genotyping assays were attempted (Table 3) to generate the Phase I data on more than 1 million SNPs. The 0.3 million SNPs not part of the Phase I data set include 73,652 that passed QC filters but were monomorphic in all 269 samples. The remaining SNPs failed the QC filters in one or more analysis panels mostly because of inadequate completeness, non-mendelian inheritance, deviations from Hardy–Weinberg equilibrium, discrepant genotypes among duplicates, and data transmission discrepancies.

Table 3. HapMap Phase I genotyping success measures

SNP categories	Analysis panel		
	YRI	CEU	CHB+JPT
Assays submitted	1,273,716	1,302,849	1,273,703
Passed QC filters	1,123,296 (88%)	1,157,650 (89%)	1,134,726 (89%)
Did not pass QC filters*	150,420 (12%)	145,199 (11%)	138,977 (11%)
> 20% missing data	98,116 (65%)	107,626 (74%)	93,710 (67%)
> 1 duplicate inconsistent	7,575 (5%)	6,254 (4%)	10,725 (8%)
> 1 mendelian error	22,815 (15%)	13,600 (9%)	0 (0%)
< 0.001 Hardy–Weinberg *P*-value	12,052 (8%)	9,721 (7%)	16,176 (12%)
Other failures†	23,478 (16%)	17,692 (12%)	23,722 (17%)
Non-redundant (unique) SNPs	1,076,392	1,104,980	1,087,305
Monomorphic	156,290 (15%)	234,482 (21%)	268,325 (25%)
Polymorphic	920,102 (85%)	870,498 (79%)	818,980 (75%)
	All analysis panels		
Unique QC-passed SNPs	1,156,772		
Passed in one analysis panel	52,204 (5%)		
Passed in two analysis panels	97,231 (8%)		
Passed in three analysis panels	1,007,337 (87%)		
Monomorphic across three analysis panels	75,997		
Polymorphic in all three analysis panels	682,397		
MAF ⩾ 0.05 in at least one of three analysis panels	877,351		

* Out of 95 samples in CEU, YRI; 94 samples in CHB+JPT.

† "Other failures" includes SNPs with discrepancies during the data transmission process. Some SNPs failed in more than one way, so these percentages add up to more than 100%.

SNPs on the Phase I map are evenly spaced, except on Y and mtDNA. The Phase I data

在 10 个 ENCODE 区域中(表 2)，SNP 的密度大约是全基因组图谱的十倍：5 Mb 区域中有 17,944 个 SNP(每 279 bp 一个)。

我们对超过 130 万个 SNP 进行了基因分型(表 3)，得到了第 I 阶段的超过 100 万个 SNP 的数据。30 万个不在第 I 阶段数据集的 SNP 包括 73,652 个通过了 QC 过滤但是在所有 269 个样本中都是单态性的 SNP。剩下的 SNP 则是在一个或更多分析小组中没有通过 QC 过滤的，这大多是因为完整性不足、非孟德尔遗传、偏离哈迪-温伯格平衡、重复间基因型不一致以及数据传送不一致。

表 3. 第 I 阶段 HapMap 成功的基因分型

SNP 类别	分析小组		
	YRI	CEU	CHB+JPT
提交的试验	1,273,716	1,302,849	1,273,703
通过 QC 过滤	1,123,296 (88%)	1,157,650 (89%)	1,134,726 (89%)
没有通过 QC 过滤 *	150,420 (12%)	145,199 (11%)	138,977 (11%)
缺失的数据 ＞20%	98,116 (65%)	107,626 (74%)	93,710 (67%)
＞1 个重复不一致	7,575 (5%)	6,254 (4%)	10,725 (8%)
＞1 个孟德尔错误	22,815 (15%)	13,600 (9%)	0 (0%)
哈迪-温伯格平衡 P 值 ＜0.001	12,052 (8%)	9,721 (7%)	16,176 (12%)
其他失败 †	23,478 (16%)	17,692 (12%)	23,722 (17%)
非冗余(唯一) SNP	1,076,392	1,104,980	1,087,305
单态性的	156,290 (15%)	234,482 (21%)	268,325 (25%)
多态性的	920,102 (85%)	870,498 (79%)	818,980 (75%)
	所有分析小组		
特定的通过 QC 的 SNP	1,156,772		
在一个分析小组中通过	52,204 (5%)		
在两个分析小组中通过	97,231 (8%)		
在三个分析小组中通过	1,007,337 (87%)		
在三个分析小组中的单态性	75,997		
在所有三个分析小组中的多态性	682,397		
在三个分析小组的至少一个中 MAF ≥ 0.05	877,351		

* 得自 95 个 CEU 和 YRI 样本；94 个 CHB+JPT 样本。

† “其他失败”包括在数据传输过程中不一致的 SNP。一些 SNP 存在不止一种失败，所以这些比例总计超过 100%。

除 Y 染色体和线粒体 DNA 上，第 I 阶段图谱中的 SNP 是均匀分布的 第 I 阶

include a successful, common SNP every 5 kb across most of the genome in each analysis panel (Supplementary Fig. 1): only 3.3% of inter-SNP distances are longer than 10 kb, spanning 11.9% of the genome (Fig. 2; see also Supplementary Fig. 2). One exception is the X chromosome (Supplementary Fig. 1), where a much higher proportion of attempted SNPs were rare or monomorphic, and thus the density of common SNPs is lower.

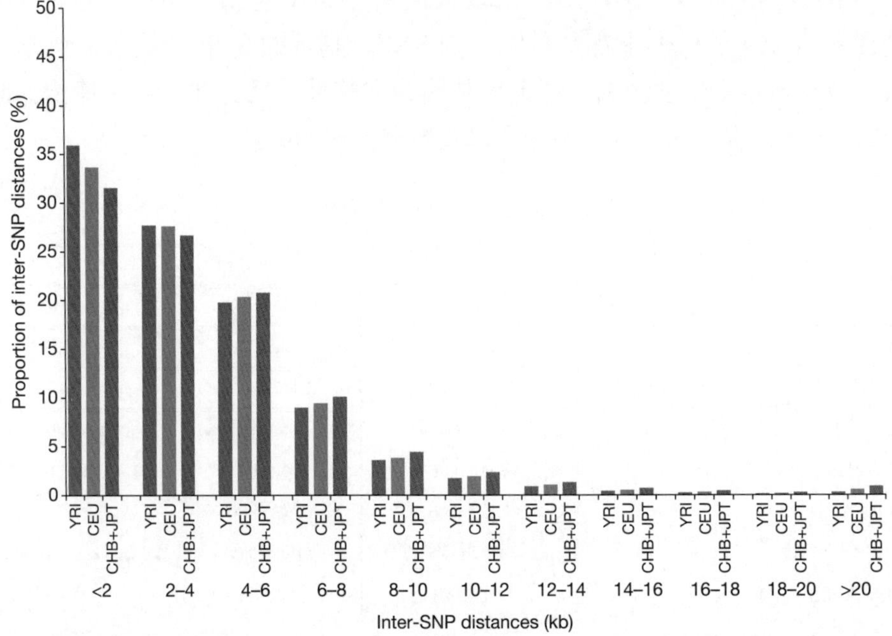

Fig. 2. Distribution of inter-SNP distances. The distributions are shown for each analysis panel for the HapMappable genome (defined in the Methods), for all common SNPs (with MAF ⩾ 0.05).

Two intentional exceptions to the regular spacing of SNPs on the physical map were the mitochondrial chromosome (mtDNA), which does not undergo recombination, and the non-recombining portion of chromosome Y. On the basis of the 168 successful, polymorphic SNPs, each HapMap sample fell into one of 15 (of the 18 known) mtDNA haplogroups[34] (Table 4). A total of 84 SNPs that characterize the unique branches of the reference Y genealogical tree[35-37] were genotyped on the HapMap samples. These SNPs assigned each Y chromosome to 8 (of the 18 major) Y haplogroups previously described (Table 4).

Table 4. mtDNA and Y chromosome haplogroups

MtDNA haplogroup	DNA sample*			
	YRI (60)	CEU (60)	CHB (45)	JPT (44)
L1	0.22	–	–	–
L2	0.35	–	–	–
L3	0.43	–	–	–
A	–	–	0.13	0.04

段的数据在各个分析小组中大多数的基因组范围内每 5 kb 有一个成功分型的常见 SNP(补充信息图 1)：只有 3.3% 的 SNP 间距超过 10 kb，占基因组的 11.9%(图 2；也可详见补充信息图 2)。X 染色体是一个例外(补充信息图 1)，该染色体上罕见的或者单多态性的 SNP 的比例高得多，因而常见 SNP 的密度较低。

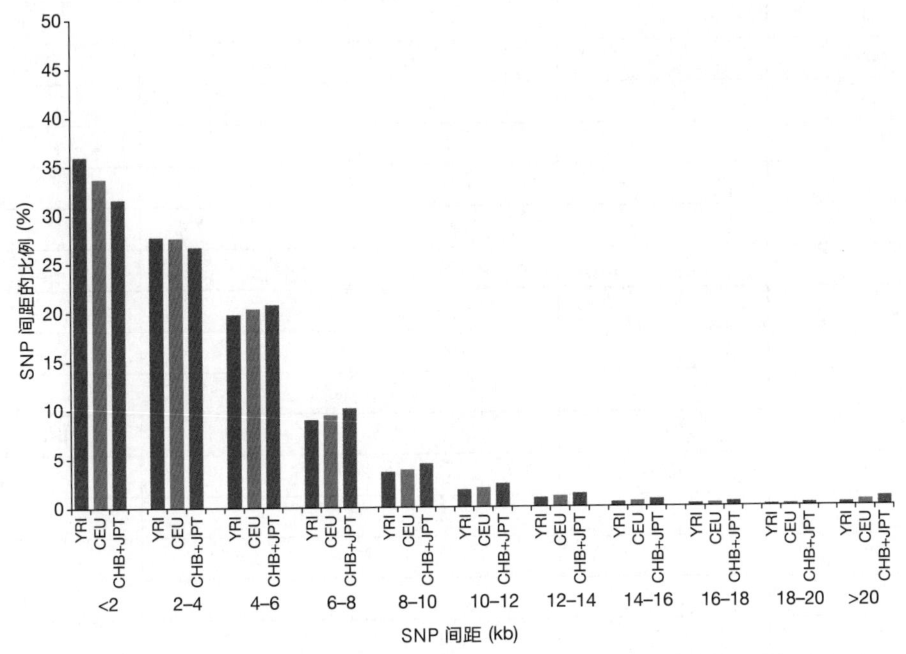

图 2. SNP 间距的分布。展示了各个分析小组中所有常见 SNP(MAF ≥ 0.05) 在可被 HapMap 比对的基因组(在方法中有定义)中的分布。

物理图谱上 SNP 均匀间距的两个例外是不进行重组的线粒体染色体(mtDNA)和 Y 染色体的非重组部分。基于 168 个成功分型的多态性的 SNP，各个 HapMap 样本分别归属于 18 个已知 mtDNA 单倍体型中的 15 个 [34](表 4)。总共 84 个表征参考 Y 染色体系统树独特分支的 SNP[35-37] 在 HapMap 样本中进行了基因分型。这些 SNP 将各个样本的 Y 染色体分配到 18 个主要的 Y 单体型中之前描述过的 8 个中(表 4)。

表 4. mtDNA 和 Y 染色体单体型

MtDNA 单体型	DNA 样本 *			
	YRI (60)	CEU (60)	CHB (45)	JPT (44)
L1	0.22	–	–	–
L2	0.35	–	–	–
L3	0.43	–	–	–
A	–	–	0.13	0.04

Continued

MtDNA haplogroup	DNA sample*			
	YRI (60)	CEU (60)	CHB (45)	JPT (44)
B	–	–	0.33	0.30
C	–	–	0.09	0.07
D	–	–	0.22	0.34
M/E	–	–	0.22	0.25
H	–	0.45	–	–
V	–	0.07	–	–
J	–	0.08	–	–
T	–	0.12	–	–
K	–	0.03	–	–
U	–	0.23	–	–
W	–	0.02	–	–
Y chromosome haplogroup	DNA sample*			
	YRI (30)	CEU (30)	CHB (22)	JPT (22)
E1	0.07	–	–	–
E3a	0.93	–	–	–
F, H, K	–	0.03	0.23	0.14
I	–	0.27	–	–
R1	–	0.70	–	–
C	–	–	0.09	0.09
D	–	–	–	0.45
NO	–	–	0.68	0.32

* Number of chromosomes sampled is given in parentheses.

Highly accurate phasing of long-range chromosomal haplotypes. Despite having collected data in diploid individuals, the inclusion of parent–offspring trios and the use of computational methods made it possible to determine long-range phased haplotypes of extremely high quality for each individual. These computational algorithms take advantage of the observation that because of LD, relatively few of the large number of possible haplotypes consistent with the genotype data actually occur in population samples.

The project compared a variety of algorithms for phasing haplotypes from unrelated individuals and trios[38], and applied the algorithm that proved most accurate (an updated version of PHASE[39]) separately to each analysis panel. (Phased haplotypes are available for download at the Project website.) We estimate that "switch" errors—where a segment of the maternal haplotype is incorrectly joined to the paternal—occur extraordinarily rarely in the trio samples (every 8 Mb in CEU; 3.6 Mb in YRI). The switch rate is higher in the CHB+JPT samples (one per 0.34 Mb) due to the lack of information from parent–

续表

MtDNA 单体型	DNA 样本 *			
	YRI (60)	CEU (60)	CHB (45)	JPT (44)
B	–	–	0.33	0.30
C	–	–	0.09	0.07
D	–	–	0.22	0.34
M/E	–	–	0.22	0.25
H	–	0.45	–	–
V	–	0.07	–	–
J	–	0.08	–	–
T	–	0.12	–	–
K	–	0.03	–	–
U	–	0.23	–	–
W	–	0.02	–	–
Y 染色体单体型	DNA 样本 *			
	YRI (30)	CEU (30)	CHB (22)	JPT (22)
E1	0.07	–	–	–
E3a	0.93	–	–	–
F, H, K	–	0.03	0.23	0.14
I	–	0.27	–	–
R1	–	0.70	–	–
C	–	–	0.09	0.09
D	–	–	–	0.45
NO	–	–	0.68	0.32

* 染色体样本数在括号中给出。

远距离染色体单体型的精准确定 尽管收集的是二倍体个体的数据，亲本–子代三联家系的纳入和计算方法的使用使我们能够非常高质量地确定各个个体远距离的单体型。这些计算算法利用了这样一个现象：由于 LD 现象的存在，可能与基因型数据一致的大量单体型中，只有很少会实际存在于种群样本中。

本计划比较了各种来自不相关个体和三联家系 [38] 的单体型确定算法，并且分别对各个分析小组应用了被证明最准确的算法（PHASE 的一个最新版本 [39]）。（确定的单体型数据可在本计划的网站上下载。）我们估计“转换”错误——母本单体型的一个片段错误地连到父本上——在三联家系样本中特别罕见（在 CEU 样本中每 8 Mb 一个；在 YRI 样本中每 3.6 Mb 一个）。由于缺少亲本–子代三联家系的信息，转换错

offspring trios, but even for the unrelated individuals, statistical reconstruction of haplotypes is remarkably accurate.

Estimating properties of SNP discovery and dbSNP. Extensive sequencing and genotyping in the ENCODE regions characterized the false-positive and false-negative rates for dbSNP, as well as polymerase chain reaction (PCR)-based resequencing (see Methods). These data reveal two important conclusions: first, that PCR-based sequencing of diploid samples may be biased against very rare variants (that is, those seen only as a single heterozygote), and second, that the vast majority of common variants are either represented in dbSNP, or show tight correlation to other SNPs that are in dbSNP (Fig. 3).

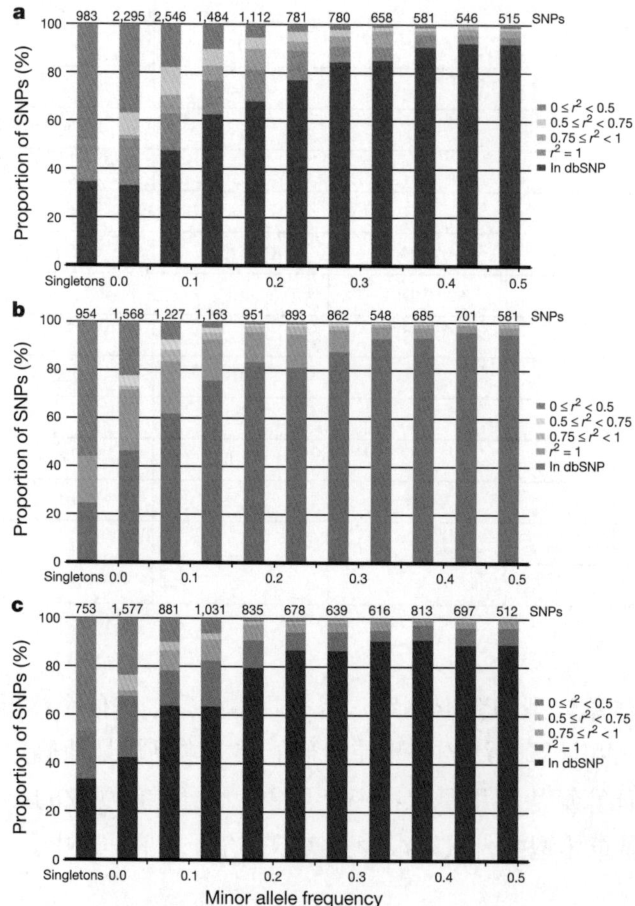

Fig. 3. Allele frequency and completeness of dbSNP for the ENCODE regions. **a**–**c**, The fraction of SNPs in dbSNP, or with a proxy in dbSNP, are shown as a function of minor allele frequency for each analysis panel (**a**, YRI; **b**, CEU; **c**, CHB+JPT). Singletons refer to heterozygotes observed in a single individual, and are broken out from other SNPs with MAF < 0.05. Because all ENCODE SNPs have been deposited in dbSNP, for this figure we define a SNP as "in dbSNP" if it would be in dbSNP build 125 independent of the HapMap ENCODE resequencing project. All remaining SNPs (not in dbSNP) were discovered only by ENCODE resequencing; they are categorized by their correlation (r^2) to those in dbSNP. Note that the number of SNPs in each frequency bin differs among analysis panels, because not all SNPs are polymorphic in all analysis panels.

误率在 CHB+JPT 样本中较高（每 0.34 Mb 一个），不过即使对于不相关的个体，单体型的统计学重建也是非常准确的。

SNP 发现和 dbSNP 数据库的特征估计 对 ENCODE 区域广泛的测序和基因分型，以及基于聚合酶链式反应（PCR）的重测序（详见方法），确定了 dbSNP 数据库的假阳性率和假阴性率。这些数据反映了两个重要的结论：第一，对二倍体样本进行基于 PCR 的测序，可能会在检测非常罕见的变异（即那些仅在单个杂合子中发现的变异）时产生偏倚；第二，大多数的常见变异要么收录在 dbSNP 中，要么与 dbSNP 中的其他 SNP 有紧密关联（图 3）。

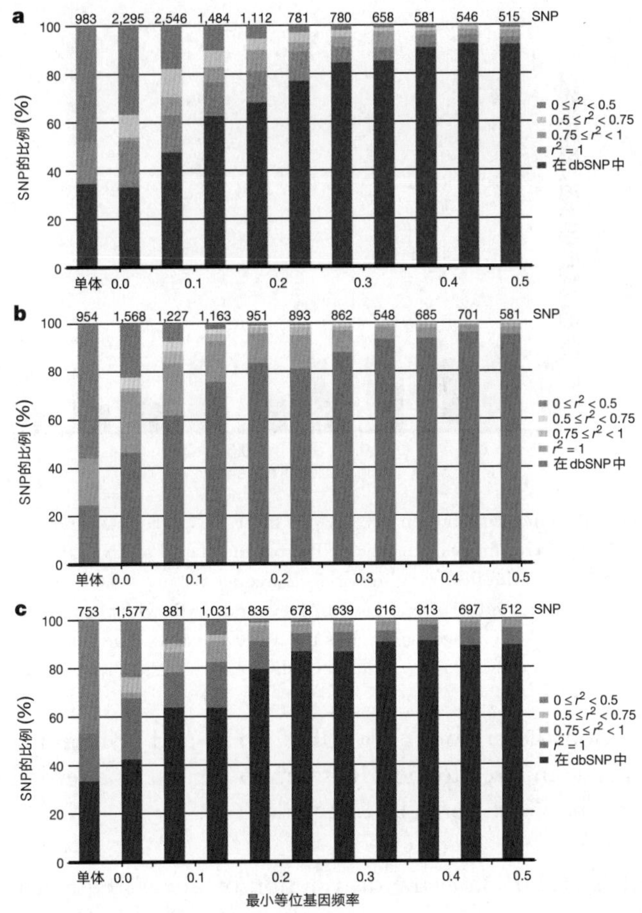

图 3. ENCODE 区域的 dbSNP 等位基因频率和完整性。**a**～**c**，根据各个分析小组中最小等位基因频率表示 dbSNP 中或者在 dbSNP 中有代表物的 SNP 所占的比例（**a**，YRI；**b**，CEU；**c**，CHB+JPT）。单体指的是仅在单个个体中观察到的杂合子，是从 MAF < 0.05 的 SNP 中单独分出去的。因为所有的 ENCODE SNP 已经被收录到 dbSNP 中，所以对于这张图，如果一个 SNP 不是基于 HapMap ENCODE 重测序计划而被收录在 dbSNP 版本 125 中，我们就定义它为“在 dbSNP 中”。所有剩下的 SNP（不在 dbSNP 中）只通过 ENCODE 重测序而被发现，并根据与 dbSNP 中的 SNP 的关联（r^2）而进一步分类。注意，各个频率块中 SNP 的数量在不同分析小组中是不同的，因为不是所有 SNP 在所有分析小组中都是多态性的。

Allele frequency distributions within population samples. The underlying allele frequency distributions for these samples are best estimated from the ENCODE data, where deep sequencing reduces bias due to SNP ascertainment. Consistent with previous studies, most SNPs observed in the ENCODE regions are rare: 46% had MAF < 0.05, and 9% were seen in only a single individual (Fig. 4). Although most varying sites in the population are rare, most heterozygous sites within any individual are due to common SNPs. Specifically, in the ENCODE data, 90% of heterozygous sites in each individual were due to common variants (Fig. 4). With ever-deeper sequencing of DNA samples the number of rare variants will rise linearly, but the vast majority of heterozygous sites in each person will be explained by a limited set of common SNPs now contained (or captured through LD) in existing databases (Fig. 3).

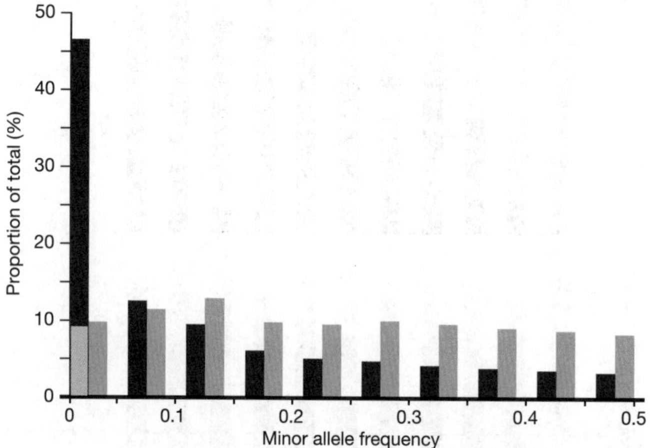

Fig. 4. Minor allele frequency distribution of SNPs in the ENCODE data, and their contribution to heterozygosity. This figure shows the polymorphic SNPs from the HapMap ENCODE regions according to minor allele frequency (blue), with the lowest minor allele frequency bin (< 0.05) separated into singletons (SNPs heterozygous in one individual only, shown in grey) and SNPs with more than one heterozygous individual. For this analysis, MAF is averaged across the analysis panels. The sum of the contribution of each MAF bin to the overall heterozygosity of the ENCODE regions is also shown (orange).

Consistent with previous descriptions, the CEU, CHB and JPT samples show fewer low frequency alleles when compared to the YRI samples (Fig. 5), a pattern thought to be due to bottlenecks in the history of the non-YRI populations.

In contrast to the ENCODE data, the distribution of allele frequencies for the genome-wide data is flat (Fig. 5), with much more similarity in the distributions observed in the three analysis panels. These patterns are well explained by the inherent and intentional bias in the rules used for SNP selection: we prioritized using validated SNPs in order to focus resources on common (rather than rare or false positive) candidate SNPs from the public databases. For a fuller discussion of ascertainment issues, including a shift in frequencies over time and an excess of high-frequency derived alleles due to inclusion of chimpanzee data in determination of double-hit status, see the Supplementary Information (Supplementary Fig. 3).

种群样本中的等位基因频率分布　对这些样本的潜在等位基因频率分布的最佳估计来自 ENCODE 数据，因为深度测序减少了确定 SNP 过程的偏倚。与之前的研究一致，ENCODE 区域内观察到的 SNP 大多数是罕见的：46% 的 MAF < 0.05，9% 仅在单个个体中存在(图 4)。尽管在种群中大多数变异位点是罕见的，但是在任何一个个体中，大多数杂合位点是常见 SNP。特别地，在 ENCODE 数据中，各个个体中 90% 的杂合位点是常见变异(图 4)。随着对 DNA 样本更加深度的测序，罕见变异的数量将会线性增长，但是每个个体中绝大多数的杂合位点仍可以用有限的一套已包含(或者通过 LD 而捕获)在已有数据库中的常见变异来解释(图 3)。

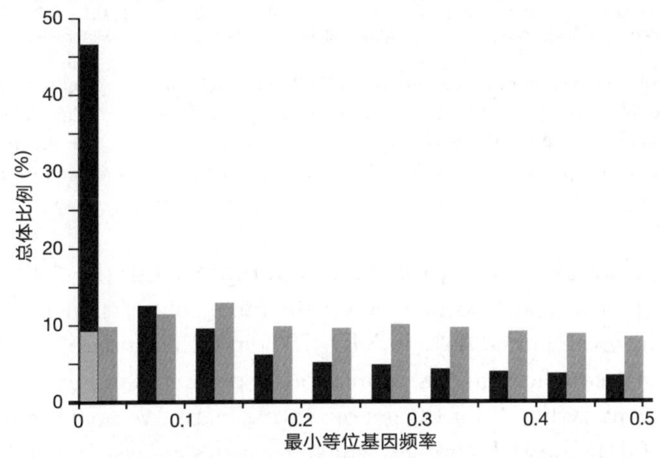

图 4. ENCODE 数据中 SNP 的最小等位基因频率分布和它们对杂合性的贡献。根据最小等位基因频率(蓝色)展示了来自 HapMap ENCODE 区域的多态性 SNP，最低的最小等位基因频率块(< 0.05)被分为单体(仅在一个个体中杂合的 SNP，用灰色表示)和在多于一个个体中杂合的 SNP。本分析中，MAF 是各分析小组的平均值。另外还展示了各个 MAF 块对 ENCODE 区域总体杂合性的贡献总和(橘色)。

与之前的描述一致，跟 YRI 样本相比，CEU、CHB 和 JPT 样本的低频等位基因更少(图 5)，这种现象被认为是由于非 YRI 种群历史中的瓶颈效应。

与 ENCODE 数据相比，全基因组范围的等位基因频率分布是“平”的(图 5)，在三个分析小组中的分布的相似性要高得多。这种模式可以用 SNP 筛选规则中固有的和人为的偏倚很好地进行解释：我们优先使用验证的 SNP，以便将资源集中于来自公共数据的常见的(而不是罕见或假阳性的)候选 SNP。对于确定 SNP 的问题的更加完整的讨论，包括频率随着时间的改变，以及在确定“双打击”状态时纳入黑猩猩数据而导致的高频率等位基因过量，详见补充信息(补充信息图 3)。

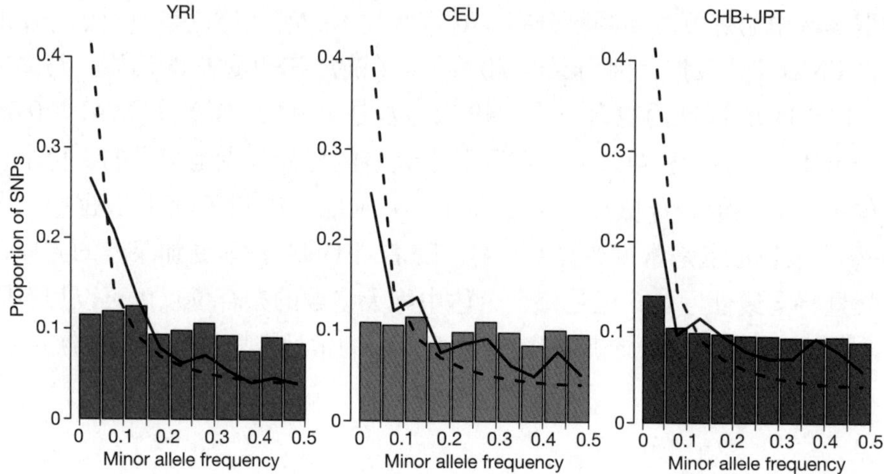

Fig. 5. Allele frequency distributions for autosomal SNPs. For each analysis panel we plotted (bars) the MAF distribution of all the Phase I SNPs with a frequency greater than zero. The solid line shows the MAF distribution for the ENCODE SNPs, and the dashed line shows the MAF distribution expected for the standard neutral population model with constant population size and random mating without ascertainment bias.

SNP allele frequencies across population samples. Of the 1.007 million SNPs successfully genotyped and polymorphic across the three analysis panels, only a subset were polymorphic in any given panel: 85% in YRI, 79% in CEU, and 75% in CHB+JPT. The joint distribution of frequencies across populations is presented in Fig. 6 (for the ENCODE data) and Supplementary Fig. 4 (for the genome-wide map). We note the similarity of allele frequencies in the CHB and JPT samples, which motivates analysing them jointly as a single analysis panel in the remainder of this report.

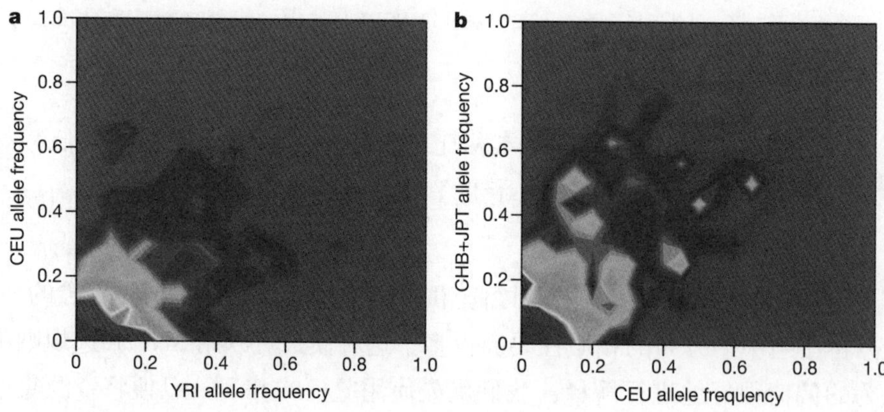

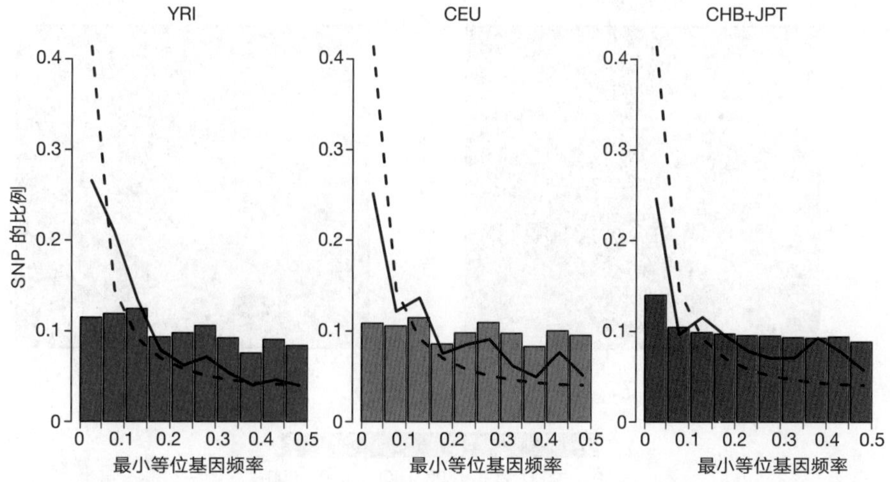

图 5. 常染色体 SNP 的等位基因频率分布。对于各个分析小组中所有第 I 阶段 HapMap 得到的频率大于零的 SNP，我们对其 MAF 分布作图（条形图）。实线表示 ENCODE SNP 的 MAF 分布，虚线表示基于恒定种群大小、随机交配以及无检测偏倚的标准中性种群模型得到的预期 MAF 分布。

种群样本间的 SNP 等位基因频率 三个分析小组共有 100.7 万成功基因分型并呈多态性的 SNP，其中只有一部分在三个分析小组中都是多态性的：YRI 中有 85%，CEU 中有 79%，CHB+JPT 中有 75%。种群间等位基因频率的联合分布见图 6（ENCODE 数据）和补充信息图 4（全基因组范围图谱）。我们注意到 CHB 和 JPT 样本等位基因频率相似，所以在这份报告的剩余部分将它们作为一个分析小组联合分析。

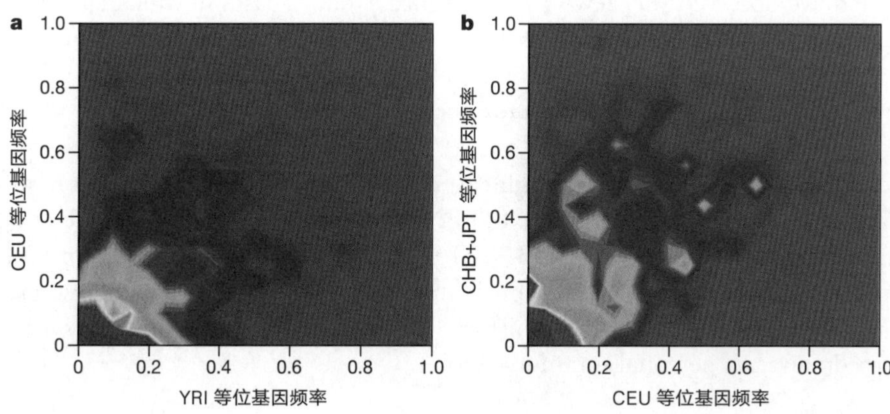

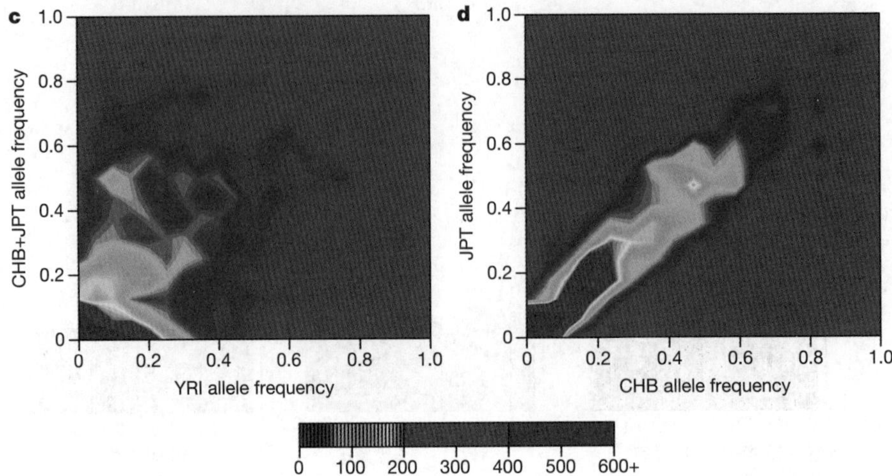

Fig. 6. Comparison of allele frequencies in the ENCODE data for all pairs of analysis panels and between the CHB and JPT sample sets. For each polymorphic SNP we identified the minor allele across all panels (**a**–**d**) and then calculated the frequency of this allele in each analysis panel/sample set. The colour in each bin represents the number of SNPs that display each given set of allele frequencies. The purple regions show that very few SNPs are common in one panel but rare in another. The red regions show that there are many SNPs that have similar low frequencies in each pair of analysis panels/sample sets.

A simple measure of population differentiation is Wright's F_{ST}, which measures the fraction of total genetic variation due to between-population differences[40]. Across the autosomes, F_{ST} estimated from the full set of Phase I data is 0.12, with CEU and CHB+JPT showing the lowest level of differentiation ($F_{ST} = 0.07$), and YRI and CHB+JPT the highest ($F_{ST} = 0.12$). These values are slightly higher than previous reports[41], but differences in the types of variants (SNPs versus microsatellites) and the samples studied make comparisons difficult.

As expected, we observed very few fixed differences (that is, cases in which alternate alleles are seen exclusively in different analysis panels). Across the 1 million SNPs genotyped, only 11 have fixed differences between CEU and YRI, 21 between CEU and CHB+JPT, and 5 between YRI and CHB+JPT, for the autosomes.

The extent of differentiation is similar across the autosomes, but higher on the X chromosome ($F_{ST} = 0.21$). Interestingly, 123 SNPs on the X chromosome were completely differentiated between YRI and CHB+JPT, but only two between CEU and YRI and one between CEU and CHB+JPT. This seems to be largely due to a single region near the centromere, possibly indicating a history of natural selection at this locus (see below; M. L. Freedman *et al.*, personal communication).

Haplotype sharing across populations. We next examined the extent to which haplotypes are shared across populations. We used a hidden Markov model in which each haplotype is modelled in turn as an imperfect mosaic of other haplotypes (see Supplementary Information)[42]. In essence, the method infers probabilistically which other haplotype in the sample is the closest relative (nearest neighbour) at each position along the chromosome.

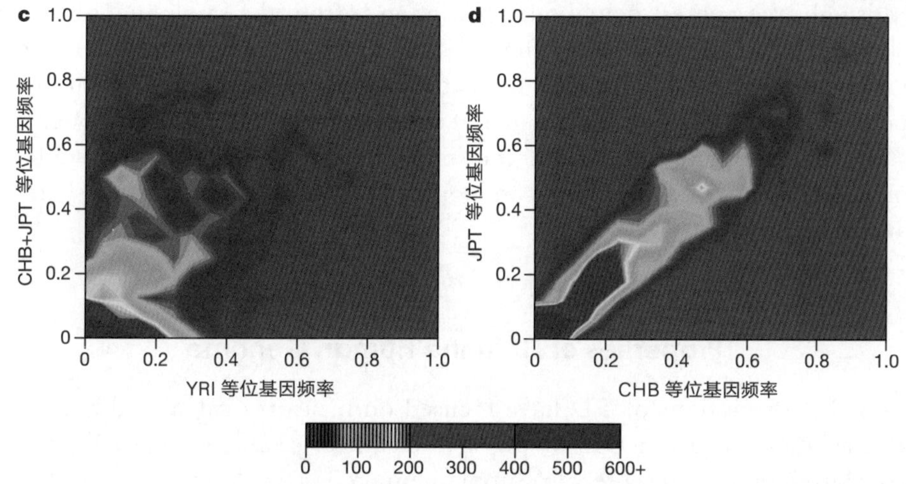

图 6. ENCODE 数据中的等位基因频率在所有分析小组间以及 CHB 和 JPT 样本集间的比较。对于各个多态性 SNP，我们鉴定其在所有分析小组的最小等位基因(**a** ~ **d**)，然后计算这个等位基因在各个分析小组/样本集的频率。各个块中的颜色代表显示特定等位基因频率集的 SNP 的数量。紫色区域表明非常少的 SNP 是在一个分析小组中常见而在另一个分析小组中罕见的。红色区域表明很多 SNP 在各对分析小组/样品集中有类似的低频率。

种群差异的一个简单的衡量方法是赖特的 F_{ST}，它测量总的遗传差异中由种群间差异导致的比例[40]。对于常染色体，从全套第 I 阶段数据估计的 F_{ST} 是 0.12，其中 CEU 和 CHB+JPT 表现出最低水平的差异($F_{ST}=0.07$)，而 YRI 和 CHB+JPT 则表现出最高的差异($F_{ST}=0.12$)。这些数值略微高于之前的报道[41]，但是变异类型的不同(SNP 相对于微卫星)和研究样本的差异使得比较变得困难。

正如预期的，我们观察到非常少的固定的差异(即不同的等位基因特定地出现在不同的分析小组中的情况)。在进行了基因分型的 100 万 SNP 中，对于常染色体，只有 11 个 SNP 在 CEU 和 YRI 间有固定的差异，CEU 和 CHB+JPT 间有 21 个，YRI 和 CHB+JPT 间有 5 个。

种群差异的程度在常染色体间是类似的，但是在 X 染色体上更高($F_{ST}=0.21$)。有趣的是，X 染色体上有 123 个 SNP 在 YRI 和 CHB+JPT 间完全不同，而 CEU 和 YRI 间仅有两个，CEU 和 CHB+JPT 间仅有一个。这或许在很大程度上是由于着丝粒附近的一个区域，可能暗示着该基因座被自然选择的历史(见下文；弗里德曼等，个人交流)。

种群间共同的单体型 我们接下来检查了单体型在种群间一致的程度。我们使用的是隐马尔可夫模型，各个单体型依次被模型化为其他单体型的一个不完美镶嵌(详见补充信息)[42]。本质上，这个方法概率地推断在染色体的各个位置上，样本中的哪一个其他单体型是最近的近亲(近邻)。

Unsurprisingly, the nearest neighbour most often is from the same analysis panel, but about 10% of haplotypes were found most closely to match a haplotype in another panel (Supplementary Fig. 5). All individuals have at least some segments over which the nearest neighbour is in a different analysis panel. These results indicate that although analysis panels are characterized both by different haplotype frequencies and, to some extent, different combinations of alleles, both common and rare haplotypes are often shared across populations.

Properties of LD in the Human Genome

Traditionally, descriptions of LD have focused on measures calculated between pairs of SNPs, averaged as a function of physical distance. Examples of such analyses for the HapMap data are presented in Supplementary Fig. 6. After adjusting for known confounders such as sample size, allele frequency distribution, marker density, and length of sampled regions, these data are highly similar to previously published surveys[43].

Because LD varies markedly on scales of 1–100 kb, and is often discontinuous rather than declining smoothly with distance, averages obscure important aspects of LD structure. A fuller exploration of the fine-scale structure of LD offers both insight into the causes of LD and understanding of its application to disease research.

LD patterns are simple in the absence of recombination. The most natural path to understanding LD structure is first to consider the simplest case in which there is no recombination (or gene conversion), and then to add recombination to the model. (For simplicity we ignore genotyping error and recurrent mutation in this discussion, both of which seem to be rare in these data.)

In the absence of recombination, diversity arises solely through mutation. Because each SNP arose on a particular branch of the genealogical tree relating the chromosomes in the current populations, multiple haplotypes are observed. SNPs that arose on the same branch of the genealogy are perfectly correlated in the sample, whereas SNPs that occurred on different branches have imperfect correlations, or no correlation at all.

We illustrate these concepts using empirical genotype data from 36 adjacent SNPs in an ENCODE region (ENr131.2q37), selected because no obligate recombination events were detectable among them in CEU (Fig. 7). (We note that the lack of obligate recombination events in a small sample does not guarantee that no recombinants have occurred, but it provides a good approximation for illustration.)

毫不奇怪，最靠近的单体型大多来自同一分析小组，不过大约有 10% 的单体型与其他分析小组的单体型最匹配(补充信息图 5)。所有个体都有一些片段，其最靠近的单体型存在于其他分析小组中。这些结果说明，尽管分析小组具有不同的单体型频率，在某种程度上还具有不同的等位基因组合，但是常见的以及罕见的单体型在不同种群间经常是一致的。

人类基因组中 LD 的特征

传统上，LD 的描述集中于 SNP 对之间计算得到的测度，结果作为物理距离的函数取平均值。对 HapMap 数据的这种分析的例子在补充信息图 6 中有展示。对已知的混杂因素如样本量大小、等位基因频率分布、标志物密度和检测的区域长度进行校正后，这些数据跟之前发表的研究高度相似[43]。

因为 LD 在 1 ~ 100 kb 尺度上变化较大，而且经常是不连续的而非随距离平滑减少，因此取平均值会模糊 LD 结构的重要方面。对 LD 精细尺度结构更全面的探索，将不仅为 LD 的产生原因提供线索，也为它在疾病研究中的应用提供知识。

没有重组发生时 LD 模式是简单的 了解 LD 结构最自然的途径是首先考虑最简单的情况，即没有重组(或者基因转变)的情况，然后向模型中加入重组。(为了简便，在这部分分析中我们忽略了基因分型错误和频发突变，在这些数据中这两种情况看起来是罕见的。)

没有重组时，多样性只通过突变产生。因为各个 SNP 产生于系统树特定的分支，而该系统树与目前种群中的染色体相关，所以可观察到多个单体型。系统树相同分支上产生的 SNP 在样本中完美相关，而在不同分支上产生的 SNP 则不完美相关或者完全不相关。

我们利用实验观察到的基因型数据对这些概念进行了说明，这些数据来自一个 ENCODE 区域(ENr131.2q37)的 36 个相邻 SNP，选择它们是因为在 CEU 中没有检测到它们有专性重组事件(图 7)。(我们注意到，在小样本中没有专性重组事件并不能保证没有重组发生，但是它提供了一个很好的近似的例子。)

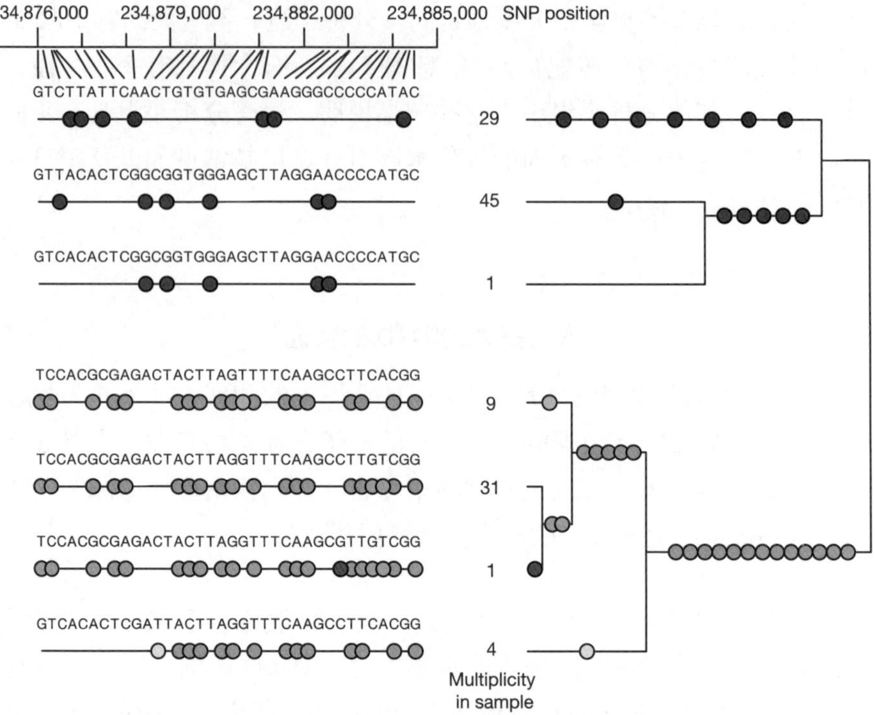

Fig. 7. Genealogical relationships among haplotypes and r^2 values in a region without obligate recombination events. The region of chromosome 2 (234,876,004–234,884,481 bp; NCBI build 34) within ENr131.2q37 contains 36 SNPs, with zero obligate recombination events in the CEU samples. The left part of the plot shows the seven different haplotypes observed over this region (alleles are indicated only at SNPs), with their respective counts in the data. Underneath each of these haplotypes is a binary representation of the same data, with coloured circles at SNP positions where a haplotype has the less common allele at that site. Groups of SNPs all captured by a single tag SNP (with $r^2 \geqslant 0.8$) using a pairwise tagging algorithm[53,54] have the same colour. Seven tag SNPs corresponding to the seven different colours capture all the SNPs in this region. On the right these SNPs are mapped to the genealogical tree relating the seven haplotypes for the data in this region.

In principle, 36 such SNPs could give rise to 2^{36} different haplotypes. Even with no recombination, gene conversion or recurrent mutation, up to 37 different haplotypes could be formed. Despite this great potential diversity, only seven haplotypes are observed (five seen more than once) among the 120 parental CEU chromosomes studied, reflecting shared ancestry since their most recent common ancestor among apparently unrelated individuals.

In such a setting, it is easy to interpret the two most common pairwise measures of LD: D' and r^2. (See the Supplementary Information for fuller definitions of these measures.) D' is defined to be 1 in the absence of obligate recombination, declining only due to recombination or recurrent mutation[27]. In contrast, r^2 is simply the squared correlation coefficient between the two SNPs. Thus, r^2 is 1 when two SNPs arose on the same branch of the genealogy and remain undisrupted by recombination, but has a value less than 1 when SNPs arose on different branches, or if an initially strong correlation has been disrupted by crossing over.

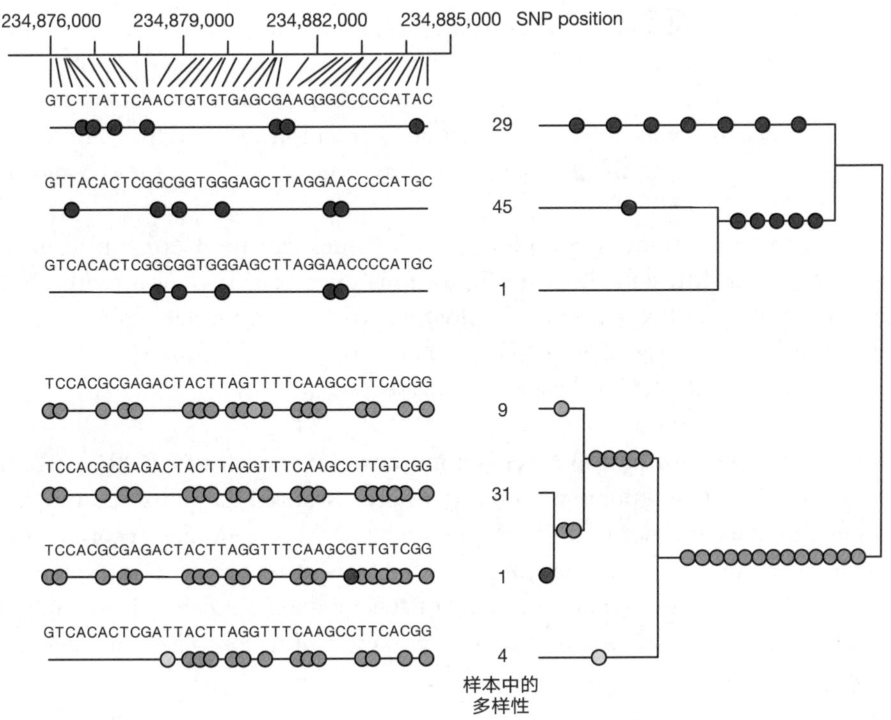

图 7. 没有专性重组事件的区域内单体型间的系谱关系和 r^2 值。2 号染色体 ENr131.2q37 区域 (234,876,004 ~ 234,884,481 bp；NCBI 版本 34) 包含 36 个 SNP，在 CEU 样本中没有专性重组事件。图的左边部分表示在这个区域中观察到的七个不同的单体型（仅在 SNP 处标示等位基因），以及它们在数据中各自的数量。在这些单体型各自的下面是相同数据的二进制表示，有颜色的圆圈表示单体型在该位点含有相对不常见等位基因的 SNP 位置。相同的颜色表示的是利用成对标签算法 [53,54] 通过单个标签 SNP 捕获 ($r^2 \geqslant 0.8$) 的各组 SNP。对应于七种不同颜色的七个标签 SNP 捕获了该区域所有的 SNP。在图的右边部分，这些 SNP 比对到系统树上，该系统树将该区域数据得到的七个单体型关联起来。

原则上，36 个这样的 SNP 能够产生 2^{36} 个不同的单体型。即使没有重组、基因转变或者频发突变，也能够形成多达 37 个不同的单体型。尽管存在这种巨大的潜在多样性，在研究的 120 个亲本 CEU 染色体中，只观察到七个单体型（其中五个观察到超过一次），这反映了自它们最近的共同祖先之后明显不相关的个体也有着共同的血统。

在这样的设定中，很容易解释这一对最常见的度量 LD 的方法：D' 和 r^2。（对于这些度量方法更全面的定义见补充信息。）D' 在不存在专性重组时被定义为 1，只由于重组或频发突变而减小 [27]。相反，r^2 就是简单的两个 SNP 间相关系数的平方。因而，当两个 SNP 在系统树的同一分支上产生并且不受重组的破坏时，r^2 为 1；但是当 SNP 在不同分支上产生，或者一个一开始很强的相关被互换所破坏时，r^2 值小于 1。

In this region, $D' = 1$ for all marker pairs, as there is no evidence of historical recombination. In contrast, and despite great simplicity of haplotype structure, r^2 values display a complex pattern, varying from 0.0003 to 1.0, with no relationship to physical distance. This makes sense, however, because without recombination, correlations among SNPs depend on the historical order in which they arose, not the physical order of SNPs on the chromosome.

Most importantly, the seeming complexity of r^2 values can be deconvolved in a simple manner: only seven different SNP configurations exist in this region, with all but two chromosomes matching five common haplotypes, which can be distinguished from each other by typing a specific set of four SNPs. That is, only a small minority of sites need be examined to capture fully the information in this region.

Variation in local recombination rates is a major determinant of LD. Recombination in the ancestors of the current population has typically disrupted the simple picture presented above. In the human genome, as in yeast[44], mouse[45] and other genomes, recombination rates typically vary dramatically on a fine scale, with hotspots of recombination explaining much crossing over in each region[28]. The generality of this model has recently been demonstrated through computational methods that allow estimation of recombination rates (including hotspots and coldspots) from genotype data[46,47].

The availability of nearly complete information about common DNA variation in the ENCODE regions allowed a more precise estimation of recombination rates across large regions than in any previous study. We estimated recombination rates and identified recombination hotspots in the ENCODE data, using methods previously described[46] (see Supplementary Information for details). Hotspots are short regions (typically spanning about 2 kb) over which recombination rates rise dramatically over local background rates.

Whereas the average recombination rate over 500 kb across the human genome is about 0.5 cM[48], the estimated recombination rate across the 500-kb ENCODE regions varied nearly tenfold, from a minimum of 0.19 cM (ENm013.7q21.13) to a maximum of 1.25 cM (ENr232.9q34.11). Even this tenfold variation obscures much more dramatic variation over a finer scale: 88 hotspots of recombination were identified (Fig. 8; see also Supplementary Fig. 7)—that is, one per 57 kb—with hotspots detected in each of the ten regions (from 4 in 12q12 to 14 in 2q37.1). Across the 5 Mb, we estimate that about 80% of all recombination has taken place in about 15% of the sequence (Fig. 9, see also refs 46, 49).

在这个区域，由于没有证据表明历史上有过重组，对于所有的标志物对而言，$D'=1$。相反，尽管单体型结构特别简单，r^2 值却呈现复杂的模式，在 0.0003 到 1.0 之间变化，与物理距离无关。这一点可以得到解释，因为在没有重组的情况下，SNP 间的相关依赖于它们出现的历史顺序，而不是它们在染色体上的物理顺序。

最重要的是，这种看起来复杂的 r^2 值能够以一种简单的方式去卷积：只有七种不同的 SNP 组合存在于这个区域，除了两条染色体外所有的染色体与五种常见的单体型相匹配，这些单体型能够通过输入特异的一组四个 SNP 而互相区分开来。也就是说，要完全捕获这个区域的信息，只需要检查少量的位点。

局部重组率的差异是 LD 的一个主要决定因素 现有种群的祖先中的重组通常已经破坏了上面呈现的简单图景。人类基因组与酵母 [44]、小鼠 [45] 和其他基因组一样，重组率一般在精细尺度上差异巨大，而重组热点解释了各个区域的很多互换 [28]。这个模型的普遍性最近已经通过计算方法得到证实，这种计算方法能够从基因型数据估计重组率（包括热点和冷点）[46,47]。

ENCODE 区域中常见 DNA 变异的接近完整的信息使我们能够比任何之前的研究在大的区域中更加精确地估计重组率。利用之前描述的方法 [46]，我们在 ENCODE 数据中估计了重组率，并且鉴定了重组热点（详见补充信息）。热点是短区域（一般长约 2 kb），其上的重组率比局部背景率高得多。

人类基因组中 500 kb 尺度上的平均重组率大约是 0.5 cM[48]。在 500 kb 的 ENCODE 区域中，重组率的变化幅度估计将近 10 倍，从最小的 0.19 cM（ENm013.7q21.13）到最大的 1.25 cM（ENr232.9q34.11）。然而这种 10 倍的差别仍然模糊了在更精细的尺度上更加巨大的差别：鉴定到 88 个重组热点（图 8；也可见补充信息图 7）——也就是每 57 kb 一个，并且存在于所有 10 个区域（从 12q12 区域的 4 个到 2q37.1 的 14 个）。在这个 5 Mb 的区域中，我们估计大约所有重组的 80% 发生在约 15% 的序列中（图 9，也可见参考文献 46、49）。

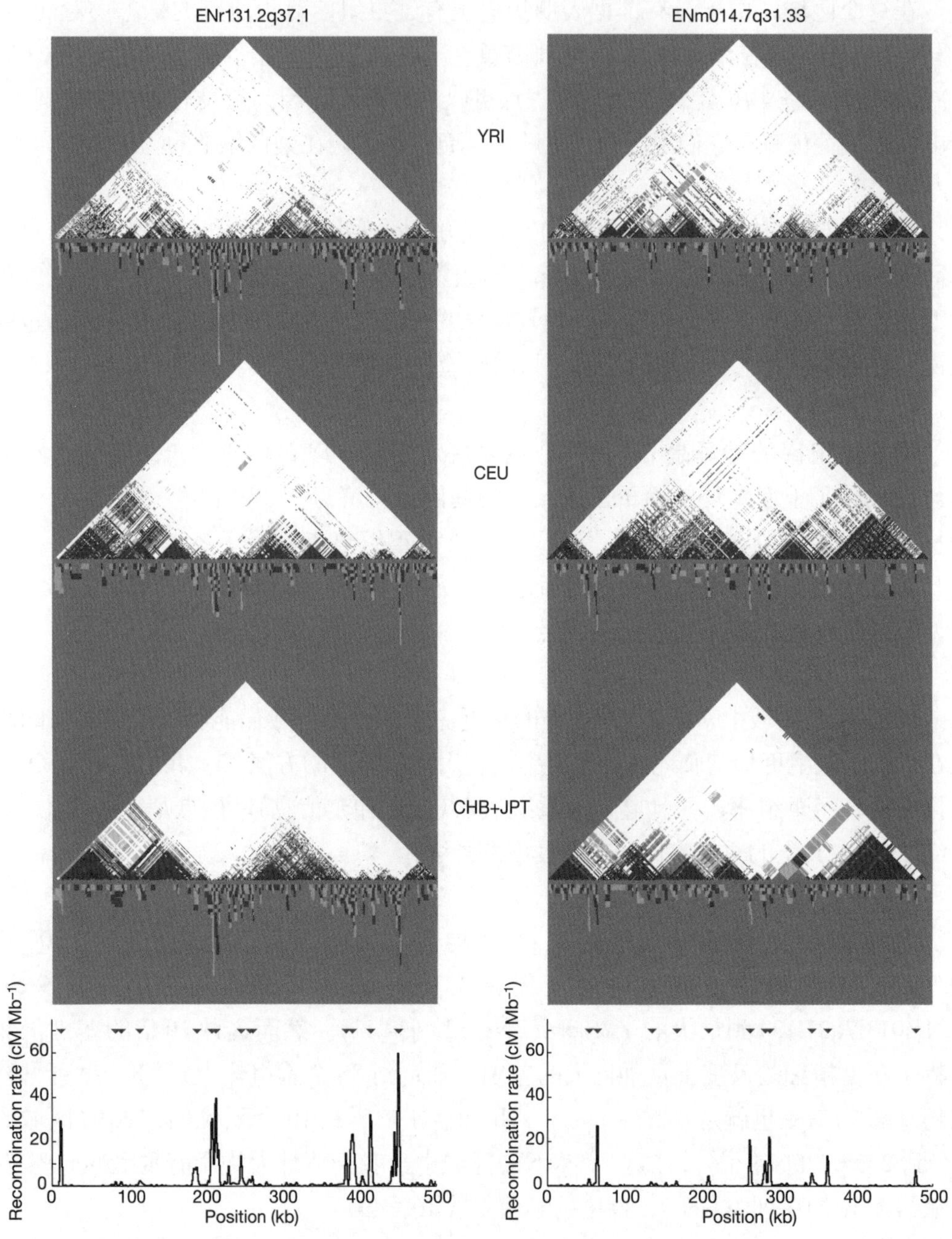

Fig. 8. Comparison of linkage disequilibrium and recombination for two ENCODE regions. For each region (ENr131.2q37.1 and ENm014.7q31.33), D' plots for the YRI, CEU and CHB+JPT analysis panels are shown: white, $D' < 1$ and LOD < 2; blue, $D' = 1$ and LOD < 2; pink, $D' < 1$ and LOD $\geqslant 2$; red, $D' = 1$ and LOD $\geqslant 2$. Below each of these plots is shown the intervals where distinct obligate recombination events must have occurred (blue and green indicate adjacent intervals). Stacked intervals represent regions where there are multiple recombination events in the sample history. The bottom plot shows estimated recombination rates, with hotspots shown as red triangles[46].

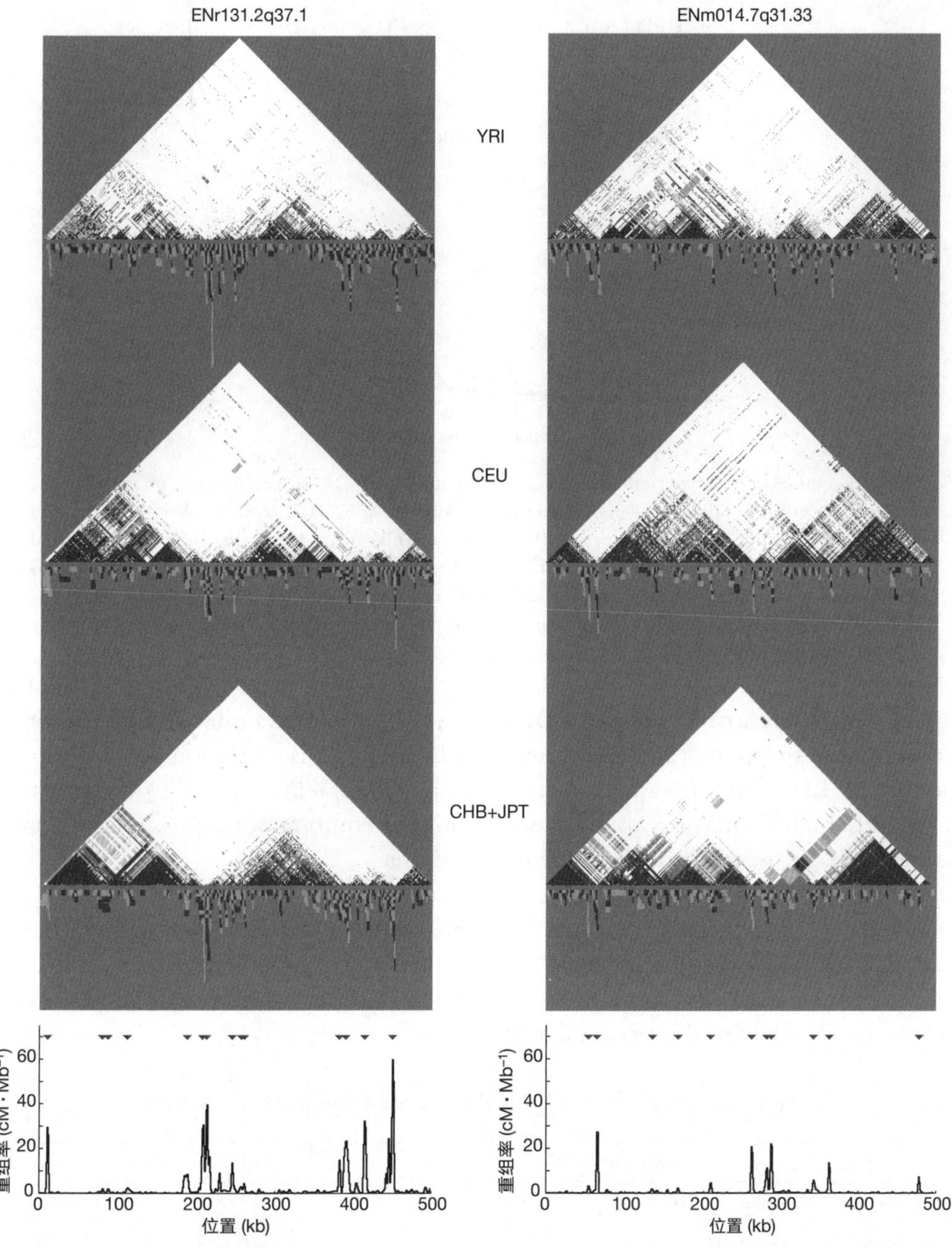

图 8. 两个 ENCODE 区域的 LD 和重组的比较。每个区域（ENr131.2q37.1 和 ENm014.7q31.33）的 YRI、CEU 和 CHB+JPT 分析小组的 D' 图：白色，$D'<1$ 且 LOD <2；蓝色，$D'=1$ 且 LOD <2；粉色，$D'<1$ 且 LOD $\geqslant 2$；红色，$D'=1$ 且 LOD $\geqslant 2$。各个图的下方展示了不同专性重组事件发生的区间（蓝色和绿色表示邻近的区间）。堆积的区间代表在样本历史中有多个重组事件发生的区域。底部的图展示估计的重组率，热点以红色三角形表示 [46]。

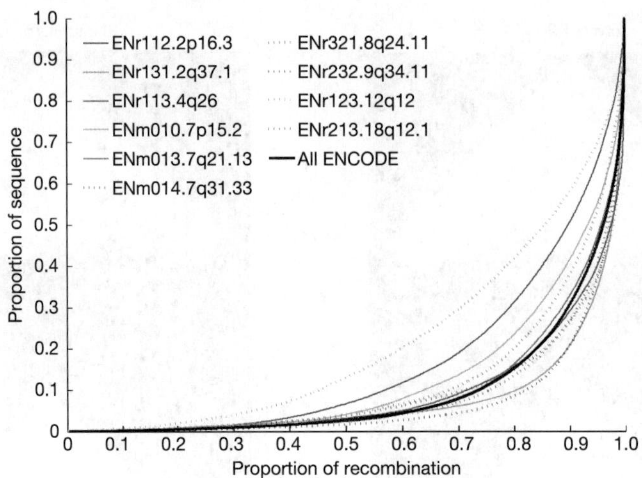

Fig. 9. The distribution of recombination events over the ENCODE regions. Proportion of sequence containing a given fraction of all recombination for the ten ENCODE regions (coloured lines) and combined (black line). For each line, SNP intervals are placed in decreasing order of estimated recombination rate[46], combined across analysis panels, and the cumulative recombination fraction is plotted against the cumulative proportion of sequence. If recombination rates were constant, each line would lie exactly along the diagonal, and so lines further to the right reveal the fraction of regions where recombination is more strongly locally concentrated.

A block-like structure of human LD. With most human recombination occurring in recombination hotspots, the breakdown of LD is often discontinuous. A "block-like" structure of LD is visually apparent in Fig. 8 and Supplementary Fig. 7: segments of consistently high D' that break down where high recombination rates, recombination hotspots and obligate recombination events[50] all cluster.

When haplotype blocks are more formally defined in the ENCODE data (using a method based on a composite of local D' values[30], or another based on the four gamete test[51]), most of the sequence falls into long segments of strong LD that contain many SNPs and yet display limited haplotype diversity (Table 5).

Table 5. Haplotype blocks in ENCODE regions, according to two methods

Parameter	YRI	CEU	CHB+JPT
Method based on a composite of local D' values[30]			
Average number of SNPs per block	30.3	70.1	54.4
Average length per block (kb)	7.3	16.3	13.2
Fraction of genome spanned by blocks (%)	67	87	81
Average number of haplotypes (MAF $\geqslant$ 0.05) per block	5.57	4.66	4.01
Fraction of chromosomes due to haplotypes with MAF $\geqslant$ 0.05 (%)	94	93	95
Method based on the four gamete test[51]			
Average number of SNPs per block	19.9	24.3	24.3

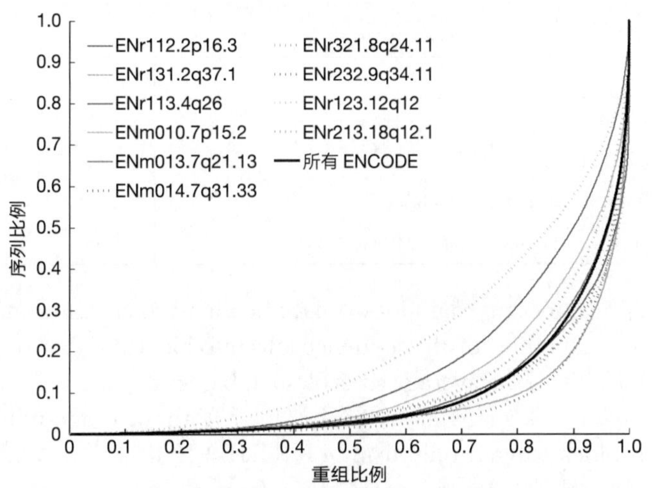

图 9. ENCODE 区域重组事件的分布。对于 10 个 ENCODE 区域（彩色线）和合并的全部区域（黑色线），含有特定比例重组的序列所占的比例。对于各条线，SNP 区间按估计的重组率逐渐减小的次序放置 [46]，该重组率是各分析小组合计的，x 轴是累积的重组比例，y 轴是累积的序列比例。如果重组率是恒定的，各条线将会正好沿着对角线，所以偏向右边的线反映的是重组更加局部集中的区域所占的比例。

人类 LD 的区块样结构 大部分人类重组发生在重组热点处，LD 的分解常常是不连续的。LD 的“区块样”结构在图 8 和补充信息图 7 中看起来很明显：在高重组率、重组热点和专性重组事件 [50] 聚集的区域分成一段一段的高 D' 的片段。

当单体型区块在 ENCODE 数据中更正规地定义时（利用基于局部 D' 值组合的方法 [30]，或者基于四配子检验的另一方法 [51]），大部分的序列落入强 LD 的长片段，这些片段含有很多 SNP 但是表现出有限的单体型多样性（表 5）。

表 5. 根据两种方法得到的 ENCODE 区域的单体型块

参数	YRI	CEU	CHB+JPT
基于局部 D' 值组合的方法 [30]			
每个区块的 SNP 平均数量	30.3	70.1	54.4
每个区块平均长度（kb）	7.3	16.3	13.2
被区块横跨的基因组的比例（%）	67	87	81
每个区块的单体型（MAF ≥ 0.05）平均数量	5.57	4.66	4.01
MAF ≥ 0.05 的单体型占染色体的比例（%）	94	93	95
基于四配子检验的方法 [51]			
每个区块的 SNP 平均数量	19.9	24.3	24.3

Continued

Parameter	YRI	CEU	CHB+JPT
Average length per block (kb)	4.8	5.9	5.9
Fraction of genome spanned by blocks (%)	86	84	84
Average number of haplotypes (MAF ⩾ 0.05) per block	5.12	3.63	3.63
Fraction of chromosomes due to haplotypes with MAF ⩾ 0.05 (%)	91	95	95

Specifically, addressing concerns that blocks might be an artefact of low marker density[52], in these nearly complete data most of the sequence falls into blocks of four or more SNPs (67% in YRI to 87% in CEU) and the average sizes of such blocks are similar to initial estimates[30]. Although the average block spans many SNPs (30–70), the average number of common haplotypes in each block ranged only from 4.0 (CHB+JPT) to 5.6 (YRI), with nearly all haplotypes in each block matching one of these few common haplotypes. These results confirm the generality of inferences drawn from disease-mapping studies[27] and genomic surveys with smaller sample sizes[29] and less complete data[30].

Long-range haplotypes and local patterns of recombination. Although haplotypes often break at recombination hotspots (and block boundaries), this tendency is not invariant. We identified all unique haplotypes with frequency more than 0.05 across the 269 individuals in the phased data, and compared them to the fine-scale recombination map. Figure 10 shows a region of chromosome 19 over which many such haplotypes break at identified recombination hotspots, but others continue. Thus, the tendency towards colocalization of recombination sites does not imply that all haplotypes break at each recombination site.

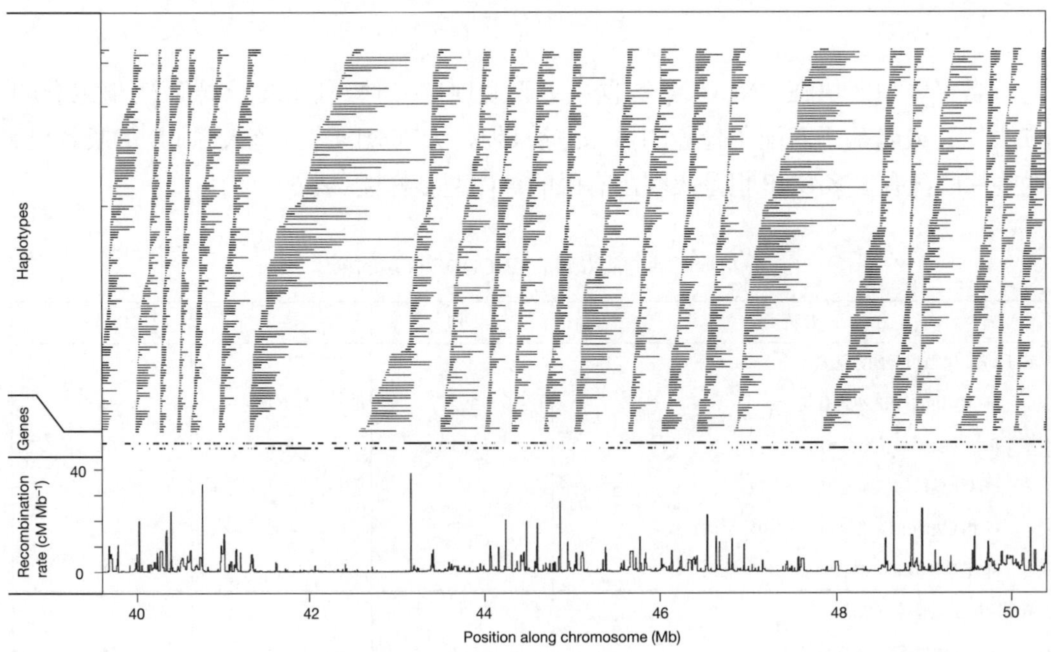

Fig. 10. The relationship among recombination rates, haplotype lengths and gene locations. Recombination

续表

参数	YRI	CEU	CHB+JPT
每个区块的平均长度(kb)	4.8	5.9	5.9
被区块横跨的基因组的比例(%)	86	84	84
每个区块的单体型(MAF ≥ 0.05)平均数量	5.12	3.63	3.63
MAF ≥ 0.05 的单体型占染色体的比例(%)	91	95	95

特别地，在这些接近完整的数据中，大部分的序列落入含有四个或更多 SNP 的区块中(从 YRI 中的 67% 到 CEU 中的 87%)，而且这些区块的平均大小与之前的估计类似[30]，这解决了关于区块可能是低标签密度的假象的担忧[52]。尽管一般的区块横跨很多 SNP(30～70 个)，但是在各个区块中，常见单体型的平均数量只有 4.0(CHB+JPT)到 5.6 个(YRI)，几乎所有的单体型都是不常见的单体型。这些结果证实了从样本量更小[29]、数据更不完整[30]的疾病遗传定位研究[27]以及基因组调查所得出的推断的普遍性。

远距离单体型和局部重组模式 尽管单体型经常在重组热点(和区块的边界)中断，这个趋势并不是一成不变的。我们鉴定了阶段性数据中所有在 269 个个体中频率均超过 0.05 的独特单体型，并将它们与精细尺度的重组图谱相比较。图 10 展示了 19 号染色体的一个区域，这个区域上很多这样的单体型在鉴定的重组热点位置中断，但是其他的则是连续的。因而，重组位点共定位的趋势并不意味着所有单体型在各个重组位点中断。

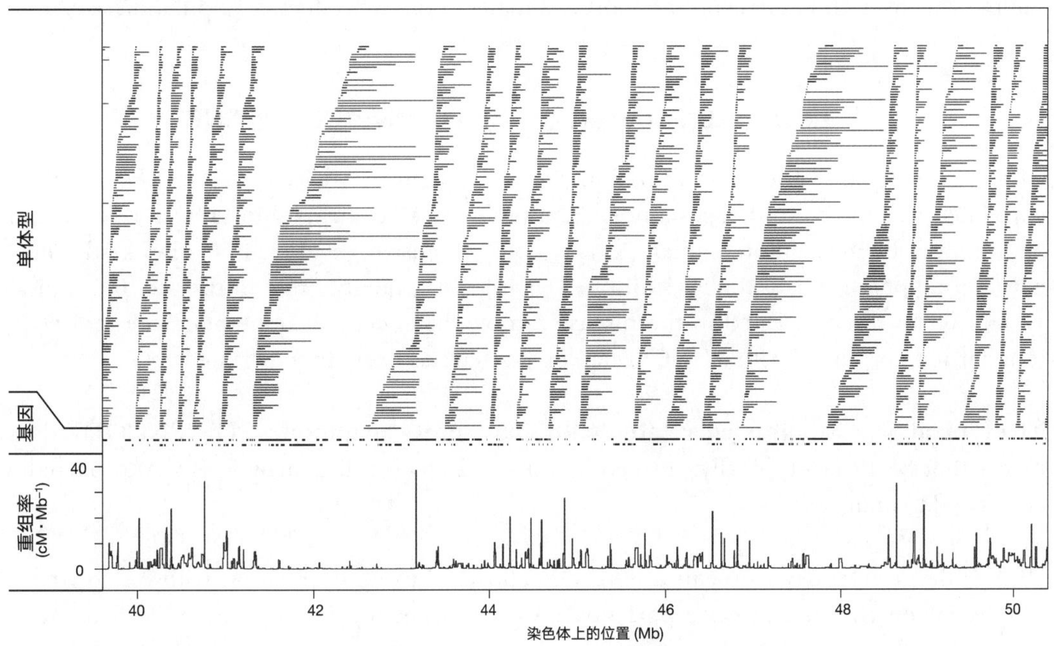

图 10. 重组率、单体型长度和基因位置之间的关系。重组率单位为 cM · Mb^{-1}(蓝色)。在 19 号染色体的

rates in cM Mb^{-1} (blue). Non-redundant haplotypes with frequency of at least 5% in the combined sample (bars) and genes (black segments) are shown in an example gene-dense region of chromosome 19 (19q13). Haplotypes are coloured by the number of detectable recombination events they span, with red indicating many events and blue few.

Some regions display remarkably extended haplotype structure based on a lack of recombination (Supplementary Fig. 8a, b). Most striking, if unsurprising, are centromeric regions, which lack recombination: haplotypes defined by more than 100 SNPs span several megabases across the centromeres. The X chromosome has multiple regions with very extensive haplotypes, whereas other chromosomes typically have a few such domains.

Most global measures of LD become more consistent when measured in genetic rather than physical distance. For example, when plotted against physical distance, the extent of pairwise LD varies by chromosome; when plotted against average recombination rate on each chromosome (estimated from pedigree-based genetic maps) these differences largely disappear (Supplementary Fig. 6). Similarly, the distribution of haplotype length across chromosomes is less variable when measured in genetic rather than physical distance. For example, the median length of haplotypes is 54.4 kb on chromosome 1 compared to 34.8 kb on chromosome 21. When measured in genetic distance, however, haplotype length is much more similar: 0.104 cM on chromosome 1 compared to 0.111 cM on chromosome 21 (Supplementary Fig. 9).

The exception is again the X chromosome, which has more extensive haplotype structure after accounting for recombination rate (median haplotype length = 0.135 cM). Multiple factors could explain different patterns on the X chromosome: lower SNP density, smaller sample size, restriction of recombination to females and lower effective population size.

A View of LD Focused on the Putative Causal SNP

Although genealogy and recombination provide insight into why nearby SNPs are often correlated, it is the redundancies among SNPs that are of central importance for the design and analysis of association studies. A truly comprehensive genetic association study must consider all putative causal alleles and test each for its potential role in disease. If a causal variant is not directly tested in the disease sample, its effect can nonetheless be indirectly tested if it is correlated with a SNP or haplotype that has been directly tested.

The typical SNP is highly correlated with many of its neighbours. The ENCODE data reveal that SNPs are typically perfectly correlated to several nearby SNPs, and partially correlated to many others.

We use the term proxy to mean a SNP that shows a strong correlation with one or more others. When two variants are perfectly correlated, testing one is exactly equivalent to testing the other; we refer to such collections of SNPs (with pairwise $r^2 = 1.0$ in the HapMap samples) as "perfect proxy sets".

一个高密度基因示例区域（19q13）内展示了在合并样本中频率至少为 5% 的非冗余单体型（棒状）和基因（黑色片段）。根据单体型区域内可检测的重组事件的数量，单体型标记为不同的颜色，红色代表很多的重组事件，而蓝色代表很少的重组事件。

由于缺少重组，一些区域表现出明显扩展的单体型结构（补充信息图 8a、8b）。最值得注意的是缺少重组的着丝粒区域：超过 100 个 SNP 组成的单体型横跨着丝粒前后数 Mb 的范围。X 染色体的多个区域含有范围非常大的单体型，而其他染色体则通常仅含有少量这样的区域。

当以遗传学距离而不是物理距离测量时，大多数对 LD 的全局性测量变得更加一致。例如，当针对物理距离作图时，成对 LD 的程度随染色体的不同而变化；当针对各条染色体上的平均重组率（估计自基于家系的遗传图谱）作图时，这些差异大部分消失了（补充信息图 6）。类似地，当以遗传学距离而不是物理距离测量时，单体型长度在染色体上的分布差异也更小。例如，单体型长度的中位数在 1 号染色体上是 54.4 kb，而在 21 号染色体上则是 34.8 kb。但是，当以遗传学距离测量时，单体型长度则接近得多：1 号染色体上是 0.104 cM，21 号染色体上是 0.111 cM（补充信息图 9）。

X 染色体同样是例外，扣除重组率的影响后，它含有范围更广的单体型结构（单体型长度中位数 = 0.135 cM）。有多个因素可以解释 X 染色体上呈现的不同模式：更低的 SNP 密度、更小的样本量、限于女性的重组以及更小的有效种群尺寸。

集中于潜在致病 SNP 的 LD 概览

尽管家系和重组为解释为什么邻近的 SNP 常常关联提供了线索，但是对于关联研究的设计和分析，SNP 中的冗余才是重中之重。一个真正全面的遗传关联研究必须考虑所有可能致病的等位基因，并且检验它们在疾病中潜在的作用。如果一个导致疾病的变异没有在疾病样本中被直接检验，但是它与一个已经被直接检验的 SNP 或单体型相关联，则它的作用就可以被间接检验。

一个典型的 SNP 与很多邻近 SNP 高度相关 ENCODE 数据反映了 SNP 通常与多个邻近的 SNP 完美相关，与很多其他的 SNP 部分相关。

我们使用术语“代理者”来表示与一个或者更多其他 SNP 高度相关的一个 SNP。当两个变异完美相关时，检验一个就完全等同于检验另一个；我们称这样的 SNP 集（在 HapMap 样本中成对的 $r^2 = 1.0$）为“完美的代理者集”。

Considering only common SNPs (the target of study for the HapMap Project) in CEU in the ENCODE data, one in five SNPs has 20 or more perfect proxies, and three in five have five or more. In contrast, one in five has no perfect proxies. As expected, perfect proxy sets are smaller in YRI, with twice as many SNPs (two in five) having no perfect proxy, and a quarter as many (5%) having 20 or more (Figs 11 and 12). These patterns are largely consistent across the range of frequencies studied by the project, with a trend towards fewer proxies at MAF < 0.10 (Fig. 11). Put another way, the average common SNP in ENCODE is perfectly redundant with three other SNPs in the YRI samples, and nine to ten other SNPs in the other sample sets (Fig. 13).

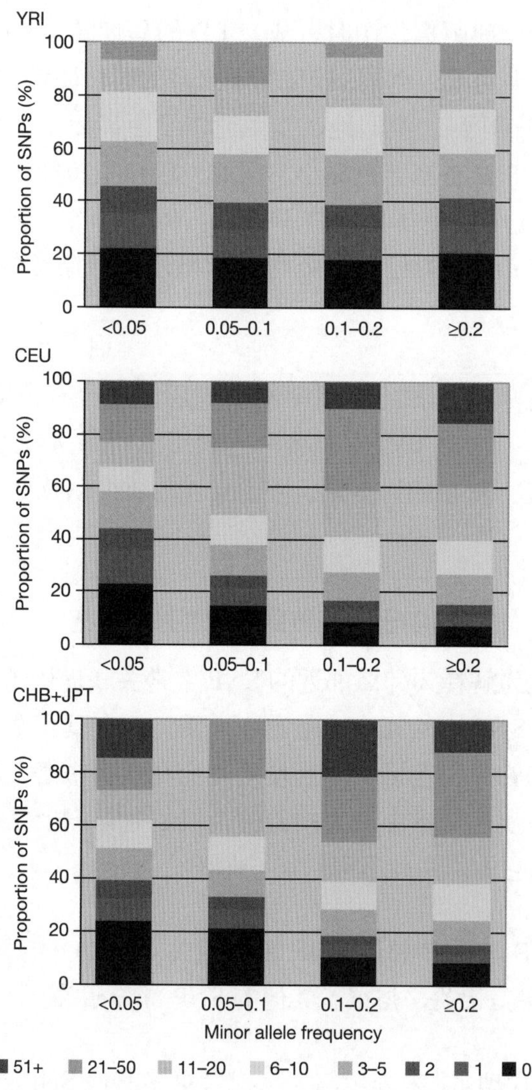

Fig. 11. The number of proxy SNPs ($r^2 \geqslant 0.8$) as a function of MAF in the ENCODE data.

只考虑 ENCODE 数据 CEU 中常见的 SNP(HapMap 计划的研究目标)，则每五个 SNP 中有一个有 20 个或更多的完美代理者，有三个含有五个或更多的完美代理者，只有一个没有完美代理者。正如预期的，完美代理者集在 YRI 中更小，没有完美代理者的 SNP(每五个中有两个)是另两个小组的两倍，含有 20 个或更多代理者的 SNP(5%)是另两个小组的四分之一(图 11 和 12)。在计划所研究的频率范围内，这些模式大部分是一致的，并且表现出 MAF < 0.10 时代理者更少的趋势(图 11)。换句话说，ENCODE 中常见的 SNP，在 YRI 样本中平均跟三个其他 SNP 完全冗余，在其他样本集中平均跟 9 到 10 个 SNP 完全冗余(图 13)。

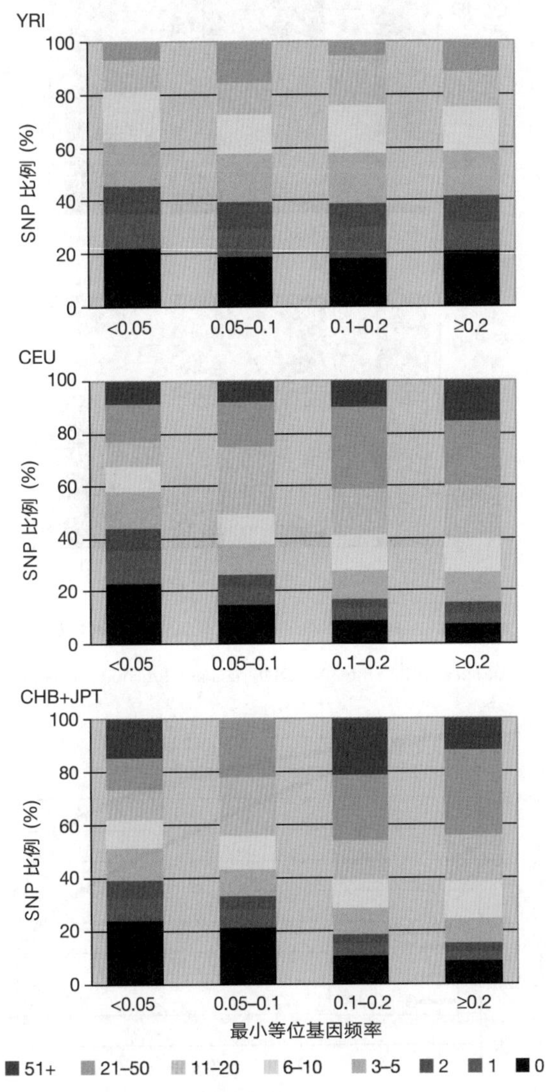

图 11. ENCODE 数据中不同 MAF 的 SNP 的代理者 SNP($r^2 \geqslant 0.8$)数量

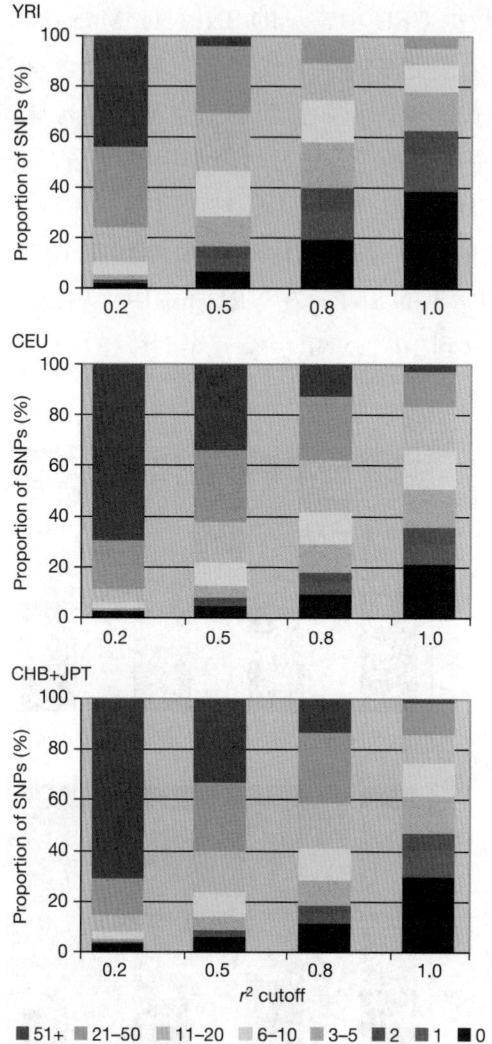

Fig. 12. The number of proxies per SNP in the ENCODE data as a function of the threshold for correlation (r^2).

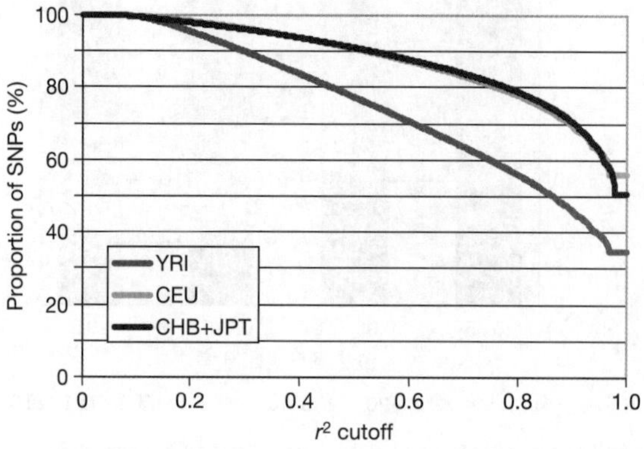

Fig. 13. Relationship in the Phase I HapMap between the threshold for declaring correlation between proxies and the proportion of all SNPs captured.

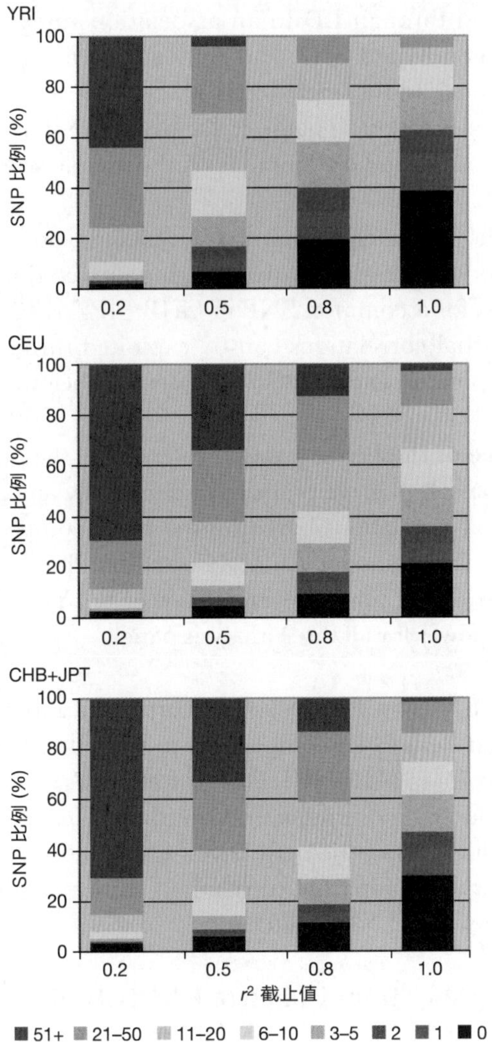

图 12. ENCODE 数据中不同相关性阈值 (r^2) 的每个 SNP 的代理者数量

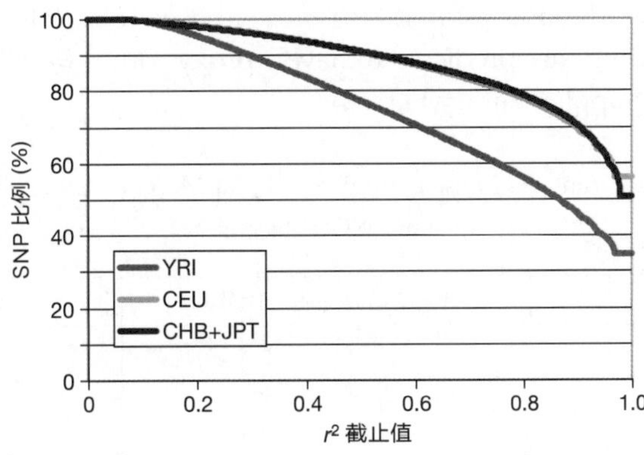

图 13. 第 I 阶段 HapMap 中用于判定代理者间相关的阈值和所有捕获到的 SNP 的比例之间的关系

Of course, to be detected through LD in an association study, correlation need not be complete between the genotyped SNP and the causal variant. For example, under a multiplicative disease model and a single-locus χ^2 test, the sample size required to detect association to an allele scales as $1/r^2$. That is, if the causal SNP has an $r^2 = 0.5$ to one tested in the disease study, full power can be maintained if the sample size is doubled.

The number of SNPs showing such substantial but incomplete correlation is much larger. For example, using a looser threshold for declaring correlation ($r^2 \geqslant 0.5$), the average number of proxies found for a common SNP in CHB+JPT is 43, and the average in YRI is 16 (Fig. 12). These partial correlations can be exploited through haplotype analysis to increase power to detect putative causal alleles, as discussed below.

Evaluating performance of the Phase I map. To estimate the proportion of all common SNPs captured by the Phase I map, we evaluated redundancy among SNPs on the genome-wide map, and performed simulations based on the more complete ENCODE data. The two methods give highly similar answers, and indicate that Phase I should provide excellent power for CEU, CHB and JPT, and substantial power for YRI. Phase II, moreover, will provide nearly complete power for all three analysis panels.

Redundancies among SNPs in Phase I HapMap. Redundancy offers one measure that Phase I has sampled densely in comparison to the underlying scale of correlation. Specifically, 50% (YRI) to 75% (CHB+JPT, CEU) of all SNPs on the Phase I map are highly correlated ($r^2 \geqslant 0.8$) to one or more others on the map (Fig. 13; see also Supplementary Fig. 10). Over 90% of all SNPs on the map have highly statistically significant correlation to one or more neighbours. These partial correlations can be combined to form haplotypes that are even better proxies for a SNP of interest.

Modelling Phase I HapMap from complete ENCODE data. A second approach to evaluating the completeness of the Phase I data involves thinning the more complete ENCODE data to match Phase I for allele frequency and SNP density. Simulated Phase I HapMaps were used to evaluate coverage in relation to the full set of common SNPs (Table 6), and provided nearly identical estimates to those above: 45% (YRI) to 74% (CHB+JPT, CEU) of all common SNPs are predicted to have a proxy with $r^2 \geqslant 0.8$ to a SNP included in the Phase I HapMap (Supplementary Fig. 11).

Table 6. Coverage of simulated Phase I and Phase II HapMap to capture all common SNPs in the ten ENCODE regions

Analysis panel	Per cent maximum $r^2 \geqslant 0.8$	Mean maximum r^2
Phase I HapMap		
YRI	45	0.67
CEU	74	0.85
CHB+JPT	72	0.83

当然，为了能在关联研究中通过 LD 被检测到，基因分型的 SNP 和导致表型的变异间的相关性不需要是完全的。例如，在乘性疾病模型和单基因座 χ^2 检验下，检测一个等位基因的相关性所要求的样本大小是 $1/r^2$。也就是说，当在疾病研究中导致表型的 SNP 与被检验的 SNP 的 $r^2 = 0.5$ 时，如果样本大小翻倍就能够保持完全的效能。

表现出这种相当程度但不完全相关的 SNP 的数量大得多。例如，利用一个宽松的阈值来判定相关性（$r^2 \geqslant 0.5$），在 CHB+JPT 中一个常见 SNP 的代理者的平均数量是 43 个，在 YRI 中是 16 个（图 12）。正如下文讨论的，这些部分相关可以通过单体型分析而被挖掘，进而来增加检测可能导致表型的等位基因的能力。

第 I 阶段图谱质量的评估　为了估计第 I 阶段图谱捕获的所有常见 SNP 的比例，我们评估了全基因组范围图谱上 SNP 间的冗余，并且基于更完整的 ENCODE 数据进行了模拟。这两种方法给出了高度相似的答案，说明第 I 阶段图谱应该为 CEU、CHB 和 JPT 提供了非常好的效能，为 YRI 提供了相当程度的效能。并且，第 II 阶段将会为所有三个分析小组提供接近完全的效能。

第 I 阶段 HapMap 中 SNP 间的冗余　与潜在的相关性指标相比，冗余提供了一种测量手段，第 I 阶段图谱已经对其进行了密集取样。特别地，第 I 阶段图谱上所有 SNP 中的 50%（YRI）到 75%（CHB+JPT、CEU）与一个或更多其他 SNP 高度相关（$r^2 \geqslant 0.8$）（图 13；也可见补充信息图 10）。图谱上所有 SNP 中超过 90% 与一个或更多邻近 SNP 具有统计学意义上的高度相关。这些部分相关能被联合起来形成单体型，这些单体型对于感兴趣的 SNP 甚至是更好的代理者。

基于完整 ENCODE 数据的第 I 阶段 HapMap 建模　评估第 I 阶段数据完整性的第二个方法包括缩减更加完整的 ENCODE 数据来匹配第 I 阶段数据的等位基因频率和 SNP 密度。模拟的第 I 阶段 HapMap 被用于评估其相对全套常见 SNP 的覆盖度（表 6），它提供了跟上面几乎一致的估计值：所有常见 SNP 的 45%（YRI）到 74%（CHB+JPT、CEU）被预测在第 I 阶段的 HapMap 中有一个 $r^2 \geqslant 0.8$ 的代理者（补充信息图 11）。

表 6. 模拟的 HapMap 第 I 和第 II 阶段捕获 10 个 ENCODE 区域中所有常见 SNP 的覆盖度

分析小组	最大 r^2 值 ≥0.8 的比例	平均最大 r^2 值
第 I 阶段 HapMap		
YRI	45	0.67
CEU	74	0.85
CHB+JPT	72	0.83

Continued

Analysis panel	Per cent maximum $r^2 \geq 0.8$	Mean maximum r^2
Phase II HapMap		
YRI	81	0.90
CEU	94	0.97
CHB+JPT	94	0.97

Simulated Phase I HapMaps were generated from the phased ENCODE data (release 16c1) by randomly picking SNPs that appear in dbSNP build 121 (excluding "non-rs" SNPs in release 16a) for every 5-kb bin until a common SNP was picked (allowing up to three attempts per bin). The Phase II HapMap was simulated by picking SNPs at random to achieve an overall density of 1 SNP per 1 kb. These numbers are averages over 20 independent iterations for all ENCODE regions in all three analysis panels.

Statistical power in association studies may be more closely approximated by the average (maximal) correlation value between a SNP and its best proxy on the map, rather than by the proportion exceeding an arbitrary (and stringent) threshold. The average values for maximal r^2 to a nearby SNP range from 0.67 (YRI) to 0.85 (CEU and CHB+JPT).

Modelling Phase II HapMap from complete ENCODE data. A similar procedure was used to generate simulated Phase II HapMaps from ENCODE data (Table 6). Phase II is predicted to capture the majority of common variation in YRI: 81% of all common SNPs should have a near perfect proxy ($r^2 \geq 0.8$) to a SNP on the map, with the mean maximal r^2 value of 0.90. Unsurprisingly, the CEU, CHB and JPT samples, already well served by Phase I, are nearly perfectly captured: 94% of all common sites have a proxy on the map with $r^2 \geq 0.8$, with an average maximal r^2 value of 0.97.

These analyses indicate that the Phase I and Phase II HapMap resources should provide excellent coverage for common variation in these population samples.

Selection of Tag SNPs for Association Studies

A major impetus for developing the HapMap was to guide the design and prioritization of SNP genotyping assays for disease association studies. We refer to the set of SNPs genotyped in a disease study as tags. A given set of tags can be analysed for association with a phenotype using a variety of statistical methods which we term tests, based either on the genotypes of single SNPs or combinations of multiple SNPs.

The shared goal of all tag selection methods is to exploit redundancy among SNPs, maximizing efficiency in the laboratory while minimizing loss of information[24,27]. This literature is extensive and varied, despite its youth. Some methods require that a single SNP serve as a proxy for other, untyped variants, whereas other methods allow combinations of alleles (haplotypes) to serve as proxies; some make explicit use of LD blocks whereas others are agnostic to such descriptions. Although it is not practical to implement all such

续表

分析小组	最大 r^2 值 ≥0.8 的比例	平均最大 r^2 值
第 II 阶段 HapMap		
YRI	81	0.90
CEU	94	0.97
CHB+JPT	94	0.97

模拟的第 I 阶段 HapMap 是从区分染色体的 ENCODE 数据（版本 16c1）产生的，方法是每 5 kb 随机选取出现在 dbSNP 版本 121（不包括版本 16a 中的"non-rs"SNP）中的 SNP，直到取到一个常见 SNP（每个 5 kb 区间允许三次尝试）。第 II 阶段 HapMap 的模拟是通过随机取 SNP 以达到总体每 1 kb 1 个 SNP 的密度。这些数值是三个分析小组中所有 ENCODE 区域 20 次独立迭代的平均值。

通过 SNP 和它在图谱上的最佳代理者的平均（最大）相关值，而不是通过超过一个主观设定的（并且严格的）阈值的比例，可以对关联研究中的统计效能进行更接近的估算。SNP 与附近 SNP 的最大 r^2 的平均值在 0.67（YRI）到 0.85（CEU 和 CHB+JPT）之间变化。

基于完整 ENCODE 数据的第 II 阶段 HapMap 建模 类似的步骤被用于从 ENCODE 数据产生模拟的第 II 阶段 HapMap（表 6）。第 II 阶段预计能捕获 YRI 中绝大多数的常见变异：所有常见 SNP 的 81% 应该在图谱上有一个接近完美的代理者（$r^2 \geqslant 0.8$），平均最大 r^2 值是 0.90。与预期相符，CEU、CHB 和 JPT 样本已经在第 I 阶段满足要求，被近乎完美地捕获：所有常见位点的 94% 在图谱上有一个 $r^2 \geqslant 0.8$ 的代理者，平均最大 r^2 值为 0.97。

这些分析说明第 I 和第 II 阶段 HapMap 的资源应该能够非常好地覆盖这些种群样本中的常见变异。

用于关联研究的标签 SNP 的选择

开展 HapMap 研究的一个主要的动力是指导疾病关联研究中 SNP 基因分型实验的设计和优先次序。我们称疾病研究中被基因分型的 SNP 集为标签。基于单个 SNP 或多个 SNP 的组合的基因型，利用我们称之为检验的各种统计学方法，可以分析一组给定的标签与某种表型的关联。

所有标签选择方法的共同目标是利用 SNP 间的冗余，最大化实验效率，同时最小化信息的丢失[24,27]。尽管刚起步，但这样的文献广泛而多样。一些方法要求单个 SNP 作为其他未分型变异的代理者，而其他方法允许等位基因的组合（单体型）作为代理者；一些人明确使用 LD 区块，而其他人则没有这样的描述。尽管在本计划的

methods at the project website, the HapMap genotypes are freely available and investigators can apply their method of choice to the data. To assist users, both a single-marker tagging method and a more efficient multimarker method have been implemented at http://www.hapmap.org.

Tagging using a simple pairwise method. To illustrate general principles of tagging, we first applied a simple and widely used pairwise algorithm[53,54]: SNPs are selected for genotyping until all common SNPs are highly correlated ($r^2 \geqslant 0.8$) to one or more members of the tag set.

Starting from the substantially complete ENCODE data, the density of common SNPs can be reduced by 75–90% with essentially no loss of information (Fig. 14). That is, the genotyping burden can be reduced from one common SNP every 500 bp to one SNP every 2 kb (YRI) to 5 kb (CEU and CHB+JPT). Because LD often extends for long distances, studies of short gene segments tend to underestimate the redundancy across the genome[43].

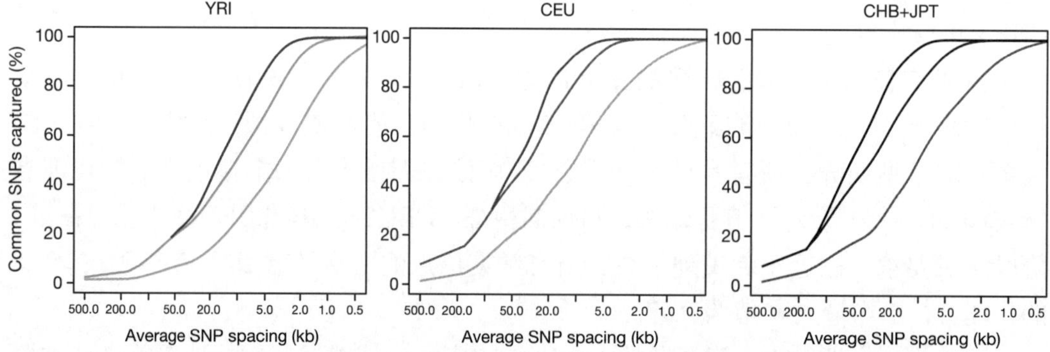

Fig. 14. Tag SNP information capture. The proportion of common SNPs captured with $r^2 \geqslant 0.8$ as a function of the average tag SNP spacing is shown for the phased ENCODE data, plotted (left to right) for tag SNPs prioritized by Tagger (multimarker and pairwise) and for tag SNPs picked at random. Results were averaged over all the ENCODE regions.

Although tags selected based on LD offer the greatest improvements in efficiency and information capture, even randomly chosen subsets of SNPs offer considerable efficiencies (Fig. 14).

The data also reveal a rule of diminishing returns: a small set of highly informative tags captures a large fraction of all variation, with additional tags each capturing only one or a few proxies. For example, in CHB+JPT the most informative 1% of all SNPs (one per 50 kb) is able to proxy (at $r^2 \geqslant 0.8$) for 40% of all common SNPs, whereas a substantial proportion of SNPs have no proxies at all.

These observations are encouraging with respect to genome-wide association studies. A set of SNPs typed every 5–10 kb across the genome (within the range of current technology) can capture nearly all common variation in the genome in the CEU and CHB+JPT samples, with more SNPs required in the YRI samples.

网站上实现所有这些方法并不现实，但是 HapMap 的基因型数据是自由获取的，研究者可以用他们选择的方法分析这些数据。为了帮助使用者，单个标志物标签法和更有效的多标志物法都已经在 http://www.hapmap.org 网站上实现了。

使用简单成对方法进行的标记 为了说明标记的一般原则，我们首先应用了一个简单而广泛使用的成对算法[53,54]：选择 SNP 进行基因分型，直到所有常见 SNP 与一个或更多的标签集成员高度连锁（$r^2 \geqslant 0.8$）。

先以相当完整的 ENCODE 数据为例，常见 SNP 的密度能被减少 75% ~ 90% 而基本不丢失信息（图 14）。也就是说，基因分型的负担能从每 500 bp 一个常见 SNP 减少到每 2 kb（YRI）或每 5 kb（CEU 和 CHB+JPT）一个 SNP。因为 LD 常常延伸很长距离，短基因片段的研究通常低估基因组中的冗余度[43]。

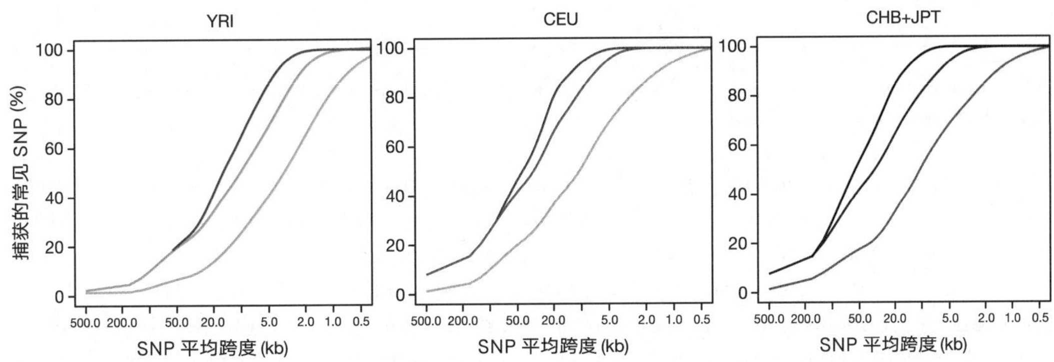

图 14. 标签 SNP 信息捕获。图中展示的是 ENCODE 数据中，被不同平均跨度的标签 SNP 捕获（$r^2 \geqslant 0.8$）的常见 SNP 的比例，从左到右依次为由 Tagger 软件（多标志物和成对标志物）定为优先级的标签 SNP 和随机选取的标签 SNP。结果是所有 ENCODE 区域的平均值。

基于 LD 选择的标签在效率和信息的捕获上提供了最大的改进，而即使是随机选择的 SNP 子集也提供了相当高的效率（图 14）。

这个数据还反映了一个回报不断减少的规律：一个小的高度富含信息的标签集可以捕获所有变异中的大部分，而增加的标签仅捕获一个或很少的代理者。例如，在 CHB+JPT 中，所有 SNP（每 50 kb 一个）中最富有信息的 1% 能够代理（$r^2 \geqslant 0.8$）所有常见 SNP 的 40%，而相当大比例的 SNP 完全没有代理者。

对于全基因组关联研究，这些现象是鼓舞人心的。对于 CEU 和 CHB+JPT 样本，在全基因组中每 5 ~ 10 kb 被分型的一组 SNP（现有技术的能力范围之内）就能够捕获基因组中几乎所有的常见变异，对于 YRI 样本则需要更多的 SNP。

Tagging from the genome-wide map. Whereas analysis of the complete ENCODE data set reveals the maximal efficiency likely to be possible with this tag selection strategy, analysis of the Phase I map illuminates the extent to which the current resource can be used for near-term studies. Specifically, using the same pairwise tagging approach above, 260,000 (CHB+JPT) to 474,000 (YRI) SNPs are required to capture all common SNPs in the Phase I data set (Table 7). That is, being incomplete and thus less redundant, the Phase I data are much less compressible by tag SNP selection than are the ENCODE data. Nevertheless, even at this level a half to a third of all SNPs can be selected as proxies for the remainder (and, by inference, the bulk of other common SNPs in the genome).

Table 7. Number of selected tag SNPs to capture all observed common SNPs in the Phase I HapMap

r^2 threshold*	YRI	CEU	CHB+JPT
$r^2 \geqslant 0.5$	324,865	178,501	159,029
$r^2 \geqslant 0.8$	474.409	293,835	259,779
$r^2 = 1.0$	604,886	447,579	434,476

Tag SNPs were picked to capture common SNPs in HapMap release 16c1 using the software program Haploview.
* Pairwise tagging at different r^2 thresholds.

Increasing the efficiency of tag SNPs. Although the pairwise method is simple, complete and straightforward, efficiency can be improved with a number of simple changes. First, relaxing the threshold on r^2 for tag SNP selection substantially reduces the number of tag SNPs selected, with only a modest decrease in the correlations among SNPs (Table 7). For example, reducing the r^2 threshold from 0.8 to 0.5 decreases the number of tag SNPs selected from the HapMap by 39% in CHB+JPT (260,000 to 159,000) and 32% in YRI (474,000 to 325,000). The average r^2 value between tags and other (unselected) SNPs falls much less dramatically than the number of tags selected, increasing efficiency. Whether such a loss of power is justified by the disproportionate reduction in work is a choice each investigator will need to make.

A second enhancement exploits multimarker haplotypes. Many investigators have discussed using multiple SNPs (in haplotypes and regression models) to serve as proxies for untyped sites[55-58], which may reduce the number of tags required and increase the power of analyses performed. Figure 14 illustrates the point with one such method[55], showing that a multimarker method allows greater coverage for a fixed set of markers (or, alternatively, fewer markers to achieve the same coverage). Although a full consideration of this issue is beyond the scope of this paper, the availability of these and other data should allow the comparison and application of such methods.

A third approach to increasing efficiency is to prioritize tags based on the number of other SNPs captured. Whereas 260,000 SNPs are required to provide $r^2 \geqslant 0.8$ for all SNPs in the Phase I HapMap (CHB+JPT), the best 10,000 such SNPs (4%) capture 22% of all common

从全基因组图谱进行的标签选取 尽管对完整 ENCODE 数据集的分析反映了利用这种标签选择策略可能具有的最大效率，对第 I 阶段图谱的分析则阐明了现有资源能够在多大程度上被用于近期的研究。特别地，利用相同的上述成对标签方法，为捕获第 I 阶段数据集中所有常见 SNP，需要 260,000(CHB+JPT) 到 474,000(YRI) 个 SNP(表 7)。也就是说，第 I 阶段数据不够完整因而冗余更少，与 ENCODE 数据相比，能够通过标签 SNP 选择而被压缩的程度要小很多。但是，即使在这种水平下，SNP 中的一半或三分之一可以被选作剩余 SNP(推断可知也是基因组中其他常见 SNP 的大部分)的代理者。

表 7. 为捕获第 I 阶段 HapMap 中观察到的所有常见 SNP 所选择的标签 SNP 的数量

r^2 阈值 *	YRI	CEU	CHB+JPT
$r^2 \geq 0.5$	324,865	178,501	159,029
$r^2 \geq 0.8$	474,409	293,835	259,779
$r^2 = 1.0$	604,886	447,579	434,476

标签 SNP 是利用软件 Haploview 选取的，以捕获 HapMap 版本 16c1 中常见的 SNP。
* 在不同 r^2 阈值下取成对的标签。

增加标签 SNP 的效率 尽管成对的方法简单、完整而直接，我们还是可以通过一些简单的改变提高效率。首先，放宽用于标签 SNP 选择的 r^2 的阈值，可以大大减少选择的标签 SNP 的数量，而在 SNP 间的相关性上仅稍有下降(表 7)。例如，r^2 阈值从 0.8 减少到 0.5，在 CHB+JPT 中减少了 39% 从 HapMap 选择的标签 SNP 的数量(从 260,000 到 159,000)，而在 YRI 中减少了 32%(从 474,000 到 325,000)。标签与其他(未选择)SNP 间平均 r^2 值下降的水平要比选择的标签数量的下降少得多，因而提高了效率。是否这样一种效能的损失会因为不成比例地减少工作量变得合理，是每个研究者需要作出的一个选择。

第二个提高的方法是利用多标志物单体型。很多研究者已经讨论了利用多个 SNP(在单体型和回归模型中)作为未分型位点的代理者 [55-58]，这可能能够减少需要的标签的数量，并且增加分析的效能。图 14 利用一个这样的方法 [55] 说明了这一点，它表明，对于固定的一组标志物，多标志物方法能够达到更高的覆盖度(换言之，更少的标志物可以达到相同的覆盖度)。尽管全面考虑这个问题超出了本文的范畴，已有的这些以及其他数据足以对这样的方法进行比较和应用。

第三个增加效率的方法是基于捕获的其他 SNP 的数量对标签划分优先级。尽管要求 260,000 个 SNP 才能以 $r^2 \geq 0.8$ 捕获第 I 期 HapMap 里所有的 SNP(CHB+JPT)，其中最佳的 10,000 个(4%)就能以 $r^2 \geq 0.8$ 捕获所有常见变异位点的

variable sites with $r^2 \geqslant 0.8$ (Table 8). Such prioritization can be applied using different weights for SNPs based on genomic annotation (for example, non-synonymous coding SNPs, SNPs in conserved non-coding sequence, and candidate genes of biological interest).

Table 8. Proportion of common SNPs in Phase I captured by sets of tag SNPs

Tag SNP set size	Common SNPs captured (%)		
	YRI	CEU	CHB+JPT
10,000	12.3	20.4	21.9
20,000	19.1	30.9	33.2
50,000	32.7	50.4	53.6
100,000	47.2	68.5	72.2
250,000	70.1	94.1	98.5

As in Table 7, tag SNPs were picked to capture common SNPs in HapMap release 16c1 using Haploview, selecting SNPs in order of the fraction of sites captured. Common SNPs were captured by fixed-size sets of pairwise tags at $r^2 \geqslant 0.8$.

Tag transferability across populations. The most complete set of tags would be those based on all 269 samples; however, many studies may be performed in individuals more closely related to one particular HapMap population, and efficiency may be gained by selecting tags only from that population sample. (Selecting tags in a HapMap population sample that is known to be more distantly related than is another, for example, using CEU to pick tags for a study of Japanese, seems inefficient.)

An important question is how tags selected in one or more analysis panels will transfer to disease studies performed in these or other populations. Our data do not address this question directly, although the known similarity of allele and haplotype frequencies across populations within continents[41] is encouraging. More data are clearly needed, however.

Tag selection based on initial genotyping. Whereas the discussions above assume *de novo* selection of SNPs, many investigators will have already performed initial studies, and wish to design follow-on experiments. The HapMap data can be used to highlight SNPs that might potentially explain a positive association signal, or those that were poorly captured (and thus still need to be tested) after a negative scan. In cases where multiple SNPs are both associated with the trait and with each other, the HapMap data can be queried to identify whether samples from any other analysis panel show a breakdown of LD in that region, and thus the possibility of narrowing the span over which the causal variant may reside.

Applications to the Analysis of Association Data

Beyond guiding selection of tag SNPs, HapMap data can inform the subsequent analysis and interpretation in disease association studies.

22%(表 8)。这样的划分优先级法可以通过根据基因组注释对 SNP 使用不同的权重来应用(例如，非同义编码 SNP、在保守的非编码序列中的 SNP，以及有生物学意义的候选基因)。

表 8. 通过不同标签 SNP 集捕获的第 I 阶段中常见 SNP 的比例

标签 SNP 集大小	捕获的常见 SNP(%)		
	YRI	CEU	CHB+JPT
10,000	12.3	20.4	21.9
20,000	19.1	30.9	33.2
50,000	32.7	50.4	53.6
100,000	47.2	68.5	72.2
250,000	70.1	94.1	98.5

如表 7 一样，标签 SNP 是使用 Haploview 选取的，以捕获 HapMap 版本 16c1 中常见的 SNP，SNP 按捕获的位点的比例顺序选择。常见 SNP 被固定大小的一组成对标签以 $r^2 \geqslant 0.8$ 捕获。

标签在不同种群间的可转移性　最完整的一组标签是基于所有 269 个样本的标签；但是，很多研究可能是在与一个特定的 HapMap 种群亲缘关系更近的个体中进行的，因此效率仅在从那个种群样本中选择标签时可能会增加。(在亲缘关系更远的 HapMap 种群样本中选择标签看起来是低效的，例如使用 CEU 种群选取用于研究日本人的标签。)

一个重要的问题是：在一个或更多个分析小组中选择的标签如何转移到在这些或者其他种群中所开展的疾病研究。尽管已知的同一大陆上各个种群中等位基因和单体型频率的相似性 [41] 是令人鼓舞的，但是我们的数据并没有直接回答这个问题。很明显，要回答这个问题还需要更多的数据。

基于初步基因分型的标签选择　上述的讨论是假定从头选择 SNP 的，但是很多研究者已经进行了初步研究，希望在此基础上设计后续实验。HapMap 数据可以用来突出有望解释阳性关联信号的 SNP，或者经过阴性扫描不容易被捕获的(因而仍需要被检验)SNP。在多个 SNP 都跟性状关联并且 SNP 也彼此关联的情况下，可以查询 HapMap 数据来鉴定任一其他分析小组的样本是否在那个区域表现为 LD 的断裂，以及缩小导致性状的变异可能存在的区域范围的可能性。

在关联数据分析中的应用

在疾病关联研究中，HapMap 数据除了指导标签 SNP 的选择，还能够影响后续的分析和解释。

Analysis of an existing genotype data set. The HapMap can be used to inform association testing, regardless of how tags were selected. Specifically, as long as the SNPs genotyped in a disease study have also been typed in the HapMap samples, it is possible to identify which SNPs are well captured by the genotyped SNPs (either singly, or in haplotype combinations), and which are not[55].

This is of particular importance for genome-wide association studies performed using array-based, standardized genotyping reagents, which do not allow investigators to choose their own sets of tag SNPs. The Affymetrix 120K SNP array data included in Phase I of the HapMap provides a simple example: in CEU 48% of HapMap SNPs have substantial pairwise correlation ($r^2 \geqslant 0.5$) to one or more of the 120K SNPs on the array. An additional 13%, however, are not correlated to a single SNP, but are to a specific haplotype of two members of the 120K panel. By identifying such haplotype predictors in the HapMap, and testing them (in addition to the single SNPs) in a disease study, it is likely that power will be increased (I. Pe'er *et al.*, manuscript in preparation).

Evaluating statistical significance and interpreting results. An important challenge in genome-wide association testing is to develop statistical procedures that minimize false positives without greatly sacrificing true positives. The challenge is amplified by the correlated nature of polymorphism data, which makes simple frequentist approaches that assume independence (such as Bonferonni correction) highly conservative. To illustrate this point, we used the ENCODE data to estimate the "effective number of independent tests" (the statistical burden of testing all common ($MAF \geqslant 0.05$) variation) across large genomic regions. Specifically, we re-sampled from the phased ENCODE chromosomes to create mock case-control panels in which all common SNPs were observed, but there was not a causal allele. The resulting χ^2 distribution for association indicates that complete testing of common variation in each 500-kb region is equivalent to performing about 150 independent statistical tests (in CEU and CHB+JPT) and about 350 tests (in YRI). Although it will probably be desirable to perform such empirical estimates of significance within each disease study, these results illustrate how Bonferonni correction overestimates the statistical penalty of performing many correlated tests.

Study of less common alleles. We have focused primarily on the hypothesis that a single, common causal allele exists, and needs to be tested for association to disease. Of course, in many cases the causal allele(s) will be less common, and might be missed by such an approach.

It is possible to perform additional haplotype tests, beyond those that capture known polymorphisms, in the hope of capturing less common or unrepresented alleles[56]. Such haplotype analysis has a long history and proven value in mendelian genetics; the causal mutation is generally rare and unexamined during initial genotyping, but is frequently recognized by its presence on a long, unique haplotype of common alleles[18,19,59-62].

Admixture mapping. Although not designed specifically to enable admixture mapping[63],

对一个已有的基因型数据集的分析 HapMap能够用于关联检验，而不受标签如何选择的影响。特别地，只要在疾病研究中被基因分型的SNP也在HapMap样本中被分型，就有可能鉴定哪些SNP被已进行基因分型的SNP较好地捕获（单一地或者在单体型组合中），哪些没有被捕获[55]。

这对于使用基于芯片的、标准化的基因分型试剂所进行的全基因组关联研究特别重要，这些研究不允许研究者选择他们自己的标签SNP集。包含在第I阶段的HapMap的Affymetrix 120K SNP芯片数据提供了一个简单的例子：在CEU中，48%的HapMap SNP与芯片上的120K SNP中的一个或者更多有着相当程度的成对相关性（$r^2 \geqslant 0.5$）。另外还有13%的SNP不与单个SNP相关，但是与120K组中两个成员组成的特定单体型相关。通过鉴定HapMap中这样的单体型预测者，并且在疾病研究中对它们（以及单个SNP）进行检验，效能有可能会被提高（佩尔等，稿件准备中）。

统计学意义评估和结果解释 在全基因组关联检验中一个重要的挑战是开发最小化假阳性且不大幅牺牲真正阳性数据的统计学方法。多态性数据的关联特性将这个挑战进一步放大，使得假定独立的简单频率论方法（如邦费罗尼校正）高度保守。为了阐释这一点，我们使用ENCODE数据来估计在大的基因组区域内“独立检验的有效数量”（检验所有常见（MAF ≥ 0.05）变异的统计学负担）。特别地，我们从区分染色体的ENCODE染色体中重新取样，进而产生模拟的样本-对照组，其中可观察到所有常见SNP，但是没有引起疾病的等位基因。对于关联分析，得到的χ^2分布说明对各个500 kb区域内常见变异的完整检验等同于进行大约150次独立统计学检验（在CEU和CHB+JPT中）和大约350次检验（在YRI中）。尽管可能需要在各个疾病研究中都进行这样的对统计学意义的经验估计，这些结果阐释了邦费罗尼校正是如何高估了进行大量相关性检验的统计惩罚的。

不常见等位基因的研究 我们之前主要聚焦于这样的假说，即单个常见的引起疾病的等位基因是存在的，需要检验其与疾病的关联。当然，在很多情况下，导致疾病的等位基因是不常见的，可能被这样的方法所遗漏。

除了捕获已知的多态性的检验，还可以进行额外的单体型检验，以尝试捕获不常见或者无代表的等位基因[56]。这种单体型分析有着很长的历史，已经证明在孟德尔遗传研究中具有价值；引起疾病的突变通常是罕见的，并且在一开始的基因分型中未被检测到，但是经常由于其存在于常见等位基因的长的唯一的单体型上而被识别[18,19,59-62]。

混合作图 尽管没有特别设计来实现混合作图[63]，但是HapMap已经为这种方

the HapMap has helped lay the groundwork for this approach. Admixture mapping requires a map of SNPs that are highly differentiated in frequency across population groups. By typing many SNPs in samples from multiple geographical regions, the data have helped to identify such SNPs for the design of genome-wide admixture mapping panels[64,65] and can be further used to identify candidate SNPs with large allele frequency differences for follow-up of positive admixture scan results[66].

Loss of heterozygosity in tumours. Loss of heterozygosity (LOH) in tumour tissue can be a powerful indicator of the location of tumour suppressor genes, and genome-wide, fine-scale LOH analysis has been empowered by genome-wide SNP arrays[67]. Germline DNA is not always available from the same subjects, however, and even if available, typing of germline DNA doubles project costs. In lower density scans for LOH (with markers far apart relative to the scale of LD), long runs of homozygosity in tumours are nearly always indicative of LOH. However, at higher densities runs of homozygosity can be due to haplotype homozygosity in the inherited germline DNA, rather than LOH.

The HapMap data can help minimize this difficulty; previous probabilities for homozygosity based on known frequencies of haplotypes in the HapMap data can be used to distinguish homozygosity due to haplotype sharing rather than LOH[68].

Identifying Structural Variants in HapMap Data

Structural variations—segments where DNA is deleted, duplicated, or rearranged—are common[69,70] and have an important role in diseases[71-73]. The HapMap can provide some insight into structural variation because, in many cases, structural variants reveal themselves through signatures in SNP genotype data. In particular, polymorphic deletions are important to discover, because loss of genetic material is of obvious functional relevance, and results in aberrant patterns of SNP genotypes. These include apparent non-mendelian inheritance of SNP alleles, null genotypes and deviations from Hardy–Weinberg equilibrium. However, such SNPs are routinely discarded as technical failures of genotyping.

Thus, we scanned the unfiltered Phase I HapMap data using an approach developed and validated to identify polymorphic deletions from clusters of SNPs with aberrant genotype patterns (calibrated across the multiple centres and genotyping platforms[74]). In total, 541 candidate deletion polymorphisms were identified, of which 150 were common enough to be observed as homozygotes.

The properties of these candidate deletions, including experimental validation of 90 candidates, are described in ref. 74. Validated polymorphisms include 10 that remove coding exons of genes, such that in many cases individuals are homozygous null for the encoded transcript. Analysis of confirmed deletions often shows strong LD with nearby SNPs, indicating that LD-based approaches can be useful for detecting disease associations due to structural (as well as SNP) variants.

法打下了基础。混合作图需要在不同种群中频率高度不同的一群 SNP 的图谱。通过对来自多个地理区域的样本的很多 SNP 进行分型，得到的数据已经帮助鉴定到可用于设计全基因组混合作图的 SNP[64,65]，并且能被进一步用于鉴定后续阳性混合扫描结果所需的存在较大等位基因频率差异的候选 SNP[66]。

肿瘤中的杂合性缺失 肿瘤组织中的杂合性缺失（LOH）可以用作抑癌基因位置的强有力的指示物，全基因组 SNP 芯片使得全基因组范围的精细 LOH 分析成为可能[67]。然而，对于同一个个体，种系 DNA 并不总是能够获得的，而且即使获得了，对种系 DNA 的基因型分型也会使项目的花费加倍。在低密度扫描 LOH 时（相对于 LD 的尺度，标志物之间相距较远），肿瘤中纯合性的大量出现几乎总是提示 LOH 的存在。不过，在高密度扫描 LOH 时，纯合性的大量出现可能是由于遗传的种系 DNA 中单体型的纯合性，而不是 LOH。

HapMap 数据能够帮助将这种困难最小化；前期基于 HapMap 数据中单体型的已知频率得到的纯合性可能性可以用于区分单体型共享而非 LOH 导致的纯合性[68]。

HapMap 数据中结构变异的鉴定

结构变异——片段上发生 DNA 缺失、重复或重排——是常见的[69,70]，并且在疾病中有着重要的作用[71-73]。HapMap 能够提供关于结构变异的信息，因为在很多情况下，结构变异可通过 SNP 基因型数据中的特征反映出来。发现多态性的缺失尤其重要，因为遗传材料的缺失是明显与功能相关的，并且会导致 SNP 基因型的异常模式。这包括 SNP 等位基因明显的非孟德尔遗传、频率为零的基因型和哈迪–温伯格平衡的偏离。然而，这样的 SNP 通常作为基因分型的技术问题而被弃掉。

因而，在扫描未过滤的第 I 阶段 HapMap 数据时，我们采用了一种为从带有异常基因型模式的 SNP 簇（在多个中心和基因分型平台进行了校准[74]）中鉴定多态性缺失而开发和验证的方法。总共 541 个候选的缺失多态性被鉴定出来，其中 150 个较为常见，可观察到是纯合的。

这些候选缺失的组成，包括对 90 个候选者的实验验证，都在参考文献 74 中有所描述。得到验证的多态性中有 10 个使基因缺失外显子，在很多情况下，这种个体纯合缺少编码的转录本。对已证实的缺失的分析经常发现这种缺失与邻近 SNP 存在强 LD，表明基于 LD 的方法可以用于检测结构（以及 SNP）变异导致的疾病关联。

Polymorphic inversions may also be reflected in the HapMap data as long regions where multiple SNPs are perfectly correlated: because recombination between an inverted and non-inverted copy is lethal, the inverted and non-inverted copies of the region evolve independently. A striking example corresponds to the known inversion polymorphism on chromosome 17, present in 20% of the CEU chromosomes, that has been associated with fertility and total recombination rate in females among Icelanders[75]. Long LD may also arise, however, due to a low recombination rate or certain forms of natural selection, as discussed below.

Insights into Recombination and Natural Selection

In addition to its intended function as a resource for disease studies, the HapMap data provide clues about the biology of recombination and history of natural selection.

A genome-wide map of recombination rates at a fine scale. On the basis of the HapMap data, we created a fine-scale genetic map spanning the human genome (Supplementary Fig. 12), including 21,617 identified recombination hotspots (one per 122 kb).

Both the number and intensity of hotspots contribute to overall variation in recombination rate. For example, we selected 25 regions of 5 Mb as having the highest (>2.75 cM Mb^{-1}) and lowest (<0.5 cM Mb^{-1}) rates of recombination in the deCODE (pedigree-based) genetic map[48]. We detected recombination hotspots in all regions, even the lowest. But in the high cM Mb^{-1} regions hotspots are more closely spaced (one per 84 kb) and have a higher average intensity (0.124 cM) as compared to the low cM Mb^{-1} regions (one every 208 kb, and 0.051 cM, respectively).

Estimates of recombination rates and identified hotspots are robust to the specific markers and samples studied. Specifically, we compared these results to a similar analysis[76] of the data of ref. 77 (with about 1.6 million SNPs genotyped in 71 individuals). We find nearly complete correlation in rate estimates at a coarse scale (5 Mb) between these two surveys ($r^2 = 0.99$) and to the pedigree map ($r^2 = 0.95$). Very substantial correlation is found at finer scales: $r^2 = 0.8$ at 50 kb and $r^2 = 0.59$ at 5 kb. Moreover, of the 21,617 hotspots identified using the HapMap data, 78% (16,923) were also identified using the data of ref. 77.

The ability to detect events depends on marker density, with the larger number of SNPs studied by ref. 77 increasing power to detect hotspots, and presumably precision of rate estimates. There are, however, substantial genomic regions where the HapMap data have a higher SNP density. For example, more hotspots are detected on chromosomes 9 and 19 from the HapMap data. We expect that Phase II of HapMap will provide a genome-wide recombination map of substantially greater precision than either ref. 77, or Phase I, at fine scales.

多态性的倒位可能也在 HapMap 数据中有所反映，反映为多个 SNP 完美相关的长区域：因为倒位和非倒位拷贝间的重组是致死的，所以区域的倒位和非倒位拷贝各自独立进化。一个明显的例子是 17 号染色体上的一个已知的倒位多态性，它存在于 20% 的 CEU 染色体中，在冰岛女性中与生育力和总重组率相关[75]。不过，长 LD 也可能由于低重组率或者特定形式的自然选择而增加，详见下文的讨论。

对重组和自然选择的认识

除了原本预期的作为疾病研究的资源这一功能外，HapMap 数据还为研究重组的机理和自然选择的历史提供了线索。

重组率的全基因组精细图谱 基于 HapMap 数据，我们绘制了整个人类基因组的精细遗传学图谱（补充信息图 12），包括了鉴定到的 21,617 个重组热点（每 122 kb 一个）。

重组热点的数量和强度都对重组率的整体变化有所贡献。例如，我们选择了 deCODE（基于家系）遗传图谱[48]中含有最高重组率（$> 2.75\ cM \cdot Mb^{-1}$）和最低重组率（$< 0.5\ cM \cdot Mb^{-1}$）的 25 个 5 Mb 的区域。我们检测了所有区域中的重组热点，甚至是最低重组率的区域。不过，与低 $cM \cdot Mb^{-1}$ 区域相比，在高 $cM \cdot Mb^{-1}$ 区域中重组热点靠得更近（每 84 kb 一个），平均强度更高（0.124 cM）（低 $cM \cdot Mb^{-1}$ 区域分别是每 208 kb 一个和 0.051 cM）。

对重组率和鉴定到的重组热点的估计，对于特定标志物和研究样本而言是稳健的。特别地，我们将这些结果与参考文献 77 的数据（含有 71 个个体的大约 160 万个基因分型的 SNP）的一个类似的分析[76]进行了比较。我们发现，在粗略的尺度（5 Mb）上，这两项研究之间相比（$r^2 = 0.99$）以及与家系图谱相比（$r^2 = 0.95$），重组率的估计几乎完全相关。在更加精细的尺度上，我们也发现了相当程度的相关性：在 50 kb 上 $r^2 = 0.8$，在 5 kb 上 $r^2 = 0.59$。并且，利用 HapMap 数据鉴定到的 21,617 个重组热点中，有 78%（16,923 个）也在参考文献 77 的数据中鉴定到。

检测重组事件的能力取决于标志物的密度，参考文献 77 所研究的 SNP 数量更多，这增加了其检测重组热点的效能，并且很可能提高了重组率估计的准确性。不过，也有很大一部分基因组区域，其中 HapMap 数据有着更高的 SNP 密度。例如，在 9 号和 19 号染色体上，在 HapMap 数据中检测到了更多的重组热点。我们预期，第 II 阶段的 HapMap 将会提供一个在精细尺度上与参考文献 77 或者第 I 阶段 HapMap 相比准确度大大提高的全基因组范围的重组图谱。

Little is yet known about the molecular determinants of recombination hotspots. In an analysis of the data of ref. 77, another study (ref. 76) found significant evidence for an excess of the THE1A/B retrotransposon-like elements within recombination hotspots, and more strikingly for a sixfold increase of a particular motif (CCTCCCT) within copies of the element in hotspots, compared to copies of the element outside hotspots. In analysing the HapMap data, we confirmed these findings (Supplementary Fig. 13). Furthermore, THE1B elements with the motif are particularly enriched within 1.5 kb of the centre of the hotspots compared to flanking sequence ($P < 10^{-16}$).

Correlations of LD with genomic features. Variation in recombination rate is important, in large part, because of its impact on LD. We thus examined genome-wide LD for correlation to recombination rates, sequence composition and gene features.

We confirmed previous observations that LD is generally low near telomeres, elevated near centromeres, and correlated with chromosome length (Fig. 15; see also Supplementary Figs 8b and 14)[48,78-80]. These patterns are due to recombination rate variation as discussed above. We also confirmed previously described relationships between LD and G+C content[78,81,82], sequence polymorphism[83] and repeat composition[78,82].

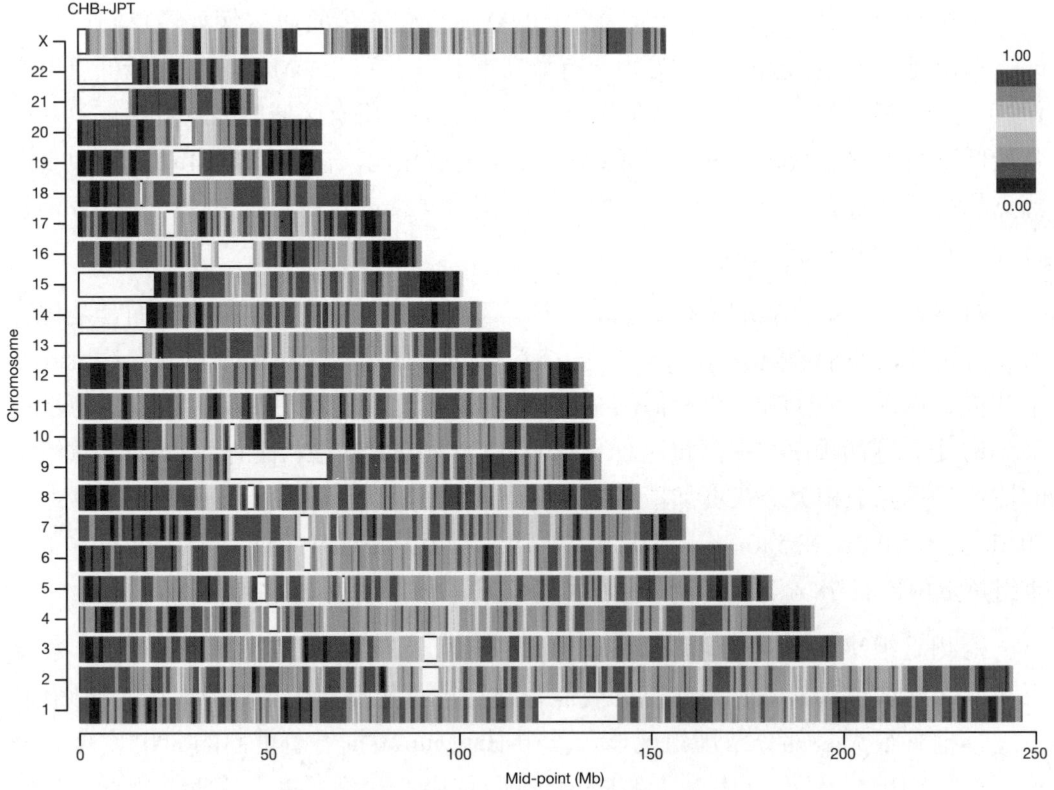

Fig. 15. Length of LD spans. We fitted a simple model for the decay of linkage disequilibrium[103] to windows of 1 million bases distributed throughout the genome. The results of model fitting are summarized for the CHB+JPT analysis panel, by plotting the fitted r^2 value for SNPs separated by 30 kb. The overall pattern

关于重组热点的分子机制目前还知之甚少。在分析参考文献 77 的数据时，另一项研究（参考文献 76）发现在重组热点中有明显的 THE1A/B 逆转录转座子样元件的富集，并且更明显地是，相对于重组热点外的元件，重组热点中的元件中一个特殊的模体（CCTCCCT）有六倍的增加。在分析 HapMap 数据的过程中，我们证实了这些发现（补充信息图 13）。并且，与旁侧序列相比，含有这个模体的 THE1B 元件在重组热点中心的 1.5 kb 范围内特别富集（$P < 10^{-16}$）。

LD 与基因组特征的相关性 重组率的变化的重要性很大程度上是由于它对 LD 的影响。因此我们研究了全基因组的 LD 与重组率、序列组成和基因特征的相关性。

我们证实了之前的发现，即 LD 普遍在端粒附近低，在着丝粒附近较高，并且与染色体的长度相关（图 15；也可见补充信息图 8b 和 14）[48,78-80]。正如上文的讨论，这种模式是由于重组率的变化。我们也证实了之前描述的 LD 与 G+C 含量 [78,81,82]、序列多态性 [83] 和重复序列组成 [78,82] 的关系。

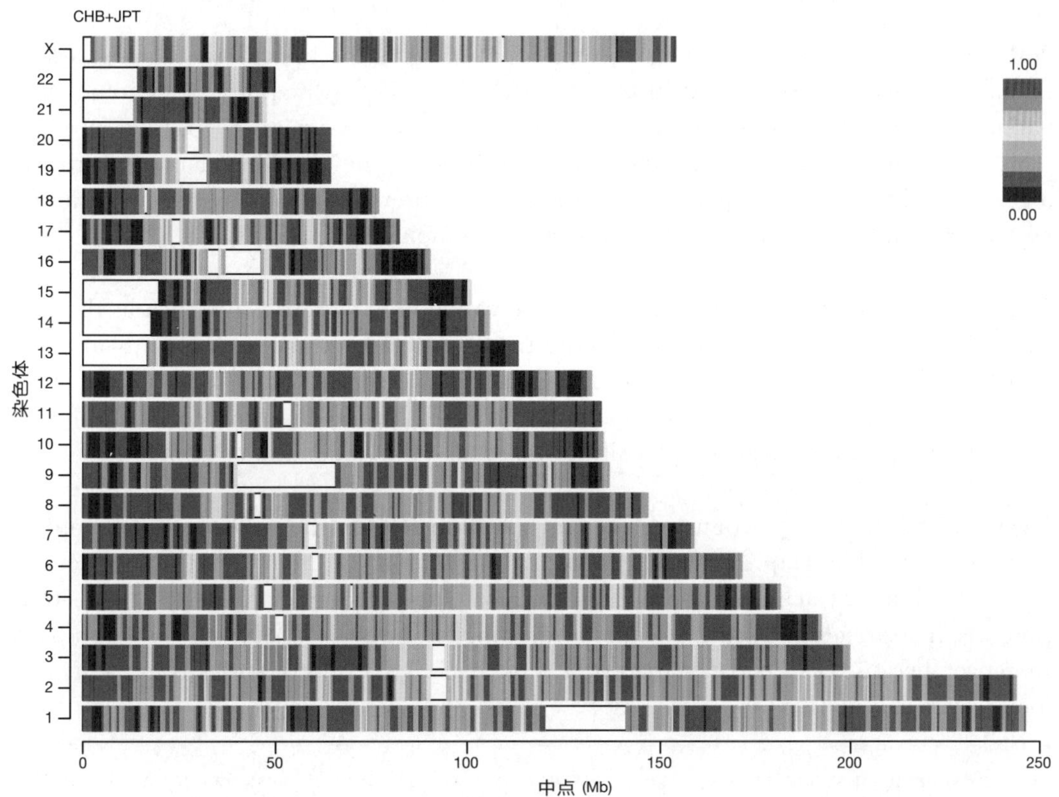

图 15. LD 横跨的长度。对于连锁不平衡的减弱 [103]，我们以全基因组范围内的 100 万碱基为窗口拟合了一个简单的模型。通过画出相距 30 kb 的 SNP 的拟合 r^2 值，我们总结了 CHB+JPT 分析小组模型拟合

of variation was very similar in the other analysis panels[84] (see Supplementary Information).

We observe, for the first time, that LD tracks with both the density and functional classification of genes. We examined quartiles of the genome based on extent of LD, and looked for correlations to gene density. Surprisingly, we find that both the top and bottom quartiles of the genome have greater gene density as compared to the middle quartiles (6.7 as compared to 6.1 genes per Mb), as well as percentage of bases in codons (1.24% as compared to 1.08%). We have no explanation for this observation.

Although the majority of gene classes are equally divided between these two extreme quartiles of the genome, some classes of genes show a marked skew in their distribution[64,84,85]. Genes involved in immune responses and neurophysiological processes are more often located in regions of low LD, whereas genes involved in DNA and RNA metabolism, response to DNA damage and the cell cycle are preferentially located in regions of strong linkage disequilibrium. It is intriguing to speculate that the extent of LD (and sequence diversity) might track with gene function due to natural selection, with increased diversity being favoured in genes involved in interface with the environment such as the immune response[86], and disadvantageous for core cell biological processes such as DNA repair and packaging[87,88].

Natural selection. The preceding observation highlights the hypothesis that signatures of natural selection are present in the HapMap data. The availability of genome-wide variation data makes it possible to scan the genome for such signatures to discover genes that were subject to selection during human evolution[89]; the HapMap data also provide a genome-wide empirical distribution against which previous claims of selection can be evaluated (rather than relying solely on theoretical computer simulations).

Natural selection influences patterns of genetic variation in various ways, such as through the removal of deleterious mutations, the fixation of advantageous variants, and the maintenance of multiple alleles through balancing selection. Each form of selection may have occurred uniformly across the world (and thus be represented in all human populations) or have been geographically localized (and thus differ among populations).

Nearly all methods for recognizing natural selection rely on the collection of complete sequence data. The HapMap Project's focus on common variation—and the process of SNP selection that achieved a preponderance of high-frequency alleles (Fig. 5)—thus prevents their straightforward application. Adjusting for the effect of SNP choice is complex, moreover, because SNP choice varied over time as dbSNP evolved, and was implemented locally at each centre.

For these reasons, we focus here on two types of analysis. First, we examined the distributions of signatures of selection across the genome. Although the absolute value of these measures is difficult to interpret (owing to SNP ascertainment), the most extreme cases in a genome-wide distribution are important candidates to evaluate for selection. Second,

的结果。变异的总体模式在其他分析小组中是非常类似的[84](详见补充信息)。

我们第一次观察到LD与基因的密度和功能分类均同步。我们基于LD的含量研究了基因组的四分位数，寻找其与基因密度的相关性。令人惊奇的是，我们发现与中间四分位数相比，基因组的上四分位数和下四分位数均含有更高的基因密度(中间四分位数每Mb 6.1个基因，上、下四分位数每Mb 6.7个)，位于编码区的碱基比例也是如此(前者为1.08%，后者为1.24%)。我们对这种现象还没有解释。

尽管大部分基因分类在基因组的两个极端四分位间是均等分布的，也有一些基因分类在分布上呈现明显的倾向性[64,84,85]。参与免疫反应和神经生理过程的基因更多位于低LD区域，而参与DNA和RNA代谢、DNA损伤反应和细胞周期的基因更多位于强连锁不平衡的区域。一种有趣的猜测是，由于自然选择，LD的含量(以及序列的多样性)可能与基因的功能同步，基因多样性的增加在环境界面如免疫反应相关的基因中是有利的[86]，而在核心细胞生物学过程如DNA修复和包装相关的基因中是不利的[87,88]。

自然选择 前面的发现突出了自然选择的特征存在于HapMap数据中这一假说。全基因组变异数据的获得，使得对这些特征进行全基因组扫描以发现在人类进化过程中被选择的基因成为可能[89]；HapMap数据还提供了一个全基因组的经验分布，以此可以评估之前声称的自然选择(而不是仅仅依赖于理论上的计算机模拟)。

自然选择通过多种方式对遗传变异的模式产生影响，这些方式如去除有害突变，固定有利变异，以及通过平衡选择对多个等位基因进行保持。各种形式的选择可能在全世界一致地发生(因而在所有人类种群中都表现出来)或者局限于特定的地理位置(因而在不同种群中有差异)。

几乎所有识别自然选择的方法都依赖于收集完整的序列数据。HapMap计划聚焦于常见变异——以及实现了高频等位基因的数量优势的SNP选择过程(图5)，因而使这些遗传变异不能直接应用。并且，调整SNP选择的影响是复杂的，因为随着dbSNP的演化，SNP选择也随时间变化，并在各个中心中局部执行。

因为这些原因，我们这里聚焦于两种类型的分析。首先，我们研究了自然选择的特征在全基因组的分布。尽管这些测量的绝对值难以解释(原因在于SNP检测方法)，但是对于自然选择，在基因组范围的分布中最极端的例子是重要的待评估的候

we compared across functional categories, because SNP choice was largely agnostic to such features, and thus systematic differences may be a sign of selection.

The outcomes of these analyses confirm a number of previous hypotheses about selection and identify new loci as candidates for selection.

Evidence for selective sweeps in particular genomic regions. First we consider population differentiation, generally accepted as a clue to past selection in one of the populations. The HapMap data reveal 926 SNPs with allele frequencies that differ across the analysis panels in a manner as extreme as the well-accepted example of selection at the Duffy (*FY*) locus (Supplementary Fig. 8c). Of these 926 SNPs, 32 are non-synonymous coding SNPs and many others occur in transcribed regions, making them strong candidates for functional polymorphisms that have experienced geographically restricted selection pressures (see Table 9 and Supplementary Information for details). In particular, the *ALMS1* gene on chromosome 2 has six amino acid polymorphisms that show very strong population differentiation.

Table 9. High-differentiation non-synonymous SNPs

Chromosome	Position (base number)	Gene*	SNP
1	54,772,383	*THEA*	rs1702003
1	156,000,000	*FY*	rs12075
1	244,000,000	Q8NGY8_human†	rs7555046
2	3,184,917	*COLEC11*	rs7567833
2	73,563,622	*ALMS1*	rs3813227
2	73,589,553	*ALMS1*	rs6546837
2	73,591,645	*ALMS1*	rs6724782
2	73,592,163	*ALMS1*	rs6546839
2	73,629,222	*ALMS1*	rs2056486
2	73,629,311	*ALMS1*	rs10193972
2	109,000,000	*EDAR*	rs3827760
3	182,000,000	*FXR1*	rs11499
3	185,000,000	*MCF2L2*	rs7639705
4	41,844,599	*SLC30A9*	rs1047626
4	46,567,077	ENSG00000172895.1	rs5825
4	101,000,000	*ADH1B*	rs1229984
8	10,517,787	*RP1L1*	rs6601495
8	146,000,000	*SLC39A4*	rs1871534
10	50,402,145	*ERCC6*	rs4253047
10	71,002,210	*NEUROG3*	rs4536103
11	46,701,579	*F2*	rs5896

选者。其次，我们比较了各个功能分类，因为 SNP 选择在很大程度上对于这些特征是不可知的，因而系统的差异可能是选择的一个标志。

这些分析的结果证实了大量的之前关于自然选择的假说，并且鉴定到了新的自然选择的候选基因座。

特定基因组区域内选择性清理的证据 首先，我们考虑了种群差异，它通常被认为是其中一个种群中以前的自然选择的线索。HapMap 数据发现了 926 个 SNP，其等位基因频率在各个分析小组中存在差异，且这种差异跟公认的在 Duffy(*FY*) 基因座的自然选择的例子一样极端(补充信息图 8c)。在这 926 个 SNP 中，有 32 个是非同义编码 SNP，并且很多其他的 SNP 位于转录区域中，使得它们成为经历过地理位置限制的选择压力的功能多态性的有力候选者(详见表 9 和补充信息)。特别地，2 号染色体上的 *ALMS1* 基因有六个表现出非常强的种群差异的氨基酸多态性。

表 9. 高度分化的非同义 SNP

染色体	位置(碱基数量)	基因*	SNP
1	54,772,383	*THEA*	rs1702003
1	156,000,000	*FY*	rs12075
1	244,000,000	Q8NGY8_human†	rs7555046
2	3,184,917	*COLEC11*	rs7567833
2	73,563,622	*ALMS1*	rs3813227
2	73,589,553	*ALMS1*	rs6546837
2	73,591,645	*ALMS1*	rs6724782
2	73,592,163	*ALMS1*	rs6546839
2	73,629,222	*ALMS1*	rs2056486
2	73,629,311	*ALMS1*	rs10193972
2	109,000,000	*EDAR*	rs3827760
3	182,000,000	*FXR1*	rs11499
3	185,000,000	*MCF2L2*	rs7639705
4	41,844,599	*SLC30A9*	rs1047626
4	46,567,077	ENSG00000172895.1	rs5825
4	101,000,000	*ADH1B*	rs1229984
8	10,517,787	*RP1L1*	rs6601495
8	146,000,000	*SLC39A4*	rs1871534
10	50,402,145	*ERCC6*	rs4253047
10	71,002,210	*NEUROG3*	rs4536103
11	46,701,579	*F2*	rs5896

Continued

Chromosome	Position (base number)	Gene*	SNP
15	46,213,776	*SLC24A5*	rs1426654
15	61,724,262	*HERC1*	rs7162473
16	30,996,126	*ZNF646*	rs749670
16	46,815,699	*ABCC11*	rs17822931
17	26,322,430	*RNF135*	rs7225888
17	26,399,303	ENSG00000184253.2	rs6505228
18	66,022,323	*RTTN*	rs3911730
19	5,782,891	*FUT6*	rs364637
19	47,723,209	*CEACAM1*	rs8110904
22	18,164,095	*GNB1L*	rs2073770
X	65,608,007	*EDA2R*	rs1385699

* Where no standard gene abbreviation exists, the ENSEMBL gene ID has been given.
† It is unclear from current annotations whether this is a pseudogene.

Another signature of an allele having risen to fixation through selection is that all other diversity in the region is eliminated (known as a selective sweep). We identified extreme outliers in the joint distribution of heterozygosity (as assessed from shotgun sequencing SNP discovery projects) and either population differentiation or skewing of allele frequency towards rare alleles in each analysis panel (Supplementary Fig. 15). We identified 19 such genomic regions (13 on autosomes, 6 on the X chromosome) as candidates for future study (Supplementary Table 4); these include candidates for population-specific sweeps and sweeps in the ancestral population. Encouragingly, this analysis includes among its top-scoring results the *LCT* gene, which influences the ability to digest dairy products[90] and has been shown to be subject to past natural selection[91].

Long haplotypes as candidates for natural selection. Selective sweeps that fail to fix in the population, as well as balancing selection, lead to haplotypes that are relatively high in frequency and long in duration. In the *HLA* region (which is widely believed to have been influenced by balancing selection) multiple haplotypes of 500 SNPs that extend more than 1 cM in length are observed with a frequency in the HapMap samples of more than 1%. We identified other such occurrences of long haplotypes across the genome (Supplementary Fig. 8 and Supplementary Tables 5 and 6).

An approach to long haplotypes designed specifically to identify regions having undergone partial selective sweeps is the long range haplotype (LRH) test[91,92], which compares the length of each haplotype to that of others at the locus, matched across the genome based on frequency. Previously identified outliers to the genome-wide distribution for the LRH test (Fig. 16) that have been identified as candidates for selection include the *LCT* gene in the CEU sample (empirical P-value = 1.3×10^{-9}), which was an outlier for the

续表

染色体	位置(碱基数量)	基因 *	SNP
15	46,213,776	*SLC24A5*	rs1426654
15	61,724,262	*HERC1*	rs7162473
16	30,996,126	*ZNF646*	rs749670
16	46,815,699	*ABCC11*	rs17822931
17	26,322,430	*RNF135*	rs7225888
17	26,399,303	ENSG00000184253.2	rs6505228
18	66,022,323	*RTTN*	rs3911730
19	5,782,891	*FUT6*	rs364637
19	47,723,209	*CEACAM1*	rs8110904
22	18,164,095	*GNB1L*	rs2073770
X	65,608,007	*EDA2R*	rs1385699

* 没有标准基因缩写的使用 ENSEMBL 基因 ID。

† 根据现有的注释不清楚它是否是一个假基因。

一个等位基因通过自然选择而被固定的另一个特征是区域内所有其他的多样性都被消除(这被称为选择性清除)。我们在杂合性(SNP 发现计划进行鸟枪法测序所检测的)与种群差异或各分析小组中等位基因频率向罕见等位基因的偏斜的联合分布中鉴定到了极端的离群值(补充信息图 15)。我们鉴定到了 19 个这样的基因组区域(13 个位于常染色体上，6 个位于 X 染色体上)，将其作为下一步研究的候选者(补充信息表 4)；它们包括种群特异性的清理以及古老种群中的清理的候选者。鼓舞人心的是，这个分析在高分结果中包括了 *LCT* 基因，该基因影响消化奶制品的能力[90]，并且已经被发现经历过以前的自然选择[91]。

作为自然选择候选者的长单体型 没有在种群中固定下来的选择性清理，以及平衡选择，产生了频率相对较高且跨度相对较长的单体型。在 *HLA* 区域(普遍认为它被平衡选择所影响过)中，多个单体型含有 500 个 SNP，长度上超过 1 cM，在 HapMap 样本中频率超过 1%。我们还在全基因组中鉴定到了其他这样的长单体型情况(补充信息图 8 和补充信息表 5 和 6)。

长距离单体型(LRH)检验[91,92]是一种专门设计的用来鉴定经历过部分选择性清理的长单体型区域的方法，它是将各个单体型的长度与全基因中频率匹配的其他单体型的长度相比较。之前通过 LRH 检验鉴定到的相对于全基因组分布的(图 16)，已经被认为是选择候选者的离群值，包括 CEU 样本中的 *LCT* 基因(经验 P 值 $= 1.3 \times 10^{-9}$)——上面提到的杂合性/等位基因频率检验的一个离群值，以及 YRI 样

heterozygosity/allele frequency test above, and the *HBB* gene (empirical *P*-value = 1.39×10^{-5}) in the YRI sample. However, most of the strongest signals in the LRH test (Table 10) were not previously hypothesized as undergoing selection.

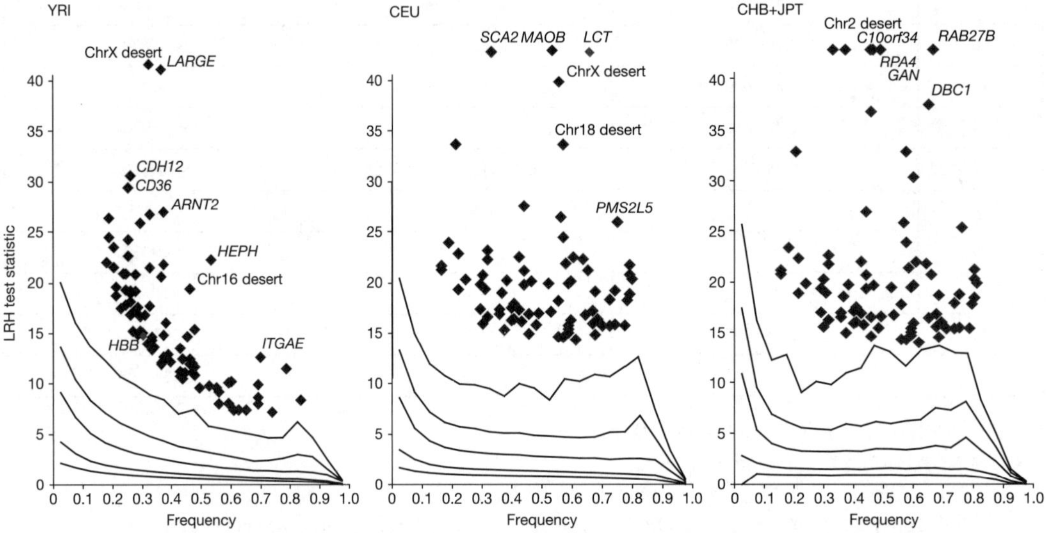

Fig. 16. The distribution of the long range haplotype (LRH[92]) test statistic for natural selection. In the YRI analysis panel, diversity around the *HBB* gene is highlighted by the red point. In the CEU analysis panel, diversity within the LCT gene region is similarly highlighted.

Table 10. Candidate loci in which selection occurred

Chromosome	Position (base number) at centre	Genes in region	Population	Haplotype frequency	Empirical *P*-value
2	137,224,699	*LCT*	CEU	0.65	1.25×10^{-9}
5	22,296,347	*CDH12, PMCHL1*	YRI	0.25	5.77×10^{-8}
7	79,904,387	*CD36*	YRI	0.24	2.72×10^{-6}
7	73,747,934	*PMS2L5, WBSCR16*	CEU	0.76	3.37×10^{-6}
12	109,892,896	*CUTL2*	CEU	0.36	7.95×10^{-9}
15	78,558,508	*ARNT2*	YRI	0.32	6.92×10^{-7}
16	75,661,011	Desert	YRI	0.46	5.01×10^{-7}
17	3,945,580	*ITGAE, GSG2,* HSA277841, *CAMKK1, P2RX1*	YRI	0.70	9.26×10^{-7}
18	24,502,756	Desert	CEU	0.57	2.23×10^{-7}
22	32,459,471	*LARGE*	YRI	0.36	7.82×10^{-9}
X	20,171,291	Desert	YRI	0.33	5.02×10^{-9}
X	64,323,320	*HEPH*	YRI	0.55	3.02×10^{-8}
X	42,763,073	*MAOB*	CEU	0.53	4.21×10^{-9}
X	34,399,948	Desert	CEU	0.57	8.85×10^{-8}

本中的 *HBB* 基因(经验 *P* 值 $=1.39\times10^{-5}$)。不过,LRH 检验中大部分的最强信号(表 10)之前并不认为经历过自然选择。

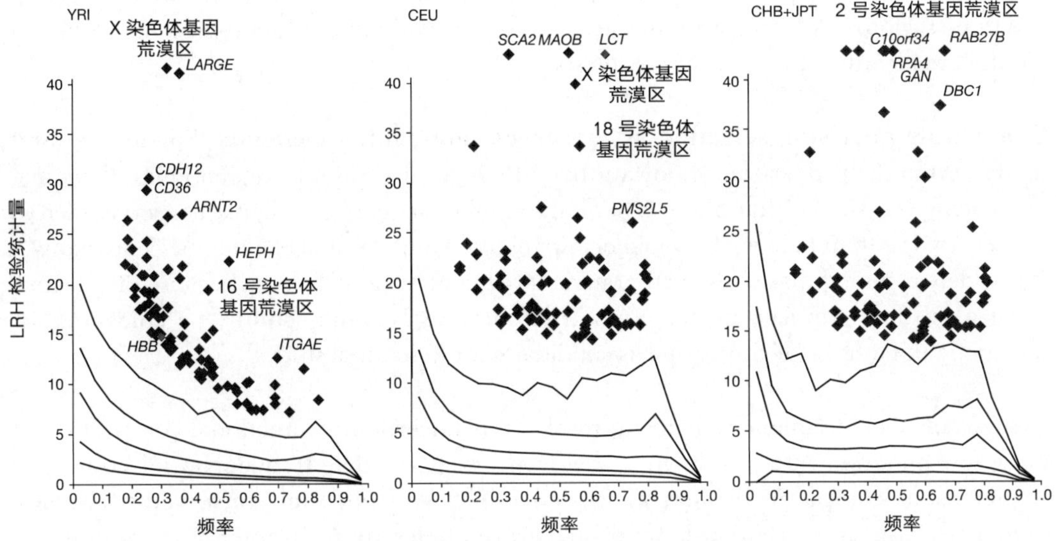

图 16. 自然选择的长距离单体型(LRH[92])检验统计量的分布。在 YRI 分析小组,*HBB* 基因周围的多样性通过红点突出显示。在 CEU 分析小组中,*LCT* 基因区域内的多样性以类似的方式突出显示。

表 10. 发生选择的候选基因座

染色体	中心位置(碱基数量)	区域内基因	种群	单体型频率	经验 *P* 值
2	137,224,699	*LCT*	CEU	0.65	1.25×10^{-9}
5	22,296,347	*CDH12*、*PMCHL1*	YRI	0.25	5.77×10^{-8}
7	79,904,387	*CD36*	YRI	0.24	2.72×10^{-6}
7	73,747,934	*PMS2L5*、*WBSCR16*	CEU	0.76	3.37×10^{-6}
12	109,892,896	*CUTL2*	CEU	0.36	7.95×10^{-9}
15	78,558,508	*ARNT2*	YRI	0.32	6.92×10^{-7}
16	75,661,011	Desert	YRI	0.46	5.01×10^{-7}
17	3,945,580	*ITGAE*、*GSG2*、HSA277841、*CAMKK1*、*P2RX1*	YRI	0.70	9.26×10^{-7}
18	24,502,756	Desert	CEU	0.57	2.23×10^{-7}
22	32,459,471	*LARGE*	YRI	0.36	7.82×10^{-9}
X	20,171,291	Desert	YRI	0.33	5.02×10^{-9}
X	64,323,320	*HEPH*	YRI	0.55	3.02×10^{-8}
X	42,763,073	*MAOB*	CEU	0.53	4.21×10^{-9}
X	34,399,948	Desert	CEU	0.57	8.85×10^{-8}

These four tests overlap only partially in the hypotheses they address—heterozygosity, for example, is sensitive to older sweeps, whereas the haplotype tests are most powerful for partial sweeps—but encouragingly some candidate regions are found by more than one test. In particular, six regions are identified both by long haplotypes and by low heterozygosity, and three regions (*LCT* on chromosome 2, and two regions on the X chromosome at 20 and 65 Mb) are identified by three different tests.

Confirming purifying selection at conserved non-coding elements. Finally, we used the HapMap data to test an important hypothesis from comparative genomics. Genomic sequencing has shown that about 5% of the human sequence is highly conserved across species, yet less than half of this sequence spans known functional elements such as exons[45]. It is widely assumed that conserved non-genic sequences lack diversity because of selective constraint (that is, purifying selection), but such regions may simply be coldspots for mutation, and thus be of little value as candidates for functional study.

Analysis of allele frequencies helps to resolve this uncertainty. Functional constraint, but not a low mutation rate, results in a downward skew in allele frequencies for conserved sequences as compared to neutral sequences[93,94]. We find that conserved non-genic sequences display a greater skew towards rare alleles than do intergenic regions, as predicted under purifying selection. This skew is less extreme than that observed for exons (Supplementary Fig. 16), reflecting either stronger purifying selection or the prioritization of coding SNPs for genotyping by the HapMap centres regardless of validation status. This novel evidence for ongoing constraint shows that conserved non-genic sequences are not mutational coldspots, and thus remain of high interest for functional study.

Conclusions

The International HapMap Project set out to create a resource that would accelerate the identification of genetic factors that influence medical traits. Analyses reported here confirm the generality of hotspots of recombination, long segments of strong LD, and limited haplotype diversity. Most important is the extensive redundancy among nearby SNPs, providing (1) the potential to extract extensive information about genomic variation without complete resequencing, and (2) efficiencies through selection of tag SNPs and optimized association analyses. Beyond the biomedical context, these data have made it possible to identify deletion variants in the genome, explore the nature of fine-scale recombination and identify regions that may have been subject to natural selection.

The HapMap Project (along with a previous genome-wide assessment of LD[77]) is a natural extension of the Human Genome Project. Where the reference sequence constructed by the Human Genome Project is informative about the vast majority of bases that are invariant across individuals, the HapMap focuses on DNA sequence differences among individuals. Our understanding of SNP variation and LD around common variants in the sampled populations is reasonably complete; the current picture is unlikely to change with additional

这四种检验提出的假设仅仅部分重合——例如，杂合性对更古老的清理敏感，而单体型检验对于部分清理最有效；但是令人鼓舞的是，一些候选区域在超过一种检验中发现。特别地，有六个区域同时被长单体型和低杂合性鉴定出来，有三个区域(2 号染色体上的 *LCT*，以及 X 染色体在 20 和 65 Mb 处的两个区域)被三种不同的检验同时鉴定出来。

在保守的非编码元件中验证纯化选择 最后，我们使用 HapMap 数据来检验一个来自比较基因组学的重要假说。基因组测序已经发现，大约 5% 的人类序列是在物种间高度保守的，但是其中横跨已知的功能元件如外显子 [45] 的只有不到一半。普遍认为，保守的非基因序列由于选择性限制(即纯化选择)而缺少多样性，但是，这样的区域也可能单纯是突变的冷点而几乎没有作为功能研究候选者的价值。

分析等位基因频率能帮助解决这种不确定性。功能限制，而非低突变率，会导致相比于中性序列，保守序列的等位基因频率向下偏斜 [93,94]。我们发现，保守的非基因序列比基因间区域表现出更大的向罕见等位基因的偏斜，这与在纯化选择压力下预测的一致。这种偏斜没有外显子中观察到的那么极端(补充信息图 16)，不管验证状态如何，这反映了更强的纯化选择，或是 HapMap 中心对基因分型的编码 SNP 的优先选择。这个正在进行的限制的新证据表明，保守的非基因序列并不是突变冷点，因而在功能研究中仍然很有价值。

结　论

国际 HapMap 计划的建立是为了创造一种能够加快对影响医学特性的遗传因子的鉴定的资源。本文报道的分析结果验证了重组热点、强 LD 的长片段和有限的单体型多样性的普遍性。最重要的是邻近 SNP 间广泛的冗余性，这提供了(1)不需要全基因组重测序就可以提取大量基因组变异信息的可能，以及(2)通过标签 SNP 的选择和优化后的关联分析而提高的效率。除了生物医学方面，这些数据还使得鉴定基因组中的缺失突变，探索精细尺度上重组的本质以及鉴定可能经历过自然选择的区域成为可能。

HapMap 计划(以及之前的一项全基因组范围的对 LD 的评估 [77])是人类基因组计划的自然的延伸。人类基因组计划搭建的参考序列表明绝大多数碱基在人与人之间是不变的，而 HapMap 聚焦于个体间存在差异的 DNA 序列。我们对于取样种群中的 SNP 变异和常见变异附近的 LD 的了解是相当完整的；即使再增加数据，目前

data. In other aspects—such as the fine details of local correlation among SNPs, rarer alleles, structural variants, and interpopulation differences—these resources are only a first step on the path towards a complete characterization of genetic variation of the human population. Planned extensions of the Phase I map include Phase II of HapMap, with genotyping of another 4.6 million SNPs attempted in the HapMap samples, and detailed genotyping of the HapMap ENCODE regions in additional members of each HapMap population sampled, as well as in samples from additional populations. These results should guide understanding of the robustness and transferability of LD inferences and tag SNPs selected from the current set of HapMap samples.

An important application of the HapMap data is to help make possible comprehensive, genome-wide association studies. There are now laboratory tools that make it practical to undertake such studies, and initial results are encouraging[13]. Given the low prior probability of causality for each SNP in the genome, however, rigorous standards of statistical significance will be needed to avoid a flood of false-positive results. Multiple replications in large samples provide the most straightforward path to identifying robust and broadly relevant associations. Given the potential for confusion if associations of uncertain validity are widely reported (and a persistent tendency towards genetic determinism in public discourse), we urge conservatism and restraint in the public dissemination and interpretation of such studies, especially if non-medical phenotypes are explored. It is time to create mechanisms by which all results of association studies, positive and negative, are reported and discussed without bias.

The success of the HapMap will be measured in terms of the genetic discoveries enabled, and improved knowledge of disease aetiology. Specifically, identifying which genes and pathways are causal in humans has the potential to provide a new and solid foundation for biomedical research. This is equally true whether the variants that lead to the discovery of those genes are themselves rare or common, or of large or small effect. The impact on diagnostics and targeted prevention, however, will depend on how predictive each given allele may be. Where genetic mechanisms underlie treatment responses, both more efficient trials and individualized preventive and treatment strategies may become practical[95].

Success identifying alleles conferring susceptibility or resistance to common diseases will also provide a deeper understanding of the architecture of disease: how many genes are involved in each case, whether and how alleles interact with one another[96] and with environmental exposures to shape clinical phenotypes. In this regard, it will be important to invest heavily in the discovery and characterization of relevant lifestyle factors, environmental exposures, detailed characterization of clinical phenotypes, and the ability to obtain such information in longitudinal studies of adequate size. Where environmental and behavioural factors vary across studies, replication will be hard to come by (as will clinical utility) unless we can learn to capture these variables with the same precision and completeness as genotypic variation. Technological innovation and international collaboration in these realms will probably be required (as they have been in the Human Genome Project and the HapMap) to advance the shared goal of understanding, and ultimately preventing, common human diseases.

的图景也基本不会改变。对于其他方面——比如 SNP 间局部相关性、更罕见的等位基因、结构变异和种群间差异等细节，这些资源仅仅是通往完整表征人类种群遗传变异的道路的第一步。第 I 阶段图谱的延伸计划包括 HapMap 的第 II 阶段，即在 HapMap 样本中对另外 460 万个 SNP 进行基因分型，以及在已取样的各 HapMap 种群的更多成员中和另外的种群的样本中，对 HapMap ENCODE 区域进行详细的基因分型。这些结果会对理解 LD 推断和从现有 HapMap 样本集中选择的标签 SNP 的稳健性和可转移性提供指导。

HapMap 数据的一个重要应用是帮助我们实现全面的全基因组关联研究。现有的实验工具已经使进行这样的研究切实可行，并且初步结果是令人鼓舞的 [13]。但是，鉴于基因组中各个 SNP 导致疾病的先验概率偏低，统计显著性将需要严格的标准来避免大量的假阳性结果。大量样本中的多个重复提供了最直接的鉴定稳健且广泛相关的关联的方法。鉴于可能存在的对不确定有效的关联是否被广泛报道的困惑(以及公共讨论中朝向遗传决定论的持续倾向)，我们呼吁在公开宣传和说明这样的研究时保持保守和克制，特别是研究非医学表型时。现在需要建立一种机制，以使所有关联研究的结果，无论阳性或阴性，都得到无偏倚的报道和讨论。

HapMap 的成功会通过它带来的遗传学发现和提高的病因知识来衡量。特别地，对人类中导致疾病的基因和通路的鉴定有望为生物医学研究提供新的坚实基础。不管使我们发现这些致病基因的遗传变异是罕见的还是常见的，效应是大还是小，这一点是始终成立的。不过，对诊断和靶向预防的影响将取决于各个给定的等位基因可以在多大程度上预测疾病发生。一旦清楚了治疗响应下的遗传机制，更加有效的试验和个性化预防与治疗策略都可能成为现实 [95]。

成功鉴定影响常见疾病易感性或抵抗性的等位基因还将会帮助我们更加深入地理解疾病的机制：多少基因参与其中，等位基因间 [96] 以及等位基因与环境是否和如何相互作用进而影响临床表型。在这一点上，大力投入以发现和表征相关生活方式因子、环境暴露、详细的临床表型，以及在足够大的纵向研究中获取这种信息，将会十分重要。各个研究的环境和行为因子不同，除非我们能学会以与遗传学变异相同的准确度和完整性捕获这些变化，否则将难以进行重复(临床应用也是如此)。这些领域有可能需要技术的革新和国际合作(像人类基因组计划和 HapMap 那样)，以推进理解和最终预防常见人类疾病这一共同的目标。

Methods

The project was undertaken by investigators from Japan, the United Kingdom, Canada, China, Nigeria and the United States, and from multiple disciplines: sample collection, sequencing and genotyping, bioinformatics, population genetics, statistics, and the ethical, legal, and social implications of genetic research. The Supplementary Information contains information about project participants and organization.

Choice of DNA samples. Any choice of DNA samples represents a compromise: a single population offers simplicity, but cannot be representative, whereas grid-sampling is representative of the current worldwide population, but is neither practical nor captures historical genetic diversity. The project chose to include DNA samples based on well-known patterns of allele frequencies across populations[41], reflecting historical genetic diversity[31,32].

For practical reasons, the project focused on SNPs present at a minor allele frequency (MAF) $\geqslant 0.05$ in each analysis panel, and thus studied a sufficient number of individuals to provide good power for this frequency range[31]. Cell lines and DNA are available at the Coriell Institute for Medical Research (http://locus.umdnj.edu/nigms/products/hapmap.html).

Community engagement was employed to explain the project, and to learn how the project was viewed, in the communities where samples were collected[31,32]. Papers describing the community engagement processes are being prepared.

One JPT sample was replaced for technical reasons, but not in time for inclusion in this report. We surveyed cryptic relatedness among the study participants, and identified a small number of pairs with unexpectedly high allele sharing (Supplementary Information). As the total level of sharing is not great, and as a subset of analyses performed without these individuals were unchanged, we include these individuals in the data and analyses presented here.

Genome-wide SNP discovery. The project required a dense map of SNPs, ideally containing information about validation and frequency of each candidate SNP. When the project started, the public SNP database (dbSNP) contained 2.6 million candidate SNPs, few of which were annotated with the required information.

To generate more SNPs and obtain validation information, shotgun sequencing of DNA from whole-genome libraries and flow-sorted chromosomes was performed[31], augmented by analysis of sequence traces produced by Applied Biosystems[97,98], and information on 1.6 million SNPs genotyped by Perlegen Sciences[77], including 425,000 not in dbSNP when released (Supplementary Table 7). The HapMap Project contributed about 6 million new SNPs to dbSNP.

At the time of writing (October 2005) dbSNP (http://www.ncbi.nlm.nih.gov/projects/SNP/) contains 9.2 million candidate human SNPs, of which 3.6 million have been validated by both alleles having been seen two or more times during discovery ("double-hit" SNPs), and 2.4 million have genotype

方　法

本计划是由来自日本、英国、加拿大、中国、尼日利亚和美国的研究者实施的，这些研究者来自多个方向：样本收集，测序和基因分型，生物信息学，种群遗传学，统计学，以及遗传学研究的伦理、法律和社会影响。补充信息包括了参与本计划的人员和组织的信息。

DNA 样本的选择　任何一种 DNA 样本选择方法都代表着一种妥协：单一种群简单，但是不够有代表性，而网格采样代表了目前全球的种群，但是既不现实也不能捕获历史上的遗传多样性。本计划选择基于种群间等位基因频率的已知模式来纳入 DNA 样本[41]，进而反映历史上的遗传多样性[31,32]。

出于实际的原因，本计划聚焦于在各个分析小组中最小等位基因频率（MAF）≥0.05 的 SNP，因而研究了足量的个体来为这个频率段提供较好的效能[31]。细胞系和 DNA 可在科里尔医学研究所获取（http://locus.umdnj.edu/nigms/products/hapmap.html）。

本研究在样本采集的社区开展了参与活动，包括解释本计划，以及了解本计划是如何被看待的[31,32]。介绍社区参与过程的文章正在准备中。

只有一个 JPT 样本由于技术原因被替换掉，但是没有及时包含进本报道中。我们还调查了样本之间的隐性亲缘关系，鉴定到少量几对含有超预期的高等位基因共享性（详见补充信息）。由于总体的共享性水平并不高，以及不含这些个体的分析结果不变，我们在本文的数据和分析中还是包含了这些个体。

全基因组范围的 SNP 发现　本计划要求高密度的 SNP 图谱，最好包含各个候选 SNP 的验证和频率的信息。在本计划启动时，公共的 SNP 数据库（dbSNP）包含 260 万个候选 SNP，其中很少有所要求的注释信息。

为了产生更多的 SNP 并获得验证信息，我们对来自全基因组文库和流式筛选的染色体的 DNA 进行了鸟枪法测序[31]，并辅以 Applied Biosystems 公司的序列踪迹分析[97,98]，以及 Perlegen Sciences 公司基因分型的 160 万个 SNP 的信息[77]，这其中有 425,000 个在数据释放时并不在 dbSNP 中（补充信息表格 7）。HapMap 计划为 dbSNP 贡献了大约 600 万个新的 SNP。

在写作本文时（2005 年 10 月），dbSNP（http://www.ncbi.nlm.nih.gov/projects/SNP/）包含 920 万个候选人类 SNP，其中 360 万个 SNP 的两个等位基因都在发现过程中被观察到两次或更多

data (Fig. 1).

Comprehensive study of common variation across 5 Mb of DNA. To study patterns of genetic variation as comprehensively as possible, we selected ten 500-kb regions from the ENCODE Project[33]. These ten regions were chosen in aggregate to approximate the genome-wide average for G+C content, recombination rate, percentage of sequence conserved relative to mouse sequence, and gene density (Table 2).

In each such region additional sequencing and genotyping were performed to obtain a much more complete inventory of common variation. Specifically, bidirectional PCR-based sequencing was performed across each 500-kb region in 48 individuals (16 YRI, 16 CEU, 8 CHB, 8 JPT). Although the intent was for these same DNA samples to be included in Phase I, eight Yoruba and one Han Chinese sample used in sequencing were not among the 269 samples genotyped. (The nine samples are available from Coriell.)

All variants found by sequencing, and any others in dbSNP (build 121) not found by sequencing, were genotyped in all 269 HapMap samples. If the first attempt at genotyping was unsuccessful, a second platform was tried for each SNP.

False-positive and false-negative rates in PCR-based SNP discovery. The false-positive rate of SNP discovery by PCR-based resequencing was estimated at 7–11% (for the two sequencing centres), based on genotyping of each candidate SNP in the same samples used for discovery.

The false-negative rate of SNP discovery by PCR resequencing was estimated at 6%, using as the denominator a set of SNPs previously in dbSNP and confirmed by genotyping in the specific individuals sequenced. The false-negative rate was considerably higher, however, for singletons (SNPs seen only as a single heterozygote): 15% of singletons covered by high-quality sequence data were not detected by the trace analysis, and another 25% were missed due to a failure to obtain a high-quality sequence over the relevant base in the one heterozygous individual (D. J. Richter *et al.*, personal communication).

False-positive and false-negative rates in dbSNP. The false positive rate (candidate SNPs that cannot be confirmed as variable sites) estimated for dbSNP was 17%. This represents an upper bound, because dbSNP entries that are monomorphic in the 269 HapMap samples could be rare variants, or polymorphic in other samples. We note that as the catalogue of dbSNP gets deeper, the rate at which candidate SNPs are monomorphic in any given sample is observed to rise (Supplementary Table 8). This is expected because the number of rare SNPs and false positives scales with depth of sequencing, whereas the number of true common variants will plateau.

SNP genotyping for the genome-wide map. Genotyping assays were designed from dbSNP, with priority given to SNPs validated by previous genotyping data or both alleles having been seen more than once in discovery. Data from the Chimpanzee Genome Sequencing Project[99] were used in SNP validation if they confirmed the ancestral status of a human allele seen only once in discovery (Supplementary Information). Non-synonymous coding SNPs were also prioritized for genotyping.

次而得到验证（“双打击”的 SNP），240 万个含有基因型数据（图 1）。

对 5 Mb DNA 区域内常见变异的全面研究 为了尽可能全面地研究遗传变异的模式，我们从 ENCODE 计划中选择了 10 个 500 kb 区域[33]。这 10 个区域合起来可近似代表全基因组平均的 G+C 含量、重组率、相对于小鼠基因组保守序列的百分比，以及基因密度（表 2）。

在各个这样的区域中，我们进行了额外的测序和基因分型，以获得更加完整的常见变异清单。特别地，对 48 个个体（16 个 YRI、16 个 CEU、8 个 CHB、8 个 JPT）在各个 500 kb 区域上进行了基于双向 PCR 的测序。尽管目的是将这些用在测序中的 DNA 样本都包含进第 I 阶段 HapMap，但是其中有 8 个约鲁巴人和 1 个汉族中国人样本并没有包含在基因分型的 269 个样本当中。（这 9 个样本可从科里尔医学研究所获得。）

通过测序发现的所有变异，以及任何其他未被测序发现但在 dbSNP（版本 121）中的变异，都在所有 269 个 HapMap 样本中进行了基因分型。对于各个 SNP，如果第一次基因分型尝试不成功，则尝试第二个基因分型平台。

基于 PCR 的 SNP 发现中的假阳性率和假阴性率 通过基于 PCR 的重测序进行 SNP 发现，其假阳性率估计为 7% ~ 11%（对于两个测序中心），这是基于在相同的样本中对各个候选 SNP 进行基因分型得出来的。

通过 PCR 重测序进行 SNP 发现，其假阴性率估计为 6%，这是使用之前在 dbSNP 中的并在被测序的特定个体中通过基因分型而被验证的 SNP 集作为分母得出来的。不过，对于单体 SNP（仅作为单个杂合子而被发现），假阴性率则高得多：被高质量的测序数据所覆盖的单体 SNP 中有 15% 没有被踪迹分析检测到，另外还有 25% 由于不能在唯一的杂合个体中获得相关碱基的高质量测序而被漏掉（里克特等，个人交流）。

dbSNP 中的假阳性率和假阴性率 对于 dbSNP，假阳性率（不能被证实在人群中存在变异的候选 SNP）估计为 17%。这代表着上限，因为在 269 个 HapMap 样本中呈单态性的 dbSNP 条目有可能是罕见变异，或者在其他样本中是多态性的。我们注意到，随着 dbSNP 目录的深入，候选 SNP 在任一给定样本中是单态性的比率是增加的（补充信息表 8）。这是符合预期的，因为罕见 SNP 和假阳性的数量与测序的深度成比例，而真正的常见变异的数量则将会保持稳定。

全基因组图谱的 SNP 基因分型 基因分型实验是根据 dbSNP 而设计的，首先针对由之前的基因分型数据验证过的 SNP，或者两个等位基因在发现过程中都被观察到超过一次的 SNP。对于在发现过程中仅被看到一次的人类等位基因，如果来自黑猩猩基因组测序计划[99]的数据确认了该等位基因的祖先状态，这些数据就被用于 SNP 验证。非同义编码 SNP 也被

Two whole-genome, array-based genotyping reagents were used efficiently to increase SNP density: 40,000 SNPs from Illumina, and 120,000 SNPs from Affymetrix[100].

To monitor progress, the genome was partitioned into 5-kb bins, with genotyping continuing through iterative rounds until a set of predetermined "stopping rules" was satisfied in each analysis panel. (1) Minor allele frequency: in each analysis panel a common SNP (MAF $\geqslant 0.05$) was obtained in each 5-kb bin. (2) Spacing: the distance between adjacent SNPs was 2–8 kb, with at least 9 SNPs across 50 kb. (3) "HapMappable" genome: with available technologies it is challenging to study centromeres, telomeres, gaps in genome sequence, and segmental duplications. The project identified such regions[101] (Supplementary Table 9), spanning 4.4% of the finished human genome sequence, in which only a single attempt to develop a genotyping assay was required. (4) Three strikes, you're out: if the above rules were not satisfied after three attempts to develop an assay in a given 5-kb region, or if all available SNPs in dbSNP had been tried, genotyping was considered complete for Phase I. Two attempts were considered sufficient if one attempt was of a SNP previously shown to have MAF $\geqslant 0.05$ in the appropriate population sample in a previous genome-wide survey[77]. (5) Quality control: ongoing and standardized quality control (QC) filters and three rounds of quality assessment (QA) were used to ensure and document the high quality of the genotype data.

QC filters were systematically performed, with each SNP tested for completeness ($>80\%$), consistency across five duplicate genotypes ($\leqslant 1$ discrepancy), mendelian inheritance in 60 trios ($\leqslant 1$ discrepancy in each of YRI and CEU), and Hardy–Weinberg equilibrium ($P > 0.001$[102]). SNPs in the Phase I data set passed all the QC filters in all the analysis panels and were polymorphic in the HapMap samples. Failing SNPs were released (with a special flag), as they can help to identify polymorphisms under primers, insertions/deletions, paralogous loci and natural selection.

Three QA exercises were carried out. First, a calibration exercise to "benchmark" each platform and laboratory protocol. Second, a mid-project evaluation of each genotyping centre. Third, a blind analysis of a random sample of the complete Phase I data set. A number of SNPs were genotyped more than once during the project, or by other investigators, providing additional information about data quality. See the Supplementary Information for full information about the QA exercises.

An exhaustive approach was taken to mtDNA. Alignment of more than 1,000 publicly available mtDNA sequences of African (n = 87), European (n = 928) and Asian (n = 238) geographical origin[34] was used to identify 210 common variants (MAF $\geqslant 0.05$ in at least one continental region) that were attempted in the samples.

Data release. Data deposited at the Data Coordination Center and released at http://www.hapmap.org, a Japanese mirror site http://hapmap.jst.go.jp/ and dbSNP include ascertainment status of each SNP at the time of selection, primer sequences, protocols for genotyping, genotypes for each sample, allele frequencies, and, for SNPs that failed QC filters, a code indicating the mode(s) of failure.

Initially, because of concern that third parties might seek patents on HapMap data, users were required to agree to a web-based "click-wrap license", assenting that they would not prevent others from using the data (http://www.hapmap.org/cgi-perl/registration). In December 2004 this license

优先进行基因分型。两个全基因组的基于芯片的基因分型试剂被高效用于增加SNP密度：40,000个SNP来自Illumina，120,000个SNP来自Affymetrix[100]。

为了监控进度，基因组被分成5 kb的块，基因分型通过迭代循环持续进行，直到满足在各个分析小组中预先确定的一套“停止规则”。(1)最小等位基因频率：在各个分析小组中每个5 kb块均获得一个常见SNP(MAF ≥ 0.05)。(2)间隔：相邻SNP间的距离是2 ~ 8 kb，50 kb区域内至少有9个SNP。(3)“可被HapMap的”基因组：使用现有技术研究着丝粒、端粒、基因组序列缺口和片段重复还比较有挑战性。本计划鉴定了占已完成的人类基因组序列的4.4%的区域[101](补充信息表9)，在这些区域内设计基因分型实验只需要尝试一次。(4)三振出局：如果在一个给定的5 kb区域内设计一个基因分型实验时经过三次尝试而上述规则都不满足，或者如果dbSNP中所有SNP都已经被尝试了，则对于第I阶段HapMap，基因分型被认为是完整的。如果其中一次尝试针对的是一个在之前的全基因组调查中在合适的种群样本中MAF ≥ 0.05的SNP[77]，则两次尝试也被认为是足够的。(5)质量控制：不间断的标准化质量控制(QC)过滤和三轮质量评估(QA)被用来确保和记录基因型数据的高质量。

我们系统实施了QC过滤，检验各个SNP的完整性(> 80%)、五个重复基因型间的一致性(≤ 1个不一致)、60个三联家系中的孟德尔遗传(在YRI和CEU中均 ≤ 1个不一致)，以及哈迪–温伯格平衡($P > 0.001$[102])。第I阶段数据集中的SNP在所有分析小组中都通过了所有的QC过滤，并且在HapMap样本中是多态性的。失败的SNP也被释放(含有一个特殊的标记)，因为它们能够帮助鉴定在引物、插入/缺失、同源基因座和自然选择下的多态性。

我们实施了三项QA。第一是“基准”各个平台和实验室操作手册的一项校准。第二是在计划的中期对各个基因分型中心进行的评估。第三是对完整的第I阶段数据集的随机样本进行的盲分析。大量的SNP在计划中被基因分型超过一次，或者被其他的研究者进行过分型，这提供了额外的关于数据质量的信息。关于QA的完整信息见补充信息。

对于mtDNA，我们采取了穷举的方法。超过1,000个可公开获得的来自非洲(n = 87)、欧洲(n = 928)和亚洲(n = 238)的mtDNA序列[34]被用来鉴定在样本中尝试过的210个常见变异(在至少一个大陆区域中MAF ≥ 0.05)。

数据释放 储存在数据协调中心，并在http://www.hapmap.org、一个日本镜像网站http://hapmap.jst.go.jp/和dbSNP上释放的数据包括各个SNP在选择时的确定状态、引物序列、基因分型的操作手册、各个样本的基因型、等位基因频率，此外，对于未通过QC过滤的SNP，数据还包含一个表示失败模式的编码。

最初，由于考虑到第三方可能会寻求HapMap数据的专利，所有的使用者都被要求同意一项基于网络的“点击同意许可”，表示其同意不会阻止其他人使用数据(http://www.hapmap.

was dropped, and all data were released without restriction into the public domain.

(**437**, 1299-1320; 2005)

Received 11 August; accepted 12 September 2005.

References:

1. Lechler, R. & Warrens, A. *HLA in Health and Disease* 2nd edn (Academic Press, San Diego, California, 2005).
2. Strittmatter, W. J. & Roses, A. D. Apolipoprotein E and Alzheimer's disease. *Annu. Rev. Neurosci.* **19**, 53-77 (1996).
3. Dahlbäck, B. Resistance to activated protein C caused by the factor V $R^{506}Q$ mutation is a common risk factor for venous thrombosis. *Thromb. Haemost.* **78**, 483-488 (1997).
4. Altshuler, D. *et al.* The common PPARγ Pro12Ala polymorphism is associated with decreased risk of type 2 diabetes. *Nature Genet.* **26**, 76-80 (2000).
5. Deeb, S. S. *et al.* A Pro12Ala substitution in PPARγ 2 associated with decreased receptor activity, lower body mass index and improved insulin sensitivity. *Nature Genet.* **20**, 284-287 (1998).
6. Florez, J. C., Hirschhorn, J. & Altshuler, D. The inherited basis of diabetes mellitus: implications for the genetic analysis of complex traits. *Annu. Rev. Genomics Hum. Genet.* **4**, 257-291 (2003).
7. Begovich, A. B. *et al.* A missense single-nucleotide polymorphism in a gene encoding a protein tyrosine phosphatase (*PTPN22*) is associated with rheumatoid arthritis. *Am. J. Hum. Genet.* **75**, 330-337 (2004).
8. Bottini, N. *et al.* A functional variant of lymphoid tyrosine phosphatase is associated with type I diabetes. *Nature Genet.* **36**, 337-338 (2004).
9. Bell, G. I., Horita, S. & Karam, J. H. A polymorphic locus near the human insulin gene is associated with insulin-dependent diabetes mellitus. *Diabetes* **33**, 176-183 (1984).
10. Ueda, H. *et al.* Assocation of the T-cell regulatory gene *CTLA4* with susceptibility to autoimmune disease. *Nature* **423**, 506-511 (2003).
11. Ogura, Y. *et al.* A frameshift mutation in *NOD2* associated with susceptibility to Crohn's disease. *Nature* **411**, 603-606 (2001).
12. Hugot, J. P. *et al.* Association of NOD2 leucine-rich repeat variants with susceptibility to Crohn's disease. *Nature* **411**, 599-603 (2001).
13. Klein, R. J. *et al.* Complement factor H polymorphism in age-related macular degeneration. *Science* **308**, 385-389 (2005).
14. Haines, J. L. *et al.* Complement factor H variant increases the risk of age-related macular degeneration. *Science* **308**, 419-421 (2005).
15. Edwards, A. O. *et al.* Complement factor H polymorphism and age-related macular degeneration. *Science* **308**, 421-424 (2005).
16. Puffenberger, E. G. *et al.* A missense mutation of the endothelin-B receptor gene in multigenic Hirschsprung's disease. *Cell* **79**, 1257-1266 (1994).
17. Emison, E. S. *et al.* A common sex-dependent mutation in a *RET* enhancer underlies Hirschsprung disease risk. *Nature* **434**, 857-863 (2005).
18. Kerem, B. *et al.* Identification of the cystic fibrosis gene: genetic analysis. *Science* **245**, 1073-1080 (1989).
19. Hästbacka, J. *et al.* Linkage disequilibrium mapping in isolated founder populations: diastrophic dysplasia in Finland. *Nature Genet.* **2**, 204-211 (1992).
20. Pritchard, J. K. & Przeworski, M. Linkage disequilibrium in humans: models and data. *Am. J. Hum. Genet.* **69**, 1-14 (2001).
21. Jorde, L. B. Linkage disequilibrium and the search for complex disease genes. *Genome Res.* **10**, 1435-1444 (2000).
22. Reich, D. E. *et al.* Linkage disequilibrium in the human genome. *Nature* **411**, 199-204 (2001).
23. Kruglyak, L. Prospects for whole-genome linkage disequilibrium mapping of common disease genes. *Nature Genet.* **22**, 139-144 (1999).
24. Johnson, G. C. *et al.* Haplotype tagging for the identification of common disease genes. *Nature Genet.* **29,** 233-237 (2001).
25. Nickerson, D. A. *et al.* DNA sequence diversity in a 9.7-kb region of the human lipoprotein lipase gene. *Nature Genet.* **19**, 233-240 (1998).
26. Zhu, X. *et al.* Localization of a small genomic region associated with elevated ACE. *Am. J. Hum. Genet.* **67**, 1144-1153 (2000).
27. Daly, M. J., Rioux, J. D., Schaffner, S. F., Hudson, T. J. & Lander, E. S. High-resolution haplotype structure in the human genome. *Nature Genet.* **29**, 229-232 (2001).
28. Jeffreys, A. J., Kauppi, L. & Neumann, R. Intensely punctate meiotic recombination in the class II region of the major histocompatibility complex. *Nature Genet.* **29**, 217-222 (2001).
29. Patil, N. *et al.* Blocks of limited haplotype diversity revealed by high-resolution scanning of human chromosome 21. *Science* **294**, 1719-1723 (2001).
30. Gabriel, S. B. *et al.* The structure of haplotype blocks in the human genome. *Science* **296**, 2225-2229 (2002).
31. The International HapMap Consortium. The International HapMap Project. *Nature* **426**, 789-796 (2003).
32. The International HapMap Consortium. Integrating ethics and science in the International HapMap Project. *Nature Rev. Genet.* **5**, 467-475 (2004).
33. The ENCODE Project Consortium. The ENCODE (ENCyclopedia Of DNA Elements) Project. *Science* **306**, 636-640 (2004).
34. Herrnstadt, C. *et al.* Reduced-median-network analysis of complete mitochondrial DNA coding-region sequences for the major African, Asian, and European haplogroups. *Am. J. Hum. Genet.* **70**, 1152-1171 (2002).
35. Jobling, M. A. & Tyler-Smith, C. The human Y chromosome: an evolutionary marker comes of age. *Nature Rev. Genet.* **4**, 598-612 (2003).
36. The Y Chromosome Consortium. A nomenclature system for the tree of human Y-chromosomal binary haplogroups. *Genome Res.* **12**, 339-348 (2002).
37. Underhill, P. A. *et al.* The phylogeography of Y chromosome binary haplotypes and the origins of modern human populations. *Ann. Hum. Genet.* **65**, 43-62 (2001).
38. Marchini, J. *et al.* A comparison of phasing algorithms for trios and unrelated individuals. *Am. J. Hum. Genet.* (in the press).

org/cgi-perl/registration)。2004 年 12 月，这一许可被终止，所有数据面向公众无限制释放。

(李平 翻译；常江 审稿)

39. Stephens, M. & Donnelly, P. A comparison of Bayesian methods for haplotype reconstruction from population genotype data. *Am. J. Hum. Genet.* **73**, 1162-1169 (2003).

40. Wright, S. *Evolution and the Genetics of Populations Volume 2: the Theory of Gene Frequencies* 294-295 (Univ. of Chicago Press, Chicago, 1969).

41. Rosenberg, N. A. *et al.* Genetic structure of human populations. *Science* **298**, 2381-2385 (2002).

42. Li, N. & Stephens, M. Modeling linkage disequilibrium and identifying recombination hotspots using single-nucleotide polymorphism data. *Genetics* **165**, 2213-2233 (2003).

43. Pe'er, I. *et al.* Reconciling estimates of linkage disequilibrium in the human genome. *Genome Res.* (submitted).

44. Lichten, M. & Goldman, A. S. Meiotic recombination hotspots. *Annu. Rev. Genet.* **29**, 423-444 (1995).

45. Mouse Genome Sequencing Consortium. Initial sequencing and comparative analysis of the mouse genome. *Nature* **420**, 520-562 (2002).

46. McVean, G. A. *et al.* The fine-scale structure of recombination rate variation in the human genome. *Science* **304**, 581-584 (2004).

47. Crawford, D. C. *et al.* Evidence for substantial fine-scale variation in recombination rates across the human genome. *Nature Genet.* **36**, 700-706 (2004).

48. Kong, A. *et al.* A high-resolution recombination map of the human genome. *Nature Genet.* **31**, 241-247 (2002).

49. Winckler, W. *et al.* Comparison of fine-scale recombination rates in humans and chimpanzees. *Science* **308**, 107-111 (2005).

50. Myers, S. R. & Griffiths, R. C. Bounds on the minimum number of recombination events in a sample history. *Genetics* **163**, 375-394 (2003).

51. Hudson, R. R. & Kaplan, N. L. Statistical properties of the number of recombination events in the history of a sample of DNA sequences. *Genetics* **111**, 147-164 (1985).

52. Phillips, M. S. *et al.* Chromosome-wide distribution of haplotype blocks and the role of recombination hot spots. *Nature Genet.* **33**, 382-387 (2003).

53. Chapman, J. M., Cooper, J. D., Todd, J. A. & Clayton, D. G. Detecting disease associations due to linkage disequilibrium using haplotype tags: a class of tests and the determinants of statistical power. *Hum. Hered.* **56**, 18-31 (2003).

54. Carlson, C. S. *et al.* Selecting a maximally informative set of single-nucleotide polymorphisms for association analyses using linkage disequilibrium. *Am. J. Hum. Genet.* **74**, 106-120 (2004).

55. de Bakker, P. I. W. *et al.* Efficiency and power in genetic association studies. *Nature Genet.* Advance online publication, 23 October 2005 (doi:10.1038/ng1669).

56. Lin, S., Chakravarti, A. & Cutler, D. J. Exhaustive allelic transmission disequilibrium tests as a new approach to genome-wide association studies. *Nature Genet.* **36**, 1181-1188 (2004).

57. Weale, M. E. *et al.* Selection and evaluation of tagging SNPs in the neuronal-sodium-channel gene *SCN1A*: implications for linkage-disequilibrium gene mapping. *Am. J. Hum. Genet.* **73**, 551-565 (2003).

58. Stram, D. O. *et al.* Choosing haplotype-tagging SNPs based on unphased genotype data using a preliminary sample of unrelated subjects with an example from the Multiethnic Cohort Study. *Hum. Hered.* **55**, 27-36 (2003).

59. de la Chapelle, A. & Wright, F. A. Linkage disequilibrium mapping in isolated populations: the example of Finland revisited. *Proc. Natl Acad. Sci. USA* **95**, 12416-12423 (1998).

60. Mootha, V. K. *et al.* Identification of a gene causing human cytochrome c oxidase deficiency by integrative genomics. *Proc. Natl Acad. Sci. USA* **100**, 605-610 (2003).

61. Engert, J. C. *et al.* ARSACS, a spastic ataxia common in northeastern Québec, is caused by mutations in a new gene encoding an 11.5-kb ORF. *Nature Genet.* **24**, 120-125 (2000).

62. Richter, A. *et al.* Location score and haplotype analyses of the locus for autosomal recessive spastic ataxia of Charlevoix-Saguenay, in chromosome region 13q11. *Am. J. Hum. Genet.* **64**, 768-775 (1999).

63. Chakraborty, R. & Weiss, K. M. Admixture as a tool for finding linked genes and detecting that difference from allelic association between loci. *Proc. Natl Acad. Sci. USA* **85**, 9119-9123 (1988).

64. Smith, M. W. & O'Brien, S. J. Mapping by admixture linkage disequilibrium: advances, limitations and guidelines. *Nature Rev. Genet.* **6**, 623-632 (2005).

65. Smith, M. W. *et al.* A high-density admixture map for disease gene discovery in African Americans. *Am. J. Hum. Genet.* **74**, 1001-1013 (2004).

66. Zhu, X. *et al.* Admixture mapping for hypertension loci with genome-scan markers. *Nature Genet.* **37**, 177-181 (2005).

67. Zhao, X. *et al.* An integrated view of copy number and allelic alterations in the cancer genome using single nucleotide polymorphism arrays. *Cancer Res.* **64**, 3060-3071 (2004).

68. Huang, J. *et al.* Whole genome DNA copy number changes identified by high density oligonucleotide arrays. *Hum. Genomics* **1**, 287-299 (2004).

69. Iafrate, A. J. *et al.* Detection of large-scale variation in the human genome. *Nature Genet.* **36**, 949-951 (2004).

70. Sebat, J. *et al.* Large-scale copy number polymorphism in the human genome. *Science* **305**, 525-528 (2004).

71. Stankiewicz, P. & Lupski, J. R. Genome architecture, rearrangements and genomic disorders. *Trends Genet.* **18**, 74-82 (2002).

72. Gonzalez, E. *et al.* The influence of *CCL3L1* gene-containing segmental duplications on HIV-1/AIDS susceptibility. *Science* **307**, 1434-1440 (2005).

73. Singleton, A. B. *et al.* α-Synuclein locus triplication causes Parkinson's disease. *Science* **302**, 841 (2003).

74. McCarroll, S. *et al.* Common deletion variants in the human genome. *Nature Genet.* (in the press).

75. Stefansson, H. *et al.* A common inversion under selection in Europeans. *Nature Genet.* **37**, 129-137 (2005).

76. Myers, S., Bottolo, L., Freeman, C., McVean, G. & Donnelly, P. A fine-scale map of recombination rates and recombination hotspots in the human genome. *Science* **310**, 321-324 (2005).

77. Hinds, D. A. *et al.* Whole-genome patterns of common DNA variation in three human populations. *Science* **307**, 1072-1079 (2005).

78. Yu, A. *et al.* Comparison of human genetic and sequence-based physical maps. *Nature* **409**, 951-953 (2001).

79. Broman, K. W., Murray, J. C., Sheffield, V. C., White, R. L. & Weber, J. L. Comprehensive human genetic maps: individual and sex-specific variation in

recombination. *Am. J. Hum. Genet.* **63**, 861-869 (1998).

80. Weissenbach, J. *et al.* A second-generation linkage map of the human genome. *Nature* **359**, 794-801 (1992).

81. Fullerton, S. M., Bernardo Carvalho, A. & Clark, A. G. Local rates of recombination are positively correlated with GC content in the human genome. *Mol. Biol. Evol.* **18**, 1139-1142 (2001).

82. Dawson, E. *et al.* A first-generation linkage disequilibrium map of human chromosome 22. *Nature* **418**, 544-548 (2002).

83. Begun, D. J. & Aquadro, C. F. Levels of naturally occurring DNA polymorphism correlate with recombination rates in *D. melanogaster*. *Nature* **356**, 519-520 (1992).

84. Smith, A. V., Thomas, D. J., Munro, H. M. & Abecasis, G. R. Sequence features in regions of weak and strong linkage disequilibrium. *Genome Res.* **15**, 1519-1534 (2005).

85. The Gene Ontology Consortium. Gene ontology: tool for the unification of biology. *Nature Genet.* **25**, 25-29 (2000).

86. Trachtenberg, E. *et al.* Advantage of rare HLA supertype in HIV disease progression. *Nature Med.* **9**, 928-935 (2003).

87. Pehrson, J. R. & Fuji, R. N. Evolutionary conservation of histone macroH2A subtypes and domains. *Nucleic Acids Res.* **26**, 2837-2842 (1998).

88. Modrich, P. & Lahue, R. Mismatch repair in replication fidelity, genetic recombination, and cancer biology. *Annu. Rev. Biochem.* **65**, 101-133 (1996).

89. Nielsen, R. Human genomics: disclosure of variation. *Nature* **434**, 288-289 (2005).

90. Enattah, N. S. *et al.* Identification of a variant associated with adult-type hypolactasia. *Nature Genet.* **30**, 233-237 (2002).

91. Bersaglieri, T. *et al.* Genetic signatures of strong recent positive selection at the lactase gene. *Am. J. Hum. Genet.* **74**, 1111-1120 (2004).

92. Sabeti, P. C. *et al.* Detecting recent positive selection in the human genome from haplotype structure. *Nature* **419**, 832-837 (2002).

93. Dermitzakis, E. T. *et al.* Numerous potentially functional but non-genic conserved sequences on human chromosome 21. *Nature* **420**, 578-582 (2002).

94. Margulies, E. H., Blanchette, M., Haussler, D. & Green, E. D. Identification and characterization of multi-species conserved sequences. *Genome Res.* **13**, 2507-2518 (2003).

95. Need, A. C., Motulsky, A. G. & Goldstein, D. B. Priorities and standards in pharmacogenetic research. *Nature Genet.* **37**, 671-681 (2005).

96. Brem, R. B., Storey, J. D., Whittle, J. & Kruglyak, L. Genetic interactions between polymorphisms that affect gene expression in yeast. *Nature* **436**, 701-703 (2005).

97. Istrail, S. *et al.* Whole-genome shotgun assembly and comparison of human genome assemblies. *Proc. Natl Acad. Sci. USA* **101**, 1916-1921 (2004).

98. Venter, J. C. *et al.* The sequence of the human genome. *Science* **291**, 1304-1351 (2001).

99. The Chimpanzee Sequencing and Analysis Consortium. Initial sequence of the chimpanzee genome and comparison with the human genome. *Nature* **437**, 69-87 (2005).

100. Matsuzaki, H. *et al.* Genotyping over 100,000 SNPs on a pair of oligonucleotide arrays. *Nature Methods* **1**, 109-111 (2004).

101. Bailey, J. A. *et al.* Recent segmental duplications in the human genome. *Science* **297**, 1003-1007 (2002).

102. Wigginton, J. E., Cutler, D. J. & Abecasis, G. R. A note on exact tests of Hardy–Weinberg equilibrium. *Am. J. Hum. Genet.* **76**, 887-893 (2005).

103. Hill, W. G. & Weir, B. S. Maximum-likelihood estimation of gene location by linkage disequilibrium. *Am. J. Hum. Genet.* **54**, 705-714 (1994).

Supplementary Information is linked to the online version of the paper at www.nature.com/nature.

Acknowledgements. We thank many people who contributed to this project: J. Beck, C. Beiswanger, D. Coppock, A. Leach, J. Mintzer and L. Toji (Coriell Institute for Medical Research) for transforming the Yoruba, Japanese and Han Chinese samples, distributing the DNA and cell lines, storing the samples for use in future research, and producing the community newsletters and reports; J. Greenberg and R. Anderson (NIH National Institute of General Medical Sciences) for providing funding and support for cell line transformation and storage in the NIGMS Human Genetic Cell Repository at the Coriell Institute; T. Dibling, T. Ishikura, S. Kanazawa, S. Mizusawa and S. Saito (SNP Research Center, RIKEN) for help with genotyping; C. Hind and A. Moghadam for technical support in genotyping and all members of the subcloning and sequencing teams at the Wellcome Trust Sanger Institute; X. Ke (Wellcome Trust Centre for Human Genetics at the University of Oxford) for help with data analysis; Oxford E-Science Centre for provision of high-performance computing resources; H. Chen, W. Chen, L. Deng, Y. Dong, C. Fu, L. Gao, H. Geng, J. Geng, M. He, H. Li, H. Li, S. Li, X. Li, B. Liu, Z. Liu, F. Lu, F. Lu, G. Lu, C. Luo, X. Wang, Z. Wang, C. Ye and X. Yu (Beijing Genomics Institute) for help with genotyping and sample collection; X. Feng, Y. Li, J. Ren and X. Zhou (Beijing Normal University) for help with sample collection; J. Fan, W. Gu, W. Guan, S. Hu, H. Jiang, R. Lei, Y. Lin, Z. Niu, B. Wang, L. Yang, W. Yang, Y. Wang, Z. Wang, S. Xu, W. Yan, H. Yang, W. Yuan, C. Zhang, J. Zhang, K. Zhang and G. Zhao (Chinese National Human Genome Center at Shanghai) for help with genotyping; P. Fong, C. Lai, C. Lau, T. Leung, L. Luk and W. Tong (University of Hong Kong, Genome Research Centre) for help with genotyping; C. Pang (Chinese University of Hong Kong) for help with genotyping; K. Ding, B. Qiang, J. Zhang, X. Zhang and K. Zhou (Chinese National Human Genome Center at Beijing) for help with genotyping; Q. Fu, S. Ghose, X. Lu, D. Nelson, A. Perez, S. Poole, R. Vega and H. Yonath (Baylor College of Medicine); C. Bruckner, T. Brundage, S. Chow, O. Iartchouk, M. Jain, M. Moorhead and K. Tran (ParAllele Bioscience Inc.); N. Addleman, J. Atilano, T. Chan, C. Chu, C. Ha, T. Nguyen, M. Minton and A. Phong (UCSF) for help with genotyping, and D. Lind (UCSF) for help with quality control and experimental design; R. Donaldson and S. Duan (Washington University) for help with genotyping, and J. Rice and N. Saccone (Washington University) for help with experimental design; J. Wigginton (University of Michigan) for help with implementing and testing QA/QC software; A. Clark, B. Keats, R. Myers, D. Nickerson and A. Williamson for providing advice to NIH; J. Melone, M. Weiss and E. DeHaut-Combs (NHGRI) for help with project management; M. Gray for organizing phone calls and meetings; D. Leja for help with figures; the Yoruba people of Ibadan, Nigeria, the people of Tokyo, Japan, and the community at Beijing Normal University, who participated in public consultations and community engagements; the people in these communities who were generous in donating their blood samples; and the people in the Utah CEPH community who allowed the samples they donated earlier to be used for the Project. We also thank A. Clark, E. Lander, C. Langley and R. Lifton for comments on earlier drafts of the manuscript. This work was supported by the Japanese Ministry of Education, Culture,

Sports, Science, and Technology, the Wellcome Trust, Nuffield Trust, Wolfson Foundation, UK EPSRC, Genome Canada, Génome Québec, the Chinese Academy of Sciences, the Ministry of Science and Technology of the People's Republic of China, the National Natural Science Foundation of China, the Hong Kong Innovation and Technology Commission, the University Grants Committee of Hong Kong, the SNP Consortium, the US National Institutes of Health (FIC, NCI, NCRR, NEI, NHGRI, NIA, NIAAA, NIAID, NIAMS, NIBIB, NIDA, NIDCD, NIDCR, NIDDK, NIEHS, NIGMS, NIMH, NINDS, NLM, OD), the W.M. Keck Foundation, and the Delores Dore Eccles Foundation.

Author Contributions. David Altshuler, Lisa D. Brooks, Aravinda Chakravarti, Francis S. Collins, Mark J. Daly and Peter Donnelly are members of the writing group responsible for this manuscript.

Author Information. Reprints and permissions information is available at npg.nature.com/reprintsandpermissions. The authors declare competing financial interests: details accompany the paper at www.nature.com/nature. Correspondence and requests for materials should be addressed to D.A. (altshuler@molbio.mgh.harvard.edu) or P.D. (donnelly@stats.ox.ac.uk).

Analysis of One Million Base Pairs of Neanderthal DNA

R. E. Green *et al.*

Editor's Note

Neanderthals were the closest relatives of modern humans. They lived in Europe and Western Asia about 400,000–24,000 years ago. As modern humans were present in Europe at least 41,000 years ago, there is much speculation about whether Neanderthals and modern humans interbred. The only way to decide this is from genome analysis. In this paper Svante Pääbo and colleagues describe the sequencing of a million base pairs of Neanderthal nuclear DNA from a bone from Croatia. The full genome sequence took a further four years to complete, and the analysis of this and other Neanderthal genomes suggests that around 4% of the DNA of modern Europeans is of Neanderthal origin.

Neanderthals are the extinct hominid group most closely related to contemporary humans, so their genome offers a unique opportunity to identify genetic changes specific to anatomically fully modern humans. We have identified a 38,000-year-old Neanderthal fossil that is exceptionally free of contamination from modern human DNA. Direct high-throughput sequencing of a DNA extract from this fossil has thus far yielded over one million base pairs of hominoid nuclear DNA sequences. Comparison with the human and chimpanzee genomes reveals that modern human and Neanderthal DNA sequences diverged on average about 500,000 years ago. Existing technology and fossil resources are now sufficient to initiate a Neanderthal genome-sequencing effort.

NEANDERTHALS were first recognized as a distinct group of hominids from fossil remains discovered 150 years ago at Feldhofer in Neander Valley, outside Düsseldorf, Germany. Subsequent Neanderthal finds in Europe and western Asia showed that fossils with Neanderthal traits appear in the fossil record of Europe and western Asia about 400,000 years ago and vanish about 30,000 years ago. Over this period they evolved morphological traits that made them progressively more distinct from the ancestors of modern humans that were evolving in Africa[1,2]. For example, the crania of late Neanderthals have protruding mid-faces, brain cases that bulge outward at the sides, and features of the base of the skull, jaw and inner ears that set them apart from modern humans[3].

The nature of the interaction between Neanderthals and modern humans, who expanded out of Africa around 40,000–50,000 years ago and eventually replaced Neanderthals

对尼安德特人 DNA 的一百万碱基对的分析

格林等

编者按

尼安德特人是现代人类最近的亲戚。它们大约在 400,000 到 24,000 年前生活在欧洲和西亚。由于现代人类至少在 41,000 年前出现在欧洲，人们对尼安德特人和现代人类是否杂交有很多猜测。确定这一点的唯一方法是从基因组进行分析。在本文中，斯凡特·帕波及其同事描述了来自克罗地亚骨骼的尼安德特人核 DNA 的一百万个碱性对的测序。完整的基因组序列又用了四年时间才完成。对这个基因组序列和其他尼安德特人基因组的分析表明，现代欧洲人中约有 4%的 DNA 来自尼安德特人。

尼安德特人是与当代人类亲缘关系最近的已经灭绝的人类类群，所以他们的基因组能为我们提供一个独特的机遇来了解解剖学意义上完全的现代人所特有的遗传变化。我们已经甄别选定了一件罕见不受现代人 DNA 污染的有着 38,000 年历史的尼安德特人化石。对该化石提取出的 DNA 进行的直接高通量测序，目前已经得到了 100 多万碱基对的类人猿核 DNA 序列。该序列与人和黑猩猩的基因组对比表明，现代人和尼安德特人的 DNA 序列平均大约在 500,000 年前发生分歧。现有的技术和化石资源足以支撑着手开展尼安德特人基因组测序工作。

基于 150 年前发现于德国杜塞尔多夫外的尼安德谷费尔德霍费尔山洞的化石遗存，尼安德特人最初被认为是一个独特的人类类群。后来欧洲和西亚的尼安德特人的发现显示具有尼安德特人特征的化石在欧洲和西亚的化石记录中出现于约 400,000 年前，消失于约 30,000 年前。在这期间，他们进化出了与当时正在非洲进化的现代人祖先渐行渐远的形态特征 [1,2]。例如，晚期尼安德特人的头骨有着突出的中面部，两侧向外突出的脑颅以及将他们与现代人区别开来的颅底、颌骨和内耳特征 [3]。

现代人大约于 40,000 到 50,000 年前走出非洲，并最终取代了尼安德特人和其他生活在旧大陆的古老型人类。这些现代人与尼安德特人间相互交流的本质至今仍是

as well as other archaic hominids across the Old World is still a matter of some debate. Although there is no evidence of contemporaneous cohabitation at any single site, there is evidence of geographical and temporal overlap in their ranges before the disappearance of Neanderthals. Additionally, late in their history, some Neanderthal groups adopted cultural traits such as body decorations, potentially through cultural interactions with incoming modern humans[4].

In 1997, a segment of the hypervariable control region of the maternally inherited mitochondrial DNA (mtDNA) of the Neanderthal type specimen found at Feldhofer was sequenced. Phylogenetic analysis showed that it falls outside the variation of contemporary humans and shares a common ancestor with mtDNAs of present-day humans approximately half a million years ago[5,6]. Subsequently, mtDNA sequences have been retrieved from eleven additional Neanderthal specimens: Feldhofer 2 in Germany[7], Mezmaiskaya in Russia[8], Vindija 75, 77 and 80 in Croatia[9,10], Engis 2 in Belgium, La Chapelle-aux-Saints and Rochers de Villeneuve in France[10], Scladina in Belgium[11], Monte Lessini in Italy[12], and El Sidron 441 in Spain[13]. Although some of these sequences are extremely short, they are all more closely related to one another than to modern human mtDNAs[9,11].

This fact, in conjunction with the absence of any related mtDNA sequences in currently living humans or in a small number of early modern human fossils[5,10] strongly suggests that Neanderthals contributed no mtDNA to present-day humans. On the basis of various population models, it has been estimated that a maximal overall genetic contribution of Neanderthals to the contemporary human gene pool is between 25% and 0.1% (refs 10, 14). Because the latter conclusions are based on mtDNA, a single maternally inherited locus, they are limited in their ability to detect a Neanderthal contribution to the current human gene pool both by the vagaries of genetic drift and by the possibility of a sex bias in reproduction. However, both morphological evidence[4,15] and the variation in the modern human gene pool[16] support the conclusion that if any genetic contribution of Neanderthals to modern human occurred, it was of limited magnitude.

Neanderthals are the hominid group most closely related to currently living humans, so a Neanderthal nuclear genome sequence would be an invaluable resource for annotating the human genome. Roughly 35 million nucleotide differences exist between the genomes of humans and chimpanzees, our closest living relatives[17]. Soon, genome sequences from other primates such as the orang-utan and the macaque will allow such differences to be assigned to the human and chimpanzee lineages. However, temporal resolution of the genetic changes along the human lineage, where remarkable morphological, behavioural and cognitive changes occurred, are limited without a more closely related genome sequence for comparison. In particular, comparison to the Neanderthal would enable the identification of genetic changes that occurred during the last few hundred thousand years, when fully anatomically and behaviourally modern humans appeared.

一个存在争议的问题。尽管没有证据显示他们在任何一个地点同时生存过，但是有证据表明在尼安德特人消失之前，他们彼此存在地理和时代分布的交叠。此外，在他们的历史的晚期，有些尼安德特人群体接纳了身体装饰等文化特征，可能是他们通过与迁徙而来的现代人之间进行文化交流而实现的[4]。

1997 年，对发现于费尔德霍费尔的尼安德特人模式标本的母系遗传的线粒体 DNA(mtDNA)的高变控制区的一个片段进行了测序。系统发育分析的结果表明，该序列落于当代人的变异范围之外，并且于大概 500,000 年前与当代人的 mtDNA 共有一个祖先[5,6]。接着，从另外 11 个尼安德特人标本得到了 mtDNA 序列，包括德国的费尔德霍费尔 2[7]，俄罗斯的梅兹迈斯卡亚[8]，克罗地亚的温迪迦 75、77 和 80[9,10]，比利时的英格斯 2，法国的拉沙佩勒–欧赛恩茨和维伦纽夫岩[10]，比利时的斯卡拉迪纳[11]，意大利的蒙蒂莱西尼[12]，以及西班牙的埃尔西德隆 441[13]。尽管其中有些序列非常短，但是比起与现代人 mtDNA 的关系，它们彼此的亲缘关系要近得多[9,11]。

这一事实与现存的人类或少量早期现代人化石[5,10]中缺少任何相关的 mtDNA 序列一起强烈表明尼安德特人对当代人的 mtDNA 没有贡献。基于各种人类群体模型，已估算出尼安德特人对当代人基因库的最大整体遗传贡献在 25% 到 0.1% 之间(参考文献 10 和 14)。因为后面的推论是根据 mtDNA 推导出来的，而 mtDNA 只是一个单一的母系遗传位点，所以遗传漂移的不确定性及繁殖过程中性别差异的可能性导致它们在检测尼安德特人对当前人类基因库的贡献度方面的能力是有限的。然而，形态学证据[4,15]和现代人类基因库的变异[16]都支持如下结论，即如果尼安德特人对现代人有过任何的遗传贡献的话，那么这种贡献程度也是有限的。

尼安德特人是与现存的人类亲缘关系最密切的人类类群，所以尼安德特人的核基因组序列对于注释人类基因组来说将是无价的资源。我们现存的至亲黑猩猩与人类的基因组之间大概存在 35,000,000 个核苷酸的差异[17]。不久的将来，其他灵长类(如猩猩和猕猴)的基因组序列也会支持这种差异添加到人类和黑猩猩的演化谱系中。但是，由于所发生的形态、行为和认知变化非常显著，如果没有与人类亲缘关系更加接近的基因组序列能够用来做比较，发生在人类演化谱系中遗传变化的年代分辨率会受到局限。特别是，通过与尼安德特人进行比较，就可以对过去几十万年里出现的解剖学和行为学上完全的现代人发生的遗传变化进行鉴别。

Identification of a Neanderthal Fossil for DNA Sequencing

Although it is possible to recover mtDNA[18] and occasionally even nuclear DNA sequences[19-22] from well-preserved remains of organisms that are less than a few hundred thousand years old, determination of ancient hominid sequences is fraught with special difficulties and pitfalls[18]. In addition to degradation and chemical damage to the DNA that can cause any ancient DNA to be irretrievable or misread, contamination of specimens, laboratory reagents and instruments with traces of DNA from modern humans must be avoided. In fact, when sensitive polymerase chain reaction (PCR) is used, human mtDNA sequences can be retrieved from almost every ancient specimen[23,24]. This problem is especially severe when Neanderthal remains are studied because Neanderthal and human are so closely related that one expects to find few or no differences between Neanderthals and modern humans within many regions[25], making it impossible to rely on the sequence information itself to distinguish endogenous from contaminating DNA sequences. A necessary first step for sequencing nuclear DNA from Neanderthals is therefore to identify a Neanderthal specimen that is free or almost free of modern human DNA.

We tested more than 70 Neanderthal bone and tooth samples from different sites in Europe and western Asia for bio-molecular preservation by removing samples of a few milligrams for amino acid analysis. The vast majority of these samples had low overall contents of amino acids and/or high levels of amino acid racemization, a stereoisomeric structural change that affects amino acids in fossils, indicating that they are unlikely to contain retrievable endogenous DNA[26]. However, some of the samples are better preserved in that they contain high levels of amino acids (more than 20,000 p.p.m.), low levels of racemization of amino acids such as aspartate that racemize rapidly, as well as amino acid compositions that suggest that the majority of the preserved protein stems from collagen.

From 100–200 mg of bone from six of these specimens we extracted DNA and analysed the relative abundance of Neanderthal-like mtDNA sequences and modern human-like mtDNA sequences by performing PCR with primer pairs that amplify both human and Neanderthal mtDNA with equal efficiency. The amplification products span segments of the hypervariable region of the mtDNA in which all Neanderthals sequenced to date differ from all contemporary humans. From subsequent cloning into a plasmid vector and sequencing of more than a hundred clones from each product, we determined the ratio of Neanderthal-like to modern human-like mtDNA in each extract. We used two different primer pairs that amplify fragments of 63 base pairs and 119 base pair to gauge the contamination levels for different lengths of DNA molecules.

Figure 1 shows that the level of contamination differs drastically among the samples. Whereas only around 1% of the mtDNA present in three samples from France, Russia and Uzbekistan was Neanderthal-like, one sample from Croatia and one from Spain contained around 5% and 75% Neanderthal-like mtDNA, respectively. One bone (Vi-80) from Vindija Cave, Croatia, stood out in that ~99% of the 63-base-pair mtDNA segments and

用于 DNA 测序的尼安德特人化石的甄选

尽管从保存良好且年龄小于几十万年的生物遗骸中得到 mtDNA[18]，极少数情况下甚至得到核 DNA 序列 [19-22] 也是可能的，但是古老人类序列的确定充满特殊的困难和陷阱 [18]。除了导致古 DNA 不可获取或误读的 DNA 降解和化学损伤外，标本的污染、带有现代人 DNA 残留的实验室试剂和仪器也是必须要避免的。事实上，当使用敏感的聚合酶链式反应（PCR）时，几乎可以从每件古老的标本中得到人类 mtDNA 序列 [23,24]。当研究尼安德特人时，这一问题尤为严重，因为尼安德特人与现代人的亲缘关系如此之近，所以研究者会预期在许多区域上现代人和尼安德特人之间都很少有或根本没有差异 [25]，这使得仅仅依靠序列本身不可能将内源 DNA 与污染源 DNA 序列区别开。因此如果要对尼安德特人的核 DNA 进行测序，第一步必须要做的就是甄选一件没有或几乎没有受到现代人 DNA 污染的尼安德特人标本。

通过去除几毫克样品进行氨基酸分析，我们检测了欧洲和西亚的不同地点采集的 70 多个尼安德特人骨骼和牙齿样品以确认其生物分子保存状况。绝大部分样品的总氨基酸含量很低并且（或者）氨基酸外消旋化（一种对化石中的氨基酸产生影响的立体异构变化）水平很高，表明这些样本不可能含有可获取的内源 DNA[26]。然而，有些样本的保存情况较好，因为它们的氨基酸含量水平很高（大于 20,000 ppm）；而通常快速外消旋化的天冬氨酸等氨基酸的外消旋化水平很低；另外这些样本的氨基酸构成意味着大部分保存下来的蛋白质来源于胶原蛋白。

从其中六件标本的 100 mg 到 200 mg 骨骼样品中我们提取到了 DNA。通过执行等效引物对 PCR 扩增现代人和尼安德特人的 mtDNA，我们分析了似尼安德特人的 mtDNA 序列及似现代人的 mtDNA 序列的相对丰度。扩增产物涵盖了 mtDNA 高变区的片段。迄今为止，在这一区域中所有经过测序的尼安德特人与所有现代人都不一样。随后通过将 PCR 产物克隆到质粒载体中并对每个产物的一百多个克隆进行测序，我们确定了每个提取物中似尼安德特人 mtDNA 与似现代人 mtDNA 的比率。我们使用了两个不同引物对分别扩增 63 个碱基对（bp）和 119 个碱基对的片段，用以计量不同长度的 DNA 分子的污染水平。

图 1 显示样品间的污染水平显著不同。尽管来自法国、俄罗斯和乌兹别克斯坦的三个样品中只有 1% 左右的 mtDNA 是似尼安德特人的，而一个来自克罗地亚的样品和一个来自西班牙的样品分别含有 5% 和 75% 左右的似尼安德特人 mtDNA。来自克罗地亚温迪迦洞穴的一件骨骼标本（Vi-80）的数据异常突出，约 99% 的 63 bp

~94% of the 119-base pair segments are of Neanderthal origin. Assuming that the ratio of Neanderthal to contaminating modern human DNA is the same for mtDNA as it is for nuclear DNA, the Vi-80 bone therefore yields DNA fragments that are predominantly of Neanderthal origin and provided that the contamination rate was not increased during the downstream sequencing process, the extent of contamination in the final analyses is below ~6%.

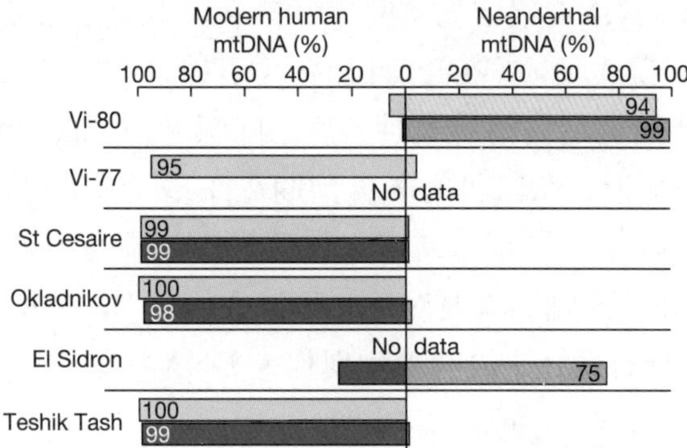

Fig. 1. Ratio of Neanderthal to modern human mtDNA in six hominid fossils. For each fossil, primer pairs that amplify a long (119 base pairs; upper lighter bars) and short (63 base pairs; lower darker bars) product were used to amplify segments of the mtDNA hypervariable region. The products were sequenced and determined to be either of Neanderthal (yellow) or modern human (blue) type.

The Vi-80 bone was discovered by M. Malez and co-workers in layer G3 of Vindija Cave in 1980. It has been dated by carbon-14 accelerator mass spectrometry to 38,310 ± 2,130 years before present and its entire mtDNA hypervariable region I has been sequenced[10]. Out of 14 Neanderthal remains from layer G3 that we have analysed, this bone is one of six samples that show good bio-molecular preservation, while the other eight bones show intermediate to bad states of preservation that do not suggest the presence of amplifiable DNA. Preservation conditions in Vindija Cave thus vary drastically from bone to bone, a situation that may be due to different extents of water percolation in different parts of the cave.

Direct Large-scale DNA Sequencing from the Vindija Neanderthal

Because the Vi-80 Neanderthal bone extract is largely free of contaminating modern human mtDNA, we chose this extract to perform large-scale parallel 454 sequencing[27]. In this technology, single-stranded libraries, flanked by common adapters, are created from the DNA sample and individual library molecules are amplified through bead-based emulsion PCR, resulting in beads carrying millions of clonal copies of the DNA fragments from the samples. These are subsequently sequenced by pyrosequencing on the GS20 454 sequencing system.

For several reasons, the 454 sequencing platform is extremely well suited for analyses of bulk

的 mtDNA 片段和约 94% 的 119 bp 的 mtDNA 片段源自尼安德特人。假设尼安德特人的核 DNA 与现代人污染源的核 DNA 的比率与 mtDNA 的情况一样，那么从该 Vi-80 骨骼标本得到的 DNA 片段应该主要来自尼安德特人。那么，如果在下游测序过程中没有增加污染率的话，那么最终分析时的污染率应该低于大约 6%。

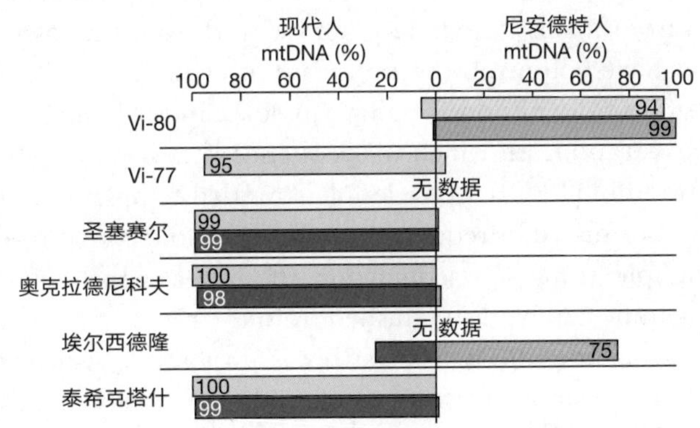

图 1. 六件人类化石标本中尼安德特人与现代人 mtDNA 的比例。对于每件化石标本，都使用了两个引物对来扩增 mtDNA 的高变区片段，其中一对扩增的产物较长（119 bp；上面的浅颜色横杠），另一对扩增的产物较短（63 bp；下面的深颜色横杠）。对产物进行了测序并确定其属于尼安德特人（黄色）类型还是现代人（蓝色）类型。

梅勒兹及其同事于 1980 年在温迪迦洞穴的 G3 层中发现了 Vi-80 骨骼标本。通过碳–14 加速器质谱法测定其年代为距今 38,310±2,130 年，其 mtDNA 高变区 I 已经完全测序[10]。我们已经对 G3 层发现的 14 个尼安德特人样本进行了分析，该骨骼标本是生物分子保存状况良好的六个样品之一，而其余八件骨骼标本则显示出中度至较差的保存状况，表明其不存在可扩增的 DNA。因此温迪迦洞穴中各骨骼标本彼此的保存状况非常不同，可能是由洞穴中不同部位的水渗透情况不同造成的。

温迪迦尼安德特人的直接大范围 DNA 测序

由于 Vi-80 尼安德特骨骼标本提取物基本排除了现代人 mtDNA 污染的可能，所以我们选择这一提取物来进行大范围平行 454 测序[27]。这一技术中，由 DNA 样本创建出两侧连有通用衔接子的单链 DNA 库，各个库中的分子通过磁珠乳液 PCR 进行扩增，于是就可以从样本中得到携带有数以百万计的 DNA 片段的克隆拷贝的磁珠。然后在 GS20 454 测序系统上通过焦磷酸测序法对这些 DNA 片段进行测序。

由于以下几个原因，454 测序平台非常适合于分析古代残骸中提取到的大量

DNA extracted from ancient remains[28]. First, it circumvents bacterial cloning, in which the vast majority of initial template molecules are lost during transformation and establishment of clones. Second, because each molecule is amplified in isolation from other molecules it also precludes template competition, which frequently occurs when large numbers of different DNA fragments are amplified together. Third, its current read length of 100–200 nucleotides covers the average length of the DNA preserved in most fossils[29]. Fourth, it generates hundreds of thousands of reads per run, which is crucial because the majority of the DNA recovered from fossils is generally not derived from the fossil species, but rather from organisms that have colonized the organism after its death[20,30]. Fifth, because each sequenced product stems from just one original single-stranded template molecule of known orientation, the DNA strand from which the sequence is derived is known[28]. This provides an advantage over traditional PCR from double-stranded templates, in which the template strand is not known, because the frequency of different nucleotide misincorporations can be deduced. For example, using 454 sequencing, the rate at which cytosine is converted to uracil and read as thymine can be distinguished from the rate at which guanine is converted to xanthine and read as adenine, whereas this is impossible using traditional PCR or bacterial cloning. This is important since nucleotide conversions and misincorporations in ancient DNA are caused by damage that affects different bases differently[28,31] and this pattern of false substitutions can be used to estimate the relative probability that a particular substitution (that is, the observation of a nucleotide difference between DNA sequences) represents the authentic DNA sequence of the organism versus an artefact from DNA degradation.

We recovered a total of 254,933 unique sequences from the Vi-80 bone (see Supplementary Methods). These were aligned to the human (build 36.1)[32], chimpanzee (build 1)[17] and mouse (build 34.1)[33] complete genome sequences, to environmental sample sequences in the GenBank *env* database (version 3, September 2005), and to the complete set of redundant nucleotide sequences in GenBank *nt* (version 3, September 2005, excluding EST, STS, GSS, environmental and HTGS sequences)[34] using the program BLASTN (NCBI version 2.2.12)[35]. The most similar database sequence for each query was identified and classified by its taxonomic order (Fig. 2) (see Supplementary Methods). No significant nucleotide sequence similarity in the databases was found for 79% of the fossil extract sequence reads. This is typical of large-scale sequencing both from other ancient bones[20,22,28] and from environmental samples[36,37], although some permafrost-preserved specimens can yield high amounts of endogenous DNA[22]. Sequences with similarity to a database sequence were classified by the taxonomic order of their most significant alignment. Actinomycetales, a bacterial order with many soil-living species, was the most populous order and accounted for 6.8% of the sequences. The second most populous order, to which 15,701 unique sequences or 6.2% of the sequence reads were most similar, was that of primates. All other individual orders were substantially less frequent. Notably, the average percentage identity for the primate sequence alignments was 98.8%, whereas it was 92–98% for the other frequently occurring orders. Thus, the primate reads, unlike many of the prokaryotic reads, are aligned to a very closely related species.

DNA[28]。首先，它绕过了细菌克隆，因为在转化和建立克隆的过程中会丢失绝大多数的初始模板分子。其次，由于每个分子都是同其他分子隔离开进行扩增的，所以可以预先排除模板竞争(模板竞争经常发生在同时扩增大量不同的 DNA 片段时)。第三，454 测序目前的读取长度是 100 到 200 个核苷酸，这一长度可以涵盖大部分化石中保存的 DNA 的平均长度[29]。第四，454 测序技术每次运行可以产生数十万个读取，这一特点是很重要的，因为大部分从化石中复原的 DNA 都不是来自化石物种本身，而是来自那些在化石物种死后聚居其上的生物[20,30]。第五，由于每个测序产物仅由一个已知方向的初始单链模板分子扩增而来，所以产生该序列的 DNA 链是已知的[28]。传统 PCR 中，我们不知道模板链是哪条，所以 454 测序技术比传统的从双链模板得到 PCR 产物的技术更具优越性，因为同时可以推导出不同核苷酸错配率。举例而言，使用 454 测序技术，可以将胞嘧啶被转变为尿嘧啶而被读为胸腺嘧啶的比率与鸟嘌呤被转变为黄嘌呤而被读成腺嘌呤的比率区别开来，而这在使用传统 PCR 或者细菌克隆时是不可能办到的。由于古 DNA 中的损伤会不同程度地影响不同碱基而发生核苷酸转变和错配，而且这种虚假的碱基替换模式可以用来估计代表生物真实 DNA 序列的特定碱基替换(即观察到的 DNA 序列间的核苷酸差异)与 DNA 降解所导致人为替代的相对概率，所以该测序技术的这一特点非常重要[28,31]。

我们总共从 Vi-80 骨骼标本中复原了 254,933 条不同序列(见补充方法)。使用 BLASTN 程序(NCBI 2.2.12 版本)[35] 将这些序列与人(版本 36.1)[32]、黑猩猩(版本 1)[17] 和小鼠(版本 34.1)[33] 的全基因组序列进行了对齐，与 GenBank *env* 数据库(2005 年 9 月第 3 版)中的环境样本序列进行了对齐，以及与 GenBank *nt*(2005 年 9 月第 3 版，不包括 EST、STS、GSS、环境和 HTGS 序列)[34] 中全套冗余核苷酸序列进行了对齐。根据“目”一级分类学单元将每次查询到的最相似的数据库序列进行确认并分类(图 2)(见补充方法)。其中 79% 的化石提取物的序列读取没有在数据库中找到有意义的核苷酸序列相似性。虽然一些永久冻土中保存的标本可以产生大量的内源 DNA[22]，但是对于其他古代骨骼[20,22,28] 和环境样品[36,37] 的大范围测序来说，这是很典型的现象。与数据库序列具有相似性的序列根据其最有意义的对齐的“目”一级分类学单元进行分类。放线菌目是一个含有许多生活在土壤中的种类的细菌目，是序列数目最多的细菌目，占序列的 6.8%。第二大目是灵长目，有 15,701 条不同序列或者说 6.2% 的序列读取与其最相似。所有其他的各目的频率都低得多。值得注意的是，灵长目序列对齐的平均百分率一致性达到了 98.8%，而其余经常出现的目是 92% 到 98%。因此，与许多原核生物的读取不同，灵长目的序列读取被对齐到了亲缘关系很近的物种上。

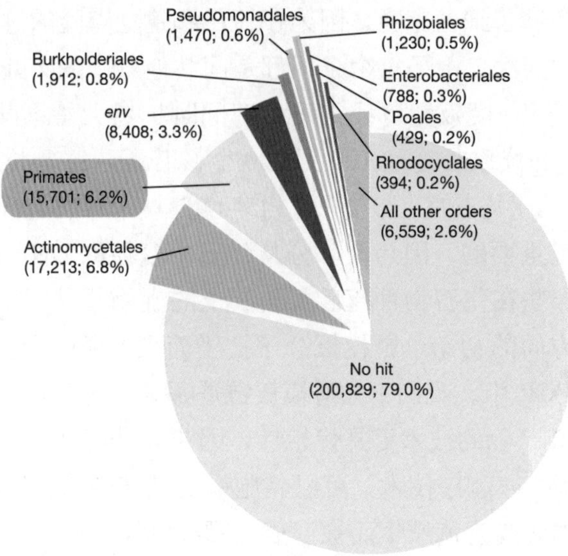

Fig. 2. Taxonomic distribution of DNA sequences from the Vi-80 extract. The taxonomic order of the database sequence giving the best alignment for each unique sequence read was determined. The most populous taxonomic orders are shown.

Neanderthal mtDNA Sequences

Among the 15,701 sequences of primate origin, we first identified all mtDNA in order to investigate whether their evolutionary relationship to the current human mtDNA pool is similar to what is known from previous analyses of Neanderthal mtDNA. A total of 41 unique DNA sequences from the Vi-80 fossil had their closest hits to different parts of the human mtDNA, and comprised, in total, 2,705 base pairs of unique mtDNA sequence. None of the putative Neanderthal mtDNA sequences map to the two hypervariable regions that have been previously sequenced in Neanderthals. We aligned these mtDNA sequences to the complete mtDNA sequences of 311 modern humans from different populations[38] as well as to the complete mtDNA sequences of three chimpanzees and two bonobos (Supplementary Information). A schematic neighbour-joining tree estimated from this alignment is shown in Fig. 3. In agreement with previous results, the Neanderthal mtDNA falls outside the variation among modern humans. However, the length of the branch leading to the Neanderthal mtDNA is 2.5 times as long as the branch leading to modern human mtDNAs. This is likely to be due to errors in our Neanderthal sequences derived from substitution artefacts from damaged, ancient DNA and from sequencing errors[28].

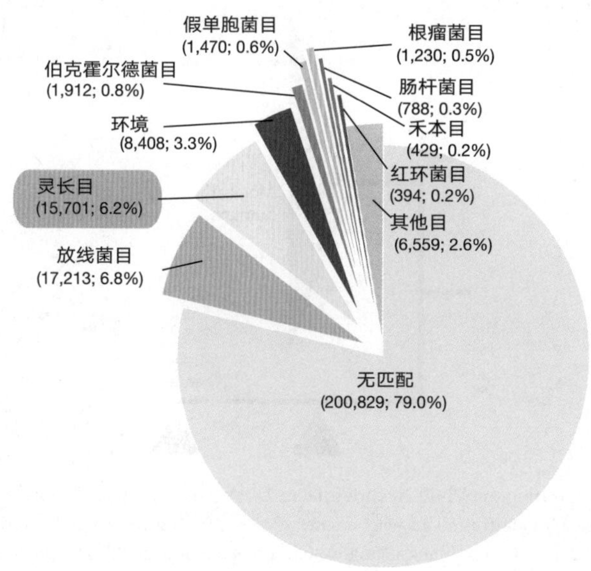

图 2. Vi-80 提取物的 DNA 序列的分类学分布情况。确定了与每条不同序列读取最佳对齐的数据库序列所属的分类学目。图中展示了序列数量最多的几个分类学目。

尼安德特人的 mtDNA 序列

在源于灵长目的 15,701 条序列中，我们首先鉴定了所有 mtDNA，以研究它们同现代人 mtDNA 库的演化关系是否与之前对尼安德特人 mtDNA 的分析结果相似。从 Vi-80 化石共得到 41 条与人类 mtDNA 的不同部位最为匹配的不同序列，总共包含 2,705 bp 的独特 mtDNA 序列。这些推定的尼安德特人 mtDNA 序列不能够映射到之前已测序的尼安德特人两个高变区。我们将这些 mtDNA 序列与来自不同人类群体的 311 个现代人的 mtDNA 全序列 [38] 及三只黑猩猩和两只倭黑猩猩的 mtDNA 全序列进行了对齐（见补充信息）。由该对齐估算出的邻接树的示意图如图 3 所示。得到的结果与以前的研究一致，尼安德特人 mtDNA 也落在现代人的变异范围之外。然而，尼安德特人 mtDNA 分支的长度是现代人 mtDNA 分支长度的 2.5 倍。导致这种情况的原因可能是尼安德特人序列中存在的由受损古 DNA 导致的人为碱基替换错误以及测序错误 [28]。

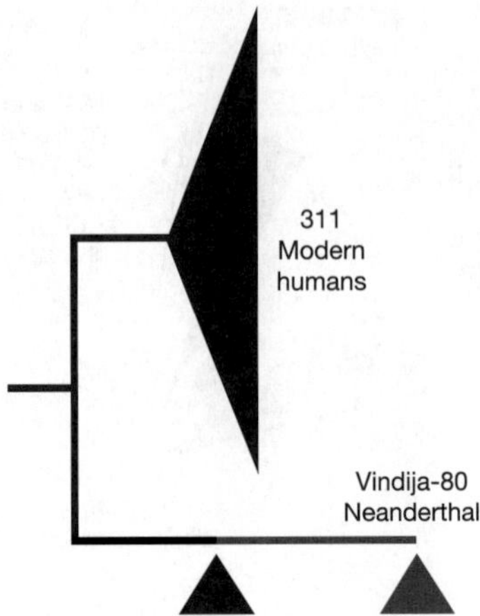

Fig. 3. Schematic tree relating the Vi-80 Neanderthal mtDNA sequences to 311 human mtDNA sequences. The Neanderthal branch length is given with uncorrected sequences (red triangle) and after correction of sequences via independent PCRs (black triangle). Chimpanzee and bonobo sequences (not shown) were used to root the neighbour-joining tree. Several substitution models (Kimura 2-parameter, Tajima-Nei, and Tamura 3-parameter with uniform or gamma-distributed ($\gamma = 0.5–1.1$) rates) yielded bootstrap support values for the human branch from 72–83%.

To analyse the extent to which errors occur in the Neanderthal mtDNA reads, we designed 29 primer pairs (Supplementary Methods) flanking all 39 positions at which the Vi-80 Neanderthal mtDNA sequences differed by substitutions from the consensus bases seen among the 311 human mtDNA sequences. These primer pairs, which are designed to yield amplification products that vary in length between 50 and 98 base pairs (including primers), were used in a multiplex two-step PCR[39] from the same Neanderthal extract that had been used for large-scale 454 sequencing. Twenty five of the PCR products, containing 34 of the positions where the Neanderthal differs from humans, were successfully amplified and cloned, and then six or more clones of each product were sequenced. The consensus sequence seen among these clones revealed the same nucleotides seen by the 454 sequencing at 20 of the 34 positions and no additional differences. Of the 14 positions found to represent errors in the sequence reads, seven were C to T transitions, four were G to A, two were G to T and one was T to C. This pattern of change is typical for ancient DNA, where deamination of cytosine residues[31] and, to a lesser extent, modifications of guanosine residues[28] have been found to account for the majority of nucleotide misincorporations during PCR.

These results also show that the likelihood of observing errors in the sequencing reads is drastically different depending on whether one considers nucleotide positions where a base in the Neanderthal mtDNA sequence differs from both the human and chimpanzee sequences, or positions where the Neanderthal differ from the humans but is identical to the chimpanzee mtDNA sequences. Among the mtDNA sequences analysed, there are 14 positions where the Neanderthal carries a base identical to the chimpanzee, and 13 of those were confirmed by PCR. In contrast, among the remaining 20 positions, where the Neanderthal sequences differed from both humans and chimpanzees, only seven were

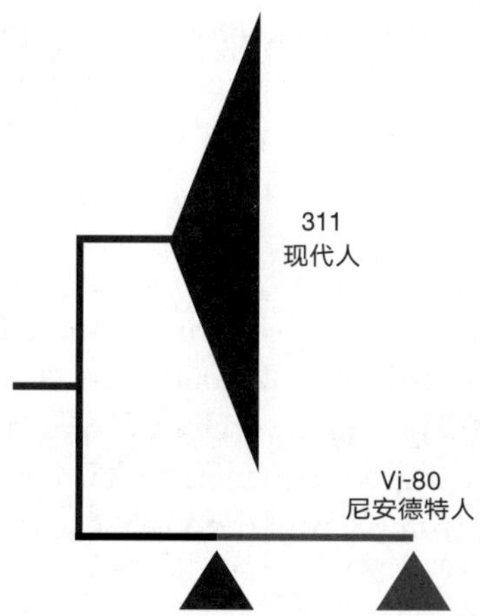

图 3. Vi-80 尼安德特人 mtDNA 序列和 311 个人类的 mtDNA 序列的关系树示意图。尼安德特人分支的长度根据未修正序列（红色三角形）及通过独立 PCR 修正过的序列（黑色三角形）分别给出。黑猩猩和倭黑猩猩的序列（图中未示出）用来作为邻接树的根部。几种替换模型（替代速率为均匀分布或伽马分布（γ = 0.5 ~ 1.1）的 Kimura 双参数模型、Tajima-Nei 模型以及 Tamura 三参数模型）对人类分支的自举支持值达到了 72% 到 83%。

为了分析在尼安德特人 mtDNA 序列读取中发生错误的程度，我们设计了 29 个引物对（见补充方法），从两侧囊括了 39 个 Vi-80 尼安德特人 mtDNA 序列由于替换而与 311 个人类 mtDNA 序列中所见的合意碱基不同的位置。这些被设计为获取 50 bp 到 98 bp（包括引物）长度不等的扩增产物的引物对被用于大范围 454 测序的尼安德特人相同提取物的多重两步 PCR[39]。最终成功扩增及克隆了 25 个包含尼安德特人不同于人类的 34 个位置的 PCR 产物。随后对每个产物挑选了六个或者更多克隆进行了测序。这些克隆的合意序列与 454 测序法测得的序列相比，这 34 个位置中的 20 个位置的核苷酸是相同的，而且不存在此外的差异。在这 14 个被发现是错误的序列读取中，其中七个是 C → T 转换，四个 G → A 转换，两个 G → T 转换及一个 T → C 转换。这种变化模式是古 DNA 的典型变化类型。这当中，胞嘧啶残基的脱氨基作用 [31] 以及程度较弱的鸟嘌呤核苷残基的修改 [28] 导致了 PCR 过程中发生的大多数核苷酸错参。

这些结果也表明在测序序列读取当中观察到错误发生的可能性的显著差异取决于一个人是否考虑尼安德特人 mtDNA 中某些核苷酸位置上一个碱基与人和黑猩猩序列都不同，或者在这些位置上尼安德特人与现代人不同但与黑猩猩 mtDNA 序列完全相同。在所分析的 mtDNA 序列中，在 14 个位置上，尼安德特人所带的碱基与黑猩猩相同，其中 13 个通过 PCR 得到了确认。相比之下，余下的 20 个位置上，尼安德特人与人类和黑猩猩都不相同，其中只有 7 处得到了确认。当只使用经过 PCR

confirmed. When only PCR-confirmed sequence data are used to estimate the mtDNA tree (Fig. 3), the Neanderthal branch has a length comparable to that of contemporary humans. This suggests that no large source of errors other than what is detected by the PCR analysis affects the sequences.

Using these PCR-confirmed substitutions and a divergence time between humans and chimpanzees of 4.7–8.4 million years[40-42], we estimate the divergence time for the mtDNA fragments determined here to be 461,000–825,000 years. This is in general agreement with previous estimates of Neanderthal–human mtDNA divergence of 317,000–741,000 years[6] based on mtDNA hypervariable region sequences and is compatible with our presumption that the mtDNA sequences determined from the Vi-80 extract are of Neanderthal origin.

Nuclear DNA Sequences

We next analysed the sequence reads whose closest matches are to the human or chimpanzee nuclear genomes and that are at least 30 base pairs long. Figure 4 shows where they map to the human karyotype (see Supplementary Methods). Overall, 0.04% of the autosomal genome sequence is covered by the Neanderthal reads—on average 3.61 bases per 10,000 bases. Both X and Y chromosomes are represented, with a lower coverage of 2.18 and 1.62 bases per 10,000, respectively, showing that the Vi-80 bone is derived from a male individual.

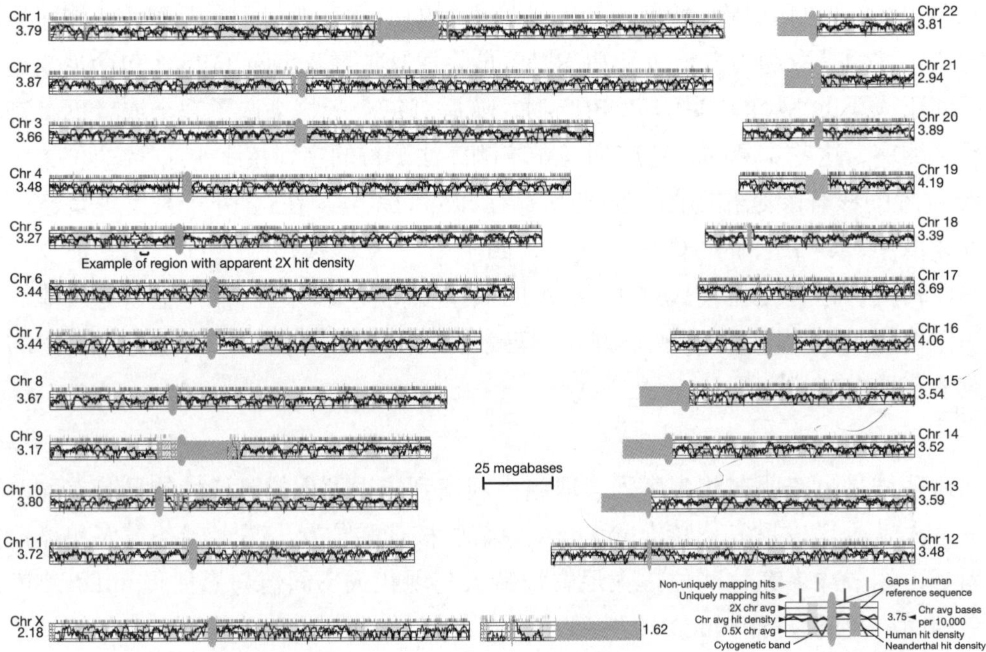

Fig. 4. Location on the human karyotype of Neanderthal DNA sequences. All sequences longer than 30 nucleotides whose best alignments were to the human genome are shown. The blue lines above each chromosome mark the position of all alignments that are unique in terms of bit-score within the human

确认过的序列数据来估算 mtDNA 树（图 3）时，尼安德特人分支的长度基本与现代人的分支长度相当。这表明除了 PCR 分析检测出的错误外，没有其他大的错误来源影响到序列的测定。

我们使用这些 PCR 确认过的碱基替换并采纳人类和黑猩猩间的分歧时间为距今 470 万年到 840 万年 [40-42]，由此估算出这里所确定的 mtDNA 片段的分歧时间是距今 461,000 年到 825,000 年。这与以前基于 mtDNA 高变区序列的尼安德特人–现代人 mtDNA 分歧时间为距今 317,000 年到 741,000 年 [6] 基本一致，也与我们的假定是相符的，即假定 Vi-80 提取物确定的 mtDNA 序列是源于尼安德特人的。

核 DNA 序列

接着我们分析了与现代人或黑猩猩核基因组最匹配的序列读取，这些序列至少 30 bp 长。图 4 展示了将它们映射到人类染色体核型上的情况（见补充方法）。总的来说，这些尼安德特人的序列读取涵盖了 0.04% 的常染色体基因组序列——平均每 10,000 碱基中有 3.61 个。其中也展示了 X 染色体和 Y 染色体的情况，涵盖范围比常染色体的稍低，分别是每 10,000 碱基有 2.18 个和 1.62 个，表明 Vi-80 骨骼标本是一个男性个体的。

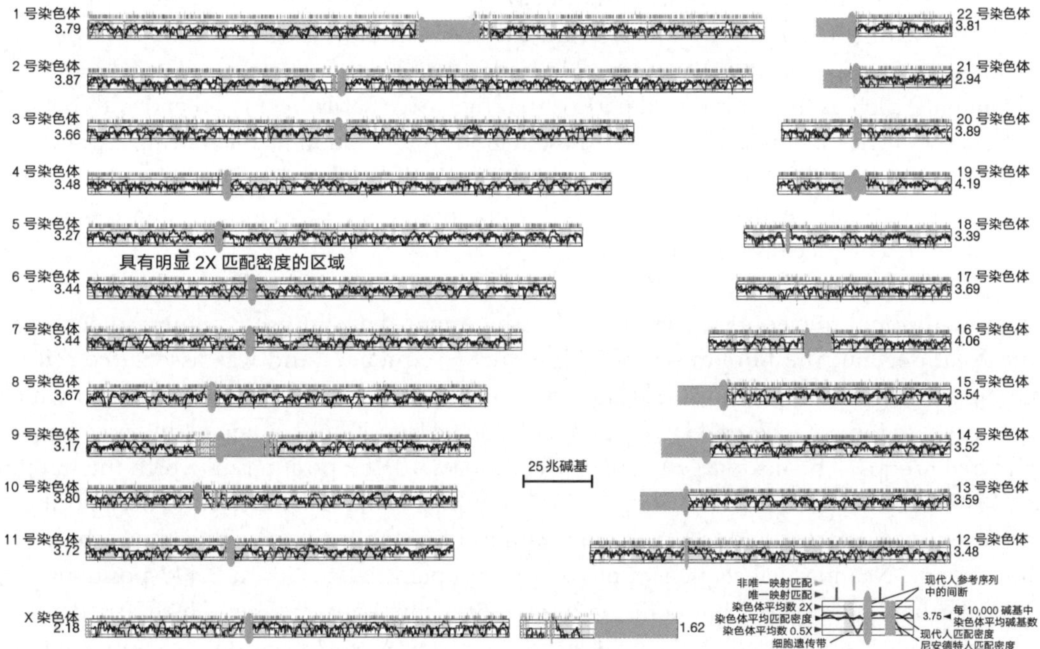

图 4. 尼安德特人 DNA 序列在人类染色体核型上的位置。所有长度大于 30 个核苷酸的序列在现代人基因组中的最佳对齐位置都在此示出。每条染色体上部的蓝线表示在现代人基因组中具有唯一位分值的最

genome. Orange lines are alignments that have more than one alignment of equal bit-score. To the left of each chromosome, the average number of Neanderthal bases per 10,000 is given. Lines (Neanderthal, blue; human, red) within each chromosome show the hit density, on a log-base 2 scale, within sliding windows of 3 megabases along each chromosome. The centre black lines indicate the average hit-density for the chromosomes. The purple lines above and below indicate hit densities of 2X and 1/2X the chromosome average, respectively. On chromosome 5, an example of a region of increased sequence density is highlighted. Sequence gaps in the human reference sequence are indicated by dark grey regions. Chromosomal banding pattern is indicated by light grey regions.

The data presented in Fig. 4 show that when the hit density for sequences that have a single best hit in the human genome is plotted along the chromosomes, several suggestive local deviations from the average hit density are seen, which may represent copy-number differences in the Neanderthal relative to the human reference genome. For comparison, we generated 454 sequence data from a DNA sample from a modern human. Interestingly, some of the deviations seen in the Neanderthal are present also in the modern human, whereas others are not. The latter group of sequences may indicate copy-number differences that are unique to the Neanderthal relative to the modern human genome sequence. Thus, when more Neanderthal sequence is generated in the future, it may be possible to determine copy number differences between the Neanderthal, the chimpanzee and the human genomes.

Patterns of Nucleotide Change on Lineages

We generated three-way alignments between all Neanderthal sequences that map uniquely within the human genome and the corresponding human and chimpanzee genome sequences (see Supplementary Methods). An important artefact of local sequence alignments, such as those produced here, is that they necessarily begin and end with regions of exact sequence identity. The size of these regions is a function of the scoring parameters for the alignment. In this case, five bases at both ends of the alignments, amounting to ~14% of all data, needed to be removed (Supplementary Fig. 1) to eliminate biases in estimates of sequence divergence.

Each autosomal nucleotide position in the alignment that did not contain a deletion in the Neanderthal, the human or the chimpanzee sequences and was associated with a chimpanzee genome position with quality score $\geqslant 30$ was classified according to which species share the same bases (Fig. 5). A total of 736,941 positions contained the same base in all three groups. The next largest category comprises 10,167 positions in which the human and Neanderthal base are identical, but the chimpanzee base is different. These positions are likely to have changed either on the hominid lineage before the divergence between human and Neanderthal sequences or on the chimpanzee lineage. At 3,447 positions, the Neanderthal base differs from both the human and chimpanzee bases, which are identical to each other. As suggested by the analysis of the mtDNA sequences, this category contains positions that have changed on the Neanderthal lineage, as well as a large proportion of errors that derive both from base damage that have accumulated in the ancient DNA and

佳对齐位置。橙色线表示具有相等位分值的超过一个的对齐位置。每条染色体的左侧给出了每 10,000 个碱基中尼安德特人碱基的平均数目。每条染色体内部的曲线(蓝色代表尼安德特人，红色代表现代人)表示沿每条染色体在三百万碱基滑动窗口内以 2 为底取对数的匹配密度。中间的黑线表示该染色体的平均匹配密度。上面和下面的紫色线分别表示相当于染色体平均数 2X 和 1/2X 的匹配密度。5 号染色体上，标出了一个序列密度增加的区域的例子。现代人参考序列中的序列间断用深灰色区域标出。染色体带型用浅灰色区域表示。

图 4 列出的数据表明当在现代人基因组中只有一个最佳匹配的序列的匹配密度沿着染色体被描绘的时候，能够看到几个平均匹配密度的提示性局部偏差，可能代表了尼安德特人相对于现代人参考基因组的拷贝数差异。为了进行比较，我们由一个现代人的 DNA 样本生成了 454 序列数据。有趣的是，某些在尼安德特人序列中看到的偏差同样也存在于现代人中，而其他却不存在。后一组序列可能暗示相对于现代人基因组序列而言，尼安德特人所特有的拷贝数差异。因此，如果将来可以生成更多的尼安德特人序列，那么就有可能确定尼安德特人、黑猩猩和现代人基因组间的拷贝数差异。

各谱系中的核苷酸变化模式

我们对所有在现代人基因组内有唯一映射的尼安德特人序列与相应的现代人和黑猩猩基因组序列进行了三重对齐(见补充方法)。就像这里的对齐一样，局部的序列对齐一个重要的人为因素就是它们必须以序列完全相同的区域开始和结束。这些区域的大小是对齐的评分参数的一个函数。这里，对齐两端各有五个碱基(补充图 1)，占所有数据的约 14%，需要予以剔除，以消除估算序列分化时的偏差。

尼安德特人、现代人或者黑猩猩序列中不包含缺失，且在黑猩猩基因组中位置质量得分 ≥ 30 那些对齐中，每个常染色体核苷酸位置都根据共有相同碱基的物种进行分类(图 5)。所有三组中含有相同碱基的总共有 736,941 个位置。第二大类中，现代人和尼安德特人具有相同碱基但与黑猩猩不同的位置有 10,167 个。这些位置可能在现代人和尼安德特人序列发生分歧之前或者是黑猩猩谱系中就发生了变化。尼安德特人在 3,447 个位置的碱基与人类和黑猩猩都不同，但后两者完全相同。正如 mtDNA 序列分析所表明的那样，这一类别包含了尼安德特人谱系中发生改变的位置，也包含了由古 DNA 中累积的碱基损伤以及测序错误导致的大部分错误。现代

from sequencing errors. At 434 positions, the human base differs from both the Neanderthal and chimpanzee bases, which are identical to each other. These positions are likely to have changed on the human lineage after the divergence from Neanderthal. Finally, a total of 51 positions contain different bases in all three groups.

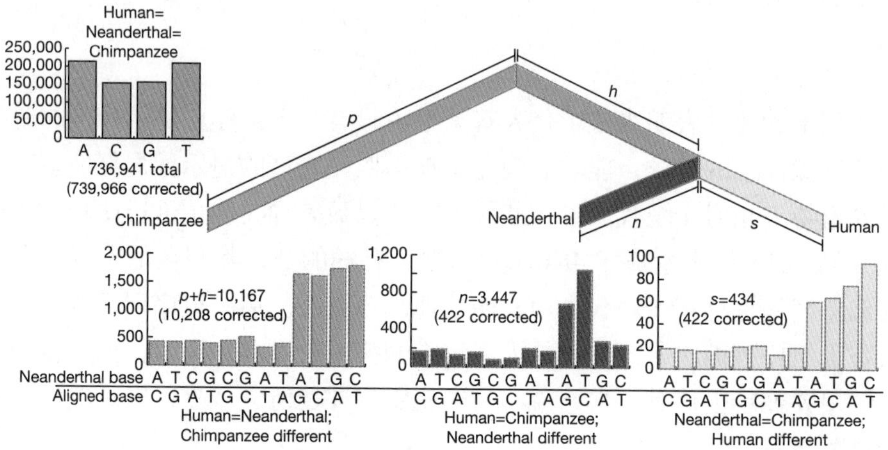

Fig. 5. Schematic tree illustrating the number of nucleotide changes inferred to have occurred on hominoid lineages. In blue is the distribution of all aligned positions that did not change on any lineage. In brown are the changes that occurred either on the chimpanzee lineage (p) or on the hominid lineage (h) before the human and Neanderthal lineages diverged. In red are the changes that are unique to the Neanderthal lineage (n), including all changes due to base-damage and base-calling errors. In yellow are changes unique to the human lineage. The distributions of types of changes in each category are also given. The numbers of changes in each category, corrected for base-calling errors in the Neanderthal sequence (see Supplementary Methods), are shown within parentheses.

Because the 454 sequencing technology allows the base in a base pair from which a sequence is derived to be determined, the relative frequencies of each of the 12 possible categories of base changes can be estimated for each evolutionary lineage. As seen in Fig. 5, the patterns of the chimpanzee-specific and human-specific changes are similar to each other in that the eight transversional changes are of approximately equal frequency and about fourfold less frequent than each of the four transitional changes, yielding a transition to transversion ratio of 2.04, typical of closely related mammalian genomes[43]. For the Neanderthal-specific changes the pattern is very different in that mismatches are dominated by C to T and G to A differences. Thus, the pattern of change seen among the Neanderthal-specific alignment mismatches is typical of the nucleotide substitution pattern observed in PCR of ancient DNA.

Consistent with this, modern human sequences determined by 454 sequencing show no excess amount of C to T or G to A differences (Supplementary Fig. 2), indicating that lesions in the ancient DNA rather than sequencing errors account for the majority of the errors in the Neanderthal sequences. Assuming that the evolutionary rate of DNA change was the same on the Neanderthal and human lineages, the majority of observed differences specific to the Neanderthal lineage are artefacts. All Neanderthal-specific changes were therefore

人在 434 个位置处的碱基与尼安德特人和黑猩猩都不同，而后两者是完全相同的。这些位置可能是人类与尼安德特人发生分歧之后才在人类谱系中发生了变化的。最后，有 51 个位置所有三组中的碱基均不同。

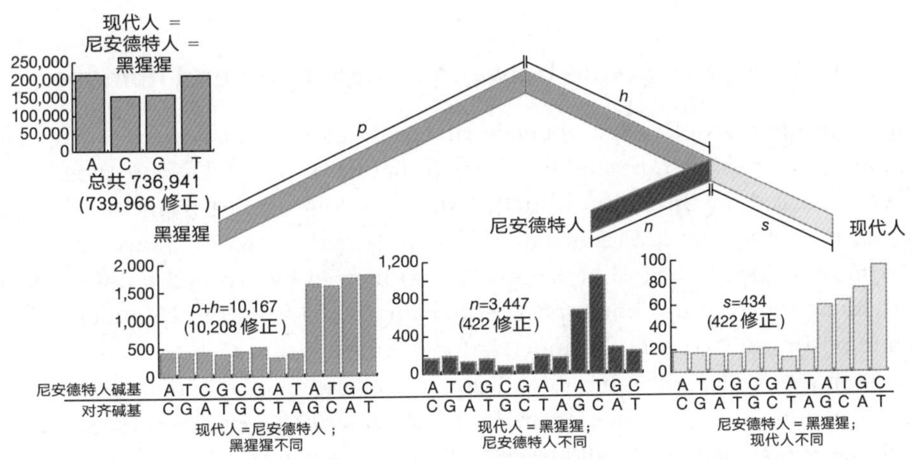

图 5. 人猿超科谱系上已发生的核苷酸变化数目的推测示意树。蓝色表示在任何谱系上都没有变化的对齐位置的分布情况。褐色表示在现代人和尼安德特人谱系分歧之前，黑猩猩谱系（p）或者人类谱系（h）上发生过的变化。红色表示尼安德特人谱系（n）独有的变化，包括碱基损伤和碱基识别错误等引起的所有变化。黄色是现代人谱系独有的变化。图中也给出了每个类别的变化类型的分布情况。圆括号中给出的是对尼安德特人序列碱基识别错误进行修正（见补充方法）后的每个类别的变化数目。

由于 454 测序技术能够确定序列来源的碱基对中的碱基，所以可以估算出每一演化谱系 12 个可能的碱基变化种类的每一种的相对发生频率。正如从图 5 中看到的，黑猩猩特有以及现代人特有的变化模式彼此很相似，因为八种颠换发生的频率基本相等，大概是四种转换中的每一种的发生频率的 1/4，产生了一个转换对颠换比率为 2.04 的结果，这对亲缘关系近的哺乳动物基因组而言是很具有代表性的 [43]。尼安德特人特有的变化主要是 C 到 T 和 G 到 A 错配的差异，所以与上述模式很不一样。因此，尼安德特人特有的对齐中的错配的变化模式，是在古 DNA 的 PCR 中典型的核苷酸替换。

与此一致，454 测序技术确定的现代人序列显示出没有过量的 C 到 T 或 G 到 A 的差异（补充图 2），这表明造成尼安德特人序列中出现的错误，大部分是由于古 DNA 中的损伤而非测序错误。假定 DNA 变化的演化速率在尼安德特人和现代人谱系中是相等的，那么大部分观察到的尼安德特谱系特有的差异就都是人为噪音了。因此在接下来的分析中我们忽略了所有尼安德特人特有的变化，尼安德特人序列只

disregarded in the subsequent analyses and the Neanderthal sequences were used solely to assign changes to the human or chimpanzee lineage where the human and chimpanzee genome sequences differ and the Neanderthal sequence carries either the human or the chimpanzee base.

Genomic Divergence between Neanderthals and Humans

Assuming that the rates of DNA sequence change along the chimpanzee lineage and the human lineage were similar, it can be estimated that 8.2% of the DNA sequence changes that have occurred on the human lineage since the divergence from the chimpanzee lineage occurred after the divergence of the Neanderthal lineage. However, although the Neanderthal-specific changes that are heavily influenced by errors are not used for this analysis, some errors in the single-pass sequencing reads from the Neanderthal extract will create positions where the Neanderthal is identical either to human or chimpanzee sequences, and thus affect the estimates of sequence change on the human and chimpanzee lineages. When the effects of such errors in the Neanderthal sequences are quantified and removed (see Supplementary Methods), ~7.9% of the sequence changes along the human lineage are estimated to have occurred after divergence from the Neanderthal. If the human–chimpanzee divergence time is set to 6,500,000 years (refs 40, 41, 44), this implies an average human–Neanderthal DNA sequence divergence time of ~516,000 years. A 95% confidence interval generated by bootstrap re-sampling of the alignment data gives a range of 465,000 to 569,000 years. Obviously, these divergence estimates are dependent on the human–chimpanzee divergence time, which is a much larger source of uncertainty.

We analysed the DNA sequences generated from a contemporary human using the same sequencing protocol as was used for the Neanderthal. Although ancient DNA is degraded and damaged, this comparison controls for many of the aspects of the analysis including sequencing and alignment methodology. In this case, ~7.1% of the divergence along the human lineage is assigned to the time subsequent to the divergence of the two human sequences. The average divergence time between alleles within humans is thus ~459,000 years with a 95% confidence interval between 419,000 and 498,000 years. As expected, this estimate of the average human diversity is less than the divergence seen between the human and the Neanderthal sequences, but constitutes a large fraction of it because much of the human sequence diversity is expected to predate the human–Neanderthal split[25]. Neanderthal genetic differences to humans must therefore be interpreted within the context of human diversity.

Ancestral Population Size

Humans differ from apes in that their effective population size is of the order of 10,000 while those of chimpanzees, gorillas and orangutans are two to four times larger[45-47]. Furthermore, the population size of the ancestor of humans and chimpanzees was found to

用来在现代人和黑猩猩基因组序列不同的位置并且在该位置尼安德特人序列或者携带现代人或者携带黑猩猩碱基的情况下，将变化分配到现代人或者黑猩猩谱系中去。

尼安德特人和现代人基因组之间的分歧

假定黑猩猩谱系和人类谱系中 DNA 序列变化速率是相似的，那么可以估算出从人类与黑猩猩谱系发生分歧后所发生的 DNA 序列变化中，有 8.2% 是在与尼安德特人谱系发生分歧之后发生的。但是，尽管本分析中并未使用那些受错误严重影响的尼安德特人特有的变化，但是有些来自尼安德特人提取物的单通道测序读取错误能告知尼安德特人与现代人或者黑猩猩序列相同的位置，因此会影响到对现代人和黑猩猩谱系中发生的序列变化的估算。当把尼安德特人序列中这种错误的影响量化并且去除后(见补充方法)，可以估算出，现代人谱系中的 7.9% 的序列变化是发生在与尼安德特人发生分歧之后。如果人类–黑猩猩分歧时间被设定在 6,500,000 年(参考文献 40、41、44)，那么就意味着现代人–尼安德特人 DNA 序列的平均分歧时间约为距今 516,000 年。通过对对齐数据进行自举重新抽样产生的 95% 置信区间给出的范围是距今 465,000 年至 569,000 年。很明显，这些分歧估值取决于人类–黑猩猩的分歧时间，这是一个更大的不确定性来源。

我们使用与尼安德特人同样的测序方案分析了一例现代人产生的 DNA 序列。尽管古 DNA 会被降解和损坏，但是这一比较在包括测序和对齐方法论在内的很多方面对分析进行了控制。这里，现代人谱系中 7.1% 的分歧被分配到两条人类序列发生分歧之后的时间里。于是，现代人中等位基因之间的平均分歧时间大约是距今 459,000 年，95% 置信区间的范围在距今 419,000 年到 498,000 年间。正如预期的那样，这一平均人类多样性的估值比现代人和尼安德特人序列间的分歧要晚，但是因为现代人序列许多多样性理应早于现代人–尼安德特人分歧时间，所以年代范围占据了其中很大一部分[25]。因此必须在现代人多样性这一背景下解读尼安德特人与现代人间的遗传差异。

祖先种群大小

人不同于猿，人的有效种群大小是 10,000，而黑猩猩、大猩猩和猩猩的有效种群大小是人类的两到四倍[45-47]。此外，研究发现人类和黑猩猩的祖先的种群大小与

be similar to those of apes, rather than to humans[42,48]. The Neanderthal sequence data now allow us to ask if the effective size of the population ancestral to humans and Neanderthals was large, as is the case for apes and the human–chimpanzee ancestor, or small, as for present-day humans.

We applied a method[42] that co-estimates the ancestral effective population size and the split time between Neanderthal and human populations (Fig. 6a; see Supplementary Methods). As seen in Fig. 6b, we recover a line describing combinations of population sizes and split times compatible with the data and lack power to be more precise (see Supplementary Methods and Results). Using this line we can estimate the ancestral population size, given estimates about the population split time from independent sources. If we use a split time of 400,000 years inferred from the fossil record (J. J. Hublin, personal communication), then our point estimate of the ancestral population size is ~3,000. Given uncertainty in both the sequence divergence time and the population split time, our estimate of the ancestral population size varies from 0 to 12,000.

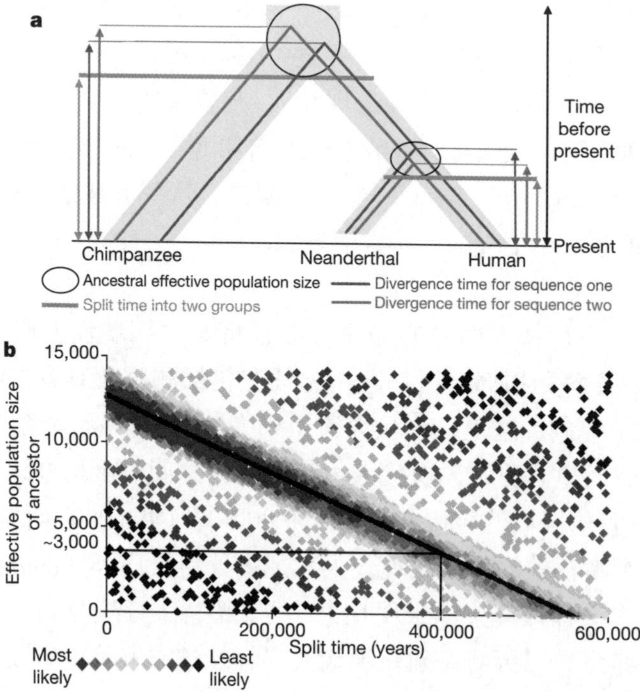

Fig. 6. Estimate of the effective population size of the ancestor of humans and Neanderthals. **a**, Schematic illustration of the model used to estimate ancestral effective population size. By split time, we mean the time, in the past, after which there was no more interbreeding between two groups. By divergence, we mean the time, in the past, at which two genetic regions separated and began to accumulate substitutions independently. Effective population size is the number of individuals needed under ideal conditions to produce the amount of observed genetic diversity within a population. **b**, The likelihood estimates of population split times and ancestral population sizes. The likelihoods are grouped by colour. The red–yellow points are statistically equivalent based on the likelihood ratio test approximation. The black line is the line of best fit to red–yellow points (see Supplementary Methods). This graph is scaled assuming a human–chimpanzee average sequence divergence time of 6,500,000 years.

猿类很相似，而非与人类相似[42,48]。现在，尼安德特人的序列数据允许我们提出如下问题，现代人和尼安德特人的祖先的有效种群大小是像猿类和人类–黑猩猩的祖先的种群一样大呢，还是像如今的现代人一样小？

我们运用了一种可以同时估算祖先有效种群大小和尼安德特人与现代人种群之间的分化时间的方法[42](图 6a；见补充方法)。如图 6b 所示，我们复原了一条同时描述种群大小和分化时间组合且与数据相吻合的拟合线，但是统计学功效欠缺无法更加精确(见补充方法和结果)。使用这条拟合线，给定独立来源的种群的分化时间估值，我们可以估算出祖先种群的大小。如果我们使用根据化石记录推断的距今 400,000 年(胡布林，个人交流)作为分化时间的话，那么我们对祖先种群大小的点估计大约是 3,000。鉴于序列分歧时间和种群分化时间二者的不确定性，我们对祖先种群大小的估计从 0 到 12,000 不等。

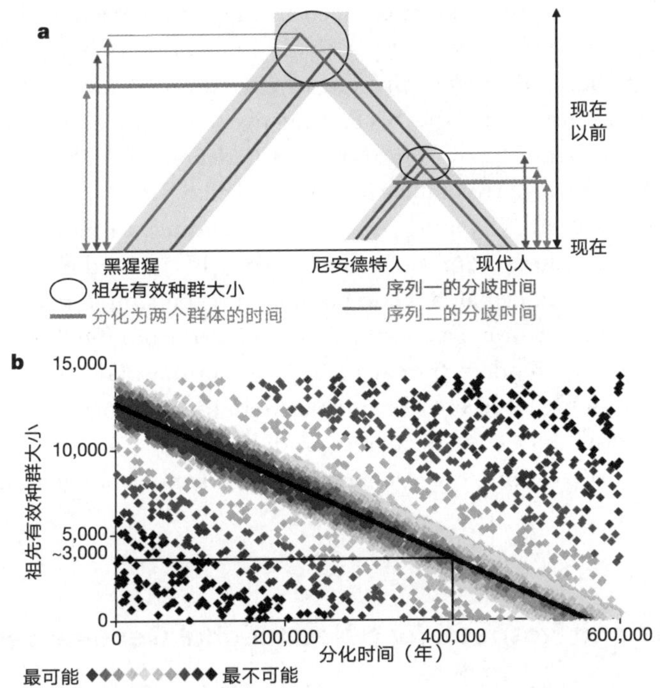

图 6. 现代人和尼安德特人的祖先的有效种群大小的估计。**a**，用来估算祖先有效种群大小的模型的示意图。我们所说的分化时间是指过去的某一时间，在该时间之后两个群体之间不再有杂交。我们所说的分歧是指过去某一时间，在该时间两个遗传区域发生分离并且开始独立积累替换。有效种群大小是指理想条件下能在一个种群内产生特定数量的可见遗传多样性所需的个体数目。**b**，种群分化时间和祖先种群大小的可能性估计。用不同颜色对可能性进行了分组。红黄点根据似然比检验逼近在统计意义上是等同的。黑线是红黄点的最佳拟合线(见补充方法)。该图比例设定的是在假定人类–黑猩猩的平均序列分歧时间为距今 6,500,000 年的前提下。

These results suggest that the population ancestral to present-day humans and Neanderthals was similar to present-day humans in having a small effective size and thus that the effective population size on the hominid lineage had already decreased before the split between humans and Neanderthals. Therefore, the small effective population size seen in present-day human samples may not be unique to modern humans, but was present also in the common ancestor of Neanderthals and modern humans. We speculate that a small effective size, perhaps associated with numerous expansions from small groups, was typical not only of modern humans but of many groups of the genus *Homo*. In fact, the origin of *Homo erectus* may have been associated with genetic or cultural adaptations that resulted in drastic population expansions as indicated by their appearance outside Africa around two million years ago.

Neanderthal Sequences and Human Polymorphisms

Another question that can be addressed with these data is how often the Neanderthal has the ancestral allele (that is, the same allele seen in the chimpanzee) versus the derived (or novel) allele at sites where humans carry a single nucleotide polymorphism (SNP). The latter case identifies SNPs that were present in the common ancestor of Neanderthals and present-day humans. Using the SNPs that overlap with our data from two large genome-wide data sets (HapMap[49], 786 SNPs and Perlegen[50], 318 SNPs), we find that the Neanderthal sample has the derived allele in ~30% of all SNPs. This number is presumably an overestimate since the SNPs analysed were ascertained to be of high frequency in present-day humans and hence are more likely to be old. Nevertheless, this high level of derived alleles in the Neanderthal is incompatible with the simple population split model estimated in the previous section, given split times inferred from the fossil record. This may suggest gene flow between modern humans and Neanderthals. Given that the Neanderthal X chromosome shows a higher level of divergence than the autosomes (R.E.G., unpublished observation), gene flow may have occurred predominantly from modern human males into Neanderthals. More extensive sequencing of the Neanderthal genome is necessary to address this possibility.

Rationale and Prospects for a Neanderthal Genome Sequence

We demonstrate here that DNA sequences can be generated from the Neanderthal nuclear genome by massive parallel sequencing on the 454 sequencing platform. It is thus feasible to determine large amounts of sequences from this extinct hominid. As a corollary, it is possible to envision the determination of a Neanderthal genome sequence. For several reasons, we believe that this would represent a valuable genomic resource.

First, a Neanderthal genome sequence would allow all nucleotide sequence differences as well as many copy-number differences between the human and chimpanzee genomes to be temporally resolved with respect to whether they occurred before the separation of

这些结果表明如今的现代人和尼安德特人的祖先种群与如今的现代人相似，有效种群大小都比较小，所以人类谱系的有效种群大小在现代人和尼安德特人发生分化之前已经有所减少。因此，我们看到的如今的现代人样本中有效种群小的现象，可能并不是现代人特有的，而是早在尼安德特人和现代人的共同祖先时就具有的。我们推测，小规模有效种群可能与小群体的扩张有关，不仅在现代人中很典型，在人属的许多其他类群中也很典型。事实上，直立人的起源可能与遗传和文化适应有关，这导致了明显的种群扩张，正如大约两百万年前他们就出现在非洲之外所表明的那样。

尼安德特人序列和人类的遗传多态性

基于这些数据可以探讨的另一个问题是尼安德特人在现代人携带的单核苷酸多态性（SNP）的位点处出现祖先型等位基因（即在黑猩猩中也见到的同样的等位基因）与衍生的（或新型）等位基因的频率。后一种情况可以识别出存在于尼安德特人和现代人的共同祖先中的 SNP。我们使用两大全基因组数据库中与我们的数据相重叠的那些 SNP（HapMap[49] 有 786 个 SNP 及 Perlegen[50] 有 318 个 SNP），发现尼安德特人样品具有的衍生等位基因约占到了所有 SNP 的 30%。这一数字很可能是过高估算，因为已经查明所有纳入分析的 SNP 在如今的现代人中都有很高的出现频率，故而其年代更有可能比较久远。然而，根据由化石记录推断的分化时间，尼安德特人中这种高水平衍生等位基因与前一部分估算的简单种群分化模型是相矛盾的。这可能表明现代人和尼安德特人之间存在基因流。考虑到尼安德特人的 X 染色体显示出比常染色体更高水平的分歧（理查德 · 格林，未发表观察结果），因此基因流可能主要是从现代人男性流向尼安德特人的。要确认这一可能性，须要对尼安德特人基因组进行更广泛的测序。

尼安德特人基因组序列的缘由和前景

我们在此证明可以通过在 454 测序平台上进行大规模平行测序从尼安德特人核基因组生成 DNA 序列。因此从这种灭绝的人类确定大量序列是可行的。作为必然的结果，确定尼安德特人基因组序列是可能的。由于如下原因，我们相信这将是一种非常有价值的基因组资源。

首先，尼安德特人基因组序列使得人类与黑猩猩基因组之间所有的核苷酸序列差异以及许多拷贝数差异在时间上得到解决，即差异究竟是发生在现代人与尼安德

humans from Neanderthals, or whether they occurred after or at the time of separation. The latter class of changes is of interest, because some of them will be associated with the emergence of modern humans. A Neanderthal genome sequence would therefore allow the research community to determine whether DNA sequence differences between humans and chimpanzees that are found to be functionally important represent recent changes on the human lineage. No data other than a Neanderthal genome sequence can provide this information.

Second, the fact that Neanderthals carry the derived allele for a substantial fraction of human SNPs suggests a method of identifying genomic regions that have experienced a selective sweep subsequent to the separation of human and Neanderthal populations. Such selective sweeps in the human genome will make the variation in these regions younger than the separation of humans and Neanderthals. As we show above, in regions not affected by sweeps a substantial proportion of polymorphic sites in humans will carry derived alleles in the Neanderthal genome sequence, whereas no sites will do so in regions affected by sweeps. This represents an approach to identifying selective sweeps in humans that is not possible from other data.

Third, once large amounts of Neanderthal genome sequence is generated, it will become possible to estimate the misincorporation probabilities for each class of nucleotide differences between the Neanderthal and chimpanzee genomes with high accuracy by analysing regions covered by many reads such as mtDNA, repeated genome regions of high sequence identity, as well as single-copy regions covered by multiple reads. Once this is done, the confidence that any particular nucleotide position where the Neanderthal differs from human as well as chimpanzee is correct can be reliably estimated. In combination with future knowledge about the function of genes and biological systems, comprehensive information from the Neanderthal genome will then allow aspects of Neanderthal biology to be deciphered that are unavailable by any other means.

Are fossil and technical resources today sufficient to imagine the determination of a Neanderthal genome sequence? The results presented here are derived from approximately one fifteenth of an extract prepared from ~100 mg of bone. To achieve one-fold coverage of the Neanderthal genome (3 gigabases) without any further improvement in technology, about twenty grams of bone and 6,000 runs on the current version of the 454 sequencing platform would be necessary. Although this is at present a daunting task, technical improvements in the procedures described here that would make the retrieval of DNA sequences of the order of ten times more efficient can easily be envisioned (our unpublished results). In view of that prospect, we have recently initiated a project that aims at achieving an initial draft version of the Neanderthal genome within two years.

(**444**, 330-336; 2006)

特人分离之前，还是发生在分离之后或分离之时。后一类变化更加应该被关注，因为其中有些与现代人的出现有关。因此尼安德特人的基因组序列将令学术界能够确定人和黑猩猩间功能上重要的 DNA 序列差异是否代表了人类谱系最近发生的变化。除了尼安德特人的基因组序列外，没有其他数据能够提供这一信息。

其次，尼安德特人携带着大量人类 SNP 的衍生等位基因这一事实指明了一种识别基因组中在现代人与尼安德特人种群分离之后经历了选择性清除的区域的方法。人类基因组中的这些选择性清除会使这些区域内的变异比现代人和尼安德特人的分离年轻。正如我们上面所陈述的，在未受选择性清除影响的区域中，对于人类的大部分多态位点来说，尼安德特人基因组序列中会携带存在的衍生等位基因，然而在受选择性清除影响的区域中，没有位点会出现这种情况。这代表一种识别人类中的选择性清除的方法，其他数据是不可能做到这点的。

再者，一旦生成了大量尼安德特人基因组序列，通过分析像 mtDNA 那样由多读取覆盖的区域、序列高度等同的重复基因组区域以及多读取覆盖的单拷贝区域，以高精度估算尼安德特人和黑猩猩基因组间每类核苷酸差异的错配概率将成为可能。一旦完成了这一步，就能够可靠地估计出尼安德特人不同于现代人及黑猩猩的任何特定核苷酸位置的正确性的可信度。加上将来对基因功能和生物系统的进一步了解，尼安德特人基因组提供的全面信息就可以对尼安德特人生物学的方方面面进行解读，这是其他方式无法做到的。

现在的化石和技术资源足够让我们奢想确定尼安德特人基因组序列吗？这里介绍的结果源自由约 100 mg 骨骼样品产生的提取物的大概 1/15。在无须任何技术改进的基础上，为了得到尼安德特人整个基因组（3 千兆碱基）的信息，在现有版本的 454 测序平台上，将需要约 20 g 骨骼，进行 6,000 次运行。尽管目前这是一项令人畏惧的任务，但是对这里描述的步骤进行技术改良之后，很轻易就能够更有效地获取十倍的 DNA 序列（我们尚未发表的结果）。鉴于这一前景，我们最近启动了一个旨在两年内完成尼安德特人基因组草图的项目。

（刘皓芳 翻译；张颖奇 审稿）

Richard E. Green[1], Johannes Krause[1], Susan E. Ptak[1], Adrian W. Briggs[1], Michael T. Ronan[2], Jan F. Simons[2], Lei Du[2], Michael Egholm[2], Jonathan M. Rothberg[2], Maja Paunovic[3‡] & Svante Pääbo[1]

[1] Max-Planck Institute for Evolutionary Anthropology, Deutscher Platz 6, D-04103 Leipzig, Germany

[2] 454 Life Sciences, 20 Commercial Street, Branford, Connecticut 06405, USA

[3] Institute of Quaternary Paleontology and Geology, Croatian Academy of Sciences and Arts, A. Kovacica 5/II, HR-10 000 Zagreb, Croatia

‡ Deceased

Received 14 July; accepted 11 October 2006.

References:

1. Bischoff, J. L. *et al.* The Sima de los Huesos hominids date to beyond U/Th equilibrium (> 350 kyr) and perhaps to 400–500 kyr: New radiometric dates. *J. Archaeol. Sci.* **30**, 275-280 (2003).
2. Hublin, J-J. (ed.) *Climatic Changes, Paleogeography, and the Evolution of the Neandertals* (Plenum Press, New York, 1998).
3. Franciscus, R. G. (ed.) *Neanderthals* (Oxford Univ. Press, Oxford, 2002).
4. Hublin, J. J., Spoor, F., Braun, M., Zonneveld, F. & Condemi, S. A late Neanderthal associated with Upper Palaeolithic artefacts. *Nature* **381**, 224-226 (1996).
5. Krings, M. *et al.* Neandertal DNA sequences and the origin of modern humans. *Cell* **90**, 19-30 (1997).
6. Krings, M., Geisert, H., Schmitz, R. W., Krainitzki, H. & Pääbo, S. DNA sequence of the mitochondrial hypervariable region II from the Neandertal type specimen. *Proc. Natl Acad. Sci. USA* **96**, 5581-5585 (1999).
7. Schmitz, R. W. *et al.* The Neandertal type site revisited: Interdisciplinary investigations of skeletal remains from the Neander Valley, Germany. *Proc. Natl Acad. Sci. USA* **99**, 13342-13347 (2002).
8. Ovchinnikov, I. V. *et al.* Molecular analysis of Neanderthal DNA from the northern Caucasus. *Nature* **404**, 490-493 (2000).
9. Krings, M. *et al.* A view of Neandertal genetic diversity. *Nature Genet.* **26**, 144-146 (2000).
10. Serre, D. *et al.* No evidence of Neandertal mtDNA contribution to early modern humans. *PLoS Biol.* **2**, 313-317 (2004).
11. Orlando, L. *et al.* Revisiting Neandertal diversity with a 100,000 year old mtDNA sequence. *Curr. Biol.* **16**, R400-R402 (2006).
12. Caramelli, D. *et al.* A highly divergent mtDNA sequence in a Neandertal individual from Italy. *Curr. Biol.* **16**, R630-R632 (2006).
13. Lalueza-Fox, C. *et al.* Neandertal evolutionary genetics: mitochondrial DNA data from the Iberian peninsula. *Mol. Biol. Evol.* **22**, 1077-1081 (2005).
14. Currat, M. & Excoffier, L. Modern humans did not admix with Neanderthals during their range expansion into Europe. *PLoS Biol.* **2**, e421 (2004).
15. Stringer, C. Modern human origins: progress and prospects. *Phil. Trans. R. Soc. Lond. B* **357**, 563-579 (2002).
16. Takahata, N., Lee, S. H. & Satta, Y. Testing multiregionality of modern human origins. *Mol. Biol. Evol.* **18**, 172-183 (2001).
17. Chimpanzee Sequencing and Analysis Consortium. Initial sequence of the chimpanzee genome and comparison with the human genome. *Nature* **437**, 69-87 (2005).
18. Pääbo, S. *et al.* Genetic analyses from ancient DNA. *Annu. Rev. Genet.* **38**, 645-679 (2004).
19. Greenwood, A. D., Capelli, C., Possnert, G. & Pääbo, S. Nuclear DNA sequences from late Pleistocene megafauna. *Mol. Biol. Evol.* **16**, 1466-1473 (1999).
20. Noonan, J. P. *et al.* Genomic sequencing of Pleistocene cave bears. *Science* **309**, 597-599 (2005).
21. Rompler, H. *et al.* Nuclear gene indicates coat-color polymorphism in mammoths. *Science* **313**, 62 (2006).
22. Poinar, H. N. *et al.* Metagenomics to paleogenomics: large-scale sequencing of mammoth DNA. *Science* **311**, 392-394 (2006).
23. Hofreiter, M., Serre, D., Poinar, H. N., Kuch, M. & Pääbo, S. Ancient DNA. *Nature Rev. Genet.* **2**, 353-359 (2001).
24. Malmstrom, H., Stora, J., Dalen, L., Holmlund, G. & Gotherstrom, A. Extensive human DNA contamination in extracts from ancient dog bones and teeth. *Mol. Biol. Evol.* **22**, 2040-2047 (2005).
25. Pääbo, S. Human evolution. *Trends Cell Biol.* **9**, M13-M16 (1999).
26. Poinar, H. N., Höss, M., Bada, J. L. & Pääbo, S. Amino acid racemization and the preservation of ancient DNA. *Science* **272**, 864-866 (1996).
27. Margulies, M. *et al.* Genome sequencing in microfabricated high-density picolitre reactors. *Nature* **437**, 376-380 (2005).
28. Stiller, M. *et al.* Patterns of nucleotide misincorporations during enzymatic amplification and direct large-scale sequencing of ancient DNA. *Proc. Natl Acad. Sci. USA* **103**, 13578-13584 (2006).
29. Pääbo, S. Ancient DNA: extraction, characterization, molecular cloning, and enzymatic amplification. *Proc. Natl Acad. Sci. USA* **86**, 1939-1943 (1989).
30. Höss, M., Dilling, A., Currant, A. & Pääbo, S. Molecular phylogeny of the extinct ground sloth *Mylodon darwinii*. *Proc. Natl Acad. Sci. USA* **93**, 181-185 (1996).
31. Hofreiter, M., Jaenicke, V., Serre, D., Haeseler Av, A. & Pääbo, S. DNA sequences from multiple amplifications reveal artifacts induced by cytosine deamination in ancient DNA. *Nucleic Acids Res.* **29**, 4793-4799 (2001).
32. International Human Genome Sequencing Consortium. Initial sequencing and analysis of the human genome. *Nature* **409**, 860-921 (2001).
33. Mouse Genome Sequencing Consortium. Initial sequencing and comparative analysis of the mouse genome. *Nature* **420**, 520-562 (2002).
34. Benson, D. A., Karsch-Mizrachi, I., Lipman, D. J., Ostell, J. & Wheeler, D. L. GenBank. *Nucleic Acids Res.* **34**, D16-D20 (2006).
35. Altschul, S. F. *et al.* Gapped BLAST and PSI-BLAST: a new generation of protein database search programs. *Nucleic Acids Res.* **25**, 3389-3402 (1997).
36. Beja, O. *et al.* Construction and analysis of bacterial artificial chromosome libraries from a marine microbial assemblage. *Environ. Microbiol.* **2**, 516-529 (2000).

37. Venter, J. C. *et al.* Environmental genome shotgun sequencing of the Sargasso Sea. *Science* **304**, 66-74 (2004).

38. Ingman, M. & Gyllensten, U. mtDB: Human Mitochondrial Genome Database, a resource for population genetics and medical sciences. *Nucleic Acids Res.* **34**, D749-D751 (2006).

39. Krause, J. *et al.* Multiplex amplification of the mammoth mitochondrial genome and the evolution of Elephantidae. *Nature* **439**, 724-727 (2006).

40. Kumar, S., Filipski, A., Swarna, V., Walker, A. & Blair Hedges, S. Placing confidence limits on the molecular age of the human-chimpanzee divergence. *Proc. Natl Acad. Sci. USA* **102**, 18842-18847 (2005).

41. Patterson, N., Richter, D. J., Gnerre, S., Lander, E. S. & Reich, D. Genetic evidence for complex speciation of humans and chimpanzees. *Nature* **441**, 1103-1108 (2006).

42. Wall, J. D. Estimating ancestral population sizes and divergence times. *Genetics* **163**, 395-404 (2003).

43. Yang, Z. & Yoder, A. D. Estimation of the transition/transversion rate bias and species sampling. *J. Mol. Evol.* **48**, 274-283 (1999).

44. Innan, H. & Watanabe, H. The effect of gene flow on the coalescent time in the human-chimpanzee ancestral population. *Mol. Biol. Evol.* **23**, 1040-1047 (2006).

45. Kaessmann, H., Wiebe, V., Weiss, G. & Pääbo, S. Great ape DNA sequences reveal a reduced diversity and an expansion in humans. *Nature Genet.* **27**, 155-156 (2001).

46. Yu, N., Jensen-Seaman, M. I., Chemnick, L., Ryder, O. & Li, W-H. Nucleotide diversity in gorillas. *Genetics* **166**, 1375-1383 (2004).

47. Fischer, A., Pollack, J., Thalmann, O., Nickel, B. & Pääbo, S. Demographic history and genetic differentiation in apes. *Curr. Biol.* **16**, 1133-1138 (2006).

48. Rannala, B. & Yang, Z. Bayes estimation of species divergence times and ancestral population sizes using DNA sequences from multiple loci. *Genetics* **164**, 1645-1656 (2003).

49. The International HapMap Consortium. A haplotype map of the human genome. *Nature* **437**, 1299-1320 (2005).

50. Hinds, D. A. *et al.* Whole genome patterns of common DNA variation in three human populations. *Science* **307**, 1072-1079 (2005).

Supplementary Information is linked to the online version of the paper at www.nature.com/nature.

Acknowledgements. We are indebted to G. Coop, W. Enard, I. Hellmann, A. Fischer, P. Johnson, S. Kudaravalli, M. Lachmann, T. Maricic, J. Pritchard, J. Noonan, D. Reich, E. Rubin, M. Slatkin, L. Vigilant and T. Weaver for discussions. We thank A. P. Derevianko, C. Lalueza-Fox, A. Rosas and B. Vandermeersch for fossil samples. We also thank the Croatian Academy of Sciences and Arts for support and the Innovation Fund of Max Planck Society for financial support. 454 Life Sciences thanks NHGRI for continued support for the development of this platform, as well as all of its employees who developed the sequencing system. R.E.G. is supported by an NSF postdoctoral fellowship in Biological Informatics.

Author Contributions. M.P. provided Neanderthal samples and palaeontological information; J.M.R. and S.P. conceived of and initiated the 454 Neanderthal sequencing approach; M.T.R. developed the library preparation method, and generated and processed the sequencing data; J.F.S. planned and coordinated library preparation and sequencing activities; L.D. processed and transferred data between 454 Life Sciences and the MPI; M.E. supervised, planned and coordinated research between MPI and 454 Life Sciences; J.K. and A.W.B. extracted ancient DNA and performed analyses in the "Identification of a Neanderthal fossil for DNA sequencing" section; J.K. and R.E.G. performed analyses in the "Neanderthal mtDNA sequences" section; R.E.G. performed the analyses in the sections "Direct large-scale DNA sequencing" to "Genomic divergence between Neanderthals and humans"; S.E.P. performed analyses in the sections "Ancestral population size" and "Neanderthal sequences and human polymorphisms"; S.P. conceived of the ideas presented in the section "Rationale and prospects for a Neanderthal genome sequence", and initiated, planned and coordinated the study; R.E.G., S.E.P., J.K. and S.P. wrote the paper.

Author Information. Neanderthal fossil extract sequences were deposited at EBI with accession numbers CAAN01000001-CAAN01369630. Reprints and permissions information is available at www.nature.com/reprints. The authors declare competing financial interests: details accompany the paper on www.nature.com/nature. Correspondence and requests for materials should be addressed to R.E.G. (green@eva.mpg.de).

Identification and Analysis of Functional Elements in 1% of the Human Genome by the ENCODE Pilot Project

The ENCODE Project Consortium*

Editor's Note

With the human genome sequence completed, the logical next step was to work out how cells interpret the genetic instructions. This paper reveals the first findings of the Encyclopedia of DNA Elements (ENCODE) project, an initiative aiming to identify the functional elements of the human genome. The pilot program, which scrutinizes just 1% of the genome, offers insights into the nature and evolution of DNA sequences. Around half the functional elements appear able to change sequence more freely than expected, challenging the idea that biologically relevant DNA resists change. The study also finds that most DNA is transcribed into RNA, undermining the common picture of our genome as a rather small number of discrete genes amidst a mass of inactive "junk DNA".

We report the generation and analysis of functional data from multiple, diverse experiments performed on a targeted 1% of the human genome as part of the pilot phase of the ENCODE Project. These data have been further integrated and augmented by a number of evolutionary and computational analyses. Together, our results advance the collective knowledge about human genome function in several major areas. First, our studies provide convincing evidence that the genome is pervasively transcribed, such that the majority of its bases can be found in primary transcripts, including non-protein-coding transcripts, and those that extensively overlap one another. Second, systematic examination of transcriptional regulation has yielded new understanding about transcription start sites, including their relationship to specific regulatory sequences and features of chromatin accessibility and histone modification. Third, a more sophisticated view of chromatin structure has emerged, including its inter-relationship with DNA replication and transcriptional regulation. Finally, integration of these new sources of information, in particular with respect to mammalian evolution based on inter- and intra-species sequence comparisons, has yielded new mechanistic and evolutionary insights concerning the functional landscape of the human genome. Together, these studies are defining a path for pursuit of a more comprehensive characterization of human genome function.

THE human genome is an elegant but cryptic store of information. The roughly three billion bases encode, either directly or indirectly, the instructions for synthesizing

---

* The full list of authors and affiliations has been removed. The original text is available in the *Nature* online archive.

通过 ENCODE 先导计划鉴定与分析人类基因组 1% 区域内的功能元件

ENCODE 计划协作组*

编者按

人类基因组测序完成后的下一步必然是阐释细胞如何执行这个遗传指南。DNA 元件百科全书 (ENCODE) 计划是一个旨在鉴定人类基因组功能元件的倡议，本文介绍了该计划的早期发现。先导项目仅仔细研究了 1% 的基因组，使人们对 DNA 序列的本质和进化有了初步了解。大约半数的功能元件看起来比预期能够更自由地改变序列，这对生物学相关的 DNA 抵抗改变的观点提出了挑战。大多数的 DNA 被转录成 RNA，这个发现颠覆了我们把基因组看作是相当少量且分散的基因位于大量不活跃的“垃圾 DNA”之间的通俗认识。

作为 ENCODE 计划先导阶段的一部分，针对人类基因组 1% 的目标区域进行了多种不同的实验，得到了功能数据，我们对这些功能数据的产生和分析进行了报道。这些数据已经通过大量的进化和计算分析被进一步整合和扩充。总的来说，我们的结果在几个主要方面增加了对人类基因组功能的认识。首先，我们的研究提供了令人信服的证据，证明了基因组是广泛转录的，以至于它的大部分碱基都可以在原始转录本中找到，这包括非蛋白质编码的转录本和广泛的互相重合的转录本。第二，转录调控的系统研究对转录起始位点产生了新的认识，这包括其与特定调控序列的关系、染色可接近性和组蛋白修饰的特征。第三，一个关于染色质结构的更加复杂精致的图景已经浮现，这包括它与 DNA 复制和转录调控的相互关系。最后，这些新的信息的整合，特别是基于物种间和物种内序列比较对哺乳动物进化的研究，已经对人类基因组的功能图景产生了机制和进化方面新的认识。总而言之，这些研究为进一步阐明人类基因组功能指明了道路。

人类基因组以简洁而隐晦的方式储存信息。大约 30 亿个碱基通过直接或间接的方式编码合成分子的指令，指导合成几乎所有构成人类细胞、组织和器官的分子。

* 作者和其他附加信息已经移除，原文可以从《自然》在线数据库中获得。

nearly all the molecules that form each human cell, tissue and organ. Sequencing the human genome[1-3] provided highly accurate DNA sequences for each of the 24 chromosomes. However, at present, we have an incomplete understanding of the protein-coding portions of the genome, and markedly less understanding of both non-protein-coding transcripts and genomic elements that temporally and spatially regulate gene expression. To understand the human genome, and by extension the biological processes it orchestrates and the ways in which its defects can give rise to disease, we need a more transparent view of the information it encodes.

The molecular mechanisms by which genomic information directs the synthesis of different biomolecules has been the focus of much of molecular biology research over the last three decades. Previous studies have typically concentrated on individual genes, with the resulting general principles then providing insights into transcription, chromatin remodelling, messenger RNA splicing, DNA replication and numerous other genomic processes. Although many such principles seem valid as additional genes are investigated, they generally have not provided genome-wide insights about biological function.

The first genome-wide analyses that shed light on human genome function made use of observing the actions of evolution. The ever-growing set of vertebrate genome sequences[4-8] is providing increasing power to reveal the genomic regions that have been most and least acted on by the forces of evolution. However, although these studies convincingly indicate the presence of numerous genomic regions under strong evolutionary constraint, they have less power in identifying the precise bases that are constrained and provide little, if any, insight into why those bases are biologically important. Furthermore, although we have good models for how protein-coding regions evolve, our present understanding about the evolution of other functional genomic regions is poorly developed. Experimental studies that augment what we learn from evolutionary analyses are key for solidifying our insights regarding genome function.

The Encyclopedia of DNA Elements (ENCODE) Project[9] aims to provide a more biologically informative representation of the human genome by using high-throughput methods to identify and catalogue the functional elements encoded. In its pilot phase, 35 groups provided more than 200 experimental and computational data sets that examined in unprecedented detail a targeted 29,998 kilobases (kb) of the human genome. These roughly 30 Mb—equivalent to ~1% of the human genome—are sufficiently large and diverse to allow for rigorous pilot testing of multiple experimental and computational methods. These 30 Mb are divided among 44 genomic regions; approximately 15 Mb reside in 14 regions for which there is already substantial biological knowledge, whereas the other 15 Mb reside in 30 regions chosen by a stratified random-sampling method (see http://www.genome.gov/10506161). The highlights of our findings to date include:

- The human genome is pervasively transcribed, such that the majority of its bases are associated with at least one primary transcript and many transcripts link distal regions to established protein-coding loci.

对人类基因组的测序[1-3]为 24 条染色体提供了高精度的 DNA 序列。但是目前，我们还不能完整理解基因组的蛋白质编码部分，而对于非蛋白质编码的转录本和在时空水平上调控基因表达的基因组元件，更是知之甚少。为了了解人类基因组，以及它调控的生物学过程和它的缺陷导致疾病的机制，我们需要对它所编码的信息有一个更清楚的认知。

在过去的 30 多年里，基因组信息指导合成不同生物分子的分子机制一直是很多分子生物学研究的焦点。之前的研究通常聚焦于单个基因，获得一般规律，进而促进对转录、染色质重塑、信使 RNA 剪接、DNA 复制和众多其他基因组学过程的认识。尽管许多这样的规律也适用于其他研究的基因，但是它们通常还未提供在全基因组范围内认识生物学功能的信息。

第一个在基因组层面阐明人类基因组功能的研究是关于进化的行为。随着越来越多的脊椎动物基因组被测序[4-8]，人们越来越能够发现被进化的力量影响最多和最少的基因组区域。不过，尽管这些研究有力地表明，存在大量的基因组在进化上高度保守，但是这些研究在鉴定精确的保守碱基上还不够有力，对于为什么那些碱基在生物学上是重要的也几乎不能提供有用的信息。并且，尽管我们对于蛋白质编码区域如何进化有很好的模型，但是我们目前对其他有功能的基因组区域的了解还很欠缺。为了巩固我们对基因组功能的理解，通过实验研究扩充我们进化分析所得是关键。

DNA 元件百科全书(ENCODE)计划[9]旨在利用高通量方法对人类基因组所编码的功能元件进行鉴定和分类，进而提供人类基因组更加具有生物学信息的表征。在该计划的先导阶段，35 个课题组贡献了超过 200 个实验和计算数据集，以前所未有的详细度研究了人类基因组特定的 29,998 千碱基(kb)区域。这大约 30 Mb——相当于人类基因组的 ~ 1%——已经足够大和足够多样，可以严格地用来先期测试多种实验和计算方法。这 30 Mb 分布在 44 个基因组区域；大约 15 Mb 位于 14 个已经有大量的生物学知识的区域，而另外 15 Mb 则位于 30 个通过分层随机抽样方法选择的区域(见 http://www.genome.gov/10506161)。目前我们研究发现的亮点包括：

- 人类基因组是广泛转录的，以至于它的大部分碱基至少和一个原始转录本相关联，并且许多转录本将远端区域与已知的蛋白质编码基因座相关联。

- Many novel non-protein-coding transcripts have been identified, with many of these overlapping protein-coding loci and others located in regions of the genome previously thought to be transcriptionally silent.

- Numerous previously unrecognized transcription start sites have been identified, many of which show chromatin structure and sequence-specific protein-binding properties similar to well-understood promoters.

- Regulatory sequences that surround transcription start sites are symmetrically distributed, with no bias towards upstream regions.

- Chromatin accessibility and histone modification patterns are highly predictive of both the presence and activity of transcription start sites.

- Distal DNaseI hypersensitive sites have characteristic histone modification patterns that reliably distinguish them from promoters; some of these distal sites show marks consistent with insulator function.

- DNA replication timing is correlated with chromatin structure.

- A total of 5% of the bases in the genome can be confidently identified as being under evolutionary constraint in mammals; for approximately 60% of these constrained bases, there is evidence of function on the basis of the results of the experimental assays performed to date.

- Although there is general overlap between genomic regions identified as functional by experimental assays and those under evolutionary constraint, not all bases within these experimentally defined regions show evidence of constraint.

- Different functional elements vary greatly in their sequence variability across the human population and in their likelihood of residing within a structurally variable region of the genome.

- Surprisingly, many functional elements are seemingly unconstrained across mammalian evolution. This suggests the possibility of a large pool of neutral elements that are biochemically active but provide no specific benefit to the organism. This pool may serve as a "warehouse" for natural selection, potentially acting as the source of lineage-specific elements and functionally conserved but non-orthologous elements between species.

Below, we first provide an overview of the experimental techniques used for our studies, after which we describe the insights gained from analysing and integrating the generated data sets. We conclude with a perspective of what we have learned to date about this 1% of the human genome and what we believe the prospects are for a broader and deeper

- 许多新的非蛋白质编码转录本被鉴定，其中很多与蛋白质编码基因座重合，其他的则位于以前被认为是不转录的基因组区域。

- 大量的以前未被识别的转录起始位点被鉴定，其中很多具有与已知的启动子类似的染色质结构和序列特异的蛋白质结合特性。

- 转录起始位点附近的调控序列是对称分布的，并没有偏向上游区域。

- 染色质可接近性和组蛋白修饰特征不仅可以准确预测转录起始位点的存在，而且可以预测转录起始位点的活性。

- 远端的 DNA 酶 I 超敏感位点含有与启动子截然不同的组蛋白修饰特征；其中一些远端位点含有的标志物与绝缘子功能一致。

- DNA 复制时相与染色质的结构相关。

- 基因组总计 5% 的碱基可以明确地鉴定为在哺乳动物中进化保守；基于目前已有的实验结果，大约 60% 的保守碱基，有证据表明是有功能的。

- 尽管实验鉴定到的功能性基因组区域与进化保守的区域有广泛的重叠，但是这些实验确定的区域内，并非所有的碱基都有进化保守的证据。

- 不同的功能元件在人群中序列变异性方面以及位于基因组结构可变区的可能性方面差别很大。

- 出人意料的是，很多功能元件看起来在哺乳动物进化中并不保守。这暗示可能存在一个巨大的中性元件库，这些中性元件在生化上活跃，但是对机体并没有特别的益处。这个中性元件库可以作为一个用于自然选择的“仓库”，作为谱系特异元件以及种间功能保守但非同源元件的潜在来源。

接下来，我们首先概述我们研究中所用到的实验技术，然后描述由分析和整合这些实验产生的数据集所得到的见解。最后我们总结了目前 1% 的人类基因组中已有的发现，并对更广更深地研究人类基因组中的功能元件进行了展望。为了帮助读

investigation of the functional elements in the human genome. To aid the reader, Box 1 provides a glossary for many of the abbreviations used throughout this paper.

Box 1. Frequently used abbreviations in this paper

AR Ancient repeat: a repeat that was inserted into the early mammalian lineage and has since become dormant; the majority of ancient repeats are thought to be neutrally evolving

CAGE tag A short sequence from the 5′end of a transcript

CDS Coding sequence: a region of a cDNA or genome that encodes proteins

ChIP-chip Chromatin immunoprecipitation followed by detection of the products using a genomic tiling array

CNV Copy number variants: regions of the genome that have large duplications in some individuals in the human population

CS Constrained sequence: a genomic region associated with evidence of negative selection (that is, rejection of mutations relative to neutral regions)

DHS DNaseI hypersensitive site: a region of the genome showing a sharply different sensitivity to DNaseI compared with its immediate locale

EST Expressed sequence tag: a short sequence of a cDNA indicative of expression at this point

FAIRE Formaldehyde-assisted isolation of regulatory elements: a method to assay open chromatin using formaldehyde crosslinking followed by detection of the products using a genomic tiling array

FDR False discovery rate: a statistical method for setting thresholds on statistical tests to correct for multiple testing

GENCODE Integrated annotation of existing cDNA and protein resources to define transcripts with both manual review and experimental testing procedures

GSC Genome structure correction: a method to adapt statistical tests to make fewer assumptions about the distribution of features on the genome sequence. This provides a conservative correction to standard tests

HMM Hidden Markov model: a machine-learning technique that can establish optimal parameters for a given model to explain the observed data

Indel An insertion or deletion; two sequences often show a length difference within alignments, but it is not always clear whether this reflects a previous insertion or a deletion

者阅读，框 1 提供了一个词汇表，里面有本论文所用的很多缩写词。

框 1. 本文常用的缩写词

AR 古老的重复：一种在早期就插入到哺乳动物谱系但自此休眠的重复；大部分古老的重复被认为正在进行中性进化

CAGE tag 转录本 5′端的短序列

CDS 编码序列：cDNA 区域或编码蛋白质的基因组区域

ChIP-chip 染色质免疫沉淀后利用基因组叠瓦式阵列检测产物

CNV 拷贝数变异：在人群中一些个体中发生大片段重复的基因组区域

CS 保守序列：有证据表明跟负选择相关的基因组区域(也就是，跟中性区域相比更排斥突变)

DHS DNA 酶 I 超敏感位点：跟附近区域相比，对 DNA 酶 I 非常敏感的基因组区域

EST 表达序列标签：表示在该点表达的 cDNA 短序列

FAIRE 基于甲醛的调控元件分离：一种检测开放染色质的方法，甲醛交联后利用基因组叠瓦式阵列检测产物

FDR 错误发现率：一种为统计学检验设定阈值以校正多重检验的统计学方法

GENCODE 通过人工审查和实验检测的方法，整合注释已知的 cDNA 和蛋白质资源，以定义转录本

GSC 基因组结构校正：一种适合统计学检验的方法，使其对基因组序列上的分布特征做出更少假设。这为标准检验提供了一种保守的校正

HMM 隐马尔可夫模型：一种机器学习技术，能为一个给定的模型建立最优的变量以解释观察到的数据

Indel 插入或缺失；在比对两个序列时经常出现长度不一的情况，但是有时并不清楚这反映的是一个插入还是缺失

PET A short sequence that contains both the 5′ and 3′ ends of a transcript

RACE Rapid amplification of cDNA ends: a technique for amplifying cDNA sequences between a known internal position in a transcript and its 5′ end

RFBR Regulatory factor binding region: a genomic region found by a ChIP-chip assay to be bound by a protein factor

RFBR-Seqsp Regulatory factor binding regions that are from sequence-specific binding factors

RT–PCR Reverse transcriptase polymerase chain reaction: a technique for amplifying a specific region of a transcript

RxFrag Fragment of a RACE reaction: a genomic region found to be present in a RACE product by an unbiased tiling-array assay

SNP Single nucleotide polymorphism: a single base pair change between two individuals in the human population

STAGE Sequence tag analysis of genomic enrichment: a method similar to ChIP-chip for detecting protein factor binding regions but using extensive short sequence determination rather than genomic tiling arrays

SVM Support vector machine: a machine-learning technique that can establish an optimal classifier on the basis of labelled training data

TR50 A measure of replication timing corresponding to the time in the cell cycle when 50% of the cells have replicated their DNA at a specific genomic position

TSS Transcription start site

TxFrag Fragment of a transcript: a genomic region found to be present in a transcript by an unbiased tiling-array assay

Un.TxFrag A TxFrag that is not associated with any other functional annotation

UTR Untranslated region: part of a cDNA either at the 5′ or 3′ end that does not encode a protein sequence

Experimental Techniques

Table 1 (expanded in Supplementary Information section 1.1) lists the major experimental techniques used for the studies reported here, relevant acronyms, and references reporting the generated data sets. These data sets reflect over 400 million experimental data points (603 million data points if one includes comparative sequencing bases). In describing the major

PET 同时包含一个转录本 5′和 3′端的短序列

RACE cDNA 末端快速扩增：一种扩增技术，扩增一个转录本中间的已知位置和它的 5′端之间的 cDNA 序列

RFBR 调控因子结合区域：ChIP-chip 实验发现的与蛋白质因子结合的基因组区域

RFBR-Seqsp 被序列特异的结合因子所结合的调控因子结合区域

RT–PCR 反转录酶聚合酶链式反应：扩增转录本特定区域的技术

RxFrag RACE 反应的片段：通过无偏性叠瓦式阵列实验在 RACE 产物中发现的基因组区域

SNP 单核苷酸多态性：人群中两个个体间单个碱基对的改变

STAGE 基因组丰度的序列标签分析：类似于 ChIP-chip 用于检测蛋白质因子结合区域的方法，不过是利用大量短序列确定，而不是基因组叠瓦式阵列实验

SVM 支持向量机：一种机器学习技术，能基于标记的训练数据建立最优的分类器

TR50 一种衡量复制时相的方法，对应于细胞周期中 50% 的细胞在特定的基因组位置复制 DNA 所用的时间

TSS 转录起始位点

TxFrag 转录本片段：通过无偏性叠瓦式阵列实验在转录本中发现基因组片段

Un.TxFrag 与任何功能注释均不关联的 TxFrag

UTR 非翻译区：位于 5′或 3′端的不编码蛋白质序列的部分 cDNA

实验技术

表 1(在补充信息 1.1 中有扩充) 列出了本文报道的研究所用的主要实验技术、相关的首字母缩略词以及报道这些实验技术所涉及的参考文献。这些数据集反映了超过 4 亿个实验数据点(如果包含比较测序的碱基则是 6.03 亿个数据点)。在描述主

results and initial conclusions, we seek to distinguish "biochemical function" from "biological role". Biochemical function reflects the direct behaviour of a molecule(s), whereas biological role is used to describe the consequence(s) of this function for the organism. Genome-analysis techniques nearly always focus on biochemical function but not necessarily on biological role. This is because the former is more amenable to large-scale data-generation methods, whereas the latter is more difficult to assay on a large scale.

Table 1. Summary of types of experimental techniques used in ENCODE

Feature class	Experimental technique(s)	Abbreviations	References	Number of experimental data points
Transcription	Tiling array, integrated annotation	TxFrag, RxFrag, GENCODE	117 118 19 119	63,348,656
5′ends of transcripts*	Tag sequencing	PET, CAGE	121 13	864,964
Histone modifications	Tiling array	Histone nomenclature†, RFBR	46	4,401,291
Chromatin‡ structure	QT-PCR, tiling array	DHS, FAIRE	42 43 44 122	15,318,324
Sequence-specific factors	Tiling array, tag sequencing, promoter assays	STAGE, ChIP-Chip, ChIP-PET, RFBR	41, 52 11, 120 123 81 34, 51 124 49 33 40	324,846,018
Replication	Tiling array	TR50	59 75	14,735,740
Computational analysis	Computational methods	CCI, RFBR cluster	80 125 10 16 126 127	NA
Comparative sequence analysis*	Genomic sequencing, multi-sequence alignments, computational analysis	CS	87 86 26	NA
Polymorphisms*	Resequencing, copy number variation	CNV	103 128	NA

* Not all data generated by the ENCODE Project.

† Histone code nomenclature follows the Brno nomenclature as described in ref. 129.

‡ Also contains histone modification.

要的结果和初始的结论上，我们试着区分“生化功能”和“生物学功能”。生化功能反映的是一个或多个分子的直接行为，而生物学功能则用来描述这个功能对机体的影响。全基因分析技术几乎都是聚焦于生化功能，并不一定关心生物学功能。这是因为前者更容易被大规模数据产生方法所检验，而后者在大通量下却更难分析。

表 1. ENCODE 中使用的实验技术类型总结

特征分类	实验技术	缩写	参考文献	实验数据点的数量
转录	叠瓦式阵列、整合注释	TxFrag、RxFrag、GENCODE	117 118 19 119	63,348,656
转录本 5′端 *	标签测序	PET、CAGE	121 13	864,964
组蛋白修饰	叠瓦式阵列	组蛋白系统命名法 †、RFBR	46	4,401,291
染色质结构 ‡	QT-PCR、叠瓦式阵列	DHS、FAIRE	42 43 44 122	15,318,324
序列特异因子	叠瓦式阵列、标签测序、启动子实验	STAGE、ChIP-chip、ChIP-PET、RFBR	41、52 11、120 123 81 34、51 124 49 33 40	324,846,018
复制	叠瓦式阵列	TR50	59 75	14,735,740
计算分析	计算方法	CCI、RFBR 簇	80 125 10 16 126 127	NA
序列比较分析 *	基因组测序、多重序列比对、计算分析	CS	87 86 26	NA
多态性 *	重测序、拷贝数目变异	CNV	103 128	NA

* 并非所有的数据都由 ENCODE 计划产生。

† 组蛋白密码系统命名法遵循参考文献 129 中描述的布尔诺系统命名法。

‡ 也包括组蛋白修饰。

The ENCODE pilot project aimed to establish redundancy with respect to the findings represented by different data sets. In some instances, this involved the intentional use of different assays that were based on a similar technique, whereas in other situations, different techniques assayed the same biochemical function. Such redundancy has allowed methods to be compared and consensus data sets to be generated, much of which is discussed in companion papers, such as the ChIP-chip platform comparison[10,11]. All ENCODE data have been released after verification but before this publication, as befits a "community resource" project (see http://www.wellcome.ac.uk/doc_wtd003208.html). Verification is defined as when the experiment is reproducibly confirmed (see Supplementary Information section 1.2). The main portal for ENCODE data is provided by the UCSC Genome Browser (http://genome.ucsc.edu/ENCODE/); this is augmented by multiple other websites (see Supplementary Information section 1.1).

A common feature of genomic analyses is the need to assess the significance of the co-occurrence of features or of other statistical tests. One confounding factor is the heterogeneity of the genome, which can produce uninteresting correlations of variables distributed across the genome. We have developed and used a statistical framework that mitigates many of these hidden correlations by adjusting the appropriate null distribution of the test statistics. We term this correction procedure genome structure correction (GSC) (see Supplementary Information section 1.3).

In the next five sections, we detail the various biological insights of the pilot phase of the ENCODE Project.

Transcription

Overview. RNA transcripts are involved in many cellular functions, either directly as biologically active molecules or indirectly by encoding other active molecules. In the conventional view of genome organization, sets of RNA transcripts (for example, messenger RNAs) are encoded by distinct loci, with each usually dedicated to a single biological role (for example, encoding a specific protein). However, this picture has substantially grown in complexity in recent years[12]. Other forms of RNA molecules (such as small nucleolar RNAs and micro (mi)RNAs) are known to exist, and often these are encoded by regions that intercalate with protein-coding genes. These observations are consistent with the well-known discrepancy between the levels of observable mRNAs and large structural RNAs compared with the total RNA in a cell, suggesting that there are numerous RNA species yet to be classified[13-15]. In addition, studies of specific loci have indicated the presence of RNA transcripts that have a role in chromatin maintenance and other regulatory control. We sought to assay and analyse transcription comprehensively across the 44 ENCODE regions in an effort to understand the repertoire of encoded RNA molecules.

Transcript maps. We used three methods to identify transcripts emanating from the ENCODE regions: hybridization of RNA (either total or polyA-selected) to unbiased tiling

ENCODE 先导计划旨在建立不同数据集发现结果的冗余。在一些情形下，这包括专门使用基于相似技术的不同实验结果，而在其他情形下，则是不同的技术测定同一生化功能。这种冗余使得不同方法可以进行比较，产生一致的数据集，这部分在关于比较的文章中有很多讨论，比如 ChIP-chip 平台的比较 [10,11]。作为一个"共享资源"计划，所有经过验证的 ENCODE 数据在本文发表前就已经发布了(详见 http://www.wellcome.ac.uk/doc_wtd003208.html)。验证的定义是实验被重复确认(详见补充信息 1.2)。UCSC 基因组浏览器提供获取 ENCODE 数据的主要入口(http://genome.ucsc.edu/ENCODE/)；除此之外还有多个其他的网站(详见补充信息 1.1)。

基因组分析的一个常见的特征是需要评估特征共发生的显著性，或者其他统计学检验的显著性。一个混淆因素是基因组的异质性，这会使得分布于基因组上的变量产生无意义的相关。我们已经开发和使用了一种统计学框架，它通过检验统计调整合适的零分布，以减少很多这样的隐含相关。我们将这种校正方法称为基因组结构校正(GSC)(详见补充信息 1.3)。

在接下来的五部分中，我们详细介绍 ENCODE 先导计划的各种生物学发现。

转　录

概况　RNA 转录本通过直接作为生物学活性分子，或者间接编码其他活性分子，参与很多细胞功能。关于基因组结构，传统上认为，不同的 RNA 转录本(如信使 RNA)由不同的基因座编码，每一种 RNA 转录本通常具有单独一种生物学功能(如编码一个特定的蛋白质)。但是，近年来这种描述逐渐变得复杂起来 [12]，我们已经发现其他形式的 RNA 分子(如小核仁 RNA 和微 RNA(miRNA))的存在，它们大多间插在编码蛋白质的基因区域。众所周知，目前观察到的 mRNA 和大的结构性 RNA 水平与细胞总 RNA 水平不符，这与上述发现是一致的，说明还有大量的 RNA 类型未被归类 [13-15]。另外，特定基因座的研究已经表明，RNA 转录本的存在在染色质维持和其他调节控制中有重要作用。为了理解作为一种分子库的编码 RNA，我们全面地检测和分析了 44 个 ENCODE 区域的转录。

转录本图谱　我们使用了三种方法来鉴定 ENCODE 区域产生的转录本：RNA 杂交(总的或多聚 A 筛选的)到无偏性叠瓦式阵列(详见补充信息 2.1)，在 5′端或

arrays (see Supplementary Information section 2.1), tag sequencing of cap-selected RNA at the 5′ or joint 5′/3′ ends (see Supplementary Information sections 2.2 and S2.3), and integrated annotation of available complementary DNA and EST sequences involving computational, manual, and experimental approaches[16] (see Supplementary Information section 2.4). We abbreviate the regions identified by unbiased tiling arrays as TxFrags, the cap-selected RNAs as CAGE or PET tags (see Box 1), and the integrated annotation as GENCODE transcripts. When a TxFrag does not overlap a GENCODE annotation, we call it an Un.TxFrag. Validation of these various studies is described in papers reporting these data sets[17] (see Supplementary Information sections 2.1.4 and 2.1.5).

These methods recapitulate previous findings, but provide enhanced resolution owing to the larger number of tissues sampled and the integration of results across the three approaches (see Table 2). To begin with, our studies show that 14.7% of the bases represented in the unbiased tiling arrays are transcribed in at least one tissue sample. Consistent with previous work[14,15], many (63%) TxFrags reside outside of GENCODE annotations, both in intronic (40.9%) and intergenic (22.6%) regions. GENCODE annotations are richer than the more-conservative RefSeq or Ensembl annotations, with 2,608 transcripts clustered into 487 loci, leading to an average of 5.4 transcripts per locus. Finally, extensive testing of predicted protein-coding sequences outside of GENCODE annotations was positive in only 2% of cases[16], suggesting that GENCODE annotations cover nearly all protein-coding sequences. The GENCODE annotations are categorized both by likely function (mainly, the presence of an open reading frame) and by classification evidence (for example, transcripts based solely on ESTs are distinguished from other scenarios); this classification is not strongly correlated with expression levels (see Supplementary Information sections 2.4.2 and 2.4.3).

Table 2. Bases detected in processed transcripts either as a GENCODE exon, a TxFrag, or as either a GENCODE exon or a TxFrag

	GENCODE exon	TxFrag	Either GENCODE exon or TxFrag
Total detectable transcripts (bases)	1,776,157 (5.9%)	1,369,611 (4.6%)	2,519,280 (8.4%)
Transcripts detected in tiled regions of arrays (bases)	1,447,192 (9.8%)	1,369,611 (9.3%)	2,163,303 (14.7%)

Percentages are of total bases in ENCODE in the first row and bases tiled in arrays in the second row.

Analyses of more biological samples have allowed a richer description of the transcription specificity (see Fig. 1 and Supplementary Information section 2.5). We found that 40% of TxFrags are present in only one sample, whereas only 2% are present in all samples. Although exon-containing TxFrags are more likely (74%) to be expressed in more than one sample, 45% of unannotated TxFrags are also expressed in multiple samples. GENCODE annotations of separate loci often (42%) overlap with respect to their genomic coordinates, in particular on opposite strands (33% of loci). Further analysis of GENCODE-annotated sequences with respect to the positions of open reading frames revealed that some component exons do not have the expected synonymous versus non-synonymous substitution patterns of protein-coding sequence (see Supplement Information section 2.6)

同时在 5′/3′端采用标签测序经帽子筛选的 RNA(详见补充信息 2.2 和 S2.3)，以及通过计算、手工和实验方法整合注释已知互补的 DNA 和 EST 序列 [16](详见补充信息 2.4)。我们把经无偏性叠瓦式阵列鉴定到的区域简称为 TxFrag，将经帽子筛选的 RNA 称为 CAGE 或 PET 标签(详见框 1)，将整合的注释称为 GENCODE 转录本。对于不与 GENCODE 注释重合的 TxFrag，我们称之为 Un.TxFrag。这些不同研究的验证在报道这些数据集的文章中有描述 [17](详见补充信息 2.1.4 和 2.1.5)。

这些方法虽然再现了之前的发现，但这次不仅有更多组织取样，还整合了三种不同的方法，从而提供了更高的分辨率(见表 2)。首先，我们的研究发现，无偏性叠瓦式阵列上的碱基有 14.7% 在至少一种组织样本中转录。与之前的研究一致 [14,15]，很多(63%) TxFrag 位于 GENCODE 注释之外，内含子区有(40.9%)，基因间区有(22.6%)。与相对保守的 RefSeq 或 Ensembl 注释相比，GENCODE 注释更丰富，2,608 个转录本聚集成 487 个基因座，平均每个基因座 5.4 个转录本。最后，对位于 GENCODE 注释之外的预测蛋白编码序列进行了广泛的检测，发现只有 2% 是阳性的 [16]，这说明 GENCODE 注释涵盖了几乎所有的蛋白质编码序列。既可通过可能的功能(主要是开放阅读框的存在)也可通过类别证据(例如，仅基于 EST 的转录本跟其他类型是不同的)对 GENCODE 注释进行分类；这种分类跟表达水平并没有很强的相关性(详见补充信息 2.4.2 和 2.4.3)。

表 2. 在处理过的转录本中检测到的是 GENCODE 外显子、TxFrag 或两者其一的碱基

	GENCODE 外显子	TxFrag	GENCODE 外显子或 TxFrag
总的检测到的转录本(碱基)	1,776,157 (5.9%)	1,369,611 (4.6%)	2,519,280 (8.4%)
在阵列的叠瓦区域检测到的转录本(碱基)	1,447,192 (9.8%)	1,369,611 (9.3%)	2,163,303 (14.7%)

比例是第一行中 ENCODE 中总的碱基和第二行阵列中叠瓦的碱基。

对更多生物学样本的分析更加丰富了对转录本特异性的描述(详见图 1 和补充信息 2.5)。我们发现 40% 的 TxFrag 仅在一种样本中存在，只有 2% 的 TxFrag 在所有样本中都存在。尽管含有外显子的 TxFrag 更倾向于(74%)在超过一种样本中表达，45% 未被注释的 TxFrag 也在多种样本中表达。不同基因座 42% 的 GENCODE 注释大多与它们的基因组位置重合，特别是与另一条链(33% 的基因座)。根据开放阅读框的位置进一步分析 GENCODE 注释的序列，发现一些组成的外显子没有预期的蛋白质编码序列中同义和非同义替换情况(详见补充信息 2.6)，一些外显子含

and some have deletions incompatible with protein structure[18]. Such exons are on average less expressed (25% versus 87% by RT–PCR; see Supplementary Information section 2.7) than exons involved in more than one transcript (see Supplementary Information section 2.4.3), but when expressed have a tissue distribution comparable to well-established genes.

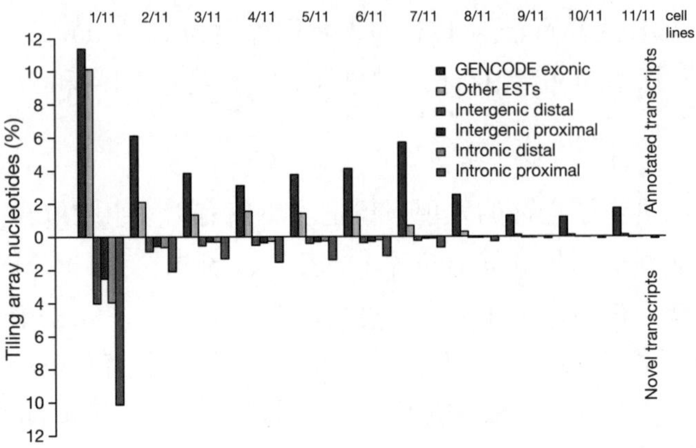

Fig. 1. Annotated and unannotated TxFrags detected in different cell lines. The proportion of different types of transcripts detected in the indicated number of cell lines (from 1/11 at the far left to 11/11 at the far right) is shown. The data for annotated and unannotated TxFrags are indicated separately, and also split into different categories based on GENCODE classification: exonic, intergenic (proximal being within 5 kb of a gene and distal being otherwise), intronic (proximal being within 5 kb of an intron and distal being otherwise), and matching other ESTs not used in the GENCODE annotation (principally because they were unspliced). The *y* axis indicates the per cent of tiling array nucleotides present in that class for that number of samples (combination of cell lines and tissues).

Critical questions are raised by the presence of a large amount of unannotated transcription with respect to how the corresponding sequences are organized in the genome—do these reflect longer transcripts that include known loci, do they link known loci, or are they completely separate from known loci? We further investigated these issues using both computational and new experimental techniques.

Unannotated transcription. Consistent with previous findings, the Un.TxFrags did not show evidence of encoding proteins (see Supplementary Information section 2.8). One might expect Un.TxFrags to be linked within transcripts that exhibit coordinated expression and have similar conservation profiles across species. To test this, we clustered Un.TxFrags using two methods. The first method[19] used expression levels in 11 cell lines or conditions, dinucleotide composition, location relative to annotated genes, and evolutionary conservation profiles to cluster TxFrags (both unannotated and annotated). By this method, 14% of Un.TxFrags could be assigned to annotated loci, and 21% could be clustered into 200 novel loci (with an average of ~7 TxFrags per locus). We experimentally examined these novel loci to study the connectivity of transcripts amongst Un.TxFrags and between Un.TxFrags and known exons. Overall, about 40% of the connections (18 out of 46) were validated by RT–PCR. The second clustering method involved analysing a time course (0, 2, 8 and 32 h) of expression changes in human HL60 cells following retinoic-acid

有与蛋白质结构不相容的缺失[18]。这种外显子（详见补充信息 2.7）比参与一种以上转录本的外显子（详见补充信息 2.4.3）平均表达水平更低（根据 RT-PCR 是 25% 比 87%），不过，当它们表达时，其组织分布与已确定的基因相比差别不大。

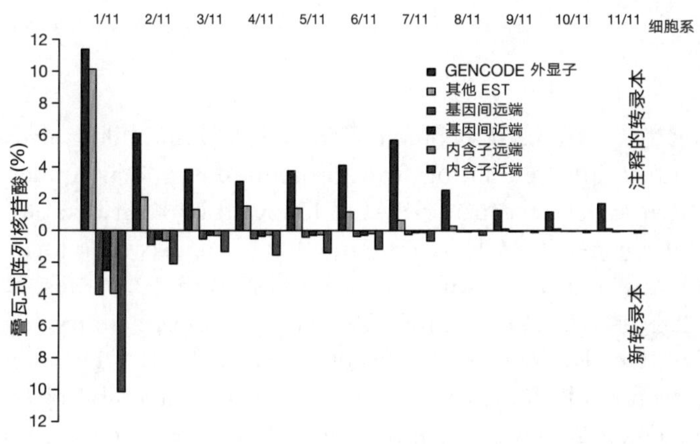

图 1. 不同细胞系中检测到的注释和未注释的 TxFrag。在指明数量的细胞系中检测到的不同类型的转录本的比例（从最左边的 1/11 到最右边的 11/11）。注释和未注释的 TxFrag 的数据分开展示，并且还基于 GENCODE 分类分成了不同的类别：外显子、基因间（距基因 5 kb 以内是近端，大于 5 kb 则是远端）、内含子（距内含子 5 kb 以内是近端，大于 5 kb 则是远端）和未在 GENCODE 注释中使用的匹配其他 EST（主要是因为它们没有剪接）。y 轴表示特定数量的样本（包括了细胞系和组织）特定分类下检测到的叠瓦式阵列核苷酸的百分比。

大量未注释转录本的存在带来了一个重要问题：它们相应的序列在基因组中是如何组织的——它们是否反映了包含已知基因座的更长转录本，它们是否与已知基因座有联系，或者它们是否完全独立于已知基因座之外？我们利用计算和新实验技术进一步探究了这些问题。

未被注释的转录　与之前的发现一致，没有证据表明 Un.TxFrag 编码蛋白质（详见补充信息 2.8）。我们期待 Un.TxFrag 在协调表达且在物种间有相似保守性的转录本中可能有关联。为检验这种猜测，我们利用两种方法将 Un.TxFrag 聚类。第一种方法[19]利用 11 种细胞系或条件下表达水平、二核苷酸构成、相对注释基因的位置和进化保守谱来聚类 TxFrag（未注释的和已注释的）。通过这种方法，14% 的 Un.TxFrag 能被分配到注释的基因座，21% 能被聚类成 200 个新的基因座（平均一个基因座大约 7 个 TxFrag）。我们利用实验检查了这些新的基因座，以研究 Un.TxFrag 内部以及 Un.TxFrag 跟已知的外显子间的转录本的关联性。总体上，大约 40% 的关联（46 个中有 18 个）被 RT-PCR 验证。第二种聚类方法是分析视黄酸刺激人 HL60 细胞后一段时间内（0、2、8 和 32 小时）的表达变化。这是一个描述注释基因座表达

stimulation. There is a coordinated program of expression changes from annotated loci, which can be shown by plotting Pearson correlation values of the expression levels of exons inside annotated loci versus unrelated exons (see Supplementary Information section 2.8.2). Similarly, there is coordinated expression of nearby Un.TxFrags, albeit lower, though still significantly different from randomized sets. Both clustering methods indicate that there is coordinated behaviour of many Un.TxFrags, consistent with them residing in connected transcripts.

Transcript connectivity. We used a combination of RACE and tiling arrays[20] to investigate the diversity of transcripts emanating from protein-coding loci. Analogous to TxFrags, we refer to transcripts detected using RACE followed by hybridization to tiling arrays as RxFrags. We performed RACE to examine 399 protein-coding loci (those loci found entirely in ENCODE regions) using RNA derived from 12 tissues, and were able to unambiguously detect 4,573 RxFrags for 359 loci (see Supplementary Information section 2.9). Almost half of these RxFrags (2,324) do not overlap a GENCODE exon, and most (90%) loci have at least one novel RxFrag, which often extends a considerable distance beyond the 5′ end of the locus. Figure 2 shows the distribution of distances between these new RACE-detected ends and the previously annotated TSS of each locus. The average distance of the extensions is between 50 kb and 100 kb, with many extensions (> 20%) being more than 200 kb. Consistent with the known presence of overlapping genes in the human genome, our findings reveal evidence for an overlapping gene at 224 loci, with transcripts from 180 of these loci (~50% of the RACE-positive loci) appearing to have incorporated at least one exon from an upstream gene.

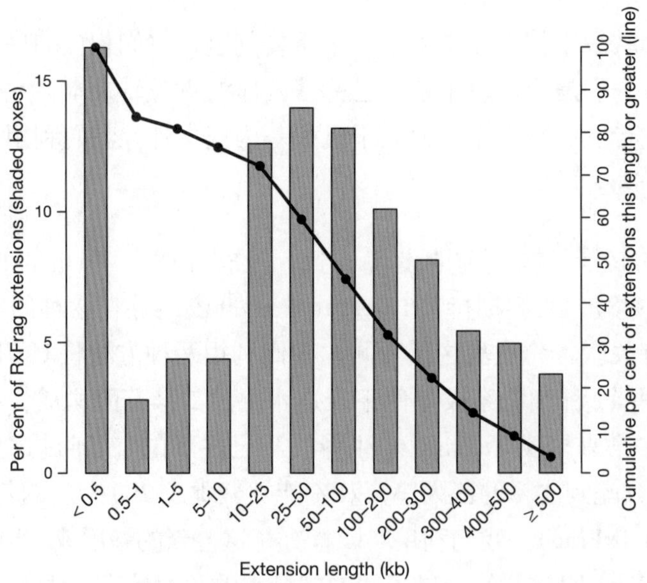

Fig. 2. Length of genomic extensions to GENCODE-annotated genes on the basis of RACE experiments followed by array hybridizations (RxFrags). The indicated bars reflect the frequency of extension lengths among different length classes. The solid line shows the cumulative frequency of extensions of that length or greater. Most of the extensions are greater than 50 kb from the annotated gene (see text for details).

变化的协调性项目，通过跟无关外显子相比较描绘注释基因座外显子表达水平的皮尔逊相关值（详见补充信息 2.8.2）。同样，靠近的 Un.TxFrag 协调表达，虽然水平更低一些，但仍然与随机组显著不同。这两种聚类方法说明很多 Un.TxFrag 存在协调关系，这与它们位于相关联转录本上一致。

转录本的关联性 我们联合使用 RACE 和叠瓦式阵列[20] 来研究蛋白质编码基因座产生的转录本的多样性。类似于 TxFrag，我们称 RACE 后杂交到叠瓦式阵列所检测到的转录本为 RxFrag。利用 12 种组织的 RNA，我们通过 RACE 检测了 399 个蛋白质编码基因座（这些基因座全部位于 ENCODE 区域），能够确定地在 359 个基因座检测到 4,573 个 RxFrag（详见补充信息 2.9）。这些 RxFrag 近乎半数（2,324 个）不与 GENCODE 外显子重合，并且大部分基因座（90%）至少有一个全新的 RxFrag，这些全新的 RxFrag 大多从基因座的 5′端延伸相当长的距离。图 2 展示了各个基因座中这些 RACE 检测到的全新的末端与已注释的 TSS 之间的距离的分布。延伸的平均距离在 50 kb 到 100 kb 之间，很多（> 20%）超过了 200 kb。与已知人类基因组中存在重合基因一致，我们发现 224 个基因座存在重合基因的证据，其中 180 个基因座（～50% 的 RACE 阳性基因座）的转录本看起来包含了上游基因的至少一个外显子。

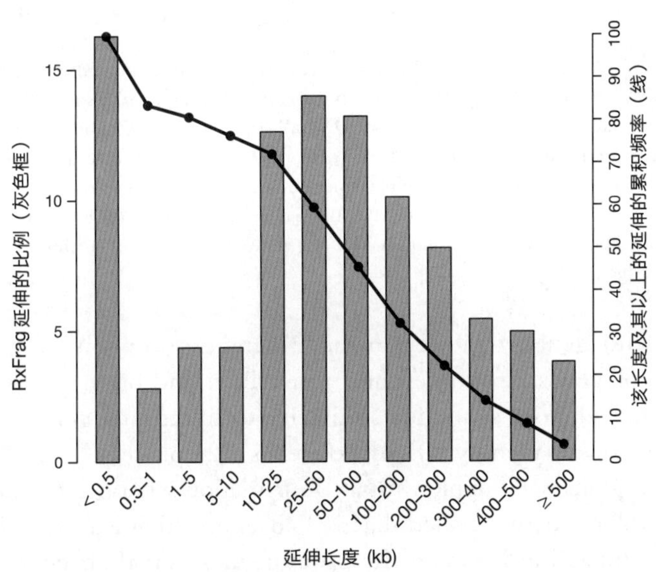

图 2. 基于 RACE 实验结合微阵列杂交（RxFrag）得到 GENCODE 注释基因的延伸长度。各个条表示在不同长度类别下延伸长度的频率。实线表示特定长度及其以上的延伸的累积频率。大部分延伸距离注释基因超过 50 kb（详见正文）。

To characterize further the 5′ RxFrag extensions, we performed RT–PCR followed by cloning and sequencing for 550 of the 5′ RxFrags (including the 261 longest extensions identified for each locus). The approach of mapping RACE products using microarrays is a combination method previously described and validated in several studies[14,17,20]. Hybridization of the RT–PCR products to tiling arrays confirmed connectivity in almost 60% of the cases. Sequenced clones confirmed transcript extensions. Longer extensions were harder to clone and sequence, but 5 out of 18 RT–PCR-positive extensions over 100 kb were verified by sequencing (see Supplementary Information section 2.9.7 and ref. 17). The detection of numerous RxFrag extensions coupled with evidence of considerable intronic transcription indicates that protein-coding loci are more transcriptionally complex than previously thought. Instead of the traditional view that many genes have one or more alternative transcripts that code for alternative proteins, our data suggest that a given gene may both encode multiple protein products and produce other transcripts that include sequences from both strands and from neighbouring loci (often without encoding a different protein). Figure 3 illustrates such a case, in which a new fusion transcript is expressed in the small intestine, and consists of at least three coding exons from the *ATP5O* gene and at least two coding exons from the *DONSON* gene, with no evidence of sequences from two intervening protein-coding genes (*ITSN1* and *CRYZL1*).

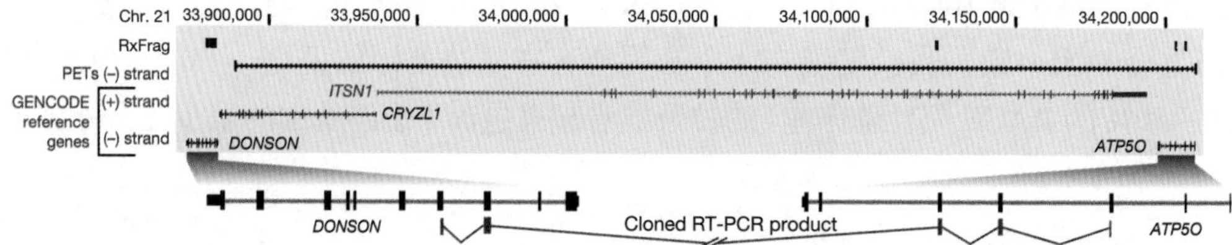

Fig. 3. Overview of RACE experiments showing a gene fusion. Transcripts emanating from the region between the *DONSON* and *ATP5O* genes. A 330-kb interval of human chromosome 21 (within ENm005) is shown, which contains four annotated genes: *DONSON*, *CRYZL1*, *ITSN1* and *ATP5O*. The 5′ RACE products generated from small intestine RNA and detected by tiling-array analyses (RxFrags) are shown along the top. Along the bottom is shown the placement of a cloned and sequenced RT–PCR product that has two exons from the *DONSON* gene followed by three exons from the *ATP5O* gene; these sequences are separated by a 300 kb intron in the genome. A PET tag shows the termini of a transcript consistent with this RT–PCR product.

Pseudogenes. Pseudogenes, reviewed in refs 21 and 22, are generally considered non-functional copies of genes, are sometimes transcribed and often complicate analysis of transcription owing to close sequence similarity to functional genes. We used various computational methods to identify 201 pseudogenes (124 processed and 77 non-processed) in the ENCODE regions (see Supplementary Information section 2.10 and ref. 23). Tiling-array analysis of 189 of these revealed that 56% overlapped at least one TxFrag. However, possible cross-hybridization between the pseudogenes and their corresponding parent genes may have confounded such analyses. To assess better the extent of pseudogene transcription, 160 pseudogenes (111 processed and 49 non-processed) were examined for expression using RACE/tiling-array analysis (see Supplementary Information section 2.9.2). Transcripts were detected for 14 pseudogenes (8 processed and 6 non-processed) in at least

为了进一步研究 5′ RxFrag 延伸的特征，我们对 550 个 5′ RxFrag(包括各个基因座延伸最长的 261 个)进行了 RT−PCR，然后进行克隆和测序。通过微阵列检测 RACE 产物的方法是一种之前报道的并且被很多研究验证过的联合方法[14,17,20]。将 RT−PCR 产物杂交到叠瓦式阵列验证了几乎 60% 的连通情况。被测序的克隆验证了转录本的延伸。越长的延伸越难克隆测序，但是 18 个超过 100 kb 的 RT−PCR 阳性延伸中，有 5 个经测序验证(详见补充信息 2.9.7 和参考文献 17)。大量的 RxFrag 延伸被检测到，并且有证据表明有相当数量的内含子转录本存在，都说明蛋白质编码基因座的转录要比之前想象的更加复杂。传统的观点认为很多基因含有一个或多个选择性转录本，编码选择性蛋白质，而我们的数据则暗示一个特定的基因可能既编码多个蛋白质产物，又产生包含两条链以及附近基因座序列的其他转录本(大多不编码其他蛋白质)。图 3 展示了这样一个例子来加以说明，在这个例子中，一个新的融合转录本在小肠中表达，它含有 *ATP5O* 基因的至少三个编码外显子和 *DONSON* 基因的至少两个编码外显子，但是并没有证据显示它含有位于这两个干扰蛋白编码基因(*ITSN1* 和 *CRYZL1*)的序列。

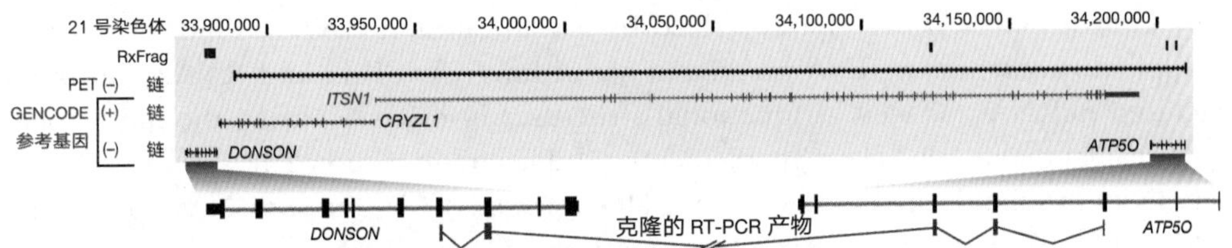

图 3. 概述 RACE 实验，展示基因融合现象。转录本产生于 *DONSON* 和 *ARP5O* 基因间区域。图中展示了人类 21 号染色体的一段 330 kb 的区域(在 ENm005 中)，它包含四个注释基因：*DONSON*、*CRYZL1*、*ITSN1* 和 *ATP5O*。顶端展示的是叠瓦式阵列检测到的来自小肠 RNA 的 5′ RACE 产物(RxFrag)。底端展示的是克隆测序的 RT−PCR 产物的位置，该 PCR 产物含有 *DONSON* 基因的两个外显子和 *ATP5O* 基因的三个外显子；这些序列在基因组上被一个 300 kb 的内含子分开。PET 标签展示的与这个 RT−PCR 产物一致的一个转录本的末端。

假基因 假基因通常被认为是没有功能的基因拷贝，有时候会被转录，因为与正常基因序列的高度相似性常使转录分析变得复杂，在参考文献 21 和 22 中有对其的综述。利用多种计算方法，我们在 ENCODE 区域鉴定到 201 个假基因(124 个加工，77 个未加工)(见补充信息 2.10 和参考文献 23)。对其中的 189 个进行叠瓦式阵列分析，发现 56% 与至少一个 TxFrag 重合。不过，假基因可能和它们对应的亲本基因交叉杂交，进而可能使得上述分析产生混乱。为了更好地检测假基因转录的范围，利用 RACE/叠瓦式阵列检测 160 个假基因(111 个加工，49 个未加工)的表达(详见补充信息 2.9.2)。14 个假基因(8 个加工，6 个未加工)在 12 个 RNA 样本中的

one of the 12 tested RNA sources, the majority (9) being in testis (see ref. 23). Additionally, there was evidence for the transcription of 25 pseudogenes on the basis of their proximity (within 100 bp of a pseudogene end) to CAGE tags (8), PETs (2), or cDNAs/ESTs (21). Overall, we estimate that at least 19% of the pseudogenes in the ENCODE regions are transcribed, which is consistent with previous estimates[24,25].

Non-protein-coding RNA. Non-protein-coding RNAs (ncRNAs) include structural RNAs (for example, transfer RNAs, ribosomal RNAs, and small nuclear RNAs) and more recently discovered regulatory RNAs (for example, miRNAs). There are only 8 well-characterized ncRNA genes within the ENCODE regions (*U70*, *ACA36*, *ACA56*, *mir-192*, *mir-194-2*, *mir-196*, *mir-483* and *H19*), whereas representatives of other classes, (for example, box C/D snoRNAs, tRNAs, and functional snRNAs) seem to be completely absent in the ENCODE regions. Tiling-array data provided evidence for transcription in at least one of the assayed RNA samples for all of these ncRNAs, with the exception of mir-483 (expression of mir-483 might be specific to fetal liver, which was not tested). There is also evidence for the transcription of 6 out of 8 pseudogenes of ncRNAs (mainly snoRNA-derived). Similar to the analysis of protein-pseudogenes, the hybridization results could also originate from the known snoRNA gene elsewhere in the genome.

Many known ncRNAs are characterized by a well-defined RNA secondary structure. We applied two *de novo* ncRNA prediction algorithms—EvoFold and RNAz—to predict structured ncRNAs (as well as functional structures in mRNAs) using the multi-species sequence alignments (see below, Supplementary Information section 2.11 and ref. 26). Using a sensitivity threshold capable of detecting all known miRNAs and snoRNAs, we identified 4,986 and 3,707 candidate ncRNA loci with EvoFold and RNAz, respectively. Only 268 loci (5% and 7%, respectively) were found with both programs, representing a 1.6-fold enrichment over that expected by chance; the lack of more extensive overlap is due to the two programs having optimal sensitivity at different levels of GC content and conservation. We experimentally examined 50 of these targets using RACE/tiling-array analysis for brain and testis tissues (see Supplementary Information sections 2.11 and 2.9.3); the predictions were validated at a 56%, 65%, and 63% rate for Evofold, RNAz and dual predictions, respectively.

Primary transcripts. The detection of numerous unannotated transcripts coupled with increasing knowledge of the general complexity of transcription prompted us to examine the extent of primary (that is, unspliced) transcripts across the ENCODE regions. Three data sources provide insight about these primary transcripts: the GENCODE annotation, PETs, and RxFrag extensions. Figure 4 summarizes the fraction of bases in the ENCODE regions that overlap transcripts identified by these technologies. Remarkably, 93% of bases are represented in a primary transcript identified by at least two independent observations (but potentially using the same technology); this figure is reduced to 74% in the case of primary transcripts detected by at least two different technologies. These increased spans are not mainly due to cell line rearrangements because they were present in multiple tissue experiments that confirmed the spans (see Supplementary Information section 2.12).

至少 1 个样本中检测到转录本，大部分（9 个）在睾丸中被检测到（见参考文献 23）。另外，有证据表明 25 个假基因在转录，这是基于它们靠近（假基因末端距离 100 bp 以内）CAGE 标签（8 个）、PET（2 个）或 cDNA/EST（21 个）。总体来说，我们估计 ENCODE 区域内至少 19% 的假基因被转录，这与之前的估计是一致的[24,25]。

非蛋白质编码 RNA　非蛋白质编码 RNA（ncRNA）包括结构 RNA（如转运 RNA、核糖体 RNA 和核内小 RNA）和最近发现的调控 RNA（如 miRNA）。在 ENCODE 区域内只有 8 种熟知的 ncRNA 基因（*U70*、*ACA36*、*ACA56*、*mir-192*、*mir-194-2*、*mir-196*、*mir-483* 和 *H19*），其他类型（如 C/D 核仁小 RNA、tRNA 和功能性 snRNA）在 ENCODE 区域似乎完全不存在。叠瓦式阵列数据为所有这些 ncRNA 在至少一种被分析 RNA 样本中的转录提供了证据，这其中 mir-483 除外（mir-483 可能在胎肝中特异表达，而胎肝没有被检测）。也有证据表明 8 个 ncRNAs 假基因（主要是 snoRNA 产生的）中有 6 个转录。类似于蛋白质假基因的分析，杂交结果也可能是来源于基因组中其他地方的已知 snoRNA 基因。

很多已知的 ncRNA 具有明确的 RNA 二级结构。我们通过两种从头预测 ncRNA 的算法——EvoFold 和 RNAz，利用多物种序列比对来预测结构 ncRNA（详见下文、补充信息 2.11 和参考文献 26）。使用能检测所有已知的 miRNA 和 snoRNA 的敏感性阈值，我们通过 EvoFold 和 RNAz 分别鉴定到 4,986 和 3,707 个候选 ncRNA 基因座。只有 268 个基因座（分别是 5% 和 7%）在两个程序中都被发现，相对偶然概率有 1.6 倍富集；没有更高的重合是因为两个程序的最佳敏感性所要求的 GC 含量和保守性水平不同。我们使用 RACE/叠瓦式阵列实验在大脑和睾丸组织中检查了其中的 50 个（详见补充信息 2.11 和 2.9.3）；Evofold、RNAz 和两种方法联合预测的被验证率分别为 56%、65% 和 63%。

初级转录本　由于大量未被注释的转录本被检测到，加上我们对转录的基本复杂性认识的深入，使我们不得不检查初级（也就是未被剪接的）转录本在 ENCODE 区域内的分布范围。三个数据源为认识这些初级转录本提供了线索：GENCODE 注释、PET 和 RxFrag 延伸。图 4 总结了 ENCODE 区域内碱基与这些技术鉴定到的转录本重合的比例。值得注意的是，93% 的碱基位于被至少两种独立观察（但是有可能使用同一种技术）鉴定到的初级转录本中；如果是被至少两种不同的技术鉴定到的初级转录本，则这个数字下降到 74%。这些增加的跨度并不主要是由于细胞系的重排，因为在多种组织的实验中都验证了这种跨度的存在（详见补充信息 2.12）。这些

These estimates assume that the presence of PETs or RxFrags defining the terminal ends of a transcript imply that the entire intervening DNA is transcribed and then processed. Other mechanisms, thought to be unlikely in the human genome, such as *trans*-splicing or polymerase jumping would also produce these long termini and potentially should be reconsidered in more detail.

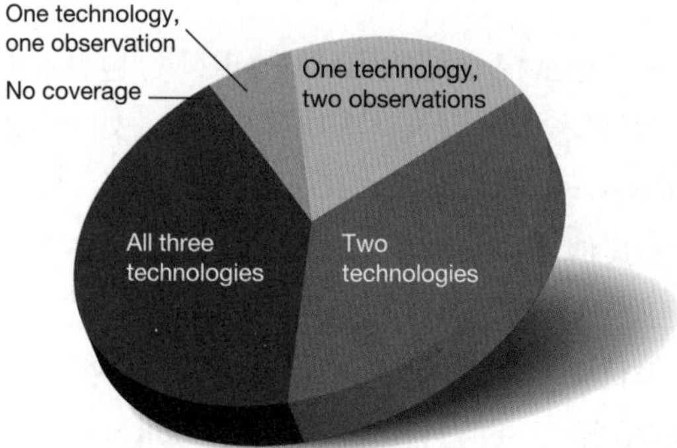

Fig. 4. Coverage of primary transcripts across ENCODE regions. Three different technologies (integrated annotation from GENCODE, RACE-array experiments (RxFrags) and PET tags) were used to assess the presence of a nucleotide in a primary transcript. Use of these technologies provided the opportunity to have multiple observations of each finding. The proportion of genomic bases detected in the ENCODE regions associated with each of the following scenarios is depicted: detected by all three technologies, by two of the three technologies, by one technology but with multiple observations, and by one technology with only one observation. Also indicated are genomic bases without any detectable coverage of primary transcripts.

Previous studies have suggested a similar broad amount of transcription across the human[14,15] and mouse[27] genomes. Our studies confirm these results, and have investigated the genesis of these transcripts in greater detail, confirming the presence of substantial intragenic and intergenic transcription. At the same time, many of the resulting transcripts are neither traditional protein-coding transcripts nor easily explained as structural non-coding RNAs. Other studies have noted complex transcription around specific loci or chimaeric-gene structures (for example refs 28–30), but these have often been considered exceptions; our data show that complex intercalated transcription is common at many loci. The results presented in the next section show extensive amounts of regulatory factors around novel TSSs, which is consistent with this extensive transcription. The biological relevance of these unannotated transcripts remains unanswered by these studies. Evolutionary information (detailed below) is mixed in this regard; for example, it indicates that unannotated transcripts show weaker evolutionary conservation than many other annotated features. As with other ENCODE-detected elements, it is difficult to identify clear biological roles for the majority of these transcripts; such experiments are challenging to perform on a large scale and, furthermore, it seems likely that many of the corresponding biochemical events may be evolutionarily neutral (see below).

估计假定如果存在定义转录本两个末端的 PET 或 RxFrag，就意味着整个中间 DNA 首先被转录，然后再被加工。被认为不可能存在于人类基因组的其他机制，例如反式剪接或者聚合酶跳跃，也会产生这种长末端，可能应该被更仔细地重新思考。

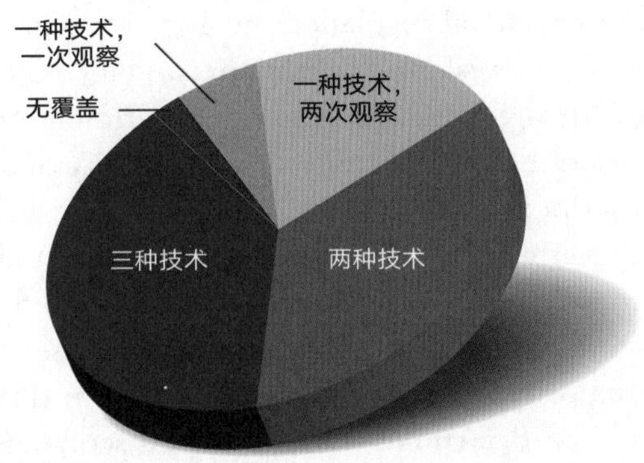

图 4. 原始转录本在 ENCODE 区域的覆盖度。三种不同的技术（来自 GENCODE、RACE 阵列实验（RxFrag）和 PET 标签的整合注释）被用来评估初级转录本中碱基的存在。这些技术的应用使人们有机会对每一个发现进行多次观察。图中展示了在 ENCODE 区域中被检测的基因组碱基与下列情形相关的各自所占的比例：被三种技术都检测到，被三种技术中的两种检测到，被一种技术但是多次观察到，以及被一种技术一次观察到。图中还展示了没有检测到初级转录本覆盖的基因组碱基所占的比例。

之前的研究已经表明人类 [14,15] 和小鼠 [27] 基因组都存在类似的大量转录。我们的研究验证了这些结果，并且还更详细地研究了这些转录本的产生，验证了大量基因内和基因间转录的存在。同时，得到的很多转录本既不是传统的蛋白质编码转录本，也不能简单地解释为结构性非编码 RNA。其他的研究已经注意到在特定基因座或者嵌合基因结构存在复杂的转录（如参考文献 28 ~ 30），但是这些常常被认为是特例，我们的数据说明复杂的有插入的转录在很多基因座是常见的。下一部分呈现的结果展示了在新的 TSS 周围有大量的调控因子，与这种广泛转录一致。这些研究还不能回答这些未注释转录本的生物学意义。进化的信息（下文有详细介绍）在这方面是混杂的，例如，它暗示未注释的转录本比很多其他注释特征表现为更弱的进化保守性。和其他 ENCODE 检测到的元件一样，这些转录本中的大多数难以鉴定明确的生物学功能。大规模地进行这种实验很有挑战性，再者，有可能很多对应的生化事件在进化上是中性的（见下文）。

Regulation of Transcription

Overview. A significant challenge in biology is to identify the transcriptional regulatory elements that control the expression of each transcript and to understand how the function of these elements is coordinated to execute complex cellular processes. A simple, commonplace view of transcriptional regulation involves five types of *cis*-acting regulatory sequences—promoters, enhancers, silencers, insulators and locus control regions[31]. Overall, transcriptional regulation involves the interplay of multiple components, whereby the availability of specific transcription factors and the accessibility of specific genomic regions determine whether a transcript is generated[31]. However, the current view of transcriptional regulation is known to be overly simplified, with many details remaining to be established. For example, the consensus sequences of transcription factor binding sites (typically 6 to 10 bases) have relatively little information content and are present numerous times in the genome, with the great majority of these not participating in transcriptional regulation. Does chromatin structure then determine whether such a sequence has a regulatory role? Are there complex inter-factor interactions that integrate the signals from multiple sites? How are signals from different distal regulatory elements coupled without affecting all neighbouring genes? Meanwhile, our understanding of the repertoire of transcriptional events is becoming more complex, with an increasing appreciation of alternative TSSs[32,33] and the presence of non-coding[27,34] and anti-sense transcripts[35,36].

To better understand transcriptional regulation, we sought to begin cataloguing the regulatory elements residing within the 44 ENCODE regions. For this pilot project, we mainly focused on the binding of regulatory proteins and chromatin structure involved in transcriptional regulation. We analysed over 150 data sets, mainly from ChIP-chip[37-39], ChIP-PET and STAGE[40,41] studies (see Supplementary Information section 3.1 and 3.2). These methods use chromatin immunoprecipitation with specific antibodies to enrich for DNA in physical contact with the targeted epitope. This enriched DNA can then be analysed using either microarrays (ChIP-chip) or high-throughput sequencing (ChIP-PET and STAGE). The assays included 18 sequence-specific transcription factors and components of the general transcription machinery (for example, RNA polymerase II (Pol II), TAF1 and TFIIB/GTF2B). In addition, we tested more than 600 potential promoter fragments for transcriptional activity by transient-transfection reporter assays that used 16 human cell lines[33]. We also examined chromatin structure by studying the ENCODE regions for DNaseI sensitivity (by quantitative PCR[42] and tiling arrays[43,44], see Supplementary Information section 3.3), histone composition[45], histone modifications (using ChIP-chip assays)[37,46], and histone displacement (using FAIRE, see Supplementary Information section 3.4). Below, we detail these analyses, starting with the efforts to define and classify the 5′ ends of transcripts with respect to their associated regulatory signals. Following that are summaries of generated data about sequence-specific transcription factor binding and clusters of regulatory elements. Finally, we describe how this information can be integrated to make predictions about transcriptional regulation.

转录的调控

概况 生物学中一个重大的挑战是鉴定控制各个转录本表达的转录调控元件，以及理解这些元件的功能是如何协调以完成复杂的细胞过程。对转录调控一个简单朴素的认识包括五种类型的顺式调控元件——启动子、增强子、沉默子、绝缘子和基因座控制区 [31]。总体上，转录调控包括多种组分的相互作用，特定转录因子的可用性和特定基因组区域的开放性决定一个转录本的产生与否 [31]。不过，目前对转录调控的认识被认为过于简单，很多细节还未完善。例如，转录因子结合位点的共有序列（通常 6 到 10 个碱基）含有相对很少的信息量，在基因组中大量存在，其中大部分并不参与转录调控。接下来，是否染色质结构决定这样一个序列是否具有调控作用？是否有复杂的因子间相互作用整合来自多个位点的信号？来自不同的远端调控元件的信号如何联系起来并且还不影响所有邻近的基因？同时，随着对可变 TSS[32,33] 认识的更加深入以及非编码 [27,34] 和反义 [35,36] 转录本的存在，我们对转录事件库的理解正变得更加复杂。

为更好地理解转录调控，我们开始为位于 44 个 ENCODE 区域内的调控元件编制目录。对于这个先导计划，我们主要关注调控蛋白的结合和参与转录调控的染色质结构。我们分析了超过 150 个数据集，这些数据集主要来自 ChIP-chip[37-39]，ChIP-PET 和 STAGE[40,41] 研究（详见补充信息 3.1 和 3.2）。这些方法使用染色质免疫沉淀结合特定的抗体从而物理靠近目标表位富集 DNA。这些富集的 DNA 可以利用芯片（ChIP-chip）或高通量测序（ChIP-PET 和 STAGE）分析。这样的实验包括 18 个序列特异的转录因子和基本转录机制的组分（如 RNA 聚合酶 II(Pol II)、TAF1 和 TFIIB/GTF2B）。另外，我们利用 16 个人细胞系，通过瞬时转染报告实验检测了超过 600 个潜在的启动子片段的转录活性 [33]。我们还通过研究 ENCODE 区域的 DNaseI 敏感性（通过定量 PCR[42] 和叠瓦式阵列 [43,44]，详见补充信息 3.3）、组蛋白构成 [45]、组蛋白修饰（利用 ChIP-chip 实验）[37,46] 和组蛋白置换（使用 FAIRE，详见补充信息 3.4）检测了染色质的结构。下面，我们详细介绍这些信息，首先根据相关的调控信号来定义和分类转录本的 5′端；然后总结产生的关于序列特异转录因子结合和调控元件簇的数据；最后我们描述这些信息如何被整合起来，从而对转录调控做出预测。

Transcription start site catalogue. We analysed two data sets to catalogue TSSs in the ENCODE regions: the 5′ ends of GENCODE-annotated transcripts and the combined results of two 5′-end-capture technologies—CAGE and PET-tagging. The initial results suggested the potential presence of 16,051 unique TSSs. However, in many cases, multiple TSSs resided within a single small segment (up to ~200 bases); this was due to some promoters containing TSSs with many very close precise initiation sites[47]. To normalize for this effect, we grouped TSSs that were 60 or fewer bases apart into a single cluster, and in each case considered the most frequent CAGE or PET tag (or the 5′-most TSS in the case of TSSs identified only from GENCODE data) as representative of that cluster for downstream analyses.

The above effort yielded 7,157 TSS clusters in the ENCODE regions. We classified these TSSs into three categories: known (present at the end of GENCODE-defined transcripts), novel (supported by other evidence) and unsupported. The novel TSSs were further subdivided on the basis of the nature of the supporting evidence (see Table 3 and Supplementary Information section 3.5), with all four of the resulting subtypes showing significant overlap with experimental evidence using the GSC statistic. Although there is a larger relative proportion of singleton tags in the novel category, when analysis is restricted to only singleton tags, the novel TSSs continue to have highly significant overlap with supporting evidence (see Supplementary Information section 3.5.1).

Table 3. Different categories of TSSs defined on the basis of support from different transcript-survey methods

Category	Transcript survey method	Number of TSS clusters (non-redundant)*	*P* value†	Singleton clusters‡ (%)
Known	GENCODE 5′ ends	1,730	2×10^{-70}	25 (74 overall)
Novel	GENCODE sense exons	1,437	6×10^{-39}	64
	GENCODE antisense exons	521	3×10^{-8}	65
	Unbiased transcription survey	639	7×10^{-63}	71
	CpG island	164	4×10^{-90}	60
Unsupported	None	2,666	–	83.4

* Number of TSS clusters with this support, excluding TSSs from higher categories.

† Probability of overlap between the transcript support and the PET/CAGE tags, as calculated by the Genome Structure Correction statistic (see Supplementary Information section 1.3).

‡ Per cent of clusters with only one tag. For the "known" category this was calculated as the per cent of GENCODE 5′ ends with tag support (25%) or overall (74%).

Correlating genomic features with chromatin structure and transcription factor binding. By measuring relative sensitivity to DNaseI digestion (see Supplementary Information section 3.3), we identified DNaseI hypersensitive sites throughout the ENCODE regions. DHSs and TSSs both reflect genomic regions thought to be enriched for regulatory information and many DHSs reside at or near TSSs. We partitioned DHSs into those within 2.5 kb of a TSS (958; 46.5%) and the remaining ones, which were classified

转录起始位点的分类 我们分析了两个数据集，进而对 ENCODE 区域的 TSS 进行分类：GENCODE 注释的转录本的 5′端和两个 5′端捕获技术——CAGE 和 PET 标签的综合结果。起初的结果表明有 16,051 个唯一的 TSS 的潜在存在。但是，在很多情况下，多个 TSS 位于同一个小的片段内(最多 ~ 200 碱基)，这是因为一些含有 TSS 的启动子含有很多非常靠近的精确的起始位点[47]。为了使这种影响正常化，我们将相距 60 或者更少碱基的 TSS 归为一个簇，并且在这种情况下把频率最高的 CAGE 或 PET 标签(或者对于仅在 GENCODE 数据鉴定到 TSS 则是最 5′的 TSS)作为这个簇的代表用于下游分析。

上述分析在 ENCODE 区域产生了 7,157 个 TSS 簇。我们将这些 TSS 分为三类：已知的(在 GENCODE 确定的转录本的末端存在)，全新的(被其他证据支持的)和不被支持的。基于支持的证据的性质，全新的 TSS 被进一步分类(详见表 3 和补充信息 3.5)，利用 GSC 统计发现，产生的 4 个亚类都与实验证据高度重合。当仅分析单一标签时，尽管单一标签的相对比例在新类里更大，但是全新的 TSS 仍然与支持证据高度重合(详见补充信息 3.5.1)。

表 3. 基于支持不同转录本鉴定方法而定义的不同类的 TSS

类别	转录本鉴定方法	TSS 簇的数量(非冗余的)*	*P* 值†	单一簇‡(%)
已知的	GENCODE 5′端	1,730	2×10^{-70}	25(总共 74)
全新的	GENCODE 正义链外显子	1,437	6×10^{-39}	64
	GENCODE 反义链外显子	521	3×10^{-8}	65
	无偏性的转录鉴定	639	7×10^{-63}	71
	CpG 岛	164	4×10^{-90}	60
不被支持的	无	2,666	–	83.4

* 基于本文 TSS 簇的数量，不包括更高层级分类的 TSS。

† 转录本支持和 PET/CAGE 标签重合的概率，通过基因组结构校正统计计算(详见补充信息 1.3)。

‡ 只有一个标签的簇的比例。对于“已知的”这一类，计算的是含有标签支持的(25%)或所有的(74%) GENCODE 5′端的比例。

将基因组特征与染色质结构和转录因子结合相关联 通过检测对 DNaseI 切割的相对敏感度(详见补充信息 3.3)，我们在整个 ENCODE 区域鉴定了 DNaseI 超敏感位点。DHS 和 TSS 都反映了被认为是富集调控信息的基因组区域，并且很多 DHS 位于 TSS 或其附近。我们将 DHS 分为在 TSS 2.5 kb 以内的(958；46.5%)和其他位于远端的(1,102；53.5%)。然后我们通过根据相对 TSS 或 DHS 的距离将信号聚集，

as distal (1,102; 53.5%). We then cross-analysed the TSSs and DHSs with data sets relating to histone modifications, chromatin accessibility and sequence-specific transcription factor binding by summarizing these signals in aggregate relative to the distance from TSSs or DHSs. Figure 5 shows representative profiles of specific histone modifications, Pol II and selected transcription factor binding for the different categories of TSSs. Further profiles and statistical analysis of these studies can be found in Supplementary Information 3.6.

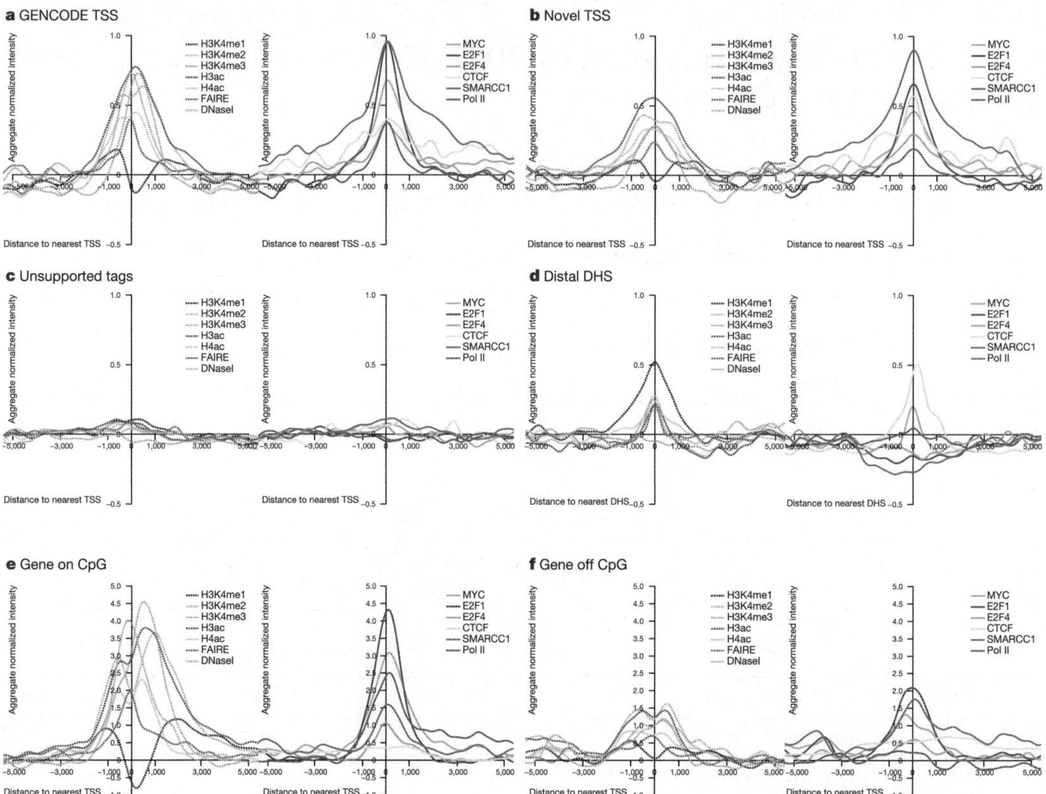

Fig. 5. Aggregate signals of tiling-array experiments from either ChIP-chip or chromatin structure assays, represented for different classes of TSSs and DHS. For each plot, the signal was first normalized with a mean of 0 and standard deviation of 1, and then the normalized scores were summed at each position for that class of TSS or DHS and smoothed using a kernel density method (see Supplementary Information section 3.6). For each class of sites there are two adjacent plots. The left plot depicts the data for general factors: FAIRE and DNaseI sensitivity as assays of chromatin accessibility and H3K4me1, H3K4me2, H3K4me3, H3ac and H4ac histone modifications (as indicated); the right plot shows the data for additional factors, namely MYC, E2F1, E2F4, CTCF, SMARCC1 and Pol II. The columns provide data for the different classes of TSS or DHS (unsmoothed data and statistical analysis shown in Supplementary Information section 3.6).

In the case of the three TSS categories (known, novel and unsupported), known and novel TSSs are both associated with similar signals for multiple factors (ranging from histone modifications through DNaseI accessibility), whereas unsupported TSSs are not. The enrichments seen with chromatin modifications and sequence-specific factors, along with the significant clustering of this evidence, indicate that the novel TSSs do not reflect false

对 TSS 和 DHS 与组蛋白修饰、染色质开放性和序列特异的转录因子结合相关的数据集进行了交叉分析。图 5 展示了特定的组蛋白修饰、Pol II 和特定转录因子结合不同类的 TSS 的代表性特征。这些研究更深入的特征和统计学分析可以在补充信息 3.6 中找到。

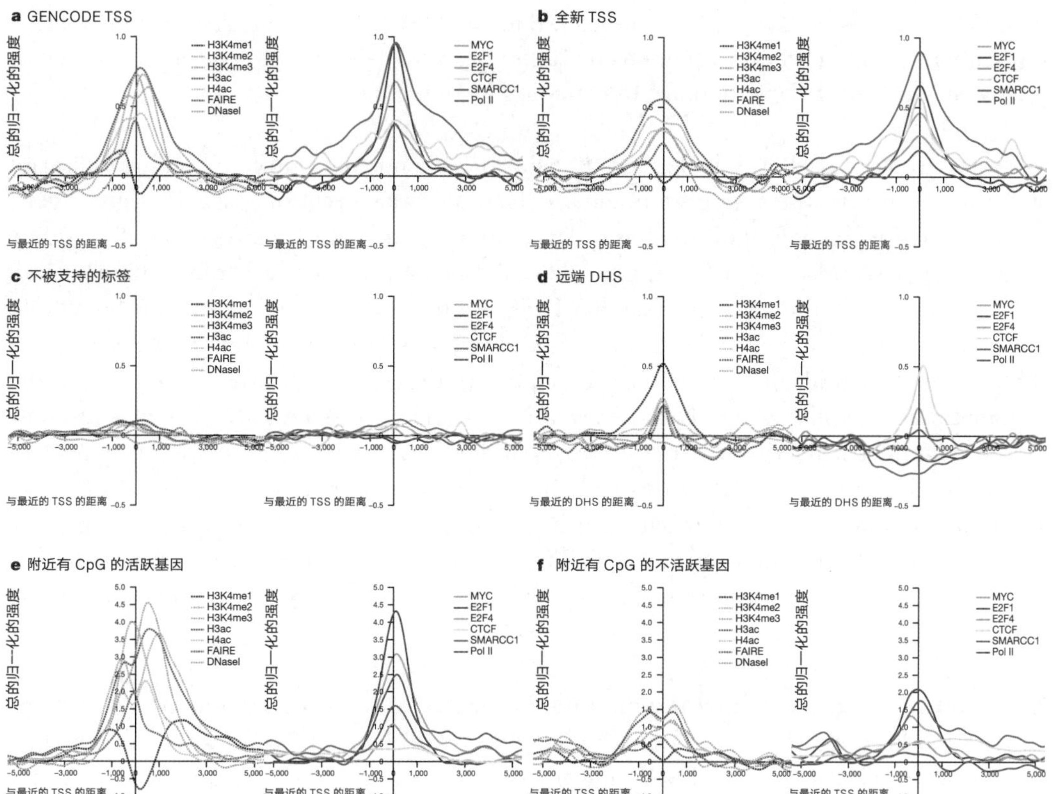

图 5. ChIP-chip 或染色质结构实验中的叠瓦式阵列实验的聚集信号，展示的是不同类型的 TSS 和 DHS 的情况。对于每一个图，信号先根据均值为 0、标准差为 1 被归一化，然后归一化后的值在相应类型的 TSS 或 DHS 的每一个位置求和，并利用核密度方法平滑化（详见补充信息 3.6）。对于各类位点有两个相邻的图。左图展示的是基本因子的数据：FAIRE 和 DNaseI 敏感性作为染色质开放性的检测方法，以及 H3K4me1、H3K4me2、H3K4me3、H3ac 和 H4ac 组蛋白修饰（如图所示）；右图展示的是另外的因子的数据，即 MYC、E2F1、E2F4、CTCF、SMARCC1 和 Pol II。各栏提供了不同类型的 TSS 或 DHS 的数据（未平滑的数据和统计学分析见补充信息 3.6）。

对于这三类 TSS（已知的、全新的和不被支持的），已知的和全新的 TSS 与多个因子类似的信号（从组蛋白修饰到 DNaseI 开放性）都有关联，而不被支持的 TSS 则不是这样。染色质修饰和序列特异的因子的富集，以及这些证据的明显聚集，说明全新的 TSS 并不是假阳性，可能使用和其他启动子一样的生物机制。序列特异的转

positives and probably use the same biological machinery as other promoters. Sequence-specific transcription factors show a marked increase in binding across the broad region that encompasses each TSS. This increase is notably symmetric, with binding equally likely upstream or downstream of a TSS (see Supplementary Information section 3.7 for an explanation of why this symmetrical signal is not an artefact of the analysis of the signals). Furthermore, there is enrichment of SMARCC1 binding (a member of the SWI/SNF chromatin-modifying complex), which persists across a broader extent than other factors. The broad signals with this factor indicate that the ChIP-chip results reflect both specific enrichment at the TSS and broader enrichments across ~5-kb regions (this is not due to technical issues, see Supplementary Information section 3.8).

We selected 577 GENCODE-defined TSSs at the 5′ ends of a protein-coding transcript with over 3 exons, to assess expression status. Each transcript was classified as: (1) "active" (gene on) or "inactive" (gene off) on the basis of the unbiased transcript surveys, and (2) residing near a "CpG island" or not ("non-CpG island") (see Supplementary Information section 3.17). As expected, the aggregate signal of histone modifications is mainly attributable to active TSSs (Fig. 5), in particular those near CpG islands. Pronounced doublet peaks at the TSS can be seen with these large signals (similar to previous work in yeast[48]) owing to the chromatin accessibility at the TSS. Many of the histone marks and Pol II signals are now clearly asymmetrical, with a persistent level of Pol II into the genic region, as expected. However, the sequence-specific factors remain largely symmetrically distributed. TSSs near CpG islands show a broader distribution of histone marks than those not near CpG islands (see Supplementary Information section 3.6). The binding of some transcription factors (E2F1, E2F4 and MYC) is extensive in the case of active genes, and is lower (or absent) in the case of inactive genes.

Chromatin signature of distal elements. Distal DHSs show characteristic patterns of histone modification that are the inverse of TSSs, with high H3K4me1 accompanied by lower levels of H3K4Me3 and H3Ac (Fig. 5). Many factors with high occupancy at TSSs (for example, E2F4) show little enrichment at distal DHSs, whereas other factors (for example, MYC) are enriched at both TSSs and distal DHSs[49]. A particularly interesting observation is the relative enrichment of the insulator-associated factor CTCF[50] at both distal DHSs and TSSs; this contrasts with SWI/SNF components SMARCC2 and SMARCC1, which are TSS-centric. Such differential behaviour of sequence-specific factors points to distinct biological differences, mediated by transcription factors, between distal regulatory sites and TSSs.

Unbiased maps of sequence-specific regulatory factor binding. The previous section focused on specific positions defined by TSSs or DHSs. We then analysed sequence-specific transcription factor binding data in an unbiased fashion. We refer to regions with enriched binding of regulatory factors as RFBRs. RFBRs were identified on the basis of ChIP-chip data in two ways: first, each investigator developed and used their own analysis method(s) to define high-enrichment regions, and second (and independently), a stringent false discovery rate (FDR) method was applied to analyse all data using three cut-offs (1%, 5% and 10%).

录因子在包含 TSS 的较宽区域内的结合明显增加。这种增加是明显对称的，在 TSS 上游或下游结合的可能性一样（详见补充信息 3.7，其中有解释为什么这种对称的信号不是一种信号分析的假象）。另外，还有 SMARCC1 结合的富集（SWI/SNF 染色质修饰复合物的一个组分），它比其他因子分布的范围更宽。这个因子较宽的信号暗示，ChIP-chip 结果不仅反映在 TSS 特异的富集，也反映在更宽的约 5 kb 区域的富集（这不是因为技术的原因，详见补充信息 3.8）。

我们选择 577 个 GENCODE 定义的 TSS 来评价表达水平，这些 TSS 都位于蛋白质编码且含有超过三个外显子的转录本的 5′端。各个转录本被分为：(1) 基于无偏性的转录本检测确定的“活跃的”（基因开启）或者“不活跃的”（基因关闭）和 (2) 位于“CpG 岛”附近或者不在附近的（“非 CpG 岛”）（详见补充信息 3.17）。正如预期的，组蛋白修饰的聚集信号主要位于活跃的 TSS（图 5），特别是那些 CpG 岛附近的。由于在 TSS 染色质的开放性，在 TSS 位置这些大的信号能看到明显的双峰（类似于之前在酵母中的发现 [48]）。正如预期的，很多组蛋白标志和 Pol II 信号现在明显是不对称的，Pol II 信号延伸到了基因区。但是，序列特异的因子仍然大部分是对称分布的。与不靠近 CpG 岛的 TSS 相比，CpG 岛附近的 TSS 呈现更宽的组蛋白标志分布（详见补充信息 3.6）。对于活跃基因，一些转录因子（E2F1、E2F4 和 MYC）广泛结合，而对于不活跃的基因，则结合水平较低（或不结合）。

远端元件的染色质标志　远端 DHS 呈现的特征性组蛋白修饰模式与 TSS 相反，高水平的 H3K4me1 伴随着低水平的 H3K4Me3 和 H3Ac（图 5）。很多在 TSS 结合水平较高的因子（如 E2F4）在远端 DHS 几乎没有富集，而其他因子（如 MYC）则在 TSS 和远端 DHS 都有富集 [49]。一个特别有意思的现象是，绝缘子相关因子 CTCF[50] 在远端 DHS 和 TSS 均有相对富集，这与 SWI/SNF 组分 SMARCC2 和 SMARCC1 明显不同，后者以 TSS 为中心。序列特异的因子的这种差异的行为表明，在远端调控位点和 TSS 之间，存在由转录因子介导的明显的生物学差异。

序列特异的调控因子结合的无偏性图谱　前面的部分聚焦于 TSS 或 DHS 定义的特定的位置。接下来我们以无偏性的方式分析序列特异的转录因子结合数据。我们称有调控因子富集结合的区域为 RFBR。基于 ChIP-chip 数据，RFBR 通过两种方式被鉴定：第一，每个研究者开发和使用他们自己的分析方法定义高富集区；第二（单独地），一种严格的错误发现率（FDR）方法被应用于分析所有的数据，它使用三个阈值（1%、5% 和 10%）。实验室特异的和基于 FDR 的方法高度相关，特别是对于

The laboratory-specific and FDR-based methods were highly correlated, particularly for regions with strong signals[10,11]. For consistency, we used the results obtained with the FDR-based method (see Supplementary Information section 3.10). These RFBRs can be used to find sequence motifs (see Supplementary Information section S3.11).

RFBRs are associated with the 5′ ends of transcripts. The distribution of RFBRs is non-random (see ref. 10) and correlates with the positions of TSSs. We examined the distribution of specific RFBRs relative to the known TSSs. Different transcription factors and histone modifications vary with respect to their association with TSSs (Fig. 6; see Supplementary Information section 3.12 for modelling of random expectation). Factors for which binding sites are most enriched at the 5′ ends of genes include histone modifications, TAF1 and RNA Pol II with a hypo-phosphorylated carboxy-terminal domain[51]—confirming previous expectations. Surprisingly, we found that E2F1, a sequence-specific factor that regulates the expression of many genes at the G1 to S transition[52], is also tightly associated with TSSs[52]; this association is as strong as that of TAF1, the well-known TATA box-binding protein associated factor 1 (ref. 53). These results suggest that E2F1 has a more general role in transcription than previously suspected, similar to that for MYC[54-56]. In contrast, the large-scale assays did not support the promoter binding that was found in smaller-scale studies (for example, on SIRT1 and SPI1 (PU1)).

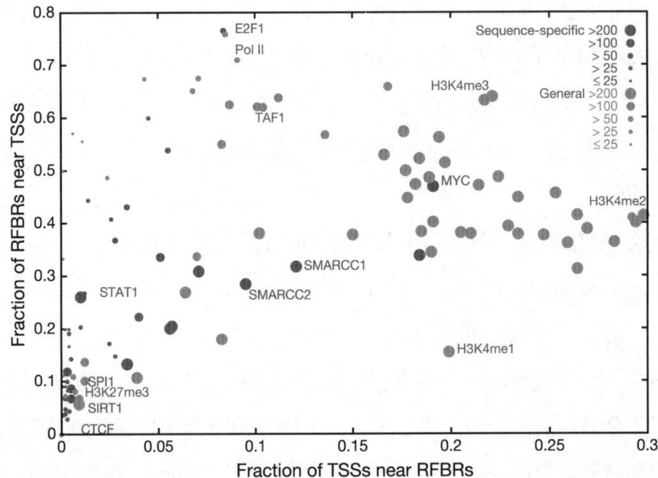

Fig 6. Distribution of RFBRs relative to GENCODE TSSs. Different RFBRs from sequence-specific factors (red) or general factors (blue) are plotted showing their relative distribution near TSSs. The x axis indicates the proportion of TSSs close (within 2.5 kb) to the specified factor. The y axis indicates the proportion of RFBRs close to TSSs. The size of the circle provides an indication of the number of RFBRs for each factor. A handful of representative factors are labelled.

Integration of data on sequence-specific factors. We expect that regulatory information is not dispersed independently across the genome, but rather is clustered into distinct regions[57]. We refer to regions that contain multiple regulatory elements as "regulatory clusters". We sought to predict the location of regulatory clusters by cross-integrating data generated using all transcription factor and histone modification assays, including

强信号区域[10,11]。为了统一性，我们使用基于 FDR 方法获得的结果(详见补充信息 3.10)，这些 RFBR 可被用于发现序列模序(详见补充信息 S3.11)。

RFBR 与转录本的 5′端关联 RFBR 的分布不是随机的(详见参考文献 10)，并且与 TSS 的位置相关。我们考察了特定 RFBR 相对已知的 TSS 的分布。不同的转录因子和组蛋白修饰在它们与 TSS 关联方面各不相同(图 6；随机期望的建模详见补充信息 3.12)。在基因 5′端最富集的与位点结合的因子包括组蛋白修饰、TAF1 和含有低磷酸化羧基端结构域的 RNA Pol II[51]——验证之前的预期。令人意外的是，我们发现 E2F1，一个在 G1 到 S 转换阶段调控很多基因表达的序列特异的因子[52]，也与 TSS 紧密关联[52]；这种关联跟 TAF1 与 TSS 的关联一样强，TAF1 即人们熟知的 TATA 框结合蛋白相关因子 1(参考文献 53)。这些结果说明相比于之前的猜测，E2F1 在转录中有着更基本的功能，这类似于 MYC[54-56] 的功能。相反，大规模的实验并不支持之前小规模研究发现的启动子的结合(如 SIRT1 和 SPI1(PU1))。

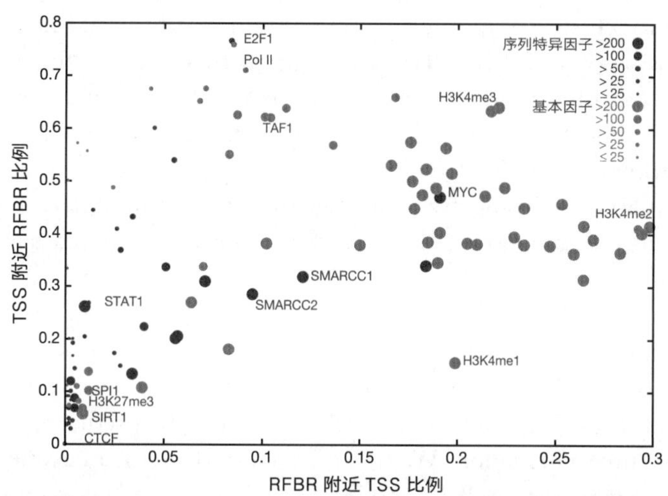

图 6. RFBR 相对 GENCODE TSS 的分布。图中展示的是源自序列特异的因子(红色)或基本因子(蓝色)的不同 RFBR 在 TSS 附近的相对分布。x 轴表示靠近(2.5 kb 以内)特定因子的 TSS 的比例。y 轴表示靠近 TSS 的 RFBR 的比例。圆圈的大小表示各个因子的 RFBR 的数量。少数几个代表性的因子被标记出来了。

序列特异的因子的数据的整合 我们预期调控信息并不是各自分散在基因组上，而是聚集在特定的区域[57]。我们称含有多个调控元件的区域为“调控簇”。利用所有转录因子和组蛋白修饰实验所产生的交叉整合的数据，包括在单个实验中低于主观阈值的结果，我们来预测调控簇的位置。特别地，我们利用了四种互为补充的

results falling below an arbitrary threshold in individual experiments. Specifically, we used four complementary methods to integrate the data from 129 ChIP-chip data sets (see Supplementary Information section 3.13 and ref. 58). These four methods detect different classes of regulatory clusters and as a whole identified 1,393 clusters. Of these, 344 were identified by all four methods, with another 500 found by three methods (see Supplementary Information section 3.13.5). 67% of the 344 regulatory clusters identified by all four methods (or 65% of the full set of 1,393) reside within 2.5 kb of a known or novel TSS (as defined above; see Table 3 and Supplementary Information section 3.14 for a breakdown by category). Restricting this analysis to previously annotated TSSs (for example, RefSeq or Ensembl) reveals that roughly 25% of the regulatory clusters are close to a previously identified TSS. These results suggest that many of the regulatory clusters identified by integrating the ChIP-chip data sets are undiscovered promoters or are somehow associated with transcription in another fashion. To test these possibilities, sets of 126 and 28 non-GENCODE-based regulatory clusters were tested for promoter activity (see Supplementary Information section 3.15) and by RACE, respectively. These studies revealed that 24.6% of the 126 tested regulatory clusters had promoter activity and that 78.6% of the 28 regulatory clusters analysed by RACE yielded products consistent with a TSS[58]. The ChIP-chip data sets were generated on a mixture of cell lines, predominantly HeLa and GM06990, and were different from the CAGE/ PET data, meaning that tissue specificity contributes to the presence of unique TSSs and regulatory clusters. The large increase in promoter proximal regulatory clusters identified by including the additional novel TSSs coupled with the positive promoter and RACE assays suggests that most of the regulatory regions identifiable by these clustering methods represent bona fide promoters (see Supplementary Information 3.16). Although the regulatory factor assays were more biased towards regions associated with promoters, many of the sites from these experiments would have previously been described as distal to promoters. This suggests that commonplace use of RefSeq- or Ensembl-based gene definition to define promoter proximity will dramatically overestimate the number of distal sites.

Predicting TSSs and transcriptional activity on the basis of chromatin structure. The strong association between TSSs and both histone modifications and DHSs prompted us to investigate whether the location and activity of TSSs could be predicted solely on the basis of chromatin structure information. We trained a support vector machine (SVM) by using histone modification data anchored around DHSs to discriminate between DHSs near TSSs and those distant from TSSs. We used a selected 2,573 DHSs, split roughly between TSS-proximal DHSs and TSS-distal DHSs, as a training set. The SVM performed well, with an accuracy of 83% (see Supplementary Information section 3.17). Using this SVM, we then predicted new TSSs using information about DHSs and histone modifications—of 110 high-scoring predicted TSSs, 81 resided within 2.5 kb of a novel TSS. As expected, these show a significant overlap to the novel TSS groups (defined above) but without a strong bias towards any particular category (see Supplementary Information section 3.17.1.5).

To investigate the relationship between chromatin structure and gene expression, we examined transcript levels in two cell lines using a transcript-tiling array. We compared

方法整合来自 129 个 ChIP-chip 数据集的数据（详见补充信息 3.13 和参考文献 58）。这四种方法检测不同类型的调控簇，合起来鉴定到 1,393 个簇。其中 344 个被四种方法都鉴定到，另外 500 个被三种方法鉴定到（详见补充信息 3.13.5）。被四种方法都鉴定到的 344 个调控簇中（或 1,393 个簇中的 65%），有 67% 位于已知的或全新的 TSS 2.5 kb 范围内（上文所定义的；按类分解情况详见表 3 和补充信息 3.14）。仅分析之前已经注释的 TSS（如 RefSeq 或者 Ensembl）发现大约 25% 的调控簇靠近之前鉴定的 TSS。这些结果说明通过整合 ChIP-chip 数据集鉴定到的调控簇中，有很多是还未被发现的启动子或者以其他的某种方式与转录存在着关联。为了检查这些可能性，启动子活性（详见补充信息 3.15）和 RACE 实验分别检测了 126 和 28 个非基于 GENCODE 的调控簇。这些研究发现 126 个被检测的调控簇中有 24.6% 具有启动子活性，28 个被 RACE 分析的调控簇中有 78.6% 产生产物，这与 TSS 一致[58]。与 CAGE/PET 数据不同，ChIP-chip 数据集产生于混合的细胞系，主要是 HeLa 和 GM06990，这意味着组织特异性也是存在独特的 TSS 和调控簇的一个原因。包括了额外全新的 TSS 以及阳性启动子和 RACE 实验后，鉴定到的启动子附近的调控簇大大增加了，这说明大部分通过这些聚类方法鉴定的调控区域代表的是真实的启动子（详见补充信息 3.16）。尽管调控因子实验更偏向启动子相关区域，很多这些实验得到的位点如果按以前的描述是远离启动子的。这说明简单地使用基于 RefSeq 或 Ensembl 定义的基因来确定是否靠近启动子会极大地高估远距离位点的数量。

基于染色质结构预测 TSS 和转录活性 TSS 与组蛋白修饰和 DHS 的高度的相关性促使我们探究是否仅基于染色质结构信息就能预测 TSS 的位置和活性。我们使用 DHS 周围的组蛋白修饰数据训练了一个支持向量机（SVM），用来区分靠近 TSS 的 DHS 和远离 TSS 的 DHS。我们选择了 2,573 个 DHS，大体分为靠近 TSS 的 DHS 和远离 TSS 的 DHS，作为一个训练集。SVM 表现良好，准确度为 83%（详见补充信息 3.17）。使用这个 SVM，然后我们利用 DHS 和组蛋白修饰信息预测新的 TSS——110 个高分预测的 TSS 中，81 个位于全新的 TSS 的 2.5 kb 范围内。正如预期的，它们与全新 TSS 组（上文定义的）有高度的重合，不过也没有特别偏向某一类（详见补充信息 3.17.1.5）。

为了研究染色质结构和基因表达的关系，我们利用转录本阵列实验在两种细胞系中检测了转录本的表达水平。我们将这个转录本数据与测定 ENCODE 区域组蛋白

this transcript data with the results of ChIP-chip experiments that measured histone modifications across the ENCODE regions. From this, we developed a variety of predictors of expression status using chromatin modifications as variables; these were derived using both decision trees and SVMs (see Supplementary Information section 3.17). The best of these correctly predicts expression status (transcribed versus non-transcribed) in 91% of cases. This success rate did not decrease dramatically when the predicting algorithm incorporated the results from one cell line to predict the expression status of another cell line. Interestingly, despite the striking difference in histone modification enrichments in TSSs residing near versus those more distal to CpG islands (see Fig. 5 and Supplementary Information section 3.6), including information about the proximity to CpG islands did not improve the predictors. This suggests that despite the marked differences in histone modifications among these TSS classes, a single predictor can be made, using the interactions between the different histone modification levels.

In summary, we have integrated many data sets to provide a more complete view of regulatory information, both around specific sites (TSSs and DHSs) and in an unbiased manner. From analysing multiple data sets, we find 4,491 known and novel TSSs in the ENCODE regions, almost tenfold more than the number of established genes. This large number of TSSs might explain the extensive transcription described above; it also begins to change our perspective about regulatory information—without such a large TSS catalogue, many of the regulatory clusters would have been classified as residing distal to promoters. In addition to this revelation about the abundance of promoter-proximal regulatory elements, we also identified a considerable number of putative distal regulatory elements, particularly on the basis of the presence of DHSs. Our study of distal regulatory elements was probably most hindered by the paucity of data generated using distal-element-associated transcription factors; nevertheless, we clearly detected a set of distal-DHS-associated segments bound by CTCF or MYC. Finally, we showed that information about chromatin structure alone could be used to make effective predictions about both the location and activity of TSSs.

Replication

Overview. DNA replication must be carefully coordinated, both across the genome and with respect to development. On a larger scale, early replication in S phase is broadly correlated with gene density and transcriptional activity[59-66]; however, this relationship is not universal, as some actively transcribed genes replicate late and vice versa[61,64-68]. Importantly, the relationship between transcription and DNA replication emerges only when the signal of transcription is averaged over a large window ($>$ 100 kb)[63], suggesting that larger-scale chromosomal architecture may be more important than the activity of specific genes[69].

The ENCODE Project provided a unique opportunity to examine whether individual histone modifications on human chromatin can be correlated with the time of replication and whether such correlations support the general relationship of active, open chromatin with early replication. Our studies also tested whether segments showing interallelic

修饰的 ChIP-chip 实验结果进行了比较。从中，我们以染色质修饰为变量开发了很多表达状态的预测器；这些是利用决策树和 SVM 派生的(详见补充信息 3.17)。其中最好的预测器正确预测了 91% 的表达状态(表达对不表达)。当预测算法吸收一个细胞系的结果去预测另一个细胞系的表达状态时，成功率并没有大幅降低。有趣的是，尽管组蛋白修饰富集情况在靠近 CpG 岛的 TSS 和和远离 CpG 岛的 TSS 上有很大不同(详见图 5 和补充信息 3.6)，包含靠近 CpG 岛的信息后并没有对预测器有所改观。这说明尽管在这些 TSS 种类中组蛋白修饰有很大不同，利用不同组蛋白修饰水平的相互作用，仅需一个预测器即可。

总结一下，我们研究了很多数据集，目的是更全面地了解调控信息，不仅是特定位点(TSS 和 DHS)附近的，并且是以无偏性的方式。从多个数据集的分析中，我们在 ENCODE 区域发现了 4,491 个已知的和全新的 TSS，这几乎比已确定的基因数量多了十倍。这么大量的 TSS 可能解释了上文提到的广泛的转录，也开始改变我们关于调控信息的认识——如果没有这么一大类 TSS，很多调控簇将会被分类为远离启动子。除了揭露了靠近启动子的调控元件的丰富性外，我们还鉴定到相当数量的假定远端调控元件，特别是基于 DHS 的存在。我们关于远端调控元件的研究最可能被远端元件相关转录因子的数据的缺乏所影响，但是，我们还是清楚地检测到了一组被 CTCF 或 MYC 结合的远端 DHS 相关片段。最后，我们表明仅靠染色质结构的信息就能对 TSS 的位置和活性做出有效的预测。

复　制

概况　无论是在基因组范围内，还是在发育过程中，DNA 复制必须被精确地协调。在更大的尺度下，在 S 期的早期复制与基因密度和转录活性有着广泛的相关性 [59-66]；但是，这种关系并不是普遍适用的，因为一些活跃转录的基因复制较晚，反之亦然 [61,64-68]。重要的是，只有当转录信号在一个较大的窗口(> 100 kb)中被平均时，转录与 DNA 复制之间的联系才会出现 [63]，这说明大尺度的染色体构建可能比特定基因的活性更重要 [69]。

ENCODE 计划提供了一个独特的机会来检验是否人类染色质上单一的组蛋白修饰可以与复制的时间相关联，以及是否这种关联支持活跃的开放染色质与早期复制

variation in the time of replication have two different types of histone modifications consistent with an interallelic variation in chromatin state.

DNA replication data set. We mapped replication timing across the ENCODE regions by analysing Brd-U-labelled fractions from synchronized HeLa cells (collected at 2 h intervals throughout S phase) on tiling arrays (see Supplementary Information section 4.1). Although the HeLa cell line has a considerably altered karyotype, correlation of these data with other cell line data (see below) suggests the results are relevant to other cell types. The results are expressed as the time at which 50% of any given genomic position is replicated (TR50), with higher values signifying later replication times. In addition to the five "activating" histone marks, we also correlated the TR50 with H3K27me3, a modification associated with polycomb-mediated transcriptional repression[70-74]. To provide a consistent comparison framework, the histone data were smoothed to 100-kb resolution, and then correlated with the TR50 data by a sliding window correlation analysis (see Supplementary Information section 4.2). The continuous profiles of the activating marks, histone H3K4 mono-, di-, and tri-methylation and histone H3 and H4 acetylation, are generally anti-correlated with the TR50 signal (Fig. 7a and Supplementary Information section 4.3). In contrast, H3K27me3 marks show a predominantly positive correlation with late-replicating segments (Fig. 7a; see Supplementary Information section 4.3 for additional analysis).

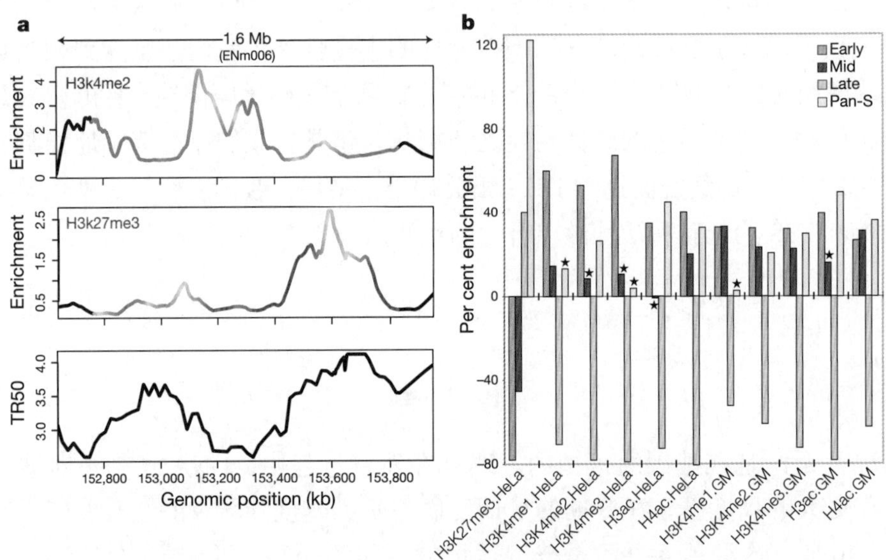

Fig. 7. Correlation between replication timing and histone modifications. **a**, Comparison of two histone modifications (H3K4me2 and H3K27me3), plotted as enrichment ratio from the Chip-chip experiments and the time for 50% of the DNA to replicate (TR50), indicated for ENCODE region ENm006. The colours on the curves reflect the correlation strength in a sliding 250-kb window. **b**, Differing levels of histone modification for different TR50 partitions. The amounts of enrichment or depletion of different histone modifications in various cell lines are depicted (indicated along the bottom as "histone mark.cell line"; GM = GM06990). Asterisks indicate enrichments/depletions that are not significant on the basis of multiple tests. Each set has four partitions on the basis of replication timing: early, mid, late and Pan-S.

的基本关系。我们的研究还检验了表现为等位基因间复制时间不同的片段是否有两种不同的组蛋白修饰，即染色质状态在等位基因间不同。

DNA 复制数据集 我们利用叠瓦式阵列分析 Brd-U 标记的同步化的 HeLa 细胞（在 S 期每隔 2h 收集），绘制 ENCODE 区域的复制时间图谱（详见补充信息 4.1）。尽管 HeLa 细胞系核型变化较大，但是利用其他细胞系数据（见下文）进行校正发现，结果跟其他细胞系是相近的。结果以任一给定基因组位置在 50% 的细胞中被复制时的时间（TR50）表示，数值越大代表复制时间越晚。除了五个“激活的”组蛋白标志，我们还将 TR50 与 H3K27me3 关联了起来，H3K27me3 是一个与多梳蛋白介导的转录抑制相关的修饰[70-74]。为了统一比较，组蛋白数据被平滑化到 100 kb 的分辨率，然后通过滑动窗口相关性分析将其与 TR50 数据关联（详见补充信息 4.2）。激活标志物组蛋白 H3K4 单、二和三甲基化以及组蛋白 H3 和 H4 乙酰化的连续图谱通常与 TR50 信号负相关（图 7a 和补充信息 4.3）。相反，H3K27me3 标志物则明显与晚期复制片段正相关（图 7a；额外的分析详见补充信息 4.3）。

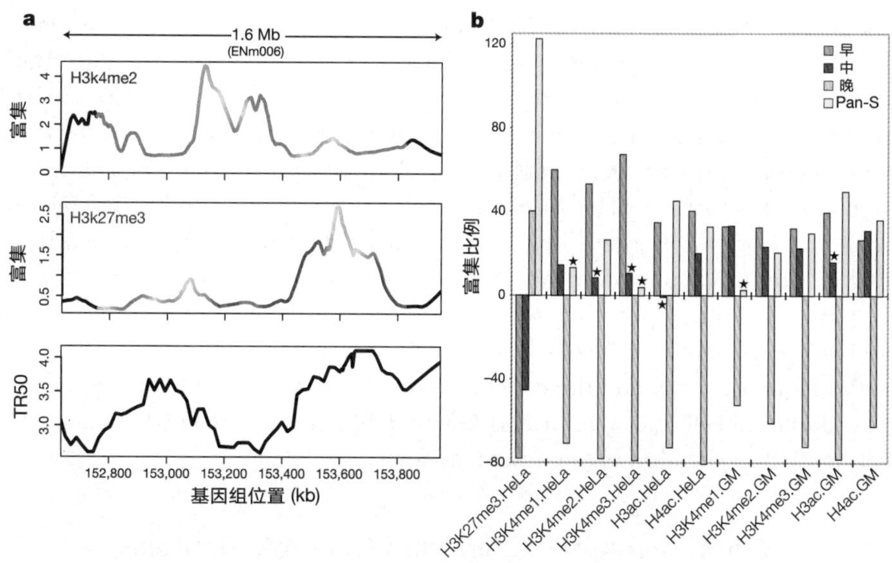

图 7. 复制时间和组蛋白修饰的相关性。**a**，比较两种组蛋白修饰（H3K4m2 和 H3K27me3），以 ChIP-chip 实验得到的富集率和 50% 的 DNA 复制的时间（TR50）画图，展示的是 ENCODE 区域 ENm006。曲线的颜色表示在一个 250 kb 的滑动窗口内的相关强度。**b**，对于不同 TR50 段组蛋白修饰的不同水平。图中描绘的是不同组蛋白修饰在不同细胞系中富集或缺失的量（在底部以“组蛋白标志 . 细胞系”表示；GM = GM06990）。星号表示基于多重检验不显著的富集/缺失。基于复制时间各组含有四段：早、中、晚和 Pan-S。

Although most genomic regions replicate in a temporally specific window in S phase, other regions demonstrate an atypical pattern of replication (Pan-S) where replication signals are seen in multiple parts of S phase. We have suggested that such a pattern of replication stems from interallelic variation in the chromatin structure[59,75]. If one allele is in active chromatin and the other in repressed chromatin, both types of modified histones are expected to be enriched in the Pan-S segments. An ENCODE region was classified as non-specific (or Pan-S) regions when > 60% of the probes in a 10-kb window replicated in multiple intervals in S phase. The remaining regions were sub-classified into early-, mid- or late-replicating based on the average TR50 of the temporally specific probes within a 10-kb window[75]. For regions of each class of replication timing, we determined the relative enrichment of various histone modification peaks in HeLa cells (Fig. 7b; Supplementary Information section 4.4). The correlations of activating and repressing histone modification peaks with TR50 are confirmed by this analysis (Fig. 7b). Intriguingly, the Pan-S segments are unique in being enriched for both activating (H3K4me2, H3ac and H4ac) and repressing (H3K27me3) histones, consistent with the suggestion that the Pan-S replication pattern arises from interallelic variation in chromatin structure and time of replication[75]. This observation is also consistent with the Pan-S replication pattern seen for the H19/IGF2 locus, a known imprinted region with differential epigenetic modifications across the two alleles[76].

The extensive rearrangements in the genome of HeLa cells led us to ask whether the detected correlations between TR50 and chromatin state are seen with other cell lines. The histone modification data with GM06990 cells allowed us to test whether the time of replication of genomic segments in HeLa cells correlated with the chromatin state in GM06990 cells. Early- and late-replicating segments in HeLa cells are enriched and depleted, respectively, for activating marks in GM06990 cells (Fig. 7b). Thus, despite the presence of genomic rearrangements (see Supplementary Information section 2.12), the TR50 and chromatin state in HeLa cells are not far from a constitutive baseline also seen with a cell line from a different lineage. The enrichment of multiple activating histone modifications and the depletion of a repressive modification from segments that replicate early in S phase extends previous work in the field at a level of detail and scale not attempted before in mammalian cells. The duality of histone modification patterns in Pan-S areas of the HeLa genome, and the concordance of chromatin marks and replication time across two disparate cell lines (HeLa and GM06990) confirm the coordination of histone modifications with replication in the human genome.

Chromatin Architecture and Genomic Domains

Overview. The packaging of genomic DNA into chromatin is intimately connected with the control of gene expression and other chromosomal processes. We next examined chromatin structure over a larger scale to ascertain its relation to transcription and other processes. Large domains (50 to > 200 kb) of generalized DNaseI sensitivity have been detected around developmentally regulated gene clusters[77], prompting speculation that the genome is organized into "open" and "closed" chromatin territories that represent higher-order functional domains. We explored how different chromatin features, particularly histone

尽管大多数的基因组区域在 S 期以一种时间特异窗口复制，也有一些区域呈现非典型的复制模式（Pan-S），即复制信号出现在 S 期的多个阶段。我们已经表明这样一种复制模式源于染色质结构在等位基因间的不同[59,75]。如果一个等位基因位于活跃的染色质而另一个位于被抑制的染色质，则这两种修饰的组蛋白都会在 Pan-S 区间富集。当一个 ENCODE 区域在 10 kb 窗口里有 ＞60% 的探针在 S 期的多个阶段被复制，则这个区域被归类为非特异的（或者 Pan-S）区域。基于在 10 kb 窗口内的时间特异探针的平均 TR50 值，剩下的区域被进一步分类为早、中或晚期复制[75]。对于每一类复制时间的区域，我们确定了各种组蛋白修饰峰在 HeLa 细胞中的相对富集（图 7b；补充信息 4.4）。这个分析证实了激活和抑制的组蛋白修饰峰与 TR50 的相关性（图 7b）。有趣的是，Pan-S 区间是独特的，激活（H3K4me2、H3ac 和 H4ac）和抑制（H3K27me3）的组蛋白都有富集，这也说明 Pan-S 复制模式是由于在染色质结构和复制时间方面等位基因的差异[75]。这种现象也与 H19/IGF2 基因座的 Pan-S 复制模式一致，该基因座是一个已知的两个等位基因不同的表观遗传学修饰印记区域。

HeLa 细胞广泛的基因组重排使我们不禁要问检测到的 TR50 和染色质状态之间的相关性是否能在其他细胞系中看到。GM06990 细胞的组蛋白修饰数据使我们能够检测是否 HeLa 细胞中基因组片段的复制时间与 GM06990 细胞中的染色质状态相关。HeLa 细胞中的早期和晚期复制片段分别在 GM06990 细胞中富集和缺失激活的标志物（图 7b）。因而，尽管存在基因组的重排（详见补充信息 2.12），HeLa 细胞中 TR50 和染色质状态差不多就是一个组成性的基准，也能在源于另一谱系的细胞系中看到。S 期早期复制片段的这种对多种激活性组蛋白修饰的富集和对抑制性修饰的缺失，以前所未有的详细度和尺度拓展了之前在哺乳动物细胞中这一领域的工作。HeLa 基因组 Pan-S 区域的组蛋白修饰模式的二元性，以及在两种不同的细胞系（HeLa 和 GM06990）中染色质标志物和复制时间的一致性，验证了人类基因组中组蛋白修饰和复制的相关性。

染色质结构和基因组结构域

概况　基因组 DNA 包装成染色质与基因表达的调控以及其他染色质过程紧密相关。接下来，我们探究更大尺度的染色质结构来确定它与转录和其他过程的关系。在发育调控的基因簇附近已经检测到广泛的 DNaseI 敏感性的大结构域（50 到 ＞200 kb）[77]，这促使我们猜测基因组被组成“开放”和“闭合”的染色质区域，代表着更高级的功能结构域。我们探索了不同的染色质特征，特别是组蛋白修饰，在短

modifications, correlate with chromatin structure, both over short and long distances.

Chromatin accessibility and histone modifications. We used histone modification studies and DNaseI sensitivity data sets (introduced above) to examine general chromatin accessibility without focusing on the specific DHS sites (see Supplementary Information sections 3.1, 3.3 and 3.4). A fundamental difficulty in analysing continuous data across large genomic regions is determining the appropriate scale for analysis (for example, 2 kb, 5 kb, 20 kb, and so on). To address this problem, we developed an approach based on wavelet analysis, a mathematical tool pioneered in the field of signal processing that has recently been applied to continuous-value genomic analyses. Wavelet analysis provides a means for consistently transforming continuous signals into different scales, enabling the correlation of different phenomena independently at differing scales in a consistent manner.

Global correlations of chromatin accessibility and histone modifications. We computed the regional correlation between DNaseI sensitivity and each histone modification at multiple scales using a wavelet approach (Fig. 8 and Supplementary Information section 4.2). To make quantitative comparisons between different histone modifications, we computed histograms of correlation values between DNaseI sensitivity and each histone modification at several scales and then tested these for significance at specific scales. Figure 8c shows the distribution of correlation values at a 16-kb scale, which is considerably larger than individual *cis*-acting regulatory elements. At this scale, H3K4me2, H3K4me3 and H3ac show similarly high correlation. However, they are significantly distinguished from H3K4me1 and H4ac modifications ($P < 1.5 \times 10^{-33}$; see Supplementary Information section 4.5), which show lower correlation with DNaseI sensitivity. These results suggest that larger-scale relationships between chromatin accessibility and histone modifications are dominated by sub-regions in which higher average DNaseI sensitivity is accompanied by high levels of H3K4me2, H3K4me3 and H3ac modifications.

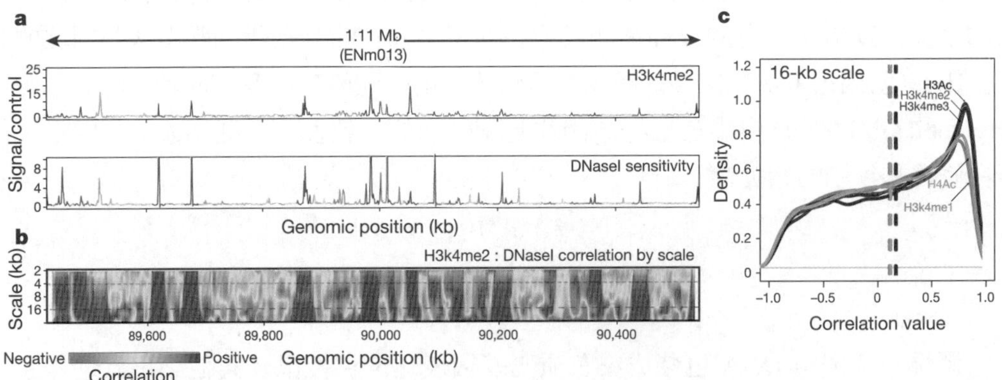

Fig. 8. Wavelet correlations of histone marks and DNaseI sensitivity. As an example, correlations between DNaseI sensitivity and H3K4me2 (both in the GM06990 cell line) over a 1.1-Mb region on chromosome 7 (ENCODE region ENm013) are shown. **a**, The relationship between histone modification H3K4me2 (upper plot) and DNaseI sensitivity (lower plot) is shown for ENCODE region ENm013. The curves are coloured with the strength of the local correlation at the 4-kb scale (top dashed line in panel **b**). **b**, The same data as in **a** are represented as a wavelet correlation. The *y* axis shows the differing scales decomposed

距离以及长距离方面，如何与染色质结构关联。

染色质开放性和组蛋白修饰 我们使用了组蛋白修饰研究和 DNaseI 敏感性数据集（上文介绍的）来探究整体的染色质开放性，而不是聚焦于特定的 DHS 位点（详见补充信息 3.1、3.3 和 3.4）。在大的基因组区域范围内的连续数据方面，一个主要的困难是确定合适的分析尺度（如 2 kb、5 kb、20 kb 等等）。为了解决这个问题，我们基于小波分析开发了一种方法，而小波分析是主要应用于信号处理领域的一种数学工具，最近已经被应用到连续值的基因组分析当中。小波分析提供了一种可以一致地将连续信号转换为不同尺度的方法，从而能以一种一致的方式将不同的现象在不同尺度下独立地关联。

染色质开放性与组蛋白修饰的整体相关性 我们利用小波的方法，在多种尺度下计算了 DNaseI 敏感性与每一种组蛋白修饰区域的相关性（图 8 和补充信息 4.2）。为了定量比较不同的组蛋白修饰，我们计算了 DNaseI 敏感性和每一种组蛋白修饰在多个尺度下的相关性值并画了直方图，然后对其进行检验以寻找在特定的尺度下的显著性。图 8c 展示了在 16 kb 尺度下相关性值的分布，它的值明显比单个顺式调控元件的大。在这个尺度下，H3K4me2、H3K4me3 和 H3ac 表现类似，都是高度相关。但是，它们跟 H3K4me1 和 H4ac 修饰显著不同（$P < 1.5 \times 10^{-33}$；详见补充信息 4.5），后者与 DNaseI 敏感性的相关性更低。这些结果表明染色质开放性和组蛋白修饰在大尺度下的关系，主要取决于更高 DNaseI 敏感性平均值伴随着较高水平 H3K4me2、H3K4me3 和 H3ac 修饰的亚区域。

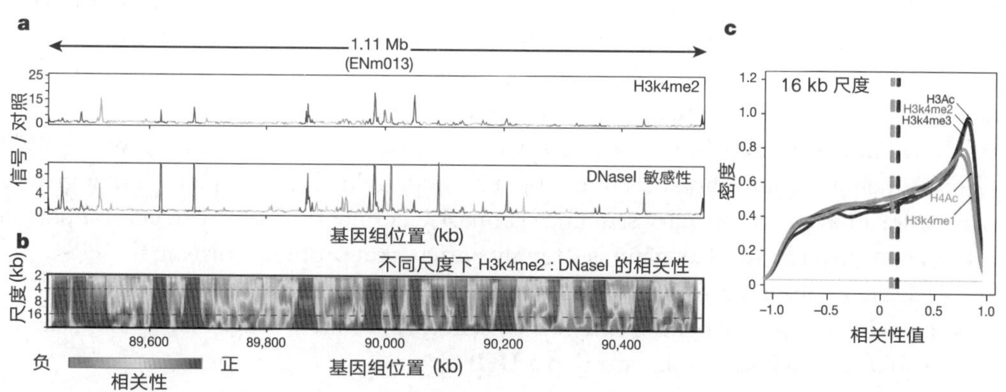

图 8. 组蛋白标志物和 DNaseI 敏感性的小波关联。作为一个例子，展示了在 7 号染色体超过 1.1 Mb 的区域内（ENCODE 区域 ENm013）DNaseI 敏感性和 H3K4me2（都是在 GM06990 细胞系）之间的相关性。**a**，展示的是 ENCODE 区域 ENm013 内组蛋白修饰 H3K4me2（上图）和 DNaseI 敏感性（下图）之间的关系。曲线的颜色是根据在 4 kb 尺度下局部相关性的强度（**b** 图中上方的虚线）。**b**，与 **a** 中同样的数据作

by the wavelet analysis from large to small scale (in kb); the colour at each point in the heatmap represents the level of correlation at the given scale, measured in a 20 kb window centred at the given position. **c**, Distribution of correlation values at the 16 kb scale between the indicated histone marks. The *y* axis is the density of these correlation values across ENCODE; all modifications show a peak at a positive-correlation value.

Local correlations of chromatin accessibility and histone modifications. Narrowing to a scale of ~2 kb revealed a more complex situation, in which H3K4me2 is the histone modification that is best correlated with DNaseI sensitivity. However, there is no clear combination of marks that correlate with DNaseI sensitivity in a way that is analogous to that seen at a larger scale (see Supplementary Information section 4.3). One explanation for the increased complexity at smaller scales is that there is a mixture of different classes of accessible chromatin regions, each having a different pattern of histone modifications. To examine this, we computed the degree to which local peaks in histone methylation or acetylation occur at DHSs (see Supplementary Information section 4.5.1). We found that 84%, 91% and 93% of significant peaks in H3K4 mono-, di- and tri-methylation, respectively, and 93% and 81% of significant peaks in H3ac and H4ac acetylation, respectively, coincided with DHSs (see Supplementary Information section 4.5). Conversely, a proportion of DHSs seemed not to be associated with significant peaks in H3K4 mono-, di- or tri-methylation (37%, 29% and 47%, respectively), nor with peaks in H3 or H4 acetylation (both 57%). Because only a limited number of histone modification marks were assayed, the possibility remains that some DHSs harbour other histone modifications. The absence of a more complete concordance between DHSs and peaks in histone acetylation is surprising given the widely accepted notion that histone acetylation has a central role in mediating chromatin accessibility by disrupting higher-order chromatin folding.

DNA structure at DHSs. The observation that distinctive hydroxyl radical cleavage patterns are associated with specific DNA structures[78] prompted us to investigate whether DHS subclasses differed with respect to their local DNA structure. Conversely, because different DNA sequences can give rise to similar hydroxyl radical cleavage patterns[79], genomic regions that adopt a particular local structure do not necessarily have the same nucleotide sequence. Using a Gibbs sampling algorithm on hydroxyl radical cleavage patterns of 3,150 DHSs[80], we discovered an 8-base segment with a conserved cleavage signature (CORCS; see Supplementary Information section 4.6). The underlying DNA sequences that give rise to this pattern have little primary sequence similarity despite this similar structural pattern. Furthermore, this structural element is strongly enriched in promoter-proximal DHSs (11.3-fold enrichment compared to the rest of the ENCODE regions) relative to promoter-distal DHSs (1.5-fold enrichment); this element is enriched 10.9-fold in CpG islands, but is higher still (26.4-fold) in CpG islands that overlap a DHS.

Large-scale domains in the ENCODE regions. The presence of extensive correlations seen between histone modifications, DNaseI sensitivity, replication, transcript density and protein factor binding led us to investigate whether all these features are organized systematically across the genome. To test this, we performed an unsupervised training of a two-state HMM with inputs from these different features (see Supplementary Information

为小波关联的代表。y 轴从大到小表示小波分析分解的不同尺度（单位 kb）；热图中每一个点的颜色代表在给定尺度下以给定位置为中心的 20 kb 窗口内测得的相关水平。**c**，在 16 kb 尺度下与相应组蛋白标志物的相关性值的分布。y 轴是这些相关性值在 ENCODE 区域的密度；所有的修饰在正相关值范围内都有一个峰。

染色质开放性和组蛋白修饰的局部相关性　将尺度缩小到约 2 kb 发现了一个更加复杂的情况，H3K4m2 是跟 DNaseI 敏感性最相关的组蛋白修饰。但是，没有明显的标志物组合与在更大尺度下看到的类似的方式与 DNaseI 敏感性相关（详见补充信息 4.3）。对于在更小尺度下增加的复杂性，一种解释是它混合了不同类型的开放染色质区域，每一类型含有不同的组蛋白修饰模式。为了研究这种可能性，我们计算了组蛋白甲基化或乙酰化中有多少局部峰是在 DHS 上（详见补充信息 4.5.1）。我们发现 H3K4 单、二和三甲基化中分别有 84%，91% 和 93% 显著的峰，以及 H3ac 和 H4ac 中分别有 93% 和 81% 的显著峰，与 DHS 是一致的（详见补充信息 4.5）。相反，一部分 DHS 则好像并不与 H3K4 单、二或三甲基化中显著的峰关联（分别为 37%、29% 和 47%），也不与 H3 或 H4 乙酰化中的峰关联（均为 57%）。因为只有有限数量的组蛋白修饰标志被检测，所以一些含有其他的组蛋白修饰的 DHS 也可能存在。鉴于广泛接受的观点是组蛋白乙酰化通过破坏更高级的染色质折叠而在介导染色质开放性方面发挥核心作用，DHS 和组蛋白乙酰化峰之间缺少更完全的一致性是出乎意料的。

DHS 处的 DNA 结构　观察到不同的羟自由基切割模式与特定的 DNA 结构[78]相关，促使我们研究 DHS 亚类是否会因它们局部的 DNA 结构而不同。反过来，因为不同的 DNA 序列能产生相似的羟自由基切割模式[79]，采用特定一种局部结构的基因组区域也未必含有相同的核苷酸序列。对 3,150 个 DHS 的羟自由基切割模式应用吉布斯采样算法，我们发现了一个 8 碱基片段含有保守的切割特征（CORCS；详见补充信息 4.6）。尽管有着这种类似的结构模式，产生这种模式的潜在 DNA 序列在一级序列上却几乎没有相似性。另外，这种结构元件在靠近启动子的 DHS(跟其他的 ENCODE 区域相比有 11.3 倍富集）比在远离启动子的 DHS 明显富集（1.5 倍富集），这种元件在 CpG 有 10.9 倍的富集，而在与 DHS 重合 CpG 到则富集倍数更高（26.4 倍）。

在 ENCODE 区域内的大尺度结构域　组蛋白修饰、DNaseI 敏感性、复制、转录本密度和蛋白质因子结合间广泛的相关性的存在，促使我们研究是否所有这些特征是在基因组范围内系统地组织起来的。为了检验这种可能性，我们输入这些不同的特征数据，无监督训练一个两态的 HMM(详见补充信息 4.7 和参考文献 81)。在

section 4.7 and ref. 81). No other information except for the experimental variables was used for the HMM training routines. We consistently found that one state ("active") generally corresponded to domains with high levels of H3ac and RNA transcription, low levels of H3K27me3 marks, and early replication timing, whereas the other state ("repressed") reflected domains with low H3ac and RNA, high H3K27me3, and late replication (see Fig. 9). In total, we identified 70 active regions spanning 11.4 Mb and 82 inactive regions spanning 17.8 Mb (median size 136 kb versus 104 kb respectively). The active domains are markedly enriched for GENCODE TSSs, CpG islands and Alu repetitive elements ($P < 0.0001$ for each), whereas repressed regions are significantly enriched for LINE1 and LTR transposons ($P < 0.001$). Taken together, these results demonstrate remarkable concordance between ENCODE functional data types and provide a view of higher-order functional domains defined by a broader range of factors at a markedly higher resolution than was previously available[82].

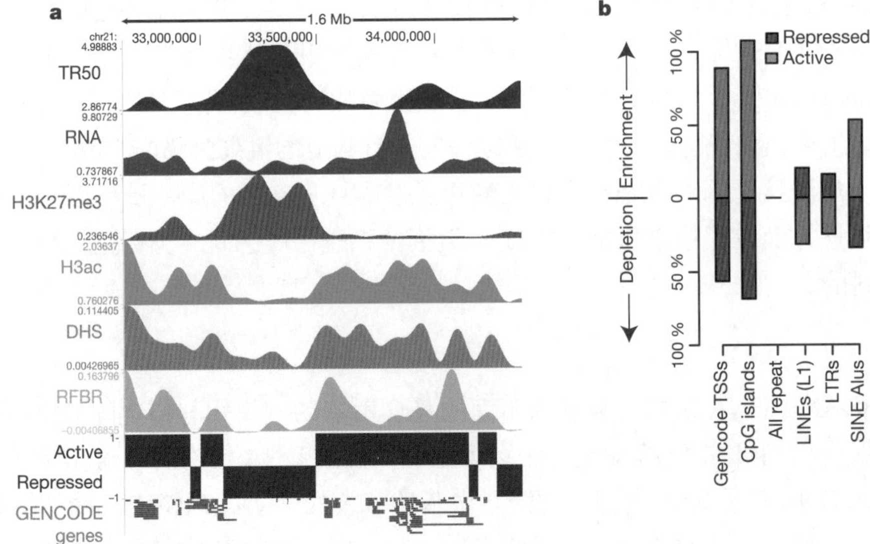

Fig. 9. Higher-order functional domains in the genome. The general concordance of multiple data types is shown for an illustrative ENCODE region (ENm005). **a**, Domains were determined by simultaneous HMM segmentation of replication time (TR50; black), bulk RNA transcription (blue), H3K27me3 (purple), H3ac (orange), DHS density (green), and RFBR density (light blue) measured continuously across the 1.6-Mb ENm005. All data were generated using HeLa cells. The histone, RNA, DHS and RFBR signals are wavelet-smoothed to an approximately 60-kb scale (see Supplementary Information section 4.7). The HMM segmentation is shown as the blocks labelled "active" and "repressed" and the structure of GENCODE genes (not used in the training) is shown at the end. **b**, Enrichment or depletion of annotated sequence features (GENCODE TSSs, CpG islands, LINE1 repeats, Alu repeats, and non-exonic constrained sequences (CSs)) in active versus repressed domains. Note the marked enrichment of TSSs, CpG islands and Alus in active domains, and the enrichment of LINE and LTRs in repressed domains.

Evolutionary Constraint and Population Variability

Overview. Functional genomic sequences can also be identified by examining evolutionary changes across multiple extant species and within the human population. Indeed, such

HMM 训练过程中，除了实验变量，没有使用其他的信息。我们不断发现一个状态（“活跃的”）通常对应于含有高水平 H3ac 和 RNA 转录、低水平 H3K27me3 标志物以及早期复制时间的结构域，而另一个状态（“抑制的”）反映的是含有低水平 H3ac 和 RNA、高水平 H3K27me3 以及晚期复制的结构域（见图 9）。总共，我们鉴定到横跨 11.4 Mb 的 70 个活跃区域和横跨 17.8 Mb 的 82 个不活跃的区域（中位数大小分别为 136 kb 和 104 kb）。活跃的结构域明显富集 GENCODE TSS、CpG 岛和 Alu 重复元件（每个都 $P < 0.0001$），而抑制的区域则明显富集 LINE1 和 LTR 转座子（$P < 0.001$）。总而言之，这些结果显示了 ENCODE 功能数据类型间显著的一致性，并且提供一个比之前已有的更高级的功能结构域图谱，该功能结构域被更广范围的因子所定义以及有着明显更高的分辨率。

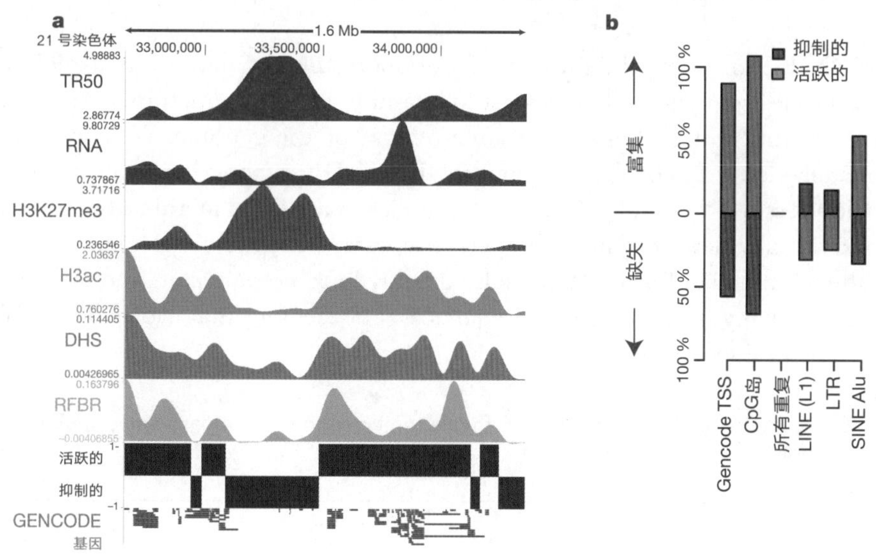

图 9. 基因组中更高级的功能结构域。本图展示的是一个用来说明 ENCODE 区域（ENm005）内多种数据类型的普遍一致性。**a**，结构域是通过对 1.6 Mb 的 ENm005 范围内对复制时间（TR50；黑色）、大量 RNA 转录（蓝色）、H3K27me3（紫色）、H3ac（橘色）、DHS 密度（绿色）和 RFBR 密度（淡蓝色）连续测量，同时进行 HMM 分段而确定的。所有的数据都是利用 HeLa 细胞产生的。组蛋白、RNA、DHS 和 RFBR 信号被小波平滑化成大约 60 kb 的尺度（详见补充信息 4.7）。HMM 分段以标为“活跃的”和“抑制的”的区块表示，最下面是 GENCODE 基因的结构（训练中没有使用）。**b**，在活跃和抑制的结构域中被注释的序列特征（GENCODE TSS、CpG 岛、LINE1 重复、Alu 重复和非外显子的限制序列（CS））的富集或缺失的比较。注意在活跃的结构域中 TSS、CpG 岛和 Alu 明显富集，以及在抑制的结构域中 LINE 和 LTR 富集。

进化的限制和群体的变异性

概况 功能性的基因组序列也能通过分析多个现存物种间和人群内部进化的变化而被鉴定。确实，这样的研究补充了鉴定特定功能元件的实验方法[83-85]。进化的

studies complement experimental assays that identify specific functional elements[83-85]. Evolutionary constraint (that is, the rejection of mutations at a particular location) can be measured by either (i) comparing observed substitutions to neutral rates calculated from multi-sequence alignments[86-88], or (ii) determining the presence and frequency of intra-species polymorphisms. Importantly, both approaches are indifferent to any specific function that the constrained sequence might confer.

Previous studies comparing the human, mouse, rat and dog genomes examined bulk evolutionary properties of all nucleotides in the genome, and provided little insight about the precise positions of constrained bases. Interestingly, these studies indicated that the majority of constrained bases reside within the non-coding portion of the human genome. Meanwhile, increasingly rich data sets of polymorphisms across the human genome have been used extensively to establish connections between genetic variants and disease, but far fewer analyses have sought to use such data for assessing functional constraint[85].

The ENCODE Project provides an excellent opportunity for more fully exploiting inter- and intra-species sequence comparisons to examine genome function in the context of extensive experimental studies on the same regions of the genome. We consolidated the experimentally derived information about the ENCODE regions and focused our analyses on 11 major classes of genomic elements. These classes are listed in Table 4 and include two non-experimentally derived data sets: ancient repeats (ARs; mobile elements that inserted early in the mammalian lineage, have subsequently become dormant, and are assumed to be neutrally evolving) and constrained sequences (CSs; regions that evolve detectably more slowly than neutral sequences).

Table 4. Eleven classes of genomic elements subjected to evolutionary and population-genetics analyses

Abbreviation	Description
CDS	Coding exons, as annotated by GENCODE
5′UTR	5′ untranslated region, as annotated by GENCODE
3′UTR	3′ untranslated region, as annotated by GENCODE
Un.TxFrag	Unannotated region detected by RNA hybridization to tiling array (that is, unannotated TxFrag)
RxFrag	Region detected by RACE and analysis on tiling array
Pseudogene	Pseudogene identified by consensus pseudogene analysis
RFBR	Regulatory factor binding region identified by ChIP-chip assays
RFBR-SeqSp	Regulatory factor binding region identified only by ChIP-chip assays for factors with known sequence-specificity
DHS	DNaseI hypersensitive sites found in multiple tissues
FAIRE	Region of open chromatin identified by the FAIRE assay
TSS	Transcription start site
AR	Ancient repeat inserted early in the mammalian lineage and presumed to be neutrally evolving
CS	Constrained sequence identified by analysing multi-sequence alignments

限制(即在特定位置排斥变异)能通过(i)比较观察到的替换与多重序列比对计算得到的中性率[86-88]，或者(ii)确定物种内多态性的存在和频率而被衡量。重要的是，两种方法无关乎限制序列可能发挥的任何特定功能。

之前的研究比较了人类、小鼠、大鼠和狗的基因组，分析了基因组中所有核苷酸大的进化特性，而对受限制的碱基的精确位置几乎没有了解。有意思的是，这些研究表明大部分受限制的碱基位于人类基因组的非编码部分。同时，人类基因组范围的多态性的数据集越来越丰富，并且已经被广泛地用于建立遗传变异与疾病的联系，但是目前利用这样的数据来评价功能限制的分析还很少[85]。

ENCODE 计划提供了一个很好的机会来更加全面地探索物种间和物种内的序列比较，进而结合在基因组相同区域广泛进行的实验研究来研究基因组的功能。我们统一了实验得到关于 ENCODE 区域的信息，将我们的分析聚焦于 11 个主要类别的基因组元件。这些类别在表 4 中被列出，其中包含两个非实验得到数据集：古老的重复(AR；一种可移动的元件，在早期就插入到哺乳动物谱系但自此休眠，被认为是中性进化)和受限制的序列(CS；可检测到比中性序列进化得更缓慢的区域)。

表 4. 十一类被用于进化和群体遗传学分析的基因组元件

缩写	描述
CDS	编码外显子，根据 GENCODE 注释
5′UTR	5′非翻译区，根据 GENCODE 注释
3′UTR	3′非翻译区，根据 GENCODE 注释
Un.TxFrag	通过 RNA 杂交到叠瓦式阵列检测到的非翻译区(即非翻译的 TxFrag)
RxFrag	通过 RACE 和叠瓦式阵列分析检测到的区域
假基因	通过共有假基因分析鉴定到的假基因
RFBR	通过 ChIP-chip 实验鉴定到的调控因子结合区域
RFBR-SeqSp	对于已知含有序列特异性的因子，仅通过 ChIP-chip 实验鉴定到的调控因子结合区域
DHS	在多种组织中发现的 DNaseI 超敏感位点
FAIRE	通过 FAIRE 实验鉴定到的开放染色质区域
TSS	转录起始位点
AR	在早期就插入到哺乳动物谱系中的古老的重复，被认为是中性进化
CS	通过多重序列比对分析鉴定到的限制序列

Comparative sequence data sets and analysis. We generated 206 Mb of genomic sequence orthologous to the ENCODE regions from 14 mammalian species using a targeted strategy that involved isolating[89] and sequencing[90] individual bacterial artificial chromosome clones. For an additional 14 vertebrate species, we used 340 Mb of orthologous genomic sequence derived from genome-wide sequencing efforts[3-8,91-93]. The orthologous sequences were aligned using three alignment programs: TBA[94], MAVID[95] and MLAGAN[96]. Four independent methods that generated highly concordant results[97] were then used to identify sequences under constraint (PhastCons[88], GERP[87], SCONE[98] and BinCons[86]). From these analyses, we developed a high-confidence set of "constrained sequences" that correspond to 4.9% of the nucleotides in the ENCODE regions. The threshold for determining constraint was set using a FDR rate of 5% (see ref. 97); this level is similar to previous estimates of the fraction of the human genome under mammalian constraint[4,86-88] but the FDR rate was not chosen to fit this result. The median length of these constrained sequences is 19 bases, with the minimum being 8 bases—roughly the size of a typical transcription factor binding site. These analyses, therefore, provide a resolution of constrained sequences that is substantially better than that currently available using only whole-genome vertebrate sequences[99-102].

Intra-species variation studies mainly used SNP data from Phases I and II, and the 10 re-sequenced regions in ENCODE regions with 48 individuals of the HapMap Project[103]; nucleotide insertion or deletion (indel) data were from the SNP Consortium and HapMap. We also examined the ENCODE regions for the presence of overlaps with known segmental duplications[104] and CNVs.

Experimentally identified functional elements and constrained sequences. We first compared the detected constrained sequences with the positions of experimentally identified functional elements. A total of 40% of the constrained bases reside within protein-coding exons and their associated untranslated regions (Fig. 10) and, in agreement with previous genome-wide estimates, the remaining constrained bases do not overlap the mature transcripts of protein-coding genes[4,5,88,105,106]. When we included the other experimental annotations, we found that an additional 20% of the constrained bases overlap experimentally identified non-coding functional regions, although far fewer of these regions overlap constrained sequences compared to coding exons (see below). Most experimental annotations are significantly different from a random expectation for both base-pair or element-level overlaps (using the GSC statistic, see Supplementary Information section 1.3), with a more striking deviation when considering elements (Fig. 11). The exceptions to this are pseudogenes, Un.TxFrags and RxFrags. The increase in significance moving from base-pair measures to the element level suggests that discrete islands of constrained sequence exist within experimentally identified functional elements, with the surrounding bases apparently not showing evolutionary constraint. This notion is discussed in greater detail in ref. 97.

可比较的序列数据集和分析。 利用包括分离[89]和测序[90]单个细菌人工染色体克隆的靶向策略，我们从 14 个哺乳动物物种中得到了 206 Mb 跟 ENCODE 区域直系同源的基因组序列。对于另外的 14 种脊椎动物，我们使用的是从全基因组测序得到的 340 Mb 直系同源的基因组序列[3-8,91-93]。我们使用三种比对程序进行直系同源序列的比对：TBA[94]、MAVID[95] 和 MLAGAN[96]。四个独立的方法(PhastCons[88]、GERP[87]、SCONE[98] 和 BinCons[86]) 得到了高度一致的结果[97]，然后被用于鉴定受限制的序列。从这些分析中，我们形成了一组高度可信的"受限制的序列"，对应于 ENCODE 区域内 4.9% 的核苷酸。使用 5% 的 FDR 率作为受限制的阈值(见参考文献 97)；该水平跟之前估计的人类基因组中受哺乳动物限制的比例类似[4,86-88]，但是 FDR 率并不是为了符合这个结果。这些受限制的序列长度的中位数是 19 个碱基，最短是 8 个碱基——差不多一个典型的转录因子结合位点的长度。因而，这些分析提供的受限制的序列的分辨率明显好于目前仅利用全基因组脊椎动物序列获得的[99-102]。

物种内的变异研究主要使用来自 HapMap 计划 I 和 II 期的 SNP 数据，以及其中 48 个个体的 10 个位于 ENCODE 区域的区域的重测序数据[103]；核苷酸的插入或者缺失(插入缺失位)数据来自 SNP 协作组和 HapMap。我们也分析了 ENCODE 区域与已知的片段复制[104] 和 CNV 重合情况。

实验鉴定的功能元件和受限制的序列 我们首先将检测到的受限制的序列跟与实验鉴定到的功能元件进行了比较。总共 40% 受限制的碱基位于蛋白质编码的外显子和它们相关的非翻译区域(图 10)，并且，与之前的全基因组估计一致，剩下的受限制的碱基不与蛋白质编码的基因的成熟转录本重合[4,5,88,105,106]。当包括了其他实验注释时，我们发现了 20% 额外受限制的碱基与实验鉴定到的非编码功能区域重合，尽管与受限制序列重合的这些区域要远远少于编码的外显子(见下文)。无论是碱基对还是元件水平的重合，大多数的实验注释显著不同于随机预期(利用 GSC 统计，详见补充信息 1.3)，当考虑元件时差异更加明显(图 11)。其中的例外是假基因、Un.TxFrag 和 RxFrag。从碱基对测量到元件水平时显著水平升高，说明受限制序列的分离的岛存在于实验鉴定到的功能元件内，而周围的碱基则明显不具有进化上的限制。这个观点在参考文献 97 中有更详细的讨论。

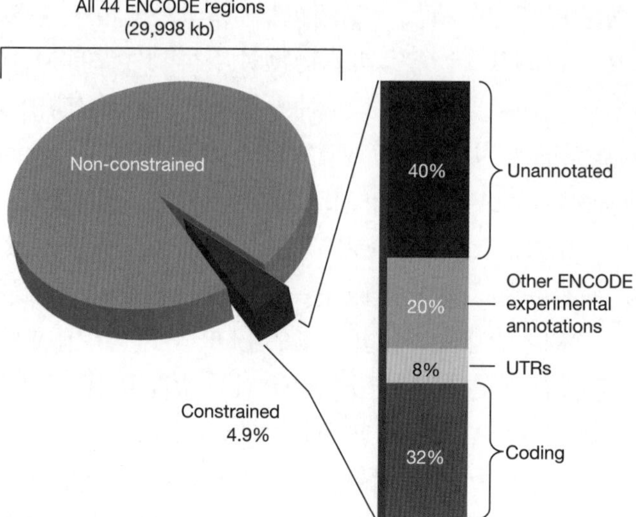

Fig. 10. Relative proportion of different annotations among constrained sequences. The 4.9% of bases in the ENCODE regions identified as constrained is subdivided into the portions that reflect known coding regions, UTRs, other experimentally annotated regions, and unannotated sequence.

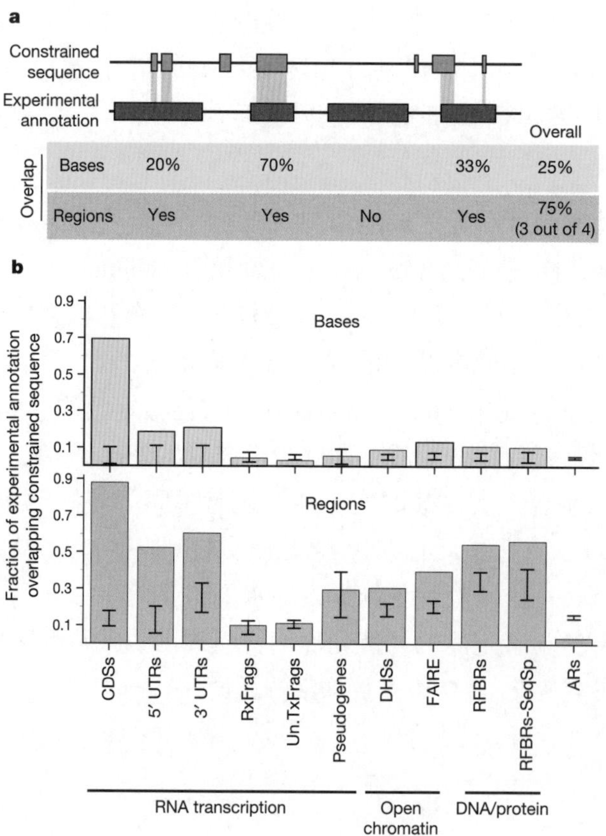

Fig. 11. Overlap of constrained sequences and various experimental annotations. **a**, A schematic depiction shows the different tests used for assessing overlap between experimental annotations and constrained

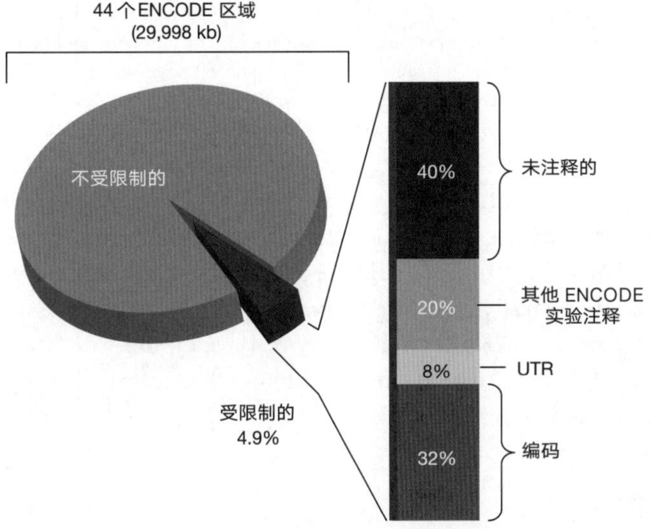

图 10. 不同注释在受限制序列中的相对比例。ENCODE 区域中有 4.9% 的碱基被鉴定为受限制的，其又被进一步分为已知的编码区域、UTR、其他实验注释的区域和未注释的序列。

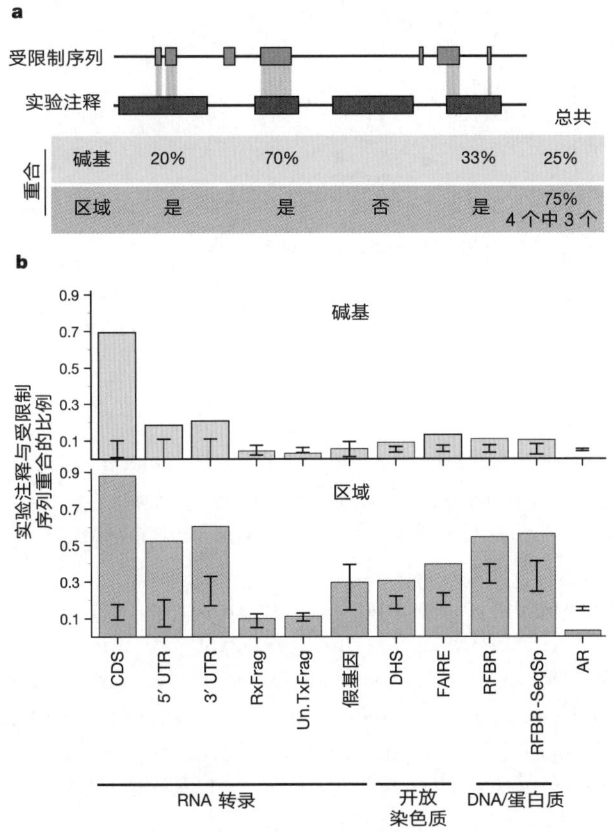

图 11. 受限制序列与各种实验注释的重合情况。**a**，示意图展示用于评价实验注释和受限制序列重合情况的不同检验方法，不只是对于单个碱基，而且针对整个区域。**b**，观察到的重合的比例，对于碱基和

sequences, both for individual bases and for entire regions. **b**, Observed fraction of overlap, depicted separately for bases and regions. The results are shown for selected experimental annotations. The internal bars indicate 95% confidence intervals of randomized placement of experimental elements using the GSC methodology to account for heterogeneity in the data sets. When the bar overlaps the observed value one cannot reject the hypothesis that these overlaps are consistent with random placements.

We also examined measures of human variation (heterozygosity, derived allele-frequency spectra and indel rates) within the sequences of the experimentally identified functional elements (Fig. 12). For these studies, ARs were used as a marker for neutrally evolving sequence. Most experimentally identified functional elements are associated with lower heterozygosity compared to ARs, and a few have lower indel rates compared with ARs. Striking outliers are 3′ UTRs, which have dramatically increased indel rates without an obvious cause. This is discussed in more depth in ref. 107.

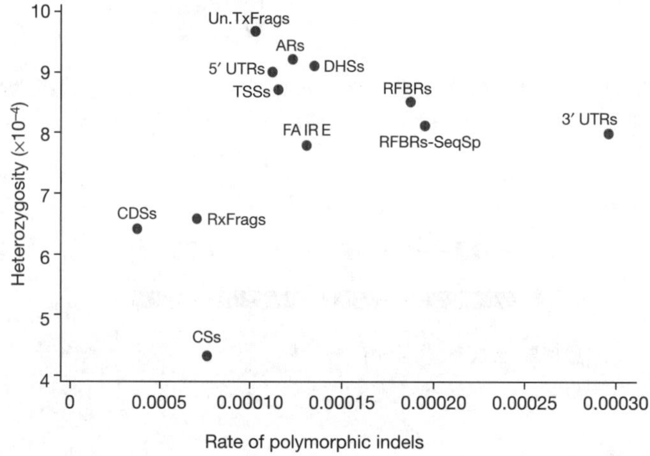

Fig. 12. Relationship between heterozygosity and polymorphic indel rate for a variety of experimental annotations. 3′ UTRs are an expected outlier for the indel measures owing to the presence of low-complexity sequence (leading to a higher indel rate).

These findings indicate that the majority of the evolutionarily constrained, experimentally identified functional elements show evidence of negative selection both across mammalian species and within the human population. Furthermore, we have assigned at least one molecular function to the majority (60%) of all constrained bases in the ENCODE regions.

Conservation of regulatory elements. The relationship between individual classes of regulatory elements and constrained sequences varies considerably, ranging from cases where there is strong evolutionary constraint (for example, pan-vertebrate ultraconserved regions[108,109]) to examples of regulatory elements that are not conserved between orthologous human and mouse genes[110]. Within the ENCODE regions, 55% of RFBRs overlap the high-confidence constrained sequences. As expected, RFBRs have many unconstrained bases, presumably owing to the small size of the specific binding site. We investigated whether the binding sites in RFBRs could be further delimited using information about evolutionary constraint. For 7 out of 17 factors with either known TRANSFAC or Jaspar

区域分开展示。展示的结果是选出来的实验注释。中间的横杠表示实验元件随机排布的 95% 的置信区间，这是利用 GSC 方法得到的，目的是为了消除数据集中的异质性。当这个横杠与观察值重合时，则不能拒绝这些重合与随机排布一致的假说。

我们也分析了实验鉴定到的功能元件中人类的变异量（杂合度、衍生的等位基因频率谱和插入缺失率）（图 12）。对于这些研究，AR 被用作中性进化的标志物。与 AR 相比，大多数实验鉴定到的功能元件与更低的杂合度相关，一小部分含有比 AR 更低的插入缺失率。显著的异常情况是 3′ UTR，它含有明显更高的插入缺失率，却没有明显的原因。在参考文献 107 中有更深入的讨论。

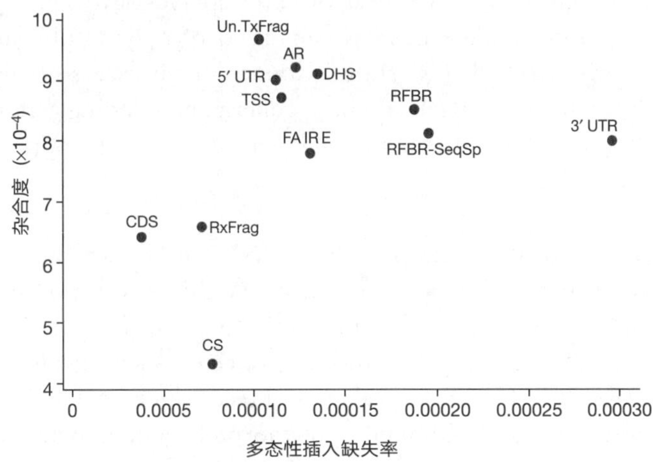

图 12. 对于各种实验注释，杂合度与多态性的插入缺失率的关系。对于插入缺失的测量，3′ UTR 是一个符合预期的离群值，原因是低复杂序列的存在（导致更高的插入缺失率）。

这些发现表明大部分进化上受限制的实验鉴定到的功能元件呈现负选择的迹象，这不仅是在哺乳动物间，在人类群体内部也是如此。再者，我们已经将 ENCODE 区域内所有受限制碱基中的大部分（60%）至少赋予了一种分子功能。

调控元件的保守性 各种调控元件与受限制序列之间的关系差别很大，有的受到强烈的进化限制（如多个脊椎动物超保守区域 [108,109]），有的调控元件则在直系同源的人和小鼠基因间不保守 [110]。在 ENCODE 区域内，55% 的 RFBR 与高可信度的受限制序列重合。正如预期的，RFBR 含有很多不受限制的碱基，可能是由于特定结合位点的长度较短。利用关于进化限制的信息，我们研究了 RFBR 中的结合位点能否被进一步划分。对于含有已知的 TRANSFAC 或 Jaspar 模序的 17 个因子中的 7 个，我们的 ChIP-chip 数据发现，相比不受限制的 RFBR，在受限制的 RFBR 中明显

motifs, our ChIP-chip data revealed a marked enrichment of the appropriate motif within the constrained versus the unconstrained portions of the RFBRs (see Supplementary Information section 5.1). This enrichment was seen for levels of stringency used for defining ChIP-chip-positive sites (1% and 5% FDR level), indicating that combining sequence constraint and ChIP-chip data may provide a highly sensitive means for detecting factor binding sites in the human genome.

Experimentally identified functional elements and genetic variation. The above studies focus on purifying (negative) selection. We used nucleotide variation to detect potential signals of adaptive (positive) selection. We modified the standard McDonald–Kreitman test (MK-test[111,112]) and the Hudson–Kreitman–Aguade (HKA)[113] test (see Supplementary Information section 5.2.1), to examine whether an entire set of sequence elements shows an excess of polymorphisms or an excess of inter-species divergence. We found that constrained sequences and coding exons have an excess of polymorphisms (consistent with purifying selection), whereas 5′ UTRs show evidence of an excess of divergence (with a portion probably reflecting positive selection). In general, non-coding genomic regions show more variation, with both a large number of segments that undergo purifying selection and regions that are fast evolving.

We also examined structural variation (that is, CNVs, inversions and translocations[114]; see Supplementary Information section 5.2.2). Within these polymorphic regions, we encountered significant over-representation of CDSs, TxFrags, and intra-species constrained sequences ($P < 10^{-3}$, Fig. 13), and also detected a statistically significant under-representation of ARs ($P = 10^{-3}$). A similar overrepresentation of CDSs and intra-species constrained sequences was found within non-polymorphic segmental duplications.

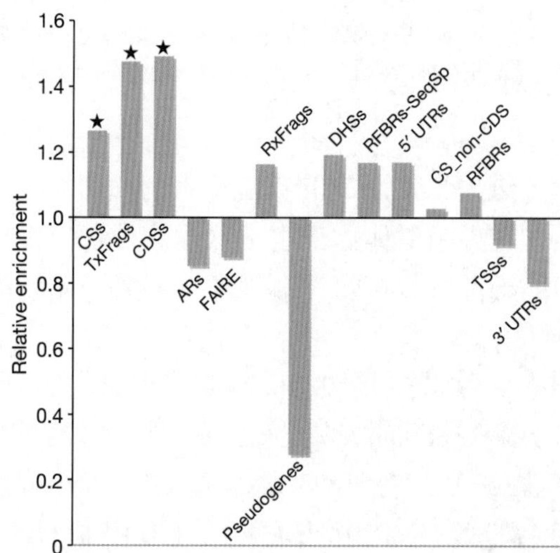

Fig. 13. CNV enrichment. The relative enrichment of different experimental annotations in the ENCODE regions associated with CNVs. CS_non-CDS are constrained sequences outside of coding regions. A value of 1 or less indicates no enrichment, and values greater than 1 show enrichment. Starred columns are cases that are significant on the basis of this enrichment being found in less than 5% of randomizations that matched each element class for length and density of features.

富集合适的模序（详见补充信息 5.1）。这种富集是在用于确定 ChIP-chip 阳性位点的阈值水平下看到的（1% 和 5%FDR 水平），说明结合序列的受限制和 ChIP-chip 数据可能提供一种高度敏感的方法来确定因子在人类基因组中的结合位点。

实验鉴定的功能元件和遗传变异 上述研究聚焦于纯化（负）选择。我们使用核苷酸的变异来检测潜在的适应性（正）选择信号。我们修改标准的麦克唐纳–克莱特曼检验（MK 检验 [111, 112]）和赫德森–克莱特曼–阿瓜德（HKA）[113] 检验（详见补充信息 5.2.1），来检查一整套序列元件是否呈现过度的多态性或者过度的物种间差异。我们发现受限制序列和编码的外显子呈现过度的多态性（与纯化选择一致），而 5′UTR 呈现过度的差异迹象（其中一部分可能反映了正选择）。总的来说，非编码的基因组区域呈现更多的变异，不仅大量的片段在进行纯化选择，而且有区域正在快速地进化。

我们还检查了结构变异（即 CNV，倒位和易位 [114]；详见补充信息 5.2.2）。在这些多态性的区域，我们发现 CDS、TxFrag 和物种内受限制序列显著富集（$P < 10^{-3}$，图 13），还检测到有统计学差异的 AR 的缺失（$P = 10^{-3}$）。CDS 和物种内受限制序列类似的富集在非多态性的片段重复中也被发现。

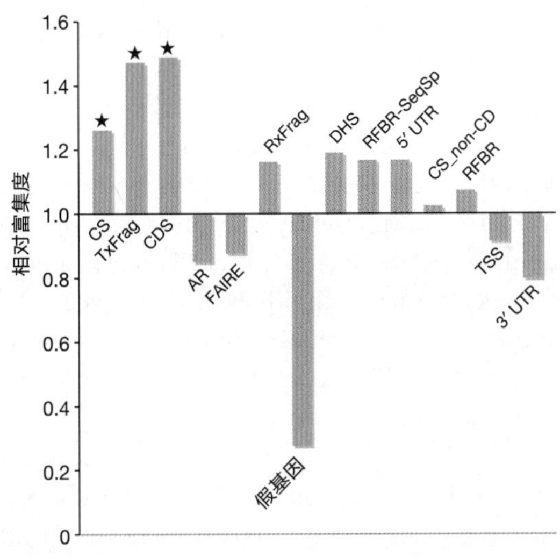

图 13. CNV 富集。在 ENCODE 区域内，与 CNV 相关的不同实验注释的相对富集程度。CS_non-CDS 是编码区域以外的受限制序列。数值为 1 或更小表示没有富集，数值大于 1 表示富集。标记星号的柱是显著的组，这种显著是基于在与相应类别元件长度和特征密度相匹配的随机组中仅有不到 5% 的富集。

Unexplained constrained sequences. Despite the wealth of complementary data, 40% of the ENCODE-region sequences identified as constrained are not associated with any experimental evidence of function. There is no evidence indicating that mutational cold spots account for this constraint; they have similar measures of constraint to experimentally identified elements and harbour equal proportions of SNPs. To characterize further the unexplained constrained sequences, we examined their clustering and phylogenetic distribution. These sequences are not uniformly distributed across most ENCODE regions, and even in most ENCODE regions the distribution is different from constrained sequences within experimentally identified functional elements (see Supplementary Information section 5.3). The large fraction of constrained sequence that does not match any experimentally identified elements is not surprising considering that only a limited set of transcription factors, cell lines and biological conditions have thus far been examined.

Unconstrained experimentally identified functional elements. In contrast, an unexpectedly large fraction of experimentally identified functional elements show no evidence of evolutionary constraint ranging from 93% for Un.TxFrags to 12% for CDS. For most types of non-coding functional elements, roughly 50% of the individual elements seemed to be unconstrained across all mammals.

There are two methodological reasons that might explain the apparent excess of unconstrained experimentally identified functional elements: the underestimation of sequence constraint or overestimation of experimentally identified functional elements. We do not believe that either of these explanations fully accounts for the large and varied levels of unconstrained experimentally functional sequences. The set of constrained bases analysed here is highly accurate and complete due to the depth of the multiple alignment. Both by bulk fitting procedures and by comparison of SNP frequencies to constraint there is clearly a proportion of constrained bases not captured in the defined 4.9% of constrained sequences, but it is small (see Supplementary Information section 5.4 and S5.5). More aggressive schemes to detect constraint only marginally increase the overlap with experimentally identified functional elements, and do so with considerably less specificity. Similarly, all experimental findings have been independently validated and, for the least constrained experimentally identified functional elements (Un.TxFrags and binding sites of sequence-specific factors), there is both internal validation and cross-validation from different experimental techniques. This suggests that there is probably not a significant overestimation of experimentally identified functional elements. Thus, these two explanations may contribute to the general observation about unconstrained functional elements, but cannot fully explain it.

Instead, we hypothesize five biological reasons to account for the presence of large amounts of unconstrained functional elements. The first two are particular to certain biological assays in which the elements being measured are connected to but do not coincide with the analysed region. An example of this is the parent transcript of an miRNA, where the current assays detect the exons (some of which are not under evolutionary selection), whereas the intronic miRNA actually harbours the constrained bases. Nevertheless, the

未解释的受限制序列　尽管有丰富的互补数据，被鉴定为受限制的 ENCODE 区域序列中仍有 40% 不与任何一项功能的实验证据有关联。没有证据表明突变冷点导致这种限制，它们与实验鉴定到的元件有类似的程度的受限制，并且含有相同比例的 SNP。为了进一步描述未解释的受限制序列的特征，我们检查了它们的聚集和系统进化分布。这些序列并非均匀分布在大多数的 ENCODE 区域，甚至在大多数的 ENCODE 区域，它们的分布跟位于实验鉴定的功能元件内的受限制序列也是不同的(详见补充信息 5.3)。考虑到目前为止仅检测了有限数量的转录因子、细胞系和生物学条件，很大一部分受限制序列不与任何实验鉴定到的元件相匹配其实并不奇怪。

不受限制的实验鉴定到的功能元件　相反，出乎意料的是很大比例实验鉴定到的功能元件中没有表现出进化受限制的证据，从 Un.TxFrag 的 93% 到 CDS 的 12%。对于大多数类型的非编码功能元件，每种元件都有大约 50% 好像在所有哺乳动物中不受限制。

有两个方法学的原因可能解释这种明显过量的不受限制的实验鉴定的功能元件：低估了受限制的序列或者高估了实验鉴定的功能元件。我们不相信这些解释中的某一个能完全解释这大量且不同水平的非受限制的实验鉴定的功能序列。由于多重比对的深度，这里分析的这组受限制的碱基是高度准确而完整的。不管是通过批量拟合方法还是通过比较 SNP 频率跟受限制性，明显有一部分受限制的碱基在 4.9% 的受限制序列中没有被捕获到，但是这个数量是很小的(详见补充信息 5.4 和 S5.5)。更激进的检测受限制的策略仅仅略微增加了与实验鉴定到的功能元件的重合度，而特异性明显降低。类似地，所有实验发现已经被独立验证，对于最不受限制的实验鉴定的功能元件(UnTxFrag 和序列特异因子的结合位点)，既有内部验证也有不同实验技术的交叉验证。这说明可能并没有显著高估实验鉴定的功能元件。因而，这两种解释可能有助于解释普遍观察到的非受限制功能元件，但是还不能完全解释它。

作为替代，我们猜测了五个生物学原因来解释大量的非受限制功能元件的存在。前两个特别针对特定的生物学实验，实验中测到的元件与分析的区域相关但是并不一致。一个这样的例子是 miRNA 的亲本转录本，现有的实验检测的是外显子(其中一些并没有受到进化的筛选)，而位于内含子的 miRNA 实际上含有受限制的碱基。尽管如此，转录本序列还是提供了被调控的启动子与 miRNA 之间关键的耦合。转

transcript sequence provides the critical coupling between the regulated promoter and the miRNA. The sliding of transcription factors (which might bind a specific sequence but then migrate along the DNA) or the processivity of histone modifications across chromatin are more exotic examples of this. A related, second hypothesis is that delocalized behaviours of the genome, such as general chromatin accessibility, may be maintained by some biochemical processes (such as transcription of intergenic regions or specific factor binding) without the requirement for specific sequence elements. These two explanations of both connected components and diffuse components related to, but not coincident with, constrained sequences are particularly relevant for the considerable amount of unannotated and unconstrained transcripts.

The other three hypotheses may be more general—the presence of neutral (or near neutral) biochemical elements, of lineage-specific functional elements, and of functionally conserved but non-orthologous elements. We believe there is a considerable proportion of neutral biochemically active elements that do not confer a selective advantage or disadvantage to the organism. This neutral pool of sequence elements may turn over during evolutionary time emerging via certain mutations and disappearing by others. The size of the neutral pool would largely be determined by the rate of emergence and extinction through chance events; low information-content elements, such as transcription factor-binding sites[110] will have larger neutral pools. Second, from this neutral pool, some elements might occasionally acquire a biological role and so come under evolutionary selection. The acquisition of a new biological role would then create a lineage-specific element. Finally, a neutral element from the general pool could also become a peer of an existing selected functional element and either of the two elements could then be removed by chance. If the older element is removed, the newer element has, in essence, been conserved without using orthologous bases, providing a conserved function in the absence of constrained sequences. For example, a common HNF4A binding site in the human and mouse genomes may not reflect orthologous human and mouse bases, though the presence of an HNF4A site in that region was evolutionarily selected for in both lineages. Note that both the neutral turnover of elements and the "functional peering" of elements has been suggested for *cis*-acting regulatory elements in *Drosophila*[115,116] and mammals[110]. Our data support these hypotheses, and we have generalized this idea over many different functional elements. The presence of conserved function encoded by conserved orthologous bases is a commonplace assumption in comparative genomics; our findings indicate that there could be a sizable set of functionally conserved but non-orthologous elements in the human genome, and that these seem unconstrained across mammals. Functional data akin to the ENCODE Project on other related species, such as mouse, would be critical to understanding the rate of such functionally conserved but non-orthologous elements.

Conclusion

The generation and analyses of over 200 experimental data sets from studies examining the 44 ENCODE regions provide a rich source of functional information for 30 Mb of

录因子的滑动（转录因子可能结合一个特定的序列但是随后沿着 DNA 移动）或者组蛋白修饰在染色质上的延伸性则是更加异乎寻常的例子。与之相关的第二个假说是基因组不受位置限制的行为，一般的染色质开放性可能通过一些生化过程而被维持（如基因间区域的转录或者特定因子的结合）不需要特定的序列元件。这两种对相连组分和扩散组分都与受限制序列相关但又不完全一致现象的解释，与相当数量未注释且未受限制的转录本特别相关。

其他三种假说——中性（或者近中性）生化元件、谱系特异的功能元件和功能保守但非直系同源元件的存在——可能更常见。我们相信存在相当大的一部分中性的生化活跃元件，并不对机体产生选择性优势或者劣势。这种中性序列元件库可能在进化过程中发生转变，通过特定突变而出现，通过其他一些突变而消失。中性库的大小很大程度上取决于元件通过随机事件而产生和消失的速率；低信息量的元件例如转录因子结合位点[110]会有更大的中性库。第二，从这个中性库中，一些元件可能偶然获得一个生物学功能，因而进入进化选择。一个新的生物学功能的获得会进而产生一个谱系特异的元件。最后，普通库中的一个中性元件也能成为一个同等的已被选择的功能元件，这两个元件中的一个会随机被去除。如果更老的元件被去除，那么更新的元件即使不使用直系同源碱基，在本质上就已经是保守的，在不存在受限制序列的情况下提供了一个保守功能。例如，在人和小鼠基因组中一个共同的 HNF4A 结合位点可能并不反映直系同源的人和小鼠碱基，尽管 HNF4A 位点在那个区域的存在在两个谱系中都是被进化选择的。需要注意的是，元件的中性转换和元件的“功能等价”在果蝇[115,116]和哺乳动物[110]中都被认为是顺式调控元件。我们的数据支持这些假说，我们已经将这个观点普遍地应用于很多不同的功能元件。由保守直系同源碱基编码的保守功能的存在，在比较基因组学中是一个很平常的假设，我们的发现提示可能在人类基因组中存在相当数量的功能保守但非直系同源的元件，这些元件在哺乳动物中似乎并不受限制。类似于 ENCODE 计划的关于其他相关物种的功能数据（例如小鼠的），对于理解这种功能保守的但非直系同源元件的速率将会至关重要。

结　论

研究针对 44 个 ENCODE 区域，产生和分析了超过 200 个实验数据集，为人类

the human genome. The first conclusion of these efforts is that these data are remarkably inform-ative. Although there will be ongoing work to enhance existing assays, invent new techniques and develop new data-analysis methods, the generation of genome-wide experimental data sets akin to the ENCODE pilot phase would provide an impressive platform for future genome exploration efforts. This now seems feasible in light of throughput improvements of many of the assays and the ever-declining costs of whole-genome tiling arrays and DNA sequencing. Such genome-wide functional data should be acquired and released openly, as has been done with other large-scale genome projects, to ensure its availability as a new foundation for all biologists studying the human genome. It is these biologists who will often provide the critical link from biochemical function to biological role for the identified elements.

The scale of the pilot phase of the ENCODE Project was also sufficiently large and unbiased to reveal important principles about the organization of functional elements in the human genome. In many cases, these principles agree with current mechanistic models. For example, we observe trimethylation of H3K4 enriched near active genes, and have improved the ability to accurately predict gene activity based on this and other histone modifications. However, we also uncovered some surprises that challenge the current dogma on biological mechanisms. The generation of numerous intercalated transcripts spanning the majority of the genome has been repeatedly suggested[13,14], but this phenomenon has been met with mixed opinions about the biological importance of these transcripts. Our analyses of numerous orthogonal data sets firmly establish the presence of these transcripts, and thus the simple view of the genome as having a defined set of isolated loci transcribed independently does not seem to be accurate. Perhaps the genome encodes a network of transcripts, many of which are linked to protein-coding transcripts and to the majority of which we cannot (yet) assign a biological role. Our perspective of transcription and genes may have to evolve and also poses some interesting mechanistic questions. For example, how are splicing signals coordinated and used when there are so many overlapping primary transcripts? Similarly, to what extent does this reflect neutral turnover of reproducible transcripts with no biological role?

We gained subtler but equally important mechanistic findings relating to transcription, replication and chromatin modification. Transcription factors previously thought to primarily bind promoters bind more generally, and those which do bind to promoters are equally likely to bind downstream of a TSS as upstream. Interestingly, many elements that previously were classified as distal enhancers are, in fact, close to one of the newly identified TSSs; only about 35% of sites showing evidence of binding by multiple transcription factors are actually distal to a TSS. This need not imply that most regulatory information is confined to classic promoters, but rather it does suggest that transcription and regulation are coordinated actions beyond just the traditional promoter sequences. Meanwhile, although distal regulatory elements could be identified in the ENCODE regions, they are currently difficult to classify, in part owing to the lack of a broad set of transcription factors to use in analysing such elements. Finally, we now have a much better appreciation of how DNA replication is coordinated with histone modifications.

基因组的这 30 Mb 区域提供了一个丰富的功能信息的资源。这些工作的第一个结论是这些数据所包含的信息量是巨大的。尽管将有进一步的工作来促进现有的实验分析、发明新的技术和开发新的数据分析方法，类似于 ENCODE 先导阶段的全基因组实验数据集的产生还是会为以后的基因组研究提供一个出色的平台。现在看来这似乎是可行的，因为很多实验的通量在提升而全基因组叠瓦式阵列和 DNA 测序费用在大幅下降。就像其他大规模基因组计划一样，这种全基因组的功能数据应该开放获取，以保证它作为一个新的基础能被所有生物学家用来研究人类基因组。正是这些生物学家将会经常为鉴定到的元件提供从生化功能到生物学功能的关键联系。

ENCODE 计划先导阶段的规模也是足够大且无偏性，进而能够反映人类基因组中功能元件的重要组织原理。在很多例子中，这些原理与现有的机制模型是一致的。例如，我们观察到 H3K4 的三甲基化在活跃基因的附近富集，基于这种以及其他组蛋白修饰，我们提高了精确预测基因活性的能力。然而，我们也发现了一些意料之外的事情，对现有的生物学机制的信条提出了挑战。已经反复发现，有大量的间插性转录本的产生，这些转录本横跨大部分的基因组 [13,14]，但是对于这种现象中这些转录本的生物学意义有不同的观点。我们通过对大量的正交数据集的分析，确凿地证实了这些转录本的存在，因而简单地认为基因组含有确定的一套分离的基因座独立地被转录好像并不是准确的。也许基因组编码一个转录本的网络，很多转录本与蛋白质编码的转录本有联系，而对于大部分的转录本，我们(还)不能确定其生物学功能。我们关于转录和基因的观点可能不得不进化，另外也产生了一些有趣的关于机制的问题。例如，当有这么多重叠的原始转录本时，剪接信号是如何协调和使用的？类似地，这在多大程度上反映了没有生物学功能的重复转录本的中性转换？

我们获得了微妙但同等重要的与转录、复制和染色质修饰相关的机制方面的发现。之前认为主要结合启动子的转录因子，其实结合更广泛，那些确定结合启动子的转录因子除了结合 TSS 上游之外，同样可能结合 TSS 的下游。有趣的是，很多元件之前被划分为远端增强子，其实靠近新鉴定到的 TSS；对于有证据表明被多个转录因子结合的位点，只有大约 35% 确实远离 TSS。这不一定说明大部分的调控信息局限于经典的启动子，但是确实暗示转录和调控并不仅是传统启动子序列的协调活动。同时，尽管远端调控元件能在 ENCODE 区域中被鉴定，但是它们目前还难以分类，部分是由于缺少一套广泛的转录因子来分析这些元件。最后，目前我们对于 DNA 复制如何与组蛋白修饰协调有了更好的认识。

At the outset of the ENCODE Project, many believed that the broad collection of experimental data would nicely dovetail with the detailed evolutionary information derived from comparing multiple mammalian sequences to provide a neat "dictionary" of conserved genomic elements, each with a growing annotation about their biochemical function(s). In one sense, this was achieved; the majority of constrained bases in the ENCODE regions are now associated with at least some experimentally derived information about function. However, we have also encountered a remarkable excess of experimentally identified functional elements lacking evolutionary constraint, and these cannot be dismissed for technical reasons. This is perhaps the biggest surprise of the pilot phase of the ENCODE Project, and suggests that we take a more "neutral" view of many of the functions conferred by the genome.

Methods

The methods are described in the Supplementary Information, with more technical details for each experiment often found in the references provided in Table 1. The Supplementary Information sections are arranged in the same order as the manuscript (with similar headings to facilitate cross-referencing). The first page of Supplementary Information also has an index to aid navigation. Raw data are available in ArrayExpress, GEO or EMBL/GenBank archives as appropriate, as detailed in Supplementary Information section 1.1. Processed data are also presented in a user-friendly manner at the UCSC Genome Browser's ENCODE portal (http://genome.ucsc.edu/ENCODE/).

(**447**, 799-816; 2007)

Received 2 March; accepted 23 April 2007.

References:

1. International Human Genome Sequencing Consortium. Initial sequencing and analysis of the human genome. *Nature* **409**, 860-921 (2001).
2. Venter, J. C. *et al.* The sequence of the human genome. *Science* **291**, 1304-1351 (2001).
3. International Human Genome Sequencing Consortium. Finishing the euchromatic sequence of the human genome. *Nature* **431**, 931-945 (2004).
4. Mouse Genome Sequencing Consortium. Initial sequencing and comparative analysis of the mouse genome. *Nature* **420**, 520-562 (2002).
5. Rat Genome Sequencing Project Consortium. Genome sequence of the Brown Norway rat yields insights into mammalian evolution. *Nature* **428**, 493-521 (2004).
6. Lindblad-Toh, K. *et al.* Genome sequence, comparative analysis and haplotype structure of the domestic dog. *Nature* **438**, 803-819 (2005).
7. International Chicken Genome Sequencing Consortium. Sequence and comparative analysis of the chicken genome provide unique perspectives on vertebrate evolution. *Nature* **432**, 695-716 (2004).
8. Chimpanzee Sequencing and Analysis Consortium. Initial sequence of the chimpanzee genome and comparison with the human genome. *Nature* **437**, 69-87 (2005).
9. ENCODE Project Consortium. The ENCODE (ENCyclopedia Of DNA Elements) Project. *Science* **306**, 636-640 (2004).
10. Zhang, Z. D. *et al.* Statistical analysis of the genomic distribution and correlation of regulatory elements in the ENCODE regions. *Genome Res.* **17**, 787-797 (2007).
11. Euskirchen, G. M. *et al.* Mapping of transcription factor binding regions in mammalian cells by ChIP: comparison of array and sequencing based technologies. *Genome Res.* **17**, 898-909 (2007).
12. Willingham, A. T. & Gingeras, T. R. TUF love for "junk" DNA. *Cell* 125, 1215-1220 (2006).
13. Carninci, P. *et al.* Genome-wide analysis of mammalian promoter architecture and evolution. *Nature Genet.* **38**, 626-635 (2006).
14. Cheng, J. *et al.* Transcriptional maps of 10 human chromosomes at 5-nucleotide resolution. *Science* **308**, 1149-1154 (2005).
15. Bertone, P. *et al.* Global identification of human transcribed sequences with genome tiling arrays. *Science* **306**, 2242-2246 (2004).
16. Guigó, R. *et al.* EGASP: the human ENCODE Genome Annotation Assessment Project. *Genome Biol.* 7, (Suppl. 1; S2) 1-31 (2006).

在 ENCODE 计划的开始时，很多人认为广泛收集的实验数据将会完美吻合来自比较多种哺乳动物序列得到的详细进化信息，进而提供一个纯粹的保守基因组元件的“词典”，每一个元件含有更多的关于它们生化功能的注释。在某种意义上，这已经达到了。在 ENCODE 区域中，大部分的受限制碱基现在至少与某些实验得到的功能信息相关联。但是，我们也遇到了大量多余的实验鉴定到的功能元件，它们缺少进化的限制，又不能因为技术原因而摒弃。这可能是 ENCODE 计划先导阶段最大的意外，提示我们要更加“中性”地看待很多基因组赋予的功能。

方　法

在补充信息中对方法进行了介绍，对于各个实验，更多的技术细节大多能在表 1 中提供的参考文献中找到。补充信息部分跟正文排列顺序相同（使用类似的标题以便于交叉参考）。补充信息的第一页也有一个索引以便于查找。原始数据可在 ArrayExpress、GEO 或者 EMBL/GenBank 档案文件获得，详见补充信息 1.1。处理过的数据也以用户友好的方式在 UCSC 基因组浏览器的 ENCODE 入口呈现（http://genome.ucsc.edu/ENCODE/）。

（李平 翻译；于军 审稿）

17. Denoeud, F. *et al.* Prominent use of distal 5′ transcription start sites and discovery of a large number of additional exons in ENCODE regions. *Genome Res.* **17**, 746-759 (2007).

18. Tress, M. L. *et al.* The implications of alternative splicing in the ENCODE protein complement. *Proc. Natl Acad. Sci. USA* **104**, 5495-5500 (2007).

19. Rozowsky, J. *et al.* The DART classification of unannotated transcription within ENCODE regions: Associating transcription with known and novel loci. *Genome Res.* **17**, 732-745 (2007).

20. Kapranov, P. *et al.* Examples of the complex architecture of the human transcriptome revealed by RACE and high-density tiling arrays. *Genome Res.* **15**, 987-997 (2005).

21. Balakirev, E. S. & Ayala, F. J. Pseudogenes: are they "junk" or functional DNA? *Annu. Rev. Genet.* **37**, 123-151 (2003).

22. Mighell, A. J., Smith, N. R., Robinson, P. A. & Markham, A. F. Vertebrate pseudogenes. *FEBS Lett.* **468**, 109-114 (2000).

23. Zheng, D. *et al.* Pseudogenes in the ENCODE regions: Consensus annotation, analysis of transcription and evolution. *Genome Res.* **17**, 839-851 (2007).

24. Zheng, D. *et al.* Integrated pseudogene annotation for human chromosome 22: evidence for transcription. *J. Mol. Biol.* **349**, 27-45 (2005).

25. Harrison, P. M., Zheng, D., Zhang, Z., Carriero, N. & Gerstein, M. Transcribed processed pseudogenes in the human genome: an intermediate form of expressed retrosequence lacking protein-coding ability. *Nucleic Acids Res.* **33**, 2374-2383 (2005).

26. Washietl, S. *et al.* Structured RNAs in the ENCODE selected regions of the human genome. *Genome Res.* **17**, 852-864 (2007).

27. Carninci, P. *et al.* The transcriptional landscape of the mammalian genome. *Science* **309**, 1559-1563 (2005).

28. Runte, M. *et al.* The IC-*SNURF–SNRPN* transcript serves as a host for multiple small nucleolar RNA species and as an antisense RNA for *UBE3A*. *Hum. Mol. Genet.* **10**, 2687-2700 (2001).

29. Seidl, C. I., Stricker, S. H. & Barlow, D. P. The imprinted *Air* ncRNA is an atypical RNAPII transcript that evades splicing and escapes nuclear export. *EMBO J.* **25**, 3565-3575 (2006).

30. Parra, G. *et al.* Tandem chimerism as a means to increase protein complexity in the human genome. *Genome Res.* **16**, 37-44 (2006).

31. Maston, G. A., Evans, S. K. & Green, M. R. Transcriptional regulatory elements in the human genome. *Annu. Rev. Genomics Hum. Genet.* **7**, 29-59 (2006).

32. Trinklein, N. D., Aldred, S. J., Saldanha, A. J. & Myers, R. M. Identification and functional analysis of human transcriptional promoters. *Genome Res.* **13**, 308-312 (2003).

33. Cooper, S. J., Trinklein, N. D., Anton, E. D., Nguyen, L. & Myers, R. M. Comprehensive analysis of transcriptional promoter structure and function in 1% of the human genome. *Genome Res.* **16**, 1-10 (2006).

34. Cawley, S. *et al.* Unbiased mapping of transcription factor binding sites along human chromosomes 21 and 22 points to widespread regulation of noncoding RNAs. *Cell* **116**, 499-509 (2004).

35. Yelin, R. *et al.* Widespread occurrence of antisense transcription in the human genome. *Nature Biotechnol.* **21**, 379-386 (2003).

36. Katayama, S. *et al.* Antisense transcription in the mammalian transcriptome. *Science* **309**, 1564-1566 (2005).

37. Ren, B. *et al.* Genome-wide location and function of DNA binding proteins. *Science* **290**, 2306-2309 (2000).

38. Iyer, V. R. *et al.* Genomic binding sites of the yeast cell-cycle transcription factors SBF and MBF. *Nature* **409**, 533-538 (2001).

39. Horak, C. E. *et al.* GATA-1 binding sites mapped in the β-globin locus by using mammalian chIp-chip analysis. *Proc. Natl Acad. Sci. USA* **99**, 2924-2929 (2002).

40. Wei, C. L. *et al.* A global map of p53 transcription-factor binding sites in the human genome. *Cell* **124**, 207-219 (2006).

41. Kim, J., Bhinge, A. A., Morgan, X. C. & Iyer, V. R. Mapping DNA–protein interactions in large genomes by sequence tag analysis of genomic enrichment. *Nature Methods* **2**, 47-53 (2005).

42. Dorschner, M. O. *et al.* High-throughput localization of functional elements by quantitative chromatin profiling. *Nature Methods* **1**, 219-225 (2004).

43. Sabo, P. J. *et al.* Genome-scale mapping of DNase I sensitivity *in vivo* using tiling DNA microarrays. *Nature Methods* **3**, 511-518 (2006).

44. Crawford, G. E. *et al.* DNase-chip: a high-resolution method to identify DNase I hypersensitive sites using tiled microarrays. *Nature Methods* **3**, 503-509 (2006).

45. Hogan, G. J., Lee, C. K. & Lieb, J. D. Cell cycle-specified fluctuation of nucleosome occupancy at gene promoters. *PLoS Genet.* **2**, e158 (2006).

46. Koch, C. M. *et al.* The landscape of histone modifications across 1% of the human genome in five human cell lines. *Genome Res.* **17**, 691-707 (2007).

47. Smale, S. T. & Kadonaga, J. T. The RNA polymerase II core promoter. *Annu. Rev. Biochem.* **72**, 449-479 (2003).

48. Mito, Y., Henikoff, J. G. & Henikoff, S. Genome-scale profiling of histone H3.3 replacement patterns. *Nature Genet.* **37**, 1090-1097 (2005).

49. Heintzman, N. D. *et al.* Distinct and predictive chromatin signatures of transcriptional promoters and enhancers in the human genome. *Nature Genet.* **39**, 311-318 (2007).

50. Yusufzai, T. M., Tagami, H., Nakatani, Y. & Felsenfeld, G. CTCF tethers an insulator to subnuclear sites, suggesting shared insulator mechanisms across species. *Mol. Cell* **13**, 291-298 (2004).

51. Kim, T. H. *et al.* Direct isolation and identification of promoters in the human genome. *Genome Res.* **15**, 830-839 (2005).

52. Bieda, M., Xu, X., Singer, M. A., Green, R. & Farnham, P. J. Unbiased location analysis of E2F1-binding sites suggests a widespread role for E2F1 in the human genome. *Genome Res.* **16**, 595-605 (2006).

53. Ruppert, S., Wang, E. H. & Tjian, R. Cloning and expression of human $TAF_{II}250$: a TBP-associated factor implicated in cell-cycle regulation. *Nature* **362**, 175-179 (1993).

54. Fernandez, P. C. *et al.* Genomic targets of the human c-Myc protein. *Genes Dev.* **17**, 1115-1129 (2003).

55. Li, Z. *et al.* A global transcriptional regulatory role for c-Myc in Burkitt's lymphoma cells. *Proc. Natl Acad. Sci. USA* **100**, 8164-8169 (2003).

56. Orian, A. *et al.* Genomic binding by the *Drosophila* Myc, Max, Mad/Mnt transcription factor network. *Genes Dev.* **17**, 1101-1114 (2003).

57. de Laat, W. & Grosveld, F. Spatial organization of gene expression: the active chromatin hub. *Chromosome Res.* **11**, 447-459 (2003).

58. Trinklein, N. D. *et al.* Integrated analysis of experimental datasets reveals many novel promoters in 1% of the human genome. *Genome Res.* **17**, 720-731 (2007).

59. Jeon, Y. *et al.* Temporal profile of replication of human chromosomes. *Proc. Natl Acad. Sci. USA* **102**, 6419-6424 (2005).

60. Woodfine, K. *et al.* Replication timing of the human genome. *Hum. Mol. Genet.* **13**, 191-202 (2004).

61. White, E. J. *et al.* DNA replication-timing analysis of human chromosome 22 at high resolution and different developmental states. *Proc. Natl Acad. Sci. USA* **101**, 17771-17776 (2004).

62. Schubeler, D. *et al.* Genome-wide DNA replication profile for *Drosophila melanogaster*: a link between transcription and replication timing. *Nature Genet.* **32**, 438-442 (2002).

63. MacAlpine, D. M., Rodriguez, H. K. & Bell, S. P. Coordination of replication and transcription along a *Drosophila* chromosome. *Genes Dev.* **18**, 3094-3105 (2004).

64. Gilbert, D. M. Replication timing and transcriptional control: beyond cause and effect. *Curr. Opin. Cell Biol.* **14**, 377-383 (2002).

65. Schwaiger, M. & Schubeler, D. A question of timing: emerging links between transcription and replication. *Curr. Opin. Genet. Dev.* **16**, 177-183 (2006).

66. Hatton, K. S. *et al.* Replication program of active and inactive multigene families in mammalian cells. *Mol. Cell. Biol.* **8**, 2149-2158 (1988).

67. Gartler, S. M., Goldstein, L., Tyler-Freer, S. E. & Hansen, R. S. The timing of *XIST* replication: dominance of the domain. *Hum. Mol. Genet.* **8**, 1085-1089 (1999).

68. Azuara, V. *et al.* Heritable gene silencing in lymphocytes delays chromatid resolution without affecting the timing of DNA replication. *Nature Cell Biol.* **5**, 668-674 (2003).

69. Cohen, S. M., Furey, T. S., Doggett, N. A. & Kaufman, D. G. Genome-wide sequence and functional analysis of early replicating DNA in normal human fibroblasts. *BMC Genomics* **7**, 301 (2006).

70. Cao, R. *et al.* Role of histone H3 lysine 27 methylation in Polycomb-group silencing. *Science* **298**, 1039-1043 (2002).

71. Muller, J. *et al.* Histone methyltransferase activity of a *Drosophila* Polycomb group repressor complex. *Cell* **111**, 197-208 (2002).

72. Bracken, A. P., Dietrich, N., Pasini, D., Hansen, K. H. & Helin, K. Genome-wide mapping of Polycomb target genes unravels their roles in cell fate transitions. *Genes Dev.* **20**, 1123-1136 (2006).

73. Kirmizis, A. *et al.* Silencing of human polycomb target genes is associated with methylation of histone H3 Lys 27. *Genes Dev.* **18**, 1592-1605 (2004).

74. Lee, T. I. *et al.* Control of developmental regulators by Polycomb in human embryonic stem cells. *Cell* **125**, 301-313 (2006).

75. Karnani, N., Taylor, C., Malhotra, A. & Dutta, A. Pan-S replication patterns and chromosomal domains defined by genome tiling arrays of human chromosomes. *Genome Res.* **17**, 865-876 (2007).

76. Delaval, K., Wagschal, A. & Feil, R. Epigenetic deregulation of imprinting in congenital diseases of aberrant growth. *Bioessays* **28**, 453-459 (2006).

77. Dillon, N. Gene regulation and large-scale chromatin organization in the nucleus. *Chromosome Res.* **14**, 117-126 (2006).

78. Burkhoff, A. M. & Tullius, T. D. Structural details of an adenine tract that does not cause DNA to bend. *Nature* **331**, 455-457 (1988).

79. Price, M. A. & Tullius, T. D. How the structure of an adenine tract depends on sequence context: a new model for the structure of T_nA_n DNA sequences. *Biochemistry* **32**, 127-136 (1993).

80. Greenbaum, J. A., Parker, S. C. J. & Tullius, T. D. Detection of DNA structural motifs in functional genomic elements. *Genome Res.* **17**, 940-946 (2007).

81. Thurman, R. E., Day, N., Noble, W. S. & Stamatoyannopoulos, J. A. Identification of higher-order functional domains in the human ENCODE regions. *Genome Res.* **17**, 917-927 (2007).

82. Gilbert, N. *et al.* Chromatin architecture of the human genome: gene-rich domains are enriched in open chromatin fibers. *Cell* **118**, 555-566 (2004).

83. Nobrega, M. A., Ovcharenko, I., Afzal, V. & Rubin, E. M. Scanning human gene deserts for long-range enhancers. *Science* **302**, 413 (2003).

84. Woolfe, A. *et al.* Highly conserved non-coding sequences are associated with vertebrate development. *PLoS Biol.* **3**, e7 (2005).

85. Drake, J. A. *et al.* Conserved noncoding sequences are selectively constrained and not mutation cold spots. *Nature Genet.* **38**, 223-227 (2006).

86. Margulies, E. H., Blanchette, M., NISC Comparative Sequencing Program, Haussler D. & Green, E. D. Identification and characterization of multi-species conserved sequences. *Genome Res.* **13**, 2507-2518 (2003).

87. Cooper, G. M. *et al.* Distribution and intensity of constraint in mammalian genomic sequence. *Genome Res.* **15**, 901-913 (2005).

88. Siepel, A. *et al.* Evolutionarily conserved elements in vertebrate, insect, worm, and yeast genomes. *Genome Res.* **15**, 1034-1050 (2005).

89. Thomas, J. W. *et al.* Parallel construction of orthologous sequence-ready clone contig maps in multiple species. *Genome Res.* **12**, 1277-1285 (2002).

90. Blakesley, R. W. *et al.* An intermediate grade of finished genomic sequence suitable for comparative analyses. *Genome Res.* **14**, 2235-2244 (2004).

91. Aparicio, S. *et al.* Whole-genome shotgun assembly and analysis of the genome of *Fugu rubripes. Science* **297**, 1301-1310 (2002).

92. Jaillon, O. *et al.* Genome duplication in the teleost fish *Tetraodon nigroviridis* reveals the early vertebrate proto-karyotype. *Nature* **431**, 946-957 (2004).

93. Margulies, E. H. *et al.* An initial strategy for the systematic identification of functional elements in the human genome by low-redundancy comparative sequencing. *Proc. Natl Acad. Sci. USA* **102**, 4795-4800 (2005).

94. Blanchette, M. *et al.* Aligning multiple genomic sequences with the threaded blockset aligner. *Genome Res.* **14**, 708-715 (2004).

95. Bray, N. & Pachter, L. MAVID: constrained ancestral alignment of multiple sequences. *Genome Res.* **14**, 693-699 (2004).

96. Brudno, M. *et al.* LAGAN and Multi-LAGAN: efficient tools for large-scale multiple alignment of genomic DNA. *Genome Res.* **13**, 721-731 (2003).

97. Margulies, E. H. *et al.* Relationship between evolutionary constraint and genome function for 1% of the human genome. *Genome Res.* **17**, 760-774 (2007).

98. Asthana, S., Roytberg, M., Stamatoyannopoulos, J. A. & Sunyaev, S. Analysis of sequence conservation at nucleotide resolution. *PLoS Comp. Biol.* (submitted).

99. Cooper, G. M., Brudno, M., Green, E. D., Batzoglou, S. & Sidow, A. Quantitative estimates of sequence divergence for comparative analyses of mammalian genomes. *Genome Res.* **13**, 813-820 (2003).

100. Eddy, S. R. A model of the statistical power of comparative genome sequence analysis. *PLoS Biol.* **3**, e10 (2005).

101. Stone, E. A., Cooper, G. M. & Sidow, A. Trade-offs in detecting evolutionarily constrained sequence by comparative genomics. *Annu. Rev. Genomics Hum. Genet.* **6**, 143-164 (2005).
102. McAuliffe, J. D., Jordan, M. I. & Pachter, L. Subtree power analysis and species selection for comparative genomics. *Proc. Natl Acad. Sci. USA* **102**, 7900-7905 (2005).
103. International HapMap Consortium. A haplotype map of the human genome. *Nature* **437**, 1299-1320 (2005).
104. Cheng, Z. *et al.* A genome-wide comparison of recent chimpanzee and human segmental duplications. *Nature* **437**, 88-93 (2005).
105. Cooper, G. M. *et al.* Characterization of evolutionary rates and constraints in three Mammalian genomes. *Genome Res.* **14**, 539-548 (2004).
106. Dermitzakis, E. T., Reymond, A. & Antonarakis, S. E. Conserved non-genic sequences - an unexpected feature of mammalian genomes. *Nature Rev. Genet.* **6**, 151-157 (2005).
107. Clark, T. G. *et al.* Small insertions/deletions and functional constraint in the ENCODE regions. *Genome Biol.* (submitted) (2007).
108. Bejerano, G. *et al.* Ultraconserved elements in the human genome. *Science* **304**, 1321-1325 (2004).
109. Woolfe, A. *et al.* Highly conserved non-coding sequences are associated with vertebrate development. *PLoS Biol.* **3**, e7 (2005).
110. Dermitzakis, E. T. & Clark, A. G. Evolution of transcription factor binding sites in Mammalian gene regulatory regions: conservation and turnover. *Mol. Biol. Evol.* **19**, 1114-1121 (2002).
111. McDonald, J. H. & Kreitman, M. Adaptive protein evolution at the *Adh* locus in *Drosophila. Nature* **351**, 652-654 (1991).
112. Andolfatto, P. Adaptive evolution of non-coding DNA in *Drosophila. Nature* **437**, 1149-1152 (2005).
113. Hudson, R. R., Kreitman, M. & Aguade, M. A test of neutral molecular evolution based on nucleotide data. *Genetics* **116**, 153-159 (1987).
114. Feuk, L., Carson, A. R. & Scherer, S. W. Structural variation in the human genome. *Nature Rev. Genet.* **7**, 85-97 (2006).
115. Ludwig, M. Z. *et al.* Functional evolution of a *cis*-regulatory module. *PLoS Biol.* **3**, e93 (2005).
116. Ludwig, M. Z. & Kreitman, M. Evolutionary dynamics of the enhancer region of even-skipped in *Drosophila. Mol. Biol. Evol.* **12**, 1002-1011 (1995).
117. Harrow, J. *et al.* GENCODE: producing a reference annotation for ENCODE. *Genome Biol.* **7**, (Suppl. 1; S4) 1-9 (2006).
118. Emanuelsson, O. *et al.* Assessing the performance of different high-density tiling microarray strategies for mapping transcribed regions of the human genome. *Genome Res.* advance online publication, doi: 10.1101/gr.5014606 (21 November 2006).
119. Kapranov, P. *et al.* Large-scale transcriptional activity in chromosomes 21 and 22. Science 296, 916-919 (2002).
120. Bhinge, A. A., Kim, J., Euskirchen, G., Snyder, M. & Iyer, V. R. Mapping the chromosomal targets of STAT1 by Sequence Tag Analysis of Genomic Enrichment (STAGE). *Genome Res.* **17**, 910-916 (2007).
121. Ng, P. *et al.* Gene identification signature (GIS) analysis for transcriptome characterization and genome annotation. *Nature Methods* **2**, 105-111 (2005).
122. Giresi, P. G., Kim, J., McDaniell, R. M., Iyer, V. R. & Lieb, J. D. FAIRE (Formaldehyde-Assisted Isolation of Regulatory Elements) isolates active regulatory elements from human chromatin. *Genome Res.* **17**, 877-885 (2006).
123. Rada-Iglesias, A. *et al.* Binding sites for metabolic disease related transcription factors inferred at base pair resolution by chromatin immunoprecipitation and genomic microarrays. *Hum. Mol. Genet.* **14**, 3435-3447 (2005).
124. Kim, T. H. *et al.* A high-resolution map of active promoters in the human genome. *Nature* **436**, 876-880 (2005).
125. Halees, A. S. & Weng, Z. PromoSer: improvements to the algorithm, visualization and accessibility. *Nucleic Acids Res.* **32**, W191-W194 (2004).
126. Bajic, V. B. *et al.* Performance assessment of promoter predictions on ENCODE regions in the EGASP experiment. *Genome Biol.* **7**, (Suppl 1; S3) 1-13 (2006).
127. Zheng, D. & Gerstein, M. B. A computational approach for identifying pseudogenes in the ENCODE regions. *Genome Biol.* **7**, S13.1-S13.10 (2006).
128. Stranger, B. E. *et al.* Genome-wide associations of gene expression variation in humans. *PLoS Genet* **1**, e78 (2005).
129. Turner, B. M. Reading signals on the nucleosome with a new nomenclature for modified histones. *Nature Struct. Mol. Biol.* **12**, 110-112 (2005).

Supplementary Information is linked to the online version of the paper at www.nature.com/nature.

Acknowledgements. We thank D. Leja for providing graphical expertise and support. Funding support is acknowledged from the following sources: National Institutes of Health, The European Union BioSapiens NoE, Affymetrix, Swiss National Science Foundation, the Spanish Ministerio de Educación y Ciencia, Spanish Ministry of Education and Science, CIBERESP, Genome Spain and Generalitat de Catalunya, Ministry of Education, Culture, Sports, Science and Technology of Japan, the NCCR Frontiers in Genetics, the Jérôme Lejeune Foundation, the Childcare Foundation, the Novartis Foundations, the Danish Research Council, the Swedish Research Council, the Knut and Alice Wallenberg Foundation, the Wellcome Trust, the Howard Hughes Medical Institute, the Bio-X Institute, the RIKEN Institute, the US Army, National Science Foundation, the Deutsche Forschungsgemeinschaft, the Austrian Gen-AU program, the BBSRC and The European Molecular Biology Laboratory. We thank the Barcelona SuperComputing Center and the NIH Biowulf cluster for computer facilities. The Consortium thanks the ENCODE Scientific Advisory Panel for their advice on the project: G. Weinstock, M. Cherry, G. Churchill, M. Eisen, S. Elgin, J. Lis, J. Rine, M. Vidal and P. Zamore.

Author Information. Reprints and permissions information is available at www.nature.com/reprints. The authors declare no competing financial interests. The list of individual authors is divided among the six main analysis groups and five organizational groups. Correspondence and requests for materials should be addressed to the co-chairs of the ENCODE analysis groups (listed in the Analysis Coordination group) E. Birney (birney@ebi.ac.uk); J. A. Stamatoyannopoulos (jstam@u.washington.edu); A. Dutta (ad8q@virginia.edu); R. Guigó (rguigo@imim.es); T. R. Gingeras (Tom_Gingeras@affymetrix.com); E. H. Margulies (elliott@nhgri.nih.gov); Z. Weng (zhiping@bu.edu); M. Snyder (michael.snyder@yale.edu); E. T. Dermitzakis (md4@sanger.ac.uk) or collectively (encode_chairs@ebi.ac.uk).

Generation of Germline-competent Induced Pluripotent Stem Cells

K. Okita *et al.*

Editor's Note

In 2006, Japanese cell biologist Shinya Yamanaka claimed to have made stem cell-like cells by inserting a handful of genes into mouse fibroblasts. The so-called induced pluripotent stem (iPS) cells displayed many of the hallmarks of true stem cells, but failed one acid test: they were unable to contribute to live animals when injected into early embryonic mice. In this paper, Yamanaka and colleagues overcome this problem, tweaking their methodology to produce iPS cells that appear completely reprogrammed and truly stem cell-like. Two other papers, published at the same time, describe a similar feat and iPS cells remain the focus of great attention, offering the potential to make therapeutically useful stem cells from a person's own cells without the need for a donated egg or embryo.

We have previously shown that pluripotent stem cells can be induced from mouse fibroblasts by retroviral introduction of Oct3/4 (also called Pou5f1), Sox2, c-Myc and Klf4, and subsequent selection for *Fbx15* (also called *Fbxo15*) expression. These induced pluripotent stem (iPS) cells (hereafter called Fbx15 iPS cells) are similar to embryonic stem (ES) cells in morphology, proliferation and teratoma formation; however, they are different with regards to gene expression and DNA methylation patterns, and fail to produce adult chimaeras. Here we show that selection for *Nanog* expression results in germline-competent iPS cells with increased ES-cell-like gene expression and DNA methylation patterns compared with Fbx15 iPS cells. The four transgenes (*Oct3/4*, *Sox2*, *c-myc* and *Klf4*) were strongly silenced in Nanog iPS cells. We obtained adult chimaeras from seven Nanog iPS cell clones, with one clone being transmitted through the germ line to the next generation. Approximately 20% of the offspring developed tumours attributable to reactivation of the c-*myc* transgene. Thus, iPS cells competent for germline chimaeras can be obtained from fibroblasts, but retroviral introduction of c-Myc should be avoided for clinical application.

ALTHOUGH ES cells are promising donor sources in cell transplantation therapies[1], they face immune rejection after transplantation and there are ethical issues regarding the usage of human embryos. These concerns may be overcome if pluripotent stem cells can be directly derived from patients' somatic cells[2]. We have previously shown that iPS cells can be generated from mouse fibroblasts by retrovirus-mediated introduction of four transcription factors (Oct3/4 (refs 3, 4), Sox2 (ref. 5), c-Myc (ref. 6) and Klf4 (ref. 7)) and by selection for *Fbx15* expression[8]. Fbx15 iPS cells, however, have different gene

具有种系功能的诱导多能干细胞的产生

冲田奎介等

编者按

在 2006 年，日本细胞生物学家山中伸弥声称通过在小鼠成纤维细胞中插入少量基因制造出了干细胞样细胞。所谓的诱导多能干细胞(iPS 细胞)显示出真正干细胞的许多特征，但尚未通过一项严格的测试：当被注射到早期胚胎小鼠时，它们无法培养成为活体动物。在本文中，山中及其同事克服了这个问题，他们调整了实验方法，产生的 iPS 细胞呈现完全重编程和真正干细胞样。同时发表的另外两篇论文描述了类似的壮举。iPS 细胞仍然是人们关注的重点，因为它可以从一个人自己的细胞中制造出可用于治疗的干细胞而不需要捐赠的卵子或胚胎。

先前我们已经发现，通过逆转录病毒导入 Oct3/4(又称 Pou5f1)、Sox2、c-Myc 和 Klf4 并随后选择 *Fbx15*(又称 *Fbxo15*)表达，可以从小鼠成纤维细胞中诱导产生多能干细胞。这些诱导多能干细胞(iPS 细胞)(以下称为 Fbx15 iPS 细胞)与胚胎干细胞(ES 细胞)在形态、增殖和形成畸胎瘤方面类似；但是，它们在基因表达和 DNA 甲基化模式上是不同的，并不能形成成熟的嵌合体。这里我们发现选择 *Nanog* 基因的表达可以产生具有种系功能的 iPS 细胞，与 Fbx15 iPS 细胞相比增加了 ES 细胞样基因的表达和 DNA 甲基化模式。这四个转基因(*Oct3/4*、*Sox2*、*c-myc* 和 *Klf4*)在 Nanog iPS 细胞中极度沉默。我们从 7 个 Nanog iPS 细胞克隆中获得了成熟的嵌合体，其中一个克隆还通过种系培养到了下一代。由于重新激活了 *c-myc* 转基因，大约 20% 的后代出现了肿瘤。总之，具有产生种系嵌合体功能的 iPS 细胞能够从成纤维细胞中诱导获得，但是通过逆转录病毒导入 c-Myc 在临床应用时应该被避免。

尽管胚胎干细胞是细胞移植治疗很有前途的材料来源 [1]，它们仍存在移植后免疫排斥的问题，而且还存在与使用人类胚胎有关的伦理学问题。如果多能干细胞能够直接来源于病人的体细胞，那么这些问题就都可以被解决 [2]。先前我们已经发现 iPS 细胞能够从小鼠成纤维细胞通过逆转录病毒导入四个转录因子(Oct3/4(文献 3，4)、Sox2(文献 5)、c-Myc(文献 6)和 Klf4(文献 7))并继以选择 *Fbx15* 表达而产生 [8]。但是，Fbx15 iPS 细胞具有与 ES 细胞不同的基因表达和 DNA 甲基化模式，

expression and DNA methylation patterns compared with ES cells and do not contribute to adult chimaeras. We proposed that the incomplete reprogramming might be due to the selection for *Fbx15* expression, and that by using better selection markers, we might be able to generate more ES-cell-like iPS cells. We decided to use *Nanog* as a candidate of such markers.

Although both *Fbx15* and *Nanog* are targets of Oct3/4 and Sox2 (refs 9–11), Nanog is more tightly associated with pluripotency. In contrast to *Fbx15*-null mice and ES cells that barely show abnormal phenotypes[9], disruption of *Nanog* in mice results in loss of the pluripotent epiblast[12]. *Nanog*-null ES cells can be established, but they tend to differentiate spontaneously[12]. Forced expression of *Nanog* renders ES cells independent of leukaemia inhibitory factor (LIF) for self-renewal[12,13] and confers increased reprogramming efficiency after fusion with somatic cells[14]. These results prompted us to propose that if we use *Nanog* as a selection marker, we might be able to obtain iPS cells displaying a greater similarity to ES cells.

Generation of Nanog iPS Cells

To establish a selection system for *Nanog* expression, we began by isolating a bacterial artificial chromosome (BAC, ~200 kilobases) containing the mouse *Nanog* gene in its centre. By using recombineering technology[15,16], we inserted a green fluorescent protein (GFP)-internal ribosome entry site (IRES)-puromycin resistance gene ($Puro^r$) cassette into the 5′ untranslated region (UTR; Fig. 1a). ES cells that had stably incorporated the modified BAC were positive for GFP, but became negative when differentiation was induced (not shown). By introducing these ES cells into blastocysts, we obtained chimaeric mice and then transgenic mice containing the Nanog-GFP-IRES-$Puro^r$ reporter construct. In transgenic mouse blastocysts, GFP was specifically observed in the inner cell mass (Fig. 1b). In 9.5 days post coitum (d.p.c.) embryos, only migrating primordial germ cells (PGCs) showed GFP signal. In 13.5 d.p.c. embryos, GFP was specifically detected in the genital ridges of both sexes. After removing the brain, visceral tissues and genital ridges, we isolated mouse embryonic fibroblasts (MEFs) from 13.5 d.p.c. male embryos. Flow cytometry analyses showed that these MEFs did not contain GFP-positive cells, whereas ~1% of cells isolated from genital ridges showed GFP signals (Fig. 1c).

在成年鼠中无嵌合。我们认为这种不完全的重编程可能是由于通过 *Fbx15* 表达来选择细胞，因此如果选用更好的选择标记物，我们可能能够产生更加像 ES 细胞的 iPS 细胞。我们计划使用 *Nanog* 作为这种标记物的候选者。

尽管 *Fbx15* 和 *Nanog* 都是 Oct3/4 和 Sox2 的靶目标（文献 9 ~ 11），但 Nanog 与多能性的关系更加密切。与几乎没有异常表型的 *Fbx15* 敲除小鼠以及 ES 细胞 [9] 相反，小鼠中 *Nanog* 基因的紊乱可以导致多能上胚层的丧失 [12]。可以建立敲除 *Nanog* 的 ES 细胞，但是它们倾向于发生自发分化 [12]。强制表达 *Nanog* 可以使 ES 细胞的自我更新不依赖于白血病抑制因子（LIF）[12,13]，而且在体细胞融合后赋予其更高的重编程效率 [14]。这些结果促使我们假设如果使用 *Nanog* 作为选择标记物，我们可能能够获得更像 ES 细胞的 iPS 细胞。

Nanog iPS 细胞的产生

为了建立 *Nanog* 表达的选择系统，我们首先分离了在中心含有小鼠 *Nanog* 基因的细菌人工染色体（BAC，约 200 个千碱基）。通过使用重组工程技术 [15,16]，我们将一个绿色荧光蛋白（GFP）–内部核糖体进入位点（IRES）–嘌呤霉素抗性基因（$Puro^r$）盒插入 5′非翻译区（UTR，图 1a）。那些和修饰过的 BAC 稳定整合的 ES 细胞就含有 GFP，但是如果诱导分化，就变为 GFP 阴性（没有显示）。通过将这些 ES 细胞导入胚泡，我们就获得了嵌合小鼠，然后获得含有 Nanog-GFP-IRES-$Puro^r$ 报告基团的转基因小鼠。在转基因的小鼠胚泡中，GFP 特有地出现在内细胞团中（图 1b）。在交配后第 9.5 天的胚胎中，只有迁移的原始生殖细胞（PGC）发出 GFP 信号。在交配后第 13.5 天，GFP 特定地出现在两种性别的生殖嵴上。去除大脑、内脏组织和生殖嵴后，我们从交配后 13.5 天的雄性胚胎中分离出小鼠胚胎成纤维细胞（MEF）。流式细胞分析显示这些 MEF 不含有 GFP 阳性的细胞，然而约 1% 从生殖嵴分离出的细胞显示 GFP 信号（图 1c）。

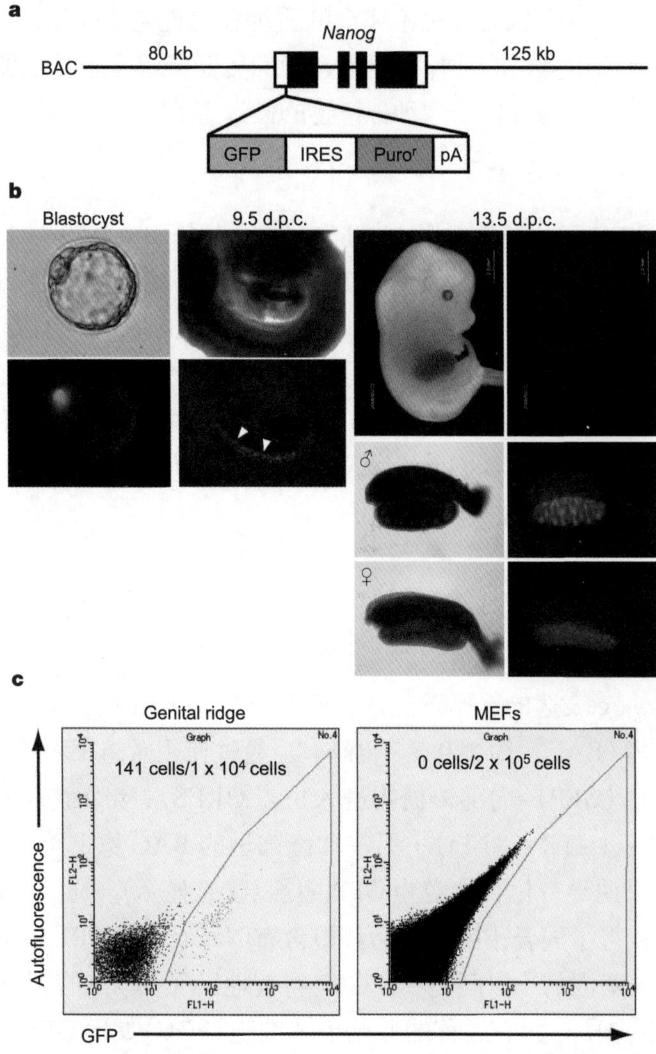

Fig. 1. Nanog-GFP-IRES-Puro^r transgenic mice. **a**, Modified BAC construct. White boxes indicate the 5′ and 3′ UTRs of the mouse *Nanog* gene. Black boxes indicate the open reading frame. **b**, GFP expression in Nanog-GFP transgenic mouse embryos. Whole embryos (top panels) and isolated genital ridges (bottom panels) from 13.5 d.p.c. mice are shown. **c**, Histogram showing GFP fluorescence in cells isolated from genital ridges of a 13.5 d.p.c. Nanog-GFP transgenic mouse embryo (left) or in MEFs isolated from the same embryo (right).

Next, we introduced the four previously described factors (Oct3/4, Sox2, Klf4 and the c-Myc mutant c-Myc(T58A)) into Nanog-GFP-IRES-Puro^r MEFs cultured on SNL feeder cells with the use of retroviral vectors. Three, five, or seven days after retroviral infection, we started puromycin selection in ES cell medium. GFP-positive cells first became apparent ~7 days after infection. Twelve days after infection, a few hundred colonies appeared, regardless of the timing of puromycin selection (Fig. 2a). By contrast, no colonies emerged from MEFs transfected with mock DNA. Among puromycin-resistant colonies, ~5% were positive for GFP (Fig. 2b). When the puromycin selection was started at 7 days after infection, we

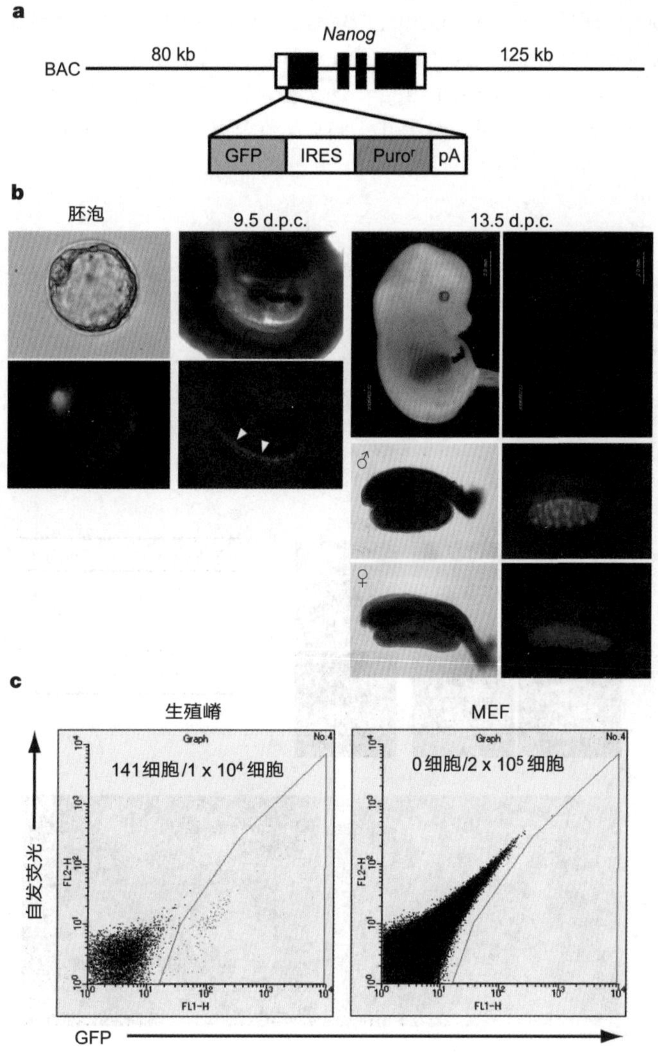

图 1. Nanog-GFP-IRES-Puro^r 转基因小鼠。**a**，修饰的 BAC 结构。白框表示小鼠 *Nanog* 基因的 5′和 3′UTR。黑框代表可读框。**b**，Nanog-GFP 转基因小鼠胚胎中的 GFP 表达。图中显示了交配后 13.5 天小鼠的整个胚胎（上部）和分离的生殖嵴（下部）。**c**，图中显示从交配后 13.5 天的 Nanog-GFP 转基因小鼠胚胎的生殖嵴中分离的细胞的 GFP 荧光（左）和从相同胚胎中分离出来的 MEF 的 GFP 荧光（右）。

然后，我们将先前所述的四种因子（Oct3/4、Sox2、Klf4 和 c-Myc 突变体 c-Myc（T58A））利用逆转录病毒载体导入到培养在 SNL 饲养细胞中的 Nanog-GFP-IRES-Puro^r MEF 内。逆转录病毒感染后的 3 天、5 天和 7 天，我们分别在 ES 细胞基质中进行嘌呤霉素选择。感染后约 7 天 GFP 阳性的细胞开始出现。感染后 12 天出现了数百个集落，并且不受嘌呤霉素选择时间的影响（图 2a）。相反地，用对照 DNA 进行转染的 MEF 没有出现集落。在嘌呤霉素抗性集落中，大约 5% 是 GFP 阳性的（图 2b）。感染 7 天后当嘌呤霉素选择开始时，我们获得了 GFP 阳性最高的集

obtained the most GFP-positive colonies. Because we used the GFP-IRES-Puro^{r} cassette, it is unclear why we obtained GFP-negative colonies. With increased concentrations of puromycin, we obtained fewer GFP-negative colonies (Fig. 2c). With any combination of three of the four factors, we did not obtain any GFP-positive colonies (Supplementary Fig. 1).

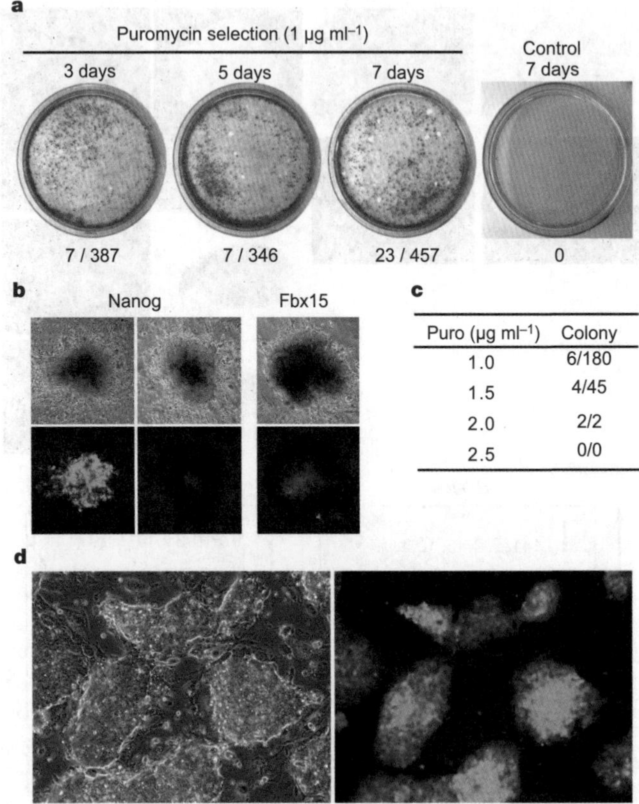

Fig. 2. Generation of iPS cells from MEFs of Nanog-GFP-IRES-Puro^{r} transgenic mice. **a**, Puromycin-resistant colonies. Puromycin selection was initiated at 3, 5, or 7 days after retroviral transduction. Numbers indicate GFP-positive colonies/total colonies. **b**, GFP fluorescence in resulting colonies. Phase contrast (top row) and fluorescence (bottom row) micrographs are shown. iPS cells were also generated from Fbx15 β-geo knockin MEFs. **c**, Effect of increasing concentrations of puromycin. Numbers of GFP-positive colonies/total colonies are shown on the right. **d**, Morphology of established Nanog iPS cells (clone 20D17). Phase contrast (left) and fluorescence (right) micrographs are shown.

By continuing cultivation of these GFP-positive colonies, we obtained cells that were morphologically indistinguishable from ES cells (Fig. 2d). These cells also demonstrated ES-like proliferation, with slightly longer doubling times than that of ES cells (Fig. 3a). Subcutaneous transplantation of these cells into nude mice resulted in tumours that consisted of various tissues of all three germ layers, indicating that these cells are pluripotent (Fig. 3b and Supplementary Fig. 2). We therefore refer to these cells as Nanog iPS cells in the remainder of this manuscript. Induced pluripotent stem cells were established from Fbx15 β-geo MEFs in parallel and are referred to as Fbx15 iPS cells.

落。因为我们插入的是 GFP-IRES-Puro^r 盒，所以不清楚为什么会出现 GFP 阴性的集落。随着嘌呤霉素浓度的增加，我们得到 GFP 阴性的集落越少（图 2c）。这四个因子中的任意三个组合都不能出现 GFP 阳性的集落（补充信息图 1）。

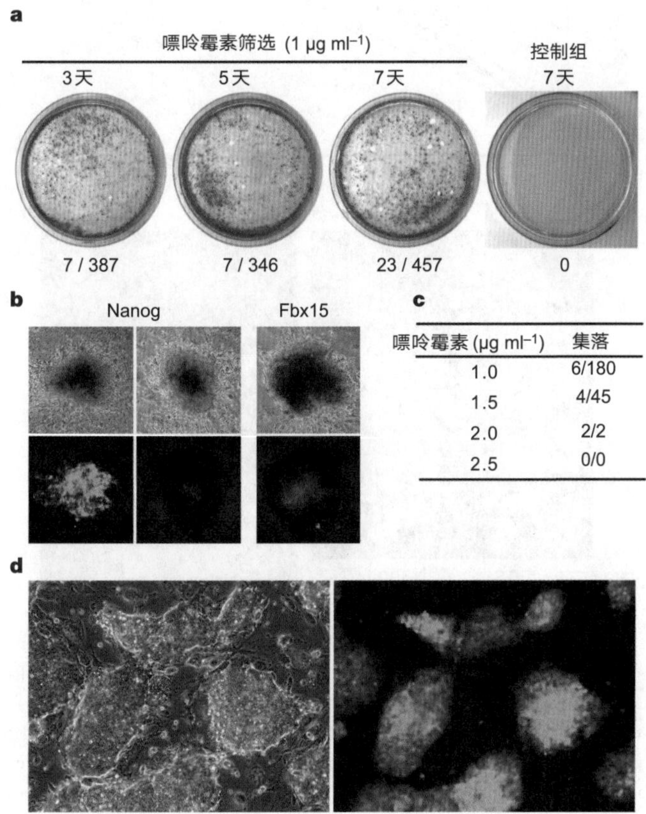

图 2. 从 Nanog-GFP-IRES-Puro^r 转基因小鼠 MEF 中生成 iPS 细胞。**a**，嘌呤霉素抗性集落。嘌呤霉素筛选开始于逆转录病毒转导后的 3、5 和 7 天。数字代表 GFP 阳性集落/总集落。**b**，集落中的 GFP 荧光。图片为相差（顶行）和荧光（底行）显微图。iPS 细胞也从 Fbx15 β-geo 基因敲入 MEF 中产生。**c**，嘌呤霉素浓度增加的影响。右边显示 GFP 阳性集落/总集落的数目。**d**，已建立的 Nanog iPS 细胞的形态（克隆 20D17）。图片是相差（左）和荧光（右）显微图。

通过继续培养这些 GFP 阳性的集落，我们获得了形态上与 ES 细胞无法区分的细胞（图 2d）。这些细胞可以出现 ES 样的增殖，倍增时间略长于 ES 细胞（图 3a）。将这些细胞皮下移植到裸鼠中可以产生含有全部三个胚层各种组织的肿瘤，提示这些细胞是多能的（图 3b 和补充信息图 2）。因此在本文的剩余部分我们称这些细胞为 Nanog iPS 细胞。我们平行建立了来源于 Fbx15 β-geo MEF 的诱导多能干细胞，称为 Fbx15 iPS 细胞。

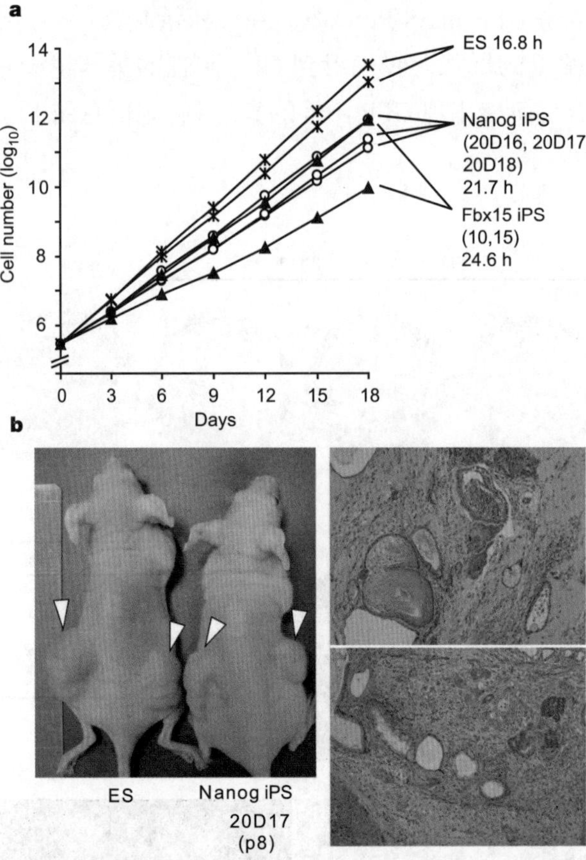

Fig. 3. Characterization of Nanog iPS cells. **a**, Proliferation. ES cells, Nanog iPS cells (clones 20D16, 20D17 and 20D18) and Fbx15 iPS cells (clones 10 and 15) were passaged every 3 days (3×10^5 cells per each well of a 6-well plate). Calculated doubling times are indicated. **b**, Teratomas. ES cells or Nanog iPS cells (clone 20D17, 1×10^6 cells) were subcutaneously transplanted into nude mice. After 8 weeks, teratomas were photographed (left) and analysed histologically with haematoxylin and eosin staining (right).

Similarity between Nanog iPS Cells and ES Cells

Polymerase chain reaction with reverse transcription (RT–PCR) showed that Nanog iPS cells expressed most ES cell marker genes, including *Nanog*, at higher and more consistent levels compared with Fbx15 iPS cells (Fig. 4a). DNA microarray analyses confirmed that Nanog iPS cells had greater ES-cell-like gene expression compared with Fbx15 iPS cells (Fig. 4b). The expression level of *Rex1* (also called *Zfp42*) in Nanog iPS cells was higher compared with Fbx15 iPS cells, but still lower than in ES cells. Thus, Nanog iPS cells show greater gene expression similarity to ES cells (without being identical) than do Fbx15 iPS cells.

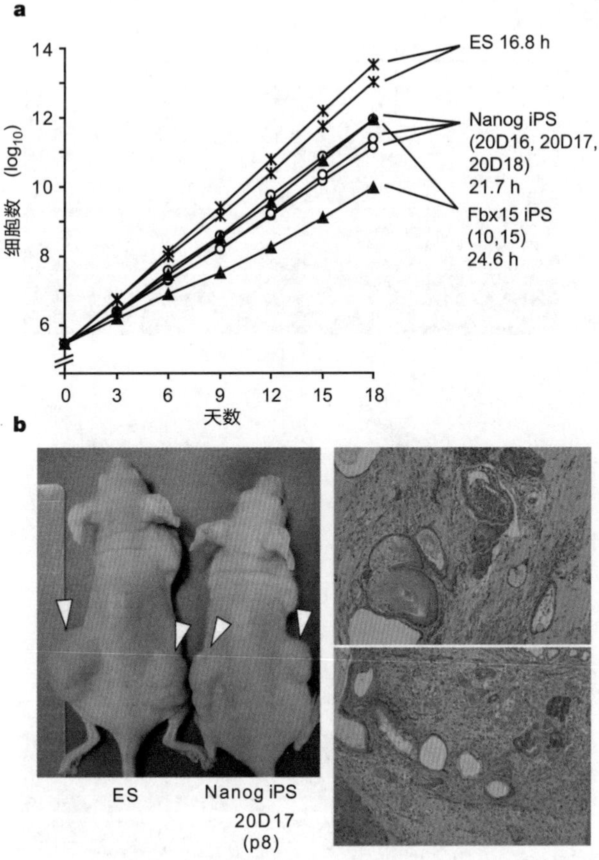

图 3. Nanog iPS 细胞的特征。**a**，增殖。ES 细胞、Nanog iPS 细胞（克隆 20D16、20D17 和 20D18）以及 Fbx15 iPS 细胞（克隆 10 和 15）每 3 天都进行传代（6 孔板上每孔 3×10^5 个细胞）。图中显示了计算出的倍增时间。**b**，畸胎瘤。ES 细胞或者 Nanog iPS 细胞（克隆 20D17，1×10^6 个细胞）皮下移植到裸鼠上。8 周以后，进行畸胎瘤照相（左图）并用苏木精-伊红染色观察组织学结构（右图）。

Nanog iPS 细胞和 ES 细胞间的相似性

反转录聚合酶链反应（RT–PCR）显示 Nanog iPS 细胞表达包括 *Nanog* 在内的大部分 ES 细胞标志物基因，而且与 Fbx15 iPS 细胞相比，Nanog iPS 细胞表达水平更高、更符合 ES 细胞（图 4a）。DNA 微阵列分析确证了 Nanog iPS 细胞与 Fbx15 iPS 细胞相比表达更多的 ES 细胞样基因（图 4b）。Nanog iPS 细胞中 *Rex1*（又称 *Zfp42*）的表达水平高于 Fbx15 iPS 细胞，但是仍然低于 ES 细胞。因此，Nanog iPS 细胞与 ES 细胞的基因表达相似程度（不是完全一致）要超过 Fbx15 iPS 细胞。

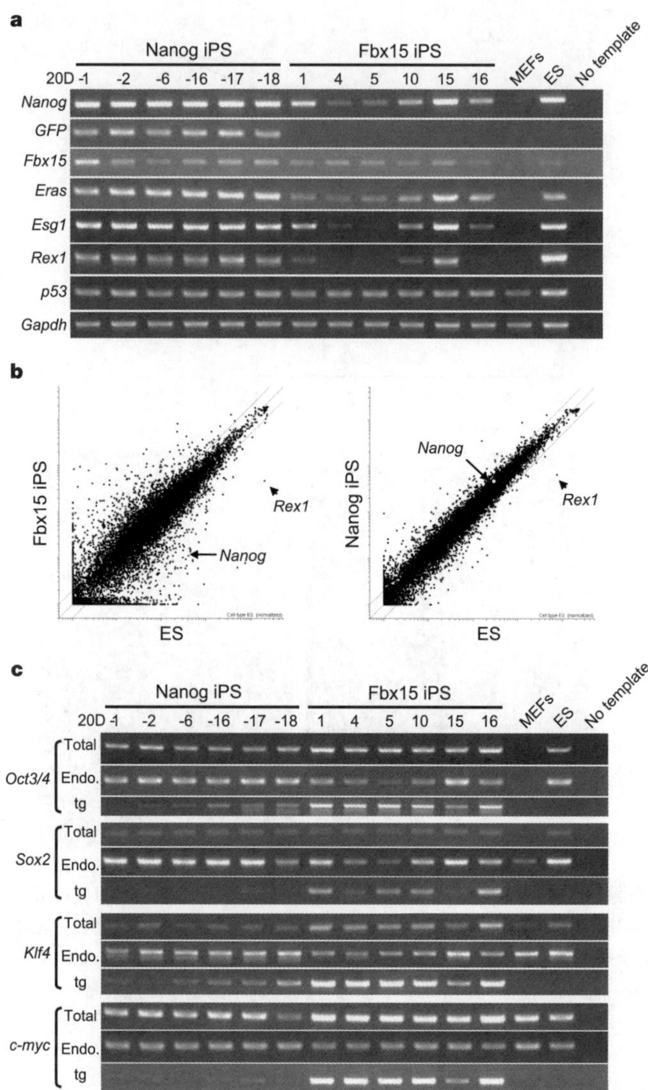

Fig. 4. Gene expression in Nanog iPS cells. **a**, RT–PCR. Total RNA was isolated from six clones of Nanog iPS cells (clones 20D1, 20D2, 20D6, 20D16, 20D17 and 20D18), six clones of Fbx15 iPS cells (clones 1, 4, 5, 10, 15 and 16), MEFs and ES cells. **b**, Scatter plots showing comparison of global gene expression between ES cells and Nanog iPS cells (right), and between ES cells and Fbx15 iPS cells (left), as determined by DNA microarrays. **c**, Expression levels of the four transcription factors. Total RNA was isolated from six clones of Nanog iPS cells (clones 20D1, 20D2, 20D6, 20D16, 20D17 and 20D18), six clones of Fbx15 iPS cells (clones 1, 4, 5, 10, 15 and 16), MEFs and ES cells. RT–PCR analyses were performed with primers that amplified the coding regions of the four factors (Total), endogenous transcripts only (Endo.), and transgene transcripts only (tg).

RT–PCR showed that Nanog iPS cells have significantly lower expression levels of the four transgenes than Fbx15 iPS cells (Fig. 4c). Real-time PCR confirmed that transgene expression was very low in Nanog iPS cells (Supplementary Fig. 4a–d). In contrast, Southern blot analyses showed similar copy numbers of retroviral integration in Nanog

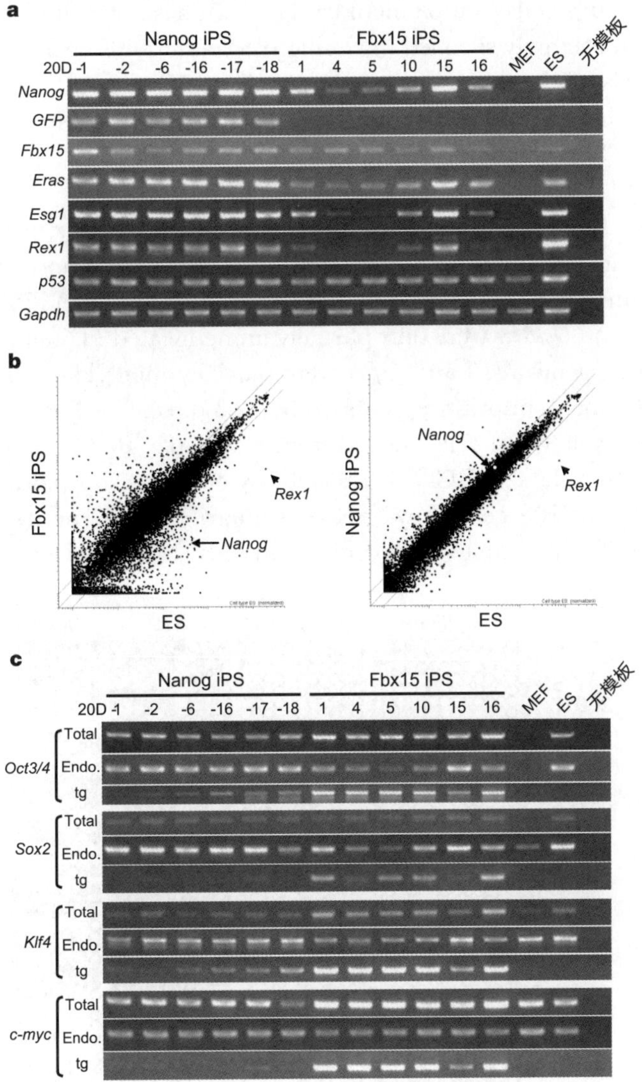

图 4. Nanog iPS 细胞的基因表达。**a**，RT–PCR。从 6 个 Nanog iPS 细胞克隆（克隆 20D1、20D2、20D6、20D16、20D17 和 20D18）、6 个 Fbx15 iPS 细胞克隆（克隆 1、4、5、10、15 和 16）、MEF 和 ES 细胞中分离出总 RNA。**b**，散点图显示了使用 DNA 微阵列得出的 ES 细胞和 Nanog iPS 细胞（右图）以及 ES 细胞和 Fbx15 iPS 细胞（左图）之间的整体基因表达的对比。**c**，四个转录因子的表达水平。从 6 个 Nanog iPS 细胞克隆（克隆 20D1、20D2、20D6、20D16、20D17 和 20D18）、6 个 Fbx15 iPS 细胞克隆（克隆 1、4、5、10、15 和 16）、MEF 和 ES 细胞中分离出总 RNA。然后进行 RT–PCR 分析，使用的引物分别扩增四个因子的编码区（Total）、仅内源转录物（Endo.）和仅转基因转录物（tg）。

RT–PCR 显示 Nanog iPS 细胞中四个转基因的表达水平远远低于 Fbx15 iPS 细胞（图 4c）。实时 PCR 确证了 Nanog iPS 细胞中转基因的表达非常低（补充信息图 4a ~ 4d）。相反，Southern 印迹分析显示 Nanog iPS 细胞和 Fbx15 iPS 细胞中逆转录

iPS cells and Fbx15 iPS cells (Supplementary Fig. 5). These data indicate that retroviral transgene expression is largely silenced in Nanog iPS cells, as has been shown in ES cells[17]. The expression levels of the transgenes are reversely correlated with *Dnmt3a2* expression, suggesting that *de novo* methyltransferase[18] may be involved in the retroviral silencing observed in iPS cells (Supplementary Fig. 6).

Bisulphite genomic sequencing analyses also revealed similarities between Nanog iPS cells and ES cells (Fig. 5). The promoter regions of *Nanog*, *Oct3/4* and *Fbx15* were largely unmethylated in Nanog iPS cells. This is in marked contrast to Fbx15 iPS cells in which the promoters of *Nanog* and *Oct3/4* were only partially unmethylated[8]. Differentially methylated regions of imprinting genes *H19* and *Igf2r* were partially methylated in Nanog iPS cells. During PGC development, imprinting is erased by 12.5 d.p.c.[19-21]. The loss of imprinting is maintained in embryonic germ cells derived from 12.5 d.p.c. PGCs[22] and cloned embryos derived from 12.5–16.5 d.p.c. PGCs[23,24]. ES cells, by contrast, showed normal imprinting patterns[25]. Thus, Nanog iPS cells show greater similarity in the methylation patterns of imprinting genes to ES cells than to embryonic germ cells.

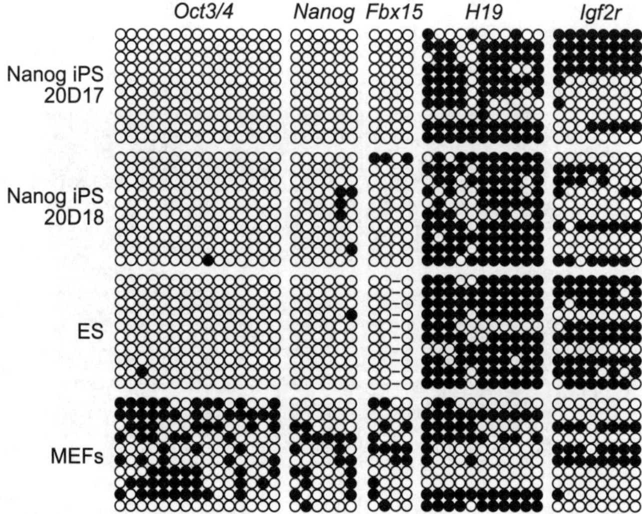

Fig. 5. DNA methylation of ES-cell-specific genes and imprinting genes. White circles indicate unmethylated CpG dinucleotides, whereas black circles indicate methylated CpG dinucleotides.

Simple sequence length polymorphism (SSLP) analyses showed that Nanog iPS cells are largely of the DBA background but also have some contribution from the C57BL/6 and 129S4 backgrounds (Supplementary Fig. 3). This result is consistent with the genetic background of the MEFs, which was 75% DBA, 12.5% C57BL/6 and 12.5% 129S4. This result also confirms that Nanog iPS cells are not a contamination of ES cells that exists in our laboratory, which are either pure 129S4 or C57BL/6.

We next compared the stability of Nanog iPS cells and Fbx15 iPS cells (Supplementary Fig. 7). Cells were cultivated in the presence of the selection drug for up to 22–26 passages.

病毒整合的拷贝数目相似（补充信息图 5）。这些数据提示 Nanog iPS 细胞中逆转录病毒转基因的表达大量地被抑制，这和 ES 细胞中出现的情况一样[17]。转基因的表达水平与 *Dnmt3a2* 的表达呈负相关，说明新生的甲基转移酶[18]可能参与了 iPS 细胞内逆转录病毒被抑制的过程（补充信息图 6）。

亚硫酸氢盐基因组测序分析也揭示了 Nanog iPS 细胞与 ES 细胞间的相似性（图 5）。Nanog iPS 细胞内的 *Nanog*、*Oct3/4* 和 *Fbx15* 的启动子区域大部分都是未甲基化的。这与 Fbx15 iPS 细胞明显相反，后者的 *Nanog* 和 *Oct3/4* 启动子仅仅是部分未甲基化[8]。印记基因 *H19* 和 *Igf2r* 的差异甲基化区域在 Nanog iPS 细胞中部分被甲基化。在 PGC 的发育过程中，印记在交配后 12.5 天消除[19-21]。这种印记消除的现象在交配后 12.5 天的 PGC 来源的胚胎生殖细胞以及交配后 12.5 ~ 16.5 天的 PGC 来源的克隆胚胎中保持[23,24]。相反，ES 细胞具有正常的印记分布[25]。因此，Nanog iPS 细胞在印记基因的甲基化分布方面与 ES 细胞的相似程度比胚胎生殖细胞更高。

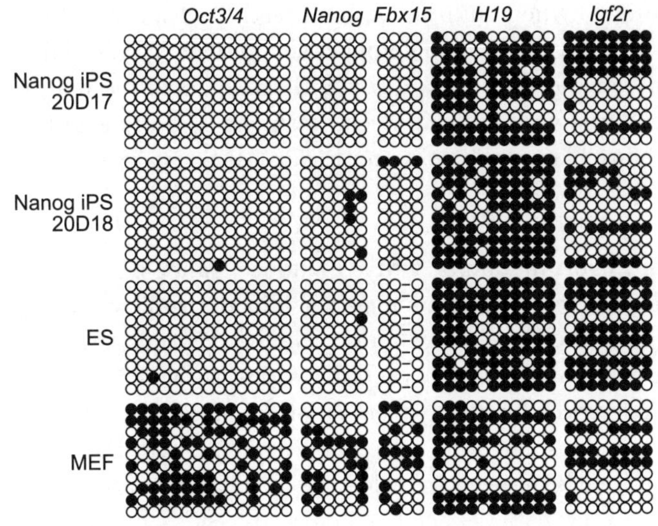

图 5. ES 细胞特异性基因和印记基因的 DNA 甲基化。白圈代表未甲基化的 CpG 二核苷酸，而黑圈代表甲基化的 CpG 二核苷酸。

简单序列长度多态性（SSLP）分析显示 Nanog iPS 细胞大部分是来源于 DBA 背景的，但是也有一部分来源于 C57BL/6 和 129S4 背景（补充信息图 3）。这个结果与 MEF 的基因背景相一致，即 75% DBA、12.5% C57BL/6 和 12.5% 129S4。该结果也证实了 Nanog iPS 细胞不是存在于我们实验室中的 ES 细胞的污染，因为那些 ES 细胞都是纯的 129S4 或者 C57BL/6 来源的。

我们接下来比较了 Nanog iPS 细胞和 Fbx15 iPS 细胞的稳定性（补充信息图 7）。

Morphologically, we did not observe significant changes over the long-term culture course. However, RT–PCR showed that Fbx15 iPS cells lost the expression of ES cell marker genes after prolonged culture. By contrast, Nanog iPS cells maintained relatively high expression levels of the ES cell marker genes. These data demonstrate that Nanog iPS cells are more stable than Fbx15 iPS cells.

We also compared the induction efficiency of Nanog iPS cells and Fbx15 iPS cells. In independent experiments, we obtained 4–125 GFP-positive colonies from 8×10^5 Nanog-reporter MEFs transfected with the four transcription factors. Because ~50% of transfected MEFs are supposed to express all four factors[8], the induction efficiency is approximately 0.001–0.03%. In contrast, from the same number of Fbx15-reporter MEFs, we obtained 47–1,800 G418-resistant colonies, with the induction efficiency of approximately 0.01–0.5%. Thus, the efficiency of Nanog iPS cell induction is approximately one-tenth that of Fbx15 iPS cells.

We then compared the responses of Nanog iPS cells and Fbx15 iPS cells to LIF or retinoic acid (Supplementary Fig. 8). As we have shown previously[8], Fbx15 iPS cells do not remain undifferentiated when cultured without feeder cells, even in the presence of LIF. Furthermore, Fbx15 iPS cells formed compact colonies when cultured without feeder cells in the presence of retinoic acid. In contrast, LIF maintained the undifferentiated state of Nanog iPS cells cultured without feeder cells. Retinoic acid induced the differentiation of Nanog iPS cells. Thus, Nanog iPS cells are similar to ES cells in their response to LIF and retinoic acid.

Initially we used the T58A mutant of c-Myc to induce Nanog iPS cells. We also tested wild-type c-Myc for Nanog iPS cell induction. We obtained a similar number of colonies with both wild-type c-Myc and the T58A mutant. Nanog iPS cells established with wild-type c-Myc were indistinguishable from those established with the T58A mutant with regards to morphology, gene expression (analysed via microarrays), teratoma formation (Supplementary Fig. 2) and stability under puromycin selection (Supplementary Fig. 9). Without puromycin selection, Nanog iPS cells induced by wild-type c-Myc were more stable (Supplementary Fig. 9).

Germline Chimaeras from Nanog iPS Cells

We next examined the ability of Nanog iPS cells to produce adult chimaeras. We injected 15–20 male Nanog iPS cells (five clones with the T58A mutant and three with wild-type c-Myc) into C57BL/6-derived blastocysts, which we then transplanted into the uteri of pseudo-pregnant mice. We obtained adult chimaeras from seven clones (four clones with the T58A mutant and three with wild-type c-Myc) as determined by coat colour (Fig. 6a and Supplementary Table 1). SSLP analyses showed that Nanog iPS cells contributed to various organs, with the level of chimaerism ranging from 10% to 90%. Chimeras from clone 20D17 showed highest iPS cell contribution in the testes. From clone 20D18, we

细胞在存在选择药物的环境中培养 22 ~ 26 代。在这段长时间的培养过程中，我们没有发现形态学上明显的改变。但是，RT–PCR 提示经过延长培养后 Fbx15 iPS 细胞丢失了 ES 细胞标志基因的表达。相反，Nanog iPS 细胞保持着相对高水平的 ES 细胞标志基因表达。这些数据说明 Nanog iPS 细胞比 Fbx15 iPS 细胞更加稳定。

我们还比较了 Nanog iPS 细胞和 Fbx15 iPS 细胞的诱导效率。在独立的实验中，我们从 8×10^5 个转染了 4 个转录因子的 Nanog 报告基因 MEF 中获得了 4 ~ 125 个 GFP 阳性的集落。因为大约 50% 的转染 MEF 被认为会表达所有的四个因子[8]，诱导效率约为 0.001% ~ 0.03%。相反，从同样数量的 Fbx15 报告基因 MEF 中，我们获得了 47 ~ 1,800 个 G418 抗性集落，诱导效率大约是 0.01% ~ 0.5%。因此，Nanog iPS 细胞的诱导效率大约是 Fbx15 iPS 细胞的十分之一。

然后我们比较了 Nanog iPS 细胞和 Fbx15 iPS 细胞对 LIF 或者视黄酸的响应（补充信息图 8）。正如我们之前所示[8]，即便是存在 LIF，Fbx15 iPS 细胞在没有饲养细胞的情况下也不会保持未分化状态。此外，Fbx15 iPS 细胞在存在视黄酸和没有饲养细胞的情况下也形成紧致的集落。相反，没有饲养细胞存在时，LIF 能保持 Nanog iPS 细胞的未分化状态。视黄酸诱导 Nanog iPS 细胞的分化。因此，Nanog iPS 细胞在对 LIF 和视黄酸的反应方面与 ES 细胞相似。

起初我们使用 c-Myc 的 T58A 突变体来诱导 Nanog iPS 细胞。我们也实验了野生型 c-Myc 诱导 Nanog iPS 细胞。两种情况下我们获得的集落数相似。在形态、基因表达（通过微阵列分析）、畸胎瘤形成（补充信息图 2）和嘌呤霉素选择下的稳定性（补充信息图 9）等方面，无法区分用野生型 c-Myc 诱导的与 T58A 突变体诱导的 Nanog iPS 细胞。没有嘌呤霉素选择的话，野生型 c-Myc 诱导的 Nanog iPS 细胞更加稳定（补充信息图 9）。

Nanog iPS 细胞来源的种系嵌合体

我们接下来检验了 Nanog iPS 细胞产生成熟嵌合体的能力。我们将 15 ~ 20 个雄性 Nanog iPS 细胞（5 个 T58A 突变体诱导的克隆和 3 个野生型 c-Myc 诱导的克隆）注射到 C57BL/6 来源的胚泡中，然后将胚泡移植到假孕小鼠的子宫内。根据毛色，我们得到了 7 个克隆的成熟嵌合体（4 个 T58A 突变体诱导的克隆和 3 个野生型 c-Myc 诱导的克隆）（图 6a 和补充信息表 1）。SSLP 分析显示 Nanog iPS 细胞参与生成多个器官，嵌合水平从 10% 到 90% 不等。克隆 20D17 来源的嵌合体在睾丸内出现最高的 iPS 细胞嵌合。对于克隆 20D18，我们仅从感染后胚泡中获得少量非嵌合

obtained only a few nonchimaeric pups from infected blastocysts; thus, whether this clone has competency for producing adult chimaeras remained to be determined. These data demonstrate that most Nanog iPS clones are competent for adult chimaeric mice.

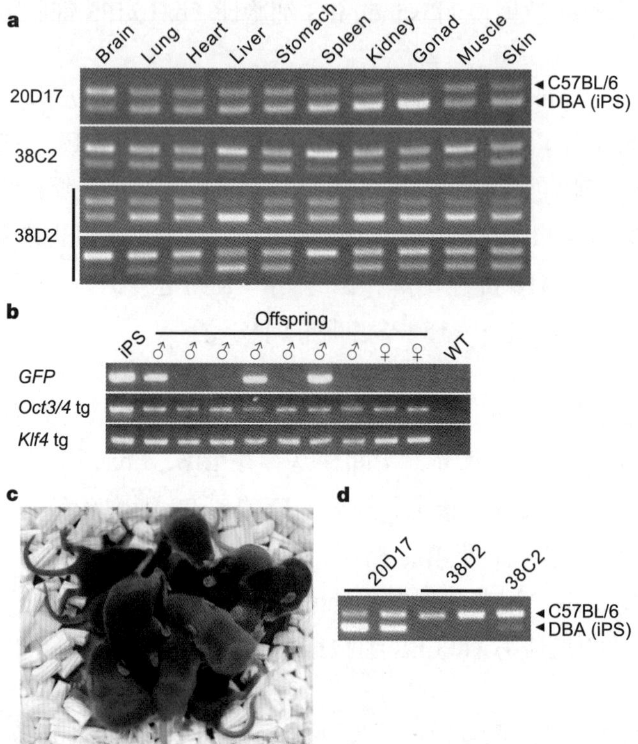

Fig. 6. Germline chimaeras from Nanog iPS cells. **a**, Tissue distribution of iPS cells in chimaeras. Genomic DNA was isolated from the indicated organs of chimaeras derived from three Nanog iPS cell clones (20D17, 38C2 and 38D2). SSLP analyses were performed for D6Mit15. **b**, PCR analyses showing the presence of the GFP cassette and retroviral transgenes in F_1 mice obtained from the intercross between a chimaeric male and a C57BL/6 female. **c**, Coat colours of F_2 mice obtained from F_1 intercrosses. **d**, Sperm contribution of iPS cells in chimaeric mice. Spermatozoa were isolated from the epididymides of chimaeric mice derived from three Nanog iPS cell clones (20D17, 38D2 and 38C2). iPS cell contribution was determined by SSLP of D6Mit15.

We then crossed three of the chimaeras from clone 20D17—for which the highest iPS cell contribution was in the testes—with C57BL/6 females. Whereas all F_1 mice showed black coat colour, all contained retroviral integration of the four transcription factors and approximately half contained the GFP-IRES-$Puro^r$ cassette (Fig. 6b), indicating germline transmission. Furthermore, approximately half of the F_2 mice born from F_1 intercrosses showed agouti coat colour, confirming germline transmission of Nanog-iPS-20D17 (Fig. 6c).

We also examined germline competency for two other clones that produced adult chimaeras. In one chimaeric mouse from Nanog-iPS-38C2 cell line, PCR analysis detected iPS cell contribution in isolated spermatozoa (Fig. 6d), suggesting that germline competency

型幼仔。因此，该克隆是否具有产生成熟嵌合体的能力还有待确定。这些数据提示大部分 Nanog iPS 细胞具有产生成熟嵌合体小鼠的能力。

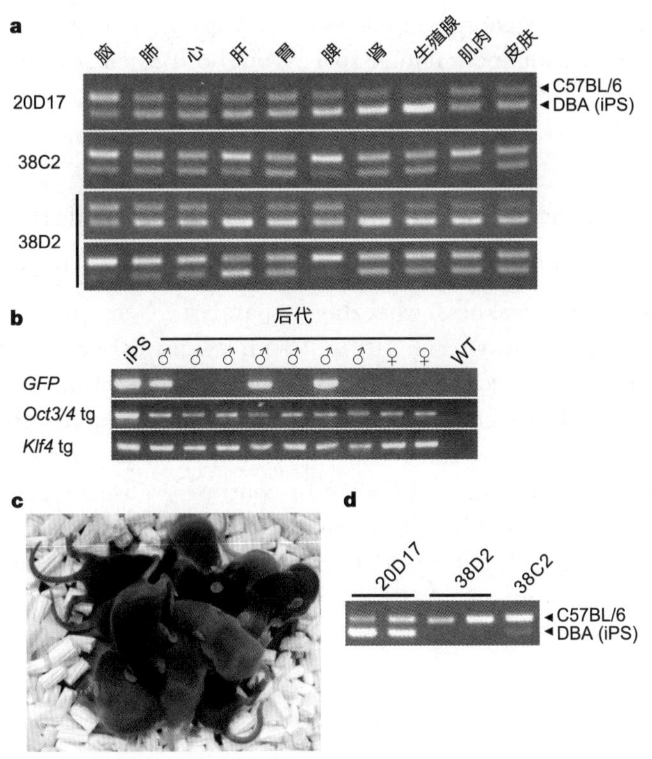

图 6. Nanog iPS 细胞来源的种系嵌合体。**a**，嵌合体中 iPS 细胞的组织分布。基因组 DNA 从三个 Nanog iPS 细胞克隆（20D17、38C2 和 38D2）来源的嵌合体的所示器官中分离出来。对 D6Mit15 进行 SSLP 分析。**b**，PCR 分析显示嵌合体雄鼠和 C57BL/6 雌鼠交配生成的 F_1 代小鼠中存在 GFP 盒和逆转录病毒转基因。**c**，F_1 代互交产生的 F_2 代小鼠的毛色。**d**，嵌合体小鼠中 iPS 细胞对精子生成的贡献。从三个 Nanog iPS 细胞克隆（20D17、38D2 和 38C2）来源的嵌合体小鼠的附睾中分离出精子。通过 D6Mit15 的 SSLP 分析确定 iPS 细胞的嵌合。

我们然后用克隆 20D17 来源的 3 只嵌合体（在这些嵌合体中，iPS 细胞在睾丸中贡献最大）与 C57BL/6 雌性小鼠杂交。所有的 F_1 代小鼠显示黑色的毛色，且都含有逆转录病毒整合的这四个转录因子，几乎一半含有 GFP-IRES-Puro^r 盒（图 6b），这标志着发生了种系的传递。此外，F_1 代互交产生的几乎一半的 F_2 代小鼠显示刺豚鼠样的毛色，证实了发生 Nanog-iPS-20D17 的种系传递（图 6c）。

我们还检测了产生成熟嵌合体的另两个克隆的种系产生能力。在一个 Nanog-iPS-38C2 细胞系来源的嵌合型小鼠中，PCR 分析检测到分离的精子中有 iPS 细胞来源（图 6d），表明其种系能力并不局限于克隆 20D17。但是，克隆 38C2 中 iPS 细

is not confined to clone 20D17. However, the iPS cell contribution to sperm of clone 38C2 is much smaller than that of clone 20D17, and no iPS-cell-derived offspring were found for 119 mice born from the cross between the 38C2 chimaera and C57BL/6 female mice. Most male mice with a high degree of chimaerism from the Nanog-iPS-38D2 cell line showed small testes and aspermatogenesis (Supplementary Fig. 13). The testes of some chimaeras from Nanog-iPS-38D2 contained mature sperm, but no iPS cell contribution was detected by PCR (Fig. 6d).

Tumour Formation by *c-myc* Reactivation

Out of 121 F_1 mice (aged 8–41 weeks) derived from the Nanog-iPS-20D17 cell line, 24 died or were killed because of weakness, wheezing or paralysis. Necropsy of 17 mice identified neck tumours (Supplementary Fig. 10) in 13 mice and other tumours in five mice, including two mice with neck tumours. Histological examination of one neck tumour showed that it was a ganglioneuroblastoma with follicular carcinoma of the thyroid gland (not shown). In these tumours, retroviral expression of *c-myc*, but not *Oct3/4*, *Sox2*, or *Klf4*, is reactivated (Supplementary Fig. 11). In contrast, transgene expression of all four transcription factors remained low in normal tissues, except for *c-myc* in muscle in one mouse (Supplementary Fig. 12). These data indicate that reactivation of *c-myc* retrovirus is attributable to tumour formation.

Discussion

Our results demonstrate that Nanog selection allows the generation of high-quality iPS cells that are comparable to ES cells in morphology, proliferation, teratoma formation, gene expression and competency for adult chimaeras. Nearly all Nanog iPS clones showed these properties, indicating that Nanog is a major determinant of quality in cellular pluripotency. However, germline competence was variable among Nanog iPS clones, indicating the existence of other important determinants of germline competency in addition to Nanog. The high quality of Nanog iPS cells underscores the possibility of using this technology to generate patient-specific pluripotent stem cells. In a separate study, we found that germline-competent iPS cells can also be obtained from adult mouse somatic cells (T. Aoi and S.Y., unpublished data). The current study, however, also reveals that reactivation of *c-myc* retrovirus may result in tumour formation. There may be ways to overcome this problem. Strong silencing of the four retroviruses in Nanog iPS cells indicates that they are only required for the induction, but not the maintenance, of pluripotency. Therefore, the retrovirus-mediated system might be eventually replaced by transient expression, such as the adenovirus-mediated system. Alternatively, high-throughput screening of chemical libraries might identify small molecules that can replace the four genes. These are crucial research areas in order to apply iPS cells to regenerative medicine.

We found that the efficiency of Nanog iPS cell induction is less than 0.1%. The low

胞产生精子的比例要远远少于克隆 20D17，而且 38C2 嵌合体与 C57BL/6 雌性小鼠间交配生出的 119 只小鼠中没有发现 iPS 细胞来源的后代。大部分具有高嵌合度 Nanog-iPS-38D2 细胞系来源的雄性小鼠都表现为小睾丸和无精子发生(补充信息图 13)。Nanog-iPS-38D2 细胞系来源的另一些嵌合型小鼠的睾丸内含有成熟的精子，但是 PCR 没有检测到其中有来源于 iPS 细胞的(图 6d)。

c-myc 再激活后的肿瘤形成

Nanog-iPS-20D17 细胞系来源的 121 只 F_1 小鼠中(年龄 8～41 周)，24 只因为虚弱、喘鸣或者瘫痪而死亡或者被杀死。对 17 只小鼠进行尸检发现其中 13 只小鼠有颈部肿瘤(补充信息图 10)，五只小鼠有其他肿瘤(其中两只有颈部肿瘤)。对一个颈部肿瘤进行组织学检查显示，它是滤泡性甲状腺癌的成神经节细胞瘤(未显示)。在这些肿瘤中，只有逆转录病毒的基因 *c-myc* 被重新激活，而 *Oct3/4*、*Sox2* 或者 *Klf4* 并没有(补充信息图 11)。相反，除了一只小鼠肌肉中的 *c-myc* 外，正常组织中所有四个转录因子的转基因表达都非常低(补充信息图 12)。这些数据提示逆转录病毒 *c-myc* 基因的再激活导致了肿瘤的生成。

讨　论

我们的结果显示 Nanog 选择可以产生高质量的 iPS 细胞，其在形态、增殖、畸胎瘤形成、基因表达和成熟嵌合体形成能力上与 ES 细胞相似。几乎所有的 Nanog iPS 克隆都有这些特性，表明 Nanog 是细胞多能性质量的主要决定因素。但是，Nanog iPS 克隆间的种系形成能力有差别，这提示除了 Nanog 外还存在其他重要的决定因素。Nanog iPS 细胞的这种高性能使得应用该技术产生病人特异性的多能干细胞成为可能。在另一个研究中，我们发现具有种系形成功能的 iPS 细胞也能从成年小鼠的体细胞中产生(青井贵之和山中伸弥，未发表的数据)。但是，目前的研究也揭示了 *c-myc* 基因的再激活可能导致肿瘤的形成。可能有解决这个问题的方法。Nanog iPS 细胞中这四个逆转录病毒基因的高度沉默说明，它们仅是诱导而非维持多能性所必需的。因此，逆转录病毒介导的系统可能最终被瞬时表达取代，比如腺病毒介导的系统。或者，高通量筛选化合物库可能会发现能够替代这四个基因的小分子。这是将 iPS 细胞应用于再生医学的关键研究领域。

我们发现 Nanog iPS 细胞诱导的效率低于 0.1%。这种低效率表明 iPS 细胞的起

efficiency suggests that the origin of iPS cells might be rare stem cells co-existing in MEF culture. Alternatively, activation of additional genes by retroviral integration might be required for iPS cell generation in addition to the four transcription factors. This is relevant to the fact that we have been able to obtain iPS cells only with retroviral transduction. Identification of such factor(s) may lead to the generation of iPS cells with higher efficiency, and without the need for retroviruses.

Methods Summary

To generate Nanog-reporter mice, we isolated a BAC clone containing the mouse *Nanog* gene in its centre. By using the RED/ET recombination technique (Gene Bridges), we inserted a GFP-IRES-$Puro^r$ cassette into the 5′ UTR of the mouse *Nanog* gene. We introduced the modified BAC into RF8 ES cells by electroporation[26]. We then microinjected transgenic ES cells into C57BL/6 blastocysts to generate Nanog-reporter mice containing the modified BAC. MEFs were isolated from 13.5 d.p.c. male embryos after removing genital ridges. Generation of Nanog iPS cells was performed as described[8], except that puromycin was used instead of G418 as a selection antibiotic. Retroviruses (pMXs) were generated with Plat-E packaging cells[27]. RF8 ES cells[26] and iPS cells were cultured on SNL feeder cells[28]. Analyses of iPS cells, such as RT–PCR, real-time PCR, bisulphite genomic sequencing, SSLP analyses, DNA microarrays, teratoma formation, and microinjection into C57BL/6 blastocysts, were performed as described[8]. Contribution of iPS cells in chimaeric mice was determined by PCR for the SSLP marker D6Mit15.

Full Methods and any associated references are available in the online version of the paper at www.nature.com/nature.

(**448**, 313-317; 2007)

Keisuke Okita[1], **Tomoko Ichisaka**[1,2] **& Shinya Yamanaka**[1,2]
[1] Department of Stem Cell Biology, Institute for Frontier Medical Sciences, Kyoto University, Kyoto 606-8507, Japan
[2] CREST, Japan Science and Technology Agency, Kawaguchi 332-0012, Japan

Received 6 February; accepted 22 May 2007. Published online 6 June 2007.

References:

1. Thomson, J. A. *et al.* Embryonic stem cell lines derived from human blastocysts. *Science* **282**, 1145-1147 (1998).
2. Hochedlinger, K. & Jaenisch, R. Nuclear reprogramming and pluripotency. *Nature* **441**, 1061-1067 (2006).
3. Niwa, H., Miyazaki, J. & Smith, A. G. Quantitative expression of Oct-3/4 defines differentiation, dedifferentiation or self-renewal of ES cells. *Nature Genet.* **24**, 372-376 (2000).
4. Nichols, J. *et al.* Formation of pluripotent stem cells in the mammalian embryo depends on the POU transcription factor Oct4. *Cell* **95**, 379-391 (1998).
5. Avilion, A. A. *et al.* Multipotent cell lineages in early mouse development depend on SOX2 function. *Genes Dev.* **17**, 126-140 (2003).
6. Cartwright, P. *et al.* LIF/STAT3 controls ES cell self-renewal and pluripotency by a Myc-dependent mechanism. *Development* **132**, 885-896 (2005).
7. Li, Y. *et al.* Murine embryonic stem cell differentiation is promoted by SOCS-3 and inhibited by the zinc finger transcription factor Klf4. *Blood* **105**, 635-637 (2005).
8. Takahashi, K. & Yamanaka, S. Induction of pluripotent stem cells from mouse embryonic and adult fibroblast cultures by defined factors. *Cell* **126**, 663-676 (2006).
9. Tokuzawa, Y. *et al.* Fbx15 is a novel target of Oct3/4 but is dispensable for embryonic stem cell self-renewal and mouse development. *Mol. Cell. Biol.* **23**, 2699-2708 (2003).

源可能是 MEF 培养物中共存的少量干细胞。或者，在 iPS 细胞的生成过程中除了这四个转录因子外，可能还需要通过逆转录病毒整合激活额外的基因。这与我们只能够通过逆转录病毒转导才能产生 iPS 细胞的事实相关。鉴定这些因子可能有助于更高效率地产生 iPS 细胞，而且可能不再需要借助逆转录病毒。

方法概述

为了产生 Nanog 报告小鼠，我们分离了一个在中心含有小鼠 *Nanog* 基因的 BAC 克隆。通过使用 RED/ET 重组技术（基因桥），我们将一个 GFP-IRES-Puro^r 盒插入小鼠 *Nanog* 基因的 5′UTR 区。我们通过电穿孔技术将修饰的 BAC 转到 RF8 ES 细胞内 [26]。然后我们将转基因的 ES 细胞微注射到 C57BL/6 胚泡中以产生含有修饰 BAC 的 Nanog 报告小鼠。在交配后 13.5 天的雄性胚胎中移除生殖嵴后分离出 MEF。Nanog iPS 细胞的产生如文献所述 [8]，但是我们用嘌呤霉素代替 G418 作为选择抗生素。逆转录病毒（pMX）用 Plat-E 包装细胞产生 [27]。RF8 ES 细胞 [26] 和 iPS 细胞都在 SNL 饲养细胞上培养 [28]。对 iPS 细胞的分析，比如 RT–PCR、实时 PCR、亚硫酸氢盐基因组测序、SSLP 分析、DNA 微阵列、畸胎瘤形成和微注射到 C57BL/6 胚泡中等，都如文献所述进行 [8]。通过 SSLP 标记物 D6Mit15 的 PCR 确定嵌合体小鼠中 iPS 细胞的嵌合。

完整的方法和相关的文献可在本文的网上版本中获得。

（毛晨晖 翻译；王宇 审稿）

10. Kuroda, T. *et al.* Octamer and Sox elements are required for transcriptional cis regulation of Nanog gene expression. *Mol. Cell. Biol.* **25**, 2475-2485 (2005).

11. Rodda, D. J. *et al.* Transcriptional regulation of Nanog by OCT4 and SOX2. *J. Biol. Chem.* **280**, 24731-24737 (2005).

12. Mitsui, K. *et al.* The homeoprotein Nanog is required for maintenance of pluripotency in mouse epiblast and ES cells. *Cell* **113**, 631-642 (2003).

13. Chambers, I. *et al.* Functional expression cloning of nanog, a pluripotency sustaining factor in embryonic stem cells. *Cell* **113**, 643-655 (2003).

14. Silva, J., Chambers, I., Pollard, S. & Smith, A. Nanog promotes transfer of pluripotency after cell fusion. *Nature* **441**, 997-1001 (2006).

15. Copeland, N. G., Jenkins, N. A. & Court, D. L. Recombineering: a powerful new tool for mouse functional genomics. *Nature Rev. Genet.* **2**, 769-779 (2001).

16. Testa, G. *et al.* Engineering the mouse genome with bacterial artificial chromosomes to create multipurpose alleles. *Nature Biotechnol.* **21**, 443-447 (2003).

17. Cherry, S. R., Biniszkiewicz, D., van Parijs, L., Baltimore, D. & Jaenisch, R. Retroviral expression in embryonic stem cells and hematopoietic stem cells. *Mol. Cell. Biol.* **20**, 7419-7426 (2000).

18. Chen, T., Ueda, Y., Xie, S. & Li, E. A novel Dnmt3a isoform produced from an alternative promoter localizes to euchromatin and its expression correlates with active de novo methylation. *J. Biol. Chem.* **277**, 38746-38754 (2002).

19. Davis, T. L., Yang, G. J., McCarrey, J. R. & Bartolomei, M. S. The H19 methylation imprint is erased and re-established differentially on the parental alleles during male germ cell development. *Hum. Mol. Genet.* **9**, 2885-2894 (2000).

20. Sato, S., Yoshimizu, T., Sato, E. & Matsui, Y. Erasure of methylation imprinting of Igf2r during mouse primordial germ-cell development. *Mol. Reprod. Dev.* **65**, 41-50 (2003).

21. Brandeis, M. *et al.* The ontogeny of allele-specific methylation associated with imprinted genes in the mouse. *EMBO J.* **12**, 3669-3677 (1993).

22. Labosky, P. A., Barlow, D. P. & Hogan, B. L. Mouse embryonic germ (EG) cell lines: transmission through the germline and differences in the methylation imprint of insulin-like growth factor 2 receptor (Igf2r) gene compared with embryonic stem (ES) cell lines. *Development* **120**, 3197-3204 (1994).

23. Kato, Y. *et al.* Developmental potential of mouse primordial germ cells. *Development* **126**, 1823-1832 (1999).

24. Lee, J. *et al.* Erasing genomic imprinting memory in mouse clone embryos produced from day 11.5 primordial germ cells. *Development* **129**, 1807-1817 (2002).

25. Geijsen, N. *et al.* Derivation of embryonic germ cells and male gametes from embryonic stem cells. *Nature* **427**, 148-154 (2004).

26. Meiner, V. L. *et al.* Disruption of the acyl-CoA:cholesterol acyltransferase gene in mice: evidence suggesting multiple cholesterol esterification enzymes in mammals. *Proc. Natl Acad. Sci. USA* **93**, 14041-14046 (1996).

27. Morita, S., Kojima, T. & Kitamura, T. Plat-E: an efficient and stable system for transient packaging of retroviruses. *Gene Ther.* **7**, 1063-1066 (2000).

28. McMahon, A. P. & Bradley, A. The Wnt-1 (int-1) proto-oncogene is required for development of a large region of the mouse brain. *Cell* **62**, 1073-1085 (1990).

Supplementary Information is linked to the online version of the paper at www.nature.com/nature.

Acknowledgements. We thank K. Takahashi, M. Nakagawa and T. Aoi for scientific discussion; M. Maeda for histological analyses; M. Narita, J. Iida, H. Miyachi and S. Kitano for technical assistance; and R. Kato, R. Iyama and Y. Ohuchi for administrative assistance. We also thank T. Kitamura for Plat-E cells and pMXs retroviral vectors, and R. Farese for RF8 ES cells. This study was supported in part by a grant from the Uehara Memorial Foundation, the Program for Promotion of Fundamental Studies in Health Sciences of NIBIO, a grant from the Leading Project of MEXT, and Grants-in-Aid for Scientific Research of JSPS and MEXT (to S.Y.). K.O. is a JSPS research fellow.

Author Contributions. K.O. conducted most of the experiments in this study. T.I. performed manipulation of mouse embryos to generate Nanog-GFP transgenic mice. T.I. also maintained the mouse lines. S.Y. designed and supervised the study, and prepared the manuscript. S.Y. also performed computer analyses of DNA microarray data.

Author Information. The microarray data are deposited in GEO under accession number GSE7841. Reprints and permissions information is available at www.nature.com/reprints. The authors declare no competing financial interests. Correspondence and requests for materials should be addressed to S.Y. (yamanaka@frontier.kyoto-u.ac.jp).

In Vitro Reprogramming of Fibroblasts into a Pluripotent ES-cell-like State

M. Wernig *et al.*

Editor's Note

Here, American cell biologist Rudolf Jaenisch and colleagues manage to convert mature mouse fibroblasts into stem-like cells called induced pluripotent stem (iPS) cells. The study builds on the work of Japanese cell biologist Shinya Yamanaka who showed, the previous year, that turning on the expression of just four genes in adult mouse cells could partially reprogram them to become stem cell-like. Jaenisch tweaked Yamanaka's egg- and embryo-free "recipe" to produce cells that could contribute to a whole organism—a key hallmark of a true stem cell. Yamanaka simultaneously published an essentially identical study, as did a third group of scientists based in America. Jaenisch's group has since made dopamine-producing neurons from iPS cells derived from patients with Parkinson's disease, highlighting the potential of these cells for reproductive medicine.

Nuclear transplantation can reprogramme a somatic genome back into an embryonic epigenetic state, and the reprogrammed nucleus can create a cloned animal or produce pluripotent embryonic stem cells. One potential use of the nuclear cloning approach is the derivation of "customized" embryonic stem (ES) cells for patient-specific cell treatment, but technical and ethical considerations impede the therapeutic application of this technology. Reprogramming of fibroblasts to a pluripotent state can be induced *in vitro* through ectopic expression of the four transcription factors Oct4 (also called Oct3/4 or Pou5f1), Sox2, c-Myc and Klf4. Here we show that DNA methylation, gene expression and chromatin state of such induced reprogrammed stem cells are similar to those of ES cells. Notably, the cells—derived from mouse fibroblasts—can form viable chimaeras, can contribute to the germ line and can generate live late-term embryos when injected into tetraploid blastocysts. Our results show that the biological potency and epigenetic state of *in-vitro*-reprogrammed induced pluripotent stem cells are indistinguishable from those of ES cells.

EPIGENETIC reprogramming of somatic cells into ES cells has attracted much attention because of the potential for customized transplantation therapy, as cellular derivatives of reprogrammed cells will not be rejected by the donor[1,2]. Thus far, somatic cell nuclear transfer and fusion of fibroblasts with ES cells have been shown to promote the epigenetic reprogramming of the donor genome to an embryonic state[3-5]. However, the therapeutic application of either approach has been hindered by technical complications as well as ethical objections[6]. Recently, a major breakthrough was reported whereby

体外重编程技术将成纤维细胞诱导到 ES 样细胞的多能性状态

沃尼格等

编者按

在本文中，美国细胞生物学家鲁道夫·耶尼施和他的同事们成功地将成熟的小鼠成纤维细胞转化为干细胞样细胞，称为“诱导多能性干细胞”(iPS)。这项研究建立在日本细胞生物学家山中伸弥工作的基础上。山中伸弥在前一年的研究中发现，在胎鼠或者成年小鼠的成纤维中，只要激活四个转录因子的表达，就能将它们部分重编程，使其变成干细胞样的细胞。耶尼施调整了山中伸弥的不直接操作卵子和胚胎的“配方”，从而产生了可以形成整个个体的细胞——这是真正干细胞的一个关键特征。山中伸弥和美国的第三组科学家同时发表了一项基本相同的研究。自那以后，耶尼施的团队从帕金森病患者的 iPS 细胞分化得到能够形成多巴胺的神经元，凸显了这些细胞在再生医学上的潜力。

核移植能够将体细胞的基因组重新编程回到胚胎表观遗传状态，而重新编程的细胞核能够形成克隆动物或者产生多能性胚胎干细胞。这种核克隆方法的一个潜在应用价值就是衍生出“定制的”胚胎干(ES)细胞，用于患者的特异性的细胞治疗，但是技术和伦理学方面的考虑限制了该技术的治疗应用。成纤维细胞能够在体外通过四个转录因子 Oct4(又称 Oct3/4 或 Pou5f1)、Sox2、c-Myc 和 Klf4 的异位表达诱导至多能性状态。在此，我们证明了这些诱导多能性干细胞在 DNA 甲基化、基因表达和染色质状态上都与 ES 细胞相似。尤为重要的是，这些来自小鼠成纤维细胞的细胞可以形成可存活的嵌合体，促进生殖系的发育，并可以在注射到四倍体囊胚时产生活的晚期胚胎。我们的结果显示了体外重编程得到的诱导多能性干细胞在生物学功能和表观遗传状态上与 ES 细胞没有明显区别。

由于重编程得到的细胞衍生物不会被供体排斥，体细胞经过表观遗传重编程形成的 ES 细胞具有定制移植治疗的潜力，因此引起了广泛关注 [1,2]。到目前为止，体细胞核转移以及成纤维细胞与 ES 细胞的融合被证实能够促进供体基因组的表观遗传重编程，使其处于胚胎状态 [3-5]。但是，这两种方法的治疗应用都受到技术并发症以及伦理问题的阻碍 [6]。最近，有一个重大的突破，即转录因子 Oct4、Sox2、

expression of the transcription factors Oct4, Sox2, c-Myc and Klf4 was shown to induce fibroblasts to become pluripotent stem cells (designated as induced pluripotent stem (iPS) cells), although with a low efficiency[7]. The iPS cells were isolated by selection for activation of *Fbx15* (also called *Fbxo15*), which is a downstream gene of *Oct4*. This important study left a number of questions unresolved: (1) although iPS cells were pluripotent they were not identical to ES cells (for example, iPS cells injected into blastocysts generated abnormal chimaeric embryos that did not survive to term); (2) gene expression profiling revealed major differences between iPS cells and ES cells; (3) because the four transcription factors were transduced by constitutively expressed retroviral vectors it was unclear why the cells could be induced to differentiate and whether continuous vector expression was required for the maintenance of the pluripotent state; and (4) the epigenetic state of the endogenous pluripotency genes *Oct4* and *Nanog* was incompletely reprogrammed, raising questions about the stability of the pluripotent state.

Here we used activation of the endogenous *Oct4* or *Nanog* genes as a more stringent selection strategy for the isolation of reprogrammed cells. We infected fibroblasts with retroviral vectors transducing the four factors, and selected for the activation of the endogenous *Oct4* or *Nanog* genes. Positive colonies resembled ES cells and assumed an epigenetic state characteristic of ES cells. When injected into blastocysts the reprogrammed cells generated viable chimaeras and contributed to the germ line. Our results establish that somatic cells can be reprogrammed to a pluripotent state that is similar, if not identical, to that of normal ES cells.

Selection of Fibroblasts for *Oct4* or *Nanog* Activation

Using homologous recombination in ES cells we generated mouse embryonic fibroblasts (MEFs) and tail-tip fibroblasts (TTFs) that carried a neomycin-resistance marker inserted into either the endogenous *Oct4* (Oct4-neo) or *Nanog* locus (Nanog-neo) (Fig. 1a). These cultures were sensitive to G418, indicating that the *Oct4* and *Nanog* loci were, as expected, silenced in somatic cells. These MEFs or TTFs were infected with Oct4-, Sox2-, c-Myc- and Klf4-expressing retroviral vectors and G418 was added to the cultures 3, 6 or 9 days later. The number of drug-resistant colonies increased substantially when analysed at day 20 (Fig. 1i). Most colonies had a flat morphology (Fig. 1h, right) and between 11% and 25% of the colonies were "ES-like" (Fig. 1h, left) when selection was applied early (Fig. 1k), a percentage that increased at later time points. At day 20, ES-like colonies were picked, dissociated and propagated in G418-containing media. They gave rise to ES-like cell lines (designated as Oct4 iPS or Nanog iPS cells, respectively) that could be propagated without drug selection, displayed homogenous Nanog, SSEA1 and alkaline phosphatase expression (Fig. 1b–g and Supplementary Figs 1 and 5), and formed undifferentiated colonies when seeded at clonal density on gelatincoated dishes (see inset in Fig. 1b). Four out of five analysed lines had a normal karyotype (Supplementary Table 1).

c-Myc和Klf4的表达可以诱导成纤维细胞成为诱导多能性干细胞，尽管其效率很低[7]。这些iPS细胞是通过选择激活*Oct4*的下游基因*Fbx15*(又称*Fbxo15*)而分离出来的。这个重要的研究遗留了一系列没有解决的问题：(1)尽管iPS细胞具有多能性，但是它们与ES细胞并不相同(比如，注射到囊胚中的iPS细胞产生异常的嵌合体胚胎，无法长久存活)；(2)基因表达分析显示iPS细胞和ES细胞之间存在较大差异；(3)由于这四个转录因子是通过组成性表达的逆转录病毒载体转导的，目前尚不清楚为什么可以诱导细胞分化，以及是否需要持续的载体表达才能维持这种多能状态；(4)内源性多能基因*Oct4*和*Nanog*的表观遗传状态并不是被完全重编程的，研究者们对多能状态的稳定性提出了质疑。

在这里，我们使用内源性基因*Oct4*或*Nanog*的激活作为更加严格的选择策略来分离重编程的细胞。我们用转导这四个因子的逆转录病毒感染成纤维细胞，并选择激活内源性基因*Oct4*或*Nanog*的细胞。阳性克隆酷似ES细胞，呈现出ES细胞的表观遗传特性。当被注射到囊胚中后，这些重编程的细胞能够形成存活的嵌合体并对生殖系的形成具有贡献。我们的研究结果证明体细胞能够被重编程成与ES细胞相似(尽管不是完全一致)的多能性状态。

选择用于*Oct4*或*Nanog*激活的成纤维细胞

通过在ES细胞中进行同源重组，我们得到了小鼠胚胎成纤维细胞(MEF)和尾尖纤维母细胞(TTF)，这些成纤维细胞携带新霉素耐药标记物，并将其插入到内源性*Oct4*(Oct4-neo)或者*Nanog*(Nanog-neo)位点中(图1a)。这些细胞对G418敏感，表明*Oct4*和*Nanog*位点在体细胞中如预期的那样沉默。分别用表达Oct4、Sox2、c-Myc和Klf4的逆转录病毒载体包装得到的病毒感染MEF或TTF，然后分别在第3、6、9天后加入G418进行培养。第20天进行分析时，耐药克隆的数量显著增加(图1i)。大部分克隆的形态都是扁平的(图1h，右侧)，当早期进行选择时(图1k)，大约11%~25%的克隆形态像ES细胞(图1h，左侧)，这个比例随着药物处理时间延长而增加。在第20天，挑出ES样的克隆，将其消化后用含有G418的培养基进行培养。它们能够形成ES样的细胞系(分别称为Oct4 iPS细胞或者Nanog iPS细胞)，而且这些细胞系能够在没有药物选择的情况下繁殖，展示出均质的Nanog、SSEA1和碱性磷酸酶的表达(图1b~g和补充信息图1和5)，当在明胶包被的培养皿上克隆密接种时，这些细胞能够形成未分化的克隆(见图1b插图)。进行分析的细胞系中有五分之四具有正常的核型(补充信息表1)。

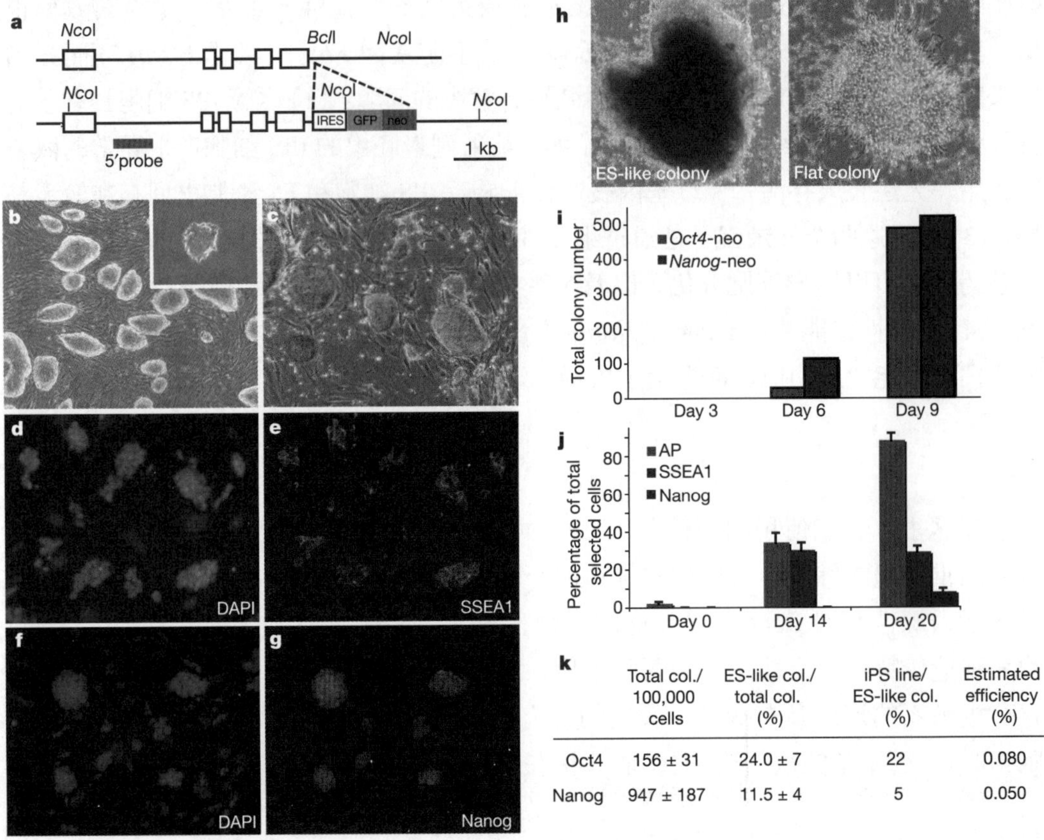

	Total col./ 100,000 cells	ES-like col./ total col. (%)	iPS line/ ES-like col. (%)	Estimated efficiency (%)
Oct4	156 ± 31	24.0 ± 7	22	0.080
Nanog	947 ± 187	11.5 ± 4	5	0.050

Fig. 1. Generation of Oct4- and Nanog-selected iPS cells. **a**, Targeting strategy to generate an Oct4-IRES-GFPneo allele. The resulting GFPneo fusion protein has sufficient neomycin-resistance activity in ES cells; GFP fluorescence, however, is not visible. **b**, Phase-contrast micrograph of Oct4 iPS cells (clone 18) grown on irradiated MEFs. Inset: an ES-cell-like colony 5 days after seeding in clonal density without feeder cells. iPS clone 18 cells exhibited strong alkaline phosphatase activity (**c**) and were homogenously labelled with antibodies against SSEA1 (**d**, **e**) and Nanog (**f**, **g**). **h**, One example of an ES-like colony 16 days after infection (left). Most G418-resistant colonies, however, consisted of flat non-ES-like cells (right): **b**, 10 ×; **c–g**, 20 ×; **h**, 4 ×. **i**, Gradual activation of the Nanog and Oct4-neo alleles. Shown are the total colony numbers of one experiment at day 20 after infection starting neomycin selection at day 3, 6 and 9. **j**, Fraction of total selected cells expressing alkaline phosphatase, SSEA1 and Nanog 0, 14, and 20 days after infection (counted were more than ten visual fields containing $n > 1{,}000$ total cells for every time point; error bars indicate s.d.). **k**, Estimated reprogramming efficiency of Oct4 selection and Nanog selection ($n = 3$ different experiments; s.e.m. is shown). Indicated are the total number of drug-resistant colonies per 100,000 plated MEFs 20 days after infection; the fraction of ES-like colonies per total number of colonies; the fraction of iPS cell lines that could be established from picked ES-like colonies as defined by homogenous alkaline phosphatase, SSEA1 and Nanog expression. After determining the fraction of *Sox2*- (83.4%), *Oct4*- (53.2%) and *c-myc*- (46.3%) infected MEFs 2 days after infection by immunofluorescence and assuming 50% were infected by *Klf4* viruses, we estimated the overall reprogramming efficiency as the ratio of quadruple-infected cells and the extrapolated total number of iPS cell lines that could be established with G418 selection starting at day 6 after infection.

Although the timing and appearance of colonies were similar between the Oct4 and Nanog selection, we noticed pronounced quantitative differences between the two

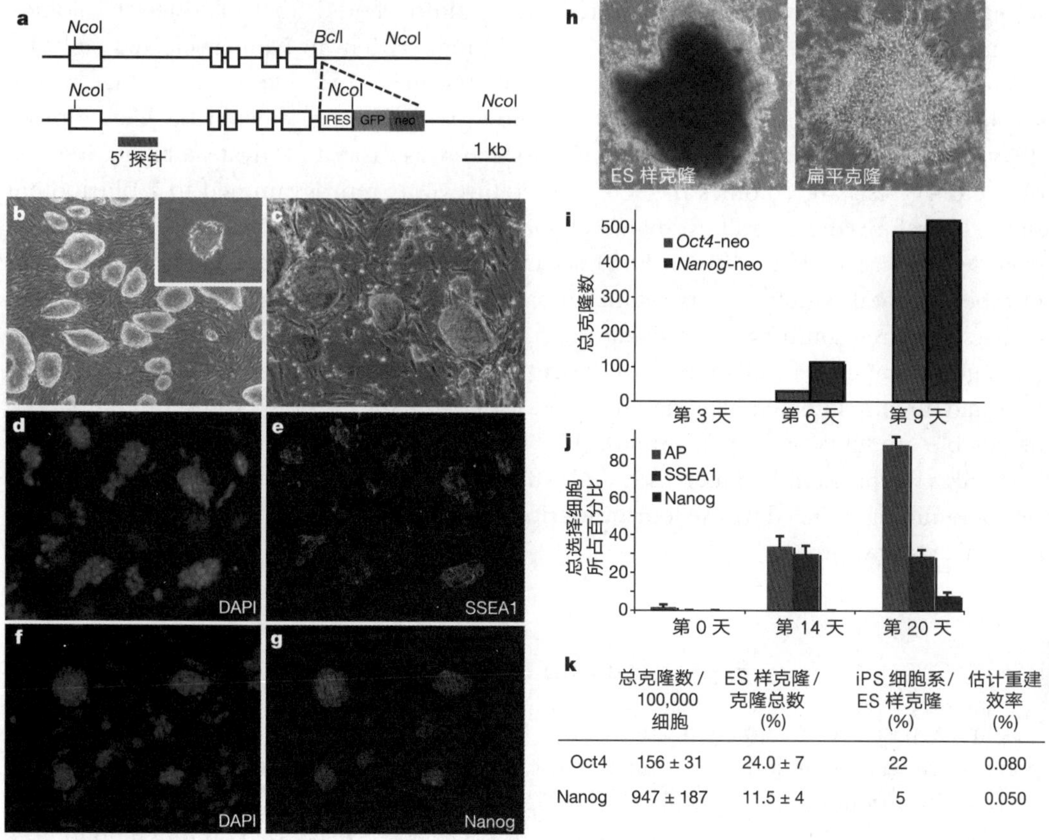

	总克隆数 / 100,000 细胞	ES 样克隆 / 克隆总数 (%)	iPS 细胞系 / ES 样克隆 (%)	估计重建效率 (%)
Oct4	156 ± 31	24.0 ± 7	22	0.080
Nanog	947 ± 187	11.5 ± 4	5	0.050

图 1. 选择性表达 *Oct4* 和 *Nanog* 的 iPS 细胞的产生。**a**，产生 Oct4-IRES-GFPneo 等位基因的靶向策略。得到的 GFPneo 融合蛋白在 ES 细胞内具有足够的新霉素抗性活性；但是看不到 GFP 荧光。**b**，生长在辐射过的 MEF 上的 Oct4 iPS 细胞（克隆 18）的相差显微镜图像。插图：在没有饲养细胞的情况下，将细胞以克隆的密度接种 5 天后形成的 ES 细胞样克隆。iPS 克隆 18 的细胞表现出强烈的碱性磷酸酶活性（**c**），并用抗 SSEA1（**d,e**）和 Nanog（**f,g**）的抗体均匀标记。**h**，感染 16 天后的 ES 样克隆的一个例子（左图）。然而，大多数 G418 耐药细胞集落都由扁平的非 ES 样细胞组成（右图）：**b**, 10×；**c**～**g**, 20×；**h**, 4×。**i**, Nanog 和 Oct4-neo 等位基因的逐渐激活。图为在第 3、6 和 9 天开始用新霉素进行筛选，病毒感染后第 20 天，一个实验的克隆总数。**j**，感染后第 0 天、第 14 天和第 20 天表达碱性磷酸酶、SSEA1 和 Nanog 的细胞的比例（每个时间点只有超过 10 个视野含有 $n > 1{,}000$ 个总细胞的才计数，误差条指代 s.d.）。**k**，估算 *Oct4* 和 *Nanog* 选择的重建效率（$n = 3$ 个不同的实验，s.e.m. 如图所示）。图中表示的是感染 20 天后每 100,000 个 MEF 重编程得到的总数；ES 样克隆占克隆总数的百分比；从挑取的 ES 样克隆中能够均质表达碱性磷酸酶，SSEA1 和 Nanog 的 iPS 克隆比例。通过免疫荧光确定了感染 2 天后 *Sox2*（83.4%）、*Oct4*（53.2%）和 *c-myc*（46.3%）感染 MEF 的比例后，并且假设 50% 受到 *Klf4* 病毒的感染，我们估计总体的重建效率为四倍感染的细胞的比例，并推算出在感染后第 6 天用 G418 选择获得的 iPS 细胞系的总数。

虽然 *Oct4* 和 *Nanog* 选择性表达的克隆出现的时间和外观相似，但我们注意到这两种选择策略之间存在明显的数量差异：*Oct4* 选择性表达的 MEF 能够形成的克

selection strategies: whereas Oct4-selected MEF cultures had 3- to 10-fold fewer colonies, the fraction of ES-like colonies was 2- to 3-fold higher than in Nanog-selected cultures. Accordingly, approximately four times more Oct4-selected ES-like colonies gave rise to stable and homogenous iPS cell lines compared with Nanog-selected ES-like colonies (Fig. 1k). This suggests that although the *Nanog* locus was easier to activate, a higher fraction of the drug-resistant colonies in Oct4-neo cultures was reprogrammed to a pluripotent state. Therefore, the overall estimated efficiency of 0.05–0.1% to establish iPS cell lines from MEFs was similar between Oct4 selection and Nanog selection, despite the larger number of total Nanog-neo resistant colonies (Fig. 1k). Next we investigated the time course of reprogramming by studying the fraction of alkaline-phosphatase-, SSEA1- and Nanog-positive cells in Oct4-selected MEF cultures. Fourteen days after infection some cells had already initiated alkaline phosphatase activity and SSEA1 expression, but lacked detectable amounts of Nanog protein (Fig. 1j), whereas by day 20, alkaline phosphatase and SSEA1 expression had increased and ~8% of the cells were Nanog-positive. Thus, the reprogramming induced by the four transcription factors (Oct4, Sox2, c-Myc and Klf4) is a gradual and slow process.

Expression and DNA Methylation

To characterize the reprogrammed cells on a molecular level we used quantitative polymerase chain reaction with reverse transcription (qRT–PCR) to measure the expression of ES-cell- and fibroblast- specific genes. Figure 2a shows that in Oct4 iPS cells the total level of Nanog and Oct4 was similar to that in ES cells but decreased on differentiation to embryoid bodies. MEFs did not express either gene. Using specific primers for endogenous or total *Sox2* transcripts showed that most *Sox2* transcripts originated from the endogenous locus rather than the viral vector (Fig. 2b). In contrast, *Hoxa9* and *Zfpm2* were highly expressed in MEFs but were expressed at very low levels in iPS or ES cells (Fig. 2c). Western blot analysis showed that multiple iPS clones expressed Nanog and Oct4 proteins at similar levels compared to ES cells (Fig. 2d). Finally, we used microarray technology to compare gene expression patterns on a global level. Figure 2f shows that the iPS cells clustered with ES cells in contrast to wild-type or donor MEFs.

To investigate the DNA methylation level of the *Oct4* and *Nanog* promoters we performed bisulphite sequencing and combined bisulphite restriction analysis (COBRA) with DNA isolated from ES cells, iPS cells and MEFs. As shown in Fig. 2g, both loci were demethylated in ES and iPS cells and fully methylated in MEFs. To assess whether the maintenance of genomic imprinting was compromised we assessed the methylation status of the four imprinted genes *H19*, *Peg1* (also called *Mest*), *Peg3* and *Snrpn*. As shown in Fig. 2e, bands corresponding to an unmethylated and methylated allele were detected for each gene in MEFs, iPS cells and TTFs. In contrast, embryonic germ cells, which have erased all imprints[8], were unmethylated. Our results indicate that the epigenetic state of the *Oct4* and *Nanog* genes was reprogrammed from a transcriptionally repressed (somatic) to an active (embryonic) state and that the pattern of somatic imprinting was maintained in iPS cells.

隆是 *Nanog* 选择性表达的 1/3 ~ 1/10，但是其能够形成 ES 样克隆的比例却是 *Nanog* 选择性表达的 MEF 的 2 ~ 3 倍。因此，与 *Nanog* 选择性表达的 ES 样克隆相比，*Oct4* 选择性表达的 ES 样克隆能够产生的稳定且均质的 iPS 细胞系的数量约是前者的 4 倍（图 1k）。这表明，尽管 *Nanog* 的基因位点更加容易被激活，但是 Oct4-neo 的细胞能够被重编程到多能性状态的比例更高。因此，尽管 Nanog-neo 的细胞总数更多（图 1k），但从 *Oct4* 选择性表达和 *Nanog* 选择性表达的 MEF 重编程获得 iPS 克隆的比例几乎都是 0.05 ~ 0.1（图 1k）。接下来，我们通过研究 *Oct4* 选择性表达的 MEF 重编程后获得的碱性磷酸酶、SSEA1 和 Nanog 阳性克隆的比例，研究重编程的时间进程。感染后 14 天，部分细胞已经出现了碱性磷酸酶活性和 SSEA1 表达，但检测不到 Nanog 的蛋白表达（图 1j），而到了第 20 天左右，碱性磷酸酶和 SSEA1 的表达增加了，大约 8% 的细胞是 Nanog 阳性的。因此，这四种转录因子（Oct4、Sox2、c-Myc 和 Klf4）诱导的重编程是一个渐进而缓慢的过程。

基因表达和 DNA 甲基化

为了在分子水平描述重编程后细胞的特征，我们使用定量反转录聚合酶链式反应（qRT–PCR）测量 ES 细胞和成纤维细胞特异性表达的基因。图 2a 显示了在 Oct4 iPS 细胞中，*Nanog* 和 *Oct4* 的总水平与 ES 细胞中相似，但分化成 EB 球后表达量就减少了。MEF 不表达这两个基因。使用内源或者总 *Sox2* 转录的特异性引物进行检测基因的表达，结果显示大部分 *Sox2* 转录起源于内源性位点而不是病毒载体（图 2b）。相反，*Hoxa9* 和 *Zfpm2* 在 MEF 中高表达，但是在 iPS 或者 ES 细胞中表达水平很低（图 2c）。Western bolt 分析显示多个 iPS 克隆表达的 Nanog 和 Oct4 蛋白水平与 ES 细胞相近（图 2d）。最后，我们使用微阵列技术在整体水平上比较了基因表达模式。图 2f 显示了 iPS 细胞与 ES 细胞聚集在一起，而不是与野生型或者供者 MEF 聚集在一起。

为了研究 *Oct4* 和 *Nanog* 启动子的 DNA 甲基化水平，我们对从 ES 细胞、iPS 细胞和 MEF 中分离出的 DNA 进行了亚硫酸氢盐测序和联合亚硫酸氢盐限制分析（COBRA）。如图 2g 所示，两个位点在 ES 和 iPS 细胞中都去甲基化，而在 MEF 中完全甲基化。为了评估基因组印迹的维持是否受到损害，我们评价了 4 个印记基因 *H19*、*Peg1*（又称为 *Mest*）、*Peg3* 和 *Snrpn* 的甲基化状态。如图 2e 所示，在 MEF、iPS 细胞和 TTF 中，每个基因都检测到与未甲基化和甲基化等位基因对应的条带。相反，已经消除所有印记的胚胎生殖细胞[8]没有甲基化。我们的结果表明，*Oct4* 和 *Nanog* 基因的表观遗传状态从转录抑制（体细胞）状态被重编辑为活跃（胚胎）状态，

Furthermore, the presence of imprints suggests a non-embryonic-germ-cell origin of iPS cells.

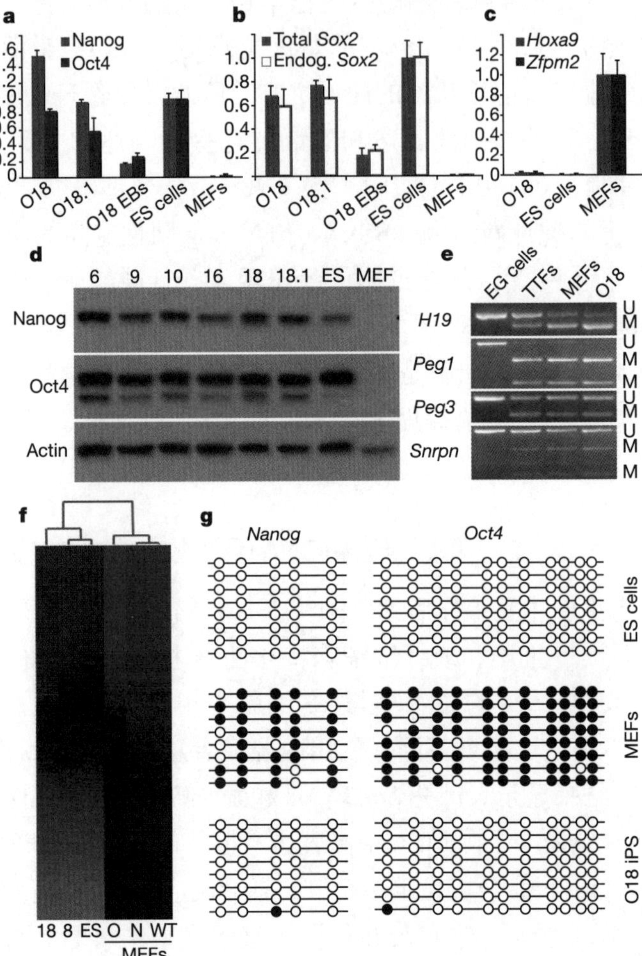

Fig. 2. Expression and promoter methylation analysis of iPS cells. **a–c**, qRT–PCR analysis ($n = 3$ independent PCR reactions; error bars indicate s.d.) of Oct4 iPS clone 18, subclone 18.1, 2-week-old embryoid bodies (EBs) derived from clone 18, V6.5 ES cells and Oct4-neo MEFs shows similar Nanog (red bars) and total Oct4 (blue bars) levels as in ES cells (**a**); slightly lower total Sox2 levels (filled red bars), mostly due to expression of endogenous Sox2 transcripts (open red bars, **b**); and strong downregulation of Hoxa9 (red) and Zfpm2 (blue) transcripts in iPS cells (**c**). Transcript levels were normalized to Gapdh expression, with expression levels in ES cells (**a**, **b**) and MEFs (**c**) set as 1. **d**, Western blot analysis for Oct4 and Nanog expression of different Oct4 iPS clones (6, 9, 10, 16, 18) and a GFP-labelled subclone of clone 18 (18.1). **e**, COBRA methylation analysis[32] of imprinted genes *H19* (maternally expressed), *Peg1* (paternally expressed), *Peg3* (paternally expressed) and *Snrpn* (paternally expressed). Upper band, unmethylated (U); lower band, methylated (M). **f**, Unsupervised hierarchical clustering of averaged global transcriptional profiles obtained from Oct4-neo iPS clone 18, Nanog-neo iPS clone 8, genetically matched ES cells (V6.5;129SvJae/C57Bl/6), Oct4-neo MEFs (O), Nanog-neo MEFs (N) and wild-type 129/B6 F1 MEFs (WT). **g**, Analysis of the methylation state of the *Oct4* and *Nanog* promoters using bisulphite sequencing. Open circles indicate unmethylated and filled circles methylated CpG dinucleotides. Shown are eight representative sequenced clones from ES cells (V6.5), Oct4-neo MEFs and Oct4-neo iPS clone 18.

并在 iPS 细胞中保持了体细胞印记的模式。此外，印记基因的存在进一步表明 iPS 细胞的非胚胎生殖细胞起源。

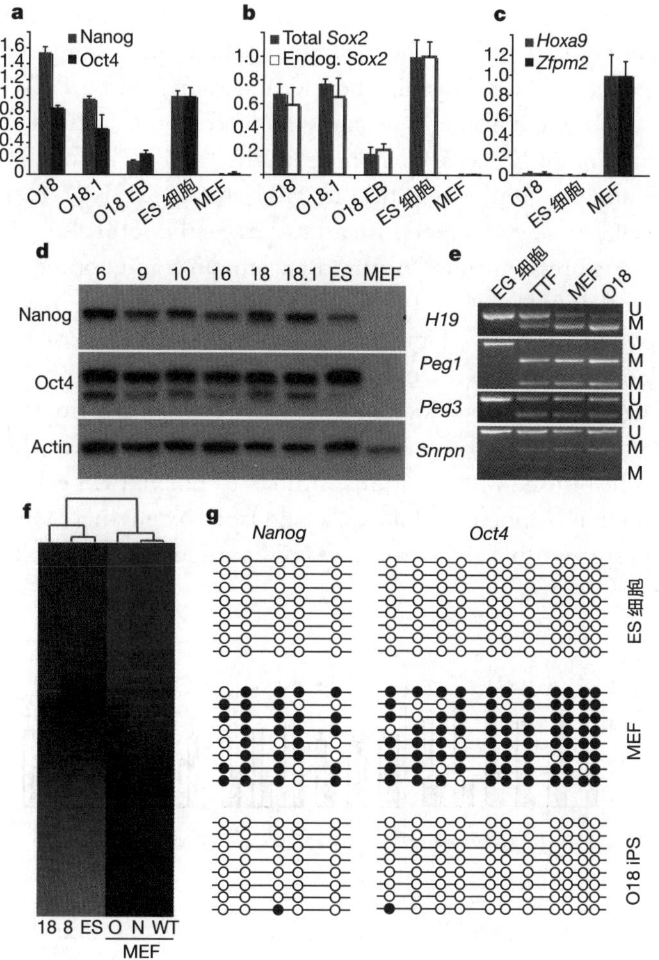

图 2. iPS 细胞的基因表达和启动子甲基化分析。**a**～**c**，对 Oct4 iPS 克隆 18、亚克隆 18.1、来源于克隆 18 的 2 周龄胚胎体、V6.5 ES 细胞和 Oct4-neo MEF 进行 qRT–PCR 分析（$n=3$ 个独立 PCR 反应，误差条表示 s.d.），显示 *Nanog*（红条）和总 *Oct4*（蓝条）水平与 ES 细胞（**a**）相似；*Sox2* 总水平略低（实心红条），主要是由于内源性 *Sox2* 转录本的表达（空心红条，**b**）；以及 iPS 细胞中 *Hoxa9*（红）和 *Zfpm2*（蓝）转录本的明显下调（**c**）。转录本水平根据 Gapdh 的表达进行了标化，即将 ES 细胞（**a**,**b**）和 MEF 细胞（**c**）内的表达水平设置为 1。**d**，不同 Oct4 iPS 克隆（6，9，10，16，18）和 GFP 标记的克隆 18 的亚克隆（18.1）中 *Oct4* 和 *Nanog* 表达的 Western 印迹分析。**e**，印记基因 *H19*（母方表达）、*Peg1*（父方表达）、*Peg3*（父方表达）和 *Snrpn*（父方表达）的 COBRA 甲基化分析[32]。上面的条带，未甲基化（U）；下面的条带，甲基化（M）。**f**，Oct4-neo iPS 克隆 18、Nanog-neo iPS 克隆 8、与之遗传信息匹配的 ES 细胞（V6.5；129SvJae/C57BL/6）、Oct4-neo MEF（O）、Nanog-neo MEF（N）和野生型 129/B6 F1 MEF（WT）中的平均整体转录情况的自发分级集合。**g**，使用亚硫酸氢盐测序分析 *Oct4* 和 *Nanog* 启动子的甲基化状态。空圈表示未甲基化，实圈表示甲基化的 CpG 二核苷酸。图中所示的是 ES 细胞（V6.5）、Oct4-neo MEF 和 Oct4-neo iPS 克隆 18 中的 8 个代表性的测序方法。

Chromatin Modifications

Recently, downstream target genes of *Oct4*, *Nanog* and *Sox2* have been defined in ES cells by genome-wide location analyses[9,10]. These targets include many important developmental regulators, a proportion of which is also bound and repressed by PcG (Polycomb-Group) complexes[11,12]. Notably, the chromatin at many of these non-expressed target genes adopts a bivalent conformation in ES cells, carrying both the "active" histone H3 lysine 4 (H3K4) methylation mark and the "repressive" histone H3 lysine 27 (H3K27) methylation mark[13,14]. In differentiated cells, those genes tend instead to carry either H3K4 or H3K27 methylation depending on their expression state. We used chromatin immunoprecipitation (ChIP) and real-time PCR to quantify H3K4 and H3K27 methylation for a set of genes reported to be bivalent in pluripotent ES cells[13]. Figure 3a shows that the fibroblast-specific genes *Zfpm2* and *Hoxa9* carried stronger H3K4 methylation than H3K27 methylation in the donor MEFs, whereas the silent genes *Nkx2.2*, *Sox1*, *Lbx1* and *Pax5* primarily carried H3K27 methylation. In contrast, in the Oct4 iPS cells, all of these genes showed comparable enrichment for both histone modifications, similar to normal ES cells (Fig. 3a). Identical results were obtained in Nanog iPS clones selected from Nanog-neo MEFs (Supplementary Fig. 2). These data suggest that the chromatin configuration of somatic cells is re-set to one that is characteristic of ES cells.

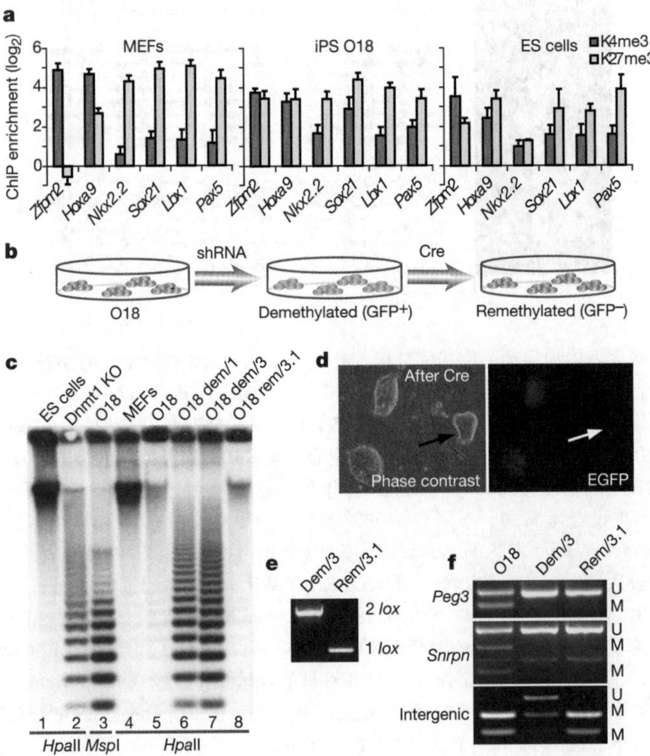

Fig. 3. Reprogrammed MEFs acquire an ES-cell-like epigenetic state. **a**, Real-time PCR after chromatin immunoprecipitation using antibodies against tri-methylated histone H3K4 and H3K27. Shown are the

染色质修饰

最近，通过全基因组的定位分析[9,10]，在 ES 细胞中发现了 *Oct4*、*Nanog* 和 *Sox2* 的下游目的基因。这些目的基因包括了许多重要的发育调节因子，其中一部分还受到 PcG(多聚硫)复合物的限制和抑制[11,12]。值得注意的是，这些未表达的目的基因中的染色质状态在 ES 细胞中形成双标构型，同时携带激活基因表达的组蛋白 3 赖氨酸 4(H3K4) 甲基化标志和抑制基因表达的组蛋白 3 赖氨酸 27(H3K27) 甲基化标志[13,14]。在分化的细胞中，这些基因根据倾向于携带 H3K4 或 H3K27 甲基化，这取决于它们的表达状态。我们用染色质免疫沉淀法(ChIP)和实时 PCR 对多能 ES 细胞内一系列被认为是双标基因的 H3K4 和 H3K27 甲基化进行了定量分析[13]。图 3a 显示了在供体 MEF 中成纤维细胞中特异性表达的基因 *Zfpm2* 和 *Hoxa9* 携带的 H3K4 甲基化水平强于 H3K27，而沉默基因 *Nkx2.2*、*Sox1*、*Lbx1* 和 *Pax5* 主要带有 H3K27 甲基化标志。相反，在 Oct4 iPS 细胞内，所有这些基因在两种组蛋白修饰上都表现出类似于正常 ES 细胞的相似富集(图 3a)。从 Nanog-neo MEF 中获得的 Nanog iPS 克隆得到了相同的结果(补充信息图 2)。这些数据表明，体细胞的染色质结构被重置成了具有 ES 细胞特征的模式。

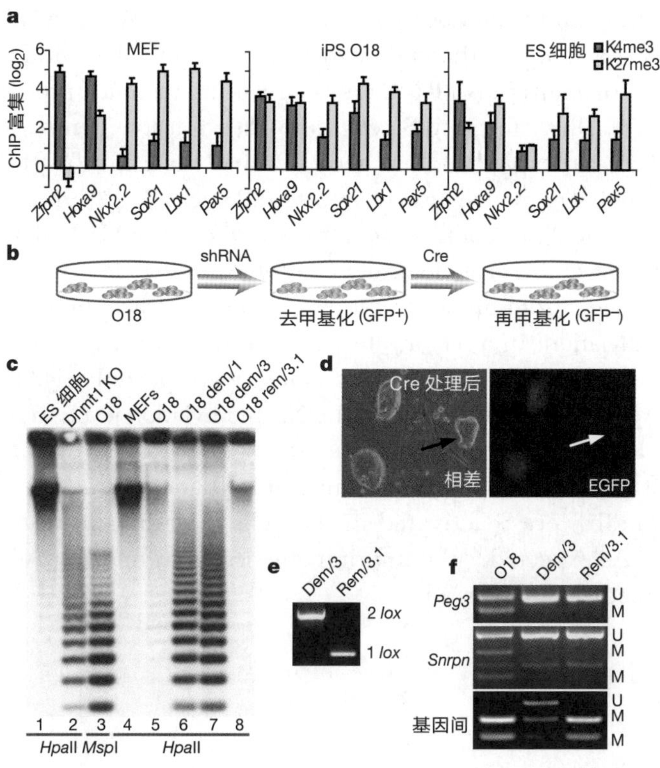

图 3. 重编程的 MEF 获得了 ES 细胞样的表观遗传状态。**a**，用三甲基化组蛋白 H3K4 和 H3K27 抗体进行染色质免疫沉淀后的实时 PCR。图中所示的是先前报道的 ES 细胞中数个“双标”位点的 log_2 富集

log_2 enrichments for several previously reported "bivalent" loci in ES cells ($n = 3$ experiments; error bars indicate s.d.). *Zfpm2* and *Hoxa9* show enrichment for the active (H3K4) mark in MEFs and are expressed (Fig. 1c and microarray data), whereas the other tested genes remain silent (microarray data). All loci tested in iPS clone O18 show enrichment for both H3K4 and H3K27 tri-methylation ("bivalent"), as seen in ES cells (V6.5). (See Supplementary Fig. 2 for H3K4 and H3K27 tri-methylation analysis of a subclone (clone O18.1) and Nanog-neo iPS clone N8.) **b**, Experimental design to de- and remethylate genomic DNA. Clone O18 was infected with the *Dnmt1*-hairpin-containing lentiviral vector pSicoR-GFP. The shRNA and GFP marker in the pSicoR vector are flanked by *loxP* sites[18]. Green colonies were expanded and passaged four times. Tat-Cre protein transduction was used to remove the shRNA[33]. **c**, Southern blot analysis of the minor satellite repeats using a methylation-sensitive restriction enzyme (*Hpa*II) and its methyl-insensitive isoschizomer (*Msp*I) as a control. Loss of methylation in two different clones (lanes 6 and 7) is comparable to Dnmt1 knockout ES cells (lane 2). After Cre-mediated recombination, complete remethylation (lane 8) of the repeats is observed within four passages. **d**, **e**, Successful loop out after Tat-Cre treatment was identified by disappearance of EGFP fluorescence (arrow) and verified by PCR analysis (**e**). **f**, COBRA assay of the imprinted genes *Peg3* and *Snrpn* and a random intergenic region close to the *Otx2* locus (Intergenic), demonstrating the expected resistance to *de novo* methylation of imprinted genes in contrast to non-imprinted intergenic sequences. U, unmethylated band; M, methylated band.

iPS Cells Tolerate Genomic Demethylation

Tolerance of genomic demethylation is a unique property of ES cells in contrast to somatic cells, which undergo rapid apoptosis on loss of the DNA methyltransferase Dnmt1 (refs 15–17). We investigated whether iPS cells would be resistant to global demethylation after Dnmt1 inhibition and would be able to re-establish global methylation patterns after restoration of Dnmt1 activity. To this end, we used a conditional lentiviral vector harbouring a *Dnmt1*-targeting short hairpin (sh)RNA and a green fluorescence protein (GFP) reporter gene (Fig. 3b and ref. 18). Infected iPS cells were plated at low density and GFP-positive colonies were picked and expanded. Southern blot analysis using *Hpa*II-digested genomic DNA showed that global demethylation of infected iPS cells (Fig. 3c, lanes 6, 7) was similar to *Dnmt1*$^{-/-}$ ES cells (lane 2). In contrast, uninfected iPS cells or MEFs (lanes 4, 5) displayed normal methylation levels. Morphologically, the GFP-positive cells were indistinguishable from the parental line or from uninfected sister subclones, indicating that iPS cells tolerate global DNA demethylation. In a second step, the *Dnmt1* shRNA was excised through Cre-mediated recombination and GFP-negative clones were picked (Fig. 3d). The cells had excised the shRNA vector (Fig. 3e) and normal DNA methylation levels were restored (Fig. 3c, lane 8) and were able to generate chimaeras (see below, Table 1), as has been reported previously for ES cells[19]. These observations imply that the *de novo* methyltransferases Dnmt3a and Dnmt3b were reactivated in iPS cells[20], leading to restoration of global methylation levels. As expected[19], the imprinted genes *Snrpn* and *Peg3* were unmethylated and resistant to remethylation (Fig. 3f).

（$n = 3$ 个实验，误差条表示 s.d.）。*Zfpm2* 和 *Hoxa9* 显示了 MEF 中活性（H3K4）标记的富集和表达（图 1c 和微阵列数据），而其他被检测的基因都保持沉默（微阵列数据）。iPS 克隆 O18 中检测的所有位点都显示 H3K4 和 H3K27 三甲基化的富集（双标性），与 ES 细胞（V6.5）一样。（亚克隆（克隆 18.1）和 Nanog-neo iPS 克隆 N8 的 H3K4 和 H3K27 三甲基化分析见补充信息图 2）。**b**，基因组 DNA 去甲基化和再甲基化的实验。克隆 O18 感染了含有 *Dnmt1* 发夹结构的慢病毒载体 pSicoR-GFP。pSicoR 载体中 shRNA 和 GFP 标记物的两侧是 *loxP* 位点[18]。绿色的克隆进行扩增并且传代四次。Tat-Cre 蛋白转导被用于去除 shRNA[33]。**c**，用甲基化敏感的限制性内切酶（*Hpa*II）进行小卫星重复序列的 Southern 印迹分析，以甲基化不敏感型异构酶（*Msp*I）作为对照。两个不同的克隆（泳道 6 和 7）中甲基化的缺失与 Dnmt1 敲除的 ES 细胞（泳道 2）相似。Cre 介导的重组之后，在四代内观察到了重复序列的完全再甲基化（泳道 8）。**d, e**，通过 EGFP 荧光的消失（箭头）验证了 Tat-Cre 处理后成功的消除作用，并用 PCR 进行验证（**e**）。**f**，印记基因 *Peg3* 和 *Snrpn* 以及邻近 *Otx2* 位点的随机基因间区域（基因间的）的 COBRA 实验，说明印记基因相对于非印记的基因间序列而言具有预期的对从头合成甲基化的耐受作用。U，未甲基化条带；M，甲基化条带。

iPS 细胞耐受基因组去甲基化

与体细胞不同，基因组去甲基化的耐受是 ES 细胞的特有性质，而体细胞在 DNA 甲基转移酶 Dnmt1 缺失后迅速凋亡（参考文献 15 ~ 17）。我们研究了在 Dnmt1 抑制后，iPS 细胞是否会对全面去甲基化产生耐药性，以及在 Dnmt1 活性恢复后能否重建甲基化模式。为此，我们使用了条件慢病毒载体，它含有一个 *Dnmt1* 靶向短发夹（sh）RNA 和一个绿色荧光蛋白（GFP）报告基因（图 3b 和参考文献 18）。将感染后的 iPS 细胞以较低的密度种下，并挑出 GFP 阳性克隆进行扩增。使用 *Hpa*II 消化的基因组 DNA 进行的 Southern 印迹分析显示，感染后 iPS 细胞的整体去甲基化状态（图 3c，6，7 道）与 $Dnmt1^{-/-}$ 的 ES 细胞相似（2 道）。相反，未感染的 iPS 细胞或者 MEF（4，5 道）显示正常的甲基化水平。形态学上，GFP 阳性的细胞与其父代或者未受感染的姐妹亚克隆无明显区别，表明 iPS 细胞能够耐受整体的 DNA 去甲基化。第二步，通过 Cre 介导的重组将 *Dnmt1* shRNA 去除，并选择 GFP 阴性克隆（图 3d）。这些细胞已经去除了 shRNA 载体（图 3e），并恢复了正常的 DNA 甲基化水平（图 3c，8 道），能够产生嵌合体（见下，表 1），正和之前报道的 ES 细胞性质相同[19]。这些观察结果表明，在 iPS 细胞中从头合成甲基转移酶 Dnmt3a 和 Dnmt3b 被重新激活[20]，导致了整体甲基化水平的恢复。如预期所示[19]，印记基因 *Snrpn* 和 *Peg3* 未发生甲基化，并且能够抵御再甲基化（图 3f）。

Table 1. Summary of blastocyst infections

Cell line	2N injections				4N injections		
	Injected blastocysts	Live chimaeras	Chimaerism (%)	Germ line	Injected blastocysts	Dead embryos (arrested)	Live embryos (analysed)
O6	ND	ND	ND	ND	13	0	2 (E12.5)
O9	30	5	30–70	Yes	90	3 (E11–13.5)	12 (E10–12.5)
O16	15	3	10–30	Yes	ND	ND	ND
O18	95	8	5–50	No	134	7 (E9–11.5)	4* (E10–12.5)
O3-2	ND	ND	ND	ND	25	2 (E8,11.5)	0
O4-16	ND	ND	ND	ND	35	4 (E11–13.5)	3 (E14.5)
N7	30	1	30	ND	ND	ND	ND
N8	90	14	5–50	No	118	9 (E9–11.5)	1* (E12.5)
N14	30	5	5–20	ND	46	2 (E8,11.5)	1 (E12.5)
TT-O25	50	2	30†	ND	39	3 (E9.5)	0
O18 rem/3.1	25	1	30	ND	ND	ND	ND

The extent of chimaerism was estimated on the basis of coat colour or EGFP expression. ND, not determined. 4N injected blastocysts were analysed between embryonic day E10.5 and E14.5.

"Analysed" indicates the day of embryonic development analysed; "arrested" indicates the estimated stage of development of dead embryos.

* Developmentally retarded or abnormal. O18 rem/3.1 is a de- and remethylated iPS clone (Fig. 3c).

† On the basis of GFP fluorescence.

Maintenance of the Pluripotent State

Southern blot analysis indicated that Oct4-neo iPS clone 18 carried four to six copies of the *Oct4*, *c-myc* and *Klf4* retroviral vectors and only one copy of the *Sox2* retroviral vector (Fig. 4a). Because these four factors were under the control of the constitutively expressed retroviral long terminal repeat, it was unclear in a previous study why iPS cells could be induced to differentiate[7]. To address this question, we designed primers specific for the four viral-encoded transcription factor transcripts and compared expression levels by qRT–PCR in MEFs 2 days after infection in iPS cells, in embryoid bodies derived from iPS cells, and in demethylated and remethylated iPS cells (Fig. 4b). Although the MEFs represented a heterogeneous population composed of uninfected and infected cells, virally encoded RNA levels of *Oct4*, *Sox2* and *Klf4* RNA were 5-fold higher and of *c-myc* more than 10-fold higher than in iPS cells. This suggests silencing of the viral long terminal repeat by *de novo* methylation during the reprogramming process. Accordingly, the total *Sox2* and *Oct4* RNA levels in iPS cells were similar to those in wild-type ES cells, and the *Sox2* transcripts in iPS cells were mostly, if not exclusively, transcribed from the endogenous gene (compare Fig. 2b). On differentiation to embryoid bodies, both viral and endogenous transcripts were downregulated. All viral *Sox2*, *Oct4* and *Klf4* transcripts were upregulated by approximately twofold in Dnmt1 knockdown iPS cells, and again downregulated on restoration of Dnmt1 activity. This is consistent with previous data that Moloney virus is efficiently *de*

表 1. 囊胚感染总结

细胞系	2N 注射				4N 注射		
	注射的囊胚	活的嵌合体	嵌合率 %	胚系	注射的囊胚	死胚胎（停止）	活胚胎（已分析）
O6	ND	ND	ND	ND	13	0	2(E12.5)
O9	30	5	30 ~ 70	Yes	90	3(E11 ~ 13.5)	12(E10 ~ 12.5)
O16	15	3	10 ~ 30	Yes	ND	ND	ND
O18	95	8	5 ~ 50	No	134	7(E9 ~ 11.5)	4*(E10 ~ 12.5)
O3-2	ND	ND	ND	ND	25	2(E8,11.5)	0
O4-16	ND	ND	ND	ND	35	4(E11 ~ 13.5)	3(E14.5)
N7	30	1	30	ND	ND	ND	ND
N8	90	14	5 ~ 50	No	118	9(E9 ~ 11.5)	1*(E12.5)
N14	30	5	5 ~ 20	ND	46	2(E8,11.5)	1(E12.5)
TT-O25	50	2	30†	ND	39	3(E9.5)	0
O18 rem/3.1	25	1	30	ND	ND	ND	ND

嵌合的程度根据毛色或者 EGFP 的表达估计。ND：不确定。4N 注射的囊胚在胚胎发育的 E10.5 到 E14.5 天之间进行了分析。

“已分析”表示已分析的胚胎发育天数；“停止”表示死亡胚胎的估计发育阶段。

* 发育迟缓或者异常。O18rem/3.1 是一个去甲基化和再甲基化的 iPS 克隆（图 3c）。

† 基于 GFP 荧光。

多能状态的保持

Southern 印迹分析显示 Oct4-neo iPS 克隆 18 带有 4 到 6 个拷贝的 *Oct4*、*c-Myc* 和 *Klf4* 逆转录病毒载体，只带有一个拷贝的 *Sox2* 逆转录病毒载体（图 4a）。因为这四个因子都是在组成性表达的逆转录病毒长末端重复的控制之下，所以在以前的研究中并不清楚 iPS 细胞能够被诱导分化的原因[7]。为了解决这个问题，我们设计了针对这四个病毒编码转录因子特异性的引物，并用 qRT–PCR 比较 iPS 细胞感染后 2 天的 MEF、iPS 细胞衍生的 EB 球、去甲基化和再甲基化的 iPS 细胞中的表达水平（图 4b）。虽然 MEF 是由未感染和感染细胞组成的异质性群体，但 *Oct4*、*Sox2* 和 *Klf4* RNA 的病毒编码 RNA 水平是 iPS 细胞的 5 倍，*c-Myc* 则是 10 倍以上。这表明在重编程的过程中发生了新的甲基化使病毒的长末端重复沉默。因此，iPS 细胞内的 *Sox2* 和 *Oct4* RNA 的总水平与野生型 ES 细胞内相似，而且 iPS 细胞内的 *Sox2* 转录本不完全是大部分来源于内源性基因（对照图 2b）。在向 EB 球分化的过程中，病毒和内源性的转录产物都发生了下调。在 Dnmt1 敲降的 iPS 细胞中，所有的病毒 *Sox2*、*Oct4* 和 *Klf4* 转录本都上调至原来的将近 2 倍，而在 Dnmt1 功能恢复时又下调回来。这与先前的数据一致，在胚胎中莫洛尼病毒能够有效地被甲基化和沉默，

novo methylated and silenced in embryonic but not in somatic cells[21,22]. Transcript levels of *c-myc* were about 20-fold lower in iPS cells than in infected MEFs, and did not change on differentiation or demethylation.

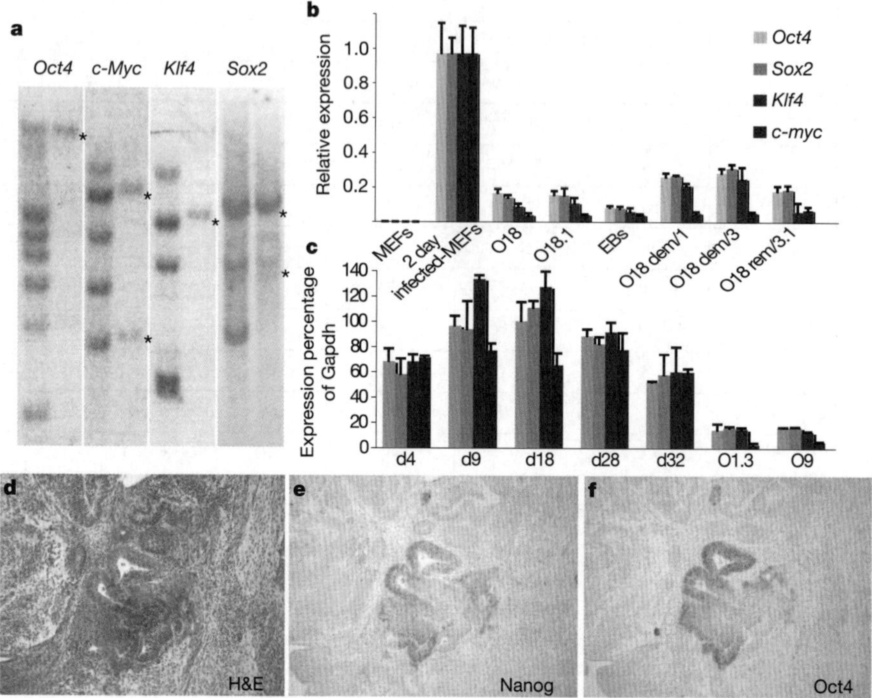

Fig. 4. Efficient silencing of retroviral transcripts in induced pluripotent cells. **a**, Southern blot analysis of proviral integrations in iPS clone O18 (left lanes) for the four retroviral vectors. Uninfected ES cells (right lanes) show only one or two bands corresponding to the endogenous gene (marked by an asterisk). **b**, Quantitative RT–PCR using primers specifically detecting the four viral transcripts. Shown are Oct4-neo iPS clone 18 and a GFP-labelled subclone, Oct4-neo MEFs, 2-week-old embryoid bodies generated from clone 18, two demethylated clones (18 dem/1 and 18 dem/3), a remethylated clone (18 rem/3.1), and Oct4-neo MEFs 2 days after infection with all four viruses but not selected with G418 ($n = 3$ independent experiments; error bars indicate s.d.). **c**, Viral transcript levels at various time points in cell populations after infection and Oct4 selection and in the two Oct4 iPS cell lines O1.3 and O9 ($n = 3$ independent experiments; error bars indicate s.d.). **d–f**, Paraffin sections of a teratoma 26 days after subcutaneous injection of Oct4 iPS clone 18 cells into SCID mice. H&E, haematoxylin and eosin. Nanog (**e**) and Oct4 (**f**) expression was confined to undifferentiated cell types as indicated an immunohistochemical analysis.

To follow the kinetics of vector inactivation during the reprogramming process, we isolated RNA from drug-resistant cell populations at different times after infection. Figure 4c shows that the viral-vector-encoded transcripts were gradually silenced during the transition from MEFs to iPS cells with a time course that corresponded to the gradual appearance of pluripotency markers (compare Fig. 1j). Finally, to visualize directly Oct4 and Nanog expression during differentiation, we injected Oct4 iPS cells into SCID mice to induce teratoma formation (Fig. 4d). Immunostaining revealed that Oct4 and Nanog were expressed in the centrally located undifferentiated cells but were silenced in the differentiated parts of the teratoma (Fig. 4e, f). Our results suggest that the retroviral vectors

而在体细胞中不行[21,22]。iPS 细胞中的 *c-Myc* 转录本水平大约为感染的 MEF 的二十分之一，而且在分化或者去甲基化过程中没有变化。

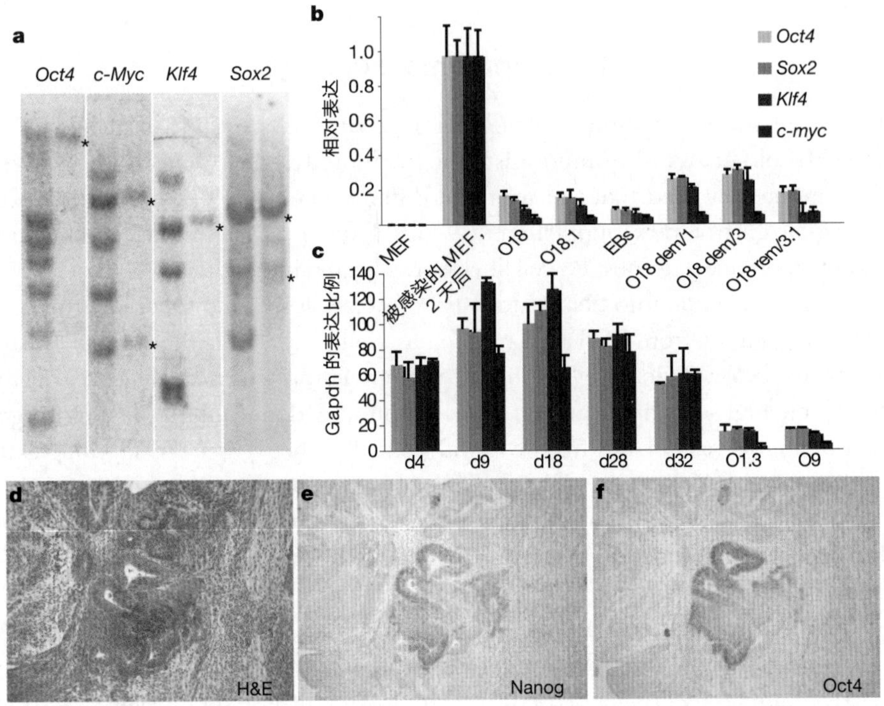

图 4. 诱导的多能细胞中逆转录病毒转录本的有效沉默。**a**，iPS 克隆 O18（左侧泳道）中四个逆转录病毒载体的前病毒整合的 Southern 印迹分析。未感染的 ES 细胞（右侧泳道）只显示与内源性基因对应的一条或者两条带（星号标记）。**b**，使用识别四个病毒转录子的引物进行的定量 RT–PCR。显示的分别是 Oct4-neo iPS 克隆 18 和 GFP 标记的亚克隆，OCT4-neo MEF，来源于克隆 18 的 2 周龄胚胎体，两个去甲基化的克隆（18dem/1 和 18dem/3），一个再甲基化的克隆（18dem/3.1）和用所有四种病毒感染 2 天后但是不用 G418 进行选择得到的 OCT4-neo MEF（$n = 3$ 独立的实验，误差条表示 s.d.）。**c**，经过感染和 Oct4 选择的细胞群以及两个 Oct4 iPS 细胞系 O1.3 和 O9 在不同时间点的病毒转录子水平（$n = 3$ 独立的实验，误差条表示 s.d.）。**d~f**，皮下注射 Oct4 iPS 克隆 18 到 SCID 小鼠中 26 天后形成的畸胎瘤的石蜡切片。H&E，苏木精和伊红。免疫组化分析显示，Nanog（**e**）和 Oct4（**f**）表达仅限于未分化的细胞类型。

为了研究重编程过程中外源基因失活的动力学，我们从感染后不同时间的耐药细胞中分离出 RNA。图 4c 显示了从 MEF 向 iPS 转化的过程中，病毒载体编码的转录本逐渐沉默，这一过程与多能性标志物的逐渐出现相对应（对照图 1j）。最后，为了直接观察 Oct4 和 Nanog 在分化过程中的表达，我们将 Oct4 iPS 细胞注射到 SCID 小鼠体内诱导畸胎瘤形成（图 4d）。免疫染色显示 Oct4 和 Nanog 在中心位置的未分化细胞中表达，但在畸胎瘤的已分化部分沉默（图 4e、f）。我们的结果表明，在重

are subject to gradual silencing by *de novo* methylation during the reprogramming process. The maintenance of the pluripotent state and induction of differentiation strictly depends on the expression and normal regulation of the endogenous *Oct4* and *Nanog* genes.

Developmental Potency

We determined the developmental potential of iPS cells by teratoma and chimaera formation. Histological and immunohistochemical analysis of Oct4- or Nanog-iPS-cell-induced teratomas revealed that the cells had differentiated into cell types representing all three embryonic germ layers (Supplementary Figs 3 and 4). To assess more stringently their developmental potential, various iPS cell lines were injected into diploid (2N) or tetraploid (4N) blastocysts. After injection into 2N blastocysts both Nanog iPS and Oct4 iPS clones derived from MEFs (Fig. 5a) or from TTFs (Fig. 5b, c), as well as iPS cells that had been subjected to a consecutive cycle of demethylation and remethylation (compare Fig. 3b, c), efficiently generated viable high-contribution chimaeras (summarized in Table 1). To test for germline transmission, chimaeras derived from two different iPS lines (Oct4 iPS O9 and O16) were mated with normal females, and blastocysts were isolated and genotyped by three different PCR reactions for the presence of the multiple viral *Oct4* and *c-myc* genes and for the single-copy GFPneo sequences inserted into the *Oct4* locus of the donor cell (Fig. 1a). Figure 5f shows that 9 out of 16 embryos from two chimaeras were positive for the viral copies. As expected, only half of the viral-positive blastocysts contained the GFPneo sequences (5 out of 9 embryos, Fig. 5f, left panel). When embryonic day (E)10 embryos derived from an Oct4 iPS line O16 chimaera were genotyped, three out of eight tested embryos were transgenic (Fig. 5f, right panel). Finally, we injected iPS cells into 4N blastocysts as this represents the most rigorous test for developmental potency, because the resulting embryos are composed only of the injected donor cells ("all ES embryo"). Figure 5d, e shows that both Oct4 and Nanog iPS cells could generate mid- and late-gestation "all iPS embryos" (summarized in Table 1). These findings indicate that iPS cells can establish all lineages of the embryo and thus have a similar developmental potential as ES cells.

建的过程中逆转录病毒载体通过从头合成甲基化逐渐沉默。保持多能状态和诱导分化严格依赖于内源性 *Oct4* 和 *Nanog* 基因的表达和正常调控。

发育潜力

我们用畸胎瘤和嵌合体的形成来确定 iPS 细胞的发育潜力。对 Oct4 或 Nanog iPS 细胞诱导的畸胎瘤进行组织学和免疫组化分析显示，这些细胞已经分化成代表三个胚层(补充信息图 3 和 4)的细胞类型。为了更加严格地评估其发育潜力，不同的 iPS 细胞系被注射到二倍体(2N)或者四倍体(4N)囊胚中。注射到 2N 囊胚后，来源于 MEF(图 5a)或者 TTF(图 5b、c)的 Nanog iPS 和 Oct4 iPS 细胞克隆以及经过连续去甲基化和再甲基化循环的 iPS 细胞(对照图 3b、c)都有效地产生高度嵌合的嵌合体(总结见表 1)。为了检验胚系传递的能力，来源于两个不同 iPS 细胞系(Oct4 iPS O9 和 O16)的嵌合体与正常雌性小鼠进行交配，分离出囊胚并用三种不同的 PCR 反应进行基因分型，验证是否存在多种病毒 *Oct4* 和 *c-Myc* 基因以及为了将单拷贝 GFPneo 序列插入到供者细胞 *Oct4* 位点内(图 1a)。图 5f 显示两个嵌合体共 16 个胚胎中的 9 个含有病毒拷贝。正如所期望的，只有一半的病毒阳性囊胚含有 GFPneo 序列(9 个胚胎中有 5 个，图 5f，左侧栏)。当对 Oct4 iPS 细胞系 O16 嵌合体的 10 个胚胎进行基因型鉴定时，8 个被测胚胎中有 3 个是转基因的(图 5f，右侧栏)。最后，我们将 iPS 细胞注射到 4N 囊胚中，这代表了发育潜力最严格的检测，因为得到的胚胎仅仅由注射的供者细胞组成(全 ES 胚胎)。图 5d、e 显示了 Oct4 和 Nanog iPS 细胞都能够产生孕中期和孕晚期的全 iPS 胚胎(总结见表 1)。这些结果表明 iPS 细胞能够建立胚胎的所有细胞系，因此具有与 ES 细胞类似的发育潜力。

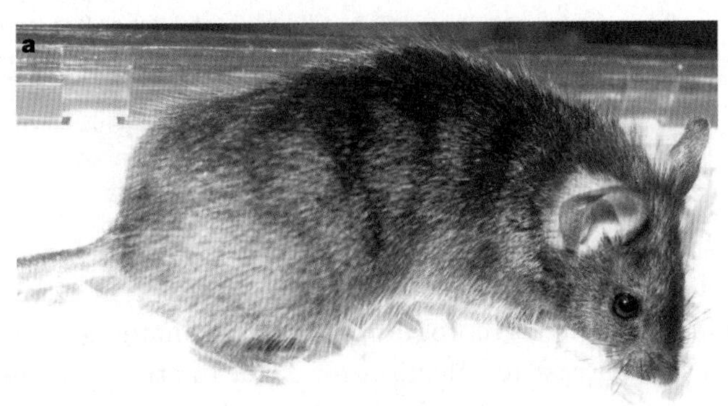

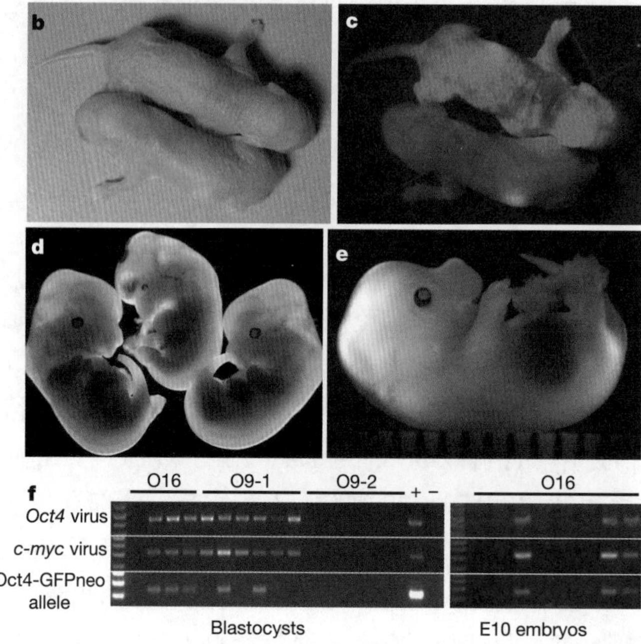

Fig. 5. Developmental pluripotency of reprogrammed fibroblasts. **a**, A 6-week-old chimaeric mouse. Agouti-coloured hairs originated from Oct4 iPS cell line O18.1. **b**, **c**, Two live pups after 2N blastocyst injection, one of which shows a high contribution (**c**) of the TTF-derived Oct4 iPS cell line TT-O25, which had been GFP-labelled with a lentiviral ubiquitin-EGFP vector. **d**, "All iPS cell embryos" were generated by injection of iPS cells into 4N blastocysts[34]. Live E12.5 embryos generated from Oct4 iPS line O6 (left), from Nanog iPS line N14 (middle) and from V.6.5 ES cells (right) are shown. **e**, A normally developed E14.5 embryo was derived from Oct4 iPS cell line O4-16 after tetraploid complementation and was isolated by screening MEFs for activation of GFP inserted into the *Oct4* locus. **f**, Germline contribution of Oct4 iPS clones O9 and O16. Genotyping of blastocysts from females mated with three chimaeric males demonstrated the presence of *Oct4* and *c-myc* virus integrations and the Oct4-IRES-GFPneo allele (left panel). Because of the multiple integrations (Fig. 4a) all embryos with iPS cell contribution are expected to be positive for proviral sequences in this assay. In contrast, the single-copy Oct4-IRES-GFPneo allele segregated into only 5 of the 9 virus-positive embryos. All six blastocysts from O9 chimaera 1 were iPS-cell-derived, suggesting that this chimaera was a pseudo-male. Additional genotyping identified 13 out of 72 tested blastocysts derived from iPS line O9 and 4 out of 13 blastocysts derived from iPS line O16 chimaeras carrying the viral transgenes. The right panel shows that 3 out of 8 tested E.10 mid-gestation embryos were sired by a chimaera derived from the donor iPS line O16. +, positive control; −, negative control.

Discussion

The results presented here demonstrate that the four transcription factors Oct4, Sox2, c-myc and Klf4 can induce epigenetic reprogramming of a somatic genome to an embryonic pluripotent state. In contrast to selection for Fbx15 activation[7], fibroblasts that had reactivated the endogenous *Oct4* (Oct4-neo) or *Nanog* (Nanog-neo) loci grew independently of feeder cells, expressed normal Oct4, Nanog and Sox2 RNA and protein levels, were epigenetically identical to ES cells by a number of criteria, and were able to generate viable chimaeras, contribute to the germ line and generate viable late-gestation embryos after

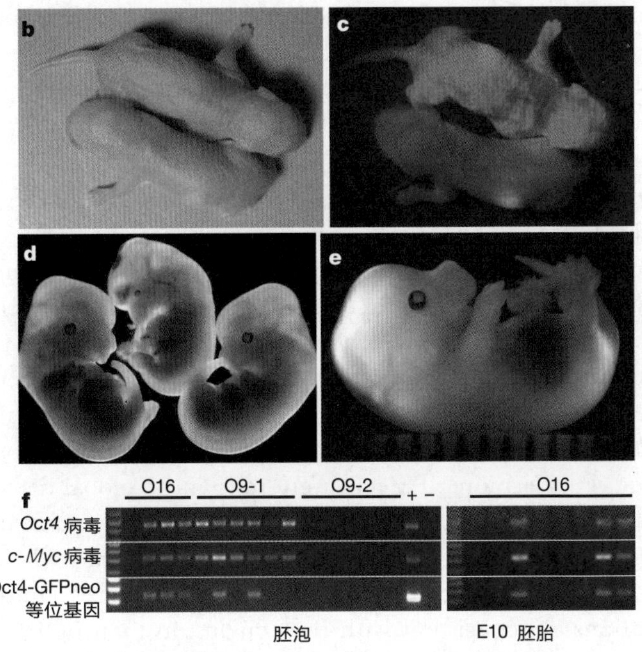

图 5. 重编程后成纤维细胞的发育多能性。**a**，一只 6 周龄的嵌合体小鼠。刺豚鼠样的毛色来源于 Oct4 iPS 细胞系 O18.1。**b, c**，2N 囊胚注射后的两只活幼鼠，其中一只显示了高比例（**c**）的 TTF 来源的 Oct4 iPS 细胞系 TT-O25，该细胞系已被慢病毒泛素化 -EGFP 进行 GFP 标记。**d**，所有的 iPS 细胞胚胎都是通过将 iPS 细胞注入 4N 囊胚生成的[34]。图中显示了来源于 Oct4 iPS 细胞系 O6（左侧）、Nanog iPS 细胞系 N14（中间）和 V.6.5 ES 细胞（右侧）的活的 E12.5 胚胎。**e**，经过四倍体互补后，从 Oct4 iPS 细胞系 O4-16 中获得一个正常发育的 E14.5 胚胎，通过筛选插入到 Oct4 位点的 MEF，激活 GFP，将其分离出来。**f**，Oct4 iPS 克隆 O9 和 O16 的生殖系嵌合贡献。对雌鼠和三只嵌合体雄鼠交配产生的囊胚进行基因型鉴定，显示存在 *Oct4* 和 *c-Myc* 病毒整合以及 Oct4-TRES-GFP 等位基因（左侧泳道）。由于多重整合（图 4a），在本实验中，所有含有 iPS 细胞成分的胚胎的前病毒序列均呈阳性。相反，9 只病毒阳性的胚胎中只有 5 只具有单拷贝 Oct4-TRES-GFPneo 等位基因。O9 嵌合体 1 的 6 个囊胚都是 iPS 细胞来源，表明该嵌合体是假雄性。另外的基因分型在 72 个测试的囊胚中鉴定出 13 个来自 iPS 细胞系 O9，在 13 个来自携带病毒转基因的 iPS 系 O16 嵌合体的囊胚中鉴定出 4 个。右侧泳道显示了 8 个检测的 E10 孕中期胚胎中有 3 个来源于供体 iPS 细胞系 O16 的嵌合体。+，阳性对照；−，阴性对照。

讨　论

本研究表明，Oct4、Sox2、c-Myc 和 Klf4 这四个转录因子能够诱导体细胞基因组的表观遗传重编程成胚胎多能状态。相对于 *Fbx15* 活化选择[7]，激活内源性 *Oct4*（Oct4-neo）或者 *Nanog*(Nanog-neo) 位点的成纤维细胞可以不依赖饲养细胞而独立生长，能够表达正常的 *Oct4*、*Nanog* 和 *Sox2* RNA 和蛋白水平。在许多标准下，成纤维细胞与 ES 细胞在表观遗传学上是相同的，它能够产生可以存活的嵌合体，在注

injection into tetraploid blastocysts. Transduction of the four factors generated significantly more drug-resistant cells from Nanog-neo than from Oct4-neo fibroblasts but a higher fraction of Oct4-selected cells had all the characteristics of pluripotent ES cells, suggesting that *Nanog* activation is a less stringent criterion for pluripotency than *Oct4* activation.

Our data suggest that the pluripotent state of Oct4 iPS and Nanog iPS cells is induced by the virally transduced factors but is largely maintained by the activity of the endogenous pluripotency factors including Oct4, Nanog and Sox2, because the viral-controlled transcripts, although expressed highly in MEFs, become mostly silenced in iPS cells. The total levels of Oct4, Nanog and Sox2 were similar in iPS and wild-type ES cells. Consistent with the conclusion that the pluripotent state is maintained by the endogenous pluripotency genes is the finding that the *Oct4* and the *Nanog* genes became hypomethylated in iPS cells as in ES cells, and that the bivalent histone modifications of developmental regulators were re-established. Furthermore, iPS cells were resistant to global demethylation induced by inactivation of Dnmt1, similar to ES cells but in contrast to somatic cells. Re-expression of Dnmt1 in the hypomethylated ES cells resulted in global remethylation, indicating that the iPS cells had also reactivated the *de novo* methyltransferases Dnmt3a and Dnmt3b. All these observations are consistent with the conclusion that the iPS cells have gained an epigenetic state that is similar to that of normal ES cells. This conclusion is further supported by the recent observation that female iPS cells, similar to ES cells, reactivate the somatically silenced X chromosome[23].

Expression of the four transcription factors proved to be a robust method to induce reprogramming of somatic cells to a pluripotent state. However, the use of retrovirus-transduced oncogenes represents a serious barrier to the eventual use of reprogrammed cells for therapeutic application. Much work is needed to understand the molecular pathways of reprogramming and to eventually find small molecules that could achieve reprogramming without gene transfer of potentially harmful genes.

Methods Summary

Cell culture, gene targeting and viral infections. ES and iPS cells were cultivated on irradiated MEFs. Using homologous recombination we generated ES cells carrying an IRES-GFPneo fusion cassette downstream of *Oct4* exon 5 (Fig. 1a). The *Nanog* gene was targeted as described[24]. Transgenic MEFs were isolated and selected from E13.5 chimaeric embryos after blastocyst injection of Oct4-IRES-GFPneo- or Nanog-neo-targeted ES cells. MEFs were infected overnight with the Moloney-based retroviral vector pLIB (Clontech) containing the murine complementary DNAs of *Oct4*, *Sox2*, *Klf4* and *c-myc*.

Southern blot, methylation and chromatin analyses. To assess the levels of DNA methylation, genomic DNA was digested with *Hpa*II and hybridized to a probe for the minor satellite repeats[25] or with an IAP probe[26]. Bisulphite treatment was performed with the Qiagen EpiTect Kit. For the methylation status of *Oct4* and *Nanog* promoters, bisulphite sequencing analysis was performed

射到四倍体囊胚中后，对生殖系的形成有促进作用，能够产生孕晚期的胚胎。这四个因子的转导使 Nanog-neo 成纤维细胞中的耐药细胞明显比 Oct4-neo 成纤维细胞中的多，但 *Oct4* 选择性表达的细胞中具有多能 ES 细胞所有的特征的比例较高，这表明 *Nanog* 活化是比 *Oct4* 活化更加不严格的多能性标准。

我们的数据表明 Oct4 iPS 和 Nanog iPS 细胞的多能状态是通过病毒转导的因子诱导的，但大部分是由内源多能性因子包括 *Oct4*、*Nanog* 和 *Sox2* 的活性来维持的，因为病毒控制的转录子尽管在 MEF 中高度表达，却在 iPS 细胞内大部分都沉默了。iPS 细胞和野生型 ES 细胞中的 *Oct4*、*Nanog* 和 *Sox2* 总水平相似。与内源性多能基因保持多能状态的结论相一致的是发现了 *Oct4* 和 *Nanog* 基因在 iPS 细胞内与在 ES 细胞一样低甲基化，而且发育调节因子的双标组蛋白修饰发生了重建。此外，iPS 细胞对 Dnmt1 失活诱导的整体去甲基化具有抗性，这与 ES 细胞相似但与体细胞相反。Dnmt1 在低甲基化 ES 细胞中的重新表达导致了整体的再甲基化，这表明 iPS 细胞也重新激活了从头合成甲基转移酶 Dnmt3a 和 Dnmt3b。所有的这些发现与 iPS 细胞已经获得了类似于正常 ES 细胞的表观遗传状态的结论相一致。最近的发现进一步支持了这一结论，即雌性 iPS 细胞，类似于 ES 细胞，能够重新激活体细胞中沉默的 X 染色体 [23]。

四个转录因子的表达被证实是一种能将体细胞重编程成多能状态的可靠的方法。但是，使用逆转录病毒转导的致癌基因意味着最终将重编程的细胞用于临床治疗存在严重的障碍。我们需要做更多的工作来了解重编程的分子机制，并最终找到小分子物质，可以在没有潜在有害基因的基因转移情况下实现重编程。

方法概述

细胞培养、基因靶向和病毒感染　在辐射过的 MEF 上进行 ES 和 iPS 细胞培养。通过同源重组，我们获得了在 *Oct4* 的 5 号外显子下游产生携带 IRES-GFPneo 融合盒的 ES 细胞系（图 1a）。*Nanog* 基因如所描述的是靶向的 [24]。将 Oct4-IRES-GFPneo 或者 Nanog-neo 靶向 ES 细胞注入囊胚后，从 E13.5 嵌合体胚胎中选择和分离出转基因 MEF。MEF 用含有 *Oct4*、*Sox2*、*c-Myc* 和 *Klf4* 的小鼠互补 DNA 的莫洛尼逆转录病毒载体 pLIB（Clontech）产生的病毒感染过夜。

Southern 印迹、甲基化和染色质分析　为了评估 DNA 甲基化水平，基因组 DNA 用 *Hpa*II 进行消化，并与小卫星重复序列探针 [25] 或者 IAP 探针 [26] 进行杂交。用 Qiagen EpiTect 试剂盒进行亚硫酸氢盐处理。如先前所述的用亚硫酸氢盐序列分析法研究 *Oct4* 和 *Nanog* 启

as described previously[27]. For imprinted genes, a COBRA assay was performed. PCR primers and conditions were as described previously[28]. The status of bivalent domains was determined by chromatin immunoprecipitation followed by quantitative PCR analysis, as described previously[12].

Expression analysis. Total RNA was reverse-transcribed and quantified using the QuantTtect SYBR green RT–PCR Kit (Qiagen) on a 7000 ABI detection system. Western blot and immunofluorescence analysis was performed as described[29,30]. Microarray targets from 2 μg total RNA were synthesized and labelled using the Low RNA Input Linear Amp Kit (Agilent), hybridized to Agilent whole-mouse genome oligonucleotide arrays (G4122F) and analysed as previously described[31].

Full Methods and any associated references are available in the online version of the paper at www.nature.com/nature.

(**448**, 318-324; 2007)

Marius Wernig[1]*, Alexander Meissner[1]*, Ruth Foreman[1,2]*, Tobias Brambrink[1]*, Manching Ku[3]*, Konrad Hochedlinger[1]†, Bradley E. Bernstein[3,4,5] & Rudolf Jaenisch[1,2]

[1]Whitehead Institute for Biomedical Research and [2]Department of Biology, Massachusetts Institute of Technology, Cambridge, Massachusetts 02142, USA

[3]Molecular Pathology Unit and Center for Cancer Research, Massachusetts General Hospital, Charlestown, Massachusetts 02129, USA

[4]Broad Institute of Harvard and MIT, Cambridge, Massachusetts 02142, USA

[5]Department of Pathology, Harvard Medical School, Boston, Massachusetts 02115, USA

†Present address: Center for Regenerative Medicine and Cancer Center, Massachusetts General Hospital, Harvard Medical School and Harvard Stem Cell Institute, Boston, Massachusetts 02414, USA

*These authors contributed equally to this work

Received 27 February; accepted 22 May 2007. Published online 6 June 2007.

References:

1. Hochedlinger, K. & Jaenisch, R. Nuclear transplantation, embryonic stem cells, and the potential for cell therapy. *N. Engl. J. Med.* **349**, 275-286 (2003).
2. Yang, X. *et al.* Nuclear reprogramming of cloned embryos and its implications for therapeutic cloning. *Nature Genet.* **39**, 295-302 (2007).
3. Hochedlinger, K. & Jaenisch, R. Nuclear reprogramming and pluripotency. *Nature* **441**, 1061-1067 (2006).
4. Tada, M., Takahama, Y., Abe, K., Nakatsuji, N. & Tada, T. Nuclear reprogramming of somatic cells by *in vitro* hybridization with ES cells. *Curr. Biol.* **11**, 1553-1558 (2001).
5. Cowan, C. A., Atienza, J., Melton, D. A. & Eggan, K. Nuclear reprogramming of somatic cells after fusion with human embryonic stem cells. *Science* **309**, 1369-1373 (2005).
6. Jaenisch, R. Human cloning—the science and ethics of nuclear transplantation. *N.Engl. J. Med.* **351**, 2787-2791 (2004).
7. Takahashi, K. & Yamanaka, S. Induction of pluripotent stem cells from mouse embryonic and adult fibroblast cultures by defined factors. *Cell* **126**, 663-676 (2006).
8. Labosky, P. A., Barlow, D. P. & Hogan, B. L. Mouse embryonic germ (EG) cell lines: transmission through the germline and differences in the methylation imprint of insulin-like growth factor 2 receptor (*Igf2r*) gene compared with embryonic stem (ES) cell lines. *Development* **120**, 3197-3204 (1994).
9. Boyer, L. A. *et al.* Core transcriptional regulatory circuitry in human embryonic stem cells. *Cell* **122**, 947-956 (2005).
10. Loh, Y. H. *et al.* The Oct4 and Nanog transcription network regulates pluripotency in mouse embryonic stem cells. *Nature Genet.* **38**, 431-440 (2006).
11. Lee, T. I. *et al.* Control of developmental regulators by Polycomb in human embryonic stem cells. *Cell* **125**, 301-313 (2006).
12. Boyer, L. A. *et al.* Polycomb complexes repress developmental regulators in murine embryonic stem cells. *Nature* **441**, 349-353 (2006).
13. Bernstein, B. E. *et al.* A bivalent chromatin structure marks key developmental genes in embryonic stem cells. *Cell* **125**, 315-326 (2006).
14. Azuara, V. *et al.* Chromatin signatures of pluripotent cell lines. *Nature Cell Biol.* **8**, 532-538 (2006).
15. Jackson-Grusby, L. *et al.* Loss of genomic methylation causes p53-dependent apoptosis and epigenetic deregulation. *Nature Genet.* **27**, 31-39 (2001).
16. Li, E., Bestor, T. H. & Jaenisch, R. Targeted mutation of the DNA methyltransferase gene results in embryonic lethality. *Cell* **69**, 915-926 (1992).

动子的甲基化水平[27]。对于印记基因，进行COBRA实验。PCR引物和条件如前所述[28]。通过染色质免疫沉淀然后通过PCR分析确定双标结构域的状态[12]。

基因表达分析 用QuantTtect SYBR green RT–PCR试剂盒(Qiagen)将总RNA进行反转录，之后再通过7000ABI检测系统完成定量检测。按照上述方法进行Western印迹和免疫荧光分析[29,30]。基因微阵列的靶标用Law RNA Input Linear Amp试剂盒，由2μg总的RNA合成，并进行标记，然后与Agilent全鼠基因组寡核苷酸阵列(G4122F)进行杂交和分析[31]。

完整的方法和相关的文献可在本文的网上版本中获得。

(毛晨晖 翻译；裴端卿 审稿)

17. Meissner, A. *et al.* Reduced representation bisulfite sequencing for comparative high-resolution DNA methylation analysis. *Nucleic Acids Res.* **33**, 5868-5877 (2005).

18. Ventura, A. *et al.* Cre-lox-regulated conditional RNA interference from transgenes. *Proc. Natl Acad. Sci. USA* **101**, 10380-10385 (2004).

19. Holm, T. M. *et al.* Global loss of imprinting leads to widespread tumorigenesis in adult mice. *Cancer Cell* **8**, 275-285 (2005).

20. Okano, M., Bell, D. W., Haber, D. A. & Li, E. DNA methyltransferases Dnmt3a and Dnmt3b are essential for *de novo* methylation and mammalian development. *Cell* **99**, 247-257 (1999).

21. Stewart, C. L., Stuhlmann, H., Jähner, D. & Jaenisch, R. *De novo* methylation, expression, and infectivity of retroviral genomes introduced into embryonal carcinoma cells. *Proc. Natl Acad. Sci. USA* **79**, 4098-4102 (1982).

22. Jähner, D. *et al. De novo* methylation and expression of retroviral genomes during mouse embryogenesis. *Nature* **298**, 623-628 (1982).

23. Maherali, N. *et al.* Global epigenetic remodeling in directly reprogrammed fibroblasts. *Cell Stem Cells* (in the press).

24. Mitsui, K. *et al.* The homeoprotein Nanog is required for maintenance of pluripotency in mouse epiblast and ES cells. *Cell* **113**, 631-642 (2003).

25. Chapman, V., Forrester, L., Sanford, J., Hastie, N. & Rossant, J. Cell lineage specific undermethylation of mouse repetitive DNA. *Nature* **307**, 284-286 (1984).

26. Walsh, C. P., Chaillet, J. R. & Bestor, T. H. Transcription of IAP endogenous retroviruses is constrained by cytosine methylation. *Nature Genet.* **20**, 116-117 (1998).

27. Blelloch, R. *et al.* Reprogramming efficiency following somatic cell nuclear transfer is influenced by the differentiation and methylation state of the donor nucleus. *Stem Cells* **24**, 2007-2013 (2006).

28. Lucifero, D., Mertineit, C., Clarke, H. J., Bestor, T. H. & Trasler, J. M. Methylation dynamics of imprinted genes in mouse germ cells. *Genomics* **79**, 530-538 (2002).

29. Hochedlinger, K., Yamada, Y., Beard, C. & Jaenisch, R. Ectopic expression of Oct-4 blocks progenitor-cell differentiation and causes dysplasia in epithelial tissues. *Cell* **121**, 465-477 (2005).

30. Wernig, M. *et al.* Functional integration of embryonic stem cell-derived neurons *in vivo*. *J. Neurosci.* **24**, 5258-5268 (2004).

31. Brambrink, T., Hochedlinger, K., Bell, G. & Jaenisch, R. ES cells derived from cloned and fertilized blastocysts are transcriptionally and functionally indistinguishable. *Proc. Natl Acad. Sci. USA* **103**, 933-938 (2006).

32. Eads, C. A. & Laird, P. W. Combined bisulfite restriction analysis (COBRA). *Methods Mol. Biol.* **200**, 71-85 (2002).

33. Peitz, M., Pfannkuche, K., Rajewsky, K. & Edenhofer, F. Ability of the hydrophobic FGF and basic TAT peptides to promote cellular uptake of recombinant Cre recombinase: a tool for efficient genetic engineering of mammalian genomes. *Proc. Natl Acad. Sci. USA* **99**, 4489-4494 (2002).

34. Eggan, K. *et al.* Hybrid vigor, fetal overgrowth, and viability of mice derived by nuclear cloning and tetraploid embryo complementation. *Proc. Natl Acad. Sci. USA* **98**, 6209-6214 (2001).

35. Naviaux, R. K., Costanzi, E., Haas, M. & Verma, I. M. The pCL vector system: rapid production of helper-free, high-titer, recombinant retroviruses. *J. Virol.* **70**, 5701-5705 (1996).

Supplementary Information is linked to the online version of the paper at www.nature.com/nature.

Acknowledgements. We thank H. Suh, D. Fu and J. Dausman for technical assistance; J. Love for help with the microarray analysis; S. Markoulaki for help with blastocyst injections; F. Edenhofer for a gift of Tat-Cre; and S. Yamanaka for the Nanog-neo construct. We acknowledge L. Zagachin in the MGH Nucleic Acid Quantitation core for assistance with real-time PCR. We also thank C. Lengner, C. Beard and M. Creyghton for constructive criticism. M.W. was supported in part by fellowships from the Human Frontiers Science Organization Program and the Ellison Foundation; B.B. by grants from the Burroughs Wellcome Fund, the Harvard Stem Cell Institute and the NIH; and R.J. by grants from the NIH.

Author Contributions. M.W., A.M. and R.J. conceived and designed the experiments and wrote the manuscript; M.W. derived all iPS lines; M.W. and A.M. performed the *in vitro* and *in vivo* characterization of the iPS lines (teratoma, 2N and 4N injections and IHC) and the conditional Dnmt1 experiment; A.M. investigated the promoter and imprinting methylation; M.K. and B.B. performed and analysed the real-time PCRs and ChIP experiments; R.F. and K.H. generated the selectable MEFs and TTFs; R.F. performed western blot and PCR analyses; and T.B. performed the microarray analysis and the proviral integration Southern blots.

Author Information. All microarray data from this study are available from Array Express at the EBI (http://www.ebi.ac.uk/arrayexpress) under the accession number E-MEXP-1037. Reprints and permissions information is available at www.nature.com/reprints. The authors declare no competing financial interests. Correspondence and requests for materials should be addressed to R.J. (jaenisch@wi.mit.edu).